FOR
REFERENCE ONLY

The GALE ENCYCLOPEDIA of SCIENCE

THIRD EDITION

The GALE ENCYCLOPEDIA of SCIENCE

THIRD EDITION

VOLUME 6
Star cluster - Zooplankton
General Index

K. Lee Lerner and
Brenda Wilmoth Lerner,
Editors

GALE®

THOMSON

GALE

Detroit • New York • San Diego • San Francisco • Cleveland • New Haven, Conn. • Waterville, Maine • London • Munich

Gale Encyclopedia of Science, Third Edition

K. Lee Lerner and Brenda Wilmoth Lerner, Editors

Project Editor
Kimberley A. McGrath

Editorial
Deirdre S. Blanchfield, Chris Jeryan, Jacqueline Longe, Mark Springer

Editorial Support Services
Andrea Lopeman

Indexing Services
Synapse

Permissions
Shalice Shah-Caldwell

Imaging and Multimedia
Leitha Etheridge-Sims, Lezlie Light, Dave Oblender, Christine O'Brien, Robyn V. Young

Product Design
Michelle DiMercurio

Manufacturing
Wendy Blurton, Evi Seoud

LIBRARY OF CONGRESS CATALOGING-IN-PUBLICATION DATA

Gale encyclopedia of science / K. Lee Lerner & Brenda Wilmoth Lerner, editors.— 3rd ed.
   p. cm.
     Includes index.
     ISBN 0-7876-7554-7 (set) — ISBN 0-7876-7555-5 (v. 1) — ISBN
     0-7876-7556-3 (v. 2) — ISBN 0-7876-7557-1 (v. 3) — ISBN 0-7876-7558-X
     (v. 4) — ISBN 0-7876-7559-8 (v. 5) — ISBN 0-7876-7560-1 (v. 6)
     1. Science—Encyclopedias. I. Lerner, K. Lee. II. Lerner, Brenda Wilmoth.

Q121.G37 2004
503—dc22
                                 2003015731

CONTENTS

TOPIC LIST

A

Aardvark
Abacus
Abrasives
Abscess
Absolute zero
Abyssal plain
Acceleration
Accelerators
Accretion disk
Accuracy
Acetic acid
Acetone
Acetylcholine
Acetylsalicylic acid
Acid rain
Acids and bases
Acne
Acorn worm
Acoustics
Actinides
Action potential
Activated complex
Active galactic nuclei
Acupressure
Acupuncture
ADA (adenosine deaminase) deficiency
Adaptation
Addiction
Addison's disease
Addition
Adenosine diphosphate
Adenosine triphosphate
Adhesive

Adrenals
Aerobic
Aerodynamics
Aerosols
Africa
Age of the universe
Agent Orange
Aging and death
Agouti
Agricultural machines
Agrochemicals
Agronomy
AIDS
AIDS therapies and vaccines
Air masses and fronts
Air pollution
Aircraft
Airship
Albatrosses
Albedo
Albinism
Alchemy
Alcohol
Alcoholism
Aldehydes
Algae
Algebra
Algorithm
Alkali metals
Alkaline earth metals
Alkaloid
Alkyl group
Alleles
Allergy
Allotrope
Alloy

Alluvial systems
Alpha particle
Alternative energy sources
Alternative medicine
Altruism
Aluminum
Aluminum hydroxide
Alzheimer disease
Amaranth family (Amaranthaceae)
Amaryllis family (Amaryllidaceae)
American Standard Code for Information Interchange
Ames test
Amicable numbers
Amides
Amino acid
Ammonia
Ammonification
Amnesia
Amniocentesis
Amoeba
Amphetamines
Amphibians
Amplifier
Amputation
Anabolism
Anaerobic
Analemma
Analgesia
Analog signals and digital signals
Analytic geometry
Anaphylaxis
Anatomy
Anatomy, comparative
Anchovy
Anemia

Anesthesia
Aneurism
Angelfish
Angiography
Angiosperm
Angle
Anglerfish
Animal
Animal breeding
Animal cancer tests
Anion
Anode
Anoles
Ant-pipits
Antarctica
Antbirds and gnat-eaters
Anteaters
Antelopes and gazelles
Antenna
Anthrax
Anthropocentrism
Anti-inflammatory agents
Antibiotics
Antibody and antigen
Anticoagulants
Anticonvulsants
Antidepressant drugs
Antihelmintics
Antihistamines
Antimatter
Antimetabolites
Antioxidants
Antiparticle
Antipsychotic drugs
Antisepsis
Antlions
Ants
Anxiety
Apes
Apgar score
Aphasia
Aphids
Approximation
Apraxia
Aqueduct
Aquifer
Arachnids
Arapaima

Arc
ARC LAMP
Archaebacteria
Archaeoastronomy
Archaeogenetics
Archaeology
Archaeometallurgy
Archaeometry
Archeological mapping
Archeological sites
Arithmetic
Armadillos
Arrow worms
Arrowgrass
Arrowroot
Arteries
Arteriosclerosis
Arthritis
Arthropods
Arthroscopic surgery
Artifacts and artifact classification
Artificial fibers
Artificial heart and heart valve
Artificial intelligence
Artificial vision
Arum family (Araceae)
Asbestos
Asexual reproduction
Asia
Assembly line
Asses
Associative property
Asteroid 2002AA29
Asthenosphere
Asthma
Astrobiology
Astroblemes
Astrolabe
Astrometry
Astronomical unit
Astronomy
Astrophysics
Atmosphere, composition and structure
Atmosphere observation
Atmospheric circulation
Atmospheric optical phenomena
Atmospheric pressure

Atmospheric temperature
Atomic clock
Atomic models
Atomic number
Atomic spectroscopy
Atomic theory
Atomic weight
Atoms
Attention-deficit/Hyperactivity disorder (ADHD)
Auks
Australia
Autism
Autoimmune disorders
Automatic pilot
Automation
Automobile
Autotroph
Avogadro's number
Aye-ayes

B

Babblers
Baboons
Bacteria
Bacteriophage
Badgers
Ball bearing
Ballistic missiles
Ballistics
Balloon
Banana
Bandicoots
Bar code
Barberry
Barbets
Barbiturates
Bariatrics
Barium
Barium sulfate
Bark
Barley
Barnacles
Barometer
Barracuda

C

Calibration

Caliper

Calorie

Calorimetry

Camels

Canal

Cancel

Cancer

Canines

Cantilever

Capacitance

Capacitor

Capillaries

Capillary action

Caprimulgids

Captive breeding and reintroduction

Capuchins

Capybaras

Carbohydrate

Carbon

Carbon cycle

Carbon dioxide

Carbon monoxide

Carbon tetrachloride

Carbonyl group

Carboxyl group

Carboxylic acids

Carcinogen

Cardiac cycle

Cardinal number

Cardinals and grosbeaks

Caribou

Carnivore

Carnivorous plants

Carp

Carpal tunnel syndrome

Carrier (genetics)

Carrot family (Apiaceae)

Carrying capacity

Cartesian coordinate plane

Cartilaginous fish

Cartography

Cashew family (Anacardiaceae)

Cassini Spacecraft

Catabolism

Catalyst and catalysis

Catastrophism

Catfish

Catheters

Cathode

Cathode ray tube

Cation

Cats

Cattails

Cattle family (Bovidae)

Cauterization

Cave

Cave fish

Celestial coordinates

Celestial mechanics

Celestial sphere: The apparent motions of the Sun, Moon, planets, and stars

Cell

Cell death

Cell division

Cell, electrochemical

Cell membrane transport

Cell staining

Cellular respiration

Cellular telephone

Cellulose

Centipedes

Centrifuge

Ceramics

Cerenkov effect

Cetaceans

Chachalacas

Chameleons

Chaos

Charge-coupled device

Chelate

Chemical bond

Chemical evolution

Chemical oxygen demand

Chemical reactions

Chemical warfare

Chemistry

Chemoreception

Chestnut

Chi-square test

Chickenpox

Childhood diseases

Chimaeras

Chimpanzees

Chinchilla

Chipmunks

Chitons

Chlordane

Chlorinated hydrocarbons

Chlorination

Chlorine

Chlorofluorocarbons (CFCs)

Chloroform

Chlorophyll

Chloroplast

Cholera

Cholesterol

Chordates

Chorionic villus sampling (CVS)

Chromatin

Chromatography

Chromosomal abnormalities

Chromosome

Chromosome mapping

Cicadas

Cigarette smoke

Circle

Circulatory system

Circumscribed and inscribed

Cirrhosis

Citric acid

Citrus trees

Civets

Climax (ecological)

Clingfish

Clone and cloning

Closed curves

Closure property

Clouds

Club mosses

Coal

Coast and beach

Coatis

Coca

Cocaine

Cockatoos

Cockroaches

Codeine

Codfishes

Codons

Coefficient

Coelacanth

Density
Dentistry
Deoxyribonucleic acid (DNA)
Deposit
Depression
Depth perception
Derivative
Desalination
Desert
Desertification
Determinants
Deuterium
Developmental processes
Dew point
Diabetes mellitus
Diagnosis
Dialysis
Diamond
Diatoms
Dielectric materials
Diesel engine
Diethylstilbestrol (DES)
Diffraction
Diffraction grating
Diffusion
Digestive system
Digital Recording
Digitalis
Dik-diks
Dinosaur
Diode
Dioxin
Diphtheria
Dipole
Direct variation
Disease
Dissociation
Distance
Distillation
Distributive property
Disturbance, ecological
Diurnal cycles
Division
DNA fingerprinting
DNA replication
DNA synthesis
DNA technology
DNA vaccine

Dobsonflies
Dogwood tree
Domain
Donkeys
Dopamine
Doppler effect
Dories
Dormouse
Double-blind study
Double helix
Down syndrome
Dragonflies
Drift net
Drongos
Drosophila melanogaster
Drought
Ducks
Duckweed
Duikers
Dune
Duplication of the cube
Dust devil
DVD
Dwarf antelopes
Dyes and pigments
Dysentery
Dyslexia
Dysplasia
Dystrophinopathies

E

e (number)
Eagles
Ear
Earth
Earth science
Earth's interior
Earth's magnetic field
Earth's rotation
Earthquake
Earwigs
Eating disorders
Ebola virus
Ebony
Echiuroid worms

Echolocation
Eclipses
Ecological economics
Ecological integrity
Ecological monitoring
Ecological productivity
Ecological pyramids
Ecology
Ecosystem
Ecotone
Ecotourism
Edema
Eel grass
El Niño and La Niña
Eland
Elapid snakes
Elasticity
Electric arc
Electric charge
Electric circuit
Electric conductor
Electric current
Electric motor
Electric vehicles
Electrical conductivity
Electrical power supply
Electrical resistance
Electricity
Electrocardiogram (ECG)
Electroencephalogram (EEG)
Electrolysis
Electrolyte
Electromagnetic field
Electromagnetic induction
Electromagnetic spectrum
Electromagnetism
Electromotive force
Electron
Electron cloud
Electronics
Electrophoresis
Electrostatic devices
Element, chemical
Element, families of
Element, transuranium
Elements, formation of
Elephant
Elephant shrews

Elephant snout fish

Elephantiasis

Elevator

Ellipse

Elm

Embiids

Embolism

Embryo and embryonic development

Embryo transfer

Embryology

Emission

Emphysema

Emulsion

Encephalitis

Endangered species

Endemic

Endocrine system

Endoprocta

Endoscopy

Endothermic

Energy

Energy budgets

Energy efficiency

Energy transfer

Engineering

Engraving and etching

Enterobacteria

Entropy

Environmental ethics

Environmental impact statement

Enzymatic engineering

Enzyme

Epidemic

Epidemiology

Epilepsy

Episomes

Epstein-Barr virus

Equation, chemical

Equilibrium, chemical

Equinox

Erosion

Error

Escherichia coli

Ester

Esterification

Ethanol

Ether

Ethnoarchaeology

Ethnobotany

Ethyl group

Ethylene glycol

Ethylenediaminetetra-acetic acid

Etiology

Eubacteria

Eugenics

Eukaryotae

Europe

Eutrophication

Evaporation

Evapotranspiration

Even and odd

Event horizon

Evolution

Evolution, convergent

Evolution, divergent

Evolution, evidence of

Evolution, parallel

Evolutionary change, rate of

Evolutionary mechanisms

Excavation methods

Exclusion principle, Pauli

Excretory system

Exercise

Exocrine glands

Explosives

Exponent

Extinction

Extrasolar planets

Eye

F

Factor

Factorial

Falcons

Faraday effect

Fat

Fatty acids

Fault

Fauna

Fax machine

Feather stars

Fermentation

Ferns

Ferrets

Fertilization

Fertilizers

Fetal alcohol syndrome

Feynman diagrams

Fiber optics

Fibonacci sequence

Field

Figurative numbers

Filtration

Finches

Firs

Fish

Flagella

Flame analysis

Flamingos

Flatfish

Flatworms

Flax

Fleas

Flies

Flightless birds

Flooding

Flora

Flower

Fluid dynamics

Fluid mechanics

Fluorescence

Fluorescence in situ hybridization (FISH)

Fluorescent light

Fluoridation

Flying fish

Focused Ion Beam (FIB)

Fog

Fold

Food chain/web

Food irradiation

Food poisoning

Food preservation

Food pyramid

Foot and mouth disease

Force

Forensic science

Forestry

Forests

Formula, chemical

Formula, structural
Fossa
Fossil and fossilization
Fossil fuels
Fractal
Fraction, common
Fraunhofer lines
Freeway
Frequency
Freshwater
Friction
Frigate birds
Frog's-bit family
Frogs
Frostbite
Fruits
Fuel cells
Function
Fundamental theorems
Fungi
Fungicide

G

Gaia hypothesis
Galaxy
Game theory
Gamete
Gametogenesis
Gamma-ray astronomy
Gamma ray burst
Gangrene
Garpike
Gases, liquefaction of
Gases, properties of
Gazelles
Gears
Geckos
Geese
Gelatin
Gene
Gene chips and microarrays
Gene mutation
Gene splicing
Gene therapy
Generator

Genetic disorders
Genetic engineering
Genetic identification of
    microorganisms
Genetic testing
Genetically modified foods and
    organisms
Genetics
Genets
Genome
Genomics (comparative)
Genotype and phenotype
Geocentric theory
Geochemical analysis
Geochemistry
Geode
Geodesic
Geodesic dome
Geographic and magnetic poles
Geologic map
Geologic time
Geology
Geometry
Geomicrobiology
Geophysics
Geotropism
Gerbils
Germ cells and the germ cell line
Germ theory
Germination
Gerontology
Gesnerias
Geyser
Gibbons and siamangs
Gila monster
Ginger
Ginkgo
Ginseng
Giraffes and okapi
GIS
Glaciers
Glands
Glass
Global climate
Global Positioning System
Global warming
Glycerol
Glycol

Glycolysis
Goats
Goatsuckers
Gobies
Goldenseal
Gophers
Gorillas
Gourd family (Cucurbitaceae)
Graft
Grand unified theory
Grapes
Graphs and graphing
Grasses
Grasshoppers
Grasslands
Gravitational lens
Gravity and gravitation
Great Barrier Reef
Greatest common factor
Grebes
Greenhouse effect
Groundhog
Groundwater
Group
Grouse
Growth and decay
Growth hormones
Guenons
Guillain-Barre syndrome
Guinea fowl
Guinea pigs and cavies
Gulls
Guppy
Gutenberg discontinuity
Gutta percha
Gymnosperm
Gynecology
Gyroscope

H

Habitat
Hagfish
Half-life
Halide, organic
Hall effect

Halley's comet
Hallucinogens
Halogenated hydrocarbons
Halogens
Halosaurs
Hamsters
Hand tools
Hantavirus infections
Hard water
Harmonics
Hartebeests
Hawks
Hazardous wastes
Hazel
Hearing
Heart
Heart diseases
Heart, embryonic development and
    changes at birth
Heart-lung machine
Heat
Heat capacity
Heat index
Heat transfer
Heath family (Ericaceae)
Hedgehogs
Heisenberg uncertainty principle
Heliocentric theory
Hematology
Hemophilia
Hemorrhagic fevers and diseases
Hemp
Henna
Hepatitis
Herb
Herbal medicine
Herbicides
Herbivore
Hermaphrodite
Hernia
Herons
Herpetology
Herrings
Hertzsprung-Russell diagram
Heterotroph
Hibernation
Himalayas, geology of
Hippopotamuses

Histamine
Historical geology
Hoatzin
Hodgkin's disease
Holly family (Aquifoliaceae)
Hologram and holography
Homeostasis
Honeycreepers
Honeyeaters
Hoopoe
Horizon
Hormones
Hornbills
Horse chestnut
Horsehair worms
Horses
Horseshoe crabs
Horsetails
Horticulture
Hot spot
Hovercraft
Hubble Space Telescope
Human artificial chromosomes
Human chorionic gonadotropin
Human cloning
Human ecology
Human evolution
Human Genome Project
Humidity
Hummingbirds
Humus
Huntington disease
Hybrid
Hydra
Hydrocarbon
Hydrocephalus
Hydrochlorofluorocarbons
Hydrofoil
Hydrogen
Hydrogen chloride
Hydrogen peroxide
Hydrogenation
Hydrologic cycle
Hydrology
Hydrolysis
Hydroponics
Hydrosphere
Hydrothermal vents

Hydrozoa
Hyena
Hyperbola
Hypertension
Hypothermia
Hyraxes

I

Ibises
Ice
Ice age refuges
Ice ages
Icebergs
Iceman
Identity element
Identity property
Igneous rocks
Iguanas
Imaginary number
Immune system
Immunology
Impact crater
Imprinting
In vitro fertilization (IVF)
In vitro and in vivo
Incandescent light
Incineration
Indicator, acid-base
Indicator species
Individual
Indoor air quality
Industrial minerals
Industrial Revolution
Inequality
Inertial guidance
Infection
Infertility
Infinity
Inflammation
Inflection point
Influenza
Infrared astronomy
Inherited disorders
Insecticides
Insectivore

Lorises
Luminescence
Lungfish
Lycophytes
Lyme disease
Lymphatic system
Lyrebirds

M

Macaques
Mach number
Machine tools
Machine vision
Machines, simple
Mackerel
Magic square
Magma
Magnesium
Magnesium sulfate
Magnetic levitation
Magnetic recording/audiocassette
Magnetic resonance imaging (MRI)
Magnetism
Magnetosphere
Magnolia
Mahogany
Maidenhair fern
Malaria
Malnutrition
Mammals
Manakins
Mangrove tree
Mania
Manic depression
Map
Maples
Marfan syndrome
Marijuana
Marlins
Marmosets and tamarins
Marmots
Mars
Mars Pathfinder
Marsupial cats
Marsupial rats and mice

Marsupials
Marten, sable, and fisher
Maser
Mass
Mass extinction
Mass number
Mass production
Mass spectrometry
Mass transportation
Mass wasting
Mathematics
Matrix
Matter
Maunder minimum
Maxima and minima
Mayflies
Mean
Median
Medical genetics
Meiosis
Membrane
Memory
Mendelian genetics
Meningitis
Menopause
Menstrual cycle
Mercurous chloride
Mercury (element)
Mercury (planet)
Mesoscopic systems
Mesozoa
Metabolic disorders
Metabolism
Metal
Metal fatigue
Metal production
Metallurgy
Metamorphic grade
Metamorphic rock
Metamorphism
Metamorphosis
Meteorology
Meteors and meteorites
Methyl group
Metric system
Mice
Michelson-Morley experiment
Microbial genetics

Microclimate
Microorganisms
Microscope
Microscopy
Microtechnology
Microwave communication
Migraine headache
Migration
Mildew
Milkweeds
Milky Way
Miller-Urey Experiment
Millipedes
Mimicry
Mineralogy
Minerals
Mining
Mink
Minnows
Minor planets
Mint family
Mir Space Station
Mirrors
Miscibility
Mistletoe
Mites
Mitosis
Mixture, chemical
Möbius strip
Mockingbirds and thrashers
Mode
Modular arithmetic
Mohs' scale
Mold
Mole
Mole-rats
Molecular biology
Molecular formula
Molecular geometry
Molecular weight
Molecule
Moles
Mollusks
Momentum
Monarch flycatchers
Mongooses
Monitor lizards
Monkeys

Organic farming
Organism
Organogenesis
Organs and organ systems
Origin of life
Orioles
Ornithology
Orthopedics
Oryx
Oscillating reactions
Oscillations
Oscilloscope
Osmosis
Osmosis (cellular)
Ossification
Osteoporosis
Otter shrews
Otters
Outcrop
Ovarian cycle and hormonal
    regulation
Ovenbirds
Oviparous
Ovoviviparous
Owls
Oxalic acid
Oxidation-reduction reaction
Oxidation state
Oxygen
Oystercatchers
Ozone
Ozone layer depletion

P

Pacemaker
Pain
Paleobotany
Paleoclimate
Paleoecology
Paleomagnetism
Paleontology
Paleopathology
Palindrome
Palms
Palynology

Pandas
Pangolins
Papaya
Paper
Parabola
Parallax
Parallel
Parallelogram
Parasites
Parity
Parkinson disease
Parrots
Parthenogenesis
Particle detectors
Partridges
Pascal's triangle
Passion flower
Paternity and parentage testing
Pathogens
Pathology
PCR
Peafowl
Peanut worms
Peccaries
Pedigree analysis
Pelicans
Penguins
Peninsula
Pentyl group
Peony
Pepper
Peptide linkage
Percent
Perception
Perch
Peregrine falcon
Perfect numbers
Periodic functions
Periodic table
Permafrost
Perpendicular
Pesticides
Pests
Petrels and shearwaters
Petroglyphs and pictographs
Petroleum
pH
Phalangers

Pharmacogenetics
Pheasants
Phenyl group
Phenylketonuria
Pheromones
Phlox
Phobias
Phonograph
Phoronids
Phosphoric acid
Phosphorus
Phosphorus cycle
Phosphorus removal
Photic zone
Photochemistry
Photocopying
Photoelectric cell
Photoelectric effect
Photography
Photography, electronic
Photon
Photosynthesis
Phototropism
Photovoltaic cell
Phylogeny
Physical therapy
Physics
Physiology
Physiology, comparative
Phytoplankton
Pi
Pigeons and doves
Pigs
Pike
Piltdown hoax
Pinecone fish
Pines
Pipefish
Placebo
Planck's constant
Plane
Plane family
Planet
Planet X
Planetary atmospheres
Planetary geology
Planetary nebulae
Planetary ring systems

Rate
Ratio
Rational number
Rationalization
Rats
Rayleigh scattering
Rays
Real numbers
Reciprocal
Recombinant DNA
Rectangle
Recycling
Red giant star
Red tide
Redshift
Reflections
Reflex
Refrigerated trucks and railway cars
Rehabilitation
Reinforcement, positive and
    negative
Relation
Relativity, general
Relativity, special
Remote sensing
Reproductive system
Reproductive toxicant
Reptiles
Resins
Resonance
Resources, natural
Respiration
Respiration, cellular
Respirator
Respiratory diseases
Respiratory system
Restoration ecology
Retrograde motion
Retrovirus
Reye's syndrome
Rh factor
Rhesus monkeys
Rheumatic fever
Rhinoceros
Rhizome
Rhubarb
Ribbon worms
Ribonuclease

Ribonucleic acid (RNA)
Ribosomes
Rice
Ricin
Rickettsia
Rivers
RNA function
RNA splicing
Robins
Robotics
Rockets and missiles
Rocks
Rodents
Rollers
Root system
Rose family (Rosaceae)
Rotation
Roundworms
Rumination
Rushes
Rusts and smuts

S

Saiga antelope
Salamanders
Salmon
Salmonella
Salt
Saltwater
Sample
Sand
Sand dollars
Sandfish
Sandpipers
Sapodilla tree
Sardines
Sarin gas
Satellite
Saturn
Savanna
Savant
Sawfish
Saxifrage family
Scalar
Scale insects

Scanners, digital
Scarlet fever
Scavenger
Schizophrenia
Scientific method
Scorpion flies
Scorpionfish
Screamers
Screwpines
Sculpins
Sea anemones
Sea cucumbers
Sea horses
Sea level
Sea lily
Sea lions
Sea moths
Sea spiders
Sea squirts and salps
Sea urchins
Seals
Seamounts
Seasonal winds
Seasons
Secondary pollutants
Secretary bird
Sedges
Sediment and sedimentation
Sedimentary environment
Sedimentary rock
Seed ferns
Seeds
Segmented worms
Seismograph
Selection
Sequences
Sequencing
Sequoia
Servomechanisms
Sesame
Set theory
SETI
Severe acute respiratory syndrome
    (SARS)
Sewage treatment
Sewing machine
Sex change
Sextant

Stroke
Stromatolite
Sturgeons
Subatomic particles
Submarine
Subsidence
Subsurface detection
Subtraction
Succession
Suckers
Sudden infant death syndrome
    (SIDS)
Sugar beet
Sugarcane
Sulfur
Sulfur cycle
Sulfur dioxide
Sulfuric acid
Sun
Sunbirds
Sunspots
Superclusters
Superconductor
Supernova
Surface tension
Surgery
Surveying instruments
Survival of the fittest
Sustainable development
Swallows and martins
Swamp cypress family
    (Taxodiaceae)
Swamp eels
Swans
Sweet gale family (Myricaceae)
Sweet potato
Swifts
Swordfish
Symbiosis
Symbol, chemical
Symbolic logic
Symmetry
Synapse
Syndrome
Synthesis, chemical
Synthesizer, music
Synthesizer, voice
Systems of equations

T

T cells
Tanagers
Taphonomy
Tapirs
Tarpons
Tarsiers
Tartaric acid
Tasmanian devil
Taste
Taxonomy
Tay-Sachs disease
Tea plant
Tectonics
Telegraph
Telemetry
Telephone
Telescope
Television
Temperature
Temperature regulation
Tenrecs
Teratogen
Term
Termites
Terns
Terracing
Territoriality
Tetanus
Tetrahedron
Textiles
Thalidomide
Theorem
Thermal expansion
Thermochemistry
Thermocouple
Thermodynamics
Thermometer
Thermostat
Thistle
Thoracic surgery
Thrips
Thrombosis
Thrushes
Thunderstorm
Tides

Time
Tinamous
Tissue
Tit family
Titanium
Toadfish
Toads
Tomato family
Tongue worms
Tonsillitis
Topology
Tornado
Torque
Torus
Total solar irradiance
Toucans
Touch
Towers of Hanoi
Toxic shock syndrome
Toxicology
Trace elements
Tragopans
Trains and railroads
Tranquilizers
Transcendental numbers
Transducer
Transformer
Transgenics
Transistor
Transitive
Translations
Transpiration
Transplant, surgical
Trapezoid
Tree
Tree shrews
Trichinosis
Triggerfish
Triglycerides
Trigonometry
Tritium
Trogons
Trophic levels
Tropic birds
Tropical cyclone
Tropical diseases
Trout-perch
True bugs

True eels
True flies
Trumpetfish
Tsunami
Tuatara lizard
Tuber
Tuberculosis
Tumbleweed
Tumor
Tuna
Tundra
Tunneling
Turacos
Turbine
Turbulence
Turkeys
Turner syndrome
Turtles
Typhoid fever
Typhus
Tyrannosaurus rex
Tyrant flycatchers

U

Ulcers
Ultracentrifuge
Ultrasonics
Ultraviolet astronomy
Unconformity
Underwater exploration
Ungulates
Uniformitarianism
Units and standards
Uplift
Upwelling
Uranium
Uranus
Urea
Urology

V

Vaccine

Vacuum
Vacuum tube
Valence
Van Allen belts
Van der Waals forces
Vapor pressure
Variable
Variable stars
Variance
Varicella zoster virus
Variola virus
Vegetables
Veins
Velocity
Venus
Verbena family (Verbenaceae)
Vertebrates
Video recording
Violet family (Violaceae)
Vipers
Viral genetics
Vireos
Virtual particles
Virtual reality
Virus
Viscosity
Vision
Vision disorders
Vitamin
Viviparity
Vivisection
Volatility
Volcano
Voles
Volume
Voyager spacecraft
Vulcanization
Vultures
VX agent

W

Wagtails and pipits
Walkingsticks
Walnut family
Walruses

Warblers
Wasps
Waste management
Waste, toxic
Water
Water bears
Water conservation
Water lilies
Water microbiology
Water pollution
Water treatment
Waterbuck
Watershed
Waterwheel
Wave motion
Waxbills
Waxwings
Weasels
Weather
Weather forecasting
Weather mapping
Weather modification
Weathering
Weaver finches
Weevils
Welding
West Nile virus
Wetlands
Wheat
Whisk fern
White dwarf
White-eyes
Whooping cough
Wild type
Wildfire
Wildlife
Wildlife trade (illegal)
Willow family (Salicaceae)
Wind
Wind chill
Wind shear
Wintergreen
Wolverine
Wombats
Wood
Woodpeckers
Woolly mammoth
Work

Wren-warblers
Wrens
Wrynecks

X

X-ray astronomy
X-ray crystallography
X rays
Xenogamy

Y

Y2K
Yak
Yam
Yeast
Yellow fever
Yew
Yttrium

Z

Zebras
Zero
Zodiacal light
Zoonoses
Zooplankton

ORGANIZATION OF THE ENCYCLOPEDIA

The *Gale Encyclopedia of Science, Third Edition* has been designed with ease of use and ready reference in mind.

- Entries are alphabetically arranged across six volumes, in a single sequence, rather than by scientific field

- Length of entries varies from short definitions of one or two paragraphs, to longer, more detailed entries on more complex subjects.

- Longer entries are arranged so that an overview of the subject appears first, followed by a detailed discussion conveniently arranged under subheadings.

- A list of key terms is provided where appropriate to define unfamiliar terms or concepts.

- Bold-faced terms direct the reader to related articles.

- Longer entries conclude with a "Resources" section, which points readers to other helpful materials (including books, periodicals, and Web sites).

- The author's name appears at the end of longer entries. His or her affiliation can be found in the "Contributors" section at the front of each volume.

- "See also" references appear at the end of entries to point readers to related entries.

- Cross references placed throughout the encyclopedia direct readers to where information on subjects without their own entries can be found.

- A comprehensive, two-level General Index guides readers to all topics, illustrations, tables, and persons mentioned in the book.

AVAILABLE IN ELECTRONIC FORMATS

Licensing. *The Gale Encyclopedia of Science, Third Edition* is available for licensing. The complete database is provided in a fielded format and is deliverable on such media as disk or CD-ROM. For more information, contact Gale's Business Development Group at 1-800-877-GALE, or visit our website at www.gale.com/bizdev.

ADVISORY BOARD

A number of experts in the scientific and libary communities provided invaluable assistance in the formulation of this encyclopedia. Our advisory board performed a myriad of duties, from defining the scope of coverage to reviewing individual entries for accuracy and accessibility, and in many cases, writing entries. We would therefore like to express our appreciation to them:

ACADEMIC ADVISORS

Marcelo Amar, M.D.
Senior Fellow, Molecular Disease Branch
National Institutes of Health (NIH)
Bethesda, Maryland

Robert G. Best, Ph.D.
Director
Divison of Genetics, Department of Obstetrics and
 Gynecology
University of South Carolina School of Medicine
Columbia, South Carolina

Bryan Bunch
Adjunct Instructor
Department of Mathematics
Pace University
New York, New York

Cynthia V. Burek, Ph.D.
Environment Research Group, Biology Department
Chester College
England, UK

David Campbell
Head
Department of Physics
University of Illinois at Urbana Champaign
Urbana, Illinois

Morris Chafetz
Health Education Foundation
Washington, DC

Brian Cobb, Ph.D.
Institute for Molecular and Human Genetics
Georgetown University
Washington, DC

Neil Cumberlidge
Professor
Department of Biology

Northern Michigan University
Marquette, Michigan

Nicholas Dittert, Ph.D.
Institut Universitaire Européen de la Mer
University of Western Brittany
France

William J. Engle. P.E.
Exxon-Mobil Oil Corporation (Rt.)
New Orleans, Louisiana

Bill Freedman
Professor
Department of Biology and School for Resource and
 Environmental Studies
Dalhousie University
Halifax, Nova Scotia, Canada

Antonio Farina, M.D., Ph.D.
Department of Embryology, Obstetrics, and
 Gynecology
University of Bologna
Bologna, Italy

G. Thomas Farmer, Ph.D., R.G.
Earth & Environmental Sciences Division
Los Alamos National Laboratory
Los Alamos, New Mexico

Jeffrey C. Hall
Lowell Observatory
Flagstaff, Arizona

Clayton Harris
Associate Professor
Department of Geography and Geology
Middle Tennessee State University
Murfreesboro, Tenness

Lyal Harris, Ph.D.
Tectonics Special Research Centre
Department of Geology & Geophysics

CONTRIBUTORS

Nasrine Adibe
Professor Emeritus
Department of Education
Long Island University
Westbury, New York

Mary D. Albanese
Department of English
University of Alaska
Juneau, Alaska

Margaret Alic
Science Writer
Eastsound, Washington

James L. Anderson
Soil Science Department
University of Minnesota
St. Paul, Minnesota

Monica Anderson
Science Writer
Hoffman Estates, Illinois

Susan Andrew
Teaching Assistant
University of Maryland
Washington, DC

John Appel
Director
Fundación Museo de Ciencia y
  Tecnología
Popayán, Colombia

David Ball
Assistant Professor
Department of Chemistry
Cleveland State University
Cleveland, Ohio

Dana M. Barry
Editor and Technical Writer
Center for Advanced Materials
  Processing
Clarkston University
Potsdam, New York

Puja Batra
Department of Zoology
Michigan State University
East Lansing, Michigan

Donald Beaty
Professor Emeritus
College of San Mateo
San Mateo, California

Eugene C. Beckham
Department of Mathematics and
  Science
Northwood Institute
Midland, Michigan

Martin Beech
Research Associate
Department of Astronomy
University of Western Ontario
London, Ontario, Canada

**Julie Berwald, Ph.D. (Ocean
  Sciences)**
Austin, Texas

Massimo D. Bezoari
Associate Professor
Department of Chemistry
Huntingdon College
Montgomery, Alabama

John M. Bishop III
Translator
New York, New York

T. Parker Bishop
Professor
Middle Grades and Secondary
  Education
Georgia Southern University
Statesboro, Georgia

Carolyn Black
Professor
Incarnate Word College
San Antonio, Texas

Larry Blaser
Science Writer
Lebanon, Tennessee

Jean F. Blashfield
Science Writer
Walworth, Wisconsin

Richard L. Branham Jr.
Director
Centro Rigional de
  Investigaciones Científicas y
  Tecnológicas
Mendoza, Argentina

Patricia Braus
Editor
American Demographics
Rochester, New York

David L. Brock
Biology Instructor
St. Louis, Missouri

Leona B. Bronstein
Chemistry Teacher (retired)
East Lansing High School
Okemos, Michigan

Brandon R. Brown
Graduate Research Assistant
Oregon State University
Corvallis, Oregon

Lenonard C. Bruno
Senior Science Specialist
Library of Congress
Chevy Chase, Maryland

Janet Buchanan, Ph.D.
Microbiologist
Independent Scholar
Toronto, Ontario, Canada.

Scott Christian Cahall
Researcher
World Precision Instruments, Inc.
Bradenton, Florida

G. Lynn Carlson
Senior Lecturer
School of Science and
   Technology
University of Wisconsin—
   Parkside
Kenosha, Wisconsin

James J. Carroll
Center for Quantum Mechanics
The University of Texas at Dallas
Dallas, Texas

Steven B. Carroll
Assistant Professor
Division of Biology
Northeast Missouri State
   University
Kirksville, Missouri

Rosalyn Carson-DeWitt
Physician and Medical Writer
Durham, North Carolina

Yvonne Carts-Powell
Editor
Laser Focus World
Belmont, Massachustts

Chris Cavette
Technical Writer
Fremont, California

Lata Cherath
Science Writer
Franklin Park, New York

Kenneth B. Chiacchia
Medical Editor
University of Pittsburgh Medical
   Center
Pittsburgh, Pennsylvania

M. L. Cohen
Science Writer
Chicago, Illinois

Robert Cohen
Reporter
KPFA Radio News
Berkeley, California

Sally Cole-Misch
Assistant Director
International Joint Commission
Detroit, Michigan

George W. Collins II
Professor Emeritus
Case Western Reserve
Chesterland, Ohio

Jeffrey R. Corney
Science Writer
Thermopolis, Wyoming

Tom Crawford
Assistant Director
Division of Publication and
   Development
University of Pittsburgh Medical
   Center
Pittsburgh, Pennsylvania

Pamela Crowe
Medical and Science Writer
Oxon, England

Clinton Crowley
On-site Geologist
Selman and Associates
Fort Worth, Texas

Edward Cruetz
Physicist
Rancho Santa Fe, California

Frederick Culp
Chairman
Department of Physics
Tennessee Technical
Cookeville, Tennessee

Neil Cumberlidge
Professor
Department of Biology
Northern Michigan University
Marquette, Michigan

Mary Ann Cunningham
Environmental Writer
St. Paul, Minnesota

Les C. Cwynar
Associate Professor
Department of Biology
University of New Brunswick
Fredericton, New Brunswick

Paul Cypher
Provisional Interpreter
Lake Erie Metropark
Trenton, Michigan

Stanley J. Czyzak
Professor Emeritus
Ohio State University
Columbus, Ohio

Rosi Dagit
Conservation Biologist
Topanga-Las Virgenes Resource
   Conservation District
Topanga, California

David Dalby
President
Bruce Tool Company, Inc.
Taylors, South Carolina

Lou D'Amore
Chemistry Teacher
Father Redmund High School
Toronto, Ontario, Canada

Douglas Darnowski
Postdoctoral Fellow
Department of Plant Biology
Cornell University
Ithaca, New York

Sreela Datta
Associate Writer
Aztec Publications
Northville, Michigan

Sarah K. Dean
Science Writer
Philadelphia, Pennsylvania

Sarah de Forest
Research Assistant
Theoretical Physical Chemistry
 Lab
University of Pittsburgh
Pittsburgh, Pennsylvania

Louise Dickerson
Medical and Science Writer
Greenbelt, Maryland

Marie Doorey
Editorial Assistant
Illinois Masonic Medical Center
Chicago, Illinois

Herndon G. Dowling
Professor Emeritus
Department of Biology
New York University
New York, New York

Marion Dresner
Natural Resources Educator
Berkeley, California

John Henry Dreyfuss
Science Writer
Brooklyn, New York

Roy Dubisch
Professor Emeritus
Department of Mathematics
New York University
New York, New York

Russel Dubisch
Department of Physics
Sienna College
Loudonville, New York

Carolyn Duckworth
Science Writer
Missoula, Montana

Laurie Duncan, Ph.D.
 (Geology)
Geologist
Austin, Texas

Peter A. Ensminger
Research Associate
Cornell University
Syracuse, New York

Bernice Essenfeld
Biology Writer
Warren, New Jersey

Mary Eubanks
Instructor of Biology
The North Carolina School of
 Science and Mathematics
Durham, North Carolina

Kathryn M. C. Evans
Science Writer
Madison, Wisconsin

William G. Fastie
Department of Astronomy and
 Physics
Bloomberg Center
Baltimore, Maryland

Barbara Finkelstein
Science Writer
Riverdale, New York

Mary Finley
Supervisor of Science Curriculum
 (retired)
Pittsburgh Secondary Schools
Clairton, Pennsylvania

Gaston Fischer
Institut de Géologie
Université de Neuchâtel
Peseux, Switzerland

Sara G. B. Fishman
Professor
Quinsigamond Community
 College
Worcester, Massachusetts

David Fontes
Senior Instructor
Lloyd Center for Environmental
 Studies
Westport, Maryland

Barry Wayne Fox
Extension Specialist,
 Marine/Aquatic Education
Virginia State University
Petersburg, Virginia

Ed Fox
Charlotte Latin School
Charlotte, North Carolina

Kenneth L. Frazier
Science Teacher (retired)
North Olmstead High School
North Olmstead, Ohio

Bill Freedman
Professor
Department of Biology and
 School for Resource and
 Environmental Studies
Dalhousie University
Halifax, Nova Scotia

T. A. Freeman
Consulting Archaeologist
Quail Valley, California

Elaine Friebele
Science Writer
Cheverly, Maryland

Randall Frost
Documentation Engineering
Pleasanton, California

Agnes Galambosi, M.S.
Climatologist
Eotvos Lorand University
Budapest, Hungary

Robert Gardner
Science Education Consultant
North Eastham, Massachusetts

Gretchen M. Gillis
Senior Geologist
Maxus Exploration
Dallas, Texas

Larry Gilman, Ph.D. (Electrical
 Engineering)
Engineer
Sharon, Vermont

Kathryn Glynn
Audiologist
Portland, Oregon

David Goings, Ph.D. (Geology)
Geologist
Las Vegas, Nevada

Natalie Goldstein
Educational Environmental
 Writing
Phoenicia, New York

David Gorish
TARDEC
U.S. Army
Warren, Michigan

Louis Gotlib
South Granville High School
Durham, North Carolina

Hans G. Graetzer
Professor
Department of Physics
South Dakota State University
Brookings, South Dakota

Jim Guinn
Assistant Professor
Department of Physics
Berea College
Berea, Kentucky

Steve Gutterman
Psychology Research Assistant
University of Michigan
Ann Arbor, Michigan

Johanna Haaxma-Jurek
Educator
Nataki Tabibah Schoolhouse of
   Detroit
Detroit, Michigan

Monica H. Halka
Research Associate
Department of Physics and
   Astronomy
University of Tennessee
Knoxville, Tennessee

Brooke Hall, Ph.D.
Professor
Department of Biology
California State University at
   Sacramento
Sacramento, California

Jeffrey C. Hall
Astronomer
Lowell Observatory
Flagstaff, Arizona

C. S. Hammen
Professor Emeritus
Department of Zoology
University of Rhode Island

Lawrence Hammar, Ph.D.
Senior Research Fellow
Institute of Medical Research
Papua, New Guinea

William Haneberg, Ph.D.
   (Geology)
Geologist
Portland, Oregon

Beth Hanson
Editor
The Amicus Journal
Brooklyn, New York

Clay Harris
Associate Professor
Department of Geography and
   Geology
Middle Tennessee State
   University
Murfreesboro, Tennessee

Clinton W. Hatchett
Director Science and Space
   Theater
Pensacola Junior College
Pensacola, Florida

Catherine Hinga Haustein
Associate Professor
Department of Chemistry
Central College
Pella, Iowa

Dean Allen Haycock
Science Writer
Salem, New York

Paul A. Heckert
Professor
Department of Chemistry and
   Physics
Western Carolina University
Cullowhee, North Carolina

Darrel B. Hoff
Department of Physics
Luther College
Calmar, Iowa

Dennis Holley
Science Educator
Shelton, Nebraska

Leonard Darr Holmes
Department of Physical Science
Pembroke State University
Pembroke, North Carolina

Rita Hoots
Instructor of Biology, Anatomy,
   Chemistry
Yuba College
Woodland, California

Selma Hughes
Department of Psychology and
   Special Education
East Texas State University
Mesquite, Texas

Mara W. Cohen Ioannides
Science Writer
Springfield, Missouri

Zafer Iqbal
Allied Signal Inc.
Morristown, New Jersey

Sophie Jakowska
Pathobiologist, Environmental
   Educator
Santo Domingo, Dominican
   Republic

Richard A. Jeryan
Senior Technical Specialist
Ford Motor Company
Dearborn, Michigan

Stephen R. Johnson
Biology Writer
Richmond, Virginia

Kathleen A. Jones
School of Medicine
Southern Illinois University
Carbondale, Illinois

Harold M. Kaplan
Professor
School of Medicine
Southern Illinois University
Carbondale, Illinois

Anthony Kelly
Science Writer
Pittsburgh, Pennsylvania

Amy Kenyon-Campbell
Ecology, Evolution and
    Organismal Biology Program
University of Michigan
Ann Arbor, Michigan

Judson Knight
Science Writer
Knight Agency
Atlanta, Georgia

Eileen M. Korenic
Institute of Optics
University of Rochester
Rochester, New York

Jennifer Kramer
Science Writer
Kearny, New Jersey

Pang-Jen Kung
Los Alamos National Laboratory
Los Alamos, New Mexico

Marc Kusinitz
Assistant Director Media
    Relations
John Hopkins Medical Institution
Towsen, Maryland

Arthur M. Last
Head
Department of Chemistry
University College of the Fraser
    Valley
Abbotsford, British Columbia

Nathan Lavenda
Zoologist
Skokie, Illinios

Jennifer LeBlanc
Environmental Consultant
London, Ontario, Canada

Nicole LeBrasseur, Ph.D.
Associate News Editor
Journal of Cell Biology
New York, New York

Benedict A. Leerburger
Science Writer
Scarsdale, New York

Betsy A. Leonard
Education Facilitator

Reuben H. Fleet Space Theater
    and Science Center
San Diego, California

Adrienne Wilmoth Lerner
Graduate School of Arts &
    Science
Vanderbilt University
Nashville, Tennessee

Lee Wilmoth Lerner
Science Writer
NASA
Kennedy Space Center, Florida

Scott Lewis
Science Writer
Chicago, Illinois

Frank Lewotsky
Aerospace Engineer (retired)
Nipomo, California

Karen Lewotsky
Director of Water Programs
Oregon Environmental Council
Portland, Oregon

Kristin Lewotsky
Editor
Laser Focus World
Nashua, New Hamphire

Stephen K. Lewotsky
Architect
Grants Pass, Oregon

Agnieszka Lichanska, Ph.D.
Department of Microbiology &
    Parasitology
University of Queensland
Brisbane, Australia

Sarah Lee Lippincott
Professor Emeritus
Swarthmore College
Swarthmore, Pennsylvania

Jill Liske, M.Ed.
Wilmington, North Carolina

David Lunney
Research Scientist
Centre de Spectrométrie
    Nucléaire et de Spectrométrie
    de Masse
Orsay, France

Steven MacKenzie
Ecologist
Spring Lake, Michigan

J. R. Maddocks
Consulting Scientist
DeSoto, Texas

Gail B. C. Marsella
Technical Writer
Allentown, Pennsylvania

Karen Marshall
Research Associate
Council of State Governments
    and Centers for Environment
    and Safety
Lexington, Kentucky

Liz Marshall
Science Writer
Columbus, Ohio

James Marti
Research Scientist
Department of Mechanical
    Engineering
University of Minnesota
Minneapolis, Minnesota

Elaine L. Martin
Science Writer
Pensacola, Florida

Lilyan Mastrolla
Professor Emeritus
San Juan Unified School
Sacramento, California

Iain A. McIntyre
Manager
Electro-optic Department
Energy Compression Research
    Corporation
Vista, California

Jennifer L. McGrath
Chemistry Teacher
Northwood High School
Nappanee, Indiana

Margaret Meyers, M.D.
Physician, Medical Writer
Fairhope, Alabama

G. H. Miller
Director
Studies on Smoking
Edinboro, Pennsylvania

J. Gordon Miller
Botanist
Corvallis, Oregon

Kelli Miller
Science Writer
NewScience
Atlanta, Georgia

Christine Miner Minderovic
Nuclear Medicine Technologist
Franklin Medical Consulters
Ann Arbor, Michigan

David Mintzer
Professor Emeritus
Department of Mechanical
  Engineering
Northwestern University
Evanston, Illinois

Christine Molinari
Science Editor
University of Chicago Press
Chicago, Illinois

Frank Mooney
Professor Emeritus
Fingerlake Community College
Canandaigua, New York

Partick Moore
Department of English
University of Arkansas at Little
  Rock
Little Rock, Arkansas

Robbin Moran
Department of Systematic Botany
Institute of Biological Sciences
University of Aarhus
Risskou, Denmark

J. Paul Moulton
Department of Mathematics
Episcopal Academy
Glenside, Pennsylvania

Otto H. Muller
Geology Department

Alfred University
Alfred, New York

Angie Mullig
Publication and Development
University of Pittsburgh Medical
  Center
Trafford, Pennsylvania

David R. Murray
Senior Associate
Sydney University
Sydney, New South Wales,
  Australia

Sutharchana Murugan
Scientist
Three Boehringer Mannheim
  Corp.
Indianapolis, Indiana

Muthena Naseri
Moorpark College
Moorpark, California

David Newton
Science Writer and Educator
Ashland, Oregon

F. C. Nicholson
Science Writer
Lynn, Massachusetts

James O'Connell
Department of Physical Sciences
Frederick Community College
Gaithersburg, Maryland

Dúnal P. O'Mathúna
Associate Professor
Mount Carmel College of
  Nursing
Columbus, Ohio

Marjorie Pannell
Managing Editor, Scientific
  Publications
Field Museum of Natural History
Chicago, Illinois

Gordon A. Parker
Lecturer
Department of Natural Sciences
University of Michigan-Dearborn
Dearborn, Michigan

David Petechuk
Science Writer
Ben Avon, Pennsylvania

Borut Peterlin, M.D.
Consultant Clinical Geneticist,
  Neurologist, Head Division of
  Medical Genetics
Department of Obstetrics and
  Gynecology
University Medical Centre
  Ljubljana
Ljubljana, Slovenia

John R. Phillips
Department of Chemistry
Purdue University, Calumet
Hammond, Indiana

Kay Marie Porterfield
Science Writer
Englewood, Colorado

Paul Poskozim
Chair
Department of Chemistry, Earth
  Science and Physics
Northeastern Illinois University
Chicago, Illinois

Andrew Poss
Senior Research Chemist
Allied Signal Inc.
Buffalo, New York

Satyam Priyadarshy
Department of Chemistry
University of Pittsburgh
Pittsburgh, Pennsylvania

Patricia V. Racenis
Science Writer
Livonia, Michigan

Cynthia Twohy Ragni
Atmospheric Scientist
National Center for Atmospheric
  Research
Westminster, Colorado

Jordan P. Richman
Science Writer
Phoenix, Arizona

Kitty Richman
Science Writer
Phoenix, Arizona

Vita Richman
Science Writer
Phoenix, Arizona

Michael G. Roepel
Researcher
Department of Chemistry
University of Pittsburgh
Pittsburgh, Pennsylvania

Perry Romanowski
Science Writer
Chicago, Illinois

Nancy Ross-Flanigan
Science Writer
Belleville, Michigan

Belinda Rowland
Science Writer
Voorheesville, New York

Gordon Rutter
Royal Botanic Gardens
Edinburgh, Great Britain

Elena V. Ryzhov
Polytechnic Institute
Troy, New York

David Sahnow
Associate Research Scientist
John Hopkins University
Baltimore, Maryland

Peter Salmansohn
Educational Consultant
New York State Parks
Cold Spring, New York

Peter K. Schoch
Instructor
Department of Physics and
    Computer Science
Sussex County Community
    College
Augusta, New Jersey

Patricia G. Schroeder
Instructor
Science, Healthcare, and Math
    Division
Johnson County Community
    College
Overland Park, Kansas

Randy Schueller
Science Writer
Chicago, Illinois

Kathleen Scogna
Science Writer
Baltimore, Maryland

William Shapbell Jr.
Launch and Flight Systems
    Manager
Kennedy Space Center
KSC, Florida

Kenneth Shepherd
Science Writer
Wyandotte, Michigan

Anwar Yuna Shiekh
International Centre for
    Theoretical Physics
Trieste, Italy

Raul A. Simon
Chile Departmento de Física
Universidad de Tarapacá
Arica, Chile

Michael G. Slaughter
Science Specialist
Ingham ISD
East Lansing, Michigan

Billy W. Sloope
Professor Emeritus
Department of Physics
Virginia Commonwealth
    University
Richmond, Virginia

Douglas Smith
Science Writer
Milton, Massachusetts

Lesley L. Smith
Department of Physics and
    Astronomy
University of Kansas
Lawrence, Kansas

Kathryn D. Snavely
Policy Analyst, Air Quality Issues
U.S. General Accounting Office
Raleigh, North Carolina

Charles H. Southwick
Professor
Environmental, Population, and
    Organismic Biology
University of Colorado at Boulder
Boulder, Colorado

John Spizzirri
Science Writer
Chicago, Illinois

Frieda A. Stahl
Professor Emeritus
Department of Physics
California State University, Los
    Angeles
Los Angeles, California

Robert L. Stearns
Department of Physics
Vassar College
Poughkeepsie, New York

Ilana Steinhorn
Science Writer
Boalsburg, Pennsylvania

David Stone
Conservation Advisory Services
Gai Soleil
Chemin Des Clyettes
Le Muids, Switzerland

Eric R. Swanson
Associate Professor
Department of Earth and Physical
    Sciences
University of Texas
San Antonio, Texas

Cheryl Taylor
Science Educator
Kailua, Hawaii

Nicholas C. Thomas
Department of Physical Sciences
Auburn University at
    Montgomery
Montgomery, Alabama

W. A. Thomasson
Science and Medical Writer
Oak Park, Illinois

Marie L. Thompson
Science Writer
Ben Avon, Pennsylvania

Laurie Toupin
Science Writer
Pepperell, Massachusetts

Melvin Tracy
Science Educator
Appleton, Wisconsin

Karen Trentelman
Research Associate
Archaeometric Laboratory
University of Toronto
Toronto, Ontario, Canada

Robert K. Tyson
Senior Scientist
W. J. Schafer Assoc.
Jupiter, Florida

James Van Allen
Professor Emeritus
Department of Physics and
  Astronomy
University of Iowa
Iowa City, Iowa

Julia M. Van Denack
Biology Instructor
Silver Lake College
Manitowoc, Wisconsin

Kurt Vandervoort
Department of Chemistry and
  Physics
West Carolina University
Cullowhee, North Carolina

Chester Vander Zee
Naturalist, Science Educator
Volga, South Dakota

Rashmi Venkateswaran
Undergraduate Lab Coordinator
Department of Chemistry
University of Ottawa
Ottawa, Ontario, Canada

R. A. Virkar
Chair
Department of Biological
  Sciences
Kean College
Iselin, New Jersey

Kurt C. Wagner
Instructor
South Carolina Governor's
  School for Science and
  Technology
Hartsville, South Carolina

Cynthia Washam
Science Writer
Jensen Beach, Florida

Terry Watkins
Science Writer
Indianapolis, Indiana

Joseph D. Wassersug
Physician
Boca Raton, Florida

Tom Watson
Environmental Writer
Seattle, Washington

Jeffrey Weld
Instructor, Science Department
  Chair
Pella High School

Pella, Iowa

Frederick R. West
Astronomer
Hanover, Pennsylvania

Glenn Whiteside
Science Writer
Wichita, Kansas

John C. Whitmer
Professor
Department of Chemistry
Western Washington University
Bellingham, Washington

Donald H. Williams
Department of Chemistry
Hope College
Holland, Michigan

Robert L. Wolke
Professor Emeritus
Department of Chemistry
University of Pittsburgh
Pittsburgh, Pennsylvania

Xiaomei Zhu, Ph.D.
Postdoctoral research associate
Immunology Department
Chicago Children's Memorial
  Hospital, Northwestern
  University Medical School
Chicago, Illinois

Jim Zurasky
Optical Physicist
Nichols Research Corporation
Huntsville, Alabama

Star cluster

Star clusters are groups of stars that occur close to each other in **space**, appear to have roughly similar ages, and therefore, seem to have had a common origin. Star clusters are typically classified into one of two large subgroups, galactic clusters and globular clusters. Galactic clusters are sometimes also known as open clusters. Astronomers have identified thousands of galactic star clusters in the **Milky Way**, but no more than about 200 globular star clusters.

The two types of star clusters found in the Milky Way differ from each other in a number of ways. First, galactic clusters occur in the **plane** of the **galaxy**, while globular clusters are found outside the galactic plane in the region known as the galactic halo. Second, globular clusters tend to be much larger than galactic clusters with an average of a few thousand to a million stars in the former and a few hundred stars in the latter. In fact, some galactic clusters contain no more than a half dozen stars. Probably the most famous of all galactic clusters is the Pleiades, or the Seven Sisters. This grouping consists of six or seven stars as seen by the naked **eye** (depending on the accuracy of one's eyesight), but of many more when viewed by **telescope**. Third, globular clusters, as their name suggests, tend to have a rather clearly defined spherical shape with a higher concentration of stars at the center of the **sphere**. In contrast, galactic clusters, as their alternative name also suggests, tend to be more open and lacking in any regular shape.

Fourth, the compositions of stars found in each kind of cluster are quite different. The stars that make up galactic clusters tend to consist primarily of **hydrogen** (more

The Pleiades open star cluster (M45), which is situated in the constellation Taurus. The Pleiades cluster is about 400 light years from Earth and is young (only about 50 million years old) on a galactic time scale. The cluster is still embedded in a cloud of cold gas and interstellar dust, material left over from its formation. *Photo Researchers, Inc. Reproduced by permission.*

than 90%) and helium (almost 10%), with small amounts of heavier elements (less than 1%). Stars in globular clusters, on the other hand, contain even smaller amounts of heavier elements. This difference suggests that the stars in galactic clusters are much younger than those in globular clusters. When the latter were formed, the universe still consisted almost entirely of hydrogen and helium, so those were the only elements used in the formation of globular cluster stars. Much later in the history of the universe, some heavier elements had been formed and were present at the formation of galactic cluster stars.

Star formation

Star formation is the process by which a cold, dark cloud of gas and dust is transformed into a brilliant star with a surface **temperature** anywhere from 3,000-50,000K (4,900–90,000°F; 2,700–50,000°C). Many regions of our **galaxy** are filled with cold **clouds** of gas that begin to contract, under certain conditions, as a result of their own gravitational attraction. As one of these clouds contracts, it heats up and tends to become spherical. The heating, however, produces **pressure** in the gas that counteracts the contraction, and eventually the contraction may stop if the gravity and gas pressure balance one another. If the cloud has become hot enough to begin thermonuclear fusion reactions at its center, it can then sustain itself against its own gravity for a long time. Such a cloud is then called a star.

The interstellar medium

When you look up on a clear night, you see stars—thousands of them—glittering against the seemingly empty backdrop of **space**. But there is something else out there; vast clouds of cold, dark gas and dust, visible only by the dimming effect they have on starlight shining through them. This is the *interstellar medium*, and it is the birthplace of the stars.

In most places the interstellar medium is almost a **vacuum**, a million trillion times less dense than air. In other places, however, there are much greater concentrations of clouds, sometimes so thick and dense that we cannot see through them at all. Such a cloud is the famous Horsehead Nebula in the **constellation** Orion. Often these clouds are enormous, thousands of times as massive as the **Sun**.

Unlike the Sun, however, these interstellar clouds have relatively weak gravity. The gravitational attraction between two particles decreases as the separation between them increases, and even in a huge cloud like the Horse-

An infrared image of the molecular cloud and region of star formation NGC 7538. The bright areas (just below center, and left of bottom center) are sites where stellar formation is taking place. Giant clouds of dust are fragmenting and forming new protostar systems; the ultraviolet radiation of the newborn stars is lighting up the surrounding dust. The larger hazy glow (right of top center) is radiation from the protostars pouring out from a gap in the dust clouds. *Photograph by Ian Gatley. Photo Researchers, Inc. Reproduced by permission.*

head Nebula, the **matter** is much more thinly distributed than in the Sun. Therefore, the matter in the cloud tends not to condense. It remains roughly the same size, slowly changing its shape over the course of millennia.

The birth of a star

Imagine a cloud, drifting along through the interstellar medium. The cloud is unthinkably cold, in excess of -400°F (-240°C). It is not very dense, but it is so large that it renders the stars behind it either invisible or as dim, red points of **light**. It is made mostly of **hydrogen**, and has had its present shape and size for thousands of years.

Then, one year, something happens. A hundred parsecs away (about 190 trillion miles), a star explodes. It is a **supernova**, the violent end to a massive star's life. An expanding, devastating blast races outward, forming a shock wave. It sweeps everything before it, clearing the space through which it passes of the interstellar medium. And eventually, it encounters the cloud.

The shock wave slams into the cloud. The cold gas and dust is violently compressed by the shock, and as the particles are squeezed together, their mutual gravitational attraction grows. So tightly are they now packed that they begin to coalesce under their own gravity. The shock has transformed the cloud: many parts are still thin and diffuse, but now there are multitudes of condensing blobs of gas. They did not contract by themselves before, but now they have been given the necessary impetus.

When a blob of gas condenses, **energy** is released, and one of the beautiful theorems of **physics** shows us that half the energy goes into heating the gas. So as the blobs in the disrupted cloud condense, they get progressively hotter. Eventually they begin to glow a dull red, much as an electric burner on your stove begins to glow when it becomes sufficiently hot.

This process of contraction cannot continue indefinitely. As the temperature in a contracting blob of gas becomes higher, the gas exerts a pressure that counteracts the inward **force** of gravity. At this point, perhaps millions of years after the shock wave slammed into the dark cloud, the contraction stops. If the blob of gas has become hot enough at its center to begin thermonuclear fusion of hydrogen into helium, it will remain in this stable configuration for millions or billions of years. It has become a star.

Nature is filled with symmetries, and this is one of the most enchanting. The death of one star triggers the birth of new stars. And what of the rest of the dead star, the expanding blast of gas and dust that encountered no interstellar clouds? Eventually it comes to a halt, cooling and fading into darkness, where it becomes part of the interstellar medium. Perhaps, millions of years in the future, a shock wave will plow into it.

Other methods of star formation

The scenario described above leads to a situation like that shown in the Great Orion Nebula. Brilliant, newly born stars blaze in the foreground, while the great cloud surrounding them glows in the background. This nebula glows because the intense **radiation** from the massive young stars near it is heating it. Contrast this with the Horsehead Nebula, which has no such sources of **heat** and therefore is dark.

These newly formed stars can themselves trigger star formation. Radiation—that is, light—exerts pressure on surrounding matter. The young stars in the Orion Nebula are huge by stellar standards, and their radiation is intense. Many of them lose **mass** continuously in a **stellar wind** that streams out into the cloud. After a few million years, the most massive of them will explode as supernovae. These effects can cause other parts of the neighboring cloud to begin contracting. Therefore, star formation might be able to bootstrap its way through an entire cloud, even if only part of the cloud is disrupted by a shock wave.

An interstellar cloud does not always have to be disrupted by a shock wave to form stars, however. Sometimes a cloud may collapse spontaneously, and the process describing this phenomenon was discovered by the astronomer James Jeans (1877-1947). Above the so-called "Jeans mass," which depends on the temperature and **density** of the cloud, a cloud will break up and contract spontaneously under its own gravity. Large clouds can break up into numerous cloudlets this way, and this process leads to the formation of *star clusters* such as the Pleiades. Often, two stars will form very close to one another, sometimes separated by a distance less than that from the **Earth** to the Sun. These *binary* systems, as well as *multiple* systems containing three to six stars, are quite common. They are more common, in fact, than single stars: most of the stars you see at night are actually binaries.

Current research on star formation

An important avenue of research involves studying the cycle of star births and deaths in the galaxy. Formation of stars depletes the interstellar medium, since some of its gas goes into making the stars. But then, as a star shines, a small part of its matter escapes its gravity and returns to the interstellar medium. More importantly, massive stars return a large fraction of their matter to the interstellar medium when they explode and die. This cycle of depletion and replenishment is critically important in understanding the types of stars we see in the galaxy, and the **evolution** of the galactic system as a whole.

The advent of powerful new telescopes like the **Hubble Space Telescope** has opened astronomer's eyes to new stars that may require new theories of formation. In 1997 the brightest star ever seen was discovered at the core of our own galaxy, the **Milky Way**. Named the Pistol Star, it has the energy of 10 million Suns and would fill the distance of the Earth's **orbit** around the Sun. The Pistol Star is about 25,000 light-years from Earth; it is so turbulent that its eruptions create a gas cloud four light-years across. It had been thought that a star so big could not have formed without blowing itself apart, and so the Pistol Star may require astronomers to reexamine their ideas about stellar formation, especially of supermassive stars near the centers of galaxies.

While some astronomers study the galactic or the interstellar medium, others study newly forming *protostars*. Protostars are hot, condensing blobs of gas that have not quite yet achieved starhood, and they are hard to observe for two reasons. First, the phase of star formation is quite short by astronomical standards, so there are not nearly as many protostars as there are fully formed stars. Second, protostars are often thickly shrouded by the remnants of the cloud from which they are forming. This makes them appear much dimmer, and so much harder to observe and study.

Fortunately, newly forming stars do have some observable characteristics. A protostar may be girdled by a disk of dust and gas, and an exciting possibility is that these disks are protoplanetary systems. Our own **solar system** is thought to have formed from such a disk that surrounded the newly forming Sun, and disks around other stars such as Beta Pictoris may be current sites of planetary formation. Additionally, a protostar with a disk may produce two "beams" of gas that stream outward from its poles along the lines of magnetic field associated with the disk. These so-called *bipolar outflows* are classic signatures of a protostar with a disk.

It is not necessary to observe only our own Milky Way Galaxy to find newly forming stars. Modern telescopes, including the Hubble Space Telescope, are used to study star-forming regions in other galaxies. High-resolution observations can detect individual stars in the Milky Way's **satellite** galaxies and in some other nearby galaxies. In more distant galaxies, the regions of heated gas produced by new stars are visible. Observations of star formation in other parts of the Universe help confirm and give us a broader perspective on our theories regarding star formation in our own celestial neighborhood.

See also Binary star; Gravity and gravitation; Interstellar matter; Star cluster; Stellar evolution.

Resources

Periodicals

Croswell, K. "Galactic Archaeology." *Astronomy* (July 1992): 28.

Meyer. "Quasars From a Complete Spectroscopic Survey." *Monthly Notices of the Royal Astronomical Society* 324, no. 2 (2001): 2001.

O' Dell, C.R., "Secrets of the Orion Nebula." *Sky & Telescope* (Dec 1994): 20.

Jeffrey C. Hall

Starburst galaxy

Billions of large, essentially independent groups of stars exist in the universe. These are called galaxies. A **galaxy** is labeled a starburst galaxy if an exceptionally high **rate** of **star formation** is found to be taking place within it. This often occurs in galaxies that are in the process of or have recently undergone a merging or collision with another galaxy. Although astronomers do not know exactly what causes starbursts in galaxies, or what creates observed ripples in gas and dust of the outer regions of some of them, both phenomena occur in collisions between galaxies. The gravitational pull of the stars of two galaxies passing close to one another seems to cause increased **star** formation activity and the rippling effect.

First identified by an excess of infrared (**heat**) **radiation** from dust within galaxies, violent star formation is usually associated with very distant galaxies. However, it is occurring in some of the nearby galaxies such as the closest starburst galaxy, NGC 253. Within a region 1,000 light-years (about six quadrillion miles) across, the **Hubble Space Telescope** (HST) high resolution camera has shown very bright star clusters, dust lanes, which trace regions of dense gas, and filaments of glowing gas in the starburst core of this galaxy. The HST has identified several regions of intense star formation, which include a bright, super-compact **star cluster**. This confirms the theory that stars are often born in dense clusters within starbursts. In NGC 253, dense gas coexists with and obscures the starburst core. Other measurements revealed unusual motions of the gas in the nucleus of NGC 253,

which seem to indicate a fast-rotating ring of cold gas as well as gas flowing outward from the nucleus. Similar features have been found in other starburst galaxies.

Ground-based telescopes have shown that the core of one starburst galaxy contains massive clumps of young stars and nebulous ripples in its outermost stellar disk. These observations were made long after any collision, but it is expected that this galaxy, and all starburst galaxies, will eventually settle down to reduced star formation levels, and may one day resemble normal **spiral** galaxies similar to the **Milky Way**.

Clint Hatchett

Starch *see* **Carbohydrate**

Starfish

Starfish are marine **invertebrates** in the phylum Echinodermata, which also includes **sea urchins**, brittle stars, sea lilies, and **sea cucumbers**. Starfish belong to the class Asteroidea, which includes 1,500 **species** inhabiting the shallow margins of all of the world's oceans. Starfish vary widely in appearance. Some species grow up to 3 ft (1 m) in diameter; others are barely 0.5 in (1.3 cm) across. Starfish come in a rainbow of colors including bright red, cobalt blue, yellows, and the familiar orange-brown.

Starfish are radially symmetrical with from 5 to 50 arms radiating from a central disk. The skin of starfish is thick with bony plates (ossicles), spines, tiny pincers on stalks (the pedicillerae which keep the animal's skin clean of debris), and bumps, between which are tiny folds of skin which function as the starfish's gills.

The **nervous system** of starfish consists of three main networks: the ectoneural (oral), the hyponeural (deep oral), and the entoneuoral (aboral) systems. There is no central ganglion, but this rather simple arrangement effectively allows the starfish to move (including the ability to right itself should it be turned over) and sense the world around it.

The eyes of starfish are extremely simple, are located at the tip of each arm, and are primarily light-sensing dots. Starfish can tell **light** from dark, but are unlikely to see much more than that. The sense of **smell**, however, is quite sensitive. Chemoreceptors on the starfish's skin can detect the faintest smell of its **prey** (clams), and even determine the direction from which it is coming. The starfish then sets off to catch its prey, slowly and deliberately, at the rate of 6 in (15.25 cm) per minute. As it moves it does not pinwheel, but follows one arm. The underside of each arm is lined with hundreds of tiny tube feet. Each tube foot ends in a suction cup, and is the terminal point of an elaborate hydraulic system within the **animal**.

This hydraulic system has as its starting point a small reddish spot on the top of the central disk, the madreporite. The madreporite is comparable to the drain of a sink, as it serves as the entry for **water** into the stone **canal**, which joins the ring canal, off which radiate the tubes that run down the starfish's arms and branch off into the tube feet. Movement is an elaborate process for a creature with so many feet, which are extended and placed on the substratum by filling each tube foot with water. To attach the tube foot, the starfish creates suction by drawing water out again and closing a tiny valve on the ampulla, a bulb at the top of the tube. The animal then contracts a muscle and draws itself forward on its tube feet, which are tremendously strong and able to keep a starfish clinging to **rocks** in all but the heaviest storms.

Starfish also use their tube feet to prey on bivalve molluscs. When a starfish encounters a clam, it attaches its tube feet and begins to pull. It can pull for hours, or even days. Eventually, the clam's adductor muscle that keeps the shell closed tires under this relentless tug, and the clam's shell opens a bit. The starfish does not need much of an opening-just enough to get part of its **digestive system** in. Starfish have two stomachs, one that remains inside the body and another than can be protruded through the starfish's mouth on the underside of the body. The starfish inserts this into the clam's shell, and releases digestive enzymes.

Because starfish lack teeth, they must convert their food to liquid form before they can ingest it. Among the other prey items this bottom-dwelling **predator** eats are sea urchins, other starfish, small **fish**, **sponges**, and carrion. Some species draw in mud as they crawl along the bottom and extract organic material from the mud. One genus, *Acanthaster*, (the crown-of-thorns starfish) has become famous for the damage it does to coral reefs, moving over the reef and stripping it clean of coral polyps. (The overabundance of the crown-of-thorns starfish can be partly attributed to the reduction in the population of its major predator, the giant triton, by humans.) Starfish have long been the bane of shellfishermen. In an effort to kill starfish, shellfishermen hack them to pieces and throw the pieces back into the sea. Unknown to humans, all that was being done was the creation of more starfish, since starfish have remarkable regenerative abilities; all species can regenerate lost arms, and some can produce a whole new starfish from an arm with a piece of the central disk attached.

An ochre sea star (*Pisaster ochraceus*) on the California coast. *Photograph by Robert J. Huffman. Field Mark Publications. Reproduced by permission.*

Although such regeneration is a form of reproduction, starfish generally reproduce by shedding eggs and sperm into the water. Once one female releases her eggs (up to 2.5 million at a time), other starfish nearby join in a kind of breeding frenzy, all releasing their sperm or eggs. The eggs float free with the **plankton** and develop into bipinnaria larvae, which remain free floating for another three weeks. They then settle to the bottom and metamorphose into the familiar **star** shapes.

See also Brittle star; Sea lily.

Resources

Books

Brusca, Richard C., and Gary J. Brusca. *Invertebrates.* Sunderland, MA: Sinaur Associates, 1990.

Whiteman, Kate. *World Encyclopedia of Fish & Shellfish.* New York: Lorenz Books, 2000.

F. C. Nicholson

Starlings

Starlings are robust, stocky **song birds** in the family Sturnidae. They have a stout beak and strong legs, and are included with other perching **birds** in the order Passeriformes. There are about 110 **species** of starlings, whose natural range includes Eurasia, **Africa**, the Pacific islands, and **Australia**. Starlings are small- to medium-sized birds, ranging in body length from about 4-17 in (10-43 cm), and are mostly found in **forests**, shrubby woodlands, and urban and suburban habitats. Starlings tend to be fast, direct fliers. Most species form flocks during the non-breeding season, and most northern species are migratory to some degree. Their songs are

A flock of starlings swarming telephone poles. *Photograph by Richard R. Hansen. Photo Researchers, Inc. Reproduced by permission.*

Golden-breasted starling perched on a branch. *Photograph by Tom & Pat Leeson. Photo Researchers, Inc. Reproduced by permission.*

usually inventive and consist of garrulous chatters of whistles, squeaks, and imitated sounds. Starlings feed widely on small **invertebrates** and **fruits**. Most species nest in cavities in trees or **rocks**, and both sexes cooperate in the feeding and rearing of the young.

Most species of starlings, including the mynah bird, are distributed in tropical regions. Some of these are extremely beautiful birds. In Africa, for example, some of the most attractive bird species are starlings, with their brilliant metallic-green, blue, purple, and violet plumage. Notable are the long-tailed glossy starling (*Lamprotornis caudtus*), the chestnut-bellied starling (*Spreo pulcher*), and the superb starling (*Spreo superbus*). The African starlings also include the oxpeckers (*Buphagus*), which glean ticks and blood-sucking **flies** off the back of large **mammals**.

Many species of starlings are endangered because of the widespread destruction of their natural **habitat** (tropical forest, **savanna**, or shrubland). For example, the beautifully white Rothschild's mynah (*Leucospar rothschildi*) of Bali, Indonesia, is endangered because its natural forest has been extensively cleared and converted into agricultural use.

Starlings in North America

One of the world's most widely introduced birds is the European or common starling (*Sturnus vulgaris*), that now occurs virtually worldwide in temperate re-

gions of Eurasia, **North America**, and Australia. This starling was first successfully introduced to North America in 1890 in Central Park, New York City, when 60 birds were released. There had been earlier releases of common starlings by homesick European immigrants, but these had failed to establish a breeding population. However, once the common starling became locally established in New York, it expanded its range explosively, and this species now occurs throughout most of temperate North America. In recent decades the European starling has consistently been the most numerous species tallied on the annual Christmas bird counts, and it may now be the most abundant species of bird in North America.

The European starling is an attractive bird, especially during the late winter to summer when it bears its dark-glossy, spotted, nuptial plumage. These short-tailed birds flock together during the non-breeding season, and they sometimes occur in huge aggregations of hundreds of thousands of birds. The European starling forages widely for invertebrates, especially **insects** living in the ground or in grass. During winter this bird mostly eats grains and other **seeds**. Although not a very accomplished singer, the renditions of the European starling are interesting, rambling assemblages of squeaks, whistles, gurgles, and imitations of the songs of other birds, and also of other sounds, such as squeaky clotheslines. Because the European starling is so common and lives in cities and most agricultural areas, it is possibly the most frequently heard and seen bird in North America, and also in much of the rest of the temperate world.

Another starling introduced to North America is the crested mynah (*Acridotheres cristatellus*), released in Vancouver in the 1890s, where it became established but did not spread more widely.

Importance of starlings

A few species of starlings are considered to be important **pests**. For example, in North America the European starling is widely regarded as a problem when it occurs in large numbers. This species has contributed to the decline of some native species of birds, by competitively displacing them from nesting cavities, which are usually in short supply. Various native birds have been affected by this **competition** with starlings, including the eastern and mountain **bluebirds** (*Sialia sialis* and *S. mexicanus*, respectively), the **tree** swallow (*Iridoprocne bicolor*), and the red-headed woodpecker (*Melanerpes erythrocephalus*). The European starling can also foul buildings with its excrement, corroding stone and metals and creating a health hazard to people through exposure to pathogenic **fungi**. In addition, the European starling sometimes causes agricultural damage, especially to certain tender fruits, such as cherries. For similar reasons, the Indian mynah (*Acridotheres tristis*) is often considered a pest in tropical regions.

However, these abundant species of starlings are also beneficial in some respects, because they eat large numbers of potentially injurious insects, such as cutworms and other beetle larvae that can damage lawns. The European starling and Indian mynah are also among the few non-human animals that can tolerate urban environments, and these birds provide an aesthetic benefit in cities.

A few species of starlings are easily bred in captivity, and are important in the pet trade. The best known example of this is the hill mynah (*Gracula religiosa*), native to South and Southeast **Asia** and widely kept as a pet. This attractive species maintains a busy and noisy chatter, and can be easily trained to mimic human words and phrases.

See also Introduced species.

Bill Freedman

States of matter

Matter includes all the material that makes up the universe. It has **mass** and it takes up space. It includes everything around us: the food we eat, the **water** we drink, the air we breathe, the ores deep within the **earth**, as well as the atmosphere above it, the substances that make up the **moon**, and the stars as well as the dust in the tail of a comet. It is fairly easy to observe that matter exists in different forms or states: solids, liquids, gases, and the less familiar **plasma** state.

Nature of matter

All matter is composed of very small, discrete particles, either **atoms**, ions, or molecules. The nature of a particular substance depends on the type and arrangement of the atoms within the **molecule**.

It is possible for these particles to assume different arrangements in space. For example, they can be arranged close together or far apart. They can be neat and orderly or **random** and disordered. Since two particles cannot occupy the same place at the same **time**, they can be pushed closer together only if there is empty space between particles. Sometimes they slip and slide past each other and sometimes they are locked rigidly into a specific position. The state in which any particular piece of matter exists depends on these properties. Under the right conditions, any substance can exist in all of the states of matter: solid, liquid, gas, or plasma.

The atoms, molecules, and ions that make up all matter are in constant **motion**, which can range from vibrating within a fairly rigid position to moving randomly at very high speeds. Movement is always in a straight line until some other **force** interferes with the motion. Like billiard balls, the moving particles can hit other particles or objects such as the walls of their container. These collisions cause the particles to change direction and, although no **energy** is lost, it can be transferred to other particles.

Various forces exist between the particles of matter. The degree of attraction or repulsion depends on such factors as whether the particles are electrically neutral or carry a charge, whether the charges are localized or balanced out, how big or small the particles are, and how far apart the particles are from each other.

Solids

Matter is said to be in the solid state when it is rigid, that is, when it retains a definite shape and **volume** against the pull of gravity. Strong attractive forces exist among the particles that make up solids, causing them to position themselves close together in an orderly and definite arrangement in space. Their motion consists mainly of vibrating in a fixed position so the shape and the volume (amount of space they occupy) are maintained. The

A piece of SEAgel sitting on soap bubbles. SEAgel (Safe Emulsion Agar gel) is a material claimed to be the first lighter-than-air solid. It is made from agar, a seaweed derivative used as a thickening agent in the food industry, and is biodegradable. SEAgel could be used as a thermal insulator instead of plastic foam or balsa wood. High density SEAgel could be used instead of plastic packaging. The substance is soluble in water above 122°F (50°C). *Lawrence Livermore National Laboratory/Science Photo Library, National Audubon Society Collection/Photo Researchers, Inc. Reproduced by permission.*

atoms, ions, or molecules cannot be pushed closer together; therefore, solids cannot be compressed.

Many solids exist in the form of crystals, which have simple geometric shapes, reflecting the regular spatial arrangement forms and shapes depending on the arrangement of the atoms, ions, or molecules of which they are made. This arrangement is called a lattice. Other solids, such as lumps of clay, seem to have no preferred shapes at all. They are said to be amorphous (without form). This is true because the individual crystals may be very tiny or because the substance consists of several kinds of crystals, randomly mixed together. Other solids, such as **glass**, contain no crystals at all.

When solids are cooled to lower temperatures, the orderly arrangement of their particles stays basically the same. The vibrations become slightly slower and weaker, however, causing the particles to move closer together, and the solid contracts slightly. But when solids are heated, the vibrations become faster and broader. This causes the particles to move slightly farther apart, and the solid expands a little. If heated enough, the particles will vibrate so vigorously that the rigid structure can no longer be maintained. The lattice begins to fall apart, first into clumps, and eventually into individual particles which can slip and slide past each other as they move about freely. At this point, the solid has become a liquid.

The **temperature** at which a solids loses its rigid form and turns into a liquid is called the melting point. Different substances have different melting points that are dependent on the sizes of the particles and the strength of the attractions between the particles. In general, heavier particles require more energy (higher temperatures) in order to vibrate vigorously enough to come apart. Also, the stronger the attractions between the particles, the more energy is required to break them apart and change the solid into a liquid. In both cases—heavier particles and stronger attractions—the melting point will be higher. Water serves as a good example. Liquid water freezes at the same temperature that **ice** melts, and the melting and freezing points are therefore identical. This is true for all substances. Ice melts at 32°F (0°C) which is uncharacteristically high for particles the weight of water molecules. This unusually high melting/freezing point is caused by the very strong attractive forces that exist between the molecules, making it very difficult for particles to move away from their neighbors and for the crystalline structure to collapse. Metals melt at much higher temperatures than ice. For example, **copper** is made into various shapes by melting it at 1,985°F (1,085°C), pouring it into molds, and cooling it. It is then usually purified further by **electrolysis** before it is commercially useful. Since pure substances have unique melting points which are usually quite easy to determine, chemists often use them as the first step in identifying unknown substances.

The amount of energy required to change a solid to a liquid varies from substance to substance. This energy is called the **heat** of fusion. Ice, for example, must absorb 80 calories per gram in order to melt into liquid water. Similarly, water releases 80 calories per gram of water to freeze into ice. Each of these changes occurs at the melting/freezing point of water, 32°F (0°C). In melting, since all the heat energy is used up in breaking the crystalline lattice, there is no change in temperature. However, once all the ice has melted, the absorbed energy causes the temperature of the liquid water to rise. This is generally true of the melting of all solids.

Liquids

The change from solid to liquid is a physical rather than chemical change because no chemical bonds have been broken. The individual particles—atoms, ions, or molecules—that made up the solid are the same individual particles that make up the liquid. What does change is the arrangement of the particles. In the liquid, the particles are at a higher temperature, having more energy than in the solid, and this allows them to move away from their nearest neighbors. The attractions between liquid particles, though less than those of solids, is still fairly strong. This keeps the particles close to each other and touching, even though they can around past one another. They cannot be pushed closer together, and so, like solids, liquids maintain their volume and cannot be compressed. Because their particles move freely around, liquids can flow, and they will assume the shape of any container.

Like solids, the particles of liquids are close to each other; therefore, the amount of space occupied by liquids is quite close to that of their corresponding solids. However, because of the disorderly arrangement, the empty space between the liquid particles is usually slightly greater than that between the particles of the solid. Therefore, liquids usually have a slightly larger volume-that is, they are less dense-than solids. A very unusual exception to this is the case of ice melting to form water, when the volume actually decreases. The crystalline lattice of ice has a cage-like structure of H_2O molecules with big, open spaces in the middle of the cages. When the ice melts and the **crystal** breaks down, the cages collapse and the molecules move closer together, taking up less space. Consequently, a given weight of water occupies more volume as ice than as liquid. In other words, ice is less dense than water. Therefore, ice floats on liquid water. Also, a full, closed container of water will break as it freezes because the ice must expand. A water pipe may break if it freezes in winter because of this unusual property of water.

Boiling

As the temperature of a liquid is increased, the particles gain more energy and move faster and faster. Jostling about and colliding increases until eventually the particles at the surface gain enough energy to overcome the attractive forces from their neighbors and break away into the surrounding space. At this point, the liquid is becoming a gas (also called a vapor). The temperature at which this happens depends on what the substance is. This temperature, known as the **boiling point**, remains constant during the entire process of boiling because the added heat is being used up to break the attraction between the particles. The reverse process, condensation, occurs at the same temperature as boiling. Like the melting point, the boiling point is unique for each pure substance, and can be used as an analytical tool for determining the identities of unknown substances.

The amount of energy required for a given amount of a liquid to vaporize or become a gas is called the heat of vaporization (or condensation). It varies from substance to substance because the particle of different substances may be heavier or lighter and may exert different attractive forces. The amount of energy absorbed when 1 gram of water completely changes to a vapor is 540 calories. Conversely, 540 calories are released when 1 gram of water vapor changes back to liquid.

When a liquid reaches the boiling point, particles on the surface actually gain enough energy to break away from the surface. But as heating continues, particles

Gallium melts at 86°F (30° C). © *Yoav Levy/Phototake NYC. Reproduced with permission.*

throughout the liquid are also increasing in energy and moving faster. In a body of the liquid, however, the particles cannot escape into the air, as those on the surface can. That is not only because they happen to be buried deep down below the surface. It is also because the atmosphere is pushing down on the entire liquid and all the particles within it, and, in order to break away, these particles deep within the liquid must acquire enough energy to overcome this additional **pressure**. (The surface particles can just fly off into the spaces between the air molecules.) When a group of interior particles finally do get enough energy-get hot enough-to overcome the **atmospheric pressure**, they can push each other away, leaving a hollow space within the liquid. This is a bubble. It is not entirely empty, however, because it contains many trapped particles, flying around inside. The light-weight bubble then rises through the liquid and breaks at the surface, releasing its trapped particles as vapor. We then say the liquid is boiling.

Since the pressure inside the bubbles must overcome atmospheric pressure in order for the bubbles to form, the boiling point of a substance depends on atmospheric pressure. Liquids will boil at lower temperatures if the atmospheric pressure is lower, as it is on a mountain. At the top of Mount Everest, 29,000 ft (8,839 m) above **sea level**, where the pressure is only about one-third that at sea level, water boils at 158°F (70°C). At 10,000 ft (3,048 m) above sea level, water boils at 192°F (89°C). It would take longer to cook an egg where the boiling point is 192°F (89°C) than at sea level where the boiling point is 212°F (100°C). The normal boiling point of a liquid is defined as its boiling point when the atmospheric pressure is exactly 760 mm Hg, or 1 atmosphere.

With the diminishing supplies of fresh water today, it is increasingly important to find ways of desalinating-

removing the **salt** from sea water in order to make it useful for human consumption, agriculture, and industry. Changes in state, both boiling and freezing, are useful for this purpose. When salt water is heated to boiling and the vapors cooled, they condense to form water again, but the salt stays behind in a very salty residue called brine. By this process, called **distillation**, **freshwater** has been recovered from salt water. Similarly, when salt water freezes, much of the salt stays behind as a very salty slush. The ice is removed from the brine and melted to produce relatively fresh water.

Gases

When a substance has reached the gaseous state, the particles are moving at relatively high speeds, and in straight lines, until they encounter other particles or some other barrier. The spaces between the particles are many times the size of the particles themselves. Generally, gas particles travel large distances through space before colliding with another particle or object. When colliding, although energy can be lost by one particle, it is gained by another and there is no net gain or loss of energy.

Because the particles are flying freely in the gaseous state, gases will fill whatever space is available to them. Thus, 100, 1000, or ten million particles of gas in a container spread out and fill the entire container.

Plasma

Plasmas are considered by some to be the fourth phase of matter. They are closely related to gases. In a plasma, the particles are neither atoms nor molecules, but electrons and positive ions. Plasmas can be formed at very high temperatures-high enough to ionize (remove electrons from) the atoms. The resulting electrons and positive ions can then move freely, like the particles in a gas. Although not found on Earth except in the outermost atmosphere, plasmas are probably more prevalent in the universe than all of the other three states of matter. The stars, comets' tails, and the aurora borealis are all plasmas. Because their particles are electrically charged, plasmas are greatly influenced by electric and magnetic fields.

Much research today involves the study of plasmas and the ability to control them. One possible method for producing enormous amounts of energy through **nuclear fusion** involves the production and control of plasmas.

See also Density; Desalination; Evaporation; Gases, properties of.

Resources

Books

Caro, Paul. *Water.* New York: McGraw Hill, 1993.

Close, Frank. *Too Hot to Handle: The Race for Cold Fusion.* Princeton: Princeton University Press, 1991.

Periodicals

Burgess, David. "Stronger than Atoms." *New Scientist* 140, (November 1993): 28-33.

Fortman, John J. "States of Matter." *Journal of Chemical Education* 70, (January 1993): 56-57.

Leona B. Bronstein

Static electricity *see* **Electrostatic devices**

Statistical mechanics

Statistical mechanics is a sub-branch of **physics** that attempts to explain the behavior of a macroscopic system based on the behavior and properties of that system's microscopic elements.

The number of these microscopic elements is usually very large, and it is impossible to accurately predict the behavior of each of these elements as they interact. However, the large number of interactions makes it theoretically possible for statistical mechanics to predict the behavior of the system as a whole.

Statistics

Statistics is that branch of **mathematics** devoted to the collection, compilation, display, and interpretation of numerical data. In general, the **field** can be divided into two major subgroups, *descriptive statistics* and *inferential statistics*. The former subject deals primarily with the accumulation and presentation of numerical data, while the latter focuses on predictions that can be made based on those data.

Some fundamental concepts

Two fundamental concepts used in statistical analysis are population and **sample**. The term *population* refers to a complete set of individuals, objects, or events that belong to some category. For example, all of the players who are employed by Major League Baseball teams make up the population of professional major league baseball players. The term *sample* refers to some subset of a population that is representative of the total population. For example, one might go down the com-

TABLE 1. STATISTICS

Number of Female African-Americans in Various Age Groups

Age	Number
0 - 19	5,382,025
20 - 29	2,982,305
30 - 39	2,587,550
40 - 49	1,567,735
50 - 59	1,335,235
60 +	1,606,335

plete list of all major league baseball players and select every tenth name. That subset of every tenth name would then make up a sample of all professional major league baseball players.

Another concept of importance in statistics is the distinction between discrete and continuous data. Discrete variables are numbers that can have only certain specific numerical value that can be clearly separated from each other. For example, the number of professional major league baseball players is a discrete **variable**. There may be 400 or 410 or 475 or 615 professional baseball players, but never 400.5, 410.75, or 615.895.

Continuous variables may take any value whatsoever. The readings on a **thermometer** are an example of a continuous variable. The **temperature** can range from 10°C to 10.1°C to 10.2°C to 10.3°C (about 50°F) and so on upward or downward. Also, if a thermometer accurate enough is available, even finer divisions, such as 10.11°C, 10.12°C, and 10.13°C, can be made. Methods for dealing with discrete and continuous variables are somewhat different from each other in statistics.

In some cases, it is useful to treat continuous variable as discrete variables, and vice versa. For example, it might be helpful in some kind of statistical analysis to assume that temperatures can assume only discrete values, such as 5°C, 10°C, 15°C (41°F, 50°F, 59°F) and so on. It is important in making use of that statistical analysis, then, to recognize that this kind of assumption has been made.

Collecting data

The first step in doing a statistical study is usually to obtain raw data. As an example, suppose that a researcher wants to know the number of female African-Americans in each of six age groups (1-19; 20-29; 30-39; 40-49; 50-59; and 60+) in the United States. One way to answer that question would be to do a population survey, that is, to in-

terview every single female African-American in the United States and ask what her age is. Quite obviously, such a study would be very difficult and very expensive to complete. In fact, it would probably be impossible to do.

A more reasonable approach is to select a sample of female African-Americans that is smaller than the total population and to interview this sample. Then, if the sample is drawn so as to be truly representative of the total population, the researcher can draw some conclusions about the total population based on the findings obtained from the smaller sample.

Descriptive statistics

Perhaps the simplest way to report the results of the study described above is to make a table. The advantage of constructing a table of data is that a reader can get a general idea about the findings of the study in a brief glance.

Graphical representation

The table shown above is one way of representing the frequency distribution of a sample or population. A *frequency distribution* is any method for summarizing data that shows the number of individuals or **individual** cases present in each given **interval** of measurement. In the table above, there are 5,382,025 female African-Americans in the age group 0-19; 2,982,305 in the age group 20-29; 2,587,550 in the age group 30-39; and so on.

A common method for expressing frequency distributions in an easy-to-read form is a graph. Among the kinds of graphs used for the display of data are histograms, bar graphs, and line graphs. A histogram is a graph that consists of solid bars without any space between them. The width of the bars corresponds to one of the variables being presented, and the height of the bars to a second variable. If we constructed a histogram based on the table shown above, the graph would have six bars, one for each of the six age groups included in the study. The height of the six bars would correspond to the frequency found for each group. The first bar (ages 0-19) would be nearly twice as high as the second (20-29) and third (30-39) bars since there are nearly twice as many individuals in the first group as in the second or third. The fourth, fifth, and six bars would be nearly the same height since there are about the same numbers of individuals in each of these three groups.

Another kind of graph that can be constructed from a histogram is a frequency polygon. A frequency polygon can be made by joining the midpoints of the top lines of each bar in a histogram to each other.

Distribution curves

Finally, think of a histogram in which the vertical bars are very narrow...and then very, very narrow. As one connects the midpoints of these bars, the frequency polygon begins to look like a smooth **curve**, perhaps like a high, smoothly shaped hill. A curve of this kind is known as a *distribution curve*.

Probably the most familiar kind of distribution curve is one with a peak in the middle of the graph that falls off equally on both sides of the peak. This kind of distribution curve is known as a "normal" curve. Normal curves result from a number of **random** events that occur in the world. For example, suppose you were to flip a penny a thousand times and count how many times heads and how many times tails came up. What you would find would be a normal distribution curve, with a peak at equal heads and tails. That means that, if you were to flip a penny many times, you would most commonly expect equal numbers of heads and tails. But the likelihood of some other distribution of heads and tails—such as 10% heads and 90% tails—would occur much less often.

Frequency distributions that are not normal are said to be skewed. In a skewed distribution curve, the number of cases on one side of the maximum is much smaller than the number of cases on the other side of the maximum. The graph might start out at **zero** and rise very sharply to its maximum point and then drop down again on a very gradual slope to zero on the other side. Depending on where the gradual slope is, the graph is said to be skewed to the left or to the right.

Other kinds of frequency distributions

Bar graphs look very much like histograms except that gaps are left between adjacent bars. This difference is based on the fact that bar graphs are usually used to represent discrete data and the space between bars is a reminder of the discrete character of the data represented.

Line graphs can also be used to represent continuous data. If one were to record the temperature once an hour all day long, a line graph could be constructed with the hours of day along the horizontal axis of the graph and the various temperatures along the vertical axis. The temperature found for each hour could then be plotted on the graph as a point and the points then connected with each other. The assumption of such a graph is that the temperature varied continuously between the observed readings and that those temperatures would fall along the continuous line drawn on the graph.

A **circle** graph, or "pie chart," can also be used to graph data. A circle graph shows how the total number of individuals, cases or events is divided up into various categories. For example, a circle graph showing the population of female African-Americans in the United States would be divided into pie-shaped segments, one (0-19) twice as large as the next two (20-20 and 30-39), and three about equal in size and smaller than the other three.

Measures of central tendency

Both statisticians and non-statisticians talk about "averages" all the **time**. But the term average can have a number of different meanings. In the field of statistics, therefore, workers prefer to use the term "measure of central tendency" for the concept of an "average." One way to understand how various measures of central tendency (different kinds of "average") differ from each other is to consider a classroom consisting of only six students. A study of the six students shows that their family incomes are as follows: $20,000; $25,000; $20,000; $30,000; $27,500; $150,000. What is the "average" income for the students in this classroom?

The measure of central tendency that most students learn in school is the *mean*. The mean for any set of numbers is found by adding all the numbers and dividing by the quantity of numbers. In this example, the mean would be equal to ($20,000 + $25,000 + $20,000 + $30,000 + $27,500 + $150,000) ÷ 6 = $45,417. But how much useful information does this answer give about the six students in the classroom? The mean that has been calculated ($45,417) is greater than the household income of five of the six students.

Another way of calculating central tendency is known as the *median*. The **median** value of a set of measurements is the middle value when the measurements are arranged in order from least to greatest. When there are an even number of measurements, the median is half way between the middle two measurements. In the above example, the measurements can be rearranged from least to greatest: $20,000; $20,000; $25,000; $27,500; $30,000; $150,000. In this case, the middle two measurements are

TABLE 2. STATISTICS			
	Improved	*Not Improved*	*Total*
Experimental Group	62	38	100
Control Group	45	55	100
Total	107	93	200

$25,000 and $27,500, and half way between them is $26,250, the median in this case. You can see that the median in this example gives a better view of the household incomes for the classroom than does the mean.

A third measure of central tendency is the mode. The **mode** is the value most frequently observed in a study. In the household income study, the mode is $20,000 since it is the value found most often in the study. Each measure of central tendency has certain advantages and disadvantages and is used, therefore, under certain special circumstances.

Measures of variability

Suppose that a teacher gave the same test to two different classes and obtained the following results: Class 1: 80%, 80%, 80%, 80%, 80% Class 2: 60%, 70%, 80%, 90%, 100% If you calculate the mean for both sets of scores, you get the same answer: 80%. But the collection of scores from which this mean was obtained was very different in the two cases. The way that statisticians have of distinguishing cases such as this is known as measuring the variability of the sample. As with measures of central tendency, there are a number of ways of measuring the variability of a sample.

Probably the simplest method is to find the range of the sample, that is, the difference between the largest and smallest observation. The range of measurements in Class 1 is 0, and the range in class 2 is 40%. Simply knowing that fact gives a much better understanding of the data obtained from the two classes. In class 1, the mean was 80%, and the range was 0, but in class 2, the mean was 80%, and the range was 40%.

Other measures of variability are based on the difference between any one measurement and the mean of the set of scores. This measure is known as the deviation. As you can imagine, the greater the difference among measurements, the greater the variability. In the case of Class 2 above, the deviation for the first measurement is 20% (80%-60%), and the deviation for the second measurement is 10% (80%-70%).

Probably the most common measures of variability used by statisticians are the **variance** and standard deviation. Variance is defined as the mean of the squared deviations of a set of measurements. Calculating the variance is a somewhat complicated task. One has to find each of the deviations in the set of measurements, **square** each one, add all the squares, and divide by the number of measurements. In the example above, the variance would be equal to $[(20)^2 + (10)^2 + (0)^2 + (10)^2 + (20)^2] 4 \div 5 = 200$.

For a number of reasons, the variance is used less often in statistics than is the standard deviation. The standard deviation is the **square root** of the variance, in this case, $\sqrt{+200} = 14.1$. The standard deviation is useful because in any normal distribution, a large fraction of the measurements (about 68%) are located within one standard deviation of the mean. Another 27% (for a total of 95% of all measurements) lie within two standard deviations of the mean.

Inferential statistics

Expressing a collection of data in some useful form, as described above, is often only the first step in a statistician's work. The next step will be to decide what conclusions, predictions, and other statements, if any, can be made based on those data. A number of sophisticated mathematical techniques have now been developed to make these judgments.

An important fundamental concept used in inferential statistics is that of the null hypothesis. A null hypothesis is a statement made by a researcher at the beginning of an experiment that says, essentially, that nothing is happening in the experiment. That is, nothing other than natural events are going on during the experiment. At the conclusion of the experiment, the researcher submits his or her data to some kind of statistical analysis to see if the null hypothesis is true, that is, if nothing other than normal statistical variability has taken place in the experiment. If the null hypothesis is shown to be true, than the experiment truly did not have any effect on the subjects.

KEY TERMS

Continuous variables—A variable that may take any value whatsoever.

Deviation—The difference between any one measurement and the mean of the set of scores.

Discrete variable—A number that can have only certain specific numerical value that can be clearly separated from each other.

Frequency polygon—A type of frequency distribution graph that is made by joining the midpoints of the top lines of each bar in a histogram to each other.

Histogram—A bar graph that shows the frequency distribution of a variable by means of solid bars without any space between them.

Mean—A measure of central tendency found by adding all the numbers in a set and dividing by the quantity of numbers.

Measure of central tendency—Average.

Measure of variability—A general term for any method of measuring the spread of measurements around some measure of central tendency.

Median—The middle value in a set of measurements when those measurements are arranged in sequence from least to greatest.

Mode—The value that occurs most frequently in any set of measurements.

Normal curve—A frequency distribution curve with a symmetrical, bellshaped appearance.

Null hypothesis—A statistical statement that nothing unusual is taking place in an experiment.

Population—A complete set of individuals, objects, or events that belong to some category.

Range—The set containing all the values of the function.

Standard deviation—The square root of the variance.

If the null hypothesis is shown to be false, then the researcher is justified in putting forth some alternative hypothesis that will explain the effects that were observed. The role of statistics in this process is to provide mathematical tests to find out whether or not the null hypothesis is true or false.

A simple example of this process is deciding on the effectiveness of a new medication. In testing such medications, researchers usually select two groups, one the control group and one the experimental group. The control group does not receive the new medication; it receives a neutral substance instead. The experimental group receives the medication. The null hypothesis in an experiment of this kind is that the medication will have no effect and that both groups will respond in exactly the same way, whether they have been given the medication or not.

Suppose that the results of one experiment of this kind was as follows, with the numbers shown being the number of individuals who improved or did not improve after taking part in the experiment.

At first glance, it would appear that the new medication was at least partially successful since the number of those who took it and improved (62) was greater than the number who took it and did not improve (38). But a statistical test is available that will give a more precise answer, one that will express the probability (90%, 75%, 50%, etc.) that the null hypothesis is true. This test,

called the **chi square test,** involves comparing the observed frequencies in the table above with a set of expected frequencies that can be calculated from the number of individuals taking the tests. The value of chi square calculated can then be compared to values in a table to see how likely the results were due to chance and how likely to some real affect of the medication.

Another example of a statistical test is called the Pearson correlation coefficient. The Pearson correlation **coefficient** is a way of determining the extent to which two variables are somehow associated, or correlated, with each other. For example, many medical studies have attempted to determine the connection between smoking and lung **cancer**. One way to do such studies is to measure the amount of smoking a person has done in her or his lifetime and compare the rate of lung cancer among those individuals. A mathematical formula allows the researcher to calculate the Pearson correlation coefficient between these two sets of data-rate of smoking and risk for lung cancer. That coefficient can range between 1.0, meaning the two are perfectly correlated, and -1.0, meaning the two have an inverse relationship (when one is high, the other is low).

The correlation test is a good example of the limitations of statistical analysis. Suppose that the Pearson correlation coefficient in the example above turned out to be 1.0. That number would mean that people who smoke the most are always the most likely to develop lung cancer. But what the correlation coefficient does not say is what the cause and effect relationship, if any, might be. It does not say that smoking causes cancer.

Chi square and correlation coefficient are only two of dozens of statistical tests now available for use by researchers. The specific kinds of data collected and the kinds of information a researcher wants to obtain from these data determine the specific test to be used.

See also Accuracy.

Resources

Books

Freund, John E., and Richard Smith. *Statistics: A First Course.* Englewood Cliffs, NJ: Prentice Hall Inc., 1986.

Hastie, T., et al. *The Elements of Stastical Learning: Data Mining, Inference, and Prediction.* New York: Springer Verlag, 2001.

Walpole, Ronald, and Raymond Myers, et al. *Probability and Statistics for Engineers and Scientists.* Englewood Cliffs, NJ: Prentice Hall, 2002.

Witte, Robert S. *Statistics.* 3rd ed. New York: Holt, Rinehart and Winston, Inc., 1989.

David E. Newton

Steady-state theory

Was there a moment of creation for the universe, or has the universe always existed? The steady-state theory is a cosmological theory for the origin of the universe that suggests the universe has always existed and did not have a moment of creation. This theory was popular during the 1950s and 1960s, but because of observations made during the 1960s, few, if any, astronomers now think that the steady-state theory is correct. The basic tenet of the steady-state theory is that the universe on a large scale does not change with time (evolve). It has always existed and will always continue to exist looking much as it does now. The universe is, however, known to be expanding. To allow for this expansion in an unchanging universe, the authors of the steady-state theory postulated that **hydrogen atoms** appeared out of empty **space**. These newly created hydrogen atoms were just enough to fill in the gaps caused by the expansion. Because hydrogen is continuously being created, the steady-state theory is sometimes called the continuous creation theory. This theory achieved great popularity for a couple of decades, but mounting observational evidence caused its demise in the late 1960s. The discovery in 1965 of the **cosmic background radiation** provided one of the most serious blows to the steady-state theory.

Cosmological assumptions

The steady-state model is based on a set of four assumptions collectively known as the perfect cosmological principle. The first assumption is that physical laws are universal. Any science experiment, if performed under identical conditions, will have the same result anywhere in the universe because physical laws are the same everywhere in the universe. Second, on a sufficiently large scale the universe is homogeneous. We know there is large scale structure in the universe, such as clusters of galaxies; so, we assume that the universe is homogenous only on scales large enough for even the largest structures to average out. Third, we assume that the universe is *isotropic*, meaning that there is no preferred direction in the universe. Fourth, we assume that over sufficiently long times the universe looks essentially the same at all times.

Collectively, the first three of the above assumptions are the cosmological principle (not to be confused with the perfect cosmological principle). In a nutshell, the cosmological principle states that the universe looks essentially the same at any location in the universe. This principle remains a largely untested assumption because we cannot travel to every location in the universe to perform experiments to test the assumption. Adding the fourth assumption, that the universe does not change on the large scale with time, gives us the perfect cosmological principle. Essentially, the universe looks the same at all times as well as at all locations within the universe. The perfect cosmological principle forms the philosophical foundation for the steady-state theory. With the addition of the fourth assumption, the universe does not evolve in the steady-state theory, so observational evidence that the universe evolves would be evidence against the steady-state theory.

Cosmological observations

There are a number of observations that astronomers have made to test cosmological theories, including both the steady-state and the **big bang theory**. Some of these cosmological observations are described below.

Evolution of the universe

When we look at the most distant objects in the universe, we are looking back in time. For example if we observe a **quasar** that is three billion **light** years away, it has taken the light three billion years to get here, because a light year is the distance light travels in one year. We are therefore seeing the quasar as it looked three billion years ago. Quasars, the most distant objects known in the universe, are thought to be very active nuclei of distant galaxies. The nearest quasar is about a billion light years away. The fact that we do not see any quasar closer than a billion light years away suggests that quasars disappeared at least a billion years ago. The universe has changed with time. Several billion years ago, quasars existed; they no longer do. This observation provides evidence that the perfect cosmological principle is untrue, and therefore that the steady-state theory is incorrect. Note, however, that when the steady-state theory and the perfect cosmological principle were first suggested, we had not yet discovered quasars.

Expansion of the universe

In his work measuring distances to galaxies, Edwin Hubble, after whom the **Hubble space telescope** was named, noticed an interesting correlation. The more distant a **galaxy** is, the faster it is moving away from us. This relationship is called the Hubble law. This relationship can be used to find the distances to additional galaxies, by measuring the speed of recession. More importantly, Hubble deduced the cause of this correlation. The universe is expanding. To visualize this expansion, draw some galaxies on an ordinary **balloon** and blow it up. Notice how the "galaxies" move farther apart as the balloon expands. Measuring distances between the drawn in galaxies at the rates at which they move apart, would give a relationship similar to Hubble's law.

The expanding universe can be consistent with either the big bang or the steady-state theory. However in the steady-state theory, new **matter** must appear to fill in the gaps left by the expansion. Normally as the universe expands, the average distance between galaxies would increase as the **density** of the universe decreases. These evolutionary changes with time would violate the fundamental assumption behind the steady-state theory. Therefore, in the steady-state theory, hydrogen atoms appear out of empty space and collect to form new galaxies. With these new galaxies, the average distance between galaxies remains the same even in an expanding universe.

The Hubble plot also provides evidence that the universe changes with time. The slope of the Hubble plot gives us the rate at which the universe is expanding. If the universe is not evolving, this slope should remain the same even for very distant galaxies. The measurements are difficult, but the Hubble plot seems to **curve** upward for the most distant galaxies. The universe was expanding faster in the distant past, contrary to the prediction of the steady-state theory that the universe is not evolving.

Cosmic background radiation

In the mid 1960s, Arno Penzias and Robert Wilson were working on a low noise (static) microwave **antenna**

when they made an accidental discovery of cosmic significance. After doing everything possible to eliminate sources of noise, including cleaning out nesting pigeons and their waste, there was still a small noise component left. This weak noise did not vary with direction or with the time of day or year, because it was cosmic in origin. It also corresponded to a **temperature** of 3K (-518°F; -270°C, three degrees above absolute zero,). This 3K cosmic background **radiation** turned out to be the leftover **heat** from the initial big bang that had been predicted by proponents of the big bang theory as early as the 1940s.

Because this cosmic background radiation was a prior prediction of the big bang theory, it provided strong evidence to support the big bang theory. Proponents of the steady-state theory have been unable to explain in detail how this background radiation could arise in a steady-state universe. The cosmic background radiation therefore gave the steady-state theory its most serious setback. Penzias and Wilson received the 1978 Nobel Prize in **physics** for their work.

Steady-state theory

The steady-state theory was inspired at least in part by a 1940s movie entitled *Dead of Night*. The movie had four parts and a circular structure such that at the end the movie was the same as at the beginning. After seeing this movie in 1946, Thomas Gold, Hermann Bondi, and Fred Hoyle wondered if the universe might not be constructed the same way. The discussion that followed led ultimately to the steady-state theory.

In 1948, Hermann Bondi and Thomas Gold proposed extending the cosmological principle to the perfect cosmological principle, so that the universe looks the same at all times as well as at all locations. They then proposed the steady-state theory based on the new perfect cosmological principle. Because Hubble had already observed that the universe is expanding, Bondi and Gold proposed the continuous creation of matter. Hydrogen atoms created from nothing combine to form galaxies. In this manner the average density of the universe remains the same as the universe expands. In the steady-state, the rate at which new matter is created must exactly balance the rate at which the universe is expanding. Otherwise, the average density of the universe will change and the universe will evolve, violating the perfect cosmological principle. To maintain the steady-state, in a cubic meter of space one hydrogen atom must appear out of nothing every five billion years. In a **volume** of space the size of the **earth** the amount of new matter created would amount to roughly a grain of dust in a million years. In the entire observable universe, roughly one new galaxy per year will form from these atoms. Bondi and Gold recognized that a new theory must be developed to explain how the hydrogen atoms formed out of nothing, but did not suggest a new theory.

In the same year, Fred Hoyle proposed a modification of Einstein's general theory of relativity. Hoyle worked independently of Bondi and Gold, but they did discuss the new theories. Hoyle's modification used a mathematical device to allow the creation of matter from nothing in general relativity. No experiments or observations have been made to justify or contradict this modification of general relativity.

Arguments for and against the steady-state theory

There are a number of problems with the steady-state theory, but at the time the theory was proposed, there were also points in its favor.

The steady-state theory rests on the foundation of the perfect cosmological principle. Hence, any evidence that the universe evolves is evidence against the steady-state theory. The existence of quasars and the change in the expansion rate of the universe a few billion years in the past, discussed earlier, are evidence against the steady-state. This evidence for the **evolution** of the universe did not exist in 1948, when the steady-state theory originated. It became part of the cumulative weight of evidence that had built up against the steady-state theory by the mid 1960s. In gallant attempts to save the steady-state theory, its proponents, chiefly Hoyle and Jayant Narlikar, have argued that the universe can change over time periods of a few billion years without violating the perfect cosmological principle. We must look at even longer time spans to see that these changes with time average out.

The cosmic background radiation is widely considered the final blow to the steady-state theory. Again, proponents of the steady-state theory have made gallant efforts to save their theory in the face of what most astronomers consider overwhelming evidence. They argue that the background radiation could be the cumulative radiation of a large number of **radio** sources that are too faint to detect individually. This scheme requires the existence of roughly 100 trillion (about 10,000 times the number of observable galaxies) such sources that about are one millionth as bright as the radio sources we do detect. Few astronomers are willing to go to such great lengths to rescue the steady-state theory.

Another objection raised against the steady-state theory is that it violates one of the fundamental laws of physics as that law is currently understood. The law of **conservation** of matter and **energy** states that matter and energy are interchangeable and can change between forms, but the total amount of matter and energy in the universe must remain constant. It can be neither created nor destroyed. The steady-state theory requires continuous creation of matter in violation of this law. However, laws of science result from our experimental evidence and are subject to change, not at our whim, but as experimental results dictate. The rate at which matter is created in the steady-state theory is small enough that we would not have noticed. Hence we would not have discovered experimentally any conditions under which matter could be created or any modifications required in this law.

Were there ever any points in favor of the steady-state theory? When the steady-state model was first suggested, our best estimate of the **age of the universe** in the context of the big bang model was about two billion years. However, the earth and **solar system** are about five billion years old. The oldest stars in our galaxy are at least 10-12 billion years old. These age estimates present the obvious problem of a universe younger than the objects it contains. This problem is no longer so severe. Modern estimates for the age of the universe range from about 10 billion years to about 20 billion years. The lower end of this range still has the problem, but the upper end of the range gives a universe old enough to contain the oldest objects we have found so far. In the steady-state theory the universe has always existed, so there are no problems presented by the ages of objects in the universe.

For some people there are philosophical or esthetic grounds for preferring the steady-state hypothesis over the big bang theory. The big bang theory has a moment of creation, which some people prefer for personal or theological reasons. Those who do not share this preference often favor the steady-state hypothesis. They prefer the grand sweep of a universe that has always existed to a universe that had a moment of creation and may, by inference, also have an end in some far distant future time.

The weight of evidence against the steady-state theory has convinced most modern astronomers that it is incorrect. The steady-state theory does, however, still stand as a major intellectual achievement and as an important part of the history of the development of **cosmology** in the twentieth century.

See also Doppler effect.

Resources

Books

Bacon, Dennis Henry, and Percy Seymour. *A Mechanical History of the Universe.* London: Philip Wilson Publishing, Ltd., 2003.

Harrison, Edward R. *Cosmology The Science of the Universe.* Cambridge: Cambridge University Press, 1981.

KEY TERMS

. . . .

Cosmic background radiation—The leftover heat radiation from the big bang.

Cosmological principle—The set of fundamental assumptions behind the big bang theory that state the universe is essentially the same at all locations.

Hubble's law—The law that states a galaxy's red-shift is directly proportional to its distance from Earth. This observation that tells us the universe is expanding: distant galaxies are receding at a speed proportional to their distance.

Perfect cosmological principle—The set of fundamental assumptions behind the steady-state theory that state the universe is essentially the same at all locations and times.

Hoyle, Fred, and Jayant Narlikar. *The Physics-Astronomy Frontier.* San Francisco: Freeman. 1980.

Hoyle, Fred. *Astronomy.* London: Rathbone Books, 1962.

Narlikar, Jayant V. *The Primeval Universe.* Oxford: Oxford University Press, 1988.

Paul A. Heckert

Stealth technology *see* **Aircraft**

Steam engine

A steam engine is a machine that converts the **heat energy** of steam into mechanical energy by means of a piston moving in a cylinder. As an external **combustion engine**—since it burns its fuel outside of the engine—a steam engine passes its steam into a cylinder where it then pushes a piston back and forth. It is with this piston movement that the engine can do mechanical work. The steam engine was the major power source of the **Industrial Revolution** and dominated industry and transportation for 150 years. It is still useful today in certain situations.

History

The earliest known steam engines were the novelties created by the Greek engineer and mathematician named Hero who lived during the first century A.D. His most famous invention was called the aeliopile. This was a small, hollow **sphere** to which two bent tubes were at-

tached. The sphere was attached to a boiler that produced steam. As the steam escaped from the sphere's hollow tubes, the sphere itself would begin to whirl and rotate. Hero of Alexandria and several other Greeks designed many other steam-powered devises, such as a steam **organ** and automatic doors, but always in the context of playfulness and seemingly without any interest in using steam in a practical way. Nonetheless, their work established the principle of steam power and their playful devices were a real demonstration of converting steam power into some kind of **motion**.

Although the Greeks established the principle of steam power, it lay ignored for over 1,500 years until the late 1600s in **Europe**. During this long period, the main sources of power were first, human muscle power or draft animals, and later, **wind** and **water** power. Windmills and waterwheels were adequate for slow, repetitive jobs like grinding corn, in which an interruption of power was of little consequence. However, for certain jobs, like pumping water from a mine shaft, a power source that could cease at any time was not always satisfactory. In fact, it was the very deepness of the English mines that spurred engineers to search for pumps that were quicker than the old water pumps. By the mid-sixteenth century, work on air pumps had established the notion of a piston working in a cylinder, and around 1680, the French physicist Denis Papin (1647-1712) put some water at the bottom of a tube, heated it, converted it to steam, and saw that the expanded steam pushed forcibly and moved a piston just ahead of it. When the tube cooled, the piston returned to its previous position. Although Papin was well aware he had created an engine that could eventually do work, he was deterred by the very real mechanical difficulties of his time and chose to work on a smaller scale-creating the world's first **pressure** cooker.

Following Papin, an English military engineer, Thomas Savery (c.1650-1715), built what most regard as the first practical steam engine. Unlike Papin's system, this had no piston since Savery wanted only to draw water from the **coal** mines deep below the **earth**. Knowing that he could use steam to produce a vacuum in a vessel, he connected such a vessel to a tube leading into the water below. The vacuum then drew water up the tube and blew it out by steam pressure. Savery's system was called the "Miner's Friend" as it raised water from the mines using the suction produced by condensing steam. A few years later, an English engineer and partner of Savery named Thomas Newcomen (1663-1729) improved the steam pump by reintroducing the piston. By 1712 he had built an engine that used steam at **atmospheric pressure** (ordinary boiling water) and which was fairly easy to build. His piston engine was very reliable and came into general use in England around 1725. His machine was called a beam engine because it had a huge rocking-arm or see-saw beam at its top whose motion transferred power from the engine's single cylinder to the water pump.

Understanding how the Newcomen engine worked provides insight into all later steam engines. First, the entire machine was contained in an engine house, about three stories high, out of whose top wall poked a long oak beam that could rock up and down. The house was constructed off to the side of the mine shaft. At the bottom of the shaft was the water pump which was connected to the engine by a long pump-rod. Below the beam inside the house was a long brass cylinder which sat atop a **brick** boiler. The boiler was fed by coal and supplied the steam. Inside the cylinder was the piston that could slide up and down and was connected to the beam above. The engine always started with the piston in the up position. Then steam filled the cylinder from an open valve. When filled, the cylinder was sprayed with water which caused the steam inside to condense into water and create a partial vacuum. With this, the pressure of the outside air would **force** the piston down, which rocked the beam and pulled up the pump rods and sucked up about 12 gal (45 l) of water. The piston then returned to its starting position (up) in the cylinder and the process was repeated. Besides being called a beam engine, Newcomen's engine was also called an atmospheric engine since it used air pressure to move the piston (down).

The most important improvement in steam engine design was brought about by the Scottish engineer James Watt (1736-1819). In 1763, Watt was asked to repair a Newcomen engine and was struck by what he considered its inefficiency. He set out to improve its performance and by 1769 had arrived at the conclusion that if the steam were condensed separately from the cylinder, the latter could always be kept hot. That year he introduced a steam engine with a separate condenser. Since this kept the heating and cooling processes separate, his machine could work constantly without any long pause at each cycle to reheat the cylinder. Watt continued to improve his engine and made three additions that were highly significant. First, he made it double-acting by allowing steam to enter alternately on either side of the piston. This allowed the engine to work rapidly and deliver power on the downward as well as on the upward piston stroke. Second, he devised his sun-and-planet gearing which was able to translate the **reciprocal**, or to-and-fro motion of the beam into rotary motion. Third, he added a centrifugal governor that maintained a constant engine speed despite varying loads. This highly innovative device marks the early beginnings of **automation**, since Watt had created a system that was essentially self-regulating. Watt also devised a pressure gauge that he added to his engine. By 1790, Watt's improved steam engines offered a powerful, reliable power source that could be located almost anywhere. This meant that factories no longer had to be sited next to water sources, but could be built closer to both their raw materials and transport systems. More than anything, it was Watt's steam engine that speeded up the Industrial Revolution both in England and the rest of the world.

Watt's steam engine was not perfect however, and did have one major limitation; it used steam at low pressure. High pressure steam meant greater power from smaller engines, but it also meant extreme danger since explosions of poorly-made boilers were common. The first to show any real success with it was the English inventor Richard Trevithick (1771-1833). By the end of the eighteenth century, metallurgical techniques were improving and Trevithick believed he could build a system that would handle steam under high pressure. By 1803, Trevithick had built a powerful, high-pressure engine that he used to power a train. His technical innovations were truly remarkable, but high-pressure engines had earned such a bad reputation in England that twenty years would pass before English inventor George Stephenson (1781-1848) would prove their worth with his own locomotives.

In the United States however, there was little bias against, or hardly any knowledge of, steam power. Toward the end of the eighteenth century, Evans began work on a high-pressure steam engine that he could use as a stationary engine for industrial purposes and for land and water transport. By 1801, he had built a stationary engine that he used to crush limestone. His major high-pressure innovation placed both the cylinder and the crankshaft at the same end of the beam instead of at opposite ends. This allowed him to use a much lighter beam.

Over the years, Evans built some 50 steam engines which were not only used in factories, but also to power an amphibious digger. High-pressure steam ran this odd-looking scow that was a dredge that could move on land as well as in water. It was the first powered road vehicle to operate in the United States.

Despite Evans' hard work and real genius, his innovative efforts with steam met with little real success during his lifetime. He was often met with indifference or simple reluctance on the part of manufacturers to change their old ways and convert to steam. His use of steam for land propulsion was set back by poor roads, vested interest in **horses**, and woefully inadequate materials. After Evans, high-pressure steam became widely used in America, unlike England where Watt's low-pressure engines took a long time to be replaced. But improvements were made nonetheless, and **iron** would eventually replace timber in engine construction, and horizontal engines came to be even more efficient than the old vertical ones.

The workings of a steam engine

Throughout all of this development and improvement of the steam engine, no one really knew the science behind it. Basically, all of this work had been accomplished on an empirical basis without reference to any theory. It was not until 1824 that this situation changed with the publication of *Reflexions sur La Puissance Motrice du Feu* by the French physicist, Nicolas Leonard Sadi Carnot (1796-1832). In his book *On the Motive Power of Fire*, Carnot founded the science of **thermodynamics** (or heat movement) and was the first to consider quantitatively the manner in which heat and work are related. Defining work as "weight lifted through a height," he attempted to determine how efficient or how much "work" a Watt engine could produce. Carnot was able to prove that there was a maximum theoretical limit to the efficiency of any engine, and that this depended upon the **temperature** difference in the engine. He showed that for high efficiency, steam must pass through a wide temperature range as it expands within the engine. Highest efficiency is achieved by using a low condenser temperature and a high boiler pressure. Steam was successfully adapted to power boats in 1802 and railways in 1829. Later, some of the first automobiles were powered by steam, and in the 1880s, the English engineer Charles A. Parsons (1854-1931) produced the first steam **turbine**. This high-powered, highly efficient turbine could produce not only mechanical energy but electrical energy as well. By 1900, the steam engine had evolved into a highly sophisticated and powerful engine that propelled huge ships in the oceans and ran turbogenerators that supplied **electricity**.

Once the dominant power source, steam engines eventually declined in popularity as other power sources became available. Although there were more than 60,000 steam cars made in the U. S. between 1897 and 1927, the steam engine eventually gave way to the **internal combustion engine** for vehicle propulsion. Today, interest in

steam has revived somewhat as improvements make it increasingly efficient and its low-pollution factors make it more attractive.

See also Diesel engine; Jet engine.

Resources

Books

Hindle, Brooke and Steven Lubar. *Engines of Change*. Washington: Smithsonian Institution Press, 1986.

Rutland, Jonathan. *The Age of Steam*. New York: Random House, 1987.

Leonard C. Bruno

Steam pressure sterilizer

Steam **pressure** sterilization requires a combination of pressure, high temperatures, and moisture, and serves as one of the most widely used methods for sterilization where these functions will not effect a load. The simplest example of a steam pressure sterilizer is a home pressure cooker, though it is not recommended for accurate sterilization. Its main component is a chamber or vessel in which items for sterilization are sealed and subjected to high temperatures for a specified length of **time**, known as a cycle.

Steam pressure sterilizer has replaced the term autoclave for all practical purposes, though autoclaving is still used to describe the process of sterilization by steam. The function of the sterilizer is to kill unwanted **microorganisms** on instruments, in cultures, and even in liquids, because the presence of foreign microbes might negatively effect the outcome of a test, or the purity of a sample. A sterilizer also acts as a test vehicle for industrial products such as **plastics** that must withstand certain pressures and temperatures.

Larger chambers are typically lined with a **metal** jacket, creating a pocket to trap pressurized steam. This method preheats the chamber to reduce condensation and cycle time. Surrounding the unit with steam-heated tubes produces the same effect. Steam is then introduced by external piping or, in smaller units, by internal means, and begins to circulate within the chamber. Because steam is lighter than air, it quickly builds enough **mass** to displace it, forcing interior air and any air-steam mixtures out of a trap or drain.

Most sterilization processes require temperatures higher than that of boiling **water** (212°F; 100°C), which is not sufficient to kill microorganisms, so pressure is increased within the chamber to increase **temperature**. For example, at 15 pounds per square inch (psi) the temperature rises to 250°F

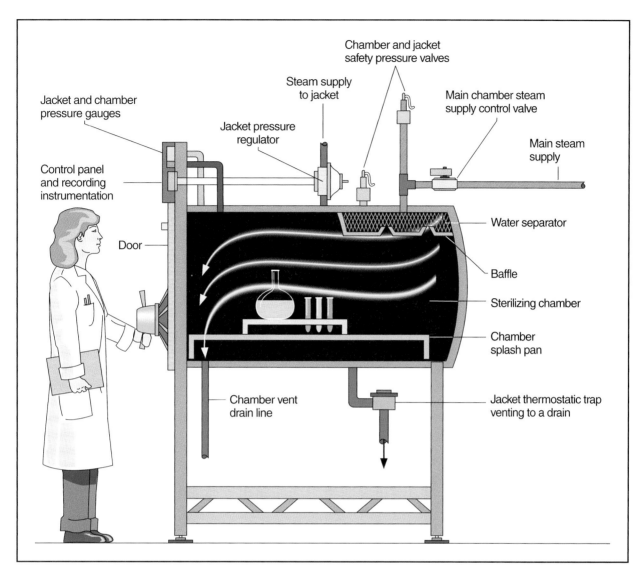

The structure of a steam pressure sterilizer. *Illustration by Hans & Cassidy. Courtesy of Gale Group.*

(121°C). Many clinical applications require a cycle of 20 minutes at this temperature for effective sterilization. Cycle variables can be adjusted to meet the requirements of a given application. The introduction of a **vacuum** can further increase temperature and reduce cycle time by quickly removing air from the chamber. The process of steam sterilization is kept in check by pressure and temperature gauges, as well as a safety valve that automatically vents the chamber should the internal pressure build beyond the unit's capacity.

Resources

Books

Gardner, Joan F., and Peel Margaret M. *Introduction to Sterilization and Disinfection.* Melbourne: Churchill Livingstone, 1986.

Reichert, Marimargaret and Young, Jack H., eds. *Sterilization Technology for the Health Care Facility.* Maryland: Aspen Publishers, 1993.

Webster, John G., ed. *Encyclopedia of Medical Devices and Instrumentation.* New York: John Wiley & Sons, 1988.

Stearic acid

Stearic acid is a chemical compound consisting of an 18 **carbon** chain whose terminal carbon is connected to an **oxygen** atom with a double bond and a hydroxyl group (OH) by a single bond. It belongs to class of materials known as **fatty acids**, produced primarily from natural fats and oils. Stearic acid is an important component in **soap** and other cosmetic and industrial preparations.

Stearic acid is derived predominantly from fats and oils. These materials contain **triglycerides** which are glycerine molecules attached to long **hydrocarbon** chains. These hydrocarbons can be removed from the glycerine backbone through a variety of techniques. When a triglyceride **molecule** is split, it yields three molecules of fatty acid and one molecule of glycerine. The major **fat** used in the production of stearic acid is beef fat, also known as tallow. Stearic acid is also obtained in lesser amounts from herring and sardine. **Plant** oils such as **cotton**, coconut, palm kernel, castor beans, rapeseed, soybeans, and sunflowers are also natural sources. In the United States, almost all stearic acid is made from tallow and coconut oil, although lesser amounts are made from palm oil. The other plant sources are more commonly used in third world countries.

Historically, stearic acid has been made by a process known as **hydrolysis**, which involves heating the fat in an alkaline **solution**. This process is also known as saponification. The alkali that is traditionally used is **sodium hydroxide**, also known as caustic soda or lye. Hence the term "lye soap." Other methods used to produce fatty acids include solvent crystallization, **hydrogenation**, and **distillation**.

Pure stearic acid is a white, waxy solid crystalline material that melts at 156°F (69°C). It is odorless and tasteless. However, because of its natural origin, pure stearic is hard to obtain. Instead, stearic acid usually includes minor amounts of other fatty acids with different carbon chain lengths, such as lauric and palmitic acids. These trace impurities can cause the acid to vary in **molecular weight**, **solubility**, melting point, **color**, odor, and other physical and chemical properties. In addition to the carbon chain distribution, the degree of **neutralization**, or the amount of free acid present, also determines the acid's properties. These are a number of physical and chemical specifications used to ensure that the stearic acid is of a consistent quality. Specifications include the acid's saponification value, iodine value, peroxide value, free fatty acids, unsaponifiables, moisture, and trace impurities.

When fatty acids are neutralized with an alkali, the resultant **salt** is known as a soap. No one knows for sure when soap was first discovered, but it was well known in the ancient world. Some legends attribute it to the Romans, others say the ancient Gauls accidentally discovered it as they tried to extract oil from **animal** fat. Soap making techniques were common in the Old World and the Phoenicians, Arabs, Turks, and the Celts were all regarded as master soap makers. It wasn't until the dawn of the nineteenth century, however, that soap **chemistry** was understood. Two key events triggered the modern soap making industry. First, in 1790, Nicholas Leblanc discovered a process to make **sodium** hydroxide from **sodium chloride**; this established an inexpensive source of lye for soap making. Then, in 1823, Michel Chevreul identified the structure of fatty acids. Chevreul discovered the first fatty acid when he analyzed a potassium soap made from pig fat. After treating the soap with various chemicals, he found that it yielded a crystalline material with acid properties. Hence the first fatty acid was isolated. Over the next decade, Chevreul decomposed a variety of soaps made from different animal soaps. He identified and named many fatty acids, including stearic and butyric.

While soaps can be made using a variety of fatty acids, stearic acid is one of the most popular. Stearic acid salts made with sodium, potassium, **calcium**, strontium, **barium**, and **magnesium** are used in a variety of applications. Sodium stearate is the most common type of soap and has been used extensively in cleansing for personal care in bar soaps. Potassium stearate is softer and more **water** soluble and has been used in water solutions for hard surface cleaning. Despite their widespread use, stearic acid soaps do have certain drawbacks. First, it is difficult to prepare concentrated solutions of these soaps because they are only marginally soluble in water. Furthermore, they can react with **minerals** present in **hard water** and form insoluble salts such as calcium stearate. These insoluble salts are responsible for bathtub ring and can leave undesirable film on hair, skin, and clothing. In the 1940s, due to wartime shortages of certain natural materials, synthetic soaps, also known as detergents, became commercially available. These detergents had the cleansing properties of soap without its negative properties.

Beside soap making, stearic acid is used to form stable creams, lotions and ointments. It is used in products like deodorants and antiperspirants, foundation creams, hand lotions, hair straighteners, and shaving creams. It is also used as a softener in chewing gum base and for suppositories. It may be further reacted to form stearyl **alcohol** which is used in a variety of industrial and cosmetic products as a thickener and lubricant. It is also used in candles to modify the melting point of the waxes.

Resources

Periodicals

"Soap Technology for the 1990s." *American Oil Chemists Society* Champaign, Il: Luis Spitz, 1990.

Randy Schueller

Steel

Steel is the most widely used of all metals, with uses ranging from **concrete** reinforcement in highways and in high-rise buildings to automobiles, **aircraft**, and vehicles in space. Steel is **iron** combined or alloyed with other metals or nonmetals such as **carbon**. Steel is more ductile (able to deform without breakage) and durable than cast iron and is generally forged, rolled, or drawn into various shapes.

Since the beginning of the Iron Age, about 1000 B.C., mankind's progress has been greatly dependent on tools and equipment made with iron. The iron tools were then used to fashion many other much needed goods. Eventually, this was followed by the **Industrial Revolution**, a period of change beginning in the middle of the eighteenth century in England where extensive mechanization of production systems resulted in a shift from home manufacturing and farms to large-scale factory production. **Machine tools** and other equipment made of iron and steel significantly changed the economy of both farm and city.

The history of iron and steel began at least 6,000 years ago. It is speculated that early mankind first learned to use iron from fallen meteorites. Many meteorites are composed of iron and nickel, which forms a much harder **metal** than pure iron. The ancients could make crude tools and weapons by hammering and chipping this metal. Because this useful metal came from the heavens, early human beings probably did not associate it with the iron found in the ground. It is likely that metallic iron was found in the ashes of fires that had been built on outcroppings of red iron **ore**, also called iron oxide. The red ore was called paint rock, and fires were built against banks of ore that had been exposed to **wind** and **weather**. Iron ore is found worldwide on each of the seven continents.

Smelting iron, a primitive direct reduction method of separating iron from its ore using a charcoal forge or furnace, probably began in China and India and then spread westward to the area around the Black Sea. Unlike **copper** ores, which yielded molten copper in these furnaces, iron would not melt at temperatures below 2,799°F (1,537°C) and the highest **temperature** that could be reached in these primitive smelters appears to have been about 2,192°F (1,200°C). Iron ore subjected to that temperature does not melt, but instead results in a spongy **mass** (called "sponge" iron) mixed with impurities called slag. The iron worker removed this spongy mass from the furnace and then squeezed the slag out of it by hammering. This "wrought" iron had less tendency to corrode and had a fibrous quality from the stringers of slag which gave it a certain toughness.

The Hittites, an ancient tribe living in **Asia** Minor and northern Syria, produced iron starting about 2500 B.C. The Chalybes, a subject tribe of the Hittites, invented a cementation process about 1400 B.C. to make the iron stronger. The iron was hammered and heated in contact with charcoal. The carbon absorbed from the charcoal produced a much harder iron. With the fall of the Hittite empire, the various tribes scattered, carrying the knowledge of smelting and the cementation process with them to Syria, Egypt, and Macedonia. Widespread use of iron for weapons and tools began about 1000 B.C., marking the beginning of the Iron Age.

The ancient Egyptians learned to increase smelting temperature in the furnace by blowing a stream of air into the fire using blowpipes and bellows. Around 500 B.C., the Greek soldiers used iron weapons which had been hardened by quenching the hot metal in cold **water**. The Romans learned to reheat the iron after quenching in a process called tempering which made the iron less brittle.

During the Middle Ages, from about A.D. 500 to A.D. 1500, the old methods of smelting and cementation continued. Early blacksmiths made chain mail, weapons, nails, horseshoes, and tools such as iron plows. The *Stückofen*, a furnace first developed by the Romans, was made larger and higher for better air draft. This was a forerunner of the modern blast furnace. Waterwheels came into use for ironmaking between A.D. 1200 and A.D. 1350. The waterwheels converted the **energy** of swift stream **currents** into work that moved air bellows, forcing blasts of air into the furnace. The resulting higher temperature melted the iron, which was then formed into "pigs" (so named because as the pig iron was cast, the runners and series of ingots resembled **pigs** suckling their mother) of cast iron. As **time** progressed, these early blast furnaces were built larger and better, reaching 30 ft (9 m) in height and able to operate continuously for weeks at a time.

About A.D. 1500, ironmakers faced **wood** shortages that affected their source of charcoal. Increased warfare and the resulting demand for more iron weapons forced ironmakers to use **coal** as an alternate source of fuel. A major problem with coal was that it contained impurities such as **sulfur** and **phosphorus** that tended to make the iron brittle. In 1709 Abraham Darby of England used "coke," the residue left after soft coal was heated to remove impurities, to successfully smelt pig iron. Crucible cast steel was invented around 1740 by Benjamin Huntsman of England. A clay crucible, or cup, of iron ore was placed in a furnace and when molten, was cast. The resulting cast steel was of very high purity since the molten steel did not come into contact with the fuel. In 1784 another improvement was made by Henry Cort, an English ironmaker, who invented the puddling of molten pig iron. Puddling involved stirring air into the liquid iron by a worker who stood near the furnace door. A reverberatory furnace was used in which the coal was separated from the iron to prevent **contamination**. After the pig iron had been converted into wrought iron, it was run through a rolling mill which used grooved **rollers** to press out the remaining slag. Cort's rolling mill was patented in 1783 and could make iron bars about 15 times faster than the old hammer method.

From 1850 to 1865, great advances were made in iron and steel processing. Steel was gaining more popularity than iron beginning around 1860 as less expensive manufacturing methods were discovered and greater quantity and quality were being produced.

William Kelly of the United States, and Henry Bessemer of England, both working independently, discovered the same method for converting iron into steel. They subjected molten pig iron to a blast of air which burned out most of the impurities and the carbon contained in the molten iron acted as its own fuel. Kelly built his first converter in 1851 and received an American patent in 1857. He also went bankrupt the same year and the method finally became known as the Bessemer process. In 1856 Bessemer completed his vertical converter, and in 1860 he patented a tilting converter which could be tilted to receive molten iron from the furnace and also to pour out its load of liquid steel. The Bessemer converter made possible the high tonnage production of steel for ships, railroads, **bridges**, and large buildings in the mid-nineteenth century. However, the steel was brittle from the many impurities which remained, especially phosphorus and sulfur, and by the **oxygen** from the air blast. An English metallurgist, Robert F. Mushet, discovered in 1856 that adding an iron **alloy** (spiegeleisen) containing manganese would remove the oxygen. Around 1875, Sidney G. Thomas and Percy Gilchrist, two English chemists, discovered that by adding limestone to the converter they could remove the phosphorus and most of the sulfur.

In England, another new furnace was introduced in 1861 by two brothers, William and Frederick Siemans. This was the open-hearth furnace, also known as the regenerative open-hearth because the outgoing hot gases were used to preheat the incoming air. Pierre Émile Martin of France improved the process in 1864 by adding scrap steel to the molten iron to speed purification. During this period hardened alloy steels came into commercial use; Mushet made a high carbon steel in 1868 which gave tools longer life in France, a chromium steel alloy was produced in 1877 and a nickel steel alloy in 1888. An Englishman, Sir Robert Hadfield, discovered in 1882 how to harden manganese tool steel by heating it to a high temperature and then quenching it in water.

Around 1879, the electric furnace was developed by William Siemans. This furnace was used very little prior to 1910 because of the high electrical costs and the poor quality of electrodes used to produce the arc for melting.

The open-hearth furnace was the most popular method of steel production until the early 1950s. Pure oxygen became more economical to produce in large quantities and in 1954 the first basic oxygen process facility opened for production in the United States. Today, most of the world's steel is made by either a basic oxygen furnace or an electric furnace.

Raw materials

The ores used in making iron and steel are iron oxides, which are compounds of iron and oxygen. The major iron oxide ores are hematite, which is the most plentiful, limonite, also called brown ore, taconite, and magnetite, a black ore. Magnetite is named for its magnetic property and has the highest iron content. Taconite, named for the Taconic Mountains in the northeastern United States, is a low-grade, but important ore, which contains both magnetite and hematite.

Ironmaking furnaces require at least a 50% iron content ore for efficient operation. Also, the cost of shipping iron ores from the mine to the smelter can be greatly reduced if the unwanted rock and other impurities can be removed prior to shipment. This requires that the ores undergo several processes called "beneficiation." These processes include crushing, screening, tumbling, floatation, and magnetic separation. The refined ore is enriched to over 60% iron by these processes and is often formed into pellets before shipping. Taconite ore powder, after beneficiation, is mixed with coal dust and a binder and rolled into small balls in a drum pelletizer where it is then baked to hardness. About two tons of unwanted material is removed for each ton of taconite pellets shipped.

The three raw materials used in making pig iron (which is the raw material needed to make steel) are the processed iron ore, coke (residue left after heating coal in the absence of air, generally containing up to 90% carbon) and limestone ($CaCO_3$) or burnt lime (CaO), which are added to the blast furnace at intervals, making the process continuous. The limestone or burnt lime is used as a fluxing material that forms a slag on top of the liquid metal. This has an oxidizing effect on the liquid metal underneath which helps to remove impurities. Approximately two tons of ore, one ton of coke, and a half ton of limestone are required to produce one ton of iron.

There are several basic elements which can be found in all commercial steels. Carbon is a very important element in steel since it allows the steel to be hardened by **heat** treatment. Only a small amount of carbon is needed to produce steel: up to 0.25% for low carbon steel, 0.25-0.50% for medium carbon steel, and 0.50-1.25% for high carbon steel. Steel can contain up to 2% carbon, but over that amount it is considered to be cast iron, in which the excess carbon forms graphite. The metal manganese is used in small amounts (0.03-1.0%) to remove unwanted oxygen and to control sulfur. Sulfur is difficult to remove from steel and the form it takes in steel (iron sulfide, FeS) allows the steel to become brittle, or *hot-short*, when forged or rolled at elevated temperatures. Sulfur content in commercial steels is usually kept below 0.05%. A small quantity of phosphorus (usually below 0.04%) is present, which tends to dissolve in the iron, slightly increasing the strength and hardness. Phosphorus in larger quantities reduces the ductility or formability of steel and can cause the material to crack when cold worked in a rolling mill, making it *cold-short*. Silicon is another element present in steel, usually between 0.5-0.3%. The silicon dissolves in the iron and increases the strength and toughness of the steel without greatly reducing ductility. The silicon also deoxidizes the molten steel through the formation of silicon dioxide (SiO_2), which makes for stronger, less porous castings. Another element that plays an important part in the processing of steel is oxygen. Some large steel mills have installed their own oxygen plants, which are located near basic oxygen furnaces. Oxygen injected into the mix or furnace "charge" improves and speeds up steel production.

Steel can be given many different and useful properties by alloying the iron with other metals such as chromium, molybdenum, nickel, **aluminum**, cobalt, tungsten, vanadium, and **titanium**, and with nonmetals such as boron and silicon.

Manufacturing processes

Most steel is produced using one of four methods: Bessemer converters, open-hearth furnaces, basic oxygen furnaces, and electric furnaces. The basic oxygen process is the most efficient, while the Bessemer and open-hearth methods have become obsolete. Electric furnaces are used to produce high quality steels from selected steel scrap, lime, and mill scale (an iron oxide that forms on the surface of hot steel and falls off as black scale).

Until 1909, most steel made in the United States came from Bessemer converters. A Bessemer converter looks like a huge pear-shaped pot and can hold anywhere from 5-25 tons. It is balanced on axles so that its open top can be tilted one way to take a charge and the other way to pour out steel. After the converter is charged with hot metal, it is swung to the upright position. Air is then blown through holes in its bottom at a typical **rate** of 30,000 cubic feet per minute. Sparks and thick, brown smoke pour from the con-

verter's mouth as the oxygen in the blow combines with the iron, silicon, and manganese to form slag. Then 30-ft (9-m) flames replace the smoke as the oxygen combines with the carbon fuel and burns. The whole process took less than 15 minutes. Unfortunately, the blowing air contained contaminants (such as **nitrogen**) and also removed some desirable elements such as carbon and manganese. This was solved by adding the necessary elements back into the converter after the blow. Because of stricter **air pollution** regulations and more efficient processes, the Bessemer converter is no longer used.

From 1909 until the 1960s, the open-hearth process was the most popular method of steel production. Open-hearth furnaces got their name from a shallow area called a hearth that is exposed to a blast of flames that alternately sweeps across the hearth from one side for a period of time and then to the side of the furnace. To make a "heat," or one batch of steel, pig iron, limestone, and scrap steel, are initially "charged," or loaded, into the hearth. These materials are heated for about two hours at temperatures 2,700–3,000°F (1,482–1,649°C) until they begin to fuse. Then the furnace is charged with many tons of molten pig iron. Scrap is placed in the furnace with a charging machine which usually serves a series of open hearth furnaces in a single building. Other elements, such as fluxing agents, carbon (usually in the form of anthracite coal pellets), and alloying materials, are then added to improve the steel. These elements can be added either in the furnace charge, the melt or "bath," ladle, or the ingot molds to meet the desired chemical composition of the finished steel or to eliminate or counteract the effect of oxides or other impurities. Fluxing agents (primarily lime, added in the form of either limestone or burnt lime and supplemented by magnesia, MgO, and lime from the furnace bottom and sides) melt and combine with the impurities to form slag at the top of the melt which is poured off into a separate slag pot. Mill scale, a form of iron oxide (Fe_3O_4), is used to reduce carbon content. Aluminum ferrosilicon is added if the steel is to be "killed." A killed steel is one that has been deoxidized to prevent gas **evolution** in the ingot **mold**, making a more uniform steel. "Rimmed" steel is steel that has not been deoxidized and gas pockets and holes from free oxygen form in the center of the ingot while the rim near the surface of the ingot is free of defects. Rolling processes are used in later operations to remove these defects. Semikilled steels are a compromise between rimmed and killed steels and are mainly limited to steels to be rolled into sheet bar, plate, and structural sections. The quantity of deoxidizers used must be closely controlled to allow a limited evolution of gas from the carbon-oxygen reaction.

When the contents of the heat are acceptable and the temperature is right, the furnace is tapped and the molten metal is poured into a ladle. An open-hearth furnace is tapped through a hole in the furnace's bottom. A heat is refined into steel during an 8-12 hour time period. Oxygen released from the ore and additional injected oxygen combine with carbon in the molten pig iron to form carbon gases. These, along with any additional gases from the burned fuel, are used to heat incoming air and this is why the open-hearth process is sometimes called the regenerative open-hearth.

The basic oxygen converter resembles a Bessemer converter. It receives materials from the top and tips to pour off the finished steel into ladles. The main element is a water-cooled oxygen lance, which is placed into the top of the converter after it is charged with scrap steel, molten pig iron, and fluxing agents. The lance, lowered to within a few feet of the charge, directs high-purity oxygen at supersonic speeds into the molten metal. This burns out the impurities and also enables the making of steel with a minimum amount of nitrogen, which can make steel brittle. The oxidation of the carbon and impurities causes a violent rolling agitation which brings all the metal into contact with the oxygen stream. The furnace ladle is first tipped to remove slag and then rotated to pour molten steel into a ladle. The speed and efficiency of the oxygen process has had a significant impact on the steel industry. An oxygen converter can produce a heat of quality steel in 30-45 minutes. An open-hearth furnace without an oxygen lance requires as much as eight hours to produce steel of a similar quality. Recent advances in refractory "brick," the insulating **ceramics** that protect vessels from the hot steel, have allowed injection of oxygen from the bottom of a vessel without a large complicated lance. This allows for a much more efficient use of the oxygen and can lower the capital costs in constructing a basic oxygen facility, especially if the building and cranes of a retired open-hearth facility is used.

High-quality carbon and alloy steels, such as tool and stainless steels, are produced in **electric arc** furnaces. These furnaces can make 150-200 tons in a single heat in as little as 90 minutes. The charge is melted by the arcing between carbon electrodes and high quality scrap steel. Some of the electrodes can be 2 ft (0.6 m) in diameter and 24 ft (7.2 m) long. The entire electric furnace is tilted during a tapping operation in which molten steel flows into a waiting ladle. Electric furnaces are the most competitive where low-cost **electricity** is available and where very little coal or iron ore is found.

After the steel in the ladle has cooled to the desired temperature, the ladle is moved by a traveling **crane** to either a pouring platform for ingot production or to a continuous caster. Ingots may be square, rectangular, or round and weigh anywhere from a few hundred pounds to 40 tons. A small amount of steel is cast directly into

the desired shape in molds of fine **sand** and fireclay. Small rail cars carrying a series of heavy cast iron ingot molds wait alongside the pouring platform. The steel is "teemed" or poured into the molds through a fire-clay nozzle in the bottom of the ladle. After the steel in the molds has solidified, the cars are pulled under a stripping crane. The crane's plunger holds down the ingot top as its jaws lift the mold from the glowing hot ingot. The ingots are then taken to soaking pits for further processing.

An underground soaking pit is used to heat the steel ingots to a uniform temperature throughout. The ingots must be the same temperature throughout so that they can be easily plastically deformed and to prevent damage to the heavy machinery of the mills. The jaws of the crane clamp onto the ingots and lower them into the open soaking pits. The roof of the pits is then closed and burning oil or gas heats the ingots to about 2,200°F (1,204°C). After "soaking" in the pits for several hours, the ingots are then lifted out by crane and transported by rail to the blooming and slabbing mills.

The mechanical working of steel, such as rolling, forging, hammering, or squeezing, improves it in several ways. Cavities and voids are closed, harmful concentrations of nonmetallic impurities are broken up and more evenly disbursed, and the grain structure is refined to produce a more homogeneous or uniform product. Some ingots are sent directly to a universal plate mill for immediate rolling of steel plates. Most ingots, however, are sent to semifinishing mills (also known as slabbing or blooming mills) for reduction and shaping into slabs, blooms, or billets. A slab is generally a large flat length of steel wider than a bloom, a bloom is a length of steel either square or rectangular with a cross-sectional area larger than 36 in (90 cm), and a billet is generally two to five inches square, although some billets can be round or rectangular. The exact sizes of slabs, blooms, and billets depend on the requirements of further processing.

In slabbing and blooming mills, the steel ingot is gradually squeezed between heavy rolls. To make billets, the steel is first shaped into blooms, then further reduced in a billet mill. Each time the ingot is forced through the rolls, it is further reduced in one dimension. Blooming mills can be classified as either two-high or three-high, depending on the number of rolls used. The two rolls of the two-high mill can be reversed so that the ingot is flattened and lengthened as it passes back and forth between the rolls. The top and bottom rolls of the three-high mill turn in one direction while the middle roll turns in the opposite direction. The ingot is flattened first between the bottom and middle rolls and ends up on a runout table. The table rises and the steel is then fed through the top and middle rolls. The continuous, or cross-country, mill is a third type of blooming mill. This mill has a series of two-high rolls. As many as 15 passes may be required to reduce an ingot 21 in^2 (135 cm^2) in **cross section** to a bloom 8 in^2 (52 cm^2) in cross section. The two- and three-high blooming mills roll the top and bottom of the steel in every pass. After one or two passes, mechanical manipulators on the runout table turn the steel to bring the side surfaces under the rolls for a more uniform material. After the steel is rolled, the uneven ends are sheared off, and the single long piece is cut into shorter lengths. The sheared off ends are reused as scrap. Most of the rolls used in these mills are horizontal, but there are also vertical rolls which squeeze the blooms or slabs from the sides. High-pressure water jets are used to remove mill scale which forms on the surface. Surface defects on the finished blooms and slabs are burned off, or scarfed, with an oxygen flame. The hot lengths of steel are moved from one station to another on a series of roller conveyors. The mill operations are automatically controlled by workers in an overhead glass-enclosed room called a "pulpit." The slabs, blooms, and billets are then taken to finishing mills where they are formed into special shapes and forms such as bars, beams, plates, and sheets. The steel is still not completely "finished" but it is closer to the form in which it will eventually be used in manufactured goods. Blooms and billets are finished into **rails**, wire rods, wires, bars, tubes, seamless pipe, and structural shapes such as I and H beams. Slabs are converted into plates, sheets, strips, and welded pipe.

After they are hot rolled, steel plates or shapes undergo further processing such as cleaning and pickling by chemicals to remove surface oxides, cold rolling to improve strength and surface finish, annealing (also known as **stress** relieving), and coating (galvanizing or aluminizing) for **corrosion** resistance.

Continuous or "strand" casting of steel eliminates the need to produce ingots and the use of soaking pits. In addition to costing less, continuously cast steels have more uniform compositions and properties than ingot cast steels. Continuous casting of steel produces an endless length of steel which is cut into long slabs or blooms that are ready for shaping in rolling mills. Molten steel is poured into the top of a continuous casting machine and is cooled by passing through a water-cooled mold. Pinch rolls draw the steel downward as it solidifies. Additional cooling is provided by water sprays along the travel path of the solidifying metal. The thickness of the steel "strand" is typically 10 in (25 cm) but new developments have reduced this thickness to 1 in (2.5 cm) or less. The thinner strand reduces the number of rolling operations required and improves the economy of the overall process. Some continuous cast machines bend the steel while it is still hot and flexible so that it comes out of the bottom in a horizontal position. Other machines cut the steel into sections while it is still in a vertical position. The continuous cast process has become the most economical method to produce large quantities of conventional steels. Small heats of alloy and specialty steels are still cast in ingots because the small size makes the continuous cast process impractical.

Some steel shapes are produced from powder. There are several chemical, electrochemical, and mechanical ways to make steel powder. One method involves improving the ore by magnetically separating the iron. A ball mill is then used to grind the ore into a powder that is then purified with hot **hydrogen**. This powder, under heat and **pressure**, is pressed into molds to form irregularly shaped objects; objects that would be hard to form any other way.

Quality control

To specify the various physical and mechanical properties of the finished product, various tests, both destructive and nondestructive, are performed. Metallurgical, hardness, hardenability, tension, ductility, compression, fatigue, impact, wear, corrosion, creep, machinability, radiography, magnetic particle, ultrasonic, and eddy current are some of the major tests that are performed by quality control personnel.

Metallurgical testing is used to determine the quality of steel by analyzing the microstructure of a sample under a **microscope**. A cross section of a sample is first highly polished and then examined at a magnification from 100-500 diameters. The microstructure of steel consists of grains of different compositions and sizes. Generally, a sample of steel with fine grains is tougher than one with large grains. Different characteristics are produced through alloying the steel with other substances. It is possible to determine grain size and the size, shape, and distribution of various phases and inclusions (nonmetallic material) which have a great effect on the mechanical properties of the metal. Some grains are made of ferrite, or pure metallic iron; graphite, a **crystal** form of carbon; pearlite, an alloy of iron of carbon; cementite, also called iron carbide, a hard compound of iron and carbon and other carbide-forming elements; austenite, a solid **solution** of carbon in gamma iron, a nonmagnetic form of iron; and martensite, an extremely hard constituent of steel produced by heat-treating. The sample can also be etched to make visible many structural characteristics of the metal or alloy by a preferential attack on the different constituents. The microstructure will reveal the mechanical and thermal treatment of the metal, and it may be possible to predict its expected behavior under a given set of conditions.

Hardness is not a fundamental property of a material, but is related to its elastic and plastic properties. The hardness value obtained in a particular test serves only as a comparison between materials or treatments. The test procedure and sample preparation are fairly simple and the results may be used in estimating other mechanical properties. Rockwell and Brinell are two popular hardness tests that are widely used for inspection and control. These tests are usually performed by impressing into the test specimen, which is resting on a rigid platform, an indenter of fixed and known **geometry**, under a known static load.

Hardenability is a property that determines the depth and distribution of hardness induced by quenching. The standardized test used is called the end-quench hardenability test, also known as the Jominy test. A 1-in (2.54 cm) round 4-in (10 cm) long sample is heated uniformly to the austenitizing temperature (this temperature depends on the material composition, ranging from 1,500–1,900°F [816–1,038°C]). The sample is removed from the furnace and placed on a fixture where a jet of water contacts the bottom face of the sample. After ten minutes on the fixture, the sample is removed and two flat parallel surfaces are ground on the sample. Rockwell hardness readings are taken along the ground surfaces at certain intervals from the quenched end. The results are expressed as a curve of hardness values versus distance from the quenched end. Plain carbon steels tend to be hard on the surface, near the quenched end, but remain relatively soft at the core, or further away from the quenched end. Alloyed steels, in general, have an increased depth of hardenability which is one of the main advantages of using alloyed steels.

Next to the hardness test, the tensile test is the most frequently performed test to determine certain mechanical properties. A specifically prepared tensile sample is placed in the heads of a testing machine and an axial load is placed on the sample through a hydraulic loading system. The tensile test is used to determine several important material properties such as yield strength, where the material starts to exhibit plastic or permanent deformation, and the ultimate tensile or breaking strength.

Ductility of a material is indicated by the amount of deformation that is possible until fracture and can be determined by measuring elongation and reduction in area of a tensile sample that has been tested to failure.

Compression tests are performed on small cylinders, blocks, or strips to determine the ability of a material to undergo large plastic deformations (a mechanical property also known as malleability) and its limits. Stress-strain relations determined from this testing are used to predict the pressures and forces arising in industrial forming operations such as rolling, forging, or extrusion. Samples are placed between anvils or pressure plates and are compressed (**friction** is also a factor to consider as the material slides sidewise over the anvils).

The fatigue test is used to determine the behavior of materials when subjected to repeated or fluctuating loads. It is used to simulate stress conditions developed in materials under service conditions. The fatigue potential, or endurance limit, is determined by counting the number of cycles of stress, applied first in one direction and then another, to which the metal can be subjected before it breaks. Fatigue tests can be used to study the material behavior under various types and ranges of fluctuating loads and also the effect of corrosion, surface conditions, temperature, size, and stress concentrations.

Impact tests are used to determine the behavior of materials when subjected to high rates of loading, usually in bending, tension, or torsion. The quantity measured is the energy absorbed in breaking the specimen in one blow, two such tests are called the Charpy and the Izod, which use notched bar specimens. A swinging pendulum of fixed weight raised to a standard height is used to strike the specimen. Some of the energy of the pendulum is used to rupture the specimen so that the pendulum rises to a lower height than the standard height. The weight of the pendulum times the difference in heights indicates the energy absorbed by the specimen, usually measured in foot-pounds.

Wear resistance is represented by few standardized tests because of its complex nature. One test is the "pin on disk" method, where a pin is moved against a disk of the test material. Usually, wear testing is application specific and the equipment is designed to simulate actual service conditions.

Corrosion involves the destruction of a material by chemical, electrochemical, or metallurgical interaction between the environment and the material. Various types of environmental exposure testing is done to simulate actual use conditions, such as **salt** bath immersion testing. Zinc coating, or *galvanizing*, is commonly applied to sheet and structural steel used for outdoor applications to protect against corrosion.

Creep tests are used to determine the continuing change in the deformation of a material at elevated temperatures when stressed below the yield strength. This is important in the design of parts exposed to elevated temperatures. Creep may be defined as a continuing slow plastic flow under constant load conditions. A creep test is a tension test run at a constant load and temperature. The **percent** elongation of the sample is measured over time.

Machinability is the ease with which a metal may be machined. Many factors are considered in arriving at machinability ratings. Some of the more important factors are the rate of metal removal, quality of the finished surface, and tool life. Machinability ratings are expressed as a percentage, in comparison with AISI 1112 steel, which is rated at 100%. Metals which are more difficult to machine have a rating of less than 100% while metals which machine easily have a rating more than 100%.

Radiography of metals involves the use of **x rays** or gamma rays. The short-wavelength electromagnetic rays are capable of going through large thickness of metal and are typically used to nondestructively test castings and welded joints for shrinkage voids and porosity.

Magnetic particle inspection (also called "Magnaflux") is a method of detecting cracks, tears, seams, inclusions, and similar discontinuities in iron and steel. This method will detect surface defects too fine to be seen by the naked **eye** and will also detect discontinuities just below the surface. The sample is magnetized and then covered with a fine iron powder. The presence of an imperfection is indicated by a pattern that assumes the approximate shape of the defect.

Ultrasonic testing utilizes **sound waves** above the audible range with a frequency of 1-5 million Hz (cycles per second). **Ultrasonics** allow for fast, reliable, nondestructive testing which employs electronically produced high-frequency sound waves to penetrate metals and other materials at speeds of several thousand feet per second. If there is a flaw in the path of the ultrasonic wave, part of the energy will be reflected and the signal received by a receiving **transducer** will be reduced. Ultrasonic inspection is used to detect and locate such defects as shrinkage voids, internal cracks, porosity, and large nonmetallic inclusions.

Eddy current inspection is used to inspect electrically conducting materials for defects and variations in composition. Eddy current testing involves placing a varying magnetic field (which is produced by connecting alternating current to a coil) near an electrically conducting sample. Eddy currents are induced in the sample which then produces a magnetic field of its own. A detection unit measures this new magnetic field and converts the signal into a voltage which can be read on a meter for comparison. Properties such as hardness, alloy composition, chemical purity, and heat treat condition influence the magnetic field and may be measured through the use of eddy current testing.

Byproducts/waste

There are a number of waste byproducts from the steel making process. Mine tailings from the ore beneficiation process are returned to the **mining** site. Growing vetches (a **species** of **plant** valuable for fodder), **grasses**, and trees on some of these barren landscapes has been a project of biologists and foresters. Gases that are given

off from the coke ovens, blast furnaces, and steel furnaces are largely recovered for reuse. After use in iron and steelmaking, most slags are used for other purposes such as railroad ballast and road fill, an ingredient in cement or blocks, insulating material, or fertilizer.

The future

In the future there will be many new developments involving computer controls and **automation** that will improve economy and quality and lower energy consumption and **pollution**. More automation will also lead to more robots replacing humans in hazardous areas. Computers can be used to control several rolling mills operating as a continuous unit. The decreasing material thickness can be maintained automatically as it passes through the various mills to produce a more uniform final sheet. Continued research and development is ongoing to connect continuous casting machines with rolling mills to provide a single continuous process from molten metal to the final product. This will produce energy and cost savings because the material would not have to be reheated for processing, and result in a higher quality end product.

The use of 100% scrap in charging electric furnaces has cut the dependence on pig iron and ores, and has resulted in the development of more small steel mills, also called mini-mills, which can be located far from natural resources to serve wider geographical areas.

More net steel shapes will be formed using powder **metallurgy** as direct reduction processes produce steel powders directly from iron ore, bypassing the blast furnace and making difficult shapes easier to form.

See also Metal production; Metallurgy.

Resources

Books

Hudson, Ray and Sadler, David. *The International Steel Industry.* Routledge, 1989.
Kalpakjian, Serope. *Manufacturing Processes for Engineering Materials.* 2nd ed. Addison-Wesley Publishing, 1991.
Walsh, Ronald A. *McGraw-Hill Machining and Metalworking Handbook.* McGraw-Hill, Inc., 1994.

Periodicals

ASM Publication. *ASM Metals Handbook Vol. 1: Properties and Selection: Irons, Steels, and High Performance Alloys*

Stellar evolution

The **mass** of a **star** determines the ultimate fate of a star. Stars that are more massive **burn** their fuel quicker and lead shorter lives. Because stars shine, they must change. The **energy** they lose by emitting **light** must come from the **matter** of which the star is made. This will lead to a change in its composition. Stars are formed from the material between stars, shine until they exhaust their fuel, and then die a predictable death based upon their initial mass. The changes that occur during a star's life are called stellar **evolution**.

From atoms to stars

Understanding of the processes of stellar evolution came as a result of twentieth century advances in both **astronomy** and atomic **physics**. Advances in quantum theory and improved models of atomic structure made it clear to astronomers that deeper understanding of the life cycle of stars and of cosmological theories explaining the vastness of **space** was to be forever tied to advances in understanding inner workings of the universe on an atomic scale. In addiiton, a complete understanding of the energetics of mass conversion in stars was provided by Albert Einstein's (1879–1955) special theory of relativity and his **relation** of mass to energy (Energy = mass times the square of the speed of light squared).

Indian-born American astrophysicist Subrahmanyan Chandrasekhar (1910–1995) first articulated the evolution of stars into **supernova**, white dwarfs, **neutron** stars and for predicting the conditions required for the formation of black holes subsequently confirmed by observation in the last years of the twentieth century.

Stellar mechanics

The material between stars occurs in **clouds** of varying mass. By processes that are still not completely clear, but involve cooling of the cloud-center with the formation of molecules, and the squeezing of the cloud by outside star light or perhaps a stellar explosion, the cloud begins to collapse under its own self-gravity. The collapse of the cloud results in the material becoming hotter simply from the squeezing of the collapse. At this point, the interior of the star churns. This churning process is called **convection**. Its rate of collapse is determined by the rate at which it can lose energy from its surface. Atomic processes keep the surface near a constant **temperature** so that a rapid collapse is slowed by the radiating surface area shrinking during the collapse. The star simply gets fainter while the interior gets progressively hotter.

Finally, the internal temperature rises to the point where **atoms** located at the center of the star, where the temperature is the hottest, are moving so fast from the **heat** that they begin to stick together. This process is called **nuclear fusion**, and it results in an additional pro-

duction of energy. Thus the star has a new source of heat. The subsequent evolution of the star will be largely determined by its mass.

If the mass of the star is about like that of the **Sun** or less, the nuclear "fires" which now provide the energy for the star to shine will determine its internal structure. A central radiative core is surrounded by a convective envelope. In the radiative core the material remains quiescent, while energy generated by nuclear fusion of **hydrogen** to helium simply diffuses through it like the light from auto headlights shines through a **fog**. It is at the very center of this radiative core that the helium "ash" of the nuclear "fires" accumulates as the star ages. Beyond the radiative core lies the churning convective envelope through which the energy is carried by blobs of hot matter rising past returning cooler blobs. At the atmospheric surface, the energy again flows as it did in the core until it physically leaves the star as starlight.

The structure of stars more than twice the mass of the sun is essentially the reverse of the low-mass stars. The cores of these stars are fully convective so that the energy produced by nuclear fusion is carried outward by the churning **motion** of the material in the core. The surrounding radiative envelope behaves much like the cores of lower-mass stars except no new energy is produced there. The churning motion of the material in the convective core causes the nuclear ash of helium to be well-mixed with the surrounding hydrogen fuel. This motion ensures that virtually all the hydrogen will be available to the nuclear fires which heat the star.

Both high- and low-mass stars respond to the depletion of hydrogen fuel in a similar manner. In order to supply the heat to oppose its own self-gravity, the star's core again responds by shrinking. In a sort of **reflex** reaction, the outer regions of the star expand, causing a great increase of its radiating surface area. Although the total energy output of the star increases during this phase, the greatly enhanced surface area results in a cooling of the surface and the star takes on a redder appearance. The size and **color** change lead to the name of red giant for these stars. If the star is very massive, it may become what is called a red supergiant.

For the low-mass stars, the expansion to the red giant phase will begin when about 90% of its hydrogen has been converted to helium. During the contraction of its core, a complicated sequence of events occurs. The shrinkage required to produce the energy radiated by the large giant causes the core to shrink to the dimensions of a **white dwarf**, while hydrogen continues to burn by nuclear fusion in a thin shell surrounding the core. It is this shell that provides most of the energy that is radiated away by the star. However, the core material, having attained the dimensions of a white dwarf, behaves very differently than the high-density gas that it was earlier in its life. No longer must it be heated to generate the **pressure** required to oppose the weight of the overlying material. When matter reaches this state it is called degenerate matter. The degenerate core just sits there, becoming hotter from the energy released by the surrounding hydrogen-burning shell and growing slowly from the helium ash generated by the shell. The hydrogen-burning shell is required to produce increasing amounts of energy from decreasing amounts of hydrogen fuel to sustain the brightening red giant. This continuing increase in the energy output from the shell heats the core, which finally reaches a temperature where the helium begins to undergo nuclear reactions, producing **carbon**. In this fusion process three helium nuclei collide yielding one carbon nucleus and additional energy for the support of the star.

The star now has a new energy source. However, the degenerate nature of the core doesn't allow it to expand and cool as would a core made of ordinary gas. Thus the onset of helium burning leads to a rapid rise in core temperature which is not balanced by a cooling expansion. The increased core temperature leads to a dramatic increase in helium burning. This sequence, known as the helium flash, continues until the degeneracy of the material making up the core is removed by the intense heat. The return of the material to its ordinary gaseous state leads to a rapid expansion which cools the core and reduces the helium burning. An equilibrium is established with the star generating progressively more energy from helium fusion, while the energy from the hydrogen burning shell is reduced, and it ultimately goes out from lack of fuel. The star continues to shine through the red giant phase by converting helium into carbon through nuclear fusion.

As the helium becomes depleted, the outer layers of the star become unstable and rather gently lift off the star to be slowly blown away by the light from the star. (Such an image was captured in 1998 by the **Hubble Space Telescope**, of a dying star known as NGC7027, located about 3,000 light-years from the sun in the direction of Cygnus the Swan.) Such shells of expanding gas are observed as greenish disks that eventually become greenish rings and the material becomes less dense. These greenish clouds are called **planetary nebulae**, even though they have nothing whatever to do with planets. Astronomers of two centuries ago gave them that name, for their telescopic appearance from **Earth** was like that of the outer planets **Uranus** and **Neptune**.

The remaining core of the red giant, now exposed, cools rapidly, again becomes degenerate, and is known as a white dwarf. A white dwarf has reached a stalemate between its own self-gravity and the nature of the degenerate stuff of which it is now composed. It may now simply cool off to become a dark stellar cinder about the size of the earth.

The evolution of a massive stars follow a somewhat different course. The churning of the convective core makes most of the hydrogen fuel available for consumption in the nuclear fires. Thus, these stars will not suffer the effects of core contraction until more than 99.9% of their hydrogen has been consumed. Even though they can consume more of their hydrogen (on a percentage basis), and they have more fuel to burn, they also shine much brighter than the low-mass stars. Thus their overall lifetimes will be far less than the low-mass stars. While the lifetime of a star like the sun may approach ten billion years, a star with ten times the mass of the sun may last less than ten million years.

The exhaustion of the hydrogen convective core leads to its contraction and the expansion of the outer layers, as was the case with the low-mass stars. However, the fate of the core is rather different from that of the low-mass stars. The core is far too massive to reach equilibrium as a degenerate structure like a white dwarf, so that contraction continues heating the core until the ignition of helium fusion is achieved. Unlike the lower-mass stars where the onset of helium burning occurs with a flash, the helium fusion in massive stars begins slowly and systematically takes over from the hydrogen-burning shell surrounding the core. Throughout the red giant or supergiant phase the role of energy production steadily shifts from hydrogen burning to helium burning. Eventually, helium becomes exhausted around a growing carbon core. While helium continues to undergo fusion in a shell surrounding the core, carbon fusion is ignited. Just as a degenerate helium core gives rise to the unstable ignition of helium, called the helium flash in low-mass stars, so the degenerate carbon core of moderate mass stars can result in an unstable ignition of carbon. However, whereas the helium flash is quickly quelled in low-mass stars, the carbon ignites explosively in the cores of these moderate-mass stars. This process is called carbon deflagration and may result in the destruction of the star.

In even more massive stars, the onset of carbon burning is a controlled process and the star develops multiple shells of energy sources involving carbon fusion, helium fusion, and even some hydrogen fusion in the outer regions of the star. The ignition by nuclear fusion of each new element yields less energy than the one before it. In addition, the increased temperature required for the nuclear fusion of these additional sources leads to an increase in the stellar luminosity. The result is an ever-increasing rate of the formation of less-efficient energy sources. When nuclear fusion in the core of the star yields **iron**, further nuclear fusion will no longer yield energy. Instead, nuclear fusion of iron will use up energy-robbing

KEY TERMS

. . . .

Convective core—The central, or surrounding regions of a star where the energy is carried by envelope convection. Convective transport of energy is the same as that found in a pan of boiling water where hot material physically rises, carrying energy, and having deposited that energy at the top of the region, descends as cooler material.

Deflagration—The explosive onset of nuclear fusion leading to the disruption of the reaction structure.

Degenerate gas—A gas whose constituents are packed to such high densities that further packing would violate the Pauli Exclusion Principle. The pressure of such a gas exhibits almost no dependence on temperature.

Ideal gas—Gas that obeys the Ideal Gas Law, where the pressure is proportional to the product of the local temperature and density.

Neutrino—A nuclear particle resulting from nuclear reactions. Neutrinos interact very weakly with ordinary matter, and can easily traverse a normal star without colliding with any of the stellar material.

Neutron—Together with protons, neutrons comprise the basic building blocks of the nuclei of the elements. They have a mass just slightly greater than that of a proton, but lack its electric charge.

Neutron star—A star with a mass similar to the Sun, but composed almost entirely of neutrons. The neutrons are packed so tightly that they are degenerate, like the electrons of a white dwarf—but the resulting density is far greater. The typical size of such a star is about 6.2 mi (10 km).

Nuclear fusion—The combining of the nuclei of two elements so as to produce a new, more massive element. The process is accompanied by the release of energy as long as the end product of the reaction is less massive than iron.

Planetary nebula—An expanding shell of gas ejected by a low-mass red giant, which may take on the appearance of one of the outer planets of the solar system when seen in a small telescope.

Radiative core—The central, or surrounding regions of a star where the energy is carried by envelope radiative diffusion. Radiative diffusion describes the flow of particles of light (photons) through a medium where there is little mechanical change to the medium.

Supernova—The final collapse stage of a supergiant star.

White dwarf—A star that has used up all of its thermonuclear energy sources and has collapsed gravitationally to the equilibrium against further collapse that is maintained by a degenerate electron gas.

thermal energy from the surrounding material. This sudden cooling of the core will bring about its collapse.

As the **density** increases in the collapsing core, there is less and less room for the free electrons that have been stripped from the atomic nuclei by the extreme temperature. These electrons must go somewhere, so they will begin to be "absorbed" in the protons of the atomic nuclei, turning them into neutrons. The process is called neutronization. This reaction generates particles called neutrinos, which interact very weakly with ordinary matter, and so normally escape directly from the star. The energy robbed from the core by the neutrinos also adds to the energy crises in the core and contributes to the core collapse.

The production of elements with masses greater than iron also produces large quantities of neutrinos, so that whichever process dominates, a great deal of energy is lost directly from the star, resulting in a catastrophic gravitational collapse of the core. This is followed

promptly by the collapse of the entire star. The rapid increase in the density of the collapsing core finally reaches the point where the material becomes opaque to the energy-robbing neutrinos, and their continued escape is stopped. The sudden deposition of the **neutrino** energy in the collapsing core reverses the collapse, bringing about an explosion of unprecedented magnitude. The infalling matter and trapped photons are hurled into space, liberating as much energy in a few minutes as the star has radiated in its lifetime of millions of years.

The remains of this titanic explosion depend on the initial mass of the collapsing star. Very-massive stars may leave a **black hole** of completely collapsed matter behind. Should the collapse involve a star of less mass, the remainder may be something called a **neutron star**, similar to that formed by the collapse of a white dwarf. In some instances, the entire star may be involved in the explosion and there will be no remains at all. While there have been recent attempts to refine the classification of these explosions, astronomers still refer to the explosion of a massive star as a supernova of type II. Supernovae of type I are thought to result from the collapse of a white dwarf which has exceeded its critical mass. Unlike the evolution of low-mass stars, in which an accommodation between the forces of gravity and degenerate structure of the star is achieved through the formation of a white dwarf, the evolution of a massive star must end in a violent stellar explosion. Gravity appears to win its struggle with nuclear physics, but at the last moment, the energy of collapse is turned to an explosion leaving either a collapsed corpse, or perhaps nothing at all.

The accession of the Hubble Space Telescope had given astronomers a valuable tool to study the evolution of stars in the universe, at the same time challenging their understanding. In 1997, Hubble detected rogue stars that belong to no **galaxy**, displaced long ago and now hanging in empty intergalactic space among star clusters like the Virgo Cluster, about 60 million light-years from Earth.

In 1996, astronomers found evidence of many isolated, dim brown dwarfs, lacking sufficient mass to start nuclear fusion. They detected light spectra from the element **lithium**, which quickly burns in true stars. These brown dwarfs, called "L dwarfs," are typically smaller then our sun but much larger than even **Jupiter**, and some may resemble Saturn's **moon** Titan.

On the opposite scale, in 1997 Hubble detected the then brightest star ever seen. Discovered at the core of our own galaxy and named the Pistol Star, it has the energy of ten million Suns and would fill the distance of the Earth's **orbit** around the Sun. The Pistol Star is about 25,000 light-years from Earth; it is so turbulent that its eruptions create a gas cloud four light-years across. It had been thought that a star so big could not have formed without blowing itself apart, and so the Pistol Star will require astronomers to re-examine their ideas about stellar formation, especially of supermassive stars near the centers of galaxies.

By 2003, other observations, including x-ray observations from the ROSAT Observatory and NASA's Chandra x-ray Observatory, allowed the identification of high intensity ultra-bright x-ray sources that many astronomers argued were evidence of black holes in star-forming galaxies. Although there are other explanations for these phenomena, the fact that they provide additional confirmation of black holes is enhanced by Hubble observations of stars rotating around stellar cores of these galaxies.

In early 2003, the Chandra x ray Observatory, provided extended observations of Sagittarius A (or Sgr A), the supermassive black hole at the center of Earth's own **Milky Way** galaxy.

See also Gravity and gravitation; Red giant star; Star formation.

Resources

Books

Collins, G.W., II, *Fundamentals of Stellar Astrophysics.* New York: W.H. Freeman, 1989.

Prialnik, Dina. *An Introduction to the Theory of Stellar Structure and Evolution.* Cambridge University Press, 2000.

Other

University of Cambridge. "Our Own Galaxy: The Milky Way." *Cambridge Cosmology*, May 16, 2002 [cited January 21 2003]. <http://www. damtp. cam.ac.uk/user/gr/public/gal_milky.htm>.

George W. Collins
K. Lee Lerner

Stellar magnetic fields

Stellar magnetic fields are an array of forces that can be observed surrounding and at the surfaces of stars like the **Sun**. They are similar in nature to the effect of the well-known dipolar magnets found in science laboratories, classrooms, and toys, but far more powerful and infinitely more complex. They are an important part of the physical makeup of stars because they affect their interiors, atmospheres, and immediate environments. Observations of the Sun show that it has a dynamic, overall magnetic field and also smaller, but often much stronger "pockets" of **magnetism** associated with **sunspots**. The influence of these more localized magnetic fields can sometimes be quite dramatic when they are involved in the creation, shaping, and size of solar prominences, flares, and some features in the solar atmosphere (the corona). The large-scale magnetic field of the Sun helps determine processes by which chemical elements are transported within and around the Sun, and even the spin, or **rotation**, of the stellar surface. The study of the sun's magnetic fields, particularly their large- and small-scale structures, helps in understanding their origins and it is assumed by astronomers that when magnetic fields around stars other than the sun are studied in detail they will show similar features and dynamics. Knowledge gained about the magnetic fields of stars can lead to an understanding of their potential impact on long-term **stellar evolution**.

Exactly how stellar magnetic fields work is somewhat of a mystery. The most widely accepted explanation for them is called the dynamo model. The dynamo principle is used in generators on **Earth**, but may be thought of as the reverse of what is happening in a **star**. In a simple emergency **generator**, a gas engine spins a magnet within a coil of wire. The interaction of the moving magnetic field within the coil generates **electricity** in the wire, which is then sent out to a connector that provides electrical power to devices outside the generator. This is how hydroelectric power is generated at **dams** like the famous Hoover Dam, but instead of a gas engine, **water** under high **pressure** provides the **motion** required to make the generator work.

In a stellar dynamo, rather than electricity being generated because of a moving magnetic field, a magnetic field seems to be generated by two major motions within the star. The first is the movement of the gases in the **convection** zone, which makes up the upper layer of the star. In this region, material at and just beneath the surface moves up as **heat** is transferred outward from lower layers to the surface by a process in which hot gas rises just as hot air does on earth. Once some of the heat of the gas is released at the surface of the sun, that gas drops down again as it is replaced by hotter gases from below.

The second motion is caused by the simple fact that the Sun is made of gas. Because of this, it does not rotate at the same speed everywhere as would a solid object like a **planet**. This is called differential rotation and it causes the material at the equator to move faster than material at the poles. While scientists have not worked out all the details, it appears that these two effects together create the basic stellar magnetic field of the Sun and other stars. However, to be able to create a full picture, it would be necessary to describe accurately all the physical processes operating on the surface of and in the interior of every area of the sun including small- and large-scale **turbulence**. In addition, the overall magnetic field and sunspot fields themselves effect the movements of the convection zone, creating a situation far more complex than the highly unpredictable **weather** patterns of Earth. A deeper understanding of the causes of stellar magnetic fields will require observations of many more stars and a more complete understanding processes within them.

On the Sun, more localized magnetic fields can be found and are made visually obvious by the appearance of sunspots, which were first recorded by ancient Chinese astronomers. They can be so large that they can indeed be observed, with proper filtering, with the naked **eye**. In the 1600s, Galileo Galilei and his contemporaries rediscovered sunspots shortly after the start of telescopic **astronomy**. Sunspots are regions on the solar surface that appear dark because they are cooler than the surrounding surface area (photosphere) by about 2,200°F (1,200°C). This means they are still at a **temperature** of about 7,600°F (4,200°C). Even though they look dark in photographs of the Sun, they are still very bright. If a piece of sunspot could be brought to Earth, it would be extremely hot and blinding to look at just as any other piece of the Sun. Sunspots develop and persist for periods ranging from hours to months, and are carried around the surface of the sun by its rotation. Sunspots usually appear in pairs or groups and consist of a dark central region called the umbra and a slightly lighter surrounding region called the penumbra. The rotation period of the Sun was first measured by tracking sunspots as they appeared to move around the Sun. Galileo used this method to deduce that the Sun had a rotational period of about a month. However, because the Sun is not a solid body, it does not have one simple rotational period. Modern measurements indicate that the rotation period of the Sun is about 25 days near its equator, 28 days at 40° latitude, and 36 days near the poles. The rotation direction is same as the motion of the planets in their orbits around the Sun.

The magnetic causes of sunspots were not known until the early years of the twentieth century, when George Ellery Hale mapped the solar magnetic field through its effect—called the Zeeman effect—on the detailed shape and polarization of **spectral lines**. Spectra show the chemical makeup of stars and are a major source of information for astronomers. They are created by spreading the **light** of a star into its component parts in the same way a **prism** creates a rainbow of colors from a light source. **Chemical reactions** in the star create lines of different intensities at predictable places along the **spectrum** allowing scientists to determine the makeup of the star. Since the chemical reactions would produce a spectral line in a given way in the absence of a magnetic field, we can see the effects of fields by comparison to the known spectrum of the reaction. The Zeeman effect is a change in the spectral lines caused by the sun's magnetic field. The sun's magnetic field has been mapped on a regular basis ever since Hale first did it, and it is now known that the 11-year sunspot cycle is just a part of an overall 22-year magnetic cycle. The shape of the sun's magnetic field changes throughout the 11-year cycle, reverses its magnetic polarity and begins the whole process over again. In addition to the differential rotation helping to cause the magnetic field of the sun, it also stretches the north-south magnetic field lines until they run east-west during the first 11 years of the magnetic cycle. Rotating convection then somehow regenerates the north-south field, but with a reversed polarity, causing the process to start again for another 11 years. During these half-cycles, the number and intensity of sunspots increases and decreases with the changes in the overall magnetic field.

Using special high-resolution spectropolarimeters combined with other techniques, the magnetic fields of stars beyond the Sun can be detected through the effect they have on the Zeeman signatures found in the shape

KEY TERMS

· ·

Corona—The outermost layer of the sun's atmosphere, seen during total solar eclipses as a glowing irregular halo.

Dipolar magnet—The common bar magnet that has opposing north and south magnetic fields.

Flare—A sudden burst of electromagnetic energy and particles from a magnetic loop in an active region of the sun. Sends material out into the solar system that can disrupt electronic devices even on Earth.

Galileo Galilei—The Italian physicist and astronomer (1564-1642) who is credited with first turning a telescope to the sky. Discovered the moons of Jupiter, providing the first observational evidence of smaller celestial bodies moving around larger ones. For stating that Earth definitely must, therefore, move around the Sun, he was placed under house arrest for the latter part of his life.

George Ellery Hale—The American astronomer

(1868-1938) best known for his contribution to the design and development of the world-famous 200-in (508 cm) telescope on Mt. Palomar in California.

Photosphere—The visible surface of the Sun. The region from which light escapes from the Sun into space.

Prominence—A cool cloud of hydrogen gas above the sun's surface in the corona. Shaped by local magnetic fields of active regions on the Sun.

Spectropolarimeter—A device that gathers information on the polarization state of individual chemical reactions from a star seen as lines in the star's spectrum.

Zeeman-Doppler imaging—The process of using a spectropolarimeter to measure the Zeeman effect, the polarization of spectral lines and a shift in frequency of the lines due to the effect of magnetic fields on the light from a star.

and polarization state of spectral lines of those stars. Zeeman-Doppler imaging (ZDI) works best for moderate to ultra-fast rotating stars, for which the polarization of individual magnetic regions match the different speeds at which the surface of the star rotates. This method was used to detect the magnetic fields in cool stars other than the Sun, showing that the same type of phenomena occur on other stars. Using Zeeman-Doppler imaging, astronomers have managed to detect and map the surface magnetic field of a few extremely active stars of about one solar **mass** (with ages ranging from a few million to more than ten billion years—twice the Sun's age). Some major differences were found between the alignment of the magnetic field lines of these stars and those of the Sun, adding to the mystery of understanding stellar magnetic fields. The conclusion of astronomers studying these results is that the entire convection zone of these active stars is involved in forming the magnetic field rather than just the upper layers as appears to be the case with the Sun.

These methods allow monitoring of the long-term **evolution** of the magnetic field shape and strength of other stars. Using them, astronomers hope to be able to detect the polarity switch of the large-scale field and observe a stellar analog of the solar magnetic cycle. If a change in magnetic field polarity is observed, it may indicate the approach of a polarity switch in the magnetic field of the star. Such observations would show that stellar magnetic fields are indeed very similar to those of the Sun.

Resources

Books

Introduction to Astronomy and Astrophysics. 4th ed. New York: Harcourt Brace, 1997.

Zelnik, Michael. *Astronomy.* 7th ed. Wiley and Sons, Inc. 1994.

Clint Hatchett

Stellar magnitudes

Magnitude is the unit used in **astronomy** to describe a star's brightness in a particular portion of the **electromagnetic spectrum**. Stars emit different amounts of **radiation** in different regions of the **spectrum**, so a star's brightness will differ from one part of the spectrum to the next. An important field of research in modern astronomy is the accurate measurement of stellar brightness in magnitudes in different parts of the spectrum.

How bright it looks: apparent magnitude

The Greek astronomer Hipparchus devised the first magnitudes in the second century B.C. He classified stars according to how bright they looked to the **eye**: the brightest stars he called "1st class" stars, the next brightest "2nd class," and so on down to "6th class." In this way all the stars visible to the ancient Greeks were neatly classified into six categories.

Modern astronomers still use Hipparchus' categories, though in considerably refined form. With modern instruments astronomers measure a quantity called V, the star's brightness in the visual portion of the spectrum. Since visual **light** is what our eyes most readily detect, V is analogous to Hipparchus' classes. For example, Hipparchus listed Aldebaran, the brightest **star** in the **constellation** Taurus (the Bull), as a 1st class star, while today we know that for Aldebaran $V = 0.85$. Astronomers often refer to a star's visual brightness as its apparent magnitude, which makes sense since this describes how bright the star appears to the eye (or the **telescope**). You will hear an astronomer say that Aldebaran has an apparent magnitude of 0.85.

Hipparchus' scheme defined from the outset one of the quirks of magnitudes: they run backwards. The fainter the star, the larger the number describing its magnitude. Therefore, the **Sun**, the brightest object in the sky, has an apparent magnitude of -26.75, while Sirius, the brightest star in the sky other than the Sun and visible on cold winter nights in the constellation Canis Major (the Big Dog), has an apparent magnitude of -1.45. The faintest star you can see without optical aid is about +5 (or +6 if your eyes are very sharp), and the faintest objects visible to the most powerful telescope on **Earth** have an apparent magnitude of about +30.

How bright it really is: absolute magnitude

More revealing than apparent magnitude is absolute magnitude, which is the apparent magnitude a star would have if it were ten parsecs from the Earth (a parsec is a unit of distance equal to 12 trillion mi [19 km]). This is important because apparent magnitude can be deceiving. You know that a lit match is not as bright as a streetlight, but if you hold the match next to your eye, it will appear brighter than a streetlight six blocks away. That's why V is called apparent magnitude: it is only how bright the star appears to be. For example, the Sun is a fainter star than Sirius! Sirius emits far more **energy** than the Sun does, yet the Sun appears brighter to us because it is so much closer. Absolute magnitude, however, reveals the truth: the Sun has an absolute magnitude of +4.8, while Sirius is +1.4 (remember, smaller numbers mean brighter, not fainter).

The nature of the magnitude scale

In 1856, the British scientist N. R. Pogson noticed that Hipparchus' 6th class stars were roughly 100 times fainter than his 1st class stars. Pogson did the sensible thing: he redefined the stars' V brightness so that a difference of five magnitudes was exactly a **factor** of 100 in brightness. This meant that a star with $V = 1.00$ appeared to be precisely 100 times brighter than a star with $V = 6.00$. One magnitude is then a factor of about 2.512 in brightness. Try it: enter 1 on a **calculator** and multiply it by 2.512 five times-you've just gone from first magnitude to sixth (or eight to thirteenth, or minus seventh to minus second, or any other combination).

Looking back to the numbers above, we see that the Sun ($V = -26.75$) has an apparent visual brightness 25 magnitudes greater than Sirius. That's five factors of five magnitudes, or $100 \times 100 \times 100 \times 100 \times 100$: ten billion! And the difference in apparent brightness between the Sun and the faintest object humans have ever seen (using the **Hubble Space Telescope**) is more than 56 magnitudes, or a factor of ten billion trillion.

Magnitudes in modern astronomy

In the 140 years since Pogson created the modern magnitudes, astronomers have developed many different brightness systems, and the most popular are less than 50 years old.

For example, in 1953, H. L. Johnson created the *UBV system* of brightness measurements. We've already met V, which is measured over that portion of the electromagnetic spectrum to which our eyes are most sensitive. B is the star's brightness in magnitudes measured in the blue part of the spectrum, while U is the brightness in the ultraviolet—a spectral region our eyes cannot detect. There are many other brightness measurement systems in use, so many that astronomers often disagree on how measurements in one system should be converted to another.

Accurate measurement of stellar brightness is important because subtracting the brightness in one part of the spectrum from the brightness in another part reveals important information about the star. For many stars the quantity $B\text{-}V$ gives a good **approximation** of the star's **temperature**. And it was established in 1978 that the quantity $V\text{-}R$, where R is the brightness in the red part of the spectrum, can be used to estimate a star's radius. This is important, because advances in our understanding of the stars require knowledge of basic parameters like temperature and radius, and careful measurements of brightness can provide some of that information.

See also Spectral classification of stars.

Resources

Books

Introduction to Astronomy and Astrophysics. 4th ed. New York: Harcourt Brace, 1997.
Kaufmann, William. *Discovering the Universe.* New York: W. H. Freeman, 1990.

KEY TERMS

. . . .

Absolute magnitude—The apparent brightness of a star, measured in units of magnitudes, at a fixed distance of 10 parsecs.

Apparent magnitude—The brightness of a star, measured in units of magnitudes, in the visual part of the electromagnetic spectrum, the region to which our eyes are most sensitive.

Brightness—The amount of energy a star radiates in a certain portion of the electromagnetic spectrum in a given amount of time, often measured in units of magnitudes.

Magnitude—The unit used in astronomy to measure the brightness of a star. One magnitude corresponds to a factor of 2.512 in brightness, so that five magnitudes = a factor of 100.

Mitton, Simon P., ed. *The Cambridge Encyclopedia of Astronomy.* Cambridge: Cambridge University Press, 1977.
Sherrod, P. Clay. *A Complete Manual of Amateur Astronomy.* New York: Dover, 2003.

Jeffrey C. Hall

Stellar populations

Stars fall into distinct groups or populations. The basic stellar populations are Population I stars and Population II stars. The **sun** and most stars near the sun are Population I stars. They are young second- to third-generation stars with compositions that include 2% of elements heavier than **hydrogen** and helium. Population II stars, on the other hand, are older stars whose compositions are just hydrogen and helium. The brightest stars in a group of Population I stars are blue and in a group of Population II stars are red. There are also additional subclassifications within this basic classification.

History

Walter Baade discovered the stellar populations during World War II, which contributed to his discovery. During the war, most scientists worked on war-related projects, such as the Manhattan Project. Baade, who was on the staff of Mount Wilson Observatory near Los Angeles, was not allowed to work on war projects because he was German-born. He did, however, conduct research at Mount Wilson Observatory, which at the time had the world's largest **telescope**. The wartime blackouts in Los Angeles contributed to darker skies on Mount Wilson. There were also few astronomers wanting to use the telescope because they were working on war-related projects. So Baade had plenty of time to use the world's largest telescope under better than normal sky conditions.

During this time Baade was able to resolve, for the first time, the stars in the Andromeda **Galaxy**. He noticed that there were two distinct populations composed of predominantly red and blue stars. He labeled them Population I and Population II. Further study since that time has produced the additional subclassifications.

Properties of populations

Population I

Population I stars have properties similar to those of the sun. They are less than ten billion years old and include newly-formed and still-forming stars. Because they are younger second- to third-generation stars, they contain heavy elements that were manufactured in previous generations of stars. (To astronomers, a heavy element is anything heavier than hydrogen or helium. These two lightest elements make up roughly 98% of the **matter** in the universe.) Population I stars contain roughly 2% heavy elements and 98% hydrogen and helium.

In a group of Population I stars, the brightest stars will be hot blue giants and supergiants. These stars, much more massive than the sun, are in the main part of their life cycles, burning hydrogen in their cores. The dominance of hot blue stars does not mean that cooler, less-massive red and yellow stars such as the sun cannot be Population I stars. Rather, the cooler stars like the sun are not as bright, so in a group of Population I stars viewed from a distance they will be less noticeable.

It turns out that the stellar populations also have dynamic properties in common. Population I stars are concentrated in the disk of the galaxy. They have circular orbits around the center of the galaxy with very little **motion** in a direction **perpendicular** to the galactic **plane**. They tend to have a patchy distribution within the disk and **spiral** arms of the galaxy. They also tend to be located in regions that have significant amounts of interstellar gas and dust, the raw materials for forming new stars.

Population II

The older Population II stars are usually over ten billion years old. Because they are first-generation stars that formed early in the history of the universe, they are devoid of heavy elements. Their composition is similar to that of the early universe. The brightest stars in a group of Population II stars are red giants. Red giants are stars in the process of dying. They have run out of hydrogen fuel in the core and swollen into cool red giants typically the size of the earth's **orbit** around the sun. Because they are so large, they are very bright and stand out in a group of Population II stars. Groups of Population I stars do not contain red giants simply because they are younger; they have not had enough time to exhaust the hydrogen fuel in their cores.

The Population II stars also have different dynamical properties. They are not confined to the plane of the galaxy. They have highly eccentric noncircular orbits that often go far above or below the plane of the galaxy to form a smoothly distributed spherical halo around the galaxy. They therefore must have significant components of their motions that are perpendicular to the plane of the galaxy. They also have much higher orbital velocities than Population I stars. Population I stars have orbital velocities that are typically about 5-6 mi (8-10 km) per second. Population II stars zip along at velocities ranging up to 45 mi (75 km) per second in the most extreme cases.

Other populations

Like most initial classifications, the division of stars into Population I and Population II stars is a bit of an oversimplification. Astronomers now classify stars into five distinct populations based on how strongly they exhibit the Population I or II characteristics. These populations are: Extreme Population I, Older Population I, Disk Population II, Intermediate Population II, and Halo Population II. Astronomers are currently arguing over whether these groups represent distinct populations, or a gradual blending from Population I properties to Population II properties.

There are also some exceptions to the rule that young Population I stars have heavy elements while older Population II stars do not. For example, the Magellanic **Clouds**, irregular companion galaxies to the **Milky Way**, are Population I stars with few heavy elements. The core of the Milky Way also contains Population II stars that do contain heavy elements. Why? Most likely, in the core of the galaxy there was a very rapid generation of very massive stars. They are gone now, but they enriched the core with heavy elements very quickly. The high concentration of stars near the core contributed to this process. The opposite occurred in the Magellanic Clouds. **Star formation** proceeded so slowly, that there was not an early generation of massive stars to produce heavy elements.

Reasons for different populations

These different populations can be understood in the context of **stellar evolution**. When the universe formed in the big bang, only hydrogen and helium were made.

KEY TERMS

Blue giant, supergiant—The most massive stars in the hydrogen burning stage.

Heavy elements—To astronomers, anything that is not hydrogen or helium.

Population I—The younger second- to third-generation stars.

Population II—The younger first-generation stars.

The heavy elements were made later in the cores of stars. Therefore, the older Population II stars are deficient in heavy elements, while the younger Population I stars contain heavy elements that were made by the massive first generation stars. Massive stars manufacture and recycle heavy elements. These stars are blue giants and supergiants most of their lives, which are very short by stellar standards. These stars are the brightest blue stars in a group of Population I stars. In a group of older Population II stars, these stars have finished their life cycles, so the brightest stars are the red giants formed near the end of stellar life cycles. The distribution of these different populations is then related to the **evolution** of the galaxy.

Stellar population studies help us understand stellar evolution, evolution of the galaxy, and the history of the universe.

Resources

Books

Bacon, Dennis Henry, and Percy Seymour. *A Mechanical History of the Universe.* London: Philip Wilson Publishing, Ltd., 2003.

Morrison, David, Wolff, Sidney, and Fraknoi, Andrew. *Abell's Exploration of the Universe.* 7th ed. Philadelphia: Saunders College Publishing, 1995.

Zeilik, Michael, Gregory, Stephen, and Smith, Elske. *Introductory Astronomy and Astrophysics.* Philadelphia: Saunders, 1992.

Paul A. Heckert

Stellar structure

It is said that Fred Hoyle once described the **evolution** of a **star** as a continual war between nuclear **physics** and gravity. The structure of a star can be characterized as a polarized battle in that war. The gravity of the stellar material pulls on all the other stellar material striving to bring about a collapse. However, the gravity is opposed by the internal **pressure** of the stellar gas which normally results from **heat** produced by nuclear reactions. This balance between the forces of gravity and the pressure forces is called hydrostatic equilibrium, and the balance must be exact or the star will quickly respond by expanding or contracting in size. So powerful are the separate forces of gravity and pressure that should such an imbalance occur in the **sun**, it would be resolved within half an hour. That fact that the sun is about five billion years old emphasizes just how exactly and continuously that balance is maintained.

In addition to its reliance on balance between gravity and pressure, the internal structure depends on the behavior of the stellar material itself. Most stars are made primarily of **hydrogen**, since it is the dominant form of **matter** in the universe. However, the behavior of hydrogen will depend on the **temperature**, pressure, and **density** of the gas. Indeed, the quantities temperature, pressure, and density, are known as state variables, since they describe the state of the material. Any equation or expression that provides a relationship between these variables is called an equation of state. The relevant equation of state and hydrostatic equilibrium go a long way toward specifying the structure of the star. However, there is one more ingredient necessary in order to determine the structure of the star.

Most of the **energy** which flows from a star originates at its center. The way in which this energy travels to the surface will influence the internal structure of the star.

There are basically three ways by which energy flows outward through a star. They are conduction, **convection**, and **radiation**. The flow of energy up an **iron** poker, which has one end in a fire, is **energy transfer** by conduction. This mode of energy flow is too inefficient to be of any interest in most stars. Most people have watched the energy flow from the bottom of a heated pot of **water** to the top. One sees large **motion** of heated material rising which is balanced by cooler material falling to the bottom where it, in turn, becomes heated and rises. This organized churning is called convection. In the interior of stars, when convection takes place it dominates the transport of energy over radiation. When the conditions are such that no convective churning takes place, no energy is carried by convection and the only remaining alternative is radiative transport.

The heat produced by an infrared heat lamp or the warmth of the sun on a clear day are good examples of energy transport by radiation. Radiative transport simply means the direct flow of radiant energy from one place to another. If the material through which the energy flows is relatively transparent, the flow takes place at virtually the speed of **light**. However, the more opaque the material is, the slower the flow of energy will be. In the sun, where

light flowing out from in the core will travel less than a centimeter before it is absorbed, it may take a million years for the light energy to make its way to the surface.

The mode of energy transport, equation of state, and hydrostatic equilibrium can be quantified and self-consistent solutions found numerically on a computer for a star of given **mass**, composition and age. Such solutions are called model stellar interiors, and supply the detailed internal structure of a particular star. For the vast majority of stars which derive their energy from the **nuclear fusion** of hydrogen into helium the internal structure is quite similar. We call such stars main sequence stars.

The sun is such a star and has a central region where the material is quiescent and the energy flows through by radiative **diffusion**. The radiative core is surrounded by a churning convective envelope which carries the energy to within a few thousand kilometers of the surface. This outer **percent** or so of the sun is called the atmosphere, and here the energy again flows primarily by radiation as it escapes into **space**. This structure is common to all main sequence stars with mass less than about one and a half time the mass of the sun. The structure of more massive stars is nearly reversed. They consist of a convective core surrounded by a radiative envelope, which in turn gives way to an atmosphere where the energy also leaves the star by radiation.

There is a class of very dense stars known as white dwarfs. These stars originate as main sequence stars, but have changed dramatically over time. As a result, their internal structure has changed drastically. The matter of a **white dwarf** no longer behaves like a normal gas, and it has a different equation of state. The structure of such a star is vastly different from main sequence star. For example, while a white dwarf has a comparable mass to that of the sun, it is not much larger than the **earth**. This means that their average density is huge. A cubic inch of white dwarf material might weigh forty tons and exert a tremendous pressure to balance the fierce gravity resulting from as much material as one finds in the sun in a **sphere** no larger than the earth. Material in such a state is said to be degenerate. In these weird stars, conduction by free moving electrons becomes extremely efficient so that any internal energy is transported almost instantly within the star. The entire interior is at the same temperature and only a thin outer blanket of more normal nondegenerate material keeps the energy from leaking quickly into interstellar space.

More than 90% of all stars fall into the categories described above. The remaining stars are called red giants or red supergiants. These names are derived from the external appearance, for they are relatively cool and therefore red, but are extremely distended. If one were to replace the sun with the typical red giant, it would occu-

KEY TERMS

. . . .

Hydrostatic equilibrium—Expresses a static balance between the forces of gravity and the forces due to the internal pressure.

Ideal gas—A term used to describe the behavior of a gas composed of simple atoms under standard conditions. An ideal gas obeys the following relation for the state variables of temperature T, number density n, and pressure P. The constant k is known as the Boltzmann constant. P = nkT.

Main sequence—A sequence of stars of differing mass, but all deriving their energy from the nuclear fusion of hydrogen into helium.

Nuclear fusion—The combining of the nuclei of two elements so as to produce an new more massive element. The process is accompanied by the release of energy as long as the end product of the reaction is less massive than iron.

py most of the inner **solar system**. Should the replacement involve a red supergiant, even the **planet Mars** would find itself within the supergiant's atmosphere and **Jupiter** itself would be heated to the point where much of its substance would evaporate away. However, so huge is the **volume** occupied by these stars that their average density would be considered a pretty good **vacuum** by most physicists. The situation is made even worse since the inner cores of these stars are about the size of the earth and contain about ninety percent of the mass of the giant. Most red giants have an object much like a white dwarf located at their center. Therefore, their structure would best be described as a white dwarf surrounded by an extensive **fog**. The energy for most of these stars is supplied by the nuclear fusion of helium into **carbon**, which is a much less energy efficient source than the fusion of hydrogen to helium. This is the basic reason why there are so many more main sequence stars than red giants and super giants. Another reason is that a star uses up its helium much more quickly and spends relatively little time in the red giant phase. By determining how the structure of a star changes with time, the evolutionary history and fate of that star can be constructed. This is called the theory of **stellar evolution**.

See also Red giant star.

Resources

Books

Abell, G.O. *Exploration of the Universe.* Philadelphia, New York: Saunders College Publishing, 1982.

Arny, T.T. *Explorations: An Introduction to Astronomy.* St. Louis, MO: Mosby, 1994.

Collins, G.W. II. *Fundamentals of Stellar Astrophysics.* New York: W.H. Freeman, 1989.

Seeds, M.A. *Horizons-Exploring the Universe.* Belmont, CA: Wadsworth, 1995.

George W. Collins

Stellar wind

The **sun** emits a constant stream of particles, mostly protons and electrons, that are known as the **solar wind**. Many stars also have a similar phenomenon, known as stellar **wind**. The solar wind is fairly gentle. Stellar winds as gentle as the solar wind are difficult to detect from the **earth** because other stars are so distant. However, many stars at certain stages in their **evolution** have very strong stellar winds. These strong winds produce effects that we can observe from Earth. They also can cause the **star** to lose significant amounts of **mass**.

Solar wind

The outermost layer of the sun is the corona, which is a very hot tenuous gas visible only during a solar eclipse. Because the coronal temperatures are typically one or two million degrees Kelvin, the individual **atoms** are ionized and moving very rapidly. The fastest ions are moving faster than the sun's escape **velocity**. So they escape, forming the solar wind.

Near Earth's **orbit**, the solar wind particles rush by at speeds of roughly 248-310 mi per second (400-500 km per second). There is considerable variation in the speed because the solar wind is gusty. The solar wind **density** is also variable, but typically runs to a few particles per cubic centimeter. The solar wind extends well beyond the orbit of **Pluto** to roughly 100 astronomical units (100 times the earth to sun distance). At this point, called the heliopause, the solar wind merges into the interstellar gas and dust.

Stellar winds

Analogous to the solar wind many stars have stellar winds. Because stars are so distant, stellar winds that are as gentle as the solar wind do not produce dramatic effects as seen from the earth. The stellar winds that we observe are therefore much stronger than the solar wind. A variety of different types of stars display interesting stellar winds.

Massive hot stars

The hottest, most massive stars are O spectral class stars, which have at least 15 times the mass of the sun. Wolf-Rayet stars have many characteristics in common with the O stars, but their nature is still not completely understood. Both O stars and Wolf-Rayet stars often have very strong stellar winds.

The surface temperatures of O stars are above 30,000 degrees Kelvin. Their stellar winds can blow as much as the sun's mass into **space** in 100,000 to one million years, the mass of the earth every year or so. The winds travel outward at speeds as high as 2,175 mi per second (3,500 km per second), almost ten times as fast as the solar wind. The Wolf-Rayet stars are hotter, 50,000K (89,541°F; 49,727°C), and their winds are more powerful than the O stars.

These powerful winds from hot stars create huge bubbles around the stars that can be as big as a few hundred **light** years across. These bubbles form when the stellar wind interacts with the surrounding interstellar medium, the gas and dust between the stars. The stellar wind slows down as it pushes into the interstellar medium. Dragging the interstellar medium along, it creates a region of higher density that is moving more slowly than the stellar wind. But the stellar wind keeps coming and slams into this region, creating a shock wave. The shock wave heats up the gas in this region and makes it glow, so that we see a bubble around the star. The bubbles are produced from the powerful stellar winds described above. There are also O stars with weaker stellar winds that have less dramatic effects.

Baby stars

The sun and similar stars form from collapsing **clouds** of gas and dust. After the star forms, the leftover material still surrounds it in a cocoon of gas and dust. How do stars like the sun shed their cocoons? One way is stellar winds. Astronomers think that shortly after the sun formed it went through a period when it had a very strong solar wind, which helped blow away the cocoon. It is difficult to know how accurate this scenario is because no one was around to witness the birth of the **solar system**.

We can however watch the birth of other stars similar to the sun. T Tauri stars are stars in the process of forming that astronomers think will eventually have properties similar to the sun's. Among other properties T Tauri stars show evidence of strong stellar winds. They also show a range of thick and thin circumstellar cocoons. Studying the stellar winds from T Tauri stars will help us understand how the Sun and similar stars shed their initial cocoons.

KEY TERMS

Interstellar medium—The matter between the stars.

O stars—The hottest most massive stars when classified on the basis of the star's spectrum..

Planetary nebula—A shell of hot gas surrounding a star in the late stages of its evolution.

Red giant—An extremely large star that is red because of its relatively cool surface.

Solar wind—A stream of charged and neutral particles that emanates from the Sun and moves into the solar system.

Stellar wind—A stream of particles blowing out from a star.

T Tauri star—An early stage in the evolution of stars like the sun.

White dwarf—A star that has used up all of its thermonuclear energy sources and has collapsed gravitationally to the equilibrium against further collapse that is maintained by a degenerate electron gas.

Wolf-Rayet star—A very hot energetic star that ejects a shell of gas.

Many young stars show bipolar outflows, which are two streams of material blowing away from the star in opposite directions. They usually occur in the birth of stars more massive than the sun. The bipolar outflow stage only lasts about 10,000 years or so, but during that time we see a strong stellar wind from the newly forming star. Why is the outflow bipolar? One theory suggests that the outflow is bipolar because an equatorial disk of material surrounding the star constrains the wind to flow out from the two polar regions. This disk may be the material that will eventually form planets around the star. Another possibility is that the star's magnetic field forces the outflow into a direction perpendicular to the equatorial disk. The study of stellar winds from newly forming stars will eventually provide us with clues to help us understand how the sun and solar system formed.

Dying stars

Old stars in the process of dying can also have very strong stellar winds. When stars like the sun exhaust the **hydrogen** fueling their nuclear fires, they expand into red giants. A typical red giant is about the size of the earth's orbit around the sun. Because red giants are so large and the gravitational **force** decreases with distance from the center, the gravitational force at the surface of a red giant is much less than at the surface of the sun. Hence only a gentle push is needed to allow **matter** to escape and form a stellar wind. This push might come from the light leaving the star. Light or other **radiation** striking an object will produce a very small but non zero force that is called radiation **pressure**. The radiation pressure on dust grains might provide the needed push. Similarly, this radiation pressure might also play a role in causing the stellar winds from the hot O stars mentioned in a previous paragraph. In addition, many red giants pulsate. They expand and contract in periods of a few years. These pulsations can also provide the needed push to cause significant stellar winds and to cause the loss of quite a bit of mass.

In some cases red giants form **planetary nebulae**, glowing shells of gas around the star. According to one model, the pulsations create a gentle wind moving at about 6 mi per second (10 km per second). This wind is gentle but can carry away the mass of the sun in as little as 10,000 years. Removing this much mass from the shell of the star exposes the more violent core and unleashes a wind blowing out thousands of miles per second. The fast wind slams into the slow wind and creates a shock wave that heats up the shell of gas until it glows as a planetary nebula. The remaining core of the star collapses into a **white dwarf** star about the size of the earth.

Mass loss

The single most important property affecting the evolution of a star is its mass. Therefore when the stellar wind causes a star to lose mass, its evolution is affected. In some cases these effects are still poorly understood. One reasonably well understood effect occurs for red giants collapsing into white dwarfs as described above. A star having more than 1.4 times the mass of the sun can not collapse into a stable white dwarf. Stars that exceed this mass limit collapse into **neutron** stars or black holes. Stellar winds can cause red giants with masses up to about 8 times the mass of the sun to lose enough mass to collapse into a white dwarf having less than 1.4 times the mass of the sun.

Roughly half of all stars occur in binary systems. When a star in a binary system has a stellar wind it loses mass. Some of this mass is transferred to the other star in the system, so it gains mass. Hence the mass of both stars in the system changes when one of the stars has a stellar wind. The evolution of both stars in the system is affected.

See also Binary star; Red giant star; Stellar evolution.

Resources

Books

Bacon, Dennis Henry, and Percy Seymour. *A Mechanical History of the Universe*. London: Philip Wilson Publishing, Ltd., 2003.

Morrison, David, Sidney Wolff, and Andrew Fraknoi. *Abell's Exploration of the Universe*. 7th ed. Philadelphia: Saunders College Publishing, 1995.

Periodicals

Frank, Adam. "Winds of Change." *Discover* (June 1994): 100-104.

Van Buren, David. "Bubbles in the Sky." *Astronomy* (January, 1993): 46-49.

Paul Heckert

Stem *see* **Plant**

Stem cells

Stem cells are undifferentiated cells that have the capability of self replication as well as being able to give rise to diverse types of differentiated or specialized **cell** lines. Stem cells are subclassified as embryonic stem cells, embryonic germ cells, or adult stem cells. Embryonic stem cells are cultured cells that were originally collected from the inner cell **mass** of an embryo at the blastocyst stage of development (four days post **fertilization**). Embryonic germ cells are derived from the fetal gonads that arise later in fetal development. Both of these stem cell types are pluripotent, that is, they are capable of producing daughter cells that can differentiate into all of the various tissues and organs of the body that are derived from the endoderm, ectoderm and mesoderm. Adult stem cells, found in both children and adults, are somewhat more limited, or multipotent, since they are associated with a single **tissue** or **organ** and function primarily in cell renewal for that tissue.

Because they are undifferentiated, stem cells have unique properties that may make them useful for new clinical applications. Initially, stem cells were considered as a potential source of tissue for transplantation. The current standard of care for many diseases that result in total tissue and/or organ destruction is transplantation of donor tissues, but the number of available organs is limited. However, bone marrow transplantation has proven to be highly successful, and studies have shown that using **blood** enriched with hematopoetic stem cells leads to a higher engraftment rate than when an equivalent bone marrow sample is used. Expanding on that idea, it was hypothesized that if adult stem cells from a specific organ could be collected and multiplied, it might be possible to use the resultant cells to replace a diseased organ or tissue. One drawback to this is that adult stem cells are very rare and although they have been isolated from bone marrow, **brain**, eyes, muscle, skin, liver, pancreas, and the digestive system, there are many tissues and organs for which it is not known if stem cells exist. Adult stem cells are also difficult to identify and isolate, and even when successfully collected, the cells often fail to survive outside of the body. However, despite the obstacles, the theory appears to be sound, so research is continuing.

Approaching the problem from another direction, researchers hypothesized that embryonic stem cells and embryonic germ cells, under the right conditions, might be induced *in vitro* to produce a broad range of different tissues that could be utilized for transplantation. Research on **Parkinson disease**, a neurodegenerative disorder that results in loss of brain function following the death of **dopamine** producing cells, underscored the potential of this approach. In the 1980s, studies on **monkeys** and **rats** showed that when fetal brain tissue rich in stem cells was implanted into the brains of diseased animals, there was a regeneration of functional brain cells and a reduction or elimination the symptoms of the **disease**. One disadvantage to this as a clinical procedure is that **random** pieces of undefined tissue are used resulting in the significant possibility of variability from one patient to the next. A better solution would be to isolate the embryonic stem cells, induce these cells to differentiate, and generate a population of dopamine producing cells. Theoretically, if these cells were transplanted back into the brains of Parkinson patients, they would replace the defective cells and reverse the course of the disease. However, the mechanisms that trigger differentiation of embryonic stem cells into various specialized tissue types are not yet well understood, so it will require additional research before transplantable tissues derived from embryonic stem cells will be a reality.

In addition to possible applications in transplantation, embryonic stem cells may be useful tools in other clinical disciplines. These cells represent a stage of development about which relatively little is known. Close observation in the laboratory could provide a better understanding of normal development versus abnormal development and what triggers fetal demise. Studies on the causes and control of childhood tumors may also be possible. Embryonic stem cell lines could aid in testing the effect of new drugs and investigating appropriate drug dosages, eliminating the need for human subjects. Similarly, such cell lines may be utilized to investigate the biological effects of toxins on human cells.

It has also been suggested that embryonic stem cells might be used in **gene therapy**. If a population of embryonic stem cells containing a known, functional **gene** can be engineered, these cells might function as vectors to transfer the gene into target tissues. Once in place, the cells would hopefully become part of the unit, begin to replicate, and restore lost function. Initial studies in **mice** confirmed the

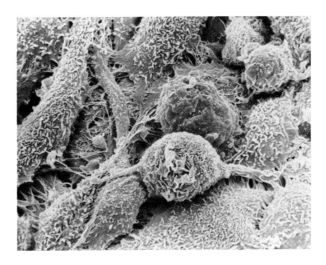

Colored scanning electron micrograph of cultured mammalian embryonic stem cells. *Yorgos Nikas. Photo Researchers. Reproduced by permission.*

idea was feasible. Investigators in Spain incorporated an **insulin** gene into mouse embryonic stem cells. After demonstrating the production of insulin *in vitro*, the cells were injected into the spleens of diabetic mice that subsequently showed evidence of disease reversal.

Although work is ongoing, research on embryonic stem/germ cells has generated some questions regarding the source of the cells. For research purposes, embryonic stem cells are primarily derived from leftover products of *in vitro* fertilization procedures. Embryonic germ cells from later gestational age fetuses have been obtained from elective termination of pregnancy or spontaneous fetal demise with appropriate parental consent. Use of such fetal tissues has posed an ethical dilemma, so new limitations on research projects have been imposed including careful review of all protocols and restriction to the use of already existing cell lines.

Although still in its infancy, stem cell research holds great promise for providing important new medical treatments in the future. There are many different diseases, ranging from **heart** disease to spinal cord injury and **autoimmune disorders** that could benefit from a better understanding of and the use of stem cells as therapeutic agents. In addition, study of these cells will impart new knowledge about human cells and early fetal development.

See also Embryo and embryonic development; Embryology; Genetic engineering; Genetics.

Resources

Books

Alberts, B., et al. *Essential Cell Biology* New York: Garland Publishing, Inc., 1998.

KEY TERMS

. . . .

Adult stem cells—Stem cells associated with a single tissue or organ that function primarily in cell renewal for that tissue.

Differentiation—The process whereby generalized cells change by acquiring a specific set of characters to perform a particular function.

Embryonic stem cells—Pluripotent cells that exist in an embryo at four days post-fertilization.

In vitro—Cells or biological processes that are present or occur outside the organism, i.e., they exist in a test tube or culture vessel.

Pluripotent—Pertaining to a cell that has the capacity to develop into any of the various tissues and organs of the body.

Stem cells—Undifferentiated cells capable of self-replication and able to give rise to diverse types of differentiated or specialized cell lines.

Transplantation—Moving cells or tissues from their point of origin in one organism to a secondary site in the same or a different organism.

Mueller, R.F., and I.D. Young. *Emery's Elements of Medical Genetics.* 11th ed. Edinburgh: Churchill Livingstone, 2001.

Nussbaum, R.L., et al. *Thompson and Thompson Genetics in Medicine.* 6th ed. Philadelphia: W. B. Saunders Co., 2001.

Organizations

American Association for the Advancement of Science. "Stem Cell Research and Applications: Monitoring the Frontiers of Biomedical Research." November 1999 [cited March 10, 2003]. <http://www. aaas.org/spp/dspp/sfrl/projects/stem/report.pdf>.

Other

Mayo Clinic. "Stem Cells: Medicine's New Frontier." August 10, 2001 [cited March 10, 2003]. <http://www.mayoclinic.com/invoke.cfm?id= CA00013>.

National Institutes of Health. "Stem Cells: A Primer". September 2002 [cited March 10, 2003]. <http://www.nih.gov/news/stemcell/primer. htm>.

Department of Health and Human Services. "Stem Cells: Scientific Progress and Future Research Directions." 2001 [cited March 10, 2003]. <http://www.nih.gov/news/stemcell/scireport.htm>.

University of Wisconsin-Madison. "Embryonic Stem Cells." 2001 [cited March 10, 2003]. <http://www.news.wisc.edu/packages/stemcells/ index.html?get=facts#>.

Constance Stein

Stereochemistry

Stereochemistry is the study of the three dimensional shape of molecules and the effects of shape upon the properties of molecules. The term stereochemistry is derived from the Greek word *stereos*, which means solid.

Historical development

Dutch chemist Jacobus Hendricus van't Hoff (1852–1911), the winner of the first Nobel Prize in **chemistry** (1901), pioneered the study of molecular structure and stereochemistry. Van't Hoff proposed that the concept of an asymmetrical **carbon** atom explained the existence of numerous isomers that had baffled the chemists of the day. Van't Hoff's work gave eventual rise to stereochemistry when he correctly described the existence of a relationship between a molecule's optical properties and the presence of an asymmetrical carbon atom.

The stereochemistry of carbon is important in all biological processes. Stereochemistry is also important in **geology**, especially **mineralogy**, with dealing with silicon based **geochemistry**.

Fundamentals of stereochemistry

Assuming that the all reactants are present, inorganic reactions are chiefly governed by **temperature**, that is, temperature is critical to determining whether or not a particular reaction will proceed. In biological reactions, however, the shape of the molecules becomes the critical factor. Small changes in the shape or alignment of molecules can determine whether or not a reaction will proceed. In fact, one of the critical roles of enzymes in **biochemistry** is to lower the temperature requirements for **chemical reactions**. Assuming the proper enzymes are present, biological temperatures are usually sufficient to allow reactions to proceed. This leaves the stereochemistry of molecules as the controlling factor in biological and organic (molecules and compounds with Carbon) reactions (assuming all the reactants are present) is the shape and alignment of the reacting molecules.

The **molecular geometry** around any atom is depends upon the number of bonds to other **atoms** and the presence or absence of lone pairs of electrons associated with the atom.

The chemical formula of a **molecule** is only a simple representation of the order of arrangement of atoms. It does not show the three-dimensional structure of the molecule. It is usually left up to the reader to translate the chemical formula into its geometric arrangement. For example, the chemical formula for methane is CH_4. This formula indicates that a central carbon atom is bonded to four **hydrogen** atoms (C-H). In order to convert this formula into the three dimensional molecular array for methane, one must know that when a carbon atom has four single bonds to four atoms, each of the bonds points towards a different corner of a **tetrahedron**, as shown in Figure 1. In the figure, the solid wedge shaped bonds are coming out of the **paper** and the dotted wedges are going into the paper.

Another way to visualize a carbon atom with four single bonds is to consider the central carbon atom at the center of a **pyramid**, also shown in Figure 1. At each point in the pyramid is located a hydrogen atom that is bonded to you. One hydrogen or pyramid point is directly above your head. One is in front of you, one point is behind you to your right, and another behind you to your left. These three hydrogen atoms or points are all on a level below the one you are on. The three dimensional arrangement of each carbon atom with four single bonds is always the same and the angle between any two bonds is 109.5°.

Stereoisomers

Some compounds differ only in their shape or orientation in **space**. Compounds that have the same **molecular formula** are called isomers. Stereoisomers are isomers (i.e., they have the same **molecular weight** and formula) but that differ in their orientation in space. No matter how a stereoisomer is rotated it presents a different picture than its stereoisomer counterpart. Most importantly, stereoisomers are not superimposable.

Enantiomers are stereoisomers that are mirror images, that is, they can map onto one another (if the molecules were two dimensional we would say that the molecules, just like human hands, could not be laid on top or superimposed upon each other).

Stereoisomers that rotate polarized **light** are called optical isomers. With the help of an instrument called a polarimeter, molecules are assigned a sign or **rotation**, either (+) for dextrorotatory molecules that rotate a **plane** of polarized light to the right, or (−) for levorotatory molecules that rotate a plane of polarized light to the left. Enantiomers differ in the direction that they rotate a plane of polarized light and in the rate that they react with other chiral molecules. Racemic mixtures of compounds contain equal amounts of enantiomers.

Symmetry and handedness

Symmetry is a term used to describes molecules with equal parts. When a molecule is symmetrical it has portions that correspond in shape, size, and structure so

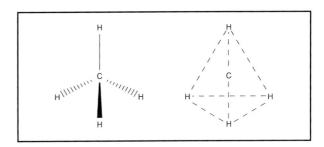

Figure 1. *Illustration by Hans & Cassidy. Courtesy of Gale Group.*

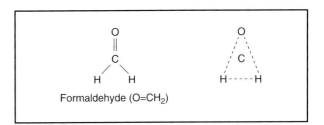

Figure 2. *Illustration by Hans & Cassidy. Courtesy of Gale Group.*

that they could be mapped or transposed on one another. Bilateral symmetry means that a molecule can be divided into two corresponding parts. Radial symmetry means that if a molecule is rotated about an axis that a certain number of degrees rotation (always less than 360°) it looks identical to the molecule prior to rotation.

A molecule is said to be symmetrical if it can be divided into equal mirror image parts by a line or a plane. Humans are roughly bilaterally symmetrical. Draw a line down the middle of the human body and the line divides the body into two mirror image halves. If a blob of ink were placed on a piece of paper, and then the paper was folded over and then unfolded again, you would find two ink spots—the original and the image—symmetrical about the fold in the paper. Molecules and complexes can have more than just two planes of symmetry.

Human hands provide an excellent example of the concept of molecular handedness. The right and left hands are normally mirror images of each other, the only major difference between them being in the direction one takes to go from the thumb to the fingers. This sense of direction is termed handedness, that is, whether a molecule or complex has a left and right orientation. Two molecules can be mirror images of each other, alike in every way except for their handedness.

Handedness can have profound implications. Some medicines are vastly more effective in their left-handed configuration than in their right-handed configuration. In some cases biological systems make only one of the forms. In some cases only one of the forms is effective in cellular chemical reactions.

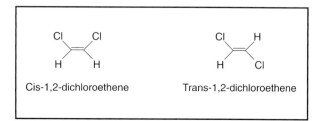

Figure 3. *Illustration by Hans & Cassidy. Courtesy of Gale Group.*

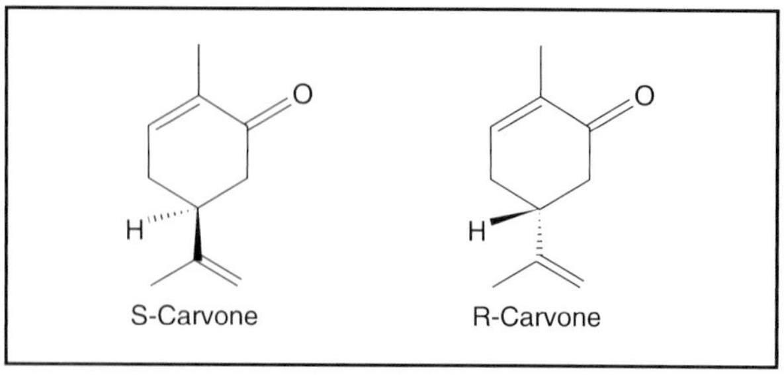

Figure 4. *Illustration by Hans & Cassidy. Courtesy of Gale Group.*

A molecule that is not symmetric, that is, a molecule without a plane of symmetry, is termed an asymmetric molecule. Asymmetric molecules can have another property termed chirality.

Chiral molecules

A molecule is said to be chiral if it lacks symmetry and its mirror images are not superimposable. To be chiral a molecule must lack symmetry, that is, a chiral molecule can not have any type or symmetry.

Carbon atoms with four sp^3 hybridized orbitals can enter into up to four different bonds about the central carbon atom. When the central carbon bonds with differing atoms or groups of atoms the carbon is termed an asymmetric carbon atom. Bromochlorofluoromethane is an example of such a molecule. The central carbon, with four sp^3 bonds oriented (pointing) to the corners of a tetrahedron, is bonded to a bromine, **chlorine**, fluorine and methane atoms. There is no symmetry to this molecule.

Chiral carbon atoms are also assigned an R and S designation. Although the rules for determining this designation can be complex, for simple molecules and compounds with chiral carbons the determination is easily accomplished with the help of a model of the molecule. The four different bonded groups are assigned a priority. When assigning priority to groups, atoms that are directly bonded to the central chiral carbon atom have their priority based upon their **atomic number**. The atom with the highest atomic number has highest priority and

atom with the lowest atomic number the lowest priority. As a result, hydrogen atoms bonded to the chiral molecule have the lowest priority. If isotopes are bonded then the **isotope** with the largest **mass** has the higher priority. The molecule is then turned so that the lowest priority group is farthest away from view. If one must take a counterclockwise path from the highest to lowest priority group the chiral configuration is said to be sinister (S). If the path from highest to lowest priority groups is clockwise then the chiral molecule is said to be rectus (R).

The compound carvone has two three-dimensional structures, one S and the other R (see Figure 4).

The compounds differ in their three-dimensional structure by the position of the indicated hydrogen atom. In S-Carvone, only the hydrogen atom is pointed into the paper, while in the R compound, the hydrogen atom is coming out of the paper. S-Carvone has a caraway flavor when tasted, whereas the R compound has the flavor of spearmint.

The rectus (R) and sinister (S) property relates to the structure of an individual molecule. In contrast, dextro (+) and levo (−) properties are based on the properties of a large collection of the molecules or complex.

Because a molecule can have more than one chiral carbon. The number of stereoisomers can be determined by the 2*n* rule, where *n* = the number of chiral carbons. Thus, if one chiral carbon is present there are two possible stereoisomers, with two chiral carbons there are four possible stereoisomers. Any chemical reaction that yields predominantly one stereoisomer out of several stereoisomer possibilities is said to be a stereoselective reaction.

Determination of stereochemical properties

Sometimes it is difficult to tell whether or not two molecules or complexes will exhibit stereochemical properties. If two molecules or complexes have the same molecular formula they are candidates for stereochemical analysis. The first step is to determine if the two molecules or complexes are superimposable. If they are then are identical structures and will not exhibit stereochemical properties. The second step is to determine if the atoms are connected to each other in the same order. If the atoms are not connected in the same order then the molecules or complexes are constitutional isomers and will not exhibit stereochemical properties. If the atoms are connected in the same order then they are stereoisomers. The next step is to see if the stereoisomers can be made identical by rotating them around a single bond in the molecule or complex then they are called conformational isomers. Stereoisomers that can not be so rotated are called configurational isomers. The last step is to analyze the configurational isomers to determine whether they are enantiomers, diastereomers, or cis-trans isomers. Those that are mirror images are enantiomers. Those stereoisomers that are not mirror images of each other are diastereomers (the prefix dia indicated opposite or across from as in diagonal) or cis-trans isomers. Stereoisomers can also be characterized as *cis* (Latin for "on this side") or *trans* (Latin for "across") when they differ in the positions of atoms or groups relative to a reference plane. They are cis-isomers if the atoms are on the same side of the plane or trans-isomers if they are on opposite sides of the reference plane.

If the molecule has a double bond in its chemical formula—for example, formaldehyde, $O=CH_2$—then the three-dimensional structure of the molecule is somewhat different. To translate formaldehyde into its geometric structure, one must know its chemical formula indicates a central carbon atom that has a double bond to an **oxygen** atom (C=O) and two single bonds to hydrogen atoms (C-H). In the geometric arrangement of a carbon atom that has a double bond to another atom, there is a 120° angle between any two bonds, and each bond points away from the central carbon atom. If the bonded atoms are connected by imaginary lines, they represent the corners of an equilateral triangle (see Figure 2). In molecules that contain two carbon atoms connected by a double bond and each of which is bonded to a hydrogen atom and another atom, then the geometric **isomer** that has both hydrogen atoms on the same side is in a *cis* configuration. The molecule with the hydrogen atoms on opposite sides of the double bond is designated as the *trans* configuration. For example, *cis*-1,2-dichloroethene has the hydrogen atoms on the same side of the double bond, where as *trans*-1,2-dichloroethene has them on opposite sides. Both of these compounds have the same chemical formula (ClHC=CHCl), but their geometric representations are different (see Figure 3).

The only other type of bond a carbon atom can have is a triple bond—that is, three bonds to the same atom. Acetylene (HCCH) is a molecule that contains a triple bond between the two carbon atoms, and each carbon atom is bonded to a hydrogen atom (C-H). A carbon atom with a triple bond to another atom is geometrically straight or linear.

$$H—C \equiv C—H$$

The importance of stereochemistry

The three-dimensional structure of a molecule determines its physical properties, such as the temperature at which it turns from a liquid to a gas (**boiling point**) and the temperature at which it changes from a solid to a liquid (melting point). The geometric structure of a molecule is also responsible for its chemical properties, such as its strength as an acid or base. The compound *trans*-1,2-

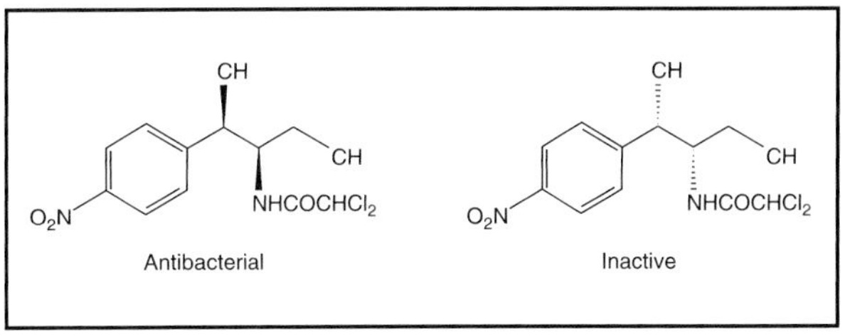

Figure 5. *Illustration by Hans & Cassidy. Courtesy of Gale Group.*

dichloroethene becomes a gas at a much higher temperature than the structurally similar *cis*-1,2-dichloroethene. The compound *cis*-3-phenylpropenoic acid is a stronger acid than *trans*-3-phenylpropenoic acid only because the hydrogen atoms are connected to the doubly bonded carbon atoms differently.

The geometric structure of a molecule can also have a dramatic effect on how that molecule tastes or how it functions as a drug. The antibacterial drug chloramphenicol is commercially produced as a mixture of the two compounds in Figure 5. One three-dimensional arrangement of atoms is an active drug, the other geometric structure is ineffective as an antibacterial agent.

In most cases the **energy** of a molecule or a compound, that is, the particular energy level of its electrons depends upon the relative **geometry** of the atoms comprising the molecule or compound. Nuclear geometry means the geometrical or spatial relationships between the nucleus of the atoms in a compound or molecule (e.g., the balls in a ball and stick model). When a molecule or compound's energy is related to its shape this is termed a stereoelectronic property.

Stereoelectronic effects arise from the different alignment of electronic orbitals with different arrangements of nuclear geometry. It is possible to control the rate or products of some chemical reactions by controlling the stereoelectronic properties of the reactants.

See also Chemical bond; Formula, chemical; Formula, structural.

Resources

Books

Boyer, Rodney. *Concepts in Biochemistry*. Pacific Grove, CA: Brooks/Cole Publishing Company, 1999.
Carroll, Felix A. *Perspectives on Structure and Mechanism in Organic Chemistry*. Pacific Grove, CA: Brooks/Cole Publishing Company, 1998.
Mislow, Kurt M. *Introduction to Stereochemistry*. Dover Publications, 2002.
Morris, David G. *Stereochemistry*. John Wiley & Sons, 2002.

Andrew J. Poss
K. Lee Lerner

Sticklebacks

Sticklebacks are small, **bony fish** in the family Gasterosteidae that rarely exceed 3 in (8 cm) in body length. Instead of scales, these **fish** have bony plates covering their body. Sticklebacks are found in **North America** and northern Eurasia. The name stickleback is derived from the sharp, thick spines arising in the first dorsal fin. The number of these spines forms part of the basis for the identification of the different **species** of sticklebacks.

Sticklebacks provide a good example of male dominance in mating and nesting **behavior**. At the beginning of the breeding season, a male stickleback selects a suitable spot in quiet **water**, where he builds a nest of **plant** parts stuck together by a sticky fluid produced by his kidneys. The fish shapes the nest by his body movements. A male three-spined stickleback is normally blue-green, but it develops a bright red **color** for the breeding season. The male ten-spined stickleback becomes brown, while the 15-spined stickleback changes to a blue color. Breeding male sticklebacks are aggressive during the breeding season and will readily fight with other males. The male performs a **courtship** dance to entice a female into the nest, but if this is not successful he attempts to chase the female into his nest, where she deposits her eggs. The male then expels sperm (milt) over the eggs to fertilize them. The male will then search for another female, repeating the process until his nest is filled with eggs. The male aerates the eggs by using his pectoral fins to set up a water current. He guards the eggs until they hatch, and then continues to guard the brood afterwards, maintaining the young in the nest until they are able to obtain their own food.

Some sticklebacks appear to be capable of rapid **evolution**, including the development of apparently separate species in different habitats within the same **lake**. For example, in Paxton Lake, British Columbia, morphologically and behaviorally distinct species of *Gasterosteus* sticklebacks utilize benthic and mid-water habitats, and these have evolved in only a few thousand years since deglaciation.

The brook stickleback (*Culaea inconstans*) is common in brooks in the United States from Pennsylvania to Kansas. It is a small fish less than 3 in (8 cm) in length, with five to six spines on its back. The 15-spine stickleback (*Spinachia spinachia*) is found in **saltwater** in the British Isles and around the North Sea. The nine-spine stickleback (*Pungitius pungitius*) is found on both sides of the Atlantic in northern latitudes. Its coloring is dark brown, but the male turns black during courtship and spawning.

Stigma *see* **Flower**

Stilts and avocets

Stilts and avocets are long-legged, long-beaked wading **birds** of the muddy shores of shallow lakes and lagoons, including both fresh and saline waters.

There are fewer than ten **species** of stilts and avocets all of which are included in the family Recurvirostridae. These birds occur in the temperate and tropical zones of all of the continents except **Antarctica**. The bill of stilts is rather straight, while that of avocets curves upwards, and that of the ibisbill curves downwards.

The **habitat** of the ibisbill (*Ibidorhyncha struthersii*) is the shores of cold, glacial streams of the **mountains** of the Himalayas. The Andean avocet (*Recurvirostra andina*) utilizes similarly cold, but shallow-lake habitat in the Andes of **South America**.

Stilts and avocets feed on small **invertebrates** on the **water** surface or at the mud-water interface. The ibisbill feeds by probing among the cobble and pebbles of the cold streams that it inhabits.

Stilts and avocets commonly construct nests as mounds of vegetation, on the boggy edges of lakes or on islands. The chicks of these species are *precocial*, meaning that they can leave the nest within a short time of being born. Remarkably, stilt and avocet chicks feed themselves, and are not fed by their parents.

Some stilts and avocets breed in semi-arid climates, where the rainfall is unpredictable. Under such conditions, these birds nest opportunistically, whenever recent rains have been sufficient to develop **wetlands** containing enough water to support the breeding effort of these birds. Stilts and avocets tend to be gregarious, forming small breeding colonies numbering as many as tens of pairs. After breeding has been accomplished, birds from

A black-winged stilt. *Photograph by Roger Wilmshurst/Photo Researchers, Inc. Reproduced by permission.*

various colonies congregate in flocks that can number hundreds of birds.

Species of stilts and avocets

The American avocet (*Recurvirostra americana*) is a chestnut-headed bird with a white body and black wings. This species is relatively abundant, and breeds on the shores of shallow lakes and marshes in the western United States and adjacent parts of Canada. The black-necked stilt (*Himantopus mexicanus*) is more southwestern in its distribution in the United States. Both of these species are migratory, mostly spending the winter months in Central America.

The Hawaiian stilt or ae'o (*Himantopus mexicanus knudseni*) is a distinct subspecies of the black-necked stilt that breeds in wetlands in the Hawaiian Islands. Fewer than 1,000 pairs of this bird remain. The subspecies is considered to be endangered, mostly by the loss of its wetland habitat through conversion to agricultural uses or for residential or tourism development.

The pied or black-winged stilt (*Himantopus himantopus*) is a very wide-ranging species, occurring in suitable habitat in Eurasia, **Africa**, Southeast **Asia**, and **Aus-**tralia. The black-winged stilt (*Himantopus melanurus*) of South America is closely related to the previous species. The banded stilt (*Cladorhynchus leucocephalus*) occurs in Australia.

The avocet (*Recurvirostra avosetta*) occurs in parts of Eurasia and Africa. Australia has a red-necked avocet (*R. novaehollandiae*), while the Andean avocet (*R. andina*) occurs in montane habitats of South America.

The ibisbill (*Ibidorhyncha struthersii*) occurs in the mountain zone of the Himalayan Mountains of south Asia.

Most ornithologists believe that the family Recurvirostridae is not a very natural grouping of birds. The several species of avocets are obviously closely related to each other, as are the various stilts. However, the ibisbill does not seem to be closely related to the stilts and avocets, and may eventually be assigned to a separate family.

See also Endangered species; Migration.

Resources

Books

Forshaw, Joseph. *Encyclopedia of Birds.* New York: Academic Press, 1998.

Bill Freedman

Stimulus

The term stimulus has many meanings; very generally, it is any occurrence (be it an external event, or anything perceived or thought) that causes a detectable response. Stimulus is often used with qualifying terms to further specify its meaning, for example, conditioned stimulus and neutral stimulus.

Various fields of study use the term stimulus in different ways. In **psychology**, it is most often used to describe **energy** forms that reach sense organs and cause a response. For example, the visual sense using the eyes responds to photic **radiation** or **light**. Because human sense organs respond to a limited number of energy forms, and even then to only limited amounts of that energy, some energy reaching the sense organs is not detected and does not cause a response. The energy reaching the sense organs but not causing a response may be deemed a stimulus to a physiologist, but for psychologists it would not be considered a stimulus unless it had been responded to or detected by the **organism**. A stimulus may also be an internal mental event that causes a response.

Stimulus is the primary term in stimulus-response theory, which refers to a number of **learning** theories that are theoretically based on conditioned bonds or associations between a stimulus and response. The associative bonds are formed through the repeated pairing of certain stimuli and certain responses. Most of these theories are also behavioristic in that they focus on behaviors and do not look at mental processes, and they see the environment as the most important determinant of human **behavior**. Indeed, these theories view the bond between stimulus and response as the basis of behavior and believe that psychology's primary goal should be to discover rules governing how stimuli and responses interact. The two dominant stimulus-response theories are classical and operant **conditioning** theories.

See also Perception.

Resources

Books

Atkinson, R.L., R.C. Atkinson, E.E. Smith, and D.J. Bem. *Introduction to Psychology*. 10th ed. New York: Harcourt Brace Jovanovich, 1990.

Masin, S.C., ed. *Foundations of Perceptual Theory*. New York: Elvesier Science, 1993.

Periodicals

Friedrich, M.J. "A Bit of Culture for Children: Probiotics May Improve Health and Fight Disease." *Journal of the American Medical Association* no. 284 (September 2000): 1365-1366.

Stone and masonry

It is possible that ever since people first came to be, stone was used in constructing something: a fence, an oven in a hole or trench, or a shelf in a **cave**. And it is possible that sometime during this era, someone coined the statement: "Leave no stone unturned." The many cairns and stone hedges erected for religious or astronomical uses were the initial attempts at masonry. However, true masonry did not begin until the Egyptians built the pyramids. Previous to this, most stone structures were constructed by placing one stone upon, or next to, another, regardless of size or shape.

In constructing the pyramids, the stone was first hewn or carved into a certain shape and then placed into a preplanned position. Other great stone undertakings were walls, the wall of China being the largest. The Roman wall in England was 10 ft (3 m) thick at the base and up to 15 ft (4.6 m) in height in some places. Although it was filled with mud and pebbles, both faces were constructed with squared stones. It ran for over 70 mi (113 km). Surviving portions are about 6 ft (1.8 m) high.

It was during the Norman period that most old stone structures were erected in England. And many of the craftsmen were of Norman descent. Therefore many of the architectural terms used in the English language are of French origin. Some of these are vault, buttress, **niche**, oriel, trefoil, fillet, and chamfer. French influence is also evident in the standardization of various building units, such as the course heights and in various moldings and carvings. Many of these were designed to fit the individual blocks of stone rather than create a regular repetitive pattern. The majority of the original stone buildings were cathedrals, churches, and castles. But as **time** went on, especially where stone was quite abundant, manor houses, farmhouses, and even barns began to be built of stone.

In America, especially in northern America, when settlers first came, many were tillers of the **soil**, or became such. The last glacier brought many stones from Canada and northern America and deposited them on the land. The settlers then used them for fences and for barns and house foundations. Today stones are used to face many buildings.

Stone types

In working with stone one should know the various types of stone. There are three main types, given their name from the manner whereby they were formed. Igneous stone is formed when **magma** from below the earth's crust comes to the surface and solidifies. The liquefied material from beneath the crust of the **earth** spews forth from a **volcano** as lava. Basalt is the most common stone to be formed volcanically. It is composed mainly of silica, as is diabase and other primordial stone. Feldspar, which contains **aluminum** and **calcium** compounds, is the other common mineral spewed from a volcano.

Metamorphic stones are made from existing materials that have undergone change. As an example, when weathered on the surface, feldspar becomes clay, which can undergo tremendous **pressure** and become metamorphic slate. Granite is often taken for granted, but comes from quartz and feldspar, sometimes at molten temperatures. It can be either igneous or metamorphic. Most metamorphic **rocks** have been crushed into their present state by a arent stone upon it. Thus limestone under pressure becomes marble. Quartzite is metamorphic quartz. Gneiss and schist are two other metamorphic rocks.

Sedimentary stones are formed from sediments on the earth's surface. Such stones cover three fourths of the earth's crust, but account for less than five **percent** of the earth's **volume**. Shale, sandstone, and limestone make up ninety-nine percent of the sedimentary stone. Shale comes from clay deposits which were formed from feldspar. Sandstone comes from river carried **erosion** of other stones and **minerals** that build up at continents' edges. It can also be formed by erosion of igneous stone. Limestone is formed mainly from the buildup of exoskeletons of tiny sea animals drifting to shore. The shell material is mainly **calcium carbonate**. Limestone is often quarried for building stone, gravel, cement or agricultural lime. Weathered stone supplies the minerals that plants and animals need for their existence.

Chemical composition

The builder of a stone structure is more concerned with its chemical composition than its geological classification. There are three main categories for stone chemical content. Siliceous stone has as its main element, silica, or silicon dioxide (SiO_2). Most stone from volcanoes is siliceous. This type of stone also includes compressed sediments of siliceous stone, like sandstone. Quartz is a very pure pressurized sandstone.

Argillaceous stone has as its main element alumina AlO_2). It, along with its compounds, comes from feldspar in the crust. When these meet up with the atmosphere, they change into clay-like compounds. Slate, a sort of petrified clay, is the most common to the mason. When clay combines with other stone in varying degrees, it can overlap into other types. If clay and **sand** mix, brownstone is formed. **Brick** is artificial argillaceous stone.

Calcareous stone is made up mostly of calcium carbonate ($CaCO_3$), or lime. Lime comes from the bodies of sea creatures, whose skeletons have accumulated at the bottom of the seas. When lime is pressurized for millennia, it becomes marble. Marble is mainly a **metamorphic rock**. But because it is still mainly lime after **metamorphosis**, it is calcareous in classification.

Certainly, there are many combinations of the above three classifications, but rocks are distinguished by the abundance of glassy (silceous), clayey (argillaceous), or limy (calcareous) material in them. It is the quantity of each basic compound found in rocks, along with the way they were formed and the presence of other minerals in smaller amounts, that give rocks their particular desirability by masons.

Construction rocks

The strongest rock is trap, a very old igneous and siliceous primeval stone that can withstand over 67,000 lb/in^2 (3,900 kg/cm^2) pressure. Gabbro and basalt are similar. But these rocks are very difficult to work, or quarry. They are not formed in layers and hence are very difficult to layer. Granite is somewhat manageable, and is used where great strength or resistance to **weather** are needed. But it is difficult to work with and therefore quite expensive. Other qualities of rock that make it more desirable for use in **constructions** are ease of quarrying, nearness to quarry, durability, resistance to absorbing **water**, and strength. However, stone masons look mainly for shape (they desire stones more or less **square**), their proximity, and the price. The stone most often used by masons is sandstone or limestone.

Stone masonry always has stood for permanence. Anything properly made of stone outlasts the same thing made of any other material, even **concrete** with **steel** mesh imbedded. But one must be certain of one's work. It can take weeks to repair a mistake. Stone masonry differs entirely from brick or block masonry, where every layer must lie in a straight line, where there must be square and level units and mortar joints that are uniform. The main reasons for building with stone are its beauty, endurance, strength, and mass.

Stone construction

Stone walls can be constructed as dry walls or, if mortar is used, as wet walls. Dry walls are used mainly for fences or retaining walls. Dry walls are usually 2-3 ft (0.6-0.9 m) thick. The base is usually a bed of sand around 5 in (12.1 cm) thick. At the beginning of the first layer of stones usually one lays a stone, the bonding stone, which is faced relatively even. This helps keep the wall together. The bottom course should consist of alternating larger and smaller stones, all with the longest side along the outside surface. Such stones are placed on either side of the wall. Then stones are put between the sides, filling in the center. The layers that are above the base layer are constructed such that stones connect two lower stones, and, if possible, are kept reasonably level. If spaces result, and stones do not seat firmly, gaps are filled by chinking, by driving narrow stones or chips in the spaces so that the wall is locked tight and weight is pressed inward to prevent collapsing.

A wet wall begins with a mortared bed. A trench is dug and filled with sand or gravel. This then is allowed to settle for a few days. An inch of mortar is then laid on the top of the slab. As above, a bonding stone is placed at the end. Other stones are added along the sides, leaving spaces for the mortar which is poured as one moves along. The inside is filled with small stones, and mortar added. After laying but not filling in the next course, a broomstick is inserted in-between stones on the first tier to make weep holes, which should pass entirely through the wall. When the wall is finished, it should be raked with a piece of **wood** to compact the mortar.

KEY TERMS

Arch—Curved path or span across an opening, constructed of voussoirs (curved arch stones).

Bed—A stratified layer of stone in sedimentary rock, laid down by nature. A bed is also the term used to describe the horizontal layer of mortar, cement, sand, etc. on which the blocks or stone are laid. It is the base for the first layer.

Bevel—An instrument to determine that a layer of stone or brick does not change height over a distance. It also is the name for a splayed angle on a worked stone.

Block stone—A large stone block, roughly squared at the quarry.

Bonder—Any stone or brick which is so laid that it increases the strength of a wall, either in thickness or in length.

Brought to course—Where random walling is brought up to a level line, such as at the top of a wall or the bottom of a window.

Buttress—A strong stone pier built against a wall to give additional strength. (Flying buttresses extend quite far out from the wall.) A buttress must be bonded to the wall.

Chamfer—A flat splayed edge between two flat plain surfaces.

Chisel—A generic term for a certain type of cutting tool.

Cladding—Nonload bearing thin stone slabs or bricks used for facing a building.

Cock's comb—A thin shaped steel plate with a serrated edge. It is used for finishing moldings and other shapes.

Corbel—A bracket of stone that projects from the face of a wall. It usually is used to give support to some feature above.

Course—A continuous layer of stone or brick of uniform height in a wall.

Diamond saw—A saw whose cutting edge has industrial diamonds inserted.

Dormer—A window that projects vertically from a sloping roof.

Dowel—A sort piece of metal (not iron) or slate fixed into a mortise or adjoining stones or tile to prevent movement.

Dressings—A generic term used to wart, chimneys, etc. made of freestone, on the elevation of a building.

Drum—A separate circular stone in a shaft or column.

Eave—The part of a roof or other structure which oversails the wall face.

Entablature—In classical architecture, the upper part of an order. It comprises architrave, frieze and cornice.

Fault—A natural fissure in a bed or stratum of stone at a quarry. Usual faults are somewhat vertical.

Fillet—A small member between moldings. It also is called a quirk.

Flagstone—Stone that is naturally stratified in slabs about two to three inches thick, used mainly for paving, copings, etc.

Float—A rectangular wooden hand trowel, used for smoothing surfaces.

Fluted stone—Stone worked with regular concave grooves, as found in columns.

Gauge—A tool used for marking parallel lines when setting out a piece of work.

Grain—A word used at times to describe the natural visible bedding planes in stone.

Most modern stone houses have stone faces only; the entire wall is usually not completely stone. The main reason is that stone is very expensive and the inner part of the wall usually is blocked by furniture, pictures, or other items. So only an outer facade of stone is added to the house. This is between 3-4 in (7.6-10 cm) thick, and can be either field or quarried stone. One of the hardest stones that is relatively easy to work with is limestone. One of the hardest limestone is quarried in Valders, Wisconsin, which is about 30 mi (48 km) south of Green Bay. Here the Valders glacier covered the area. But when one digs beneath the glaciated land, an extra dense level of the Niagara ledge is found. Valders dolomite limestone ranges in **color** from a silvery white to a buff texture. It is one of the most enduring limestone in the United States.

Marble is another stone used by masons, and is found in almost every state. Depending on color, porosity, and strength, various marbles are used in various

Grout—Mortar used for filling vertical joints (perpends).

Header—A stone which has its longest dimension built into the thickness of the wall to improve bonding and strength.

Hip—The inclined angle at which two sloping roofs meet. The converse of valley.

Hod—A device used for carrying mortar or other materials when climbing. It is placed on the shoulder.

Hydrated lime—Quicklime processed into an inert powdered lime ready for use.

Jamb—Vertical side of an archway, doorway or window.

Keystone—The central stone in an arch

Ledger—A large flat stone covering an altar tomb or grave. Often forms part of a church floor.

Level—A tool for telling if horizontal surfaces are true and level.

Lintel—The block or stone that spans the top of an opening, such as a doorway or window. Sometimes called a head.

Molding—A projecting or recessed part, used to give shadows to a wall, arch, or other surfaces.

Mortar board—A board placed near the work. It holds the mortar, allowing the mason to pick it up with a trowel.

Mortise—A recess in a block cut to receive a dowel or tenon.

Mould—A shaped pattern used to set out the work. Called a template in carpentry.

Niche—A recess in a wall, usually prepared to receive a carved figure.

Oriel window—A window that projects from an upper story.

Pilaster—A flat pier attached to a wall, with a base and a capital.

Plinth—The projecting base of a wall or column, generally molded at the top.

Plumb rule—A straightedge about four or six feet long with a spirit level recessed in it. It is used to ensure vertical with when building a wall. (Along with the level, it is one of a mason's most important tools.)

Pointing—Filling of mortar joints in masonry.

Quarry—Usually an opencast pit (but sometimes a mine) from which stone is extracted.

Rendering—A coating, usually of cement and sand, that covers rough stone.

Sill—Either the threshold stone under a doorway, if flush, that is not a step; or the stone across the base of a window opening.

Soffit—The underside of an arch, vault, cornice, lintel, or the like.

Square—A tool having a fixed right angle. It is used to set out work and keep angles truly square.

Straightedge—An unmarked ruler of any length that can be used to draw lines through two given points.

Throat—A name sometimes given to the small grove under a window sill or dripstone. Its purpose is to deflect water from the wall.

Transom—A structural part dividing a window horizontally.

Vein—A colored marking in limestone, marble, etc.

Weathered—Stones that have been exteriorly exposed in a building for many years to the elements can be weathered, well or badly.

buildings, such as churches, museums, art galleries and the like. Again, due to expense, today the most common stone used in construction (for facing mainly) is limestone, although some marble can be found.

Bricks

Bricks are manmade argillaceous rock. Brick and cement or cinder block masonry are not quite as artistic as stone masonry, but are usually quicker and cheaper. Like stone masonry, a footing is laid, and tiers of block or brick are placed upon the footing. The way a brick is positioned in a wall is given a name. The long part of the brick is termed the stretcher, the short end, the header, and the flat face of the brick is the bed. A tier or layer of brick can be a stretcher course (if all faces showing are stretchers), a headed course (ends showing are heads), a sailor course (where bricks are upright, with beds showing) or soldier course (where bricks are upright, with stretchers showing). Bricks laid in stretcher position with headers abutting are termed rowlock stretchers, and with headers upright are termed rowlock headers. There are various patterns that are used in brick laying. Strictly, in bricklaying, pattern refers to changes in arrangement or varied brick texture or color used in the face. But the different methods of joining bricks with the mortar can also be termed patterns. There are five types of bonding. They are:

- Running bond—Consists of all stretchers, but each tier is moved over one half brick.

- Block or stack bond—Each tier is placed over the previous in exactly the same position.

- American bond—A variation of running bond, with a full course of headers every fifth, sixth or seventh course.

- English bond—Alternates block bond tiers with header tiers.

- Flemish bond—Similar to English bond, but differs in that instead of alternating tiers, it alternates stretcher and header in the same tier.

Modern building codes require that masonry-bonded walls have no less that 4% headers in the wall surface. Headers bond the wall across its thickness. **Metal** ties are usually recommended for bonding an exterior wall. This allows for stretching and resistance to cracking.

The surface of the brick wall can vary, depending on how the joints are finished. There are about seven types of finished joints.

Concave is the most common. It keeps moisture out and the mortar is forced tightly between the brick and thus makes an excellent bond.

The V-joint is formed with a special jointer, a piece of wood or a trowel, and it stands out as a sharp line. V-joints direct water off.

The raked joint forms a deep recess, which does not keep water out, but produces a shading that accentuates the brick courses.

The weathered joint sheds the water better than any other. It is recessed at the top, and is level with the brick surface at the bottom of the joint.

The struck joint is the opposite of weathered—it is recessed at the bottom and slants towards the top.

The flush joint is the easiest of all to make. One simply uses the trowel to smooth the face. It is water resistant but not very strong since the mortar is not compacted.

The extruded joint, also called the "weeping" joint, is water resistant and is used mainly for garden walls or where a rustic appearance is desired. It is made by using an excess of mortar between tiers of brick. When a brick is put into place the excess mortar is then squeezed out, and left where it hangs over.

The type of stone, brick, or bonding to be used depends on the purpose and desire of the builder. But whatever type is chosen, stone and brick last the longest, besides being pleasant to the **eye**.

See also Igneous rocks; Sedimentary rock.

Melvin Tracy

Stoneflies

Stoneflies or salmonflies are a group of **insects** with aquatic nymphal stages in the order Plecoptera. Stoneflies have a simple **metamorphosis**, with three life-history stages: egg, nymph or naiad, and adult.

Adult stoneflies have two pairs of membranous wings that are folded back over the abdomen when not in use. Stoneflies are rather weak fliers, and are not usually found very far from the aquatic **habitat** of their nymphs. Adult stoneflies do not feed, their short life being devoted to procreation.

The nymphs of stoneflies are soft-bodied, rather flattened, and usually occur in well-aerated, running waters of streams and **rivers**, or sometimes along rocky **lake** shores, where they hide in gravel and organic debris. Stonefly nymphs have an abdomen with ten segments, and with two extended appendages known as cerci sticking out from the back end. Stonefly nymphs are most easily differentiated from superficially-similar mayfly nymphs (order Ephemeroptera) on the basis of their two cerci (**mayflies** almost always have three abdominal "tails"). The nymphal stage of stoneflies is long-lived, commonly lasting for a year or more. There can be more than 20 nymphal stages, each separated by a molt.

Stoneflies are an abundant component of the bottom **fauna** of many aquatic communities, especially in streams. Although their **biomass** and productivity are not usually as large as that of mayflies, stoneflies are an important food of sportfish such as trout and **salmon**, and for this reason these insects are economically important. Some of the best designs for fish-catching "flies" are based on the shape and **color** of stoneflies, especially the adult insects.

Storks

Storks are large wading **birds** of the tropics and subtropics. They belong to the order Ciconiiformes, which

A white stork and its nest in Turkey. *JLM Visuals. Reproduced by permission.*

also includes the **ibises** and spoonbills. Storks are in the family Ciconiidae.

Unlike most tall wading birds, storks will **perch** in trees. They also nest in high places, and often return to the same nesting site year after year. These tendencies have long made the white stork (*Ciconia ciconia*) a favorite among villagers of **Europe**, who look forward to the annual return of these birds after wintering in **Africa**. Folklore that children are delivered by these birds, as well as other tales about storks, were brought to America by European settlers.

Storks look somewhat like **cranes**, but their bill is considerably longer and heavier. Most **species** have a straight bill. Cranes have three front toes like most birds, but storks have an additional toe at the back of the foot that lets them cling to a branch when perching. **Herons** and egrets also have this special toe.

Storks fly with their neck outstretched (except for the adjutant storks; *Leptoptilos* spp.). Storks have broad, strong wings that allow them to soar easily for long periods on rising warm air currents. The different species are often distinguished by the **color** of their bill and the bare skin on their head and legs.

The bill of a stork can be up to 9 in (23 cm) long and 7 in (18 cm) around at the thick base. Storks are adept at using their beak, to build a nest, find food, care for the young, and in fighting. The saddle-bill stork (*Ephippi-orhynchus senegalensis*) of Africa has a vivid orange beak with a wide black band around it. It also has orange anklets at the bend in its legs (which are actually its ankles). The Asian open-bill stork (*Anastomas lamel-ligerus*) of India and Nepal has a beak in which the two mandibles do not close in the middle, which is an **adaptation** to capturing its specialized diet of **freshwater** mussels. The bill of the whale-headed stork (*Balaeniceps rex*) of freshwater swamps in Africa, from Sudan down through Zaire, is thick, rounded, spotted, and hooked on the end. However, this large gray bird is not a true stork; it is in a related family of its own, the Balaenicipitidae.

All storks (except the wood stork) study the ground or **water** intently when looking for food. Their keen eyesight and fast reaction time lets them quickly grab moving animals.

The adjutant storks of India eat carrion so they do not have to chase their meal. The greater adjutant (*Leptoptilos dubius*) and the lesser adjutant (*L. javanicus*) can be found along with **vultures** feeding at the leftovers of the kills of large predators. However, the adjutants do not tear meat off the carcass; they steal bits from the vultures. The marabou stork (*L. crumeniferus*) of the African **savanna** is also a carrion thief. Not an attractive bird, the marabou stork has a splotchy head, short neck, and a huge pink pouch hanging from its neck. The soft feathers on the tail of *Leptoptilos* storks have been used for hat decorations.

American storks

The only stork of **North America** is the wood stork (*Mycteria americana*), sometimes called the wood ibis. The wood stork has white body feathers, black flight feathers, gray legs, and a blue-gray featherless head. Its 9-in (23-cm) bill is gray, and it curves downward slightly at the end.

The wood stork lives primarily in the swamps of southern Florida, where it breeds in stands of large bald cypress trees. It also occurs more widely in Central and **South America**. In Florida, however, many of its nesting trees have been cut down, and wood stork **habitat** has also been destroyed or damaged by agricultural activities and drainage. In the 1950s, these birds were seriously threatened, until the Corkscrew Swamp Sanctuary was designated to protect a large nesting population. Their population, however, is still seriously depleted.

Wood storks feed primarily on **fish**, plus small **frogs**, **reptiles**, crustaceans, and **mollusks**. They use their feet and their bill to feel for the presence of food, which they snap up in their beak. They rarely depend on sight when hunting. Some wood storks also feed in open fields, where they hunt for **insects** in newly turned **soil**.

Wood storks do not mate so much according to the season, as to the level of water in their swampy habitat. They raise their young during the dry season, when their **wetlands** concentrate into prey-rich pools. These dense food supplies makes feeding hungry young storks easier. At the proper time, the male selects a nesting site, usually in a large bald cypress. At first, any female approaching the nest is driven away, but eventually he accepts a mate. The female holds the chosen nesting site while the male collects sticks as nest-building material. The female lays 2-5 eggs, but incubation starts as soon as the first one is laid. The eggs hatch after about 30 days of incubation by both birds. The fluffy white young remain in the nest until they are fully feathered, about 55 days after hatching.

The nest can be a dangerous place. Adolescent storks are prone to an activity that has been likened to that of juvenile gangs of humans. Groups of immature storks, younger than the age of three years, will attack nests, trying to drive the parents away. Then they tear apart the nest and kill the young. If this happens, the parents will build another nest and lay more eggs.

The only other American stork resides in tropical wetlands in South America. The jabiru (*Jabiru mycteria*) is among the largest storks, and can stand as tall as a human. Except for its head and neck, which are black and red, its plumage is white. The wings may span 7 ft (2.2 m) or more. Its 12-in (30-cm)-long bill is also black. The name *jabiru* is also sometimes given to the saddle-billed stork of Africa.

All large birds are in danger from hunters, and storks have the added problem of losing much of their wetland habitat to development. The presence of white storks on the rooftops of Europe has long been considered good luck. However, the numbers of these famous storks have dropped by about 90% in the last hundred years. Several changes explain this loss, including the effects of **pesticides**, loss of nesting and foraging habitat, changing climatic conditions, hunting, and the intensification of agricultural management (which degrades foraging habitat of the storks). Most species of storks are threatened by similar environmental changes associated with human activities.

Resources

Books

Sibley, David Allen. *The Sibley Guide to Birds.* New York: Knopf, 2000.

Jean F. Blashfield

Storm

A storm is any disturbance in the atmosphere that has noticeable effects on the earth's surface. The term suggests disagreeable **weather** conditions that may bring discomfort, inconvenience, economic disaster and loss of human lives. In spite of that fact, storms have a generally positive effect on the environment and on human societies because they are the source of most of the rain and snow on which the **planet** depends.

Among the many kinds of storms that exist are thunderstorms, hurricanes, tornadoes, blizzards, and **ice** and hail storms. As different as these types of storms may be, they possess a few common characteristics. They all result from significant atmospheric instabilities, and they all involve dramatic **convection** currents in which air masses travel upwards at high rates of speed.

The formation of a **thunderstorm** is typical of the way many storms develop. Imagine that a large **mass** of air encounters conditions that **force** it to move upwards. An approaching front or some kind of geographical barrier might produce such an effect. The air mass will continue to rise as long as it is warmer than the atmosphere

Multiple lightning strikes over Tucson Mountain, Arizona. *Photograph by Kent Wood. National Audubon Society Collection/Photo Researchers, Inc. Reproduced by permission.*

around it. The upward movement of such an air mass constitutes a convection current.

At some point, moisture within the air mass may begin to condense, releasing **heat** as it does so. When this happens the convection current may begin to move even more rapidly. The upward movement of air in a thunder cloud has been measured at more than 50 MPH (80 km/h). As the upward movement of air continues, more moisture condenses out of the air mass and a large cloud begins to form. Depending on atmospheric conditions, a thundercloud of this type may rise to a height of anywhere from 6-9 mi (10-15 km). Eventually, ice crystals within the thundercloud will begin to condense as rain, snow, or some other form of **precipitation** and a thunderstorm will occur.

Some of the most severe types of storms (tornadoes and hurricanes, for example) occur when the upward convention current receives a rotational push. This push converts a purely vertical air movement into a spiraling **motion** characteristic of these great tropical and mid-latitude storms.

See also Air masses and fronts; Cyclone and anticyclone; Tornado.

Storm surge

Storm surge, caused by very low **atmospheric pressure**, is a **volume** of oceanic **water** driven by the **wind** toward the shore where it "builds up" along the coast producing a localized increase in **sea level**.

Such low atmospheric **pressure** occurs during cyclonic storms, called typhoons in the Pacific region and hurricanes along the Atlantic seaboard. During these storms, a dome of water forms in the area of low pressure that moves across the **ocean** as the winds drive the storm. Upon reaching shallow coastal water, winds blowing toward the shore move over the domed water, and pile water along the coast, producing an elevated sea level.

Storm surges are the deadliest element of cyclonic storms. These storms form in areas with warm surface water temperatures in the zone 8° north and 15° south of the equator. Winds may reach 328 MPH (100 km/h) and move in a counterclockwise **spiral** around a calm center. Winds of nearly 200 MPH (60 km/h) associated with cyclonic storms can extend the storm's diameter to more than 2,000 mi (640k). Hurricane Hugo, the most expensive cyclonic storm in U.S. history had a storm surge of

7.9 ft (2.4 m) above normal tide levels. However, smaller storms can produce storm surge when they coincide with high **tides**.

Tides, the rhythmic rise and fall of the ocean's surface caused by the gravitational attraction of the **sun** and **moon**, can amplify the effect of storm surges. In most areas, two high tides and two low tides occur daily. Tides may rise and fall a few inches to more than 49 ft (15 m).

Normal tides are driven by three astronomical occurrences. First, when the moon is nearest the **earth**, due to its oval shaped **orbit**, the moon's gravitational attractions is at its greatest; thus tides occurring during this period are greater than normal tides. Second, when the sun, moon and earth are aligned, "spring" tides occur. These are 20% greater than normal tides. When the sun, moon and earth are at 90°, to one another "neap" tides, which are 20% weaker than normal, occur. Finally, twice a month the moon crosses the earth's equatorial **plane** resulting in higher than normal tides.

In addition to tidal influences, the shape of ocean basins, inlets and bays, and sea level rise may cause higher than normal storm surges. In February 1976, Bucksport, Maine experienced a 5.8-ft (1.76-m) storm

surge and Bangor, located 15 mi (24 km) inland from Bucksport, suffered a storm surge of nearly 11 ft (3.35 m). The difference in storm surges was due to the funnel shape of the bay. The great amount of water in the surge that was forced through a relatively narrow channel caused water levels to rise rapidly in Bangor. Damages of more than $2 million occurred.

London historically experienced recurrent tidal **flooding** from storm surges until the Thames barrier was constructed in 1986. Ten **steel** gates reaching 1,640 mi (500 m) across the river now protect the city from storm surge impacts.

Monica Anderson

Strata

Strata (singular: stratum) are the horizontal layers, or beds, present in most sedimentary **rocks**. During or immediately after the accumulation of sediments, physical, biological, and chemical processes produce sedimentary structures. Strata are probably the most com-

Visible strata in the Grand Canyon, Arizona. *Photograph by M. Woodbridge Williams, National Park Service.*

mon sedimentary structures, as almost all sedimentary rocks display some type of bedding. A rock that contains beds is stratified or displays stratification.

Strata form during sediment deposition, that is, the laying down of sediment. Meanwhile, if a change in current speed or sediment grain size occurs or perhaps the sediment supply is cut off, a bedding **plane** forms. Bedding planes are surfaces that separate one stratum from another. Bedding planes can also form when the upper part of a sediment layer is eroded away before the next episode of deposition. Strata separated by a bedding plane may have different grain sizes, grain compositions, or colors. Sometimes these other traits are better indicators of stratification as bedding planes may be very subtle.

The strata in an exposure or outcropping of **sedimentary rock** can range from layers as thin as **paper**, known as lamina (plural: laminae or laminations) to beds tens of feet thick. Generally, the more stable and consistent the environmental conditions during deposition, the thicker the strata. For example, in a river with very consistent current speeds, thick sediment layers with widely spaced bedding planes form. In a different river, where current speeds vary often, thin sediment layers and closely spaced bedding planes form instead.

The area covered by a bed, that is its areal extent, is also highly variable. Some beds can be traced for hundreds of square miles. Others may cover an area of only a few hundred square feet. Many factors influence the areal extent of beds. Among the more important factors is the setting in which the bed formed. **Rivers**, for obvious reasons, **deposit** beds shaped like a shoe string (long and thin); deposits in the open **ocean** often extend for great distances in all directions. **Erosion** of strata after deposition also affects their areal extent.

Bedding planes indicate variable environmental conditions during sediment deposition, but they may also be evidence of a gap in the geologic record. Many times a bedding plane develops because no sediment accumulates for at least a brief period of time or it is later eroded away. This represents an interval of time for which there is no sediment record. If we think of strata as a record of **geologic time** preserved in sediment (or sedimentary rock), this break or gap between sedimentation events actually is a gap in the geologic record. In other words, think of an **outcrop** of rock as being like a book; strata are pages and bedding planes are pages that have been torn from the book.

The time gap represented by a bedding plane may be very short, a fraction of a second, or perhaps a few minutes. After that interval passes, recording continues in the form of sediment deposition. However, sometimes the amount of time that is unrepresented (the gap) can be quite long, perhaps hundreds or thousands of years. These longer gaps are called unconformities. In some rock outcrops, more geologic time may be represented by the bedding planes (the gaps) than by the strata that lie between them.

See also Sediment and sedimentation; Unconformity.

Stratigraphy

Stratigraphy is the science of interpreting and describing layers and **strata** of sediments. Commonly these layers are levels of **sedimentary rock**, but stratigraphy can also include the study of non-ossified sediments, like those in stream beds and **lake** bottoms, of inclusions such as volcanic ash and lava, and even the study of different layers of human occupation. Sediment usually forms distinct strata with the most recent layers on top and, although they may be folded by **continental drift**, interrupted by inclusions and slippages, and even metamorphosed into other forms of rock, as long as these strata can be untangled and interpreted, scientists can perform stratigraphic analyses. The processes of sedimentation—including the presence of certain types of fossils—provide scientists with valuable clues about the age of the **earth** and its history. These principles are thus valuable for many different types of scientist, ranging from prospecting geologists to city planners to archaeologists and paleontologists studying human and **animal** history and prehistory.

Stratigraphic fundamentals

The basic principle of sedimentation—that in any given set of layers of material the most recent levels are closest to the top—were established as long ago as the seventeenth century. By the nineteenth century early geologists like Charles Lyell recognized that this accumulation was not necessarily regular nor was it obvious. Interruptions and inversions (known collectively as discontinuities) in the stratigraphic record can change the actual position of layers of sedimentation, but their apparent position remains evident to trained geologists in relation to other layers and to their contents—especially fossil contents. **Weathering** can also influence the stratigraphic record by introducing **trace elements** into the various layers. Patterns in the very layering of sediments, such as ripple marks and flumes, can introduce discontinuities. Changes in climate, which bring about changes in **sea level**, also create discontinuities.

Equally as important as the contents of the layers themselves, however, are the borders between them.

These separations mark discontinuities, breaks between one time and another. They can mark changes in accumulation of sediment as well as changes in time. Sediments do not **deposit** evenly—rates of sedimentation are influenced by extraordinary events as well as everyday processes. During periods of flood, for instance, **rivers** can drop tons of silt on what had been working farm land, and a single **storm** can carry away tons of beach **sand** into the **ocean** depths. The borders marked by the beginning and ends of such events can represent as little time as a single day. Because sediments generally accumulate over long periods of time, however, the borders between different layers usually represent a long-term change in local geography.

Geologists have created terminology to describe the different types of layers based on their thickness. Sediments are generally divided between laminae and beds, with the laminae represented by an accumulation of less than one centimeter, and the beds represented by accumulations ranging from 0.4-47 in (1-120 cm). The beds are subdivided into very thin, thin, thick, and very thick—respectively measuring 0.4-2 in (1-5 cm), 2-24 in (5-60 cm), 24-47 in (60-120 cm), and more than 47 in (120 cm) across. Beds are also graded on the size and type of the individual sand grains. Beds and laminae together form primary sedimentary structures, which indicate the way in which strata are laid down.

Geologists have also introduced various subgenres of stratigraphy classified by the types of layered material. Lithostratigraphers trace changes in layers of rock. This is the type of stratigraphy most commonly seen on geological survey maps. Biostratigraphy uses microscopic fossils to determine the relative ages of **rocks** and helps paleontologists trace local variations in climate. Tephrostratigraphy is the study of deposits of volcanic ash, while magnetostratigraphers trace fluctuations in the earth's magnetic field—specifically, reversals in its polarity—over millions of years. Other useful applications of stratigraphic analysis include seismic stratigraphy, which applies the principles of **acoustics** (sending shock waves through the earth) to determine the positions of pockets of **petroleum** and other substances.

Applications of stratigraphy in historical studies

Because strata are deposited in layers that scientists can interpret, they can be used to study history, both the history of the earth and, on a shorter time scale, of humankind. Anthropologists and archaeologists use stratigraphic principles to understand how and under what circumstances a site was occupied, how long the people that lived there stayed, and how they lived while they were in residence. Archaeologists regularly apply microstratigraphic principles such as observation of the process of **soil** formation and landscape development based on weathering and sediment accumulation to their sites in order to classify and date artifacts. For example, excavators working at the Paleolithic site of Mezhirich in what is now Ukraine uncovered the bones of hundreds of mammoths, some of which had been burnt, others of which had been arranged to form houses. The scientists suggested, based on stratigraphic principles— the thinness of accumulated layers of occupation debris—that the site was not occupied year- round, but instead was a seasonal dwelling-place. Paleontologists studying the border regions between the Eocene and Oligocene periods in ancient history have studied eastern Oregon's stratigraphy to draw conclusions about global climactic conditions. They suggest that the regularity of sediments in the area reflect, not a gradual change in sea level, but a cooling trend in which the area changed from a subtropical to a temperate climate.

Archaeologists have even applied stratigraphic principles to understanding the history of the famous Roman city of Pompeii, which was buried following an eruption of the **volcano** known as Vesuvius in A.D. 79. Although the historical record of the explosion itself is quite clear, scientists use stratigraphy to help unwrap the city's past before the eruption. Excavators had assumed, based on the testimony of ancient written sources, that parts of Pompeii had been built as long ago as the fifth century B.C. and had been occupied ever since. Nineteenth and early twentieth-century archaeologists had accepted this reasoning, based on analyses of building styles and construction. Beginning in the 1930s, however, scientists began to revise their thinking using observations of the microstratigraphy of the site. Modern excavations suggest that most of the buildings standing at the time the volcano erupted were built in the period of Roman occupation—in other words, no earlier than the second century B.C. Some finds of debris unrelated to the A.D. 79 eruption can be dated back to the fifth century B.C., but these are not directly connected with standing houses. Stratigraphy promises to change the history of a site historians believed they knew very well.

Resources

Books

Boggs., Sam, Jr. *Principles of Sedimentology and Stratigraphy.* 2nd edition. Englewood Cliffs, NJ: Prentice Hall, 1995.

Opdyke, Neil D., and James E.T. Channell. *Magnetic Stratigraphy.* San Diego, CA: Academic Press, 1996.

Waters, Michael. *Principles of Geoarchaeology: A North American Perspective.* Tucson, AZ: University of Arizona Press, 1992.

Periodicals

Fulford, Michael, and Andrew Wallace-Hadrill. "Unpeeling Pompeii." *Antiquity,* 72. (March 1998): 128-46.

Soffer, Olga, James M. Adovasio, Ninelj L. Kornietz, Andrei A. Velichko, Yurij N. Gribchenko, Brett R. Lenz and Valeriy Yu. Suntsov. "Cultural Stratigraphy at Mezhirich, an Upper Paleolithic Site in Ukraine with Multiple Occupations." *Antiquity,* 71. (March 1997): 48-63.

Kenneth R. Shepherd

Stratigraphy (archeology)

Stratigraphy is the study of layered materials (**strata**) that were deposited over time—their lateral and vertical relations, as well as their composition. The basic law of stratigraphy, the law of superposition, states that lower layers are older than upper layers, unless the sequence has been disturbed. Stratified deposits may include soils, sediments, and **rocks**, as well as man-made structures such as pits and postholes. The adoption of this principle by archeologists greatly improved excavation and archeological dating methods.

By digging from the top downward, the archeologist can trace the buildings and objects on a site back through time using techniques of typology (i.e., the study of how types change in time). Object types, particularly types of pottery, can be compared with those found at other sites in order to reconstruct patterns of trade and communication between ancient cultures. When combined with stratification analysis, an analysis of the stylistic changes in objects found at a site can provide a basis for recognizing **sequences** in stratigraphic layers.

Archeological stratigraphy, which focuses on stratifications produced by man, was derived largely from the observations of stratigraphic geologists, or geomorphologists. A geomorphologist studies stratigraphy in order to determine the natural processes, such as floods, that altered and formed local terrain. By comparing natural strata and man-made strata, archaeologists are often able to determine a depositional history, or stratigraphic sequence—a chronological order of various layers, interfaces, and stratigraphic disturbances. Stratigraphic data may be translated into abstract diagrams, with each deposit's diagram positioned relative to the deposits above and below it. By this method, archeologists can illustrate the stratigraphic sequence of a given site with a single diagram. Such a diagram, showing the different layers with the oldest at the bottom and the youngest at the top, may cover 3,000 years. The diagram also records finds such as pits, post holes, and burials that may have belonged to a single period. The archeologist may also document the site with notes about the relationships of stratigraphic units and **soil** composition.

History of stratigraphy

The basic principles of stratigraphy were developed primarily by geologists in the nineteenth century. Many of the fundamental ideas drew on the observations of Jens Jacob Asmussen Worsaae in Denmark, and Thomas Jefferson in Virginia.

Among the first archeologists to understand the stratigraphy of tells (artificial mounds) were William Matthew Flinders Petrie at Tell-el-Hesi in 1890, Heinrich Schliemann at Troy between 1871 and 1890, and R. Pumpelly and Hubert Schmidt at Anau in 1904. Another major force behind the acceptance of archeological stratigraphy was General Pitt-Rivers (1827–1900), who considered that material culture could be explained in terms of a typological sequence—objects that had evolved over time. In his excavations, he practiced the total excavation of sites, emphasizing the principles of stratigraphy. Giuseppe Fiorelli, who assumed responsibility for the excavation of Pompeii in 1860, also pioneered the use of stratigraphic methods in archeology.

Some early advocates of the principles of stratigraphy found opposition from many of the same traditionalists who opposed the theory of **evolution**. The French scientist Georges Cuvier (1769–1832), for example, was convinced that the history of **Earth** had been characterized by a series of catastrophic events, the last being the biblical flood of Genesis. Charles Lyell (1797–1875), a contemporary of Couvier, argued that geologic change throughout Earth's history had taken place gradually. Although many of Lyell's ideas were not new, they had tremendous influence because he presented them more clearly than had any of his predecessors. As the biblical accounts of the Flood became less convincing to many scientists in light of new scientific discoveries, the historical record of stratified rocks began to replace the story of Genesis as a basis or understanding the past.

How stratigraphy is used

In the case of societies that have left no written histories, the excavation and recording of strata, features and artifacts often provides the only method of learning about those societies. Even when recorded histories exist, stratigraphic investigations can provide an excellent complement to what is already known.

According to the law of superposition, in a given series of layers, as originally created, the upper layers are younger and the lower layers older because each layer presumably has been added to a pre-existing **deposit**.

Based on this law, archeologists have been able to assign dates, in relative sequence, to stratified layers. The law of superposition is not infallible. Sites often contain strata that have been disturbed by natural processes, such as floods, and human activities, such as digging. In these instances, several original layers may be intermixed, and the artifacts contained within may be out of chronological sequence.

In stratigraphic excavations, deposits from a site are removed in reverse order to determine when they were made. Each deposit is assigned a number, and this number is appended to all objects, including artifacts, bones, and soil samples containing organic **matter**, found in the layer. Each layer provides a unique snapshot of a past culture, the environment in which it existed, and its relative period in time. Stratigraphic dating does not require the existence of artifacts, but their presence may facilitate dating the site in absolute time. Without such clues, it can be very difficult to date the layers; a deep layer of **sand**, for example, might have been deposited very quickly in the course of a sand **storm**, while another layer of the same thickness could have taken hundreds of years or longer to form.

Problems with stratigraphy

Unfortunately for archeologists, it is not always the case that the oldest layer lays at the bottom of an excavated site. In one excavation, an archeologist found the surface of a site littered with old coins dating to the seventeenth century. Subsequent investigations, however, revealed that a bulldozer had earlier overturned the soil at the site to a depth of several feet as part of a preparation for building homes on the site.

The problems of relying entirely on stratigraphic analyses to evaluate the antiquity of a find were made even clearer in an incident known as the great **Piltdown hoax**. Between 1909 and 1915, an amateur British paleontologist made claims of having discovered the fossils of a prehistoric human being in a gravel pit in Piltdown, Sussex (England). But in 1953, tests revealed that the Piltdown man actually had the jaw of a nineteenth-century ape, and the skull of a modern human. The planting of

faked remains at a site of known stratigraphic antiquity had in this case succeeded in deceiving even the head geologist at the British Museum, who had been among many who authenticated the find.

See also Archaeology; Archeological sites.

Resources

Books

Fagan, Brian M., ed. *The Oxford Companion to Archeology.* New York: Oxford University Press, 1996.

Lyman, R. Lee, Michael J. O'Brien. *Seriation, Stratigraphy, and Index Fossils—The Backbone of Archaeological Dating.* New York: Kluwer Academic Publishers, 1999.

Maloney, Norah. *The Young Oxford Book of Archeology* New York: Oxford University Press, 1997.

Nash, Stephen Edward, ed. *It's about Time: A History of Archaeological Dating in North America.* Salt Lake City, UT: University of Utah Press, 2000.

Randall Frost

Strawberry *see* **Rose family (Rosaceae)**

Stream capacity and competence

Streams channel **water** downhill under the influence of gravity. Stream capacity is a measure of the total sediment (material other than water) a stream can carry. Stream competence reflects the ability of a stream to transport a particular size of particle (e.g., boulder, pebble, etc). With regard to calculation of stream capacity and competence, streams broadly include all channelized movement of water, including large movements of water in **rivers**.

Under normal circumstances, the major factor affecting stream capacity and stream competence is channel slope. Channel slope (also termed stream gradient) is measured as the difference in stream elevation divided by the linear distance between the two measuring points. The **velocity** of the flow of water is directly affected by channel slope, the greater the slope the greater the flow velocity. In turn, an increased velocity of water flow increases stream competence. The near level **delta** at the lower end of the Mississippi River is a result of low stream velocities and competence. In contrast, the Colorado River that courses down through the Grand Canyon (where the river drops approximately 10 ft per mile [3 m/1.6 km]) has a high stream velocity that results in a high stream capacity and competence.

Channelization of water is another critical component affecting stream capacity and stream competence. If a stream narrows, the velocity increases. An overflow or broadening of a stream channel results in decreased stream velocities, capacity, and competence.

The amount of material (other than water) transported by a stream is described as the stream load. Stream load is directly proportional to stream velocity and stream gradient and relates the amount of material transported past a point during a specified time interval. The greater the velocity, the greater the sum of the **mass** that can be transported by a stream (stream load). Components of stream load contributing to stream mass include the suspended load, dissolved load, and bed load. Broad, slow moving streams are highly depositional (low stream capacity) while high velocity streams have are capable of moving large **rocks** (high stream competence).

Alluvial fans form as stream streams channeling mountain runoff reach flatter (low, slope, low gradient) land at the base of the mountain. The stream loses capacity and a significant portion of the load can then settle out to for the alluvial fan.

The ultimate site of deposition of particular types and sizes of particles is a function of stream capacity and stream competence (along with settling velocity of particles). These factors combine to allow the formation of articulated sedimentary deposits of gypsum, limestone, clay, shale, siltstone, sandstone, and larger rock conglomerates. No **matter** how low the stream capacity, the **solution** load usually retains ions in solution until the water evaporates or the **temperature** of the water cools to allow **precipitation**.

In confined channels, stream competence can vary with seasonal runoff. A stream with low **volume** may only be able to transport ions, clays, and silt in its solution and suspension loads and transport **sand** as part of its saltation load. As stream flow increases the stream competence and during seasonal **flooding** stream may gain the competence to move pebbles, cobbles and boulders.

See also Hydrologic cycle; Hydrology; Landform; Sediment and sedimentation; Sedimentary environment; Stream valleys, channels, and floodplains.

Stream valleys, channels, and floodplains

Stream valleys, channels, and floodplains form complicated systems that evolve through time in response to changes in sediment supply, **precipitation**, **land use**, and rates of tectonic **uplift** affecting a drainage **basin**.

Stream channels serve to convey flow during normal periods, whereas floodplains accommodate flow above the bankfull stage (floods) that occurs with frequencies inversely proportional to their magnitude. Bankfull stage is defined as the discharge at which the **water** level is at the top of the channel. Any further increase in discharge will cause the water to spill out of the channel and inundate the adjacent floodplain. Flood **frequency** studies of streams throughout the world show a remarkably consistent 1.0 to 2.5 year recurrence interval for bankfull discharge in most streams, averaging about 1.5 years, meaning that small floods are relatively common events.

Stream channels are classified according to four basic variables: their slope or gradient (change in elevation per unit of distance along the stream channel), their width to depth **ratio**, their entrenchment ratio (floodplain width to bankfull width), and the predominant channel bed material (**bedrock**, gravel, cobble, **sand**, or clay). In general, the width of stream channels increases downstream more than the depth, so that large **rivers** such as the Mississippi may be a kilometer or more wide but only tens of meters deep. Channels with large bed loads of coarse-grained materials, steep gradients, and banks composed of easily eroded sediments tend to be shallow and braided, meaning that flow occurs through many anastomosing channels separated by bars or islands. Streams with low gradients, small bed loads, and stable banks tend to meander in space and time, following a pattern that resembles an exaggerated sine wave. Another characteristic of streams with beds coarser than sand is the occurrence of riffle-pool **sequences**, in which the channel is segregated into alternating deep pools and shallow riffles. In steep mountain streams, the riffles can be replaced by steep steps over boulders or bedrock outcrops to form a step-pool sequence.

Stream channels can change in form over time as a function of climate, precipitation, sediment supply, tectonic activity, and land use changes. Increased precipitation or human activities, for example, heavy grazing or clear-cut logging, can lead to increased **erosion** or **mass wasting** that subsequently increase the amount of sediment delivered to streams. As a consequence, the channel and stream gradient change to accommodate the increased sediment load, which may in term have adverse effects on aquatic **habitat**. For example, an influx of fine-grained sediment can clog the gravel beds in which **salmon** and trout spawn. The effect of urbanization is generally to increase **storm** runoff and the erosive power of streams because impervious areas (principally pavement and rooftops) decrease the amount of water that can infiltrate into the **soil** while at the same time decreasing the amount of sediment that is available for erosion before runoff enters stream channels. Tectonic uplift can

A canyon (goosenecks) formed by the San Juan River in Utah. *CORBIS/Tom Bean. Reproduced by permission.*

increase the **rate** of stream valley incision. Thus, stream channels represent the continually changing response of the stream system to changing conditions through over geologic and human time spans.

Because streams are the products of continual change, many stream valleys contain one or more generations of stream terraces that represent alternating stages of sediment deposition (valley filling) and erosion (stream incision). Each flat terrace surface, or tread, is a former floodplain. Stream terraces can often be recognized by a stair-step pattern of relatively flat surfaces of increasing elevation flanking the channel; in many cases, however, stream terraces are subtle features that can be distinguished and interpreted only with difficulty.

Floodplains form an important part of a stream system and provide a mechanism to dissipate the effects of floods. When a stream exceeds bankfull discharge, floodwater will begin to spill out onto the adjacent flat areas where its depth and **velocity** decrease significantly, causing sediment to fall out of suspension. The construction of flood control structures such as artificial levies has allowed development on many floodplains that would otherwise be subjected to regular inundation. Artificial levies, however, also increase the severity of

less frequent large floods that would have been buffered by functioning floodplains, and can thereby provide a false sense of security. Current trends in flood hazard mitigation are therefore shifting away from the construction of containment structures and towards the use of more enlightened land use practices such as the use of floodplains for parks or green belts rather than residential development.

See also Hydrologic cycle; Hydrology; Landform; Sediment and sedimentation; Sedimentary environment; Stream capacity and competence.

Strepsiptera

Also known as twisted-winged **parasites**, strepsiterans are small **insects** which are internal parasites of other insects. Measuring between 0.02-0.16 in (0.5 and 4 mm) long, the males and females lead totally different lives. Males are free, winged insects—resembling some forms of beetles—and females are wingless, shapeless insects living as parasites. Strepsipterans live all over the world, except **Antarctica**.

Belonging to the largest class of animals in the world—the class Insecta—the superclass Hexapoda contains over 750,000 **species**. There are two subclasses within Hexapoda: (1) Apterygota (insects without wings) which contains two orders, and (2) Pterygota (insects with wings, accounting for 99.9% of all insect species) which contains twenty-eight orders. Further classification of strepsipterans is continuously being revised. Sometimes, they are considered to be a suborder of the order Coleoptera, an order containing **beetles**; however, often they are given their own order—the order Strepsiptera. Currently, there are seven families within the order Strepsiptera, containing about 300 species of insects, 60 of which live in **North America**.

As mentioned before, the appearance and **behavior** of the male and female strepsipterans differ markedly. The female resembles a grub, having no wings, legs, eyes, or mouth. Also, her **nervous system** is very diminished. She generally attaches herself to another insect as a host. For instance, the female of the Stylopidae species attaches herself to the stomach of a wasp or bee. She burrows almost her entire body into her host, sticking out slightly. While she does not usually kill her host, her presence sterilizes it, causing it to have both male and female sexual organs, and alters its appearance otherwise.

The male strepsipteran is independent of any host, having antennae, wings, and large eyes. He darts about during his ten hour lifetime, continuously looking for a female to mate. She lures him with a special odor, since she is almost invisible within her host. He injects his sperm through a small hole located between her thorax and abdomen. The male dies soon after mating; when he dies, his forewings dry out and twist up like a corkscrew, giving these insects their common name.

After a few days, the female hatches about 1,500 tiny larvae that are born with eyes, mouths, and three pair of legs ending in **suckers**. Leaving their mother through an opening in her back, they continue to exist on the host bee (or other insect) until it lands on a **flower**. At this point, the offspring climb onto the flower and await another bee. Being consumed by a new bee, the young strepsipterans ride back to the hive. The bee regurgitates them when it stocks its nest with **nectar**, and the young strepsipterans are free to bore into the bee larvae and molt to their legless, inactive forms. Thus, the cycle begins again.

Stress

Stress is mental or physical tension brought about by internal or external pressures. Researchers have found significant biochemical changes that take place in the body during stress. Exaggerated, prolonged, or genetic tendencies to stress cause destructive changes which lower the body's **immune system** response and can lead to a variety of diseases and disorders. These include **depression**, cardiovascular **disease**, **stroke**, and **cancer**.

People experience stress from many different sources. It can come from having to take a test or dealing with a difficult person; from traumatic experiences such as the death of a loved one, or a serious illness. Stress can be acute—as in the face of immediate danger when the "fight-or-flight" response is triggered; or chronic—such as when a person is involved in a long-term stressful situation.

People who experience severe traumas, as do soldiers during combat, may develop a condition called post-traumatic stress disorder (PTSD). During World War I this was called shell shock; during World War II it was called battle fatigue. Since 1980, PTSD has been listed as a diagnostic category by the American Psychiatric Association. Sufferers of PTSD experience depression, feelings of guilt for having survived, nightmares, and flashbacks to the traumatic events. They may be excessively sensitive to noise, become violent, and have difficulty holding a job.

General adaptation syndrome

Dr. Hans Selye, an endocrinologist, developed a three-stage model of the body's response to stress. He called his theory the general adaptation **syndrome** (GAS). The first phase is an alarm reaction, the second stage is one of resistance or adaptation, and the final stage is one of exhaustion.

In the alarm stage the body responds to a stressor, which could be physical or psychological. Perhaps you are crossing the street and a car suddenly speeds toward you. Your **heart** begins to beat fast and the release of adrenaline makes you move quickly from the path of the oncoming car. Or another response might include **butterflies** in your stomach, a rise in your **blood pressure**, heavy breathing, dilation of your eyes, dry mouth, and the hair on your arms might even stand on end. To help you meet the sudden danger, your blood flows away from the organs not needed to confront the danger, to organs and tissues which are; for example, your heart races, your eyes dilate, your muscles tense up, and you will not be able to concentrate on any kind of problem solving outside the danger confronting you.

During the resistance stage of a stress reaction, your body remains on alert for danger. When this part of the GAS is prolonged, your immune system may become compromised and you may become susceptible to illness. Even within days of becoming stressed and maintaining a stress alertness, changes take place that weaken your body's ability to fight off disease.

The final stage of Selye's GAS is the exhaustion stage. As your body readjusts during this period, **hormones** are released to help bring your body back to normal, to the state of balance called **homeostasis**. Until balance is reached, the body continues to release hormones, ultimately suppressing your immune system.

Stress and illness

Continuously, studies are being aimed at trying to determine the relationship of illness and state of mind. During the 1980s, physicians at the University of California Medical Center in Los Angeles determined that emotional stress affected the immune system and that, conversely, the reduction of stress boosted the immune system.

Significant breakthroughs in the late 1990s found stress causes an immediate and significant increase in the release of the hormone corticotrophin (ACTH) by the anterior pituitary gland, causing many stress-related behaviors in the **nervous system**, including the fight-or-flight response. This is followed soon thereafter by drastically increased secretion of the hormone cortisol, which is intended to relieve the damaging effects of stress. However, the prolonged secretion of cortisol has the potential to cause or worsen biological and psychological diseases and disorders.

Clinical studies reported in the December 2, 1997 issue of the journal *Circulation*, published by the American Heart Association, found exaggerated response to mental stress can produce the same degree of atherosclerosis risk as does smoking and high **cholesterol**, thereby drastically increasing the risk of heart disease and stroke. Another study in 1997 showed that the stress of being diagnosed with cancer also reduces the activity and therefore effectiveness of natural killer (NK) cells—cells whose role it is to seek and kill malignant (cancer) cells. It also decreases their ability to respond to recombinant interferon gamma, a form of cancer therapy aimed to help them do their job.

Other diseases associated with stress are the onset of adult diabetes (type 2 non-insulin-dependent), **ulcers**, respiratory infections, and depression. The stress can be psychological or it can come from stressful situations such as accidents or illness.

Recognition of stress

Identifying physical and psychological responses to stress and understanding the cause of the stress itself are important factors in reducing its negative effects. Denying the existence of stress, on the other hand, is a contributing factor to intensifying those effects. Some physical signs are a dry mouth and throat; tight muscles in the neck, shoulders, and back; chronic neck and back problems; headaches; indigestion; tremors; muscle tics; **insomnia**; and fatigue. Emotional signs of stress include difficulty in concentrating, feeling tense, negative thinking, brooding, worrying, depression, **anxiety**, or feelings of worthless.

Irritability, restlessness, impulsive **behavior**, difficulty in making a decision, poor judgment, difficulty relating to—and mistrusting—people, as well as tobacco, **alcohol**, and drug use, may all be indications of stress.

Treatments for stress reduction

A form of psychotherapy called medical psychotherapy is one of the methods used to deal with stress. It is primarily a talking therapy based on the principle that when people can talk about what is troubling them, they can lessen the amount of stress they feel. It is important for the therapist to understand the nature of the illness around which the stress is involved. In this kind of supportive therapy, the goal is to help patients deal with the feelings stimulated by traumatic events, illness, or conflicts that produce stress. Medications may also be used, such as anti-anxiety or **antidepressant drugs**, which are best administered by a psychiatrist knowledgeable about stress-induced disorders. The American Academy of Family Physicians reports that two-thirds of their patients come for treatment of a stress-related condition.

Selye believed that the stress of life contributed a great deal to aging. A method of treatment for stress developed after his theories became popular was called progressive relaxation. Relaxation techniques such as yoga and creative visualization are often used successfully to reduce stress and boost the immune system. Often, life-style changes are necessary also, such as a healthier diet, smoking cessation, **aerobic exercise**, and group discussions.

Advances in biochemical research hold a promising future for the treatment of stress and its related diseases. A 1995 study published in the *Journal of Neuroscience* in 1997 indicated that researchers have identified a peptide in the **brain** and the body—prepro-TRH178-199—which significantly reduces hormonal and behavioral manifestations of stress by as much as 50%. This peptide acts by reducing levels of ACTH and prolactin—another pituitary hormone stimulated by stress—which subsequently lowers cortisol levels. By reducing levels of these hormones, anxiety-related behaviors and fear were significantly decreased. Because the overproduction of cortisol is also found in serious depression, and anxiety disorders such as anorexia nervosa, the prepro-TRH pep-

KEY TERMS

. . . .

Atherosclerosis—Abnormal narrowing of the arteries of the body that generally originates from the buildup of fatty plaque on the artery wall.

Corticotropin (ACTH)—A hormone released by the anterior pituitary gland in the brain in response to stress or strong emotions.

Cortisol—A hormone involved with reducing the damaging nature of stress.

General adaptation syndrome (GAS)—The three-phase model of stress reaction.

Homeostasis—The state of being in balance.

Peptide—A combination of two or more amino acids.

Post-traumatic stress disorder (PTSD)—A response to traumatic events such as those experienced in combat.

Prostaglandin—A fatty acid in the stomach that protects it from ulcerating.

tide may also become a valuable new approach in treating these disorders.

Resources

Books

Cotton, Dorothy H.G. *Stress Management.* New York: Bruner/Mazel, 1990.

Friedman, Howard S., ed. *Hostility, Coping and Health.* Washington, DC: American Psychological Association, 1992.

Green, Stephen A. *Feel Good Again.* Mount Vernon, New York: Consumers Union, 1990.

Moyers, Bill. *Healing and the Mind.* New York: Doubleday, 1993.

Sapolsky, Robert M. *Why Zebras Don't Get Ulcers.* New York: W.H. Freeman and Company, 1994.

Schafer, Walter E. *Stress Management for Wellness.* New York: Holt, Rinehart & Winston, 1992.

Periodicals

"Abnormal, Abusive, And Stress-Related Behaviors In Baboon Mothers." *Biological Psychiatry* 52, no. 11 (2002): 1047-1056.

Vita Richman

Stress, ecological

Environmental **stress** refers to physical, chemical, and biological constraints on the productivity of **species** and on the development of ecosystems. When the exposure to environmental stressors increases or decreases in intensity, ecological responses result. Stressors can be natural environmental factors, or they may result from the activities of humans. Some environmental stressors exert a relatively local influence, while others are regional or global in their scope. Stressors are challenges to the integrity of ecosystems and to the quality of the environment.

Species and ecosystems have some capacity to tolerate changes in the intensity of environmental stressors. This is known as resistance, but there are limits to this attribute, which represent thresholds of tolerance. When these thresholds are exceeded by further increases in the intensity of environmental stress, substantial ecological changes are caused.

Environmental stressors can be grouped into the following categories:

(1) Physical stress refers to brief but intense exposures to kinetic **energy**. This is a type of ecological disturbance because of its acute, episodic nature. Examples include volcanic eruptions, windstorms, and explosions.

(2) **Wildfire** is also a disturbance, during which much of the **biomass** of an **ecosystem** is combusted, and the dominant species may be killed.

(3) **Pollution** occurs when chemicals are present in concentrations large enough to affect organisms and thereby cause ecological changes. Toxic pollution can be caused by gases such as **sulfur dioxide** and **ozone**, by elements such as arsenic, **lead**, and mercury, and by **pesticides** such as DDT. Inputs of **nutrients** such as phosphate and nitrate can influence productivity and other ecological processes, causing a type of pollution known as **eutrophication**.

(4) Thermal stress occurs when releases of **heat** influence ecosystems, as happens in the vicinity of natural hot-water vents on the **ocean** floor, and near industrial discharges of heated **water**.

(5) **Radiation** stress is associated with excessive loads of ionizing energy. This can occur on mountain tops where there are intense exposures to ultraviolet radiation, and in places where there are exposures to radioactive materials.

(6) Climatic stress is associated with excessive or insufficient regimes of **temperature**, moisture, solar radiation, and combinations of these. **Tundra** and deserts are examples of climatically stressed ecosystems, while tropical rainforests occur under a relatively benign climatic regime.

(7) Biological stresses are associated with the diverse interactions that occur among organisms of the same or different species. Biological stresses can result

from **competition**, herbivory, predation, parasitism, and **disease**. The harvesting and management of species and ecosystems by humans is a type of biological stress. The introduction of invasive, non-native species may be regarded as a type of biological pollution.

Various types of ecological responses occur when the intensity of environmental stress causes significant changes. For example, disruption of an ecosystem by an intense disturbance causes mortality of organisms and other ecological damage, followed by recovery through **succession**.

More permanent ecological adjustments occur in response to longer-term increases in the intensity of environmental stress, associated perhaps with chronic pollution or climate change. The resulting effects can include reductions in the abundance of vulnerable species, their elimination from sites stressed over the longer term, and replacement by species that are more tolerant of the changed environmental conditions. Other commonly observed responses to longer-term increases in stress include a simplification of species richness and decreased rates of productivity, **decomposition**, and nutrient cycling. In total, these changes represent a longer-term change in the character of the ecosystem, or an ecological conversion.

See also Ecological integrity.

String theory

String theory (also termed "superstring" theory) is a mathematical attempt to describe all fundamental forces and particles as manifestations of a single, underlying entity, the "string."

String theory's predictions are consistent with all known experimental data, and it is felt by some physicists to be a candidate for the long-sought "theory of everything" (i.e., a single theory describing all fundamental physical phenomena); however, string theory has proved difficult to subject to definitive experimental tests, and therefore actually remains a speculative hypothesis.

Physics as defined before the twentieth century, "classical" physics, envisioned the fundamental particles of **matter** as tiny, solid spheres. Quantum physics, which originated during the first 40 years of the twentieth century, envisioned the fundamental particles simultaneously as particles and as waves; both mental pictures were necessary to make sense out of certain experimental results.

String theory, the first forms of which were developed in the late 1960s, proposes that the fundamental unit of everything is the "string," pictured as a bit of taut wire or string on the order of 10^{-33} cm in length (a **factor** of 10^{-20} smaller than a **proton**). These strings may be "open," like a guitar string, or form loops like rubber bands; also, they may merge with other strings or divide into substrings.

Beginning a few years after its initial formulation, string theory stagnated for a decade because of mathematical difficulties, but exploded in the 1980s with the discovery that the theory actually possesses a highly desirable mathematical feature termed $E(8) \times E(8)$ **symmetry**. Several major theoretical victories were won by string theorists in the 1990s, and intense efforts to extend string theory continue today. String theory is not the product of a single mind, like the theory of relativity, but has been produced by scores of physicists refining each other's ideas in stages.

Like the "waves" or "particles" of traditional **quantum mechanics**, of which string theory is an extension or refinement, "strings" are not objects like those found in the everyday world. A string-theory string is not made of any substance in the way that a guitar string, say, may be made of **steel**; nor is it stretched between anchor-points. If string theory is right, a fundamental string simply *is*.

Not only does string theory propose that the string is the fundamental building block of all physical reality, it makes this proposition work, mathematically, by asserting that the Universe works not merely in the four dimensions of traditional physics—three spatial dimensions plus time—but in 10 or 11 dimensions, 6 or 7 of which are "hidden" from our senses because they "curled up" to subatomic size. Experimental proof of the existence of these extra dimensions has not yet been produced.

Although the "strings" of string theory are not actual strings or wires, the "string" concept is nevertheless a useful mental picture. Just as a taut string in the everyday world is capable of vibrating in many modes and thus of producing a number of distinct notes (**harmonics**), the vibrations of an elementary string manifest, the theory proposes, as different particles: **photon**, **electron**, quark, and so forth.

The string concept also resolves the problem of the "point particle" in traditional quantum physics. This arises during the mathematical description of collisions between particles, during which particles are treated as mathematical points having zero diameter. Because the fields of **force** associated with particles, such as the electric field that produces repulsion or attraction of charges, go by $1/r$, where r is the distance to the particle, the force associated with a zero-diameter particle goes to **infinity** during a collision as $r \rightarrow 0$. The infinities in the point-particle theory have troubled quantum physicists' efforts to describe particle interactions for decades, but in the **mathematics** of string theory they do not occur at all.

In the **Standard Model**, quantum physicists' systematic list of all the fundamental particles and their properties, the graviton (the particle that mediates the gravitational force) is tacked on as an afterthought because it is hypothesized to exist, not because the equations of the Standard Model explicitly predict its existence; in string theory, however, a particle having all the properties required of the graviton is predicted as a natural consequence of the mathematical system.

In fact, when the existence of this particle was calculated by early string-theory workers, they did not recognize that it might be the graviton, for it had not occurred to them that their new theory might be powerful enough to resolve *the* biggest problem of modern fundamental physics, the split between general relativity (the large-scale theory of space, time, and gravity) and quantum mechanics (the small-scale theory of particles and of all forces except gravity).

String theory—or, rather, the string theories, as a variety of different versions of string theory have been put forward—thus not only predict all the particles and forces catalogued by the Standard Model, but may offer a theory of "quantum gravity," a long-sought goal of physics.

Doubt lingers, however, as to whether string theory may be too flexible to fulfill its promise. If it cannot be cast into a form specific enough to be tested against actual data, then its mathematical beauty may be a valuable tool for exploring new ideas, but it will fail to constitute an all-embracing theory of the real world, a "theory of everything." Excitement and skepticism about string theory both, therefore, continue to run high in the world of professional physics.

See also Cosmology; Relativity, general; Relativity, special.

Resources

Books

Barnett, Michael R., Henry Möhry, and Helen R. Quinn. *The Charm of Strange Quarks: Mysteries and Revolutions of Particle Physics.* New York: Springer-Verlag, 2000.

Kaku, Michio, and Jennifer Thompson. *Beyond Einstein.* New York: Anchor Books, 1995.

Periodicals

Taubes, Gary, "String Theorists Find a Rosetta Stone." *Science* Vol. 285, No. 5427 (23 July 1999): 512-517.

Other

Cambridge University. "Cambridge Cosmology." [cited February 14, 2003]. <http://www. damtp.cam.ac.uk/user/gr/public/cos_home.html>.

Schwarz, Patricia. "The Official String Theory Website." 2002 [cited February 13, 2003]. <http://superstringtheory.com/>.

Larry Gilman

Stroke

A stroke, also called a cerebral vascular accident or CVA, is a sudden, often crippling disturbance in **blood** circulation in the **brain**. Interruption in blood circulation may be the result of a burst artery or of an artery that has become closed off because a blood clot has lodged in it. Blood circulation to the area of the brain served by that artery stops at the point of disturbance, and the brain **tissue** beyond that is damaged or dies.

Stroke is the third leading cause of death in the United States. **Heart** attack is the leading cause of death, and all forms of **cancer** are second. Approximately 500,000 strokes, new and recurrent, are reported each year. Of these, about 150,000 will be fatal. Today approximately three million Americans who have had a stroke are alive.

The death **rate** from stroke has been steadily declining since the early 1970s. A great part of this success has been the efforts of various agencies to make physicians and the general population aware of the danger of high blood **pressure**. The importance of regular blood pressure checks has begun to infiltrate the general public's knowledge. High blood pressure, or **hypertension**, is the most important risk **factor** for stroke.

The brain requires a constant and steady infusion of blood to carry out its functions. Blood delivers the **oxygen** and **nutrients** needed by the brain cells, so the circulation to the brain is copious. The carotid **arteries** on each side of the throat and the vertebral arteries in the back of the neck carry blood to the brain. As the arteries enter the brain they begin to divide into smaller and smaller vessels until they reach the microscopic size of **capillaries**. The venous system to drain blood from the brain is equally large.

A burst blood vessel, which may occur in a weak area in the artery, or a blood vessel that becomes plugged by a floating blood clot no longer supplies blood to the brain tissue beyond the point of the occurrence. The effect of the interruption in circulation to the brain depends upon the area of the brain that is involved. Interruption of a small blood vessel may result in a **speech** impediment or difficulty in **hearing** or an unsteady gait. If a larger blood vessel is involved the result may be the total paralysis of one side of the body. Damage to the right hemisphere of the brain will result in injury to the left side of the body, and vice versa. The onset of the stroke

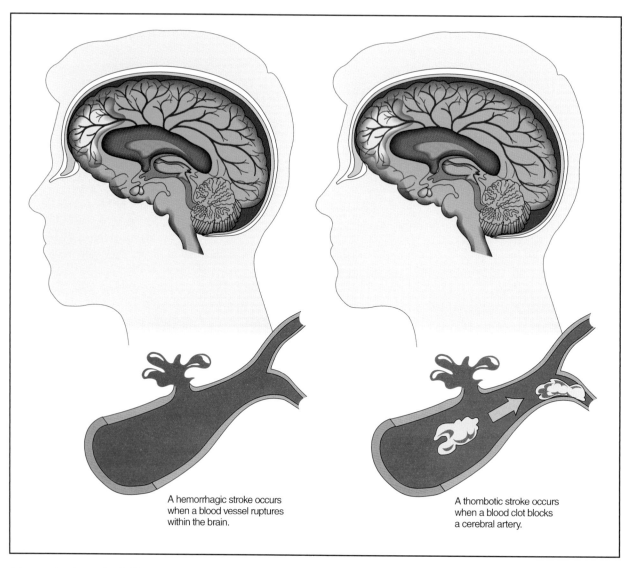

A hemorrhagic stroke occurs when a blood vessel ruptures within the brain.

A thombotic stroke occurs when a blood clot blocks a cerebral artery.

A hemorrhagic stroke (left) compared to a thrombotic stroke (right). *Illustration by Hans & Cassidy. Courtesy of Gale Group.*

may be so sudden and severe that the patient is literally struck down in his tracks. Some patients have early warnings that a stroke may be developing, however.

These people have what is called transient ischemic attacks, or TIAs. These usually result when a cerebral artery is closing because of the build-up of fatty deposits inside the vessel, a condition called atherosclerosis. Slight disturbances in circulation occur at intervals in such a case and the patient will experience momentary symptoms such as unexplained dizziness, blurring of **vision**, loss of hearing, or other event. It is important in this case to consult a physician so that proper testing can be carried out to determine the cause of the TIAs and steps taken to prevent development of a full stroke.

A stroke that results from a burst blood vessel is called a hemorrhagic stroke. In addition to the damage to

brain tissue caused by lack of blood, damage may be brought about by the blood clot formed from the bleeding. The doctor may need to surgically remove the clot to relieve pressure on the brain.

A stroke caused by a clot sealing a blood vessel is called a thrombotic stroke, a derivation of the technical name for a clot, a thrombus. The clot can be formed elsewhere in the body and be carried in the circulating blood until it reaches a vessel too small for it to pass through. At that point it will lodge and dam up the flow of blood, inflicting damage on the tissue beyond the blockage. A thrombotic stroke, if attended quickly, sometimes can be minimized in terms of damage.

People who are known to form blood clots in their **circulatory system** can be given medications to prevent it. Also, current therapy includes medications that can be given to dissolve clots (a process called thrombolysis) and remove the barrier to blood flow. Some brain function may be preserved in this way that would otherwise be lost if the clot is allowed to remain in place.

Not all vascular events bring on a stroke. Pathologists often find areas in the brain in which a small blood vessel has ruptured but the patient was not reported to have suffered a stroke. Blood vessels that form what is called collateral circulation often make up for the blood circulation lost by a plugged or ruptured vessel. Collateral circulation is carried by other vessels that help to supply blood to a given area of the brain.

Strokes can be prevented by effective treatment of high blood pressure and by taking an aspirin tablet every day, for those who can tolerate such medication. The aspirin helps to prevent clot formation and a number of clinical trials have shown it to be effective in stroke reduction.

Recovery from a stroke varies from one person to the next. Swift treatment followed by effective **physical therapy** may restore nearly full function of an affected area of the body. Some individuals have experienced severe enough damage that their recovery is minimal and they may be confined to a wheelchair or bed for the remainder of their lives.

Because stroke is a major cause of disability and death, much research effort is being put into this topic. Research is looking at a variety of ways to improve the odds of stroke patients, investigating substances which could be "neuroprotective," by protecting the cells from certain events that occur after stroke, and result in increasing damage to brain tissue; vasodilators (medications that work to increase the diameter of blood vessels, thereby allowing a greater flow of blood to areas served by those vessels), which hold the potential to increase blood flow and therefore oxygen delivery to areas of the brain threatened by or injured by stroke; **genetic engineering** which holds the future hope of allowing genes to be inserted in the brain, to provoke brain cells into producing chemicals which could be either neuroprotective or perhaps reparative to areas damaged by stroke; **animal hibernation**, to increase understanding of how a hibernating animal can have decreased blood flow to the brain without brain damage; and specialized magnets to ascertain whether application of a magnetic field to areas of the brain damaged by stroke could potentially help stroke victims recapture functioning to those areas.

See also Thrombosis.

Resources

Periodicals

Aldhous, P. "A Hearty Endorsement for Aspirin." *Science* 263 (January 7, 1994):24.

KEY TERMS

- -

Cerebral—Refers to the brain. Cerebral vascular is reference to the blood supply to the brain.

Death rate—The number of deaths from a given cause per specified number of people. The death rate from a disease may be given as 30 per 10,000 population.

Pathologist—A physician who studies the disease process. He/she learns by conducting autopsies on people who have died.

Other

National Heart, Lung, and Blood Institute. *Report of the Joint National Committee on Detection, Evaluation, and Treatment of High Blood Pressure* Bethesda, MD: 1993.

Larry Blaser

Stromatolite

A stromatolite is a preserved structure in **sedimentary rock** that is a series of thin layers of sediment. These layers formed when a colony of **algae** trapped loose sediment particles.

Stromatolites occur in **rocks** that range in age from very recent to more than 3.5 billion years old. Ancient stromatolites are the oldest evidence of life that is visible without a **microscope**. They are also the most common evidence of life found in rocks dating from about 3.5 billion years ago to 1.5 billion years ago. The algae that form them are of the blue-green variety. Also called cyanobacteria due to their primitive nature, blue-green algae carry on **photosynthesis** to produce their own food. Thin algal mats (colonies) grow today in almost any moist environment—rocky stream beds, boat piers, shower stalls, etc.

Stromatolites are not true fossils. They contain no preserved body parts (bones, for example) or impressions (for example, footprints). Stromatolites consist of only layered sediment. They form when algal mats trap sediment on a sticky "carpet" of thin, thread-like filaments. **Currents** or waves move sediment over algal mats located in shallow **water**. Fine sediment—especially particles less than 0.0025 in (1/16 mm)—sticks to the filaments. Sediment collects on the algal filaments like dirt sticking to a rug where you spilled a soft drink. Eventually the sediment limits the amount of **light** available to the algae for photosynthesis. In response, the

Stromatolites at Hamelin Pool in western Australia. *Photograph by Francois Gohier. National Audubon Society Collection/Photo Researchers, Inc. Reproduced by permission.*

algae send new filaments up through the accumulated sediment layer, creating a "clean carpet," that is, a new algal mat. Another sediment layer then begins to accumulate. In this way, many thin layers of sediment form. Later, the algal filaments usually decay so that only the sediment layers remain.

Most stromatolites occur in limestone, rather than in sandstone or shale. The sediments that form limestone are usually bound together (cemented) soon after the particles accumulate. Cementation occurs much later in most sandstone and shale. Thanks to this early cementation, stromatolites in limestone are more likely to avoid destruction by burrowing animals and other disruptive processes. Stromatolites also are less common in rocks younger than about 500 million years old. Plant-eating animals (herbivores) probably began to evolve about 850-900 million years ago. For the next few hundred million years, herbivores became increasingly common, and their grazing activities began to destroy more and more stromatolites. By 500 million years ago, preservation of stromatolites became much less likely.

Ancient stromatolites probably populated the sea bottom in much the same way as modern coral reefs,

preferring shallow, well-lit areas. However, modern stromatolites are common only in very salty, shallow **ocean** waters or in highly mineralized **lake** water. These high salinity (salty) environments are hostile to many organisms, but stromatolites thrive there because of the absence of grazing herbivores. In this setting, algal mats build dome- and column-shaped stromatolites that reach as much as 3 ft (1 m) in height.

Strong electrolyte *see* **Electrolyte**

Strontium *see* **Alkaline earth metals**

Strontium-90 *see* **Radioactive pollution**

Structural formula *see* **Formula, structural**

Strychnine *see* **Nux vomica tree**

Sturgeons

Sturgeons are large shark-like **fish**, with a heterocercal tail like that of a shark and a small ventral mouth behind an extended snout. The mouth has long barbells used for feeding on small animals on the bottom. Sturgeons feed on aquatic **insects, snails, crayfish,** small clams, and small fish.

Sturgeons have a cartilaginous skeleton with bony plates instead of scales in their skin which cover the sides of the body, and which completely cover the head.

Sturgeons are distributed throughout **North America, Asia** and northern **Europe.** A few **species** are anadromous, living in the oceans but swimming up **rivers** to spawn, and other species live only in fresh **water.** Of the anadromous sturgeons, one of the two species on the Pacific coast, *Acipenser transmontanus,* may weigh over 1,000 lb (450 kg). The two anadromous species on the Atlantic coast have been recently greatly reduced in population.

Sturgeons are found both in the rivers and the eastern coasts of North America. The most common species in North America is the **lake** sturgeon, *Acipenser fulvescens.* In Europe they are found in the rivers and the coast line from Scandinavia to the Black Sea. From time to time trawlers fishing along the coasts of Britain and Ireland may catch sturgeons. **Pollution** and the presence of weirs have been instrumental in reducing the populations to the point of **extinction** in rivers where they were once plentiful.

Spawning occurs from May to July when the sturgeons enter the rivers of the United States and continental Europe. A female sturgeon may lay up to three million eggs, each about 0.08 in (2 mm) in diameter, and covered with **gelatin.** Eggs remain on the bottom of the river, hatch within 3–7 days, and release larvae that measure about 0.4 in (9 mm). At one month the young fish may measure 4-5.5 in (10-14 cm) long. The young may not start the seaward journey until they reach two or three years of age and are 3.3 ft (1 m) long.

The flesh of the sturgeon is edible but is not prized; it is the sturgeon's eggs used to make caviar that are in great demand. The swim bladders of sturgeons are used to make isinglass, a semi-transparent whitish gelatin used in jellies, glues, and as a clarifying agent.

Since the population of sturgeons has been greatly diminished, commercial fishing of these fish is now limited. Some sturgeons may provide considerable excitement because of the battle they provide when hooked on light tackle.

The sturgeon family has the distinction of providing the largest **freshwater** fish in North America. In the last century three fish were reported to exceed 1,500 lb (680 kg). At present some specimens may weigh over 1,000 lb (450 kg), and one fish caught by gill net weighed 1,285 lb (584 kg). Although the species extends from Alaska to

A lake sturgeon (*Acipenser fulvescens*). *Photograph by Tom McHugh. Photo Researchers, Inc. Reproduced by permission.*

the middle of California, the largest fish are found mainly in the Columbia and Fraser rivers in British Columbia, Washington, and Oregon.

The white sturgeon is actually grayish brown with a white belly. When at 9 ft (2.7 m) long, it may be about 50 years of age. The 3,000,000 eggs laid by a 10–foot female may weigh almost 250 lb (113 kg). Laws protecting large sturgeons, which tend to be females with eggs, are now in effect.

The green sturgeon, *A. medirostris*, grows up to 7 ft (2.1 m) and weighs about 350 lb (159 kg). It has a greenish **color** and its barbells are located nearer to the end of the snout, and has fewer bony plates along the back.

The largest fish caught in fresh water along the Atlantic coast is the Atlantic sturgeon, *A. oxyrhynchus.* It has been verified that a 14–ft (4.3–m) fish weighing 611 lb (278 kg) has been caught. As is the case with the Pacific sturgeon, the Atlantic sturgeon populations are being depleted because of pollution and **dams,** which prevent them from reaching their breeding grounds.

Nathan Lavenda

Style *see* **Flower**

Subatomic particles

Subatomic particles are particles that are smaller than an atom. Early in the twentieth century, electrons, protons, and neutrons were thought to be the only subatomic particles; these were also thought to be elementary (i.e., inca-

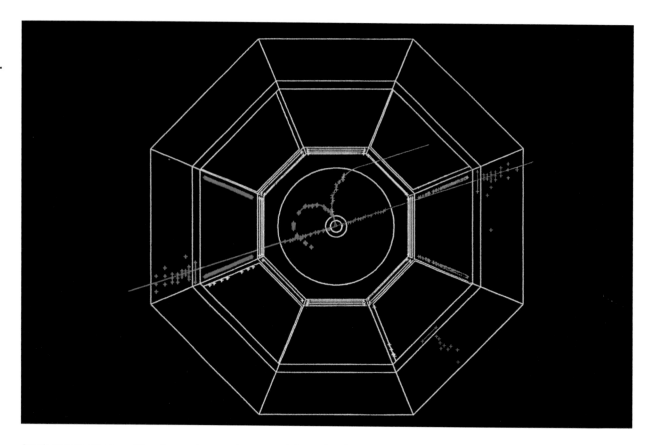

An electronic display of the decay of an upsilon, a particle made of a bottom quark and an antiquark, in the CLEO detector at the CESR collider at Cornell University. An electron and a positron have met and annihilated (center); their energy produces an excited upsilon, which lives too briefly to show. It decays into a lower energy state by emitting a photon which converts into a positron (curves away tightly to the left) and an electron (curves to the right). The lower-energy upsilon then decays to its ground state by emitting another photon (detected in the bottom right sector of the grid). The ground-state upsilon finally decays into a high energy electron-positron pair (seen as the two long paths that cut diagonally across the grid in opposite directions from the center). *Newman Laboratory of Nuclear Science, Cornell University, National Audubon Society Collection/Photo Researchers, Inc. Reproduced by permission.*

pable of being broken down into yet smaller particles). However, the list of subatomic particles has now been expanded to include a large number of elementary particles and the particles they can be combined to make.

There are two types of elementary particles. One type of makes up **matter**. Examples of these particles include **quarks** (which make up protons and neutrons) and electrons. Baryons and mesons are combinations of quarks and are considered subatomic particles. The most famous baryons are protons and neutrons.

The other elementary particles are mediators of the fundamental forces. These mediator particles enable the matter particles to interact with each other. That is, when two electrons collide, they do not simply bounce off of each other like two billiard balls: they exchange a **photon** (one of the mediator particles). All forces, including gravity, are thought to be mediated by particle exchanges.

Discovery of particles

Electrons

The first subatomic particle to be discovered was the **electron**. While others had deduced the existence of a negatively charged particle in what were called **cathode** rays (and which are now known to be beams of electrons), it was English physicist J. J. Thomson (1856–1940), who in 1897 measured the **velocity** and charge-to-mass **ratio** of these particles. The charge-to-mass ratio was found to be relatively large, and independent of the gas used in his experiments, which indicated to him that he had found a true particle. Thomson gave it the name "corpuscle," which was later changed to "electron."

The charges of all particles are traditionally measured in terms of the size of the charge of the electron. The electron has a charge, e, of 1.6×10^{-19} Coulombs.

Photons

The first mediator particle to be discovered was the photon. In 1900, German physicist Max Planck (1858–1947) reported that **light** came in little packages of **energy**, which he called "quanta." In 1905, German physicist Albert Einstein (1879–1955) studied the **photoelectric effect** and proposed that **radiation** is quantized by its nature—that is, transfers energy in minimal packets termed quanta. A photon (the name was coined by U.S. chemist Gilbert Lewis [1875–1946] in 1926) is one of these quanta, the smallest possible piece of energy in a light wave. (The word "wave" is applied by physicists to describe some observable aspects of the behavior of light, while the particle terminology of the "photon" is applied to describe others. Both words convey mental pictures that are useful in some physical applications, but neither picture is sufficient: a photon is not a "particle" in the sense of a perfectly round, hard, self-contained **sphere**, nor is light a "wave" in the sense of being a smooth undulation in some medium.)

Protons

The **proton** was one of the earliest particles known. (The word proton is Greek for "the first one.") In 1906 the first clues to the nature of the proton were seen. J. J. Thomson reported detecting positively charged **hydrogen** "atoms." These were in fact, hydrogen nuclei (protons), but atomic structure was not understood at the time. Thomson thought that protons and electrons were randomly scattered throughout the atom, the so-called "plum-pudding model." In 1909–1911, English physicist Ernest Rutherford (1871–1937) and his colleagues, German physicist Hans Wilhelm Geiger (1882–1947) and New Zealand physicist Ernest Marsden (1888–1970) did their famous scattering experiments involving alpha particles (two protons and two neutrons; a helium-atom nucleus) shot through gold foil. From their observations of the angles at which alpha particles were deflected, they deduced that **atoms** had relatively hard and small centers, thus proving the existence of the atomic nucleus and disproving the plum-pudding model.

In 1913, the **Bohr model** of the atom was introduced (named after Danish physicist Neils Bohr, 1885–1962). In this model, the hydrogen atom consists of an electron orbiting the nucleus (a single proton), much as the **Earth** orbits the **Sun**. The Bohr model also requires that the angular **momentum** (**mass** times velocity times distance from the orbital center) of the electron be limited to certain values (that is, be "quantized") in order that the electron not fall into the nucleus. Though known to have serious defects, the Bohr model still supplies the standard graphic representation of the atom: a solid nucleus around which electrons **orbit** like tiny planets.

When the principles of **quantum mechanics** were developed, the **Heisenberg uncertainty principle**, discovered by German physicist Werner Heisenberg (1901–1976) meant the Bohr atom had to be modified. The Heisenberg uncertainty principle states that it is impossible to accurately determine both the position and the momentum (mass times velocity) of a subatomic particle at the same time. Indeed, a subatomic particle cannot be thought of as *having* precise values of these quantities simultaneously, measured or not. This means that the electrons in an atom can still be thought of as orbiting, the nucleus, but their position is smeared throughout a wide region or "cloud" rather than confined to well-defined orbits.

Neutrinos

In 1930, scientists started to suspect the existence of another subatomic particle that came to be known as the **neutrino**. Neutrinos are considered matter particles, but they do not make up normal matter by themselves. In fact, neutrinos are very common–about 60 billion neutrinos from the Sun pass through every square centimeter of the Earth's surface every second–but we do not observe them because they interact only rarely with other particles.

In 1930 a problem with a process called nuclear beta decay had developed. Nuclear beta decay is when an unstable, or radioactive, nucleus decays into a lighter nucleus and an electron. Scientists observed that the energy before the beta decay was greater than the energy after the beta decay. This was puzzling because one of the most basic laws of **physics**, the law of **conservation** of energy, states that the amount of energy in any process must remain the same. To keep the idea of energy conservation intact, Austrian physicist Wolfgang Pauli (1900–1958) proposed that a hitherto-unidentified particle carried off the missing energy. In 1933 Italian physicist Enrico Fermi (1901–1954) named this hard-to-detect particle the neutrino, and used it to successfully explain the theory of beta decay.

One type of neutrino, the electron neutrino, was finally detected in 1956. Later, a second type of neutrino, the muon neutrino, was found, and a third type, called the tau neutrino, was discovered in the late 1990s. For decades physicists debated the question of whether the neutrino is a massless particle, like the photon, or has a finite mass. In 1998 physicists discovered that at least one of these types of neutrinos must have mass. Though it would have to be very tiny, it must at least be greater than 20-billionths of the mass of the electron—extremely small, but not zero.

Positrons

In 1931–1932, U.S. physicist Carl Anderson (1905–) experimentally observed the anti-electron, which he called the positron, after its positive charge. The positron is an **antiparticle** which had been predicted by English physicist Paul Dirac (1902–1984) in 1927–1930. Every particle has a corresponding antiparticle that has the same properties except for an opposite electrical properties (charge and magnetic moment). Antiparticles make up what is called **antimatter**. Matter is much more common in our universe than antimatter, though it is unknown why this is so.

Neutrons

In 1932 English physicist James Chadwick (1891–1974) discovered another matter particle, the **neutron**. The neutron is very similar to the proton except that it is electrically neutral (i.e., has no charge). Chadwick found the neutron by hitting a chemical called beryllium with alpha particles. When this occurred, highly penetrating radiation was emitted. This "radiation" turned out to be a stream of neutrons. After Chadwick's experiment, Werner Heisenberg proposed that the nucleus is made up of protons and neutrons, which was later found to be true.

Pion, muons, and kaons

The second mediator particle discovered (after the photon) was the pion. In 1935, Japanese physicist Hideki Yukawa (1907–1981) formulated the idea that protons and neutrons were held together by a nuclear **force** that was mediated by a particle called the pion. Yukawa described it in detail. In 1937 the first evidence for the pion was obtained by studying cosmic rays (high-energy particles from **space**). By 1947 it became clear that cosmic rays did contain Yukawa's pions, but also contained another particle, a heavy electron-like particle, which was given the name muon. In 1947 yet another particle was detected from cosmic rays, the kaon. The kaon is like a heavy pion, and decays into two lighter pions. The kaons are considered strange particles because they can be made fairly quickly, but it takes a long time for them to decay. Usually the time to make a particle and the time for it to decay to be about the same, but this is not true for the kaon.

Quarks

In 1980, Maurice Jacob (1933–) and Peter Lanshoff detected small, hard, objects inside the proton by firing high-energy electrons and protons at it. Most of the high-energy particles seemed to pass right through the proton. However, a few of these high-energy particles were reflected back, as if they had hit something. These and other experiments indicated that the proton contains three small, hard, solid objects. Thus protons are not elementary, but the objects inside them may be. These objects are now called quarks.

Quark model

Quarks had been postulated much earlier, in 1964, by American physicist Murray Gell-Mann (1929–) and, inde-

pendently, by American physicist George Zweig (1937–). The theory describing quarks was called the quark model. In 1964 it was thought that there should be three different quarks. These different quarks each have a unique property called flavor. These first three quarks had flavors that were whimsically named up, down, and strange. Up-flavored quarks have an **electric charge** of $(2/3)e$, where e is the fundamental quantum of charge such as that of the negatively-charged electron. Down- and strange-flavored quarks have an electric charge of $(-1/3)e$. The quark model also says that quarks must remain bound inside their particles—in nature, quarks cannot exist by themselves. This idea is called quark confinement, and is based on the experimental observation that a free quark has never been seen. Since we cannot isolate quarks, it is very difficult to determine their masses.

In 1964 physicist Oscar W. Greenberg (1932–) suggested each quark has a quality he termed **color**. The label "color" for this quark property does not refer to the usual definition of color, but is just a way to keep track of quarks. Using this idea of color, the improved quark model says only overall-colorless particles can exist in nature. There are only three different kinds of color in the quark model, usually designated red, blue, and green. Color had to be introduced when a particle called the $\Delta++$ (pronounced delta-plus-plus) baryon was discovered to avoid violating the **Pauli exclusion principle**. The Pauli exclusion principle says that each particle in a system of matter particles must have unique properties like electric charge, mass, and spin. The $\Delta++$ baryon is made of three up quarks. Without color, each of its three up quarks cannot have its own properties. Color has been proven experimentally, and a theory called the **standard model** of elementary particles has updated the quark model.

Subatomic particle classifications

Elementary matter particles

There are two kinds of elementary (indivisible) matter particles, the quarks and the leptons. The two lowest-mass leptons are the electron (e^-) and its partner the neutrino, usually called the electron-neutrino (ν_e). For unknown reasons, this lepton pairing is repeated two more times, each time with increasing mass. These leptons are called the muon (μ^-) and muon neutrino (ν_μ) and the tau (τ^-) and tau neutrino (ν_τ). There are said to be three families, or generations, of leptons.

Like the leptons, the quarks have three families. The first family of quarks are the up and down quarks, the second contains the strange and "charmed" quarks, and the third the "bottom" and "top" quarks. Though all matter we see around us contains only up, down, and strange quarks, physicists have proven the existence of all six

TABLE 1. ELEMENTARY MATTER PARTICLES

1st family			2nd family			3rd family		
particle	*charge*	*mass*	*particle*	*charge*	*mass*	*particle*	*charge*	*mass*
leptons								
ν_e	0 e	0	ν_μ	0 e	0	ν_τ	0 e	0
e-	-1 e	.511	μ-	-1 e	106	τ-	-1 e	1777
quarks								
u	2/3 e	2-8	c	2/3 e	1000-1600	t	2/3 e	176000
d	-1/3 e	5-15	s	-1/3 e	100-300	b	-1/3 e	4100-4500

TABLE 2. ELEMENTARY MEDIATOR PARTICLES

particle	charge	mass	force
γ	0 e	0	Electromagnetic
g	0 e	0	Strong
$W^\pm$	$\pm$1 e	80200	Weak
Z^0	0 e	91200	Weak

flavors of quarks, culminating with the discovery of the top quark in 1995.

Another property of elementary particles is termed "spin." Spin is akin to the **rotation** of a particle on its axis, as the earth spins on its axis to give us day and night. (In actuality elementary particles do not rotate like spheres; it is only that the particle property termed spin obeys rules that mathematically are similar to those used to describe the rotation of macroscopic bodies.) The spin of elementary particles is measured in special units called "h-bar" (h-bar is **Planck's constant** divided by 2π), and $= 1.1 \times 10^{-34}$ Joule-seconds. Using the property called spin, all matter particles are fermions which have spin one-half h-bar or three-halves h-bar. All quarks and leptons have spins of one-half h-bar. The matter particles and some of their properties are summarized in Table 1.

Masses are given in units of MeV/c^2, where c is the speed of light (three-hundred-million meters per second). The quark masses are approximate.

Elementary mediator particles

Bosons are particles defined to have spin of zero h-bar, one h-bar, or two h-bar. The elementary mediator particles are bosons with spins of one h-bar. The force we are most familiar with is the electromagnetic force. The electromagnetic force is responsible for keeping electrons and nuclei together to form atoms. The electromagnetic force is mediated by photons, which are massless. The mediators of the strong force are called gluons, because they glue quarks together to form mesons and baryons. Like the quarks, the gluons carry the color property, and as a result there are eight different types of gluons.

The weak force is more uncommon. It is responsible for radioactive decays like nuclear beta decay. The mediators of the weak force are the electrically charged W-bosons ($W\pm$), and the electrically neutral Z-bosons (Z^0), both discovered in 1983. Some properties of the mediator particles are given in Table 2.

TABLE 3. BARYONS				
baryon	quark content	spin	charge	mass
p	u u d	1/2	+1 e	938
n	u d d	1/2	0 e	939
Λ	u d s	1/2	0 e	1116
Σ^+	u u s	1/2	1 e	1189
Σ^0	u d s	1/2	0 e	1192
Σ^-	d d s	1/2	-1 e	1197
Ξ^0	u s s	1/2	0 e	1315
Ξ^-	d s s	1/2	-1 e	1321
Λ_c^+	u d c	1/2	1 e	2281
Δ^{++}	u u u	3/2	2 e	1232
Δ^+	u u d	3/2	1 e	1232
Δ^0	u d d	3/2	0 e	1232
Δ^-	d d d	3/2	-1 e	1232
Σ^{*+}	u u s	3/2	1 e	1383
Σ^{*0}	u d s	3/2	0 e	1384
Σ^{*-}	d d s	3/2	-1 e	1387
Ξ^{*0}	u s s	3/2	0 e	1532
Ξ^{*-}	d s s	3/2	-1 e	1535
Ω^-	s s s	3/2	-1 e	1672

Baryons

One of the main rules of the standard quark model is that combinations of three quarks are called baryons. Protons and neutrons are the most important baryons. Protons are made of two up quarks and one down quark. Neutrons are made of two down quarks and one up quark. Since the quark model requires that naturally-occurring particles be colorless, a baryon must be made of a red, a blue, and a green quark. These combine to make a white, or colorless particle. Spin is also important in classifying baryons. Baryons are fermions and so have spins of one-half h-bar or three-halves h-bar. Table 3 summarizes several kinds of baryons, with masses in MeV/c^2 (millions of electron-volts divided by the speed of light squared) and spin in terms of h-bar.

Mesons

The second main idea of the standard quark model is that combinations of one quark and one antiquark form mesons. Pions (π) and kaons (K) are examples of mesons. Thus now we see Yukawa's nuclear force mediator particle, the pion, is really a matter particle made of a quark and an antiquark. There are several kinds of pions. For example, the positively charged pion, π^+, is made of an up quark and a down antiquark. Similarly there are several kinds of kaons. One kind of kaon, K^+, is made of an up quark and a strange antiquark. The colorless rule requires that mesons must be made of quarks with opposite color, red and anti-red for example. All mesons are bosons and so have spins of zero h-bar or one h-bar.

Current and future research

Subatomic particles are important in all electronic, optical, and nuclear technologies. Cathode-ray tubes, for example, use beams of electrons to create the pictures. A **television antenna** first picks up the television signal—a series of radio-frequency photons—which is then processed electronically and used to control an electron gun. An electron gun shoots a beam of electrons which is steered by magnets and hits the coated inner surface of the picture tube. When electrons hit this surface, it lights up, creating the picture as the electron beam is steered rapidly across it. A common type of smoke detector that uses subatomic particles is an ionization smoke detector; in an ionization smoke detector, alpha particles ionize (strip electrons from) air molecules. These ionized air molecules cause **electric current** to flow in the detector. If there is a fire, other particles enter the detector and interfere with the flow of the electric current, and this makes the alarm go off. Proton beams are used to treat **cancer**; all technologies involving **optics** or **radio** manipulate photons; all electronic devices manipulate electrons; **nuclear weapons** and **nuclear power** depend on controlling neutrons so as to produce either an explosive or a controlled nuclear chain reaction, respectively; positron-emitting isotopes are used to image metabolic activity in the human **brain** in real time; and so on.

In recent years, particle physics has been particularly exciting, with several important experimental developments. Besides the discovery of the W and Z bosons and the top quark, scientists working in Japan in 1998 found evidence that at least some of the three types of neutrinos have a small but nonzero mass. Their experiment did not allow them to determine the exact value for the mass, but subsequent work has shown that the mass of the neutrino is too small to account for the "dark matter" which astronomers have shown must account for a significant fraction of the mass of the Universe. Previously, it seemed that the unknown mass of the neutrino might explain the "dark matter" mystery; today, suspicion centers on "dark energy" rather than on "dark matter" as an explanation of the Universe's nonvisible mass.

See also Spin of subatomic particles.

Resources

Books

Barnett, Michael R., Henry Möhry, and Helen R. Quinn. *The Charm of Strange Quarks: Mysteries and Revolutions of Particle Physics.* New York: Springer-Verlag, 2000.

Gribbin, John. *Q is for Quantum: An Encyclopedia of Particle Physics.* New York: The Free Press, 1998.

Kane, Gordon. *The Particle Garden: Our Universe as Understood by Particle Physicists.* Reading, MA: Helix Books, 1995.

Weinberg, S. *The First Three Minutes.* New York: Basic Books, 1977.

Weinberg, Steven. *Dreams of a Final Theory.* New York: Patheon Books, 1992.

Periodicals

Barnett, R. "The Charm of Strange Quarks: Mysteries and Revolutions of Particle Physics." *Physics Today* 54 (2001)50–51.

Gibbs, W. Wayt. "A Massive Discovery." *Scientific American* (August 1998).

Kalmus, P. I. P. "Particle physics at the Turn of the Century." *Contemporary Physics.* 41 (2000):129–142.

Pokrovsky, V. "Particle Physics: Russian Turmoil Rattles CERN." *Science* 292 (2001):2414B–22415.

Lesley L. Smith
Larry Gilman
K. Lee Lerner

KEY TERMS

Coulomb—The standard unit of electric charge, defined as the amount of charge flowing past a point in a wire in one second, when the current in the wire is one ampere.

Fundamental force—A basic force, which has its own elementary mediator particle(s). There are four fundamental forces: the strong force, the electromagnetic force, the weak force, and gravity.

Mega electron volt (MeV)—A unit of energy. One MeV is one million Electron Volts. An Electron Volt is the amount of energy an electron gains as it passes through one Volt of potential difference.

Quarks—Believed to be the most fundamental units of protons and neutrons.

Subduction zones *see* **Plate tectonics**

Submarine

The first known treatise on submarines was written in 1578. Published by William Bourne in his *Inventions or Devices*, the document describes a ship with two hulls, the outer made of **wood**. While no record exists concerning its manufacture, the ship, according to Bourne, could be submerged or raised by taking in or expelling **water** from between the double hulls. The first known submarine was built by Dutch inventor Cornelius Drebbel, and consisted of greased leather over a wooden framework. It was propelled either on or beneath the surface by eight oars sealed through the sides with leather flaps. During a demonstration for James I in 1620, this vessel was successfully piloted just under the surface of the Thames River. It was unable, however, to make deep descents.

During the American Revolution, David Bushnell built a one-man submarine called the Turtle. It resembled an egg squashed thin with a height of 6 ft (2 m), and had two hand-cranked screw propellers, a hand-operated control lever connected to the rudder, foot-operated pumps to let water in or send it out (to submerge or surface), and a crudely-lit control panel. As if it was not dangerous enough simply to get in the water sealed inside this device, the Turtle also had a large explosive attached to it in the hopes the operator could maneuver under an enemy ship, screw the explosive into the ship's hull, and depart before the explosive's timing device discharged it. Unfortunately, the Turtle failed to sink any ship. On its only test mission, the Turtle was assigned the task of bombing the British HMS Eagle in New York, but its pilot was unable to screw the explosive into the Eagle's **copper** hull.

Others, such as English carpenters Symons and Day, included ballast systems on their submarines to permit descents. Day's submarine resembled a sloop, and had two large bags of stones hanging from its bottom to serve as ballast. Day would sink, then jettison the **rocks** to return to the surface. After two successful tests, Day confidently decided he would test his vessel off Plymouth Sound, a site with a depth of 900 ft (274 m). Apparently his ship was crushed by high water **pressure**, for when he and his crew descended, a crowd of onlookers waited in vain for his return. Day and his crew had become the first victims of a submarine mishap.

Perhaps the most successful early submarine was designed by Robert Fulton. In an age of naval battles, Fulton, who detested war, felt that a device capable of neutralizing the effectiveness of warships would end war altogether. While living in France in 1767, he outlined plans to build a sub called the Nautilus and unsuccessfully attempted to interest the French government in his idea. By 1801, however, he had managed to complete a submarine on his own. A 21 ft (6 m) vessel with a two-bladed propeller, the Nautilus performed well in tests, even sinking a ship with an explosive charge. But he was once again rejected by the French government, so he moved to England, hoping for a better reception there.

It soon became clear that the English did not want his submarine either. In fact, Fulton had failed not because his vessel did not work, but because major naval powers feared his vessel and did not want to participate in developing a weapon that could negate their military strength. Fulton went on to produce his famous steamboats in the United States.

After the American Civil War, designers, spurred on by the invention of the self-propelled torpedo in 1866, increasingly sought alternatives to human-powered propulsion for submarines. Several systems proved unsuitable—steam engines made the craft unbearably hot and an electric **battery** could not be recharged at sea. In the late 1890s, however, Irish-born American John Holland solved the problem with the use of a new power source, the gasoline engine. Because it needed **oxygen**, the gasoline engine could not be used while a submarine was underwater, but on the surface it could not only provide propulsion but also charge the batteries used while submerged. Holland's vessels incorporated many of the features we associate with modern subs: a powerful engine, advanced control and balancing systems, and a circular-shaped hull to withstand pressure. The United States Navy accepted his submarine, the Holland, in 1900.

Around this time, two other improvements were introduced. Simon Lake (1866-1945), who also built an early gasoline-powered submarine, created the first periscope specifically for submarines: it provided a magnified view and a wide angle of **vision**. In the 1890s Rudolf Diesel invented an engine that was fired by compression rather than an electric spark. The **diesel engine** was more economical than the gasoline engine and its fumes were much less toxic and volatile. This new engine became the mainstay of all submarines until **nuclear power** was introduced as a means of propulsion in the 1950s.

Germany made good use of diesel propulsion. Unlike Britain's small, coastal subs, Germany's vessels, displacing up to 3,200 tons, were capable of crossing the Atlantic. Their U-boat (short for unterseeboot) sent more than 11 million tons of Allied shipping to the bottom and, in the process, created a new, terrifying type of warfare.

In World War II submarines played an even larger role in Germany's repeated attacks on Allied shipping,

A Trident submarine under construction in Groton, Connecticut. *Photograph by Tom Kelly. Phototake NYC. Reproduced by permission.*

eventually destroying 14 million tons of shipping. Meanwhile, American submarines crippled the Japanese by sinking nearly 1,400 merchant and naval ships. The greatest improvement came through the development of the snorkel, a set of two fixed air pipes that projected from the sub's topside. One tube brought fresh air into the vessel, and the other vented engine exhaust fumes. Now a sub could stay hidden below the surface when running on its diesel engine and recharging its batteries.

The greatest advance in submarine technology was the advent of nuclear power. With the encouragement of United States Navy Captain Hyman Rickover, American inventors Ross Gunn and Phillip Abelson designed the U.S.S. *Nautilus,* the first nuclear-powered submarine. Launched in 1955, the *Nautilus* carried a reactor in which controlled **nuclear fission** provided the **heat** that converted water into steam for turbines. With this new power source, the submarine could remain under water indefinitely and cruise at top speed for any length of time required.

For a submarine able to remain under water for longer distances at higher speeds, a needle-like shape proved inefficient. The Davis Taylor Model Basin in the United States developed a new teardrop design, first test-ed on its Albacore submarine. Vessels with this improved shape easily drove through the water at speeds of 35-40 knots per hour. The United States Navy later adopted the Albacore's shape for its submarines.

Submarines have also benefited from advances in navigation equipment. Inertial navigation systems, relying on gyroscopes, now fix their position with extreme accuracy. The U.S.S. *Skate* used this system to navigate under the polar **ice** cap at the North Pole in 1959.

See also Internal combustion engine.

Submarine canyon *see* **Continental margin**

Subsidence

The term subsidence is used in both atmospheric and geological sciences. Atmospheric subsidence refers to the sinking of air that is denser (heavier) than the air below. As it subsides, increasing air **pressure** compresses the air parcel, causing it to warm. Geologic subsidence is a form of **mass wasting** that refers to the sinking of geologic materials (**rocks** or sediments) as underlying materials are removed or change their position.

Atmospheric subsidence

Atmospheric subsidence occurs when the normal upward flow of air in the atmosphere, known as atmospheric **convection**, is disturbed. To understand what happens during convection, imagine a parcel of air located immediately above the ground during sunrise. As solar **energy** warms the **earth**, **heat** is transferred to that parcel of air. Warm air is less dense than cold air, so the heated parcel has a tendency to rise upward, or convect, into the atmosphere. As the parcel rises, it expands, causing cooling. Cooling causes **water** vapor (gas) in the air to change state to a liquid; water droplets form, producing **clouds**. An air parcel will continue to rise upward until its **density** is equal to the surrounding atmosphere, that is, until it is stable.

Convection, which creates a large area of low pressure and converging winds at the earth's surface, known as a cyclone, is not always present. Subsidence, or sinking of air, may happen instead, forming an area of high pressure, an anticyclone. Large scale subsidence occurs when air several thousands of feet overhead is denser than the surrounding air. This denser air is produced when winds aloft converge or air aloft is very cold, or warm, but unusually dry. The dense air sinks due to the pull of gravity, compressing the air, creating high pressure at the surface and diverging winds just above the surface. Warming of the air as it subsides increases **evaporation**, causing clear skies. That is why high pressure systems are usually associated with fair **weather**.

The subsiding air may settle onto a cooler air layer, creating what is known as a **temperature** inversion, or subsidence inversion. In a temperature inversion, a warm air layer, several hundred or thousand feet above the surface, is trapped between cooler layers above and below. This inversion resists convection of surface air, since the surface air is stable, and causes **air pollution** to be trapped at the surface. Subsidence inversions commonly occur at high latitudes during the winter and over the eastern United States during the late summer months. During an inversion, an urban area's air **pollution** may become a serious health hazard.

Geologic subsidence

Geological subsidence involves the settling or sinking of a body of rock or sediment. Subsidence is a type of **mass** wasting, or mass movement-transport of large volumes of earth material primarily by gravity. Subsidence may occur as the result of either natural or human-caused events.

Earthquakes are commonly associated with subsidence. When two blocks of the earth's crust slide against each other, causing an **earthquake**, ground movement may occur, raising or lowering the ground surface. During Alaska's 1964 Good Friday earthquake, an area of at least 70,000 sq mi (180,000 sq km), much of it coastline, subsided 3 ft (1 m) or more—some areas now flood at high tide.

Another way that earthquakes can cause subsidence is by rapidly decreasing the load-bearing capacity, or strength, of loose earth materials, or sediments, due to liquefaction. Liquefaction occurs when vibrations from an earthquake, or other disturbance, cause water-saturated sediments to temporarily lose their grain-to-grain contact, which is what gives them their load-bearing capacity. For just an instant, the weight of the overlying materials is supported only by the water between the grains. An instant later, when the grains begin to settle, the weight of the overlying sediment (or buildings) causes the grains to be forced closer together and the land to subside.

During the 1989 earthquake along the San Andreas Fault in California, some of the most serious damage occurred in San Francisco's Marina District. Buildings were constructed on old bay deposits, unconsolidated water-saturated sediments, which when shaken by the earthquake temporarily lost their strength due to liquefaction. In the Marina district, sediment that had previously supported large buildings quickly turned into a mud-like material that could no longer support a building's weight.

Another example of natural subsidence can be found in regions where caves are common. Caves form when underground water dissolves limestone and carries it away. The resulting void spaces grow larger and larger over **time** until they become the features we call caves.

If limestone dissolves for a long enough time, the hole (**cave**) that forms becomes too large to support the weight of its walls. In such a case, the ceiling of the cave will subside, either slowly or in a catastrophic collapse, forming a large depression at the surface, known as a sinkhole. If conditions are right, the sinkhole may eventually fill with water, forming a **lake**. A landscape containing many dry or water-filled **sinkholes** formed by limestone dissolution is called **karst topography**.

Yet another cause of subsidence is volcanic eruption. Whether molten rock, or **magma**, is lost suddenly and dramatically, as in the Mt. St. Helens eruption of May 18, 1980, or the slow flow of lava (magma flowing on the surface), as happens in the Hawaiian Islands, land subsidence is likely to follow. The material ejected from the **earth's interior** leaves an empty **space** that must be filled in one way or another. In many cases, a large section of the overlying crust—along with the geologic and human features attached to it—collapses into the earth's interior, forming what is called a caldera.

Subsidence may also result from the accumulation of large volumes of sediment at the earth's surface in what is known as a sediment **basin**. An obvious setting in which this occurs is at river deltas. Each day, the Mississippi River deposits up to 1.8 million metric tons of sediment at its mouth near New Orleans. The weight of this sediment contributes to a gradual subsidence of the land on which New Orleans resides. Basins between **mountains** also can subside due to the weight of accumulating sediments.

Human causes of subsidence

Many forms of human activities can result in subsidence. One of the most widespread of these problems involves the removal of **groundwater** for agricultural, municipal, and other purposes. In large parts of the United States, for example, farmers and ranchers depend heavily on water removed from underground aquifers to irrigate their **crops** and water their **livestock**. Such activities have now been going on with increasing intensity for at least a century.

This practice will not lead to subsidence as long as enough rainfall filters downward to replace, or recharge, the groundwater removed by humans. However, when the **rate** of removal exceeds the rate of recharge, significant decreases in the **volume** of the **aquifer** begin to occur. The pore spaces between the grains of the aquifer, previously occupied by water, are emptied. The grains then begin to compact more tightly, and they collapse. Eventually, the aquifer begins to subside.

A dramatic example of subsidence as a result of groundwater removal has taken place in a region southeast of Phoenix, Arizona. There a section of land covering 120 sq mi (310 sq km) has sunk more than 7 ft (2 m). This phenomenon has occurred at many locations above the Ogallala Aquifer, which lies beneath the High Plains region, stretching from Kansas and Nebraska to Wyoming and from Texas and New Mexico to Colorado.

The removal of **fossil fuels** is also a major human cause of subsidence. A traditional method of removing **coal**, for example, is known as room-and-pillar because vertical columns of the coal (the "pillar") are left in position while the coal around it is removed. When such mines are abandoned, however, the pillars of coal left behind are often not strong enough to hold up the overlying ground. When the pillars break, the ceilings of the mined room collapse and the overlying ground does so also.

KEY TERMS

Aquifer—A formation of soil or rock that holds water underground.

Karst topography—A region in which numerous caves, sinkholes, and other formations resulting from the dissolving of underground limestone rock are apparent.

Liquefaction (of rocks)—The process by which changes in pressure cause a rocky material that was originally strong and stable to change into a liquid-like material.

Temperature inversion—A situation in which a layer of cool air is trapped beneath a layer of warmer, less dense air.

With more than 90,000 abandoned mines in the United States, this source of subsidence is likely to be a problem into the foreseeable future.

The pumping of oil and **natural gas** from underground sources can have similar effects. Similar to removal of water from an aquifer, when these materials are removed from the reservoir, the reservoir's grains compact and the reservoir occupies a smaller volume than it did before the oil or gas was removed. As a result, overlying ground subsides as the reservoir slowly collapses.

One of the most famous of these instances occurred as far back as the late 1920s in Southern California. Oil removed from the Wilmington and Signal Hill oil fields caused unstable ground to subside by as much as 29 ft (9 m). Since this region lies along the coastline of the Pacific Ocean, drastic efforts were required to prevent ocean water from flowing into lands that were now lower in some places than **sea level**. By 1968, subsidence in the area had been stopped, but only after huge quantities of seawater had been injected into the empty oil wells in order to prop open the pores in the reservoir. This success in restoring stability to the area came at a high price, however, as docks, highways, sewer systems, and other municipal structures had, by that time, been completely rebuilt.

Effects of subsidence

Whether caused by natural or human activities, subsidence often has a number of serious consequences for human societies. Probably the most dramatic, of course, is the disappearance of whole sections of land, as occurred in Alaska's Good Friday earthquake. Today, the sudden appearance of sinkholes in Florida is no longer unusual news. In many cases, these sinkholes appear because the removal of groundwater has left limestone caves that are unable to support the land overlying them.

Even relatively modest subsidence can also damage a variety of human structures. Buildings are weakened and collapse, railway lines and roads are twisted and broken, and underground sewer, power, and water lines are torn apart. Due to its ability to destroy property on a large scale, subsidence is a very expensive type of mass wasting that also poses some risk to human lives.

See also Atmospheric circulation; Volcano.

Resources

Books

Erickson, Jon. *Quakes, Eruptions, and Other Geologic Cataclysms.* New York: Facts on File, 1994.

David E. Newton

Subsurface detection

Making inferences about the nature and structure of buried rock bodies, without access to them, is called subsurface detection. Using geophysical techniques, we obtain data at the surface that characterize the feature buried below. Then we construct models of the feature, trying to invent combinations of reasonable rock bodies which are consistent with all of the observations. Finally, using intuition, logic, and guesswork, we may select one or a few of the models as representing the most likely subsurface situation.

Seismic techniques

An **earthquake** generates seismic waves which can travel through the entire **Earth**. If you stamp on the ground you make the same kinds of waves, although obviously they are much weaker and do not travel as far. We know a great deal about how these waves travel through rock. By generating waves and then carefully timing how long it takes them to travel different distances we can learn a lot about the structures of rock units at depth.

Seismic reflection

The next time you go to a shopping mall on a bright afternoon, notice your reflection in the two sets of double doors at the entrance. You'll probably see yourself reflected twice, once from the outer set, and once from the inner set. Unlike opaque **mirrors**, **glass** doors permit some **light** to go through them, while at the same time they reflect some light back. Rock layers behave in a similar

manner with respect to seismic waves. An explosion detonated at the surface will reflect off of many layers at depth. Detectors set up in an array can time the arrival of each of these reflected waves. Just as **bats** and dolphins can use the echoes of **sound waves** to locate food, geophysicists can use the echoes of seismic waves to locate reflecting boundaries at depth. These reflecting boundaries occur where there is an abrupt contrast in some seismic wave transmission properties (usually velocities) of the material. Most often this is a result of the sedimentary layering. A seismic reflection profile, which actually shows reflection horizons, is usually interpreted as revealing the structure of the underlying layers. The oil bearing structures in many productive oil and gas fields were located using reflection seismic surveys, so this technique has been very important to the **petroleum** industry.

At first seismic reflection profiling was used only by oil companies. Because it is economical to recover oil only if it is relatively near the surface, such surveys did not seek out much information at great depths. Over the last fifteen years or so, however, scientists have adapted the technique to probe the very deepest parts of the crust. One remarkable discovery is a nearly horizontal **fault** underlying much of Georgia and the adjacent **continental shelf**.

Explosives are still used to generate seismic waves in some areas, particularly at sea, but in many places they have been supplanted by special trucks which vibrate the ground beneath them with hydraulic **jacks**. The signals generated by these vibrations reflect just like any other seismic waves, but because the **energy** is put into the ground over a period of time, instead of instantly, there is less damage to structures in the area.

Seismic refraction

If you live in a good sized city and want to travel a few blocks, you would probably take the direct route and put up with the stoplights and traffic. If you want to go further, though, you might find it takes less time to go out of your way to use an expressway. Although you have to travel a greater distance, you save time because the expressway is so much faster.

Similarly, seismic waves may go further, but reach faster layers at depth and arrive at a sensor before those taking the direct route do. When this occurs the path of the waves bends as it crosses boundaries between layers of different velocities, a phenomenon called " refraction." This technique can be used to determine how thick the **soil** is above **bedrock**. (This might be an important consideration in siting a **landfill**, for example.) Solid rock has faster seismic velocities than soil. If the depths of interest are small, the source of the seismic waves does not need to be very energetic. A sledgehammer, a dropped weight, or a blasting cap might be used. A detector located near this source will pick up the waves traveling through the soil. By plotting how long it takes them to travel different distances, their **velocity** through the soil can be determined. As the detector is moved further from the source, however, a point will be reached where the waves traveling through the bedrock start arriving first. This is equivalent to the distance you would need to go (in the city example) before it was quicker to use the expressway. Continuing to measure travel times for more distant stations permits the seismic velocity in the bedrock to be found. Knowing the **physics** of refraction, the two velocities, and the location where the bedrock path became faster yields the thickness of the soil layer.

The same principles can be applied to problems where the depths of interest are much greater, but the source of the seismic waves must be more energetic. When Andrija Mohorovicic observed a similar set of two lines for the arrival times of earthquake generated seismic waves, he realized that the earth must have a layered structure. In this case the upper layer was the crust, and it was about 18.5 mi (30 km) thick. The transition zone between the core and mantle now bears his name, the Mohorovicic discontinuity.

Potential field methods

Some properties of a material can be sensed at a distance because they generate *potential fields*. Gravitational fields are produced by any object with **mass**, and magnetic fields can be produced or distorted by objects with appropriate magnetic properties. Techniques to measure and interpret such fields are extremely useful in subsurface detection.

Gravity

Every object with mass produces a gravitational field. In theory it should be possible to detect objects which are denser than average or less dense than average lying beneath the surface. A large body of **lead ore**, for example, should cause the gravitational field above it to be somewhat greater than normal. A deep trough of loose sediments should result in a weaker gravitational field. The problem is that the gravitational attraction of the **planet** is huge, so that it is difficult to separate and measure that little modification produced by an anomalous mass.

Halite is a rock made of **salt**, NaCl. Therefore, rock salt, which is composed of halite, is much lighter than most other **rocks**, and deforms very easily. Big blobs of it sometimes develop at depth and rise toward the surface in bubble-like structures called diapirs or salt domes. As they move up, they can warp the **sedimentary rock** units they move through, producing rich, but localized, accu-

mulations of oil. Much of the oil found in the Gulf Coast states occurs in association with these salt domes, and so the petroleum industry had considerable incentive to develop techniques to detect them. As a result, the gravity meter, or gravimeter, was invented.

Essentially a very delicate bathroom scale, the gravimeter measures how much a spring is extended by the gravitational **force** acting on a mass at the end of the spring. What makes it work, and also makes it expensive, is that this extension can be measured with great precision, generally one part per million or better. That precision is equivalent to measuring a mile to within a sixteenth of an inch.

Such instruments immediately proved their worth, successfully detecting scores of salt domes beneath the nearly flat Gulf Coast states. Extending this technique to regional surveys, involving larger areas extending over greater elevation ranges, required refinements in our models for the gravitational field of the planet. This work has continued in conjunction with ever more refined gravity surveys. In the process, a major rift running down the center of **North America** has been discovered, subsurface continuations of rock units cropping out at the surface have been delimited, and even precursors for earthquakes have been detected.

Magnetism

Few phenomena seem so magical as the invisible attraction and repulsion we can feel when we play with magnets. When we stick a **paper** clip on one end of a magnet, it becomes a magnet, too, capable of holding up another paper clip. Some rock types exhibit their own magnetic fields, much like the magnets on a refrigerator door. Others distort **Earth's magnetic field**, similar to the way a paper clip temporarily becomes a magnet if it is in contact with a kitchen magnet. Sedimentary rocks rarely exhibit either magnetic behavior, and so they are effectively transparent to the magnetic signal.

Using a magnetometer we can measure the strength of the magnetic field anywhere on Earth. Often this is done with airborne surveys, which cover tremendous areas in little time. The results are mapped and the maps are used in several different ways.

The thickness of the sedimentary cover over igneous and metamorphic rocks (often called the "depth to basement") can often be inferred qualitatively from the magnetic maps. Just as a newspaper picture looks like a gray block when seen from a distance, a photograph when seen from arm's length, and a collection of printed dots when seen under a magnifying **lens**, so too the magnetic signal from the basement looks considerably different when seen from different distances. Little detail and subdued images suggest a thick blanket of sedimentary rocks, often miles.

Sweeping patterns or textures, caused by the combination of many outcrops involved in the same tectonic deformation, suggest a sedimentary cover of moderate thickness. If we can see distinct outlines, produced by the basement rock's **outcrop** patterns, we can safely infer that there is little or no sedimentary cover.

Magnetic maps are also utilized to **map** the continuation of units from places where they crop out into areas where they are buried. Much of the recent increase in our knowledge of the **geology** of the Adirondack Mountains in New York stems from this use of magnetic maps.

A third technique uses the strength and form of the magnetic signals to put limits on the **geometry** of buried units. This is similar to the situation with gravity, where models are developed and tested for consistency with the data. Often magnetic data can be used to constrain gravity models, and vice versa.

Electric techniques

The various responses of Earth materials to electric fields of different types permits additional characterization of the subsurface. Natural and artificial signals at a variety of different frequencies can be observed to travel through different parts of Earth. Measurements are made of how the signals are modified in their paths, and then models are constructed which try to emulate this behavior. Often these techniques are most useful where the extension of a geological body with distinctive electrical properties is sought in the subsurface. Metallic ore deposits or ion-rich pollutant plumes are good examples. Measurements of electrical resistivity, the resistance of a material to passing a current of **electricity**, have been used for decades in the oil industry to help locate oil- or gas-bearing rock units.

Nuclear survey methods

Nuclear survey methods are of two basic types. The more common of the two involves measurement of natural radioactivity in rocks or soils. This method might be used, for example, to identify potential **uranium** ores for **mining**. In this case a hand-held geiger counter could be employed. Another use is to measure the natural gamma ray emissions from rock formations in a drill hole when searching for oil. A gamma ray counter is lowered down the hole on a wire, and the natural gamma ray emissions of the rocks in the borehole wall are measured. Different rocks exhibit various levels of radioactivity, making remote identification of rock type possible. Geophysical surveys of boreholes that are done in this manner are called wireline surveys or wireline bogs.

The second type of nuclear survey method is stimulated radioactivity. In this method, a radioactive source is used to bombard a rock and induce **radioactive decay** in the rock. The level of induced radioactive decay is measured and the rock type is interpreted from the measurements. Wireline surveys employing this method are often used when exploring for oil.

Satellite altimeter data

Satellites can measure the elevation of the **sea level** surface with astonishing precision. Because their path crosses the same place on Earth over and over again, the effects of waves and **tides** can be accounted for, and the actual variations in the elevation of sea level can be determined. These variations are most often the result of relief on the **ocean** floor. Where there is a mountain on the ocean floor, its additional mass draws extra **water** toward it, which will elevate the level of the sea above it. Where a trench exists on the ocean floor, the absence of mass attracts less water and a depression in the elevation of sea level will occur above it. Many of the detailed maps of the sea floor were obtained indirectly in this way.

The inverse problem

Subsurface detection relies on solutions to what is often called the "inverse problem." A particular set of data are observed, and then models are developed which attempt to fit all the data. Sometimes, when a very nice match is made, it is tempting to assume that a particular model is the only model which can fit the data, although this is rarely true.

An example may illustrate this: Imagine that I am trying to figure out the value of the change you have in your pocket. Suppose I have a device which will detect how many coins you have, and it says you have seven. Then I would know that you have at least $0.07 and at most $1.75. Suppose I have another device which tells me that the coins are of two different sizes, with four of them larger than the other three. The big ones might be quarters, leaving a range from $1.03 to $1.30, or they could be nickels, leaving a range from $0.23 to $0.50. Finally, if I were to think about this more carefully, I would see that not all of the values in these ranges are possible. So you could have $1.03, $1.12, $1.15, $1.21, $1.30 if the larger coins are quarters, or $0.23, $0.32, $0.41, $0.50, if the larger coins are nickels. We have reduced the number of possibilities to nine, but there is no way we can use these "subsurface detection" techniques to constrain things further. Because each of these nine possibilities fits the data perfectly, we might find one and erroneously conclude that because it fit the data so well it must be true. Assumptions were built into our conclusions, also; we assumed it is United States currency, and no half dollars or dollar coins. Such assumptions make sense for most pockets we are

KEY TERMS

. . . .

Reflection—When some of the energy of a seismic wave bounces off a boundary between two media, instead of traveling through it.

Refraction—The bending of light that occurs when traveling from one medium to another, such as air to glass or air to water.

Seismic wave—A disturbance produced by compression or distortion on or within the earth, which propagates through Earth materials; a seismic wave may be produced by natural (e.g. earthquakes) or artificial (e.g. explosions) means.

likely to run into in this country, but are obviously not valid in other countries.

This example illustrates the nature of subsurface detection. Results are usually somewhat ambiguous, depend on assumptions, and do not directly give the answers we seek. Yet they can provide important constraints. I may be able to make some additional assumptions from other data, hunches, or guesses. For instance, I may know you well enough to figure that you would not keep pennies, which would reduce the number of options to three.

Usefulness of subsurface detection

The techniques described here are very often used together to improve our understanding of what exists below the surface. By adjusting the sensitivity of the instruments and the spacing of the measurements, the scale and the depth of interest may be varied. The same theory and principles used with magnetic techniques which delineate the rift running through the crust of North America, at a depth of 18 mi (30 km), can be used to locate buried pipes and cables at a depth of less than 10 ft (3 m).

Often, subsurface detection is the only way to study the area of interest, because it lies too deep to be reached directly. Other times it is used because it is less expensive than digging or drilling, or because it disrupts the environment less.

See also Seismograph.

Resources

Books

Press, F., and R. Siever. *Understanding Earth*. 3rd ed. New York: W.H Freeman and Company, 2001.
Telford, William Murray, L.P. Geldart, and R.E. Sheriff. *Applied Geophysics*. Cambridge; New York: Cambridge University Press, 1990.

Otto H. Muller

Subtraction

Subtraction is the operation that is the inverse of **addition**. Subtracting a number has the effect of nullifying the addition of that same number. For example, if one adds 11 to 41 then subtracts 11, the result is 41 again.

Symbolically, for any numbers a and b, $(a + b) - b = a$

Or one can subtract first, then add: $(a - b) + b = a$

Thus, one can say that subtraction and addition are "inverse operations."

The definitions of subtraction

The second of the two rules above can be taken as a definition of subtraction. That is "the difference of two numbers, a - b, is the number which must be added to the second number, b, to equal the first, a." Before the widespread use of electronic cash registers, grocery clerks and others, making change, would use this definition directly. If, for example, a customer bought groceries worth $3.70 and gave the clerk $5.00 to pay for them, the clerk would not subtract in order to compute the change. Instead he (or she) would count it out. Starting with "three seventy," he would give the customer a nickel, and say, "three seventy-five." He would give the customer a quarter and say, "four." Then he would give the customer a dollar and say, "five." This method is still in use when no **calculator** is available.

A second definition of subtraction is often given in **algebra** books after the introduction of negative numbers. There a - b is defined as a + (-b), i.e. "To subtract a number, add its opposite." This definition is mathematically efficient, but it hides a lot of the ways in which subtraction is used, as in the change-making example above.

Terminology

In a subtraction problem it is useful to have names for the various parts. In a - b the entire expression is called a "difference," and the answer to a - b is the difference, or occasionally the "remainder." These two terms arise from two ways in which subtraction is used in practical problems. The first number, a, is called the "minuend." This is the part that is diminished or made smaller when something is subtracted from it (provided that that something is positive). The second number, b, is the "subtrahend" or the part subtracted.

Properties

It matters very much which of the two parts of a difference are named first. Taking $500 from an account with $300 in it is very different from taking $300 from $500. For this reason, subtraction is *not* commutative: a - b does not equal b - a.

Subtraction is not associative either: (a - b) - c does not equal a - (b - c).

An example will demonstrate this: (20 - 10) - 3 is 7, but 20 - (10 - 3) is 13.

Often one encounters expressions such as $5x^2 - 2 - 3x^2$, with no indication of which subtraction is to be done first. Since subtraction is non-associative, it matters. To avoid this ambiguity one can agree that subtractions, unless otherwise indicated, are to be done left-to-right. This is a rather limiting agreement, therefore, it may be more convenient to use some other order. Another agreement, which is the common agreement of algebra, is to treat the minus sign as a plus-the-opposite-of sign. Thus one would interpret the example above as $5x^2 + (-2) + (-3x^2)$. In this interpretation it becomes a sum, whose terms can be combined in any order one pleases.

In certain sets subtraction is not a closed operation. The set of **natural numbers**, for instance, is not closed with respect to subtraction. If a merchant will not extend credit, one cannot buy an article whose price is greater than the amount of money one has.

Uses of subtraction

The most familiar meaning for subtraction is called "take away." To find the answer if you take 5 eggs from a dozen, you subtract: 12 - 5 = 7.

Subtraction is also used to compare two numbers. When one wants to know how much colder -7.3° is than 13.8°, one computes the difference -7.3 - 13.8 to get -21.1°.

A third use of subtraction is to figure out a missing addend. If one has a certain sum of money, say $45.20 and wants to buy an article costing $85.50, subtraction is used to compute that needed amount: 85.50 - 45.20.

J. Paul Moulton

Succession

Succession is a process of ecological change, involving the progressive replacement of earlier biotic communities with others over time. Succession usually begins with the disturbance of a pre-existing **ecosystem**, followed by recovery. In the absence of further stand-level disturbance, succession culminates in a stable climax community, the nature of which is determined by climate, **soil**, and the nature of the local biota. Primary succession occurs on bare substrates that have not been previously modified by biological processes. Secondary succession occurs on substrates that have been modified biologically, and it follows disturbances that have not been so intense as to eliminate the regenerative capacity of the vegetation.

Disturbance, stress, and succession

Disturbance is an episodic **stress** that causes substantial changes in the structure and function of ecosystems. Depending on its severity and extent, disturbance can influence individual organisms or entire stands, disrupting relationships among individuals within the community, and affecting ecological processes such as productivity, nutrient cycling, and **decomposition**.

Natural disturbances can be caused by intense events of physical disruption associated with, for example, a hurricane, **tornado**, oceanic tidal wave, **earthquake**, the blast of a volcanic eruption, or over geological timespans, the advance and retreat of **glaciers**. Natural disturbances are also associated with **wildfire**, and with biological events, such as an irruption of defoliating **insects** that can kill a mature forest. Human activities are also important sources of ecological disturbances, as is the case of agricultural practices such as plowing, the harvesting of trees from **forests**, construction activities, and explosions associated with military activities. There are numerous other examples of disturbances caused by human activities.

Note that all of these disturbances can operate at various spatial scales. In some cases, they can result in extensive disturbances over very large areas, for example, when a very severe wildfire consumes millions of hectares of forest. In other cases a disturbance may not cause a stand-level mortality, as when a **lightning** strike kills a single mature **tree** in a forest, creating a gap in the overstory, and initiating a microsuccession that culminates in occupation of the gap by another mature tree. Scale is a very important aspect of succession.

Once the intense physical, chemical, or biological stress associated with an event of disturbance is relaxed, succession begins, and this process may eventually restore an ecosystem similar to that present prior to the disturbance. However, depending on environmental circumstances, a rather different ecosystem may develop through post-disturbance succession. For example, if climate change has resulted in conditions that are no longer suitable for the establishment and growth of the same **species** of trees that dominated a particular stand of old-growth forest prior to its disturbance by a wildfire, then succession will result in the development of a forest of a different character.

Succession can also follow the alleviation of a condition of longer-term, chronic environmental stress. For example, if a local source of **pollution** by poisonous chemicals is cleaned up, succession will occur in response to the decreased exposure to toxic chemicals. This might occur, for example, if emissions of toxic **sulfur dioxide** and metals from a smelter are reduced or stopped. This abatement of longer-term environmental stress allows pollution-sensitive species to invade the vicinity of the source, so that succession could proceed. Similarly, removal of some or all of a population of cattle that are chronically overgrazing a pasture would allow the **plant** community to develop to a more advanced successional stage, characterized by greater **biomass** and possibly more **biodiversity**.

Succession can also be viewed as a process of much longer-term ecological change occurring, for example, as a **lake basin** gradually in-fills with sediment and organic debris, eventually developing into a terrestrial **habitat** such as a forest. This particular succession, which occurs through the development and replacement of a long series of community types, can take thousands of years to achieve its progression through the aquatic to terrestrial stages. The series of ecological changes is still, however, properly viewed as a successional sequence (or sere).

Primary succession

Primary successions occur after disturbances that have been intense enough to obliterate any living organisms from the site, and even to have wiped out all traces of previous biological influences, such as soil development. Natural disturbances of this intensity are associated with glaciation, lava flows, and very severe wildfires, while human-caused examples might include the abandonment of a paved parking lot, or an above-ground explosion of a nuclear weapon. In all of these cases, organisms must successfully invade the disturbed site for succession to begin.

A well-known study of primary succession after deglaciation has been conducted in Alaska. This research examined a time series of substrates and plant communities of known age (this is called a chronosequence) at various places along a fiord called Glacier Bay. In this study, historical positions of the glacial front could be dated by various means, including written records and old photographs. The primary succession begins as the first plants colonize newly deglaciated sites and, in combination with climatic influences, begin to modify the post-glacial substrate. These initial plants include mosses, **lichens**, herbaceous dicotyledonous species such as the river-beauty (*Epilobium latifolium*), and a nitrogen-fixing cushion plant known as mountain avens (*Dryas octopetala*). These pioneers are progressively replaced as

succession continues, first by short statured species of willow (*Salix* spp.), then by taller shrubs such as alder (*Alnus crispa*, another nitrogen-fixing plant), which dominates the community for about 50 years. The alder is progressively replaced by sitka **spruce** (*Picea sitchensis*), which in turn is succeeded by a forest of western hemlock (*Tsuga heterophylla*) and mountain hemlock (*T. mertensiana*). The primary succession at Glacier Bay culminates in the hemlock forest—this is the climax stage, dominated by species that are most tolerant of stresses associated with **competition**, in a mature habitat in which access to resources is almost fully allocated among the dominant biota.

The plant succession at Glacier Bay is accompanied by other ecological changes, such as soil development. The mineral substrate that is initially available for plant colonization after the meltback of glacial **ice** is a fine till, with a slightly alkaline **pH** of about 8, and as much as 7-10% carbonate **minerals**. As this primary substrate is progressively leached by percolating **water** and modified by developing vegetation, its acidity increases, reaching pH 5 after about 70 years under a spruce forest, and eventually stabilizing at about pH 4.7 under a mature hemlock forest. The acidification is accompanied by large reductions of **calcium concentration**, from initial values as large as 10%, to less than 1%, because of **leaching** and uptake by vegetation. Other important soil changes include large accumulations of organic **matter** and **nitrogen**, due to biological fixations of atmospheric **carbon dioxide** (CO_2) and dinitrogen (N_2).

Other studies of primary succession have examined chronosequences on **sand** dunes of known age. At certain places on the Great Lakes, sandy beach ridges are being slowly uplifted from beneath the lakewaters by a ponderous rebounding of the land. The process of uplifting is called **isostasy**, and it is still occurring in response to meltback of the enormously heavy glaciers that were present only 10,000 years ago. The initial plant colonists of newly exposed dunes that were studied on Lakes Michigan and Huron are short-lived species, such as sea rocket (*Cakile edentula*) and beach spurge (*Euphorbia polygonifolia*). These ephemeral plants are quickly replaced by a perennial dunegrass community, dominated by several **grasses** (*Ammophila breviligulata* and *Calamovilfa longifolia*). With time, a tall-grass **prairie** develops, dominated by other species of tall grasses and by perennial, dicotyledonous herbs. The prairie is invaded by shade-intolerant species of shrubs and trees, which form nuclei of forest. Eventually a climax forest develops, dominated by several species of **oaks** (*Quercus* spp.) and tulip-tree (*Liriodendron tulipifera*). This series of successional plant communities is accompanied by soil development, with broad characteristics similar to those observed at Glacier Bay (although, of course, the rates of change are different).

Secondary succession

Secondary succession occurs after disturbances that are not intense enough to kill all plants, so that regeneration can occur by re-sprouting and growth of surviving individuals, and by the **germination** of pre-existing **seeds** to establish new plants. This regeneration by surviving plants and seeds is supplemented by an aggressive invasion of plant seeds from elsewhere. Another characteristic of secondary succession is that the soil still retains much of its former character, including previous biological and climatic influences on its development.

Secondary successions are much more common than primary successions, because disturbances are rarely intense enough to obliterate previous ecological influences. Most natural disturbances, such as windstorms, wildfires, and insect defoliations, are followed by ecological recovery through secondary succession. The same is true of most disturbances associated with human activities, such as the abandonment of agricultural lands, and the harvesting of forests.

Secondary succession can be illustrated by an example involving natural regeneration after the clear-cutting of a mature, mixed-species forest in northeastern **North America**. In this case, the original forest was dominated by a mixture of **angiosperm** and coniferous tree species, plus various plants that are tolerant of the stressful, shaded conditions beneath a closed forest canopy. Some of the plants of the original community survive the disturbance of clear-cutting, and they immediately begin to regenerate. For example, each cut stump of red maple (*Acer rubrum*) rapidly issues hundreds of new sprouts, which grow rapidly, and eventually self-thin to only 1-3 mature stems after about 50 years. Other species regenerate from a long-lived seed bank, buried in the forest floor and stimulated to germinate by the environmental conditions occurring after disturbance of the forest. Pin cherry (*Prunus pensylvanica*) and red raspberry (*Rubus strigosus*) are especially effective at this type of regeneration, and these species are prominent during the first several decades of the secondary succession. Some of the original species do not survive the clear-cutting in large numbers, and they must re-invade the developing habitat. This is often the case of coniferous trees, such as red spruce (*Picea rubens*). Another group of species is not even present in the community prior to its disturbance, but they quickly invade the site to take advantage of the temporary conditions of resource availability and little competition immediately after disturbance. Examples of these so-called ruderal species are woody plants such as alders (in

this case, *Alnus rugosa*) and white birch (*Betula papyrifera*), and a great richness of herbaceous perennial plants, especially species in the aster family (Asteraceae), such as goldenrod (*Solidago canadensis*) and aster (*Aster umbellatus*), along with various species of grasses, **sedges**, and other monocotyledonous plants.

This secondary succession of plant communities is accompanied by a succession of **animal** communities. For example, the mature forest is dominated by various species of **warblers**, **vireos**, **thrushes**, **woodpeckers**, flycatchers, and others. After clear-cutting, this avian community is replaced by a community made up of other native species of **birds**, which specialize in utilizing the young habitats that are available after disturbance. Eventually, as the regenerating forest matures, the bird species of mature forest re-invade the stand, and their community re-assembles after about 30-40 years has passed.

Mechanisms of succession

As noted previously, succession generally begins after disturbance creates a situation of great resource availability that can be exploited by organisms, but under conditions of little competition. The classical explanation of the ecological mechanism of community change during succession is the so-called facilitation model. This theory suggests that the recently disturbed situation is first exploited by certain pioneer species that are most capable of reaching and establishing on the site. These initial species modify the site, making it more suitable for invasion by other species, for example, by carrying out the earliest stages of soil development. Once established, the later-successional species eliminate the pioneers through competition. This ecological dynamic proceeds through a progression of stages in which earlier species are eliminated by later species, until the climax stage is reached, and there is no longer any net change in the community.

Another proposed mechanism of succession is the tolerance model. This concept suggests that all species in the succession are capable of establishing on a newly disturbed site, although with varying successes in terms of the rapid attainment of a large population size and biomass. In contrast with predictions of the facilitation model, the early occupants of the site do not change environmental conditions in ways that favor the subsequent invasion of later-successional species. Rather, with increasing time, the various species sort themselves out through their differing tolerances of the successionally increasing intensity of biological stresses associated with competition. In the tolerance model, competition-intolerant species are relatively successful early in succession when site conditions are characterized by a free availability of resources. However, these species are eliminated later on because they are not as competitive as later species, which eventually develop a climax community.

A third suggested mechanism of succession is the inhibition model. As with the tolerance model, both early- and later-successional species can establish populations soon after disturbance. However, some early species make the site less suitable for the development of other species. For example, some plants are known to secrete toxic biochemicals into soil (these are called allelochemicals), which inhibit the establishment and growth of other species. Eventually, however, the inhibitory species die, and this creates opportunities that later-successional species can exploit. These gradual changes eventually culminate in development of the climax community.

All three of these models, facilitation, tolerance, and inhibition, can be supported by selected evidence from the many ecological studies that have been made of succession (especially plant succession). Although these models differ significantly in their predictions about the organizing principles of successional dynamics, it appears that none of them are correct all of the time. Facilitation seems to be most appropriate in explaining changes in many primary successions, but less so for secondary successions, when early post-disturbance site conditions are suitable for the vigorous growth of most species. The relatively vigorous development of ecological communities during secondary succession means that competition rapidly becomes an organizing **force** in the community, so there is an intensification of interactions by which organisms interfere with and inhibit each other. Aspects of these interactions are more readily explained by the tolerance and inhibition models. Overall, it appears that successions are idiosyncratic—the importance of the several, potential mechanisms of succession varies depending on environmental conditions, the particular species that are interacting, and the influence of haphazard events, such as which species arrived first, and in what numbers.

Climax—the end point of succession

The climax of succession is a relatively stable community that is in equilibrium with environmental conditions. The climax condition is characterized by slow rates of change in an old-growth community, compared with more dynamic, earlier stages of succession. The climax stage is dominated by species that are highly tolerant of the biological stresses associated with competition, because access to resources is almost completely allocated among the dominant organisms of the climax community. However, it is important to understand that the climax community is not static, because of the dynamics of within-community microsuccession, associated, for example, with gaps in a forest canopy caused by

KEY TERMS

Canopy closure—A point in succession where the sky is obscured by tree foliage, when viewed upward from the ground surface.

Chronosequence—A successional series of stands or soils of different age, originating from a similar type of disturbance.

Competition—An interaction between organisms of the same or different species associated with their need for a shared resource that is present in a supply that is smaller than the potential, biological demand.

Cyclic succession—A succession that occurs repeatedly on the landscape, as a result of a disturbance that occurs at regular intervals.

Ruderal—Refers to plants that occur on recently disturbed sites, but only until the intensification of competition related stresses associated with succession eliminates them from the community.

Sere—A successional sequence of communities, occurring under a particular circumstance of types of disturbance, vegetation, and site conditions.

Successional trajectory—The likely sequence of plant and animal communities that is predicted to occur on a site at various times after a particular type of disturbance.

the death of individual trees. Moreover, if events of stand-level disturbance occur relatively frequently, the climax or old-growth condition will not be achieved.

See also Climax (ecological); Stress, ecological.

Resources

Books

Begon, M., J. L. Harper, and C. R. Townsend. *Ecology. Individuals, Populations and Communities.* 2nd ed. London: Blackwell Sci. Pub., 1990.

Freedman, B. *Environmental Ecology.* 2nd ed. San Diego: Academic Press, 1995.

Bill Freedman

Suckers

Suckers are cylindrical **fish** with a downward-pointing suckering mouth in the family Catostomidae, which is in the large suborder Cyprinoidea, which also includes **minnows**, carps, and loaches.

Most **species** in the sucker family occur in the Americas, over a range that extends from the boreal forest of **North America** through much of Central America. A few other species occur in eastern Siberia, and there is one isolated species in eastern China.

Suckers are distinguished from other members of the Cyprinoidea by aspects of their jaw structure, the presence of a single row of teeth in their pharynx, a lack of barbels, and their round, downward-pointing, fleshy-lipped, sucking mouth. Suckers generally have a cylindrical or slightly compressed body.

Most species of suckers occur in flowing waters, such as **rivers** and streams. Some species also occur in still waters such as lakes and large ponds, but suckers living in these habitats spawn in nearby rivers.

The largest species of suckers can attain a length of 6.6 ft (2 m), but most are much smaller than this. Male suckers are smaller than the females. Both sexes become relatively brightly colored during the breeding season.

The primary food of suckers are aquatic **invertebrates**, which are mostly hunted in the sediment. Suckers that live in lakes may also eat some aquatic vegetation. The larger species of suckers are of some economic importance as food-fishes for humans, although their flesh is rather bony. Other, smaller species are important as forage species for larger, more valuable species of fishes.

Species of suckers

There are over one hundred species of suckers. The common or white sucker (*Catostomus commersoni*) is a widespread species throughout much of northern and central North America. This species has a round mouth, useful for feeding on its usual **prey** of bottom-dwelling **insects**, crustaceans, molluscs, and other invertebrates. The common sucker is a relatively large species, attaining a length of up to 10 in (45 cm), and a weight of 2.2 lb (1 kg). The common sucker is often found in lakes and ponds. These fish generally run up nearby streams to spawn in gravel beds in the springtime, but they sometimes lay their eggs in gravel along shallow lakeshores. Individuals of this species can live as long as 12 years.

The longnose or northern sucker (*Catostomus catostomus*) is also widely distributed in northern North America, and it also occurs in eastern Siberia. The longnose sucker generally inhabits cooler waters and occurs in deeper lakes and larger rivers and streams than the common sucker. This species is exploited commercially on the Great Lakes and elsewhere, although it is not considered to be a high-value species of fish. Other species of suckers

are more local in distribution, for example, the Sacramento sucker (*C. occidentalis*) of northern California.

Both the bigmouth buffalo (*Ictiobus cyprinellus*) and the smallmouth buffalo (*I. bubalus*) are widely distributed in the eastern United States. These species have also been transplanted farther to the west to establish some sportfishing populations. These fish can attain a large size, exceeding 22 lb (10 kg) in some cases, and are commonly fished as food.

The northern redhorse or redfin sucker (*Moxostoma macrolepidotum*) occurs widely in central North America. The **lake** or northern chub (*Couesius plumbeus*) is a small minnow-sized fish that occurs widely across northern North America. This is an important forage and bait fish. The lake chubsucker (*Erimyzon sucetta*) occurs in the eastern United States, including Lake Saint Clair and Lake Erie.

Resources

Books

Page, L., and Burr, B. *Field Guide to Freshwater Fishes of North America.* Boston, MA: Houghton Mifflin, 1991.
Whiteman, Kate. *World Encyclopedia of Fish & Shellfish.* New York: Lorenz Books, 2000.

Bill Freedman

Sucking lice *see* **Lice**
Sucrose *see* **Carbohydrate**

Sudden infant death syndrome (SIDS)

Sudden infant death **syndrome** (SIDS), also called crib death, is the death without apparent organic cause of an infant under the age of one year. A **diagnosis** of SIDS can only be made after experts have investigated the death scene, autopsied the dead infant, reviewed the baby's medical history, and ruled out all other possible explanations. About 7,000 babies die of SIDS each year in the United States. This baffling disorder is the leading cause of death in infants ages 1-12 months. Although SIDS cannot be prevented completely, research has shown that parents can reduce the risk by putting their baby to **sleep** on its back (supine position) or side rather than on its stomach.

For unknown clinical reasons, in the United States, African American and Native American babies are up to three times more likely to die of SIDS than Caucasian infants. In all cases and groups, the majority of SIDS victims are male infants.

The mysterious malady

The SIDS definition is purposefully vague and reflects how little actually is known about what causes the syndrome. SIDS victims seem to stop breathing in their sleep. They typically are found lifeless, limp, and blue. Often they have blood-tinged mucus coming from their mouth or nose. Ninety **percent** of SIDS victims die before six months. Most appear perfectly healthy beforehand or at most have a slight cold. There is a statistically significant correlation between SIDS deaths and respiratory infections prior to death. Although they are usually found in their cribs, babies have died of SIDS in car seats, strollers, and their mother's arms.

Although SIDS researchers have investigated hundreds of possible theories regarding the causes of SIDS, no clear answers have been found. Autopsies fail to show any abnormalities in SIDS victims; they seem to be healthy, normal babies. Scientists are not even sure whether death is caused by cardiac arrest or respiratory failure.

Some experts estimate that 1-20% of all diagnosed SIDS deaths are actually the result of other causes, including child abuse and murder. For this reason an autopsy and a thorough examination of the scene of death must be done. This suspicion adds to the parents' grief and guilt. It also confuses the public's understanding of SIDS. But until a more definitive diagnosis of SIDS exists such steps must be taken to rule out the possibility of murder.

SIDS research

The age of its victims offers an important clue towards better understanding SIDS. Almost all sudden deaths occur between one week and six months of age, a time of rapid growth and change in a baby. Neurological control of the baby's circulatory and respiratory systems is still evolving. Some scientists theorize that very subtle flaws in the baby's physical development are responsible for SIDS. Instead of breathing evenly, young babies tend to stop breathing for a few seconds and then begin again with a gasp. According to one theory, babies who die of SIDS have difficulty re-starting their breathing. Much more needs to be known about the normal respiratory processes and sleep patterns of babies in order to detect abnormalities.

Another clue may lie in the observation that many SIDS victims have a cold in the weeks before death. SIDS deaths are more common in the winter, a season when colds are frequent. This suggests that an upper respiratory **infection** might somehow trigger a series of events that leads to sudden death. Some researchers believe that no one **factor** is responsible for SIDS but that a number of events must come together to cause the syndrome.

Risk factors

By studying large groups of young infants, a few of whom eventually go on to die of SIDS, scientists have found certain factors that occur more frequently in sudden death victims. For example, a genetic defect in an **enzyme** involved in fatty acid **metabolism** has been identified as a possible cause of death in a small percentage of SIDS victims. With this defect, the infant's **brain** can become starved for **energy** and the baby enters a **coma**. Italian researchers have demonstrated a link between a particular type of irregular heartbeat and SIDS. Infants who inherit this irregularity, called "long QT syndrome," are 41 times more likely to die of SIDS. This syndrome also is a leading cause of sudden death in adults. It is possible to screen for this irregularity with **heart** monitors and to treat it with drugs.

Babies born prematurely are at greater risk for SIDS. So are twins and triplets. A twin is more than twice as likely as a non-twin to die of SIDS. Boys are more susceptible than girls. Formula-fed infants are more susceptible than breast-fed babies. SIDS also is more common in the babies of mothers who are poor, under 20 years old, have other children, and receive little medical care during pregnancy. Mothers who smoke during pregnancy and the presence of **cigarette smoke** in the home after **birth** also increases the likelihood of SIDS. Other proposed risk factors including childhood vaccines and allergies to cow's milk have failed to show any link to SIDS.

It is important to remember that more than two-thirds of SIDS cases occur in babies without known risk factors. Some scientists believe that all infants are potential victims if certain factors in their bodies and their environment interact in a particular unknown way. By studying risk factors researchers hope to gain important insights into the causes of the syndrome.

In general, SIDS does not appear to be hereditary. Siblings of SIDS victims have only a slightly higher risk of sudden death compared to the average population. Yet some parents who have a subsequent baby after suffering the loss of a baby to SIDS find it very reassuring to use a home monitor. Attached to the baby, this machine sounds an alarm if the baby's **respiration** or heart **rate** drops below normal. The National Institutes of Health has stated that home monitors have not been shown to prevent SIDS.

"Back to sleep" campaign

SIDS occurs more frequently in New Zealand and the United States than in Japan and China. Within the United States it is more common in African-American babies than Hispanic babies. These differences suggest that certain cultural factors of baby care, particularly how he or she is put to bed, may affect the incidence of SIDS. Scientists do not understand exactly why these differences matter—just that they do.

The single most important thing a parent or caregiver can do to lower the risk of SIDS is to put the baby to sleep on its back or side rather than on its stomach. In 1992 the American Academy of Pediatrics recommended placing healthy infants to sleep on their backs or sides. The group made the recommendation after reviewing several large studies done in New Zealand, England, and **Australia**. The studies demonstrated that SIDS declined as much as 50% in communities that had adopted this sleeping position for infants. In the United States, the "Back to Sleep" campaign has been very successful and appears to have contributed significantly to a sharp drop in SIDS in the past few years. However it also may have resulted in an increase in "misshapen head" syndrome, caused by infants always sleeping on their backs. This syndrome is readily treatable with **physical therapy**.

In the past the supine (or back-sleeping) position has been discouraged for fear that a sleeping infant might spit up and then suffocate on its own vomit. A careful examination of studies of infants placed prone (on their stomachs) has shown that this does not happen. Some infants with certain health problems might best be placed prone. Parents who suspect this should check with their doctor.

SIDS in history

The phenomenon of sudden death in babies has been recorded for centuries. SIDS has been described as a distinct disorder for nearly a century. In 1979 it was officially accepted as a cause of death. The current definition of the condition was developed by the National Institutes of Health in 1989.

As sleeping habits for families and babies changed over time so have the explanations offered for sudden death. Until a century ago infants and small children slept in the same beds as their mother. When babies were found dead their mothers were often blamed for rolling on top of them. In the 1700s and 1800s mothers were accused of rolling on their babies while drunk. After noting that SIDS is very rare in Asian countries where parents and babies typically sleep together, some recent researchers have theorized that sleeping with a parent might help regulate an infant's respiration and thus prevent SIDS.

In the early 1900s in the United States when co-sleeping became rare, sudden death was blamed on dressing a baby too warmly at night. Before physicians realized that all babies have large thymus **glands**, enlargements of the thymus gland were also blamed for SIDS.

Studies release in 2003 showed no correlation between immunization schedules and SIDS death.

Support groups for parents

Parents who suffer the loss of a child to SIDS typically feel immense sorrow and grief over the unexpected and mysterious death. They also may feel guilty and blame themselves for not being more vigilant although there was nothing they could have done. Parents and other relatives of SIDS victims often find it helpful to attend support groups designed to offer them a safe place to express their emotions.

Everyone affected by the death of a baby to SIDS may need special support. This includes doctors, nurses, paramedics, other health care providers, police officers, and the babysitter or friend who may have been caring for the baby when it died. Ideally counseling should be made available to these people. Health care professionals and law enforcement officers often have special training to help comfort the grieving survivors of a SIDS death.

See also Neuroscience; Sleep disorders.

Resources

Periodicals

Beckwith, J.B. "Defining the Sudden Infant Death Syndrome." *Arch Pediatr Adolesc Med.* 157(3) (2003):286-90.

Berry, P.J. "SIDS: Permissive or Privileged Diagnosis?" *Arch Pediatr Adolesc Med.* 157(3) (2003):293-4.

Bower, Bruce. "Co-sleeping Gives Babies a Boost." *Science News.* 144 (4 December 1993): 380.

Byard, R.W., H.F. Krous. "Sudden Infant Death Syndrome: Overview and Update." *Pediatr. Dev. Pathol.* (Jan 21, 2003) [epub ahead of print].

Cutz, E. "New Challenges for SIDS Research." *Arch Pediatr Adolesc Med.* 157(3) (2003):292-3.

Horchler, Joani Nelson, and Robin Rice Morris. *The SIDS Survival Guide.* Hyattsville: SIDS Educational Services, 1994.

James, C., H. Klenka, D. Manning. "Sudden Infant Death Syndrome: Bed Sharing with Mothers who Smoke." *Arch. Dis. Child.* 88 (2003) 112-113.

Matthews, T. "Sudden Infant Death Syndrome—a Defect in Circulatory Control?" *Child Care Health Dev.* 28 Suppl. 1(2002) 41-43.

Sunshine-Genova, Amy. "Sleep Position and SIDS." *Parents.* (January 1994).

Liz Marshall

Sugar *see* **Carbohydrate**

Sugar beet

The possibility of **beet** sugar was first discovered in 1605 when a French scientist found that the boiled root of garden beet (*Beta vulgaris*) yielded a syrup similar to that obtained from **sugarcane** (*Saccharum officinarum*). It was not until the mid-1700s, however, that the commercial potential of sugar beets was recognized. Once realized, sugar beets quickly became a major crop in **Europe** and elsewhere, displacing some of the sugarcane that could only be obtained from tropical plantations. The cul-

tivated sugar beet is, therefore, a variety of the common garden beet, known as *Beta vulgaris* var. *crassa*.

The sugar beet has wide, thin leaves, growing from a large, tuberous root **mass**. It is a biennial **herb** (having a two-year life cycle), storing most of its first-year production of **biomass** in its large, carbohydrate-rich root (containing 17-27% sugar).

In 1999, about 16.8 million acres (6.8 million ha) of sugar beets were grown world-wide, and total production was 286 million tons of root mass (260 million tonnes). Sugar beets are used to manufacture sucrose-sugar, as well as secondary products such as **alcohol**. The pressed remains of sugar extraction can be fed to cows and other **livestock**.

Bill Freedman

Sugarcane

The sugarcane (*Saccharum officinale*) is a 12-26 ft tall (4-8 m), perennial, tropical grass (family Poaceae). The tough, semi-woody stems of sugarcane are up to 2 in (5 cm) in diameter, with leafy nodes and a moist internal

Harvesting sugarcane. *Photo Researchers, Inc. Reproduced by permission.*

pith containing 15-20% sucrose-sugar. The sugar **concentration** is highest just before the **plant** flowers, so this is when harvesting occurs. The plants are propagated by cuttings placed into the ground, but a single planting can last several harvest rotations.

Sugarcane is thought to have originated in southern **Asia**, where it has been cultivated for at least 3,000 years. It is believed to be a cultivated **hybrid** of various **species** of *Saccharum*, which are still wild plants in South and Southeast Asia. These wild progenitors include *S. barbari*, *S. robustum*, *S. sinense*, and *S. spontaneum*.

In 1999, about 48.4 million acres (19.6 million ha) of sugarcane were cultivated worldwide, and the total production was 1.41 billion tons (1.28 billion tonnes). Sugarcane is used to manufacture sucrose-sugar, and accounts for about 60% of the global supply. The sucrose is used to manufacture many secondary products, such as molasses and **alcohol**. It is also an ingredient in innumerable prepared foodstuffs, such as candy, chocolate, carbonated beverages, **ice** cream, and other sweetened foods. The pressed remains of sugar extraction can be fed to cows and other **livestock**.

Bill Freedman

Sulfur

Sulfur is the non-metallic chemical element of **atomic number** 16. It has a symbol of S, an **atomic weight** of 32.07, and a specific gravity of 2.07 (rhombic form) or 1.96 (monoclimic form). Sulfur boils at 832.5°F (444.7°C) and consists of four stable isotopes of **mass** numbers 32 (95.0%), 33 (0.75%), 34 (4.2%) and 36 (0.015%). Sulfur **atoms** found in different locations have slightly different percentages of the four isotopes, however.

Sulfur is a bright yellow solid that can exist in many allotropic forms with slightly different melting points, all around 239°F (115°C). The two main forms are called rhombic sulfur and monoclinic sulfur. There is also a rubbery, non-crystalline form, called plastic or amorphous—without shape—sulfur. An ancient name for sulfur is brimstone, meaning "burning stone." It does indeed **burn** in air with a blue flame, producing **sulfur dioxide**:

$$S \quad + \quad O_2 \quad \rightarrow \quad SO_2$$
sulfur   oxygen   sulfur
          gas   dioxide
                           gas

Sulfur itself has no odor at all, but it has a bad reputation because it makes many smelly compounds. Sulfur dioxide is one of them; it has a sharp, choking, suffocating effect on the unfortunate breather. The "fire and brimstone" of the Bible was one of the worst punishments that its authors could think of. The fact that sulfur comes from deep under the ground and that sulfur dioxide can be smelled in the fumes of volcanoes further fueled people's imaginations of what Hell must be like.

Where sulfur is found

Sulfur makes up only 0.05% of the Earth's crust, but it is easy to get because it occurs uncombined as the element, S_8 (eight sulfur atoms tied together into each **molecule** of the element and joined in a ring).

Sulfur occurs in huge, deep underground deposits of almost-pure element, notably along the Gulf coast of the United States and in Poland and Sicily. However, miners do not have to go underground to get it. They make the sulfur come to them by a clever arrangement of three pipes within pipes, drilled down into the sulfur bed. This arrangement is called the Frasch Process. Superheated steam is shot down through the outermost pipe to melt the sulfur. (The steam has to be superheated because sulfur doesn't melt until 239°F [115°C], which is hotter than "regular" steam.) Then, compressed air is shot down the innermost pipe, which forces the liquid sulfur up and out the middle pipe.

Sulfur is also widely distributed in the form of **minerals** and ores. Many of these are in the form of sulfates, including gypsum (**calcium sulfate**, $CaSO_4 \cdot 2H_2O$), barite (**barium sulfate**, $BaSO_4$) and Epsom salts (**magnesium sulfate**, $MgSO_4 \cdot 7H_2O$). Others are **metal** sulfides, including **iron** pyrites (iron sulfide, FeS_2), galena (**lead** sulfide, PbS), cinnabar (mercuric sulfide, HgS), stibnite (antimony sulfide, Sb_2S_3) and zinc blende (zinc sulfide, ZnS). The sulfur is recovered from these metal ores by roasting them—heating them strongly in air, which converts the sulfur to sulfur dioxide. For example,

$$2PbS \quad + \quad 3O_2 \quad \rightarrow \quad 2PbO \quad + \quad 2SO_2$$
lead   oxygen   lead   sulfur
sulfide   oxide   dioxide

Then the sulfur dioxide can go directly into the manufacture **sulfuric acid**, which is where more than 90% of the world's mined sulfur winds up.

Compounds of sulfur

Some sulfur is used directly as a **fungicide** and insecticide, in matches, fireworks, and gunpowder, and in the **vulcanization** of natural rubber. Most, however, is converted into a multitude of useful compounds.

Sulfur is in group 16 of the **periodic table**, directly below **oxygen**. It can form compounds in any one of its three major oxidation states: -2, +4 or +6.

Sulfuric acid is the parent acid of all sulfate compounds. Among the most important sulfates is **calcium** sulfate ($CaSO_4$), which occurs naturally as gypsum, alabaster and selenite. **Copper** sulfate is used as an agricultural insecticide and to kill **algae** in **water** supplies. Alums are double sulfates of **aluminum** and another metal such as potassium, chromium or iron. The most common alum is **potassium aluminum sulfate**, $KAl(SO_4)_2 \bullet 12H_2O$. In water, it makes a gelatinous, goopy precipitate of **aluminum hydroxide**, $Al(OH)_3$, which, when it sinks to the bottom, carries along with it all sorts of suspended dirt, leaving behind clear water. Alum is therefore used in water purification.

Hydrogen sulfide is the parent acid of the sulfides, a family of compounds that contain nothing but sulfur and a metal. This family includes many metal ores, including iron pyrites, galena, cinnabar, stibnite and zinc blende, as mentioned above, plus sulfides of copper and silver. Other sulfides are used in leather tanning, as **pesticides**, and in depilatories—creams that remove "unwanted hair" from the hides of cattle and from people.

Hydrogen sulfide itself is a foul-smelling gas. When eggs and certain other **animal** matter go rotten and putrefy, the stench is mostly hydrogen sulfide. It is often present in flatus—expelled intestinal gas. Hydrogen sulfide is extremely poisonous, but fortunately, it smells so bad that people don't hang around long enough to be overcome by it. Other very bad-smelling compounds of sulfur are the mercaptans, a family of organic sulfur-containing compounds. A mercaptan is the major ingredient in the aroma of **skunks**. A tiny amount of a gaseous mercaptan is deliberately added to the **natural gas** that is used for home heating and cooking, so that dangerous gas leaks can be detected by smell. Natural gas itself is odorless.

Resources

Books

Emsley, John. *Nature's Building Blocks: An A-Z Guide to the Elements*. Oxford: Oxford University Press, 2002.
Greenwood, N.N., and A. Earnshaw. *Chemistry of the Elements*. 2nd ed. Oxford: Butterworth-Heinemann Press, 1997.
Hancock P.L., and B.J. Skinner, eds. *The Oxford Companion to the Earth*. Oxford: Oxford University Press, 2000.
Kirk-Othmer Encyclopedia of Chemical Technology. 4th ed. New York: John Wiley & Sons, 1998.
Lide, D.R., ed. *CRC Handbook of Chemistry and Physics*. Boca Raton: CRC Press, 2001.

Robert L. Wolke

Sulfur cycle

Sulfur is an important nutrient for organisms, being an key constituent of certain amino acids, **proteins**, and other biochemicals. Plants satisfy their nutritional needs for sulfur by assimilating simple mineral compounds from the environment. This mostly occurs as sulfate dissolved in **soil water** that is taken up by roots, or as gaseous **sulfur dioxide** that is absorbed by foliage in environments where the atmosphere is somewhat polluted with this gas. Animals obtain the sulfur they need by eating plants or other animals, and digesting and assimilating their organic forms of sulfur, which are then used to synthesize necessary sulfur-containing biochemicals.

In certain situations, particularly in intensively managed agriculture, the availability of biologically useful forms of sulfur can be a **limiting factor** to the productivity of plants, and application of a sulfate-containing fertilizer may prove to be beneficial. Sulfur compounds may also be associated with important environmental damages, as when sulfur dioxide damages vegetation, or when acidic drainages associated with sulfide **minerals** degrade ecosystems.

Chemical forms and transformations of sulfur

Sulfur (S) can occur in many chemical forms in the environment. These include organic and mineral forms, which can be chemically transformed by both biological and inorganic processes.

Sulfur dioxide (SO_2) and sulfate (SO_4^{-2})

Sulfur dioxide is a gas that can be toxic to plants at concentrations much smaller than one part per million in the atmosphere, and to animals at larger concentrations. There are many natural sources of **emission** of SO_2 to the atmosphere, such as volcanic eruptions and forest fires. Large emissions of SO_2 are also associated with human activities, especially the burning of **coal** and the processing of certain **metal** ores. In the atmosphere, SO_2 is oxidized to sulfate, an **anion** that occurs as a tiny particulate in which the negative charges are electrochemically balanced by the positive charges of cations, such as ammonium (NH_4^+), **calcium** (Ca^{2+}), or **hydrogen** ion (H^+). These fine particulates can serve as condensation nuclei for the formation of **ice** crystals, which may settle from the atmosphere as rain or snow, delivering the sulfate to terrestrial and aquatic ecosystems. If the sulfate is mostly balanced by hydrogen ion, the **precipitation** will be acidic, and can damage some types of **freshwater** ecosystems.

Hydrogen sulfide (H_2S)

Hydrogen sulfide is another gas, with a strong smell of rotten eggs. Hydrogen sulfide is emitted from situations in which organic sulfur compounds are being decomposed in the absence of **oxygen**. Once H_2S is emit-

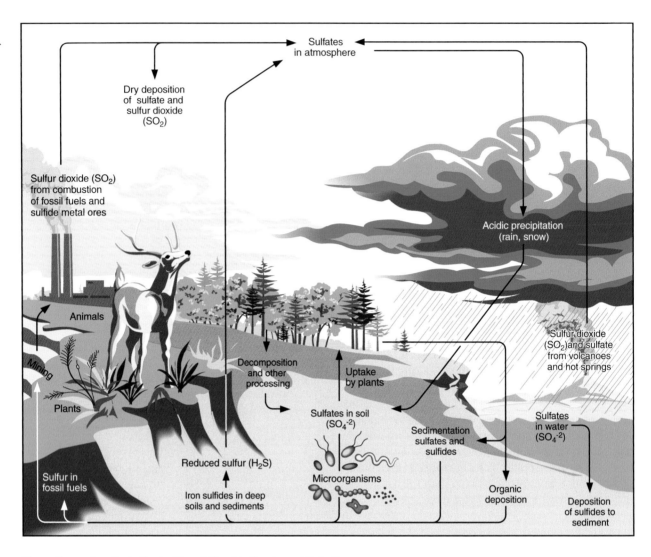

The sulfur cycle. *Illustration by Hans & Cassidy. Courtesy of Gale Group.*

ted to the atmosphere, it is slowly oxidized to sulfate, which behaves as described above.

Metal sulfides

Sulfur can occur in many chemically reduced mineral forms, or sulfides, in association with many metals. The most common metal sulfides in the environment are **iron** sulfides (such as FeS_2; these are called pyrites when they occur as cubic crystals), but all heavy metals can occur in this mineral form. Whenever metal sulfides are exposed to an oxygen-rich environment, certain **bacteria** begin to oxidize the sulfide, generating sulfate as a product, and tapping **energy** from the process which is used to sustain their own growth and reproduction. This autotrophic process is called chemosynthesis, and the bacteria involved are named *Thiobacillus thiooxidans*. When a large quantity of sulfide is oxidized in this way, an enormous amount of acidity is associated with the sulfate product.

Organic sulfur

Organic sulfur refers to an extremely diverse array of sulfur-containing organic molecules. These can range in weight and complexity from amino acids, proteins, and nucleic acids, to large molecules such as humic substances in soil or water. Organic sulfur is a chemically reduced form of sulfur. When it is exposed to an oxygen-rich atmosphere, bacteria oxidize the organic sulfur to derive energy, while liberating sulfate, usually in association with hydrogen ions and acidity.

Humans and the sulfur cycle

Human activities influence the rates and character of certain aspects of the sulfur cycle in important ways, sometimes causing substantial environmental damages.

Acid rain is a well-known environmental problem. Acid rain is ultimately associated with large emissions of

KEY TERMS

. . . .

Acid mine drainage—Surface water or groundwater that has been acidified by the oxidation of pyrite and other reduced-sulfur minerals that occur in coal and metal mines and their wastes.

Electrochemical balance—In an aqueous solution, the number of positive charges associated with cations must equal the number of negative charges of anions, so the solution does not develop an electrical charge.

sulfur dioxide to the atmosphere by human sources, such as oil- and coal-fired power plants, metal smelters, and the burning of fuel oil to **heat** homes. The SO_2 is eventually oxidized in the atmosphere to sulfate, much of which is balanced by hydrogen ions, so the precipitation **chemistry** is acidic. In addition, the vicinity of large point-sources of SO_2 emission is generally polluted by relatively large concentrations of this gas. If its **concentration** is large enough, the SO_2 can cause toxicity to plants, which may be killed, resulting in severe ecological damages. In addition, atmospheric SO_2 can be directly deposited to surfaces, especially moist soil, **plant**, or aquatic surfaces, since SO_2 can readily dissolve in water. When this happens, the SO_2 becomes oxidized to sulfate, generating acidity. This means a direct input of sulfur dioxide is called dry deposition, and is a fundamentally different process from the so-called wet deposition of sulfate and acidity with precipitation.

Acid mine drainage is another severe environmental problem that is commonly associated with coal and metal **mining**, and sometimes with construction activities such as road building. In all of these cases, physical disturbance results in the exposure of large quantities of mineral sulfides to atmospheric oxygen. This causes the sulfides to be oxidized to sulfate, a process accompanied by the generation of large amounts of acidity. Surface waters exposed to acid mine drainage can become severely acidified, to a **pH** less than 3, resulting in severe biological damages and environmental degradation.

Sulfur is also an important mineral commodity, with many industrial uses in manufacturing. Sulfur for these purposes is largely obtained by cleaning sour **natural gas** of its content of H_2S, and from **pollution control** at some metal smelters.

In a few types of intensively managed agriculture, **crops** may be well fertilized with **nitrogen**, **phosphorus**, and other **nutrients**, and in such cases there may be a deficiency of sulfate availability. Because sul-

fate is an important plant nutrient, it may have to be applied in the form of a sulfate-containing fertilizer. In **North America**, sulfate **fertilization** is most common in **prairie** agriculture.

Resources

Books

Atlas, R.M., and R. Bartha. *Microbial Ecology.* Menlo Park: Benjamin/Cummings, 1994.

Freedman, B. *Environmental Ecology.* 2nd ed. San Diego: Academic Press, 1995.

Bill Freedman

Sulfur dioxide

Sulfur dioxide, SO_2, is a colorless gas with an irritating, choking odor. It is produced naturally in volcanoes and in geysers, and is produced in ever-increasing amounts by the burning of sulfur-containing **coal**. Sulfur dioxide is a key contributor to **acid rain**. It is used in the manufacturing of **sulfuric acid**, in preserving many foods, as a disinfectant, and in the wine industry.

Sulfur dioxide is produced naturally inside the **earth** by the reaction of **sulfur** with **oxygen** gas. This accounts for the "sulfur" smell around active volcanoes and geysers. Sulfur is a natural component of most coal, and it is the burning of sulfur-containing coal that has contributed the most to the acid rain problem in the world. When burned, the sulfur in coal reacts with oxygen and makes sulfur dioxide gas. This gas may react with **water** to make sulfurous acid (H_2SO_3, a moderately strong acid), or it may react with oxygen to form sulfur trioxide, SO_3, which then reacts with water in the atmosphere to produce sulfuric acid, H_2SO_4, the major contributor to acid rain. This same series of reactions is used to make sulfuric acid, the most widely produced chemical in the world. Efforts to reduce the amounts of sulfur dioxide in the atmosphere use two approaches. Low sulfur coal can be used, reducing emissions. However, as we use up more and more of the low sulfur coal, we are forced to use the remaining higher sulfur content coal. Many factories now use *scrubbers*, which put sulfur dioxide gas into contact with **calcium oxide** (CaO) to produce solid **calcium** sulfite, thus preventing the sulfur dioxide from reaching the atmosphere. The reaction is shown below:

$$CaO + SO_2(g) \rightarrow CaSO_3$$

One drawback to this approach is that for every ton of sulfur dioxide that reacts, two tons of solid calcium sulfite are produced, and they must be disposed of.

Sulfur dioxide is used widely to prevent dried **fruits** (especially apricots and peaches) from becoming discolored. Sulfur dioxide is a strong reducing agent, and prevents the formation of discolored oxidation products in the fruits. It is also used to **bleach** vegetables and in the wine industry is used to prevent wines from discoloring and turning brown. In the food industry, sulfur dioxide is used as a disinfectant during the manufacturing process.

See also Volcano.

Sulfuric acid

Sulfuric acid, H_2SO_4, is a viscous (thick and syrupy), oily liquid which has for years been the most widely used chemical in the world. Over 100 billion lb (45 billion kg) of sulfuric acid are produced each year. It is also one of the least expensive acids, which makes it a favorite of industries around the world. It is used in the production of **fertilizers** and as an industrial catalyst (a substance which speeds up **chemical reactions**).

One of the major uses of sulfuric acid is in the production of fertilizers. Phosphate rock is treated with sulfuric acid to produce **water** soluble phosphates, which are essential for **plant** growth and survival. It is also the acid used in car batteries. **Automobile** batteries contain **lead**, lead oxide, and sulfuric acid. These lead storage batteries are used because they can not only provide the **electric current** needed to start a car, but can be recharged by the car's electrical system while the car is running.

Sulfuric acid is one of the major components of **acid rain**. **Coal** contains **sulfur** as a natural impurity and when coal is burned **sulfur dioxide** (SO_2) and sulfur trioxide (SO_3) gases are produced. Sulfur trioxide then reacts with water in the air, creating sulfuric acid. This acid rain can damage buildings, corrode **metal**, and destroy plant and **animal** life. Acid rain is an increasing problem not only in major industrialized nations, but also in neighboring countries that are downwind, since pollutants produced by a country do not stay in the air above that country.

One of the major industrial uses of sulfuric acid is as a dehydrating agent (a substance that removes water from other substances). Sulfuric acid is an extremely effective dehydrating agent. Upon contact with living **tissue** it kills cells by removing water from them.

Sumac *see* **Cashew family (Anacardiaceae)**

Sun

The Sun is the **star** at the center of our **solar system**. It has a diameter of about 420,000 mi (700,000 km) and a surface **temperature** of 9,981°F (5,527°C). Its visible "surface" is actually a thin gas, as is the rest of its atmosphere and interior.

The Sun shines as a result of thermonuclear fusion reactions in its core, and the **energy** produced by these reactions heats the gas in the Sun's interior sufficiently to prevent the weight of its own **matter** from crushing it. This energy also is the source of **heat** and life on **Earth**, and small variations in the Sun's energy output, or even in the features present in its atmosphere, may be sufficient to profoundly affect terrestrial climate. Although the Sun is by far the nearest star, the processes causing solar variability are still poorly understood and continue to challenge astronomers.

A brief history of solar observations

The Sun is about 90 million mi (150 million km) away from Earth. It is modest by stellar standards, although it is over 100 times larger than the earth and over 300,000 times more massive. Although it consumes more than half a billion tons of its nuclear fuel each second, it has been shining continuously for five billion years, and it will continue to shine for another five billion.

To the ancient Egyptians the Sun *was* a god, for anyone who felt its warmth and watched the renewal of **crops** it brought in the spring realized it was a bringer of life. Greek mythology recounts the tale of Icarus, who died in a brazen attempt to fly too close to the Sun.

Ancient dogma held that the Sun orbited Earth. It was not until 1543 that Nikolaus Copernicus (1473-1543) published the **heliocentric theory** of the Universe, in which the Sun lay at the center and Earth was relegated to a circular **orbit** around it. Copernicus refused to have this magnificent work published until near his death. This was a wise decision, since earlier publication would have resulted in his equally earlier death at the hand of Church authorities who regarded his work as heresy.

Around 1610, the newly invented **telescope** was trained on the Sun. The early solar observers, including Galileo (1564-1642), noticed that the surface of the Sun had dark spots on it. Not only that, the spots moved across the solar surface, leading Galileo to conclude that

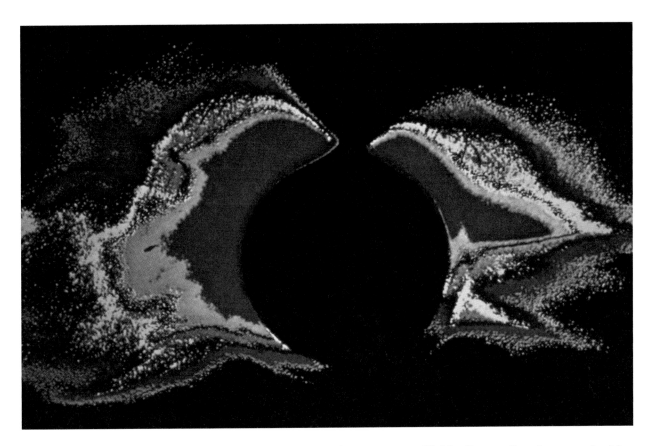

A photograph of the Sun taken with a coronagraph, a telescope that creates an artificial eclipse, so the solar corona is visible. The shape of the corona varies with time and in some areas is nearly absent. Much of its radiation is in the x-ray and extreme UV range, and the corona is many times hotter than the surface of the Sun itself. *U.S. National Aeronautics and Space Administration (NASA).*

the Sun rotated. This sent further shock waves through the religious world. The poor cardinals had hardly recovered from Copernicus's heresy, and now along came Galileo telling them that the backup embodiment of celestial perfection, the Sun, had ugly blemishes on its surface. Galileo got into serious trouble with the Church for this and other statements he made, and it would be 380 years before Pope John Paul II exonerated him.

In the nineteenth century an amateur German astronomer named Heinrich Schwabe (1789-1875) did some of the first scientific analysis of solar data. Observers had kept records over the years of the number of **sunspots** visible, and Schwabe noticed that the sunspot number rose and fell in a cyclic fashion, with a period of about 11 years. Later workers confirmed this activity cycle, and the sunspot number remains today an important piece of information about our star.

The nature of sunspots was first investigated by George Ellery Hale (1868-1938), who devoted his life to studying the Sun from the observatory he founded on Mt. Wilson near Los Angeles. Hale discovered that sunspots were cooler than the surrounding solar surface—which is why they appear dark—and that sunspots are associated with strong magnetic fields. Hale also discovered that the sun's magnetic field reverses itself with each 11-year cycle, so that there is a more fundamental 22-year magnetic activity cycle behind the sunspot cycle.

Since Hale's time, solar research has developed along two great lines of investigation. First, a huge observational database has been created. The national solar observing facilities of the United States are located at Sacramento Peak just east of Alamogordo, New Mexico, and at Kitt Peak, 30 mi (48 km) west of Tucson, Arizona. In addition to sunspots, these observatories have monitored prominences and flares, which are spectacular, eruptive events in the solar atmosphere, the solar magnetic field itself, the solar granulation, which is a result of the turbulent movement of the gas beneath the solar surface, and the total energy output of the Sun. Space-based satellites have studied the Sun in wavelength regimes, such as the ultraviolet and the x ray, not accessible from the ground. The result of these investigations is a detailed set of observations about the phenomena that occur on our star.

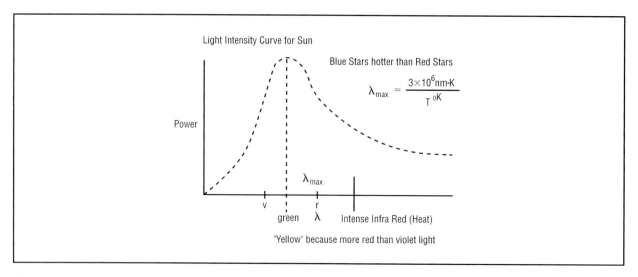

Illustration by Argosy. The Gale Group.

The other line of investigation is the theoretical analysis of processes within the Sun and in its atmosphere. Scientists have developed models to explain what is observed; if a model fails to fit the observations, it must be discarded. The goal of this work is to explain *why* the various features seen on the Sun appear. Why is there an 11-year-long **solar activity cycle**? What do the forces producing granulation have to do with the nature of the solar magnetic field? Where does the magnetic field originate? These questions, and many others, are only partially answered.

A fruitful line of research toward answering these questions involves observing not just the sun, but many other stars like the Sun. You would have a difficult time understanding what the human race was like by observing only one person; similarly, it is hopeless to try to understand the complex nature of the stars by observing only one of them. Long-term observations of solar-like stars have been carried out at the Mount Wilson Observatory near Los Angeles and at Lowell Observatory in Flagstaff, Arizona. Understanding the so-called "solar-stellar connection," derived from observations of other stars, may yield important clues about the processes at work in our own star.

Two million years after our earliest ancestors looked up and shielded their eyes from the brilliant, life-giving object in the sky, we still have only a rudimentary understanding of it.

A journey through the Sun

The solar furnace

At the sun's core, the temperature is 26,999,541°F (14,999,727°C). The matter here has a **density** of rough-

ly 2.2 lb (1 kg) per cubic centimeter—about 150 times the density of **water**. It is compressed to this degree by the crushing weight of all the matter between it and the surface—it is about 210,000 mi (350,000 km)—to the surface.

There are no **atoms** in the core. No atom (a nucleus of protons and neutrons, orbited by electrons) could exist in this inferno. There is nothing but a swirling sea of particles. We know from **physics** that the hotter a medium is, the faster its particles move. In the Sun's core the protons race around at blinding speeds, and because they are so tightly packed, they are constantly crashing into one another.

What is a **proton**? It is a **hydrogen** ion-a hydrogen atom that has had its sole **electron** stripped away. The sun is made mostly of hydrogen. In the cool regions of its atmosphere the hydrogen exists as atoms, with a single electron bound to the proton; in the core there are only the ions.

Four hydrogen nuclei smash together in a quick series of collisions, with catastrophic **force**. So violent are these collisions that the protons' natural tendency to repel one another—they all have the same positive charge—is overcome. When the various interactions are over, a new particle has emerged: a *helium* nucleus containing two protons and two neutrons. The helium nucleus gets its share of battering by the other particles, but it is larger and tightly bound together, and even the maelstrom cannot disrupt it.

There is one more product of this fusion reaction. In the series of collisions leading to the formation of the helium nucleus, two particles called *photons* are produced. A **photon** is a bundle of electromagnetic **radiation**, also

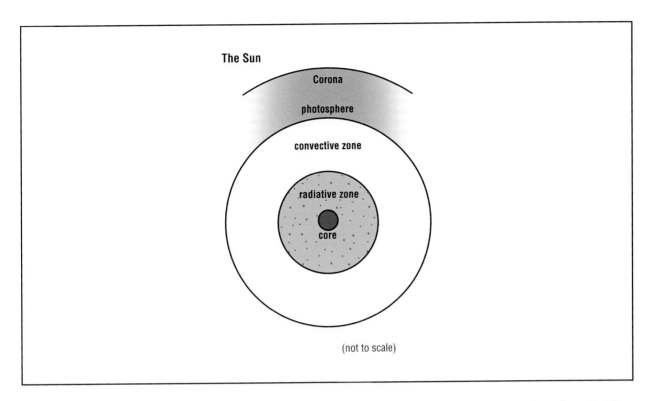

The Sun

Corona

photosphere

convective zone

radiative zone

core

(not to scale)

Nuclear fusion and the production of gamma rays, takes place in the core of the Sun. The radiative zone is so dense that it can take a million years for a photon to pass through. *Illustration by Argosy. The Gale Group.*

known as a ray of **light**. The photons race away from the Sun's core at the incredible speed of 180,000 mi (300,000 km) per second, the speed of light.

Toward the surface

Photons do not travel to the surface in a straight line. They constantly hit other particles, bouncing off them in a **random** direction. Sometimes an atom absorbs its energy, only to re-emit it a fraction of a second later in a different direction. This is the so-called random walk, and it describes how photons work out from the Sun's core.

Then toward the surface, the temperature, **pressure,** and density of the gas drop. There is not as much weight compressing the gas, so it does not need to be at as high a pressure to support the material above it. Lower pressure means lower temperature and density.

Halfway from the Sun's core to its surface, we are in the zone of *radiative* energy transport, where uncountable trillions of photons flow away from the Sun's core where they were produced. As they flow past, new photons, freshly created in the core, flow into it from below.

Into the convection zone

A little more than two-thirds of the way to the surface, gas cools to 179,492°F; (99,700°C). Instead of in-

dividual particles, atoms exist, which are capable of absorbing the photons rather than simply scattering them in a different direction. Photons have difficulty flowing through this cool gas. As they get absorbed, new photons flow into the gas from below, heating it even more. The gas begins to overheat. As a result, energy transport is now more efficient if a huge bubble of hot gas forms and begins to rise toward the surface. This is called *convection*, and we are now in the Sun's zone of convective energy transport.

A hot gas bubble rises into progressively cooler gas, releasing heat into smaller bubbles reaching the very cool region just below the Sun's surface. The bubbles release their pent-up heat. With their heat gone, they are now cooler than their surroundings, so they sink back into the Sun's interior, to pick up more heat and begin the convective cycle anew.

In the atmosphere

It takes about 30,000 years for a photon to reach the Sun's surface. Had the photon gone in a straight line, it would have reached the surface in just over one second, but 10 billion trillion interactions with matter in the Sun's interior slowed it considerably.

At the surface is a thin (300 mi/500 km) layer of matter called the photosphere. The temperature here is

9,981°F; (5,527°C), and for the first time, a photon of visual light—that is, with a wavelength that places it in the visual portion of the spectrum—has a chance of escaping directly to outer space. The density of the gas is now so low that it is nearly a **vacuum**, thousands of times less dense than air, and so little matter is left that photons escape with no further interactions.

The photosphere is a seething region of hot, rising granules and cooler, sinking ones. In places there are great, dark spots, perhaps 6,200 mi (10,000 km) across, where the temperature is only 6,692°F; (3,700°C) and where matter is constrained to flow along the intense and tangled lines of the strong magnetic fields that permeate the spots. (One phenomenon thought to contribute to the tangling of the solar magnetic field is the Sun's **rotation**. The Sun's equator rotates once every 26 days, its poles once every 36 days. This differential rotation contributes to twisting the magnetic fields and producing active features like sunspots.) The magnetic fields are invisible, but observations have revealed that they can arch high into the Sun's atmosphere, forming loops. Hot gas becomes confined in these loops, forming spectacular prominences. Violent rearrangements or eruptions in twisted magnetic fields result in flares, which spew matter and intense radiation into space. Some of this radiation may interrupt **radio** communications on Earth, while the particles will soon stream into Earth's atmosphere, causing aurorae.

Just above the photosphere, the temperature starts to climb, reaching 17,492°F; (9,700°C) a few thousand miles above the photosphere. This is the chromosphere. Most of it is ten million times less dense than air. The causes for the temperature rise are still not fully understood. One possibility is that mechanical energy from the **convection** zone—the energy associated with the **motion** of the gas—is deposited into the Sun's upper atmosphere, heating it. Because it is so thin and tenuous, the chromosphere is very faint, and under normal circumstances is invisible with the brilliant photosphere behind it. We can see the chromosphere by photographing the Sun with special filters sensitive to light that originates in the chromosphere, or during an eclipse, when the **Moon** blocks the photosphere and the chromosphere appears as a glowing ring girdling the solar limb.

Now 1,800 mi (3,000 km) above the photosphere, the temperature rises sharply—1,935,541°F; (19,727°C), then 179,541°F; (99,727°C), then 899,541°F; (499,727°C). A narrow transition region opens to the corona, an incredibly tenuous and hot—3,599,541°F; (1,999,727°C)—region extending 3,000,000 mi (5,000,000 km) above the photosphere. The corona is also very faint, and can only be observed in visible light with the photosphere blocked, as it is during an eclipse. Because the corona is so hot, it is also spectacular in x ray photographs, which can be obtained only from space-based observatories.

The solar wind

The end of the corona marks the last of the Sun's strong magnetic field regions. photons race into empty space.

There are also swarms of particles. There are only a few per cubic centimeter—an almost perfect vacuum. They are all part of the **solar wind**, a continuous stream of matter flowing away from the sun. Slowly, the Sun is losing material to space. The **rate** of this loss is very small, so it will not seriously affect the Sun's **evolution**. (Some stars, however, have powerful winds that can carry off a substantial fraction of their mass.) The solar **wind** permeates the entire solar system, and beyond.

A small blue planet

Eight minutes after a photon leaves the Sun's photosphere, it reaches Earth. Along with countless billions of other photons, it streams through Earth's atmosphere. The photon, a product of a hydrogen fusion reaction 30,000 years ago, has finished its trip. Obviously the Sun has a profound impact on doings here on Earth, but recent research suggests the connections may run deeper than initially thought.

The link may lie in the solar activity cycle, which is the periodic variation in active features such as sunspots, prominences, and flares, in the Sun's atmosphere and on its visible surface. The cause of the activity cycle is not well understood, but astronomers generally agree that the Sun's differential rotation, combined with the turbulent motions in its convection zone, create a magnetic dynamo that results in a perpetual tangling and rearrangement of the Sun's magnetic field. When magnetic field lines, which normally lie below the photosphere, become tangled and burst into the Sun's atmosphere, active features such as sunspots and prominences invariably form. When the magnetic field becomes tangled to a critical level, it rearranges and simplifies its configuration, and the amount of solar activity decreases correspondingly. The sunspot cycle typically has a length of about 11 years, but there is compelling circumstantial evidence that variations in the length of the solar activity cycle are closely related to changes in the global temperature, with shorter solar cycles corresponding to warmer temperatures on Earth.

And in the end, the sun will have its final and greatest impact on Earth. What of the core, which our photon left a million years ago? Five billion years from now, the seemingly countless hydrogen nuclei will all have been

KEY TERMS

. . . .

Chromosphere—The narrow middle layer of the Sun's atmosphere. It is about 17,492°F; (9,700°C) and is very faint relative to the photosphere.

Convection zone—The outermost third of the solar interior. Here heat is transported to the surface in giant convective bubbles of gas, which rise to the surface, release their heat, and then sink back into the interior to pick up more heat.

Core—The central region of the Sun, where thermonuclear fusion reactions take place.

Corona—The highest and hottest layer of the solar atmosphere. Matter in the corona may have a temperature of 3,599,541°F (1,999,727°C) and may be several million miles above the photosphere.

Photosphere—The lowest layer of the solar atmosphere, where most of the visible light is emitted. Because this is the layer we see in white light photographs, it is often called the solar "surface," even though it is a very thin gas.

Prominence—A large region of glowing gas suspended in magnetic fields, often arching far above the photosphere. Some prominences are *quiescent*, remaining for days, while others are *eruptive*, and dissipate violently.

Radiative zone—The central two-thirds of the solar interior. Here energy is transported by the flow of photons, or light waves, through the matter.

converted into helium "ash"—the Sun's fuel will be gone. To stave off destruction by the inexorable force of gravity, the sun's core will contract and heat to the point that the helium will ignite. In the process, the sun will expand into a **red giant star**, swallowing the innermost planet, Mercury, and turning Earth into a charred wasteland. But the helium is the last fuel reserve the sun will be able to use, and it will eject its outer layers, leaving behind only its collapsed core, a small, dying white dwarf.

See also Seasons; Solar flare; Star formation; Stellar evolution.

Resources

Books

Seeds, M.A. *Horizons, Discovering the Universe*. New York. Wiley, 1991.

Periodicals

Giampapa, Mark S. "The Solar-Stellar Connection." *Sky & Telescope* (August 1987): 142.

Pasachoff, J. "The Sun: A Star Close Up," *Mercury* (May/June 1991): 66.

Jeffrey C. Hall

Sun dog *see* **Atmospheric optical phenomena**

Sun pillar *see* **Atmospheric optical phenomena**

Sunbirds

Sunbirds are 105 **species** of small, lovely **birds** that make up the family Nectariniidae. Sunbirds occur in **Africa**, South and Southeast **Asia**, New Guinea, and **Australia**. They occupy a wide range of habitats, from **forests** and savannas to shrubby **grasslands**, and some agricultural habitats.

Sunbirds range in body length from 4-9 in (9-22 cm). The wings are short and rounded, and the tail is quite long in some species. They have a long, pointed, down-curved beak, which in some species exceeds the length of the head. The tongue is long, tubular for about two-thirds of its length, and its tip is split. The unusual bill and tongue of sunbirds are adaptations to feeding on the **nectar** of flowers. There are fine serrations near the tip of the beak, which are thought to be an **adaptation** to gripping **insects**.

Male sunbirds are brightly and garishly colored in bold patterns of green, blue, purple, red, black, or white. Many of these superb hues are due to a iridescence of the feathers, which is a prism-like, physical phenomenon, rather than the **color** of pigments. Female sunbirds are more subdued in their coloration, and are not iridescent. The sunbirds and **hummingbirds** (family Trochilidae) are not related, but they are similar in aspects of their iridescent coloration, and their feeding.

Sunbirds are active, strongly flying birds. They mostly feed on nectar, and also on insects. Sunbirds sometimes hover while feeding.

Sunbird nests are hanging, purselike structures, with a side entrance. The nest is constructed by the female of fibrous **plant** materials, partly held together using spider webs. The one to three eggs are incubated mostly by the female, but she is fed by the male while on the nest. Both parents care for the babies, which have a straight bill when born.

There are 78 species of sunbirds in the genus *Nectarinia*. Males of the superb sunbird (*Nectarinia superba*) of central Africa have a metallic-green back, a purple throat, and a burgundy breast, with black wings and tail. Clearly, this bird is appropriately named. Males of Gould's sunbird (*Aethopyga gouldiae*) of Southeast Asia have a blue head and tail, a crimson back, scarlet breast, and a lemon-yellow belly.

The spiderhunters (*Arachnothera* spp.) are greenish, less-garishly colored birds, and they have very long bills. The long-billed spiderhunter (*Arachnothera rubusta*) of Indochina and Southeast Asia is a spider- and insect-eating bird that inhabits montane forests.

Sundew *see* **Carnivorous plants**

Sunflower *see* **Composite family**

Sunspots

Sunspots are relatively dark, temporary spots that appear on the **Sun** from time to time. The largest of these spots are visible to the naked **eye** and have been noted by Chinese astronomers since antiquity, but their first mention in Western literature is in *The Starry Messenger* (1610) by Italian astronomer Galileo Galilei (1564–1642). Sunspot activity—the number of spots on the Sun at any one time—varies with a period of 11–13 years. This corresponds to the period of an overall solar-activity cycle whose other features include solar flares and prominences.

The solar cycle

At the beginning of an active period in the solar cycle, a few sunspots appear at the higher latitudes (i.e., near the poles). These are more or less stationary on the Sun's surface, but appear to us to move because of the Sun's axial **rotation**. Large spots—which may be large enough to sink many Earths in—may last for one or several solar-rotation periods of about a month each. As the solar cycle progresses, the number of spots increases and they tend to disappear at higher latitudes and appear at lower latitudes (i.e., nearer the equator). The end of the cycle is marked by a marked drop in the number of low-latitude sunspots, which is followed by the beginning of the next cycle as spots begin to appear again at high latitudes.

Sunspots, which usually occur in pairs aligned with the direction of the Sun's spin, correspond to places where intense magnetic fields emerge from or reenter the solar surface. Just as a bar magnet possesses a looping magnetic field connecting one end of the bar to the other, as revealed by **iron** filings scattered over a sheet of **paper** placed above the magnet, sunspot pairs possess a magnetic field that links them and along which charged particles align their **motion**. Scientists label the ends of a bar magnet "north" or "south" magnetic poles depending on their properties; similarly, each member of a sunspot pair corresponds to either a north or south magnetic pole. During a given solar cycle, the member of a sunspot pair that leads (i.e., is located toward the direction of the Sun's rotation) usually has the same magnetic polarity in a pair formed in a particular hemisphere. The order of polarity is reversed for sunspot pairs formed in the opposite hemisphere. The magnetic orientation spot pairs is preserved throughout one entire 11–13 year solar cycle; however, during the *next* cycle the order of leading and trailing polarities is reversed in both hemispheres. Thus the sunspot cycle actually consists of *two* 11–13 year cycles, since two cycles must pass before conditions are duplicated and the pattern can begin to repeat.

Sunspots and weather

Sunspot activity may be subtly linked to the earth's **weather**. Suggestive correlations between solar activity, global **temperature**, and rainfall have been observed, and analysis of tree-ring data spanning centuries seems to show the presence of an 11–13 year cycle. There is also geological evidence that the solar cycle may have been affecting terrestrial weather since Precambrian times. However, all these data have been disputed on statistical grounds, and there presently no consensus among scientists as to whether sunspots actually affect the earth's weather or not, or if so, how. The **energy** output of the Sun varies very little over the solar cycle (i.e., by about 0.1%), and some scientists doubt whether such slight changes can really affect the troposphere (lower atmosphere) of the **earth**, where **precipitation** occurs. A possible mechanism for amplifying the effects of the solar cycle on tropospheric weather is its influence on the *stratosphere* (the region of the atmosphere from an altitude of about 10 mi [16 km] to about 30 mi [50 km]). The stratosphere is home to the **ozone** layer, a diffuse shield of triatomic **oxygen** (O_3) that is an efficient absorber of ultraviolet **radiation**. Since the Sun's ultraviolet output varies 10 times more over the solar cycle than its overall radiation output, it is plausible—and has been confirmed by observation—that the temperature (and thus **volume**) of the stratosphere will vary significantly with the solar cycle. (Those involved in the launching and maintenance of Earth satellites are acutely aware that the upper layers of the earth's atmosphere respond to solar activity by expanding and thereby inflicting increased drag on satellites in low orbits.) However, the troposphere is many thousands of times more massive than the stratosphere, and scientists continue to investigate the question of whether temperatures in the frail film of the stratosphere can measurably affect surface weather.

One suggestion of a sunspot-weather link comes from historical records. There was a curious period of about 75 years shortly after Galileo's discovery of sunspots when few were observed. This era is called the **Maunder minimum** after the astronomer who first noted its existence. Other phenomena such as the aurora borealis (northern lights) that are associated with solar activity are also missing from European records during this period. The interval is also associated with what has long thought to have been time of unusually severe winters in both **Europe** and the **North America** that is sometimes termed the Little **ice** Age. However, there is now doubt as to whether the Little Ace Age ever happened at all, at least on a global level. Most of the evidence for its occurrence is anecdotal, and most comes from Western Europe. It is unlikely that historical evidence of this kind will resolve the scientific dispute over whether the solar cycle (or anomalies therein) significantly affect terrestrial weather.

Why sunspots are dark

The strong magnetic field in a sunspot, which is several thousand times stronger than that at the surface of the earth, accounts for the relative dimness of the spot. The hot atmosphere of the Sun contains a significant number of **atoms** having a net positive charge resulting from collisions between them (i.e., they are ionized, having each lost one or more electrons). Moving charged particles tend to **spiral** along magnetic field lines. The magnetic field lines passing vertically through a sunspot therefore tend to suppress the **convection** (heat-driven vertical circulation) that usually transports **heat** to the Sun's surface from its depths; convection requires horizontal motion so that material can reverse its direction, but the strong vertical magnetic field in the sunspot hampers cross-field (horizontal) motion of the electrically charged gas of the solar atmosphere. With less heat being supplied from below, a sunspot cools to about 3,000K (versus about 5,800K over the Sun's normal surface). Because of its lower temperature, the spot's interior is relatively dark—about 40% dimmer than the rest of the Sun's surface. Close inspection of a sunspot shows it to have a dark central region called the umbra (Latin for "shadow") surrounded by a lighter, radially structured region called the penumbra ("almost-shadow"). These regions are created by structural differences in the magnetic field responsible for the sunspot. Further, many sunspots are surrounded by bright rings. It is likely that the heat energy which cannot convect upward into the sunspot leaks up instead around its edges, superheating the material there.

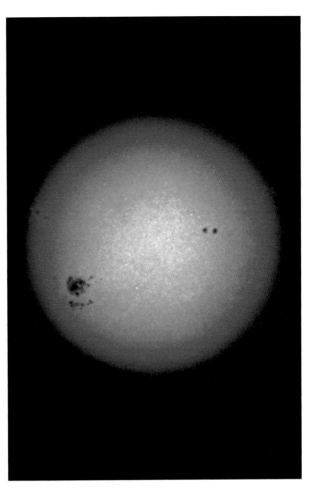

A visible light image of the Sun, showing large sunspots.
U.S. National Aeronautics and Space Administration (NASA).

Causes

In the second half of the twentieth century a mechanism of sunspot formation was proposed which accounts for much of their observed behavior. To begin with, the Sun does not rotate as a rigid body; the polar regions rotate somewhat more slowly than the equator. (The reason for this is still not known.) Because the solar material is electrically charged, the Sun's overall magnetic field is dragged along with the solar rotation; because the solar rotation is faster at the equator, the field will be dragged faster at the equator than at the poles. Although the overall magnetic field of the Sun is weak (i.e., similar to that of the earth), this differential rotation both distorts and intensifies it over time. The faster-rotating regions of the equator drag the local magnetic field so that the field lines are drawn out into long, thin tubes; the more these tubes are stretched, the more intense the magnetic field within them becomes. As the magnetic tube breaks the surface of the Sun (and returns into it, as all magnetic field lines form closed loops), it forms two spot-like structures. As the field direction is out of the solar surface at one spot and into it at the other, one of these spots will have act as north magnetic pole and the other will act as a south magnetic pole. The global nature of the general solar field is what guarantees that the stretched magnetic tubes will yield leading spots with opposite polarities in opposite hemispheres. A reversal of the Sun's general field between 11–13 year cycles would account for the observed periodic reversal of this order; however, there is no compelling explanation of why the general field should reverse after each 11–13 year solar cycle. Nevertheless, this relatively simple model does provide a beginning basis for understanding sunspots.

Resources

Books

de Pater, Imke, and Jack J. Lissauer. *Planetary Sciences.* Cambridge, UK: Cambridge University Press, 2001.

Prialnik, D. *An Introduction to the Theory of Stellar Structure and Evolution.* Cambridge: Cambridge University Press, 2000.

Periodicals

Kerr, Richard A. "Link Between Sunspots, Stratosphere Buoyed." *Science* 5412 (April 9, 1999): 234-235.

Rast, M.P., et al. "Bright Rings around Sunspots." *Nature* 401 (October 14, 1999): 678-679.

Solanki, S.K., M. Schüssler, and M. Fligge. "Evolution of the Sun's Large-scale Magnetic Field Since the Maunder Minimum." *Nature* 408 (November 23, 2000): 445-447.

George W. Collins

Superclusters

Superclusters are currently the largest structures known in the universe. Stars and clusters of stars group together into galaxies that can contain anywhere from a few million to a few trillion stars. Galaxies collect into groups known as clusters of galaxies. On a larger scale superclusters are clusters of clusters of galaxies. As clusters of galaxies group into superclusters they leave empty spaces called voids between the superclusters. Superclusters and voids typically extend for hundreds of millions of light-years.

History

In 1924 Edwin Hubble (after whom the **Hubble Space Telescope** was named) proved that there were indeed galaxies outside the **Milky Way**. The groupings of galaxies in the sky were so obvious that the existence of clusters of galaxies was accepted immediately. The existence of superclusters is less obvious. Finding the distances to galaxies, and their three dimensional distribution in space is difficult. Most astronomers accepted Hubble's word that superclusters did not exist. Clyde Tombaugh, who discovered **Pluto**, once showed Hubble his first **map** of a supercluster. Hubble refused to believe it so the idea was ignored.

Gerard deVaucouleurs was not so easily discouraged. In the early 1950s he suggested that our **galaxy** and its cluster, the Local Group, are at the edge of a much larger group. This larger group is now known as the Local Supercluster. DeVaucouleurs' suggestion was initially "received with resounding silence" (his words) but after 25 years his view became widely accepted.

By the 1970s, improvements in instrumentation allowed astronomers to measure distances to a large number of galaxies and work out the three dimensional structure of the universe. The first definite supercluster was mapped out by Stephen Gregory, Laird Thompson, and William Tifft in the 1970s. They found the Coma Supercluster extending about 100 million light-years. Two clusters, the Coma cluster and A1367 (the 1367th cluster in a catalog by Abell), were bridged together by galaxies and small clusters. The trio also unintentionally found the first void in front of the Coma supercluster. Astronomers quibble about the details of superclusters but most accept the fact that the large scale structure of the universe includes superclusters and relatively empty voids.

Large scale structures

Since the 1970s large scale structure surveys have provided a picture of the superclusters and voids in the universe. All major known clusters of galaxies and at least 95% of all galaxies are found in superclusters. The voids between superclusters may contain faint galaxies but no bright galaxies. The voids tend to be spherical but superclusters are not. Superclusters have long filamentary or sheet-like structures that provide the boundaries for the voids.

KEY TERMS

. . . .

Cluster of galaxies—A group of galaxies that is gravitationally bound.

Galactic (open) cluster—A cluster of roughly a few hundred young stars in a loose distribution. This is not the same thing as a cluster of galaxies.

Galaxy—A large collection of stars and clusters of stars containing anywhere from a few million to a few trillion stars.

Light year—The distance light travels in one year, roughly 6 trillion mi (9.5 trillion km).

Supercluster—A connected group of clusters of galaxies that may extend for hundreds of millions of light years.

Void—A region of space extending for hundreds of millions of light years that contains few, if any, galaxies.

Our galaxy is located at one end of the Local Group—a small cluster of galaxies with the Andromeda galaxy at the other end. The Local Group is near the outskirts of the Local Supercluster which has a diameter of 100 million light-years. This somewhat flattened supercluster consists of two major hot dog-shaped groups of galaxies. It contains a total of 10^{15} times the **mass** of the **Sun**, most of which is concentrated into 5% of the **volume** of the supercluster.

The biggest supercluster is the Perseus-Pegasus Filament discovered by David Batuski and Jack Burns of New Mexico State University. This filament contains the Perseus supercluster and stretches for roughly a billion light years. It is currently the largest known structure in the universe.

Margaret Geller and John Huchra have mapped a region that is 500 million light-years long but only 15 million light years thick, but this area, the Great Wall, has not been completely mapped. They may find it extends longer than 500 million light-years when the mapping is complete.

A number of other superclusters and voids are known but astronomers have only mapped the large scale structure of a small part of the universe. They need to do much more work before we completely understand the structure of superclusters and voids. There may even be larger scale structures.

One of the pressing issues related to superclusters is how such structures formed in the early universe. This question remains unanswered but observations of the cosmic background radiation indicate that its beginnings developed even before galaxy formation.

Resources

Books

Bacon, Dennis Henry, and Percy Seymour. *A Mechanical History of the Universe.* London: Philip Wilson Publishing, Ltd., 2003.

Bartusiak, Marcia. *Thursday's Universe.* Redmond, WA: Tempus Books, 1988.

Greene, Brian. *The Elegant Universe: Superstrings, Hidden Dimensions, and the Quest for the Ultimate Theory.* New York: Vintage Books, 2000.

Hodge, Paul. *Galaxies.* Cambridge, MA: Harvard University Press, 1986.

Morrison, David, Sidney Wolff, and Andrew Fraknoi. *Abell's Exploration of the Universe.* 7th ed. Philadelphia: Saunders College Publishing, 1995.

Zeilik, Michael, Stephen Gregory, and Elske Smith. *Introductory Astronomy and Astrophysics.* Philadelphia: Saunders, 1992.

Paul A. Heckert

Superconductor

A superconductor is a material that exhibits zero resistance to the flow of electrical current and becomes diamagnetic (opaque to magnetic fields) when cooled to a sufficiently low **temperature**.

An electrical current will persist indefinitely in a ring of superconducting material; also, a magnet can be levitated (suspended in space) by the magnetic field produced by a superconducting, diamagnetic object. Because of these unique properties, superconductors have found wide applications in the generation of powerful magnetic fields, magnetometry, magnetic shielding, and other technologies. Many researchers are seeking to devise "high-temperature" superconductors—materials that superconduct at or above the **boiling point** of **nitrogen** (N_2), 77 K—that can carry large amounts of current without lapsing from the superconducting state. Such materials are already increasingly useful in power transmission and other applications.

Superconductivity history and theory

Superconductivity was first discovered in 1911 by Dutch physicist Heike Kamerlingh Onnes (1853–1926). After succeeding in liquefying helium (He), Onnes observed that the **electrical resistance** of a mercury filament dropped abruptly to an experimentally undetectable value at a temperature near -451.84°F (-268.8°C, 4.2K), the boiling point of helium. Onnes wrote: "Mercury has passed into a new state, which, because of its extraordinary electrical properties, may be called the superconductive state."

The temperature below which the resistance of a material = zero is referred to as the superconducting transi-

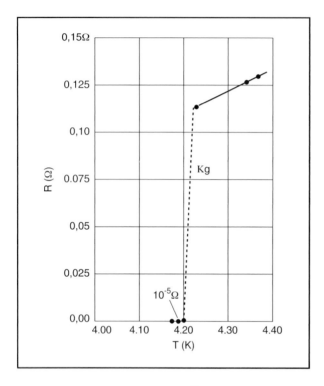

Figure 1. The curve of resistance R(Ω) versus temperature T (K) of a mercury filament (after H. Kamerlingh Onnes, 1911). *Illustration by Hans & Cassidy. Courtesy of Gale Group.*

tion temperature or the critical temperature of that material, T_c. Another unique characteristic of superconductors is their diamagnetic property, which was discovered by German physicist W. Meissner (1882–1974), working with a graduate student, in 1933. When a superconducting object is placed in a weak magnetic field, a persistent supercurrent or "screening current" is set up on its surface. This persistent current induces a magnetic field that exactly **mirrors** or cancels the external field, and the interior of the superconductor remains field-free. This phenomenon is called the Meissner effect and is the basis of the ability to of superconducting objects to levitate magnets. (Levitation only occurs when the **force** of repulsion of the magnetic field, which is a **function** of the field's intensity, exceeds the weight of the magnet itself.)

Superconductors are categorized as *type I* (soft) and *type II* (hard). For type I superconductors (e.g., most pure superconducting elements, including **lead**, tin, and mercury), diamagnetism and superconductivity break down together when the material is subjected to an external magnetic field whose strength is above a certain critical threshold H_c, the thermodynamic critical field. For type II superconductors (e.g., some superconducting alloys and compounds such as Mb_3Sn), diamagnetism (but not superconductivity) breaks down at a first threshold field strength H_{c1} and superconductivity persists until a higher threshold H_{c2} is reached. These properties arise from differences in the ways in which microscopic swirls or vortices of current tend to arise in each particular material in response to an external magnetic field.

No unified or complete theory of superconductivity yet exists. However, the basic underlying mechanism for

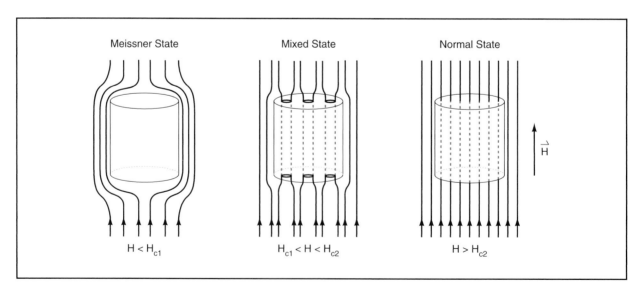

Figure 2. A type-II superconductor in various states under the application of magnetic field. *Illustration by Hans & Cassidy. Courtesy of Gale Group.*

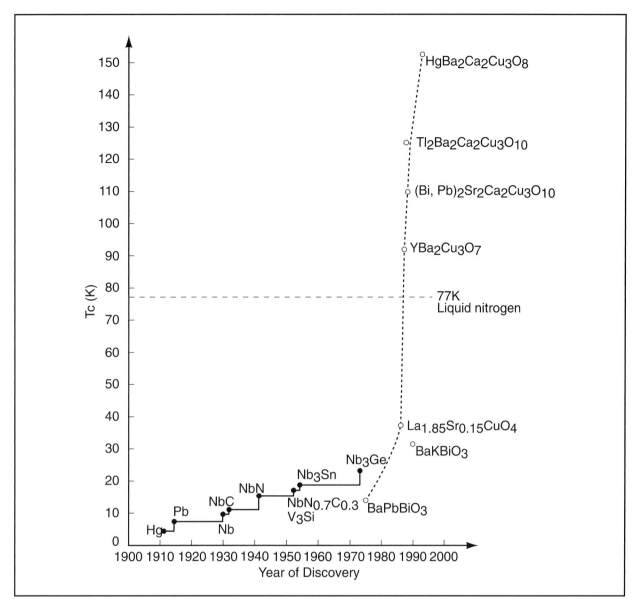

Figure 3. Increase in T_c with time since superconductivity was discovered. *Illustration by Hans & Cassidy. Courtesy of Gale Group.*

superconductivity has been suggested to be an electron-lattice interaction. U.S. physicists John Bardeen (1908–1991), Leon Cooper (1930–), and Robert Schrieffer (1931–) derived a theory (termed the BCS theory, after their initials) in 1957, proposing that in the lattice of **atoms** comprising the material, pairing occurs between electrons with opposite **momentum** and spin. These **electron** pairs are called Cooper pairs, and as described by Schrieffer, they condense into a single state and flow as a totally frictionless "fluid." BCS theory also predicts that an **energy** gap—energy levels a discrete amount below those of normal electrons—exists in superconductors. English Brian Josephson (1940–), in 1962, proposed that Cooper pairs could tunnel from one superconductor to

another through a thin insulating layer. Such a structure, called a Josephson junction, has for years been fabricated widely for superconducting electronic devices.

High-temperature superconductors

Before 1986, although a variety of superconductors had been discovered and synthesized, all had critical temperatures at or below the boiling point of He (e.g., Pb at -446.5°F [-265.8°C, 7.19K] and Nb_3Sn at -426.91°F [-254.95°C, 18.05K]). Since expensive refrigeration units are required to produce liquid He, this strictly limited the circumstances under which it was economical to apply superconductivity.

A small cube magnet hovering over a nitrogen-cooled specimen of a superconducting ceramic demonstrates the Meissner effect. Walther Meissner observed in 1933 that a magnetic field does not penetrate a superconductor. Magnetic lines of force loop around them, so the magnet is neither attracted nor repelled but instead floats above. Because the effect is so distinctive, it is used as a test for identifying superconducting materials. *Photograph by Yoav Levy. Phototake NYC. Reproduced by permission.*

The first superconductor to be discovered having T_c > -320.8°F (-196.0°C, 77K) (the boiling point of liquid N_2, which is much cheaper to produce than liquid He) was $YBa_2Cu_3O_7$ ($T_c \sim$ -294°F [-181°C, 92K]). The Y-Ba-Cu-O compound was discovered by U.S. physicist C. W. Chu (1948–), working with a graduate student, in 1987 following the 1986 discovery by German physicists G. Bednorz (1950–) and K. A. Müller (1927–) of the La-Ba-Cu-O oxide superconductor (T_c = -394.6°F [-237.0°C, 36K]). One year later, in 1988, bismuth-based (e.g., $(Bi, Pb)_2Sr_2Ca_2Cu_3O_{10}$, T_c = -261.4°F [-163.0°C, 110K) and thallium-based (e.g., $Tl_2Ba_2Ca_2Cu_3O_{10}$, T_c = -234.4°F [-148.0°C, 125K]) superconductors were successfully synthesized; their T_c's were some 20K higher than that of Y-Ba-Cu-O. Very recently, mercury-based cuprates ($HgBa_2Ca_{n-1}Cu_nO2_{n+2+\Delta}$) have been shown to have T_c values higher than 130K. These oxide superconductors are now classified as the high-temperature (or high-T_c) superconductors (HTSCs).

All HTSCs so far discovered have an atomic structure that consists of thin planes of atoms, many of which consist of the compound **copper** dioxide (CuO_2). This compound is, so far, uniquely important to producing the property of high-temperature conductivity. Ironically, CuO_2 is a Mott insulator, meaning that at temperatures approaching **absolute zero** it begins to behave as an insulator (a substance having very high resistance) rather than as a conductor: yet at higher temperatures, embedded in an appropriate **crystal** matrix, it is key to the production of *zero* resistance (superconduction).

Current flow in the CuO_2 family of HTSCs has directional properties. That is, the critical current **density**, J_c—the largest current density that a superconductor can carry without lapsing into finite resistivity—along the CuO_2 **plane** direction is orders of magnitude higher than at right angles to it. For HTSCs to carry a large amount of current, this implies that individual crystalline grains in bulk conductors (e.g., wires or tapes) of HTSC material should be well aligned with the current transport direction. Significant grain misorientation and **chemistry** inhomogeneity at grain boundaries can form weak links between neighboring grains and thus lower local J_c values. Several bulk material manufacture technologies, such as the melt-powder-melt-growth (MPMG) method for Y-Ba-Cu-O and the oxide-powder-in-tube (OPIT) process for Bi- and Tl-based superconductors, have been demonstrated to develop textured bulk structures. J_c values of 10^6–10^8 amperes per square centimeter at 77K have been achieved over small distances. For thin-film growth, dual ion beam sputtering (DIBS), molecular beam epitaxy (MBE), pulsed-laser deposition (PLD), and metal-organic chemical vapor deposition (MOCVD) have been shown to be successful methods. By optimizing processing temperature and **pressure** and using proper **buffer** layers, epitaxial HTSC films can be deposited on templates of other crystalline materials such as Si and MgO. With the integration of Si-based microelectronics processes, HTSC thin-film devices can be fabricated for a variety of applications.

Superconductivity applications

Superconductivity applications fall into two main areas—electromagnets (magnets whose **magnetism** depends on an externally-powered current passing through a winding) and **electronics**. In electromagnets, superconducting windings have much lower power consumption than do conventional copper windings, and thus are particularly attractive for high-field applications. Superconducting magnets can be used in magnetic **resonance** imaging, magnetic sorting of metals, **magnetic levitation** trains, and magnetic shielding. For power utility applications, superconductors are promising for magnetic energy storage, electrical power transmission, motors, and generators. They are also useful as the coatings for radio-frequency cavities. In electronic applications, thin-film interconnections and Josephson junctions are two key elements. Superconductors offer fast switching speeds and reduced wiring delays so that they are applicable for logic devices and memory cells. Superconducting field-effect transistors and Josephson junction integrated circuits have been demonstrated. At the temperature of liquid nitrogen, 77K, superconductors can be further integrated with semiconductors to form **hybrid** devices. For sensor operation, superconducting quantum **interference** devices (SQUIDs), based on the Josephson junction technology, are the most sensitive detector for measuring changes in magnetic field. For example, they can detect they very faint signals (on the order of 10^{-15} Tesla) produced by the human **brain** and **heart**. Also, SQUID-based gradiometry is a very powerful instrument for non-destructive evaluation of nonliving materials. The increased energy gap in HTSCs allows the fabrication of superconducting electromagnetic **radiation detectors** used for over the **spectrum** from x ray to the far infrared.

As time goes by, superconductors will find more and more applications. Recently, Y-Ba-Cu-O has been shown to be a good material for the top and bottom electrodes of oxide ferroelectric thin-film capacitors which exhibit fatigue resistance superior to that of capacitors with conventional Pt electrodes (used in dynamic random-access computer memories). This suggests that when the microstructures and the properties of HTSC materials can be well controlled and tailored, oxide superconductors are promising for many hybrid designs— designs incorporating both conventional and superconducting materials. We can also expect upcoming hybrid fabrication technologies. Processes for thin films, thick films, wires, and tapes may all be needed for the integration of a single superconductor-based instrument. Future growth in superconductors technology in electronic components, medical sensing, **geology**, military technology, transportation, and power transmission and storage is very promising, especially if, as researchers believe, transition temperatures and critical current densities can be significantly increased.

See also Electric circuit; Electromagnetism; Nanotechnology.

Resources

Books

Bennemann, K.H., J.B. Ketterson, and Joachim A. Kohler, eds. *Handbook of Superconductivity.* Springer-Verlag, 2003.
Ford, P., and George Saunders. *The Rise of Superconductors.* Taylor & Francis, 2003.

Periodicals

Orenstein, J., and A.J. Millis. "Advances in the Physics of High-Temperature Superconductivity." *Science* 280 (April 21, 2000): 468-474.

Pang-Jen Kung

KEY TERMS

· ·

Critical current density, J_c—The largest current density that a superconductor can carry nonresistively.

Diamagnetic—The material with a magnetization vector opposite to the applied magnetic field, which gives rise to a negative magnetic susceptibility.

Josephson junction—A structure with two superconductors separated by a thin insulating barrier.

Meissner effect—The expulsion of magnetic flux lines from a superconductor when the superconductor is cooled to a temperature below T_c.

Superconducting transition temperature, T_c—The highest temperature at which a given superconducting material remains superconducting.

Thermodynamic critical field, H_c—Minimum value of externally applied magnetic field that will cause breakdown of diamagnetism in a superconductor.

Superheavy element *see* **Element, transuranium**

Supernova

A *supernova* is the massive explosion of a **star**, and is one of the most violent events in the Universe. There are two types of supernovae. A Type I supernova happens when a dead star called a **white dwarf** accretes so much **matter** from a companion star that it becomes unstable and explodes. A Type II supernova occurs when a high-mass star runs out of thermonuclear fuel. In this case, the star's core collapses and becomes tremendously hard and rigid. The collapsing outer layers of the star bounce off the core and are flung outward with a burst of **energy** that can rival the output of an entire **galaxy**.

Guest stars

In A.D. 1054, a brilliant new star blazed into view into the **constellation** of Taurus, the Bull. The Chinese astronomers who observed it called it a "guest star," and at its peak brightness they could see it even during the day. Over the following months it gradually faded and disappeared. When we train our telescopes on the location of the former guest star, we see an angry-looking cloud of gas called the Crab Nebula. Several other "guest stars" appear in the historical record, usually separated by intervals of a few hundred years. They are therefore quite uncommon. Two appeared in A.D. 1572 and A.D. 1604, but after then astronomers would have to wait more than 380 years before the next one.

Studies of the Crab Nebula show that it is expanding, as if the enormous cloud of gas had been flung outward from a central point. Modern **stellar evolution** theory has provided a reason for this: the "guest stars" were the final acts in the lives of a massive star. These stars end their lives as supernovae, massive explosions that blast the stars' outer layers into **space**. For a short time they can rival the brightness of a small galaxy; later, like a dying **coal** in a fire, they fade away.

Types of supernovae

A *supernova* is the explosion of a star. In a single cataclysm, a massive star may blow itself to bits, releasing as much energy, for a brief time, as an entire galaxy. There are two types of supernovae.

A Type I supernova is the explosion and complete destruction of a dead star called a white dwarf. (The **Sun**, after it dies, will become a white dwarf.) If the white dwarf is made of **carbon** (the end product of the thermonuclear reactions that took place during the star's life), and if it is a member of a binary system, a Type I supernova can potentially occur.

Two important effects contribute to a Type I supernova. First, a white dwarf cannot be more massive than about 1.4 solar masses and remain stable. Second, if the white dwarf's companion star expands to become a red giant, some of its matter may be drawn away and sucked onto the surface of the white dwarf. If you could hover in a spacecraft at a safe distance from such a system, you would see a giant stream of matter flowing from the large, bloated star to its tiny companion, swirling into an **accretion disk** which then trickles onto its surface. A white dwarf that is almost 1.4 solar masses may be pushed over the critical **mass** limit by the constant influx of material. If this happens, the white dwarf will explode in an instantaneous nuclear reaction that involves all the mass of the star.

The popular image of a massive, supergiant star ending its life in one final, dazzling, swan song is what astronomers classify as a Type II supernova. In a Type II supernova, a massive star runs out of thermonuclear fuel and can no longer sustain itself against the inward pull of its own gravity. In a matter of seconds, the star collapses. The core is crushed into a tiny object called a **neutron star**, which may be no more than 6 mi (10 km) across. The outer layers collapse as well, but when they encounter the extremely hard, rigid, collapsed core, they

bounce off it. An immense cloud of glowing gas **rushes** outward, and some of the nebulae visible in small telescopes are these dispersed outer layers of stars.

Astronomers can tell the type of a supernova by observing its total brightness as well as its **spectrum**. Type I supernovae release more energy and therefore have a lower absolute magnitude (about -19 at peak brightness, which is as bright as a small galaxy). Since a Type I supernova is the explosion of a dead star made largely of carbon, there is little evidence in its spectrum for the element **hydrogen**. Type II supernovae, however, have prominent hydrogen lines in their spectra, for hydrogen is the primary element in the exploding star.

Type I supernovae are useful to astronomers trying to determine the distances to other galaxies, which is a very difficult task. Since all Type I supernovae have about the same absolute brightness, astronomers can calculate how far away a Type I supernova is by measuring its apparent brightness and then calculating how far away it must be to appear that bright. Type I supernovae therefore serve as one of several kinds of *distance indicators* that help us determine the size of the Universe.

Why a supernova explodes

If a Type II supernova is the collapse of a massive star, why does a huge explosion result? We do not have the answer, partly because of the extreme physical conditions that exist in the temperatures (about a trillion degrees) and pressures in a collapsing stellar core, and partly because everything happens very rapidly. The more rapidly the situation is changing, the more difficult it is to simulate it on a computer.

The usual explanation is that the outer layers "bounce" off the collapsed core. Try this experiment: hold two superballs, one larger than the other, about 5 ft (1.5 m) off the floor. Hold the smaller superball so that it is on top of and touching the larger one. Drop them simultaneously, so they fall with the little ball just above the big one (this takes some practice). If you do it right, the large ball (the "core") will stop dead on the floor, and the little ball (the "outer layers") will be flung high in the air. Something akin to this is thought to happen in a supernova. As the core collapses, a shock wave develops as the material gets jammed together. The incredible energy involved blasts the star's outer layers far into space.

Supernova 1987A

Supernovae ought to happen in our galaxy about once every 30 years. Until 1987, however, no bright supernova had been seen since 1604. By no means were we "due for one"—that is an all-too-frequent abuse of statistics—but

KEY TERMS

. . . .

Core collapse—The sudden collapse of a star's central region when the star's last fuel reserves are exhausted. With no energy being produced to sustain it against its own gravity, the core collapses in a fraction of a second, triggering a supernova explosion.

Type I supernova—The explosion of a white dwarf in a binary system. Often the white dwarf, which is the dead remnant of a star originally about as massive as the Sun, has a stream of matter being dumped onto its surface by its companion. The white dwarf may be pushed past the limit of 1.4 solar masses, after which it will become unstable and explode.

Type II supernova—The explosion of a massive star that has run out of thermonuclear fuel.

nevertheless a supernova did explode on February 23, 1987. (To be precise, that was the date that **light** from the explosion first reached us; the actual explosion took place 170,000 years earlier.) It was not in our galaxy, but in one of the small **satellite** galaxies orbiting it, the Large Magellanic Cloud, visible in the southern hemisphere. It was a distance of 170,000 light-years from **Earth**, and became known as SN 1987A. The star that exploded was a blue supergiant, probably 20 times more massive than the Sun, and when it exploded it was visible to the naked **eye**. Supernova 1987A became one of the most studied events in astronomical history, as observations of the expanding blast revealed numerous exciting results.

As a star's core collapses, the protons and electrons in it are smashed together to form a **neutron**, and every time such a reaction happens, an evanescent particle called a *neutrino* is created. The neutrinos travel outward from the core at the speed of light, and they are created in such vast numbers that they carry off most of the energy produced during the supernova. Just before SN 1987A became visible, surges of neutrinos were indeed detected here on Earth. Since the neutrinos escape from the star before the visible explosion, the timing of the event provided important observational support for the idea that core collapse and the subsequent bounce of infalling outer layers drives some supernova explosions.

There are several stars in Earth's celestial neighborhood that are prime supernova candidates. Betelgeuse, the red supergiant that marks Orion's right shoulder, and Antares, the brightest star in Scorpio, are two notable examples.

Resources

Books

Mark, Hans, Maureen Salkin, and Ahmed Yousef, eds. *Encyclopedia of Space Science & Technology.* New York: John Wiley & Sons, 2001.

Seeds, M.A. *Horizons: Discovering the Universe.* Chap. 10. Wiley, 1991.

Periodicals

Filippenko, A. "A Supernova with an Identity Crisis." *Sky & Telescope* (December 1993): 30.

Naeye, R. "Supernova 1987A Revisited." *Sky & Telescope* (February 1993): 39.

Thorpe, A. "Giving Birth to Supernovae." *Astronomy* (December 1992): 47.

Jeffrey C. Hall

Surface tension

Surface tension is the result of the cohesive forces that attract **water** molecules to one another. This surface **force** keeps objects which are more dense than water (meaning they should not float) from sinking into it. The surface tension of water makes it puddle on the ground and keeps it in a droplet shape when it falls.

If you use a table fork to carefully place a **paper** clip on the surface of some clean water, you will find that the paper clip, although more dense than water, will remain on the water's surface. If you look closely, you will see that the surface is bent by the weight of the paper clip much as your skin bends when you push on it with your finger.

A **molecule** inside a **volume** of water is pulled equally in all directions by the other molecules of water that surround it. A molecule on the water's surface, on the other hand, is pulled by the molecules below it and to its sides. The net force on this surface molecule is inward. The result is a surface that behaves as if it were under tension. If a glob of water with an irregular shape is created, the inward forces acting on the molecules at its surface quickly pull it into the smallest possible volume it can have, which is a **sphere**.

A simple apparatus can be used to measure the forces on a liquid's surface. A force is applied to a wire of known length which forms a **circle** parallel to the surface of the water. The force balances the surface forces acting on each side of the wire. The surface tension of the liquid, g, is defined as the **ratio** of the surface force to the length of the wire (the length along which the force acts). For this kind of measurement, $g = F/2L$. The force, F, applied to the wire is that required to balance the surface forces;

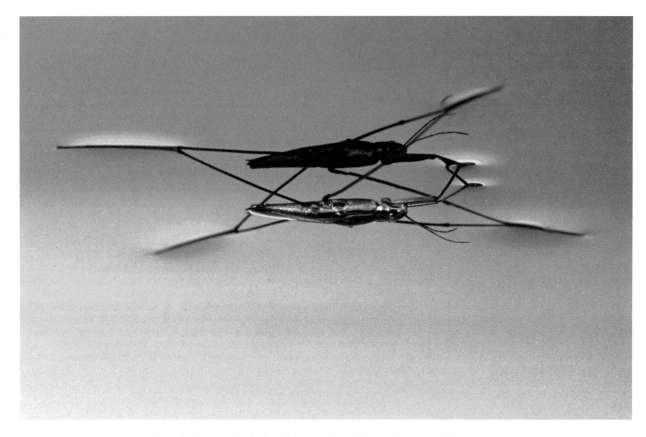

The long legs of a waterstrider (*Gerris paludum*) allow it to distribute its weight across the water and be supported by surface tension. © *Hermann Eisenbeiss, National Audubon Society Collection/Photo Researchers, Inc. Reproduced by permission.*

L is the length of the wire. The 2 appears in the denominator because there is a surface film on each side of the wire. Thus, surface tension has units of force per length—dynes/cm, newtons/m, ounces/inch, etc.

Water has a relatively high surface tension that, not surprisingly, decreases with increasing **temperature**. The increased kinetic **energy** of the molecules at higher temperature would oppose the forces of cohesion. In fact, at the boiling temperature, the kinetic energy of molecules is sufficient to overcome their cohesive forces of attraction and the molecules separate to form a gas.

Surgery

Surgery is the part of medicine which employs operative or manual treatment of **disease** or injury. Although surgery was practiced in ancient times, modern **anesthesia** was not developed until the nineteenth century. For centuries, most types of operative surgery involved high risk to patients due to **infection**. With the development of antiseptic surgical methods in the nineteenth century, the risks linked to surgery diminished. Some types of

surgery remain risky, but many have high rates of success. Advances in technological knowledge offer new horizons in surgery.

Ancient surgeons

Traditionally, wars have been the proving ground of surgeons and new types of surgery. Early surgeons developed methods to anesthetize their patients and tools to operate effectively. Because there was no global communication, many surgical advances remained geographically isolated. The Western tradition of medicine developed independently of traditions in India, **South America**, and elsewhere, although traders and others reported some medical advances.

One of the earliest operative practices was *trepanation*, or the making of a hole in the head. Evidence of trepanation is present in skulls of individuals who lived as long ago as 10,000 B.C. in areas ranging from Northern **Africa** to Tahiti, France, and Peru. Ancient surgeons appear to have performed trepanation for reasons ranging from the relief of **pressure** on the **brain**, due to head injury, to the release of demons. The practice is still used in some cases to relieve pressure on the brain.

Ancient Egyptians also practiced surgery. According to the Smith Papyrus, a document written about 1600 B.C., Egyptians were well acquainted with a variety of surgical problems. They set broken collarbones using splints, treated medical problems ranging from tumors to fractures, and treated wounds with an **adhesive** plaster.

Surgery was also mentioned in the Code of Hammurabi, the code of law developed in 1760 B.C. during the reign of the Mesopotamian King Hammurabi. The code called for strict discipline of surgeons. Surgeons who caused the death of a free man in surgery were to have their right hand amputated as punishment, while surgeons who damaged a slave had to repay the owner for the value of the slave.

Greeks and Romans used surgery as a last resort. The Greeks used herbs mixed in wine or **water** to dress and clean wounds, and performed surgical procedures ranging from trepanation to the **cauterization** of **blood** vessels.

The Greeks used a number of natural anesthetics, including opium and the mandrake, a **plant** which can have a **narcotic** affect. A first century Greek physician in Nero's army wrote that after using mandrake, patients receiving surgery "do not apprehend the **pain**, because they are overborn with dead sleep." The Greeks also developed anatomical studies based on the dissection of human bodies. Dissection for the advance of medical knowledge was forbidden during the Roman Era and during the Middle Ages, and was not revived until the fifteenth century.

The Romans developed many new surgical procedures. Roman surgeons used dilators to take barbed arrows out of wounds, amputated limbs, performed **plastic surgery**, and developed techniques for removing bladder stones, hernias, and cataracts.

Surgery also flourished in the first century A.D. in India. Medical texts document the practice of operations ranging from tonsillectomy to plastic surgery. Certain practices sound familiar, such as the use of an operating room, the use of various types of forceps, and the use of anesthesia to reduce pain. One anesthetic used was burning **hemp**, the plant used today in the drug **marijuana**. Indians also fumigated the area where surgery was to take place to reduce infection.

Other practices suggest a far different place and **time**, including the use of large black **ants** as clips for tears in the intestines. As India continued to develop its medical tradition, the Middle Ages descended on Western **Europe**, casting a cloud of religious fervor over surgery, science, and the development of medicine as a whole.

The sponge of sleep

The Roman Catholic Church was the overwhelming authority in Medieval life, dictating everything from worship to medical care. Medical teaching was seen as less important than theology. While the Greeks had idealized good health, Christian doctrine in the Middle Ages considered suffering a potentially valuable entity, which could test one's faith. As a result, the idea of healing the sick was controversial, particularly in the early Middle Ages. Some religious authorities suggested that medical treatment could be sinful. This climate was not conducive to an expansion of knowledge about surgery.

During the early Middle Ages, traveling surgeons often wandered the countryside, operating on individuals for hernias, stones, and cataracts, then frequently leaving town before complications developed. By this time, surgery was separate from medicine, and surgeons, whose work was often of low quality, had little prestige.

As the Middle Ages progressed in Europe, medical training changed. Medical doctors moved from an education based on apprenticeship to an education which included formal instruction at a university. With the founding of universities in Europe in the thirteenth and fourteenth centuries, the idea of a central well of medical knowledge expanded. Surgeons were left out of this formal educational world, with most learning their trade as apprentices. Eventually, colleges specifically for surgeons were developed. Those with the least education were called barber-surgeons.

By the thirteenth century, technical guides to surgery started to emerge, complete with records of innovative techniques. These techniques included the use of the soporific sponge, a sponge soaked in a mixture of ingredients to promote **sleep**. The sponge could include opium, mandrake, and other ingredients. Such ingredients were difficult to regulate, and could cause drug dependence, death, or other less serious side effects.

During the Middle Ages in Europe, the golden era of Islamic medicine transformed surgery. An Arab text from the eleventh century documented the use of cauterization to stop hemorrhaging and as therapy for chronic migraine headaches and **epilepsy**.

Beyond boiling oil

The Scientific Revolution in the sixteenth and seventeenth centuries revolutionized medicine. William Harvey (1578-1657) advanced all branches of medicine with his discoveries about circulation of the blood. A fascination with the human body, and the renewed study of **anatomy** helped forge dramatic advances in the understanding of how the body worked. The era also saw the innovations of great surgeons such as Ambroise Pare (1510-1590).

Trained as a barber-surgeon, Pare's writings on surgery influenced the profession for centuries. His books were translated into Japanese, Latin, German, English and Dutch. Pare, a military surgeon for four kings of France, made his first great innovation at the Siege of Turin in 1537. At the time, conventional treatment of gunshot wounds called for treatment that today would be considered sadistic torture: the cleansing of wounds by boiling oil.

Stationed at the battle field where many soldiers were wounded, Pare used up the supply of boiling oil before all the men's wounds could be treated. To treat the others, he developed a milder mixture consisting of egg yolk, turpentine, and oil of roses. The soldiers who had received Pare's mixture looked better the next morning, while the other soldiers' wounds remained unchanged. The success of the milder treatment lead to abandonment of the boiling oil dressing.

Pare's books on surgery related everything from effective treatment of gunpowder wounds to methods for removing arrows and treatment of urinary-tract infections. He also discussed the use of ligature to repair damaged blood vessels. Pare and other surgeons enhanced the reputation of the profession. But until the nineteenth century, surgeons and their patients were limited by their inability to fend off infection or control pain. Their failure to do so meant that surgery was generally painful and life-threatening.

The limits of anesthesia in the pre-modern era shaped the way doctors operated on patients. One measure of the need to limit pain was the value placed on speed in surgery. Reportedly, Scottish surgeon Robert Liston could amputate a limb in 29 seconds. Long operations were impossible, because patients died on the operating table due to shock, pain, and **heart** failure.

Surgeons from ancient times developed anesthetics drawn from plants or plant products, such as opium from the poppy plant and wine drawn from **grapes**. But all of these substances had flaws, such as dangerous side effects. All were difficult to control for dosage, and none provided surgeons with what they most needed—a patient who would lie still and not feel pain during surgery, then awaken after the procedure was over.

This changed with the development of effective anesthesia in the eighteenth century. There was no single inventor of anesthesia. Indeed, controversy marked the early use of anesthesia, with many different individuals claiming credit. **Ether**, the first gas to be used widely, was described as early as 1540 as a solvent which could also be used to aid patients with pleurisy and **pneumo-**

nia. The gas gained new life in the early 1800s, when American and British pleasure-seekers marveled at the changes in behavior and **perception** induced by inhaling ether or nitrous oxide.

The story of the first public display of anesthesia by dentist William Thomas Green Morton is a story of claims, counterclaims, and frustration. The first to have his claims of innovation overlooked was Crawford Williamson Long, a physician in Jefferson, Georgia. Long reported the safe removal of a **tumor** in the neck of a patient anesthetized with ether in 1842. His failure to promptly report his use of the substance to medical journals resulted in the eclipse of his achievement by Morton in 1846.

Another dentist, Horace Wells, also was frustrated in his claim to be the first to successfully use a gas anesthetic. Wells used nitrous oxide in 1844 to anesthetize patients while their teeth were being pulled. Wells attempted to perform a dental procedure under anesthesia before a class of Harvard Medical students. The 1845 demonstration was a failure. The patient cried out in pain, and Wells was ridiculed. Eventually he committed suicide, after a bitter campaign to gain credit for his discovery.

Morton did not know about the use of nitrous oxide as an anesthetic until after Wells, a former teacher, traveled to Boston to display the new technology in 1845. Taking a more careful and politic route than Wells, Morton conducted a series of experiments using ether with dogs, goldfish, and other animals. He applied for a patent, and in 1846 performed a successful tooth extraction on a patient who had inhaled ether. He also associated himself with a number of prestigious physicians, then scheduled a public display of ether anesthetic. The 1846 display, during which a prominent surgeon operated on a vascular tumor, was a success.

Though Morton received credit for the first successful use of ether anesthesia, his credit was challenged a second time after the procedure was deemed successful. This time, he was accused by a chemist who had advised him, Charles T. Jackson, who said he was the true inventor.

A sanitary leap forward

The development of anesthesia cleared the way for more ambitious types of surgery and more careful surgical endeavors. Without the need to operate so quickly, surgeons could focus on operating more carefully. Yet surgery still had not entered the modern era, for infection continued to make recovery treacherous.

Patients who survived surgery in the middle nineteenth century continued to face frightening odds of dying from infection. An 1867 report noted that 60% of patients who received a major **amputation** in a Paris hospital died. Physicians routinely plunged their unwashed hands inside the body and often wore the same outfit for repeated surgical operating, unknowingly passing one patient's germs on to the next.

Ignorance about the nature of germs lead to drastic surgical measures. For example, individuals with compound fractures, in which the bone protruded through the skin, faced serious risk of infection and death in the early nineteenth century. To avoid infection, surgeons often amputated the limb.

Joseph Lister (1827-1912), the British surgeon who is credited with bringing antiseptic surgery to medicine, drew upon the work of his French contemporary, Louis Pasteur (1822-1895). Pasteur observed that germs could be present in the air and on certain surfaces. Lister was the first to apply Pasteur's observation to surgery.

Lister shook the medical world with his use of carbolic acid on compound fracture wounds after surgery. Because surgery for such fractures had a high **rate** of infection, Lister's report of success with nine out of 11 patients in 1867 was a dramatic finding. Cheered by the success of this effort, Lister continued his drive to remove germs from surgery. His efforts included immersing surgical instruments in a carbolic acid **solution**, requiring surgical assistants to wash their hands in **soap** and water, and careful washing of the area to be operated upon.

The success of the sterile surgical technique transformed surgery, over time, from a risky endeavor to one which carried a low risk for most procedures. The new safety of surgery made the practice more familiar, less risky, and far more open to innovation.

A third discovery helped clear the way for the dramatic surgical developments of the twentieth century. This was the finding by immunologist Karl Landsteiner (1868-1943) that blood varied by blood group types. Landsteiner, who received the Nobel Prize in 1930 for his work, described four different blood groups. Knowledge about the compatibility of certain blood types and the incompatibility of others, enabled the development of safe blood transfusions during surgery.

The modern era

By the late nineteenth century, surgery was still performed rarely. For example, in 1867, only 3.2% of the hospital admissions involved surgery at the Charity hospital in New Orleans. By 1939, surgery was involved in about 40% of admissions.

But surgeons of the nineteenth century broke many barriers. As recently as the 1880s, most surgeons would not intentionally operate on the head, the chest, or the abdomen unless an injury already existed. Over the next

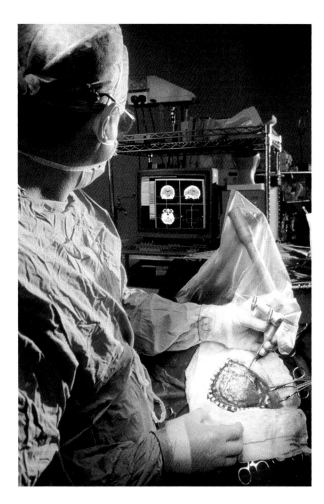

A neurosurgeon using a computer-controlled robot arm (called a "magic-wand") during an operation to remove a brain tumor. In the background, a screen shows a number of computer tomography (CT) scans of the patient's head. With the CT scan data being fed into the robot's computer, even a small tumor can be found and removed with minimal damage to surrounding tissue. *Photograph by Geoff Tompkincon. National Audubon Society Collection/Photo Researchers, Inc. Reproduced by permission.*

few years, surgeons mastered numerous abdominal operations, including surgery for appendicitis, new types of **hernia** operations, and surgery for ovarian disorders.

During the twentieth century, surgeons created successful techniques to operate on the brain and the heart and even techniques for **organ** transplantation. Surgeons became elite members of the medical establishment, earning more money than general practitioners and gaining celebrity as medical trailblazers.

Once a last resort, surgery is now performed frequently. Each year, a total of 25 million surgical operations calling for anesthesia are conducted in the United States. These operations cost a total of about $125 billion annually.

Heart of hearts

The emergence of heart surgery in the twentieth century defied earlier beliefs that the heart was inviolate and untouchable. Contemporary surgeons replace hearts in heart transplant operations, create new pathways for the blood using **tissue** from other parts of the body in coronary bypass operations, and clear out the blood vessels of the heart using special tools in coronary angioplasty. But the story of the development of heart surgery makes clear the high cost in human lives of medical advancement in this area.

As early as 1882, the German physician M.H. Block wrote of his successful suturing of rabbit hearts and suggested that human hearts could also be sutured. Surgeons in the early twentieth century who attempted surgery on heart valves found limited success. A 1928 survey of the 12 known cases of surgery on heart valves reported that 83% of those who received the procedure had died.

Due to overwhelming death rates, heart valve surgery waned until the 1940s and 1950s, when the procedure was reintroduced and surgeons could take advantage of several research advances. These included procedures which enabled surgeons to maintain circulation during surgery and to slow the beating of the heart. In 1950, the first electric **pacemaker** was developed. The race to add new ways to preserve the precious heart intensified.

In 1967, the first successful human heart transplant was performed by South African surgeon Christian Barnard. Although the patient died in 18 days, Barnard was celebrated internationally for his surgical skill. The thrill of transplanting hearts, and the great potential for saving lives, encouraged surgeons around the world to try the operation. The results were disappointing, as patient after patient died in a number of days or, sometimes, months. Most died after their bodies rejected the donated organ. Enthusiasm for the operation faded. While 99 transplants had been conducted in 1968, nine were performed in 1971.

As a small number of surgeons continued performing transplants, success rates for the operation improved. In addition, effective drugs to fight organ rejection were developed. By the early 1980s, heart transplants had regained popularity and in 1984, about 300 heart transplants were performed in the United States. In 1994, a total of 83% of individuals receiving heart transplants were expected to survive one or more years, making the surgery far safer than life with a failing heart.

A more common, and safer, procedure for individuals whose hearts are weak but not in need of complete replacement is the coronary artery bypass, a procedure developed in the late 1960s. The procedure takes tissue from elsewhere in the body to form new, clear pathways for blood to flow through the heart. A total of 309,000 coronary artery bypass grafts were performed in the United States in 1992. The procedure is not without risk, and the death rate linked to bypass surgery is from one to two individuals out of 100. Yet the prospect of a limited life without the surgery makes coronary bypass a common procedure.

One vivid measure of the extent to which surgery has become routine is the high number of babies born using caesarian section, the surgical delivery of a baby. In 1992, a total of 921,000 caesarian sections were performed, making the procedure the most common type of major inpatient surgery. The procedure is used when labor fails to progress, when a child being born is in distress, and in other situations where normal **birth** is considered unwise or dangerous. The **frequency** of the procedure is controversial, with many critics saying that cesareans are used too frequently, placing mothers at risk and adding to the high cost of health care.

Many contemporary experts challenge the long-told story that cesarean birth was named after the birth of Roman statesman Julius Caesar. They say that had Caesar been delivered using caesarian section, his mother probably would have died, due to the high risk of infection. Historical evidence suggests that she survived the birth of her son. Cesarean sections were mentioned in ancient Indian medical texts, although the outcomes are unclear. In the modern era, the first cesarean births in which mother and child were expected to survive were performed in the early 1800s. Death due to infection still occurred frequently following the surgery. Cesarean births did not become popular until the late 1920s and they did not approach their current popularity until recently.

A prettier face

Another type of surgery which has recently flourished is plastic surgery. Evidence of the reconstruction of noses, ears, and lips can be found as early as the first century A.D., in ancient Indian medical texts. In the contemporary era, surgery to repair injuries or to treat disease, particularly **cancer**, continues to be the most common type of plastic surgery.

But cosmetic plastic surgery has also become popular, and procedures to make breasts larger, noses smaller, and buttocks less saggy have gained in popularity. For example, in 1992, a total of 50,175 procedures for nose reshaping were performed, and 32,607 procedures to increase the size of the breast were performed. Nearly nine out of 10 individuals who received plastic surgery in 2002 were women, but men were also recipients of facelifts, buttock lifts and so-called "tummy tucks." The popularity of cosmetic plastic surgery testifies to the public perception of surgery as a safe, predictable activity.

Fetal surgery

Contemporary surgeons have taken the concept of surgery well beyond what their counterparts as little as 100 years ago believed to be their domain. One of the most dramatic types of surgery is fetal therapy on the unborn. The first successful effort to address fetal problems in the womb took place in the early 1960s, with the first prenatal blood transfusion. Effective techniques for human fetal surgical therapy were first performed in the 1980s and are still being developed. Because such surgery presents a risk to the fetus and the mother, surgery on the fetus is performed rarely. Such procedures are only performed if the potential benefit to the fetus is deemed to be great. Successful surgery has been performed on fetuses suffering from certain types of life-threatening hernias and urinary-tract obstructions.

The procedure involves placing the mother under anesthesia and opening the uterus to expose the fetus. Following the surgery, amniotic fluid is replaced with a liquid solution containing **antibiotics**. The uterus is then stitched closed.

Surgery of the future

Technological advances in **robotics** and imaging devices suggest dramatic changes in the operating room of the future. Robots have already performed certain procedures in clinical trials. One such trial involves using a robot in surgery to help replace non-functioning hips with a prosthesis. Hip-replacement procedures are commonly performed and have a high rate of success. But surgeons have long been concerned about the procedures that do not succeed due, in part, to the difficulty of creating an exact fit for the prosthesis.

Developers of the robot used in the clinical trials hope the robot will be able to do a better job. The robot is linked to a CT scanner which measures the exact dimensions of the thigh bone. These measurements are used to help the robot hollow out a cavity in the thigh bone which will be an exact fit for the prosthesis. Successful use of such robots could change the role of the surgeon, whose tasks may include robot supervision in the future.

New imaging devices should also help surgeons operate more safely and efficiently in the future. Researchers are currently working on imaging devices that combine powerful computer technology with existing imaging technology to produce images of the body during surgery. Such images could be used for surgeons in training, to see how the body changes during surgery. They may also be used during surgery to limit the amount of potentially harmful cutting that is done. As technology enables surgeons to be more precise about

KEY TERMS

Cauterization—The use of heat, electricity, or other substance to destroy tissue.

Ligature—Material used to tie a blood vessel or body part.

Prosthesis—An artificial body part.

Sutures—Substance used to stitch closed a surgical wound.

Trepanation—The operation in which a circular area is removed, usually in the skull.

Uterus—Organ in female mammals in which embryo and fetus grow to maturity.

where they cut, surgery could become more effective and less invasive.

One possible result of this could be what has been dubbed "trackless" surgery, procedures performed without cutting the patient. If images of harmful tissue, such as a breast tumor, could be sharpened, surgeons may be able to operate by simply focusing ultrasound waves on the tissue. Mastery of such techniques could make scalpels, sutures, and other conventional surgical tools obsolete.

Surgery of the future may bear little resemblance to the bloody, messy surgery of the present. But certain elements of surgery are unlikely to change. Surgical innovation will continue to require patients willing to take a risk and surgeons willing to challenge convention. Surgical skill will continue to require the traits Hippocrates urged surgeons to honor in ancient Greece: "ability, grace, speed, painlessness, elegance, and readiness." Finally, surgical progress will continue to depend on the ability of skilled surgeons to use their physical ability and their medical knowledge to heal patients.

See also Laser surgery; Neurosurgery; Prenatal surgery; Psychosurgery; Thoracic surgery; Transplant, surgical.

Resources

Books

Bollinger, Randal R., and Delford L. Stickel. "Transplantation." *Textbook of Surgery.* 14th ed. David C. Sabiston. Jr., Philadelphia: W.B. Saunders Co, 1991.

Brieger, Gert H. "The Development of Surgery." In *Textbook of Surgery.* 14th ed. ed. David C. Sabiston Jr., Philadelphia: W.B. Saunders Co, 1991.

Gray, Henry, Lawrence H. Bannister, Martin M. Berry, and Peter L. Williams, eds. *Gray's Anatomy: The Anatomical Basis of Medicine & Surgery.* London: Churchill Livingstone, 1995.

Magner, Lois N. *A History of Medicine.* New York: Marcel Dekker, Inc., 1992.

Nuland, Sherwin B. *Doctors: The Biography of Medicine.* New York: Alfred A. Knopf, 1988.

Periodicals

Houkin, K. "Digital recording in Microsurgery." *Journal of Neurosurgery* 92, no. 1 (2000): 176-180.

Leparc. G.F. "Nucleic Acid Testing for Screening Donor Blood." *Infectious Medicine* no. 17 (May 2000): 310-333.

Other

Current Science and Technology Center. "Robotic Surgery." [cited April 2003]. <http://www.mos.org/cst/article/1623/>.

Kobus, Nancy. *1992 Plastic Surgery Statistics.* The American Society for Aesthetic Plastic Surgery, Inc., 1992.

Patricia Braus

Surveying instruments

Surveying is the apportionment of land by measuring and mapping. It is employed to determine boundaries and property lines, and to plan construction projects.

Any civilization that had any degree of sophistication in construction methods required surveys to ensure that work came out according to plan. Surveying is thought to have originated in ancient Egypt as early as 2700 B.C., with the construction of the Great Pyramid of Khufu at Giza, though the first recorded evidence of boundary surveying dates from 1400 B.C. in the Nile River valley.

The classic surveyors were the Romans. In order to forge an Empire that stretched from the Scottish border to the Persian Gulf, a large system of roads, **bridges**, aqueducts, and canals was built, binding the country economically and militarily.

Surveying was a major part of Roman public works projects. It also was used to divide the land among the citizens. Roman land surveying was referred to as centuriation after the century, which was a common rectangular unit of land area. These land parcels can still be seen in aerial photographs taken over France and other parts of **Europe**, the work of the Roman agrimensores, or measurers of land. The property lines were usually marked by stone walls and boundary markers. With the advent of new methods of **trigonometry** and **calculus**, new surveying instruments emerged. The theodolite was invented in the sixteenth century. Its precise origin is unclear, but one version was invented by English mathematician Leonard Digges in 1571, who gave it its name. A great theodolite was invented by Jesse Ramsden more

than 200 years later in 1787. Its use led to the establishment of the British Ordnance Survey.

Made up of a **telescope** mounted on a compass or of a quadrant plus a circle and a compass, the theodolite is used to measure horizontal and vertical angles. The modern theodolite is usually equipped with a micrometer, which gives magnified readings up to 1/3600°, or one second of arc. The micrometer is derived from the vernier scale, which was invented by French engineer and soldier Pierre Vernier (1584-1638) in 1631 to measure in fractions.

The transit is a theodolite capable of measuring horizontal and vertical angles, as well as prolonging a straight line or determining a level sight line. A telescope atop a tripod assembly is clamped in position to measure either horizontal or vertical angles. The transit employs a plumb bob hanging from the center of the tripod to mark the exact location of the surveyor.

The practice of triangulation was introduced by Gemma Frisius in 1533. By measuring the **distance** of two sides of a triangle in a ground survey, the third side and the triangle's area can be calculated. Triangulation was aided by the inventions of the prismatic **astrolabe** and the heliotrope. The latter was invented by German mathematician Johann Gauss (1777-1855), who is considered the father of geodesy, the science of **Earth** measurement. Both instruments used a beam of sunlight to signal the positions of distant participants in a land survey.

Other survey instruments include the surveyor's compass, which is used for less precise surveying. The surveyor's level is used to measure heights of points above **sea level** or above local base points. **Metal** tapes, first introduced by English mathematician Edmund Gunter in 1620, are used for shorter measurements.

In the late twentieth century, surveying has been aided greatly by **remote sensing**: Photogrammetry employs aerial **photography** to survey large areas for topographic mapping and land assessment purposes, **satellite** imagery has increased the aerial coverage of surveys, and **laser** technology has increased the precision of survey sightings.

Survival of the fittest

The term "survival of the fittest" was first used by the Victorian naturalist Herbert Spencer as a metaphor to help explain natural **selection**, the central element of Charles Darwin's revolutionary theory of evolutionary change, first published in 1859 in his famous book, *The Origin of Species by Means of Natural Selection.*

In this extremely influential and important book, Darwin reasoned that all **species** are capable of producing an enormously larger number of offspring than actually survive. He believed that the survival of progeny was not a **random** process. (In fact, he described it as a "struggle for existence.") Rather, Darwin suggested that those progeny which were better adapted to coping with the opportunities and risks presented by environmental circumstances would have a better chance of surviving, and of passing on their favorable traits to subsequent generations. These better-adapted individuals, which contribute disproportionately to the genetic complement of subsequent generations of their population, are said to have greater reproductive "fitness." Hence the use, and popularization, of the phrase: "survival of the fittest." (Darwin also used another, more awkward expression to explain the same thing: the "preservation of favored races in the struggle for life." In fact, this is the subtitle that he used for *The Origin of Species by Means of Natural Selection*.)

Darwin's theory of **evolution** by natural selection is one of the most important concepts and organizing principles of modern **biology**. The differential survival of individuals that are more-fit, for reasons that are genetically heritable, is believed to be one of the most important mechanisms of evolution. And because of its clarity, the phrase "survival of the fittest" is still widely used to explain natural selection to people interested in understanding the evolution of life on **Earth**.

Sustainable development

Sustainable development is the management of renewable resources for the good of the entire human and natural community. Built into this concept is an awareness of the **animal** and **plant** life of the surrounding environment, as well as inorganic components such as **water** and the atmosphere. The goal of sustainable development is to provide resources for the use of present populations without compromising the availability of those resources for future generations, and without causing environmental damage that challenges the survival of other **species** and natural ecosystems.

The notion of sustainable development recognizes that individual humans and their larger economic systems can only be sustained through the exploitation of natural resources. By definition, the stocks of non-renewable resources, such as metals, **coal**, and **petroleum**, can only be diminished by use. Consequently, sustainable economies cannot be based on the use of non-renewable resources. Ultimately, sustainable economies must be supported by the use of renewable resources

such as biological productivity, and solar, **wind**, geothermal, and **biomass energy** sources.

However, even renewable resources may be subjected to overexploitation and other types of environmental degradation. Central to the notion of sustainable development is the requirement that renewable resources are utilized in ways that do not diminish their capacity for renewal, so that they will always be present to sustain future generations of humans.

To be truly sustainable, systems of resource use must not significantly degrade any aspects of environmental quality, including those not assigned value in the marketplace. **Biodiversity** is one example of a so-called non-valuated resource, as are many ecological services such as the cleansing of air, water, and land of pollutants by ecosystems, the provision of **oxygen** by vegetation, and the maintenance of agricultural **soil** capability. These are all important values, but their importance is rarely measured in terms of dollars.

A system of sustainable development must be capable of yielding a flow of resources for use by humans, but that flow must be maintainable over the long term. In addition, an ecologically sustainable economy must be capable of supporting viable populations of native species, viable areas of natural ecosystems, and acceptable levels of other environmental qualities that are not conventionally valued as resources for direct use by humans.

Natural resources

In economic terms, resources (or capital) are regarded as actual or potential wealth that can be applied toward the creation of additional wealth. There are three broad types of capital. First, manufactured capital is industrial infrastructure that can be applied to the production of goods and services. Examples include factories, mines, harvesting equipment, buildings, tools and machinery, computers, and information networks. Second is human capital, or the cultural means of production, encompassing a workforce with particular types of knowledge and skills. And third, natural capital refers to quantities of raw, natural resources that can be harvested, processed, used in manufacturing, and otherwise utilized to produce goods and services for an economy.

There are two types of natural resources: non-renewable and renewable. Non-renewable resources are present in a finite quantity on **Earth**. Therefore, their stock diminishes as they are mined from the environment. Non-renewable resources can only be used in an non-sustainable manner. The lifetime of a non-renewable resource is determined by the size of its recoverable stocks in the environment, and the **rate** of **mining**. However, some non-renewable resources can be reused and recycled to some degree, which extends the effective lifetime of the resource. Common examples of non-renewable resources include **metal** ores, coal, and petroleum.

Potentially, renewable resources can be sustained and harvested indefinitely. However, sustainable use requires that the rate of harvesting does not exceed the rate of renewal of the resource. Most renewable resources are biological and include trees, hunted animals such as **fish**, waterfowl, and **deer**, and the products of agriculture. Flowing surface water is an example of a non-biological resource that can potentially be sustainably used for **irrigation**, to generate hydroelectricity, and as a means of transportation.

It is important to recognize that potentially renewable resources can easily be "over-harvested," or exploited at a rate exceeding that of renewal, resulting in degradation of the resource. During over-harvesting, the resource is essentially being "mined"—that is, it is managed in the same way as a non-renewable resource. Regrettably, this is all too often the case, resulting in collapses of stocks of hunted fish, **mammals**, and **birds**; **deforestation**; declines of agricultural soil capability; and diminished river flows due to excessive withdrawals for use by humans.

Another important characteristic of renewable resources is that they can provide meaningful ecological services even when they are in their natural, unharvested state. For example, intact, natural **forests** provide biological productivity; cycling of **nutrients** and water; a sink for atmospheric **carbon**; control of **erosion**; cleansing of pollutants emitted into the environment by humans; **habitat** for diverse elements of biodiversity; aesthetics; and other important ecological services. Some of these services are of potential value in providing resources that humans require, an example being the biomass and productivity of trees and hunted animals. However, most of these are not recognized by the conventional marketplace, although they are certainly important to **ecological integrity** and environmental health.

The undeniable ecological reality is that humans have an absolute dependence on a continuous flow of natural resources to sustain their societies and economies. Over the longer term, this is particularly true of renewable resources because sustainable economies cannot be supported only by non-renewable resources. Therefore, the only way to achieve a condition of sustainable development is to build an economy that is supported by the wise harvesting and management of renewable resources.

Economics

A goal of economic systems is to maximize the utility of goods and services to society. Usually, these products are assigned value in units of currency. Some examples of valuated goods and services include the following: manufactured products, such as automobiles, computers, highways, and buildings; harvested natural resources, such as **wood**, hunted animals, and the products of agriculture; and the services provided by farmers, industrial workers, and others.

Conventional economics does not seriously consider non-valuated resources, or goods and services that are not assigned value in the marketplace. Examples of non-valuated ecological resources include the aesthetics of natural landscapes, services such as nutrient and water cycling, and rare species and natural ecosystems. Consequently, the merits of non-valuated ecological resources cannot be easily compared with those of valuated goods and services. This in turn means that degradations of non-valuated resources are not usually considered to be true "costs" by conventional economists, and they do not have a strong influence in cost-benefit calculations.

In conventional accounting, large profits can often be made by undertaking activities that cause substantial environmental damage, including the exhaustion of potentially renewable resources. Clearly, this is an ecologically false accounting, but it has been rationalized by considering degradations of environmental quality to be externalities, or costs that are not directly paid by the individuals or companies that are causing the damage. However, the costs of resource and environmental degradation are very real, and they are borne by society at large—which of course includes the individuals or institutions responsible for the degradation.

Ecological economics is a new, actively developing sub-discipline within economics. The principal distinction of ecological economics is that it attempts to find a non-anthropocentric system of valuation. This is different from conventional economics, in which valuations are based almost entirely in terms of the importance of good and services to humans, as determined in the marketplace.

Accountings in ecological economics include the important social and environmental costs that may be associated with the depletion of resources and the degradation of environmental quality. These costs are critical to achieving and measuring sustainable development, but they are not seriously considered during accountings in conventional economics.

Sustainable development and sustained growth

The notion of sustainable development refers to an economic system that is ultimately based on the wise utilization of renewable natural resources in a manner that does not threaten the availability of the resources for use by future generations of people. It is also important that damages to non-valuated resources be kept within acceptable limits.

Clearly, the existing human economy is grossly unsustainable in these respects. Modern economies are characterized by resolute economic growth, which is achieved by the vigorous mining of non-renewable resources, potentially renewable resources, and environmental quality in general.

Since the mid-1980s, when the notion was first introduced, "sustainable development" has been enthusiastically advocated by many politicians, economists, businesspeople, and resource managers. However, many of these have confused sustainable development with "sustained economic growth," which by definition is impossible because resources eventually become limiting. The first popularization of the phrase "sustainable development" was in the widely applauded report of the World Commission of Environment and Development, also known as the "Brundtland Report" after the chairperson of the commission, Gro Harlem Brundtland of Norway. However, this report appears to confuse some of the fundamental differences between economic growth and sustainable development.

Although the Brundtland Report supports the need for sustainable development, it also calls for a large increase in the size of the global economy. The Brundtland Report suggests that a period of strong economic growth is needed, in concert with a redistribution of some of the existing wealth, if the living standards of people in poorer countries are to be improved. It is believed that once this has been accomplished, social and economic conditions will favor an end to population growth and the over-exploitation of natural resources, so an equilibrium condition of a non-growing economy can be achieved.

However, the sorts of economic growth rates recommended in the Brundtland Report are equivalent to an increase of per-capita, global income of 3% per year, sufficient to double per-capita income every 23 years. The economic growth must also compensate for growth of the human population, which amounts to about 2% per year. Therefore, the adjusted rate of economic growth would have to be about 5% per year (that is, 3% + 2%), which would result in a doubling of the global economy every 14 years. Of course, in poorer countries with even higher rates of population growth, including most of **Africa**, **Asia**, and Latin America, the rate of economic growth would have to compensate and be correspondingly larger. In total, the Brundtland Report suggested that an expansion of the global economy by a **factor** of five to 10 was needed to create conditions appropriate to achieve a condition of sustainable development.

The Brundtland Report not only recommends a great deal of economic growth; it also recommends the development of technologies that would allow a more efficient economic growth, which would consume fewer resources of material and energy per unit of growth achieved. Additionally, the report advocates a redistribution of wealth from richer to poorer people and countries, as well as greater efforts towards the elimination of population growth.

The Brundtland Report, like other champions of "sustainable development," actually promotes economic growth as a cure for the present ills of human economies. However, there are profound doubts that a five-to-10-times increase in the size of the human economy could be sustained by the environment and its ecosystems. Economic growth may, in fact, be more of a cause of the environmental crisis than a cure.

Resolution of the environmental crisis and achievement of sustainable economies may require the immediate, aggressive pursuit of more difficult and unpopular solutions than those recommended by the Brundtland Report. These would include much less use of resources by richer peoples of the world, immediate redistribution of some of the existing wealth to poorer peoples, vigorous population control, and an overall focus on preventing further deterioration of ecological integrity and environmental quality more generally.

Sustainable development

A truly sustainable economic system recognizes that the human economy must be limited within the **carrying capacity** of Earth's remaining natural resources. In fact, many resource economists, environmental scientists, and ecologists believe that the human economy is already too large to be sustained by Earth's resources and ecosystems. If these specialists are correct, then not only is further economic growth undesirable, it may have to be reversed.

Non-sustainable economic growth occurs through a crude maximization of the flow of resources through an economy. In large part, economic growth is achieved by mining resources and environmental quality.

In contrast, sustainable development is ultimately based on the efficient use of renewable resources, which are not degraded over time. Moreover, this use occurs under conditions in which environmental quality is also protected. A sustainable economic system would have the following characteristics:

First, renewable resources must be exploited at or below their capability for renewal. Present economies are greatly dependent on the use of non-renewable resources, but these are being rapidly diminished by use.

As non-renewable resources become exhausted, renewable resources will become increasingly more important in the economic system. Ultimately, sustainable economic systems must be based on the wise use of renewable resources.

Second, non-renewable resources can also be utilized in a sustainable economy. However, the rates at which non-renewable resources are utilized must be balanced by the rate at which renewable substitutes are created, that is, by growth of a renewable resource. For example, **fossil fuels** can only be used in a truly sustainable economy if their utilization is compensated by net growth of a renewable energy substitute—for example, by an increase in forest biomass. To discourage the use of non-renewable resources and the unsustainable mining of potentially renewable resources, it might be possible to implement a system of natural-resource depletion taxes.

Third, there must be a markedly increased efficiency of the use and **recycling** of non-renewable resources, aimed at extending their useful lifetime in the economic system. Information systems and new technologies will be important in achieving this increased efficiency. There must also be well-designed systems of use and management of renewable resources to ensure that these are sustainably utilized over the longer term.

Fourth, it is critical that ecological resources that are not conventionally valuated also be sustained. The use and management of natural resources for human benefits will inevitably cause declines of some species and natural ecosystems, as well as other environmental damage. However, viable populations of native species, viable areas of natural ecosystems, and other aspects of environmental quality must be preserved in an ecologically sustainable economic system. Some of these ecological values cannot be accommodated on landscapes that are primarily managed for thee harvesting and management of economic resources, and they will therefore have to be preserved in ecological reserves. These ecological values must be accommodated if an economic system is to be considered truly sustainable.

Sustainable economic systems represent a very different way of doing business, in comparison with the manner in which economies are now conducted. Sustainable development requires the implementation of a sustainable economy. To achieve this would be difficult on the short term, although the longer-term benefits to society and ecosystems would be enormous. The longer-term benefit would be achievement of an economic system that could sustain humans, other species, and natural ecosystems for a long time. However, there would be short-term **pain** in implementing such a system, largely associated with substantially less use of natural re-

KEY TERMS

. . . .

Anthropocentric—Considering the implications of everything from the perspective of utility to humans, and to human welfare.

Natural resource—Any naturally occurring commodity that can be used by people. Non-renewable resources are of a finite quantity, and they can only be mined. Renewable resources can potentially by exploited indefinitely, but only if they are not degraded by overexploitation, i.e., used at a rate that exceeds renewal.

Valuation—The assignment of economic worth, for example, in dollars.

sources, abandonment of the ambition of economic growth, and rapid stabilization of the human population.

As a result of these short-term inconveniences, truly sustainable development would not be initially popular among much of the public, politicians, government bureaucrats, and industry. This is because individual humans and their societies are self-interested, and they think on the shorter-term. However, for the sake of future generations of humans, and for that of other species and natural ecosystems, it remains absolutely necessary that sustainable economic systems be designed and implemented.

See also Population, human.

Resources

Books

Bueler, W.M. *An Agenda for Sustainability: Fairness in a World of Limits.* Cross Cultural Publications, 1998.

Daly, H.E. *Beyond Growth: The Economics of Sustainable Development.* Beacon Press, 1997

Dudley, William. *Biodiversity.* San Diego: Greenhaven Press, 2002.

Dunn, Seth. *Hydrogen Futures: Toward a Sustainable Energy System.* Washington, DC: Worldwatch Institute, 2001.

French, Hilary F. *Vanishing Borders: Protecting the Planet in the Age of Globalization.* New York: W. W. Norton, 2000.

Rao, P.K. *Sustainable Development: Economics and Policy.* Blackwell Publishers, 1999.

Sheehan, Molly O'Meara, and Jane A. Peterson. *City Limits: Putting the Brakes on Sprawl.* Washington, DC: Worldwatch Institute, 2001.

Weinberg, Adam S., David N. Pellow, and Allan Schnaiberg. *Urban Recycling and the Search for Sustainable Community Development.* Princeton, NJ: Princeton University Press, 2000.

Bill Freedman
Judson Knight

Swallows and martins

Swallows and martins are small fast-flying agile **birds** in the family Hirundinidae. There are 74 **species** in this family worldwide, mostly found in open habitats, where they forage aerially for their **prey** of flying **insects**.

There is no particular biological difference between swallows and martins. Sometimes these names are used interchangeably, as in the case of *Riparia riparia*, known as the **sand** martin in Western **Europe**, and as the bank swallow in **North America**.

Biology of swallows and martins

Swallows and martins have relatively long, pointed wings, and they are swift and agile fliers. The feet of these birds tend to be small and weak, and are used for little more than perching. These birds often rest on wires and exposed branches of trees. During **migration**, large numbers of swallows and martins may roost together on these sorts of perches, often in mixed-species flocks. Swallows may also forage in large, mixed-species flocks during migration.

Many species of swallows and martins are rather plainly marked with dark brown or black backs and wings and a white breast. Other species are more boldly patterned, and may be brightly colored with red, yellow, and iridescent green and purple. Most species have a notched tail, and some have a deeply forked tail.

Swallows have a short but broad mouth, which can open with a very wide gape, an **adaptation** for catching insects on the wing. This food of flying insects is sometimes referred to as *aeroplankton*.

Many species of swallows and martins nest in colonies of various size, and most species are gregarious during the non-breeding season. Some swallows nest in natural or artificial cavities. Other species nest in tunnels that they dig in earthen banks. Many species construct an urn-like cavity of hardened mud or clay, or they make cup-shaped nests of these materials. The natural substrate for attachment of the constructed nests is cliffs and other sheer surfaces. However, some species use **bridges** and buildings as substrata upon which to build their clay nests.

All species of swallows and martins migrate between their breeding and non-breeding habitats. Species that breed at high latitudes can migrate great distances, for example, about 6,831 mi (11,000 km) in the case of the European swallow (*Hirundo rustica*), some of which breed in northern Europe and winter in South **Africa**.

North American swallows and martins

The most familiar swallow in North America is the barn swallow (*Hirundo rustica*). This is a cosmopolitan species that also occurs under other common names in Eurasia, Africa, Southeast **Asia**, and **Australia**. The barn swallow is an attractive bird, with a deeply forked tail, an iridescent purple back, and a brick-red breast. Barn swallows often nest in small colonies. The natural nesting **habitat** is cliffs and caves, where these birds build their cup-nests of mud bonded with grass. Barn swallows also commonly build their nests on the sides of buildings and bridges. The barn swallow breeds south of the **tundra** over most of North America and Mexico, and winters in **South America**.

The cliff swallow (*Petrochelidon pyrrhonota*) looks rather similar to the barn swallow, but it does not have a forked tail, and it has a white patch on its forehead, a buff rump, and a white breast. The cliff swallow is a colonial nester, building its roofed, mud nests on cliffs, and also on structures made by humans, such as bridges, **dams**, and buildings. The cliff swallow breeds locally over a wide range, from the subarctic tundra to southern Mexico, and winters in South America.

The bank swallow (*Riparia riparia*) also breeds over much of North America south of the tundra, as well as in Eurasia. North American populations of bank swallows winter in South America. The bank swallow is a brown-backed species, with a brown band across the chest. The bank swallow nests in colonies, typically excavating tunnels for its nests in the earthen banks of **rivers**. This species also uses the sides of gravel pits as a place to nest.

The rough-winged swallow (*Stelgidopteryx ruficollis*) is similarly marked to the bank swallow, but it lacks the brown breast-band, and has a darker breast. This is a non-colonial nester, which digs burrows in riverbanks, or sometimes nests in holes in cement bridges and other built structures. This species breeds widely south of the boreal forest of North America and into northern South America. The rough-winged swallow winters from the southern United States to South America.

The **tree** swallow (*Tachycineta bicolor*) has an iridescent, dark-blue or green back, and a white breast. This species breeds south of the low-arctic tundra to the northern states, and winters from the southern United States to Central America. The tree swallow nests in natural cavities in trees, or in cavities previously excavated and used by **woodpeckers**. This species will also take readily to nest boxes.

The violet-green swallow (*Tachycineta thalassina*) has a superficial resemblance to the tree swallow, with an iridescent violet-green back. This species breeds in

Barn swallows (*Hirundo rustica*) at Point Pelee National Park, Ontario. *Photograph by Robert J. Huffman. Field Mark Publications. Reproduced by permission.*

the western United States, and winters in northern Central America and Mexico. The violet-green swallow generally nests in cavities in trees or in crevices of cliffs.

The purple martin (*Progne subis*) is the largest swallow in North America. Male purple martins are a uniformly iridescent purple, while females are brown. The natural nesting sites of this colonial species are hollow cavities in trees, but purple martins also utilize multi-celled nesting boxes provided by humans. This species breeds in an area from southern Canada to Mexico, and winters in South America.

Interactions with humans

Swallows are boisterous and active birds, which maintain a stream of cheerful twitterings that many people find pleasing. The fact that some species nest in the vicinity of human habitations means that people can easily watch the comings and goings of these small, charismatic birds. Observers can gain an impression of the daily life of the swallow, from the building of nests, through the rearing of nestlings, to the trials and tribulations by which fledglings learn to fly and to hunt their own food of insects.

Swallows of all species eat enormous numbers of flying insects. Some of the prey insects (such as mosquitos and blackflies) are regarded as **pests** by humans. As a result, swallows nesting near homes are often considered to be beneficial. Barn and cliff swallows construct their mud nests on buildings. These species do not need encouragement, only tolerance of the small annoyances that some people might perceive about the defecations of these birds. The nesting of tree swallows can be assisted by providing simple nest boxes, while purple martins can be attracted by providing multiple-unit apartment boxes. Because of the relatively large number of birds that can

be supported, a colony of purple martins can have a significant effect on the abundance of biting **flies** in its vicinity.

Resources

Books

Forshaw, Joseph. *Encyclopedia of Birds.* New York: Academic Press, 1998.

Turner, A. *The Swallow.* U.K.: Hamlyn, 1994.

See also Insectivore.

Bill Freedman

Swamp cypress family (Taxodiaceae)

The swamp cypress family is more formally called the Taxodiaceae. This is a family of coniferous trees within the Gymnosperms, that is, plants which produce naked **seeds** (not in a fruit) borne on scales. These scales are usually arranged to form a cone.

Within the Taxodiaceae some **species** are evergreen, and some deciduous. There are nine genera which contain 16 species. These can be found in temperate and subtropical regions, in both the Old and New Worlds. Only one genus is represented in the Southern Hemisphere. Most of these are fast growing trees which can achieve a large size and an impressive age. Some members of this group have been known to be 3,000 years old.

Characteristics

All of the trees in this family produce resin when their branches are damaged. They all have one main trunk with a fibrous **bark**, usually of a reddish **color**. As the **tree** grows older the basal branches are lost, leaving a clear trunk. The leaves are usually dark green, needle like structures.

All of the Taxodiaceae are **wind** pollinated, with male and female reproductive structures on the same individual **plant**, but physically separate in different canes. The male part is a cluster of catkin-like cones which release pollen, which is then blown by the wind. The female cones are larger and occur singly or in groups, a drop of fluid is exuded from them on which pollen grains land. Once the pollen has been captured in this manner, the fluid is taken back in, bringing the pollen and ovule into close association so that **pollination** may occur. Some species are capable of reproducing very early in their life (e.g., *Taxodium*) while others must first achieve a few centuries of age (e.g., *Taiwania*).

The seeds are produced in mature cones, and when they are released from their high branches they can disperse over long distances due to the presence of wing-like, aerodynamic structures. Some, such as *Sequoiadendron,* only release their seeds after a forest fire, allowing a relatively competition-free start in life. The adult is little affected by the fire due to a protective layer of fibrous bark that is burnt off. However, the living **wood** underneath is not damaged, except by exceptionally severe fires.

Recognizable fossils of some of these species have been found from the Cretaceous and Tertiary eras.

The modern Taxodiaceae are found in warm temperate regions of the world, mostly in eastern **Asia**, east and west **North America**, and with a single genus in Tasmania. These plants favor areas of high rainfall with rich **soil**, tending to occur in local groves. One species, the swamp cypress (Taxodium distichum), is tolerant of wet conditions and will grow in swampy places.

Members of the Taxodiaceae

Athrotaxis comprises three species found only on the islands of Taiwan in east Asia. It is possible that one species is a **hybrid** between the other two, due to its intermediate characteristics. These are moderately tall, pine-like trees.

Cunninghamia is a tall tree of great commercial importance in its native China and Taiwan. Two, rather similar species are known.

Another genus, with two or three species, is *Taiwania*, from China, Taiwan, and north Burma. These are felled for their timber, but they are rather scarce in their native habitats, provoking much concern among conservationists.

Cryptomeria is similar to the above groups, but the several species are more-widely spread in their native Japan and China. Due to the fast growing nature and height of this group, they have been widely introduced to other areas where they are important timber producers.

The big-tree or giant redwood (*Sequoiadendron giganteum*) produces the world's most massive trees, with the tallest measured example being 344.5 ft (105 m) tall with a trunk diameter of 39 ft (12 m). These massive trees are native to California; they will not survive an intense frost and are intolerant of **pollution**. They are, however, remarkably fire tolerant due to their fibrous bark. In fact, disposal of their seeds is aided by fire, which makes the cones open, and also reduces **competition** for the new seedlings.

A closely related, and physically slightly taller genus is the coast redwood (*Sequoia sempivirens*), found

Redwoods in Jediah Smith State Park, California. *JLM Visuals. Reproduced with permission.*

naturally from southwest Oregon to California in small numbered groups.

All of the above genera are evergreens, that is, they do not lose all of their leaves in the autumn. The remaining members of the Taxodiaceae are all seasonally deciduous.

Perhaps the most endangered member of the whole group is the dawn redwood (*Metasequoia glyptostroboides*), which was found for the first time in 1941, in a small valley in Central China. No other wild populations have ever been discovered. The specific name is due to the initial physical similarity to the previous genus. While endangered in the wild, the dawn redwood can be cultivated and is widely grown.

Another **endangered species** endemic to China is *Glyptostrobus pensilis.* The swamp or bold cypress (*Taxodium distichum*) is a moderately hardy tree native to the southeastern United States. This species is not as massive as some of its relatives, but it is more hardy, and is a cultivated species. Two other *Taxodium* species are known.

See also Conifer; Gymnosperm; Sequoia.

Resources

Books

Kubitzki, K., K.U. Kramer, P.S. Green, and C.N. Page. "Pteridophytes and Gymnosperms," *The Families and Genera of Vascular Plants.* vol. 1. Springer, 1990.

Vidakovic, M. *Conifers: Morphology and Variation.* Groficki Zavod Hrvatske, 1991.

Welch, H. and Haddow, G. *The World Checklist of Conifers.* Landsmons Bookshop, 1993.

Gordon Rutter

Swamp eels

Belonging to the order Synbranchiformes, swamp eels are very slim **fish** with elongated bodies and reduced fins. Their gill system, which is very small, is linked to other organs to help them breathe air. Swamp eels live in tropical and subtropical habitats. They usually are found in stagnant fresh or **brackish** water; only one **species** lives in the sea. These fish are found in Central and **South America**, **Asia**, and **Africa**.

According to some publications, the order Synbranchiformes is made up of only one family, Synbrachidae, which contains four genera of swamp eels: the *Macrotrema*, the *Ophisternon*, the *Synbranchus*, and the *Monopterus*. Other sources report that there are three separate families within the Synbranchiform Order: the swamp eels, the singleslit eels, and the cuchias. Whichever way these fish are categorized, in total, there are about 15 distinct species.

Fish that look like **snakes** and live in dark, murky **water** are commonly called eels. Swamp eels are a great example of this type of fish. Indeed, they lack pectoral and pelvic fins, and their dorsal and anal fins are very small. Furthermore, while all species have small eyes, some are functionally blind with their eyes sunken below the skin. Some species live in caves, and many others burrow in mud. The maximum length these fish can attain is almost 3 ft (1 m). While swamp eels look a lot like eels, they are in no way related to them. Swamp eels are significantly different from eels internally and can breathe air. Furthermore, some of them estivate, which means that they **sleep** through the hot, summer months.

All 15 species of swamp eels have one or two gill openings at their throats which are designed to absorb **oxygen** from water. However, several species that live in water with small amounts of oxygen—like stagnant pools—are able to breathe air from the surface through their open mouths. One such species is the **rice** eel. Fish of this species live in **rivers**, ditches, and swamps of Southeast Asia, Indonesia, and the Philippines. During the extensive dry season, these fish burrow into the mud and can remain alive until the rainy season starts, as long as their skin remains damp. Interestingly, when rice eels reproduce, which is during the summer months, the males form the nests using air bubbles and mucus. As the female rice eels lay the eggs, the males pick them up, one by one, and spit them into the floating nests. The males guard their nests and eggs as they float freely in the water and stay with their young until they are independent. Rice eels have successfully adapted to living in rice patties and are an important source of food for people in some regions.

Swans

Swans are large **birds** in the waterfowl family, *Anatidae*, which also includes **ducks** and **geese**. There are seven **species** of swans, occurring on all continents except **Antarctica**. Three species of swan, the mute swan (*Cygnus olor*), the **tundra** swan (*C. colombianus*), and the trumpeter swan (*C. buccinator*) breed regularly in **North America**.

Swans have a very long neck, and all North American species have a white body. Swans are well adapted to the aquatic environment, having fully webbed feet for swimming. However, these birds are not capable of submerging; instead they feed by tipping up and using their long neck to reach for their food of aquatic plants and occasional **invertebrates**. The plumage and external morphology of the sexes is similar. Swans are big birds, the largest species being the mute swan, which weighs as much as 33 lb (15 kg). Swans are flightless during the molt of late summer. They generally spend this time on large lakes, where they are relatively safe from terrestrial predators.

Most species of swans undertake substantial migrations between their breeding grounds (on large ponds and lakes) and wintering grounds (lakes or estuaries), sometimes traveling thousand of miles back and forth each year. Swans are highly territorial on the breeding grounds, trumpeting loudly and often engaging in fights with other swans. They even sometimes drive other species of waterfowl from their breeding **lake**. Swans typically mate for life, the pair staying together until one of the spouses dies.

Swans of North America

Three species of swan are regular breeders in North America. Mute swans breed in ponds and **rivers** in urban parks and other disturbed places, as well as more natural and wild habitats. The graceful and charismatic mute swan is familiar to most people. This species is native to Eurasia and was introduced to North America, where it has become the commonest of our swans.

The most abundant of the truly native species is the tundra or whistling swan, closely related to the Bewick's swan (*Cygnus columbianus bewickii*) of Eurasia. The tundra swan breeds widely on the low-arctic tundra of mainland Canada and southern Baffin Island, and mostly winters on estuaries of the Pacific and Atlantic coasts. During **migration** the tundra swan occurs on lakes and rivers, and sometimes in agricultural fields near **water**. This species usually nests beside marsh-fringed ponds or lakes. Soon after hatching, the brood of 3-5 young swans (or cygnets) accompany the parents, feeding on invertebrates during their first month or so, as well as on terrestrial **grasses** and **sedges** and aquatic vegetation.

The trumpeter swan is somewhat larger than the whistling swan. It breeds in isolated populations in Alaska, Yukon, British Columbia, and as far south as Oregon and Wyoming. This species once had a much more widespread breeding range, probably extending east to Ontario and Quebec. Unfortunately, because of agricultural conversions of its **habitat** and overhunting, this species was taken perilously close to **extinction**. However, the population of trumpeter swans has recently increased to more than 5,000 individuals, and appears to be slowly expanding. Nevertheless, the trumpeter swan remains vulnerable to population decline. The trumpeter swan winters on Pacific estuaries and offshore islands, and on a few inland lakes. Some efforts are being made to expand the range of this species through captive-breeding and release, for example, in Ontario.

Conservation of swans

Although swans are not abundant birds, during migration and winter these birds may congregate in large numbers. Moreover, swans are large animals, and they are relatively easy to approach on land when they are husbanding their young, or on water when they are in molt. Because of these factors, wild swans have long been hunted for subsistence or sport. However, because of their low reproductive potential, populations of swans are easily depleted and extirpated by overhunting. This occurred during the heyday of sport and market hunting of the nineteenth century, a population decline that was exacerbated by the widespread loss of breeding, migrating, and wintering habitats. Some species of swans, including the trumpeter swan of North America, became an **endangered species**, and were almost made extinct. More recently, however, the populations of most species of swans have generally been stable or have increased

A tundra swan (*Cygnus columbianus*) at the Kellogg Bird Sanctuary, Michigan. *Photograph by Robert J. Huffman. Field Mark Publications. Reproduced by permission.*

due to **conservation** efforts by government, especially during the latter half of the twentieth century. Today, there is no legal hunt of swans in North America, although there is some poaching of these birds, and a small aboriginal harvest of tundra swans occurs on their breeding grounds.

Swans are also at risk from other environmental stresses. The occurrence of **lead** pellets from spent shotgun ammunition in the surface sediment of their wintering habitat poses a toxic risk for wild swans. Waterfowl are poisoned when they retain the lead shot in their gizzard for use in grinding their food of **seeds**. This results in the abrasion and dissolving of the lead, which enters into the bloodstream and can cause a range of toxic problems, including death (see entry on ducks). During the 1980s and 1990s, it is likely that hundreds or thousands of swans have died each year in North America from poisoning by lead shot, and probably more than a million ducks and geese. Fortunately, the use of lead shot has been, or soon will be banned over most of North America, so this will be less of a problem in the future.

Swans are also vulnerable to certain types of infectious diseases, especially avian **cholera**. This **disease** is caused by a food- and water-borne pathogen, and it can occur as a local **epidemic** that kills tens of thousands of waterfowl on particular lakes. Potentially, the trumpeter swan is especially at risk from the effects of this sort of epidemic disease, because large numbers congregate on wintering grounds in only a few lakes in Montana, Idaho, and Wyoming. Birds wintering on coastal estuaries are at less risk from avian cholera.

Because of their positive aesthetics, swans have been widely cultivated in waterfowl collections and in public parks. The most commonly kept species is the mute swan, but other species are also bred, including the unusual black swan (*Cygnus atratus*) of **Australia**. Viewings of wild swans are also widely sought after by birders and other naturalists. Both the cultivation of swans and their non-consumptive use in **ecotourism** have economic benefits, and do not endanger populations of these birds.

Resources

Books

Godfrey, W.E. *The Birds of Canada.* Toronto: University of Toronto Press, 1986.

Johnsgard, P.A. *Ducks in the Wild: Conserving Waterfowl and Their Habitats.* Swan Hill Press, 1992.

Owen, M., and J. M. Black. *Waterfowl Ecology.* London: Blackie Pub., 1990.

Sibley, David Allen. *The Sibley Guide to Birds.* New York: Knopf, 2000.

Bill Freedman

Sweet gale family (Myricaceae)

The sweet gale or bayberry family (Myricaceae) is made up of about 50 **species** of shrubs and trees. Minor economic uses of some species involve the extraction of a fragrant wax from their **fruits** and cultivation as ornamental shrubs.

The foliage of plants in the Myricaceae can be deciduous or evergreen, and the leaves are commonly fragrant when crushed. The flowers are small and occur in catkin-like inflorescences, which are made up of either male (staminate) or female (pistillate) flowers. The fruits are a one-seeded berry or drupe, often with a thick, waxy coating.

One of the most interesting characteristics of species in the sweet gale family is their ability to fix atmospheric **nitrogen** gas (N_2-sometimes known as dinitrogen) into **ammonia** (NH_3), an inorganic form of nitrogen that can be utilized by plants as a nutrient. The **nitrogen fixation** is carried out inside of specialized nodules on the roots and rhizomes of these plants. This is done by the **enzyme** nitrogenase, which is synthesized by **bacteria** that live in a **mutualism** with the vascular **plant**. The ability to fix atmospheric dinitrogen into ammonium is an extremely useful trait, because it allows species in the sweet gale family to be relatively successful in nutrient deficient habitats.

The most diverse genus in the Myricaceae is *Myrica*, of which several species occur widely in **North America**. The sweet gale (*Myrica gale*) is a common shrub along lakeshores and in other moist places over much of the temperate and boreal zones of North America and Eurasia. The sweet fern or fern-gale (*M. asplenifolia*) is widespread in eastern North America. Both the wax-myrtle (*M. cerifera*) and the bayberry or waxberry (*M. carolinensis*) of eastern North America produce fruits with thick, whitish, waxy coats. These fruits can be collected, the wax extracted using boiling **water** and then used to fragrantly scent candles.

See also Symbiosis.

Bill Freedman

Sweet potato

The sweet potato (*Ipomoea batatas*) is a creeping, vine-like **plant** that is cultivated in tropical climates for its starchy, nutritious tubers. The sweet potato is in the morning glory family (Convolvulaceae). The sweet potato is sometimes referred to as a **yam**, but it is quite different from the true yam (*Dioscorea batatas*), which is another **species** of tropical root-crop.

The sweet potato is a perennial, trailing vine that develops tubers at the ends of adventitious roots. Sweet potatoes are easily propagated vegetatively, by planting small pieces of roots or vine cuttings.

Although sweet potatoes are not consumed in large quantities in **North America** or **Europe**, they are a very important crop in the tropics. The global production of sweet potatoes ranks this plant with the eight most important foods grown by people in terms of the annual production of edible **biomass** (about 127,000,000 tons [140,000,000 mt] per year).

Sweet potatoes have an ancient history of cultivation as a food crop. Archaeological evidence suggests that sweet potatoes may have been grown in Peru for as long as 5,000 years. This crop is known to have been cultivated in the tropical Americas, Southeast **Asia**, and Polynesia at the time of the first visits to these regions by Europeans. Many botanists believe that the sweet potato originated in Central America or northern **South America**, and was then somehow transported to Asia during prehistoric times. This dispersal may have happened by now-forgotten, trans-Pacific trade or **migration** voyages, or possibly by seed capsules floating across the **ocean**.

There are many varieties of sweet potatoes, but these can be categorized into two broad groups. One group has moist, sweet tubers with an orange-colored interior—this is the type most commonly found in markets in North America. The other, more diverse group of sweet potatoes has a drier and mealier **tuber**, with a yellow, white, or red-colored interior, and is most extensively grown and eaten throughout the tropics.

Sweet potatoes are a valuable source of **carbohydrate**, beta-carotene, and fiber, containing as much as 5% protein. However, a diet rich in sweet potatoes must be balanced with other foods, mostly to ensure an adequate intake of **proteins**, and a balance of essential amino acids. Sweet potatoes can be eaten after boiling or roasting, or they can be manufactured into a flour, or used to make a product known as Brazilian **arrowroot**, and sometimes fermented into **alcohol**. Sweet potatoes are also fed to **livestock**, as are the vines and foliage of the plant. Because the tubers are rather moist, sweet potatoes spoil relatively easily, and **crops** do not always store well.

Swifts

Swifts are the fastest fliers of all of the small **birds**, reaching speeds of 172-218 mph (275-349 kph), although 35-80 mph (56-128 kph) is more common. They belong to the family Apodidae, a name meaning "without feet" and a reference to the fact that a swift in flight appears to have no legs or feet. Indeed, the legs of swifts are small and weak so that a swift that lands on the ground may have difficulty taking off again.

Swifts gather food, drink, bathe, court, and mate all while on the wing. They are more closely related to the **hummingbirds** than to the swallows, which they resemble. There are about 80 **species** of swifts, and they are found throughout the world except in the Arctic and Antarctic. Swifts are generally small birds, ranging from the size of a sparrow to the size of a small hawk. They are generally gray or brown, although some species are marked with white.

The wings of swifts are slender and pointed, the ideal shape for a rapid flier. The tail contains 10 feathers, and may be forked or short and stiff. Although their legs are weak, swifts' claws are strong and ideally suited to clinging to chimneys or rock walls on which they roost. The bill of swifts is tiny, but the gape is large, and well-suited for catching **insects** while in flight. Among the insects eaten by swifts are spiders, **aphids**, **ants**, **bees**, **wasps**, midges, **mayflies**, **beetles**, and **termites**. One Alpine swift was found to have a ball (called a bolus) of 600 insects in its throat, which it was taking back to feed its young.

Swifts can be distinguished from the superficially-similar swallows by their style of flight. Swifts alternate a short glide with a series of shallow, rapid wingbeats, and in flight can be mistaken for **bats**. Swallows, on the other hand, are more fluid fliers. It was thought that swifts alternated their wing beats in flight—using first the left then the right, but stroboscopic studies have confirmed that the wings in fact beat in unison (if jerkily) because of the swifts' short, massive wing bones.

Not only are swifts rapid flyers, but they also have great endurance. One bird was estimated to have flown more than million miles in its nine-year life span. Since swifts are long-lived birds (some have been known to live for nearly 20 years), it is not unreasonable to assume that some individuals may have flown far more miles than that. Swifts migrate seasonally with species from **North America** generally wintering in **South America**. However, one North American species, the white-throated swift, becomes torpid during periods of cold **weather**, much like its relative the hummingbird.

Swifts are gregarious. A large colony of chimney swifts nested in an air shaft at Kent State University in Ohio. When chimney swifts prepare to roost for the night, a flock will circle about a chimney for as long as an hour. The bird closest to the chimney will be the first to enter, and the rest follow, looking like a puff of smoke going down the chimney.

Nesting is usually a solitary affair. Swifts are loyal to both their traditional nesting site and to their mate. The male and female gather twigs for the nest by grabbing the twig while in flight; if it breaks off, the bird adds it to the nest. Nests are cup-shaped agglomerations of twigs, mud, **moss**, and saliva (building materials vary by species), usually in a spot secluded and inaccessible to predators, such as a crevice in a high cliff face. Because the birds need so much saliva during breeding season, their salivary **glands** increase in size accordingly, swelling to as much as 12 times the normal size.

The glutaneous saliva is considered by some people a delicacy. In Southeast **Asia**, swift nests are harvested for use in bird's-nest soup, which some believe keeps people young. The nests of these Asian species are constructed entirely of fast-drying saliva. Some swift species have become rare because of the harvesting of the nests. And, predictably, as the nests have become more rare, they have become more expensive, commanding as much as $1,000 per pound. The escalating price has not reduced demand, and nest collectors are traveling farther afield in China, Thailand, Vietnam, and other Asian countries in search of nests.

Depending on her species, the female lays between one and six white eggs. Both the male and female incubate the eggs, which hatch 17-28 days after being laid. The young swifts fledge at about 30 days old.

The chimney swift (*Chaetura pelagica*) is the only North American species generally found east of the Mississippi River. Other species include Vaux's swift (*Chaetura vauxi*), the smallest North American species, which weights just 0.66 oz (18.9 g), and the black swift (*Cypseloides niger*), the largest North American swift, which is rarely seen because of its high mountain **habitat**. The swift is not a songbird, and its voice consists of high, piercing screams.

See also Swallows and martins.

Resources

Books

Forshaw, Joseph. *Encyclopedia of Birds.* New York: Academic Press, 1998.
Terres, John K. *The Audubon Society Encyclopedia of North American Birds.* New York: Wing Books, 1991.

F. C. Nicholson

Swordfish

The swordfish (*Xiphias gladius*), also known as the broadbill, or the forktail, is the only **species** in the **bony fish** family Xiphiidae. The swordfish is highly prized as a food **fish**, and as a game fish. Its most distinguishing characteristic is the remarkable elongation of the upper jaw, which resembles a long, flattened, serrated sword and can extend up to one-third of the body length. The sword is used as an offensive weapon, and to spear **prey**, which may be other fish or large **mollusks**. Swordfish often attack schools of **mackerel**, gashing several fish with their sword before devouring them.

Swordfish are dark in **color**, most often brownish black or black, with a lighter brown below. Adult swordfish are devoid of scales and of teeth. Commercially, swordfish are caught by harpooning, and fish in excess of 1,200 lb (545 kg) have been taken by this method. Swordfish are spotted by their curved dorsal fin exposed above the **water** surface. Swordfish are also caught by rod-and-reel using heavy tackle, and are baited by trolling **squid** or mackerel. When hooked, a swordfish will make extreme efforts to free itself by leaping out of the water several times before eventually tiring. The angler may have to patiently play the fish for three or four hours before it is subdued. The record catch for a swordfish by rod-and-reel is 1,182 lb (537 kg).

When injured or hooked, a swordfish may thrust itself out of the water, squirm violently and attack anyone or anything in its path. When approached by a boat, the fish may pretend to be exhausted and then suddenly ram its sword into the side of the boat. The **force** of the thrust can be sufficient to pierce a 2-in (5 cm) thick, solid-wood side of a boat. If the boat hull is wooden, the sword may go in too deep to be removed by the fish, and may be broken off in order to escape.

The sail fish (*Istiophorus* spp.), spearfish (*Tetrapturus* spp.), and marlin (*Makaira* spp.) are relatives of the swordfish, but are placed in the family Istiophoridae.

Symbiosis

Symbiosis is a word used to refer to intimate relationships among **species**. Symbioses can involve interactions of individuals of different species, or associations of populations of one or more species. Symbiosis can involve obligate relationships, in which the symbionts cannot live apart in nature, but usually the association is more flexible than this.

Various types of symbiosis

Mutualism is a symbiosis between species in which both partners benefit. Mutualism is considered by some biologists to be the archetypal form of symbiosis. The examples of symbiosis that are discussed in the next section are all mutualisms.

Parasitism is another type of symbiotic association, in which one **organism** obtains nourishment from a host, usually without killing it. In most parasitisms, the parasite has a close and sometimes obligate relationship with the host. However, to be healthy, the host by no means needs the parasite, and in fact usually suffers a detriment from the symbiosis. **Commensalism** is a relationship in which one symbiont benefits from the interaction, while the host species is neither positively or negatively affected. For example, small epiphytic plants derive a substantial ecological benefit from living on larger plants, but the latter are not usually affected to a meaningful degree.

Examples of natural symbioses

Most biologists, when confronted by the need to illustrate the concept of symbiotic mutualism, describe the case of **lichens**. Lichens are an obligate association between a fungus (the mycobiont) and an alga or blue-green bacterium (the phycobiont). Lichen mutualisms are very distinctive, and they can be identified on the basis of the size, shape, **color**, and **biochemistry** of their **biomass**. Lichenologists have developed systematic and taxonomic treatments of lichens, even though these mutualisms are not true "organisms" in the con-

A cape buffalo with an oxpecker on its back in Kenya. The relationship between the oxpecker and the buffalo is a type of symbiosis called mutualism; the oxpecker feeds from the supply of ticks on the buffalo, which in turn benefits from tick removal. *JLM Visuals. Reproduced by permission.*

ventional meaning of the word. The fungus benefits from the lichen mutualism through access to photosynthetic products of the alga or blue-green bacterium, while the phycobiont benefits from provision of a relatively moist **habitat** and enhanced access to inorganic **nutrients**.

Certain species of **fungi** also occur in intimate associations with the roots of vascular plants, in a mutualism referred to as mycorrhizae. The **plant** benefits from the **mycorrhiza** through increased access to inorganic nutrients, especially phosphate, while the fungus gains an advantage through access to nutritious exudates from the roots of the plant. This is a very widespread mutualism—most vascular plants have mycorrhizae.

Some vascular plants live in a mutualism with particular **microorganisms** that have the ability to fix atmospheric dinitrogen into **ammonia**, a form of inorganic **nitrogen** that the plant can utilize in its **nutrition** (see entry on **nitrogen cycle**). The best known examples of this mutualism involve various species of plants in the legume family (Fabaceae) and a bacterium known as *Rhizobium japonicum*. In this mutualism, the plant benefits from increased access to an important

nutrient, while the bacterium gains an advantage through the provision of an appropriate habitat in the form of root nodules, as well as fixed **energy** provided by the host plant.

Another common mutualism occurs in the guts of animals that eat plant **matter**. Many animals consume plant biomass, but most are not very effective at digesting polymeric biochemicals such as **cellulose** and lignin. Often, these animals live in a symbiosis with microorganisms, which inhabit part of the gut and secrete specialized enzymes, such as cellulases, which digest cellulose. The herbivorous **animal** benefits from access to a large source of fixed energy, while the microorganisms benefit from access to a safe and appropriate habitat, and to nutritious chemicals available in the animal gut. This sort of mutualism occurs, for example, between **termites** and symbiotic **bacteria** and protozoans. In the case of the termite *Eutermes*, protozoans in the gut may account for 60% of the insect's weight. Many herbivorous **mammals** also live in a cellulose-digesting symbiosis with bacteria and protozoans, as is the case of ruminants such as the domestic **sheep** (*Ovis aries*) and cow (*Bos taurus*).

Symbioses between humans and other species

Humans live in symbioses of various intensities with a number of domesticated animals and plants. To varying degrees, these cultural symbioses are mutualistic, with both humans and the other species benefitting.

For example, all important agricultural plants exist in tight mutualisms with humans. Agricultural varieties of corn or maize (*Zea mays*), for example, are no longer capable of reproducing independently of human management. This is because over time the fruiting structure of maize has been selected to be enclosed in a leafy sheath that does not open, and to have **seeds** that do not easily separate (or shatter) from the supporting **tissue** (the cob). If humans did not plant the seeds of maize, the species would rapidly become extinct, because it no longer occurs as wild populations. The same is substantially true for most agricultural plants that have become extensively modified through cultural **selection** by humans. Humans, of course, benefit greatly from their mutualisms with agricultural plants, through the provision of **crops** of food, fiber, and other products.

Similarly, agricultural animals live in a symbiotic mutualism with humans. Cows (*Bos taurus*), for example, benefit from their human-managed access to fodder, veterinary services, and protection from predators, while humans benefit from access to milk and meat.

Even the keeping of animals as pets represents a type of mutualism. Pet dogs (*Canis familiaris*) and **cats** (*Felis catus*) are fed and kept safe in domestication, while humans benefit from the companionship of these animals, and sometimes from other services, as when cats kill pest **rodents**.

Symbiosis and evolution

Ideas about symbiosis have made some important contributions to theories that help explain the **evolution** of complex life forms on **Earth**. The first organisms on Earth were prokaryotic viruses and blue-green bacteria, which do not have an organized nucleus. Eukaryotic cells are more complex, having their nuclear material bounded within a nucleus, as well as other cellular organelles such as **ribosomes**, chloroplasts, cilia, flagellae, and other important structures. An exciting theory of the origin of eukaryotic cellular organization postulates the occurrence of a series of symbioses, in which prokaryotic cells became intimately associated with each other, with certain cellular functions being mutualistically divided amongst the symbionts. For example, certain tiny symbionts might have become responsible for the most

---

KEY TERMS

Mutualism—A mutually beneficial relationship between species.

Symbiosis—A biological relationship between two or more organisms that is mutually beneficial. The relationship is obligate, meaning that the partners cannot successfully live apart in nature.

---

of the respiratory **function** of the mutualism, and could then have evolved into mitochondria. Other symbionts, such as blue-green bacteria, could have been responsible for **photosynthesis** by the **cell**, and may have evolved into chloroplasts. To a degree these ideas are supported by the observation that both ribosomes and chloroplasts contain small amounts of genetic material (DNA, or deoxyribonucleic acid), which may be relict from an earlier, independent existence.

Another recent and highly controversial theory, called the **Gaia hypothesis**, suggests that Earth may represent an enormous, quasi-organismic entity, in which all species comprise a global, symbiotic, physiological system that maintains environmental conditions within a range that life can tolerate. Supporting evidence for this hypothesis includes the suggestion that the **oxygen** in Earth's atmosphere is ultimately of biological origin, having been emitted by photosynthetic organisms. Without oxygen, of course, most species could not survive. In addition, some ecologists suggest that the **concentration** of **carbon dioxide** in Earth's atmosphere is to a large degree regulated by a complex of integrated biological and physical processes by which CO_2 is emitted and absorbed. This gas is well known to be important in the planet's **greenhouse effect**, which is critical in maintaining the average surface **temperature** within a range that organisms can tolerate.

See also Parasites.

Resources

Books

Begon, M., J.L. Harper, and C.R. Townsend. *Ecology: Individuals, Populations and Communities.* 2nd ed. London: Blackwell Sci. Pub., 1990.

Brewer, R. *The Science of Ecology.* 2nd ed. Fort Worth: Saunders, 1994.

Margulis, L., and L. Olendzenski, eds. *Environmental Evolution: Effects of the Origin and Evolution of Life on Planet Earth.* Cambridge: MIT Press, 1992.

Bill Freedman

Symbol, chemical

Chemical symbols are shorthand abbreviations of the names of the 109 known elements. Each element has its own unique symbol. Since science is an international enterprise, chemical symbols are determined by international agreement.

The use of symbols for the chemical elements existed long before a systematic method was developed. The alchemists associated the symbols of the planets not only with the days of the week, but also with the seven metals known at the time: gold, silver, **iron**, mercury, tin, **copper, and lead**. By the beginning of the nineteenth century, there were about 26 known elements, but by the beginning of the twentieth century, there were more than 81. As more elements were discovered, the need for symbolic representations for these elements became more evident.

During the first half of the nineteenth century, an outstanding Swedish chemist, Jöns Jacob Berzelius, systematically assigned letters as symbols for the elements. This method soon became accepted by chemists everywhere. Today, the International Union of Pure and Applied Chemistry (IUPAC) is the organization that makes the final decision on the names and symbols of the element.

Chemical symbols are composed of one or two letters. The first letter is always capitalized and the second, if there is one, is always lowercase. Often these are the first two letters of the element's name but this is not always possible, because it would sometimes cause duplication. For example, the symbol C represents **carbon**, Ca represents **calcium**, Cd represents cadium, and Cf represents californium.

Most of the elements have symbols derived from their English name, but a few symbols stem from other languages. Notable among these are ten common elements whose names and symbols are derived from Latin. These are: antimony, Sb, from stibium; copper, Cu, from cuprum; gold, Au, from aurum; iron, Fe, from ferrum; potassium, K, from kalium; lead, Pb, from plumbum; mercury, Hg, from hydrargyrum; silver, Ag, from argentum; **sodium**, Na, from natrium; and tin, Sn, from stannum.

Chemical symbols are used by chemists everywhere in writing chemical formulas, in which the symbols represent the **atoms** of the elements present in a compound.

Some strange-looking element names with even stranger-looking three-letter symbols have been in general use recently for the heaviest chemical elements. Because they may still be seen in periodic tables published between 1980 and 1994, it is important to know what they mean. For example the element of **atomic number** 104 was referred to as unnilquadium, with the symbol Unq, and element 108 was referred to as unniloctium, with the symbol Uno. These names and symbols are based entirely on the atomic numbers themselves: un means 1, nil means 0, and quad means 4; therefore unnilquad means 104. These names and symbols were recommended by the International Union of Pure and Applied Chemistry to be used temporarily, until certain disputes about who discovered these elements could be resolved.

The discoverers of a new element have historically been given the right to suggest a name. But three groups of nuclear chemists all claim to have been the first to discover some of the transuranium elements: an American group at the Lawrence Berkeley Laboratory in California, a Russian group at the Joint Institute for Nuclear Research in Dubna, and a German group at the Gesellschaft für Schwerionenforschung in Darmstadt. In 1994, the International Union of Pure and Applied Chemistry attempted to settle the issues by recommending "official" names and symbols for elements 101 through 109, but they are still being hotly debated.

See also Element, chemical; Periodic table.

Symbolic logic

Logic is the study of the rules which underlie plausible reasoning in **mathematics**, science, law, and other disciplines.

Symbolic logic is a system for expressing logical rules in an abstract, easily manipulated form.

Symbols

In **algebra**, a letter such as x represents a number. Although the symbol gives no clue as to the value of the number, it can be used nevertheless in the formation of sums, products, etc. Similarly P, in **geometry**, stands for a point and can be used in describing segments, intersections, and the like.

In symbolic logic, a letter such as p stands for an entire statement. It may, for example, represent the statement, "A triangle has three sides." In algebra, the plus sign joins two numbers to form a third number. In symbolic logic, a sign such as ∨ connects two statements to form a third statement. For example, ∨ replaces the word "or" and ∧ replaces the word "and." The following is a list of the symbols commonly encountered:

p, q, r,…	statements	
v	"or"	
Λ	"and"	
~	"it is not the case that"	
=>	"implies" or "If…, then…"	
⟷	"implies and is implied by" or "…. if and only if…"	

Statements

Logic deals with statements, and statements vary extensively in the precision with which they may be made. If someone says, "That is a good book," that is a statement. It is far less precise, however, than a statement such as "Albany is the capital of New York." A good book could be good because it is well printed and bound. It could be good because it is written in good style. It could tell a good story. It could be good in the opinion of one person but mediocre in the opinion of another.

The statements that logic handles with the greatest certainty are those that obey the law of the excluded middle, i.e., which are unambiguously true or false, not somewhere in between. It doesn't offer much help in areas such as literary criticism or history where statements simple enough to be unequivocally true or false tend also to be of little significance. As an antidote to illogical thinking, however, logic can be of value in any discipline.

By a "statement" in logic one means an assertion which is true or false. One may not know whether the statement is true or false, but it must be one or the other. For example, the Goldbach conjecture, "Every even number greater than two is the sum of two primes," is either true or false, but no one knows which. It is a suitable statement for logical analysis.

Other words that are synonyms for "statement" are "sentence," "premise," and "proposition."

Conjunctions

If p stands for the statement, "All right angles are equal," and q the statement, "Parallel lines never meet," one can make a single statement by joining them with "and": "All right angles are equal and parallel lines never meet." This can be symbolized p Λ q, using the inverted V-shaped symbol to stand for the conjunction "and." Both the combined statement and the word "and" itself are called "conjunctions." In ordinary English, there are several words in addition to "and" which can used for joining two statements conjunctively, for example, "but." "But" is the preferred conjunction when one wants to alert the reader to a relationship which otherwise might seem contradictory. For example, "He is 6 ft (1.8 m) tall, but he weighs 120 lb (54 kg)." In logic the only conjunctive term is "and."

Negation

Negation is another logical "operation." Unlike conjunction and disjunction, however, it is applied to a single statement. If one were to say, "She is friendly," the negation of that statement would be, "She is not friendly." The symbol for negation is "~." It is placed in front of the statement to be negated, as in ~ (pΛq) or ~ p. If p were the statement, "She is friendly," ~ p means "She is not friendly," or more formally, "It is not the case that she is friendly." Prefacing the statement with, "It is not the case that…," avoids embedding the negation in the middle of the statement to be negated. The symbol lips is read "not p."

The statement ~ p is true when p is false, and false when p is true. For example, if p is the statement "x < 4," ~ p is the statement "x ≥ 4." Replacing x with S makes p false but ~ p true. If a boy, snubbed by the girl in "She is friendly," were to hear the statement, he would say that it was false. He would say, "She is not friendly," and mean it.

Truth tables

The fact that someone says something doesn't make it true.

Statements can be false as well as true. In logic, they must be one or the other, but not both and not neither. They must have a "truth value," true or false, abbreviated T or F.

p	q	~p	pΛq	pVq
T	T	F	T	T
T	F	F	F	T
F	T	T	F	T
F	F	T	F	F

Whether a conjunction is true depends on the statements which make it up. If both of them are true, then the conjunction is true. If either one or both of them are false, the conjunction is false. For example, the familiar expression $3 < x < 7$, which means "x > 3 and x < 7" is true only when both conditions are satisfied simultaneously, that is for numbers between 3 and 7.

Disjunctions

Another word used in both ordinary English and in logic is "or." Someone who says, "Either he did not hear me, or he is being rude," is saying that at least one of those two possibilities is true. By connecting the two possibilities about which he or she is unsure, the speaker can make a statement of which he or she is sure.

In logic, "or" means "and/or." If p and q are statements, p V q is the statement, called a "disjunction," formed by connecting p and q with "or," symbolized by "V."

For example if p is the statement, "Mary Doe may draw money from this account," and q is the statement, "John Doe may draw money from this account," then p V q is the statement, "Mary Doe may draw money from this account, or John Doe may draw money from this account."

The disjunction p V q is true when p, q, or both are true. In the example above, for instance, an account set up in the name of Mary or John Doe may be drawn on by both while they are alive and by the survivor if one of them should die. Had their account been set up in the name Mary and John Doe, both of them would have to sign the withdrawal slip, and the death of either one would freeze the account. Bankers, who tend to be careful about money, use "and" and "or" as one does in logic.

One use of truth tables is to test the equivalence of two symbolic expressions. Two expressions such as p and q are equivalent if whenever one is true the other is true, and whenevet one is false the other is false. One can test the equivalence of ~(p V q) and ~p∧~q (as with the minus sign in algebra, "~" applies only to the statement which immediately follows it. If it is to apply to more than a single statement, parentheses must be used to indicate it):

p	q	~p	~q	pVq	~(pVq)	~p∧~q
T	T	F	F	T	F	F
T	F	F	T	T	F	F
F	T	T	F	T	F	F
F	F	T	T	F	T	T

The expressions have the same truth values for all the possible values of p and q, and are therefore equivalent.

For instance, if p is the statement "x > 2" and q the statement "x < 2," p V q is true when x is any number except 2. Then (p V q) is true only when x = 2. The negations p and q are "x 2" and "x 2" respectively. The only number for which ~ p ∧ ~ q is true is also 2.

Algebra of statements

Equivalent propositions or statements can be symbolized with the two-headed arrow "⟷." In the preceeding section we showed the first of De Morgan's rules:

1. ~(p V q) ⟷ ~p ∧ ~q

2. ~(p ∧ q) ⟷ ~p V ~q

Rules such as these are useful for simplifying and clarifying complicated expressions. Other useful rules are

3. p ⟷ ~(~p)

4. p ∧(q V r) ⟷ (p ∧ q) V (p ∧ r) (a distributive law for "and" applied to a disjunction)

5. p ∧ q ⟷ q ∧ p p V q ⟷ q V p

6. (p ∧ q) V r ⟷ p V (q V r) (p V q) V r ⟷ p V (q V r)

Each of these rules can be verified by writing out its truth table.

A truth table traces each of the various possibilities. To check rule 4 with its three different statements, p, q, and r, would require a truth table with eight lines. On occasion one may want to know the truth value of an expression such as ((T V F) L ∧ (F V T)) V ~ F where the truth values of particular statements have been entered in place of p1 q, etc. The steps in evaluating such an expression are as follows:

((T V F) ∧(F V T)) V ~ F Given

(T ∧ T) v T Truth tables for V,~

T V T Truth table for ∧

T Truth table for V

Such a compound expression might come from the run-on sentence, "Roses are red or daisies are blue, and February has 30 days or March has 31 days; or it is not the case that May is in the fall." Admittedly, one is not likely to encounter such a sentence in ordinary conversation, but it illustrates how the rules of symbolic logic can be used to determine the ultimate truth of a complex statement. It also illustrates the process of replacing statements with known truth values instead of filling out a truth table for all the possible truth values. Since this example incorporates five different statements, a truth table of 32 lines would have been needed to run down every possibility.

Implication

In any discipline one seeks to establish facts and to draw conclusions based on observations and theories. One can do so deductively or inductively. In inductive reasoning, one starts with many observations and formulates an explanation that seems to fit. In deductive reasoning, one starts with premises and, using the rules of logical inference, draws conclusions from them. In disciplines such as mathematics, deductive reasoning is the predominant means of drawing conclusions. In fields such as **psychology**, inductive reasoning predominates, but once a theory has been formulated, it is both tested and applied through the processes of deductive thinking. It is in this that logic plays a role.

Basic to deductive thinking is the word "implies," symbolized by "=>." A statement p=> q means that whenever p is true, q is true also. For example, if p is the statement, "x is in Illinois," and q is the statement "x is in the United States," then p=> q is the statement, "If x is in Illinois, then x is in the United States."

In logic as well as in ordinary English, there are many ways of translating p=> q into words: "If p is true, then q is true" "q is implied by p;" "p is true only if q is true;" "q is a necessary condition for p;" "p is a sufficient condition for q."

The implication p => q has a truth table some find a little perplexing:

p	q	p=>q
T	T	T
T	F	F
F	T	T
F	F	T

The perplexing part occurs in the next to last line where a false value of p seems to imply a true value of q. The fact that p is false does not imply anything at all. The imolication says only that q is true whenever p is. It doesn't say what happens when p is false. In the example given earlier, replacing x with Montreal makes both p and q false, but the implication itself is still true.

Implication has two properties which resemble the reflexive and **transitive** properties of equality. One, p=> p, is called a "tautology." Tautologies, although widely used, do not add much to understanding. "Why is the **water** salty?" asks the little boy.

"Because **ocean** water is salty," says his father.

The other property, "If p=> q and q=> r, then p=>r," is also widely used. In connecting two implications to form a third, it characterizes a lot of reasoning, formal and informal. "If we take our vacation in January, there will be snow. If there is snow, we can go skiing. Let's take it in January." This property is called a "syllogism."

A third property of equality 1 "If a = b, then b = a," called **symmetry** may or may not be shared by the implication p=>q. When it is, it is symbolized by the two-headed arrow used earlier, "p $\longleftrightarrow$ q." p $\longleftrightarrow$ q means (p=> q) Λ (q=> p). It can be read "p and q are equivalent;" p is true if and only if q is true;" "p implies and is implied by q;" "p is a necessary and sufficient condition for q;" and "p implies q, and conversely."

In p=> q, p is called the "antecedent" and q the "consequent." If the antecedent and consequent are inter-

changed, the resulting implication, q=> p, is called the "converse." If one is talking about triangles, for example, there is a **theorem**, "If two sides are equal, then the angles opposite the sides are equal." The converse is, "If two angles are equal, then the sides opposite the angles are equal."

If an implication is true, it is never safe to assume that the converse is true as well. For example, "If x lives in Illinois, then x lives in the United States," is a true implication, but its converse is obviously false. In fact, assuming that the converse of an implication is true is a significant source of fallacious reasoning. "If the **battery** is shot, then the car won't start." True enough, but it is a good idea to check the battery itself instead of assuming the coverse and buying a new one.

Implications are involved in three powerful lines of reasoning. One, known as the Rule of Detachment or by the Latin *modus ponendo ponens,* states simply "If p=> q and p are both true, then q is true." This rule shows up in all sorts of areas. "If x dies, then y is to receive $100,000." When x dies and proof is submitted to the insurance company, y gets a check for the money. The statements p=> q and p are called the "premises" and q the "conclusion."

A second rule, known as *modus tollendo tollens,* says if p=>q is true and q is false, then p is false. "If x ate the cake, then x was home." If x was not at home, then someone else ate the cake.

A third rule, *modus tollerdo ponens,* says that if p V q and p are true, then q is true. Mary or Ann broke the pitcher.

Ann did not; so Mary did. Of course, the validity of the argument depends upon establishing that both premises are true.

It may have been the cat.

Another type of argument is known as *reductio ad absurdum*, again from the Latin. Here, if one can show that ~p=> (q Λ ~q), then p must be true. That is, if assuming the negation of p leads to the absurdity of a statement which is both true and false at the same time, then p itself must be true.

Resources

Books

Carroll, Lewis. *Symbolic Logic.* New York: Dover Publications Inc. 1958.

Christian, Robert R. *Logic and Sets.* Waltham, Massachusetts: Blaisdell Publishing Co., 1965.

Suppes, Patrick, and Shirley Hill. *First Course in Mathematical Logic.* Waltham, Massachusetts: Blaisdell Publishing Co., 1964.

J. Paul Moulton

Symmetry

Symmetry is a property of some images, objects, and mathematical equations whereby **reflections**, rotations, or substitutions cause no change in properties or appearance. For example, the letter M is symmetrical across a line drawn down its center, a ball is symmetrical under all possible rotations, and the equation $y = x^2$ (a **parabola**) is symmetrical under the substitution of $-x$ for x. This equation's mathematical symmetry is equivalent to its graph's physical symmetry. The ability of mathematical symmetries to reflect the physical symmetries of the real world is of great importance in **physics**, especially particle physics.

Many real objects and forces at all size scales—subatomic particles, **atoms**, crystals, organisms, stars, and galaxies—exhibit symmetry, of which there are many kinds. *Line* or *bilateral* symmetry, the simplest and most familiar, is the symmetry of by any figure or object that can be divided along a central line and then restored (geometrically) to wholeness by reflecting its remaining half in a mirror.

Symmetries are not only defined in terms of reflection across a line. A **sphere**, for example, can be rotated through any angle without changing its appearance, and in **mathematics** is said to possess O(3) symmetry. The quantum field equations whose solutions describe the **electron**, which is, like a sphere, the same viewed from any direction, also have O(3) symmetry.

In particle physics, the mathematics of symmetry is an essential tool for producing an organized account of the confusing plethora of particles and forces observed in Nature and for making predictions based on that account. An extension of the parabola example shows how it is possible for mathematical symmetry to lead to the prediction of new phenomena. Consider a system of two equations, $y = x^2$ and $y = 4$. There are two values of x that allow both equations to be true at once, $x = 2$ and $x = -2$. The two (x, y) pairs (2, 4) and (-2, 4) are termed the *solutions* to this system of two equations, because

KEY TERMS

· ·

Noether's theorem—Mathematical theorem stating that every conservation law in Nature must have a symmetrical mathematical description.

both both equations are simultaneously true if and only if x and y have these values. (The two solutions correspond to the points where a horizontal line, $y = 4$, would intersect the two rising arms of the parabola.) If this system two equations constitued an extremely simple theory of **matter**, and if one of its two solutions corresponded to a known particle, say with "spin" = x = 2 and "mass" = y = 4, then one might predict, based on the symmetry of the two solutions, that a particle with "spin" = -2 and "mass" = 4 should also exist. An analogous (though more complex) process has actually led physicists to predict, seek, and find certain fundamental particles, including the Ω^- baryon and the η^0 muon).

Symmetry, however, not only is a useful tool in mathematical physics, but has a profound connection to the laws of Nature. In 1915, German mathematician Emmy Noether (1882–1835) proved that every **conservation** law corresponds to a mathematical symmetry. A conservation law is a statement that says that the total amount of some quantity remains unchanged (i.e., is conserved) in any physical process.

Momentum, for example, is conserved when objects exert **force** on each other; **electric charge** is also conserved. The laws (mathematical equations) that describe momentum and charge must, therefore, display certain symmetries.

Noether's **theorem** works both ways: in the 1960s, a conserved quantum-mechanical quantity (unitary spin) was newly defined based on symmetries observed in the equations describing a class of fundamental particles termed hadrons, and has since become an accepted aspect of particle physics. As physicists struggle today to determine whether the potentially all-embracing theory of "strings" can truly account for all known physical phenomena, from **quarks** to gravity and the Big Bang, string theory's designers actively manipulate its symmetries in seeking to explore its implications.

See also Cosmology; Relativity, general; Relativity, special.

Resources

Books

Barnett, R. Michael, Henry Mühry, and Helen R. Quinn. *The Charm of Strange Quarks.* New York: Springer-Verlag, 2000.

Elliot, J.P., and P.G. Dawber. *Symmetry in Physics.* New York: Oxford University Press, 1979.

Silverman, Mark. *Probing the Atom* Princeton, NJ: Princeton University Press, 2000.

Other

Cambridge University. "Cambridge Cosmology." [cited February 14, 2003]. <http://www. damtp.cam.ac.uk/user/gr/public/cos_home.html>.

Larry Gilman

Synapse

Nerve impulses are transmitted through a functional gap or intercellular space between neural cells (neurons) termed the synapse (also termed the synaptic gap). Although nerve impulses are conducted electrically within the **neuron**, in the synapse they are continued (propogated) via a special group of chemicals termed neurotransmitters.

The synapse is more properly described in structural terms as a synaptic cleft. The cleft is filled with extra cellular fluid and free neurotransmitters.

The neural synapse is bound by the presynaptic terminal end of one neuron, and the dendrite of the postsynaptic neuron. Neuromuscular synapses are created when neurons terminate on a muscle. Neuroglandular synapses occur when neurons terminate on a gland. The major types of neural synapses include axodendritic synapses, axosomatic synapses, and axoaxonic synapses—each corresponding to the termination point of the presynaptic neuron.

The arrival of an **action potential** (a moving wave of electrical changes resulting from rapid exchanges of ions across the neural **cell membrane**) at the presynaptic terminus of a neuron, expels synaptic vesicles into the synaptic gap.

The four major neurotransmitters found in synaptic vesicles are noradrenaline, actylcholine, **dopamine**, and serotoin. Acetylchomine is derived from **acetic acid** and is found in both the central **nervous system** and the peripheral nervous system. Dopamine, epinephrine, and norepinephrine are catecholamines derived from tyrosine. Dopamine, epinephrine, and norepinephrine are also found in both the central nervous system and the peripheral nervous systems. Serotonin and **histamine** neurotransmitters are indolamines that primarily function in the central nervous system. Other amino acids, including gama-aminobutyric acid (GABA), aspartate, glutamate, and glycine along with neuropeptides containing bound amino acids also serve as neurotransmitters. Specialized neuropeptides include tachykinins and endorphins (including enkephalins) that function as natural painkillers.

Neurotransmitters diffuse across the synaptic gap and bind to **neurotransmitter** specific receptor sites on the dendrites of the postsynaptic neurons. When neurotransmitters bind to the dendrites of neurons across the synaptic gap they can, depending on the specific neurotransmitter, type of neuron, and timing of binding, excite or inhibit postsynaptic neurons.

After binding, the neurotransmitter may be degraded by enzymes or be released back into the synaptic cleft where in some cases it is subject to reuptake by a presynaptic neuron.

A number of neurons may contribute neurotransmitter molecules to a synaptic space. Neural transmission across the synapse is rarely a one-to-one direct **diffusion** across a synapse that separates individual presynaptic-postsynaptic neurons. Many neurons can converge on a postsynaptic neuron and, accordingly, presynaptic neurons are often able to affect the many other postsynaptic neurons. In some cases, one neuron may be able to communicate with hundreds of thousands of postsynaptic neurons through the synaptic gap.

Excitatory neurotransmitters work by causing ion shifts across the postsynaptic neural cell membrane. If sufficient excitatory neurotransmitter binds to dendrite receptors and the postsynaptic neuron is not in a refractory period, the postsynaptic neuron reaches threshold potential and fires off an electrical action potential that sweeps down the post synaptic neuron.

A summation of chemical neurotransmitters released from several presynaptic neurons can also excite or inhibit a particular postsynaptic neuron. Because neurotransmitters remain bound to their receptors for a time, excitation or inhibition can also result from an increased **rate** of release of neurotransmitter from the presynaptic neuron or delayed reuptake of neurotransmitter by the presynaptic neuron.

Bridge junctions composed of tubular **proteins** capable of carrying the action potential are found in the early embryo. During development, the **bridges** degrade and the synapses become the traditional chemical synapse.

See also Adenosine triphosphate; Nerve impulses and conduction of impulses; Neuromuscular diseases.

Resources

Books

Cooper, Geoffrey M. *The Cell—A Molecular Approach.* 2nd ed. Sunderland, MA: Sinauer Associates, Inc., 2000.

Gilbert, Scott F. *Developmental Biology.* 6th ed. Sunderland, MA: Sinauer Associates, Inc., 2000.

Guyton, Arthur C., and John E. Hall. *Textbook of Medical Physiology.* 10th ed. Philadelphia: W.B. Saunders Co., 2000.

Kandel, E.R., J.H. Schwartz, and T.M. Jessell., eds. *Principles of Neural Science.* 4th ed. New York: Elsevier, 2000.

Lodish, H., et. al. *Molecular Cell Biology.* 4th ed. New York: W. H. Freeman & Co., 2000.

Thibodeau, Gary A., and Patton, Kevin T. *Anatomy & Physiology.* 5th ed. Mosby, 2002.

Periodicals

Cowan, W.M., D.H. Harter, and E.R. Kandel. "The Emergence of Modern Neuroscience: Some Implications for Neurology and Psychiatry." *Annual Review of Neuroscience* 23: 343–39.

Abbas L. "Synapse Formation: Let's Stick Together." *Current Biology* 8 13 (January 2003): R25–7.

K. Lee Lerner

Syndrome

A syndrome is a collection of signs, symptoms, and other indications which, taken together, characterize a particular **disease** or abnormal medical condition. Medical workers place a heavy reliance on the collection of such indicators in the **diagnosis** of health problems and disorders.

The usual approach is to question patients about the nature of their complaints and then to conduct examinations and tests suggested by these reports. The collection of data resulting from the verbal report and clinical examination may then fall into a pattern—a syndrome—that makes it possible for the physician to predict the disorder responsible for the patient's problems. That diagnosis, in turn, may lead to a course of therapy designed to deal with the problem.

As an example, a patient may complain to a physician of headaches, visual problems, and difficulty in breathing when lying down. Clinical examination may then indicate a dilation of **veins** in the upper part of the chest and neck with collection of fluid in the region around the eyes. An experienced physician may recognize this collection of symptoms as an indication of superior vena cava (SVC) syndrome, an obstruction of venous drainage.

Hundreds of syndromes are now recognized by medical authorities. Indeed, a specific dictionary (*Dictionary of Medical Syndromes*, by Sergio I. Magalini, J. B. Lippincott Company) has been published to summarize and describe the named syndromes.

Synthesis, chemical

Chemical synthesis is the preparation of a compound, usually an organic compound, from easily available or inexpensive commercial chemicals. Compounds are prepared or synthesized by performing various **chemical reactions** using an inexpensive starting material and changing its molecular structure, by reactions with other chemicals. The best chemical syntheses are those that use cheap starting materials, require only a few steps, and have a good output of product based on the amounts of starting chemicals. The starting materials for organic synthesis can be simple compounds removed from oil and **natural gas** or more complex chemicals isolated in large amounts from **plant** and **animal** sources. The goal of chemical synthesis is to make a particular product that can be used commercially; for example as a drug, a fragrance, a **polymer** coating, a food or cloth dye, a herbicide, or some other commercial or industrial use. Compounds are also synthesized to test a chemical theory, to make a new or better chemical, or to confirm the structure of a material isolated from a natural source. Chemical synthesis can also be used to supplement the supply of a drug that is commonly isolated in small amounts from natural sources.

Chemical synthesis has played an important role in eradicating one of the major infectious diseases associated with the tropical regions of the world. **Malaria** is a **disease** that affects millions of people and is spread by mosquito bites. It causes a person to experience chills followed by sweating and intense fever, and in some cases can cause death. In about 1633, the Inca Indians told the Jesuit priests that the **bark** from the cinchona or quina-quina **tree** could be used to cure malaria. The cinchona tree is an evergreen tree that grows in the **mountains** of Peru. The healing properties of the bark were quickly introduced in **Europe** and used by physicians to treat the disease. The chemical responsible for the healing properties of the cinchona bark was isolated in 1820 and named **quinine**, after the quina-quina tree. By 1850, the demand for cinchona bark was so great that the trees in Peru were near **extinction**. To supplement the supply of quinine, plantations of cinchona trees were started in India, Java, and Ceylon, but by 1932, they were only able to supply 13% of the world's demand for the antimalarial drug. Chemical synthesis was used by German scientists from 1928 to 1933 to make thousands of new compounds that could be used to treat malaria and make up for the deficiency of natural quinine. They were able to identify two new antimalarial drugs and one of them, quinacrine, was used as a substitute for quinine until 1945. During World War II, the supply of cinchona bark to the Allied Forces was constantly interrupted and a new drug had to be found. British and American scientists began to use chemical synthesis to make compounds to be tested against the disease. Over 150 different laboratories cooperated in synthesizing 12,400 new substances by various and often long, involved chemical **sequences**. In just four years, they were able to identify the new antimalarial chloroquine and large quantities were quickly prepared by chemical synthesis for use by the Allied Forces in malarial regions. Today, there are more than half a dozen different drugs available to treat malaria and they are all prepared in large quantities by chemical synthesis from easily available chemicals.

Taxol is an anticancer drug that was isolated in the 1960s from the Pacific **yew** tree. In 1993, taxol was approved by the Food and Drug Administration (FDA) for treatment of ovarian **cancer** and is also active against various other cancers. The demand for this compound is expected to be very large, but only small amounts of the drug can be obtained from the yew bark, so rather than destroy all the Pacific yew trees in the world, chemists set out to use chemical synthesis to make the compound from more accessible substances. One chemical company found that they could convert 10-deacetylbaccatin III, a compound isolated from yew twigs and needles, into taxol by a series of chemical reactions. Furthermore, in 1994, two research groups at different universities devised a chemical synthesis to synthesize taxol from inexpensive starting materials.

Chemical synthesis can also be used to prove the chemical structure of a compound. In the early nineteenth century, the structure of a compound isolated from natural sources was deduced by chemical reactions that converted the original compound into substances of known, smaller molecular arrangements.

In 1979, chemical synthesis was used as a tool to determine the molecular structure of periplanone B, the sex excitant of the female American cockroach. In the Netherlands in 1974, C. J. Persons isolated 200 micrograms of periplanone B from the droppings of 75,000 virgin female **cockroaches**. He was able to deduce the gross chemical structure of the compound by modern analytical methods, but not its exact three dimensional structure. Without knowing the **stereochemistry** or three dimensional arrangement of the **carbon atoms**, larger quantities of the excitant could not be prepared and tested. In 1979, W. Clark Still at Columbia University in New York, set out to determine the structure of periplanone B. He noted that four compounds had to be made by chemical synthesis in order to determine the structure of the cockroach excitant. He chose an easily prepared starting material and by a series of chemical reactions was able to make three of the four compounds he needed to determine the chemical structure. One of the

new substances matched all the analytical data from the natural material. When it was sent to the Netherlands for testing against the natural product isolated from cockroaches, it was found to be the active periplanone B.

Resources

Books

McMurry, J. *Organic Chemistry*. 5th ed. Pacific Grove, CA: Brooks/Cole Publishing Company, 1999.

Mundy, B.P. *Concepts of Organic Synthesis*. New York: Marcel Dekker Inc, 1980.

Warren, S. *Organic Synthesis: The Disconnection Approach.* New York: John Wiley & Sons, 1983.

Periodicals

Borman, S. *Chemical and Engineering News* (February 21, 1994), 32.

Stinson, S.C. *Chemical and Engineering News* (April 30, 1979): 24.

Andrew Poss

Synthesizer, music

The virtuoso demands that composers placed on musicians at the end of the 1800s were but a foretaste of things to come in the twentieth century. Members of the orchestra were complaining that the music of contemporary composers was unplayable because of the enormous difficulty of complex orchestral writing styles. With the Paris premiere of Igor Stravinsky's "Le Sacre Du Printemps" in 1913, it seemed that the limits of performability had been reached and that the music world was about to go over the brink.

After a break in compositional flow during World War I, composers explored new, uncharted musical domains. In the 1920s and 1930s, some composers created intricate and complex avant garde music, demonstrating the ultimate limitations of human musicians. They did not know it at the time, but these pioneers were attempting to write electronic music before the needed technology had been invented.

After World War II, European composers began to experiment with a new invention, the tape recorder. Here was a medium in which an artist could actually hold sounds in her own hands. Chopping up a tape recording and reassembling the pieces in a different order opened up a new world of sounds for composers to explore. It also required artists to come to grips with the phenomenon of sound itself, and questions like what it was made of and how sounds differed from each other. These problems were eventually solved on **paper** but a real tool was required to give composers the ability to actually manipulate the building blocks of sound. In the 1950s, **electronics** technology had been developed to the point where it was finally able to meet this demand, leading ultimately to the development of the first music synthesizer by Harry Olson and Herbert Belar in the laboratories and studios of RCA.

The synthesizer is a device that creates sounds electronically and allows the user to change and manipulate the sounds. All sounds in nature consist of waves of varying air **pressure**. A synthesizer creates sounds by generating an electrical signal in the form of a waveform, usually either a sine wave or other simple mathematical wave, which is amplified and used to drive an acoustic speaker. Unfortunately, the sound quality of a simple waveform is somewhat raw and unmusical, at least to most people. The waveform is usually altered in numerous ways, using filters to create the interesting timbres, or colors, of sound that are usually associated with certain acoustical instruments. Changing the frequency of the waveform raises or lowers the pitch of the sound. A synthesizer can control and change the beginning attack of a sound, its duration, and its decay, in addition to controlling the waveform itself.

Synthesizers can receive information from numerous sources about how to set the different parameters of its output sound. Any electronic device, such as a computer program, or person can control the synthesizer. An obvious way to accomplish this is to build the synthesizer in such a manner that it resembles an already-existing musical instrument, such as a piano. A piano-like keyboard is often used to generate signals that control the pitch of the synthesizer, although a keyboard is not required, or even necessarily desirable, to do the job. One of the first commercially available keyboard-based synthesizers marketed to the general public was built by Robert Moog in the 1960s. Other early competitors of the Moog Synthesizer were built by Don Buchla and Al Perlemon.

All of the early synthesizers were built using **analog computer** technology. Since the late 1970s, however, digital synthesis has developed as the premiere technology in synthesizer design. In the process of digitally recording a sound, called sampling, any sound recording can be converted into a series of numbers that a computer can analyze. The computer takes snapshots of the sampled sound in very short increments, about forty thousand times a second. Mathematical techniques, such as Fourier analysis, are then used to calculate the complex waveform of the original sound. The sound can then be easily reproduced in real-time from a synthesizer keyboard. This technique for creating sounds, and others, form the design basis of most digital synthesizers such as the New

England Digital Synclavier and the Kurzweil music synthesizer. The same technique can be applied to synthesize drums, voices or any other kind of sound. Digital instruments can also receive input not just from a keyboard, but from the actual breath of the performer, for instance. Digital flutes and other wind-instrument synthesizers convert the **force** of the musicians breath into a signal that can modify any desired parameter of the output sound.

Synthesizers have shown themselves capable of creating a wide variety of new and interesting sounds. Their one limitation, of course, is that they sound only as good as the speaker that amplifies their signal. Most humans can hear sounds far beyond the range that even the best speakers can reproduce, sounds that acoustical instruments have always been capable of generating. Because of this limitation, and others, synthesizers are not viewed as replacements of traditional instruments, but rather as a creative tool that enhances the musician's already rich palette of musical possibilities.

See also Computer, digital; Synthesizer, voice.

Synthesizer, voice

The earliest known talking machine was developed in 1778 by Wolfgang von Kempelen. Eyewitnesses reported that it could speak several words in a timid, child-like voice. While the talking machine's success appears genuine, Baron von Kempelen's accomplishments are not above suspicion. Nine years earlier, he had built a chess-playing machine, which defeated many players, including Napoleon (who, incidentally, made several unsuccessful attempts to cheat). Eventually, it was discovered that the machine was a fraud—its cabinet concealed a hidden, human chess player, who controlled the game.

In 1830, Professor Joseph Faber of Vienna, Austria, produced his own speaking automaton. Faber's machine, dubbed Euphonia, had taken 25 years to construct. Designed to look like a bearded Turk, the creation could recite the alphabet, whisper, laugh, and ask "How do you do?" **Speech** was produced by its inner workings—double bellows, levers, **gears**, and keys located inside the mannequin. Strangely enough, Euphonia spoke English with a German accent.

The first talking machines employing electronic technology were developed in the 1930s. The Voice Operation Demonstrator, or Voder, invented by Dudley in 1933, could imitate human speech and even utter complete sentences as its operator pressed keys on a board. Speech-synthesis technology evolved further with the rapid development of computer technology in the 1950s.

During the late 1960s, the MITalk System was developed at the Massachusetts Institute of Technology. Although originally designed as a reading machine for the blind, once completed, the system could convert virtually any type of text into speech-synthesized output.

Raymond Kurzweil also developed speech-synthesis technology to aid the blind. In 1976, he produced the Kurzweil reading machine which could read everything from a phone bill to a full-length novel and provided un-limited-vocabulary synthesized output. Sometimes called a set of eyes for the blind, the reading machine has proved very popular.

Today, speech synthesis is a useful way to convey in-formation in public places. Cars, appliances, and even games are being equipped with voice-synthesizer chips.

Syphilis *see* **Sexually transmitted diseases**

Systematics *see* **Taxonomy**

Systems of equations

Systems of equations are a **group** of relationships be-tween various unknown variables which can be expressed in terms of algebraic expressions. The solutions for a sim-ple system of equation can be obtained by graphing, sub-stitution, and elimination addition. These methods became too cumbersome to be used for more complex systems however, and a method involving matrices is used to find solutions. Systems of equations have played an important part in the development of business and quicker methods for solutions continue to be explored.

Unknowns and linear equations

Many times, mathematical problems involve relation-ships between two variables. For example, the distance that a car moving 55 mph travels in a unit of **time** can be described by the equation $y = 55x$. In this case, y is the distance traveled, x is the time and the equation is known as a linear equation in two variables. Note that for every value of x, there is a value of y which makes the equation true. For instance, when x is 1, y is 55. Similarly, when x is 4, y is 220. Any pair of values, or ordered pair, which make the equation true are known as the **solution** of the equation. The set of all ordered pairs which make the equation true are called the solution set. Linear equations are more generally written as $ax + by = c$ where a, b and c represent constants and x and y represent unknowns.

Often, two unknowns can be related to each other by more than one equation. A system of equations includes all of the linear equations which relate the unknowns. An example of a system of equations can be described by the following problem involving the ages of two people. Suppose Lynn is twice as old as Ruthie, but two years ago, Lynn was three times as old as Ruthie. Two equa-tions can be written for this problem. If we let x = Lynn's age and y = Ruthie's age, then the two equations relating the unknown ages would be $x = 2y$ and $x - 2 = 3(y - 2)$. The relationships can be rewritten in the general format for linear equations to obtain,

(Eq. 1) $x - 2y = 0$

(Eq. 2) $2x - 3y = -4$

The solution of this system of equations will be any ordered pair which makes both equations true. This sys-tem has only solution, the ordered pair of x = 8 and y = 4, and is thus called consistent.

Solutions of linear equations

Since the previous age problem represents a system with two equations and two unknowns, it is called a sys-tem in two variables. Typically, three methods are used for determining the solutions for a system in two vari-ables, including graphical, substitution and elimination.

By graphing the lines formed by each of the linear equations in the system, the solution to the age problem could have been obtained. The coordinates of any **point** in which the graphs intersect, or meet, represent a solu-tion to the system because they must satisfy both equa-tions. From a graph of these equations, it is obvious that there is only one solution to the system. In general, straight lines on a coordinate system are related in only three ways. First, they can be parallel lines which never cross and thus represent an inconsistent system without a solution. Second, they can intersect at one point, as in the previous example, representing a consistent system with one solution. And third, they can coincide, or intersect at all points indicating a dependent system that has an infi-nite number of solutions. Although it can provide some useful information, the graphical method is often difficult to use because it usually provides us with only approxi-mate values for the solution of a system of equations.

The methods of substitution and elimination by ad-dition give results with a good degree of **accuracy**. The substitution method involves using one of the equations in the system to solve for one **variable** in terms of the other. This value is then substituted into the first equa-tion and a solution is obtained. Applying this method to the system of equations in the age problem, we would first rearrange the equation 1 in terms of x so it would become $x = 2y$. This value for x could then be substituted into equation 2 which would become $2y - 3y = -4$, or simply $y = 4$. The value for x is then obtained by substituting $y = 4$ into either equation.

Probably the most important method of solution of a system of equations is the elimination method because it can be used for higher order systems. The method of elimination by addition involves replacing systems of equations with simpler equations, called equivalent systems. Consider the system with the following equations; equation 1: $x - y = 1$; and equation 2: $x + y = 5$. By the method of elimination, one of the variables is eliminated by adding together both equations and obtaining a simpler form. Thus equation 1 + equation 2 results in the simpler equation $2x = 6$ or $x = 3$. This value is then put back into the first equation to get $y = 2$.

Often, it is necessary to multiply equations by other variables or numbers to use the method of elimination. This can be illustrated by the system represented by the following equations:

equation 1: $2x - y = 2$

equation 2: $x + 2y = 10$

In this case, addition of the equations will not result in a single equation with a single variable. However, by multiplying both sides of equation 2 by -2, it is transformed into $-2x - 4y = -20$. Now, this equivalent equation can be added to the first equation to obtain the simple equation, $-3y = -18$ or $y = 6$.

Systems in three or more variables

Systems of equations with more than two variables are possible. A linear equation in three variables could be represented by the equation $ax + by + cz = k$, where a, b, c, and k are constants and x, y, and z are variables. For these systems, the solution set would contain all the number triplets which make the equation true. To obtain the solution to any system of equations, the number of unknowns must be equal to the number of equations available. Thus, to solve a system in three variables, there must exist three different equations which relate the unknowns.

The methods for solving a system of equations in three variables is analogous to the methods used to solve a two variable system and include graphical, substitution, and elimination. It should be noted that the graphs of these systems are represented by geometric planes instead of lines. The solutions by substitution and elimination, though more complex, are similar to the two variable system counterparts.

For systems of equations with more than three equations and three unknowns, the methods of graphing and substitution are not practical for determining a solution. Solutions for these types of systems are determined by using a mathematical invention known as a **matrix**. A

KEY TERMS

. . . .

Consistent system—A set of equations whose solution set is represented by only one ordered pair.

Dependent system—A set of equations whose solution set has an infinite amount of ordered pairs.

Elimination—A method for solving systems of equations which involves combining equations and reducing them to a simpler form.

Graphical solution—A method for finding the solution to a system of equations which involves graphing the equations and determining the points of intersection.

Inconsistent system—A set of equations which does not have a solution.

Linear equation—An algebraic expression which relates two variables and whose graph is a line.

Matrix—A rectangular array of numbers written in brackets and used to find solutions for complex systems of equations.

Ordered pair—A pair of values which can represent variables in a system of equations.

Solution set—The set of all ordered pairs which make a system of equations true.

Substitution—A method of determining the solutions to a system of equation which involves defining one variable in terms of another and substituting it into one of the equations.

matrix is represented by a **rectangle** array of numbers written in brackets. Each number in a matrix is known as an element. Matrices are categorized by their number of rows and columns.

By letting the elements in a matrix represent the constants in a system of equation, values for the variables which solve the equations can be obtained.

Systems of equations have played an important part in the development of business, industry and the military since the time of World War II. In these fields, solutions for systems of equations are obtained using computers and a method of maximizing parameters of the system called linear programming.

See also Graphs and graphing; Solution of equation

Resources

Books

Barnett, Raymond, and Michael Ziegler. *College Mathematics.* San Francisco: Dellen Publishing Co, 1984.

Bittinger, Marvin L. and Davic Ellenbogen. *Intermediate Algebra: Concepts and Applications.* 6th ed. Reading, MA: Addison-Wesley Publishing, 2001.

Paulos, John Allen. *Beyond Numeracy.* New York: Alfred A. Knopf Inc., 1991.

Perry Romanowski

T cells

When a vertebrate encounters substances that are capable of causing it harm, a protective system known as the "immune system" comes into play. This system is a network of many different organs that work together to recognize foreign substances and destroy them. The **immune system** can respond to the presence of a **disease** causing agent (pathogen) in two ways. Immune cells called the B cells can produce soluble **proteins** (antibodies) that can accurately target and kill the pathogen. This branch of immunity is called "humoral immunity." In cell-mediated immunity, immune cells known as the T cells produce special chemicals that can specifically isolate the pathogen and destroy it.

The T cells and the B cells together are called the lymphocytes. The precursors of both types of cells are produced in the bone marrow. While the B cells mature in the bone marrow, the precursor to the T cells leaves the bone marrow and matures in the thymus. Hence the name, "T cells" for thymus-derived cells.

The role of the T cells in the immune response is to specifically recognize the **pathogens** that enter the body and to destroy them. They do this either by directly killing the cells that have been invaded by the pathogen, or by releasing soluble chemicals called "cytokines," which can stimulate other killer cells specifically capable of destroying the pathogen.

During the process of maturation in the thymus, the T cells are taught to discriminate between "self" (an individual's own body cells) and "non-self" (foreign cells or pathogens). The immature T cells, while developing and differentiating in the thymus, are exposed to the different thymic cells. Only those T cells that are "self-tolerant," that is to say, they will not interact with the molecules normally expressed on the different body cells are allowed to leave the thymus. Cells that react with the body's own proteins are eliminated by a process known as "clonal deletion." The process of clonal deletion en-

sures that the mature T cells, which circulate in the **blood**, will not interact with or destroy an individual's own tissues and organs. The mature T cells can be divided into two subsets, the T-4 cells (that have the accessory **molecule** CD4) or the T-8 (that have CD8 as the accessory molecule).

There are millions of T cells in the body. Each **T cell** has a unique protein structure on its surface known as the "T cell receptor" (TCR), which is made before the cells ever encounter an antigen. The TCR can recognize and bind only to a molecule that has a complementary structure. It is kind of like a "lock-and-key" arrangement. Each TCR has a unique binding site that can attach to a specific portion of the antigen called the epitope. As stated before, the binding depends on the complementarity of the surface of the receptor and the surface of the epitope. If the binding surfaces are complementary, and the T cells can effectively bind to the antigen, then it can set into **motion** the immunological cascade which eventually results in the destruction of the pathogen.

The first step in the destruction of the pathogen is the activation of the T cells. Once the T lymphocytes are activated, they are stimulated to multiply. Special cytokines called interleukins that are produced by the T-4 lymphocytes mediate this proliferation. It results in the production of thousands of identical cells, all of which are sp ecific for the original antigen. This process of clonal proliferation ensures that enough cells are produced to mount a successful immune response. The large clone of identical lymphocytes then differentiates into different cells that can destroy the original antigen.

The T-8 lymphocytes differentiate into cytotoxic T-lymphocytes (CTLs) that can destroy the body cells that have the original antigenic epitope on its surface, e.g., bacterial infected cells, viral infected cells, and **tumor** cells. Some of the T lymphocytes become **memory** cells. These cells are capable of rememberin g the original antigen. If the individual is exposed to the same **bacteria** or **virus** again, these memory cells will initiate a rapid and strong immune response against it. This is the reason

Scarlet tanager (*Piranga olivacea*). *Photograph by Robert J. Huffman. Field Mark Publications. Reproduced by permission.*

why the body develops a permanent immunity after an infectious disease.

Certain other cells known as the T-8 suppressor cells play a role in turning off the immune response once the antigen has been removed. This is one of the ways by which the immune response is regulated.

Tanagers

Tanagers are 239 **species** of extremely colorful, perching **birds** that make up the family Thraupidae. The evolutionary history and phylogenetic relationships of the tanagers and related birds are not well understood. Recent taxonomic treatments have included the tanagers as a subfamily (Thraupinae) of a large family of New World birds, the Emberizidae, which also includes the **wood warblers**, cardinals, sparrows, buntings, **black-**

birds, and **orioles**. Whether the tanagers are treated as a family or as a sub-family, they nevertheless constitute a distinct group of birds.

Tanagers are birds of **forests**, forest edges, and shrublands. Species of tanagers occur from temperate southern Alaska and Canada south to Brazil and northern Argentina. Almost all species, however, breed or winter in the tropics. Tanagers that breed in temperate habitats are all migratory, spending their non-breeding season in tropical forests. In the ecological sense, these migratory tanagers should be viewed as tropical birds that venture to the temperate zone for a few months each year for the purpose of breeding.

Tanagers are small birds, ranging in body length from 3-12 in (8-30 cm), and are generally smaller than 8 in (20 cm). Their body shape is not particularly distinctive, but their brilliant colors are. The plumage of tanagers can be quite spectacularly and extravagantly colored in rich hues of red, crimson, yellow, blue, or black, making these among the most beautiful of all the birds. In most species, the female has a more subdued coloration than the male.

Tanagers feed in trees and shrubs on fruit, **seeds**, and **nectar**; their diet may also include **insects** and other **invertebrates**.

Tanagers defend a territory during the breeding season. An important element of the defense is song, and some species are accomplished vocalists, although tanagers are not renowned in this respect. The cup-shaped or domed nest is built by the female. She also incubates the 1-5 eggs. In general, tropical tanagers lay fewer eggs than species that breed in the temperate zone. The female tanager is fed by the male during her egg-incubating seclusion. Both parents tend and feed their young.

North American tanagers

Four species of tanagers are native breeders in **North America**, and a fifth species has been introduced. Mature males of all of the native species are brightly colored, while the females and immature males are a more subdued olive-green or yellow **color**.

The western tanager (*Piranga ludovicianus*) breeds in **conifer** and aspen forests of western North America, as far north as southern Alaska. The male has a red head and a yellow body, with black wings and tail.

The scarlet tanager (*P. olivacea*) breeds in mature hardwood and mixed-wood forests, and also in well-treed suburbs of the eastern United States and southeastern Canada. The male has a brilliantly scarlet body with black wings and tail. The scarlet tanager winters in tropical forest in northwestern **South America**. This species is sometimes kept as a cagebird.

The summer tanager (*P. rubra*) occurs in oak and oak-pine forests and riparian woodlands of the eastern and southwestern United States. Male summer tanagers have an bright red body with slightly darker wings.

The hepatic tanager (*P. flava*) is a bird of oak-pine, oak, and related montane forests in the southwestern United States, as well as suitable habitats in Central and South America. The male of this species is a bright brick-red color, rather similar to the summer tanager. However, the hepatic tanager has a heavier, dark bill and a dark patch on the cheeks.

The blue-gray tanager (*Thraupis virens*) is a native species from Mexico to Brazil, but it has been introduced to the vicinity of Miami, Florida. This species has a brightly hued, all-blue body, with the wings and tail a darker hue. The male and the female are similarly colored in the blue-grey tanager.

Tanagers elsewhere

There are hundreds of species of tanagers in the American tropics. All of them are brilliantly and boldly colored, and are extremely attractive birds. Because of their habitat—usually areas with thick vegetation, their beauty is not always easily seen. One of the showiest species is the paradise tanager (*Tangara chilensis*) of southern South America, a brilliant bird with a crimson rump, purple throat, bright blue belly, and black back.

Tanagers are not known for their singing ability. One of the more prominent singers, however, is the blue-hooded euphonia (*Euphonia musica*), a wide-ranging species that occurs from Mexico to Argentina. This attractive bird has a bright yellow belly and rump, a sky blue cap, and a black throat, wings, back, and tail.

Some species of tanagers of deep, old-growth tropical forests are still being discovered. *Hemispingus parodii* was only discovered in a Peruvian jungle in 1974.

Resources

Books

Farrand, J., ed. *The Audubon Society Master Guide to Birding*. New York: Knopf, 1983.
Forshaw, Joseph. *Encyclopedia of Birds*. New York: Academic Press, 1998.

Bill Freedman

Tantalum *see* **Element, chemical**

Tapeworm *see* **Flatworms**

Taphonomy

Taphonomy is the study of how organisms are preserved in the fossil record (the term is derived from the Greek word *taphos*, which means *grave*). Taphonomists seek to understand how an **organism** died and what happened to its body before and during burial. They also try to determine what factors may have contributed to unequal representation of certain groups in the fossil record due to differences in their rates of preservation.

One of the challenges taphonomists are often presented with is the interpretation of fossilized broken bones. The scientists must determine if the damage occurred while the **animal** was alive, after it died, or even after the bone was buried. If the bone broke while the animal was alive, there may be new growth showing that the injury was healing. Bones broken after the animal died will not show any signs of healing. Unfortunately, it is sometimes difficult to determine whether the damage occurred before burial, perhaps by gnawing or trampling, or afterwards.

The study of preservation processes is central to taphonomy. There is much to be learned from the nature of a fossil's preservation, but if a fossil is not well preserved, then there will be little evidence to study.

Rapid burial is crucial to good preservation. One reason is that buried remains are much less likely to be disturbed by scavengers or swept away by **rivers**. Scavengers pull skeletons apart, scatter bones, and eat some parts preferentially to others. Rivers transport animal and **plant** remains far from the site of death. Indeed, most organisms are not buried where they die, but are first shuffled around either in part or in whole. This movement may be small, as in the sinking of a **fish** to the bottom of a **lake**, or extensive, as when an animal's body floats hundreds of miles in **ocean currents**. A second reason is that burial, especially in fine-grained muds, may slow down bacterial decay so tissues are better preserved.

Of the organisms that eventually become fossilized, few are preserved intact. This is due to differences in the ability of various parts to resist decay. On the one hand, teeth and bones are very hard, and fossilized specimens are relatively common. On the other, it is quite a find indeed to recover a fossilized **heart** or other soft part. Since soft parts nearly always decay, even if deeply buried, their pattern can only be preserved if **minerals** such as carbonate, pyrite, or phosphate replace the soft tissues. Such tissues will not decay if the organism happens to be buried in a sterile environment such as amber or a peat bog.

The hard parts of an organism are held together by soft parts, and when these decay, the organism's **anato-**

my often becomes jumbled and scattered. This may make it difficult to determine what type of organism a part came from, and sometimes different parts of the same organism are accidentally classified as different **species**. A case in point is *Anomalocaris canadensis*. Until intact specimens of this prehistoric animal were found, the body was thought to be a sea cucumber, the mouth a **jellyfish**, and the feeding appendages shrimp-like creatures. By helping to solve puzzles like this, taphonomists increase our understanding of the fossil record.

See also Fossil and fossilization.

Tapirs

Tapirs, of the family Tapiridae, are large, forest-dwelling **mammals** with a long flexible snout. They are found in tropical South and Central America, and in southeast **Asia**. There are four **species** of tapirs in the single genus, *Tapirus*. Tapirs are grouped with **horses** and rhinoceroses in the order Perissodactyla, which are the odd-toed, hoofed mammals. Tapirs have a fourth toe on their front feet. Each toe is covered by a small hoof.

Tapirs have been called "living fossils" because they have existed, apparently with little change, for perhaps 40 million years. The fact that they are found in two widely separated locations (the Neotropics and tropical Asia) is interpreted to have resulted from the breakup of an ancient landmass called Gondwanaland, which split into continents that drifted apart.

The muscular snout of tapirs is developed from the fusion of the upper lip and the nose. It has nostrils at the tip. The end of the snout often dangles down and over the mouth, and is flexible enough to grasp vegetation and pull it toward the mouth. Tapirs browse on leaves and **plant** shoots. Tapirs are solitary animals, and frequent riverbanks. They are also sometimes found in farm fields, where they can damage **crops**. Male tapirs mark their territory with sprays of urine.

The largest species is the Malayan or Asian tapir (*Tapirus indicus*), which occurs in rainforests of Malaysia, Myanmar, and Thailand. This species stands 40 in (1 m) tall at the shoulder and may be more than 8 ft (2.4 m) long, weighing up to 1,200 lb (550 kg). The Malayan tapir has distinctive coloring, with a solid black body and a bright white saddle around the middle. This pattern helps to conceal the tapir in patches of **sun** and shade in its **rainforest** habitat.

Baird's tapir, or the Central American tapir (*T. bairdi*), is found in swampy land in rainforest from Mexico to as far south as Ecuador. The Brazilian tapir (*T. ter-*

A mountain tapir. *Photograph by Tom McHugh/Photo Researchers, Inc. Reproduced by permission.*

restris) is the most abundant of the tapirs and is found in Amazonian **South America** as far south as Brazil and Peru. Both Baird's and Brazilian tapirs have a bristly mane that may help deflect the attacks of jaguars, which seize the neck of their **prey**. The fourth species is the mountain tapir (*T. pinchaqua*), found in montane **forests** of the northern Andes Mountains. It is reddish brown in **color**, and has thick hair to insulate it against the cold temperatures found at such high altitudes. The mountain tapir is the smallest of the four species, measuring less than 6 ft (2 m) long and weighing less than 500 lb (227 kg); it is also the rarest of the tapirs.

Tapirs spend a lot of time in streams or lakes, which both cools them and may help dislodge ectoparasites. The Malayan tapir walks on the bottom of **rivers** rather like a hippopotamus. Female tapirs protect their young from the attacks of caymans and other predators when in the **water**. On land the main defense of tapirs is to run away, but a cornered tapir will also turn and bite.

Tapirs breed at any time of the year, and their **courtship** and mating **behavior** is often accompanied by much noise and squealing. After mating, both sexes resume a solitary life. Female tapirs usually produce a single offspring about every 18 months. The young tapir is born after a gestation period of 390-400 days, and has light-colored stripes and spots on a dark reddish brown background. This protective coloration conceals young tapirs while they lie motionless on the forest floor as their mother feeds elsewhere. The coloring of the young begins to disappear at about two months and is gone by six months. Young tapirs reach sexual maturity at about two or three years of age, and can live for up to 30 years.

Tapirs are often hunted for their meat and hide, but the greatest danger they face today is the loss of their forest habitat. The habitat of all four species is being deforested rapidly, and the populations of all tapirs are declining and becoming fragmented. The mountain tapir is considered an **endangered species**, while the Baird's and Malayan tapirs are vulnerable, and the Brazilian tapir is near threatened (these are designations of the World Conservation Union, or IUCN). Although the mountain tapir is now a protected species, its remote mountain forests are rarely supervised, so it is impossible to control local hunting.

Resources

Books

Knight, Linsay. *The Sierra Club Book of Small Mammals.* San Francisco: Sierra Club Books for Children, 1993.

Patent, Dorothy Hinshaw. *Horses and Their Wild Relatives.* New York: Holiday House, 1981.

Stidworthy, John. "Mammals: The Large Plant-Eaters." *Encyclopedia of the Animal World.* New York: Facts On File, 1988.

Jean F. Blashfield

Taro *see* **Arum family (Araceae)**

Tarpons

Tarpons are large silvery **fish**, measuring 4-8 ft (1.3-2.5 m) in length, with large scales, a compressed body, a deeply forked caudal fin, and a long ray extending from the dorsal fin. The mouth is large, and contains rows of sharp, fine teeth, and the lower jaw protrudes outward. Tarpon are among the best known and most impressive of the sport-fish. They can live in both **freshwater** and **saltwater**.

Taxonomy

Tarpons are the most primitive (i.e., evolutionarily ancient) **species** classified among the 30 orders of spiny-rayed fish designated as "true" **bony fish** (superorder Teleosti). Tarpons belong to the Order Elopiformes, which is subdivided into two suborders: the Elopoidei and the Albuloidei, which include three families. These are Elopidae (tenpounders, or ladyfish), the Megalopidae (the tarpons), and the Albulidae (bonefishes). The larva (leptocephalus) of fish in the order Elopiformes resembles the larva of eels.

The family Megalopidae includes only one genus, *Megalops*, and two species of tarpon: *Megalops cyprinoides* and *M. atlanticus*. The *Megalops cyprinoides* lives in the Indian and West Pacific Oceans, from east **Africa** to the Society Islands. The *Megalops atlanticus*

lives in the western Atlantic Ocean, from North Carolina to Brazil, and also off tropical West Africa.

Physical characteristics and distribution

Tarpons are large fish, measuring up to 8 ft (2.5 m) in length and weighing up to 350 lb (160 kg). Their dorsal and anal fins have an elongated ray, which forms a threadlike projection that trails behind the fish. Tarpons are recognized by their silvery **color**, forked caudal fin, and underbite caused by an extended jawbone that juts out in front of the upper jaw, giving the fish a turned-down, frowning mouth. Both the upper and lower jaw bones extend well behind the large eyes, and the large mouth, jaws, roof of the mouth, and tongue all have rows of sharp, needlelike teeth.

Tarpons are widely distributed in oceans within their range, and are also found in fresh and **brackish** water. The swim bladder of tarpons has an open connection with the gullet, and is an **adaptation** for taking in air. Tarpons are often seen swimming in **rivers**, canals, and mangrove estuaries. In warm months, tarpons swim north, then return to tropical waters when the **weather** gets cooler.

Development

Tarpons do not mature sexually until they are six or seven years old. A large female tarpon weighing about 140 lb (64 kg) may contain over 12 million eggs. Tarpon spawning sites have not been located, and fresh tarpon eggs have not been seen. However, the females are thought to lay their eggs in shallow seas, or on the ocean floor, starting at the end of June and continuing throughout July and August.

Tarpon larvae (leptocephali) are transparent and shaped like leaves or ribbons, closely resembling eel larvae. Other than having a forked caudal fin, the larvae bear little resemblance to adult tarpons. The larvae of tarpons can be differentiated from eel larvae in that the leptocephali of eels have a rounded tail. Ocean **currents** carry the tarpon larvae close to shore, to shallow habitats such as marshes, swamps, estuaries, and ponds, where they begin their **metamorphosis** into young tarpons. They eat smaller fish, crustaceans, and **insects**.

When tarpons grow large enough to survive without the protection of these shallow-water habitats, they move out of the lagoons into the open sea. Adult tarpons eat fish and crustaceans; one of their favorite foods is the grunion-like silverside. Tarpons do not form schools, but they are often found together in bays and canals, since these are the areas where they primarily feed, usually at night. Additionally, groups of tarpons have been seen swimming into schools of small fish and attacking them in unison. The natural enemies of tarpons are **sharks** and dolphins, and young tarpons have been found in the stomachs of many other species of fish.

Tarpons as gamefish

Tarpons are the original big-game fish and are well known to sport fishers throughout the world. Long before fishers began to catch sailfish, marlin, or bluefin **tuna** for sport, they angled for tarpon. To catch this fish, the angler must be strong, skilled, and have a great deal of endurance.

Fishermen who specialize in catching gamefish look for tarpon as far north as Long Island, but they concentrate their efforts in locations where tarpons are known to live more regularly. The best places to catch *Megalops atlanticus* are in waters surrounding the Florida Keys, off the west coast of Florida, in the Rio Encantado in Cuba, and in the Rio Panuca in Mexico. *Megalops cyprinoides* is also coveted by sport fishers off the east coast of Africa.

Tarpons are exciting fish to catch because of their tenacious fighting. The instant that the fish is hooked, it hurls itself upward into the air and goes through an astounding series of leaps in an effort to free itself. In fact, this fish can jump as high as 10 ft (3 m) and as far as 20 ft (6 m). These leaps are so violent that, after a while, the tarpon exhausts itself; at this point, it can be brought to the boat.

Tarpons are not often prized as a food fish in **North America**, because the adults are relatively tough and full of bones. There are exceptions, however. In Mexico and in **South America**, people eat tarpon salted or smoked. Smaller tarpon are also eaten in Africa. The silvery scales of tarpon are sometimes used to make folk jewelry.

Resources

Books

Grzimek, H.C. Bernard, ed. *Grzimek's Animal Life Encyclopedia.* New York: Van Nostrand Reinhold Company, 1993.

Lythgoe, John, and Gillian Lythgoe. *Fishes of the Sea.* Cambridge, MA: Blandford Press, 1991.

Nelson, Joseph S. *Fishes of the World.* 3rd ed. New York: Wiley, 1994.

Whiteman, Kate. *World Encyclopedia of Fish & Shellfish.* New York: Lorenz Books, 2000.

Kathryn Snavely

Tarsiers

Tarsiers are **prosimians**, or primitive **primates**, in the family Tarsiidae, found the islands of Southeast **Asia**. Tarsiers have only 34 teeth, unlike their closest prosimian relatives, the **lemurs** and **lorises**, which have 36 teeth. Also, the upper lip of tarsiers is not fastened to the gum underneath, so that the face can be mobile, rather like the more advanced primates, **monkeys** and **apes**. Tarsiers are the only prosimians with a nose that does not stay moist. A moist nose usually indicates that a mammal depends heavily on its sense of **smell**.

Tarsiers are often called "living fossils" because they most resemble fossil primates from about 40 million years ago. Thus, instead of being grouped with other prosimians, tarsiers are placed in a separate order, the Tarsioidea.

The head and body of the tarsier measure about 5 in (12.5 cm) in length, and its long, thin, naked tail is an additional 8 or 9 in (20-22 cm). Their average weight is only slightly over 4 oz (114 g). Tarsiers have soft, brown, olive, or buff fur on their head and back, which is a lighter buff or gray below. The tail may have a small puff of fur on the end. Both the second and third toes of the hind feet have toilet claws, which are long claws used for grooming and digging for **insects**. When grooming, tarsiers make a foot fist, from which these claws protrude.

Locomotion

Moving somewhat like a small, furry frog, a tarsier can leap from small branch to small branch. In order to do this efficiently, the tibia and the fibula (the two lower leg bones) are fused about halfway down their length, giving the leg more strength. Tarsiers also have elongated ankle bones, which helps them leap, and which gives them their name, *tarsier,* a reference to the tarsal, or ankle, region. The legs are much longer than their arms.

These curious little nocturnal creatures dart around the undergrowth and low trees, keeping out the

A tarsier clinging to a narrow tree branch. *Photograph by Alan G. Nelson. The National Audubon Society Collection/Photo Researchers, Inc. Reproduced by permission.*

realm of the larger animals until they want to leap across the ground to gather up **prey**. Tarsiers are carnivorous, eating insects and small lizards. They have **fat** pads on the tips of their thin fingers and toes that help them cling to trees. These primates probably do not build nests.

Tarsiers have large bulging eyes, which close quickly for protection if large insect prey comes near. The eyes also face forward, providing **binocular vision**, an aid in catching insects at night. The animal's large ears can also be folded for protection. Their eyes do not move in their head, but they can turn their heads in a full half circle, like an owl. This fact accounts for the belief, recorded in Borneo, that tarsiers have detachable heads. The **brain** of some tarsier **species** weighs less than a single **eye**. Their big ears constantly move, listening for sounds of danger.

Tarsiers form family groups consisting of the male, the female, and their young. Each family stays in its own territory, and fusses loudly if another tarsier enters it. After a 180-day gestation, the female produces a single, fairly mature infant. The offspring rides either under its mother's abdomen or in her mouth. When she is off hunting, she may leave it in a safe place. The young can hunt on its own by the age of one month old, when it is also ready to leap.

There are only three species of tarsier in a single genus, *Tarsius*. All are endangered to some degree, and their ranges do not overlap.

The Mindanao or Philippine tarsier *(T. syrichta)* lives on several Philippine islands, where its forest **habitat** is being destroyed. The Western, or Horsfield's tarsier, *T. bancanus,* lives on Sumatra, Borneo, Java, and the nearby islands, and has been protected in Indonesia since 1931. The middle finger of its hand is amazingly long, almost as long as its upper arm.

The spectral, eastern, or Sulawesi tarsier *(T. spectrum)*, lives in three areas of Sulawesi (formerly Celebes) and nearby islands. Unlike the other species, the spectral tarsier has scales on a skinny tail, rather like a mouse. There is the possibility that another species, *T. pumilus,* still exists in the **mountains** of central Celebes.

Over the years, many attempts have been made to domesticate tarsiers. However, without a continuous source of live food, these primates quickly die.

Resources

Books

Knight, Linsay. *The Sierra Club Book of Small Mammals.* San Francisco: Sierra Club Books for Children, 1993.

Napier, J.R., and P.H. Napier. *The Natural History of the Primates.* Cambridge: MIT Press, 1985.

Napier, Prue. *Monkeys and Apes.* New York: Grosset & Dunlap, 1972.

Peterson, Dale. *The Deluge and the Ark: A Journey into Primate Worlds.* Boston: Houghton-Mifflin, 1989.

Preston-Mafham, Rod, and Ken Preston-Mafham. *Primates of the World.* New York: Facts on File, 1992.

Jean F. Blashfield

Tartaric acid

Tartaric acid is an organic (**carbon** based) compound of the chemical formula $C_4H_6O_6$, and has the official name 2,3-dihydroxybutanedioic acid. In this name, the *2,3-dihydroxy* refers to the two OH groups on the second and third carbon **atoms**, and the butane portion of the name refers to a four-carbon **molecule**. The *dioic acid* portion communicates the existence of two organic acid (COOH) groups on the molecule. Tartaric acid is found throughout nature, especially in many **fruits** and in wine. In addition to existing freely, it is also found as a **salt** (salts are the products of **acids and bases**), the most common of which are **calcium** tartrate, potassium tartrate, and **sodium** tartrate.

Tartaric acid is used making silver **mirrors**, in the manufacturing of soft drinks, to provide tartness to foods, in tanning leather and in making blueprints. Tartaric acid is used in cream of tartar (for cooking) and as an emetic (a substance used to induce vomiting). It readily dissolves in **water** and is used in making blueprints. Tartaric acid is a molecule that demonstrates properties of optical activity, where a molecule can cause the **rotation** of plane-polarized **light**. Tartaric acid exists in four forms (isomers are molecular rearrangements of the same atoms), each of which affects plane-polarized light differently.

The chemistry of tartaric acid

Tartaric acid is a white solid, possessing two **alcohol** groups and two acid groups. The second and third carbons of the molecule are asymmetrical (these are called chiral centers). The naturally occurring form of tartaric acid is the L-isomer, which rotates light to the left. The D-form of the acid, which rotates plane-polarized light to the right (the *D* refers to dextro, or right hand direc-

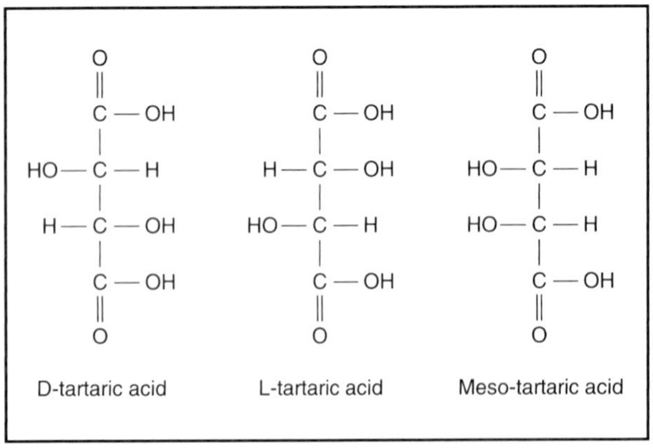

Figure 1. The D, L, and meso forms of tartaric acid.
Illustration by Hans & Cassidy. Courtesy of Gale Group.

KEY TERMS

Chiral center—A carbon atom with four different atoms or groups of atoms attached to it (sometimes called an asymmetrical carbon). Chiral centers can cause the rotation of polarized light.

Emetic—A substance used to induce vomiting, usually to remove a poison from the body.

Isomers—Two molecules in which the number of atoms and the types of atoms are identical, but their arrangement in space is different, resulting in different chemical and physical properties. Isomers based on chiral centers (such as tartaric acid) are sometimes called stereoisomers.

Polarized light—Light in which the waves vibrate in only one plane, as opposed to the normal vibration of light in all planes.

Salts—Compounds that are the products of the reaction of acids and bases. Sodium tartrate is the product of the reaction between sodium hydroxide and tartaric acid.

tion) is far less common in nature and has almost no practical uses. In general, where biological molecules have optical isomers, only one of the isomers or forms will be active biologically. The other will be unaffected by the enzymes in living cells. The *meso* form of the molecule does not affect polarized light. Figure 1 shows the D, L, and meso forms of tartaric acid. The fourth form—the DL mixture—is not a single molecule, but a mixture of equal amounts of D and L isomers. It does not rotate polarized light either (like the meso form) because the rotation of light by the D and L forms is equal in amount but opposite in direction. It is possible to separate the DL mixture into the two isomers, each of which does rotate light. In the 1840s Louis Pasteur determined that each of the two isomers of tartaric acid rotated light in opposite directions, and the meso form was inactive in this respect. He also separated by hand crystals of the racemic mixture to show that it was made of equal amounts of the D and L forms, making it different than the meso form of tartaric acid.

Uses of tartaric acid

Tartaric acid is found in cream of tartar, which is used in cooking candies and frostings for cakes. Tartaric acid is also found in baking powder, where it serves as the source of acid that reacts with **sodium bicarbonate** (baking soda). This reaction produces **carbon dioxide** gas, and lets products "rise," but does so without any "yeasty"

taste, that can result from using active **yeast** cultures as a source of the carbon dioxide gas. Tartaric acid is used in silvering mirrors, tanning leather, and in Rochelle Salt, which is sometimes used as a laxative. Blue prints are made with ferric tartarte as the source of the blue ink. In medical analysis, tartaric acid is used to make solutions for the determination of glucose. Common esters of tartaric acid are diethyl tartrate and dibutyl tartrate, which are made by reacting tartaric acid with **ethanol** and butanol. In this reaction, the H of the COOH acid group is replaced with a CH_3CH_2 (ethyl) group or a **butyl group** ($CH_3CH_2CH_2CH_2-$). These esters are used in manufacturing lacquer and in dyeing **textiles**.

Resources

Periodicals

Hunter, Beatrice. "Technological vs. Biological Needs." *Consumer Research Magazine* (August 1988): 8.

Louis Gotlib

Tasmanian devil

The Tasmanian devil (*Sarcophilus harrisii*) is the largest surviving marsupial **predator**, occurring only on the island of Tasmania in dense thickets and **forests**. The Tasmanian devil is one of about 45 **species** of marsupial predators that make up the family Dasyuridae.

The Tasmanian devil once occurred widely in **Australia** and Tasmania. However, the Tasmanian devil became extirpated from Australia following the prehistoric introduction of the dingo (*Canis dingo*; this is a placental, wild dog) by aboriginal people, and the species is now confined to the island of Tasmania.

Male Tasmanian devils can attain a body length of 32 in (80 cm) and a tail of 12 in (30 cm), and can weigh more than 20 lb (9 kg). Their pelage is colored dark brown or black, with several white spots on the rump and sides and a pinkish snout. The body is stout and badger-like, and the jaws and teeth are strong.

As is the case with all **marsupials**, young Tasmanian devils are born in an early stage of embryonic development. The tiny babies crawl slowly to a belly pouch (or marsupium) on their mother, where they suckle until they are almost fully grown and ready for an independent life.

Tasmanian devils **sleep** in a den during the day, located in a hollow log, **cave**, or another cavity. This species is a fierce, nocturnal predator of smaller animals, and a **scavenger** of dead bodies, filling a **niche** similar to those of such placental carnivores as foxes, **cats**,

A Tasmanian devil. *JLM Visuals. Reproduced by permission.*

badgers, and wild dogs. Tasmanian devils feed on a wide range of species, including domestic chickens and **sheep**.

Sometimes, individual Tasmanian devils will invade a chicken coop and create havoc there. Unfortunately, this and sheep-killing often turn out badly for the Tasmanian devil in the end. Because many people consider the Tasmanian devil a pest, this extraordinary and uncommon **animal** is still persecuted over much of its remaining native range. This is highly unfortunate, because the Tasmanian devil is the last of the large marsupial predators, and it is essential that this species survives the human onslaught on its **habitat**. Although not yet listed as an **endangered species**, the Tasmanian devil is much reduced in abundance.

Bill Freedman

Taste

Taste is one of the five senses (the others being **smell**, **touch**, **vision**, and **hearing**) through which all animals interpret the world around them. Specifically, taste is the sense for determining the flavor of food and other substances. One of the two chemical senses (the other being smell), taste is stimulated through the contact of certain chemicals in substances with clusters of taste bud cells found primarily on the tongue. However, taste is a complex sensing mechanism that is also influenced by the smell and texture of substances. An individual's

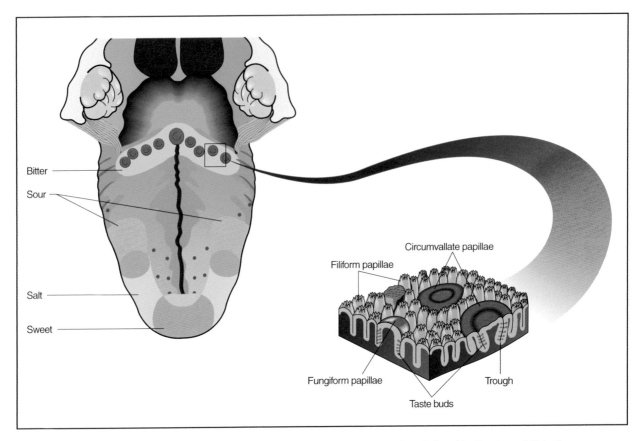

Taste regions of the tongue (left) and taste bud anatomy (right). *Illustration by Hans & Cassidy. Courtesy of Gale Group.*

unique sense of taste is partially inherited, but factors such as culture and familiarity can help determine why one person's favorite food made be hot and spicy while another just can't get enough chocolate.

The biology of taste

The primary **organ** for tasting is the mouth. Clusters of cells called taste buds (because under the **microscope** they look similar to **plant** buds) cover the tongue and are also found to a lesser extent on the cheek, throat, and the roof of the mouth. First discovered in the nineteenth century by German scientists Georg Meissner and Rudolf Wagner, taste buds lie on the elevated or ridged surface of the tongue (called the papillae) and have hairlike extensions (microvilli) to increase the receptor surface of the cells. For most foods and substances, saliva breaks down the chemical components which travel through the pores in the papillae to reach the taste buds which specialize primarily in processing one of the four major taste groups: sweet, sour, salty, and bitter.

Taste occurs when specific **proteins** in the food bind to receptors on the taste buds. These receptors, in turn, send messages to the brain's cerebral cortex, which inter-

prets the flavor. The actual chemical processes involved for each major taste group vary. For example, salty and sour flavors occur when saliva breaks down **sodium** or acids, respectively. The chemical constituents of foods that give bitter and sweet tastes, however, are much harder to specify because many chemical components are involved.

Although certain taste buds seemed to have an affinity for one of the four major flavors, continued research into this intricate biological process has revealed a complex neural and chemical network that precludes simple black and white explanations. For example, each taste bud actually has receptors for sweet, sour, salty, and bitter sensations, indicating that taste buds are sensitive to a complex flavor **spectrum** just like vision is sensitive to a broad **color** spectrum grouped into the four major colors of red, orange, yellow, and green. Particular proteins of taste are also under study, like gustducin, which may set off the plethora of **chemical reactions** that causes something to taste bitter.

Taste buds for all four taste groups can be found throughout the mouth, but specific kinds of buds are clustered together in certain areas. Think about licking an **ice** cream cone; taste buds for sweetness are grouped on the tip of our tongue. The buds for sour tastes are on the sides of the tongue and salty on the front. Bitter taste buds on the back of the tongue can make people gag, a natural defense mechanism to help prevent poisoning.

People constantly regenerate new taste buds every three to 10 days to replace the ones worn out by scalding soup, frozen yogurt, and the like. Unfortunately, as people grow older, their taste buds lose their fine tuning because they are replaced at a slower **rate**. As a result, middle-aged and older people require more of a substance to produce the same sensations of sweetness or spiciness, for example, than would be needed by a child eating the same food.

Scientists have also discovered that genetic makeup partially accounts for **individual** tasting abilities and preferences for specific foods. According to researchers at Yale University, some people are genetically programmed to have more taste buds and, as a result, taste more flavors in a particular food. (The number of taste buds varies in different **animal species**. For example cows have 25,000 taste buds, rabbits 17,000, and adult people approximately 10,000.) In general, a person's ability to taste can lie anywhere in a spectrum from poor to exceptional, with the ability to sense tastes increasing in proportion to the number of taste buds present. The difference in the number of taste buds can be extreme. Researchers have found anywhere from 11 to 1,100 taste buds per square inch in various young people tested. They have also found that women tend to have more taste buds than men and, as a result, are often better tasters. How well people taste greatly affects what they like. Studies at Yale, for example, revealed that children with fewer taste buds who are classified as poor tasters liked cheese more often than exceptional tasters, who experienced a more bitter sensation, probably because of increased sensitivity to the combination of **calcium** and the milk protein casein found in cheese.

Despite the important role that taste buds play in recognizing flavors, they do not work alone in providing the experience of taste. For example, the amount of naturally occurring **salt** in saliva varies; the result being that those with less saliva can better taste the saltiness of certain foods than others, who may end up adding salt to get a similar flavor. The smell and texture of foods are also important contributing factors to how people perceive a food to taste and whether or not they like it. Food in the mouth produces an odor that reaches the nose through the nasopharynx (the opening that links the mouth and the nose). Since smell is much more sensitive to odors than taste is to flavors, people often first experience the flavor of a food by its odor. A cold or flu is probably the most common example of how important smell is to taste. People with congestion often experience a diminished ability to taste. The taste buds, however, are working fine; it's the lack of smell that hinders the brain's ability to process

flavor. The texture and **temperature** of food also influences how it tastes. For example, many people would not think of drinking cold coffee, while others will not eat pears because of a dislike for the fruit's gritty texture.

The predilection for certain foods and tastes is not determined merely by **biology**. Culture and familiarity with foods greatly influence taste preferences. The Japanese have long considered raw **fish**, or sushi, to be a savory delicacy. But only a decade or so ago, few Americans would have enjoyed such a repast. But as the number of Japanese restaurants with sushi bars grew, so did American's familiarity with this delicacy and, as a result, their taste for it.

Taste disorders

The inability to taste is so intricately linked with smell that it is often difficult to tell whether the problem lies in tasting or smelling. An estimated two to four million people in the United States suffer from some sort of taste or smell disorder. The inability to taste or smell not only robs an individual of certain sensory pleasures, it can also be dangerous. Without smell or taste, for example, people cannot determine whether food is spoiled, making them vulnerable to **food poisoning**. Also, some psychiatrists believe that the lack of taste and smell can have a profoundly negative affect on a person's quality of life, leading to **depression** or other psychological problems.

There are a variety of causes for taste and smell disorders, from a biological breakdown to the effects of environmental toxins. In addition to cold or flu, common physical ailments that can assault the sense of taste and smell include allergies and various viral or bacterial infections that produce swollen mucous membranes. Fortunately, most of these problems are temporary and treatable. However, neurological disorders due to **brain** injury or diseases like **Parkinson disease** or **Alzheimer disease** can cause more permanent damage to the intricate neural network that processes the sense of taste and smell. Some drugs can also cause these disorders by inhibiting certain enzymes, affecting the body's **metabolism**, and interfering with the neural network and receptors needed to taste and smell. Exposure to environmental toxins like **lead**, mercury, **insecticides**, and solvents can also wreak havoc on the ability to smell and taste by causing damage to taste buds and sensory cells in the nose or brain.

See also Perception.

Resources

Books

Ackerman, Diane. *A Natural History of the Senses.* New York: Vintage Books, 1991.

Moller, Aage R. *Sensory Systems: Anatomy and Physiology.* New York: Academic Press, 2002.

KEY TERMS
. . . .

Casein—The primary protein found in cow's milk and a major component of cheese.

Cerebral cortex—The external gray matter surrounding the brain and made up of layers of nerve cells and fibers. Thought to process sensory information and impulses.

Microvilli—Hair or fingerlike projections found on cell membranes that increase surface area to better receive outside stimuli.

Papillae—Nipplelike projections found on tissue which constitute the ridge-like surfaces on the tongue.

Protein—Macromolecules that constitute three-fourths of cell matter's dry weight and which play an important role in a number of life functions, such as sensory interpretation, muscle contraction, and immunological response.

Taste buds—Cells found primarily on the tongue that are the primary biological components for interpreting the flavor of foods and other substances.

Periodicals

Lewis, Ricki. "When Smell and Taste Go Awry." *FDA Consumer* (November 1991): 29-33.

Tyler, Aubin. "Disorders That Rob You of Taste & Smell." *Good Housekeeping* (October 1992): 257-258.

Willoughby, John. "Taste? Bud to Bud, Tongues May Differ." *New York Times* (December 7, 1994): C1, C11.

David Petechuk

Taxonomy

Taxonomy is the field of **biology** which deals with the nomenclature, identification, and classification of organisms. There are over one million known **species** on **Earth** and probably several million more not yet identified. Taxonomists are responsible for identifying, naming, and classifying all these different species. Systematics is a discipline of biology that explicitly examines the natural variation and relationships of organisms, and which includes the field of taxonomy. Systematics also deals with the relationships of different groups of organisms, as most systematicists strive to construct natural classification systems reflecting evolutionary relationships. Many biologists use the terms taxonomy and systematics interchangeably.

Definition of species

Before taxonomists can identify, name, and classify organisms, they need to agree on a definition of the concept of species. The definition of species is not as simple as it may seem, and has been debated by biologists and philosophers alike for many years.

Most modern biologists agree that species, unlike higher taxa (genus, family, order, and so on), are authentic taxonomic units. In other words, species really do exist in nature, and are not merely artificial human constructs. Some of the most persuasive evidence in support of this is the close correspondence between the species identified by western taxonomists and those identified in the "folk taxonomy" of relatively isolated non-western societies. For example, Ernst Mayr, a noted bird taxonomist, identified 137 species of **birds** in his field work in the **mountains** of New Guinea; the folk taxonomy of the native New Guineans identifies 136 species. Similar correlations occur with plants. For example, an extensive interdisciplinary study of the native people and plants of the Chiapas highlands of Mexico showed a very close correspondence between species identified by western botanists and those identified by Chiapas natives.

Many modern biologists, particularly zoologists, define species according to the "biological species concept," a definition that has been forcibly advocated by Ernst Mayr. According to this definition, a species is a group of organisms reproductively isolated from other organisms. In other words, a species is a group of organisms that interbreed and produce fertile offspring only with one another.

Some microbiologists and botanists are dissatisfied with the biological species concept. Microbiologists have noted that single-celled organisms do not reproduce sexually in nature (although they may undergo genetic recombination), yet they can still be segregated into discrete natural groups considered species. Botanists have noted that many species of related plants can hybridize with one another in nature. For example, the different species in the sub-genus *Leucobalanus* (white oak) can hybridize and produce fertile offspring. Thus, rigorous application of the biological species concept would require that all of these separate species of oak be considered a single species, even though they have very different morphologies.

Most biologists basically accept the biological species concept, but argue that it is really an idealized concept, since only rarely is an actual test for reproductive compatibility performed. Thus, in practice, nearly all biologists think of species as morphologically distinct groups of organisms.

Nomenclature

Nomenclature (from the Latin term *nomenclatura*, indicating the procedure of assigning names) is the naming of organisms. For many centuries, naturalists used Latinized names to refer to different species of plants and animals in their scientific writings. Following the lead of the Swedish naturalist Carl von Linné (1707-1778) (Carolus Linnaeus) in the mid-1700s, scientists began using a binomial (two word) Latinized name for all species. The first word is the genus name and the second word is the specific epithet, also called the trivial name.

Modern taxonomists have devised formal rules of nomenclature so that scientists throughout the world can more easily communicate with one another. For example, biologists use the term *Acer* to refer to a genus of trees commonly called *maple* in England and America, *érable* in France, *acre* in Spain, and *Ahorn* in Germany. The rules of nomenclature in **botany** and zoology were established by different governing bodies, and are very similar, though not identical. In both systems the genus, species, and subspecies are italicized.

In the modern scientific literature, the complete name for a species has three parts: the genus, a name which is usually Latinized, is written in italics, and begins with a capital letter; the specific epithet, a name which is usually Latinized, is written in italics, but begins with a lowercase letter; and the author, the name of the person who gave the **organism** its Latinized name, which is capitalized and written in normal typeface. For example, the sugar maple is named *Acer saccharinum* L., where *Acer* is the genus, *saccharinum* is the specific epithet, and *L.* stands for Linneaus, who first named the species.

If subsequent studies indicate that a **plant** species should be transferred to a new genus, the original author's name is given in parenthesis and this is followed by the name of the botanist who changed the name. For example, a shrub known as inkberry is named *Ilex glabra* (L.) Gray, indicating that Linneaus first named this plant, but Gray gave it its current name. For animals, the convention is different, in that a renamed species only lists the first author's name, but places it in parentheses. For example, the red squirrel is named *Tamiasciurus hudsonicus* (Banks), indicating that a biologist named Banks was the original author, but someone else gave it its current name. The year of publication of the original description of the species is often included with the name of the original describer.

In addition to being given a genus name, specific epithet, author and year of publication, all species are assigned broader taxonomic categories. Thus, related genera are placed in the same family; related families are placed in the same order; related orders are placed in the same class; related classes are placed in the same phylum (botanists prefer the term *division* over *phylum*); and related phyla are placed in the same kingdom. Occasionally, additional groups are added to this classification scheme, such as subphylum, subclass, or suborder.

For example, human beings (*Homo sapiens* L.) are placed into the following taxonomic categories:

Kingdom-Animalia, Phylum-Chordata, Subphylum-Vertebrata, Class-Mammalia, Order-Primata, Family-Hominoidea, Genus-*Homo*, Species-*sapiens* L.

Identification

Taxonomic identification is the recognition of the identity or essential character of an organism. Taxonomists often present organized written descriptions of the characteristics of similar species so that other biologists can identify unknown organisms. These organized descriptions are referred to as taxonomic keys. A taxonomic key is often published with pictures of the species it describes. However, written descriptions are usually preferred over pictures, since pictures cannot convey the natural variation in the morphology of a species, nor the small, yet characteristic, morphological features of a species. In addition, matching an unidentified organism to one picture in a book of hundreds or thousands of pictures can be very time-consuming.

The dichotomous key is the best and most-used format for taxonomic keys. A dichotomous key sequentially presents pairs of alternative morphological features (dichotomies) and requires the user to decide which alternative best describes the unknown organism. A very simple hypothetical key for identifying four species of flowering plants illustrates the dichotomous key:

a. Flowers white; stem woody

b. Sepals present; **bark** deep-furrowed—species A

bb. Sepals absent; bark not deep-furrowed—species B

aa. Flowers not white; stem herbaceous

c. Flowers blue; flowers with five petals—species C

cc. Flowers red; flowers with three petals—species D

The first dichotomy asks the user to choose between *a* and *aa*. If alternative *a* is chosen, the user must choose between *b* and *bb*. If alternative *aa* is chosen, the user must choose between *c* and *cc*.

Taxonomic keys often require knowledge of the morphology of the taxa in question and consequently rely upon technical terminology. In the above example, it is assumed the user knows that sepals constitute the leaf-like, outermost whorl of a **flower**.

Interestingly, taxonomic keys often use the sexual organs of animals (genitalia), plants (flowers, **fruits**, and cones), and **fungi** (spore-bearing structures), because these differ significantly among closely related species. This indicates that sexual organs of animals, plants, and fungi tend to evolve rapidly and divergently, presumably because they are subjected to great evolutionary **selection** pressures. Differences in reproductive structures also indicate that mating may not be successful, and that two organisms may be reproductively isolated—that is, different species.

Classification

Taxonomic classification is the grouping together of different organisms into different taxa (taxonomic categories), such as family, genus, and species. In the time of Linneaus (mid-1700s), taxonomists and naturalists did not generally acknowledge that all species have evolutionary affinities to other species. Thus, early classification systems were artificial, in that they did not represent the lines of evolutionary descent. Most early taxonomists classified species according to their overall similarity because they thought this revealed the divine origin of nature.

Evolution and classification

The theory of organic **evolution**, developed by Charles Darwin in the mid-1800s, revolutionized taxonomy. In *Origin of Species*, Darwin proposed that "community of descent—the one known cause of close similarity in organic beings" be used as the basis for taxonomic classification systems. Ever since Darwin, taxonomists have tried to represent **phylogeny** (lines of evolutionary descent) in their classification systems. Thus, modern taxonomists who study plants or animals do not merely name and catalog species, they also try to construct evolutionary trees showing the relationships of different species.

The early evolutionary taxonomists relied on morphological features to classify organisms. Many early **animal** taxonomists found similarities in the embryos of related organisms and proposed that these similarities indicate evolutionary affinities. For example, early in development, all animals exist as a ball of cells, referred to as a blastula. The blastula of higher animals, called the coelomates, forms an invagination, called a blastopore, which later develops into the digestive cavity. Early taxonomists further divided coelomate animals into two large taxonomic groups, the protosomes, in which the blastopore opening develops into the mouth, and the deuterostomes, in which the blastopore opening develops into the anus. In contrast to the coelomates, the blastula of lower animals does not form a blastopore and these animals are called the acoelomates. When acoelomate animals are fully grown, they lack true digestive systems.

Thus, the early evolutionary taxonomists concluded humans are more closely related to **segmented worms** than to **flatworms**, since humans and segmented worms both have coelomate embryos, whereas flat worms have an acoelomate embryo. In addition, humans are more closely related to **sea urchins** than to segmented worms, since humans and sea urchins are both coelomate deuterostomes whereas round worms are coelomate protostomes.

Modern trends

Modern biologists accept many of the conclusions of the early evolutionary taxonomists, such as the relationships of the major animal phyla, as determined by the comparative morphology of their embryos. However, these early classification systems all had a weakness in that they were based, at least in part, on the intuition of taxonomists. Thus, all early classification systems had a strong element of subjectivity.

In the 1950s, R. R. Sokal and P. H. A. Sneath advocated the use of numerical taxonomy to remove the subjectivity of classification. In their method, all relevant characters (morphological features) of a group of organisms are numerically coded and the overall similarity is calculated by use of a mathematical **algorithm**. Initially, many traditional taxonomists rejected numerical taxonomy, since its results sometimes contradicted their own decade-long studies of comparative morphology. However, nearly all modern taxonomists currently use numerical methods in taxonomy, although there is often very contentious debate about which particular algorithms should be used.

In recent years, advances in **molecular biology** have had a profound impact on the field of taxonomy. In particular, biologists can now clone and sequence the DNA (deoxyribonucleic acid) in the genes of many different organisms and compare these DNA **sequences** to estimate relationships and construct classification systems. The use of such molecular data in taxonomy has several advantages. First, classification schemes for groups such as the fungi, whose phylogeny has long confounded the many taxonomists who rely upon more traditional morphological characters, can now be determined more easily. Second, organisms typically have many thousands of different genes, so there is a potential database of characters which is virtually unlimited in size. Third, since changes in DNA form the basis for all other evolutionary changes, such as changes in morphology, comparison of **gene** sequences allows study of evolution at its most basal level. Comparative studies of morphology will continue to play an important role in taxonomy, but gene sequences are becoming more widely used as gene **sequencing** becomes easier.

Methods of classification

Most modern taxonomists agree that classification systems should reflect evolutionary relationships. Thus, they agree that one should distinguish between homologous features and analogous features in constructing a classification system.

Homologous features have a common evolutionary origin, although they may differ in superficial morphology. For example, the arms of a human, fins of a whale, and front legs of a dog are all homologous. The underlying **skeletal system** of these appendages is similar and these three species had a common mammalian ancestor having appendages with similar skeletal structure.

Analogous features are superficially similar, but the similarity is not a result of common evolutionary origin. For example, the wings of a bird, bat, and butterfly are all analogous. Although they all have a similar superficial morphology and function, their underlying structure is very different. In other words, birds, **bats**, and **butterflies** did not share a common ancestor that had wings with a common design. The wings of birds, bats, and butterflies are similar merely because the laws of **physics** places certain restrictions on the shape of an appendage that can support flight.

Although taxonomists agree on the distinction between analogous and homologous features, they sometimes disagree about which specific method should be used to classify species. The three most commonly used methods are phenetics, cladistics, and evolutionary taxonomy. Some taxonomists use a combination of several of these different methods.

Phenetics

Phenetics, also known as numerical taxonomy, was proposed by Sokal and Sneath in the 1950s. Although very few modern taxonomists currently use phenetics, Sokal and Sneath's methods clearly revolutionized taxonomy by introducing computer-based numerical algorithms, now an essential tool of all modern taxonomists.

Phenetics classifies organisms based on their overall similarity. First, many different characteristics of a group of organisms are measured. These measurements are then used to calculate similarity coefficients between all pairs of organisms. The similarity **coefficient** is a number between 0 and 1, where 1 indicates absolute identity, and 0 indicates absolute dissimilarity. Finally, the similarity coefficients are used to develop a classification system.

Critics of phenetic classification have argued that it tends to classify unrelated organisms together, because it is based on overall morphological similarity, and does not distinguish between analogous and homologous features. Pheneticists have responded that they ignore the distinction between analogous and homologous features because analogous features are usually numerically overwhelmed by the larger number of homologous features. Most evolutionary biologists would consider this response questionable, at best.

Cladistics

Cladistics is a method that classifies organisms based on the order in which different evolutionary lines branch off from one another. It was first proposed in the 1950s by Willi Hennig, a German entomologist. Subsequently, many other scientists have made Hennig's original method more practicable by developing various cladistic numerical algorithms, some of which are very sophisticated. Cladistics is currently the most widely used method of classification.

One might reasonably ask: how can cladistics possibly determine the branching points of different evolutionary lines, given that we never have a complete fossil record and often have only living species for constructing a classification system? The answer is that cladistics relies upon the fact that all new species evolve by descent with modification. In other words, cladistics determines the evolutionary branching order on the basis of shared derived characteristics. It does not use shared primitive characteristics as a basis for classification, since these may be lost or modified through the course of evolution. To establish which characters are primitive and which are derived, cladistic classification generally relies upon one or more outgroups, species hypothesized to be primitive ancestors of all the organisms under study.

An example illustrates the distinction between shared derived and shared primitive characteristics in cladistic classification. The common mammalian ancestor of humans, **cats**, and **seals** had five digits on each hand and foot. Thus, the presence of five digits is a shared primitive characteristic and cladistics does not segregate humans and cats, which have five digits on their hands and feet, from seals, which have flippers instead of distinct digits. Instead, cladistics classifies seals and cats in the order Carnivora, based on certain shared derived characteristics of the Carnivora, and humans in the order Primata, based on other derived characteristics of the **Primates**.

It is important to note that a cladistic classification is not based on the amount of evolutionary change after the branching off of an evolutionary line. For example, although chimps and orangutans appear more similar to one another then either does to humans, cladistic classification places humans and chimps together since they share a more recent common ancestor than chimps and orangutans. Such a classification may seem counterintuitive;

however, cladistic taxonomists would argue that such classifications should be considered the starting point for subsequent comparative studies. Such comparative studies might seek to discover why human morphology evolved so rapidly, relative to that of chimps and orangutans.

Some botanists have noted a limitation of cladistics, in that it does not recognize the role of interspecific hybridization in the evolution of new species. In interspecific hybridization, two different species mate and produce offspring that constitute a new species. Certain derived features of the parent species revert to primitive features in the new **hybrid** species. Hybrid species appear to be more common among plants than animals, but taxonomists who study any group of organisms clearly need to account for the possibility of interspecific hybridization in the evolution of new species.

W. H. Wagner, a noted plant taxonomist, has shown that interspecific hybridization has had a particularly important role in the evolution of **ferns**. He has advocated the development of a new methodology he calls *reticulistics*, to account for the evolution of hybrid species. In reticulate phylogeny, one evolutionary line can split to form two new species or two evolutionary lines can join together to form one new species. Unfortunately, there are not yet any numerical algorithms that can be used to reconstruct a reticulate phylogeny.

Evolutionary taxonomy

Evolutionary taxonomy can be considered a mixture of phenetics and cladistics. It classifies organisms partly according to their evolutionary branching pattern and partly according to the overall morphological similarity. Evolutionary taxonomy is basically the method used by the early evolutionary taxonomists and is also called classical taxonomy.

The major limitation of evolutionary taxonomy is that it requires a highly arbitrary judgment about how much information to use for overall similarity and how much information about branching pattern to use. This judgment is always highly subjective, and makes evolutionary taxonomy a very poor method of classification, albeit one that survives in the hands of certain older taxonomists.

The kingdoms of organisms

In the past, biologists classified all organisms into the plant kingdom or animal kingdom. Fungi were placed into the plant kingdom and single-celled organisms were rather arbitrarily placed into one kingdom or the other. Very few biologists currently use this simple two kingdom system. In recent years, biologists have proposed a great variety of classification systems, based

on as many as 13 separate kingdoms of organisms. The five kingdom system, based on proposals by ecologist Robert Whittaker in 1959 and 1969, is widely used as a framework for discussing the diversity of life in most modern biology textbooks.

Five kingdom system

According to the five kingdom system designated by L. Margulis and K. V. Schwartz, all organisms are classified into one of five kingdoms: Monera, single-celled prokaryotes (**bacteria**); **Protista**, single-celled eukaryotes (**algae**, **water** molds and various other protozoans); Fungi, multicellular eukaryotic organisms which decompose organic **matter** (molds and **mushrooms**); Plantae, multicellular eukaryotic photosynthetic organisms (seed plants, mosses, ferns, and fern allies); and Animalia, multicellular eukaryotic organisms which eat other organisms (animals).

Clearly, the five kingdom system does not segregate organisms according to their evolutionary ancestry. The Monera and Protista are single-celled organisms only distinguished by their intracellular organization. The Animalia, Fungi, and Plantae are distinguished by their mode of **nutrition**, an ecological, not a phylogenetic, characteristic. Plants are considered producers, in that they use **photosynthesis** to make complex organic molecules from simple precursors and sunlight. Fungi are considered decomposers, in that they break down the dead cells of other organisms. Animals are considered consumers, in that they primarily eat other organisms, such as plants, fungi, or other animals.

More recently, a six kingdom model has been widely, although not universally, accepted. In this system, the former kingdom Monera is divided into two kingdoms: **Eubacteria** and **Archaebacteria**. The eubacteria (or true bacteria) are more common species of bacteria. Free-living decomposing bacteria, pathogenic (or **disease** causing) bacteria, and photosynthesizing cyanobacteria belong to this group. Some familiar members of Eubacteria, then, would be the bacteria found on human skin that can cause **acne**, or the bacteria found in a compost pile, which facilitate **decomposition** of organic material.

The Archaebacteria (or ancient bacteria) are quite different. Members of this group of bacteria live in very hostile environments. Examples are those living in extremely warm environments (called thermophiles) and bacteria living in extremely salty environments (called halopohiles). Archaebacteria are believed to be representative "living ancestors" of bacteria that inhabited the earth eons ago. The Archaebacteria are so fundamentally different from other bacteria that the new taxonomy reflects this difference by assigning them their own kingdom. Archaebacteria are believed to be the most primitive organisms found on earth.

Alternative systems

Many cladistic taxonomists have criticized Whittakers five kingdom classification system because it is not based on the branching pattern of evolutionary lineages. Cladistic classification systems seek to place organisms into monophyletic taxa. A monophyletic taxon is one that includes all species descended from a single common ancestor. For example, since biologists believe different groups of multicellular animals evolved from different single-celled eukaryotic ancestors, the kingdom Animalia is clearly not a monophyletic group.

Several cladistic taxonomists advocate a classification system that groups all organisms into three apparently monophyletic kingdoms, the Eukaryota, Eubacteria, and Archaebacteria (alternatively called Eukarya, Bacteria, and Archaea). The Eukaryota kingdom includes eukaryotic organisms, and all organisms in the Plantae, Animalia, Fungi, and Protista kingdoms. The kingdoms Eubacteria and Archaebacteria both consist of single-celled prokaryotes, all of which Whittaker placed into the Monera kingdom. The Archaebacteria (*ancient bacteria*) were originally considered more primitive than the Eubacteria. The Archaebacteria includes the methanogens (methane-producing bacteria), halophiles (salt-loving bacteria), and thermophiles (heat-loving bacteria), all rather unusual prokaryotes which live in very unusual habitats.

Many cladistic taxonomists are currently studying the relationships of the Eubacteria, Archaebacteria, and Eukaryota, and have proposed different classification schemes based on comparisons of different gene sequences of these organisms. Ironically, although Archaebacteria acquired their name because they were considered primitive to Eubacteria, many taxonomists now believe that Archaebacteria are in fact more modern and more closely related to the Eukaryota than are the Eubacteria. Some taxonomists have proposed a fourth kingdom called the Eocytes, a group of hyperthermophilic bacteria which other biologists include within the Archaebacteria. One of the most intriguing recent studies found that Eukaryota have some genes that are most like those in Archaebacteria and other genes most like those in Eubacteria. This suggests that the Eukaryota may have evolved as a chimera of some primitive Eubacterium and Archaebacterium.

Taxonomists who are intrigued by this controversy are encouraged in knowing there are many undiscovered species of bacteria, perhaps millions, and only a very small portion of Eubacterial, Archaebacterial, and Eukaryotal DNA sequences have been studied to date.

See also Adaptation.

KEY TERMS

Algorithm—Method for solving a numerical problem consisting of a series of mathematical steps.

Analogous features—Characteristics of organisms that are superficially similar, but have different evolutionary origins.

Chimera—Organism or part of an organism consisting of two or more genetically distinct types of cells.

Eukaryote—A cell whose genetic material is carried on chromosomes inside a nucleus encased in a membrane. Eukaryotic cells also have organelles that perform specific metabolic tasks and are supported by a cytoskeleton which runs through the cytoplasm, giving the cell form and shape.

Homologous features—Characteristics of organisms that have a common evolutionary origin, but are not necessarily similar in form.

Monophyletic—Group including all species or other taxa descended from a single common ancestor.

Outgroup—Primitive ancestor of all organisms being classified by cladistics, used to identify primitive and derived characteristics.

Phylogeny—Evolutionary history or lineage of an organism or group of related organisms.

Prokaryote—Cell without a nucleus, considered more primitive than a eukaryote.

**Taxon (plural, *taxa)*—Taxonomic group, such as species, genus, or family.

Taxonomic key—Descriptions of the characteristics of species or other taxa organized in a manner to assist in identification of unknown organisms.

Resources

Books

Gould, S. J. *The Pandas Thumb.* New York, W.W. Norton, 1980.

Luria, S.E., S.J. Gould, and S. Singer. *A View of Life.* Redwood City, CA: Benjamin-Cummings, 1989.

Margulis, L., and K.V. Schwartz. *Five Kingdoms.* New York: W.H. Freeman, 1988.

Mayr, E. *Principles of Animal Taxonomy.* New York: Columbia University Press, 1997.

Periodicals

Lewis, R. "A New Place for Fungi?" *Bioscience* 44 (1994): 389-391.

Peter A. Ensminger

Tay-Sachs disease

Background

Tay-Sachs disease, in its classical form, is a genetically inherited, progressive, neurodegenerative disorder with affected individuals having abnormal **brain** development. The outcome is a life expectancy of no more than five years of age due to complications related to the disorder. The **disease** is named after the British ophthalmologist Warren Tay (1843–1927) for his description of a patient having a cherry-red spot located on the retina of the **eye** and after a New York neurologist, Bernard Sachs (1858–19444), who demonstrated the cellular alterations associated with the disease. Sachs also observed the genetic component of the disorder, while recognizing that most affected babies were of eastern European Jewish descent.

Genetic basis and clinical manifestations of Tay-Sachs disease

Tay–Sachs disease is inherited from asymptomatic carrier parents and is, therefore, an autosomal recessive disorder. The defective **gene** produces a protein, specifically an **enzyme**, that is important for speeding up a specific biochemical reaction. The enzyme, hexosaminidase A (Hex-A) and the defective activity leads to the accumulation of the substrate (precursor) it breaksdown called GM2 gangliosides, a fatty substance found enriched in nerve cells of the brain. Gangliosides are synthesized and rapidly degraded during brain development. Accumulation of GM2 ganglioside leads to the deterioration of both mental and physical development and children become blind, deaf, and unable to swallow food, and eventually paralysis due to muscular degeneration. The cherry-red spot on the retina is a typical finding caused by accumulation of the fatty acid material in lipid-rich cells around the retina. This fatty substance can accumulate even during pregnancy, where the first destructive effects take place, although the clinical significance does not appear until the first few months after **birth** The symptom vary in severity and time at which they are evident. Failure to develop, loss of **vision**, and a characteristic abnormal startle reaction to sounds are often helpful considerations in making a **diagnosis**. There is also another hexosaminidase, or Hex-B, and if it is also defective, it leads to a distinct disorder called Sandhoff's disease.

Although Tay–Sachs disease primarily affects infants, juvenile and adult forms of Tay-Sachs disease also exist. A rarer form occurs in affected individuals that are in their late twenties or early thirties of life and is char-acterized by progressive neurological degeneration with the first recognized abnormality being an unsteady gait.

Tay-Sachs is a homozygous recessive genetic disorder caused by a defective gene on **chromosome** 15. In homozygous recessive **genetic disorders**, two defective **alleles**, or copies of the gene, one from each parent, must be passed on to produce the disease. If two people who each carry a defective copy of the gene have a child, the chances are one in four that the child will have Tay-Sachs disease. Certain populations are known to be at a higher risk for carrying a defective Hex-A gene, however, anyone in a population can be a carrier of Tay-Sachs disease. The defective Tay-Sachs allele is prevalent in Jews of Eastern European descent. About 1 in 27 people of this descent are thought to be carriers, and about 1 in 3600 Jewish infants are born with this disease, accounting for approximately 90% of all Tay-Sachs cases worldwide. Among non-Jews, about one in 300 people are carriers of the defective allele. Carriers do not have symptoms of Tay-Sachs, although their levels of the hexosaminidase A enzyme may be reduced as much as 50%. This reduction, however, is not sufficient to produce symptoms.

Testing and diagnosis

In 1969, a nationwide screening program to identify carriers was initiated in the United States. Patients and carriers can be indentified by a simple **blood** test that measures Hex-A activity. The results of screening and educational programs were extraordinarily successful. In 1970, the number of children born in the United States with Tay-Sachs disease dropped from 100 to 50; in 1980, the number dropped even further, to 13. Many carriers identified in this screening program opted not to have children. However, carriers could also reduce their risk of having children with the disease if they made sure their potential partners were also tested for carrier status.

A newer, better screening test is now available. Carrier testing involves detection of this protein or enzyme deficiency, the accuracy of which is sometimes affected by pregnancy and other factors. More recently, researchers have developed a genetic test that detects the defective gene in a blood sample. Blood samples can be analyzed for either DNA mutations or enzyme activity. The enzyme assay is a test that measures Hex-A activity with carriers having reduced activity levels than non-carriers, while affected babies have absent enzyme activity levels. This test can be used from any **individual** from any ethnic background.

In 1989, researchers identified three genetic changes, or mutations, responsible for Tay-Sachs disease. Two of these mutations cause the infantile form of the disease. By testing for the existence of these mutations in a person's blood, carriers are more accurately identified. The test is more specific than the enzyme test and is easier to perform. Over 80 mutations have been identified for the gene that encodes Hex-A. The more prevalent mutations are used for detection, especially mutations that are associated with the later-onset form. However, since all the mutations are not screened, there is a number of carriers that will not test positive. Currently, testing detects about 95% of carriers in the Ashkenazi Jewish populations and about 60% of non-Jewish individuals. This test can assist parents in making reproductive decisions. DNA testing is currently the accepted approach to test individuals of confirmed Ashkenazi Jewish descent, whereas newborn screening programs can use dried blood spots on filter **paper** to measure Hex-A activity. For prenatal diagnosis, Hex-A levels can be measured in amniotic fluid by a procedure called **amniocentesis**. As in all **genetic testing** services, the proper genetic counseling and followup procedures should always accompany any genetic test.

Treatment

Unfortunately, a clinically available treatment regimen currently does not exist. Some researchers have attempted to develop novel treatments for Tay-Sachs disease, but these attempts have not proven successful in most cases. Many laboratories around the world are investigating the utility of enzyme replacement therapy to deliver Hex-A to the brain. This technique provides only temporary therapy and the protective blood-brain barrier blocks larger molecules like enzymes from entering into the brains bloodstream. Bone marrow transplation has been unsuccessful at reducing the damaging effects. In the future, researchers hope that **gene therapy** may cure Tay-Sachs disease. Since the defective gene responsible for Tay-Sachs is known, gene therapy is currently being investigated. In gene therapy, cells that have been infected with viruses carrying normal genes are injected into the body. The healthy genes would then produce enough hexosaminidase A to breakdown the accumulating gangliosides. Researchers, however, must solve many technical difficulties before gene therapy for Tay-Sachs can be used. Information regarding Tay-Sachs disease can be obtained from the National Tay-Sachs and Allied Diseases Association.

See also Gene mutation; Genetic engineering; Genetics.

Resources

Books

Nussbaum, R.L., Roderick R. McInnes, and Huntington F. Willard. *Genetics in Medicine*. Philadelphia: Saunders, 2001.

Rimoin, David, L. *Emery and Rimoin's Principles and Practice of Medical Genetics*. London; New York: Churchill Livingstone, 2002.

Periodicals

Bach, G., J. Tomczak. N. Risch, J. Ekstein, Sheldon L. Glashow, and Leon M. Lederman. "Tay-Sachs Screening in the Jewish Ashkenazi Population: DNA Testing is the Preferred Procedure." *American Journal of Medical Genetics* 99 (2001): 70-75.

Chamoles, N.A., M. Blanco, D. Gaggioli, and C. Casentini. "Tay-Sachs and Sandhoff Diseases: Enzymatic Diagnosis in Dried Blood Spots on Filter Paper: Retrospective Diagnoses in Newborn-screening Cards." Clinica Chimica *Acta* 318 (2002): 133-137.

Kaback, M., J. Miles, M. Yaffe, H. Itabashi, H. McIntyre, M. Goldberg, and T. Mohandas. " Hexosaminidase-A (Hex A) Deficiency in Early Adulthood: A New Type of GM-2 Gangliosidosis." *American Journal & Human Genetics* 30 (1978):31A

Other

National Tay-Sachs & Allied Diseases Association. "Tay Sachs Disease" June 17, 2002 [cited December 30, 2002]. <http://www.ntsad.org/pages/t-sachs.htm>.

University of Cambridge. "Our Own Galaxy: The Milky Way" National Institute of Neurological Disorders and Stroke. August 21, 2000 [cited December 30, 2002]. <http://www.ninds.nih.gov/health_and_ medical/disorders/taysachs_doc.htm>.

National Center for Biotechnology Information. "Tay-Sachs Disease." December 13, 2002 [cited December 30, 2002]. <http://www.ncbi.nlm.nih. gov/disease/Tay_Sachs.html >.

Bryan Cobb

Tea plant

The tea plant *Camellia sinensis*, which is a member of the **plant** family Theaceae, is a small evergreen **tree** that is related to the camellias commonly grown in gardens. Although there are more than 3,000 different types, or grades, of true tea that are produced, nearly all are derived from this single **species**. Other plants such as peppermint and jasmine, which are also often steeped to yield a hot drink, are not true teas. Next to **water**, tea is said to be the most common drink in the world, with production exceeding two million tons per year worldwide.

Tea is thought to be native to China and/or Tibet, but not to India or Sri Lanka, where most commercial tea is grown today. The practice of drinking hot tea likely originated in China, and was later taken up in Japan, where elaborate tea drinking ceremonies developed. It was only much later, during the 1600s, that the export of tea to **Europe** began. Today, tea is consumed worldwide, and a great variety of practices has evolved regarding how and when tea is served and consumed. For example, Americans often drink tea with or after meals, adding milk, sugar, or lemon, while in Great Britain, it has long been

A tea plantation in east central China. *JLM Visuals. Reproduced by permission.*

the practice to serve tea in the late afternoon, often accompanied by cakes or biscuits.

The tea plant is grown almost exclusively for the drink that is made by **brewing** the **flower** buds and leaves. Before reaching the consumer, either as loose tea, tea bags, or bricks, a rather complex series of steps takes place. Dried tea is prepared by harvesting steps that usually include drying, rolling, crushing, fermenting, and heating. The result is the drink that is consumed, both hot and cold, primarily because of the stimulatory effect of the **caffeine** and the astringent effect of the tannins that the plant contains. Depending on where and how the tea is grown, when and how much of the plant is harvested, and the details of the processing, the resulting tea can take on an astounding range of tastes and colors. Commercial teas are generally of two forms—black or green—with black by far the more common. Black tea differs from green primarily because of an additional **fermentation** step caused by **microorganisms** on the dried, crushed leaves. In addition to black and green teas, other major classes of tea are produced as a result of alterations of the basic production process.

Tea was originally steeped from the loose leaves, which could be added to boiling water, or through which boiling water could be poured. In **North America**, and increasingly elsewhere, most tea is now brewed from tea bags rather than from loose tea. By steeping tea bags for shorter or longer periods of **time**, the **color** and the strength of the tea can be controlled—yet one more example of how the final drink can be manipulated. Regardless of the type of tea, virtually all commercial teas are blends of several to many different types that have been combined in such a way—and often based on secret recipes—to achieve a particular **taste** and color. In addition, dried leaves or oils from other plants are sometimes added to achieve a particular taste, as in the case of Earl

Grey tea. Although purists sometimes scoff at such practices, there are many teas in which interesting flavors have been achieved in this way.

In thinking about tea, students of United States history recall the Boston Tea Party of 1773. In this uprising, the tea cargo on three British ships was thrown into Boston Harbor by small group of colonists disguised as Native Americans to protest a tax levied by Britain's King George III. The resulting closing of Boston Harbor by the British can be interpreted as but one of the many events of history leading to the writing of the Declaration of Independence.

Resources

Books

Harrison, S.G., G.B. Masefield, and M. Wallis. *The Oxford Book of Food Plants.* Oxford: Oxford University Press, 1969.

Hobhouse, H. *Seeds of Change.* New York: Harper and Row, 1986.

Klein, R.M. *The Green World: An Introduction to Plants and People.* New York: HarperCollins, 1987.

Lewington, A. *Plants for People.* Oxford: Oxford University Press, 1990.

Simpson, B.B., and M. C. Ogorzaly. *Economic Botany: Plants in Our World.* New York: McGraw-Hill, 1986.

Steven B. Carroll

Teak *see* **Verbena family (Verbenaceae)**

Technetium *see* **Element, chemical**

Tectonics

Tectonics is the study of the deformation of Earth's lithosphere—both the causes of deformation and its effects. Tectonics focuses primarily on mountain-building, but involves other unrelated activities as well. Since the development of the theory of **plate tectonics**, tectonics has become an especially active area of research in **geology**.

Deformation of **rocks**, known as tectonism or diastrophism, is a product of the release and redistribution of **energy** from **Earth's interior**. This energy is provided by a combination of **heat**, produced by **radioactive decay** within the **earth** and by gravity. Uneven heating (or unequal distribution of **mass**) within the earth's interior creates **pressure** differences, or gradients. These pressure gradients cause rocks to experience compression, extension, **uplift**, or **subsidence**. As a result, rocks deform and shift about on the earth's surface. When deformed, rocks may break, bend, warp, slide, flow (as solids), or even melt. Structures that indicate past episodes of tectonism include faults, folds, and volcanoes. Among activities that suggest active tectonism are earthquakes and volcanic eruptions.

Tectonism is generally divided into two categories of deformation: orogenesis and epeirogenesis. Orogenesis (derived from the Greek words *oros*, meaning mountain, and *genesis*, meaning origin), or mountain-building, involves the formation of mountain ranges by folding, faulting, and volcanism. Most mountain ranges form where lithospheric plates converge (that is, along plate margins) and are a product of plate tectonics. Epeirogenesis (*epeiros*, Greek for mainland) involves vertical displacement (uplift or subsidence) of large regions of the earth's **lithosphere**, as opposed to the elongated belts of lithosphere involved in orogenesis. There are a number of other important differences between epeirogenesis and orogenesis as well. Folding and faulting is less significant in epeirogenesis and volcanism is not often involved. Epeirogenesis is also not as closely related to plate tectonics and is common throughout all areas of plates, not just at plate margins.

The products of orogenesis are obvious—mountains. Epeirogenesis is usually more subtle. **Erosion** of material from a region could cause uplift (epeirogenesis) in response to the decrease in crustal mass. Conversely, sediments eroded from one area of the earth's crust may accumulate elsewhere. The weight of these accumulated

Rift valleys like this one in Thingvellir, Iceland, are formed when an area of the earth's crust is warped upward by pressure from magma below and cracked into sections. The sections sink back lower than before as the adjacent areas of intact crust are drawn away from the fractures. *JLM Visuals. Reproduced by permission.*

sediments could cause subsidence (epeirogenesis) of the crust. The vertical movements in these examples are driven by gravity and result from changes in crustal mass within an area. Such movements are known as isostatic adjustments; they are among the most common, but subtle, epeirogenic activities.

Prior to the development of the plate tectonic theory in the 1960s, many theories had been proposed that attempted to explain regional tectonic activity, especially mountain building. None of these theories could adequately account for the range of activities and structures observed. One of the main reasons plate tectonic theory was so quickly and widely accepted is that it provides a workable explanation for a wide variety of tectonic processes and products.

In fact, plate tectonics has been called the "unifying theory of geology" because it explains and relates so many different aspects of Earth's geology. Many Earth processes, which previously were not even believed to be connected, are now known to be closely interrelated. As a result, tectonic research is now highly interdisciplinary. Thorough interpretation of the causes and effects of a single episode of deformation often requires input from a large number of geologic subdisciplines (structural geology, sedimentology, geomorphology, **geochemistry**, etc.) And to further complicate matters, most areas of the earth have been involved in multiple tectonic episodes.

See also Earthquake; Volcano.

Telegraph

A telegraph is any system that transmits encoded information by signal across a distance. Although it is associated with sending messages via an **electric current**, the word telegraph was coined to describe an optical system of sending coded messages. From its invention until the **telephone** became a viable system, the telegraph was the standard means of communicating both between and within metropolitan areas in both **Europe** and the United States. Telephones did not make the telegraph obsolete but rather complemented it for many decades. Telegrams and telexes used telegraphy but are rapidly being replaced by facsimile (fax) transmissions through telephone lines. **Satellite** transmission and high-frequency **radio** bands are used for international telegraphy.

History

The earliest forms of sending messages over distances were probably both visual and acoustic. Smoke signals by day and beacon fires by night were used by the ancient people of China, Egypt, and Greece. Drum beats extended the range of the human voice and are known to have sent messages as have reed pipes and the ram's horn. The Greek poet Aeschylus (c. 525-c. 455 B.C.) described nine beacon fires used on natural hills that could communicate over 500 mi (805 km), and the Greek historian, Polybius (c. 200-c. 118 B.C.), recounted a visual code that was used to signal the 24-letter Greek alphabet. It is also known that Native Americans used signal fires before colonial times and later. Visual systems had a greater range than ones that depended on being heard, and they were greatly stimulated by the seventeenth century invention of the **telescope**.

In 1791, the French engineer Claude Chappe (1763-1805) and his brother Ignace (1760-1829) invented the semaphore, an optical telegraph system that relayed messages from hilltop to hilltop using telescopes. The Chappes built a series of two-arm towers between cities. Each tower was equipped with telescopes pointing in either direction and a cross at its top whose extended arms could each assume seven easily-seen angular positions. Together, they could signal all the letters of the French alphabet as well as some numbers. Their system was successful and soon was duplicated elsewhere in Europe. It was Chappe who coined the word telegraph. He combined the Greek words *tele* meaning distance and *graphien* meaning to write, to define it as "writing at a distance." Its shortcomings however were its dependence on good **weather** and its need for a large operating staff. Advances in **electricity** would soon put this system out of business.

It was the invention of the **battery** and the resultant availability of electric charges moving at 186,000 mi (299,460 km) a second that accomplished this. Prior to this invention by the Italian physicist Alessandro Giuseppe A. A. Volta (1745-1827) in 1800, attempts to use electricity to communicate had failed because a dependable source of electricity was not available and the long, **iron** wires needed did not conduct electricity well and could not be properly insulated. Volta's new battery meant that experimenters had for the first **time** a reliable current of sufficient strength to transmit signals. The next major development was in 1819 when the Danish physicist Hans Christian Oersted (1777-1851) demonstrated that he could use an electric current to deflect a magnetic needle. Further, he showed that the direction of the movement depended on the direction of the flow of the current. This pointed the way to the true telegraph. While several researchers in different countries were attempting to exploit the communications aspects of this discovery, two Englishmen, William Fothergill Cooke (1806-1879) and Charles Wheatstone (1802-1875), formed a partnership and designed a five-needle telegraph system in 1837. Their system used needles to point to letters of the alphabet and numbers that were arranged on a panel. Their electric telegraph was immediately put to use on the British railway system. This system was used primarily for railroad signalling until 1845 when an event raised the public's awareness of the potential of the telegraph. On New Year's Day, 1845, the telegraph was used to catch a murderer who had been seen boarding a train bound for London. The information was telegraphed ahead and the murderer was arrested, tried, and hanged.

Although Cooke and Wheatstone built the first successful telegraph based on electricity, it was an American artist and inventor, Samuel Finley Breese Morse (1791-1872), who would devise a telegraph method that would eventually become universally adopted. Morse had begun investigating telegraphy at about the same time as his English rivals, but he had no scientific background and was getting nowhere until he was informed about the 1825 invention of the electromagnet that had been made by the English physicist William Sturgeon (1783-1850). Fortunately for Morse, he took his inquiries to the American physicist Joseph Henry (1797-1878), who had built in 1831 an extremely powerful electromagnet (it could lift 750 lb [341 kg] compared to Sturgeon's 9 lb [4.1 kg]). More importantly, Henry had successfully experimented with using the electromagnet to transmit signals and clearly understood what would become the fundamental principle of the telegraph—the opening and closing of an **electric circuit** supplied by a battery. Henry gladly enlightened Morse on the mysteries of **electromagnetism**, and the determined Morse took it from there. He enlisted the aid of a young mechanic, Alfred Vail, and together they improved on the work Morse had already started. These early attempts using an electromagnet resulted in a pen touching a moving piece of **paper** to record a series of dots and dashes. This system presumes a coded message, and Morse had created his own system which, when he collaborated with Vail, resulted in the now-famous Morse code. Vail contributed significantly to the code, having visited a printer to determine which letters were most and least often used. Their code was then based on the most common letters having the simplest, shortest of symbols (dots and dashes). By 1837, they had put together a system which used a single, simple operator key which, when depressed, completed an electric circuit and sent a signal to a distant receiver over a wire. Their first public demonstration was made at Vail's shop in Morristown, New Jersey, and in 1843, the U. S. Government appropriated funds to build a pole line spanning the 37 mi (59.5 km) between Baltimore, Maryland and Washington, D.C. On May 24, 1844, the historic message, "What

hath God wrought?" was sent and received. Once the system became practiced, it was found that skilled operators could "read" a message without looking at the dots and dashes on the paper by simply listening to the sound of the electromagnet's clicking. This led to the elimination of the paper and an even simpler electric telegraph system that used only a key, battery, pole line, and a new sounder to make the dot or dash clicking sound clear. Using such simple equipment and a single, insulated **copper** wire, Morse's telegraph system spread quickly across the United States and eventually replaced the older, English versions in Europe.

As the telegraph system grew and spread across the world, improvements followed fairly quickly. One of the first was Morse's development of a relay system to cover longer distances. His relay used a series of electromagnet receivers working on low current, each of which opened and shut the switch of a successive electric circuit supplied by its own battery. Telegraph use increased with the invention in Germany of the duplex circuit, allowing messages to travel simultaneously in opposite directions on the same line. In 1874, American inventor Thomas Alva Edison (1847-1931) designed a double duplex called a quadruplex. This higher-capacity system needed eight operators who handled four messages at one time, two in each direction. A high-speed automatic Morse system also had been invented by Wheatstone in 1858, whose punched-paper tape idea offered a means by which a message could be stored and sent by a high speed transmitter that could read the holes in the tape. This system could transmit up to 600 words per minute. The most revolutionary and innovative improvement however was a time-division, multiplex-printing telegraph system devised in 1872 by the French engineer, Jean Maurice Emile Baudot (1845-1903). His system was based on his new code which replaced the Morse code. It employed a five-unit code whose every character contained five symbol elements. The heart of his system was a distributor consisting of a stationary faceplate of concentric copper rings that were swept by brushes mounted on a rotating assembly. This logical system greatly increased the traffic capacity of each line and was so far ahead of its time that it contained many elements from which modern systems have evolved.

By the end of the nineteenth century, most of the world was connected by telegraph lines, including several cables that crossed the Atlantic Ocean. The first underwater conductor was laid by Morse in New York Harbor in 1842. Insulated with India rubber, it did not last long. After the German-English inventor, William Siemans (1823-1883) devised a machine to apply gutta-percha as insulation in 1847, **submarine** cables were laid across the English Channel from Dover, England to Calais,

France in 1850-51. Unsuccessful attempts to span the Atlantic were made in 1857, 1858, and 1865, all under the guidance of American entrepreneur, Cyrus West Field (1819-1892). On July 27, 1866, Field was successful in his fourth attempt, and having connected the United States to Europe, he immediately returned to sea, recovered the lost 1865 cable, and had a second transatlantic telegraph cable working that same year. By 1940 there were 40 transatlantic cables in operation. Ten years later, some of these began to fail and were not repaired for economic reasons. In 1956, transatlantic telephone cables were first laid, and in 1966, the last of the exclusively telegraph cables were abandoned.

Throughout its history, the telegraph proved especially useful to the military. It was first used for these purposes in 1854 by the Allied Army in Bulgaria during the Crimean War. A transcontinental telegraph line had been completed in the United States just as the Civil War began, and the telegraph proved enormously useful to both sides. During the Spanish-American War in 1898, undersea telegraph cables were cut as an act of belligerency for the first time, and in World War I, teleprinters with secret codes were heavily used by all combatants.

The earliest teleprinter was invented by an American, Royal E. House, in 1846, only two years after Morse's first success. The transmitter had 28 character keys and employed a fairly crude system that even had a hand crank. Although it was used for only a few years, it was the forerunner of both the teleprinter and the stock ticker. At the turn of the century, a Nova Scotia inventor, Frederick G. Creed (1871-1957), experimented in Scotland with using a typewriter to send printed messages without using the Morse Code. His teleprinter system did not catch on in England, and in 1907, Charles L. Krumm of the United States designed the prototype version of the modern teleprinter. This system was subsequently improved, and during the 1920s became known by the American Telephone and Telegraph trade name, Teletype. Commercial teleprinter exchange services called TRX and Telex were developed during the next decade that were capable of **printing** up to 500 characters per minute. By 1964, this was up to 900 characters per minute. By then, technical improvements in the telephone had made an entire new range of technology available to telegraphy, and today, the telegraph has evolved into a modern digital data-transmission system. Today's modern systems use **television** coaxial cables, microwave, optical fiber, and satellite links to achieve an extremely high transmission **rate**.

The invention of the telegraph could in some ways be seen as the real beginning of our modern age, given the way in which it so interconnected the entire world. Almost coincidental with its **birth** there was the emergence of a new kind of journalism that made currency its stock in

KEY TERMS

Code—A system of symbols arbitrarily used to represent words.

Electromagnet—A coil of wire surrounding an iron core that becomes magnetized when electric current flows through the wire.

Gutta-percha—A yellowish to brownish, somewhat leathery solid that is prepared from the latex of a South Sea Island tree. On heating, it becomes plastic and very resistant to water.

Semaphore—A signalling device that uses moving arms, human or mechanical, whose position indicates letters or numbers.

Sounder—The receiving device used in the aural type of telegraph reception that consists of an electromagnet constructed to give slightly different sounds to dots or dashes.

Teleprinter—A device also called a teletypewriter that sends and receives written messages along lines or via satellites.

trade. Reporting events that had only just occurred took precedence over a newspaper's traditional editorial role, and news was reported almost as soon as it happened. Corporations also could become larger and more far-flung, and nations became necessarily more interdependent. With the telegraph, information—in all its aspects and forms—began to assume the critical role it plays today.

Resources

Books

Coe, Lewis. *The Telegraph: A History of Morse's Invention and Its Predecessors in the United States.* Jefferson, NC: McFarland, 1993.

Holzmann, Gerald J., and Bjorn Pehrson. *The Early History of Data Networks.* Los Alamitos, CA: IEEE Computer Society Press, 1995.

Israel, Paul. *From Machine Shop to Industrial Laboratory: Telegraphy and the Changing Context of American Invention.* Baltimore: Johns Hopkins University Press, 1992.

Leonard C. Bruno

Telemetry

Telemetry is the science of obtaining quantities or making measurements from a distant location and transmitting them to receiving equipment where they are recorded, monitored, or displayed. A basic telemetry system consists of a measuring instrument or detector, a medium of transmission (sending), a receiver, and an output device that records and displays data. Today, telemetric systems are mainly used for monitoring manned and unmanned **space** flights, obtaining meteorological, oceanographic, and medical data, and for monitoring power-generating stations.

History

The word telemetry did not come into use until some medium of transmission had been invented. Since it is defined as the communication of measurements that were taken from some distant **point**, the earliest telemetry system was one based on a electrical wire. Following the invention of the **telegraph** and later the **telephone**, one of the earliest known United States patents for a telemetry system was granted in 1885. These first telemetry systems were used by electric power companies to monitor the distribution and use of **electricity** throughout their systems. They were called supervisory systems because of their monitoring abilities. One of these original telemetry systems was installed in Chicago, Illinois, in 1912. This network used the city's telephone lines to transmit data on its several electric power generating plants to a central control station. Following World War I, this data was transmitted by the electric power lines themselves. Although electrical telemetry systems are still in use, most modern telemetry systems use **radio** transmissions to span substantial distances.

System components

Most of today's telemetry systems consist of an input device known as a **transducer**, the radio wave medium of transmission, an instrument to receive and process the signal, and some type of recording or display instrumentation. The transducer obtains whatever it is that is being measured or monitored, like **temperature** or **pressure**, and converts that value into an electrical impulse. Transducers can have their own power source or they can be externally powered. Some examples of today's transducers are those used in **weather** balloons to obtain and convert measurements of temperature, barometric pressure, and **humidity**. Transducers are also used in manned space flights to measure an astronaut's heartbeat, **blood** pressure, and temperature as well as other biomedical indices. Transducers can also be employed for such mundane tasks as measuring the flow **rate** in a pipe.

Once something is measured and converted into an electrical signal by the transducer, this data must then be transmitted. Simple forms of telemetry, like a remote

metering system, use wire links to a central control room. While radio is the preferred medium for use over long distances, other more specialized alternates are available, such as beams of **light** or sonic signals. Manned and unmanned space systems use radio communications, and we have become more familiar with the notion of telemetry as it applies to a distant, unmanned spacecraft taking measurements as it approaches another **planet** and sending them back to **Earth**. These radio telemetry systems use what is called a modulated signal since the data is varied by a radio link. This means that the data is varied or modulated by a subcarrier signal that actually carries it. This radio signal may be either a single data channel or it may be carry several types of information. Called multiplexing, this system combines several types of information into a single signal and is used for reasons of cost and efficiency. To be effective, various forms of separating out the data from the single signal are employed. One method is called the time division multiplexing system in which data is sent and received in a certain, set order or pattern. Time division multiplexing that involves a sequential action are common and very efficient. An alternative to time division in multiplexing is a system called frequency division. Where time division combines channels sequentially, frequency division system assign each channel its own frequency band. Although the frequency bands are individually allocated, they are still combined for simultaneous transmission.

Another communication link is called the address-replay system. This special program sends data only after receiving a command signal. Modulation by a subcarrier has already been noted, and there are mainly two different methods available. The first is an AM or FM system similar to commercial radio. The other is one of several types of pulse-based methods in which data is coded digitally into pulse groups and then transmitted. User needs and preferences usually depend which system is chosen. At the receiving end of the entire telemetry system, signals are separated and can be displayed in real time and/or stored by computers. Telemetry technology is being extended to the point where we can obtain astronomical information from the farthest planet, or biological data from within a person's body via micro-miniature transmitters.

Leonard C. Bruno

Telephone

The term telephone (from Greek *tele*, afar, and *phone*, sound) in a broad sense means a specific type of

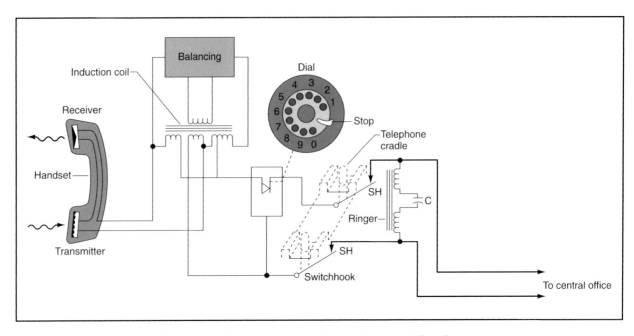

Figure 1. Telephone set simplified circuit. *Illustration by Hans & Cassidy. Courtesy of Gale Group.*

telecommunications which permits people to carry direct conversations over almost any distance. The articulate **speech** is transmitted in a form of either modulated **electric current** propagating along conductors or **radio waves**. The apparatus used for converting the verbal sounds into electromagnetic signals and vice versa is called a telephone set and is also commonly referred to as a telephone. Telephone communication is one of the most widespread and expeditious forms of communication. It has proved to be invaluable for an exchange of information in all areas of human life, whether it is business, government, science, public health, personal services, or social contacts. There were about 200,000 telephones in the world in 1887, approximately 30 million in 1928, 118 million in 1958, and more than 330 million in 1975.

The general concept of the telephone as a system involves a number of elements. First comes a telephone apparatus itself. Next, there is a variety of means for conveying the electromagnetic signals over distances (transmission paths). Third, the transmission paths are arranged in multi-leveled structures (networks) in a way that permits to interconnect (switch) any desired telephone sets upon request. And finally, there are signalling facilities for directing the switching operations, alerting the person called, and clearing the circuits upon the completion of the call.

A scientific and technological discipline which studies and develops all of the telephone-system's constituents is called telephony. Many problems in telephony draw on the principles and techniques of the electro-magnetic theory, the theory of linear and nonlinear circuits, the **probability theory**, and the queuing theory.

Invention and historical development of the telephone

Electrical telecommunication originated in the nineteenth century, with the invention of the telegraph—a method of transferring intelligence between distant places through metallic wires in a form of spaced bursts of **electricity**. A sender, using a special letter code, produced regulated electrical pulses in a circuit. These signals were converted at the receiving end into a pattern of sound clicks, which was decoded by an operator or an automatic device. The next logical step beyond sending of non-articulate messages was the instantaneous transmission of conversation over wires. The intricate and very individual nature of sounds produced by the human vocal cords with the participation of the lips, as well as of the oral and nasal cavities, made this task difficult to accomplish. Articulate (spoken) speech involves not only the elementary sounds and their assembly into syllables, words, and sentences, but also infinite variations in accent, emphasis, intonation, and voice timbre—characteristics of no importance for other types communication.

By no chance, the invention of a way to transmit the human voice electrically was made by a man deeply involved in the study of vocal **physiology** and the mechanics of speech. A native of Scotland, Alexander Graham Bell (1847-1922) inherited from his father, who was a

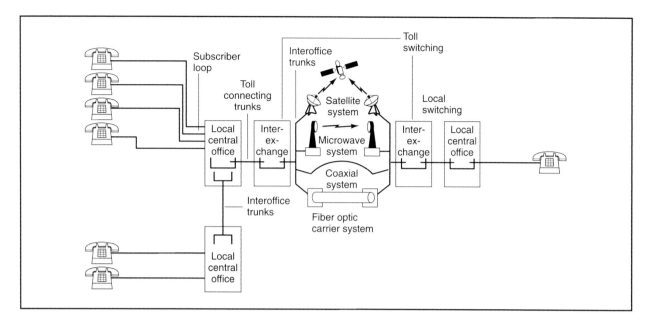

Figure 2. Telephone network. *Illustration by Hans & Cassidy. Courtesy of Gale Group.*

famous professor of elocution and an author of several textbooks on speech correction, a profound interest in speech and **hearing** theory, and put his knowledge primarily for the teaching of the deaf. Being interested in electricity, Bell set up a little laboratory and experimented on transmission of sound tones over an **electric circuit** in an attempt to make a "harmonic telegraph"—a device capable of sending multiple **telegraph** messages simultaneously over the same wire. On the basis of his experimental observations, Bell gradually came to the conclusion that **oscillations** of air **pressure (sound waves)** could be used to modulate the intensity of electric current in a circuit. Using his knowledge of **ear anatomy**, Bell attached one end of a metallic **organ** reed to a thin diaphragm intended to simulate the eardrum. In presence of even weak air-pressure variations caused by the human voice, the diaphragm forced the metallic reed to oscillate in front of the electromagnet and, therefore, to undulate the electric current in the circuit.

Telephones successfully served to the public for almost 23 years before scientists could explain theoretically why the telephone worked. Bell's **perception** of an "analog" relationship between sound pressure and **electromotive force** was one of the most fascinating examples of the ingenious intuition in the history of technological discoveries.

Bell first started testing his device for voice transmission in June, 1875, and patented it nine months later. The demonstration of the working apparatus at the Philadelphia Centennial Exposition attracted substantial public interest, which helped to raise some additional monetary funds. On 1 August 1877, with Bell's four patents as a tangible basis for a wide manufacturing of a speaking telephone, the Bell Telephone Company was formed. Bell outlined a general concept of a telephone system as a universal means of voice communication between individuals, at any time, from any location, and without any special skill requirements. This concept might seem obvious today, but it was far in advance of any existing techniques of its time. Giant technological and organizational efforts were required to make Bell's **vision** of a telephone communication a reality.

The number of existing telephones at that time counted 778, but rapidly growing demand presented the challenge of operating the existing telephone while producing new ones. The first telephone subscribers were directly connected to each other. Under such arrangement, a community of 100 subscribers used 9,900 separate wire connections. For 1,000 subscribers the number of needed wire connections was more than 100 times bigger. The evident impracticality of full-time point-to-point connections stimulated the idea of a central office, or exchange. All the telephone sets in a particular area, instead of being wired permanently to each other, were connected to the same central office, which could establish a temporary link between any two of them on demand. The process of line connecting and disconnecting (switching) was initially performed by trained operators on a switchboard. However, with more and more telephones coming into use the manual switching was becoming increasingly complex. In 1891, the first automatic switching system was introduced.

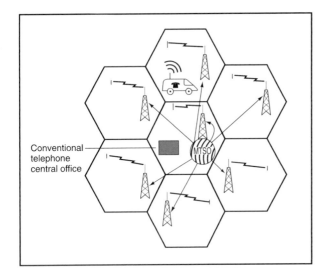

Conventional telephone central office

MTSO

Figure 3. Cellular network. *Illustration by Hans & Cassidy. Courtesy of Gale Group.*

As local telephone lines emerged and expanded, the idea of the creation and commercial operation of long-distance telephone communication became a necessity. Unfortunately, the already existing telegraph lines were of little use for this purpose. Feeble electric currents used in telephony were rapidly dissipated (attenuated) on their way through **iron** wires. Compensating for attenuation was one of the main problems of telephony during its first 50 years. Considerable success in reducing the attenuation was achieved by using **copper** in place of iron wire. The innovation extended the limits of the voice transmission through bare wires to a distance between Boston and Chicago—1180 mi (1,900 km)—in 1893. Later, the effectiveness of wire lines was crucially increased by inserting inductances, or "loading coils," at regular distances in a circuit. For crossing small bodies of **water** telephone lines were laid in a form of insulated cables instead of bare wires. The invention of **radio** transmission at the beginning of the twentieth century led to the development of wireless telephone links and accelerated creation of the worldwide telephone system.

Long-haul facilities grew swiftly. In 1900, toll lines totalled 621,000 mi (1 million km) of wire. It took only 25 years to increase this number by a **factor** of 10. In 1925, almost one-half of the long-distance circuits was cable. The first transatlantic telephone cable (TAT-1) became operational in 1956. Nine years later, the first transatlantic telephone **satellite** "Earlybird" went into service.

Telephone set

A typical telephone instrument (see Figure 1) includes a number of main blocks (a switchhook, a ringer, a dial, a receiver, a transmitter, and a balancing network), which make it capable of numerous operations. 1) It requests the use of the telephone system when the handset is lifted, and signals the system that a call is finished when a caller returns the handset to the cradle. 2) It announces the incoming call by sending audible signals. 3) It sends to the system the telephone number which is specified by the caller, who rotates a dial or presses number keys. 4) It converts the speech of the caller to electrical signals for transmission through the system and changes received electrical signals to speech. 5) It controls the level of unwanted sounds.

The switchhook interrupts the connection with the central office when the telephone set is idle (the handset is on the cradle) and completes the circuit whenever the handset is lifted off. The ringer is initiated by the system each time when an incoming call is waiting. Alternating current passing through a coil drives a permanent magnet, which makes the attached clapper periodically strike the bell.

The rotary dial is an electric contact that interrupts the circuit periodically sending to the central office a series of current pulses, corresponding to the identification number of the telephone called. When the tone dialing is used, as in touch-tone service, pressing of a number key results in sending a signal of two frequencies to the central office circuit, which identifies it and accordingly sets up the switching for the communication path. The use of tones speeds up the dialing operation and allows sending commands and control information directly to the called location.

The telephone transmitter responds to acoustic frequencies from 250 to 1,000 Hz. It consists of an electrical **capacitor** adjacent to a diaphragm, so that air-pressure vibrations caused by sound make the **membrane** produce changes in the value of **capacitance** of the circuit and, therefore, the variable electric voltage. As a result, a pulsating direct current occurs.

The typical receiver consists of a lightweight diaphragm attached to a permanent magnet associated with a coil of wire. When the alternating current of the telephone signal passes through the winding of the coil, changes are produced in the magnetic forces acting on a permanent magnet. The diaphragm moves in response to these variations, creating an acoustic pressure, more or less exactly reproducing the original sound wave from the distant telephone transmitter.

Sidetone is the sound of the speaker's voice heard in his own receiver. To keep the level of this unwanted sound in acceptable range, the balancing network is used.

Special signals informing users on the status of their calls originate from the central office. When a caller lifts off the handset from the cradle the direct current flows through a complete circuit to the central office, which immediately registers the line initiating a call and places a dial tone on it. This tone signals to the subscriber that he may proceed with dialing the number of the called party. If the connection cannot be established, the tone **generator** in the central office produces one of three possible busy signals. A signal interrupted 60 times per minute means that the called party is busy. A signal interrupted only 30 times per minute indicates that the toll line between the central offices is busy, while a signal with 120 interruptions per minute means that all the intra-office paths are busy.

Telephone network

The telephone network's structure may be defined as an entire plant of existing connections between telephone exchanges. It consists of three broad categories: local, exchange area, and long-haul networks.

The local network (see Figure 2) links telephones in residences and businesses to a central office serving a particular geographical area. The size of the area may vary from 11.5 sq mi (30 km²) in cities to 123.5 sq mi (320 km²) in the country. The telephone lines connecting a subscriber to the central office are called local lines or loops. Central offices are interconnected through the exchange area network, and all of the above are interconnected with toll (long-distance) exchanges. The telephone lines connecting one telephone exchange with another are called trunks in **North America** and junctions in **Europe**.

Each telephone is assigned a number indicating its location in the system. The switching network recognizes which telephone initiates the call and which telephone is to receive the call. From this information, it sets up the circuit connection for a signal path. Modern transmission facilities for conveying the electric analog of speech between telephone stations use diverse transmission paths, such as wire or cable circuits and microwave-radio or infrared-optical channels.

Quality of telephone communication

The intelligibility, naturalness, and audibility of transmitted speech are the main requirements for high quality telephone transmission. In technical terms, these requirements mean that: all the harmonic components of the human voice in the frequency range from 300 to 3,400 Hz pass through the communication channel; the loss of a signal during passing the channel does not exceed 30 dB; and the level of noise arising from all types of **interference** be at least 35 dB lower than the level of the main signals.

Alexander Graham Bell on the telephone at the formal opening of telephone service between New York and Chicago in 1892. *U.S. National Aeronautics and Space Administration.*

The quality of the telephone service is greatly dependent upon the structure of the networks by which subscribers are connected. The number of subscribers making telephone calls at one time is always substantially less than the total amount of subscribers. That is why the number of channels in the commercial telephone system is considerably less than the number of subscribers served by a central office (usually by a factor of 7-10 in a local office and 200-250 in toll exchanges). Because of such a design, a connection may be blocked when the telephone traffic is high. The quality of the automatic telephone service can be defined as the percentage of blocked (refused) calls during the hours of the heaviest telephone traffic. To reduce the incidence of channel overload, good planning based on statistical analysis is used.

Another way to improve network quality is to increase the amount of channels that can be carried by a single underlying medium. For example, in 1940, a coaxial cable could carry 600 voice channels. By the early 1950s, this amount increased to 1,860 voice channels and reached the number 13,200 in the 1980s. The microwave radio system experienced similar bandwidth growth, with 2,400 channels in the 1950s increased to 42,000 channels in the 1980s. The potential of the fiber optic technology promises even faster progress in this direction.

Wireless telephone systems

In wireless communication the information is superimposed on a carrier radio signal, which is sent through the air to a receiving location, where the original information is detected and isolated. Cordless, mobile, and cellular telephones perform all functions of the conventional telephone but partially use a radio link instead of wires.

The cordless telephone uses low-power radio transmissions only between the portable handset and its base. The telephone base is wired in a regular way to the telephone line completing the local loop to the central office. An internal **antenna** in the handset receives the transmission from the base unit over a range from 49 to 948.5 ft (15-300 m). The handset is powered by a **battery** which is automatically recharged when placed in a receptacle in the base unit. When the user dials the number for the outgoing calls, the dial tones are transmitted to the base unit which sends the tones to the regular telephone line.

The mobile telephone has no direct hardware connection with the conventional telephone line. It uses a high-power transmitter and an elevated antenna to establish a wireless link with the base station antenna serving a circular area of up to 31 mi (50 km) in radius. The base station receives and transmits on several different frequencies simultaneously, providing clear reliable communications. The control terminal of the base station directs telephone calls to and from the conventional telephone system, just like calls that are carried out entirely over wires. When the user in his moving vehicle lifts the handset of the mobile telephone to place a call, the control terminal automatically selects an available channel. If a channel is found, the user hears the normal dial tone and proceeds as usual.

The **cellular telephone** concept, first developed and implemented in the United States in the late 1970s, is a method of providing high quality telephone service to subscribers when they move beyond the boundaries of the home area. This concept suggests dividing a service area into a number of small cells (see Figure 3). Each **cell** is served by a control terminal, which, like a local central office, can switch, transmit, and receive calls to/from any mobile telephone located in the cell. Each cell transmitter and receiver operates on a designated channel.

There are two essential features of the cellular concept: frequency reuse and cell splitting. Frequency reuse means that the same channel may be used for conversations in cells located far apart enough to keep interference low. This is possible, because each cell uses relatively low-power transmitter covering only limited area, so cells located sufficiently far from each other may use the same frequency. Cell splitting is based on the notion that cell sizes do not have to be fixed. A system with a relatively small number of subscribers uses large cells, which can be divided into smaller ones as demand grows.

The cells are interconnected and controlled by a central Mobile Telecommunications Switching Office (MTSO), which connects the system to the conventional telephone network and keeps track of all call information for billing purposes. During the call, the terminal at the serving cell site examines the signal strength once every few seconds. If the signal level becomes too low, the MTSO looks for a closest to the active user cell site to handle the call. The actual "handoff" from one cell to the next occurs without getting noticed by a user. Decision to hand off is made by the computer, based on the location analysis, the quality of the signal and potential interference.

The convenience and efficiency of wireless telephone communication is the reason for the impressive growth of this service. Recent market data indicate that there are currently more than 100 million cellular subscribers in the United States, comparing with approximately 4.4 million in 1990, and 90,000 subscribers in 1984. Currently, cellular telephone service is available mainly in urban areas. The future expansion of cellular network on a global scale will be based on employing low altitude low weight satellites.

Modern developments in telephony

Along with the **television** and the telegraph, the telephone is only a part of a large family of complex telecommunication systems. The past few years have seen the digitalization of the networks. Digital technology uses a simple binary code to represent any signal as a sequence of ones and zeros, somewhat similar to the Morse code, which assigns dot-dash combinations to the letters of the alphabet. The smallest unit of information for the digital transmission system is a bit, which is either 1 or 0. Binary codes are very stable against different distortions, and are therefore extremely effective for long-distance transmissions. Conversion to digital form allows for the integration of various modes of information management, opening new possibilities for information display, transmission, processing, and storage.

The introduction of the new transmission media (optic fibers) has extended the capabilities and quality of telecommunication services as a whole and telephone service in particular. The fiber optic cable has an almost infinite bandwidth, and is suited especially well for digital transmission. Current tendencies in optical fiber transmission technology include the development of new fiber

KEY TERMS
. . . .

Attenuation—Loss of energy in a signal as it passes through the transmission medium.

Bandwidth—The total range of frequencies that can be transmitted through a transmission path; the bigger bandwidth means bigger information carrying capacity of the path. As a rule, transmission path carries multiple channels.

Bel—Unit of measurement associated with a tenfold increase in sound energy. Decibel (dB) is one-tenth of a bel, and represents the smallest difference in sound intensity that the trained ear can normally perceive at a frequency of 1,000 Hz. In telephony, 0 dB level is prescribed to a reference point, which is usually the sending end of the transmission line.

Hertz—A unit of measurement for frequency, abbreviated Hz. One hertz is one cycle per second.

Optical fiber—Glass strands which act as "light pipes" for light beams created by lasers.

Switching—Process whereby the temporary point to point connection between telephone sets is established.

Transmission channel—The range of frequencies needed for a transmission of a particular type of signals (for example, voice channel has the frequency range of 4,000 Hz).

materials producing a signal loss of no more than 0.001 dB per km as compared to 0.15 dB per km in existing silica-based fibers; the development of effective amplifiers; and the use of the fiber's special transmission modes, called solitons, which are so stable that they preserve their shape even after having traveled thousands of miles.

For short distances a cable-free, easily deployed, and cost-effective communication for voice, data, and video is offered by an infrared optical communication system, employing air as the transmission medium. Infrared signal is another type of electromagnetic **emission** which is characterized by a considerably higher frequency than microwave. Due to higher frequency, an infrared signal is considerably more directional than a microwave one. The current commercial infrared systems are capable of providing reliable telephone links over a range of a half-mile and are ideal for local inter-building communications.

Both voice communications and data communications today exist separately. As technologies become more advanced, the best of both worlds will be integrated

into a multimedia telecommunication network. Multimedia conferences or one-to-one multimedia calls will be set up as easily as voice calls from a desktop device. The same device will be used to access FAX and e-mail messaging. Multimedia will enable people to combine any media they need to send, receive, or share information in the form of speech, music, messages, text, data, images, video, animation, or even varieties of **virtual reality**.

Conventional networks can already accommodate some of the multimedia communication services, such as the direct participation of about 20 conferees in an audio/video/data/image work session. Technology, economics, and even environmental factors stimulate people's readiness to rely on networked contacts. Meeting somebody no longer implies being in the same place together. People separated geographically can still communicate face to face and collaborate productively. The emerging capabilities offered by the unified, intelligent telecommunication network gradually transform the way people interact, work, and learn.

Resources
Books

Brooks J. *Telephone: The First Hundred Years.* Harper & Row, 1976.
Freeman, R.L. *Telecommunication System Engineering.* New York: Wiley, 1989.

Periodicals

Chang, J.J.C., R.A. Miska, and R.A. Shober. "Wireless Systems and Technologies: An Overview." *A&T; Technical Journal* 72 (July- August 1993).
Udell, J. "Computer Telephony." *Byte* 19 (July 1994).

Elena V. Ryzhov

Telescope

The telescope is an instrument which collects and analyzes the **radiation** emitted by distant sources. The most common type is the optical telescope, a collection of lenses and/or **mirrors** that is used to allow the viewer to see distant objects more clearly by magnifying them or to increase the effective brightness of a faint object. In a broader sense, telescopes can operate at most frequencies of the **electromagnetic spectrum**, from **radio waves** to gamma rays. The one characteristic all telescopes have in common is the ability to make distant objects appear to be closer (from the Greek *tele* meaning far, and *skopein* meaning to view).

The first optical telescope was probably constructed by a Dutch lens-grinder, Hans Lippershey, in 1608. The

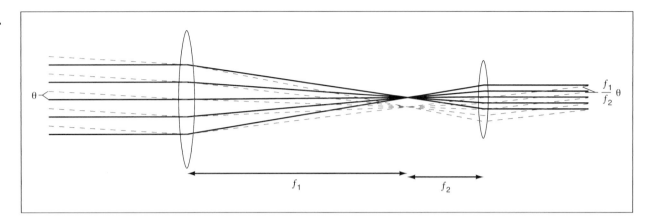

Figure 1. A refracting telescope. *Illustration by Hans & Cassidy. Courtesy of Gale Group.*

following year Galileo Galilei built the first astronomical telescope, from a tube containing two lenses of different focal lengths aligned on a single axis (the elements of this telescope are still on display in Florence, Italy). With this telescope and several following versions, Galileo made the first telescopic observations of the sky and discovered lunar **mountains**, four of Jupiter's moons, **sunspots**, and the starry nature of the **Milky Way**. Since then, telescopes have increased in size and improved in image quality. Computers are now used to aid in the design of large, complex telescope systems.

Operation of a telescope

Light gathering

The primary function of a telescope is that of **light** gathering. As will be seen below, resolution limits on telescopes would not call for an aperture much larger than about 30 in (76 cm). However, there are many telescopes around the world with diameters several times this. The reason for this is that larger telescopes can see further because they can collect more light. The 200 in (508 cm) diameter reflecting telescope at Mt. Palomar, California, for instance can gather 25 times more light than the 40 in (102 cm) Yerkes telescope at Williams Bay, Wisconsin, the largest refracting telescope in the world. The light gathering power grows as the area of the objective increases, or the square of its diameter if it is circular. The more light a telescope can gather, the more distant the objects it can detect, and therefore larger telescopes increase the size of the observable universe.

Resolution

The resolution, or resolving power, of a telescope is defined as being the minimum angular separation between two different objects which can be detected. The

angular resolution limit, q, of a telescope operating under ideal conditions is given by the simple formula:

$$\theta = 2.5 \times 10^5 \frac{\lambda}{D}$$

where λ is the wavelength of radiation being detected and D is the limiting aperture of the telescope, usually the diameter of the objective, or primary optic. Unfortunately, we are not able to increase the resolution of a telescope simply by increasing the size of the light gathering aperture to as large a size as we need. Disturbances and nonuniformities in the atmosphere limit the resolution of telescopes to somewhere in the range 0.5-2 arc seconds, depending on the location of the telescope. Telescope sights on top of mountains are popular since the light reaching the instrument has to travel through less air, and consequently the image has a higher resolution. However, a limit of 0.5 arc seconds corresponds to an aperture of only 12 in (30 cm) for visible light: larger telescopes do not provide increased resolution but only gather more light.

Magnification

Magnification is not the most important characteristic of telescopes as is commonly thought. The magnifying power of a telescope is dependent on the type and quality of eyepiece being used. The magnification is given simply by the **ratio** of the focal lengths of the objective and eyepiece. Thus a 0.8 in (2 cm) focal length eyepiece used in conjunction with a 39 in (100 cm) focal length objective will give a magnification of 50. If the field of view of the eyepiece is 20°, the true field of view will be 0.4°.

Types of telescope

Most large telescopes built before the twentieth century were refracting telescopes because techniques were

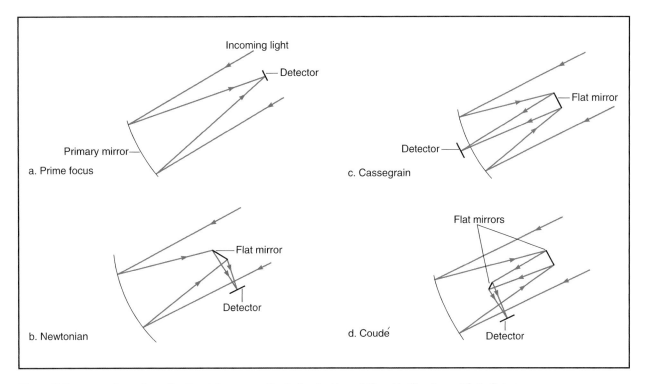

Figure 2. Focus systems for reflecting telescopes *Illustration by Hans & Cassidy. Courtesy of Gale Group.*

readily available to polish lenses. Not until the latter part of the nineteenth century were techniques developed to coat large mirrors which allowed the construction of large reflecting telescopes.

Refracting telescopes

A simple, uncorrected refracting telescope is shown in Figure 1.

The parallel light from a distant object enters the objective, of focal length f1, from the left. The light then comes to a focus at a distance f1 from the objective. The eyepiece, with focal length f2, is situated a distance f1+f2 from the objective such that the light exiting the eyepiece is parallel. Light coming from a second object (dashed lines) exits the eyepiece at an angle equal to f1/f2 times the angle of the light entering.

Refracting telescopes, i.e. telescopes which use lenses, can suffer from problems of chromatic and other aberrations, which reduce the quality of the image. In order to correct for these, multiple lenses are required, much like the multiple **lens** systems in a camera lens unit. The advantages of the refracting telescope include having no central "stop" or other diffracting element in the path of light as it enters the telescope, and the alignment and transmission characteristics are stable over long periods of time. However the refracting telescope can have low overall transmission due to reflection at the surface of all the optical elements and the largest refractor ever built has a diameter of only 40 in (102 cm): lenses of a larger diameter will tend to distort under their own weight and give a poor image. Additionally, each lens needs to have both sides polished perfectly and be made from material which is of highly uniform optical quality throughout its entire **volume**.

Reflecting telescopes

All large telescopes, both existing and planned, are of the reflecting variety. Reflecting telescopes have several advantages over refracting designs. First, the reflecting material (usually **aluminum**), deposited on a polished surface, has no chromatic aberration. Second, the whole system can be kept relatively short by folding the light path, as shown in the Newtonian and Cassegrain designs below. Third, the objectives can be made very large, since there is only one optical surface to be polished to high tolerance, the optical quality of the mirror substrate is unimportant and the mirror can be supported from the back to prevent bending. The disadvantages of reflecting systems are 1) alignment is more critical than in refracting systems, resulting in the use of complex adjustments for aligning the mirrors and the use of **temperature** insensitive mirror substrates and 2) the secondary or other auxiliary mirrors are mounted on a support structure which occludes part of the primary mirror and causes **diffraction**.

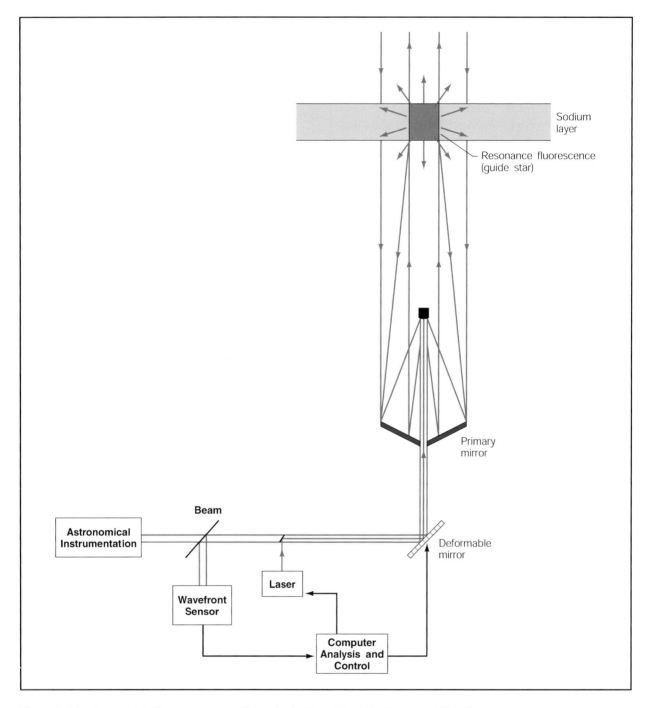

Sodium layer

Resonance fluorescence (guide star)

Primary mirror

Beam

Astronomical Instrumentation

Deformable mirror

Wavefront Sensor

Laser

Computer Analysis and Control

Figure 3. Adaptive optical telescope system. *Illustration by Hans & Cassidy. Courtesy of Gale Group.*

Figure 2 shows four different focusing systems for reflecting telescopes.

These are a) the prime focus, where the detector is simply placed at the prime focus of the mirror; b) the Newtonian, where a small, flat mirror reflects the light out to the side of the telescope; c) the Cassegrain, where the focus is located behind the **plane** of the primary mirror through a hole in its center and d) the Coudé,

where the two flat mirrors provide a long focal length path as shown.

Catadioptric telescopes

Catadioptric telescopes use a combination of lenses and mirrors in order to obtain some of the advantages of both. The best known type of catadioptric is the Schmidt telescope or camera, which is usually used to image a wide field of view for large area searches. The lens in this system is very weak and is commonly referred to as a corrector-plate.

Overcoming resolution limitations

The limits to the resolution of a telescope are, as described above, a result of the passage of the light from the distant body through the atmosphere which is optically nonuniform. Stars appear to twinkle because of constantly fluctuating optical paths through the atmosphere, which results in a variation in both brightness and apparent position. Consequently, much information is lost to astronomers simply because they do not have sufficient resolution from their measurements. There are three ways of overcoming this limitation, namely setting the telescope out in **space** in order to avoid the atmosphere altogether, compensating for the distortion on a ground-based telescope and/or stellar **interferometry**. The first two methods are innovations of the 1990s and are expected to lead to a new era in observational **astronomy**.

Space telescopes

The best known and biggest orbiting optical telescope is the **Hubble Space Telescope** (HST), which has an 8 ft (2.4 m) primary mirror and five major instruments for examining various characteristics of distant bodies. After a much publicized problem with the focusing of the telescope and the installation of a package of corrective **optics** in 1993, the HST has proved to be the finest of all telescopes ever produced. The data collected from HST is of such a high quality that researchers can solve problems that have been in question for years, often with a single photograph. The resolution of the HST is 0.02 arc seconds, close to the theoretical limit since there is no atmospheric distortion, and a **factor** of around twenty times better than was previously possible. An example of the significant improvement in imaging that space-based systems have given is the Doradus 30 nebula, which prior to the HST was thought to have consisted of a small number of very bright stars. In a photograph taken by the HST it now appears that the central region has over 3,000 stars.

Another advantage of using a telescope in **orbit** is that the telescope can detect wavelengths such as the ultraviolet and various portions of the infrared, which are absorbed by the atmosphere and not detectable by ground-based telescopes.

Adaptive optics

In 1991, the United States government declassified adaptive optics systems (systems that remove atmospheric effects), which had been developed under the Strategic Defense Initiative for ensuring that a **laser** beam could

The Clark telescope at Lowell Observatory in Arizona.
Photo Researchers, Inc. Reproduced by permission.

penetrate the atmosphere without significant distortion. The principle behind adaptive optical telescope systems is illustrated in Figure 3..

A laser beam is transmitted from the telescope into a layer of mesospheric **sodium** at 56-62 mi (90-100 km) altitude. The laser beam is resonantly backscattered from the volume of excited sodium **atoms** and acts as a guide-star whose position and shape are well defined except for the atmospheric distortion. The light from the guide-star is collected by the telescope and a wavefront sensor determines the distortion caused by the atmosphere. This information is then fed back to a deformable mirror, or an array of many small mirrors, which compensates for the distortion. As a result, stars that are located close to the guide-star come into a focus, which is many times better than can be achieved without compensation. Telescopes have operated at the theoretical resolution limit for infrared wavelengths and have shown an improvement in the visible region of more than ten times. Atmospheric distortions are constantly changing, so the deformable

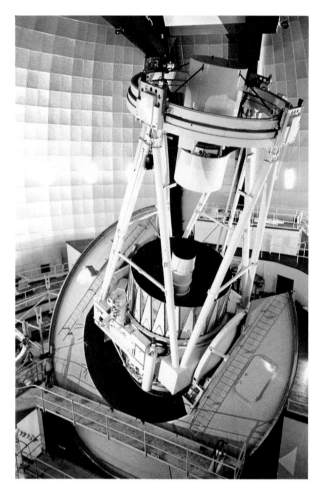

The Anglo-Australian telescope (AAT) in New South Wales, Australia. The principal mirror of this optical telescope has a diameter of 12.8 ft (3.9 m) and lies at the bottom of the white frame. During the day it is protected by an array of steel petals that are seen here in open position. The larger horseshoe shaped frame is part of the supporting structure. AAT was one of the first electronically controlled telescopes. *Royal Observatory, Edinburgh/National Audubon Society Collection/Photo Researchers, Inc. Reproduced by permission.*

mirror has to be updated every five milliseconds, which is easily achieved with modern computer technology.

Recording telescope data

Telescopes collect light largely for two types of analysis, imaging and spectrometry. The better known is imaging, the goal of which is simply to produce an accurate picture of the objects which are being examined. In past years, the only means of recording an image was to take a photograph. For long exposure times, the telescope had to track the sky by rotating at the same speed as **Earth**, but in the opposite direction. This is still the case today, but the modern telescope no longer uses photographic film but a charged coupled device (CCD) array.

The CCD is a semiconductor light detector, which is fifty times more sensitive than photographic film, and is able to detect single photons. Being fabricated using semiconductor techniques, the CCD can be made to be very small, and an array typically has a spacing of 15 microns between CCD pixels. A typical array for imaging in telescopes will have a few million pixels. There are many advantages of using the CCD over photographic film or plates, including the lack of a developing stage and the output from the CCD can be read directly into a computer and the data analyzed and manipulated with relative ease.

The second type of analysis is spectrometry, which means that the researcher wants to know what wavelengths of light are being emitted by a particular object. The reason behind this is that different atoms and molecules emit different wavelengths of light; measuring the **spectrum** of light emitted by an object can yield information as to its constituents. When performing spectrometry, the output of the telescope is directed to a spectrometer, which is usually an instrument containing a **diffraction grating** for separating the wavelengths of light. The diffracted light at the output is commonly detected by a CCD array and the data read into a computer.

Modern optical telescopes

For almost 40 years the Hale telescope at Mt. Palomar was the world's largest with a primary mirror diameter of 200 in (5.1 m). During that time improvements were made primarily in detection techniques, which reached fundamental limits of sensitivity in the late 1980s. In order to observe fainter objects, it became imperative to build larger telescopes, and so a new generation of telescopes is being developed for the 1990s and beyond. These telescopes use revolutionary designs in order to increase the collecting area; 2,260 ft^2 (210 m^2) is planned for the European Southern Observatory. This new generation of telescopes will not use the solid, heavy primary mirror of previous designs, whose thickness was between 1/6 and 1/8 of the mirror diameter, but will use a variety of approaches to reduce the mirror weight and improve its thermal and mechanical stability, including using 1) many hexagonal mirror elements forming a coherent array, 2) a single large meniscus mirror (with a thickness 1/40 of the diameter), with many active support points which bend the mirror into the correct shape and 3) a single large mirror formed from a honeycomb sandwich.

These new telescopes, combined with quantum-limited detectors, distortion reduction techniques, and coherent array operation will allow astronomers to see objects more distant than have been observed before.

One of this new generation, the Keck telescope located on Mauna Loa in Hawaii, is currently the largest oper-

TABLE 1. MAJOR GROUND-BASED OPTICAL TELESCOPES FOR THE 1990s		
Name	*Collector Area*	*Design Type*
Multi-mirror Telescope Conversion Kitt Peak, Arizona	33 m²	6.5 m honeycomb glass
Magellan Las Campanas, Chile	50 m²	8 m honeycomb glass
Keck Telescope Mauna Kea, Hawaii	76 m²	36 x 1.8 m hexagonal array
Keck I and II Mauna Key, Hawaii mirror	152 m²	two 36 x 1.8 m hexagonal arrays, spaced by ~ 75 m
Columbus, Arizona	110 m²	2 x 8.4 m honeycomb glass
Very Large Telescope Cerro Paranal, Chile	210 m²	4 x 8.2 m diameter meniscus

ating telescope, using a 32 ft (10 m) effective diameter hyperbolic primary mirror constructed from 36 6 ft (1.8 m) hexagonal mirrors. The mirrors are held to relative positions of less than 50 nm using active sensors and actuators in order to maintain a clear image at the detector.

Because of its location at over 14,000 ft (4,270 m), the Keck is useful for collecting light over the range 300 nm-30 æm. In the late 1990s, this telescope was joined by an identical twin, Keck II, which resulted in an effective mirror diameter of 279 ft (85 m) through the use of interferometry.

Alternative wavelengths

Most of the discussion so far has been concerned with optical telescopes operating in the range 300 nm-1100 nm. However, valuable information is contained in the radiation reaching us at different wavelengths and telescopes have been built to cover wide ranges of operation, including **radio** and millimeter waves, infrared, ultraviolet, **x rays,** and gamma rays.

Infrared telescopes

Infrared telescopes (operating from 1-1000 æm) are particularly useful for examining the emissions from gas **clouds**. Since **water** vapor in the atmosphere can absorb some of this radiation, it is especially important to locate infrared telescopes in high altitudes or in space. In 1983, NASA launched the highly successful Infrared Astronomical Satellite which performed an all-sky survey, revealing a wide variety of sources and opening up new avenues of astrophysical discovery. With the improve-

KEY TERMS

· ·

Chromatic aberration—The reduction in image quality arising from the fact that the refractive index in varies across the spectrum.

Objective—The large light collecting lens used in a refracting telescope.

Reflecting telescope—A telescope which uses only reflecting elements, i.e., mirrors.

Refracting telescope—A telescope which uses only refracting elements, i.e., lenses.

Spectrometry—The measurement of the relative strengths of different wavelength components which make up a light signal.

ment in infrared detection technology in the 1980s, the 1990s will see several new infrared telescopes, including the Infrared Optimized Telescope, an 26 ft (8 m) diameter facility, on Mauna Kea, Hawaii.

Several methods are used to reduce the large thermal background which makes viewing infrared difficult, including the use of cooled detectors and dithering the secondary mirror. This latter technique involves pointing the secondary mirror alternatively at the object in question and then at a patch of empty sky. Subtracting the second signal from the first results in the removal of most of the background thermal (infrared) noise received from the sky and the telescope itself, thus allowing the construction of a clear signal.

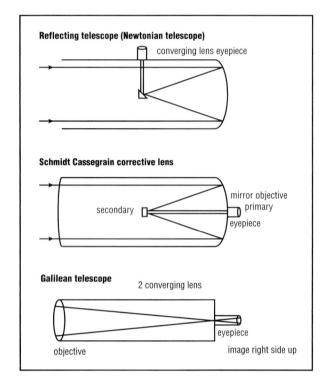

Reflecting telescope (Newtonian telescope)
converging lens eyepiece

Schmidt Cassegrain corrective lens
mirror objective
primary
secondary
eyepiece

Galilean telescope
2 converging lens
eyepiece
objective
image right side up

Regardless of design (i.e., the most common basic conbinations of lens and mirrors are shown above), the most important function of telescope is not to magnify but rather to gather light. *Illustration by Argosy. The Gale Group.*

Radio telescopes

Radio astronomy was developed following World War II, using the recently developed radio technology to look at radio emissions from the sky. The first radio telescopes were very simple, using an array of wires as the **antenna**. In the 1950s, the now familiar collecting dish was introduced and has been widely used ever since.

Radio waves are not susceptible to atmospheric disturbances like optical waves are, and so the development of radio telescopes over the past forty years has seen a continued improvement in both the detection of faint sources as well as in resolution. Despite the fact that radio waves can have wavelengths which are meters long, the resolution achieved has been to the sub-arc second level through the use of many radio telescopes working together in an interferometer array, the largest of which stretches from Hawaii to the United States Virgin Islands (known as the Very Long Baseline Array).

See also Spectroscopy.

Resources

Books

Consolmagno, Guy, and Dun M. Davis. *Turn Left at Orion.* Cambridge, UK: Cambridge University Press, 1989.

Field, George, and Donald Goldsmith. *The Space Telescope.* Chicago: Contemporary Books, 1989.

Malin, David. *A View of the Universe.* Cambridge: Sky Publishing, 1993.

Mark, Hans, Maureen Salkin, and Ahmed Yousef, eds. *Encyclopedia of Space Science & Technology.* New York: John Wiley & Sons, 2001.

Parker, Barry. *Stairway to the Stars.* New York: Plenum, 1994.

Tucker, Wallace, and Tucker, Karen. *The Cosmic Inquirers.* Cambridge: Harvard University Press, 1986.

Periodicals

Martin, Buddy, Hill, John M., and Angel, Robert. "The New Ground-Based Optical Telescopes." *Physics Today* (March 1991).

Iain A. McIntyre

Television

The invention of the **cathode ray tube** in 1897 by Ferdinand Braun quickly made possible the technology that we call television. Indeed, by 1907, the **cathode** ray tube was supplying television images. Within 50 years, television had become a dominant form of entertainment and an important way to acquire information. This remains true today, as the average American spends between two and five hours each day watching television.

The name television means distance seeing. Television, or TV, is the technology used to transmit pictures with sound using **radio** frequency and microwave signals or closed-circuit connections. Television operates on two principles that underlie how the human **brain** perceives the visual world. First, if an image is divided into a group of very small colored dots (called pixels), the brain is able to reassemble the individual dots to produce a meaningful image. Second, if a moving image is divided into a series of pictures, with each picture displaying a successive part of the overall sequence, the brain can put all the images together to form a single flowing image. The technology of the television (as well as computers) utilizes these two features of the brain to present images. The dominant basis of the technology is still the cathode ray tube.

Operation of the cathode ray tube

A cathode ray tube contains a positively charged region (the **anode**) and a negatively charged region (the cathode). The cathode is located at the back of the tube. As electrons exit the cathode, they are attracted to the anode. The electrons are also focused electronically into a tight beam, which passes into the central area of the television screen. The central region is almost free of air, so that there are few air molecules to deflect the electrons from their path. The electrons travel to the far end of the tube where they encounter a flat screen. The screen is coated with a **molecule** called phosphor. When an **electron** hits a phosphor, the phosphor glows. The electron beam can be focused in a coordinated way on different part of the phosphor screen, effectively painting the screen (a raster pattern). This process occurs very quickly—about 30 times each second—producing multiple images each second. The resulting pattern of glowing and dark phosphors is what is interpreted by the brain as a moving image.

Black and white television was the first to be developed, as it utilized the simplest technology. In this technology, the phosphor is white. **Color** television followed, as the medium became more popular, and demands for a more realistic image increased. In a color television, three electron beams are present. They are called the red, green, and blue beams. Additionally, the phosphor coating is not just white. Rather, the screen is coated with red, green, and blue phosphors that are arranged in stripes. Depending on which electron beam is firing and which color phosphor dots are being hit, a **spectrum** of colors is produced. As with the black and white television, the brain reassembles the information to produce a recognizable image.

High definition television

High definition television (HDTV) produces a much crisper and lifelike image than is possible using the cathode ray tube technology. This is because more information can be packed into the area of the television screen. Conventional cathode ray tube screens typically have 525 horizontal lines of dots on the screen. Each line contains approximately 500 dots (or pixels). Put another way, the information possible is 525 x 500 pixels. In contrast, HDTV contains 720 to 1080 x 500 pixels. The added level of detail produces a visually-richer image.

Televisions of the 1950s and 1960s utilized an analog signal. The signals were beamed out into the air from the television station, to be collected by an **antenna** positioned on a building or directly on the television ("rabbit ears"). Nowadays, the signal is digitized. This allows the electronic pulses to be sent through cable wire to the television, or to a **satellite**, which then beams the signal to a receiving dish in a format known as MPEG-2 (exactly like the video files that can be loaded on to a computer).

The digital signal is less subject to deterioration that is the analog signal. Thus, a better quality image reaches the television.

Cable television

Television signals are transmitted on frequencies that are limited in range. Only persons residing within a few-

An HDTV (high-definition television). *AP/Wide World Photos. Reproduced by permission.*

dozen miles of a TV transmitter can usually receive clear and interference-free reception. Community Antenna Television systems, often referred to as CATV, or simply cable, developed to provide a few television signals for subscribers far beyond the service area of big-city transmitters. As time passed cable moved to the big cities. Cable's appeal, even to subscribers able to receive local TV signals without an outdoor antenna, is based on the tremendous variety of programs offered. Some systems provide subscribers with a choice of hundreds of channels.

Cable systems prevent viewers from watching programs they have not contracted to buy by scrambling or by placing special traps in the subscriber's service drop that remove selected channels. The special tuner box that descrambles the signals can often be programmed by a digital code sent from the cable system office, adding or subtracting channels as desired by the subscriber.

Wired cable systems generally send their programming from a central site called a head end. All TV signals are combined at the head end, then sent down one or more coaxial-cable trunk lines. Signals for various neighborhoods along the trunk split away to serve individual neighborhoods from shorter branches called spurs.

Coaxial cable, even the special type used for CATV trunk lines, is made of material that dissipates the electrochemicals passing through. Signals must be boosted in power periodically along the trunk line, usually every time the signal level has fallen by approximately 20 decibels, the equivalent of the signal having fallen to 1/100 of its original power. The line amplifiers used must be very sophisticated to handle the wide bandwidth required for many programs without degrading the pictures or adding noise. The amplifiers must adjust for changes in the coaxial cable due primarily to **temperature** changes. The amplifiers used are very much improved over those used by the first primitive community antenna systems, but even today trunk lines are limited in length to about a dozen miles. Not much more than about one hundred line amplifiers can be used along a trunk line before problems become unmanageable.

Cable's program offerings are entirely confined within the shielded system. The signals provided to subscribers must not interfere with over-the-air radio and television transmissions using the same frequencies. Because the cable system's offerings are confined within the shielded system, pay-per-view programs can be offered on cable.

Cable is potentially able to import TV signals from a great distance using satellite or terrestrial-microwave relays. Cable systems are required to comply with a rule called Syndex, for syndication exclusivity, where an over-the-air broadcaster can require that imported signals be blocked when the imported stations carry programs they have paid to broadcast.

The next step in CATV technology is the replacement of wire-based coaxial systems with fiberoptic service. Fiberoptics is the technology where electrical signals are converted to **light** signals by solid-state **laser** diodes. The light waves are transmitted through very-fine **glass** fibers so transparent that a beam of light will travel through this glass fiber for miles.

Cable's conversion to fiberoptics results in an enormous increase in system bandwidth. Virtually the entire radio and TV spectrum is duplicated in a fiberoptic system. Every radio and TV station transmitting "over the air" can be carried in a single thread of fiberglass; 500 separate television channels on a single cable system are not beyond reach. A fiberoptic CATV system can be used for two-way communication more easily than can a wire-cable plant with electronic amplifiers. Fiberoptic cable service has great potential to support interactive television services.

Television of the future

Plasma television

Plasma television has been available commercially since the late 1990s. It is currently expensive, and as so

is not yet popular. As with other technologies, however, refinements over time will drive down the price into the affordable range for many people. Plasma televisions do not have a cathode ray tube. Thus, the screen can be very thin. Typically televisions screens are about 6 in (15 cm) thick. This allows the screen to be hung from a wall.

In a plasma television, fluorescent lights are present instead of phosphors. Red, green, and blue fluorescent lights enable a spectrum of colors to be produced, in much the same way as with conventional television. Each **fluorescent light** contains a gas called plasma. Plasma consists of electrically charged **atoms** (ions) and electrons (negative in charge). When an electrical signal encounters plasma, the added **energy** starts a process where the particles bump into one another. This bumping releases a form of energy called a **photon**. The release of ultraviolet photons causes a reaction with phosphor material, which then glows.

ATV

ATV stands for advanced television, the television system expected to replace the current system in the United States. Television technology is rapidly moving toward the ATV digital system planned to replace the aging analog process. ATV is a digital-television system, where the aspects that produce the image are processed as computer-like data. Digitally processed TV offers several tremendous advantages over analog TV methods. In addition to sharper pictures with less noise, a digital system can be much more frugal in the use of spectrum space.

Most TV frames are filled with information that has not changed from the previous frame. A digital TV system can update only the information that has changed since the last frame. The resulting picture looks to be as normal as the pictures seen for years, but many more images can be transmitted within the same band of frequencies.

TV audiences have been viewing images processed digitally in this way for years, but the final product has been converted to a wasteful analog signal before it leaves the television transmitter. Satellite relays have relayed TV as digitally compressed signals to maximize the utilization of the expensive transponder equipment in **orbit**. The small satellite "dishes" offered for home reception receive digitally-encoded television signals.

ATV will not be compatible with current analog receivers, but it will be phased in gradually in a carefully-considered plan that will allow older analog receivers to retire gracefully over time.

See also Digital recording; Electronics.

KEY TERMS

. . . .

Chrominance—Color information added to a video signal.

Coaxial cable—A concentric cable, in which the inner conductor is shielded from the outer conductor; used to carry complex signals.

Compact disc—Digital recording with extraordinary fidelity.

Field—Half a TV frame, a top to bottom sweep of alternate lines.

Frame—Full TV frame composed of two interlaced fields.

Parallax—Shift in apparent alignment of objects at different distances.

Phosphor—Chemical that gives off colored light when struck by electron.

Resources

Other

Ovadia, Schlomo. *Broadband Cable TV Access Networks: From Technologies to Applications.* Upper Saddle River, NJ: Prentice Hall, 2001.

Thomas, Jeffrey L., and Francis M. Edgington. *Digital Basis for Cable Television Systems.* Upper Saddle River, NJ: Prentice Hall, 1999.

Donald Beaty

Tellurium *see* **Element, chemical**

Temperature

Temperature is intuitively associated with the sense of hot and cold. Put your finger in a pan of hot **water** and **energy** flows as **heat** from the water to your finger; you say that the water is at a higher temperature than your finger. Now put your finger in a **glass** of **ice** water and energy flows as heat in the other direction. The direction of energy flow as heat is the basis of our definition of temperature. Temperature is the property of objects—or more generally, systems—that determines the direction of energy flow as heat when the objects are put in direct contact with each other. Energy flows as heat from objects at higher temperature to ones at lower temperature. When energy as heat ceases to flow, the objects are at the same temperature and are said to be in thermal equilibrium.

Molecular interpretation

At the molecular level, temperature is related to the **random** motions of the particles (**atoms** and molecules) in **matter**. Because there are different types of **motion**, the particles' kinetic energy (energy of motion) can take different forms, and each form contributes to the total kinetic energy of the particles.

For example, when water squirts from a hose, part of the kinetic energy of the water is due to the movement of the molecules as a collection in a single direction out the nozzle. But the individual water molecules are also moving about with random, constantly changing, speeds and directions relative to each other. This kind of kinetic energy is called molecular translational energy. This energy remains even after the squirted water becomes a quiet puddle. The temperature of the puddle, or of any object, is a measure of the average of the individual translational energies of all of its atoms or molecules.

If a swimming pool were filled from this same hose, the total molecular translational energy of the molecules in the pool would be much greater than those in the puddle because there are many more molecules in the pool. The temperatures of the puddle and the pool, however, would be the same because temperature is a measure of the average molecular translational energies.

The molecules in a kettle of boiling water have a higher average molecular translational energy—a higher temperature—than those in the swimming pool. Place the kettle on the surface of the pool and the direction of energy flow is obvious: from hotter to cooler, from higher temperature to lower, from greater average molecular translational energy to lesser. These are three ways of saying the same thing.

The reason that heat flows from an object of higher temperature to one of lower temperature is that once they are in contact, the molecular agitation is contagious. Fast-moving molecules will collide with slower-moving ones, kicking them up to higher speed and thereby raising their translational energy.

Thermometers and temperature scales

There is no easy way to measure directly the average molecular translational energies in an object. Therefore, temperature is determined indirectly by measuring a temperature-dependent property of a device that is in thermal equilibrium with the object to be measured. We call such a device a **thermometer**. One of the earliest kinds of thermometers, still in use today, has a liquid in a glass bulb attached to a glass capillary tube—a tube with a very narrow bore. The tube is sealed to make the thermometer independent of **atmospheric pressure**. The expansion and contraction of the liquid **volume** as the temperature goes up and down results in a rise and fall of the thin column of liquid in the tube, which can be calibrated in terms of temperature.

Early thermometry used water or water-alcohol mixtures as the liquid. Water turned out to be an especially poor choice, however, because the volume of water does not change uniformly with changes in temperature. When cooled down from room temperature, liquid water at first contracts. Then, as the water approaches its freezing point, it expands. This unusual property of expansion upon cooling means that a water thermometer not only goes down as the temperature falls, but it sometimes goes up. This is certainly not a desirable characteristic of a thermometer liquid.

The Fahrenheit scale of temperature

In 1714, the German physicist Daniel Gabriel Fahrenheit (1686-1736) made a better choice by selecting liquid mercury. Mercury has a uniform volume change with temperature, a lower freezing point and higher **boiling point** than water, and does not wet glass. Mercury thermometers made possible the development of reproducible temperature scales and quantitative temperature measurement. Fahrenheit first chose the name " degree" (Grad, in German) for his unit of temperature. Then, to fix the size of a degree (°), he decided that it should be of such size that there are exactly 180° between the temperature at which water freezes and the temperature at which water boils. (180 is a "good" number because it is divisible by one and by 16 other whole numbers. That is why 360, or 2×180, which is even better, was originally chosen as the number of "degrees" into which to divide a circle.) Fahrenheit now had a size for his degree of temperature, but no standard reference values. Should the freezing and boiling points of water be called **zero** and 180? Or 180 and 360, or something else? In other words, where shall 0° fall on the scale? He eventually decided to fix zero at the coldest temperature that he could make in his laboratory by mixing ice with various salts that make it colder. (Salts, when mixed with cold ice, lower the melting point of ice, so that when it is melting it is at a lower temperature than usual.) When he set his zero at that point, the normal freezing point of water turned out to be 32° higher. Adding 180 to 32 gave 212° for the normal boiling point of water. Thus, freezing water falls at 32° and boiling water falls at 212° on the Fahrenheit scale. And the normal temperature of a human being turns out to be about 99°.

The Celsius scale

In 1742, the noted Swedish astronomer Anders Celsius (1701-1744), professor of **astronomy** at the Universi-

ty of Uppsala, proposed the temperature scale which now bears his name, although for many years it was called the centigrade scale. As with the Fahrenheit scale, the reference points were the normal freezing and normal boiling points of water, but he set them to be 100° apart instead of 180. Because the boiling point and, to a lesser extent, freezing point of a liquid depend on the atmospheric **pressure**, the pressure must be specified: "normal" means the freezing and boiling points when the atmospheric pressure is exactly one atmosphere. These points are convenient because they are easily attained and highly reproducible. Interestingly, Celsius at first set boiling as zero and freezing as 100, but this was reversed in 1750 by the physicist Martin Strömer, Celsius's successor at Uppsala.

Defined in this way, a Celsius degree (°C) is 1/100 of the temperature difference between the normal boiling and freezing points of water. Because the difference between these two points on the Fahrenheit scale is 180°F, a Celsius degree is 1.8 times (or 9/5) larger than a Fahrenheit degree. You cannot convert between Fahrenheit and Celsius temperatures simply by multiplying by 1.8, however, because their zeroes are at different places. That would be like trying to measure a table in both yards and meters, when the left-hand ends (the zero marks) of the yardstick and meter stick are not starting at the same place.

To convert a temperature from Fahrenheit to Celsius or vice versa, you first have to account for the differences in their zero points. This can be done very simply by adding 40 to the temperature you want to convert. That is because -40° (40 below zero) happens to come out at the same temperature on both scales, so adding 40 gets them both up to a comparable point: zero. Then you can multiply or divide by 9/5 to account for the difference in degree size, and finally remove the 40° that you added.

Thus, to convert from Celsius to Fahrenheit, (1) Multiply by 9/5 (2) Add 32. To convert from Fahrenheit to Celsius, (1) Subtract 32 (2) Multiply by 5/9.

The Kelvin scale

About 1787 the French physicist, Jacques Charles (1746-1823) noted that a sample of gas at constant pressure regularly contracted by about 1/273 of its volume at 0°C for each Celsius degree drop in temperature. This suggests an interesting question: If a gas were cooled to 273° below zero, would its volume drop to zero? Would it just disappear? The answer is no, because most gases will condense to liquids long before such a low temperature is reached, and liquids behave quite differently from gases.

In 1848 William Thomson (1824-1907), later Lord Kelvin, suggested that it was not the volume, but the molecular translational energy that would become zero at about -273°C, and that this temperature was therefore the lowest possible temperature. Thomson suggested a new and more sensible temperature scale that would have the lowest possible temperature—absolute zero—set as zero on this scale. He set the temperature units as identical in size to the Celsius degrees. Temperature units on Kelvin's scale are now known as Kelvins (abbreviation, K); the term, degree, and its symbol, °, are not used anymore. Lord Kelvin's scale is called either the Kelvin scale or the absolute temperature scale. The normal freezing and boiling points of water on the Kelvin scale, then, are 273K and 373K, respectively, or, more accurately, 273.16K and 273.16K. To convert a Celsius temperature to Kelvin, just add 273.16.

The Kelvin scale is not the only absolute temperature scale. The Rankine scale, named for the Scottish engineer William Rankine (1820-1872), also has the lowest possible temperature set at zero. The size of the Rankine degree, however, is the same as that of the Fahrenheit degree. The Rankin temperature scale is rarely used today.

Absolute temperature scales have the advantage that the temperature on such a scale is directly proportional to the actual average molecular translational energy, the property that is measured by temperature. For example, if one object has twice the Kelvin temperature of another object, the molecules, or atoms, of the first object actually have twice the average molecular translational energy of the second. This is not true for the Celsius or Fahrenheit scales, because their zeroes do not represent zero energy. For this reason, the Kelvin scale is the only one that is used in scientific calculations.

Temperature extremes

The highest recorded **weather** temperature on **Earth** was 136°F (57.8°C), observed in North **Africa** in 1922. The record low temperature is -129°F (-89.2°C), observed in the Antarctic in 1983. Elsewhere in the universe, temperature extremes are much greater. The average surface temperatures of the most distant planets in our **solar system** (**Uranus**, **Neptune**, **Pluto**) are about 53K (-364°F; -220°C;). Although temperatures on and within the **sun** vary, the core is about 27 million° F (15 million° C). (At very high temperatures Celsius and Kelvin temperatures are virtually identical; the 273 is negligible).

Temperatures produced in laboratories can be even more extreme. The study of very low temperatures, called **cryogenics**, is an active field of scientific research because of the unusual properties of many substances when they are close to **absolute zero**. Using magnetic techniques, temperatures below one microkelvin have been achieved. (A microkelvin, μK, is 0.000001K.) Absolute zero itself, however, has not yet been reached.

Less extreme low temperatures can be obtained relatively easily with dry ice or liquid **nitrogen**. Dry ice, solid **carbon dioxide**, vaporizes (sublimes) under normal pressures rather than melting to a liquid. The sublimation temperature of dry ice is 195K (-198°F; 78°C). Liquid nitrogen can be used to obtain even lower temperatures. It is used at its normal boiling point, which is of liquid nitrogen is 77K (-321°F; -196°C).

Scientific interest in very high temperatures is largely due to the hope of achieving controlled nuclear fusion—the energy producing process in the sun and stars. By the use of powerful lasers, temperatures over 400 million kelvins have been achieved for short periods of **time**.

See also Gases, properties of; States of matter.

Resources

Books

Atkins, Peter W. *The Second Law.* New York: Freeman, 1984.
Klein, Herbert A. *The Science of Measurement.* New York: Dover, 1974.

Periodicals

Kikoyin, A. "Temperature, Heat, and Thermometers." *Quantum* (May 1990): 16-21.

"Temperature And Rainfall Tables: July 2002." *Journal of Meteorology* 27, no. 273 (2002): 362.

"Weather Extremes: July 2002." *Journal Of Meteorology* 27 no. 273 (2002): 361.

John C. Whitmer

Temperature regulation

An **organism** maintains **homeostasis**, a steady internal state, only if its body **temperature** stays within prescribed limits. Cellular activities require an optimum amount of **heat**. They depend on **enzyme** action, and enzymes function within a narrow range of temperature. For this reason, living things can only tolerate a limited rise or drop in temperature from the optimum. Mechanisms exist that regulate body temperature to stay within its survival limits.

Cells require a source of heat for their activities. The heat source can be external or internal. The **Sun** is an external source of heat for living things. Cellular **metabolism** is an internal heat source. During cellular respiration, the chemical **energy** in food is converted to high-energy phosphate groups in **adenosine triphosphate**, ATP. In the process, cells lose about 60% of the released energy as heat. Plants and most animals give up this heat to the environment. **Birds**, **mammals**, and some **fish**, however, make use of some of the heat that metabolism yields.

Organisms exchange heat with the environment by four major physical processes. First, there is conduction, the direct transfer of heat from a warmer to a cooler object. We cool off during a swim because the heat of the body is conducted to the cool **water**. During another process called **convection**, a breeze passes over a surface and brings about heat loss. This is why a fan cools us on a hot day. **Radiation**, the **emission** of electromagnetic waves, transfers heat between objects that are not in direct contact. Animals absorb heat that the sun radiates. Finally, **evaporation**, the physical change of liquid to gas removes heat from a surface. Sweating and panting are cooling processes. Perspiration evaporates from the skin, and panting increases evaporation from the **respiratory system**. Both procedures take heat away from the body.

Externally heated animals

Ectotherms are animals that warm their bodies by absorbing heat from their surroundings. In most ectotherms, the body temperature fluctuates with changes in the surrounding temperature. The body temperature of **snakes**, for example, cools in cold **weather** and warms up in hot weather. Most marine fish and **invertebrates**, however, live in water that stays the same temperature. Their body temperature, therefore, does not change. In everyday language, we say that these animals are cold-blooded. **Amphibians**, **reptiles**, most fish, and invertebrates are considered cold-blooded. This term is misleading, however, because many lizards have high body temperatures. External temperature plays a major role in the activity **rate** of ectothermic animals. When the weather is warm, they become active. They slow down when the temperature drops.

Certain ectotherm behaviors help regulate body temperature. To warm up, reptiles find sunny places, and stretch out for maximum exposure. If it gets too warm, lizards alternate between sun and shade. Amphibians warm up by moving into the sun or diving into warm water. They cool off by entering the shade. In cold weather, honeybees huddle together to retain heat. **Bees** and large **moths** build up heat before takeoff by contracting their flight muscles without moving their wings.

In addition to behaviors, physiological adaptations help ectotherms regulate temperature. Diving reptiles conserve heat because their **blood** circulates inward toward the body core during a dive. The skin of bullfrogs secretes more mucus when it is hot, allowing more cooling by evaporation. Many ectotherms exist at a lower temperature during torpor, a state of slowed metabolism. This helps them survive a food shortage. If the food supply increases, they come out of torpor in a few hours.

Internally heated animals

Endotherms are animals that warm their bodies mainly from their own metabolism. Mostly birds and mammals, they maintain a constant body temperature regardless of changes in the surrounding temperature. We commonly call them warm-blooded, because the internal temperature they generate is usually higher than the environmental temperature. Endotherms have certain advantages. Their consistent higher temperature allows them to be active at all times. It also gives them the freedom to live in varied habitats. It helps them survive on land, where the air temperature changes more drastically than water temperature.

Although most fish are ectothermic, mackerals, **tuna**, **swordfish**, **marlins**, and some **sharks** are endotherms; although in swordfish and some mackeral, warming occurs only in the central **nervous system** and the retina of the **eye**. **Endothermic** fish are more closely related to their ectothermic relatives than they are to each other. Thus, endothermy evolved independently in

Canines, like this North American timber wolf, use panting as a means of temperature regulation. *Photograph by Frank Rossotto. Stock Market. Reproduced by permission.*

these different groups, as an **adaptation** that allowed the animals to expand their habitats into waters of varying temperatures.

Certain adaptations accompany the maintenance of a constant body temperature. A high metabolic rate helps supply heat. An efficient **circulatory system** conducts and distributes heat around the body. **Fat** layers, fur, and feathers insulate the body and retain heat. Shivering muscles contract to increase body heat, and **hormones** such as epinephrine and thyroxin increase the rate of metabolism. Under unfavorable conditions some endotherms experience torpor, for example, **hibernation**, in which the body temperature drops. Hibernation enables animals to survive long periods of cold and lack of food.

Receptors in various parts of the body, such as the skin, deliver information to the hypothalamus of the **brain** about the body's temperature. The hypothalamus acts as a **thermostat**. It contains two thermoregulating areas that stimulate nerve impulses. One area, the heating center, sends out impulses that raise the body temperature. It causes blood vessels near the surface of the body to contract, thus preventing heat loss. It also causes fur to stand erect and become more insulating. It causes shivering by muscles that produce heat when they contract, and it stimulates hormonal production. The other area, the cooling center, sends out impulses that bring about a temperature drop. It causes surface blood vessels to expand thereby releasing heat. It brings about sweating or panting.

Resources

Books

Coping with Change. Films for the Humanities and Sciences. Princeton, 1994-95.

Guyton & Hall. *Textbook of Medical Physiology.* 10th ed. New York: W. B. Saunders Company, 2000.

KEY TERMS
. . . .

Cooling center—Thermoregulating area of the hypothalamus whose impulses result in temperature-lowering activities of various parts of the body.

Ectotherm—A cold-blooded animal, whose internal body temperature is similar to that of its environment. Ectotherms produce little body heat, and are dependent on external sources (such as the sun) to keep their body temperature high enough to function efficiently.

Endotherm—An animal that uses its metabolism as a primary source of body heat and uses physiological mechanisms to hold its body temperature nearly constant.

Heating center—Thermoregulating area of the hypothalamus whose impulses result in temperature-raising activities of various parts of the body.

Hibernation—A type of torpor in which an animal's body temperature drops, thereby enabling it to withstand prolonged cold and little food.

Periodicals

Abass, Mageed, and C.B. Rajashekar. "Abscisic Acid Accumulation in Leaves and Cultured Cells During Heat Acclimation in Grapes." *HortScience* (January 1993).

Bookspan, Jolie. "Exposing the Myths of Heat Loss and Gain for Divers." *Underwater USA* (July 1993).

Comis, Don. "Resetting a Plant's Thermostat." *Agricultural Research* (July 1992).

Heinrich, Bernd. "Kinglets' Realm of Cold." *Natural History* (February 1993).

Luck-Baker. "Taking the Temperature of *T. Rex.*" *New Scientist* (July 23, 1994).

Monastersky, Richard. "The Pulse of *T. Rex.*" *Science News* (May 14, 1994).

Bernice Essenfeld

Tenrecs

Tenrecs are four-legged nocturnal **mammals** belonging to the order Insectivora. Tenrecs have evolved into more distinct forms than any other family of animals within the order. Tenrecs can resemble **hedgehogs**, **moles**, **shrews**, or muskrats, depending on the **species**. Some species of tenrecs have a long tail and long hind legs, while others have a stumpy tail and short hind legs.

A stripped tenrec. *Photograph by H. Uible/Photo Researchers, Inc. Reproduced by permission.*

Furthermore, some species of tenrec have a spiny coat, similar to that of hedgehog, while others have velvety fur.

Within the order Insectivora, there are six general types: shrew type, rat type, hedgehog type, mole type, jeroba type, and otter type. Tenrecs are grouped together with the shrews, and are characterized by an elongated body, and long, pointed snout. Tenrecs belong to the family Tenrecidae.

Evolution of insectivores

Insectivores are the most primitive of all higher mammals, and insectivore-like animals predated all of the orders of today's mammals. The fossil remains of insectivores indicated that they lived during the Cretaceous period. Rat-sized ancestors of today's insectivores date even back further—to the Jurassic and Lower Cretaceous periods. These fossils also indicate that tenrec-like animals and golden moles are related to each other.

Family Tenrecidae

The family Tenrecidae includes between 30-34 species, all but three of which live on the **island** of Madagascar. The remaining three species (all **otter shrews**) live in Central and West equatorial **Africa**.

In general, tenrecs have poor **vision**, but their senses of **smell**, **hearing**, and **touch** are acute. Like other insectivores, they have an elongated snout, a small, primitive **brain**, and their skull is relatively small, long, and narrow. Their fur can be soft, coarse, or even spiny. Tenrecs have retained some reptilian characteristics, such as a cloaca-like common chamber into which their digestive, reproductive, and urinary systems empty, and from which these substances leave their body.

The family Tenrecidae family includes three subfamilies: the Tenrecinae (tenrecs), the Oryzorictinae (rice tenrecs), and the Potamogalinae (otter shrews). The best known tenrecs are similar in appearance to hedgehogs, and belong to the Tenrecinae subfamily. This subfamily includes three genera and three species. The rice tenrecs have a long tail and are closely related to the spiny hedgehogs; this subfamily includes three genera and 24 species. There are two genera and three species of otter shrews.

Tailless or common tenrec *(Tenrec ecaudatus)*

The tailless tenrec is, perhaps, the best known species of tenrec. Vaguely resembling a hedgehog, it measures about 12.5 in (32 cm) long, with a stubby tail measuring 0.4-0.6 in (1-1.6 cm) in length. It has a long snout and coarse, bristly fur interspersed with spines. These tenrecs prefer sandier environments, such as highland plateaus and cliffs along riverbanks. During the day, these animals rest in crevices or burrows; at night, they forage for food by digging with their claws and snout. Tailless tenrecs mainly eat **insects**, lizards, eggs, roots, and fruit. In the dry season, these tenrecs hibernate in deep underground burrows. At the beginning of October, after a long **hibernation**, they mate. Female tailless tenrecs commonly have as many as 16 surviving offspring.

If threatened, tailless tenrecs stand on their hind legs and bristle their coats, and try to push the bristles into the intruder, while snorting, grunting, and hissing. Although tailless tenrecs are protected by law, the people of Madagascar hunt these animals for their fatty meat.

Rice tenrecs *(Oryzorictes spp.)*

Rice tenrecs are spinier than tailless tenrecs and, are more closely related to hedgehogs. Rice tenrecs are classified with the shrews because of similarities in their skeletons. The body of rice tenrecs measures between 1.5-5 in (4-13 cm), and their tail measures between 1-6.5 in (3-16 cm). These tenrecs acquired their name because they live on and within the banks of **rice** paddies, as well as in warm, moist **forests**, swamps, and meadows. Their front limbs are well adapted for digging and these tenrecs spend a great deal of time underground. Rice tenrecs are only seen above ground at night, but it is assumed that they are very active during the day underground, eating **invertebrates** and crustaceans. Rice growers in Madagascar consider rice tenrecs to be **pests**.

Reproduction

Some species of tenrec are prolific. Tenrecs produce the most young per litter (averaging between 12 and 15) of all the world's mammals. The number of offspring can be even higher; for example, common tenrecs have been known to have as may as 32 young in one litter. Because female tenrecs almost always have multiple births, they have numerous nipples to feed their young. In fact, common tenrec females have 29 nipples.

Female tenrecs must sometimes forage for food in daylight, when the danger presented by predators is the highest, to adequately nourish themselves, so that they are able to meet the huge demand for milk. In some tenrec species, the young have extra camouflage to protect themselves from predators, which they lose as they mature. Young tenrecs are fairly independent creatures; they are able to run soon after they are born. By the time they are four weeks old, they are completely independent.

Temperature regulation

The body **temperature** of tenrecs is maintained between 78.8–86°F (26–30°C). The activity levels of streaked tenrecs vary with the surrounding temperature. Increases in physical activity generate the extra body **heat** that they need to survive in colder conditions. On a normal day, with a daytime temperature of about 68°F (20°C), streaked tenrecs rest inside their burrows; by early evening, their activity level increases. At midnight, they start taking more frequent rests and, by dawn, they crawl back inside their shelters. However, when the outside temperature goes down to 60.8–64.4°F (16–18°C), tenrecs become much more active, both day and night. If the temperature gets colder than 60.8°F (16°C), even increased activity is insufficient to keep them warm, and they perish. Because the streaked tenrec inhabits moist areas with little temperature change, these animals rarely die of cold.

The **habitat** of tailless tenrecs, however, is much more variable than that of streaked tenrecs. On Madagascar, there is little rain in the winter, the land becomes very dry, and temperatures fall to as 50°F (10°C). At this point, tailless tenrecs, which have been accumulating **fat** all summer, roll into a ball and hibernate in their deep underground burrows for about six months. During hibernation, they are cold to the touch and breath about once every three minutes. During this time, they neither eat nor defecate.

Resources

Books

Grzimek, H.C. Bernard, ed. *Grzimek's Animal Life Encyclopedia.* New York: Van Nostrand Reinhold Company, 1993.
Gunderson, Harvey L. *Mammalogy.* New York: McGraw-Hill, 1976.
Lawlor, Timothy E., *Handbook to the Orders and Families of Living Mammals.* Eureka, CA: Mad River Press, 1979.

Kathryn Snavely

Teratogen

A teratogen is an environmental agent that can cause abnormalities in a developing **organism** resulting in either fetal death or **congenital** abnormality. The human fetus is separated from the mother by the placental barrier, but the barrier is imperfect and permits a number of chemical and infectious agents to pass to the fetus.

Well known teratogens include (but are not limited to) **alcohol**, excess **vitamin** A and retinoic acid, the rubella **virus,** and high levels of **ionizing radiation**. Perhaps the best known teratogenic agent is the drug **thalidomide**, which induced severe limb abnormalities known as phocomelia in children whose mothers took the drug.

See also Birth defects; Embryo and embryonic development; Fetal alcohol syndrome.

Terbium *see* **Lanthanides**

Term

A term is an algebraic expression which can form a separable part of another expression such as an algebraic equation or a sequence. Terms are a specific part of the symbolic language of **algebra**. The symbols of this language were primarily developed during the sixteenth and seventeenth centuries and are used to represent otherwise lengthy expressions. They can be as simple as using the single character, +, to mean addition, or as complicated as $y = 4x^2 + 2x - 3$ to represent an algebraic polynomial equation.

In general, there are three types of algebraic expressions which can be classified as terms. These include expressions made up of a single **variable** or constant, ones that are the product or quotient of two or more variables and/or constants, and those that are the product or quotient of other expressions. For example, the number 4 and the variable x are both terms because they consist of a single symbol. The expression 2z is also a term because it represents the product of two symbols. It should be noted that terms like 2z, in which a number and a variable are written together, are indicated products because **multiplication** is implied. Therefore, the symbol 2z means $2 \times z$. Finally, an expression like $2pq(a + 5)n$ is a term because it represents a quotient (the result of **division**) of two expressions.

The symbols that make up a term are known as coefficients. In the term 4x, the number 4 is known as a numerical **coefficient** and the letter x is known as the literal coefficient. For this expression, we could say that 4 is the coefficient of x or x is the coefficient of 4.

Terms should be thought of as a single unit that represents the value of a particular number. This is particularly useful when discussing the terms of a larger expression such as an equation. In the expression $5x^3 + 2x^2 + 4x - 7$, there are four terms. Numbering them from left to right, the first term is $5x^3$, the second is $2x^2$, the third is $4x$, and the fourth is -7. Notice that the sign in front of a term is actually part of it.

Some expressions contain terms which can be combined to form a single term. These "like terms" contain the same variable raised to the same power. For example, the like terms in the expression $3x + 2x$ can be added and the equation simplifies to $5x$. Similarly, the expression $7y^2 - 3y^2$ can be simplified to $4y^2$. Expressions containing unlike terms can not be simplified. Therefore, $4x^2 - 2x$ is in its simplest form because the differences in the power of x prevents these terms from being combined.

Termites

Termites are slender, social **insects**, ranging in size from 0.007-0.072 in (2-22 mm) long. The reproductive members of the **species** have wing spans of 0.03-0.3 in (10-90 mm). Inhabiting nests of their own construction, they live in permanent and often highly developed communities. While termites thrive in warm, humid environments, some species have also adapted to open savannas and temperate zones. They are most commonly found in the tropical parts of **Africa**, Southeast **Asia**, **Australia**, and the Americas.

Termites belong to the order Isoptera, which includes seven families of termites, containing about 200 genera and about 2,300 species. The families are: (1) the Mastotermitidae, (2) the Kalotermitidae, (3) the Termopsidae, (4) the Hodotermitidae, (5) the Rhinotermitidae, (6) the Serritermitidae, and (7) the Termitidae. The first five families are referred to as the lower termites, while the last two are known as the higher termites.

The classification into lower and higher termites refers to the termites' level of **evolution**, both in terms of **behavior** and **anatomy**. The higher termites have a more complex and developed social structure and build a more complex and varied nest. The most advanced nests start below the ground and form a mound above. A termite nest in Australia measured almost 20 ft (6.66 m) high and 98 ft (33 m) around. Three-quarters of all species of termites belong to the higher termite groups.

Caste system

The individuals making up a termite community are anatomically and functionally distinct from each other, depending upon their role within the community. The particular duty that each termite performs determines the specific caste to which each termite belongs. Each caste can perform only a certain task, and individual termites cannot stay alive without the rest of the colony. In total, there are four basic kinds of adult termite castes: the king and queen (primary reproductives), the supplementary reproductives, the soldiers, and the workers.

Reproductives

There are two reproductive castes providing for the continuation of the species. The primary reproductive caste—known as alates—consists of potential kings and queens. At a specific time each year, depending on the species and environment, a swarm of these alates leaves the original nest to start a new colony. The vast majority of swarming alates fall **prey** to **birds**, **bats**, lizards, **snakes**, and **ants**. When the male and female alates land, they shed their wings and look for mates. Once the pairing occurs, they run on the ground in tandem, the female ahead of the male, looking for the site on which to build their new nest.

The supplementary reproductives develop functional reproductive organs but never leave the parent colony. Their purpose is to act as substitutes for the king or queen if either or both dies. Further, female reproductives can supplement the queen's egg-laying volume, should her egg-laying capacity prove insufficient to maintain the colony.

The nesting location chosen by the alates is highly dependent upon the particular species of termite. Some species prefer to nest in the ground, while others prefer to nest in **wood**. Regardless of the nest, the pair digs a chamber, closes it off from the outside world, and never leaves it again. The pair mates in the chamber, and the female lays the eggs. Her first laying contains relatively few eggs. The young king and queen care for their larvae for four to six months. During this period, wood termites get the sustenance they need from the wood in their nests, while **soil** termites feed their young with the excess **fat** stored in their own bodies and by digesting their now-useless wing muscles.

Newly hatched termite larvae are tiny, measuring about 0. 08 in (2 mm) in length. Their coats are colorless and their bodies are soft. Furthermore, they are blind; their main sensory organs are their antennae, but they also have a series of small receptors situated on their heads.

After hatching, the young termites—or nymphs—have the general appearance of adult termites. They continue to grow and molt at intervals. After several stages of development, some finally become sexually mature individuals with wings. Others either become workers or soldiers. Both the workers and the soldiers—which are usually blind, and always infertile and wingless—are the result of specialized, incomplete development.

Workers

The workers form by far the largest caste and live exclusively underground; they are pale, have large toughened heads, and lack eyes. When the first workers are old enough to leave the nest, they begin to undertake their multiple duties. First, they search for food, leaving the queen free to devote all of her time to egg-laying. Eventually, the workers provide food for the king, queen, larvae, and soldiers. Dead wood is the main part of a worker termite's diet. Often, workers eat wood before returning to the colony with it. In such cases, they partially digest it and transform it into a milky liquid. When they reach the colony, the workers feed the other termites through regurgitation; this form of food is called stomodeal food.

Most species of termite eat wood, but their digestive enzymes are not able to break down the nutritious part of the wood, called **cellulose**. To do this, termites have the help of **microorganisms** residing in their bodies that break down the cellulose into a substance that they can digest.

The workers also take part in building the colony. In the majority of species, the workers dig underground systems of tunnels. They create complex structures, which protect the inhabitants from their enemies. Because some nests contain as many as three million termites, the nests need the right level of **oxygen** and **carbon dioxide** for the members to survive. Therefore, workers build nests with ventilation systems regulating the colony's environment. It is thought that intestinal gases, including methane, rise to the top of the nest and diffuse through the walls and out of venting shafts.

Another task that the workers must carry out is caring for the colony's young. When the queen lays her second series of eggs—about one month after her initial laying—the workers take the larvae into the brooding chamber where they care for them. As part of their care, the workers turn over the eggs periodically and check them for signs of parasite damage.

Soldiers

The soldier termites' main responsibility is defending the colony. Unlike other termites, soldiers have large armor-plated heads, a tough outer skeleton, and sometimes have **glands** for squirting poison at their enemies. Furthermore, the soldiers are equipped with biting mandibles, which include cutting edge teeth and even hooks.

Although termites are often preyed upon by aardvarks, **reptiles**, and other **amphibians**, their main predators are army and chirping ants. Even though the soldiers are blind, they can sense intruders with their antennae. To warn the colony, the soldiers beat their heads against the ground. Soldiers are never able to feed themselves.

The mature colony

Deep within the nest, protected by soldiers and cared for by workers, the queen begins to grow. While her head and thorax do not change, her abdomen increases in size as her ovaries enlarge and her egg-laying capacity increases. In some highly evolved species with extraordinarily large colonies, the queen can swell to more than 5.5 in (77 cm) in length and 1.5 inches (3.81 cm) in diameter; thus, her abdomen swells to about 200 to 300 times it original size. In such a state, the queen is incapacitated and cannot feed herself.

By her third year, the queen reaches the peak of her egg production, when she typically lays over 30,000 eggs per day. The most fertile queens belong to the species *Odontotermes obesus*, which can lay about one egg per second or about 86,400 eggs each day. In all species, the king, her constant companion, changes very little throughout his life.

Resources

Books

Grzimek, HC. Bernard, Dr., ed. *Grzimek's Animal Life Encyclopedia.* New York: Van Nostrand Reinhold Company, 1993.

Harris, W. Victor. *Termites: Their Recognition and Control.* London: Longmans, Green and Co. Ltd., 1964.

Lee, K.E., and T.G. Wood. *Termites and Soils.* New York: Academic Press, 1971.

The New Larousse Encyclopedia of Animal Life. New York: Bonanza Books, 1987.

Pearl, Mary Corliss, Ph.D. Consultant. *The Illustrated Encyclopedia of Wildlife.* London: Grey Castle Press, 1991.

Kathryn Snavely

Terns

Terns are fast-flying coastal **birds** in the family Sternidae, which includes some 42 **species**. Most species of terns are found in the tropics and subtropics, but these birds occur on all continents. They range from the limits

Royal terns (*Thalasseus maximus*) on Estero Island, Florida. *Photograph by Robert J. Huffman. Field Mark Publications. Reproduced by permission.*

of land in the highest Arctic, to the fringes of **Antarctica**. Most terns breed and occur in coastal marine environments, or in the vicinity of inland lakes, **rivers**, and marshes.

Biology of terns

Terns are slender birds with long, pointed wings, and are adept fliers. Their tail is usually forked to some degree, and their bill is sharply pointed. The usual coloration is some combination of white, gray, and/or black.

The smallest species is the little tern (*Sterna albifrons*), which is only 9 in (23 cm) in body length and 1.8 oz (50 g) in weight. The largest species is the Caspian tern (*Hydroprogne caspia*), which is 20 in (50 cm) long, and weighs 25 oz (700 g).

Most terns feed on **fish**, small **squid**, or large **invertebrates**. Species occurring in **freshwater** habitats may also eat **amphibians** and large **insects**. Terns typically hunt their aquatic **prey** by plunging head-first into **water**, often after having located their quarry by briefly hovering.

Terns typically nest in colonies, some of which are large. The usual nesting locale is a gravel shore, generally on an **island** or relatively isolated **peninsula**. The typical nest built by terns is a simple scrape, but tropical terns known as noddies (*Anous* spp.) build a more substantial nest in a **tree** or on cliff ledges. Some fairy terns do not build a nest at all—they lay a single egg, wedged into the fork between two branches of a tree.

Terns of North America

Fourteen species of terns breed regularly in **North America**. The most abundant species is the common tern (*Sterna hirundo*), which also breeds widely in Eurasia. The breeding range of this species is from the subarctic, to the

Great Lakes and temperate regions of the Atlantic coast. The common tern winters from southern parts of coastal North America through to southern **South America**. This tern has a black cap, a grey mantle (the back of the wings), a white breast, and a red beak with a blackish tip.

The arctic tern (*S. paradisaea*) is an abundant species that breeds from subarctic regions to the very limit of land in the Arctic of North America and Eurasia. It winters in the waters of the Southern Ocean. The arctic tern undertakes extraordinarily long migrations between its breeding and wintering habitats, with some populations traversing a distance of more than 22,000 mi (36,000 km) each year. Because it spends so much time in high latitudes of both hemispheres, where day length is long during the summer, the arctic tern may see more hours of daylight each year than any other creature. The arctic tern has similar coloration to the common tern, but it has an all-red beak and shorter, red legs.

Forster's tern (*S. forsteri*) breeds in **salt** and freshwater marshes of the northern prairies, and to a lesser degree along the southern coasts of the Pacific and Atlantic Oceans. The roseate tern (*S. dougallii*) is locally common along the Atlantic coast of the eastern United States and as far north as Nova Scotia. The roseate tern also breeds in coastal places in western **Europe**, the West Indies, Venezuela, **Africa**, the Indian Ocean, south and southeast **Asia**, **Australia**, and many south Pacific islands.

The royal tern (*Thalasseus maximus*) is a relatively large, crested species that breeds on the Atlantic and Pacific coasts of North America, and also in Eurasia. This species winters on the coasts of south Florida, the Gulf of Mexico, and parts of the Caribbean.

The Caspian tern is the largest species of tern. This species breeds on large lakes and rivers and at a few places along the subarctic seacoast of North America. The Caspian tern is a wide-ranging species, also breeding in Eurasia, Africa, Australia, and New Zealand. This species winters along the coasts of southern California, Baha California, the Gulf of Mexico, and Caribbean islands.

The black tern (*Chlidonias niger*) is a dark-grey locally abundant species breeding on lakes and freshwater marshes in both North America and Eurasia. North American birds winter in Central America and northern South America. The sooty tern (*S. fuscata*) and noddy tern (*Anous stolidas*) only breed in the Dry Tortugas, small U.S. islands south of West Florida.

Conservation of terns

During the nineteenth century, many species of terns were rapaciously hunted for their plumage, which was valuable at the time for decorating the clothing of fashionable ladies. Sometimes, an artistic statement was made by mounting an entire, stuffed tern onto a broad-brimmed, lady's hat. Fortunately, the plumage of terns or other birds is not much used for these purposes any more.

In many places, terns have been deprived of important nesting **habitat**, as beaches and other coastal places have been appropriated and developed for use by humans. Frequent disturbances by pedestrians, all-terrain vehicles, boats, and other agents also disrupt the breeding of terns, usually by causing brooding adults to fly, which exposes their eggs or young to predation by other birds, especially **gulls**.

In many parts of their breeding range, tern eggs and chicks are taken by a number of the larger species of gulls (*Larus* spp.). The populations of many gull species have increased enormously in most of the world, because these birds have benefited greatly from the availability of fish waste discarded by fishing boats and processing plants, and from other foods available at garbage dumps. Gulls are highly opportunistic feeders, and will predate tern chicks, and sometimes adults, whenever it is easy to do so. The negative effects of gulls on terns are an important, indirect consequence of the fact that gulls have benefited so tremendously from the activities of humans.

Some species of terns are threatened, such as the black-fronted tern (*Chlidonias albostriatus*) of New Zealand, the black-bellied tern (*Sterna acuticauda*) of South and Southeast Asia, the Chinese crested-tern (*S. bernsteini*) of Southeast Asia, the fairy tern (*S. nereis*) of Australia, and the Kerguelen tern (*S. virgata*) of southern Africa.

See also Gulls; Migration.

Resources

Books

Forshaw, Joseph. *Encyclopedia of Birds.* New York: Academic Press, 1998.

Hay, J. *The Bird of Light.* New York: Norton, 1991.

Bill Freedman

Terracing

The word terrace is applied to geological formations, architecture such as a housing complex built on a slope, or an **island** between two paved roads. However,

Rice terraces in Bali, Indonesia. *JLM Visuals. Reproduced by permission.*

KEY TERMS

....

Agroecosystem—A agricultural ecosystem, comprised of crop species, noncrop plants and animals, and their environment.

Arable—An agricultural term describing fertile ground or topsoil, which can be cultivated as cropland.

Contour farming—The modern term for horizontal plowing or contour plowing, often used in conjunction with terracing to further prevent erosion.

Erosion—Damage caused to topsoil by rainwater runoff. There are various special terms for patterns of erosion caused in the soil, like sheet, rill, or gully erosion.

Point rows—These crop areas are "dead ends" in a field, which cannot be cultivated with modern heavy farm equipment without requiring the machines to turn and pass over them a second time, in order to reach the rest of the crops. These areas are prevented during terracing by moving arable turf to a more easily farmed area and smoothing.

Topographic map—A map illustrating the elevation or depth of the land surface using lines of equal elevation; also known as a contour map.

the act of terracing specifies an agricultural method of cultivating on steeply graded land. This form of **conservation** tillage breaks a hill into a series of steplike benches. These individual flat structures prevent rainwater from taking arable topsoil downhill with it. The spacing of terraces is figured mathematically by comparing the gradient of the land, the average rainfall, and the amount of topsoil which must be preserved.

Different forms of terracing are required, depending upon how steep the ground is that is intended for cultivation. The bench is the oldest type, used on very steep territory. A little dam called a riser marks off each bench, and can slow down rainwater runoff on slopes as extreme as 30%. Just the way a steplike or "switchback" layout of railroad tracks prevent trains from having to go up one steep grade, the effect of gravity is lessened by bench terracing. Climate as well as **soil** condition and farming methods must be taken into account, so the Zingg conservation bench is a type of flat-channel terrace constructed in semiarid climates. Slopes between each bench store runoff **water** after each rain.

Newer formats were developed to accommodate for mechanized farm equipment, so now variations such as

the narrow-base ridge terrace and the broadbase terrace are used for less extreme gradients. Two approaches to broadbase terracing are used, depending upon the conditions of the topsoil and its vulnerability. The Nichols or channel variation is a graded broadbase terrace for which the soil is cultivated from above. Water is channeled off by this construction at a steady **rate**. The Mangum or ridge type is a level broadbase terrace used in permeable soil. This type shores up the topsoil from above and below. A less ambitious form than the broadbase is the steep-backslope terrace, which takes soil from the downhill side of a ridge, but this backslope cannot be used as cropland.

Modern practices

Parallel terraces are a recent innovation in conservation farming. This method incorporates land forming or landscaping by moving **earth** to make the terraces more uniform. The resulting formation allows room for the use of heavy machinery, and prevents "point rows," which are areas that cannot be efficiently cultivated without doubling back over the same area. Modern terrace planning incorporates the use of topological maps, known more simply as contour maps, which take into account the surface variations of an area slated for cultivation. Otherwise, there is no need for special equipment for terracing, which can be done with an old-fashioned moldboard plow or with mechanized rigs like the bulldozer.

Worldwide methods

In **Africa** a certain method called "fanya juu" comes from the Swahili phrase meaning "make it up." It began in Kenya during the 1950s, when standard Western practices could not control the fierce **erosion** in the area and also took too much arable land out of circulation. Fanya juu produces embankments by carving out ditches and depositing the soil uphill to form embankments. The ditches can be used to grow **banana** plants while another crop is planted on the embankments. A variation involving Western channel terracing is called "fanya chini," but this is less popular because the ditches must be desilted of churned-up topsoil on a regular basis. Additionally, in very steep areas only bench terracing can be truly effective.

Yemeni mountain land was once cultivated widely by farmers, but it made for a difficult living. So when oil became a bigger economy than agriculture in the surrounding countries, the farmers slowly migrated to places like Saudi Arabia in order to seek greater fortunes in a new business. The agroecosystem left behind began to slowly contribute to soil erosion, because the arrangement of bench terraces, small **dams** and **irrigation** or runoff conduits was decaying. By 1987, one researcher found that thin or shoestring rills were deepening into gullies on the mountainsides.

The benches of Lebanon, some of which have existed for over two and a half thousand years after being instituted by the Phoenicians, were threatened by the battles of civil conflict in the area. Farmers were driven away to safer and more convenient living conditions in cities or in other countries. An investigation in 1994 warned of long-term damage to untended terraced lands including an increased possibility of landslides, and the chance that the land may be rendered eventually unfarmable.

Resources

Books

Moldenhauer, et al., eds. *Development of Conservation Farming on Hillslopes*. The Soil and Water Conservation Society, 1991.
Turner, B.L. *Once Beneath the Forest*. Boulder, CO: Westview Press, 1983.

Periodicals

"Bench Terracing in the Kerinci Uplands of Sumatra." *Journal of Soil & Water Conservation* (September/October 1990).
"Rehabilitating the Ancient Terraced Lands of Lebanon." *Journal of Soil & Water Conservation* (March/April 1994).

"Terrace Channel Design and Evaluation." *Transactions of the ASAE* (September/October 1992).

Jennifer Kramer

Territoriality

Territoriality is the **behavior** by which an **animal** lays claim to and defends an area against others of its **species**, and occasionally members of other species as well. The territory defended could be hundreds of **square** miles in size, or only slightly larger than the animal itself. It may be occupied by a single animal, a pair, family, or entire herd or swarm of animals. Some animals hold and defend a territory year-round, and use the territory as a source of food and shelter. Other animals establish a territory only at certain times of the year, when it is needed for attracting a mate, breeding, and/or raising a family.

The advantages of territoriality

Many different species exhibit territorial behavior, because it offers several advantages to the territorial ani-

Walrus battling over a hauling-out spot, Bristol Bay, Alaska. *JLM Visuals. Reproduced by permission.*

mal. An animal which has a "home ground" can develop reflexes based on its surroundings. Thus it can react quickly to dangerous situations without having to actively seek hiding places or defensible ground. By spacing out potential competitors, territoriality also prevents the depletion of an area's natural resources. This regulation of population **density** may also slow down the spread of **disease**. In addition, territorial behavior exposes weaker animals (which are unable to defend their territory) to predation, thereby promoting a healthy population.

Types of territories

Some animals will establish a territory solely for the purpose of having a place to rest. Such a territory is known as a roost, and may be established in a different area every night. Roosts are often occupied and defended by large groups of animals, for the protection offered in numbers. Individual personal spaces within the roost may be fought over as well. Roosting spots nearer the interior of a group of animals are often the safest, and therefore the most highly prized.

Several species of **birds** and a few **mammals** are known to establish specialized territories during the breeding season, which are used only to attract mates through breeding displays. This type of territory is known as a lek, and the associated behavior is called lekking. Leks are among the most strongly defended of all territories, since holding a good lek increases the chances of attracting a mate. Leks are generally of little use for feeding or for bringing up young, and the animals will abandon its lek once it attracts a mate or mates, or if it becomes too weak to defend it.

Defending a territory

Some animals will defend their territory by fighting with those who try to invade it. Fighting, however, is not often the best option, since it uses up a large amount of **energy**, and can result in injury or even death. Most animals rely on various threats, either through vocalizations, smells, or visual displays. The songs of birds, the drumming of **woodpeckers**, and the loud calls of **monkeys** are all warnings that carry for long distances, advertising to potential intruders that someone else's territory is being approached. Many animals rely on smells to mark their territories, spraying urine, leaving droppings or rubbing scent **glands** around the territories' borders. Approaching animals will be warned off the territory without ever encountering the territory's defender.

On occasion, these warnings may be ignored, and an intruder may stray into a neighboring territory, or two animals may meet near the border of their adjacent terri-

tories. When two individuals of a territorial species meet, they will generally threaten each other with visual displays. These displays often will often exaggerate an animal's size by the fluffing up of feathers or fur, or will show off the animals weapons. The animals may go through all the motions of fighting without ever actually touching each other, a behavior known as ritual fighting. The displays are generally performed best near the center of an animal's territory, where it is more likely to attack an intruder, and become more fragmented closer to the edges, where retreating becomes more of an option. This **spectrum** of performances results in territorial boundaries, where displays of neighbors are about equal in intensity, or where the tendency to attack and the tendency to retreat are balanced.

Actual fighting usually only happens in overcrowded conditions, when resources are scarce. Serious injury can result, and old or sick animals may die, leading to a more balanced population size. Under most natural conditions, territoriality is an effective way of maintaining a healthy population. The study of social behaviors such as territoriality in animals may help us also to understand human society, and to learn how individual behavior affects human populations.

See also Competition.

Resources

Books

Parker, Steve, Jane Parker, and Terry Jennings. *Territories.* New York: Gloucester Press, 1991.

David Fontes

Testing, genetic *see* **Genetic testing**

Tetanus

Tetanus, also known as lockjaw, is a **disease** caused by a type of **bacteria** that lives in the **soil** and the intestines of people and animals. When these bacteria get into the body, the poisons they produce affect the **nervous system**, causing muscle spasms and, in many cases, death. Tetanus is not contagious and can be prevented with a **vaccine**.

Tetanus is caused by the bacteria *Clostridium tetani*. Tetanus bacteria can enter the body through an open wound, such as a puncture or a cut; the disease can also be transmitted via improperly sterilized hypodermic needles and practices such as tattooing. Since the bacteria live in the intestines of animals, **animal** bites can also cause tetanus.

Once the bacteria enter the body, it generally takes anywhere from three days to three weeks for symptoms to develop. The poison, or toxin, produced by the tetanus bacteria enters the central nervous system, affecting the body's nerve cells and causing muscle spasms. When these spasms occur in the muscles involved in chewing, the condition is commonly known as lockjaw. If the muscles of the throat and chest go into spasms, tetanus can be fatal. It is estimated that 40% of the incidences are fatal. Tetanus can be treated with **antibiotics** and antitoxin medication.

Tetanus is preventable through immunization. In the United States, infants are vaccinated against the disease at 2 months, 4 months and 6 months. This vaccination is known as the DTP shot; it protects against **diphtheria**, tetanus, and pertussis (**whooping cough**). In order to insure immunity, it is necessary to get a booster shot every ten years.

Tetanus can also be prevented through the proper cleaning and disinfection of wounds. If the nature of the wound indicates the possibility of tetanus **infection** (for example, puncture wounds), treatment may include a booster shot.

In the United States, tetanus often occurs among senior citizens, who may not be up to date on their immunizations. In countries where immunization against tetanus is not routine, however, the disease is common among infants. Generally, babies are infected during childbirth or through the newly cut umbilical cord. In infants, the disease is often fatal. However, due to improving immunization programs, the incidence of tetanus worldwide has been declining in recent decades.

See also Childhood diseases.

Tetrahedron

A tetrahedron is a **polyhedron** with four triangular faces. It is determined by four points (the *vertices*) that are not all in the same **plane**. A *regular* tetrahedron is one where all of the faces are congruent equilateral triangles.

A tetrahedron is the same as a **pyramid** with a triangular base.

See also Polygons.

Textiles

Textiles are generally considered to be woven fabrics. They may be woven from any natural or synthetic fibers, filaments, or yarns that are suitable for being spun and woven into cloth.

History of textiles

The earliest known textiles were recovered from a neolithic village in southern Turkey. Rope, netting, matting, and cloth have been found in the Judean **desert** (dating from 7160 to 6150 B.C.).

Flax was the most common **plant** fiber used in antiquity. **Hemp**, rush, palm, and papyrus were also employed as textile fibers in ancient times. Early evidence of the use of flax has been found in north Syria (c. 6000 B.C.), Iraq (c. 5000 B.C.), and Egypt (c. 6000 B.C.). Evidence that fibers from **sheep**, **goats**, and dogs were used in textiles as early as the eighth century B.C. has been found in northwest Iran; early use of these fibers has also been traced to Palestine and southern Turkey. **Cotton**, native to India, was used in Assyria around 700 B.C. The use of silk, originally exclusive to China, appears to have spread to Germany by around 500 B.C.

The ancient Egyptians and later the Israelites preferred garments of white linen and wool. But Kushites, Nubians, and Libyans apparently preferred dyed fabrics. The principal dyes and mordants such as alum were probably known since earliest times. Royal purple was produced by the Phoenicians, who had become major traders in dyes and wools by around 1700 B.C., from murex. Evidence exists of trade in textiles as early as 6000 B.C., in which wool and cloth were important trade goods.

Textile techniques

Spinning

Spinning is the process of making yarn or thread by the twisting of vegetable fibers, **animal** hairs, or manmade fibers, i.e., filament-like elements only a few inches in length. In the spinning mill, the raw material is first disentangled and cleaned. Various grades or types of fibers may then be blended together to produce yarn hav-

A computer controlled textile mill. *Photograph by Mark Antman. Phototake NYC. Reproduced by permission.*

ing the desired properties. The fibers are next spread out parallel to one another in a thin web, from which a yarn-like material is formed.

Weaving

In its simplest form, i.e., basketry, weaving probably pre-dated spinning, as in early cultures individuals presumably interlaced long fibrous stems with their fingers before they learned to convert short fibers into continuous yarn. Woven structures consist of two sets of threads, the warp and the weft, which are interlaced to form cloth. Warp threads are held parallel to each other under tension; the weft is worked over and under them, row by row. Looms have been used since the time of ancient Egyptians to keep the warp threads evenly spaced and under tension.

The following techniques are used to prepare the warp and the weft prior to weaving: 1) In doubling, two or more yarns are wound on a bobbin without undergoing any twisting (as distinct from spinning, in which fibers are twisted together to give them the requisite strength). In twisting doubling, two or more yarns are twisted around each other; 2) Sizing is done to make the

warp threads smooth, to reduce the **friction** of the threads. The threads are coated or saturated with an **adhesive** paste (size); 3) Twisting joins the ends of a new warp with those of the one already in the loom. It is done by twisting the ends together, either by hand or with the aid of a special device.

Simple weaves are of three types: tabby, twill, and satin. Tabby weave, or plain weave, is produced by passing the weft across the warp twice. Twill weaves are characterized by a diagonal movement caused by starting the weave sequence one place to the right or left on each successive passage of the weft. Satin weave is distinguished by the spacing of the binding points, the normal sequence being over one warp, and under four or more. Compound weaves are based on the three basic weaves with the addition of extra warp, weft, or both. Velvet is a compound weave that starts as a basic weave.

Finger weaving techniques include twining and braiding. Twining uses two sets of yarns. In weft twining, the warp is stretched between two bars and the weft worked across in pairs. One thread passes over a warp and the other under, with the two yarns making a half turn around each other between each warp.

Finishing

Most fabrics produced by weaving or knitting have to undergo further processing before they are ready for sale. In finishing, the fabric is subjected to mechanical and chemical treatment in which its quality and appearance are improved and its commercial value enhanced. Each type of fabric has its own particular finishing operations. Textiles produced from vegetable fibers require different treatment than textiles produced from animal fibers or synthetic fibers.

Woven cloth is usually boiled with dilute caustic soda to remove natural oils and other impurities. It is then rinsed, scoured in an acid bath, further processed, and bleached with sodium chlorite. Singeing may be done to remove any fibers on cotton or rayon materials, especially if they have to be printed.

In the course of spinning, weaving, and finishing, the fabric is subjected to much pull and stretch. When the material gets wet, the material reverts to its original shape. Sanforizing mechanically shortens the fibers, so that they will not shrink in the wash.

Raising (or napping) is a process in which small **steel** hooks tear some of the fibers or ends of fibers out of the weft yarn, so the fibers acquire a wooly surface (called the "nap"). This improves **heat** retention and absorptive properties (as in flannel fabrics), making them softer to the touch. Raising is chiefly used for cotton, rayon, or woolen fabrics. It can be applied to one or both sides.

Types of textiles

Tapestries

Although the term tapestry usually conjures up images of large pictorial wall-hangings of the sort used in medieval and post-medieval **Europe**, tapestries are in fact distinctive woven structures consisting specifically of a weft-faced plain weave with discontinuous wefts. This means that the weft crosses the warp only where its particular **color** is need for the fabric design. The technique has been used in many cultures to produce fabrics ranging from heavy, durable floor coverings to delicate Chinese silk. Compared to other weaving techniques, tapestry allows the weaver much more freedom of expression.

Woven rugs

Rugs can be made by a number of techniques, including tapestry, brocade (in which a plain weave foundation is supplemented with supplementary wefts), and pile weaving. Pile rugs are most commonly associated with rug weaving, however. These rugs are made of row after row of tiny knots tied on the warps of a foundation weave which together form a thick pile.

Embroidery

Embroidery is a method of decorating an already existing structure, usually a woven foundation fabric, with a needle. Embroideries have also been done on other media such as parchment or **bark**. For the past 100 years, it has been possible to produce embroidery by machine as well as by hand. Embroidery yarns are woven into a fabric after it has come off the loom, unlike brocade, which in which yarns are placed in the fabric during the weaving process.

Lace

Lace is essentially an openwork fabric constructed by the looping, plaiting, or twisting of threads using either a needle or a set of bobbins. It is not woven. Needle lace is made with one thread at a time. Bobbin lace is constructed with many different threads, each wound on its own bobbin. These are manipulated in a manner similar to that used in braiding. Machine-made lace was first produced around 1840.

Printed and dyed textiles

Aside from exploiting the effects achieved by using **natural fibers**, the only ways to introduce color into textiles are by **printing** or dyeing.

In textile printing, the dyes are dissolved in **water**. Thickening agents (e.g., starch) are added to the solutions to increase **viscosity**. The oldest method is block printing, usually from wooden blocks in which a design is carved. Stencil printing is done with the aid of **paper** or thin **metal** stencils. In silk screen printing, the design is formed on a silk screen, which then serves as a stencil. Most cloth is printed by roller printing. The printing area is engraved on a **copper** roller to form a recessed pattern (intaglio), which is coated with a color paste. The roller transfers the paste to the cloth.

Pattern dyeing uses two principal procedures: resist dyeing and mordant dyeing. In resist dyeing, a resist substance such as hot wax, **rice** paste, or clay is applied to those areas chosen to resist the dye and remain white. The cloth is then dyed and the resist is later removed. The technique is widely known in China, Japan, and West **Africa**, but is most often identified with Javanese batik. Tie dyeing is a resist technique in which parts of the cloth are tied with bast or waxed cord before dyeing. The dyeing is done quickly so the wrappings are not penetrated, and a negative pattern emerges. Many dyestuffs are not directly absorbed by fibers, so mordants or fixing agents are used that combine with the dye and fibers to make the color insoluble.

There are two main types of textile dyeing machines: in one type the dye **solution** is circulated through the fabric, which remains at rest; in the other, the fabric is passed through a stationary bath of the dye solution.

Knits

Knitting is a looped fabric made from a continuous supply of yarn. The yarn need not be continuous throughout the piece, however. Different colors or qualities or yarn may be introduced into a single knitted piece.

Knitting is used for the production of underwear and outer garments, curtain fabrics, etc. The materials used are yarn and threads of cotton, wool, and man-made fibers; as well as blended yarns and paper yarns. The products are either flat fabrics (later made into garments) or ready-fashioned garments. In weft fabric, the threads extend crosswise across the fabric; in warp fabric, the threads extend lengthwise. Warp fabric has less **elasticity** than weft fabric and is not used as much for socks and stockings.

Netting, knotting, and crochet

Netted fabrics have been known since antiquity when they were probably used for fishing nets. In knotting, knots are first made with a small shuttle at intervals in lengths of string, linen, silk, or wool. The knotted thread is then applied to a suitable ground fabric, forming patterns or covering it completely. Twentieth century macrame is a form of knotting. Crochet is a looped fabric made with a continuous thread that forms horizontal rows, with loops locked laterally as well as vertically.

Felt and bark cloth

Felt is a fabric formed by applying **pressure** to hot, wet fibers (usually wool). The fibers become interlocked, and the process cannot be reversed. In Polynesia, bark cloth was traditionally obtained from bark stripped from trees of the mulberry family. The bark was first soaked for several days to make it soft and flexible, then the rough outer bark was scraped from the inner bark. The inner bark was next beaten with mallets to form sheets of cloth, which were treated in a variety of ways before use.

Significance of textiles

Textiles serve the everyday needs of people, but they may also serve to distinguish individuals and groups of individuals in terms of social class, gender, occupation, and status with the group. Traditional societies associated special meaning with textile designs. These meanings

tended to have specific meanings for particular ethnic groups alone. It was assumed that everyone in the group knew them. However, once the meanings have become lost, it is almost impossible to reconstruct them. The patterns in Javanese batiks, for example, originally had meaning to the wearer, but these meanings are now largely lost. Textiles also have real as well as symbolic value. Under Byzantine emperors, silk was a powerful political tool: foreign governments out of favor were denied trading privileges; those in favor were rewarded with silks.

Textiles have played major roles in the social, economic, and religious lives of communities. In many parts of Europe and **Asia**, young girls spent many months preparing clothing and furnishing textiles for their wedding trousseaus as a demonstration of their skills and wealth. Traditionally, women have played a far larger role than men in producing textiles. In many parts of Africa, however, men produce both woven and dyed textiles, and in many urban or courtly textile traditions, men were the main producers (e.g., Asian rug weaving, European tapestry).

Textiles are thus a major component of material culture. They may be viewed as the products of technology, as cultural symbols, as works of art, or as items of trade.

The textile arts are a fundamental human activity, expressing symbolically much of what is valuable in any culture.

See also Dyes and pigments.

Resources

Books

Harris, Jennifer. *Textiles—5000 Years.* New York: H. N. Abrams, 1993.

Simmons, Paula. *Spinning and Weaving Wool.* Seattle: Pacific Search Press, 1987.

Periodicals

Akin, D.E. "Enzyme-Retting Of Flax And Characterization Of Processed Fibers." *Journal Of Biotechnology* 89, no. 2-3 (2001): 193-203.

"Nontraditionally Retted Flax For Dry Cotton Blend Spinning." " *Textiler Research Journal* 71, no. 5 (2001): 375-380.

Randall Frost

Thalidomide

Thalidomide is a drug that was marketed in the late 1950s and early 1960s in Great Britain and **Europe**. It was used both as a sleeping pill and as an antidote to morning sickness in pregnant women.

In 1962, a host of usually rare limb abnormalities suddenly became much more common. About 10,000 babies were born displaying, in particular, a shortening of the arms and/or legs called phocomelia. In phocomelia, for example, the baby may be born with the upper arm and forearm completely absent, and the hand attached to the trunk of the body by a short little bone. This same situation can occur with the legs, so that the majority of the leg is totally absent, and the foot is attached to the trunk of the body. Other babies were born with amelia, the complete absence of one or more limbs. Other **birth defects** involving the eyes, teeth, **heart**, intestine, and anus were similarly noted to be significantly more common.

These **birth** defects were soon traced to the drug thalidomide, which was proven to be a potent **teratogen** (substance which interferes with normal development in the embryo). In fact, studies showed that fully 20% of all babies exposed to thalidomide during their first eight weeks in the uterus were born with the characteristic abnormalities described above. Because the **skeletal system** begins to be formed as early as the third week of development, and small buds which will become the limbs appear around week five, anything which interferes with development at this very early stage has devastating effects. Once the marketing of thalidomide was halted, these birth defects again became rare occurrences.

Currently, thalidomide has been shown to have some use in the treatment of two different illnesses. In **leprosy**, also known as Hansen's disease, a bacterial **disease** causing both skin and **nervous system** problems, thalidomide has some effect against **inflammation** and **pain**. The FDA approved thalidomide for use in certain forms of leprosy, with very strict safeguards in place against use of the drug in pregnant women. A special program, called System for Thalidomide Education and Prescribing Safety (STEPS) approves only specified practitioners to prescribe thalidomide. Mandatory monthly pregnancy tests for all women of childbearing age taking the drug, as well as a requirement that two reliable forms of birth control be used by such women who are given thalidomide, are part of the STEPS program. Men are required to sign a statement that they will use a latex condom during all sexual encounters, even if they have previously undergone a vasectomy. Required video education, strict patient registries, small prescription amounts (thalidomide is only to be prescribed for one month at a time), and careful monitoring are also part of the STEPS program. Studies are also underway to explore using thalidomide to help guard against **immune system** rejection of bone marrow transplants, **ulcers** and severe weight loss in **AIDS**, systemic lupus erythematosus, breast **cancer**, and Kaposi's sarcoma, multiple myeloma, kidney cancer, **brain** tumors, and prostate cancer.

See also Embryo and embryonic development.

Resources

Books

Behrman, Richard E., et al. *Nelson Textbook of Pediatrics.* Philadelphia: W.B. Saunders Company, 1992.

Berkow, Robert, and Andrew J. Fletcher. *The Merck Manual of Diagnosis and Therapy.* Rahway, NJ: Merck Research Laboratories, 1992.

Fanaroff, Avroy A,. and Richard J. Martin. *Neonatal-Perinatal Medicine.* St Louis: Mosby Year Book, Inc., 1992.

Sadler, T.W. *Langman's Medical Embryology.* Baltimore: Williams & Wilkins, 1985.

Sanford, Louis, and Alfred Gilman. *Goodman and Gilman's Pharmacological Basis of Therapeutics.* New York: Pergamon Press, Inc., 1990.

Taeusch, H. William, et al. *Schaffer and Avery's Diseases of the Newborn.* 6th ed. Philadelphia: W.B. Saunders Company, 1991.

Thallium *see* **Element, chemical**

Theorem

A theorem (the term is derived from the Greek *theoreo*, which means *I look at*) denotes either a proposition yet to be proven, or a proposition proven correct on the basis of accepted results from some area of **mathematics**. Since the time of the ancient Greeks, proven theorems have represented the foundation of mathematics. Perhaps the most famous of all theorems is the **Pythagorean theorem**.

Mathematicians develop new theorems by suggesting a proposition based on experience and observation which seems to be true. These original statements are only given the status of a theorem when they are proven correct by logical deduction. Consequently, many propositions exist which are believed to be correct, but are not theorems because they can not be proven using deductive reasoning alone.

Historical background

The concept of a theorem was first used by the ancient Greeks. To derive new theorems, Greek mathematicians used logical deduction from premises they believed to be self-evident truths. Since theorems were a direct result of deductive reasoning, which yields unquestionably true conclusions, they believed their theorems were undoubtedly true. The early mathematician and philosopher Thales (640-546 B.C.) suggested many early theorems, and is typically credited with beginning the tradition of a rigorous, logical **proof** before the general acceptance of a theorem. The first major collection of mathematical theorems was developed by Euclid around 300 B.C. in a book called *The Elements*.

The absolute truth of theorems was readily accepted up until the eighteenth century. At this time mathematicians, such as Karl Friedrich Gauss (1777-1855), began to realize that all of the theorems suggested by Euclid could be derived by using a set of different premises, and that a consistent non-Euclidean structure of theorems could be derived from Euclidean premises. It then became obvious that the starting premises used to develop theorems were not self-evident truths. They were in fact, conclusions based on experience and observation, and not necessarily true. In **light** of this evidence, theorems are no longer thought of as absolutely true. They are only described as correct or incorrect based on the initial assumptions.

Characteristics of a theorem

The initial premises on which all theorems are based are called axioms. An axiom, or **postulate**, is a basic fact which is not subject to formal proof. For example, the statement that there is an infinite number of even **integers** is a simple axiom. Another is that two points can be joined to form a line. When developing a theorem, mathematicians choose axioms, which seem most reliable based on their experience. In this way, they can be cer-

KEY TERMS

Axiom—A basic statement of fact that is stipulated as true without being subject to proof.

Deductive reasoning—A type of logical reasoning that leads to conclusions which are undeniably true if the beginning assumptions are true.

Definition—A single word or phrase that states a lengthy concept.

Pythagorean theorem—An idea suggesting that the sum of the squares of the sides of a right triangle is equal to the square of the hypotenuse. It is used to find the distance between two points.

tain that the theorems are proved as near to the truth as possible. However, absolute truth is not possible because axioms are not absolutely true.

To develop theorems, mathematicians also use definitions. Definitions state the meaning of lengthy concepts in a single word or phrase. In this way, when we talk about a figure made by the set of all points which are a certain **distance** from a central point, we can just use the word *circle*.

See also Symbolic logic.

Resources

Books

Dunham, William. *Journey Through Genius.* New York: Wiley, 1990.

Kline, Morris. *Mathematics for the Nonmathematician.* New York: Dover, 1967.

Lloyd, G.E R. *Early Greek Science: Thales to Aristotle.* New York: W. W. Norton, 1970.

Newman, James R., ed. *The World of Mathematics.* New York: Simon and Schuster, 1956.

Paulos, John Allen. *Beyond Numeracy.* New York: Knopf, 1991.

Perry Romanowski

Theory *see* **Scientific method**

Thermal expansion

The most easily observed examples of thermal expansion are size changes of materials as they are heated

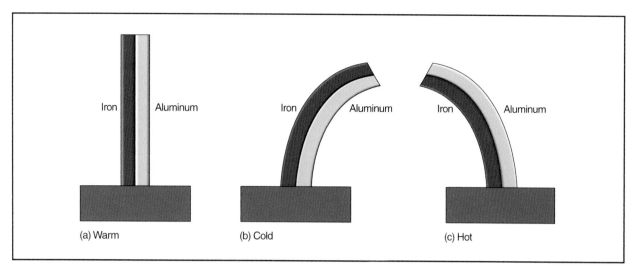

Figure 1. Representation of thermally induced change in a bimetallic strip made of iron and aluminum. The strip bends because the two materials do not expand or contract equally. *Illustration by Hans & Cassidy. Courtesy of Gale Group.*

or cooled. Almost all materials (solids, liquids, and gases) expand when they are heated, and contract when they are cooled. Increased **temperature** increases the **frequency** and magnitude of the molecular **motion** of the material and produces more energetic collisions. Increasing the **energy** of the collisions forces the molecules further apart and causes the material to expand.

Common observations

Different materials expand or contract at different rates. In general, gases expand more than liquids, and liquids expand more than solids. Observation of thermal expansion in a solid object requires careful scrutiny. Several everyday examples are: 1) The sag in outdoor electrical lines is much larger on hot summer days than it is on cold winter days. 2) The **rails** for trains are installed during warm **weather** and have small gaps between the ends to allow for further expansion during very hot summer days. 3) Because the **metal** expands more than **glass** a stuck metal lid on a glass container can be loosened by running hot **water** over the joint between the lid and the container.

Liquids generally expand by larger amounts than solids. This difference in expansion **rate** is sometimes observed when the gas tank of a car is filled on a hot day. Gasoline pumped from the underground container is cold and it gradually heats to the temperature of the car as it sits in the gas tank. The gasoline expands in **volume** faster than the gas tank and overflows onto the ground.

Gases expand even more than liquids when heated. The expansion difference between a gas and a solid can be observed by filling a plastic air mattress in a cool room and then using it on a hot beach. The difference in

thermal expansion between the container and the gas could unexpectedly over inflate the mattress and blow a hole in the plastic.

Practical applications and problems associated with thermal expansion

Sometimes man's ingenuity has led him to find practical applications for these differences in thermal expansion between different materials. In other cases, he has developed technologies or applications that overcome the problems caused by the difference in thermal expansion between different materials.

The thermally induced change in the length of a thin strip of metal differs for each material. For example, when heated, a strip of **steel** would expand by half as much as an equal length piece of **aluminum**. **Welding** together a thin piece of each of these materials produces a bimetallic strip (see Figure 1).

The difference in expansion causes the bimetallic strip to bend when the temperature is changed. This movement has many common uses including: thermostats to control temperature, oven thermometers to measure temperature, and switches to regulate toasters. Some practical solutions to everyday thermal expansion problems in solids are: 1) The material developed for filling teeth has the same expansion as the natural enamel of the tooth. 2) The steel developed to reinforce **concrete** has the same expansion as the concrete. 3) Concrete roads are poured with expansion joints between the slabs to allow for thermal expansion (these joints are the cause of the thumping noise commonly experienced when traveling on a concrete highway).

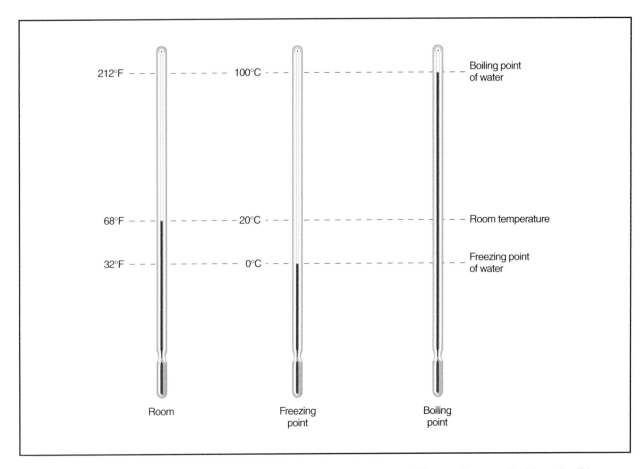

Figure 2. Mercury and alcohol thermometers function because of the expansion difference between liquids and solids. Because the liquid expands at a different rate than the tube, it rises as the temperature increases and drops as the temperature decreases. *Illustration by Hans & Cassidy. Courtesy of Gale Group.*

The manufacture of mercury and **alcohol** thermometers is based upon the expansion difference between solids and liquids (see Figure 2). **Thermometer** fabrication consists of capturing a small amount of liquid (mercury or alcohol) inside an empty tube made of glass or clear plastic.

Because the liquid expands at a faster rate than the tube, it rises as the temperature increases and drops as the temperature decreases. The first step in producing a thermometer scale is to record the height of the liquid at two known temperatures (i.e., the **boiling point** and freezing point of water). The difference in fluid height between these point is divided into equal increments to indicate the temperature at heights between these extremes.

Automobile engine coolant systems provide a practical example of a liquid-thermal expansion problem. If the radiator is filled with coolant when the engine is cold, it will overflow when the engine heats during operation. In older car models, the excess fluid produced by the hot temperatures was released onto the ground. Periodic replacement was required to avoid overheating.

Newer cars have an overflow container that collects the released fluid during thermal expansion and returns it to the radiator as the engine cools after operation. This improvement in the coolant system reduces the number of times the coolant fluid level must be checked and avoids the expense of replacing costly antifreeze material mixed with the radiator fluid.

Hot-air balloons are an obvious example of the practical use of the thermal expansion difference between a gas and a solid. Because the hot air inside the **balloon** bag increases in size faster than the container it stretches the bag so that it expands and displaces the colder (heavier) air outside the bag. The difference between the lower **density** of the air inside the bag compared to the lower density of the air outside the bag causes the balloon to rise. Cooling the air inside the bag causes the balloon to descend.

Water, like most other liquids, expands when heated and contracts when cooled, except in the temperature region between 32°F (0°C) and 39.2°F (4°C). A given **mass** of fresh water decreases in volume until the temperature is decreased to 39.2°F (4°C). Below this temperature, the

Joints like this one are used in bridges to accomodate thermal expansion. *JLM Visuals. Reproduced by permission.*

volume per unit mass increases until the water freezes. This unusual behavior is important to **freshwater** plants and animals that exist in climates where water freezes in the colder **seasons** of the year. As the water surface cools to 39.2°F (4°C), it becomes more dense and sinks to the bottom, pushing the warmer water to the surface.

This mixing action continues until all of the water has reached this temperature. The upper layer of water then becomes colder and less compact and stays near the surface where it freezes. When an **ice** layer forms, it provides an insulation barrier that retards cooling of the remaining water. Without this process, **freshwater animal** and **plant** life could not survive the winter.

See also Gases, properties of; States of matter; Thermostat.

Resources

Books

Hewitt, Paul. *Conceptual Physics.* Englewood Cliffs, NJ: Prentice Hall, 2001.
Merkin, Melvin. *Physical Science.* W. B. Sanders Company, 1976.
Serway, Raymond, Jerry S. Faughn, and Clement J. Moses. *College Physics.* 6th ed. Pacific Grove, CA: Brooks/Cole, 2002.

Jim Zurasky

Thermochemistry

Thermodynamics and thermochemistry

The word **thermodynamics** is derived from the Greek words that mean "heat" and "power." Thermodynamics is studied and has applications in all the sciences. Thermochemistry is the part of thermodynamics that studies the relationship between **heat** and **chemical reactions**. Thermochemistry is a very important field of study because it helps to determine if a particular reaction will occur and if it will release or absorb **energy** as it occurs. It is also possible to calculate how much energy a reaction will release or absorb and this information can be used to determine if it is economically viable to use a particular chemical process. Thermochemistry, however, does not predict how fast a reaction will occur.

In order to understand the terminology of thermochemistry it is first necessary to define the world as viewed by thermodynamics. The chemical reaction being studied is considered to be the "system." For instance, if an acid is being mixed with a base, the acid, the base, any **water** used to dissolve them and the beaker in which they are all held are considered the system. Everything else that is not part of the system is considered to be the "surroundings." This includes everything from the countertop on which the beaker is held to the planets in outer **space**. The system and surroundings together form the "universe." From this wide set of definitions, it is easy to understand why the system is the only part of any interest to us. The surroundings are too complex to be considered.

Change

Any process that involves a chemical reaction involves change. Sometimes the change occurs on its own. Such a process is called spontaneous. If a change does not occur on its own, it is called non-spontaneous. A spontaneous change may not occur immediately. For example, if a barrel of fuel is left alone, it will remain as fuel indefinitely. However, if a match is used to ignite the fuel, it will **burn** spontaneously until all the reactants (air, fuel) are completely consumed. In this instance, the spontaneous process required a small amount of energy to be added to the system before a much larger amount of energy could be released. However, once started, it proceeded without assistance. An **electrolysis** reaction, in which **electricity** is passed through water to dissociate it into **hydrogen** and **oxygen**, is not considered spontaneous because the reaction stops if the electricity is removed. An electrolysis reaction is a non-spontaneous process. How is it possible to determine if a process is spontaneous or non-spontaneous without actually mixing the chemicals together? There are two factors whose combination determines whether a process occurs spontaneously or not. These factors are energy and disorder.

Energy

Energy is a state function. There are a number of different forms of energy, which is the ability to do **work**. Work is done anytime a **force** is applied to make an object move. There is energy of **motion**, called kinetic energy and energy of position or stored energy, called potential energy. Potential and kinetic energy are interconvertible; that is, one form can change to the other. Different types of energy include thermal energy, electrical energy, radiant energy, chemical energy, mechanical energy and nuclear energy. One type of energy can be converted to another. However, energy can neither be created nor destroyed. It is always conserved. For example, passing electrical energy through a tungsten filament converts it to **light** energy. All the electrical energy is not converted to light however. Some of it is converted to thermal energy, which is why a light bulb becomes hot after some time.

In most chemical reactions, chemical energy is converted to some other, more useful form of energy. For example, in a flashlight, chemical energy from the batteries is converted to electrical energy. In a car, chemical energy from the **combustion** of the fuel is converted into mechanical energy. Thermochemistry concerns itself with the relation between chemical reactions and thermal energy. Thermal energy is the energy of motion of particles such as **atoms**, molecules or ions. Thermal energy depends on the quantity of a substance present and is thus known as an extensive property. The thermal energy provided by a drop of water is much less than that provided by a pot full of water. **Temperature**, however, is a property that is not dependent on the quantity of substance. The temperature of a drop of boiling water is the same as that of a pot of boiling water. Heat is the transfer of thermal energy that occurs between two objects when they are at different temperatures. If the two objects are at the same temperature, no thermal energy is transferred and no heat is felt. That is how we can tell if an object is hot by touching it. When heat is released from the system in a chemical reaction, the reaction is said to be exothermic. When heat is absorbed by the system, the reaction is said to be **endothermic**. In an endothermic reaction, the surroundings provide the heat for the reaction while in an exothermic reaction, the surroundings are heated by the reaction. For this reason it is accepted that exothermic quantities are **negative** quantities, since the system is losing energy and endothermic quantities are positive quantities since the system is gaining energy.

Measurement of thermal energy

How can thermal energy be measured? One way is to measure a quantity called the specific heat. The specific heat of a substance is the amount of thermal energy required to heat one gram of that substance by one degree. Once again, the specific heat of a substance is an intensive property, meaning that it does not depend on the amount of substance present. The specific heat of a drop of water and a pan of water are the same. When we multiply the **mass** of an object by its specific heat, it is possible to calculate the **heat capacity** of that object. Heat capacity is an extensive property, meaning that it is dependent on the quantity of substance present. The heat capacity of a drop of water is much, much less than that of a **lake**. The specific heat of water is unusually high compared to many other substances. This fact has an important impact on us. Cities located near huge bodies of water tend to have more moderate climates. Such cities are cooler in the summer as large water bodies take a long time to absorb the heat of the summer **sun**, and these cities are warmer in the winter, as the water slowly releases the heat it had absorbed during the summer. Since our bodies are composed largely of water, we are able to maintain a fairly constant body temperature, in spite of outside temperature fluctuations. Even so, one important fact arises from all the information so far. We cannot measure an absolute value for energy. We can, however, measure energy differences.

Enthalpy

The mathematical representation for thermal energy contains many terms. We can, however, simplify it based on how we perform the experiments. Most chemical reactions take place under **atmospheric pressure**, which is (for the most part) constant. When a thermal energy change is measured under constant **pressure** conditions, it is called a change in "enthalpy." The symbol for enthalpy is H. Since only a difference in enthalpy can be measured, the difference is called "delta H." When describing different kinds of changes, we can indicate the difference as part of the name by using subscripts after the H. However, reactions can be done under different pressure and temperature conditions. For the sake of uniformity, a standard state is defined as the state of a substance at one atmosphere pressure. For solids and liquids, the standard state is the pure solid or liquid at one atmosphere pressure. For gases, the standard state is the ideal gas at a partial pressure of one atmosphere. Once the standard state is defined, some simplifications can be made. The enthalpy of formation of an element in its standard state is zero. If there is more than one form of the element under the defined conditions, the most stable form is given an enthalpy of formation of zero. **Carbon**, for example, has two forms at one atmosphere, graphite and **diamond**. Graphite is more stable and is assigned an enthalpy of formation of zero. Diamond does not have an enthalpy of formation of zero.

Enthalpy has a special property. Its value is determined based on the initial state of the system and the final state of the system. It does not depend on how the system gets from the initial state to the final state. A **function** that has this property is called a "state function." The fact that enthalpy is a state function makes it possible to calculate enthalpies for some compounds without having to measure them experimentally. By combining different reactions with known enthalpies, it is possible to calculate the unknown enthalpy. Hess' Law summarizes this observation by stating that the thermal energy absorbed or released in a change is the same whether the change occurs in a single step or in multiple steps.

Those enthalpies that cannot be calculated using Hess' law can be measured experimentally. An apparatus called a calorimeter is used to measure the quantity of thermal energy gained or lost in a chemical change. A simple calorimeter can be constructed using two nested styrofoam cups with lids and a **thermometer**. A more complex type of calorimeter is the bomb calorimeter, which measures thermal energy changes under constant **volume** conditions.

Entropy

As mentioned much earlier, two quantities determine whether a reaction will be spontaneous or not, the thermal energy and disorder. Disorder is also known as **entropy**. Entropy is given the symbol S. In general, entropy always has a tendency to increase. In other words, the universe has a tendency to move towards disorder. When disorder increases, we say entropy increases. An increase in entropy is assigned a positive sign. When order increases, we say entropy decreases. A decrease in entropy is assigned a negative sign. Entropy only has a zero value if we consider a perfect **crystal** at **absolute zero**. Since it is not possible to reach absolute zero, no substance has a value of zero entropy.

Gibbs' free energy

Certain processes that release a great deal of energy are not spontaneous, even though it would seem that they should be. Similarly, certain processes that greatly increase disorder are not spontaneous, although it would seem that they should be. If we mathematically manipulate the expressions for enthalpy and entropy, it is possible to define a new quantity called the Gibbs' free energy. The Gibbs' free energy, sometimes simply called free energy, equals the change in the enthalpy minus the product of the temperature and the change in the entropy. The term free energy should not be misunderstood. As stated earlier, energy can neither be created nor destroyed. This energy does not come free of cost. The

---

KEY TERMS

Enthalpy—The measurement of thermal energy under constant pressure conditions.

Entropy—The measurement of a tendency towards increased randomness and disorder.

Equilibrium—The conditions under which a system shows no tendency for a change in its state. At equilibrium the net rate of reaction becomes zero.

Gibbs' free energy—Mathematically equal to the change in the enthalpy minus the product of the temperature and the change in the entropy. Used to determine if a process is spontaneous or not.

Heat—The transfer of thermal energy that occurs between two objects when they are at different temperatures.

Surroundings—Everything that is not part of the system.

System—The materials pertinent to the reaction being studied.

Thermal energy—The total amount of energy contained within any body as a consequence of the motion of its particles.

---

term 'free' in "free energy" is better interpreted as available. The free energy can be used to predict if a process is spontaneous or not. If the free energy is negative, the process is spontaneous. If the free energy is positive, the process is not spontaneous. A non-spontaneous process can sometimes be made spontaneous by varying the temperature. If the free energy is zero, the process is at equilibrium, meaning that the forward **rate** of the reaction equals the reverse rate of the reaction.

Resources

Books

Oxtoby, Gillis, and Nachtrieb. *Principles of Modern Chemistry.* 4th ed. Saunders College Publishing, 1999.

Silberberg. *Chemistry, the Molecular Nature of Matter and Change.* 2nd Ed., McGraw-Hill, 2000.

Rashmi Venkateswaran

Thermocouple

Accurately measuring temperatures over a wide range is a challenge to engineers, physicists, and other scientists. Many techniques have been devised to deal with a wide range of conditions and temperatures. One such technique is a thermocouple. A thermocouple makes use of one aspect of the thermoelectric effect to measure temperatures, the voltage produced between two different wires with junctions at different temperatures. Depending on the types of wires chosen, a thermocouple can be used for temperatures ranging from -454°F to 4,172°F (-270°C to 2,300°C).

A thermocouple must consist of two wires of different compositions. A popular combination is **copper** and constantan. Constantan is an **alloy** of copper and nickel. The different wires are joined at the ends to make two junctions. One of the wires is then cut so that a voltmeter can be placed in the circuit to measure the voltage between the two junctions. This voltage will depend on the **temperature** difference between the two junctions. A scientist wanting to use a thermocouple will then place one of the junctions in the object whose temperature is to be measured. Because the voltage depends on the temperature difference between the two junctions, the other junction must be maintained at an accurately known temperature. One way to maintain a known temperature is to place the junction in an **ice** water bath that will be at the freezing point of **water.** To find the unknown temperature the scientist must know what temperature difference corresponds to the measured voltage. These figures are determined by careful experiments and then tabulated, so the scientist uses the table to find the unknown temperature.

What causes this voltage difference? The two different types of **metal**, having different compositions, will have different densities of electrons. The electrons will tend to diffuse from the higher to the lower densities. These **electron** densities both depend on the temperature, so if the two junctions are at different temperatures the **diffusion** of electrons will proceed at different rates at each junction. The net result is a **motion** of the electrons, so there is a voltage between the two junctions.

Thermocouples have the advantage of being accurate over a wide temperature range and of being able to accurately follow rapid temperature changes. They can however be cumbersome to use. The need to keep one junction at an accurately known temperature limits their portability.

Thermodynamics

Thermodynamics is the science that deals with **work** and **heat**, and the transformation of one into the other. It is a macroscopic theory, dealing with **matter** in bulk, disregarding the molecular nature of materials. The corresponding microscopic theory, based on the fact that materials are made up of a vast number of molecules, is called **statistical mechanics**.

Historical background

Benjamin Thompson, Count von Rumford (1753-1814) recognized from observing the boring of cannon that the work (or mechanical **energy**) involved in the boring process was being converted to heat by **friction**, causing the **temperature** of the cannon to rise. With the experiments of James Joule (1818-1889), it was recognized that heat is a form of energy that is transferred from one object to another, and that work can be converted to heat without limit. However, the opposite is found not to be true: that is, there are limiting factors in the conversion of heat to work. The research of Sadi Carnot (1796-1832), of Lord Kelvin (1824-1907), and of Rudolph Clausius (1822-1888), among others, has led to an understanding of these limitations.

Temperature

The idea of temperature is well known to everyone, but the need to define it so that it can be used for measurements is far more complex than the simple concepts of "hot" and "cold." If a rod of **metal** is placed in an ice-water bath and the length is measured, and then placed in a steam bath and the length again measured, it will be found that the rod has lengthened. This is an illustration of the fact that, in general, materials expand when heated, and contract when cooled (however, under some conditions rubber can do the opposite, while **water** is a very special case and is treated below). One could therefore use the length of a rod as a measure of temperature, but that would not be useful, since different materials expand different amounts for the same increase in temperature, so that everyone would need to have exactly the same type of rod to make certain that they obtained the same value of temperature under the same conditions.

However, it turns out that practically all gases, at sufficiently low pressures, expand in **volume** exactly the same amount with a given increase in temperature. This has given rise to the constant volume gas **thermometer**, which consists of a flask to hold the gas, attached to a system of **glass** and rubber tubes containing mercury. A small amount of any gas is introduced into the (otherwise empty) flask, and the top of the mercury in the glass column on the left is placed at some mark on the glass (by moving the right hand glass column up or down). The difference between the heights of the two mercury columns gives the difference between **atmospheric pressure** and the **pressure** of the gas in the flask. The gas pressure changes with a change in temperature of the flask, and can be used as a definition of the temperature by taking the temperature to be proportional to the pressure; the proportionality **factor** can be found in the following manner. If the temperature at the freezing point of water is assigned the value 0° and that at the **boiling point** is called 100°, the temperature scale is called the Celsius scale (formerly called Centigrade); if those points are taken at 32° and 212°, it is known as the Fahrenheit scale. The relationship between them can be found as follows. If the temperature in the Celsius scale is T(°C), and that in the Fahrenheit scale is T(°F), they are related by T(°F)=(9/5)T(°C)+32°. The importance of using the constant volume gas thermometer to define the temperature is that it gives the same value for the temperature no matter what gas is used (as long as the gas is used at a very low pressure), so that anyone at any laboratory would be able to find the same temperature under the same conditions. Of course, a variety of other types of thermometers are used in practice (mercury-in-glass, or the change in the **electrical resistance** of a wire, for example), but they all must be calibrated against a constant volume gas thermometer as the standard.

Expansion coefficients

An important characteristic of a material is how much it expands for a given increase in temperature. The amount that a rod of material lengthens is given by $L=L_0[1+ \alpha (T-T_0)]$, where L_0 is the length of the rod at some temperature T_0, and L is the length at some other temperature T; α (Greek alpha) is called the **coefficient** of linear expansion. Some typical values for $\alpha \times 10^6$ (per °C) are: **aluminum**, 24.0; **copper**, 16.8; glass, 8.5; **steel**, 29.0 (this notation means that, for example, aluminum expands at a **rate** of 24.0/1,000,000 for each degree Celsius change in temperature). Volumes, of course, also expand with a rise in temperature, obeying a law similar to that for linear expansion; coefficients of volume expansion are approximately three times as large as that for linear expansion for the same material. It is interesting to note that, if a hole is cut in a piece of material, the hole expands just as if there were the same material filling it!

Thermostats

Since various metals expand at different rates, a **thermostat** can be made to measure changes in temperature by securely fastening together two strips of metal with different expansion coefficients. If they are straight at one temperature, they will be bent at any other temperature, since one will have expanded or contracted more than the other. These are used in many homes to regulate the temperature by causing an electrical contact to be made or broken as temperature changes cause the end of the strips to move.

Water

Water has the usual property of contracting when the temperature decreases, but only down to 39.2°F (4°C); below that temperature it expands until it reaches 32°F (0°C). It then forms **ice** at 0°C, expanding considerably in the process; the ice then behaves "normally," contracting as the temperature decreases. Since the **density** of a substance varies inversely to the volume (as a given **mass** of a substance expands, its density decreases), this means that the density of water increases as the temperature decreases until 4°C, when it reaches its maximum density. The density of the water then decreases from 4°C to 0°C; the formation of the ice also involves a decrease in density. The ice then increases its density as its temperature falls below 0°C. Thus, as a **lake** gets colder, the water at the top cools off and, since its density is increasing, this colder water sinks to the bottom. However, when the temperature of the water at the top becomes lower than 4°C, it remains at the top since its density is lower than that of the water below it. The pond then ices over, with the ice remaining at the top, while the water below remains at 4°C (until, if ever, the entire lake freezes). **Fish** are thus able to live in lakes even when ice forms at the top, since they have the 4°C water below it to live in.

Conservation of energy

The **conservation** of energy is well known from mechanics, where energy does not disappear but only changes its form. For example, the potential energy of an object at some height is converted to the kinetic energy of its **motion** as it falls. Thermodynamics is concerned with the internal energy of an object and those things that affect it; conservation of energy applies in this case, as well.

Heat

As noted in the introduction, doing work on an object (for example, by drilling a hole in a piece of metal, or by repeatedly bending it) causes its temperature to rise. If this object is placed in contact with a cooler object it is found that they eventually come to the same temperature, and remain that way as long as there are no outside influences (this is known as thermal equilibrium). This series of events is viewed as follows. Consistent with the concept of the conservation of energy, the energy due to the work done on the object is considered to be "stored" in the object as (what may be called) internal energy. In the particular example above, the increase in the internal energy of the object is recognized by the increase in temperature, but there are processes where the internal energy increases without a change in temperature. By then placing it in contact with an object of lower temperature, energy flows from the hotter to the colder one in the form of heat, until the temperatures become the same. Thus heat should be viewed as a type of energy which can flow from one object to another by virtue of a temperature difference. It makes no sense to talk of an object having a certain amount of heat in it; whenever it is placed in contact with a lower-temperature object, heat will flow from the hotter to the cooler one.

The first law of thermodynamics

These considerations may be summarized in the first law of thermodynamics: the internal energy of an object is increased by the amount of work done on it, and by the amount of heat added to it. Mathematically, if U_f is the internal energy of an object at the end of some process, and U_i is the internal energy at the beginning of the process, then $U_f - U_i = W + Q$, where W is the amount of work done on the object, and Q is the amount of heat added to the object (**negative** values are used if work is done by the object, or heat is transferred from the object). As is usual for an equation, all quantities must be expressed in the same units; the usual mechanical unit for energy (in the International System of Units-formerly the MKS system) is the joule, where 1 joule equals 1 kg- m^2/s^2.

Specific heats; the calorie

An important characteristic of materials is how much energy in the form of heat it takes to raise the temperature of some material by one degree. It depends upon the type of material being heated as well as its amount. The traditional basic unit, the **calorie**, is defined as the amount of heat that is needed to raise one gram of water by one degree Celsius. In terms of mechanical energy units, one calorie equals 4.186 joules (J).

The corresponding amount of heat necessary to raise the temperature of other materials is given by the specific **heat capacity** of a material, usually denoted by c. It is the number of kilojoules (kJ) needed to raise 1 kg of the material by 1°C. By definition, the value for water is 4.186 kilojoules. Typical values for c in kilojoules per kg (kJ/kg), at 0°C, are: ice, 2.11; aluminum, 0.88; copper, 0.38; **iron**, 0.45. It should be noted that water needs more heat to bring about a given rise in temperature than most other common substances.

Change of phase

The process of water changing to ice or to steam is a familiar one, and each is an example of a change in phase. Suppose a piece of ice were placed in a container and heated at a uniform rate, that is, a constant amount of

heat per second is transferred to the material in the container. The ice (the solid phase of water) first rises in temperature at a uniform rate until its temperature reaches 32°F (0°C), when it begins to melt, that is, some of the ice changes to water (in its liquid phase); this temperature is called the melting point. It is important to note that the temperature of the ice-water mixture remains at 32°F (0°C) until all the ice has turned to water. The water temperature then rises until it reaches 212°F (100°C), when it begins to vaporize, that is, turns to steam (the gaseous phase of water); this temperature is called the boiling point. Again, the water-steam mixture remains at 212°F (100°C) until all the liquid water turns into steam. Thereafter, the temperature of the steam rises as more heat is transferred to the container. It is important to recognize that during a change in phase the temperature of the mixture remains constant. (The energy being transferred to the mixture goes into breaking molecular bonds rather than in increasing the temperature.) Many substances undergo similar changes in phase as heat is applied, going from solid to liquid to gas, with the temperature remaining constant during each phase change. (Some substances, such as glass, do not have such a well-defined melting point.) The amount of heat needed to melt a gram of a material is known as the heat of fusion; that to vaporize it is the heat of vaporization. On the other hand, if steam is cooled at a uniform rate, it would turn to liquid water at the condensation temperature (equal to the boiling point, 212°F [100°C]), and then turn to ice at the solidification temperature (equal to the melting point, 32°F [0°C]). The heat of condensation is the amount of heat needed to be taken from a gram of a gas to change it to its liquid phase; it is equal to the heat of vaporization. Similarly, there is a heat of solidification which is equal to the heat of fusion. Some typical values are shown in Table 1.

It is interesting to note that water has much larger heats of fusion and of vaporization than many other usual substances. The melting and boiling points depend upon the pressure (the values given in the table are for atmospheric pressure). It is for this reason that water boils at a lower temperature in high-altitude Denver than at **sea level**.

Finally, below certain pressures it is possible for a substance to change directly from the solid phase to the gaseous one; this case of sublimation is best illustrated by the "disappearance" of dry ice when it is exposed to the atmosphere.

Equations of state; work

When an object of interest (usually called the system) is left alone for a sufficiently long **time**, and is subject to no outside influences from the surroundings, measurements of the properties of the object do not change

	TABLE 1. THERMODYNAMICS			
Material	**Melting Point °C**	**Heat of Fusion cal/gm**	**Boiling Point °C**	**Heat of Vaporization cal/gm**
Water	0	79.7	100	539
Ethyl alcohol	-114	24.9	78	204
Oxygen	-219	3.3	-183	51
Nitrogen	-210	6.1	-196	48
Mercury	-39	2.8	357	65

with time; it is in a state of thermal equilibrium. It is found experimentally that there are certain measurable quantities that give complete information about the state of the system in thermal equilibrium (this is similar to the idea that measurements of the **velocity** and **acceleration** of an object give complete information about the mechanical state of a system). For each such state relationships can be found which hold true over a wide range of values of the quantities. These relationships are known as equations of state.

Equations of state

Thermodynamics applies to many different types of systems; gases, elastic solids (solids that can be stretched and return to their original form when the stretching **force** is removed), and mixtures of chemicals are all examples of such systems. Each system has its own equation of state, which depends upon the variables that need to be measured in order to describe its internal state. The relevant variables for a system can only be determined by experiment, but one of those variables will always be the temperature.

The system usually given as an example is a gas, where the relevant thermodynamic variables are the pressure of the gas (P), its volume (V), and, of course, the temperature. (These variables are the relevant ones for any simple chemical system, e.g., water, in any of its phases.) The amount of gas may be specified in grams or kilograms, but the usual way of measuring mass in thermodynamics (as well as in some other fields) is in terms of the number of **moles**. One kilomole (kmol) is defined as equal to M kilograms, where M is the **molecular weight** of the substance, with carbon-12 being taken as M = 12. (One **mole** of any substance contains 6.02×10^{23} molecules, known as Avogadro's number.) Thus one

kilomole of **oxygen** has a mass of 70.56 lb (32 kg); of **nitrogen**, 61.76 lb (28.01 kg); the molar mass of air (which is, of course, actually a mixture of gases) is commonly taken as 63.87 lb (28.97 kg). It is found, by experiment, that most gases at sufficiently low pressures have an equation of state of the form: PV = NRT, where P is in Newtons/m^2, V is in m^3, N is the number of kilomoles of the gas, T is the temperature in K, and R = 8.31 kJ/kmol-K is known as the universal gas constant. The temperature is in degrees Kelvin (K), which is given in terms of the Celsius temperature as T(K) = T(°C)+273.15°C. It should be noted that real gases obey this ideal gas equation of state to within a few **percent** accuracy at atmospheric pressure and below.

The equation of state of substances other than gases is more complicated than the above ideal gas law. For example, an elastic solid has an equation of state which involves the length of the stretched material, the stretching force, and the temperature, in a relationship somewhat more complex than the ideal gas law.

Work

Work is defined in mechanics in terms of force acting over a distance; that definition is exactly the same in thermodynamics. This is best illustrated by calculating the work done by a force F in compressing a volume of gas. If a volume of gas V is contained in a cylinder at pressure P, the force needed on the piston is (by the definition of pressure) equal to PA, where A is the area of the piston. Let the gas now be compressed in a manner which keeps the pressure constant (by letting heat flow out, so that the temperature also decreases); suppose the piston moves a distance d. Then the work done is W = Fd = PAd. But Ad is the amount that the volume has decreased, $V_i - V_f$, where V_i is the initial volume and V_f is

KEY TERMS

Adiabatic process—A process during which no heat is transferred between the system and surroundings is described as "adiabatic."

Avogadro's number—The number of molecules present in one mole of whatever the compound is always equal to 6.0229×10^{23}. It was named for the Italian physicist Amedeo Avogadro.

Boiling point—The boiling point of a liquid is the temperature at which it boils, also the temperature at which its vapor condenses.

Calorie—The amount of heat necessary to increase the temperature of water by one degree Celsius.

Celsius temperature (°C)—The temperature scale on which the freezing point of water is 0° and the boiling point is 100°.

Change in phase—Change in the form and characteristics of a substance, e.g., changes from gas to liquid, or liquid to solid.

Coefficient of linear expansion ($\beta.\alpha$)—The fractional rate of change in length of an object with a change in temperature.

Condensation temperature—The temperature at which a gas changes into a liquid (equal to the boiling point).

Equation of state—Relationship among the (experimentally determined) variables, which give complete information about the state of a system.

Fahrenheit temperature (°F)—The temperature scale on which the freezing point of water is 32° and the boiling point is 212°.

First law of thermodynamics—The internal energy of a system is increased by the amount of work done on the system and the heat flow to the system (conservation of energy).

Heat of condensation—The amount of heat needed to be removed from a gas to change it to its liquid phase (equal to the heat of vaporization).

Heat of fusion—The amount of heat needed to be added to a solid to change it to its liquid phase.

Heat of solidification—The amount of heat needed to be removed from a liquid to change it to its solid phase (equal to the heat of fusion).

Heat of vaporization—The amount of heat needed to be added to a liquid to change it to its gaseous phase.

Ideal gas—A gas obeying the ideal gas equation of state, $pV = nRT$, where, e.g., p is in Newtons/meter^2, V is in m^3, n is the number of kilomoles of the gas, T

the final volume. (Note that this volume difference gives a positive value for the distance, in keeping with the fact that work done on a gas is taken as positive.) Therefore, the work done on a gas during a compression at constant pressure is $P(V_i - V_f)$.

The first law thus gives a straightforward means to determine changes in the internal energy of an object (and it is only changes in the internal energy that can be measured), since the change in internal energy is just equal to the work done on the object in the absence of any heat flow. Heat flow to or from the object can be minimized by using insulating materials, such as fiberglass or, even better, styrofoam. The idealized process where there is zero heat flow is called an adiabatic process.

The second law of thermodynamics

One of the most remarkable facts of nature is that certain processes take place in only one direction. For example, if a high temperature object is placed in contact with one of lower temperature, heat flows from the hotter to the cooler until the temperatures become equal. In this case

(where there is no work done), the first law simply requires that the energy lost by one object should be equal to that gained by the other object (through the mechanism of heat flow), but does not prescribe the direction of the energy flow. Yet, in a situation like this, heat never flows from the cooler to the hotter object. Similarly, when a drop of ink is placed in a glass of water which is then stirred, the ink distributes itself throughout the water. Yet no amount of stirring will make the uniformly-distributed ink go back into a single drop. An open bottle of perfume placed in the corner of a room will soon fill the room with its scent, yet a room filled with perfume scent will never become scent- free with the perfume having gone back into the bottle. These are all examples of the second law of thermodynamics, which is usually stated in two different ways. Although the two statements appear quite different, it can be shown that they are equivalent and that each one implies the other.

Clausius statement of the second law

The Clausius statement of the second law is: No process is possible whose only result is the transfer of heat from a cooler to a hotter object. The most common

Thermodynamics

is the temperature in K, and R = 8.31 kJ/kmolK.

Internal energy—The change in the internal energy of a system is equal to the amount of adiabatic work done on the system.

Kelvin temperature (K)—The Celsius temperature plus 273.15°C.

Kilomole (kmol)—A quantity of matter equal to M kilograms, where M is the molecular weight of the substance, with carbon-12 being taken as M = 12 (one kilomole equals 1,000 moles).

Macroscopic theory—A theory which ignores the molecular nature of matter.

Melting point—The temperature at which a solid changes into a liquid.

Microscopic theory—A theory which is based on the molecular nature of matter.

Second law of thermodynamics—No process is possible whose only result is the transfer of heat from a cooler to a hotter object (Clausius statement). No process is possible whose only result is the conversion of heat into an equivalent amount of work (Kelvin Planck statement).

Solidification temperature—The temperature at

which a liquid changes into a solid (equal to the melting point).

Specific heat—The amount of heat needed to increase the temperature of a mass of material by one degree.

Statistical mechanics—The microscopic theory of matter for which the macroscopic theory is thermodynamics, or the molecular basis of thermodynamics.

Sublimation—The change of a material from its solid phase directly to its gaseous phase.

Temperature (T)—The (experimentally determined) variable which determines the direction of heat flow; the variable which is common to all equations of state.

Thermal equilibrium—A condition between two or more objects in direct thermal contact in which no energy as heat flows from one to the other. The temperatures of such objects are identical.

Universal gas constant (R)—The constant in the ideal gas equation of state (as well as elsewhere); equal to 8.31 kJ/kmolK.

---

example of the transfer of heat from a cooler object to a hotter one is the refrigerator (air conditioners and heat pumps work the same way). When, for example, a bottle of milk is placed in a refrigerator, the refrigerator takes the heat from the bottle of milk and transfers it to the warmer kitchen. (Similarly, a heat pump takes heat from the cool ground and transfers it to the warmer interior of a house.) An idealized view of the refrigerator is as follows. The **heat transfer** is accomplished by having a motor, driven by an electrical current, run a compressor. A gas is compressed to a liquid, a phase change which generates heat (heat is taken from the gas to turn it into its liquid state). This heat is dissipated to the kitchen by passing through tubes (the condenser) in the back of (or underneath) the refrigerator. The liquid passes through a valve into a low pressure region, where it expands and becomes a gas, and flows through tubes inside the refrigerator. This change in phase from a liquid to a gas is a process which absorbs heat, thus cooling whatever is in the refrigerator. The gas then returns to the compressor where it is again turned into a liquid. The Clausius statement of the Second Law asserts that the process can only take place by doing work on the system; this work is pro-

vided by the motor which drives the compressor. However, the process can be quite efficient, and considerably more energy in the form of heat can be taken from the cold object than the work required to do it.

Kelvin-Planck statement of the second law

Another statement of the second law is due to Lord Kelvin and Max Planck (1858-1947): No process is possible whose only result is the conversion of heat into an equivalent amount of work. Suppose that a cylinder of gas fitted with a piston had heat added, which caused the gas to expand. Such an expansion could, for example, raise a weight, resulting in work being done. However, at the end of that process the gas would be in a different state (expanded) than the one in which it started, so that this conversion of all the heat into work had the additional result of expanding the "working fluid" (in this case, the gas). If the gas were, on the other hand, then made to return to its original volume, it could do so in three possible ways: (a) the same amount of work could be used to compress the gas, and the same amount of heat as was originally added would then be released from the cylinder; (b) if the cylinder were insulated so that no heat could escape, then the end result would be that the gas is at a higher temperature than originally; (c) something in-between. In the first case, there is no net work output or heat input. In the second, all the work was used to increase the internal energy of the gas, so that there is no net work and the gas is in a different state from which it started. Finally, in the third case, the gas could be returned to its original state by allowing some heat to be transferred from the cylinder. In this case the amount of heat originally added to the gas would equal the work done by the gas plus the heat removed (the first law requires this). Thus, the only way in which heat could be (partially) turned into work and the working fluid returned to its original state is if some heat were rejected to an object having a temperature lower than the heating object (so that the change of heat into work is not the only result). This is the principle of the heat engine (an **internal combustion engine** or a **steam engine** are examples).

Heat engines

The working fluid (say, water for a steam engine) of the heat engine receives heat Q_h from the burning fuel (diesel oil, for example) which converts it to steam. The steam expands, pushing on the piston so that it does work W; as it expands, it cools and the pressure decreases. It then traverses a condenser, where it loses an amount of heat Q_c to the coolant (cooling water or the atmosphere, for example), which returns it to the liquid state. The second law says that, if the working fluid (in this case the water) is to be returned to its original state so that the heat-work process could begin all over again, then some heat must be rejected to the coolant. Since the working fluid is returned to its original state, there is no change in its internal energy, so that the first law demands that $Q_h - Q_c = W$. The efficiency of the process is the amount of work obtained for a given cost in heat input: $E = W/Q_h$. Thus, combining the two laws, $E = (Q_h - Q_c)/Q_h$. It can be seen therefore that a heat engine can never run at 100% efficiency.

It is important to note that the laws of thermodynamics are of very great generality, and are of importance in understanding such diverse subjects as **chemical reactions**, very low temperature phenomena, and the changes in the internal structure of solids with changes in temperature, as well as engines of various kinds.

See also Gases, properties of.

Resources

Books

DiLavore, Philip, *Energy: Insights from Physics.* New York: Wiley, 1984.)

Goldstein, Martin, and Inge F. Goldstein. *The Refrigerator and the Universe.* Cambridge: Harvard University Press, Cambridge, 1993.

David Mintzer

Thermometer

A thermometer is a device that registers the **temperature** of a substance relative to some agreed upon standard. Thermometers use changes in the physical or electronic properties of the device to detect temperature variations. For example, the most common thermometer consists of some sort of liquid sealed into a narrow tube, or capillary, with a calibrated scale attached. The liquid, typically mercury or **alcohol**, has a high **coefficient** of **thermal expansion**, that is to say the **volume** changes significantly with changes in temperature. Combined with the narrowness of the tube, this means that the height of the column of liquid changes significantly with small temperature variations.

The oldest thermometers were not sealed, which means that air **pressure** caused inaccurate readings. The first sealed thermometers were manufactured in the seventeenth century. A further improvement took place in 1714, when the German physicist Daniel Fahrenheit (1686-1736) started using mercury instead of alcohol as the measuring liquid. The Fahrenheit thermometer set a standard for accuracy that was accepted by scientists.

All material exhibits a certain resistance to **electric current** that changes as a function of temperature; this is the basis of both the resistance thermometer and the thermistor. The resistance thermometer consists of fine wire wrapped around an insulator. With a change in temperature, the resistance of the wire changes. This can be detected electronically and used to calculate temperature change from some reference resistance/temperature. Thermistors are semiconductor devices that operate on the same principle.

A **thermocouple** is another temperature sensor based on electrical properties. When two wires of different materials are connected, a small voltage is established that varies as a function of temperature. Two junctions are used in a typical thermocouple.

One junction is the measurement junction, the other is the reference junction kept at some constant temperature. The voltage generated by the temperature difference is detected by a meter connected to the system, and as with the thermistor, this information is converted to temperature.

A digital thermometer measuring the temperature of boiling water. Pure water boils at 212°F (100°C) in standard atmospheric conditions, but the boiling point may be elevated by increased atmospheric pressure and the presence of impurities in the water. *Photograph by Adam Hart-Davis. National Audubon Society Collection/Photo Researchers, Inc. Reproduced by permission.*

A pyrometer is a temperature sensor that detects visible and infrared **radiation** and converts it to temperature. There is a direct **relation** between the **color** of **light** emitted by a hot body and its temperature; it's no accident that we speak of things as "red hot" or "white hot." All surfaces (including animals) emit or reflect radiation whose wavelength is proportional to their temperature. Pyrometers essentially compare the brightness and color of a reference filament to the radiation being emitted or reflected by the surface under test. They are excellent devices for non-contact measurements.

A wide variety of devices exist for measuring temperature; it is up to the user to choose the best thermometer for the job. For contact measurements requiring only moderate accuracy, a capillary thermometer is appropriate. Thermocouples measure temperature over a very wide range with good precision. A more accurate thermal sensor is a thermistor, which boasts the added advantages of being easy to use and inexpensive. Extremely precise contact measurements can be made with a resistance thermometer, but the devices are costly, and pyrometers are useful for non-contact measurements.

Thermostat

A thermostat is a device for controlling heating and cooling systems. It consists of a circuit controlled by a **temperature** sensitive device and connected to the environmental system. The most common thermostat, such as the one seen in homes and offices, is based on a bimetallic strip. As its name suggests, a bimetallic strip consists of thin strips of two different metals bonded together. One **metal** of the strip expands significantly with changes in temperature, the other metal changes very little with temperature increase or decrease. When the temperature increases, for example, one side of the strip will expand more than the other side, causing the strip to bend to one side. When it bends far enough, it closes the circuit that directs the air conditioner to turn on. The thermostat adjustment knob varies the distance that the bimetallic strip must bend to close the circuit, allowing **selection** of temperature level. As the air in the room gets cooler, the metal that expanded with **heat** will now contract, causing the bimetallic strip to straighten out until it no longer completes the circuit that allows the air conditioner to operate. The air conditioner will turn off until the air becomes warm enough to cause the strip to deform and close the circuit once again.

A number of thermostats that are variations on this theme have been developed. Some are based on a brass bellows filled with a thermally sensitive vapor. When the vapor is heated, it expands, pushing out the bellows until it closes the circuit and triggers the heater/air conditioner. Another thermostat design is the bulb type, which includes a bulb and capillary, similar to a **thermometer**, and a diaphragm. When the bulb is heated, the material expands and travels down the capillary to the diaphragm. The diaphragm in turn moves, moving a lever or spring post and eventually controlling the heating and cooling system.

Electronic thermostats have become increasingly popular in the past few years, offering the dual benefits of low cost and minimal moving parts. The active element is a thermistor, a semiconductor device whose resistance to electrical current changes with temperature. Temperature change is signified by a change in measured voltage, which can be used to pass information to the control systems of the heating/cooling units.

Thistle

Thistle is the common name given to some plants in several genera of the Cynareae tribe, family Asteraceae. These genera include *Cirsium, Carduus, Echinops, Onopordum, Silybum, Centaurea,* and *Cnicus.* The name thistle most often refers to the weedy, prickly plants belonging to the genera *Cirsium* and *Carduus.* Thistles are composite flowers, which means their **flower** is actually a group of small flowers that give the appearance of one, larger flower. The flowers, which are of the disc type, are surrounded by bracts. Many people group *Carduus* and *Cirsium* together, but a closer look at these two genera would reveal a distinct difference. While both genera have hairs or bristles mixed with the flowers and a hairy

Prairie thistle (*Cirsium undulatum*). *Photograph by Robert J. Huffman. Field Mark Publications. Reproduced by permission.*

pappus, the hairs on *Cirsium* **species** are plumose (they are feathery—the hairs have branch hairs) and the *Carduus* hairs are not.

The most notable characteristic of thistles are the prickly stems, leaves, and the bracts around the flower head. Among the thistle genera, there are various **leaf** shapes and colors. The flowers are most often purple, but some species have pink, white, or yellow flowers. Thistles can be found in temperate regions, can grow in many climates and soils, and their height can range from 1 foot (0.3 m) to 12 ft (3.5 m). Some well known *Cirsium* thistles are *C. vulgare* or bull thistle, *C . arvense* or Canada thistle, and *C. eriophorum* or woolly thistle. *C. nutans* or nodding thistle is the best known of the *Carduus* species. Globe thistles of the *Echinops* genus are often grown for ornamental purposes. *Onopordum acanthium* or Scottish thistle is well known for its large purple flower. *Silybum marianum* or milk thistle, *Centaurea calcitrapa* or **star** thistle, and *Cnicus benedictus* or holy thistle, have been used and cultivated for medicinal purposes.

Cynara scolymus or globe artichoke is closely related to thistle. What is eaten of the artichoke that shows up on our dinner plate is the unopened flower head and bracts of the **plant.** Some plants such as sow thistle (genus *Sonchus* of the Chicory tribe-Lactuceaea) and Russian thistle (*Salsola kali* of the Goosefoot family), are called thistle because of their prickliness but are not true thistles.

Thoracic surgery

Thoracic **surgery** refers to surgery performed in the thorax or chest. The **anatomy** and **physiology** of the thorax require special procedures to be carried out for the surgery to be done.

The thorax is the bony cage consisting of the ribs, the spine, and the breastbone or sternum. The floor of the thorax is formed by the diaphragm.

Within the thorax lie the two lungs and the **heart,** the organs whose function it is to oxygenate and pump **blood.** The lungs are large, relatively cone-shaped, spongy organs that lie with their narrow ends at the top and the broad ends at the bottom of the thorax. Between the paired lungs in an area called the mediastinum lies the heart. The inside of the chest cavity is lined with a sheet of elastic **tissue,** the pleura, which also covers each lung. The pleura on each side of the chest is independent from the other side; that is, the lining on the right covers the right lung and the right half of the thorax. A fluid called the pleural fluid fills the area between the two layers of pleura so that the **membrane** slides easily as the lungs work.

A constant **negative pressure** or **vacuum** is maintained in the chest to keep the lungs inflated. The diaphragm controls **respiration**. In its relaxed state it is dome shaped and projects into the chest. When it is tensed the diaphragm flattens and pulls air into the lungs. **Carbon dioxide** and **oxygen** are exchanged in the blood circulating through the lungs, the diaphragm relaxes and forces the air out of the thorax, and the oxygenated blood is returned to the heart for circulation.

Thoracic surgery may be needed as a result of a heart or lung **disease**, or to correct an abnormality of one of the large blood vessels. The heart may have a faulty valve that needs replacing, or a partially occluded coronary artery for which a bypass **graft** is needed. A hole in the wall separating the right and left sides of the heart may require patching. A lung **tumor** or a foreign object may need to be removed. A more serious procedure such as a heart transplant or heart-lung transplant may be needed. Any lung surgery will disrupt the negative pressure in the chest and render the lungs inoperable.

Thoracic surgery did not advance as rapidly as did surgery on other areas of the body because the means could not be found to maintain lung function during surgery and restore it after the surgery is completed. Not until early in the twentieth century did Samuel Meltzer and John Auer describe successful lung surgery carried out under positive-pressure air forced into the lungs. With this method the lungs remained inflated and the surgery could be completed without the lungs collapsing. Ironically, Andreas Vesalius (1514-1564) had described this methodology centuries earlier.

Heart surgery is carried out by placing the patient's heart and lung functions on a **heart-lung machine**, or cardiopulmonary bypass machine. First, the thorax is opened by cutting through the superficial tissue and using a saw to cut the sternum. A device called a retractor spreads the cut sternum to allow the surgeon to have full view of the heart.

The patient is connected to the bypass machine by tubes, or cannulas, attached to the large **veins** returning blood to the right side of the heart, the superior and inferior vena cavae. The cannula to return blood to the patient is implanted in the aorta, the large blood vessel leading from the heart to the body, or to a major artery such as the femoral artery in the thigh. When the machine is turned on blood is drawn from the vena cavae into the machine where it is cooled, oxygenated, and filtered to remove any unwanted particles and bubbles. The newly oxygenated blood is returned to the aorta which takes it to the body. Cooling the blood in turn cools the body **temperature** of the patient which reduces the amount of oxygen the tissues need. A third tube gathers blood at the point of surgery and shunts it into the machine to reduce blood loss.

KEY TERMS

Coronary arteries—The arteries that supply blood to the heart muscle.

Donor organs—Organs, such as a heart, kidney, or lung, removed from a person who has died to be implanted in a living person to replace a diseased organ.

Negative pressure—A pressure maintained inside the chest that is lower than the air pressure outside the body. This allows the lungs to remain inflated. Loss of negative pressure may result in a collapsed lung.

Valve—A device that controls the flow of blood between the chambers of the heart and blood entering and leaving the heart. All the valves are oneway, allowing blood to pass in one direction and not in the reverse direction.

With the patient safely on the heart-lung machine, the surgeon can stop the heart and lungs and carry out whatever procedure is needed. Attempting surgery on the beating heart or surgery while the lungs inflate and deflate would be difficult. Stopping the heart by cooling it and stopping lung action by giving a muscle relaxant that quiets the diaphragm provides an immobile field for the procedure. A vein taken from the leg, usually, can be grafted in place to bypass one or more blocked areas in the coronary **arteries**. A diseased heart valve can be removed from the heart and an artificial valve made from plastic and **steel**, pig valve, or monkey valve can be implanted. The entire heart or the heart and both lungs can be removed, emptying the chest cavity, and replaced with donor organs.

At the conclusion of the operation the chest is closed, the heart-lung machine warms the blood to restore normal body temperature, and the cannulae are removed from the vena cavae and aorta.

See also Heart diseases.

Resources

Books

Larson, David E., ed. *Mayo Clinic Family Health Book*. New York: William Morrow, 1996.

Periodicals

Sezai, Y. "Coronary Artery Surgery Results 2000." *Annals of Thoracic And Cardiovascular Surgery* 8, no. 4 (2002): 241-247.

Larry Blaser

Thorium *see* **Actinides**

Thrips

Thrips are minute (less than 0.20 in or 5 mm) slender-bodied **insects** of the order Thysanoptera, characterized by two pairs of veinless, bristle-fringed wings, which are narrow and held over the back when at rest. Although thrips have wings, they do not fly. There are 4,500 **species** of thrips worldwide. In **North America**, there are 694 species of thrips in a number of families in two suborders, the Terebrantia and the Tubulifia. Greenhouse thrips (*Heliothrips haemorrhoidalis*) in the suborder Terebrantia deposit their eggs into slits made on **plant** tissues. Black hunter thrips (*Haplothrips mali*) in the suborder Tubulifera lay their eggs on protected plant surfaces. Red-banded thrips (*Heliothrips rubrocinctus*) are widespread in North America, sucking plant juices and possibly spreading plant viruses.

Several species of predatory thrips feed on **mites**, **aphids**, and other thrips. The majority of thrips are fond of trees and shrubs, roses, onions, and many other herbaceous plants. Pale or white flowers are most attractive to thrips. Their rough mouthparts break open plant tissues, allowing them to suck up plant juices. The most obvious signs of thrip damage are tiny black specks of feces, silvering foliage, distorted flowers and leaves, or buds that fail to open. The insects themselves are rarely seen without a hand **lens**.

In warm **weather**, thrips eggs hatch in days, although cold weather may delay hatching for weeks to months. Thrips undergo incomplete **metamorphosis**, with eggs hatching into a series of nymphs (which resemble adults), which finally metamorphose into adult thrips. The wingless thrip nymphs eat their way through two larval stages, then find a protected crevice (either on the plant or in **soil**) to complete the transition to a winged adult. Female thrips can reproduce parthenogenetically—that is, without mating with males. This strategy allows rapid population explosions when conditions are optimal. The complete life cycle can occur in as little as two weeks. Adults are relatively long-lived.

Adult thrips are attracted to yellow and blue sticky traps hung a few feet above plants. Conventional **pesticides** (such as Malathion) used to control aphids and whiteflies also kill thrips. Less toxic control can be accomplished by maintaining plant health, for water-stressed plants and weedy areas are especially attractive to thrips. Treatment of infested plants involves removing infested plant tissues and mulching with plastic or newspaper to prevent successful development, followed by surrounding the plant stems with **aluminum** foil to prevent reinfestation. Predatory mites (*Neoseiulus mackenziei, Amblyseius cucumeris*) eat thrip eggs and larvae. Larval **lacewings** (*Chrysoperta carnea*) will also control greenhouse thrips as long as aphids, the preferred **prey** of lacewings, are not abundant. Predatory nematodes also consume larval thrips found in soils. Other control measures such as tobacco **water** spray, garlic/hot **pepper** spray, pyrethrin, and insecticidal **soap** sprays can be effective in controlling thrips.

Rosi Dagit

Thrombosis

Thrombosis is the formation of a **blood** clot, or thrombus, in a blood vessel. The process is an exaggeration of a normal and useful event by which the body prevents the loss of blood from **veins** and **arteries** as a result of an injury. During thrombosis, the blood clotting process goes beyond the construction of a blockage for a damaged blood vessel and actually produces a solid clump (the clot) that reduces or even interrupts the flow of blood in the blood vessel.

Thrombosis most commonly occurs for one of three reasons. First, the blood's normal clothing system may become more active with the result that clots within a vein or artery are more likely to form. After childbirth, for example, a woman's clotting mechanism becomes more active to reduce her loss of blood. If the system becomes too efficient, however, clots may form within a blood vessel.

A reduction in blood circulation can also result in the production of a clot. Primitive clots that might otherwise be swept away by the normal flow of blood may be left undisturbed on a vessel wall, allowing them to develop in size.

Finally, the normal aging of blood vessels may contribute to the formation of clots. The first signs of atheroma, or degeneration of blood vessels, can be observed as early as childhood. The more advanced stage—hardening of the arteries—usually does not become a medical problem until old age, however. At this state, scarring of the interior walls of veins and arteries provides sites on which clots can begin to form.

A variety of medical problems can result from thrombosis depending on the location of the thrombus. When thrombosis occurs in a coronary artery of the **heart**, the result is a myocardial infarction, otherwise

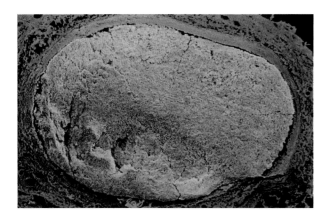

A scanning electron micrograph (SEM) of a thrombus (clot) completely occluding a major pulmonary blood vessel in a human lung. At center is the blood clot. The thick wall of the pulmonary vessel surrounds it. On the outside is the fibrous tissue and alveoli of the lung. *Moredun Animal Health LTD/National Audubon Society Collection/Photo Researchers, Inc. Reproduced by permission.*

known as a heart attack. If a thrombus forms in the **brain**, a **stroke** occurs. Thrombosis in a leg vein, especially the long saphenous vein, causes the condition known as venous thrombosis or phlebitis.

A secondary consideration in thrombosis is the possibility that a piece of the thrombus may become detached and carried away by the blood. The embolus thus formed may lodge in another part of the body, such as the lungs, where it may cause further medical problems.

Researchers have identified a number of conditions that may lead to thrombosis. These include one's genetic make-up, bodily injury, **surgery**, pregnancy, smoking, **stress**, **hypertension**, and **arteriosclerosis**.

The most suitable form of treatment depends on the form of thrombosis that occurs. With phlebitis, for example, bed rest and elevation of the limb is recommended. In some instances, by-pass surgery may be required. For many cases of thrombosis, anticoagulant drugs— drugs that reduce the **rate** of blood clotting—can also be effective.

See also Circulatory system.

Thrushes

Thrushes are a diverse group of about 305 **species** of common, medium-sized, perching, **song birds** ranging from 4.5-13 in (11-33 cm) in body length in the family Turdidae. Thrushes occur world-wide in a wide range of habitats, but mostly in **forests**.

Some species of thrushes are common in parks and gardens containing shrubs and trees not intensively treated with **insecticides**. Examples of these familiar species of thrushes include **robins**, **bluebirds**, and the European song thrush and blackbird.

Biology of thrushes

The generalized body plan of thrushes includes relatively short wings with rounded tips, a longish, weakly forked or rounded tail, stout legs and feet, and a slender beak. Coloration ranges from all-black, through various combinations of brown, blue, red, and white. The juveniles of most species have spotted breasts, as do the adults of some species.

Many thrushes are accomplished singers, with relatively loud, melodious songs that carry over a rather long distance. The rich songs of thrushes provide a pleasing component of the ambience of spring and early summer in the temperate zones of the world, when these migratory **birds** are actively establishing and defending their breeding territories.

Most species of thrushes occur in forested or shrubby habitats; others occur in **grasslands**, **tundra**, and semi-desert. A wide range of food is eaten by thrushes. Most species feed upon small **invertebrates (insects**, caterpillars, and earthworms) of diverse types, especially when these birds are raising young, which require a high-protein food. Some species feed on berries during at least part of the year.

Most thrushes, usually the female, build mud-lined, cup-shaped nests of twigs, herbaceous stems, or leaves. The nests are usually located in relatively protected places in trees or shrubs, or on the ground. The young birds are naked and helpless for the first weeks of their life and are fed and tended by both adults. Most species of thrushes raise several broods each year, using the same or different nests each time.

Species of thrushes

Robins are among the world's better known thrushes. The American robin (*Turdus migratorius*) is probably the native bird with which North Americans are most commonly familiar. The American robin has a brick-red breast and slate-grey back and is very widespread, breeding from the northern **limit** of trees and tall shrubs, through to southern Mexico. The American robin utilizes a wide range of natural habitats, and it also breeds in parks and gardens. As is implied by the scientific name of the species, the American robin is migratory, spending the non-breeding season in the more southern regions of its breeding range, as far south as Guatemala. The varied thrush (*Ixoreus naevius*) is a robin-like bird of mature and old-growth **conifer** forests of western **North America**.

The European robin (*Erithacus rubecula*) is the original robin red-breast, after which other, superficially similar species of thrushes were named, such as the American robin. The European robin is common in much of **Europe** and western Russia, breeding in open forests, shrubby habitats, hedgerows, and parks and gardens.

Another common thrush of Europe and North **Africa** is the European blackbird (*Turdus merula*). This species occurs in a wide range of forest types, and also in parks and gardens.

Bluebirds are a familiar group of thrushes in North America. The eastern bluebird (*Sialia sialis*) occurs in open, shrubby habitats of various types in eastern and central North America and south to Nicaragua. Male eastern bluebirds have a blue back and a red breast, while the **color** of the female is more subdued. The western bluebird (*S. mexicana*) and mountain bluebird (*S. currucoides*) are found in western North America.

Five species of thrushes occur in the forests of North America. These thrushes have a basic coloration of grey-brown to brown-red backs, with a white, spotted breast. All of these birds have pleasing, flute-like songs. These species include the **wood** thrush (*Hylocichla mustelina*), hermit thrush (*H. guttata*), olive-backed thrush (*H. ustulata*), gray-cheeked thrush (*H. minima*), and veery (*H. fuscescens*). Numbers of some of the more southern populations of these forest thrushes appear to be declining substantially, in part because of excessive nest parasitism by the brown-headed cowbird (*Molothrus ater*). The breeding range of the cowbird has expanded tremendously because of fragmentation of forests by human activities, especially the conversion of forests to agricultural and residential land-use. The cowbird poses a risk for many species of birds, in addition to forest thrushes.

The wheatear (*Oenanthe oenanthe*) is an arctic species of thrush, breeding as far north as the limits of land in both North America and Eurasia.

Thrushes and people

Because species of thrushes are among the more familiar and well appreciated native birds, they are an important component of the aesthetic environment. This is true of both natural habitats and managed habitats, such as gardens and parks. As such, the activities and songs of a number of species of thrushes provide people with a meaningful link to the more natural aspects of the human experience.

The European blackbird and European song thrush (*Turdus philomelos*) were introduced to New Zealand and **Australia** by British colonists. This was done as part of a larger attempt to make their new, foreign surroundings more familiar in the context of west European culture.

In some places, populations of thrushes have suffered badly as a result of poisoning caused by their exposure to insecticides used in agriculture or for other purposes. For example, during the 1950s and 1960s, populations of American robins declined greatly in places where the insecticide DDT was used to kill the beetle vectors that were spreading Dutch elm disease. The latter is an introduced pathogen that is still killing native **elm** trees over large areas, including large and valuable trees in cities and suburbs. The disappearance of robins and their prominent, melodious songs was an important component of the so-called "silent spring" that became a metaphor for the ecological damage associated with some types of pesticide uses. As such, the health of robin populations became a very important indicator of the broader health of the urban and suburban environment.

Thrushes seen only occasionally in North America include

- White-throated robin (*Turdus assimillis*). A southwestern stray. Normally a resident of the tropical **mountains**, this bird has been seen in southern Texas during particularly harsh winters.

- Clay-colored robin (*Turdus grayi*). A southwestern stray. Normally resides from eastern Mexico to northern Columbia, this bird has become a frequent visitor to southern Texas (especially in winter) in recent years.

- Redwing (*Turdus illacus*). An eastern stray. Normally residing in Europe and northern **Asia**, this bird has been observed several times in North America (mostly in winter).

- Dusky thrush (*Turdus naumanni*). An Alaskan stray. Normally a resident of Asia, this bird has been observed in Alaska on several occasions.

- Eyebrowed thrush (*Turdus obscurus*). An Alaskan stray. Native of Asia. Seen as a rare migrant in the western Aleutian Islands. Has also been observed on the Pribilofs, St. Lawrence Island, and the Alaskan mainland.

- Fieldfare (*Turdus pilaris*). An eastern stray. Common in Europe and parts of northern Asia, this bird is sometimes seen in eastern Canada and the northeastern United States. There have also been sightings in Alaska.

- Red-legged thrush (*Turdus plumbeus*). A native of the West Indies, this bird was once seen in Miami, Florida.

- Rufous-backed robin (*Turdus rufopalliatus*). A southwestern stray. A resident of Mexico, this bird occasionally strays north to southern Arizona. Sightings have also occurred in Texas, New Mexico, and California.

Status of North American thrushes

- Eastern bluebird (*Sialia sialis*) and Western bluebird (*Sialia mexicana*). Have suffered from the felling of dead trees and the removal of dead branches, which increases **competition** with other species such as House sparrows and European **starlings** for nesting cavities. Nest boxes are now used to maintain populations; this has proven more successful with the Eastern Bluebird, but the provision of nesting boxes for Western Bluebirds does not appear to have kept pace with the loss of natural sites. In recent decades, the numbers of Western bluebirds have declined over much of this bird's range. It is estimated that at one point, the Eastern bluebird population had declined by 90% since 1900; today the population seems to be increasing. Rarely, cowbirds may lay eggs in this bird's nests, with the result that fewer young of their own are hatched.

- Mountain bluebird (*Sialia currucoides*). The population of this species has suffered drastic decline since 1900. This bird competes with flickers, swallows, house sparrows, and starlings for nesting locations. But today their numbers appear stable.

- Townsend's solitaire (*Myadestes townsendi*). Rarely, cowbirds may parasitize.

- Wood thrush (*Hylocichla mustelina*). Serious population decline in recent years. May be losing winter **habitat** in tropics. Cowbirds may parasitize.

- Veery (*Catharus fuscescens*). Brown-headed Cowbirds may parasitize. Surveys suggest that this bird's numbers are declining.

- Swainson's thrush (*Catharus ustulatus*). This bird no longer breeds along the Pacific Coast as it once did, although the overall population appears stable. It may be vulnerable to loss of breeding grounds. Rarely, cowbirds may parasitize.

- Gray-cheeked thrush (*Catharus minimus*). The southern breeding populations may be in decline.

- Bicknell's thrush (*Catharus bicknelli*). The Bicknell's thrush of the Northeast, which winters mostly on the island of Hispaniola in the West Indies, is currently being watched by conservationists. Although the Bicknell's Thrush looks and sounds very much like the Gray-cheeked Thrush, it is now considered to be a separate species. Unlike most Gray-cheeked thrushes, which breed in northern **spruce** forests and in thickets of willow and alder on the tundra, the Bicknell's thrush breeds in stunted conifers on the tops of mountains and in dense second-growth woods containing many young conifers.

- Hermit thrush (*Catharus guttatus*). Numbers appear to be holding up. Because it winters farther north than other brown thrushes, it less threatened by ecological damage to the tropics. Rarely, cowbirds may parasitize.

- Varied thrush (*Ixoreus naevius*). Although still common, this bird may be vulnerable to loss of habitat due to cutting of forests of the Northwest.

- American robin (*Turdus migratorius*). Rarely, cowbirds may parasitize. Has expanded into Great plains and drier lowlands with the planting of trees and the extension of **irrigation** (creating nesting sites and moist grassland for foraging). Although this bird was once widely hunted for food, it is today abundant and widespread.

- Northern wheatear (*Oenanthe oenanthe*). The population in North America appears stable, but this bird may be increasingly seeking out breeding grounds in northeastern Canada.

- Bluethroat (*Luscinia svecica*). This small Alaskan population is probably stable, and possibly even increasing. Widespread and common in Eurasia.

See also Blackbirds.

Resources

Books

Brooke, M., and T. Birkhead, eds. *The Cambridge Encyclopedia of Ornithology.* Cambridge: Cambridge University Press, 1991.

Ehrlich, Paul R., David S. Dobkin, and Darryl Wheye. *The Birder's Handbook.* New York: Simon & Schuster Inc., 1988.

Forshaw, Joseph. *Encyclopedia of Birds.* New York: Academic Press, 1998.

Peterson, Roger Tory. *North American Birds.* Houghton Miflin Interactive (CD-ROM), Somerville, MA: Houghton Miflin, 1995.

Bill Freedman
Randall Frost

Thulium *see* **Lanthanides**

Thunderstorm

A thunderstorm is a strong disturbance in the atmosphere bringing heavy rain, **lightning**, and thunder to areas from one to hundreds of kilometers across. Thunderstorms are formed when humid air near the surface begins rising and cooling. The rising air forms **clouds**.

Lightning over Tamworth, New South Wales, Australia. *Photograph by Gordon Garrado. Science Photo Library, National Audubon Society Collection/Photo Researchers, Inc. Reproduced by permission.*

Storms develop when the clouds cool enough to bring about the growth of rain droplets or **ice** crystals. Eventually the growing drops or crystals fall out of the cloud as **precipitation**. Strong updrafts and downdrafts are inside a thunderstorm which cause static charges to build up in the cloud. Charges of opposite sign accumulate in different parts of the cloud until a spark occurs between them, resulting in the jagged bolts of lightning associated with thunderstorms. Severe thunderstorms may include hail, tornadoes, and damaging straight line winds, making these storms among nature's most destructive.

Thunderstorm development

Thunderstorms develop in the same process that forms the puffy clouds of summer skies, cumulus clouds. These clouds form when humid air (that is, air with an abundance of **water** vapor) near the surface is pushed up by being forced over a mountain range, a front, strong solar heating of the surface, or some other means. As the air rises through the atmosphere, it expands and cools. Eventually the rising air cools to the point where its water vapor condenses to form droplets of liquid water. A huge collection of these tiny suspended droplets forms a cloud. At this stage the rising air is visible as a cumulus cloud, called a convective cloud since it forms by **convection** (vertical air movement). During fair **weather** the convective clouds stop their vertical growth at this point and do not bring rain.

To form a thunderstorm from a convective cloud several conditions are necessary. Most importantly the atmosphere must be unstable. In an unstable atmosphere the air **temperature** drops rapidly with height, meaning any bubble of air that begins rising and cooling will remain warmer than its surroundings. At every point in its ascent the rising air acts like a hot air **balloon**: since it is warmer and less dense than the surrounding air it continues to rise.

A second requirement for a strong thunderstorm is plenty of humid air. This condition supports the growth of cloud droplets and actually fuels the rising air through latent **heat**. The water vapor in the air comes from the **evaporation** of liquid water somewhere—most likely the oceans. To evaporate the water into vapor, **energy** is required just as heat must be added to a kettle to make its water boil. This energy carried with the water vapor wherever it goes is latent or hidden heat. If and when the vapor condenses to form liquid water, the latent heat will be released back into the environment. Thus when the water vapor in rising air condenses to form water droplets a significant amount of heat is released to the surrounding air. Heating the air makes it less dense and increases the tendency of the air bubble, now a cloud, to rise.

As the air continues to rise and cool, droplets within the cloud begin to grow by coalescence (sticking together). In the clouds of colder climates droplets may freeze to form ice crystals, which grow as more and more water vapor condenses on them. The droplets or ice crystals, known as precipitation particles, only grow as long as they can be supported by the updrafts. When they grow too large, they begin to fall out of the cloud as drizzle or raindrops. If the updrafts in the cloud are vigorous enough, much larger precipitation will be formed. In a thunderstorm the **uplift** process is so strong that the cloud grows to the height of the entire lower atmosphere (about 40,000 ft [12 km] above the surface) allowing large raindrops and hailstones to form.

At least two distinct types of thunderstorms can be observed. Over warm humid areas such as the Gulf of Mexico the air-mass thunderstorm is the most common. These thunderstorms grow from converging cumulus clouds that rise and cool as described above. As the **storm** matures, rain begins to fall from the upper part of the cloud. The falling precipitation causes downdrafts. This downward moving air eventually overwhelms the rising air. The downdrafts effectively shut off the uplift necessary for the storm to grow, so the storm dissipates as the air sinks and no more rain is formed. These types of thunderstorms are common over the Florida **peninsula** bringing showers and lightning strikes but rarely any hail or damaging winds unless frontal action is nearby.

Potentially more severe thunderstorms form in temperate regions such as the central and eastern United States. Called frontal thunderstorms these storms often form ahead of the advancing edge of a cold air mass (a cold front). In the summer months the air ahead of the cold front is usually warm humid air that is highly unstable. The denser cold air forces the warmer lighter air ahead of it to rise forming convective clouds which eventually rain. As in an air mass thunderstorm, the falling rain causes downdrafts in the cloud. Unlike the air mass

A nighttime thunderstorm. © Keith Kent/Photo Researchers. Reproduced by permission.

storm, a frontal thunderstorm is arranged so that it is intensified by the downdrafts. The downdrafts become strong gusts of down-flowing air. When they reach the ground the downdrafts spread out and force more warm humid air to begin rising into the thunderstorm. This provides the storm with more latent heat, strengthening the cloud's updrafts, increasing its **wind** speeds, and improving the chances of heavy rain and hail. The storm advances into the warm air, vacuuming up humid air, and transforming it into a very organized system of powerful updrafts and downdrafts. After the storm and the front passes, the affected area is often affected by the cold air behind the front where temperatures and humidities are usually much lower.

Hail, lightning, and tornadoes

Strong updrafts in a thunderstorm support the growth of large rain drops and ice crystals. In a severe storm some of the ice crystals may be dragged down by the downdrafts then swept up again by updrafts. Ice particles may be circulated several times through the storm cloud in this manner picking up water with each cycle. In a process called riming, rain water freezes onto the ice particles and eventually grows to be large hailstones. Hailstones continue to be recirculated through the cloud until they grow large enough to fall out under their own weight, falling against the strong updrafts. If located in the right part of the storm, hailstones can grow to impressive sizes. Hail as large as 5.5 in (14 cm) in diameter has been recorded.

Another product of the vigorous up and down drafts in the storm cloud is lightning. Lightning is a giant spark caused by a buildup of static electrical charges, a larger version of the spark one gets by touching a **metal** doorknob after walking across a cloth carpet. By processes that still are not understood fully, thunderstorm clouds build up a large separation of **electric charge** with positive charges located near the top of the cloud and **negative** charges concentrated near the middle. Usually the cloud base has a smaller pocket of positive charge. Separating charges results in huge voltage differences within the cloud and between the cloud base and the ground. The voltage difference is equalized suddenly by a bolt of lightning between these areas. The spark heats the air in the lightning channel to over 54,000°F (30,000°C) causing a rapid expansion. The resulting sound is heard as thunder.

Severe thunderstorms also may form tornadoes, columns of air spinning at extremely high wind speeds. Tornadoes pack wind speeds of 220 mph (over 100 m/second) in a small area, making them capable of great destruction.

See also Air masses and fronts; Atmospheric circulation; Tornado.

Resources

Books

Battan, Louis J. *Weather.* Engelwood Cliffs: Prentice-Hall Inc., 1985.

Battan, Louis J. *Weather in Your Life.* New York: W.H. Freeman & Co., 1983.

Danielson, Eric W., James Levin, and Elliot Abrams. *Meteorology.* 2nd ed. with CD-ROM. Columbus: McGraw-Hill Science/Engineering/Math, 2002.

Hardy, Ralph, Peter Wright, John Kington, and John Gribben. *The Weather Book.* Boston: Little, Brown and Co., 1982.

McNeill, Robert. *Understanding the Weather.* Las Vegas: Arbor Publishers, 1991.

Mogil, H. Michael, and Barbara G. Levine. *The Amateur Meteorologist.* New York: Franklin Watts, 1993.

Wallace, John M., and Peter V. Hobbs. *Atmospheric Science: An Introductory Survey.* New York: Academic Press, 1977.

James Marti

Ticks *see* **Arachnids**

Tidal energy *see* **Alternative energy sources**

Tidal waves *see* **Tsunami**

Tides

Tides are deformations in the shape of a body caused by the gravitational **force** of one or more other bodies. All bodies in the universe exert tidal forces on each other, although the effects are generally too small to observe. As far **Earth** is concerned, the most important tidal phenomena are the **ocean** and ground tides that occur as a result of the Moon's and the Sun's gravity.

History

One of the earliest careful observers of ocean tides was the Greek geographer Pytheas of Massalia. In about the third century B.C., Pytheas traveled outside the Straits of Gibraltar and observed tidal action in the Atlantic Ocean. (Tides were essentially absent from the Mediterranean Sea, which was more familiar to the Greeks.) Pytheas proposed an explanation for tidal action: the pull of the **Moon** on the Earth's oceans, he said, caused the tides. This explanation is now known to be basically correct, although it was not widely accepted by scientists until the eighteenth century. English physicist Sir Isaac Newton (1642–1727) first succeeded in mathematically describing the tides in what is known as the equilibrium tidal theory.

Theories of tidal action

The side of the Earth that faces the Moon experiences a larger gravitational pull, due to the Moon's closer proximity, than other parts of the Earth. This force causes the Earth itself to bulge slightly toward the Moon. This bulge is termed an Earth tide. Since **water** is free to move horizontally, the oceans tend to flow over the Earth's surface and collect in a secondary bulge on top of the Earth tide; this water bulge is termed an ocean tide.

At the same time, an Earth tide and an ocean tide form on the opposite side of the Earth, directly away from the Moon. This second bulge forms as follows (focusing on the ocean tide alone, for clarity): The Moon and the Earth, like all pairs of bodies orbiting each other in **space**, actually **orbit** around their common center of mass (that is, the point where, if their individual centers were attached to opposite ends of a rigid stick, the stick could be suspended from a string and remain in balance).

In the case of the Earth-Moon system, the common center of mass happens to be inside Earth, about 1068 miles (663 km) beneath the surface along a line connecting the center of the Earth to the center of the Moon. As the Earth and Moon revolve around this point like dancers spinning with linked hands, all points on both bodies experience a centrifugal force. This centrifugal force has the same magnitude and direction at every point on and in the Earth (i.e., away from the Moon **parallel** to a line connecting the center of the Earth to the center of the Moon). Where Earth's surface is at any **angle** other than 90° to the line connecting the center of the Earth to the center of the Moon, water experiences a horizontal component of this centrifugal force. On the half of Earth's surface facing away from the Moon, this horizontal force overcomes the pull of the Moon's gravity and causes water to flow over the Earth's surface to a point on the side of the Earth directly opposite the Moon-facing tidal bulge. A second tidal bulge thus form on the side of the Earth facing directly away from the Moon. This bulge is slightly smaller than the Moon-facing bulge because the imbalance between the Moon's gravitation and centrifugal force is smaller at this point. (The Moon is closer to the Moon-facing bulge, making its gravitation stronger there, whereas the centrifugal force considered here is the same everywhere on the Earth.) The larger, Moon-facing tide is termed the direct tide; the tide on the opposite side of the Earth is termed the opposite tide.

These two tidal bulges—one Moon-facing or direct, the other on the opposite side of the Earth—are the high tides. Because the Earth is spherical, these bulges are actually arcs, meeting at the poles to form a globe-girdling ring or belt of high tide aligned with the Moon. (Centrifugal force and the Moon's gravity **cancel** exactly at the poles, so the high tide is, in this simplified model, highest at the equator and diminishes to zero toward the poles.) Movement of water to this high-tide belt causes a complementary belt of low water to form around the Earth at 90° to the line connecting the centers of the Earth and Moon. This belt produces the phenomenon known as low tide.

The high-tide belt always lies along the line connecting the centers of the Earth and Moon; however, as the Earth rotates daily on its axis, land areas approach this belt, pass through it, and leave it behind. Thus, from the point of view of an observer fixed to the surface of the rotating Earth, the ocean tides are continually sloshing up against some coastlines and draining away from others. As a result, most coastal areas experience two high tides and two low tides each day. One high tide corresponds to the high-tide **arc** facing the Moon, and the other to the high-tide arc facing away from the Moon.

The **Sun** forms similar tidal bulges in the Earth and its oceans, one set due to gravitation and the other to centrifugal force. However, the Sun's tidal effect is slightly less than one half that of the Moon. (It is both more massive than the Moon and more distant; distance wins.) As the Moon orbits the Earth every 28 days, it twice comes into alignment with the Earth and Sun—once when it is directly between the Earth and the Sun (i.e., when observers on Earth see the shadowed side of the Moon) and once when the Earth is directly between itself and the Sun (i.e., when observers on Earth see the illuminated or "full" side of the Moon). When the Moon and Sun are aligned, their tidal forces add up to produce a maximum tidal change. These maximal tides are termed spring tides because the waters of the ocean "spring up" higher (and sink lower) at these times. When the Moon and Sun are at right angles to each other (i.e., when the Moon is half-illuminated as seen from Earth), the solar and lunar tidal bulges do not add, and the least dramatic tides of the month are observed. These are termed neap tides.

Variables affecting tidal forces

Physicists have derived precise mathematical expressions to describe the gravitational effects of the Moon and the Sun on the Earth. In theory, therefore, it should be possible to make precise predictions of the timing and sizes of all of ocean tides. In fact, absolutely precise pre-

Low tide at Big Pine Key, Florida. *JLM Visuals. Reproduced by permission.*

High tide at Big Pine Key, Florida. *JLM Visuals. Reproduced by permission.*

dictions are not possible because a large number of factors contribute to the tides at a particular location. Primary among these is that the shape of ocean basins is so irregular that the water in them cannot behave in a simple way. Other variables also complicate the situation. These include variations in the Earth's axial **rotation** and variations in Earth-Moon-Sun positioning, including variations in orbital distance and inclination. Estimates of tidal **behavior** are therefore still based primarily on previous tidal observations, continuous monitoring of coastal water levels, and astronomical tables. (Comparison of predicted with measured tides along U.S. coasts can be checked at <http://tidesonline.nos.noaa.gov/>, a Website maintained by the U.S. government.)

Tide tables

Data about tidal patterns accumulated in various parts of the world are used to produce tide tables. Tide tables are constructed by examining records to find out, for any given location, the times at which high and low tides have occurred in the past and the levels which those tides have reached. These records are then used to predict the most likely times and levels to be expected for tides at various times in the future for the same locations. Because of differences in ocean bottoms, coastline shape, and other factors, unique tide tables must be constructed for each specific coastline in the world.

Semidiurnal and diurnal tides

In most places, tides are semidiurnal (twice-daily), meaning that there are two tidal cycles (with one high tide and one low apiece) each day. In other words, during a typical day the tides reach their highest point along the shore and their lowest point twice each day. The high-water level reached during one of the high tide stages is usually higher than the other high point, and the low

water level reached during one of the low tide stages is usually lower than the other low tide point. This difference is called the diurnal inequality of the tides.

In a few locations, tides occur only once a day, with a single high tide and a single low tide. Such tidal cycles are known as diurnal (daily) tides. In both diurnal and semidiurnal settings, a rising tide is termed a flood tide and a falling tide is termed an ebb tide. The moment when the water reaches its highest point at high tide (or its lowest point at low tide) is called the slack tide, since the water level is then static, neither rising nor falling, at least for a short time.

Effect of the Moon's motion around Earth

As the Earth spins on its axis, completing one turn every 24 hours, the Moon orbits around Earth, completing one orbit every 28 days. Consequently, during the time Earth rotates once (24 hours), the Moon has moved across about 1/28th of the sky. Earth, conseqently, must rotate one day plus 1/28th of a day (about 50 minutes) to bring the Moon into the same position overhead. This time period—24 hours, 50 minutes—is termed a lunar day. Since tides are caused by the Moon, they recur on a lunar-daily schedule, rather than a 24-hour schedule, and consequently shift their times of occurrence with respect to the 24-hour clock. As a result, on a coast with diurnal tides, each day the high tide (or low tide) will occur 50 minutes later than the day before; on a semidiurnal coast, each high tide (or low tide) will occur 12 hours, 25 minutes later than the previous high tide.

Tidal currents

Any movement of ocean water caused by tidal action is known as a tidal current. In open water, tidal **currents** are relatively weak and tend to change direction slowly and regularly throughout the day. They form a kind of rotary pattern that sweeps around the ocean like the minute hand on a clock. Closer to land, however, tidal currents tend to change direction rather quickly, flowing toward land during high tide and away from land during low tide. In many cases, this onshore and offshore tidal current flows up the mouth of a river or some other narrow opening. The tidal current may then attain velocities as great as 9 mi (15 km) an hour with crests as high as 10 ft (3 m) or more.

All coastal locations (as well as very large lakes) experience some variation in tidal range during each lunar cycle, due to the affects of neap versus spring tides. Most tides attain less than 10 ft in size; 3–10 ft (1–3 m) is common. In some places, however, the tides may be much greater. These locations are characterized by ocean bottoms that act as funnels through which ocean waters rush upward towards or downward away from the shore. In the Bay of Fundy, Canada, for the tidal range may be as great as 46 ft (14 m). At the opposite extreme, the Mediterranean, Baltic, and Caribbean Seas have tides of less than 1 ft (0.3 m).

Deep-ocean tidal currents also occur. In fact, it has recently been discovered that about half of the **energy** put by Earth-Moon-Sun system into dragging tides around the Earth is dissipated in deep-ocean currents, and the rest in shallow-ocean currents. Some 3×10^{12} watts of energy are dissipated through **friction** in deep-ocean circulation alone, with profound long-term effects on Earth's climate.

Some side effects of the tides

(1) Because the Earth's comparatively rapid rotation is continually dragging the tidal bulges away from their ideal locations, the Moon-facing bulge is always slightly ahead of the Moon in the direction of its orbit. (The Moon orbits in the same sense as the Earth spins.) The gravitational pull of the bulge thus tends to accelerate the Moon slightly, which causes it to orbit slightly higher above the Earth. The Moon thus moves about 1.2 inches (3 cm) farther from the Earth every year. As it moves away, the magnitude of the tides slowly decreases. (2) Tidal friction is slowing the Earth's axial spin. During the Jurassic period, for example, approximately 180 million years ago, the day was less than 23 hours long; when life first appeared on Earth, it was only about 20 hours long. Tidal friction long ago caused the Moon to slow its own axial spin to the point where it now always keeps the same side facing Earth. This is a common effect in bodies orbiting more massive ones, as for example the inner moons of **Saturn** and **Jupiter**. (3) Stresses in Earth's crust caused by the tides tend to trigger earthquakes. **Frequency** analysis of all recorded earthquakes

> ## KEY TERMS
>
> **Diurnal**—Occurring once per day.
>
> **Ebb tide**—The period when the water level is falling; the period after high tide and before low tide.
>
> **Flood tide**—The period when the water level is rising; the period after low tide and before high tide.
>
> **High tide**—The event corresponding to the largest increase in water level in an area that is induced by tidal forces.
>
> **Low tide**—The event corresponding to the largest decrease in water level in an area that is induced by tidal forces.
>
> **Neap tides**—Period of minimum tidal range that occurs about every two weeks when the Moon and Sun are at 90° to each other, that is, at the first and third quarter moons.
>
> **Period**—The interval of time between two recurring events, such as the high tides in an area.
>
> **Semidiurnal**—Occuring twice per day.
>
> **Slack tide**—Period during which the water level is neither rising nor falling.
>
> **Spring tides**—Period of maximum tidal range; occurs about every two weeks, when the Moon and Sun are in line with each other, i.e., at the new and full moons.
>
> **Tidal current**—Horizontal movement of water due to tidal forces.
>
> **Tidal range**—Vertical distance between high tide and low tide during a single tidal cycle.

shows that they have a strong tendency to occur on a semidiurnal basis, in accord with tidal stress.

Tidal power plants

Engineers have long recognized that the movement of tidal currents might be an inexpensive, efficient, and environmentally safe source of power for human use. In general, the plan would be to construct a dam across the entrance of an estuary through which tidal currents flow with significant speed. Then, as tidal currents flow into and out of the estuary twice each day, they could be used to drive turbines which, in turn, could be used to operate electrical generators.

One of the few commercial tidal power stations in operation is located at the mouth of the La Rance River in

France. Tides at this location reach a maximum of 44 ft (13.5 m). Each time the tide comes in, a dam at the La Rance station holds water back until it reaches its maximum depth. At that point, gates in the dam are opened and water is forced to flow into the La Rance River, driving a **turbine** and **generator** in the process. Gates in the dam are then closed, trapping the water inside the dam. At low tide, the gates open once again, allowing water to flow out of the river, back into the ocean. Again the power of moving water is used to drive a turbine and generator.

The plant is able to produce **electricity** only four times each day, during each of two high tides and each of two low tides. It generates a modest 250 megawatts in this way with an efficiency about equal to that of a fossil-fuel plant, 25%. With present technology, few other sites exist where tidal power generation is currently considered economically feasible.

See also Alternative energy sources; Gravity and gravitation.

Resources

Periodicals

Kasahara, Junzo. "Tides, Earthquakes, and Volcanoes." *Science* 297 (July 19, 2002): 348-349.

Wunsch, Carl. "Moon, Tides and Climate." *Nature* 405 (June 15, 2000): 743-744.

Other

National Oceanic and Atmospheric Administration, U.S. Department of Commerce. "Our Restless Tides: A Brief Explanation of the Basic Astronomical Factors which Produce Tides and Tidal Currents." Center for Operational Oceanographic Products and Services. February 1998 [December 29, 2002]. <http://www.co-ops.nos.noaa.gov/restles1.html>.

K. Lee Lerner
Larry Gilman
David E. Newton

Tiger *see* **Cats**

Time

Time is a measurement to determine the duration of an event, or to determine when an event occurred. Time has different incremental scales (year, day, second, etc.), and it has different ways by which it is reported (Greenwich Mean Time, Universal Time, Ephemeris Time, etc.).

Time measurement systems

Our time scale is based upon the movement of the **Sun**. Since the Sun appears to move from east to west in the sky, when it is 12 noon in New Jersey, people in Seattle, Washington, say that the Sun has not yet reached its apex, or noontime position. In fact, if you say that the Sun has attained its noontime position, someone who is west of you by one degree of longitude will not see the Sun in its noontime position for approximately four minutes.

In October of 1884, an international agreement divided the **planet** into 24 time zones (longitudinally into 15° segments), allowing us to standardize time keeping. The starting point, or **zero** meridian, was the longitude that ran through the Royal Observatory in Greenwich, England. This is the common position from which we measure our time and is generally referred to as Greenwich Mean Time (GMT).

As you move westward from Greenwich, you add one hour for every 15° meridian you cross. As you move eastward from Greenwich, you subtract one hour for every 15° meridian you cross. (On land the time zones are not always straight lines; this is sometimes done to keep countries, states, cities, etc., all within the same time zone.) It should also be noted that these time zones are also listed by letters of the alphabet as well as 15° increments: Greenwich is designated Z for this purpose, the zones to the East are designated A through M, omitting the letter J, and the zones to the West are N through Y.

In **North America**, most people measure time based upon two 12 hour intervals. To distinguish the first 12 hours from the second, we label the time. The first 12 hours are denoted by the letters A.M. (for the Latin *ante meridiem*, meaning *before noon*–before the Sun has reached its highest point in the sky). The second 12 hours is denoted by P.M. (for *post meridiem*, which means *after noon*).

The International Date Line is near the 180° longitude in the Pacific Ocean. When you cross it going west you add 24 hours (one day), and when you cross it going east you subtract 24 hours (one day).

If you ask someone from **Europe**, or someone in the armed services, what time it is, he or she will answer you using a 24–hour system. Their time starts at midnight of every day, which they call 0000 hours (or 2400 hours of the previous day). Their time then proceeds forward for 24 hours, at which point it starts again. If someone tells you it is 0255 Zulu time, it means that it is 2:55 A.M. in Greenwich, England. This type of time reckoning is extremely important to synchronizing clocks for navigation, etc.

Astronomers found GMT to be awkward to work with in their calculations, because its measurements were based upon the Sun's actual **motion** and period between successive occurrences at its noontime position. Therefore, in 1972 they defined a Universal Time, which starts at midnight. They also realized that the solar day, the time from noon to noon, is not necessarily constant because of the irregular motion of the **earth**. To compensate for this, they have adopted a mean (average) solar day, time from noon to noon, upon which to base their measurements.

Both GMT and Universal Time are solar-based time systems, which means that they are based upon the apparent movement of the Sun. This method is not accurate enough for all scientific measurements because the earth's **orbit** around the Sun is not circular (it is an **ellipse**), **earth's rotation rate** (how fast it spins on its axis) is not constant, and its **rotation** axis is nutating (wobbling). Astronomers compensate for these "imperfections" by using a Sidereal Time measurement, a measurement system based upon the repetitive motion of the stars. This motion is, in fact, due to the earth's motion. Ephemeris time is a Sidereal Time based on the apparent repetitious motion of the **Moon** and the planets.

During World War II a "war time" was instituted to save **electricity**. Later, this became daylight saving time. However, some states and cities no longer use daylight saving time because they see it as a needless complication.

Einstein's Theory of Relativity prompted scientists to revise the idea of time as an absolute. For example, the German mathematician Hermann Minkowski (1864-1909) described time as the fourth dimension of **space**, postulating a space-time continuum, which scientists, including Einstein, accepted. Essentially, time cannot be a basis for absolute measurement. Because we are all in motion, in one form or another, our time measurements are dependent upon how we are moving. Relativistic time measurements depend upon who is measuring it, how fast the person is moving, and whether or not they are undergoing an **acceleration**.

Time measuring devices

We measure time, in general, with a clock. However, not all clocks are based on the same time scale. Measurement systems and types of time differ widely. We have just seen that clocks in different places are calibrated to different schemes, and so are likely to tell different times.

The first clock was natural—the motion of the Sun through the sky. This led to the invention of the sundial to measure time, as well as the evening version using the Moon's position. Later came the hourglass, candles, and clocks which burned wax, or oil, at a specified rate, and **water** clocks, which allowed water to flow at a specified rate. Early clocks of greater **accuracy** used a pendulum arrangement invented in the 1656 by the Dutch astronomer Christiaan Huygens (1629-1695). Balance wheels then replaced pendulums in some clocks, which allowed clocks to become portable. In recent times, cou-

A sundial in Death Valley, California. Sundials were the first devices used to measure time. *JLM Visuals. Reproduced by permission.*

pled pendulum clocks were used to keep extremely accurate times.

Most of today's clocks and watches use a quartz **crystal** to keep the time. The specially manufactured quartz crystal, with a specific **frequency** voltage passing through it, will vibrate at a constant (characteristic) frequency. This frequency, when amplified, drives an **electric motor** which makes the hands of the clock turn, or makes the digital display change accordingly.

Atomic clocks are the most accurate clocks, and are used by the United States Naval Observatory. They vibrate at a constant sustainable frequency. The process is analogous to setting a pendulum in motion, or a quartz crystal vibrating, except the **atoms** "vibrate" between two different energies. This "vibration" between **energy** levels is associated with a specific frequency, just like the oscillating quartz crystal. This atomic frequency is what drives the clock. The current standard for atomic clocks is one that uses an **isotope** of Cesium, ^{133}Cs. In 1967, the second was redefined to allow for this accuracy. The second is now defined as 9,192,631,770 times one period, the time for one complete oscillation, of ^{133}Cs.

Research continues into ion clocks such as the **hydrogen** maserclock. This type of clock is more accurate than a ^{133}Cs clock; however, after several days its accuracy drops to, or below, that of a ^{133}Cs clock.

Time reversal

We perceive time as always proceeding forward into the future; there is no way we know of to travel into the past (except through memories). Time reversal refers the attempt to understand whether a process is moving forward or backward in time. For example, if you watch a movie of two isolated billiard balls colliding on a billiard table, can you tell if the movie is being shown forward or backward? If the two balls are exactly the same, and you do not see one of them being hit to start the process, probably not. If we were to film the process of an egg being dropped and hitting the ground, and then show the film, you could definitely determine in which direction the process continued. The ability to distinguish between forward and backward processes, between past and future on all scales, is crucial for scientific research.

To see if there really is an "arrow of time," or preferred direction in which time flows, researchers have been conducting many different types of experiments for more than 40 years. According to theory, we should be able to tell forward time processes from backward time processes. Experimental verifications of this theory delineate how difficult some experiments are to design and perform. The latest are designed to observe atomic particles as they undergo different processes. If the processes are then "run backward;" any differences that show up mean that we can tell a forward process from a backward running process.

See also Relativity, special.

Resources

Books

Ellis, G.F.R., and R. M. Williams. *Flat and Curved Space-Times.* Oxford: Clarendon Press, 2000.

Feynman, Leighton, and Sands. *The Feynman Lectures on Physics.* New York: Addison-Wesley, 1989.

Flaherty, Michael G. *A Watched Pot: How We Experience Time* New York: New York University Press, 2001.

Lineman, Rose, and Jan Popelka. *Compendium of Astrology* West Chester, PA: Schiffer, 1984.

Michelson, Neil F. (compiler). *The American Ephemeris 1931 to 1980 & Book of Tables.* San Diego: Astro Computing Services, 1982.

Periodicals

Cleere, Gail S. "Got a Second?" *Natural History* (June 1992).

Hunter, L.R. "Tests of Time-Reversal Invariance in Atoms, Molecules, and the Neutron." *Science* 252 (April 1991).

Itano, Wayne A., and Norman F. Ramsey. "Accurate Measurement of Time." *Scientific American* 269 (July 1993).

Moyer, Michael. "The Physics of Time Travel." *Popular Science* 260, no. 3 (2002): 52-54.

Tudge, Colin. "How Long is a Piece of Time?" *New Scientist* 129 (January 1991).

Westerhout, Gart, and Gernout M.R. Winkler. "Astrometry and Precise Time." *Oceanus* 33 (Winter 1990/1991).

Zee, A. "Time Reversal." *Discover* 13 (October 1992).

Peter K. Schoch

Timothy *see* **Grasses**

Tin *see* **Element, chemical**

Tinamous

Tinamous are about 45-50 **species** of ground-dwelling **birds** that comprise the family Tinamidae, the only member of the order Tinamiformes. Tinamous have a plump, partridge-like body, but they are not related to the "true" **partridges**, which are species in the pheasant family (Phasianidae). The evolutionary relationships of tinamous are not well understood, but their closest living relatives may be the rheas, which are large, **flightless birds** of **South America** that make up the order Rheiformes.

Tinamous occur from Mexico in Central America, to Patagonia in South America. Species of tinamous breed in a wide range of habitats, from lush tropical rain-forest, to savannah, grassland, and alpine **tundra**. Tinamous are resident in their habitats—they do not migrate.

Tinamous range in body length from 8–20 in (20–53 cm). These birds have a stout body, short, rounded wings, and a short tail. The legs of tinamous are short but robust, and the strong feet have three or four toes. The sternum, or breastbone, is strongly keeled for the attachment of the large flight muscles. The small head is placed at the end of a rather long neck, and the beak is downward-curved, hooked at the end, and generally fowl-like in appearance.

Tinamous are brown colored, with streaky, barred, or mottled patterns. This coloration is highly cryptic, and helps tinamous to blend in well with their surroundings, thereby avoiding predators to some degree. The sexes are colored alike, but females of some species are slightly larger than males.

Tinamous are running, terrestrial birds. They have sonorous, whistling calls, which may function to proclaim their territory. Tinamous can fly rapidly, but they tire quickly and can only fly over a short distance. Tinamous often prostrate themselves in thick vegetation to hide from predators. Tinamous eat roots, **seeds**, **fruits**, buds, and other **plant** materials, and **insects** when available.

The eggs of tinamous are brightly colored in glossy, solid hues, and are considered to be among the most beautiful of all birds' eggs. Depending on the species, the egg **color** can be brown, purple, black, gray, olive, or green. The clutch size is highly variable among species, ranging from 1-12, although the larger numbers may represent the output of several females.

Tinamous nest on the ground. Only the male incubates the eggs, and only he cares for the brood of young birds. This is, of course, a reversal of the usual role of the sexes in most groups of birds, in which females play a more prominent role. Newly hatched tinamous are highly precocious, and can leave the nest and run well within a few days of hatching. Soon after **birth**, the chicks follow their male parent about, feeding themselves.

Species of tinamous

Tinamous are obviously different from all other living birds, and this is the reason why they are assigned to their own order, the Tinamiformes. Although some avian systematists believe the Tinamiformes is most closely related to the order Rheiformes, there are important differences between these groups, especially the presence of a keeled sternum in tinamous. Rheas and other ratites, such as ostriches, emus, and cassowaries, are characterized by a nonkeeled sternum and are incapable of flight.

While they may be different from all other birds, the various species of tinamous are all rather similar to each other in their color, size, and shape. The entire group is believed to be composed of rather closely related taxa, and consequently, the **taxonomy** of tinamous is not well established. Although 45-50 species are named, further study may result in some of these taxa being joined together as subspecies of the same bird, while other species may be split into several.

Two of the more common species are the variegated tinamou (*Crypturellus variegatus*) and the Martineta or crested tinamou (*Eudromia elegans*). Both of these are birds of open **grasslands** of South America, known as pampas. The Martineta tinamou is one of the few social tinamous, occurring in flocks with as many as one-hundred individuals.

The rufescent tinamou (*Nothocercus julius*) occurs in Venezuela, Ecuador, and Columbia. The thicket tinamou (*Crypturellus cinnamomeus*) breeds in brushy habitats from Mexico to Columbia and Venezuela.

The Chilean tinamou (*Nothoprocta perdicaria*) is native to tundra-like habitats of southern South America. This species was introduced to Easter Island in the South Pacific in the late nineteenth century, and it still breeds there.

Wherever they are abundant, tinamous are hunted as a source of wild meat, or for sport. Excessive hunting, coupled with **habitat** loss, has resulted in numerous species being threatened with **extinction**. The World **Conservation** Union has listed 14 species of tinamous as being endangered.

Resources

Books

Forshaw, Joseph. *Encyclopedia of Birds.* New York: Academic Press, 1998.
Sick, H. *Birds in Brazil: A Natural History.* Princeton, N.J: Princeton University Press,1993.

Bill Freedman

Tissue

A tissue is a collection of similar cells grouped to perform a common function. Different tissues are made of their own specialized cells that are adapted for a given function.

All **animal** cells are basically similar. Each **cell** has a cell wall or **plasma membrane** that surrounds the cell and contains various receptors that interact with the outside area. A nucleus, Golgi apparatus, mitochondria, and other structures are contained in each cell. Beyond that, cells are specialized in structure for a given function.

The study of tissues is called histology. Studying the structure of tissue is done by staining a thin specimen of the tissue and placing it under a **microscope**. An experienced histologist can look at a specimen and immediately determine from which **organ** it was taken. A histologist can also see and diagnose a **disease** if it is present in the tissue. The study of disease processes is called **pathology**.

Tissues are divided by function into a number of categories. Muscle tissue, for example, makes up the muscles of the body. A muscle belongs to one of three categories-voluntary muscle which can be controlled for movement or lifting, involuntary muscle which is not under conscious control (such as the muscle tissue in the digestive organs), and cardiac muscle which forms the **heart**.

Connective tissues compose the bones, tendons, and ligaments that make up the support of the body. The body also consists of adipose tissue or **fat**.

Nervous tissue forms the **brain**, spinal cord, and the nerves that extend through the body. Digestive tissue is found in the **digestive system** including the stomach, intestines, liver, pancreas, and other organs involved in digestion. Vascular tissue comprises the blood-forming portion of the bone marrow and the **blood** cells themselves. Epithelial tissue forms sheets that cover or line other tissues. The skin and the lining of the stomach are both examples of epithelial tissue. Various reproductive tissues form the ovaries, testes and the resulting gametes (ova and sperm).

Combined, these tissues form the human body and carry out its functions.

Tit family

The tit family, Paridae, consists of 46 **species** of small **birds**, variously known as tits, titmice, and chickadees. These are all **song birds**, in the order Passeriformes. All are rather small birds, ranging in body length from about 4–8 in (11–20 cm), and mostly occurring in **forests**, shrubby woodlands, and in urban and suburban habitats. This family of birds is widespread, and its representatives are found in **North America**, **Africa**, and Eurasia.

Tits are all active feeders, constantly searching foliage and gleaning **bark** and branches for **insects**, spiders, and other **arthropods**. These birds are highly acrobatic when feeding, often hanging upside down from small twigs to gain access to their **prey**. During winter, when arthropods are scarce, these birds will eat **seeds**, and they may be among the most common species at bird feeders.

Tits actively defend territories, proclaiming their ownership by a series of loud whistles and songs. Tits nest in cavities in trees, either naturally occurring or a hole excavated by the tits in soft, rotted **wood**. Sometimes, tits will utilize a cavity that has previously been excavated and used by smaller species of **woodpeckers**, and some species of tits will use nest boxes. The clutch size is large, with sometimes more than ten eggs laid at one time. Once the young are fledged, the family of tits stays together during autumn and winter, and often joins with other families to form large foraging flocks, sometimes mixed with other species of similar sized birds.

There are seven species of chickadees in North America. These are rather tame and familiar birds, with a dark cap, and a bib on the throat, a whitish breast, and a brownish or gray back. The most widespread species is the black-capped chickadee (*Parus atricapillus*), occurring throughout the southern boreal and northern temperate forests. The territorial song of this species is a loudly whistled teea-deee, but more familiar to most people is the alarm call, chicka-dee-dee-dee, given when potentially dangerous intruders are near. The Carolina chickadee (*P. carolinensis*) occurs in southeastern North America. The boreal chickadee (*P. hudsonicus*) is common in northern coniferous forests. The mountain chickadee (*P. gambeli*) occurs in montane coniferous forests of western North America. The chestnut-backed chickadee (*P. rufescens*) is abundant in forests of the Pacific coast.

There are four species of titmice in North America. These birds have small crests on the top of their head, and are slightly larger than chickadees, with which they often flock during the non-breeding season. The most widespread species is the tufted titmouse (*P. bicolor*) of southeastern North America. The most common species in the southwest is the plain titmouse (*P. inornatus*).

Tits, titmice, and chickadees are all familiar and friendly birds, and many species can be successfully attracted to the vicinity of homes using a well-maintained feeder. Some species will quickly learn to feed directly at a steadily held hand, lured by crushed peanuts or sunflower seeds.

Bill Freedman

Titanium

Titanium is a transition **metal**, one of the elements found in Rows 4, 5, and 6 of the **periodic table**. It has an **atomic number** of 22, an atomic **mass** of 47.88, and a chemical symbol of Ti.

Properties

Titanium exists in two allotropic forms, one of which is a dark gray, shiny metal. The other **allotrope** is a dark gray amorphous powder. The metal has a melting point of 3,051°F (1,677°C), a **boiling point** of 5,931°F (3,277°C), and a **density** of 4.6 g/cm^3. At room **temperature**, titanium tends to be brittle, although it becomes malleable and ductile at higher temperatures. Chemically, titanium is relatively inactive. At moderate temperatures, it resists attack by **oxygen**, most acids, **chlorine**, and other corrosive agents.

Occurrence and extraction

Titanium is the ninth most abundant element in the earth's crust with an abundance estimated at about 0.63%. The most common sources of titanium are ilmenite, rutile, and titanite. The metal is often obtained commercially as a byproduct of the refining of **iron ore**. It can be produced from its ores by electrolyzing molten titanium chloride ($TiCl_4$): $TiCl_4$ —electric current → Ti + $2Cl_2$, or by treating hot titanium chloride with **magnesium** metal: $2Mg + TiCl_4 \rightarrow Ti + 2MgCl_2$.

Discovery and naming

Titanium was discovered in 1791 by the English clergyman William Gregor (1761-1817). Gregor was not a professional scientist, but studied **minerals** as a hobby. On one occasion, he attempted a chemical analysis of the mineral ilmenite and found a portion that he was unable to classify as one of the existing elements. He wrote a report on his work, suggesting that the unidentified material was a new element. But he went no further with his own research. It was not until four years later that German chemist Martin Heinrich Klaproth returned to an investigation of ilmenite and isolated the new element. He suggested the name of titanium for the element in honor of the Titans, mythical giants who ruled the **Earth** until they were overthrown by the Greek gods.

Uses

By far the most important use of titanium is in making alloys. It is the element most commonly added to **steel** because it increases the strength and resistance to corrosion of steel. Titanium provides another desirable property to alloys: lightness. Its density is less than half that of steel, so a titanium-steel **alloy** weighs less than pure steel and is more durable and stronger.

These properties make titanium-steel alloys particularly useful in spacecraft and **aircraft** applications, which account for about 65% of all titanium sold. These alloys are used in airframes and engines and in a host of other applications, including armored vehicles, armored vests and helmets; in jewelry and eyeglasses; in bicycles, golf clubs, and other sports equipment; in specialized dental implants; in power-generating plants and other types of factories; and in roofs, faces, columns, walls, ceilings and other parts of buildings. Titanium alloys have also become popular in body implants, such as artificial hips and knees, because they are light, strong, long-lasting, and compatible with body tissues and fluids.

The most important compound of titanium commercially is titanium dioxide (TiO_2), whose primary application is in the manufacture of white paint. About half the titanium dioxide made in the United States annually goes to this application. Another 40% of all titanium dioxide produced is used in the manufacture of various types of **paper** and plastic materials. The compound gives "body" to paper and makes it opaque. Other uses for the compound are in floor coverings, fabrics and **textiles**, **ceramics**, ink, roofing materials, and catalysts used in industrial operations.

Yet another titanium compound of interest is titanium tetrachloride ($TiCl_4$), a clear colorless liquid when kept in a sealed container. When the compound is exposed to air, it combines with **water** vapor to form a dense white cloud. This property make it useful for skywriting, in the production of smokescreens, and in **motion** picture and **television** programs where smoke effects are needed.

Toadfish

Toadfish are a poorly known group of marine fishes, the vast majority of which live close to the shoreline but remain close to the sea bed. Unlike most **fish**, these animals are extremely vocal, with some authorities even reporting that their loud calls can be heard out of **water**. One North American genus, *Porichthys*, is more commonly known as the singing midshipman. The calls are produced by muscular contractions of the swimming bladder; the intensity of the sound is related to the degree of stimulation by the muscles. Toadfish are renowned for their territorial **behavior**, and it is likely that these loud calls have developed to warn off potential rivals through a series of grunts and postures. As an additional deterrent to potential attackers, toadfish have a number of venomous spines which can inflict an irritating, if not lethal, injection of toxins to a would-be **predator**.

Another peculiar **adaptation** of toadfish is the large number of light-emitting cells, or photophores, that are scattered around the body. A toadfish can have as many as 700 of these specialized **glands**. They are thought to contain large numbers of luminescent **bacteria**, but their purpose is not yet fully understood. In some **species** these lights, which may flash on and off, serve to attract **prey** or mates. However, these features are most often found in deep sea fishes and often those that inhabit the deepest and darkest reaches of the oceans.

Toads

The true toads are **amphibians** in the order Anura, family Bufonidae. There are 355 **species** of toads in 25 genera. The largest group is the genus *Bufo*, which includes 205 species.

Toads are characterized by thick, dry, warty skin, with large poison **glands** on the side of the head and a relatively terrestrial habit as adults. In contrast, adult **frogs** have a smooth, slimy skin, and a more aquatic **habitat**.

Toads have a wide natural distribution, occurring on all major land masses except Australasia, Madagascar, and **Antarctica**. In some regions toads range as far north as the boreal forest, but most species occur in temperate or tropical climatic regimes.

The largest toads are from Central and **South America**, and include *Bufo marinus*, which can reach a body length (excluding the outstretched legs) of 11 in (29 cm), and *B. blombergi* at 10 in (25 cm). The smallest species of toad is a species of *Opeophrynella*, with a body length of only 0.75 in (20 mm).

Biology of toads

Toads are amphibious animals, breeding in **water**, but able to use terrestrial habitats as adults. As adults, toads potentially can live a long time. One of the longest-lived toads was a European common toad (*Bufo bufo*), which survived for 36 years in captivity.

Toads have a complex life cycle, similar to that of frogs. Toads lay their eggs in water, typically inside of long, intertwined strings of a gelatinous material (toad spawn). The eggs hatch into small, dark-colored aquatic larvae (called tadpoles) which have a large head that is not distinctly separated from the rest of the body, internal gills, and a large, flattened tail which undulates sideways to achieve locomotion. Toad larvae are herbivores that feed on **algae**, **bacteria**, and other materials occurring on the surface of vegetation, stones, and sediment.

The larval stage ends with a **metamorphosis** into the adult life form. Adult toads are characterized by internal lungs used for the exchange of respiratory gases, a tailless body, a pelvis that is fused with many of the vertebra to form a bulky structure known as the urostyle, large hind legs adapted for hopping and walking, and small forelegs used to achieve stability, and to help with eating by feeding certain types of **prey**, such as long earthworms, into the mouth. Adult toads also have a relatively thick, dry skin that is heavily cornified with the protein keratin, and provides these terrestrial animals with some degree of protection against dehydration and abrasion. Adult toads are exclusively carnivorous, feeding on a wide range of **invertebrates** and other small animals.

During the breeding season, toads congregate in their breeding ponds, sometimes in large numbers. The males have species-specific songs that are used to attract females. The vocalizations range from long, trilling songs, to shorter, barking sounds. Singing is enhanced by the inflatable vocal sac of male toads, which amplifies their sounds by serving as a **resonance** chamber. Mating involves the male sitting atop the female in amplexus, gripping tightly using his forelegs, and fertilizing the ova as they are laid by the female. Once laid, the eggs are abandoned to develop by themselves.

Toads are very fertile. As many as 30,000 eggs may be laid by the cane toad (*Bufo marinus*), and even the much smaller American toad (*B. americanus*) may lay as many as 25,000 eggs.

The skin of all amphibians contains glands that secrete toxic chemicals, used to deter predators. These glands are quite large and well developed in the warty skin of adult toads, especially in the paired structures at the sides of the head, known as the parotoid glands. In large toads, such as the cane toad, the quantity and toxicity of the contained poison is sufficient to kill a naive, large **predator** that foolishly takes a toad into its mouth.

Although toads are relatively resistant to dehydration, they are by no means immune to this stress. Adult toads tend to stay in a cool, moist, secluded place during the day, emerging to feed at night when the risk of dehydration is less. When the opportunity arises, toads will sit in a pool of water to rehydrate their body.

Adult toads use their tongues when feeding. The tongue is attached near the front of the mouth, and can be flipped forward to use its sticky surface to catch a prey **animal**, which is then retrieved into the mouth. This method of feeding is known as the "lingual flip," a maneuver that can be executed in less than 0.2 seconds, and is especially useful for capturing relative small, flying prey, such as **insects**.

Toads will attempt to eat any moving object of the right size. However, some prey are considered distasteful by toads, and are spat out.

Toads of North America

The only genus of true toads in **North America** is *Bufo*, of which 17 species occur north of Mexico. One of the most widespread species is the American toad (*Bufo americanus*), an abundant animal in eastern North America. The common or Woodhouse toad (*B. woodhousei*) is most abundant in the eastern and central United States, but occurs as far west as California. The great plains toad (*B. cognatus*) occurs in moist places throughout the **prairie** region. The western toad (*B. boreas*) is a widespread, western species.

Some other toad-like animals also bear mentioning, even though they are not "true" toads, that is, they are not in the family Bufonidae.

There are three species of narrow-mouthed toads (family Microhylidae) in southern North America. These are burrowing, nocturnal amphibians that specialize in eating **ants**, and are rarely seen outside of their breeding season. The eastern narrow-mouthed toad (*Gastrophryne carolinensis*) occurs in the southeastern states, while the western narrow-mouthed toad (*G. olivacea*) has a southwestern distribution.

The narrow-toed toads (family Leptodactylidae) lack webbing between their toes. These animals lay their eggs on land, with the tadpole developing within the egg, and a small toad hatching directly from the egg. The only exception is the white-lipped toad (*Leptodactylus labialis*) of Mexico, which lays its eggs near water, into which the tadpoles wriggle as soon as they hatch. All of the seven North American species of narrow-toed toads have restricted distributions. The greenhouse toad (*Eleutherodactylus planirostris*) occurs over much of peninsular Florida, while the barking toad (*Hylactophryne augusti*) occurs locally in Texas.

There is one species of bell toad (family Leiopelmatidae), the tailed toad (*Ascaphus truei*) of the northwestern states and adjacent Canada. The "tail" of the males is actually a copulatory **organ**, used to achieve the unusual, internal **fertilization** of this species.

There are five species of spadefoot toads (family Pelobatidae) in North America. These are nocturnal animals of relatively dry habitats, which breed opportunistically in the springtime if recent rains have provided them with appropriate, temporary ponds. The eastern spadefoot (*Scaphiopus holbrooki*) occurs in the southeast of the **continent**, while the plains spadefoot (*Spea bombifrons*) is widespread in the shortgrass prairie of the interior, and the western spadefoot (*Spea hammondi*) occurs in the southwest.

An American toad, *Bufo americanus*, perched in the grass. Photograph by Kenneth H. Thomas. The National Audubon Society Collection/Photo Researchers, Inc. Reproduced by permission.

Toads and humans

Like most creatures of **wetlands** and other relatively specialized habitats, some species of toads have suffered large population declines through losses of their natural habitats. This has been an especially important problem for species whose original distribution was quite restricted, for example, the rare and endangered Houston toad (*Bufo houstonensis*).

One species of toad has become an important pest in some parts of the world. The cane or marine toad (*Bufo marinus*) is a large species, reaching a body length of 11.5 in (29 cm) and weighing as much as 2 lb (1 kg). The cane toad is native to parts of Central America, but it has been introduced to many other subtropical and tropical regions in misguided attempts to achieve a measure of biological control over some insects that are agricultural **pests**.

The cane toad now occurs in **Australia**, New Guinea, Florida, and many other places, especially tropical islands. Cane toads have become serious pests in

KEY TERMS

Complex life cycle—A life marked by several radical transformations in anatomy, physiology, and ecology.

Metamorphosis—A marked anatomical and physiological transformation occurring during the life of an individual organism. This term is used to describe the changes occurring when a larval amphibian transforms into the adult form.

many of their introduced habitats, in part because of great damages that are caused to the breeding populations of large species of **birds**, lizards, and other predators, including domestic **cats** and dogs. These animals are often naive to the dangers of attempting to eat the poisonous cane toad, and they can therefore be killed in large numbers. The parotoid glands of the cane toad can emit a large quantity of a frothy substance containing noxious chemicals that block neurotransmission, and can lead to death by paralysis.

In small doses, the cane toad secretions can cause a hallucinogenic effect in humans, and people are known to lick the parotoid glands of cane toads in order to achieve this effect. However, it is easy to receive too large a dose of this chemical, and people have been made significantly sick, and have even died, while trying to get high in this manner.

Resources

Books

Conant, Roger, et al. *A Field Guide to Reptiles & Amphibians of Eastern & Central North America (Peterson Field Guide Series).* Boston: Houghton Mifflin, 1998

Harris, C.L. *Concepts in Zoology.* New York: HarperCollins, 1992.

Zug, George R., Laurie J. Vitt, and Janalee P. Caldwell. *Herpetology: An Introductory Biology of Amphibians and Reptiles.* 2nd ed. New York: Academic Press, 2001.

Bill Freedman

Tobacco plant *see* **Nightshade**

Tomato family

The tomato, or **nightshade** family (Solanaceae), contains about 85 genera and 2,300 **species**. Most of the species are tropical or subtropical in distribution, and a

Tomatoes in the field. *Photograph by Robert J. Huffman. Field Mark Publications. Reproduced by permission.*

few are of great economic importance. The region of greatest species richness is Central and **South America**, but representatives occur on all the habitable continents.

Plants in the Solanaceae family can be herbaceous annuals or perennials, shrubs, lianas (vines), or trees. They have alternate, simple leaves. The flowers are bisexual, meaning they contain both male (stamens) and female (pistil) organs. The flowers of most species have radial **symmetry**, and are pentamerous, meaning their **flower** parts occur in groups of five: there are normally five petals, sepals, pistils, and stamens. The individual flowers are aggregated into an inflorescence known as a cyme (a somewhat flat-topped cluster, in which the topmost flowers on the axis bloom first). The fruit is a many-seeded berry or a capsule.

Several species of Solanaceae are economically important food **crops**. The **potato** (*Solanum tuberosum*) is indigenous to the high plateau of the Andes region of South America, where it has been cultivated for at least 2,500 years. Wild potatoes have small, rounded, starchy, underground rhizomes known as tubers; these can weigh as much as several pounds (1 kg) in some domesticated varieties. Potatoes are a basic, carbohydrate-rich food, and are important in the diet of many people around the world. Another important cultivated species is the tomato (*Lycopersicum esculentum*), native to Central America and prized for its red, fleshy, vitamin-rich **fruits** (botanically, these are a berry). Less commonly eaten is the aubergine or eggplant (*Solanum melongena*), whose large, purplish, edible berry is usually eaten cooked. The numerous varieties of chili, or red **pepper** (*Capsicum annuum*), used to give a hot, spicy flavor to many foods, are also part of the Solanaceae family.

Other species of Solanaceae are used as medicinals and recreational drugs. Various species of thorn-apple (*Datura* species, including *Datura stramonium* of **Europe** and **North America**) are used to manufacture the **alkaloid** drugs scopolamine and hyoscine. The deadly nightshade (*Atropa belladonna*) provides similar alkaloids. The **seeds** and leaves of these plants are very poisonous if eaten. Tobacco (*Nicotiana tabacum*) is a widely used recreational drug, containing the extremely addictive alkaloid, **nicotine**.

Many species from the tomato family are used as ornamental plants because of their beautiful, colorful flowers. The most commonly cultivated plants are petunia hybrids (*Petunia x hybrida*). Also grown are the apple-of-Peru (*Nicandra physalodes*) and flowering tobacco (various species of *Nicotiana*).

Bill Freedman

Tongue worms

Tongue worms are bloodsucking endoparasites with a flattened, tongue-like body, and are in the phylum Linguatulida. The final host of these **parasites** is a predaceous vertebrate, usually a reptile, but sometimes a mammal or a bird. The intermediate host (in which the parasite lives as a larva) can be any of a number of animals, including **insects**, **fish**, **amphibians**, **reptiles**, and **mammals**. The 70 **species** of tongue worms are primarily tropical and subtropical in distribution.

Adult tongue worms live in the lungs or nasal passages of the host. Mature females reach a length of up to 15 cm (6 in) and a width of 0.4 in (1 cm), while males are 0.8-1.6 in (2-4 cm) long and 3-4 mm wide. The mouth is sited at the anterior end of the worm, and has a set of hooks for attachment to the tissues of the host. In most species of the tongue worm, the body has superficial ringlike markings, giving it the appearance of seg-

mentation. The only internal body structures are the digestive and reproductive organs, and certain **glands**. Respiratory, circulatory, and excretory organs are absent.

An typical life cycle of a tongue worm is that of *Porocephalus crotali*, whose adult stage lives and reproduces in the lungs of the rattlesnake, and whose larval stage is found in the **muskrat** and other hosts which constitute the rattlesnake's food. Following copulation, the female tongue worm releases her eggs into the host's lungs. The eggs leave the host's body in its respiratory secretions, and might be accidentally ingested by the intermediate host. In the digestive tract of the intermediate host, the eggs hatch to produce four-legged larvae which superficially resemble the immature stage of a mite. The larva penetrates the wall of the host's digestive tract, migrates to another **tissue** such as the liver, and encysts. If the intermediate host is now eaten by a rattlesnake, the parasite is transferred to the snake's digestive tract. Here it emerges from the cyst and migrates up the host's esophagus and into its lungs, where it eventually matures. *Armillifer* and *Linguatula* are two other well known genera of tongue worms, each with different final and intermediate hosts, but with an essentially similar life cycle.

Zoologists consider linguatulids to be closely related to the arthropoda, and in some classifications the tongue worms are included in that phylum.

Tonsillitis

Tonsillitis is an **inflammation** or **infection** of the tonsils, caused by either **bacteria** or viruses. The tonsils usually become swollen and very painful, making swallowing difficult. Sometimes the tonsils have spots of exudates (usually leukocytes) or pus on the surface. The surface of the tonsils is covered by recesses or crypts (*cryptae tonsillares*) that may branch and extend deep into the tonsil. The crypts often hold a variety of bacteria, some of which can cause tonsillitis.

The tonsils are conglomerations of lymphoid nodules, covered by a mucous **membrane**. They help to provide immune responses by, for example, providing lymphocytes and antibodies. There are three types of tonsils that form an irregular, circular band (Waldeyer's ring) around the throat, at the very back of the mouth. Tonsils are strategically located, guarding the opening to the digestive and respiratory tracts. The pharyngeal tonsils (adenoids) are embedded in the posterior wall of the pharynx. The lingual tonsils are located at the base of the tongue, and the palatine tonsils are the oval shaped masses situated on both sides of the throat.

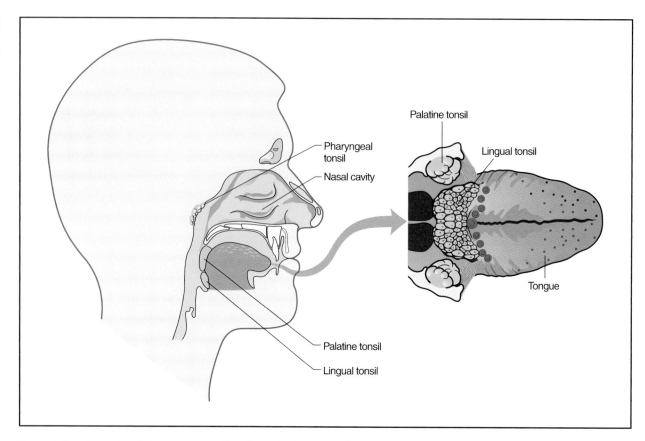

The palatine, lingual, and pharyngeal tonsils. *Illustration by Hans & Cassidy. Courtesy of Gale Group.*

It is usually the prominent, palatine tonsils that become infected or inflamed, and these are the tonsils that are removed surgically (tonsillectomy) when medically indicated, to treat chronic tonsillitis or repeated episodes of acute tonsillitis. Occasionally the pharyngeal tonsils or adenoids are also removed surgically (adenotonsillectomy) if they cause **ear** infections, or block the Eustachian tubes or nasal passages. Adenoviruses (upper respiratory infection or "common cold") are the most common cause of viral tonsillitis. The **Epstein-Barr virus** (infectious mononucleosis) can also cause tonsillitis. The most common bacteria to cause tonsillitis are any of the group A streptococcus bacteria (*Streptococcus pyogenes*). Bacterial tonsillitis is treated with **antibiotics**. However, antibiotics are ineffective with viral tonsillitis. Symptoms can be relieved with aspirin (or other anti-inflammatories), gargling with **saltwater**, and rest.

Topology

Topology, which is often described as "rubber-sheet geometry," is a branch of **geometry** that focuses on dis-

tortion. Topology describes mathematically the features of a geometric shape that do not change when the shape is twisted, stretched, or squeezed. Tearing, cutting, and combining shapes do not apply to topology. Topology helps to solve problems about determining the number of colors necessary to illustrate maps, about distinguishing the characteristics of knots, and about understanding the structure and behavior of DNA molecules.

Topological equivalency

The crucial problem in topology is deciding when two shapes are equivalent. Unlike Euclidean geometry, which focuses on the measurement of distances between points on a shape, topology focuses on the similarity and **continuity** of certain features of geometrical shapes. For example, in Figure 1, each of the two shapes has five points: a through e. The sequence of the points does not change from shape 1 to shape 2, even though the distance between the points, for example, between points b and d, changes significantly because shape 2 has been stretched. Thus the two shapes in Figure 1 are topologically equivalent, even if their measurements are different.

Similarly, in Figure 2, each of the closed shapes is curved, but shape 3 is more circular, and shape 4 is a

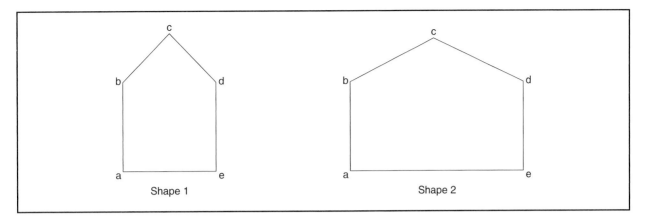

Figure 1. Topologically equivalent shapes. *Illustration by Hans & Cassidy. Courtesy of Gale Group.*

flattened **circle**, or **ellipse**. However, every point on shape 3 can be mapped or transposed onto shape 4.

Shapes 1 and 2 are both topologically equivalent to each other, as are shapes 3 and 4. That is, if each were a rubber band, it could be stretched or twisted into the same shape as the other without connecting or disconnecting any of its points. However, if either of the shapes in each pair is torn or cut, or if any of the points in each pair join together, then the shapes are not topologically equivalent. In Figure 3, neither of the shapes is topologically equivalent to any of the shapes in Figures 1 or 2, nor are shapes 5 and 6 equivalent to each other. The circles in shape 5 are fused; and the triangle in shape 6 has a broken line hanging from its apex.

Famous topologists

Topological ideas can be traced back to Gottfried Wilhelm Leibniz (1646-1716), but three of the most famous figures in the development of topology are Möbius, Riemann, and Klein.

Augustus Ferdinand Möbius (1790-1868) is best known for his invention of the **Möbius strip**, which is a simple strip of **paper** that is twisted and connected so that it has only one side. Normally, cutting a strip of paper into a long, narrow **rectangle** and connecting the ends will result in a belt-like loop with two sides. A person cannot draw a single line with a pencil on both sides of the belt-like loop without crossing an edge. In constructing the Möbius strip, however, the strip of paper is twisted as it is looped, and the result is a one-sided object.

At first, this one-sided construction seems impossible, but if a person draws a straight, continuous line on the Möbius strip, the line will cover the entire length of both sides of the strip without ever crossing an edge, and it will return to its starting point in one long stroke.

Georg Friedrich Bernhard Riemann (1826-1866) developed some of the most important topological ideas about the stretching, bending, and twisting of surfaces, but he died prematurely at the age of 39 before he could expand significantly upon his ideas.

Felix Klein (1849-1925) is best known for the paradoxical figure which was named after him: the Klein bottle.

The Klein bottle is a one-sided object that has no edge. It is a tapered tube whose neck is bent around to enter the side of the bottle. The neck continues into the base of the bottle where it flares out and rounds off to form the outer surface of the bottle. Like the Möbius strip, any two points on the bottle can be joined by a continuous line without crossing an edge, which gives the impression that the inside and outside of the Klein bottle are continuous.

Classifications

Topological shapes are classified according to how many holes they have. Shapes with no holes at all-spheres, eggs, and convex or concave shapes like bowls—are regarded as genus (or type) 0 shapes.

Genus 1 shapes have one hole in them: a donut (or **torus**), a wedding band, a pipe, or anything with a looped handle (a teacup). Genus 2 shapes have two holes in them, for example, a figure eight. Genus 3 shapes (most pretzels) have three holes in them. And so on.

The determining feature in classifying the topological genus of a shape is deciding if every point in one shape can be transposed or mapped onto a point in the other. Sometimes this process is easy, as in the case of a wedding ring and a donut, which are genus 1 topological shapes. But with complex genus shapes of 4 and above, the determination can be difficult.

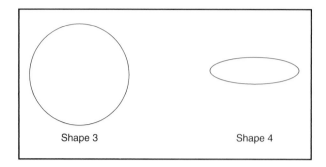

Figure 2. Topologically inequivalent shapes. *Illustration by Hans & Cassidy. Courtesy of Gale Group.*

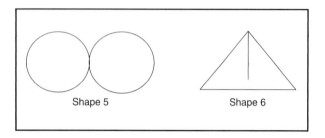

Figure 3. Topologically inequivalent shapes. *Illustration by Hans & Cassidy. Courtesy of Gale Group.*

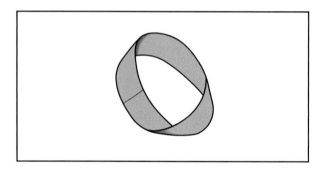

Figure 4. A Möbius strip. *Illustration by Hans & Cassidy. Courtesy of Gale Group.*

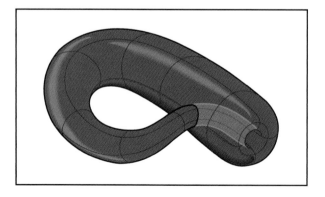

Figure 5. A Klein bottle. *Illustration by Hans & Cassidy. Courtesy of Gale Group.*

KEY TERMS

Dendrimer molecules—Branching, treelike polymers synthesized in laboratories by attaching the polymers to a core molecule.

DNA—Deoxyribonucleic acid, a thin, ladderlike molecular structure which contains the genetic code that defines life.

Euclidean geometry—Geometry based upon the postulates of the Greek educator, Euclid who lived roughly 300 B.C. Euclid's parallel postulate that only one line may be drawn through a given point such that it is parallel to another line, was overthrown in the nineteenth century and gave way to the nonEuclidean geometries created by Gauss and Riemann.

Geometry—The area of mathematics that deals with deductions concerning the measurements and relationships between points, lines, angles, and figures in physical space.

Polymers—Strings of smaller molecules.

Current research

Topology has a number of interesting applications, including **molecular biology** and synthesizing new chemical compounds to help in **gene therapy**. For example, strands of DNA (deoxyribonucleic acid, which contains the genetic code that defines life) often become knotted. Researchers need to know if the knotted mass of DNA is just one strand of DNA that has wound back upon itself, or if it is several strands of DNA which have become entangled. Topology, especially knot theory, helps molecular biologists solve such problems of equivalency.

Topology also has applications in synthesizing new molecules, called dendrimers, which may be used in **gene** therapy. Dendrimers are branching, tree-like polymers (strings of smaller molecules) synthesized in laboratories by chemically attaching the polymers to a core **molecule**. Dendrimers are approximately the same size as many naturally occurring molecules, including DNA and **proteins**. In gene therapy, new genetic material needs to be transported to the nucleus of a **cell** to replace damaged or defective genes in human cells. Then the repaired cells can reproduce more healthy cells. However, there must be a way to transport the healthy DNA into cells, which is where dendrimers come in. Normally, DNA wraps itself around clusters of proteins called histones. Gene therapists can fool the **immune system** of the human body by substituting dendrimers for naturally occurring clusters of histones and thus transport healthy DNA into damaged cells. Topology is useful in this process, because its purpose is to decide mathematically which shapes are equivalent. The closer that synthetic dendrimers mimic the naturally-occurring histones, then the greater are the chances that the DNA will be transported to create healthy cells.

See also Polymer.

Resources

Books

Ball, W.W. Rouse. *A Short Account of the History of Mathematics.* London: Sterling Publications, 2002.

Kline, Morris. *Mathematical Thought: From Ancient to Modern Times.* 3 vols. New York: Oxford University Press, 1972.

Peterson, Ivars. *The Mathematical Tourist: Snapshots of Modern Mathematics.* New York: W. H. Freeman, 1988.

Periodicals

Tomalia, Donald A. "Dendrimer Molecules." *Scientific American* 272 (May 1995): 62-66.

Patrick Moore

Torino scale *see* **NEAR-Earth Object Hazard Index**

Tornado

A tornado is a rapidly spinning column of air formed in severe thunderstorms. The rotating column, or vortex, forms inside the **storm** cloud then grows downward until it touches the ground. Although a tornado is not as large as its parent **thunderstorm**, it is capable of extreme damage because it packs very high **wind** speeds into a compact area. Tornadoes have been known to shatter buildings, drive straws through solid **wood**, lift locomotives from their tracks, and pull the **water** out of small streams. Due to a combination of geography and **meteorology**, the United States experiences most of the world's tornadoes. An average of 800 tornadoes strike the United States each year. Based on **statistics** kept since 1953, Texas, Oklahoma, and Kansas are the top three tornado states. Tornadoes are responsible for about 80 deaths, 1500 injuries, and many millions of dollars in property damage annually. While it is still impossible to predict exactly when and where tornadoes will strike, progress has been made in predicting tornado development and detecting tornadoes with Doppler **radar**.

Tornado formation

Most tornadoes form in the northern hemisphere during the months of March through June. These are months when conditions are right for the development of severe thunderstorms. To understand why tornadoes form, consider the formation and growth of a thunderstorm. Thunderstorms are most likely to develop when the atmosphere is unstable; that is when **atmospheric temperature** drops rapidly with height. Under unstable conditions, air near the surface that begins rising will expand and cool, but remains warmer (and less dense) than its surroundings. The rising air acts like a hot air **balloon**; since it is less dense than the surrounding air it continues to rise. At some point the rising air cools to the **dew point** where the water vapor in the air condenses to form liquid water droplets. The rising column of air is now a visible cloud. If the rising air, or updraft, is sustained long enough water droplets will begin to fall out of the rising air column, making it a rain cloud.

This cloud will become a severe storm capable of producing tornadoes only under certain circumstances. Severe storms are often associated with a very unstable atmosphere and moving low **pressure** systems that bring cold air into contact with warmer, more humid air masses. Such **weather** situations commonly occurs in the eastern and Midwestern United States during the spring and summer months. Large scale weather systems often sweep moist warm air from the Gulf of Mexico over these regions in a layer 1.2-1.9 mi (2-3 km) deep. At the same time winds aloft (above about 2.5 mi [4 km] in altitude) from the southwest bring cool dry air over the region. Cool air overlying humid air creates very unstable atmospheric conditions and sets the stage for the growth of strong thunderstorms.

The warm surface air is separated from colder air lying farther north by a fairly sharp **temperature** boundary called a front. A low pressure center near the earth's surface causes the cold air to advance into the warmer air. The edge of the advancing cold air, called a cold front, forces the warmer air ahead of the front to rise and cool. Since the atmosphere is so unstable the displaced air keeps rising and a cloud quickly forms. Rain that begins to fall from the cloud causes downdrafts (sinking air) in the rear of the cloud. Meanwhile the advancing edge of the storm has strong updrafts and humid air is pulled into the storm. The water vapor in this air condenses to form more water droplets as it rises and cools. When water vapor condenses it releases latent **heat**. This warms the air and forces it to rise more vigorously, strengthening the storm.

The exact mechanism of tornado formation inside severe thunderstorms is still a matter of dispute but it appears that tornadoes grow in a similar fashion to the small vortices that form in draining bathtubs. When the plug is pulled in a bathtub, water from other parts of the tub **rushes** in to replace that going down the drain. If the water has any swirl in it, the drain soon has a little vortex.

Tornadoes appear to be upside down versions of this phenomenon. As updrafts in a severe thunderstorm cloud get stronger, more air is pulled into the base of the cloud to replace the rising air. Some of this air may be rotating slightly since the air around the base of a thunderstorm usually contains some **rotation**, or vorticity. As the air converges into a smaller area it begins to rotate faster due to a law of **physics** known as the **conservation** of angular **momentum**. This effect can be seen when an **ice** skater begins spinning with arms outstretched. As the skater brings his or her arms inward, his or her rotational speed increases. In the same way air moving into a severe storm begins in a tighter column and increases its rotational speed. A wide vortex is created, called the mesocyclone. The mesocyclone begins to build vertically, extending itself upward throughout the entire height of the cloud. The rapid air movement causes the surrounding air pressure to drop, pulling more air into growing vortex. The lowered pressure causes the incoming air to cool quickly and form cloud droplets before they rise to the cloud base. This forms the wall cloud, a curtain-shaped cloud that is often seen before a tornado forms. The mesocyclone continues to contract while growing from the base of the storm cloud all the way up to 6.2 mi (10 km) above the surface. When the mesocyclone dips below the wall cloud it is called a funnel cloud because of its distinctive funnel shape. This storm is on its way to producing a tornado.

Tornado characteristics

A funnel cloud may form in a severe storm and never reach the ground. If and when it does, the funnel officially becomes a tornado. The central vortex of a tornado is typically about 328.1 ft (100 m) in diameter. Wind speeds in the vortex have been measured at greater than 220 mph (138 km/h). These high winds make incredible feats of destruction possible. They also cause the air pressure in the tornado to drop below normal **atmospheric pressure** by over 100 millibars (the normal day-to-day pressure variations we experience are about 15 millibars). The air around the vortex is pulled into this low pressure zone where it expands and cools rapidly. This causes water droplets to condense from the air, making the outlines of the vortex visible as the characteristic funnel shaped cloud. The low pressure inside the vortex picks up debris such as **soil** particles, which may give the tornado an ominous dark **color**. A tornado can act as a giant vacuum cleaner sweeping over anything unlucky enough to be in its path. The damage path of a tornado may range from 900 ft (300 m) to over 0.5 mi (1 km) wide.

Tornadoes move with the thunderstorm that they are attached to, traveling at average speeds of about 10-30 mph (15-45 kph), although some tornadoes have been seen to stand still, while other tornadoes have been clocked at 60 mph (90 kph). Since a typical tornado has a lifetime of about five to 10 minutes, it may stay on the ground for 5-10 mi (8-16km). Occasionally, a severe tornado may cut a path of destruction over 200 mi (320 km) long. Witnesses to an approaching tornado often describe a loud roaring noise made by the storm similar to jet engines at takeoff. There is no generally accepted explanation for this phenomenon although it has been suggested that supersonic winds inside the vortex cause it.

The destructive path of tornadoes appears **random**. One house may be flattened while its neighbor remains untouched. This has been explained by the tornado "skipping"—lifting up off the surface briefly then descending again to resume its destructive path. Studies made of these destructive paths after the storm suggest another possible explanation: tornadoes may have two to three smaller tornado-like vortices circling around the main vortex like **horses** on a merry-go-round. According to this theory, these "suction vortices" may be responsible for much of the actual damage associated with tornadoes. As they rotate around the main tornado core they may hit or miss objects directly in the tornado's path depending on their position. Thus if two houses were in the tornado path one may be destroyed by a suction vortex but the vortex had moved into a different position (and the next vortex had not yet taken its place) by the time it reached the next house. The tornado's skipping behavior is still not completely understood.

When houses or other structures are destroyed by a tornado, they are not simply blown down by the high winds: they appear to explode. For many years it was believed that the low pressure of the tornado vortex caused such explosions. According to this theory, if the pressure outside a building drops very quickly the air inside may not escape fast enough (through cracks, holes, and the like) to equalize the pressure difference. The higher pressure inside the building then pushes out windows or whole walls, and the structure looks like it had exploded. Studies of tornado damage have shown that buildings do not actually explode in this manner. Instead, high wind passing over a house roof acts like the air moving over an airplane wing: it gives the roof an upward **force** or lift which tends to raise the roof vertically off the house. Winds also enter the building through broken windows or doors pressurizing the house as one would blow up a balloon. The combination of these forces tends to blow

An approaching tornado with its distinctive funnel visible. *Photograph by Howard Bluestein. Photo Researchers, Inc. Reproduced by permission.*

the walls and roof off the structure from the inside out giving the appearance of an explosion.

Tornado strength is classified by the Fujita scale, which uses a scale of one to six to denote tornado wind speed. Since direct measurements of the vortex are not possible the observed destruction of the storm is used to estimate its "F scale" rating.

Tornado history

Prior to 2003, the single most violent tornado in United States history was the Tri-State tornado on March 18, 1925. Beginning in Missouri, the tornado stayed on the ground for over 220 mi (350 km), crossing Illinois, moving into Indiana, and leaving a trail of damage over one mile (1.6 km) wide in places. Tornado damage often is limited since they usually strike unpopulated areas, but the Tri-State tornado plowed through nine towns and destroyed thousands of homes. When the storm was over, 689 people had lost their lives and over 2,000 were injured making the Tri-State the deadliest tornado on record.

On May 3, 1999, a storm started in southwestern Oklahoma, near the town of Lawton. By late in the day,

it had grown into a violent storm system with 76 reported tornadoes. As the storm system tore across central Oklahoma and into Kansas, over 43 people were killed, over 500 injured and more than 1,500 buildings were destroyed. One of the tornadoes, classed as a F-5, was as much as a mile wide at times and stayed on the ground for over four hours.

Another historic storm was the severe tornado outbreak of April 3-4, 1974. As a strong low pressure system moved over the Midwest, an advancing cold front ran into warm Gulf air over the southern states. The resulting storm triggered 148 tornadoes over 13 states in the next 24 hours, some reaching F4 and F5 in strength. As severe as this outbreak was, the death toll was less than half of that from the Tri-State tornado because of advances in tornado forecasting and warnings.

Prediction and tracking of tornadoes

The precise tracking and prediction of tornadoes is not yet a reality. Meteorologists can identify conditions that are likely to lead to severe storms. They can issue warnings when atmospheric conditions are right for the development of tornadoes. They can use radar to track the path of thunderstorms that might produce tornadoes. It is still not possible, however, to detect a funnel cloud by radar and predict its path, touchdown point, and other important details. Much progress has recently been made in the detection of tornadoes using Doppler radar.

Doppler radar can measure not just the distance to an object, but also its **velocity** by using the **Doppler effect**: if an object is moving toward an observer, radar waves bounced off the object will have a higher **frequency** than if the object were moving away. This effect can be demonstrated with **sound waves**. If a car is approaching with its horn sounding, the pitch of the horn (that is, the frequency of the sound waves) seems to rise. It reaches a peak just as the car passes, then falls as the car speeds away from the listener.

Doppler radar is used to detect the **motion** of raindrops and hail in a thunderstorm, which gives an indication of the motion of the winds. With present technology it is possible to detect the overall storm circulation and even a developing mesocyclone. The relatively small size of a tornado makes direct detection very difficult with the current generation of Doppler radar. In addition any radar is limited by the curvature of **Earth**. Radar waves go in straight lines, which means distant storms that are "below the horizon" from the radar cannot be probed with this technique.

Tornadoes, which have long fascinated people with their sudden appearance and awesome destructive power, are still subjects of intense scientific study. Research continues on the formation **life history** and detection of these most impressive storms.

See also Cyclone and anticyclone; Tropical cyclone.

Resources

Books

Battan, Louis J. *Weather.* Engelwood Cliffs: Prentice-Hall Inc., 1985.

Battan, Louis J. *Weather in Your Life.* New York: W.H. Freeman & Co., 1983.

Danielson, Eric W., James Levin, and Elliot Abrams. *Meteorology.* 2nd ed. with CD-ROM. Columbus: McGraw-Hill Science/Engineering/Math, 2002.

Hardy, Ralph, Peter Wright, John Kington, and John Gribben. *The Weather Book.* Boston: Little, Brown and Co., 1982.

Lewellen, W.S. "Tornado Vortex Theory." In *The Tornado: Its Structure, Dynamics and Hazards.* Washington, DC: American Geophysical Union, 1993.

McNeill, Robert. *Understanding the Weather.* Las Vegas: Arbor Publishers, 1991.

Mogil, H. Michael, and Barbara G. Levine. *The Amateur Meteorologist.* New York: Franklin Watts, 1993.

KEY TERMS

Dew point—The temperature at which water vapor in the air condenses to form liquid water droplets.

Doppler radar—A type of radar that measures both the position and the velocity of an object.

Front—A fairly sharp temperature boundary in the lower atmosphere.

Fujita scale—A scale of one to six which rates tornado wind speed based upon the observed destruction of the storm.

Funnel cloud—A fully developed tornado vortex before it has touched the ground.

Latent heat—The heat released when water vapor condenses to form liquid water.

Skipping—The tendency of tornado damage to be random as if the tornado skips along in its path.

Suction vortices—Secondary vortices that are theorized to be part of a tornado vortex. They may be responsible for the "skipping" behavior of tornadoes.

Unstable atmosphere—The condition of the atmosphere when air temperature drops rapidly with height. Such conditions support rising air and contribute to strong thunderstorms.

Vortex—A rotating column of a fluid such as air or water.

Vorticity—The tendency of an air mass to rotate.

Wall cloud—The characteristic cloud that forms at the base of a thunderstorm before a funnel cloud appears.

Periodicals

Schmidlin, Thomas. "Unsafe At Any (Wind) Speed." *Bulletin of the American Meteorological Society* 83, no. 12 (2002): 1821-1830.

James Marti

Torque

According to Isaac Newton, an object at rest will remain at rest, and an object will remain in **motion** unless acted upon by an outside **force**. A force, therefore, is what causes any object to move. Any force which causes an object to rotate, turn, or twist is called a torque.

Torque is equal to the amount of force being exerted on the object times the object's **rotation** point to the location where the force is being applied on the object.

Seesaws are a good example of torque. Many people have had the experience of a large person sitting on one end of the seesaw, and a small person on the other end. If the larger person is sitting closer to the pivot point of the seesaw, the smaller person can lift them with little or no problem. The reason this is possible comes from the difference in torque experienced by each person. Even though the smaller person exerts a smaller force, their distance from the pivot point (lever arm) is longer, hence a large torque. The larger person exerts a larger force; however, because he/she is closer to the pivot point their lever arm is shorter, hence a smaller torque.

Wrenches also work by torque. (Some wrenches are even calibrated to display the amount of torque you are applying to a **nut**; they are called torque wrenches.) The nut (or bolt) is the point of rotation because we either want to tighten or loosen it by having it turn. The force is being exerted by your hand and arm. Since we try to pull or push (exert a force) at right angles on the wrench's handle, the lever arm is then the length of the wrench's handle. To increase the torque on the nut, we must either increase how hard we pull or push on the lever arm; or, increase the length of the lever arm by placing a pipe over the end of the wrench.

Torus

A torus is a doughnut-shaped, three-dimensional figure formed when a **circle** is rotated through 360° about a line in its **plane**, but not passing through the circle itself. Imagine, for example, that the circle lies in space such that its diameter is **parallel** to a straight line. The figure that is formed is a hollow, circular tube, a torus. A torus is sometimes referred to as an anchor ring.

The surface area and **volume** of a torus can be calculated if one knows the radius of the circle and the radius of the torus itself, that is, the distance from the furthest part of the circle from the line about which it is rotated. If the former dimension is represented by the letter r, and the latter dimension by R, then the surface area of the torus is given by $4\pi^2 Rr$, and the volume is given by $2\pi^2 Rr_2$.

Problems involving the torus were well known to and studied by the ancient Greeks. For example, the formula for determining the surface area and volume of the torus came about as the result of the work of the Greek mathematician Pappus of Alexandria, who lived around the third century A.D. Today, problems involving the torus are of special interest to topologists.

See also Topology.

Total solar irradiance

Total solar irradiance is defined as the amount of radiant **energy** emitted by the **Sun** over all wavelengths that fall each second on 11 sq ft (1 sq m) outside the earth's atmosphere.

By way of further definition, irradiance is defined as the amount of electromagnetic energy incident on a surface per unit time per unit area. *Solar* refers to electromagnetic **radiation** in the spectral range of approximately 1-9 ft (0.30-3 m), where the shortest wavelengths are in the ultraviolet region of the **spectrum**, the intermediate wavelengths in the visible region, and the longer wavelengths are in the near infrared. *Total* means that the solar flux has been integrated over all wavelengths to include the contributions from ultraviolet, visible, and infrared radiation.

By convention, the surface features of the Sun are classified into three regions: the photosphere, the chromosphere, and the corona. The photosphere corresponds to the bright region normally visible to the naked **eye**. About 3,100 mi (5,000 km) above the photosphere lies the chromosphere, from which short-lived, needle-like projections may extend upward for several thousands of kilometers. The corona is the outermost layer of the Sun; this region extends into the region of the planets. Most of the surface features of the Sun lie within the photosphere, though a few extend into the chromosphere or even the corona.

The average amount of energy from the Sun per unit area that reaches the upper regions of the earth's atmosphere is known as the solar constant; its value is approximately 1,367 watts per square meter. As earth-based measurements of this quantity are of doubtful accuracy due to variations in the earth's atmosphere, scientists have come to rely on satellites to make these measurements.

Although referred to as the solar constant, this quantity actually has been found to vary since careful measurements started being made in 1978. In 1980, a satellite-based measurement yielded the value of 1,368.2 watts per square meter. Over the next few years, the value was found to decrease by about 0.04% per year. Such variations have now been linked to several physical processes known to occur in the Sun's interior, as will be described below.

From the **earth**, it is only possible to observe the radiant energy emitted by the Sun in the direction of our **planet**; this quantity is referred to as the solar irradiance. This radiant solar energy is known to influence the earth's **weather** and climate, although the exact relationships between solar irradiance and long-term climatological changes, such as **global warming**, are not well understood.

The total radiant energy emitted from the Sun in all directions is a quantity known as solar luminosity. The luminosity of the Sun has been estimated to be 3.8478×10^{26} watts. Some scientists believe that long-term variations in the solar luminosity may be a better correlate to environmental conditions on Earth than solar irradiance, including global warming. Variations in solar luminosity are also of interest to scientists who wish to gain a better understanding of stellar **rotation**, **convection**, and **magnetism**.

Because short-term variations of certain regions of the solar spectrum may not accurately reflect changes in the true luminosity of the Sun, measurements of total solar irradiance, which by definition take into account the solar flux contributions over all wavelengths, provide a better representation of the total luminosity of the Sun.

Short-term variations in solar irradiation vary significantly with the position of the observer, so such variations may not provide a very accurate picture of changes in the solar luminosity. But the total solar irradiance at any given position gives a better representation because it includes contributions over the spectrum of wavelengths represented in the solar radiation.

Variations in the solar irradiance are at a level that can be detected by ground-based astronomical measurements of **light**. Such variations have been found to be about 0.1% of the average solar irradiance. Starting in 1978, space-based instruments aboard the *Nimbus 7, Solar Maximum Mission*, and other satellites began making the sort of measurements (reproducible to within a few parts per million each year) that allowed scientists to acquire a better understanding of variations in the total solar irradiance.

Variations in solar irradiance have been attributed to the following solar phenomena: **oscillations**, granulation, **sunspots**, faculae, and solar cycle.

Oscillations, which cause variations in the solar irradiance lasting about five minutes, arise from the action of resonant waves trapped in the Sun's interior. At any given time, there are tens of millions of frequencies represented by the resonant waves, but only certain oscillations contribute to variations in the solar constant.

Granulation, which produces solar irradiance variations lasting about 10 minutes, is closely related to the convective energy flow in the outer part of the Sun's interior. To the observer on Earth, the surface of the Sun appears to be made up of finely divided regions known as granules, each from 311-1,864 mi (500-3000 km) across, separated by dark regions. Each of these granules makes its appearance for about 10 minutes and then disappears. Granulation apparently results from convection effects that appear to cease several hundred kilometers below the visible surface, but in fact extend out into the photosphere, i.e., the region of the Sun visible to the naked eye. These granules are believed to be the centers of rising convection cells.

Sunspots give rise to variations that may last for several days, and sometimes as long as 200 days. They actually correspond to regions of intense magnetic activity where the solar atmosphere is slightly cooler than the surroundings. Sunspots appear as dark regions on the Sun's surface to observers on Earth. They are formed when the magnetic **field** lines just below the Sun's surface become twisted, and then poke though the solar photosphere. Solar irradiance measurements have also shown that the presence of large groups of sunspots on the Sun's surface produce dips ranging in amplitude from 0.1-0.25% of the solar constant. This reduction in the total solar irradiance has been attributed both to the presence of these sunspots and to the temporary storage of solar energy over times longer than the sunspot's lifetime. Another key observation has been that the largest decreases in total solar irradiance frequently coincide with the formation of newly formed active regions associated with large sunspots, or with rapidly evolving, complex sunspots. Sunspots are especially noteworthy for their 11-year activity cycle.

Faculae, producing variations that may last for tens of days, are bright regions in the photosphere where high-temperature interior regions of the Sun radiate energy. They tend to congregate in bright regions near sunspots, forming solar active regions. Faculae, which have sizes on the order of 620 mi (1000 km) or less, appear to be tube-like regions defined by magnetic field lines. These regions are less dense than surrounding areas. Because radiation from hotter layers below the photosphere can leak through the walls of the faculae, an atmosphere is produced that appears hotter, and brighter, than others.

The solar cycle is responsible for variations in the solar irradiance that have a period of about 11 years. This 11-year activity cycle of sunspot **frequency** is actually half of a 22-year magnetic cycle, which arises from the reversal of the poles of the Sun's magnetic field. From one activity cycle to the next, the north magnetic pole becomes the south magnetic pole, and vice versa. Solar luminosity has been found to achieve a maximum value at the very time that sunspot activity is highest during the 11-year sunspot cycle. Scientists have confirmed the length of the solar cycle by examining **tree** rings for variations in deuterium-to-hydrogen ratios. This **ratio** is temperature-dependent because **deuterium** molecules, which are a heavy form of the **hydrogen molecule**, are less mobile than the lighter hydrogen molecules, and therefore less responsive to thermal **motion** induced by increases in the solar irradiance.

Surprisingly, the Sun's rotation, with a rotational period of about 27 days, does not give rise to significant variations in the total solar irradiance. This is because its effects are overridden by the contributions of sunspots and faculae.

Global warming

Scientists have speculated that long-term solar irradiance variations might contribute to global warming over decades or hundreds of years. More recently, there has been speculation that changes in total solar irradiation have amplified the **greenhouse effect**, i.e., the retention of solar radiation and gradual warming of the earth's atmosphere. Some of these changes, particularly small shifts in the length of the activity cycle, seem to correlate rather closely with climatic conditions in pre- and post industrial times. Whether variations in solar irradiance can account for a substantial fraction of global warming over the past 150 years, however, remains a highly controversial point of scientific discussion.

Some researchers are convinced solar irradiance has increased between 1986-1996 (the years of the twentieth century's last two solar minima) and this increase is consistent with the conclusion that long term solar irradiance changes are occurring. But other scientists disagree, citing data inconsistent with such a conclusion. In particular, they have reported that solar irradiance was at similar levels in the years 1986 and 1996, but the global surface **temperature** of Earth had increased by about 0.2°C during the same decade. Although researchers disagree about whether recent changes in the total solar irradiance can account for global warming between 1986-1996, most agree that long-term solar irradiance measurements will help elucidate the role the Sun actually plays in driving **global climate** changes.

Measuring solar irradiance

Measurements of solar irradiance can be characterized by the range of wavelengths (or frequencies) they are sensitive to. The three types of measurements are broadband, wideband, and narrowband.

Broadband measurements

Broadband measurements typically record the complete solar spectrum. Quantities typically obtained in these types of measurements include:

- *Direct solar irradiance*, defined as the solar radiation that passes directly though the atmosphere from the Sun without being scattered or absorbed by the atmosphere. Scientists usually use pyrheliometers to measure this quantity, though more accurate measurements can be obtained using absolute cavity radiometers.

- *Diffuse sky solar irradiance* is the solar irradiance that reaches the ground after being scattered by particles in the atmosphere, including air molecules, dust, or cloud particles. To measure this quantity, scientists use a pyranometer that does not register the effects of the direct solar irradiance.

- *Downward total solar irradiance* is the total amount of solar irradiance that reaches an upward-facing horizontal surface. It is the sum of the vertical component of the direct solar irradiance and the diffuse sky irradiance. It is measured either with a pyranometer, or alternatively by summing the direct and diffuse horizontal irradiance.

- *Upward solar irradiance* is the solar irradiance that reaches a downward-facing surface. The source of this quantity is the downward solar irradiance that is reflected off the earth's surface. This quantity is measured with an inverted pyranometer.

- *Downward longwave irradiance* is thermal irradiance emitted in all directions by the atmosphere, e.g., gases, **aerosols**, and **clouds**, as received by an horizontal upward facing surface. It is measured with a pyrgeometer.

- *Upward longwave irradiance* is the thermal irradiance emitted from the earth's surface that passes through a horizontal surface at a representative distance above the ground. It is measured with an inverted pyrgeometer.

Wideband measurements

Wideband measurements typically focus on a region of the solar spectrum on the order of 10% that seen in broadband studies.

- Direct solar irradiance can be measured with a pyrheliometer equipped with suitable filters.

- Downward solar irradiance can be measured with a pyranometer equipped with an appropriate filter.

Narrowband measurements

Narrowband measurements cover a very narrow range of the solar spectrum.

- Direct, diffuse, and total solar irradiance measurements can be made using a radiometer.

- Direct solar irradiance measurements can be made using a Sunphotometer.

See also Global climate; Solar flare.

KEY TERMS

Near infrared radiation—Electromagnetic radiation typically produced by molecules that have been excited with low levels of energy. Near infrared radiation has a range of wavelengths about 2.5-0.75 μm. Such radiation can be detected by photoelectric cells.

Pyranometer—Instrument used to measure the combined intensity of incident direct solar radiation and diffuse sky radiation. It operates by comparing the heat produced by the radiation on blackened metal strips with that produced by a known electric current.

Pyrgeometer—Instrument that measures radiation from the earth's surface transmitted into space.

Pyrheliometer—Instrument that measures the total intensity of direct solar radiation received by the earth.

Radiometer—Instrument that measures radiant energy. An example is the bolometer, which measures the energy of electromagnetic radiation at

certain wavelengths by measuring the change in electrical resistance of a thin conductor due to heat accompanying the radiation.

Sunphotometer—A type of photometer used to observe a narrow range of solar wavelengths. Most instruments produce an output signal proportional to the solar irradiance within the range of wavelengths. Some instruments determine spectral atmospheric transmission, which allows the contributions of various atmospheric constituents, e.g., aerosols, water vapor, and ozone, to be calculated.

Ultraviolet radiation—Radiation similar to visible light but of shorter wavelength, and thus higher energy.

Visible radiation—Also known as light, visible radiation, like all radiation, is produced by acceleration of charged particles, often by excited electrons in atoms or molecules as they lose energy in the process of returning to their normal, or unexcited, state. Range of wavelengths in solar radiation: approximately 0.78-0.4 μm.

Resources

Books

Kitchin, C. R. *Stars, Nebulae and the Interstellar Medium.* Adam Hilger, Bristol and Boston, 1987.

Maran, Stephen P., ed. *The Astronomy and Astrophysics Encyclopedia.* Van Nostrand Reinhold (New York) and Cambridge University Press (Cambridge, England), 1992.

Serway, Raymond, Jerry S. Faughn, and Clement J. Moses. *College Physics.* 6th ed. Pacific Grove, CA: Brooks/Cole, 2002.

Randall Frost

Toucans

Toucans are 42 **species** of familiar, brilliantly colored arboreal **birds** that make up the family Ramphastidae. Toucans are in the order Piciformes, which also includes the **woodpeckers**.

Toucans range from southern Mexico to northern Argentina and Paraguay. Their usual **habitat** is tropical and subtropical **forests** and woodlands, and sometimes more open **savanna** with clumps of trees. Most species occur in lowlands, but some live in higher elevation forests.

Toucans are not migratory, although some species may undertake local movements in search of food.

Toucans are relatively large birds, with a body length of 12-24 in (30-61 cm). Their wings are short and rounded, the tail is long and wide, and the legs and feet are stout and strong, with the toes arranged in a zygodactyl pattern (i.e., two facing forward, and two backward).

The most distinctive feature of toucans is their tremendous, seemingly oversized, laterally compressed bill, which in some species is as long as the body. The bill is stout but well-chambered with air-containing cavities, and therefore surprisingly light in weight. The bill curves slightly downward, and often has a serrated inner margin. There is a small hook at the tip of the upper mandible, and the nostrils are placed at the base of the beak, near the head.

The function of the unusually enormous bill of toucans has not been conclusively determined. It may be helpful in plucking **fruits** that are far away from branches large enough for a toucan to **perch** on. Alternatively, the large bill may be used to frighten away potential predators, or to intimidate the parents of nestlings or eggs that toucans are predating upon. The bill may also have important social functions, for example, in **courtship** displays.

A toucan at Lake Arenal, Costa Rica. *Photograph by Roy Morsch. Stock Market. Reproduced by permission.*

The body plumage of toucans is soft, and usually black or dark green, with conspicuous patches of white, yellow, orange, red, or blue. Toucans have bare, brightly colored skin around the **eye**. The large bill of toucans is brightly colored in most species, and is likely important in species recognition. The sexes of most species do not differ in size or coloration.

Toucans live in small, harshly noisy, often family related groups. They feed together, with an attentive sentry posted to warn against possible intrusions by predators. Toucans mostly eat fruits, **insects**, and sometimes the eggs or nestlings of other species of birds. Toucans can manipulate their foods quite dexterously, in spite of their huge bill.

Toucans have a repertoire of harsh croaks, yelps, mews, rattles, and other sounds. Some of the larger species give a series of almost musical renditions at dawn, which may function as territorial songs.

Toucans roost and nest in cavities in trees, commonly using abandoned holes excavated by woodpeckers, although these may be further enlarged by the toucans. During roosting, several birds may crowd into the same cavity, and when doing this they commonly fold their tail up over the back to save space. Toucans lay two to four eggs in an unlined nesting cavity as high up a **tree** as possible. The eggs are incubated by both parents, who also share the chick-rearing duties, which takes about 40-50 days until fledging occurs.

Species of toucans

The largest toucans are those in the genus *Ramphastos*, which live in lowland, tropical forests. The toco toucan (*R. toco*) occurs in tropical Amazonian forests, while the chestnut-mandibled toucan (*R. swainsonii*) occurs from Honduras to Venezuela.

Species of aracari toucans (*Pteroglossus* spp.) are smaller and relatively gregarious. The green aracari (*Pteroglossus viridis*) occurs in tropical forests from Venezuela through Brazil.

The toucanets are even smaller. The spot-billed toucanet (*Selenidera maculirostris*) occurs in forests from Venezuela and Guyana to northern Argentina.

Most species of toucans are less abundant today than they used to be, mostly because of loss of habitat through **deforestation**. However, the World Conservation Union (IUCN) does not yet list any species as being threatened or endangered.

KEY TERMS

. . . .

Zygodactyl—The specific arrangement of toes of certain birds, in which two toes point forward, and two backward. This is the characteristic arrangement in the Piciformes, including the toucans.

Resources

Books

Forshaw, Joseph. *Encyclopedia of Birds.* New York: Academic Press, 1998.

Sick, H. *Birds in Brazil: A Natural History.* Princeton, NJ: Princeton University Press, 1993.

Bill Freedman

Touch

Touch is one of the five senses (the others being **smell**, **taste**, **vision**, and **hearing**) through which animals and people interpret the world around them. While the other senses are localized primarily in a single area (such as vision in the eyes or taste in the tongue), the sensation of touch (or contact with the outside world) can be experienced anywhere on the body, from the top of the head to the tip of the toe. Touch is based on nerve receptors in the skin which send electrical messages through the central **nervous system** to the cerebral cortex in the **brain**, which interprets these electrical codes. For the most part, the touch receptors specialize in experiencing either hot, cold, **pain**, or **pressure**. Arguably, touch is the most important of all the senses; without it animals would not be able to recognize pain (such as scalding **water**), which would greatly decrease their chances for survival. Research has also shown that touch has tremendous psychological ramifications in areas like child development, persuasion, healing, and reducing **anxiety** and tension.

How we feel the outside world

Our sense of touch is based primarily in the outer layer of skin called the epidermis. Nerve endings that lie in or just below the epidermis cells respond to various outside stimuli, which are categorized into four basic stimuli: pressure, pain, hot, and cold. Animals experience one or a combination of these sensations through a complex neural network that sends electrical impulses through the spinal cord to the cerebral cortex in the brain.

The cerebral cortex, in turn, contains brain cells (neurons) arranged in columns that specialize in interpreting specific types of stimuli on certain parts of the body.

The sensation of touch begins with various receptors in the skin. Although these receptors appear to specialize in reacting to certain sensations, there is some debate concerning this specificity since most touch stimuli are a combination of some or all of the four major categories.

Scientists have identified several types of touch receptors. Free nerve ending receptors, located throughout the body at the bases of hair, are associated primarily with light pressure (such as **wind**) and pain. Meissner corpuscles are nerve endings contained in tiny capsules and are found primarily in the fingertips and areas especially sensitive to touch (in the form of low-frequency vibrations), like the soles of the feet and the tongue. The Pacinian corpuscles look like the **cross section** of an onion and are found in deep tissues in the joints, the genitals, and the mammary **glands**. They are extremely sensitive to pressure and are also stimulated by rapid movement of the tissues and vibrating sensations. Ruffini endings, which are also located in the deeper layers of the skin, respond to continuous stimulation, like steady pressure or tension within the skin. Merkel disks, are found near the base of the epidermis and also respond to continuous stimulation or pressure. The skin also contains specific thermoreceptors for sensing hot and cold and nociceptors that identify high intensity stimulation in the form of pain.

Most, if not all of these receptors, are designed to adapt or become accustomed to the specific stimulation they interpret. In other words, the receptor does not continue to register a constant "feeling" with the same intensity as when it first begins and may even shut off the tactile experience. Imagine, for example, putting on a wool sweater over bare skin. The initial prickly sensation eventually abates, allowing the wearer to become accustomed to the feeling. Other examples include wearing jewelry such as rings, necklaces, and watches.

These receptors are also found in greater numbers on different parts of the body. For example, peoples' backs are the least sensitive to touch, while their lips, tongue, and fingertips are most sensitive to tactile activity. Most receptors for cold are found on the surface of the face while thermoreceptors for warmth usually lie deeper in the skin and are fewer in number. A light breeze on the arm or head is felt because there tend to be more sense receptors at the base of the hairs than anywhere else.

Touch and health

Touch has a tremendous impact on most animals' physical and psychological well being. Numerous studies of humans and other animals have shown that touch greatly impacts how we develop physically and respond to the world mentally. For example, premature babies that receive regular massages will gain weight more rapidly and develop faster mentally than those who do not receive the same attention. When baby **rats** are separated from their mothers for only 45 minutes, they undergo physiological or biochemical changes, specifically a reduction in a growth hormone. Touching of premature babies can also stimulate **growth hormones** (such as the hormone needed to absorb food) that occur naturally in healthy babies.

A baby does not have to be premature or sickly to benefit from touch. Even healthy babies show benefits from touch in terms of emotional stability. Difficult children often have a history of abuse and neglect. The reason is that touch serves as a type of reassurance to infants that they are loved and safe, which translates into emotional well being. In general, babies who are held and touched more tend to develop better alertness and cognitive abilities over the long run.

Touch continues to have a great psychological impact throughout peoples' lives. Even adults who are hospitalized or sick at home seem to have less anxiety and tension headaches when they are regularly touched or caressed by caretakers or loved ones. Numerous studies have shown that touch also has a healing power. Researchers have found that touch reduces rapid **heart** beats and irregular heart beats (arrhythmias). Another study showed that baby rats who are touched often during infancy develop more receptors to control the production of biochemicals called glucocorticoids, which are known as **stress** chemicals because of their ability to cause muscle shrinkage, high **blood** pressure, elevated **cholesterol**, and more.

Touch's psychological impact goes beyond physical and mental health. Researchers have shown that touch is a powerful persuasive force. For example, studies have shown that touch can have a big impact in marketing and sales. Salespeople often use touch to establish a camaraderie and friendship that can result in better sales. In general, people are more likely to respond positively to a request if it is accompanied by a slight touch on the arm or hand. In a study of waiters and waitresses, for example, those that lightly touched a patron often received better tips.

See also Perception.

Resources

Books

Ackerman, Diane. *A Natural History of the Senses.* New York: Vintage Books, 1991.

> ## KEY TERMS
> . . . .
>
> **Biochemical**—The biological or physiological chemicals of living organisms.
>
> **Central nervous system**—The brain and spinal cord components of the nervous system, which controls activities of internal organs, movements, perceptions, thoughts, and emotions.
>
> **Cerebral cortex**—The external gray matter surrounding the brain and made up of layers of nerve cells and fibers, thought to process sensory information and impulses.
>
> **Epidermis**—The outer layer of the skin consisting of dead cells. It is the primary protective barrier against sunlight, chemicals, and other possible harmful agents. The epidermal cells are constantly being shed and replenished.
>
> **Free nerve endings**—Touch receptors in the skin that detect light pressure.
>
> **Glucocorticoids**—A steroid or hormone like compound that affects metabolism and can have an anti-inflammatory effect.
>
> **Meissner corpuscles**—Touch receptors in the skin that are sensitive to touch. Named after German histologist Georg Meissner.
>
> **Neurons**—Nervous system unit that includes the nerve cell, dendrites, and axons.
>
> **Pacinian corpuscles**—Touch receptors in the skin that sense pressure and rapid or vibrating movement of the tissues, named after Italian anatomist Filippo Pacini.

Bennet, Thomas L. *The Sensory World. An Introduction to Sensation and Perception.* Monterey, CA: Brooks/Cole Publishing Company, 1978.

Moller, Aage R. *Sensory Systems: Anatomy and Physiology.* New York: Academic Press, 2002.

Periodicals

Ponte, Lowell. "The Sense That Shapes Our Future." *Readers Digest* (January 1992): 21-26.

Weider, Betty. "Therapeutic Touch." *Shape* (May 1992): 32.

David Petechuk

Towers of Hanoi

The towers of Hanoi is an ancient mathematical puzzle likely to have originated in India. It consists of three poles, in which one is surrounded by a certain number of discs with a decreasing diameter. The object of the puzzle is to move all of the discs from one pole onto another pole. The movement of any disc is restricted by two rules. First, discs can only be moved one at a time. Second, a larger disc can not be placed on top of a smaller disc.

Toxic shock syndrome

Toxic shock syndrome (TSS) is a serious, life-threatening **disease** caused by **bacteria** called *Staphylococcus aureus* and less typically, *Streptococcus pyrogenes*. Common causes of skin and mucous **membrane** infections, some strains of *Staph. aureus* and *Strep. pyrogenes* secrete toxins that initiate a cascade of immune reactions. These immune reactions lead to overproduction of immune **proteins**, and the abnormally high production of these proteins leads to TSS. As its name implies, the most prominent symptom of TSS is shock. During shock, the **blood pressure** drops to dangerously low levels. If TSS is not quickly diagnosed and treated with **antibiotics**, it can be fatal.

Toxic shock syndrome toxins

A toxin is a chemical that acts as a poison within the body. TSS is caused by toxins released from certain strains of *Staph. aureus* and *Strep. pyrogenes*. Not all strains of these bacteria release these toxins. About 30% of *Staph aureus* strains and less than 10% of *Strep pyrogenes* strains are TSS-toxin-producing.

Toxins that cause TSS are called "superantigens" because of their effects on the **immune system**. An antigen is the protein on a bacterial **cell** or viral coat that certain immune cells, called helper **T cells**, identify as "foreign." Helper T cells recognize antigens by binding to them. When this recognition occurs, the immune system swings into action against the invader and makes specific proteins, called antibodies, which tag the invader for destruction by other immune cells. The TSS toxins are superantigens because the immune reaction they incite is nonspecific and aggressive. Helper T cells binds to the toxins, but instead of activating one small part of the immune system—the antibody production mechanism—the helper T cells-toxin binding "turns on" *all* of the immune system.

This nonspecific activation of the immune system has devastating effects on the body. As a result of TSS-toxins binding to helper T cells, several immune proteins are overproduced. Monokines and lymphokines, proteins that promote the further proliferation of helper T cells,

are produced in large quantities. **Histamine**, a protein that functions in allergic reactions and the inflammatory response, is released from immune cells. These proteins, in turn, exert several physiological effects. Histamine causes blood vessels to dilate, increasing blood circulation. Monokines and lymphokines contribute to the body's fever response, in which the internal **temperature** of the body increases in response to an immune signal. The combination of all these effects leads to TSS.

Symptoms of toxic shock syndrome

A "syndrome" is a group of different symptoms and conditions that are traced to one specific cause. Acquired Immune Deficiency Syndrome (**AIDS**), for example, is a cluster of different diseases that stem from **infection** of helper T cells with Human Immunodeficiency **Virus** (HIV). Similarly, TSS is a spectrum of symptoms caused by infection with toxin-producing strains of *S. aureus* and *S. pyrogenes*.

The early stage of TSS is characterized by flu-like symptoms such as sudden fever, fatigue, diarrhea, and dizziness. In a matter of a few hours or days, the blood pressure drops dangerously and a sunburn-like rash forms on the body. The drastic drop in blood pressure is potentially fatal. Circulatory problems develop as a result of low blood pressure, and some extremities—such as the fingers and toes—are deprived of blood as the body tries to shunt blood to vital organs. If the syndrome is severe enough, fingers and toes may become gangrenous due to lack of circulation. TSS can be treated with antibiotics, but these drugs kill only the bacteria that release the toxins: they do not neutralize the toxin that is already in the bloodstream. For treatment to be effective, antibiotics must be given early in the illness, before a large amount of toxin has built up in the bloodstream.

Risk factors for toxic shock syndrome

In 1980, several women in the United States were diagnosed with TSS; a few of these women died. When researchers investigated these cases, they found that all the women had been menstruating and using high-absorbency tampons. Since that time, toxic shock has been associated with the use of tampons in menstruating women, who comprise about 50% of the cases of TSS per year. Researchers speculate that tampons provide a suitable environment for bacteria such as *S. aureus* to grow.

To reduce the risk of TSS, experts recommend that women who use tampons change them frequently (about every two to four hours) and use the lowest-absorbency tampon that's practical. To avoid all tampon-associated risk of TSS, avoid using tampons altogether. Since insti-

KEY TERMS

· ·

Antigen—A molecule, usually a protein, that the body identifies as foreign and toward which it directs an immune response.

Helper T cell—The "lynch pin" of specific immune responses; helper T cells bind to APCs (antigen-presenting cells), activating both the antibody and cell-mediated immune responses.

Syndrome—A set of diseases or symptoms that are traced to one specific cause; examples include Acquired Immune Deficiency Syndrome (AIDS) and toxic shock syndrome (TSS).

Toxin—A poisonous substance.

tuting these guidelines, the incidence of toxic shock has fallen significantly over the past ten years. Currently, the incidence of toxic shock syndrome in menstruating women is between 1 and 17 cases per 100,000. However, some cases of TSS in women have been associated with the use of contraceptive **sponges** and diaphragms. Like tampons, these devices should be worn for the shortest time recommended on the package directions, and for no longer than eight hours.

Cases of TSS are also found in people with pre-existing skin infections, such as boils and wound infections. Prompt treatment of these conditions can usually prevent TSS.

Resources

Periodicals

Berkley, Seth F., et al. "The Relationship of Tampon Characteristics to Menstrual Toxic Shock Syndrome." *Journal of the American Medical Association* 258 (August 21, 1987): 917-20.

Bryner, Charles L., Jr. "Recurrent Toxic Shock Syndrome." *American Family Physician* 39 (March 1989): 157-64.

Kleinman, Leanne. "Toxic Shock Revisited." *Health* 20 (April 1988): 8.

Other

Toxic Shock Syndrome and Tampons. Rockville, MD: United States Department of Health and Human Services, Food and Drug Administration, 1990.

Toxic Shock Syndrome: Assessment of Current Information and Future Research Needs: Report of a Study. Institute of Medicine (United States), Division of Health Sciences Policy. Washington, DC: National Academy Press, 1982.

Kathleen Scogna

Toxicology

Toxicology is the scientific study of poisons (or toxins). Major topics in toxicology include the detection and chemical analysis of poisons, the study of the metabolic effects of these substances on organisms, and the investigation of methods for treatment of poisoning.

The Swiss physician and alchemist Philippus Aureolus, also known as Paracelsus (1493-1541) and said to be the father of the modern science of toxicology, wrote "All things are poison, and nothing is without poison, the dose alone makes a thing not a poison." In other words, if poisoning is to be caused, an exposure to a potentially toxic chemical must result in a dose that exceeds a physiologically determined threshold of tolerance. Smaller exposures do not cause poisoning.

Physiology is the study how organisms function, and the disruption of biochemical pathways by poisons is a key aspect of toxicology. Poisons affect normal physiology in many ways, but some of the more common mechanisms involve the disabling of **enzyme** systems, induction of cancers, interference with the regulation of **blood** chemistry, and disruption of genetic processes.

Organisms vary greatly in their tolerance of exposure to chemicals, and even within populations of the same **species** there can be great variations in sensitivity. In rare cases, some individuals may be extremely sensitive to particular chemicals or groups of similar chemicals, a phenomenon known as hypersensitivity.

Organisms are often exposed to a wide variety of potentially toxic chemicals through medicine, food, **water**, and the atmosphere. Humans are exposed to complex mixtures of chemicals, many of which are synthetic and have been either deliberately or accidentally released into the environment.

In some cases, people actively expose themselves to chemicals that are known to be toxic, such as when smoking cigarettes, drinking **alcohol**, or taking recreational drugs. Voluntary exposure to chemicals also occurs when people take medicines to deal with illness, or when they choose to work in an occupation that involves routinely dealing with dangerous chemicals. However, most exposures to potentially toxic chemicals are inadvertent, and involve living in an environment that is contaminated with small concentrations of pollutants, for example, those associated with pesticide residues in food, **lead** from gasoline **combustion**, or **sulfur dioxide** and **ozone** in the urban atmosphere.

Traditionally, the discipline of toxicology has only dealt with the direct effects of poisonous chemicals on organisms, and particularly on humans. Recently, however, ecologists have broadened the scope of toxicological investigations to include the indirect effects of chemicals in the environment, a **field** known as ecotoxicology. Ecotoxicology could be defined as the study of the ecological effects of toxic chemicals, including the direct effects, but also the indirect consequences caused by changes in the structure of habitats, or in the abundance of food. A herbicide used in **forestry** may not cause direct, toxic effects to animals at the doses given during the pesticide application, but the pesticide may change the vegetation, and thereby change the ecological conditions upon which animals depend.

Toxicology in practice

The toxicologist employs the tools and methods of science to better understand the consequences of exposure to toxic chemicals. Toxicologists typically assess the relationship between toxic chemicals and environmental health by evaluating such factors as:

- Risk. To assess the risk associated with exposure to a toxic substance, the toxicologist first measures the exposure characteristics and then computes the doses that enter the human body. He or she then compares these numbers to derive an estimate of risk, sometimes based on **animal** studies. In cases where human data exist for a toxic substance, such as **benzene**, more straightforward correlations with the humans risk of illness or death are possible.

- Precautionary strategies. Given recommendations from toxicologists, government agencies sometimes decide to regulate a chemical based on limited evidence from animal and human epidemiological studies that the chemical is toxic. Such decisions may have both ethical and political ramifications; to fail to issue warnings about a "suspect" chemical could leave vulnerable members of the population at risk of contracting an avoidable illness; on the other hand, any restrictions placed on the use of the chemical could place burdensome cleanup costs on private industry.

- Clinical data. Some toxicologists devise new techniques and develop new applications of existing methods to monitor changes in the health of individuals exposed to toxic substances. For example, one academic research group in the United States has spent many years developing new methods for monitoring the effects of exposure to oxidants (e.g., free radicals) in healthy and diseased humans.

- Epidemiological evidence. Another way to understand the environmental factors contributing to human illness is to study large populations that have been exposed to substances suspected of being toxic. Scientists then at-

tempt to tie these observations to clinical data. Ecologic studies seek to correlate exposure patterns with a specific outcome. Case-control studies compare groups of persons with a particular illness with similar healthy groups, and seek to identify the degree of exposure required to bring about the illness. Other studies may refine the scope of environmental factor studies, or examine a small group of individuals in which there is a high incidence of a rare **disease** and a history of exposure to a particular chemical.

- Evidence of bio-accumulation. When a chemical is nonbiodegradable, it may accumulate in biosystems, with the result that very high concentrations may accumulate in animals at the top of food chains. Chlorinated **pesticides** such as dieldrin and DDT, for example, have been found in **fish** in much greater concentrations than in the seawater where they swim.

Common toxic materials

Toxicologists have ranked the most commonly encountered toxic chemicals in the United States. In descending order of **frequency** of encounter, they are as follows:

- Arsenic. Toxic exposure occurs mainly in the workplace, near hazardous waste sites, or in areas with high natural levels. A powerful poison, arsenic can, at high levels of exposure, cause death or illness.

- Lead. Toxic exposure usually results from breathing workplace air or dust, or from eating contaminated foods. Children may be exposed to lead from eating lead-based paint chips, or playing in **contaminated soil**. Lead damages the **nervous system**, kidneys, and the immune systems.

- Mercury. Toxic exposure results from breathing contaminated air, ingesting contaminated water and food, and possibly having dental and medical treatments. At high levels, mercury damages the **brain**, kidneys, and developing fetuses.

- Vinyl chloride. Toxic exposure occurs mainly in the workplace. Breathing high levels of vinyl chloride for short periods can produce dizziness, sleepiness, unconsciousness, and, at very high levels, death. Breathing vinyl chloride for long periods of time can give rise to permanent liver damage, immune reactions, nerve damage, and liver **cancer**.

- Benzene. Benzene is formed in both natural processes and human activities. Breathing benzene can produce drowsiness, dizziness, and unconsciousness. Long-term exposure affects the bone marrow and can produce **anemia** and leukemia.

- Polychlorinated biphenyls (PCBs). PCBs are mixtures of chemicals. They are no longer produced in the United

States, but remain in the environment. They can irritate the nose and throat, and cause **acne** and rashes. They have been shown to cause cancer in animal studies.

- Cadmium. Toxic exposure to cadmium occurs mainly in workplaces where cadmium products are made. Other sources of exposure include **cigarette smoke** and cadmium-contaminated foods. Cadmium can damage the lungs, cause kidney disease, and irritate the digestive tract.

Toxicology and the private citizen

Whenever a group of citizens becomes concerned about the toxicity of a chemical substance, they typically want immediate answers to the following questions:

- Is this chemical harmful?
- At what level of exposure is this chemical harmful?
- What are the symptoms of exposure to this chemical?

Unfortunately, from the point of view of the toxicologist, the answers to these questions are seldom simple. This problem is compounded by the fact that most chemicals of unknown toxicity did not exist until at most a few decades ago, so data for long-term exposure may not exist.

Even when the question is as apparently straightforward, such as "Is such-and-such a chemical harmful?", the answer will seldom be as simple as, "Yes, in amounts exceeding 16 parts per million." This is because a complete scientific answer must take into account other factors such as the age of the individual exposed, the duration of exposure, and whether other environmental pollutants are present that could interact with and magnify the substance's toxicity. In the same way, workplace guidelines for adults who spend an eight-hour day in a factory may be difficult to apply to a residential setting where children and homemakers spend most of their time.

Because modern science and technology developed many of the chemicals scientists are now evaluating for toxicity, it is perhaps not surprising that many laypersons have grown skeptical about the ability of scientists to identify toxic materials. Because toxicologists report their findings in the form of **statistics** and probabilities, separate government agencies have been known to issue conflicting regulations governing exposure to substances based on different interpretations of the same data.

See also Contamination; Poisons and toxins.

Bill Freedman
Randall Frost

Trace elements

Trace elements are chemicals that are required by organisms in very small quantities for proper physiological and biochemical functioning. Trace elements commonly occur in organisms in concentrations smaller than about 0.001% of the dry weight (less than 10 parts per million, or ppm). Listed in alphabetical order, the most commonly required trace elements for healthy **animal** or **plant nutrition** are: boron (B), **chlorine** (Cl), chromium (Cr), cobalt (Co), **copper** (Cu), fluorine (F), iodine (I), **iron** (Fe), manganese (Mn), molybdenum (Mo), selenium (Se), silicon (Si), tin (Sn), vanadium (V), and zinc (Zn). Some organisms also appear to require **aluminum** (Al) and nickel (Ni).

All of the 92 naturally occurring elements occur ubiquitously in the environment, in at least trace concentrations. In other words, there is a universal **contamination** of **soil**, **water**, air, and biota with all of the natural elements. As long as the methodology of analytical **chemistry** has detection limits that are small enough, this contamination will always be demonstrable. However, the mere presence of an element in organisms does not mean that it is indispensable for healthy biological functioning. To be considered an essential element, three criteria must be satisfied: (1) the element must be demonstrated as essential to normal development and **physiology** in several **species**, (2) the element must not be replaceable in this role by another element, and (3) the beneficial function of the element must be through a direct physiological role, and not related to correction of a deficiency of some other element or indirect correction of a toxic condition.

Research into the physiological roles of trace elements is very difficult, because it involves growing plants or animals under conditions in which the chemical concentrations of food and water are regulated within extremely strict standards, particularly for the trace element in question. In such research, even the slightest contamination of food with the trace element being examined could invalidate the studies. Because of the difficulties of this sort of research, the specific physiological functions of some trace elements are not known. However, it has been demonstrated that most trace elements are required for the synthesis of particular enzymes, or as co-factors that allow the proper functioning of specific **enzyme** systems.

A principle of **toxicology** is that all chemicals are potentially toxic. All that is required to cause toxicity is that organisms are exposed to a sufficiently large dose. The physiological effect of any particular dose of a chemical is related to the specific susceptibility of an organism or species, as well as to environmental conditions that influence toxicity. This principle suggests that, although trace elements are essential micronutrients, which benefit organisms that are exposed within certain therapeutic ranges, at larger doses they may cause biological damages. There are many cases of biological and ecological damages being caused by both naturally occurring and human caused pollutions with trace elements. Such occurrences may involve the natural, surface occurrences or metal-rich **minerals** such as **ore** bodies, or emissions associated with certain industries, such as **metal** smelting or refining.

See also Element, chemical; Nutrients; Toxicology.

Tragopans

Tragopans are members of the attractive bird family Phasianidae, which also includes **pheasants**, **peafowl**, **partridges**, **guinea fowl**, and **turkeys**.

There are various **species** of tragopans in Afghanistan, eastward as far as Tibet, and in the Himalayas, in the same area as *Ithaginus cruentus*, a related short-tailed species. Tragopans, however, are more colorful. Tragopans spend much of their time in the crowns of trees where they find most of their food. Their diet consists primarily of **insects**, leaves, **fruits**, and **seeds**.

Satyr Tragopan, *Tragopan satyra*, of the Himalayas, is a deep cherry red, varied with white and brown. The male bird has large areas of bare blue skin on its cheeks and two finger-like wattles of the same **color** behind the eyes. During courting these areas become very noticeable as they swell up with **blood**.

Sophie Jakowska

Trains and railroads

Trains were developed during the **Industrial Revolution** and were arguably that period's most important product. In many ways railroads made the Industrial Revolution possible. Factories could not run without a constant supply of raw materials, or without a method of moving goods to market. More than anything, the progress of the railroads depended on the development of motive power, which was, in turn, being driven by technology. If the story of the Industrial Revolution is the story of the railroads, then, the story of the railroads is the story of technology.

Like so much else in western culture, railroads had their roots in ancient Greece. Farmers and merchants transporting goods realized that their wagons could travel more quickly on a smooth, hard surface with its reduced friction than on soft dirt roads. Where possible, they cut ruts into the rock to guide the wagon wheels. These rutways were limited to areas where the rock was near the surface, but the efficiency of the approach was demonstrated.

The rutway technology was submerged in the full-width Roman roads and lost in the eventual fall of the empire. In the late Middle Ages, however, a variation of the idea surfaced. In sixteenth and seventeenth century Germany and England, primitive railway systems were developed in which wood-wheeled carts ran on wooden rails. These early lines were developed primarily for heavy industry such as **coal mining**, to make large **volume** transport viable. The ascents were made using horsepower and the descents were made with the benefit of gravity, brakes, and a few prayers. The reduced friction of the wagonways allowed **horses** to haul several times the load they could manage on a normal road, and the rails guided the wagons along.

These wooden rail systems had a number of disadvantages. When wet they were extremely slippery, causing the carts to slide out of control on grades. They were not particularly strong or durable. In particular, carts with **iron** wheels quickly wore out the soft wooden tracks. In 1767, Richard Reynolds of Coalbrookdale, England, fabricated the first iron rails. The **metal** rails reduced the rolling friction of the wheels while lasting longer than the wooden alternatives. The way was clear for motive power.

The steam locomotive

In its simplest form, a steam locomotive consists of a firebox, a boiler, a cylinder or cylinders, and wheels, all of which are mounted on a rigid frame. The flames in the firebox **heat** water in the boiler to create steam. The steam is directed into a cylinder where its **force** is used to push a plunger attached by a connector rod or **gears** to the driving wheel of the engine. These connecting elements force the wheels to turn, which moves the engine along the track.

Wheels are classified as drive wheels, which provide motive power, and carrying wheels, which distribute the weight of the engine and add stability. Carrying wheels are further divided into leading wheels, i.e., those ahead of the drivers, and trailing wheels, or those behind the drivers. A common classification scheme for steam locomotives gives the number of leading wheels, the number of driving wheels, and the number of trailing wheels.

The "choo-choo" style engine of the American West, for instance, would be classified as a 4-4-0: four leading wheels, four drivers, and no trailing wheels.

The first locomotives

The first self-propelled steam vehicle was built by Frenchman Nicolas Cugnot in 1769, followed by William Murdoch's model experimental locomotive in 1784. In 1802, Richard Trevithick built the first full-size locomotive to run on rails, thus winning a wager for his employer. A horizontal cylinder sat in the boiler and drove a piston, which drove a connecting rod that connected to a crank/flywheel. A set of gears transferred **energy** from the crankshaft to the drive wheels that moved the engine. To meet the terms of the bet, the locomotive successfully pulled a series of cars loaded with ten tons of iron and 70 people.

Trevithick used an artificial draft through the firebox to fan the flames of the coals, an important innovation. This increased the heat of the fire, generating larger amounts of high **pressure** steam. He dispensed with any additional traction mechanism to keep the engine from slipping, convinced that the friction between the iron rails and the wheels was enough to drive the vehicle forward. His invention worked admirably. At several tons in weight, however, it was far too heavy for the brittle iron plateway and left a string of broken rails in its wake.

Traction, or wheel to rail adhesion, is fundamental to the operation of a locomotive. In order to move a string of cars, the locomotive drive wheels must grip the track. If traction is insufficient, the wheels simply spin without pulling the train forward, just as car wheels can spin uselessly in mud or on **ice**. This was a special concern for early locomotive designers who, unlike Trevithick, were not convinced that wheel-to-rail adhesion was sufficient to move the train down the track. Because frictional force between wheels and rail is proportional to the downward force or weight on the driving wheels, lighter engines were more likely to encounter adhesion problems. Heavier engines had better adhesion but their weight tended to break or split the brittle cast iron tracks, and locomotive builders were under continual pressure to reduce engine weight.

A variety of solutions that balanced the issues were proposed and built. In 1812, John Blenkinsop built a substantially lighter engine than Trevithick's, compensating for any loss of adhesion by using a rack and pinion drive. The drive wheels were cogged and rails were toothed on the outside face. The teeth on the drive wheels meshed with the teeth on the rails, driving the locomotive forward with no chance for slippage. Costly and complicated, the rack and pinion drive soon proved

A steam locomotive. *JLM Visuals. Reproduced by permission.*

to be unnecessary for conventional railroads and never became popular. Other high traction methods such as chain drive or external pushing legs were simply impractical, and most of the locomotives that followed reverted to adhesion drive.

A second Blenkinsop innovation became a standard feature of almost all subsequent steam locomotives. Blenkinsop designed a two cylinder engine, with a cylinder to power the drive wheel on each side. This eliminated the use of an overhead flywheel to transfer mechanical energy from the cylinder to the drive wheels. Unlike Trevithick's design, however, the cylinders on Blenkinsop's engine were vertical. At high speeds, the rapid bounce of the pistons added a great deal of chop to the engine movement, exacerbated by the fact that the engine had no springs to absorb the **motion**. Given the weight of the engines, the chop placed a significant amount of stress on the rails, resulting in more splits and fractures.

The next groundbreaking design was an adhesion drive, two cylinder locomotive called the *Puffing Billy*. It was the first steam locomotive to feature cylinders outside of the boiler where they were easily accessible. Designed by William Hedley, the *Puffing Billy* distributed its weight over eight drive wheels, putting less concentrated load on the track and causing less wear.

One of the major locomotive companies during the nineteenth centuries was run by the father-son team of George and Robert Stephenson. The Stephensons were responsible for some of the most important technical innovations in locomotive operation. George Stephenson replaced the cylinder-to-wheel gear interface by coupling and connecting rods, streamlining the design and bringing it closer to the style of locomotive that we know. He also introduced the locomotive steam spring, which cushioned the action of the engine. The spring consisted of a vertical cylinder with a piston that carried the engine weight. Steam forced the piston to the upper end of the cylinder, applying upward force to counter the downward force of the weight of the engine. As a result of the shock-absorbing effect of the spring, the heavy engine rode more easily on its wheels and caused fewer cracks in the iron rails.

Locomotive wheels also cracked frequently, requiring costly replacements and engine down-time. Robert Stephenson and Timothy Hackworth replaced the solid cast-iron wheels of early engines with a combination design that featured durable, replaceable wrought-iron tires mounted on cast-iron hubs. In his 1827 locomotive the *Rocket*, Robert Stephenson also introduced the multitube boiler. Frenchman Marc Seguin developed a similar design at around the same time. In the multitube boiler, hot gases from the firebox move through tubes that run the length of the boiler. Heat is exchanged over a much greater surface area, making the design far more efficient than the single chamber type. Around this same time, locomotive designers abandoned the vertical cylinder for the smoother horizontally mounted type, though the cylinders on the *Rocket* compromised with a slanted orientation.

The American standard

Steam locomotives were introduced in the United States in 1829. They were initially supplied by British builders, but the development of American locomotives moved in a different direction from British and European locomotives almost immediately. Britain was a prosperous, settled country and British tracks were sturdy and well-built, with flat roadbeds and low grades. The Americans, on the other hand, were still pushing the frontier west across a vast landscape. Railroad companies were minimally financed, while striving to web the country with rails. Consequently, American tracks were built hastily, with minimal roadbed preparation. Often they consisted of just flat-topped rails spiked onto roughcut ties. Curves were tighter, grades were steeper, and because the roadbeds were poorly graded if at all, the tracks were uneven. The high performance British locomotives with their fixed, four-wheel suspension did not fare well on U.S. tracks, derailing and breaking axles on the twisting, uneven rails.

The *Experiment*, a 4-2-0 engine built by John B. Jervis in 1831, was the first locomotive designed specifically for use on the American railroads. To modify the British fixed-wheel suspension, Jervis added a four-wheeled truck mounted on a center pivot to the front of the *Experiment*. Like the front wheels of a car, this truck could shift and turn with the track, compensating for sharp curves and unevenness. The two drive wheels were

at the back of the engine, and the *Experiment* also boasted the novelty of an enclosed cab.

The design was a success and notion of a leading four-wheel truck was widely adopted in the United States. In 1836, Henry R. Campbell patented the 4-4-0 locomotive. Robust and economical, locomotives of this design could soon be ordered from a number of manufacturers. Add a cowcatcher to sweep away **livestock** from tracks running through open **prairie**, and the spark suppressing smokestack that kept glowing cinders from flying out to start fires, and you have the "American Standard," a rugged, powerful locomotive that was ubiquitous in the nineteenth century United States.

Additional accoutrements were added to these engines. Matthias Baldwin was the first to equip a locomotive with warning bells, and George Whistler added the first steam whistle. Night travel was at first accomplished by maintaining a fire on a small car hooked to the front of the engine, but was soon superceded by a headlamp.

Meanwhile, the focus in Britain was on speed. Whereas the average speed of the American Standard was around 25 MPH (40 km/h), British engines were routinely clocking speeds of 60 MPH (97 km/h) as early as 1860. The tracks were flat and smooth with few curves and low grades, and the swift engines were designed with compact, rigid frames and enormous driving wheels. In 1832, Robert Stephenson built the *Patentee*, a 2-2-0 engine whose design was to dominate in Britain and **Europe** for many years to come.

Further improvements in steam locomotive technology led to increases in speed and power. To maximize efficiency, double cylinders were constructed in which the steam from the first cylinder was let into a second cylinder to completely exhaust its pushing capabilities. More complete **combustion** was achieved by installing a firebrick arch in the firebox that routed air around prior to introducing it to the boiler. To improve power, multiwheel behemoths were built. Locomotives with six and eight wheels were commissioned, as well as the less common 10 to 12 wheelers.

Superheated steam was another method of increasing efficiency. In most early locomotives, steam included a significant portion of what was merely hot **water.** It was unable to do useful work and took up space in the cylinder where steam could normally expand to do useful work. To address this issue, engineers in Germany and Belgium developed the method of superheated steam. Steam headed toward the cylinders was heated a second time to dry it out, minimizing liquid water content. In tandem with improved cylinder valve gearing, the use of superheated steam increased engine efficiency so much that compound cylinders were eventually phased out as unnecessary.

Steam locomotives reached their peak in the middle of the twentieth century. 4-8-4s and 4-6-4s capable of speeds as high as 95 MPH (153 km/h) were built in the mid-1940s, when rail travel dominated overland passenger travel. Even as these streamliners were capturing the imagination of the public, however, diesel and electric locomotives were beginning to take over rail transportation. By the mid-1950s, the numbers of steam locomotives were dwindling rapidly, and today they exist only as sentimental reminders of a bygone era.

Diesel and electric locomotives

Diesel engines are internal combustion engines in which fuel oil is injected directly into the cylinder head and ignited by pressure. They power the wheels by direct gearing rather than the connecting rods of the steam locomotive, providing continual power. Railway diesels have been designed with electric, hydraulic, mechanical, and pneumatic transmissions; today the diesel-electric engine is most common.

When they were introduced early in the twentieth century, diesels offered unprecedented efficiency and performance over steam locomotives. Diesel engines could be operated round the clock, without timeouts to take on water for the boiler or clean out ashes from the firebox. They could carry enough fuel for a day or two of continuous operation, and running them was almost absurdly simple. Crewmen for the first diesel locomotive in the United States, for example, were trained to operate it in just 15 minutes. Initial capital outlay was high, but operating costs were only a fraction of the cost of steam locomotives.

Electric trains are the other major type of motive rail power. Particularly in Europe, passenger traffic is dominated by electric power.

Electric trains run on both direct and alternating current, with voltage in the 50 to 100 kV range. The choice of current type is driven as much by economics as by performance, involving as it does a tradeoff of cost and efficiency. Alternating current (AC) offers an economical current supply at the expense of motor complexity. Motors for the more expensive direct current (DC) supplies are very simple. Current is fed to the motors from overhead wires, as with trolleys, or from an electrified "third rail" along the ground, commonly seen in subways.

Track

From the beginning of railroad development, British and European line surveyors were extremely careful to lay flat, even track, minimizing curves and grades. A track set down by George Stephenson, for instance, was laid on stone blocks with very compact foundations. By contrast, most early American tracks were laid hastily on wooden ties. The flimsy rails were easily deformed by the repeated weight of trains, sagging where not supported by the ties. Eventually the ends of the rails would rise up due to this sagging action. When they rose high enough, these "snake heads" would be pushed up in front of the wheels and either derail the train or punch through the floorboards, impaling those unlucky enough to be sitting above them. To avoid this, American railroads began placing the ties very close to one another, a practice still followed today.

The early **wood** rails were followed by brittle cast-iron rails. It was only later that more ductile wrought iron was used. **Steel** rail came into play in the 1870s, as a byproduct of the Bessemer process, a method for economical **mass production** of steel. The steel rail was more durable, capable of supporting harder wheels and heavier loads. In recent years, rails have become heavier, weighing as much as 100 lb (45 kg) per yard. Much of it is continuously welded rail. To simplify maintenance over miles of rail, special machines have been built that detect flaws in track, raise and align track, or clean and resettle track ballast.

Track gauge, or the width between the rails, varied tremendously in the early years of railroading. Gauges ranged from 3 ft (0.9 m) called "narrow gauge" lines to 6 ft (1.8 m) called "wide gauge" lines. Wide gauges were first believed to be more stable than narrow gauges, able to support broader cars without tipping over on curves. In mountainous areas or the highly populated urban regions of Britain, however, there was not sufficient room for wide gauge tracks, and rails were laid closer together. When it came time for the tracks of different railroads to merge into one enormous net, gauge discrepancies were a major problem.

The standard gauge was a 4 ft 8.5 in (1.7 m) spacing. Common in Britain, it was quickly passed along to other countries. In the United States, standard gauge became the official gauge of the American Railway Association toward the end of the nineteenth century. This led to changes in the rail spacing of narrow and wide gauge railroads, necessitating massively coordinated efforts in which the gauge of entire railway lines, as much as 500 or 600 mi (804-965 km), would be changed in a single day.

Brakes and couplers

Early cars were coupled together using a link and pin system. Given that the pins had to be put into place and removed manually while the cars were moved by a distant engine, coupling cars was a dangerous job that all too often led to the loss of fingers and hands. Alternate coupler designs were proposed, and in 1887 a coupler designed by Eli H. Janney was approved by the Master Car Builders' Association. Resembling the curled fingers of two hands, Janney's coupler allowed cars to hook together without the use of pins.

Brakes, too, were a problem with early trains. They had to be applied on each car by hand, a time-consuming process. Brakemen on freight trains had the added difficulty of applying the brakes from a precarious **perch** on the top of the car while hoping that the train did not go under any low **bridges**. In 1869, George Westinghouse patented an air brake that used compressed air to force the brakeshoes against the wheels. Each car had a reservoir of high pressure air (70-100 lb [32-45 kg]/sq in). A control pipe filled with compressed air ran the length of the train. If the pressure in the control pipe dropped, the compressed air in the reservoir applied the brakes. This could occur when the brakes were applied or when a car became detached from the train—an added safety measure.

When diesel and electric locomotives came into use, a different approach to braking was possible. Both types of motors can be reversed such that the motor is working against the motion of the train. This dynamic braking system allows minimal use of air brakes, with longer wear on the brake shoes. Some high speed trains have computerized braking systems. If the engineer exceeds permitted speed on a section of line, the brakes are automatically applied.

Locomotive brakes took much longer to catch on than railcar brakes. Robert Stephenson's 1833 *Patentee* design included specifications for a steam brake, but the earliest recorded use of a brake on the driving wheels of an American locomotive was in 1848. Development and implementation was sporadic throughout the 1860s and 1870s, but by 1889 about half of all American locomotives were equipped with driving wheel brakes. By the end of the century, locomotives were routinely equipped with brakes, a necessity given the increased power and speeds of the twentieth century engines.

Switches and signals

In the early days of railroading, switches were set by hand and signal systems consisted of flags during the day and lamps at night. In 1856, an interlocking signal was designed to prevent signalmen from setting signals and switches in conflict with one another. In 1865, Ashbel Welch of the Camden and Amboy RR developed a new type of signal known as the manual block-signal. Trains were spaced apart by a prescribed distance or "block," and new trains could not enter this block until the first train left. The electric **telegraph** was used to pass the word that track was clear and the train ahead had reached the station.

Switch and train location information was conveyed to engineers by stationary signals such as flags, patterned disks that rotated between full view and edge view, or fixtures with semaphore arms. In 1871, electrical control of block-signals was introduced. In 1890, a compressed air switch with electronic control was installed on the Pennsylvania-Baltimore & Ohio railroad crossing. Fully automated switches soon followed in which the train wheels and axels made a complete conducting circuit, operating relays that ran the signals. The colored electronic lights used today are modern versions of these early signals.

Modern switching yards are largely automated. Car speeds are computer controlled and switching is automatic. Meanwhile, sensors and detectors check for loose wheels, damaged flanges, or other faulty equipment.

Ultrafast trains, the modern travel alternative

In 1964, the Japanese inaugurated the Shinkansen train, initially capable of going an unprecedented 100 MPH (161 km/h). They have since built a net of high speed railroads across Japan. These trains run on special tracks and have had no serious accidents since the opening of the system. Europe has a number of high speed trains, from the Swedish X2000, capable of running at 136 MPH (219 km/h) average speed, to the German Intercity Express. The French, however, are the kings of high speed rail transport. The TGV trains run regularly over the countryside at nearly 200 MPH (322 km/h). A special TGV train running over new track has reached an astounding 319 MPH (514 km/h).

At such speeds, the technology of railroads must be rethought. Locomotive and passenger car suspension must be redesigned, and most trains must run on specially graded and built track. Engineers require in-cab signaling, as with average speeds ranging between 150 and 200 MPH (241 and 322 km/h), there is not enough time to read the external signals. Brakes must be reconsidered. A train running at 155 mph capable of braking at 1.6 ft/s2 requires about 3 mi (4.8 km) to stop fully after the brakes are first applied.

A high speed option that is the topic of hot research is a non-contact, magnetically levitated (maglev) train. Strong magnetic forces would hold the train above the track. Such a train would have virtually no rolling resistance or friction because there would be no wheels and nothing rolling. The only impedance would be that of air, making it extremely efficient. Maglev trains are still in the research phase, and more development is required before this possibility can be realized.

Railroads were a significant factor in the development of industry and are still a significant mode of trans-

KEY TERMS

. . . .

Adhesion—Frictional contact that makes the wheels of an engine grip the rails.

Boiler—A single tube or multitube vessel in which water is heated to steam.

Cylinder—A cylindrical tank into which steam is introduced to push a piston head back and forth, creating mechanical motion to drive the locomotive wheels.

Maglev—A noncontact (frictionless) method of train suspension in which strong magnetic forces are used to levitate the train above the track.

Rolling stock—Railroad cars. Rolling stock can contain such specialized cars as tank cars, refrigerator cars, and piggyback cars.

portation for goods and passengers. In terms of efficiency, they cannot be rivaled. While a diesel truck can haul a single or tandem tractor-trailer, a diesel locomotive can haul a string of loaded boxcars. As we become more concerned about the environment and internal combustion engines are phased out, railroads are poised to assume an even larger role in transportation, and a new Age of the Iron Horse may well be upon us.

See also Mass transportation.

Resources

Books

Drury, G.H. *The Historical Guide to North American Railroads.* Waukesha, WI: Kalmbach Books, 1988.

Hiscox, G.D. *Mechanical Movements, Powers, and Devices.* New York: Norman W. Henley Publishing Co., 1927.

Lorie, P., and C. Garratt. *Iron Horse: Steam Trains of the World.* Garden City, NY: Doubleday & Company Inc., 1987.

Nock, O.S. *Locomotion.* New York: Charles Scribner's Sons, 1975.

Snell, J.B. *Early Railways.* London, England: Octopus Books, 1964.

Stover, J.F. *The Life and Decline of the American Railroad.* New York: Oxford University Press, 1970.

Kristin Lewotsky

Tranquilizers

The medical use of drugs to reduce or relieve **anxiety** has given rise to a group of medications called an-

tianxiety agents. These agents include anxiolytics, tranquilizers, and sedatives.

Tranquilizers were formerly grouped as either "minor" tranquilizers or "major" tranquilizers. The word major stands for "major psychiatric illness," not heavily sedating or tranquilizing. Today, major tranquilizers are more often referred to as neuroleptics or antipsychotic agents and they are used in the treatment of **schizophrenia**, **depression**, and bipolar illness. Examples of antipsychotic agents are chlorpromazine (Thorazine), synthesized in France in 1950, and the phenothiazines.

Presently, the common use of the term tranquilizer refers to the "minor" tranquilizers mainly of the benzodiazepine family. This newer group of anti-anxiety agents has tended to replace the use of **barbiturates** and meprobamate and certain **antihistamines** which were used as sedatives and anti-anxiety agents. It is these drugs that are prescribed as tranquilizers in the non-psychiatric setting of general medicine which treats anxiety brought on by **stress** rather than some disorder in the central **nervous system**.

Anxiety

While anxiety is usually an accompanying state of mind of most psychiatric disorders, it is also a special disorder of its own. Anxiety disorder or reaction is characterized by a chronic state of anxiety that does not have an immediate or visible basis. That is, the individual feels alarmed or uneasy but cannot point to any outside or realistic basis for the fear. There is a general state of unease that may become expressed by acute attacks of anxiety or panic, called panic disorder.

The emotional stress of anxiety may be triggered by impulses and mental images that in turn **lead** to a number of complex physiological responses. The autonomic nervous system may react to signals from the emotional side of the mind which call forth defense reactions of either fight or flight. An excess of adrenalin may be released which cannot be adequately discharged, thus leading to the symptoms of anxiety.

The **psychology** of anxiety often entails the repression of certain drives and needs. Sexual feelings, aggression at work, in school, in the family, and dependency on a spouse or other social relationship that is being threatened, or that the anxious person feels apprehensive toward are all examples of the circumstances that can unleash a chronic state of anxiety. The loss of a job or the sudden onslaught of an illness may, likewise, be responsible for anxiety states as the individual attempts to cope with these new conditions.

Acute anxiety

Acute anxiety panic attacks have been described as one of the most painful of life experiences. The condition can last for a few minutes to one or two hours. The individual is cast into a state of terror by some nameless imminent catastrophe. All rational thought processes cease during this time.

There are a number of cardiovascular responses to this state such as palpitations, tachycardia (elevated **heart rate**), arrhythmias of the heart, and sharp chest **pain**. Breathing becomes very difficult, almost impossible. The term given for this condition is hyperventilation. The extremities (hands and feet) feel cold, numb, and tingle with a feeling of pins and needles being present in the skin which may turn blue in places.

Other symptoms include fine trembling of the hands when they are stretched out, a feeling of "butterflies" in the stomach, sweating, a general sense of weakness, dizziness, nausea, and sometimes diarrhea. People and the environment and surrounding objects seem remote and unreal. All these symptoms reinforce the anxious patient's belief that either loss of consciousness or death are nearby.

Chronic anxiety

Many of the symptoms of chronic anxiety are similar to acute anxiety, but they are less intense and more prolonged. They can last for days, weeks, or months. There is a considerable amount of tension and expectation of conflict. There is fear about the future and an inability to deal effectively with other people, especially at home, school, and work. The condition is also characterized by chronic fatigue, **insomnia**, and headaches along with difficulty in concentration. With chronic anxiety the individual is still able to function on some level, but the ability to deal with life situations is substantially compromised.

Treatment for anxiety

Treatment for anxiety can include psychotherapy for those who are responsive for unearthing unconscious conflicts. Supportive psychotherapy is given by physicians, social workers, and therapists to reassure the individual. Relaxation techniques, meditation, and hypnosis also help to alleviate the condition.

Tranquilizers play a role in the pharmacologic treatment of anxiety. Medications, however, are usually not sufficient to deal with the root causes of anxiety, and it is not certain to what extent they play a **placebo** role in alleviating feelings of anxiety. The attitude of the taker of the medication along with the belief in the medical authority

figure administering the drug are further factors in determining the effectiveness of pharmacologic intervention.

Benzodiazepines

There are about 20 tranquilizers in the benzodiazepine family of tranquilizers. Some of the popular ones are diazepam (Valium), Fluorazepam (Dalmane), oxazepam (Serax), and chlordiazepoxide (Librium). In addition to being prescribed for anxiety, they are also used as **muscle relaxants**, sedatives, anesthetics, and as supportive medication for withdrawal from **alcohol**. These drugs were introduced in the 1960s, and they quickly replaced other drugs that were then being used as tranquilizers. Their popularity seems to be now in decline, partly due to the more cautious attitude physicians have in prescribing them.

The amount of adverse effects is low. There is only a slight depression in respiratory rate and the amount needed for an overdose is very high, at a **ratio** of 200 (excess) to 1. Used in suicide attempts the results lead more to confusion and drowsiness without damage of a permanent nature. Librium (chlordiazepoxide) has some record of causing **coma** when taken in high dosages. Some people may react to benzodiazepines by losing inhibitions and expressing hostile or aggressive **behavior** that is not characteristic of their personalities, especially if they have been experiencing a high rate of frustration.

Since benzodiazepines produce less euphoria than other types of tranquilizers there is less chance of a dependency reaction that leads to abuse. Minor tranquilizers are generally not sought after by the illegal drug abuse market, nor are any of the other neuroleptic medications used to treat such mental illnesses as schizophrenia, **manic depression**, or depression. Some withdrawal symptoms can develop with benzodiazepines if they are used for an extended period of time, such as a month or more. While there may be such symptoms as increased anxiety, sensitivity to bright lights, twitching of the muscles, nausea, and even convulsions, there is not a craving for the drug itself. Symptoms can be reduced by withdrawing from the drug gradually and Dilantin can be used for convulsions.

Action

Tranquilizers act as anti-anxiety agents by depressing the central nervous system without leading to sedation. Barbiturates are seldom used now for managing anxiety or dysphoria because of their addictive potential. The molecules of barbiturate drugs pass through the membranes of the cells in the **brain**. They are then able to block nerve signals that pass from **cell** to cell, thus inhibiting the stimulation and conduction of chemical neurotransmitters between the cells. In addition, barbiturates are able to reduce the effect of abnormal electrical activity in the brain which cause seizures, such as in the case of **epilepsy**. Phenobarbital is a barbiturate that performs this function exceptionally well, therefore it is still useful as an anticonvulsant drug.

Depressant drugs, like alcohol and barbiturates, just as stimulant drugs, like **cocaine** and **amphetamines**, all appear to have the ability to stimulate the brain's reward circuit. The behavioral effect of this action is to increase the need for more of the drug, usually to get the same effect (drug tolerance). If it is being taken for its effect as a euphoric, more of the drug is needed each time it is taken to produce a high or for **sleep** if it is being used as a sedative. Drugs that stimulate the brain reward centers also have the effect of excluding other types of reward sensations like those from food or sex.

Drugs that stimulate the brain reward centers seem to enhance the presence of a chemical found in the brain called gamma aminobutyric acid (GABA). GABA has the effect of quieting the neurons where the GABA receptors are found.

The newer benzodiazepine tranquilizers reduce **neuron** sensitivity only for cells that do have the GABA receptor sites, but the barbiturates are able to work the sedating effect elsewhere as well, wherever there are chloride channels. That difference in action may account for the higher degree of sedation afforded by the barbiturates over the benzodiazepines. Both types of drugs are able to affect the brain's reward center by increasing the amount of **dopamine** released into the limbic system, the part of the brain that regulates certain biological functions such as sleep and the emotions.

Choice of tranquilizers

Tranquilizers are the most commonly used prescription drugs in the United States. The three major groups of tranquilizers are the benzodiazepines with the brand names of Valium, Librium, and Alprazolam. The second major group are the dephenylmethanes prescribed under the brand names of Vistaril and Atarax. The third group are the older alcohol-like propanediols that came out in the 1950s, such as tybamate and meprobamate under the brand names of Equanil and Miltown.

The physician chooses among the various tranquilizers the one that will best serve the patient's need. Stress is a normal part of daily living for most people and it will produce in each individual a certain range of anxiety. Tranquilizers are prescribed on a non-psychiatric basis by the physician in general practice when the level of anxiety experienced by the individual interferes with the abili-

KEY TERMS

Arrhythmia—Any abnormal rhythm of the heart, which can be too rapid, too slow, or irregular in pace; one of the symptoms of anxiety disorder.

Autonomic nervous system—The part of the nervous system that controls involuntary processes, such as heart beat, digestion, and breathing.

Chronic anxiety—A prolonged period of an abnormal level of anxiety symptoms.

Euphoria—A feeling of intense well being; the opposite of dysphoria.

Gamma aminobutyric acid (GABA)—A chemical in the brain that quiets neuronal activity.

Hyperventilation—An autonomic reaction to anxiety which increases the breathing rate, thereby altering the ratio of the exchange of gases in the lung. That change makes the act of breathing difficult to perform.

Minor tranquilizers—As opposed to major tranquilizers generally used to treat psychoses, minor tranquilizers are used to treat anxiety, irritability, and tension.

Panic disorder—An acute anxiety attack that can last for several minutes to several hours.

Psychotherapy—A broad term that usually refers to interpersonal verbal treatment of disease or disorder that addresses psychological and social factors.

Tachycardia—An elevated heart rate due to exercise or some other physiological response to a particular condition such as an anxiety attack.

ty to cope with everyday stressful events or when anxiety symptoms have reached clinical proportions.

Panic attacks respond well with the treatment of alprazolam. Certain **antidepressant drugs** have also been found useful in treating panic disorder. Other symptoms, such as rapid heart rate, palpitations, involuntary motor reactions, insomnia or other **sleep disorders**, diarrhea, bandlike headaches, increased urination rate, and gastric discomfort can be temporarily relieved by the use of other tranquilizers.

It is, however, necessary for the physician to point out to the patient that the tranquilizers are covering up the symptoms rather than curing them. The hazard in such palliative treatment is that the underlying condition may get worse without the conflict resolution necessary for the nervous system to readjust itself to the demands of reality.

Tranquilizers are not suited for long term use and over a period of time higher dosages may be needed, especially for the mild euphoria that some of them produce. While they do not pose the degree of dependency of other psychoactive drugs, some have been limited for general use because of the potential of overdependence. Valium is an example of a benzodiazepine that now is prescribed more cautiously.

Buspirone (BuSpar) appears to avoid the problem of possible dependency as well as that of drowsiness. This drug appeared in the mid-1980s. It is reported to be a "true" tranquilizer in that it does not produce either the slight euphoria of other tranquilizers or the drowsiness which is also characteristic of the sedative effect of other tranquilizers.

Resources

Books

Longenecker, Gesina L. *How Drugs Work.* Emeryville, CA: Ziff-Davis Press, 1994.
Morgan, Robert. *The Emotional Pharmacy.* Los Angeles, CA: The Body Press, 1988.
Nicholi, Armand M., Jr. *The New Harvard Guide to Psychiatry.* Cambridge, MA: Harvard University Press, 1988.
Oppenheim, Mike. *100 Drugs That Work.* Los Angeles: Lowell House, 1994.

Jordan P. Richman

Transcendental numbers

Transcendental numbers, named after the Latin expression meaning *to climb beyond,* are numbers which exist beyond the realm of algebraic numbers. Mathematicians have defined algebraic numbers as those which can function as a solution to polynomial equations consisting of x and powers of x. In 1744, the Swiss mathematician Leonhard Euler (1707-1783) established that a large variety of numbers (for example, whole numbers, fractions, imaginary numbers, irrational numbers, **negative** numbers, etc.) can function as a solution to a polynomial equation, thereby earning the attribute *algebraic*. However, Euler pointed to the existence of certain irrational numbers which cannot be defined as algebraic. Thus, $\sqrt{2}$, π, and *e* are all irrationals, but they are nevertheless divided into two fundamentally different classes. The first number is algebraic, which means that it can be a solution to a polynomial equation. For example, $\sqrt{2}$ is the solution of $x^2 - 2 = 0$. But π and *e* cannot solve a polynomial equation, and are therefore defined as transcendental. While π, which represents the **ratio** of the circumference of a **circle** to its diameter, had been known since antiqui-

ty, its transcendence took many centuries to prove: in 1882, Ferdinand Lindemann (1852-1939) finally solved the problem of "squaring the circle" by establishing that there was no solution. There are infinitely many transcendental numbers, as there are infinitely many algebraic numbers. However, in 1874, Georg Cantor (1845-1919) showed that the former are more numerous than the latter, suggesting that there is more than one kind of **infinity**.

See also e (number); Irrational number; Pi; Polynomials.

Transducer

A transducer is a device which converts one form of **energy** to another. Typically, one of these forms is electrical while the other is either mechanical, optical, or thermal. Transducers are usually classified as either input or output devices, depending on the direction in which they transfer energy or information. Input transducers convert some property or effect into an electrical signal, while output transducers start with **electricity** and generate a mechanical or other effect.

Transducers play such fundamental roles in modern technology that examples of them abound. In virtually every electronic device or instrument, transducers act as translators between **electron** flow and the physical world. Loudspeakers are perhaps the most well-known transducers, as they are used in nearly every audio system to convert electrical signals into acoustic ones. Like loudspeakers, audio microphones are transducers. Both devices have a small diaphragm which is free to move in either a magnetic or electric **field**. In speakers, electricity pushes this diaphragm to generate a sound. In microphones, the opposite happens, and sound pushes the diaphragm to generate an electric signal. Another common transducer in audio systems is the photodetector; in a **compact disc** player this input device combines with a photoemitter to optically sense encoded information on the disc and convert it to music.

A tremendous collection of transducers can be found in the human body. The senses convert complex sights, sounds, smells, and other experiences to electrical signals, which are then sent to the **brain** for interpretation.

See also Amplifier; Electronics.

Transformer

A transformer is an electrical device which changes or transforms an alternating current (AC) signal from

changing magnetic field is essential for the operation of a transformer.

Because the two coils of a transformer are very close to each other, an **electric current** through the primary coil generates a magnetic field which is also around the secondary coil. When this magnetic field varies with time (as it does when AC is applied), it teams-up with the secondary coil to form a type of **generator**. (Recall that a generator produces electrical power by moving coils of wire through a stationary magnetic field, the converse of the transformer situation.) At any **rate**, electrical power in the primary coil is converted into a magnetic field which then generates electrical power in the secondary coil. The beauty of a transformer is that, although the power is neither increased nor decreased in this transfer (except for minor losses), the voltage level can be changed through the conversion. The **ratio** of the voltages between the two coils is equal to the ratio of number of loops in the two coils. Changing the number of windings allows a transformer to step-up or step-down voltages easily. This is extremely useful as the voltage level is converted many times between a power station, through transmission lines, into a home, and then into a household appliance.

A high voltage transformer. *Photograph by Tardos Camesi. Stock Market. Reproduced by permission.*

one level to another. The device typically consists of two sets of insulated wire, coiled around a common **iron** core. Electrical power is applied to one of these coils, called the primary coil, and is then magnetically transferred to the other coil, called the secondary. This magnetic coupling of electrical power occurs without any direct electrical contact, and allows transformers to change AC voltage level and to completely isolate two electrical circuits from one another.

When a voltage is applied to a coil of wire, an electrical current flows (just as **water** flows through a pipe when **pressure** is applied.) The flowing electrical current, however, creates a magnetic **field** about the coil. This principle can be demonstrated by simply wrapping insulated wire around a nail, and attaching a **battery** to the ends of that wire. A sufficient number of loops and ample electrical power will enable this electromagnet to lift small **metal** objects, just as an ordinary magnet can. If, however, the battery is replaced by a varying power source such as AC, the magnetic field also varies. This

Transgenics

The term transgenics refers to the process of transferring genetic information from one **organism** to another. By introducing new genetic material into a **cell** or **individual**, a transgenic organism is created that has new characteristics it did not have before. The genes transferred from one organism or cell to another are called transgenes. The development of biotechnological techniques has led to the creation of transgenic **bacteria**, plants, and animals that have great advantages over their natural counterparts and sometimes act as living machines to create therapeutics for the treatment of **disease**. Despite the advantages of transgenics, some people have great concern regarding the use of transgenic plants as food, and with the possibility of transgenic organisms escaping into the environment where they may upset **ecosystem** balance.

All of the cells of every living thing on **Earth** contain DNA (deoxyribonucleic acid). DNA is a complex and long **molecule** composed of a sequence of smaller molecules, called nucleotides, linked together. Nucleotides are nitrogen-containing molecules, called bases that are combined with sugar and phosphate. There are four different kinds of nucleotides in DNA. Each nucleotide has a unique base component. The sequence of nucleotides, and therefore of bases, within an organism's DNA is unique.

In other words, no two organisms have exactly the same sequence of nucleotides in their DNA, even if they belong to the same **species** or are related. DNA holds within its nucleotide sequence information that directs the activities of the cell. Groups, or sets of nucleotide **sequences** that instruct a single function are called genes.

Much of the genetic material, or DNA, of organisms is coiled into compact forms called chromosomes. Chromosomes are highly organized compilations of DNA and protein that make the long molecules of DNA more manageable during **cell division**. In many organisms, including human beings, chromosomes are found within the nucleus of a cell. The nucleus is the central compartment of the cell that houses genetic information and acts as a control center for the cell. In other organisms, such as bacteria, DNA is not found within a nucleus. Instead, the DNA (usually in the form of a circular **chromosome**) is free within the cell. Additionally, many cells have extrachromosomal DNA that is not found within chromosomes. The mitochondria of cells, and the chloroplasts of **plant** cells have extrachromosomal DNA that help direct the activities of these organelles independent from the activities of the nucleus where the chromosomes are found. Plasmids are circular pieces of extrachromosomal DNA found in bacteria that are extensively used in transgenics.

DNA, whether in chromosomes or in extrachromosomal molecules, uses the same code to direct cell activities. The genetic code is the sequence of nucleotides in genes that is defined by sets of three nucleotides. The genetic code itself is universal, meaning it is interpreted the same way in all living things. Therefore, all cells use the same code to store information in DNA, but have different amounts and kinds of information. The entire set of DNA found within a cell (and all of the identical cells of a multicellular organism) is called the **genome** of that cell or organism.

The DNA of chromosomes within the cellular genome is responsible for the production of **proteins**. Proteins have many varied and important functions, and in fact help determine the major characteristics of cells and whole organisms. As enzymes, proteins carry out thousands of kinds of **chemical reactions** that make life possible. Proteins also act as cell receptors and signal molecules, which enable cells to communicate with one another, to coordinate growth and other activities important for wound healing and development. Thus, many of the vital activities and characteristics that define a cell are really the result of the proteins that are present. The proteins, in turn, are determined by the genome of the organism.

Because the genetic code is universal (same for all known organisms), and because genes determine characteristics of organisms, the characteristics of one kind of organism can be transferred to another. If genes from an insect, for example, are placed into a plant in such a way that they are functional, the plant will gain characteristics of the insect. The insect's DNA provides information on how to make insect proteins within the plant because the genetic code is interpreted in the same way. That is, the insect genes give new characteristics to the plant. This very process has already been performed with firefly genes and tobacco plants. Firefly genes were transferred into tobacco plants, which created new tobacco plants that could glow in the dark. This amazing artificial genetic mixing, called recombinant **biotechnology**, is the crux of transgenics. The organisms that are created from mixing genes from different sources are transgenic. The glow-in-the-dark tobacco plants in the previous example, then, are transgenic tobacco plants.

DNA transfer

One of the major obstacles in the creation of transgenic organisms is the problem of physically transferring DNA from one organism or cell into another. It was observed early on that bacteria resistant to **antibiotics** transferred the resistance characteristic to other nearby bacterial cells that were not previously resistant. It was eventually discovered that the resistant bacterial cells were actually exchanging plasmid DNA carrying resistance genes. The plasmids traveled between resistant and susceptible cells. In this way, susceptible bacterial cells were transformed into resistant cells.

The permanent modification of a genome by the external application of DNA from a cell of a different genotype is called transformation (in bacteria) or transfection (in plant or **animal** cells). Transformed cells can pass on the new characteristics to new cells when they reproduce because copies of the foreign transgenes are replicated during cell division. Transformation can be either naturally occurring or the result of transgenic technology. Scientists mimic the natural uptake of plasmids by bacterial cells for use in creating transgenic cells. Chemical, physical, and biological methods are used to introduce DNA into the cells.

Cells can be pre-treated with chemicals in order to more willingly take-up genetically engineered plasmids. Also DNA can be mixed with chemicals such as liposomes to introduce transgenes into cells. Liposomes are microscopic spheres filled with DNA that fuse to cells. When liposomes merge with host cells, they deliver the transgenes to the new cell. Liposomes are composed of lipids very similar to the lipids that make up cell membranes, which gives them the ability to fuse with cells.

Physical methods for DNA transfer include electroporation (bacterial and animal cells), microinjection of

DNA and **gene** gun. Electroporation is a process where cells are induced by an **electric current** to take up pieces of foreign DNA. DNA can also be introduced into cells by microinjection using microscopic needles. Plant tissues are difficult to penetrate due to the presence of a cell wall so a gene gun shooting pellets covered with DNA is used to transfer DNA to plants.

Biological methods used in gene transfer include viruses, **fungi,** and bacteria that have been genetically modified. Viruses that infect bacterial cells are used to inject the foreign pieces of DNA.

Use of transgenics

The use of transgenics depends on the type of organism being modified. Transgenic bacteria are used to produce antibiotics on an industrial scale, new protein drugs and to metabolize **petroleum** products, or **plastics** for cleaning up the environment.

By creating transgenic plants, food **crops** have enhanced productivity and quality. Transgenic corn, **wheat,** and soy with herbicide resistance, for example, are able to grow in areas treated with herbicide that kills weeds. In 2000, a list of 52 transgenic plants were approved for **field** trials in the United States alone, and plants included **fruits** (cranberries or papayas), **vegetables** (potatoes or carrots), industrial plants (**cotton** or wheat), and ornamental plants. Although the majority of the improvements remain confidential, it is known that scientists try to improve sugar **metabolism**, resistance to **drought** or cold, and yields by modifying photosynthetic abilities of plants. Additionally, tests are on the way to establish feasibility of edible vaccines using lettuce, corn, tomatoes and potatoes. More recent studies suggests that plants can also be used to produce other pharmaceuticals, for example growth hormone, erythropoietin or **interferons**, however, the amounts produced are too low to be of commercial value as yet.

Transgenic animals are useful in basic research for determining gene function. They are also important for creating disease models and in searching for therapeutics. Recent developments in transgenic technology allow researchers to study the effects of gene deletion, over-expression, and inhibition in specific tissues. Such studies can allow identification of the drug targets in individual tissues or evaluate other **gene therapy** only in tissues of interest. Commercially transgenic animals are used for production of monoclonal antibodies, pharmaceuticals, xenotransplantation and meat production. New areas in large animal transgenics is expression of a bacterial **enzyme** phytase in **pigs** allowing reduction of **phosphorus** excreted into the environment. In general,

KEY TERMS

Electroporation—The induction of transient pores in the plasmalemma by pulses of high voltage electricity, in order to incorporate transgenes from an external source.

Genome—All of the genetic information for a cell or organism.

Liposomes—Lipid bubbles used to deliver transgenes to host cells. Liposomes fuse with the lipids of cell membranes, passing transgenes to the inside of host cells in the process.

Photosynthesis—The process in which plants combine water and carbon dioxide to build carbohydrates using light energy.

Plasmids—Circular pieces of extrachromosomal DNA in bacteria that are engineered and used by scientists to hold transgenes.

Transfection—The transgenic infection of host cells using viruses that infect bacterial cells.

Transgenes—Genes transferred from one organism to another in transgenic experiments. Transgenes are often engineered by scientists to possess certain desirable characteristics.

Transgenics—The process of transferring genetic material from one organism to another, or one cell to another.

Xenotransplantation—Transplantation of tissue or an organ from one species to another, for example from pig to human.

by using transgenics scientists can accomplish the results similar as with selective breeding.

Despite their incredible utility, there are concerns regarding transgenics. The **Human Genome Project** is a large collaborative effort among scientists worldwide that announced the determination of the sequence of the entire human genome in 2000. In doing this, the creation of transgenic humans could become more of a reality, which could lead to serious ramifications. Also, transgenic plants used as genetically modified food is a topic of debate. For a variety of reasons, not all scientifically based, some people argue that transgenic food is a consumer safety issue because not all of the effects of transgenic foods have been fully explored. Also of great debate are the environmental protection issues as the transgenic plants can cross-pollinate with wild varieties, which in turn can lead to unforeseen consequences.

See also Clone and cloning; Photosynthesis.

Resources

Books

Houdebine, Louis M. *Transgenic Animals.* Harwood Academic Publishing, 1997.

Rissler, Jane, and Margaret Mellon. *The Ecological Risks of Engineered Crops.* MIT Press, 1996.

Ticciati, L., and R. Ticciati. *Genetically Engineered Foods: Are They Safe? You Decide.* Keats Publishing, 1998.

Periodicals

Daniell, Henry, Stephen J. Streatfield, and Keith Wycoff. "Medical Molecular Farming: Production of Antibodies, Biopharmaceuticals and Edible Vaccines in Plants." *Trends in Plant Science* (May 2001): 219–226.

Golovan, Serguei P., et al. "Pigs Expressing Salivary Phytase Produce Low-phosphorus Manure." *Nature Biotechnology* (August 2001): 741–45.

"Hunting Down Genes that Say 'No' to Disease." *Business Week* (December 13, 1993): 113.

Terry Watkins

Transistor

A transistor is a small, solid device capable of amplifying and switching electrical signals. A transistor can be manufactured using a wide variety of materials; most transistors utilize the unique semiconducting properties of silicon or germanium that has been selectively contaminated with other elements (e.g., arsenic, phosophorus, boron, gallium). A transistor is controlled by voltages communicated to it through three or more metallic contacts. Transistors are active devices, meaning that they must be supplied with power to function. Virtually all electronic devices contain transistors, from a handful to many millions; to the extent that our civilization has come to depend on computers and electronic communications, therefore, it depends on the transistor. The term *transistor* is a shortening of TRANSfer ResISTOR.

A transistor can perform a variety of useful electrical tasks because its resistance (the ease with which an electrical current flows through it) can be adjusted using a low-power control signal applied to one of a transistor's three metallic terminals. The resulting change in resistance between the other two terminals of the transistor—through one of which current enters the transistor, leaving through the other—changes the current passed through the transistor. This current can in turn be converted into a voltage by passing it through a resistor (a passive or unpowered circuit device that simply dissipates **energy**); the change in voltage across this "load"

resistor can be many times greater than the change in voltage that was used to alter the resistance of the transistor itself. This increase in amplitude or strength is called amplification, one of the most basic processes in **electronics**.

Transistor amplification can be compared to controlling a powerful flow of **water** through a pipe by turning a valve: in this analogy, the **force** applied to the valve represents the transistor's control voltage, while the water flowing through the pipe represents its output current. A small, varying force applied to the valve—a back-and-forth wiggle—causes matched variations in the greater force carried by the water passing through the pipe; a small signal thus generates another that varies identically in time but is larger in amplitude. **Vacuum** tubes, which were developed before transistors but perform the same functions, are termed "valves" in British usage in reference to this analogy.

When a transistor acts as an **amplifier** it does not create the additional energy appearing in its output signal, just as a valve does not create water it dispenses. Rather, a transistor modulates the energy flowing from a **battery** or power supply in a way that is similar to a valve adjusting the flow **rate** from a source of pressurized water.

It is also clear from the valve analogy that instead of wiggling the valve one might choose instead to operate it in only two positions, open and shut, avoiding partial settings completely. Vacuum tubes and transistors can also be operated in this way, switching an electrical current on and off in response to a two-valued control signal. The on-off signals generated by this technique, *switching*, are the basis of digital electronics; the constantly-varying signals involved in *amplification* are the basis of analog electronics. Both rely on transistors, which are cheaper, smaller, and more reliable than other devices that can perform these functions.

The history of the transistor

Discovery of the transistor was publicly announced in 1948. Before this time, electronics had depended almost exclusively upon vacuum tubes for amplification and switching actions. Vacuum tubes are relatively bulky, short-lived, and wasteful of power; transistors are small—from peanut-size to only a few molecules across—long-lived, and dissipate far less power. Transistors are also resistant to mechanical shocks and can be manufactured by the millions on tiny semiconductor crystals (chips) using optical techniques. Transistors and related solid-state (i.e., entirely solid) devices have replaced vacuum tubes except for specialized applications, especially those involving high power.

Silicon and germanium

The first transistors were made from germanium, but now most transistors are made from silicon. Silicon (Si) and germanium (Ge) form similar **crystal** structures with similar physical properties, but silicon is preferred over germanium because of silicon's superior thermal characteristics. Crystals of Si and Ge are neither good electrical insulators nor good electrical conductors, but conduct **electricity** at a level midway between metallic conductors (which have very low resistance to **electric current**) and nonmetallic insulators such as **glass** (which have a very high resistance to electric current). Transistor action is made possible by semiconduction.

Each atom in a silicon or germanium crystal lattice has four **atoms** as close neighbors. That is, each atom is held in its place in the crystal's orderly structure because each atom shares its four outermost electrons with the outermost electrons of four nearby atoms. This sharing holds the atoms together by the process termed covalent bonding. Covalent bonding also prevents these outermost electrons from moving through the crystal (i.e., flowing as an electric current) as easily as do the conduction electrons in metals. They are not bound too tightly to break loose if given a small amount of extra energy, but cannot wander easily through the crystal. **Heat**, **light**, or **ionizing radiation** may all increase the semiconductor's **electrical conductivity** by liberating these electrons to support current. Usually these effects are unwanted, because one does not want the properties of a circuit to vary with **temperature**. Ionizing **radiation**, furthermore, may cause transistors to fail by permanently altering the crystal structure. The first active communications **satellite** placed in **orbit** by the United States, Telstar (1962), failed when its transistors were exposed to unexpected levels of ionizing radiation. (Early satellites such as Echo had been passive **radio** reflectors, containing no amplifying transistor circuits to be affected by radiation.)

Doping

An pure or "intrinsic" silicon crystal contains about one non-silicon impurity atom for every 100 million or so silicon atoms. These impurity atoms are implanted in the crystal by a process termed doping. They are located in the crystal lattice as if they were themselves silicon atoms, but change the properties of the lattice radically because of their distinct properties.

When doping adds impurity atoms with five electrons in their outermost (**valence**) orbit, the result is termed an n-type semiconductor. Arsenic, for example, has five valence electrons and is often used to produce n-type semiconductor. Pentavalent impurity atoms share only four of their five valance electrons with their four closest silicon neighbors; the fifth is free to move through the crystal in response to any electric **field** that may be present, almost like a conduction **electron** in an ordinary **metal**. An n-type semiconductor thus conducts electricity more easily than an intrinsic semiconductor.

If an impurity with only three valence electrons (e.g., boron, **aluminum**, gallium, and indium) is used, p-type semiconductor results. These atoms are short one of the electrons needed to establish a covalent bond with all four of its silicon neighbors and so introduce a defect into the crystal lattice, a positively-charged location where a negatively-charged electron would be found if a silicon atom had not been displaced by the impurity atom. This defect, termed a hole, can move when a neighboring electron slips into the hole, leaving a new hole behind. The hole will have moved from one location to another within the crystal, behaving much like a positive counterpart of an electron.

Holes travel somewhat more slowly than electrons within a an electrical field of given strength, but this difference in speed is usually not important in practice. Both the excess electrons donated in n-type semiconductor by pentavalent impurity atoms and the holes created in p-type semiconductor by trivalent impurity atoms increase the conductivity of the semiconductor; for example, at 86°F (30°C) the conductivity of n-type silicon with one pentavalent impurity atom per 100 million silicon atoms is 24,100 times greater than that of intrinsic silicon.

p-n junction diodes

A useful electrical property results at a boundary where p-type material abuts on n-type material in the same semiconductor crystal. The result is termed a p-n junction **diode** (or simply junction diode). A junction diode may be thought of as a one-way valve for electricity; it will carry current in one direction much more easily than in the opposite direction. Understanding of the transistor—especially that type termed the bipolar junction transistor or BJT—begins with knowledge of the p-n junction diode. A BJT is, in effect, a back-to-back pair of p-n diodes within a single crystal.

In either p-type or n-type semiconductors, there are two types of charge carriers that carry current: majority carriers and minority carriers. Electrons are the majority carriers in n-type material, due to the extra electrons donated by pentavalent impurity atoms, and holes are the majority carriers in p-type semiconductors. The minority carriers are the relatively few, oppositely charged carriers, electrons in p type and holes in n-type semiconductor, which cannot be eliminated entirely. Heat, ionizing radiation, and unintended impurities in the original intrinsic crystal produce minority carriers. Practical diodes do not behave ideally because minority carriers allow a small reverse current, that is, a trickle of charges leaking backward through a diode, whereas an ideal diode would present a total block to current in that direction. Leakage current occurs in transistors as well as diodes, and these **currents** can have important consequences for circuit performance.

If voltage is applied across a p-n junction diode with polarity that causes the p region of the diode to be more positive than the n region, the majority carriers in both p and n regions will be pushed toward each other, meeting at the boundary. A diode polarized in this way is said to be forward biased. A forward-biased diode conducts quite well. If the voltage polarity is reversed causing the n-type material to be more positive than the p-type material, the two types of majority carriers will be pulled away from each other. This condition is called reverse bias or back bias. The small current leak through a back-biased diode is the result of minority carriers moving in the opposite direction compared to majority carriers.

There is a very thin **volume** at the boundary where n-type semiconductor materials interfaces with p-type material, termed the depletion region. In the depletion region electrons tend to fill adjacent holes, depleting the crystal of carriers. When majority carriers from each region are pushed toward each other, hole-electron pairs continually annihilate each other. As each hole is filled by an electron, a new hole and a new electron will be injected into the crystal at the ohmic connections to the crystal (i.e., those places where metal contacts are applied). In this way current can continue to flow through the diode as long as the circuit is energized.

If the reverse-biasing voltage across a diode increases above a critical threshold the diode will suddenly break into heavy conduction when the electric field in the depletion region between the n and p materials is so strong that electrons are torn from their bonding roles. This condition is called Zener breakdown. Usually transistors are operated at voltages low enough so that this type of breakdown doesn't take place. Unless the breakdown current is limited by the external circuitry the transistor or diode may easily destroyed when excess voltages or voltages with the wrong polarity are applied.

Bipolar junction transistors

If the same crystal is doped so that each end is n-type and the very thin slice in the center is p-type, the resulting sandwich forms a bipolar junction transistor or n-p-n transistor. In an n-p-n transistor one of the n-type regions is termed the collector, the other n-type region is termed the emitter. (The emitter emits majority charge carriers, and the collector collects them.) The very thin slice of p-type material in the center is termed the base of the transistor. In a p-n-p transistor the collector and emitter regions are made from p-type semiconductor and the base has the characteristics of n-type material. Both n-p-n and p-n-p transistors are in common use but these two transistor types are not directly interchangeable since they require different power-supply polarities. Many circuits employ both n-p-n and p-n-p transistors, but the circuitry must supply the correct voltages. It is common to connect the two types of transistors together in an arrangement that is called complementary **symmetry**.

Transistor action

Transistors are able to amplify signals because their design permits the supply of charge carriers to be adjusted electrically. A transistor will have a high **electrical resistance** when it is starved for charge carriers but it will conduct quite well when a control signal injects extra carriers that can be used to support increased current.

Common base, common emitter, and common-collector configurations

There are three ways to connect a bipolar junction transistor into a working circuit, depending upon which of the three transistor elements is chosen as the common reference for the other two elements. These variations, called common base, common emitter, and common collector, produce different circuit actions each with unique characteristics. An n-p-n transistor configured as a common-emitter amplifier, where both the base and the collector circuits are referenced to the emitter, is normally connected with a positive voltage on the collector, as referenced to the emitter. The collector-base diode and the base-emitter diode appear to be in series, connected back-to-back. The collector-base diode is reverse biased so that almost no current will flow unless the base-emitter diode is forward biased. The very small current in the collector circuit under these conditions is because the p-type material in the base is starved for the n-type majority carriers that the collector circuit requires if it is to conduct a significant current. When the base-emitter junction is forward biased, the carriers needed for current in the collector circuit find their way into the collector.

The base-emitter diode in the transistor offers a very low resistance to current flow when it is forward biased. It is therefore very easy to cause current in the transistor's input circuit. Since the base region is made very thin, most of the majority carriers that flow from the emitter will be caught by the strong electric field in the collector base junction before they can exit through the base connection. It takes only a small amount of power to cause current in the transistor's forward-biased base-emitter input circuit yet almost all this easily forced input current appears in the collector circuit. A low-powered signal becomes a higher-powered signal when the input current caused by a low voltage appears almost undiminished in the collector circuit, but at a higher voltage.

Field-effect transistors (FETs)

Field-effect transistors (FETs) are solid-state active devices based on a different principle than BJTs but producing much the same result. FETs are three-terminal devices, just as are BJTs. The input terminal of an FET is termed its gate and constitutes one of the electrodes of a reverse-biased diode. FETs achieve current control by channeling current through a narrow n-type or p-type pathway whose conductivity is adjusted by the input signal. The output current controlled by an FET passes between the two remaining terminals called a source and a drain. The current through an FET must find its way through a narrow channel formed by the input-diode junction. Since this input diode is reverse biased, this channel tends to have few charge carriers. The input signal to the FET can deplete or enhance the number of available charge carriers in this channel, regulating the current in the drain circuit. Because the input diode is reverse biased, the FET demands almost no current from the signal source, therefore almost no power must be supplied. The power gain commonly achieved in an FET amplifier is very high.

A particular type of FET called a MOSFET (metal oxide semiconductor field-effect Transistor) can have an input resistance as high as 10^{18} ohms. Because of their very high input resistance, FETs are instantly destroyed if they receive even a small static-electric charge from careless handling. Sliding across a plastic chair may impart enough charge to a technician's body to destroy a field-effect transistor's input diode at the first **touch**. FETs must be handled only by persons who ground themselves before touching these devices to first dissipate static charges.

FETs are particularly useful as amplifiers of very weak signals such as those produced by high-quality microphones. FETs have more desirable overload characteristics than BJTs, so that FETs are able to handle many signals simultaneously, some strong and some weak, without suffering troublesome distortion. Before FETs were used in **automobile** receivers, these radios were easily overloaded by strong signals; the introduction of FETs made a tremendous improvement in automobile-radio receiver performance.

Integrated circuits

One of technology's most significant breakthroughs has been the discovery that many intercon-nected transistors can be created simultaneously on a single, small chip of semiconductor material. The techniques used to create individual transistors could also connect these devices to form microscopic circuits that are unified or integrated into a single solid object. The first integrated circuits (ICs) were primitive arrangements utilizing just a few transistors or diodes, but now it is common for single chips to contain millions of transistors.

Application-specific integrated circuits

Special integrated circuits are developed as application-specific integrated circuits, or ASICs, which are single chips performing a specific set of tasks, useful for only one job. For example, almost all the circuitry needed for an AM-FM radio receiver is routinely produced on a single ASIC chip that replaces the hundreds of individual components. TV receivers have also become increasingly dependent upon ASICs. ICs may also take the form of generalized modules intended to be used in a broad range of applications. Many of the first generalized ICs were designed as operational amplifiers, a general-purpose amplifier made from many transistors.

Complementary metal-oxide semiconductors

Complementary metal-oxide semiconductors or CMOS devices are coupled complementary MOSFETS in series, configured so that either MOSFET will conduct when the other is turned off. CMOS devices are frequently used as on-off switches in computer logic and memory circuits. CMOS devices use so little power that they allow electronic watches to operate for five years without battery replacement. CMOS devices only require a significant current when the two complementary FETs in series change their conduction state, i.e., a quick pulse of current flows only during each switching action. CMOS chips are commonly used in computer and **calculator** circuits, and CMOS is equally as common in consumer-type entertainment equipment.

The significance and future of the transistor

Perhaps the principal contribution of transistors has been the feasibility of highly complex yet miniature electronic equipment; they have made it possible to hold more electronics in one's hand than could be contained in a large building in the days when vacuum tubes were the only active devices available. This in turn has made it possible to pack complex functionality into packages of manageable size—computers, cell phones, automobile engine controllers, and a host of other tools. Transistors

KEY TERMS

Capacitor—Passive circuit component used to introduce capacitance.

Covalent bond—A chemical bond formed when two atoms share a pair of electrons with each other.

Crystal—Ordered three dimensional array atoms.

Dopant—A chemical impurity which is added to a pure substance in minute quantities in order to alter its properties.

Doping—Adding impurities to change semiconductor properties.

Inductor—A component designed to introduce inductance.

Intrinsic—Semiconductor material containing very few impurities.

Pentavalent—An element with five valence electrons.

Photoconducive—A better conductor of electricity when illuminated.

Resistor—An electric circuit component that opposes the flow of current.

Tetravalent—Element with four valence electrons.

Trivalent—Element with four valence electrons.

Valence electrons—The electrons in the outermost shell of an atom that determine an element's chemical properties.

continue to be intensively researched around the world, for decreasing the size and power consumption of the individual transistor on a chip offers immediate profits. Researchers have already demonstrated, in the laboratory, extremely small transistors made out of only a few molecules—even a transistor employing only a single electron. They have also demonstrated the practicality of transistors made out of plastic, which could be even cheaper and more shock-resistant than conventional devices. However, it may be years before such exotic developments see commercial application. In the near future, increasing transistor densities on chips (which are greatly desired by manufacturers) are likely to be achieved by improving fabrication techniques for traditional semiconductor devices.

Resources

Periodicals

Bachtold, Adrian, et al. "Logic Circuits with Carbon Nanotube Transistors." *Science* (November 9, 2001): 1317-1320.

Markoff, John. "Xerox Says New Material Will Allow Plastic Transistors." *New York Times* December 3, 2002.

Donald Beaty
Larry Gilman

Transition metals *see* **Periodic table**

Transitive

The concept of transitivity goes back at least 2,300 years. In the *Elements,* Euclid includes it as one of his "common notions." He says, "Things which are equal to the same thing are also equal to one another." As Euclid puts it, if a = b and c = b, then a = c, which is equivalent to the modern version, which has "b = c" rather than "c = b."

Transitivity is a property of any **relation** between numbers, geometric figures, or other mathematical elements. A relation R is said to be transitive if a R b and b R c imply that a R c. For example, 6/4 = 3/2 and 3/2 = 1.5, therefore 6/4 = 1.5.

Of course, one would not be likely to make use of the transitive property to establish such an obvious fact, but there are cases where the transitive property is very useful. If one were given the two equations

$$y = x^2$$
$$x = z + 1$$

one could use transitivity (after squaring both sides of the second equation) to eliminate x.

$$y = z^2 + 2z + 1$$

Transitivity is one of three properties which together make up an "equivalence relation."

Transitive law	If a R b and b R c, then a R c
Reflexive law	a R a
Symmetric law	If a R b, then b R a

To be an equivalence relation R must obey all three laws.

A particularly interesting relation is "wins over" in the game scissors-paper-rock. If a player chooses "paper," he or she wins over "rock;" and if the player chooses "rock," that wins over "scissors;" but "paper" does not win over "scissors." In fact, it loses. Although the various choices are placed in a winning-losing order, it is a non-transitive game. If it were transitive, of course, no one would play it.

In *Wheels, Life, and Other Mathematical Amusements,* Gardner describes a set of non-transitive dice. Die A has its faces marked 0, 0, 4, 4, 4, and 4. Each face of

KEY TERMS

. . . .

Reflexive—A relation R is reflexive if for all elements a, a R a.

Symmetric—A relation R is symmetric if for all elements a and b, a R b implies that b R a.

Transitive—A relation R is transitive if for all elements a, b, and c, a R b and b R c implies that a R c.

die B is marked with a 3. Die C is marked 2, 2, 2, 2, 6, 6. Die D is marked 1, 1, 1, 5, 5, 5. Each player chooses a die and rolls it. The player with the higher number wins. The probability that A will win over B is 2/3. The probability that B will win over C or that C will win over D is also 2/3. And, paradoxically, the probability that D will win over A is also 2/3. Regardless of the die which the first player chooses, the second player can choose one which gives him a better chance of winning. He need only pick the next die in the unending sequence... < A < B < C < D < A < B <...

There are many relations in life which are theoretically transitive, but which in practice are not. One such is the relation "like better than." One can like apples better than bananas because they are juicier, pomegranates better than apples because they have more flavor, and still like bananas better than pomegranates. They are, after all, a lot easier to eat. Transitivity, reflexivity, and **symmetry** are properties of very simple, one-dimensional relations such as one finds in **mathematics** but not in much of ordinary life.

Resources

Books

Birkhoff, Garrett, and MacLane, Saunders. *A Survey of Modern Algebra.* New York: The Macmillan Co. 1947.

Dantzig, Tobias. *Number, the Language of Science.* Garden City, N. Y.: Doubleday and Co., 1954.

Gardner, Martin. *Wheels, Life, and Other Mathematical Amusements.* New York: W. H. Freeman and Co., 1983.

J. Paul Moulton

Translations

A translation is one of the three transformations that move a figure in the **plane** without changing its size or shape. (The other two are rotations and reflections.) In a translation, the figure is moved in a single direction without turning it or flipping it over.

A translation can, of course, be combined with the two other rigid motions (as transformations which preserve a figure's size and shape are called), and it can in particular be combined with another translation. The product of two translations is also a translation, as illustrated in Figure 2.

If a set of points is drawn on a coordinate plane, it is a simple matter to write equations which will connect a point (x,y) with its translated "image" (x^1, y^1). If a point has been moved a units to the right or left and b units up or down, a will be added to its x-coordinate and b to its y-coordinate. (If a < 0, the **motion** will be to the left; and if b < 0, down.) Therefore

$$x^1 = x + a \quad \text{or} \quad x = x^1 - a$$
$$y^1 = y + b \quad \quad y = y^1 - b$$

In these equations, the axes are fixed and the points are moved. If one wishes, the points can be kept fixed and the axes moved. This is called a translation of axes. If the axes are moved so that the new origin is at the former point (a,b), then the new coordinates, (x^1, y^1) of a point (x, y) will be (x - a, y - b).

If one has two translations:

$$x^1 = x + 3 \quad y^1 = y - 6$$
$$x^{11} = x' - 2 \quad y^{11} = y' + 1$$

they can be combined into a single transformation. By substitution one has

$$x^{11} = x + 1 \quad y^{11} = y - 5,$$

which is another translation. This illustrates that the "product" of two translations is itself a translation, as claimed earlier.

If one has two translations

The idea of a translation is a very common one in the practical world. Many machines are translational in their operation. The machinist who cranks the cutting-tool holder up and down the bed of the lathe, is "translating" it. The piston of an **automobile** engine is translated up and down in its cylinder. The chain of a bicycle is translated from one sprocket wheel to another as the cyclist pedals, and so on.

The bicycle chain is not only translated, it works because it has translational **symmetry**. After a translation of one link, it looks exactly as it did before. Because of this symmetry, it continues to fit over the teeth of the sprocket wheel (which itself has rotational symmetry) and to turn it.

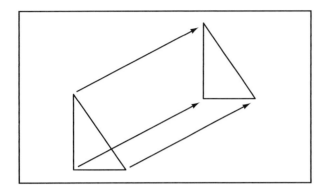

Figure 1. *Illustration by Hans & Cassidy. Courtesy of Gale Group.*

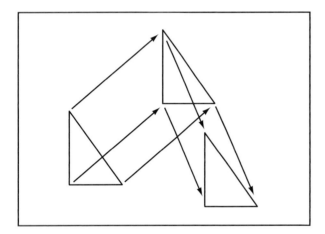

Figure 2. *Illustration by Hans & Cassidy. Courtesy of Gale Group.*

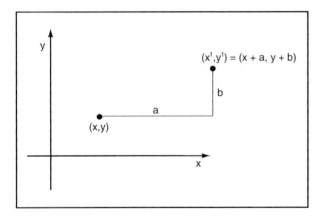

Figure 3. *Illustration by Hans & Cassidy. Courtesy of Gale Group.*

One important use of translations is to simplify an equation which represents a set of points. The equation $xy - 2x + 3y - 13 = 0$ can be written in factored form $(x + 3)(y - 2) = 7$. Then, letting $x^1 = x + 3$ and $y^1 = y - 2$, the equation is simply $x^1y^1 = 7$, which is a much simpler and more easily recognized form.

KEY TERMS
. . . .

Rigid motion—A transformation of a plane figure which does not alter the size or shape of the figure.

Such transformations are useful in drawing graphs where many points have to be plotted. The graph of $x^1y^1 = 7$ is a **hyperbola** whose branches lie entirely in the first and third quadrants with the axes as asymptotes. It is readily sketched. The graph of the original equation is also a hyperbola, but that fact may not be immediately apparent, and it will have points in all four quadrants. Many points may have to be plotted before the shape takes form.

If one has an equation of the form $ax^2 + by^2 + cx + dy + e = 0$ it is always possible to find a translation which will simplify it to an equation of the form $ax^2 + by^2 + E = 0$.

For example, the transformation $x = x^1 - 2$ and $y = y^1 + 1$ will transform $x^2 + 3y^2 + 4x - 6y - 2 = 0$

into $x^2 + 3y^2 - 9 = 0$ which is recognizable as an **ellipse** with its center at the origin.

Transformations are particularly helpful in integrating functions such as intergral $(x + 5)^4$ dx because intergral x^4 dx is very easy to integrate, while the original is not. After the translated **integral** has been figured out, the result can be translated back, substituting $x + 5$ for x.

Translational symmetry is sometimes the result of the way in which things are made; it is sometimes the goal. Newspapers, coming off a web press, have translational symmetry because the press prints the same page over and over again. Picket fences have translational symmetry because they are made from pickets all cut in the same shape. Ornamental borders, however, have translational symmetry because such symmetry adds to their attractiveness. The gardener could as easily space the plants irregularly, or use **random** varieties, as to make the border symmetric, but a symmetric border is often viewed as esthetically pleasing.

See also Rotation.

Resources

Books

Coxeter, H.S.M., and S.L. Greitzer. *Geometry Revisited.* Washington, DC: The Mathematical Association of America, 1967.

Hilbert, D., and S. Cohn-Vossen. *Geometry and the Imagination.* New York: Chelsea, 1952.

Newman, James, ed. *The World of Mathematics.* New York: Simon and Schuster, 1956.

Pettofrezzo, Anthony. *Matrices and Transformations.* New York: Dover, 1966.
Yaglom, I.M. *Geometric Transformations.* Washington, DC: The Mathematical Association of America, 1962.

J. Paul Moulton

Transpiration

Transpiration refers to the **evaporation** of **water** from a biological surface, such as leaves, skin, or lungs. In its most common usage, however, transpiration refers to the loss of water from **plant** foliage, occurring through microscopic pores known as stomata. Transpiration is a component of a larger process known as **evapotranspiration**, which is the evaporation of water from a landscape, including both inorganic surfaces such as **soil** and bodies of water, and biological surface such as foliage.

Why do organisms transpire water?

Most transpiration by plants involves water that evaporates from the moist membranes of a **tissue** known as spongy mesophyll, which occur in the minute cavities beneath the tiny **leaf** pores called stomata. Stomata can be closed tightly using bordering cells known as guard cells. However, in most plants stomata are kept open much of the time. This is done so that **carbon dioxide**, which is needed for **photosynthesis**, can diffuse into the leaf, and **oxygen**, a waste product, can diffuse out. Therefore, transpiration by plants can be viewed as an unavoidable consequence of having moist mesophyll membranes exposed to the atmosphere. It is important to understand that, in general, any area of vegetation with a multilayered canopy of foliage will evaporate much larger quantities of water than can an equivalent non-foliated area, such as the surface of a **lake** or moist soil.

A similar explanation can be developed for animals, who must also maintain moist respiratory surfaces open to the atmosphere, for example in the lungs, from which water can freely evaporate. Transpiration can be an important problem for both plants and animals, because it may be difficult to replenish their water losses in relatively dry environments, so that dehydration can occur. If severe, dehydration can kill plants and animals. In some respects, then, transpiration can be regarded as a necessary evil that organisms must endure in order to exchange oxygen and **carbon** dioxide with the atmosphere.

However, there can sometimes be important benefits of transpiration. It takes about 540 calories of thermal **energy** to evaporate 1 gallon of water at an ambient **temperature** of 68°F (20°C). This is a rather large amount of energy, and as a result transpiration can be an important way by which some animals and plants cool themselves. For example, when humans are hot, they sweat to distribute water onto their exposed skin, which is then cooled by the subsequent evaporation.

Transpiration as an ecological process

Transpiration is an ecologically important process. In areas where **forests** are common, evapotranspiration almost entirely occurs as transpiration, and this process can account for a substantial part of the manner by which the landscape deals with water inputs through rain and snow. For example, in typical forested landscapes of northeastern **North America**, evapotranspiration accounts for about 15-40% of the annual inputs of water with **precipitation**, the remainder draining to **groundwater**, or flushing from the system as streamflow.

In any regions with a seasonal climate, rates of evapotranspiration vary greatly during the year. Consider, for example, the case of a landscape in a temperate climate, covered with a seasonally deciduous, **angiosperm** forest. During winter, very little transpiration occurs because plant tissues are frozen. However, there can still be some physical evaporation of water from the surface of snow and **ice**, occurring by the direct vaporization of solid water, a process known as sublimation. During the springtime, unfrozen water is abundant, but the trees do not yet have foliage and this greatly reduces the rates of transpiration. During the growing season, air temperatures are warm and the trees are fully foliated, so transpiration occurs in large rates. During this time of the year, so much water is pumped into the atmosphere through foliage that the **rate** of evapotranspiration typically exceeds water inputs by rainfall. As a result, the soil is dried by the demands of plant roots for water, to the extent that streams may cease to flow by late summer. Once the trees drop their leaves in the autumn, transpiration rates decrease greatly, the water-depleted soil becomes recharged by rainfall, and streams again flow.

Effects of human activities on transpiration

The influence of an intact forest on evapotranspiration is affected by ecological disturbances, such as clear-cutting and **wildfire**. In general, these sorts of stand-level disturbances greatly reduce the transpiration component of evapotranspiration for several years, a change that influences other hydrological processes, such as the timing and amounts of stream flow, which may then have effects on **flooding** and **erosion**. In addition, on sites that do not drain well, substantial decreases in transpiration can increase the height of the water table.

Evapotranspiration—The evaporation of water from both inorganic and biological surfaces on a landscape.

Stomata—Pores in plant leaves which function in exchange of carbon dioxide, oxygen, and water during photosynthesis.

Watershed—The expanse of terrain from which water flows into a wetland, waterbody, or stream.

An ecological study done at Hubbard Brook, New Hampshire, involved the clear-felling of all trees on a 39.5-acre (16-hectare) **watershed**, which was then kept clear of regenerating plants for two additional years. The subsequent disruption of transpiration increased the amount of streamwater flow by an average of 31% over the three-year period. Many other studies of **forestry** have come to similar conclusions—clear-cutting increases streamflow by decreasing transpiration. However, because disturbed forests regenerate quickly through ecological **succession**, the amount of foliage on the site quickly recovers, and the effect of disturbance on streamflow tends to be rather short lived. In general, the pre-cutting stream flow volumes are substantially re-attained after three to five years of revegetation has occurred, because of the recovery of the transpirational surface area of plant foliage. (Of course, the regenerated foliage occurs on relatively short **grasses**, herbs, and shrubs, rather than on taller trees as in the initial forest.) The largest increases in stream flow usually occur in the first year after the forest is harvested, with progressively smaller effects afterwards.

Resources

Books

Freedman, B. *Environmental Ecology.* 2nd ed. San Diego: Academic Press, 1995.

Kimmins, J. P. *Forest Ecology.* New York: Macmillan, 1987.

Bill Freedman

Transplant, surgical

A surgical transplant involves the removal of body parts, organs, or tissues from one person and implanting them into the body of another person. Although the idea of transplantation to cure **disease** dates back several centuries, transplantation has been considered a viable ther-

apy for only a few decades. The relatively recent growth in transplantation stems primarily from expanding knowledge about the body's **immune system** and the ability to suppress its natural response to attack foreign **tissue**. Another advance has been the ability to preserve organs out of the body for longer periods of time until they can be transplanted. Although most transplants involve the transference of tissues or organs between two humans, research is rapidly advancing toward using **animal** organs for transplantation into humans. Unfortunately donor organs are scarce, which has led to ethical questions concerning what patients would most benefit from receiving a transplant. For example, some believe that alcoholics suffering from **cirrhosis** of the liver should not receive a liver transplant that they may eventually destroy again because of **alcohol** abuse.

The history of transplants

The idea of transplanting animal or human parts dates back for many centuries. The mythical Chimera, animals made up of different animal parts, was believed to be the work of the gods. Perhaps the most famous chimera is the Sphinx in Egypt, which has the head of a man and the body of a lion.

The first viable transplantation occurred in the sixteenth century when a technique was developed for replacing noses lost during battle or due to syphilis. The technique involved using skin from the upper inner arm and then grafting and shaping it onto the nose area. In the eighteenth century, Scottish surgeon John Hunter successfully transplanted a cock's claw to the animal's comb. Corneal transplants between two **gazelles** were also successfully performed in the latter part of the century.

During the nineteenth century, advances in **surgery** (like the development of antiseptic surgery to prevent **infection** and anesthetics to the lessen the **pain**) increased the success rates of most surgical procedures. However, transplantation of organs languished as surgeons had no knowledge of how to "reconnect" the **organ** to the new body. It was not until techniques for vascular anastomosis, or the ability to reconnect **blood** vessels, was developed near the twentieth century that transplant surgery began to move ahead. The first long-lasting renal transplant was performed in Germany in 1902, when a dog kidney was transplanted into another dog by using tube stents (a slender tube used to support the structural integrity of the artery during an operation) and ligatures (wires used to tie off the blood vessels) to make the vascular connections.

The next major advance in transplantation would not occur for more than 40 years. Although renal and kidney transplants were attempted, the transplant recipient's body always rejected the organ. In 1944, Peter Medawar

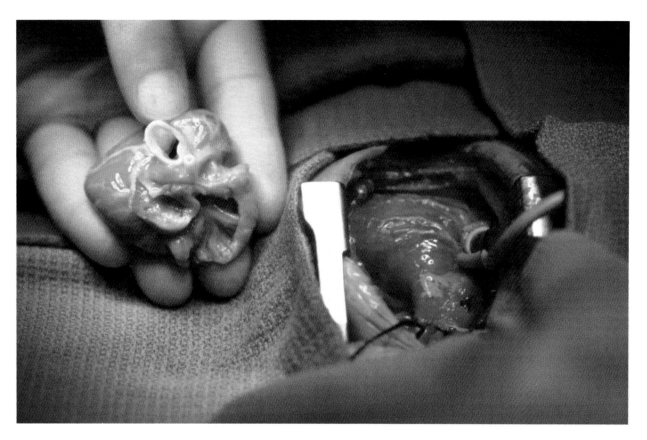

A comparison of the old and new hearts of Dylan Stork, the smallest heart transplant recipient in the world. Dylan weighed 5.5 lb (2.5 kg) at the time of the operation. *Photograph by Alexander Tsiaras. National Audubon Society Collection/Photo Researchers, Inc. Reproduced by permission.*

showed that the rejection was due to the immune system, which attacked the foreign tissues or organs as foreign invaders, much the same way it works to ward of viruses and other disease. Although short-term successful renal transplants from donor to recipient were achieved in the early 1950s, these transplants usually were rejected by the patient's immune system. As a result, scientists began to focus on manipulating the immune system so it would accept the transplant.

By the early 1960s **radiation** and drugs were being used to suppress the immune system, in effect, shutting it down to prevent rejection. As a result, that decade saw the first bone marrow transplant, kidney transplant, and kidney/pancreas transplant. In 1967, Christiaan Barnard, a South African surgeon, received world wide notoriety for achieving the first successful **heart** transplant.

The next major advance came in 1978 with the development of the extremely effective antirejection (also called immunosuppressant) drug cyclosporin. Continued research on how to selectively control the immune system has grown to the point that transplantation is now a relatively common operation with more than 35,000 sur-

gical transplants performed in medical centers throughout the world each year.

Transplantation and the immune system

Once surgeons acquired the ability to sever and re-connect **arteries**, the ability to control the immune system has been the primary concern in transplant surgery. The immune system is a complex biological network intricately linked to cells found in the blood. Many individuals share similar antigens, (small molecular **proteins** found on the surface of tissues that can stimulate an immune response,) and one of the first goals is to match antigens between organ donors and the recipients. The idea is to increase the likelihood that the patient's immune system will not attack because it recognizes a foreign antigen in the donor tissue.

Scientists have identified six primary types of antigens, called histocompatibility-locus antigens. Transplantation between people whose tissues and organs share all of these antigens generally is successful. For example, family members, particularly identical twins, usually have the same antigens. (In fact, the first successful kidney transplant was performed in 1955 when a young man received one of the kidneys of his twin brother.) As a result, relatives are often the source of donor tissues for patients who need kidneys and bone marrow transplants because one kidney or some bone marrow can be donated without causing harm to the donor.

As the number of matching antigens between donor and recipient decreases, the need to suppress the immune system with drugs becomes increasingly important. Unfortunately, many of these drugs have side effects and can suppress the immune system to the point that the patient becomes susceptible to other infections and diseases, which can cause the transplant to be rejected and lead to the patient's death. The advantage of cyclosporin is that it can be targeted somewhat to work in specific tissues and organs of the body, thus keeping the patient from becoming entirely immunocompromised, which can lead to death from any number of diseases and infections not directly related to the transplanted organ's functioning.

The immune responses

Although there are probably many types of immune responses, scientists have only identified a few associated with organ rejection. Hyperacute rejection occurs when the patient's body has already produced a large number of antibodies (proteins manufactured by the immune system to battle disease and infection) that immediately recognize the antigens from the donor organ. This type of rejection, which usually occurs because of incompatible blood types, is often instantaneous, sometimes causing the patient to die even before the surgery is completed.

Acute rejection usually takes several days to occur because the immune system's white blood cells called lymphocytes, which are integral to its functioning because they initiate the production of antibodies, are lying dormant until the patient receives the organ. As the immune system initiates a response to the antigens in the donor organ, specific immune cells in the blood begin to attack. Acute rejection is combated by the use of immunosuppressive drugs. The longer the patient survives without a immune response attacking the organ, the greater the chances for keeping the immune system under control and for long-term survival.

However, chronic rejection can occur several months or years later. Although this type of rejection is rare, when it occurs, the donor organ slowly deteriorates despite all efforts at immunosuppression. Scientists do not fully understand why the immune response may kick in months or years after an organ has been successfully transplanted.

Types of transplants

There are many types of transplant surgeries, from specific **cell** and tissue transplants to entire organs. Cornea tissue transplants for the **eye** were one of the first successful transplants routinely performed. A thin transparent **membrane** found in front of the eye, the cornea can cause blindness or clouded **vision** if it is scarred by injury or infection. The transplant procedure involves cutting out part of the damaged cornea and replacing it with cornea from a donated eye. Cornea transplants are relatively simple procedures because they require no reconnection of blood supplying arteries, thus removing the danger of an adverse immune response. These procedures are successful 90% of the time.

The transplantation of bone tissue was first performed in the 1890s and involved replacing diseased or injured bone with pieces of bone from donors. The modern procedure involves using donor bone as a type of scaffolding built over **metal** nails that immobilizes and connects the patient's remaining bone sections.

Unlike bones and the cornea, most other transplants are more difficult to achieve because they involve circulating blood and the immune system. Skin grafts for **burn** patients, for example, are often achieved using autografts (meaning from the self), in this case, segments of skin from other areas of the patient's own body. A flap **graft** begins with the partial separation of skin from its original site until adequate blood circulation is achieved, then it is completely severed and grafted onto the transplant site. This technique has a high success **rate** for curing scars or deformities. A full thickness autograft entails removing small pieces of all the layers of skin for transplantation; these types of skin grafts are particularly suitable for the face to achieve the least amount of scarring. A split-thickness autograft removes tissue-thin layers of skin that heal easily when transplanted. The one drawback to this type of skin graft is that the transplanted tissue usually appears reddish, which makes them more noticeable. Some skin grafts come from the skin of human donors or other animals, like **pigs**, and are used when large areas must be covered. Unlike the cosmetic goals of other grafts, these grafts are used primarily to stop the patient from losing fluids through the burned area and to prevent infection. Research is also underway to grow skin for grafting from a few donated skin cells.

Bone marrow transplants are often used to treat patients with blood diseases, like **leukemia**. The transplant procedure involves taking the tissue from the center of bones through a needle, a technique called bone marrow aspiration. The marrow may be an autograft taken from a patient while the disease is in remission and stored until they need the healthy blood cells. Bone marrow taken from other donors of the same **species** are called allografts. Unlike most immune responses in transplantation in which the donor's body rejects the transplant, in bone marrow allografts it is often the transplanted marrow that reacts against the host. This is known as graft-versus-host disease.

Organ transplants

By far, organ transplantation has received the most attention in the popular press. People who would be severely debilitated or die because of failing organs are the recipients of organ transplants. The primary organ transplants are heart, liver, pancreas, lung, and kidney transplants. Under certain circumstances, a patient may receive more than one transplant, the most common multiple transplant procedure is the dual heart/lung transplant.

Kidney transplants are designed to treat patients whose kidneys are failing, making them unable to process body waste products. These transplants have approached a 90% success rate. If the transplant fails, the patient may be put on **dialysis** until a new donor kidney can be found. Dialysis uses a machine to artificially remove blood from the body, clean the waste products by filtering the blood, and then return the blood to the body.

Other transplantable organs include the liver, the heart, and the lungs. With the development of immunosuppressive drugs, liver transplants have been growing in their success rates, although many patients still die because of organ rejection or from infections. Heart and lung transplants are often performed simultaneously since they share an interrelated vascular system and survival rates are higher for lung transplant patients when they also receive a new heart.

Organ transplants are delicate and complex procedures that require a well trained staff and usually take several hours or much longer to perform. As the surgeon removes the organ, sutures (wire or some other material) are used to close or tie off arteries and other connections to the organ. The new organ is then placed in position and the painstaking task of reconnecting the organ begins. Although the techniques, like vascular anastomosis, for making these connections are well established, they can be difficult procedures. For example, many of the connections to the liver lie underneath the liver and out of the surgeons' immediate view.

Organ preservation

Unlike tissues or cells, which can be maintained in functioning condition by placing them in life sustaining cultures (specially prepared nourishing fluids places in a dish or a bowl), organs that have been donated for transplantation require a more complex approach for their preservation. Perfusion is the process of using a pump or other mechanical device to circulate specific **electrolyte** solutions through the organ's vascular system.

These solutions may support the organ's metabolic (or chemical and physical) functioning. Some solutions are used, in effect, to freeze the organ. For instance, in cornea, skin, kidney, liver, heart, and pancreas transplants, some solutions are designed to cause **hypothermia** (subnormal body temperatures) in the organ. Organ tissues maintain their ability to function 10 or more times longer when kept at a **temperature** of 32–39°F (0–4°C), and kidneys have been preserved for more than 50 hours using this approach.

Donor organ and tissue networks

Most organs and tissues used for transplantation come from nationwide networks designed to provide quick access to organs when they become available. For example, the United Network for Organ Sharing (UNOS) provides access to organs and tissues throughout **North America**. The more than 70 organ procurement agencies in the United States are crucial to the success of transplants since they facilitate access to and transportation of organs and tissues that remain viable for transplanting for only a limited amount of time after they are removed from the donor's body.

The procurement begins when a hospital notifies a local organ bank that a seriously ill or dying patient (who is willing to donate his or her organ or whose family has given permission) is under their care. Organ bank staff will go to the hospital to assess the patient to determine whether their organs are healthy enough for transplantation. For example, most donor organs are harvested from people under age 65, and organs from someone who was an intravenous drug user may also be precluded from use since the organs may be infected with the **AIDS** disease or **hepatitis**. Once the patient is pronounced dead, a team of surgeons begin to remove the eyes, heart, liver, and kidneys. The next step is to determine who receives the organ. This is usually done on a geographic basis with those patients in the same general geographic area as the donor receiving first consideration to receive the organ.

However, since these organs are a precious commodity, a nationwide computer list of potential transplant recipients is also maintained. The National Organ Procurement and Transplantation Network was established by the National Organ Transplant Act of 1984. Potential donor recipients are prioritized to receive organs according to the length of time they have been on the UNOS list and on the compatibility to the donor organ in terms of blood types, body sizes, and genetic similarities.

The future of transplantation

Although many transplants, especially organ transplants, would not be needed if people took better care of their bodies by exercising and not smoking or drinking alcohol, the number of organ transplants performed each year is likely to continue to grow as long as donor organs can be obtained. In addition, new kinds of transplants are being pioneered.

One of the more exciting advances is the transplantation of tissues and cells from the central **nervous system** (CNS). These cells have been transplanted into the brains of people suffering from neurological (or nervous system) diseases, like **Parkinson disease.** CNS cells and tissues have the unique ability to regenerate, or grow back. In the case of Parkinson's patients, the substantia nigra area of the **brain** can no longer produce the **neurotransmitter** (chemical messenger) **dopamine**, which usually results in the progressive loss of many motor skills to the point where patients can no longer walk or feed themselves. One of the surgical treatments under investigation is to transplant adrenal medullary tissue in hopes that the transplanted tissue will permanently regenerate new cells in the brain.

Artificial organs and xenografts

Due to the lack of available organs for transplantation, researchers are constantly experimenting with developing new sources of organs and tissues for transplantation. Artificial organs, for example, are being developed. These include electronic devices to make the heart beat and pumps implanted into the body that can supply necessary substances like **insulin**. Many of these devices are used as "bridges" to transplantation. In other words, they are temporary therapies to keep the patient alive while a suitable donor organ can be found. Artificial skin is also under development as are mechanical implants to help cure deafness.

Although current transplantation deals primarily in allografts, or transplantation within the same species, transplantation between species, called xenografts, is also a rapidly advancing area of study. The first partially successful xenografts were performed over a quarter of a century ago when surgeons transplanted organs from **chimpanzees** to humans.

But since the use of **primates**, which are closer to humans in nature, raises certain ethical questions, recent research has focused on non-primate animals. Pigs, for example, are relatively easy to breed and have large litters. They also share with humans many similar anatomical structures, like the heart, that function in similar ways in the body. Although pig organ transplants have, to date, been unsuccessful, advances in the development of immunosuppressive drugs have led some transplant scientists to think that success in this area can be achieved. Another advance in medical science, the ability to manipulate genes (hereditary components found in cells), has led to **genetic engineering**. This process involves manipulating and combining specific genetic components to achieve desirable traits or effects. In the case of pig organs, the goal is to make them more compatible to humans. However, the primary obstacle to xenografts remains rejection by the host's immune system. Even using immunosuppressant drugs does not guarantee the immune system will not eventually recognize and destroy the non-human organ.

Ethical issues surrounding transplantation

The primary ethical issue associated with transplanted is the extreme shortage of available donors. Nearly 20,000 people die in the United States each year who would have been suitable organ donors. But only about 3,000 of these organs are ever donated and harvested. Questions surrounding the limited supply of donor organs include who should get the donor. For example, should the organ go to someone who is poor and on welfare or someone who can afford to pay for the operation. There is also concern over buying organs from people before they have died or from their families after the person has died. Some organs, like kidneys, corneas, bones, and bone marrow could even be bought and removed before death since they would not fatally harm the donor if removed.

Another ethical issue surrounding transplantation is the high costs associated with obtaining and transplanting an organ. Some transplant procedures can cost more than $200,000. Since most people cannot afford such operations, the burden falls on society in the form of higher insurance premiums and government subsidies. Even after the surgery is over, it can cost tens of thousands of dollars each year to keep the person alive because of the high cost of antirejection drugs.

Despite the difficulty of obtaining organs, the high costs, and the many ethical concerns, transplantation will continue to thrive since it is the only hope for many terminally ill patients. Society's mandate is to develop ethical regulations to ensure that growing demands are met fairly and humanely.

See also Antibody and antigen; Cyclosporine.

Resources

Books

Gutkind, L. *Many Sleepless Nights.* Pittsburgh: The University of Pittsburgh Press, 1990.

Keyes, C.D. *New Harvest: Transplanting Body Parts and Reaping the Benefits.* Clifton, NJ: Humana Press, 1991.

KEY TERMS

Allografts—Tissues and organs used for transplantation that come from donors of the same species.

Autografts—Tissues and organs used for transplantation that come from the patients themselves.

Bone marrow—A spongy tissue located in the hollow centers of certain bones, such as the skull and hip bones. Bone marrow is the site of blood cell generation.

Graft—Bone, skin, or other tissue that is taken from one place on the body (or, in some cases, from another body), and then transplanted to another place where it begins to grow again.

Immunocompromised—A condition in which the immune system suppressed so that it is not functioning completely.

Immunosuppressant—Something used to reduce the immune system's ability to function, like certain drugs or radiation.

Lymphocytes—White blood cells that play a role in the functioning of the immune system.

Vascular anastomosis—A technique for reconnecting blood vessels.

Xenografts—Tissues and organs used for transplantation that come from different animal species, like pigs or baboons.

Periodicals

Fox, Mark D. "The Transplantation Success Story." *Journal of the American Medical Association* (December 7, 1994): 1704.

"National Registry Creation Helps Assess Donor Risk." *Transplant & Tissue Weekly* (June 4, 2000).

Pace, Brian. "Suppressing the Immune System for Organ Transplants." *JAMA* 283 no. 18 (May 10, 2000): 2484.

Starzl, T.E. "The Early Days of Transplantation." *Journal of the American Medical Association* (December 7, 1994): 1705.

White, D., and J. Wallork. "Xenografting: Probability, Possibility, or Pipe Dream?" *The Lancet* (October 9, 1993): 879-880.

Other

Current Science and Technology Center. "Robotic Surgery" [cited April 2003]. <http://www.mos.org/cst/article/1623/>.

David Petechuk

Transuranium element *see* **Element, transuranium**

Trapezoid

A trapezoid is a four-sided, two-dimensional polygon.

With four sides, a trapezoid is a **quadrilateral**, just as a **square** or **rectangle** or **parallelogram**. Unlike those forms, however, a trapezoid does not necessarily have to have **parallel** sides. In other words, all rectangles are trapezoids, but not all trapezoids are rectangles.

A trapezium is a subset of trapezoids in which at least two sides are parallel; a parallelogram is one example of a trapezium. The most common image of a trapezium, often confused with a trapezoid, is a figure with two parallel faces, one longer than the other. The two parallel sides of the trapezium are called the base lines, with the longer of the two called the base. If the two non-parallel sides are the same length, the trapezium is known as an isosceles trapezium.

One important mathematical use of the trapezoid is in the discipline of **calculus**. At its most fundamental, calculus can be used to determine the area under a **curve**. We can approximate this area by a series of trapezia—one side along the x-axis, two sides rising parallel to the y-axis, and the final side slanted to approximate the slope of the curve. As the trapezia get more and more narrow, the **approximation** grows more accurate. The calculus **integral** assumes that the trapezia have become increasingly narrow so as to yield the exact area under the curve.

Kristin Lewotsky

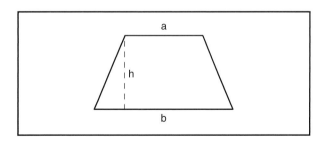

Figure 1. *Illustration by Hans & Cassidy. Courtesy of Gale Group.*

KEY TERMS

Parallelogram—A quadrilateral in which opposite sides are parallel.

Trapezium—A trapezoid in which two opposite sides are parallel.

Tree

A tree is a woody **plant** which has three principle characteristics: (a) the potential to grow to 20 ft (6.1 m) or more in height; (b) the formation of one or more trunks arising from the ground; and (c) the ability to stand on its own without support. Trees provide many products which are important to humans, such as timber, **fruits**, and nuts. They are also the dominant plants in the world's **forests**, and thus provide critical habitats for the other **species** which live there.

Tree taxonomy

Taxonomy is the identification and classification of organisms. Dendrology is the identification and classification of trees, shrubs, and vines, and is a subdiscipline of taxonomy. Shrubs and vines are also woody plants. Shrubs are shorter than trees and have multiple stems arising from the ground. Vines generally rely upon another plant or other structure for physical support.

Some plant families consist entirely of trees and other types of woody plants. Other plant families have some species that are woody, and some that are herbaceous. This indicates that arborescence, the character of being tree-like, is not a reliable character for distinguishing families and higher taxonomic groups of plants.

History of taxonomy

Carl von Linné of Sweden began the modern study of taxonomy in the mid-1700s. He classified trees and other plants according to the morphology of their reproductive structures, such as the flowers and fruits of Angiosperms, and the cones of Gymnosperms. Many religious leaders of his time considered it immoral to study the reproductive structures of plants. However, modern taxonomists still rely upon plants' reproductive structures for conclusive identification of species.

Carl Von Linné also advocated that all scientists refer to trees and other organisms by a Latinized name. He even Latinized his own name to Carolus Linneaus. Modern biologists continue to follow this convention. Thus, the tree which Americans call the white pine is known to biologists throughout the world as *Pinus strobus*, where "Pinus" is the pine genus and "strobus" is the specific epithet. There are about 90 other species of pine in the world, including red pine (*Pinus resinosa*), sugar pine (*P. lambertiana*), and pitch pine (*P. rigida*). Charles Darwin's studies of **evolution** in the mid-1800s led taxonomists to group organisms hierarchically, according to their evolutionary relationships. For example, the **pines** (genus, *Pinus*), spruces (genus, *Picea*), and about seven other genera are grouped together in the family Pinaceae, because they are evolutionarily related. In turn, the Pineaceae and about six other families are grouped together in the order Coniferales (the conifers, or cone-bearing plants) because they are evolutionarily related.

Modern taxonomy

Traditionally, plant taxonomists have relied upon the morphology of plants' reproductive structures to determine their evolutionary relationships. More recently, they have also used biochemical characteristics, DNA sequences, and additional features. Occasionally, there are disagreements about the relationships of different plant species. These disagreements are important in stimulating further research.

In identifying a tree in nature, dendrologists do not always rely upon its reproductive structures, because these are often only available for a brief time of the year. In practice, they typically rely upon features of a tree's leaves, twigs, **bark**, **wood**, habit (general shape and appearance), and **habitat** as clues for identification.

Secondary growth

Most trees increase in thickness due to **cell division** in two special layers of undifferentiated tissues near the outside of their stems. This is known as secondary growth. The two tissues are referred to as the vascular cambium and the **cork** cambium. In contrast, herbs do not have secondary growth, and they stop growing once their primary tissues have matured.

Cell layers in a tree trunk

In a typical, sawed-off sector of a tree trunk, one encounters layers of different cells and a series of concentric, annual growth rings going from the outside toward the inside.

The cork of bark is on the external surface of the trunk, and consists of dead cells which are impregnated with suberin, a waxy substance which inhibits **evaporation** of **water** through the bark. The cork cambium lies just inside the cork. It produces cork cells on its outside face and secondary cortex on its inside face. Growth and division of the cork cambium differs among tree species, and this gives the bark of each species its own characteristic appearance.

Phloem cells lie just inside the secondary cortex. Phloem cells are elongated cells specialized for the transport of plant **nutrients**, such as the carbohydrates made during **photosynthesis**. The vascular cambium lies just inside the phloem cells. It produces phloem cells on its outside face, and xylem cells on its inside surface. Xylem cells are elongated cells specialized for transport of water and dissolved ions throughout the tree. Trees growing in places with a strongly seasonal climate typically contain thick layers of xylem cells with readily apparent, concentric growth rings. This thick layer of functional xylem cells is referred to as the sapwood.

Finally, the heartwood is in the center of the tree. This layer is typically darker than the sapwood and consists of dead cells that are very stiff and serve to strengthen the tree. The heartwood may also have readily apparent growth rings.

Growth rings

In most trees growing in a strongly seasonal climate, the vascular cambium produces wide, thin-walled cells in the spring, narrow thick-walled cells in the summer, and few or no cells in the autumn and winter. This seasonal regularity of **cell** production results in the formation of annual growth rings. The bristle-cone pine (*Pinus aristata*) is the world's longest-lived tree species, and one specimen of this species has about 5,000 growth rings, indicating it is at least 5,000 years old.

Within a given growth ring, the large cells of spring-wood and the small cells of summerwood are often readily discernible with the naked **eye**. **Light**, **temperature**, **soil** moisture and other environmental factors affect the growth of trees, and therefore the width of their growth rings.

Evolution

Most botanists believe that the first land plants were herbaceous. The first woody plants were probably Lycopsids, free-sporing plants which had narrow, tubular grass-like leaves. Numerous Lycopsid fossils have been dated to the middle of the Upper Devonian period, more than 370 million years ago. The first known plant with a vascular cambium which exhibited true secondary growth was a species of Protopteridium, a free-sporing plant dated to about 370 million years ago.

Very few modern, free-sporing plants are arborescent. The only living relatives of the Lycopsids are the **club mosses** (Lycopodophyta), a group of simple, herbaceous, free-sporing plants.

Forests

Trees are the dominant organisms of forests. Climate and other factors determine which tree species grow in a forest. Forest ecologists have classified the forests of the world according to the species of trees that grow there. One classification scheme is described below.

Coniferous forests are characteristic of the boreal forests of cold regions of the northern hemisphere. The boreal forest is found in Canada, southern Alaska, northwestern America, northern **Europe**, and northern Russia. The dominant trees are conifers (cone-bearing plants) such as pines, spruces, **firs**, and larches. Many of these trees are important sources of wood used in construction, and of pulp for **paper**.

Broad-leaf forests are characteristic of temperate zones of central and eastern **North America**, central and southern Europe, and central **Asia**. Its dominant trees have broad leaves and are deciduous, in that they shed their leaves once a year. **Oaks**, hickories, **maples**, and sycamores are some of the trees commonly found in this forest type. Many broad-leaf trees are hardwoods, and are used for making some of our finest furniture.

Mediterranean forests are characteristic of regions with hot, dry summers and warm, humid winters. This forest type is found in southern California, northwestern Mexico, southern Europe, northern **Africa**, and southern **Australia**. The trees which grown in Mediterranean forests are adapted to minimizing water loss. Evergreen oaks and pines are well-known trees of this forest type. In Australia, about 600 different species of Eucalyptus grow in the Mediterranean forest type along the southern coast.

Savannas are characteristic of tropical regions with a very seasonal rainfall. The trees in a **savanna** are sparsely distributed and are specially adapted to minimize water loss. Savannas are found in northern Mexico, equatorial Asia, central Africa, and northern Australia. Acacias, Dracaenas, and the Baobab are well-known savanna trees.

Tropical rainforests are characteristic of regions with a relatively warm climate and a great deal of rainfall. This forest type is found in Central America, northern and central **South America**, central and western Africa, and the south Pacific region. Tropical rainforests have a great diversity of plant and **animal** species. The tropical rainforests of the world are currently threatened by people who are cutting down their trees at an increasing **rate**. At the same time, conservationists and tropical ecologists devote much effort to the preservation of tropical rainforests with their rich diversity of species.

Ascent of sap

It is vitally important for a tree to transport water from the soil to its upper-most leaves. This process has long fascinated plant physiologists, and has been studied for centuries. How does water move from the roots to the uppermost leaves of tall trees? The significance of this problem is best appreciated by considering the height of some of the world's tallest trees. The world's tallest tree was a *Eucalyptus regnans* of Australia, which was measured at 470 ft (143 m) in 1880. The tallest living tree is a coastal redwood (*Sequoia sempervirens*) of California, which is about 365 ft (111.3 m) in height. Its relative, the giant **sequoia** (*Sequoidendron giganteum*), is not quite as tall, but is the world's most massive tree species.

Cohesion-tension theory

Most plant physiologists now accept the "cohesion-tension theory" as an explanation for the ascent of sap. According to this theory, water moves up the trunk of a tree in narrow, elongated cells near the periphery of the trunk, referred to as the xylem, and does not require the expenditure of metabolic **energy**. The movement of water only depends upon three important physical-chemical properties of water.

The first important property of water is that it always moves from a region with a more positive water potential, to a region with a more **negative** potential. Water potential is a measure of the energy available in a **solution** of water. Thus, water moves out of the leaves and into the air because the water potential of the air is more negative; water moves out of the tree trunk and into the leaves because the water potential of the leaves is more negative; water moves out of the roots and into the trunk because the water potential of the trunk is more negative; and water moves out of the soil and into the roots because the water potential of the roots is more negative.

The second important physical-chemical property of water is that it is a cohesive **molecule**. In other words, water molecules tend to bind to one another through the formation of **hydrogen** bonds. The cohesiveness of water molecules gives the thin water columns in a tree trunk a very great tensile strength. This prevents breakage of the water column when great longitudinal stresses are placed upon it as it is pulled out of the leaves and into the air.

The third important property of water is that it adheres very tightly to the walls of xylem cells in the tree's transport pathway. Adhesion of water to these cell walls maintains the full hydration of the pathway for water transport. This prevents breakage of the water column, and allows water transport even when a tree is water-stressed in a dry environment.

Economic significance

Trees have great economic significance to humans as a source of food, building materials, and paper. Almond, coconut, cherry, prune, peach, pear, and many other tree species are grown in orchards for their fruits and nuts. The apple tree is the orchard tree of greatest economic significance, and there are several hundred dif-

KEY TERMS

Adhesion—Physical attraction between different types of molecules.

Cohesion—Physical attraction between molecules of the same type.

Cork cambium—Undifferentiated plant tissue which gives rise to cork cells and secondary cortex.

Dendrology—Identification and classification of woody plants.

Phloem—Plant tissue consisting of elongated cells which function in the transport of carbohydrates and other nutrients.

Vascular cambium—Undifferentiated plant tissue which gives rise to phloem and xylem.

Xylem—Plant tissue that transports water and minerals upward from the roots.

ferent varieties of apples. (Many of North America's best apples grow in New York state and Washington state.) Many trees are also useful for the wood they produce. Wood is used as a construction material and to make furniture. Wood is a valuable construction material because it is relatively inexpensive, easy to cut, and very strong relative to its weight. Many species of pines and other conifers are important sources of softwoods, and many broadleaf trees are important sources of hardwoods.

There are frequent conflicts between conservationists and loggers. In the United States, the coniferous **rainforest** of the Pacific northwest has been the site of one such conflict. In this region, loggers want to harvest coniferous trees from the forest just as they have done for many years, whereas conservationists seek to preserve the forest because it provides a habitat for the northern spotted owl, an **endangered species**.

See also Basswood; Citrus trees; Conifer; Deforestation; Dogwood tree; Ebony; Forestry; Horse chestnut; Magnolia; Mahogany; Mangrove tree; Nux vomica tree; Palms; Sapodilla tree; Screwpines; Spruce; Walnut family; Yew.

Resources

Books

Audubon Society and Staff. *Familiar Trees of North America: Eastern Region.* New York: Knopf, 1987.

Platt, R. *One Thousand-and-One Questions Answered about Trees.* New York: Dover Inc., 1992.

Raven, Peter, R.F. Evert, and Susan Eichhorn. *Biology of Plants.* 6th ed. New York: Worth Publishers Inc., 1998.

White, John, and David More. *Illustrated Encyclopedia of Trees*. Portland, OR: Timber Press, 2001.

Other

Chaw, S. M., et al. "Seed Plant Phylogeny Inferred From All Three Plant Genomes: Monophyly of Extant Gymnosperms and Origin of Gnetales from Conifers." *Proceedings of the National Academy of Sciences of the United States of America* 97 (2000): 4086-4091.

Peter A. Ensminger

Tree-ring dating *see* **Dating techniques**

Tree shrews

The tree shrews (order Scandentia, family Tupaiidae) comprise a small number of **species** that are only found in South and Southeast **Asia**. Five genera (19 species) are recognized. All occur in forested areas, ranging from India and Southwest China eastward through Malaysia, Indonesia (west of Wallace's Line), and the Philippines. Three genera and 10 species occur on the **island** of Borneo alone.

In appearance, tree shrews resemble long-snouted **squirrels** (the Malay word *tupai* means squirrel). All tree shrews are of a slender build; adults generally weigh 2.5-3.5 oz (70-100 g.) The length of the head and body ranges from 3.9-8.6 in (100-220 mm), while tail length varies from 3.5-8.8 in (90-225 mm). Generally a russet-brown **color**, they have a long, pointed muzzle with 38 sharp teeth. One unusual species, the pen-tailed tree shrew (*Ptilocercus lowi*), can be identified by its tail, which is naked except for a whitish feather-shaped arrangement of the hairs near the end. In all species, the fur consists of long, straight guard hairs and shorter, softer underfur. Some forms have pale shoulder stripes and others have facial markings. The ears are squirrel-like; that is, they are comparatively small and cartilaginous, except in the pen-tailed tree shrew in which they are larger and more membranous. The feet of tree shrews are naked beneath; the soles are adorned with tubercle-like pads which assist with climbing. The long and supple digits bear sharp, moderately curved claws. Tree shrews have well developed senses of **vision**, **hearing**, and **smell**.

In **addition** to their external resemblance to squirrels, tree shrews, like squirrels, are mostly diurnal, and some of their actions and movements are similar. *Ptilocercus* differs in being mainly nocturnal and, when on the ground, progresses in a series of hops. Other tree shrews are swift runners. All tupaiids are capable climbers, seeking their food in trees as well as on the ground. Their diet consists mainly of **insects** and fruit but occasionally includes other **animal** food and various types of **plant** matter. They are generally fond of **water** for both drinking and bathing. Most species nest in holes in tree trunks or branches 65.7-98.4 ft (20-30 m) high, the nest consisting of a simple structure of dried leaves, twigs, and fibers of soft **wood**.

Few other mammalian families have proved as difficult to classify as the tree shrews. Historically, the Tupaiidae have been grouped with the Macroscelididae (**elephant shrews**), Insectivora (insectivores), and even the **primates**. More recent studies of their **behavior** and reproduction, however, has led to them being placed in a distinct order, Scandentia.

Tree shrews display a wide range of social behaviors. The common tree shrew *Tupaia glis*, for example, lives in permanent pairs, with males occupying an average home range of 12,168 sq yd (10,174 sq m) and females a smaller area of about 1,052 sq yd (880 sq m). Although they basically share the same range, they remain largely solitary and defend the area from possible intruders. In contrast, the pen-tailed tree shrew generally moves about in pairs, but as many as four animals have been found together in a single nest. **Field** observations also suggest that many species exist at relatively low densities: data for *T. glis* show a varied pattern of one to two per acre (two to five per hectare) in Malaysia or two to five per acre (five to 13 per hectare) in Thailand, while *T. palawanensis* occurs at a local **density** of 0.5-1 per acre (1.5-2.5 per hectare).

Scent marking plays an important role in communications among tree shrews, all of which have specialized **glands** that secrete a wide array of chemical compounds, many of which are unique to the individual animal. The different compounds in these secretions covey a range of messages about the animal, particularly information on its sex, age, and breeding status. Such information is thought to play an important role in defending individual territories. Tree shrews seem to be creatures of habit and regularly use the same paths along the ground or on branches to reach their favorite feeding or resting areas. As they move around their ranges, their scent is liberally deposited, usually at strategic places where other animals have a chance of finding it. In this way, an intruding animal will be able to determine that the area is already occupied and defended by a resident tree shrew and should be able to withdraw without further conflict. If the intruder is a male, hoping to mate or take over the territory of a neighboring animal, it may move further into the territory, but on doing so risks attack and possible injury from the resident animals.

The reproductive behavior of most species is still poorly known. Most species apparently breed throughout the year. Following a gestation period of about 45-50 days, females give **birth** to one to four young. Some species, such as *Urogale everetti*, are receptive to breeding again shortly after giving birth.

Surprisingly, in view of their large size and diurnal habits, tree shrews have attracted relatively little biological attention. As a result there has been little study of their **ecology** or behavior in the wild. All tree shrews are forest-dwelling species, occupying a wide range of niches within different forest habitats. Many species seem capable of adapting to living in secondary forest and some even occur in rural gardens and well-established plantations.

Tropical **forests** are, however, their preferred domain, and because of their dependence on forest **habitat**, many of these species are susceptible to excessive levels of clearance and disturbance. Some species, particularly *Tupaia nicobarica* and *T. longipes*, are thought to be endangered, but other species with restricted ranges, particularly those on islands or where agricultural encroachment and/or logging is a major activity, are also seriously threatened. The most urgent among these are the golden-bellied (*Tupaia chrysogaster*), Palawan (*T. palawensis*), Philippine (*Urogale everetti*), and Madras (*Anathana elliotti*) tree shrews. Although loss of habitat is the main threat to these species, additional pressures such as hunting for food and sport can add to the **pressure** on these and other native **endemic** species. Habitat **conservation** is therefore vital to preserving tree shrews.

The forests that are important to tree shrews are also known to be of great importance to a wide range of plants, **mammals**, **birds**, and **invertebrates**, many of which are endemic. By developing and implementing conservation strategies which include tree shrew conservation requirements in these forests, the overall biological diversity of these unique ecosystems could be far better protected. These same forests also fulfill many other essential roles that benefit humans, particularly through **watershed** protection and as an important source of fruit, medicinal and herbal plants, and a wide range of timber products of economic importance.

In view of the many uncertainties surrounding the ecological status, distribution, and ability of these species to adapt to secondary habitats, there is an urgent need to promote further action for a large number of these species. Particular aspects which need attention include details of the basic ecology, home range size, feeding priorities, habitat preferences, breeding behavior, and population density of most species.

See also Shrews.

Trichinosis

Trichinosis is a **disease** caused by the roundworm (nematode) called *Trichinella spiralis* (*T. spiralis*). It is readily avoided by proper handling and cooking of certain meats, particularly pork products.

Life cycle of *Trichinella spiralis*

The life cycle of *Trichinella spiralis* includes several different stages. The adult worm lives in the intestinal lining of such carnivorous animals as swine, **bears**, walrus, and rodents. After mating, the male worm dies, while the female worm goes on to develop the offspring.

The embryonic stage, a stage often occurring after **birth** in many nematode **species**, occurs within the uterus of the female *T. spiralis,* so that the offspring which are ultimately discharged from the female are in the larval second stage of life. These larvae (about 1500 from each female worm) travel through the **circulatory system**, to the **heart**, then through the **blood** vessels leading to striated muscle (the muscle of the **skeletal system** and the heart). Any larvae not arriving in striated muscle will die.

Those larvae that reach striated muscle will grow to a length of about one millimeter, coil, and enclose themselves within a protective wall called a cyst. These cysts continue to live for up to 10 years in this form.

Trichinella spiralis in humans

A pig that has been infected with *T. spiralis* may have thousands of cysts waiting dormant within its muscles—the very muscles that humans look forward to dining on in the form of pork chops, ham, barbecued ribs, etc. When humans sit down to a meal of undercooked, *T. spiralis*-infected pig dinner, they are ingesting viable *T. spiralis* cysts. The cyst walls are broken down by the usual process of food digestion in the stomach, allowing the larvae to continue on to the new host's intestine, where the larvae mature to become adult worms, capable of reproducing a new crop of larvae. When these new larvae are born, they begin their **migration** throughout the human host's bloodstream to the human host's muscles, where they live for a short while before encysting.

Symptoms of trichinosis

A human host who eats *T. spiralis* infected meat may experience symptoms to a varying degree. If the meat ingested has only a few cysts, then the human host's "worm burden" is said to be relatively small, so that symptomatology will be relatively moderate. In fact, many infections are subclinical (exhibiting such mild symptoms as to remain undiagnosed).

In a host with a greater worm burden, the initial symptoms will be caused by the presence of the adult worms in the intestine, and usually include fever, diarrhea, abdominal **pain**, and perhaps vomiting. These symptoms begin about one to two days after eating the offending meat, and may last for a week or so.

When larvae begin their migration through the blood vessels, the host will begin to experience systemic symptoms (affecting the whole body), such as fever, swelling of the face and the area around the eyes, rash, bleeding into the nail beds, retina, and whites of the eyes, and cough. In very severe cases, **inflammation** of the heart muscle (myocarditis), lungs (pneumonitis), or **brain** (**encephalitis**) may occur, which can lead to the few deaths attributable to trichinosis.

The larvae begin to burrow into the host's muscles and form cysts within about two to three weeks of the initial **infection**. This produces signs of muscle inflammation (myositis), including swelling of the affected muscle groups, pain, and weakness.

The most frequently affected muscles are the muscles outside of the **eye**, which control eye movements (extraocular muscles), as well as the muscles of the jaw, neck, upper arm (biceps muscle), lower back (lumbar region), and diaphragm (the muscle which separates the abdominal and chest cavities and aids in inspiration).

Symptoms are at their most severe at about three weeks after infection and decrease very slowly in their severity. Recovery is extremely gradual, and symptoms may be present for as long as three months. Fatigue and myalgia (muscle pain) may take several more months to subside.

Diagnosis

An initial **diagnosis** of trichinosis relies heavily on the presence of the most classic symptoms of trichinosis (including swelling around the eyes, muscle inflammation, fever, and high levels of the blood cells called eosinophils), coupled with the patient's report of having eaten undercooked meat of a species known to potentially carry trichinosis.

The most common food culprit in the United States has been pork sausage, while outbreaks in **Europe** have been attributed to wild boar and horse meat, outbreaks in **Asia** and **Africa** have been due to dog meat, and outbreaks in Northern Canada have occurred due to consumption of walrus and bear meat.

Treatment

Treatment of trichinosis is primarily aimed at decreasing the severity of the symptomatology—bed rest and medications to relieve fever and muscle pain (aspirin, acetaminophen, or ibuprofen). Steroids (such as prednisone) are reserved for the most severe cases of muscle inflammation, or for complicated cases which include myocarditis.

Two related anti-worm medications (mebendazole and thiabendazole) have been reported to work against intestinal larvae, but not against larvae encysted in the muscles. In particular, thiabendazole has been effective when given to patients who knew within 24 hours that they had eaten infected meat.

Prevention

Prevention of trichinosis is relatively simple. Swine should be fed only grain or cooked garbage; uncooked garbage may contain contaminated pork scraps. Meats of animals prone to trichinosis infection should be cooked thoroughly, so that no pink is present. Freezing at an adequately low **temperature** 5°F (-15°C) for one week or 0°F (-18°C) for three weeks can kill most encysted larvae, except those species which infect arctic **mammals** such as walrus or bear.

See also Roundworms.

Resources

Books

Andreoli, Thomas E., et al. *Cecil Essentials of Medicine*. Philadelphia: W.B. Saunders Company, 1993.

Berkow, Robert, and Andrew J. Fletcher. *The Merck Manual of Diagnosis and Therapy*. Rahway, NJ: Merck Research Laboratories, 1992.

Cormican, M.G., and M. A. Pfaller. "Molecular Pathology of Infectious Diseases." In *Clinical Diagnosis and Management by Laboratory Methods*. 20th ed. Philadelphia: W. B. Saunders, 2001.

Francis, Frederick. *Wiley Encyclopedia of Food Science and Technology*. New York: Wiley, 1999.

Isselbacher, Kurt J., et al. *Harrison's Principles of Internal Medicine*. New York: McGraw Hill, 1994.

Kobayashi, G., Patrick R. Murray, Ken Rosenthal, and Michael Pfaller. *Medical Microbiology*. St. Louis: Mosby, 2003.

Mandell, Douglas et al. *Principles and Practice of Infectious Diseases*. New York: Churchill Livingstone Inc., 1995.

Periodicals

Hwang, Deng Fwu. "Tetrodotoxin In Gastropods (Snails) Implicated In Food Poisoning." *Journal of Food Protection* 65, no. 8 (2002): 1341-1344.

"Preventing Food Poisoning." *Professional Nurse* 18, no. 4 (2002): 185-186.

Rosalyn Carson-DeWitt

Triggerfish

Triggerfishes are members of the family Balistidae of the order Tetradontiformes. They derive their name from a unique feature of their dorsal fin. The triggerfish can lock the large dorsal spine in an upright position by supporting it with its smaller secondary spine. This protects the **fish** from predation by larger fish because the erect spine makes the fish hard to swallow or extract from small crevices. The locked dorsal spine can be "unlocked" by depressing the third spine or "trigger" which is connected to the second spine.

A distinctive characteristic of triggerfish is similarity in size and shape of the second dorsal (back) and anal fins. The belly in front of the anal fin is the widest circumference of the fish. The body of the triggerfish is protected by bony plates.

Triggerfishes are moderately large fish generally found on coral reefs widely distributed throughout the world, in all about 36 **species**. The gray triggerfish, *Balistes capriscus*, averages under 1 ft (0.3 m) in length but may grow to 2 ft (0.6 m) with a weight of 3 lb (1.4 kg). It is found in the warm waters of the Gulf of Mexico and the Mediterranean. These grayish fish often appear spotted or splotchy when swimming in among seaweeds.

Among the largest in the family is the **ocean** triggerfish, *Canthidermis sufflamen*, which grows to 2 ft (0.6 m) and weighs about 10 lb (4.5 kg). It ranges in distribution from the coast of Florida to the Caribbean. The ocean triggerfish is capable of making sounds like some of its relatives. One of the ways it makes sound is by vibrating some muscles that are attached to its swim bladder.

Several species, such as the redtail triggerfish, are found in the western Pacific. The redtail *Xanthichthys mento* is a 10 in (25 cm) fish with a purplish blue **color** and a red tail. In contrast to the redtail triggerfish, the *Abalistes stellaris* may grow to 24 in (60 cm) in length.

Triglycerides

Fats exist in foods—and are usually stored in the body—as Triglycerides. Recent research relating levels of triglycerides in the **blood** stream to **heart** attacks in human presents a sometime confusing picture but a mounting level evidence suggests that, along with other indicators, triglyceride levels can be used to predict heart attack risk, especially in women and diabetics.

Although the exact mechanisms are not fully known, elevated triglycerides allow increased blood clot formation and may slow the natural deterioration of clots once formed.

Fat molecules are generally made up of four parts: a **molecule** of **glycerol** and three molecules of **fatty acids**. Each fatty acid consists of a **hydrocarbon** chain with a **carboxyl group** at one end. The glycerol molecule has three hydroxyl groups, each able to interact with the carboxyl group of a fatty acid. Removal of a **water** molecule at each of the three positions forms a triglyceride. The three fatty acids in a single fat molecule may be all alike or they may be different. They may contain as few as four **carbon atoms** or as many as 24. Because fatty acids are synthesized from fragments containing two carbon atoms, the number of carbon atoms in the chain is almost always an even number. In **animal** fats, 16-carbon, for example, palmitic acid and 18-carbon, for example, **stearic acid** fatty acids are the most common.

Some fatty acids comprising a given triglyceride have one or more double bonds between their carbon atoms. They are then said to be unsaturated because they can hold more **hydrogen** atoms than they do. Mono-unsaturated fats have a single double bond in their fatty acids while polyunsaturated fats, such as trilinolein, have two or more. Additionally, there are trans-fats, which are only partially hydrogenated having fewer double bonds in a *trans* (as opposed to the usual *cis*) chemical configuration, and also omega-3 fats, which have at least one double bond, three carbon atoms in from the end of the fatty acid molecule. Linolenic acid is an example and **fish** oils are generally a rich source of omega-3 fatty acids.

Double bonds are rigid and those in natural fats introduce a kink into the molecule. This prevents the fatty acids from packing close together and as a result, unsaturated fats have a lower melting point than saturated fats. Because most of them are liquid at room **temperature**, they are called oils. Corn oil, canola oil, cottonseed oil, peanut oil, and olive oil are common examples. As this list suggests, **plant** fats tend to be unsaturated while fats from such animals as cattle tend to be saturated.

Ingested fats provide the precursors from which fat as well as **cholesterol** and various phospholipids are created (synthesized). In humans, fat provides the concentrated form of **energy**. The energy content of fat (9 kcal/gram) is more than twice as great as carbohydrates and **proteins** (4 kcal/gram).

Humans can synthesize fat from carbohydrates. However, there are two essential fatty acids that cannot be synthesized this way and must be incorporated into the diet. These are linoleic acid (an omega-6 fat, with the endmost double bond 6 carbons from the methyl end) and alpha-linolenic acid (an omega-3 fat, with the endmost double bond 3 carbons from the methyl end). Many studies have

KEY TERMS

. . . .

Polyunsaturated fat—A fat missing two or more hydrogen atoms from the maximum number of hydrogen atoms that can be bonded to each carbon in the carbon chain of the compound. These fats can remain liquid at room temperatures.

Saturated fats—Fats containing the maximum number of hydrogen atoms that can be bonded to each carbon in the carbon chain of the compound.

Triglycerides—A molecule containing three fatty acids chemically bonded to a glycol molecule.

examined the relationship between fat in the diet and cardiovascular **disease**. There is still no consensus, but the evidence seems to indicate that a diet high in fat is harmful and that mono- and poly-unsaturated fats are less harmful than saturated fats, with the exception of *trans* unsaturated fats which, according to some, are more harmful than saturated fats. It is also been suggested that ingestion of omega-3 unsaturated fats may be protective for the human body.

See also Biochemistry; Compound, chemical; Heart diseases; Heat.

Resources

Books

Campbell, N., J. Reece, and L. Mitchell. *Biology.* 5th ed. Menlo Park: Benjamin Cummings, Inc. 2000.

Periodicals

Austin, M.A., B. McKnight, K.L. Edwards, et al. "Cardiovascular Disease Mortality in Familial Forms of Hypertriglyceridemia: A 20-year Prospective Study." *Circulation* (June 2000): 2777-82.

Avins, A.L. and J.M. Neuhaus. "Do Triglycerides Provide Meaningful Information about Heart Disease Risk?" *Archives Internal Medicine.* (July 2000): 1937-44.

Cullen, P. "Evidence that Triglycerides are an Independent Coronary Heart Disease Risk Factor." *Am J Cardiol.* (November 2000): 943-949.

Matsubara, M., S. Maruoka, and S. Katayose. "Decreased Plasma Adiponectin Concentrations in Women with Dyslipidemia." *J. Clin. Endocrinol. Metab.* 87, (no. 6) (2002): 2764-2769.

Judyth Sassoon

Trigonometry

Trigonometry is a branch of applied **mathematics** concerned with the relationship between angles and their sides and the calculations based on them. First developed as a branch of **geometry** focusing on triangles during the third century B.C., trigonometry was used extensively for astronomical measurements. The major trigonometric functions, including sine, cosine, and tangent, were first defined as ratios of sides in a right triangle. Since trigonometric functions are intrinsically related, they can be used to determine the dimensions of any triangle given limited information. In the eighteenth century, the definitions of trigonometric functions were broadened by being defined as points on a unit **circle**. This allowed the development of graphs of functions related to the angles they represent which were periodic. Today, using the periodic nature of trigonometric functions, mathematicians and scientists have developed mathematical models to predict many natural periodic phenomena.

Historic development of trigonometry

The word trigonometry stems from the Greek words *trigonon*, which means triangle, and *metrein*, which means to measure. It began as a branch of geometry and was utilized extensively by early Greek mathematicians to determine unknown distances. The most notable examples are the use by Aristarchus (310-250 B.C.) to determine the **distance** to the **Moon** and **Sun**, and by Eratosthenes (c. 276-195 B.C.) to calculate the Earth's circumference. The general principles of trigonometry were formulated by the Greek astronomer, Hipparchus of Nicaea (162-127 B.C.), who is generally credited as the founder of trigonometry. His ideas were worked out by Ptolemy of Alexandria (A.D. c. 90-168), who used them to develop the influential Ptolemaic theory of **astronomy**. Much of the information we know about the work of Hipparchus and Ptolemy comes from Ptolemy's compendium, *The Almagest*, written around 150.

Trigonometry was initially considered a **field** of the science of astronomy. It was later established as a separate branch of mathematics—largely through the work of the mathematicians Johann Bernoulli (1667-1748) and Leonhard Euler (1707-1783).

Angles

Central to the study of trigonometry is the concept of an **angle**. An angle is defined as a geometric figure created by two lines drawn from the same point, known as the vertex. The lines are called the sides of an angle and their length is one defining characteristic of an angle. Another characteristic of an angle is its measurement or magnitude, which is determined by the amount of **rotation**, around the vertex, required to transpose one side on top of the other. If one side is rotated completely around the point, the distance travelled is known as a revolution and the path it traces is a circle.

Angle measurements are typically given in units of degrees or radians. The unit of degrees, invented by the ancient Babylonians, divides one revolution into 360° (degrees). Angles which are greater than 360° represent a magnitude greater than one revolution. Radian units, which relate angle size to the radius of the circle formed by one revolution, divide a revolution into 2π units. For most theoretical trigonometric work, the radian is the primary unit of angle measurement.

Triangles and their properties

The principles of trigonometry were originally developed around the relationship between the sides of a triangle and its angles. The idea was that the unknown length of a side or size of an angle could be determined if the length or magnitude of some of the other sides or angles were known. Recall that a triangle is a geometric figure made up of three sides and three angles, whose sum is equal to 180°. The three points of a triangle, known as its vertices, are usually denoted by capital letters.

Triangles can be classified by the lengths of their sides or magnitude of their angles. Isosceles triangles have two equal sides and two congruent (equal) angles. Equilateral, or equiangular, triangles have three equal sides and angles. If no sides are equal, the triangle is a scalene triangle. All of the angles in an acute triangle are less than 90° and at least one of the angles in an obtuse triangle is greater than 90°. Triangles, such as these, which do not contain a 90° angle, are generally known as oblique triangles. Right triangles, the most important ones to trigonometry, are those which contain one 90° angle.

Triangles which have proportional sides and congruent angles are called similar triangles. The concept of similar triangles, one of the basic insights in trigonometry, allows us to determine the length of a side of one triangle if we know the length of certain sides of the other triangle. For example, if we wanted to know the height of a **tree**, we could use the idea of similar triangles to find it without actually having to measure it. Suppose a person is 6 ft (183 cm) tall and casts an 8 ft (2.44 m) long shadow. The tree, whose height is unknown, casts a shadow that is 20 ft (6.1 m) long. The triangles that could be drawn using the shadows and objects as sides are similar. Since the sides of a similar triangles are proportional, the height of the tree is determined by setting up the mathematical equality

$$\frac{\text{height of tree}}{\substack{\text{length of}\\\text{tree shadow}}} = \frac{\text{height of person}}{\substack{\text{length of}\\\text{person shadow}}} = \frac{x}{20} = \frac{6}{8}$$

By solving this equation, the height of the tree is found to be 15 ft (4.57 m).

Right triangles and trigonometric functions

The triangles used in the previous example were right triangles. During the development of trigonometry, the parts of a right triangle were given certain names. The longest side of the triangle, which is directly across from the right angle, is known as the hypotenuse. The sides that form the right angle, denoted by a box in the diagram, are the legs of the triangle. For either acute angle in the triangle, the leg that forms the angle with the hypotenuse is known as the adjacent side. The side across from this angle is known as the opposite side. Typically, the length of each side is denoted by a lower case letter. In the diagram of triangle ABC, the length of the hypotenuse is indicated by c, the adjacent side is represented by b, and the opposite side by a. The angle of interest is usually represented by θ.

The ratios of the sides of a right triangle to each other are dependent on the magnitude of its acute angles. In mathematics, whenever one value depends on some other value, the relationship is known as a **function**. Therefore, the ratios in a right triangle are trigonometric functions of its acute angles. Since these relationships are of most importance in trigonometry, they are given special names. The **ratio** or number obtained by dividing the length of the opposite side by the hypotenuse is known as the sine of the angle θ (abbreviated sin θ). The ratio of the adjacent side to the hypotenuse is called the cosine of the angle θ (abbreviated cos θ). Finally, the ratio of the opposite side to the adjacent side is called the tangent of θ or tan θ. In the triangle ABC, the trigonometric functions are represented by the following equations.

$$\sin \theta = \frac{a}{c} \quad \cos \theta = \frac{b}{c} \quad \tan \theta = \frac{a}{b} \quad \frac{\sin \theta}{\cos \theta}$$

These ratios represent the fundamental functions of trigonometry and should be committed to **memory**. Many mnemonic devices have been developed to help people remember the names of the functions and the ratios they represent. One of the easiest is the phrase "SOH-CAH-TOA." This means: sine is the opposite over the hypotenuse, cosine is adjacent over hypotenuse, and tangent is opposite over adjacent.

In addition to the three fundamental functions, three **reciprocal** functions are also defined. The inverse of sin θ, or 1/sin θ, is known as the secant of the angle or sec θ. The inverse of the cos θ is the cosecant or csc θ. Finally, the inverse of the tangent is called the cotangent of cot θ. These functions are typically used in special instances.

The values of the trigonometric functions can be found in various ways. They can often be looked up in tables, which have been compiled over the years. They can also be determined by using infinite series formulas. Conveniently, most calculators and computers have the values of trigonometric functions preprogrammed in.

Application of the trigonometric functions

One immediate application for trigonometric functions is the simple determination of the dimensions of a right triangle, also known as the solution of a triangle, when only a few are known. For example, if the sides of a right triangle are known, then the magnitude of both acute angles can be found. Suppose we have a right triangle whose sides are 2 in (5 cm) and 4.7 in (12 cm), and whose hypotenuse is 5.1 in (13 cm). The unknown angles could be found by using any trigonometric function. Since the sine of one of the angles is equal to the length of the opposite side divided by the hypotenuse, this angle can be determined. The sine of one angle is 5/13, or 0.385. With the help of a trigonometric function table or **calculator**, it will be found that the angle which has a sine of 0.385 is 22.6°. Using the fact that the sum of the angles in a triangle is 180°, we can establish that the other angle is 180° - 90° - 22.6° = 67.4°.

In addition to solving a right triangle, trigonometric functions can also be used in the determination of the area when given only limited information. The standard method of finding the area of a triangle is by using the formula, area = 1/2b (base) × h (altitude). Often, the altitude of a triangle is not known, but the sides and an angle are known. Using the side-angle-side (SAS) **theorem**, the formula for the area of a triangle then becomes, area = 1/2 (one side) × (another side) × (sine of the included angle). For a triangle with sides of 5 cm and 3 cm respectively and an included angle of 60°, the area of the triangle would be equal to $1/2 \times 5 \times 3 \times \sin 60° = 13 \text{ cm}^2$.

The formula for the area of a triangle leads to an important concept in trigonometry known as the Law of Sines which says that for any triangle, the sine of each angle is proportional to the opposite its opposite side, symbolically written in triangle ABC as,

$$\frac{\sin A}{a} = \frac{\sin B}{b} = \frac{\sin C}{c}$$

Using the Law of Sines, we can solve any triangle if we know the length of one side and magnitude of two angles, or two sides and one angle. Suppose we have a triangle with angles of 45° and 70°, and an included side of 15.7 in (40 cm). The third angle is found to be 180° - 45° - 70° = 65°. The unknown sides, x and y, are found with the Law of Sines because

$$\frac{\sin 45}{x} = \frac{\sin 70}{y} = \frac{\sin 65}{40}$$

The lengths of the unknown sides are then x = 12.29 in (31.2 cm) and y = 16.35 in (41.5 cm).

The Law of Sines can not be used to solve a triangle unless at least one angle is known. However, a triangle can be solved if only the sides are known by using the Law of Cosines which is stated in triangle ABC, $c^2 = a^2 + b^2 - 2ab \cos C$, or can be written

$$\cos C = \frac{a^2 + b^2 - c^2}{2ab}$$

which is more convenient when using only the sides to solve a triangle. As an example, consider a triangle with sides equal to 2 in, 3.5 in, and 3.9 in (5 cm, 9 cm, and 10 cm). The cosine of one angle would be equal to $(5^2 + 9^2 - 10^2)/(2 \times 59) = 0.067$, which corresponds to the angle 86.2°. Similarly, the other two angles are found to be 29.9° and 63.9°.

Relationships between trigonometric functions

In addition to the reciprocal relationships of certain trigonometric functions, two other types of relationships exist. These relationships, known as trigonometric identities, include cofunctional relationships and Pythagorean relationships. Cofunctional relationships relate functions by their complementary angles. Pythagorean relationships relate functions by application of the **Pythagorean theorem**.

The sine and cosine of an angle are considered cofunctions, as are the secant and cosecant, and the tangent and cotangent.

The Pythagorean theorem states that the sum of the squares of the sides of a right triangle is equal to the **square** of the hypotenuse. For a triangle with sides of x and y and a hypotenuse of z, the equation for the Pythagorean Theorem is $x^2 + y^2 = z^2$. Applying this theorem to the trigonometric functions of an angle, we find that $\sin^2 \theta + \cos^2 \theta = 1$. Similarly, $1 + \tan^2 \theta = \sec^2 \theta$ and $1 + \cot^2 \theta = \csc^2 \theta$. The terms such as $\sin^2 \theta$ or $\tan^2 \theta$ traditionally have meant $(\sin \theta) \times (\sin \theta)$ or $(\tan \theta) \times (\tan \theta)$.

In some instances, it is desirable to know the trigonometric function of the sum or difference of two angles. If we have two unknown angles, θ and φ, then $\sin (\theta + \phi)$ is equal to $\sin\theta\cos\phi + \cos\theta\sin\phi$. In a similar manner, their difference, $\sin(\theta-\phi)$ is $\sin\theta\cos\phi - \cos\theta\sin\phi$. Equations for determining the sum or differences of the cosine and tangent also exist and can be stated as follows:

KEY TERMS

. . . .

Adjacent side—The side of a right triangle which forms one side of the angle in question.

Amplitude—A characteristic of a periodic graph represented by half the distance between its maximum and minimum.

Angle—A geometric figure created by two lines drawn from the same point.

Cosine—A trigonometric function that relates the ratio of the adjacent side of a right triangle to its hypotenuse, or the x coordinate of a point on a unit circle.

Degree—A unit of measurement used to describe the amount of revolution of an angle denoted by the symbol °. There are 360° in a complete revolution.

Hypotenuse—The longest side of a right triangle which is opposite the right angle.

Law of cosines—A relationship between the cosine of an angle of a triangle and its sides which can be used to determine the dimensions of a triangle.

Law of sines—A relationship between the sine of an angle of a triangle and its side which can be used to determine the dimensions of a triangle.

Opposite side—The side of a right triangle which is opposite the angle in question.

Period—A value at which a periodic function begins to repeat.

Pythagorean theorem—An idea suggesting that the sum of the squares of the sides of a right triangle is equal to the square of the hypotenuse. It is used to find the distance between two points.

Radian—A unit of angular measurement that relates the radius of a circle to the amount of rotation of an angle. One complete revolution is equal to 2π radians.

Right triangle—A triangle which contains a 90° or right angle

Similar triangles—Triangles which have congruent angles and proportional sides.

Sine—A trigonometric function which represents the ratio of the opposite side of a right triangle to its hypotenuse, or the y coordinate of a point on a unit circle.

Tangent—A trigonometric function which represents the ratio of the opposite side of right triangle to its adjacent side.

Trigonometric functions—Angular functions which can be described as ratios of the sides of a right triangle to each other.

$\cos(\theta \pm \phi) = \cos\theta\cos\phi \pm \sin\theta\sin\phi$ $\tan(\theta \pm \Phi) = (\tan\theta \pm \tan\phi)/(1 \pm \tan\theta\tan\phi)$

These relationships can be used to develop formulas for double angles and half angles. Therefore, the $\sin 2\theta = 2\sin\theta\cos\theta$ and $\cos 2\theta = 2\cos^2\theta - 1$ which could also be written $\cos\theta 2 = 1 - 2\sin^2\theta$.

Trigonometry using circles

For hundreds of years, trigonometry was only considered useful for determining sides and angles of a triangle. However, when mathematicians developed more general definitions for sine, cosine and tangent, trigonometry became much more important in mathematics and science alike. The general definitions for the trigonometric functions were developed by considering these values as points on a unit circle.

A unit circle is one which has a radius of one unit which means $x^2 + y^2 = 1$. If we consider the circle to represent the rotation of a side of an angle, then the trigonometric functions can be defined by the x and y coordinates of the point of rotation. For example, coordinates of point P(x,y) can be used to define a right triangle with

a hypotenuse of length r. The trigonometric functions could then be represented by the following equations.

$$\sin\theta = \frac{y}{r}; \quad \cos\theta = \frac{x}{r}; \quad \tan\theta = \frac{y}{x}$$

With the trigonometric functions defined as such, a graph of each can be developed by plotting its value versus the magnitude of the angle it represents.

Since the value for x and y can never be greater than one on a unit circle, the range for the sine and cosine graphs is between 1 and -1. The magnitude of an angle can be any real number, so the **domain** of the graphs is all **real numbers**. (Angles which are greater than 360° or 2π radians represent an angle with more than one revolution of rotation). The sine and cosine graphs are periodic because they repeat their values, or have a period, every 360° or 2π radians. They also have an amplitude of one which is defined as half the difference between the maximum (1) and minimum (-1) values.

Graphs of the other trigonometric functions are possible. Of these, the most important is the graph of the tangent function. Like the sine and cosine graphs, the tangent function is periodic, but it has a period of 180° or π radians. Since the tangent is equal to y/x, its range is -∞ to ∞ and its amplitude is ∞.

The periodicity of trigonometric functions is more important to modern trigonometry than the ratios they represent. Mathematicians and scientists are now able to describe many types of natural phenomena which reoccur periodically with trigonometric functions. For example, the times of sunsets, sunrises, and **comets** can all be calculated thanks to trigonometric functions. Also, they can be used to describe seasonal **temperature** changes, the movement of waves in the **ocean**, and even the quality of a musical sound.

Resources

Books

Barnett, Raymond A., Michael Zeigler, Karl Byleen, and Steven Heath. *Analytic Trigonometry with Applications.* 7th ed. New York: John Wiley & Sons, 1998.

Blitzer, Robert et al. *Algebra and Trigonometry.* 2nd ed. Englewood Cliffs, NJ: Prentice Hall, 2003.

Larson, Ron. *Calculus With Analytic Geometry.* Boston: Houghton Mifflin College, 2002.

Stewart, James, et al. *Trigonometry* Pacific Grove, CA: Brooks/Cole, 2003.

Weisstein, Eric W. *The CRC Concise Encyclopedia of Mathematics.* New York: CRC Press, 1998.

Perry Romanowski

Triple bond *see* **Chemical bond**

Tritium

Tritium is an **isotope** of the chemical element **hydrogen**. It has not only a single **proton** but also two neutrons in the nucleus of its **atoms**. Although technically it is still the element hydrogen, it has its own chemical symbol, T. Chemically, tritium reacts in exactly the same manner as hydrogen, although slightly slower because of its greater **atomic weight**. A tritium atom has almost three times the **mass** of a regular hydrogen atom: the atomic weight of tritium is 3.016 whereas the atomic weight of hydrogen is 1.008. Tritium is radioactive, with a **half-life** of 12.26 years. Its nucleus emits a low-energy beta particle, leaving behind an isotope of helium, helium-3, that has a single **neutron** in its atomic nucleus. (The common isotope of helium, helium-4, contains two neutrons in its atomic nucleus.) No gamma rays, which

are high-energy electromagnetic **radiation**, are emitted in the decay of tritium, so the **radioactive decay** of tritium is of little hazard to humans.

The heavier atomic weight of tritium has an effect on the physical properties of this hydrogen isotope. For example, tritium has a **boiling point** of 25K (-415°F; -248°C), compared with ordinary hydrogen's boiling point of 20.4K (-423°F; -252.8°C). Molecules containing tritium show similar variances. For example, **water** made with tritium and having the formula T_2O has a melting point of 40°F (4.5°C), compared with 32°F (0°C) for normal water.

Tritium was present in nature at very low levels, about 1 atom every 10^{18} atoms of hydrogen, before atmospheric nuclear bomb testing. It is produced in the upper atmosphere, as highly energetic neutrons in cosmic rays bombard **nitrogen** atoms, making a tritium atom and an atom of carbon-12.

$$N + neutron \rightarrow T + C$$

Industrially, tritium is prepared by bombarding **deuterium** with other deuterium atoms to make a tritium atom and a regular hydrogen atom:

$$D + D \rightarrow T + H$$

The resulting two types of hydrogen can be separated by distillation. Another way to make tritium is to bombard lithium-6 atoms (the less-abundant isotope of **lithium**) with neutrons, which produces a helium atom and a tritium atom:

$$Li + neutron \rightarrow T + He$$

Due to the testing of **nuclear weapons** in the atmosphere (before such testing was banned), the tritium content of the atmosphere rose to ~500 atoms per 10^{18}, declining steadily even since the ban due to radioactive decay.

Tritium is used in **nuclear fusion** processes because it is easier to fuse tritium nuclei than either of the other isotopes of hydrogen. However, because of its scarcity, it is commonly used with deuterium in fusion reactions:

$$T + D \rightarrow He + neutron + \textbf{energy}$$

This is the nuclear reaction that occurs in fusion bombs, or hydrogen bombs. Such weapons must be recharged periodically due to the radioactive decay of the tritium. Fusion reactions are also being used in experimental fusion reactors as scientists and engineers try to develop controllable nuclear fusion for peaceful power.

Tritium is used as a tracer because it is relatively easy to detect due to its radioactivity. In **groundwater** studies, tritium-labeled water can be released into the ground at one point, and the amount of tritium-labeled water that appears at other points can be monitored. In this way, the flow of water through the ground can be mapped. Such information is important when drilling oil fields, for example. Tritium can also be substituted for ordinary hydrogen in organic compounds and used to study biological reactions. Because of its radioactivity, it is easy to follow the tritium as it participates in biochemical reactions. In this way, specific metabolic processes at the cellular level can be monitored. Tritium is also used to make "glow-in-the-dark" objects by mixing tritium-containing compounds with compounds like zinc sulfide, which emit **light** when struck by alpha or beta particles from nuclear decay.

See also Radioactive tracers.

Resources

Books

Evans, E.A. *Tritium and its compounds.* New York: Wiley, Inc., 1974.

Herman, R. *Fusion: The Search for Endless Energy.* Oxford: Cambridge University Press, 1990.

Parker, Sybil, ed. *McGraw-Hill Encyclopedia of Chemistry.* 2nd ed. New York: McGraw Hill, 1999.

Romer, A. *Radioactivity and the Discovery of Isotopes.* New York: Dover, 1970.

David W. Ball

Trogons

Trogons are about 35 **species** of beautiful arboreal **birds** that constitute the family Trogonidae. Trogons have a number of peculiar features in their morphology, and are not thought to be closely related to any other groups of living birds. This is why their family is the only one in the order Trogoniformes.

Species of trogons occur throughout the tropical and subtropical parts of the world, a biogeographic distribution known as pan-tropical. Most species of trogons occur in Central and especially **South America**, with three species in **Africa**, and 11 species in **Asia**. Most trogons live in dense **forests** and woodlands, and are unobtrusive birds that tend to sit and fly quietly and are not often seen. Trogons do not undertake long-distance migrations, although some species may make local, seasonal movements.

Trogons range in body length from 9-14 in (23-34 cm). The bill of trogons is short and wide, and the upper mandible is hooked at the tip. These birds have short, rounded wings, and a long, broad, square-ended tail. Their legs and feet are small and weak, and are used only for perching. The toe arrangement of trogons is of an unusual and distinctive pattern known as heterodactylous. The first and second digits point backwards, and the third and fourth forwards. This arrangement of the toes occurs in no other group of birds.

Trogons are colorful birds, with bold patterns of bright red, green, blue, yellow, black, or white. There is bare, brightly colored skin around the **eye**. The tail has two rows of large white spots underneath. The sexes are dimorphic in most species, having differing plumage. Female trogons are beautiful birds, but somewhat less so than the males.

Trogons feed largely on **insects** and spiders, although species of the Americas also eat large quantities of **fruits**. Some species also eat **snails**, small lizards, and **frogs**. Trogons spend much of their time perched in a stiffly erect stance on mid-canopy branches, making occasional sallies to catch insects or pluck fruits, often using a hovering flight.

Trogons are solitary birds. They defend a breeding territory, which is proclaimed by simple calls. Their nests are located in a cavity excavated in rotten **wood** in a **tree**, or dug into a termite nest or paper-wasp nest. Trogons lay two to four pale-colored, unspotted eggs that are incubated by both parents. The nestlings are born naked and almost helpless, and are fed with insects regurgitated by the parents. Both sexes share in the care and feeding of the young.

Species of trogons

Most species of trogons are tropical in their distribution. One species, however, breeds as far north as the United States. This is the coppery-tailed or elegant trogon (*Trogon elegans*), which breeds in mountain forests of southern Arizona, and south into Central America. One of the best known of the Central American trogons is the resplendent **quetzal** (*Pharomachrus mocino*), which ranges from Mexico to Nicaragua. This species is brilliantly colored, with the male having greenish hues on the back, breast, tail, and crested head, a vivid red belly, black around the eyes and wings, and yellow bill and feet. The tail of the resplendent quetzal is extended by 24-inch-long (60 cm) plumes, several times longer than the body. This impressive tail is the origin of an alternative common name, the train-bearer.

Dimorphic—This refers to species in which the sexes differ in size, shape, or coloration.

Heterodactyly—An arrangement of the toes, in which the first two point backwards, and the third and fourth forwards. This only occurs in the trogons.

The resplendent quetzal was sacred to the Maya and Aztecs, and represented the god Quetzalcoatl. This bird is still a culturally important symbol within its range. Stylized renditions of quetzals are prominent in much of the folk art in Central America, particularly in Guatemala, where it is the national bird. Unfortunately, the quetzal has become extremely rare over much of its range, because of **deforestation** and hunting.

The red-headed trogon (*Harpactes erythrocephalus*) is a relatively widespread Asian species, occurring from Nepal and south China through Indochina to Sumatra in Indonesia. The male has a bright red head and breast, a cinnamon back and tail, and black wings, while the female lacks the red head. The Narina trogon (*Apaloderma narina*) occurs over much of sub-Saharan Africa. The blue-crowned trogon (*Trogon curucui*) is a widespread species of South American tropical forests, breeding from Columbia to northern Argentina.

The Cuban trogon (*Priotelus temnurus*) only occurs on that **island**, while the Hispaniolan trogon (*Temnotrogon roseigaster*) only occurs on Hispaniola (Haiti and the Dominican Republic). These are both monotypic genera, each containing one species, both of which are rare and endangered, mostly because of **habitat** loss. Other threatened species include the eared quetzal (*Euptilotis neoxenus*) of Mexico and the nearby United States, and the Ward's trogon (*Harpactes wardi*) of Southeast Asia.

Resources

Books

Forshaw, Joseph. *Encyclopedia of Birds*. New York: Academic Press, 1998.

Bill Freedman

Trophic levels

Trophic levels describe the various stages within ecological food chains or webs. Examples of trophic levels, all of which will be described below, are primary producers, primary consumers or herbivores, and secondary and higher-level consumers, or predators.

Food webs are based on the productivity of photosynthetic organisms, such as blue-green **bacteria**, **algae**, and plants. These are autotrophic organisms, which are capable of fixing some of the diffuse **energy** of solar **radiation** into simple organic compounds. This fixed energy can then be utilized by the primary producers to metabolically synthesize a diverse array of biochemicals and to support the growth of these organisms.

The solar energy fixed by **photosynthesis** is the energetic base that all heterotrophic organisms utilize to achieve their own productivity. Heterotrophic organisms include any animals and **microorganisms** that feed on the living or dead **biomass** of plants, or that of other heterotrophs. Heterotrophs include herbivores that feed directly on autotrophs, carnivores that feed on other animals, detritivores that feed on dead biomass, and omnivores that feed on any or all of the above.

Therefore, the food web is a diverse assembly of organisms that are ecologically linked through their feeding relationships, and is ultimately based on the fixation of solar radiation through photosynthesis.

Primary producers

Primary producers are autotrophic organisms that are capable of fixing solar radiation into biochemical energy, through the process of photosynthesis. Photosynthesis is comprised of a series of enzyme-mediated **chemical reactions** which result in the combination of **carbon dioxide** and **water** into glucose, a simple sugar. This chemical reaction requires an input of energy to proceed, and this energy is provided by red and blue wavelengths of solar radiation, which are captured by the photosynthetic pigment **chlorophyll**. The fixed-energy content of the glucose can then be utilized to drive a great diversity of other metabolic reactions, which are used to synthesize the myriad other biochemicals that are found in the tissues of primary producers.

Primary producers include green plants, algae, and blue-green bacteria. If the **rate** of photosynthesis by these organisms exceeds their metabolic requirements, then they are able to grow, and their biomass increases.

Primary consumers

The accumulating biomass of primary producers is a source of fixed energy that can be utilized by heterotrophic organisms by directly feeding on the autotrophic biomass. The primary consumers of autotrophic biomass are also known as herbivores and include the tiny crustacean **zooplankton** that filter microscopic algal cells out of the surface waters of lakes, ponds, and oceans, as well as much larger, mammalian herbivores, such as **mice**, **deer**, cows, and elephants. Herbivores utilize the fixed energy and **nutrients** in their food of autotrophic biomass to drive their own metabolic processes and to achieve their own growth.

Secondary and higher-order consumers

Herbivores may be fed upon by other heterotrophs, which are known as secondary consumers. If the **herbivore** must be killed before it is eaten, the secondary consumer is known as a **predator**. However, if the herbivore does not have to be killed to be eaten, the secondary consumer is known as a parasite. Predators of the tiny zooplankton described in the previous section include somewhat larger, carnivorous zooplankton, as well as small **fish**. In terrestrial ecosystems, herbivorous mice may be fed upon by predatory **weasels** and **hawks**, while deer are killed and eaten by coyotes and cougars.

If the resource base of the **ecosystem** is large enough, the secondary consumers may be killed and eaten by higher-order consumers, which will generally be the top predators in the system. For example, mature **lake** trout may be at the top of the food web of a temperate-lake ecosystem, in which the trophic structure is organized as: algae...herbivorous zooplankton...predatory zooplankton and small fish...and the largest predatory fish, such as lake trout. If bald **eagles** or humans subsequently predate on the lake trout, they would be considered the top predators in the system as well as a trophic linkage to the terrestrial part of the larger ecosystem.

Detritivores

All organisms eventually die, and detritivores are a class of organisms that feed on their dead bodies. Actually, detritivores can themselves be divided into a food web, based on the feeding relationships among the **species**. In this sense, primary detritivores feed directly on the dead biomass, while secondary detritivores feed on these direct consumers of detritus.

Omnivores

Omnivores are animals that feed at various places within the food web and are therefore difficult to classify in terms of trophic level. For example, grizzly **bears** are highly opportunistic animals that feed quite widely, on **sedges** and berries, small **mammals**, fish, and dead animals (or carrion). Of course, humans are the most omnivorous of all species (we eat just about anything that is not acutely poisonous), and in turn are not eaten by many other creatures, except, eventually, by detritivores.

KEY TERMS

. . . .

Autotroph—This refers to organisms that can synthesize their biochemical constituents using inorganic precursors and an external source of energy.

Heterotrophic—Organism that requires food from the environment since it is unable to synthesize nutrients from inorganic raw materials.

Trophic—Pertaining to the means of nutrition.

See also Autotroph; Carnivore; Food chain/web; Omnivore.

Resources

Books

Odum, E.P. *Ecology and Our Endangered Life Support Systems.* New York: Sinauer, 1993.

Ricklefs, R.E. *Ecology.* New York: W.H. Freeman and Co., 1990.

Bill Freedman

Tropic birds

Tropic **birds** are three **species** of pan-tropical seabirds that make up the family Phaethontidae, in the order Pelecaniformes, which also includes the **pelicans**, anhingas, **cormorants**, gannets, and boobies.

Tropic birds are medium-sized seabirds, weighing about 0.9 lb (0.4 kg), and having a body length of 16-18.9 in (41-48 cm). This length does not include their greatly elongated tail feathers, which are shaped as streamers that themselves can be 21 in (51-cm) long. The wings of tropic birds are short, stout, and pointed, and their legs are short and the feet small, and the toes completely webbed. Their beak is stout, pointed, and slightly down-curved, with serrated edges to help hold on to their slippery **prey** of **fish** or **squid**.

The usual coloration of the body is white, with black markings on the head and wings. The bill is colored either bright red or yellow, and the tail-streamer is prized in many species. These streamers are either white or red, depending on the ornaments in Polynesian cultures. The sexes are alike in size and coloration.

Tropic birds have a loud, shrill, piercing scream, which is the origin of one of the common names of these birds, the bosun bird. (Bosuns or boatswains are petty ship's-officers, responsible for maintenance of the ship and its gear, and equipped with a loud, shrill whistle, used to catch the attention of sailors.) Tropic birds are strong and graceful fliers, typically exhibiting bouts of fluttering, pigeon-like wing-strokes, punctuated by short, straight glides and soaring flights. Tropic birds do not swim well, and they float with their tail held in a cocked, erect position. These birds are extremely awkward on land, and can barely walk.

Tropic birds feed using partially closed-winged, aerial plunge-dives to catch their food of small fish or squid near the surface. They can even catch **flying fish**, while the fish are in the air. Tropic birds generally occur as solitary birds in off-shore, pelagic waters outside of their breeding season. They may, however, mix with large, feeding flocks of other species, such as shearwaters and **terns**.

The **courtship** of tropic birds involves the potential pair engaging in graceful, aerial wheelings and glides, with loud cries. Occasionally one bird will hover over the other, touching it with the tip of its tail-streamer. Tropic birds lay their single egg in a simple scrape in a hidden cavity on a rocky, near-shore cliff, or sometimes under a bush. Both sexes incubate the egg, and they share in guarding and feeding their white-downy, young chick. Juvenile birds leave their **birth** island as soon as they can fly, and return when they reach sexual maturity. Tropic birds may breed throughout the year.

Species of tropic birds

The red-billed tropic bird (*P. aethereus*) occurs in the tropical Caribbean, Atlantic, eastern Pacific, and Indian Oceans.

The tail-streamer of this species is white, and the upper parts of its body are barred with white and brown.

The red-tailed tropic bird (*P. rubricauda*) occurs in the tropical Indian and Pacific Oceans, but in different regions of those oceans than the red-billed tropic bird. The red-tailed tropic bird has a bright-white body, and a red streamer.

The white-tailed or yellow-billed tropic bird (*P. sepeurus*) occurs in all of the tropical oceans. This smallest tropic bird has a white body, with black patches on its wings and back, a relatively small, yellow beak, and a white streamer.

None of the tropic birds breed in **North America**. However, all species are rare visitors to coastal waters of the southern United States, especially after a heavy wind-storm. The red-billed tropic bird is a regular but rare bird off southern California, while the white-tailed tropic bird occurs off the southeastern states in the warm Atlantic, Gulf of Mexico, and Caribbean. The red-tailed tropic bird is a very rare visitor to southern California.

Resources

Books

Forshaw, Joseph. *Encyclopedia of Birds.* New York: Academic Press, 1998.
Sibley, David Allen. *The Sibley Guide to Birds.* New York: Knopf, 2000.

Bill Freedman

Tropical cyclone

Tropical cyclones are large circulating **storm** systems consisting of multiple bands of intense showers and thunderstorms and extremely high winds. These storm systems develop over warm **ocean** waters in the tropical regions that lie within about 25° latitude of the equator. Tropical cyclones may begin as isolated thunderstorms. If conditions are right they grow and intensify to form the storm systems known as hurricanes in the Americas, typhoons in East **Asia**, willy-willy in **Australia**, cyclones in Australia and India, and baguios in the Philippines. A fully developed tropical cyclone is a circular complex of thunderstorms about 403 mi (650 km) in diameter and over 7.5 mi (12 km) high. Winds near the core of the cyclone can exceed 110 MPH (50 km/h). At the center of the storm is a region about 9-12.5 mi (15-20 km) across called the **eye**, where the winds are light and skies are often clear. After forming and reaching peak strength over tropical seas, tropical cyclones may blow inshore causing significant damage and loss of life. The storm destruction occurs by very high winds and forcing rapid rises in **sea level** that flood low lying coastal areas. Better forecasting and emergency planning has lowered the death tolls in recent years from these extremely powerful storms.

Tropical cyclone geography and season

Several ocean areas adjacent to the equator possess all the necessary conditions for forming tropical cyclones. These spots are: the West Indies/Caribbean Sea where most hurricanes develop between August and November; the Pacific Ocean off the west coast of Mexico with a peak hurricane season of June through October; the western Pacific/South China Sea where most typhoons, baguios, and cyclones form between June and December; and south of the equator in the southern Indian Ocean and the south Pacific near Australia where the peak cyclone months are January to March. Note that in each area the peak season is during late summer (in the southern hemisphere summer runs from December to March). Tropical cyclones require warm surface waters at least 80°F (27°C). During the late summer months the sea surface temperatures reach their highest levels and provide tropical cyclones with the **energy** they need to develop into major storms.

The annual number of tropical cyclones reported varies widely between regions and from year to year. The West Indies recorded 658 tropical cyclones between 1886-1966, an average of about eight per year. Of these, 389, or about five per year, grew to be of hurricane strength. The Atlantic hurricane **basin** has a 50-year average of 10 tropical storms and six hurricanes annually.

In the United States, the National Weather Service names hurricanes from an alphabetic list of alternating male and female first names. New lists are drawn up each year to name the hurricanes of western Pacific and the West Indies. Other naming systems are used for the typhoons and cyclones of the eastern Pacific and Indian oceans.

Structure and behavior

In some ways tropical cyclones are similar to the low **pressure** systems that cause weather changes at higher latitudes in places like the United States and **Europe**. These systems are called extratropical cyclones and are marked with an "L" on **weather** maps. These weather systems are large masses of air circulating cyclonically (counterclockwise in the northern hemisphere and clockwise in the southern hemisphere). Cyclonic circulation is caused by two forces acting on the air: the pressure gradient and the Coriolis force.

In both cyclone types air rises at the center, creating a region of lower air (barometric) pressure. Since air is a fluid, it will rush in from elsewhere to fill the void left by air that is rising off the surface. The effect is the same as when a plug is pulled out of a full bathtub: **water** going down the drain is replaced by water rushing in from other parts of the tub. This is called the pressure gradient force because air moves from regions of high pressure to lower pressure. Pressure gradient forces are responsible for most of our day-to-day winds. As the air moves toward low pressure, the Coriolis force turns the air to the right of its straight line **motion** (when viewed from above). In the Southern Hemisphere the reverse is true: the Coriolis force pushes the moving air to the left. The air, formerly going straight toward a low pressure region, is forced to turn away from it. The two forces are in balance when the air circles around the low pressure zone with a constant radius creating a stable cyclone rotating counterclockwise in the northern hemisphere and clockwise in the southern hemisphere.

All large scale air movements such as hurricanes, typhoons, extratropical cyclones, and large thunderstorms tend to set up a cyclonic circulation in this manner.

(Smaller scale circulations such as the vortex that forms in a bathtub drain are not cyclonic because the Coriolis force is overwhelmed by other forces. You can make a bathtub drain vortex rotate clockwise or counterclockwise simply by stirring the water the right way. The larger a system is the more likely that the Coriolis force will prevail and the **rotation** will be cyclonic.) The Coriolis force is a consequence of the rotation of **Earth**. Moving air masses, like any other physical body, tend to move in a straight line. However, we observe them moving over Earth's surface, which is rotating underneath the moving air. From our perspective the air appears to be turning even though it is actually going in a straight line, and it is we who are moving.

The **Coriolis effect** can be demonstrated by two people riding across from each other on a merry-go-round. If one person throws a ball straight at his friend she will rotate out of position while the ball is moving and will be unable to catch it. To the two observers the ball seemed to curve away from the catcher as if some force pushed it. Of course the ball actually went perfectly straight but the observer's rotating frame of reference made it appear that a force was at work. On the surface of the rotating Earth this apparent force—the Coriolis force—makes moving air masses curve with respect to the surface and sets up cyclonic circulation.

In both tropical and extratropical cyclones, the rising air at the cyclone center causes **clouds** and **precipitation** to form. A fully developed hurricane consists of bands of thunderstorms that grow larger and more intense as they move closer to the cyclone center. The area of strongest updrafts can be found along the inner wall of the hurricane. Inside this inner wall lies the eye, a region where air is descending. Descending air is associated with clearing skies, therefore, in the eye the torrential rain of the hurricane ends, the skies clear, and winds drop to nearly calm. If you are in the eye of a hurricane the eye wall clouds appear as just that: towering vertical walls of **thunderstorm** clouds, stretching up to 7.5 mi (12 km) in height, and usually completely surrounding the eye. Hurricanes and other tropical cyclones move at the speed of the prevailing winds, typically 10-20 MPH (16-32 km/h) in the tropics. A hurricane eye passes over an observer in less than an hour, replaced by the high winds and heavy rain of the intense inner thunderstorms.

Life history of a tropical cyclone

Several conditions are necessary to create a tropical cyclone. Warm sea surface temperatures, which reach a peak in late summer, are required to create and maintain the very warm, humid air **mass** in which tropical cyclones grow. This provides energy for storm

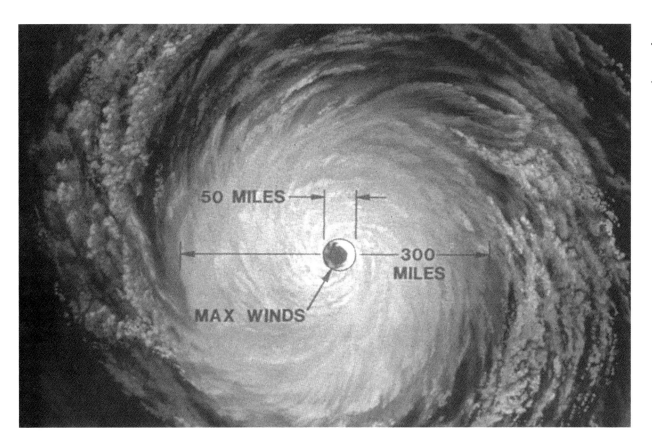

Overhead view of the eye of a tropical cyclone. *National Oceanic and Atmospheric Administration.*

development through the **heat** stored in humid air called latent heat. It takes energy to change water into vapor; that is why one must add heat to boil a kettle of water. The reverse is also true: when vapor condenses back to form liquid water, heat is released that may heat up the surrounding air. In a storm such as a hurricane, many hundreds of tons of humid air are forced to rise and cool, condensing out tons of water droplets and liberating a vast quantity of heat. This warms the surrounding air causing it to expand and become even more buoyant, that is, more like a hot air **balloon**. More air begins rising, causing even more humid air to be drawn into the cyclone. This process feeds on itself until it forms a cyclonic storm of huge proportions. The more humid air available to a tropical cyclone the greater its upward growth will be and the more intense it will become.

For storm growth to get started some air needs to begin rising. Because tropical air masses are so uniformly warm and humid, the atmosphere over much of the tropics is fairly stable; that is, it does not support rising air and the development of storms. Thunderstorms occasionally develop but tend to be short-lived and small in scale, unlike the severe thunderstorms in the middle lati-

tudes. During the late summer this peaceful picture changes. Tropical disturbances begin to appear. These can take the form of a cluster of particularly strong thunderstorms or perhaps a storm system moving westward off of the African **continent** and out to sea. Tropical disturbances are regions of lower pressure at the surface. As we have seen, this can lead to air rushing into the low pressure zone and setting up a vortex, or rotating air column, with rising air at its core.

An additional element is needed for tropical cyclone development: a constant **wind** direction with height throughout the lower atmosphere. This allows the growing vortex to stretch upward throughout the atmosphere without being sheared apart. Even with all these elements present only a few of the many tropical disturbances observed each year become hurricanes or typhoons. Some sort of extra kick is necessary to start the growth of a hurricane. This often comes when tropical disturbance near the surface encounters a similar disturbance in the air flow at higher levels such as a region of low pressure at about the 3 mi (5 km) level (called an upper low). These upper lows sometimes wander toward the equator from higher latitudes where they were part of a decaying weather system.

A computer enhanced image of Hurricane Diana at its strongest on September 11, 1984. The hurricane was just off the coast of South and North Carolina at the time, and winds within it were 130 MPH (209 km/h). *U.S. National Aeronautics and Space Administration (NASA).*

Once a tropical disturbance has begun to intensify a chain reaction occurs. The disturbance draws in humid air and begins rising. Eventually it condenses to form water droplets. This releases latent heat, which warms the air, making it less dense and more buoyant. The air rises more quickly off of the surface. As a result, the pressure in the disturbance drops and more humid air moves toward the storm. Meanwhile, the disturbance starts its cyclonic rotation and surface winds begin to increase. Soon the tropical disturbance forms a circular ring of low air pressure and becomes known as a tropical depression. As more heat energy is liberated and updrafts increase inside the vortex, the internal barometric pressure continues to drop and the incoming winds increase. When wind speeds increase beyond 37 MPH (60 km/h) the depression is upgraded to a tropical storm. If the winds reach 75 MPH (120 km/h) the tropical storm is officially classified as a hurricane (or typhoon, cyclone, etc., depending on location). The chain reaction driving this storm growth is very efficient. About 50-70% of tropical storms intensify to hurricanes.

A mature tropical cyclone is a giant low pressure system pulling in humid air, releasing its heat, and transforming it into powerful winds. The storm can range in diameter from 60-600 mi (100-1000 km) with wind speeds greater than 200 MPH (320 km/h). The central barometric pressure of the hurricane drops 60 millibars (mb) below the normal sea level pressure of 1013 mb. By comparison, the passage of a strong storm front in the middle latitudes may cause a drop of about 20-30 mb. The size and strength of the storm is limited only by the air's **humidity**, which is determined by ocean **temperature**. It is estimated that for every 1.8°F (1°C) increase in sea surface temperature the central pressure of a tropical cyclone can drop 12 mb. With such low central pressure, winds are directed inward, but near the center of

KEY TERMS

. . . .

Coriolis force—An apparent force that seems to push moving air masses into curving paths. The Coriolis effect is not a true force but is due to our observing air motion on the surface of the rotating Earth.

Extratropical cyclone—Circulating columns of air which may bring storms to areas in the middle latitudes. Often called low pressure systems.

Eye—A calm, rainfree region at the very center of a tropical cyclone.

Hurricane (typhoon, cyclone, etc.)—A tropical cyclone with winds that have reached the speed of 75 MPH (119 km/h).

Latent heat—The heat given off when water vapor condenses to form liquid water.

Midlatitudes—The portion of the earth's surface midway between the tropics and the polar regions lying about 3565° north or south of the equator.

Pressure gradient force—The force that pushes air from regions of higher pressure to regions of lower pressure.

Swell—The rise of sea level near coastal areas due to the low barometric pressure; winds and wave activity of a tropical cyclone. Also called surge.

Tropical depression—An early stage in the development of a hurricane, typhoon, or cyclone.

Tropical storm—A tropical cyclone with wind speeds 37–75 MPH (60–120 km/h).

Tropics—The region around Earth's equator spanning 23.5° north latitude to 23.5° south latitude.

the storm the winds are rotating so rapidly the Coriolis force prevents any further inward movement. This inner boundary creates the eye of the tropical cyclone. Unable to go in, the air is forced to move upward then spread out at an altitude of about 7.5 mi (12 km). Viewed from above by a **satellite**, the tropical cyclone appears as a mass of clouds diverging away from the central eye.

The tropical cyclone on land

All of the cyclone development described thus far takes place at sea, but the entire cyclone also is blown along with the prevailing winds. Often this movement brings the storm toward land. As tropical cyclones approach land they begin affecting the coastal areas with sea swells, large waves caused by the storm's high winds. Swells often reach 33 ft (10 m) in height and can travel thousands of kilometers from the storm. Coastal areas are at risk of severe damage from these swells that destroy piers, beach houses, and harbor structures every hurricane season. Particularly high swells may cause **flooding** farther inland.

Perhaps more dangerous than the gradually rising swells are the sudden rises in sea level known as storm surges. Storm surges occur when the low barometric pressure near the center of a cyclone causes the water surface below to rise. Then strong winds blowing toward the coast push this "bulge" of water out ahead of the storm. The water piles up against the coast, quickly raising sea level as much as 16 ft (5 m) or more. The highest **storm surge** generally occurs to the right of the storm's path. When storm-tossed waves 23-33 ft (7-10 m) high

are added to this wall of water land areas may be inundated. In 1900 the city of Galveston, Texas, was hit with a storm surge during a hurricane. One eyewitness reported that the sea rose 4 ft (1.3m) in a matter of seconds. Over 5,000 people lost their lives in the Galveston Hurricane and resulting flooding, making it the deadliest storm ever recorded in the United States.

Tropical cyclones that travel onto the land immediately begin to weaken since humid air, their source of energy, is cut off. The winds at the base of the cyclone encounter greater **friction** as they drag across uneven terrain that slows them. Nevertheless tropical cyclones at this stage are still capable of producing heavy rains, thunderstorms, and even tornadoes. Occasionally, the remnants of a tropical cyclone that has begun to weaken over land will unite with an extratropical low pressure system, forming a very potent rain-making storm front that may bring flooding to areas far from the coast.

Until relatively recently, people in the path of a tropical cyclone had little warning of approaching storms. Usually their only warning signs were the appearance of high clouds and a gradual increase in winds. Hurricane watch services were established beginning in the early years of the twentieth century. By the 1930s hurricanes were detected with weather balloons and ship reports while the 1940s saw the introduction of airplanes as hurricane spotters. **Radar** became available after World War II and has remained a powerful tool for storm detection in the years since. Today a global network of weather satellites allows meteorologists to identify and track tropical cyclones from their earliest appearance as disturbances over the remote ocean. This improved ability to watch storms develop anywhere in the world has meant that warnings and evacuation orders can be issued well in advance of a tropical cyclone reaching land. Even though coastal areas have more people living near them today than ever before and tropical cyclones remain just as powerful as they have always been far fewer storm related deaths are reported each year than 60 years ago thanks to advances in storm detection and forecasting.

See also Atmosphere observation; Atmospheric circulation; Atmospheric pressure; Cyclone and anticyclone; Weather forecasting.

Resources

Books

Battan, Louis J. *Weather.* Engelwood Cliffs: Prentice-Hall Inc., 1985.

Battan, Louis J. *Weather in Your Life.* New York: W.H. Freeman & Co., 1983.

Fisher, David E. *The Scariest Place on Earth: Eye to Eye with Hurricanes.* New York: Random House, 1994.

Gedzelman, Stanley D. *The Science and Wonders of the Atmosphere.* New York: John Wiley & Sons, 1980.

Hardy, Ralph, Peter Wright, John Kington, and John Gribben. *The Weather Book.* Boston: Little, Brown and Co., 1982.

McNeill, Robert. *Understanding the Weather.* Las Vegas: Arbor Publishers, 1991.

Wallace, John M., and Hobbs, Peter V. *Atmospheric Science: An Introductory Survey.* New York: Academic Press, 1977.

Periodicals

"Cyclolysis: A Diagnosis Of Two Extratropical Cyclones." *Monthly Weather Review* 129, no. 11 (2001): 2714-2729.

Leroux, M. "The Meteorology And Climate Of Tropical Africa." *Journal of Meteorology* 27, no. 271 (2002): 274.

Rodgers, Edward B. "Contribution of Tropical Cyclones to the North Atlantic Climatology." *Journal of Applied Meteorology* 40, no. 11 (2001): 1785-1800.

James Marti

Tropical diseases

Nowhere is the prevalence of certain illnesses more striking than in areas where tropical diseases flourish. In many parts of **Africa**, **South America,** and **Asia**, diseases exist that are rarely seen in the United States. These include **malaria**, which infects from 300 to 500 million people annually and kills up to 2.7 million people every year, and leishmaniasis, which affects some 12 million people internationally. Other ailments, such as measles and diarrhea, are well-known in developed areas such as the United States but are rarely fatal.

Young people are highly susceptible to death from **disease** in these areas. Children under five account for 40-60% of all deaths in Africa, Asia, and South America, although they make up only about 15% of the population. Such high death rates reflect the fact that tropical diseases are most prevalent in poor areas where health care is limited. Efforts to improve community health in areas where tropical diseases thrive include the establishment of health clinics and the development of new vaccines.

Battles against malaria

Malaria, a parasitic disease spread by **mosquitoes**, is the best-known tropical disease and infects the largest number of people internationally. It has been called the "most devastating disease in history" based on the number of people it has attacked or killed. Currently, malaria has been identified in about 100 countries, although 80% of the clinical cases are reported in Africa.

The history of malaria is different than the history of other tropical diseases, due to its immense reach and the concerted effort over time to defeat the disease. But in as much as it has been eliminated or controlled in developed countries and remains a major killer in those that are poor and less developed, its history is also similar to that of other tropical diseases.

The ancient Chinese wrote about malaria as early as 2700 B.C., describing the symptomatic fever and characteristic enlargement of the spleen. Ancient Greeks and Romans also suffered from malaria. Hippocrates (460-375 B.C.), offered an accurate account of malarial symptoms and theorized that the disease was caused by a miasma, or poisonous cloud, rising up from marshy land.

The first effective treatment for malaria was developed in the seventeenth century and utilized a traditional Peruvian treatment for fever, the cinchona **tree bark**. In 1630, Jesuit missionaries in Peru introduced the bark of the cinchona tree to Europeans. The active substance in the bark, **quinine**, was isolated in 1820. Though there were common, serious side effects to the drug, such as ringing of the ears and **hearing** loss, the substance was effective in treating malaria, which was common in **Europe** at that time.

Development of a treatment for malaria cleared the way for large scale exploration of tropical areas by Europeans. Development of a synthetic quinine, called chloroquine, in the 1940s, offered effective treatment with fewer common side effects. In the 1960s, malarial **parasites** became resistant to chloroquine, and the substance no longer worked in many areas.

The most recent effort to control malaria has been effective in many areas, including the southern United

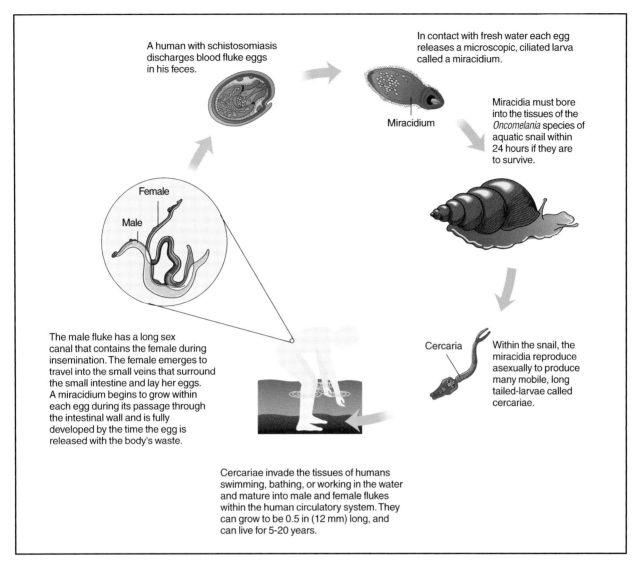

A human with schistosomiasis discharges blood fluke eggs in his feces.

In contact with fresh water each egg releases a microscopic, ciliated larva called a miracidium.

Miracidium

Miracidia must bore into the tissues of the *Oncomelania* species of aquatic snail within 24 hours if they are to survive.

Female

Male

The male fluke has a long sex canal that contains the female during insemination. The female emerges to travel into the small veins that surround the small intestine and lay her eggs. A miracidium begins to grow within each egg during its passage through the intestinal wall and is fully developed by the time the egg is released with the body's waste.

Cercaria

Within the snail, the miracidia reproduce asexually to produce many mobile, long tailed-larvae called cercariae.

Cercariae invade the tissues of humans swimming, bathing, or working in the water and mature into male and female flukes within the human circulatory system. They can grow to be 0.5 in (12 mm) long, and can live for 5-20 years.

The life cycle of *Schistosoma japonicum*, one of three species of parasitic blood flukes that cause schistosomiasis.
Illustration by Hans & Cassidy. Courtesy of Gale Group.

States. The pesticide DDT was used, successfully, in the Tennessee River Valley, Greece, Puerto Rico and other locations to kill the mosquitoes carrying malaria and eliminate the disease. A world campaign to eliminate malaria internationally used DDT from 1956-1969. However, mosquitoes became resistant to DDT and the effort failed.

There are four different types of malaria **protozoa**. One of the four, *P. falciparum,* is responsible for almost all deaths from the disease. Symptoms of the disease do not occur for one to two weeks. They often include chills, muscle aches, fatigue, and abdominal discomfort. They may include tremors and high fever, which comes and goes at regularly spaced intervals. Complications from severe malaria include renal failure, pulmonary **edema**, **coma**, and hypoglycemia.

The delayed appearance of malaria's symptoms reflects the measured course the disease takes in the body. The activity of the malaria parasite in the body is aggressive and thorough. The disease is spread by female anopheline mosquitos that inject malaria parasites into the bloodstream with their bite. The parasites, called sporozoites, travel to the liver, where they enter cells of the liver **tissue**. Once in the liver, the sporozoite changes to a **spore**, which replicates itself until there are thousands of spores in a cyst-like structure which has replaced the **cell**. Malaria manifests no symptoms while this process occurs.

Symptoms develop when the cyst bursts and the spores, called merozoites, are released into the **blood** stream. At this point, sweating and high fever can occur

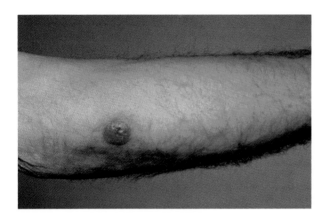

A skin lesion from cutaneous leishmaniasis. *Biophoto Associates, National Audubon Society Collection/Photo Researchers, Inc. Reproduced by permission.*

as the spores enter the host's red blood cells. The parasite consumes hemoglobin from within the red blood cells.

As the parasite grows from consuming the hemoglobin, its nucleus divides into from six to 32 parts, each of which becomes a spore of its own. The red cell bursts, and the spore moves to another cell.

After the merozoites have launched their process of reproducing in the red blood cells of the host, some of the merozoites are transformed to male or female gametocytes. These sexual cells form the base for more parasites if the host is bitten again by the female anopheles mosquito. When this occurs, the gametocytes develop into a number of differentiated cells, called gametes, zygotes, ookinetes and oocysts, within the gut of the mosquito. Eventually the parasite reproduces and sporozoites that travel in the mosquito are formed. These sporozoites move up to the mosquito's salivary gland, where they are ready to infect another host.

Treatment of malaria is usually effective. Because many malaria parasites are resistant to chloroquine, physicians have returned to using quinine in many instances. Travelers and others at risk of malaria receive chloroquine or other drugs, including mefloquine, if journeying to areas where the drugs still work. However, there is no totally effective preventive measure against malaria.

The deadly sandfly and leishmaniasis

Another disease spread by **insects** is leishmaniasis, a sickness caused by several types of protozoa carried by **sand flies**. In the mid 1990s, the disease was infecting about 12 million people worldwide, killing about 1,000 annually. It is spread when sand flies draw blood from individuals who already are infected. Hosts can be humans, dogs, or other **mammals**.

Each of the four major clinical syndromes of the disease have a long incubation period that ranges from three months to 24 months. The disease is marked by one or more skin lesions and varies in severity depending on the type. Kala Azar, one of the types, is commonly found in East Africa and the Sahara. Kala Azar causes fever, diarrhea, enlarged liver, and **anemia**. Another type, Old World leishmaniasis, usually heals on its own and is marked by multiple lesions.

Treatment depends on the type of leishmaniasis and may include the use of drug therapy and transfusions for several weeks. The need for extended medical care makes treatment of leishmaniasis impossible for many poor individuals in developing countries, raising the mortality **rate** of the disease.

Dangerous worms

Ascariasis and hookworm

A common disease in tropical countries is ascariasis. The annual death rate due to ascariasis is about 20,000, most due to complications within the intestine, where the worms settle as adults. An estimated 700 million individuals are affected at any given time by the disease.

Ascariasis is spread through dirt. The worm which grows in the intestine, the roundworm *Ascaris lumbricoides,* is the largest such parasite, growing to a length of 19.3 in (49 cm). It can live about a year in the human body.

The disease can cause difficulties, such as bronchial **asthma**, in the lungs, when the parasite settles in that part of the body. The greatest threat to the human host is the loss of food **nutrients** which go to the worm instead of the host. Individuals whose diet is already sparse may suffer from **malnutrition** after **infection** with ascariasis.

The life cycle of the creature is similar to the life cycle of the blood fluke, though there are some differences. The disease is spread when Ascaris eggs leave the human host in feces. The new host ingests the eggs through **soil** or by soil-contaminated hands or food. The eggs then travel through the liver, lungs, and throat, before ending up in the intestine. Once they are in the intestine, more eggs are laid, clearing the way for further spread of the disease.

Another parasite that is spread through dirt and grows in the human body is hookworm, which affects approximately 800 million individuals in the developing world and kills about 50,000 annually. Hookworm was at one point common in the American South, but is now seldom reported in the area.

The hookworm, which grows up to 0.5 in (13 mm) in size, sucks blood and normally lives in the small intestine. Establishment of the worm in the host generally results in iron-deficiency anemia because of blood loss, and may cause **bronchitis**, peptic **ulcers**, or even **heart** problems.

Hookworms enter the body as larvae through the skin, taking advantage of hair follicles or other openings to help the creature pierce through the skin with its boring movement. The parasite enters the **circulatory system**, moving first to the lungs then to the pharynx, where it is swallowed. The final larvae stage typically occurs in the small intestine, where adult worms emerge and ultimately lay eggs. Adult worms normally live from one to 5 years if the host is not treated. They cannot multiply in the body.

Successful control of the hookworm depends on improvements in plumbing and the use of footwear, a challenge in countries where shoes are a luxury. Medication is available to destroy hookworm, but providing it in areas where hookworm infestation is severe does little to alter the problem, as those who are medicated can easily be reinfected.

Schistosomiasis

One of the most common tropical diseases is schistosomiasis, a disease caused by a worm called a blood fluke. The worm, which can be about 0.47 in (12 mm) long, can live in the human body from five to 20 years. About 200 million people are infected by the worm in developing countries, with as many as 500,000 deaths due to the disease each year.

Symptoms include abdominal **pain**, diarrhea, and weight loss. If untreated, the disease can cause enlargement of the liver, bleeding from blood vessels in the esophagus, and problems to the central **nervous system**.

The worm, which is usually hosted by Biomphalaria **snails**, depends on fresh **water** to survive. The disease is common in areas where bathing occurs in **freshwater** contaminated by human feces and where many individuals have the disease.

The life cycle of the blood fluke entails travel throughout the human body and through the body of its other host, the snail. The worm enters the human body in a larval stage in fresh water through contact with human skin. The larva enter a blood or lymph vessel and move to the heart and lungs, where they grow for several days. Then they move to the liver and the portal circulation, where they grow for several weeks.

Ultimately, the worms move to the intestinal wall, where they settle. The creatures lay eggs, which are released in feces and become new larva after contact with fresh water. These larva enter the host snail, where they eventually produce cercariae, the final larval stage. These are the creatures that enter the human body after contact with skin.

Drug therapy is an effective treatment for the disease. Prevention efforts depend on controlling infection through drug therapy and on convincing individuals to steer clear of fresh water which may be infected.

Microscopic hazard

Amebiasis, a disease caused by a microscopic protozoa spread through dirty water, infects as many as 500 million individuals in the developing world. The disease kills about 70,000 people annually.

The protozoa can live in the large intestine and not cause damage to the host. But it is also capable of causing ulceration of the colonic wall and damage to the liver and other organs, including the **brain**.

The disease is spread through water contaminated by fecal **matter**, by contaminated flies, or by other contaminated substances. Hosts swallow cysts, which divide in the small intestine and again in the large intestine to form amoebae. These amoebae divide and form trophozoites, which feed on fecal **bacteria**. These trophozoites form cysts and are passed out in the feces of the host. This entire cycle takes from 48 hours to four months.

Diarrhea and cholera

Best known in the United States as a benign but unpleasant condition, diarrhea is one of the leading causes of death in developing countries. Indeed, diarrhea causes about 4.3 million deaths each year, more than malaria or many of the other serious tropical illnesses. Though most of the approximately 28 billion individuals who develop diarrhea survive, the health consequences of diarrhea are great in tropical areas.

Infants and small children are most likely to die of diarrhea. This is because loss of even a small amount of water can be life-threatening to small children. Death due to diarrhea stems from loss of fluid, loss of **plasma**, and the collapse of the cardiovascular system. Infants being weaned from the mother's breast are particularly susceptible to malnutrition and dehydration. Diarrhea can be caused by bacteria or viruses.

Residents of tropical disease areas are less likely than residents of developed areas to have access to emergency treatment for extreme diarrhea. They are also more likely to be exposed to **pathogens** that cause intense diarrhea in contaminated food or water. **Cholera**, a life-threatening condition whose major symptom is watery diarrhea, is spread through dirty water.

Oral rehydration therapy, an approach which uses a **solution** of glucose or sucrose and **salt** to hydrate the body, is an effective treatment for life-threatening diarrhea. One problem with the approach is that it must be made available immediately after a child gets sick, as diarrhea can cause enough fluid loss within the first 12 to 24 hours to cause death. Many individuals in tropical disease areas live many hours by foot from medical care. Therefore, health experts are working to make oral rehydration therapy, along with education about the therapy, available at home to residents of tropical-disease areas.

The most infamous type of diarrheal disease is cholera, which is also marked by vomiting, intense thirst, and abdominal cramping. Cholera is so common it is **endemic** in some parts of the world, such as India. In other parts of the world, it occurs as an **epidemic** or a pandemic.

The disease is caused by the *Vibrio cholerae* bacteria, which typically spreads through dirty water, food, or seafood from areas with **contamination**. The bacteria that causes cholera multiplies in the human intestinal tract, but has also been known to multiply in water.

Symptoms of cholera are a direct result of the cholera enterotoxin, the toxic substance elaborated by cholera bacteria that affects the cells of the small intestine. The substance forces the mucosal cells to secrete large quantities of fluid, resulting in diarrhea.

Control of cholera calls for good sanitation measures, specifically clean water and good food hygiene. Boiling water eliminates the bacteria. A **vaccine** that offers limited effectiveness—30-80% protection for up to six months—is available.

While treatment is effective, involving **antibiotics**, nutritional support, and treatment for dehydration, the death rate in epidemics may reach 30%.

Malnutrition

Malnutrition is the underlying cause of death for about two million people annually, and the disease most commonly affects children under five and the elderly. As many as 40% of all children under five in developing areas have suffered from malnutrition.

The disease stems for eating inadequate amounts of food or eating foods without enough protein. Different syndromes reflect different types of food deficits. Kwashiorkor reflects a protein deficiency, and marasmus is caused by an **energy** deficiency linked to failure to consume enough carbohydrates and fats.

Signs of malnutrition are loss of energy, wasting of muscle, and loss of **fat**. The condition can result in severe anemia, coma, or heart failure. Treatment requires efforts to rehydrate the individual and to replace deficient eating patterns with adequate consumption of healthy food. Education of some families may help improve future eating patterns. Yet, because the condition

most often occurs among impoverished families in areas with inadequate food availability, changing eating habits can be difficult or impossible.

Infectious disease killers

The most common killers in tropical areas are not exotic diseases. They are infectious diseases, many of which are not considered life-threatening in developed countries such as the United States. In the least industrialized countries, 40% or more of all deaths are caused by infectious disease. In the United States, about 1% of all deaths are caused by such illnesses. The huge difference in death rates stems from the fact that most infectious diseases are the easiest diseases to treat using modern medical care.

Approximately 10 million people die every year from **respiratory diseases** in the developing world. Such diseases are caused by the **influenza virus** or bacteria such as *pneumococcus* and *Hemophilus influenza B*. Measles, which can be prevented by immunization of infants, kills two million people in the developing world annually. The course of the disease in developing countries often includes complications such as diarrhea, **ear** infection, **pneumonia**, and weight loss. For some, weight loss leads to malnutrition. Aggressive efforts to curb measles deaths through immunization have reduced the number of potential deaths in recent years.

Whooping cough is another early childhood disease which is seen rarely in the United States but is common in developing countries. About 600,000 people die of whooping cough annually in the developing world, with many of these deaths occurring among small children. About 55 million people develop the disease. Whooping cough is also prevented by immunization.

Another infectious disease which is a common killer in developing countries is **tuberculosis**, which kills about 900,000 people every year and infects about seven million people annually. Tuberculosis can be treated through immunization and drug therapy, although drug resistant strains are becoming increasingly prevalent. Though deaths have been reduced internationally over the last 30 years, the disease is still very common in tropical areas.

Tropical-disease areas have also been hit hard by **AIDS** and HIV infection. An estimated 15 million people internationally are infected by the HIV virus, the World Health Organization estimated in 1994. Of those about one-fifth have developed AIDS, a disease which is fatal in most instances.

A key emphasis of prevention programs in Africa has been promoting the use of condoms during sexual intercourse, a practice which reduces the risk of AIDS. In Africa, condom use increased from two million per year in 1986 to about 70 million in 1993. But problems remain, particularly in places where poverty limits the safety of the **blood supply**, where condoms are in short supply, where access to accurate information about the disease is limited, and where health providers are far away.

Future trends

A hopeful new development in tropical medicine is the vaccine for malaria, which was in **field** trials in the middle 1990s. The vaccine, developed by Colombian Manuel Patarroyo, is made up of three peptides synthesized chemically and connected using a fourth peptide from the malaria parasite.

The vaccine is designed to protect against falciparum, the most deadly strain of malaria. Results of clinical tests in Colombia, Ecuador, Venezuela, and Tanzania reported in 1994 show some success against the disease. The Tanzania trial of about 600 children reduced the incidence of malaria by 31%. This shows that the vaccine has some promise. Experts note that while the vaccine did not provide 100% protection, it still would be valuable to cut sickness and death from malaria by 31%. The next step is to see if the vaccine can be improved and to conduct larger field trials.

Other experimental vaccines are also being tested. A therapeutic vaccine for tuberculosis is currently being tested which reduces the time it takes to treat the disease, which currently is about six months. Shortening the treatment time would improve the odds that treatment is completed by the patient.

Another vaccine which could reduce the sickness linked to tropical diseases is a new vaccine for leishmaniasis. The vaccine, which is in clinical trials, has fewer side affects and costs about one-tenth less than drugs currently used to treat the disease. However, because the drug is not in production for distribution yet, the ultimate cost of the medicine is not clear.

Other strategies are also essential to stem the tremendous cost in illness and death from tropical diseases. While World Health Organization efforts have boosted the rate of immunization for many diseases, there are still millions of children who do not receive adequate immunization for common early **childhood diseases** such as measles and whooping cough. Efforts to provide such proven preventive measures to more children will yield great benefits in health and longer life.

Efforts to boost access to health providers and educational information also stand to benefit residents of tropical disease areas. Knowing how to care for children with diarrhea, adults with hookworm, or many other diseases could make a vast difference. Finally, the prevalence of diseases linked to dirty water and dirty soil drive home

KEY TERMS

. . . .

Cyst—Refers to either a closed cavity or sac or the stage of life of some parasites during which they live inside a walled in area.

Dehydration—The condition of having lost too much fluid from the body.

Larva—Immature developmental stage of various species.

Lesion—An injury or wound.

Parasite—Species which depends on another species to survive.

Peptide—A class of chemical compounds that form proteins.

Protozoa—One-celled creatures and the simplest forms of animal life.

Spore—A dormant form assumed by some bacteria, such as anthrax, that enable the bacterium to survive high temperatures, dryness, and lack of nourishment for long periods of time. Under proper conditions, the spore may revert to the actively multiplying form of the bacteria.

the importance of better sewage systems and basic public health measures which western countries generally adopted years ago. Diseases carried by insects are tackled through **insecticides**, clearing brush and standing water where the carriers may proliferate; **sexually transmitted diseases** are addressed by educating communities about safer sex practices, and providing condoms; immunization programs may cut down on some diseases; efforts to create vaccines for other diseases are underway; donor blood is undergoing increasingly more stringent screening to avoid transmitting disease through this route.

Resources

Books

Desowitz, Robert S. *The Malaria Capers.* New York: W.W. Norton and Company, 1991.

Goldsmith, Robert, and Donald Hayneman, eds. *Tropical Medicine and Parasitology.* Norwalk, CT, 1989.

Najera, Jose A., Bernhard H. Liese, and Jeffrey Hammer. *Malaria: New Patterns and Perspectives.* World Bank Technical Paper Number 183, 1992.

Warren, Kenneth S., and Adel A.F. Mahmoud. *Tropical and Geographical Medicine.* 2nd ed. New York: McGraw-Hill Information Services Co., 1990.

Periodicals

Aldhous, Peter. "Vaccine Shows Promise in Tanzania Test." *Science* 266, (November 4, 1994): 724.

"No Shortcuts on International Commitment to Combat AIDS." *AIDS Weekly* (May 16, 1994): 7.

Patricia Braus

Trout-perch

The trout-perch belongs to the family Percopsidae, which includes only one genus—*Percopsis*—with only two **species**. The **fish** is found only in the fresh waters of **North America**. One species (*Percopsis omiscomaycus*) is found mainly on the eastern side; the other species, the sandroller (*P. Transmontana*), is native to the west in the regions around the Columbia River **Basin**.

Both species are small, with the eastern species averaging 3-5 in (7.6-12.7 cm) in length, with a few reaching 8 in (20.3 cm). The sandroller is a bit smaller.

As the name implies, the trout-perch has characteristics of both the trout and the **perch**. Like the trout, it has an adipose fin-a fatty fin which projects between the dorsal and caudal fins. It also has a lateral line (a row of sensory pores on its sides) and ctenoid scales (fish scales that have a comb-like projection at their margin).

The trout-perch appears to prefer deep **water** but may enter shallow water in spring to spawn. It may spawn on sandbars in lakes or up **rivers**, selecting bottoms of **sand** or gravel. It appears to be sensitive to rises in **temperature**, and, in some lakes, a considerable number of these fish die off in the summer as the water temperature rises.

Trout-perches are an important source of food for larger fishes.

True bugs

The true bugs are a large and diverse group of about 35,000 **species** of **insects** in the order Hemiptera. About 44 families of bugs occur in **North America**. Bugs typically have a flattened body, and their folded wings cross over their thorax and abdomen, giving a distinctive, cross-like pattern.

Some species of true bugs are of great economic importance as **pests** of agricultural plants. A few species of bugs are vectors of important diseases of humans.

In popular usage, the word "bug" is often used to refer to non-hemipteran insects, and not only to "true bugs." When used to refer to a species in the Hemiptera, the "bug" part of the name should be written separately, as in: stink bug, or milkweed bug. When used to refer to non-hemipteran insects, the "bug" part of the name should be used to form a single word, as in: ladybug (a family of **beetles**, order Coleoptera), or mealybug (**scale insects**, order Homoptera), or sowbug (Crustaceans in the order Isopoda, which are not even insects).

Biology of true bugs

The true bugs have an incomplete **metamorphosis**, characterized by three life-history stages: egg, nymph, and adult. The nymphs resemble the adults somewhat in form, but they are not capable of reproduction. Most North American bugs overwinter at the adult stage.

Bugs have two sets of wings. The forewings of most species of true bugs are rather tough and leathery towards their base, and membranous farther away. This unusual forewing structure is the origin of the Latin roots of the name for the order: Hemiptera, or "half wing." The diagnostic, crossed-wing appearance of bugs at rest is also due to this unusual structure of the forewings, which form a well-defined "X" when held flat over the back of the abdomen with the membranous tips overlapping. The hind wings are fully membranous, and are used for flying. Some types of true bugs have greatly reduced wings, and cannot fly.

The mouthparts of bugs are adapted for piercing and sucking. The mouthparts comprise a pointed, elongate structure known as a beak or rostrum that arises at the front of the head, and folds backwards, quite far underneath the body in some species. The beak is itself made up of specialized stylets used for piercing, and others develop channels used for actual feeding. Most species of bugs feed on **plant** juices, but a few are **parasites** of **vertebrates**, living on the animal's surface and feeding on **blood**.

Most bugs have long, segmented antennae. Bugs have well developed, compound eyes, and adult bugs may have several simple eyes (or ocelli) as well. Many species of bugs have **glands** that give off a strongly scented, distasteful odor when the insect is disturbed.

Some species are quite brightly and boldly colored. Usually these bugs feed on plants that contain poisonous chemicals, which also occur in the bugs and render them distasteful or even toxic to potential predators. This type of boldly warning **color** scheme is known as aposematic coloration.

Most species of bugs are terrestrial, living on vegetation or in organic debris on the surface. Some species of bugs are specialized for living in aquatic habitats, occurring in the **water** column, or on the surface.

Common families of terrestrial bugs in North America

The most diverse family of bugs is the plant or **leaf** bugs (family Miridae), species of which can be found in terrestrial habitats world-wide. Almost all plant bugs feed on the juices of plants, some species causing important damages to agricultural **crops**. Important agricultural pests include the tarnished plant bug (*Lygus lineolaris*), which feeds on a wide range of crop plants, the apple red bug (*Lygidea mendax*), the **cotton** fleahopper (*Psallus seriatus*), and the four-lined plant bug (*Poecilocapsus lineatus*), a pest of currants and gooseberries. The garden fleahopper (*Halticus bractatus*) is a common, jumping species in gardens and fields, which sometimes causes significant damages.

The assassin bugs (family Reduviidae) are mostly predators of other insects, although many will also give humans a painful bite if they are not handled with care, and a few are blood-sucking parasites. The blood-sucking conenose (*Triatoma sanduisuga*) sometimes occurs in houses in North America, and can inflict a particularly painful bite. In **South America**, other species in the genus *Triatoma* are the vectors of Chagas' disease, a deadly **disease** of humans.

The ambush bugs (family Phymatidae) are also predators of other insects. Yellow species of ambush bugs are common hide-and-wait predators on species of goldenrod (*Solidago* spp.) throughout North America.

Seed bugs (family Lygaeidae) are a diverse group of mostly herbivorous bugs. Two common, attractive, red-and-black marked species are the small milkweed bug (*Lygaeus kalmii*) and the large milkweed bug (*Oncopeltus fasciatus*). The bright, aposematic coloration of these milkweed bugs is meant to deter potential predators, because these insects are distasteful due to **alkaloid** chemicals accumulated from their food of milkweed (*Asclepias* spp.). The chinch bug (*Blissus leucopterus*) is a serious agricultural pest, especially of **wheat**, corn, and other grains in the grass family, as well as urban lawn-grasses.

The lace bugs (family Tingidae) are herbivorous insects with distinctive, very attractive, finely reticulated patterns on their head, thorax, and wings. The chrysanthemum lace bug (*Corythucha marmorata*) is common in much of North America, feeding on various species in the aster family, and sometimes occurring in greenhouses.

The leaf-footed or coreid bugs (family Coreidae) are common, herbivorous insects. The squash bug (*Anasa tristis*) is a pest of pumpkin and squash crops, feeding on the leaves of these plants and causing them to droop and turn black. The box-elder bug (*Leptocoris trivittatus*) is an attractive, red-and-black colored insect that feeds on species of **maples**, but does not cause important damages.

The stink bugs (family Pentatomidae) are relatively large and common bugs that produce malodorous smells as a defensive response when they are roughly handled. Many of the stink bugs are brightly colored, for example, the harlequin bug (*Murgantia histrionica*), an important pest of crops in the mustard family, such as cabbages, turnip, and radish.

Common families of aquatic bugs in North America

The water boatmen (family Corixidae) are common aquatic bugs that swim in the water column of lakes and ponds. The hind legs of water boatmen are oar-like in appearance, being long and flattened, and are used for underwater locomotion. Water boatmen do not have gills for the exchange of respiratory gases—they must breath head-first at the surface, although most species can carry a small bubble of air as they swim underwater. Most species of water boatmen are herbivorous, but some others are predators of other aquatic **invertebrates**. Water boatmen are an important food source for some species of **wildlife**, such as **ducks**. The water boatman (*Arctocorixa alternata*) is a common and widespread species in North America.

The backswimmers (family Notonectidae) are also aquatic bugs. The especially elongate hind legs of these insects are used for swimming, which the backswimmers accomplish while in an upside-down position—hence, their common name. Backswimmers must breath at the surface, but unlike the water boatmen these insects must break the surface abdomen first in order to obtain air. Backswimmers are predators of other aquatic invertebrates, and are themselves an important food for larger species. The backswimmer (*Notonecta undulata*) is a common and widespread species.

The giant water bugs (family Belostomatidae) include the world's largest bugs, one species of which can attain a most-impressive length of 3.9 in (10 cm). Giant water bugs are oval in shape, with a rather flattened body, and are often a shiny brown color. The front legs are large and strong and are used to grasp their **prey**, which can include other aquatic insects, as well as small **fish**, tadpoles, and even **frogs** and **salamanders**. Giant water bugs sometimes leave their aquatic **habitat** and fly about, possibly for the purposes of dispersal. At such times these insects are attracted to lights, where they are sometimes known as electric **light** bugs. Giant water bugs can inflict a painful bite, and should be handled with care—these insects are sometimes known as toe-biters. The giant water bug (*Lethocerus americanus*) is a widespread species in North America, and can reach a body length of more than 2.4 in (6 cm).

Water scorpions (family Nepidae) are another group of predacious aquatic bugs, with long, scissor-like front legs adapted for fiercely grasping their prey of insects and other creatures. Water scorpions can inflict a painful bite.

Water striders (family Gerridae) are semi-aquatic insects, living on the water surface. The long-legged body of water striders is suspended aloft by **surface tension**, made possible by the structure of their "feet," which are covered with fine hairs that are not easily wetted. In parts of the southern United States, these insects are known as "Jesus bugs" because of their ability to walk on water. Water striders run and skate over the surface of ponds and lakeshores, hunting terrestrial **arthropods** that fall onto their two-dimensional habitat, and aquatic insects as they come to the surface to breathe. Water striders have odoriferous scent glands, which may be a deterrent against predation by fish. The water strider (*Gerris remigis*) occurs commonly throughout much of North America.

Bugs as a health hazard

Bed bugs (family Cimicidae) are wingless bugs with a body length of about 0.2 in (6 mm). Bed bugs feed by sucking the blood of **birds** or **mammals**. Various species will bite humans, and they can be serious pests in homes, hotels, and other places, especially the common bed bug (*Cimex lectularius*). Bed bugs come out at night, hiding during the day in cracks and crevices in walls and furniture. The bites of these insects are very irritating, but bed bugs are not known to be a vector of human diseases.

Chagas' disease is common in parts of Central and South America, and is spread by blood-sucking bugs in the genus *Triatoma*, especially *T. infestans*. These are sometimes known as kissing bugs because of their tendency to bite near the mouth of their victims. Chagas' disease is a debilitating malady of humans, and is caused by a pathogenic trypanosome, *Trypanosoma cruzi*, a type of parasitic protozoan that lives in the blood. Chagas' disease is characterized by a recurring fever, and often a enfeebling **inflammation** of the **heart** muscles.

Resources

Books

Arnett, Ross H. *American Insects.* New York: CRC Publishing, 2000.

Carde, Ring, and Vincent H. Resh, eds. *Encyclopedia of Insects.* San Diego: Academic Press, 2003.

Lane, R., and R. Crosskey. *Medical Insects and Arachnids.* New York: Chapman and Hall, 1993.

Lehane, M. *Biology of Blood-sucking Insects.* London: Unwin-Hyman, 1991.

KEY TERMS

Aposematic—Refers to a bright coloration of an animal, intended to draw the notice of a potential predator, and to warn of the dangers of toxicity or foul taste.

Incomplete metamorphosis—This is characterized by three life history stages: egg, nymph, and adult. The nymphs of true bugs resemble the adults in form, but they are not capable of reproduction.

Vector—Any agent, living or otherwise, that carries and transmits parasites and diseases.

Metcalf, R. L., and R. A. Metcalf. *Destructive and Useful Insects.* New York: McGraw-Hill, 1992.

Bill Freedman

True eels

The true eels are elongate **bony fish** with a snake-like slimy body in the order Anguilliformes. There is one family of **freshwater** eels (the Anguillidae), and 25 families of exclusively marine eels. The freshwater eels must return to the oceans to spawn.

The general characteristics of eels include soft-rayed fins and elongate dorsal and anal fins which merge with the caudal fin. Eels lack pelvic fins and have small pectoral fins commonly situated immediately behind the head. The jaws of eels are relatively small, but are strong, with numerous small teeth. Most ocean-living eels do not have scales, although the freshwater eels have small, oval-shaped scales embedded in their skin.

All eels are predators, feeding on a wide range of **prey**, including small **fish**, crustaceans, **mollusks**, and worms.

Freshwater eels

The freshwater eels, comprising about 16 **species**, are the most familiar family of eels to most people. These fish have an unusual characteristic in their **life history**, known as catadromy, in which the fish spend most of their lives in fresh waters, but run to the **ocean** to spawn.

The common freshwater eel of **North America** is the American eel (*Anguilla rostrata*). The European eel (*A. anguilla*) of western **Europe** and the Japanese eel (*Anguilla japonica*) of the north Pacific coast are closely related to the American eel. Relatively large numbers of eel species, about 12, occur in the Indo-Pacific region.

A green moray eel (*Gymnothorax funebris*). The eel has blue skin covered with a yellow mucus that renders its apparent color green. It can reach a maximum length of 6 ft (1.8 m). *Photograph by J.W. Mowbray. The National Audubon Society Collection/Photo Researchers, Inc. Reproduced by permission.*

The American eel can reach a length of about 4 ft (1.2 m). This species is abundant in freshwaters that link directly to the ocean. This range includes coastal **rivers**, and the larger drainage of the Saint Lawrence River, extending inland to Lake Ontario and to a lesser degree, the upper Great Lakes. There are no freshwater eels on the west coast of the America, or on the east coast of **South America**.

American eels spends almost all of their life in fresh waters. However, to breed, this species of eel runs to the sea and migrates to a warm-water region between the West Indies and Bermuda that is known as the Sargasso Sea. European eels also commonly spawn in the Sargasso Sea, following a **migration** of 4,000 or more miles (6,440 km). After spawning, the adult eels die. The baby eels, or elvers, are leaf-shaped and transparent, and migrate to fresh waters, where they transform into miniature but transparent replicas of the adult body form,

sometimes known as "glass eels." These small fish then run up rivers, and take up residence in still waters of large rivers and lakes, where they live as adults for as long as 15 years.

This unusual breeding strategy was discovered only relatively recently. For many centuries, naturalists pondered the fact that they could never find spawning or larval eels, and the breeding habits of these fish were a mystery. It was not until 1922 that newly hatched eel larvae were observed, in the Sargasso Sea.

Freshwater eels are an economically important species of fish, and are particularly appreciated as a food in France, Belgium, The Netherlands, and Great Britain.

Other families of eels

The Moringuidae is a family of 20 species of marine eels found in tropical waters, which includes the genera

Moringua and *Stilbiscus.* These fish typically occur in shallow, soft-bottomed or gravelly habitats, where they burrow in the substratum during the day. The eel *Stilbiscus edwardsi* is a common species of the western Atlantic, while *Moringua macrochir* occurs in coastal waters of the Hawaiian Islands.

The moray eels (Muraenidae) include about 120 species that live in shallow, tropical and subtropical waters. These impressive, snake-like fish have a large mouth, well-armed with teeth and poison fangs. Morays are often brilliantly colored and marked, and are very attractive fish. Morays generally occur in rocky or coral habitats, where they hide during the day in crevices or in burrows in sediment. Many unsuspecting divers have received a nasty surprise in the form of a painful moray bite, when reaching into a rocky crevice after marine animals or other interesting things.

The zebra moray (*Echidna zebra*) is an attractive species with a brown-yellowish body and white stripes. The zebra moray occurs in the Indian and Pacific Oceans, and is sometimes seen in aquaria. The largest moray, and the largest species of living eel, is *Thyrsoidea macrurus* of coastal regions of the Indian and Pacific Oceans. This species can achieve a length greater than 9 ft (2.7 m). Morays are often caught as food fish, especially in Asiatic waters and, to a lesser degree, in the Mediterranean region.

The conger eels (Congeridae) also occur in tropical and subtropical waters. The best known species is the conger eel (*Conger conger*), with an almost world-wide distribution in suitable habitats (except for the eastern Pacific), and achieving a length of almost 9 ft (2.7 m). Some species occur in relatively deep waters, for example, *Ariosoma balearica* and *Promyllantor latedorsalis,* off the Azores of the tropical Atlantic Ocean.

See also Spiny eels.

Resources

Books

Grzimek, H.C. Bernard, Dr., ed. *Grzimek's Animal Life Encyclopedia.* New York: Van Nostrand Reinhold Company, 1993.
Whiteman, Kate. *World Encyclopedia of Fish & Shellfish.* New York: Lorenz Books, 2000.

Bill Freedman

True flies

The true **flies** are a large and diverse group of commonly observed **insects** in the order Diptera, comprising more than 100,000 **species**. About 107 families of flies occur in **North America**.

Flies have distinctive, knob-like structures known as halteres on the back of their thorax. Halteres are highly modified from the hind wings of true flies, while the fore wings are membranous and used for flying. The two-winged character of the true flues is reflected in the Latin roots of the scientific name of their order, the Diptera, which means "two wings."

Some species of true flies are of great economic importance as **pests** of agricultural plants. Other species of flies are great nuisances because they bite humans and domestic animals in order to obtain **blood** meals, as is the case of **mosquitoes**, black flies, horse flies, and others. Some of these parasitic, blood-sucking species are also vectors of deadly diseases of humans, as are some of the blow flies and house flies that feed by scavenging dead organic **matter**. However, many other species of flies provide very useful ecological services, by helping to safely dispose of decaying carcasses and other organic debris, and by serving as predators or **parasites** of other, injurious insects.

In its popular usage, the word "fly" is often used in reference to insects that are not in the order Diptera, and are therefore not "true flies." The "fly" part of the name of a dipteran should be written separately, as in: house fly, horse fly, or black fly. However, when used to refer to non-dipteran insects, the "fly" portion of a species name should be appended to form a single word, as in: sawfly (species in the family Tenthredinidae, order Hymenoptera), dragonfly and damselfly (order Odonata), mayfly (order Ephemeroptera), stonefly (order Plecoptera), caddisfly (order Trichoptera), and butterfly (order Lepidoptera).

Biology of true flies

The true flies have a complete **metamorphosis**, characterized by four stages in their **life history**: egg, larva or maggot, pupa, and adult. Fly maggots are soft-bodied, legless, and worm-like. Most flies are terrestrial animals, but many species spend their larval stages in aquatic habitats, with the adults emerging to the terrestrial environment.

The hind wings of flies are modified into small, distinctive structures known as halteres, which resemble a tiny, stalked knob. Halteres are thought to be used as an aid in achieving a sense of balance and direction. The front wings of flies are membranous and functional in more usual ways, and are used for flying. The smaller flies have very rapid wingbeats, typically 200-400 strokes per second, and as great as 1,000 per second in tiny midges in the genus *Forcipmyia*.

Other characteristic features of true flies include the division of their tarsus (that is, the leg segment immediately below the tibia) into five segments. Most flies have mouth parts of a haustellate form, that is, adapted for sucking, rather than for chewing. Flies are typically small and soft bodied, and some are minute in size.

There are many feeding strategies among the flies. Most species feed on soft foods and organic debris, while others eat **nectar**, some scavenge dead bodies of animals, or are predators of smaller **arthropods**, or are blood-sucking parasites. These various species have mouth parts and behavioral adaptations to their specific modes of feeding and living. Mosquitoes, for example, have piercing and sucking feeding structures, while flies that feed on soft organic materials have sponging or lapping mouth parts.

Some species of flies that feed on plants inject a growth-regulating chemical into the stem, which causes an abnormal **tissue**, called a gall, to develop. The gall provides **habitat** for the feeding and development of the larvae of the fly. Gall-inducing species occur in the families Agromyzidae, Cecidomyiidae, and Tephritidae, as well as in other orders of insects.

There is a great diversity of species of flies, with more than 100,000 species being identified so far, and more than 100 families. The families are mostly distinguished using characters related to the morphology of the antennae, legs, wing venation, body bristles, and other anatomical features. Habitat and other ecological information may also be useful. Aspects of body **chemistry** may also be used in the identification of closely related species, particularly the chemistry of enzymes and nucleic acids. The diversity of flies is too enormous to discuss in much detail in this entry—only a few prominent examples will be described in the following sections.

Common families of terrestrial flies in North America

The muscid flies (family Muscidae) include more than 700 species in North America, including some important pests. The house fly (*Musca domestica*) is one of the most familiar flies to most people, because it breeds readily in garbage and other organic debris, and can be very abundant in dirty places around homes, villages, and cities. The house fly does not bite, but it can be a vector of some diseases of humans, spreading the **pathogens** by contact, for example, by walking on food

that is later eaten by people. The face fly (*Musca autumnalis*) tends to cluster around the face of cows, where it feeds on mucous secretions around the nostrils and eyes, causing great irritation to the **livestock**. The stable fly (*Stomoxys calcitrans*) and horn fly (*Haematobia irritans*) are biting flies that greatly irritate livestock. Species of tsetse flies (*Glossina* spp.) occur in **Africa**, and are the vectors of **sleeping sickness** and related diseases of people, livestock, and wild large **mammals**.

Blow flies (family Calliphoridae) are scavengers, whose larvae feed on dead animals, excrement, and similar, rotting debris. Although a carcass teeming with writhing maggots is a rather disgusting spectacle, it must be remembered that blow flies provide a very useful ecological service by helping to safely dispose of unsanitary **animal** carcasses. A few species of blowflies lay their eggs in wounds on living animals, and the larvae may then attack living tissues, causing considerable damage. The screw-worm (*Cochliomyia hominivorax*) is an important pest in this regard, causing severe damages to cattle populations in some areas.

Flower flies or syrphids (family Syrphidae) include about 1,000 species in North America. Adult syrphids can be quite common in some habitats, where they are typically seen hovering in the vicinity of **flowers**. Many species of syrphids are brightly colored, sometimes with a black and yellow banding that is an obvious **mimicry** of **bees** and **wasps**.

The fruit fly family (Tephritidae) is made up of several hundred North American species, some of which are important pests in agriculture. The apple maggot (*Rhagoletis pomonella*) burrows in the **fruits** of apples and other fruits, while the Mediterranean fruit fly or medfly (*Ceratitis capitata*) is a serious pest of citrus fruits.

The small fruit flies (family Drosophilidae) are common in the vicinity of decaying vegetation and fruit. The species ***Drosophila melanogaster*** has been commonly used in biological laboratories for studies of **genetics**, because it is easily and quickly bred, and has giant chromosomes that can be readily studied using a **microscope**.

The warble and bot flies (family Oestridae) are large, stout, fast-flying flies that have a superficial resemblance to bees. The larvae of these flies are parasitic on large mammals, living in the flesh just beneath the skin. Some species are serious pests of agricultural animals, for example, the **sheep** bot fly (*Oestrus ovis*) and the ox warble fly (*Hypoderma bovis*). Warble flies are extremely irritating to cattle and to wild ungulates—there are reports of **caribou** being driven to distraction by warble flies, and jumping off cliffs in desperate attempts to escape these nasty pests.

Tachinid flies (family Tachinidae) include more than 1300 species in North America. The larvae of tachinids are parasitic on other species of insects, including some economically important pests, which are essentially eaten alive by the tachinid. As such, some species of tachinids provide a useful service to humans.

The robber flies (family Asilidae) are a diverse group, with more than 800 species in North America. Robber flies are predators of other insects, which are captured in flight.

Seaweed or wrack flies (family Coelopidae) are dark-colored flies that can be very abundant along marine shores, where they breed in natural piles of seaweed compost, with the adults swarming abundantly above, often attracting large numbers of shorebirds and swallows.

Common families of aquatic flies in North America

There are many families of flies that have aquatic larval and pupal stages, but are terrestrial as adults. Some of the more prominent of these flies are described briefly in the following paragraphs.

The mosquitoes (family Culicidae) are a diverse, well known, and important group of biting flies. Larval mosquitoes, also known as "wrigglers," are aquatic and feed on **algae** and organic debris, and the adult males feed on flower nectar. Female mosquitoes, however, require a blood meal from a bird or mammal before they can develop eggs. Many species of mosquitoes bite humans, and they can be an enormous cause of annoyance, as well as the means of spreading some important, even deadly, diseases. About 150 species of mosquitoes occur in North America, and in some habitats, for example, in northern **forests** during the summer, these blood-suckers can be enormously abundant and bothersome.

The black flies or buffalo gnats (family Simulidae) are small, dark-colored, hunch-backed flies. Female black flies require a blood meal to develop their eggs, and they obtain this food by biting the skin of a victim, and then sucking the blood that emerges from the wound. Black flies breed in cool streams, and they can be very abundant in some northern habitats. Unprotected animals have actually been killed as a result of the enormous numbers of bites that can be delivered during periods when black flies are abundant.

The horse flies and deer flies (family Tabanidae) are another group of fierce, biting flies with aquatic larvae and pupae, and terrestrial adults. Only the females require a blood meal—the males feed on nectar and **plant** juices. The eyes of deer flies are often extremely bright-colored, even iridescent.

The biting midges or no-see-ums (family Ceratopogonidae) are very small, blood-sucking flies with aquatic larval and pupal stages, but terrestrial adults. These diminutive pests can be quite abundant along the shores of lakes and oceans. These tiny flies can easily penetrate through fly screens and many types of clothing, and deliver bites that are much more painful than might be expected on the basis of the diminutive, less-than one millimeter size of these insects.

The crane flies (family Tipulidae) have a superficial resemblance to gigantic mosquitoes with extremely long and delicate legs. The larvae of crane flies occur in aquatic or moist terrestrial habitats and mostly feed on decaying organic matter, while the adults are terrestrial and feed on nectar.

The phantom midges (family Chaoboridae) are mosquito-like insects, but they do not bite. Larval phantom midges are predators of other bottom-dwelling arthropods, and have almost transparent bodies—hence their common name.

Flies as a health hazard

Some of the biting and blood-sucking flies are the vectors of important diseases of humans, domestic animals, and wild animals. The **microorganisms** that cause **malaria**, **yellow fever**, **encephalitis**, filariasis, **dengue fever**, sleeping sickness, **typhoid fever**, **dysentery**, and some other important diseases are all spread to humans by species of Diptera.

Malaria is one of the best-known cases of a **disease** that is spread by biting flies. Malaria is caused by the protozoan *Plasmodium falciparum*, and is spread to humans by a mosquito vector, especially species of *Anopheles*, which infect people when they bite them to obtain a blood meal. Malaria is an important disease in the tropics and subtropics. During the 1950s, about 5% of the world's population was infected with malaria, and during the early 1960s two to five million children died of malaria each year in Africa alone. The incidence of malaria has been greatly reduced by the use of **insecticides** to decrease the abundance of the *Anopheles* vectors, and by the use of prophylactic drugs that help to prevent infections in people exposed to the *Plasmodium* parasite. However, some of the pesticide-based control programs are becoming less effective, because many populations of *Anopheles* mosquitoes have developed a tolerance of the toxic effects of some insecticides.

The house fly (*Musca domestica*) is a non-biting fly, but it can carry some pathogens on its feet and body, and when it walks over food intended for consumption by humans, **contamination** can result. The house fly is known to be a contact vector of some deadly diseases,

KEY TERMS

. . . .

Complete metamorphosis—This is an insect life history characterized by four stages: egg, larva, pupa, and adult.

Parasite—An animal that derives its livelihood by feeding on another, usually much larger animal, but generally without causing the death of the host. For example, many bloodsucking species of flies are parasites of mammals or birds.

Vector—Any agent, living or otherwise, that carries and transmits parasites and diseases.

including typhoid fever, dysentery, yaws, **anthrax**, and conjunctivitis. House flies can be controlled using insecticides, although some populations of this insect have developed resistance to insecticides, rendering these chemicals increasing less effective as agents of control.

Resources

Books

Arnett, Ross H. *American Insects.* New York: CRC Publishing, 2000.

Borror, D.J., C.J. Triplehorn, and N. Johnson. *An Introduction to the Study of Insects.* New York: Saunders, 1989.

Carde, Ring, and Vincent H. Resh, eds. *Encyclopedia of Insects.* San Diego: Academic Press, 2003.

Lehane, M. *Biology of Blood-sucking Insects.* London: Unwin-Hyman, 1991.

Metcalf, R.L., and R.A. Metcalf. *Destructive and Useful Insects.* New York: McGraw-Hill, 1992.

Bill Freedman

Trumpetfish

Trumpetfish (*Aulostomus maculatus*) are **bony fish** in the family Aulostomidae, found from the Caribbean Sea to as far north as Bermuda. They are named for the trumpet-like shape they exhibit when their mouths are open. Trumpetfish measure up to 2 ft (0.6 m) in length, and have dorsal spines which support separate fins or finlets. They are brownish in **color** with widespread streaks and spots.

Trumpetfish are poor swimmers, but are well camouflaged, which provides protection from predators and affords a means of obtaining food. They lie motionless, practically unseen, aligned with gorgonian corals (whip corals). They appear to drift about in the **water**, since

they make no visible effort to swim and the rapid movement of their fins cannot be readily detected.

The American trumpetfish has close relatives in the Pacific—*A. valentini* in the Indo-Pacific region and *A. chinesis* along the Asian coast.

Resources

Books

Whiteman, Kate. *World Encyclopedia of Fish & Shellfish.* New York: Lorenz Books, 2000.

Tsunami

Tsunami, or seismic sea waves, are a series of very long wavelength **ocean** waves generated by the sudden displacement of large volumes of **water**. The generation of tsunami waves is similar to the effect of dropping a solid object, such as a stone, into a pool of water. Waves ripple out from where the stone entered, and thus displaced, the water. In a tsunami, the "stone" comes from underneath the ocean or very close to shore, and the waves, usually only three or four, are spaced about 15 minutes apart.

Tsunami can be caused by underwater (**submarine**) earthquakes, submarine **volcano** eruptions, falling (slumping) of large volumes of ocean sediment, coastal landslides, or even by meteor impacts. All of these events cause some sort of land mass to enter the ocean and the ocean adjusts itself to accommodate this new mass. This adjustment creates the tsunami, which can circle around the world. Tsunami is a Japanese word meaning "large waves in harbors." It can be used in the singular or plural sense. Tsunami are sometimes mistakenly called tidal waves but scientists avoid using that term since they are not at all related to **tides**.

Tsunami are classified by oceanographers as shallow water surface waves. Surface waves exist only on the surface of liquids. Shallow water waves are defined as surface waves occurring in water depths that are less than one half their wavelength. Wavelength is the distance between two adjacent crests (tops) or troughs (bottoms) of the wave. Wave height is the vertical distance from the top of a crest to the bottom of the adjacent trough. Tsunami have wave heights that are very small as compared to their wavelengths. In fact, no matter how deep the water, a tsunami will always be a shallow water wave because its wavelength (up to 150 mi [240 km]) is so much greater than its wave height (usually no more than 65 ft [20 m]).

Shallow water waves are different than deep water waves because their speed is controlled only by water depth. In the open ocean, tsunami travel quickly (up to 470 MPH [760 km/h]), but because of their low height (typically less than 3 ft [1 m]) and long wavelength, ships rarely notice them as they pass underneath. However, when a tsunami moves into shore, its speed and wavelength decrease due to the increasing **friction** caused by the shallow sea floor.

Wave **energy** must be redistributed, however, so wave height increases, just as the height of small waves increases as they approach the beach and eventually break. The increasing tsunami wave height produces a "wall" of water that, if high enough, can be incredibly destructive. Some tsunami are reportedly up to 200 ft (65 m) tall. The impact of such a tsunami can range miles inland if the land is relatively flat.

Tsunami may occur along any shoreline and are affected by local conditions such as the coastline shape, ocean floor characteristics, and the nature of the waves and tides already in the area. These local conditions can create substantial differences in the size and impact of the tsunami waves even in areas that are very close geographically.

Types of tsunami

Tsunami researchers classify tsunami according to their area of effect. They can be local, regional, or ocean-wide. Local tsunami are often caused by submarine volcanoes, submarine sediment slumping, or coastal landslides. These can often be the most dangerous because there is often little warning between the triggering event and the arrival of the tsunami.

Seventy-five **percent** of tsunami are considered to be regional events. Japan, Hawaii, and Alaska are commonly hit by regional tsunami. Hawaii, for example has been hit repeatedly during this century, about every five to 10 years. One of the worst was the April 1, 1946, tsunami that destroyed the city of Hilo.

Pacific-wide tsunami are the least common as only 3.5% of tsunami are this large, but they can cause tremendous destruction due to the massive size of the waves. In 1940 and 1960, destructive Pacific-wide tsunami occurred. More recently, there was a Pacific-wide tsunami on October 4, 1994, which caused substantial damage in Japan with 11.5 ft (3.5 m) waves. **North America** was lucky that time. Waves of only 6 in (15 cm) over the normal height were recorded in British Columbia.

Tsunami in history

Tsunami are not only a modern phenomenon. The decline of the Minoan civilization is believed to have been triggered by a powerful tsunami that hit the area in 1480 B.C. and destroyed its coastal settlements. Japan has had 65 destructive tsunami between A.D. 684 and 1960. Chile was hit in 1562 and Hawaii has a written history of tsunami since 1821. The Indian and Atlantic Oceans also have long tsunami histories. Researchers are concerned that the impact of future tsunami, as well as hurricanes, will be worse because of intensive development of coastal areas in the last 30 years.

The destructive 1946 tsunami at Hilo, Hawaii, caused researchers to think about the problem of tsunami prediction. It became clear that if we could predict when the waves are going to hit, we could take steps to minimize the impact of the great waves.

Predicting tsunami—The International Tsunami Warning System

In 1965, the Intergovernmental Oceanographic Commission of the United Nations Educational, Scientific, and Cultural Organization agreed to expand the United States' existing tsunami warning center at Ewa Beach, Hawaii. This marked the formation of the Pacific Tsunami Warning Center (PTWC) which is now operated under the U.S. Weather Service. The objectives of the PTWC are to "detect and locate major earthquakes in the Pacific **basin**; determine whether or not tsunami have been generated; and to provide timely and effective information and warnings to minimize tsunami effects."

The PTWC is the administrative center for all the associated centers, committees, and commissions of the International Tsunami Warning System (ITWS). Japan, the Russian Federation, and Canada also have tsunami warning systems and centers and they coordinate with the PTWC. In total, 27 countries now belong to the ITWS.

The ITWS is based on a world-wide network of seismic and tidal data and information dissemination stations, and specially trained people. Seismic stations measure movement of the earth's crust and are the foundation of the system. These stations indicate that some disturbance has occurred that may be powerful enough to generate tsunami. To confirm the tsunami following a seismic event, there are specially trained people, called tide observers, with monitoring equipment that enables them to detect differences in the wave patterns of the ocean. **Pressure** gauges deployed on the ocean can detect changes of less than 0.4 in (1 cm) in the height of the ocean, which indicates wave height. Also, there are accelerometers set inside moored buoys that measure the rise and fall of the ocean, which will indicate the wave speed. These data are used together to help researchers confirm that a tsunami has been generated. Tsunami can also be detected by **satellite** monitoring methods such as **radar** and photographic images.

The warning system in action

The ITWS is activated when earthquakes greater than 6.75 on the Richter scale are detected. The PTWC then collects all the data, determines the magnitude of the quake and its epicenter. Then they wait for the reports from the nearest tide stations and their tide observers. If a tsunami wave is reported, warnings are sent to the information dissemination centers.

The information dissemination centers then coordinate the emergency response plan to minimize the impact of the tsunami. In areas where tsunami **frequency** is high, such as Japan, the Russian Federation, Alaska, and Hawaii, there are also Regional Warning Systems to coordinate the flow of information. These information dissemination centers then decide whether to issue a "Tsunami Watch," which indicates that a tsunami may occur in the area, or a more serious "Tsunami Warning," which indicates that a tsunami will occur. The entire coastline of a region is broken down into smaller sections at predetermined locations known as "breakpoints" to allow the emergency personnel to customize the warnings to account for local changes in the **behavior** of the tsunami. The public is kept informed through local **radio** broadcasts. If the waves have not hit within two hours of the estimated time of arrival, or, the waves arrived but were not damaging, the tsunami threat is assumed to be over and all Watches and Warnings are canceled.

Current and future research

One of the more recent changes in the ITWS is that the Regional Centers will be taking on greater responsibility for tsunami detection and warning procedures. This is being done because there have been occasions when the warning from Hawaii came after the tsunami hit the area. This can occur with local and regional tsunami that tend to be smaller in their area of effect. Some seismically active areas need to have the warning system and equipment closer than Hawaii if they are to protect their citizens. For example, the Aleutian Islands near Alaska have two to three moderate earthquakes per week. As of May 1995, centers such as the Alaska Tsunami Warning Center located in Palmer, Alaska, have assumed a larger role in the management of tsunami warnings.

In terms of basic research, one of the biggest areas of investigation is the calculation of return rates. Return rates, or recurrence intervals, are the predicted frequency with which tsunami will occur in a given area and are useful information, especially for highly sensitive buildings such as **nuclear power** stations, offshore oil drilling platforms, and hospitals. The 1929 tsunami in Newfoundland has been studied extensively by North American researchers as a model for return rates and there has

---

KEY TERMS

. . . .

Breakpoints—Points at which the Regional Tsunami Warning Centers arbitrarily split their coastline to create smaller sections that receive customized information in the event of a tsunami.

Return rates—The predicted frequency at which a tsunami will hit a certain area.

Seismic activity/event—An earthquake or disruption of the earth's crust.

Shallow surface wave—A wave that exists only on the surface of a liquid and has a wavelength that is greater than the water depth.

Slumping—A sudden falling of unstable sediments, usually used to refer to underwater environments.

Submarine—Below the surface of the ocean.

Wave height—The vertical change in height between the top, or "crest," of the wave and the bottom, or "trough," of the wave.

Wavelength—The distance between two consecutive crests or troughs in a wave.

---

been some dispute. Columbia University researchers predict a reoccurrence in Newfoundland in 1,000-35,000 years. However, a marine geologist in Halifax believes that it may reoccur as soon as 100-1,000 years. His calculations are based on evidence from mild earthquakes and tsunami in the area. He also suggests that the 1929 tsunami left a sedimentary record that is evident in the **soil** profile, and that such records can be dated and used to calculate return rates. He is currently working with an American scientist to test this theory.

See also Earthquake.

Resources

Periodicals

Whelan, M. "The Night the Sea Smashed Lord's Cove." *Canadian Geographic* (November/December 1994): 70-73.

Jennifer LeBlanc

Tuatara lizard

Tuataras are unusual, lizard-like animals that are the only living representatives of the order Sphenodonta of the vertebrate class Reptilia. The lineage of the sphenodonts is an ancient one, with a fossil record extending back 200 million years, prior even to the **evolution** of dinosaurs and the lizards. Until the discovery of tuataras in New Zealand, biologists had believed that this reptilian lineage had been extinct for 100 million years. Hence, the tuatara became celebrated as a "living fossil."

Tuataras have a number of unusual anatomical features. The arrangement of bones and cavities in the skull of Tuataras is considered to be a primitive character. Tuataras also have a relatively well developed, **light** sensitive **organ** on the top of their head, called the *pineal organ* or *median eye*, capable of sensing light and darkness. This primitive structure has similarities to a true **eye**, and even has a **lens** and retinal **tissue**. The function of the pineal eye is not understood by physiologists, but it appears to regulate activity of the pineal gland. This organ occurs in the forebrain of **vertebrates**, and secretes the hormone melatonin into the **blood**, which changes skin coloration in some animals, regulates diurnal or 24-hour **biological rhythms**, and possibly influences seasonal reproductive cycles. A unique characteristic of tuataras, not shared with other **reptiles**, is the fusion of teeth into the jawbones, rather than being set in sockets in the bone.

Tuataras are rather long-lived animals, which can exceed 75 years in age, and reach about 28 in (70 cm) in body length. Tuataras are slow moving, sluggish animals, which mostly eat large **insects**. These reptiles live in burrows, dug by themselves, which they often share with nesting petrels (seabirds).

The two surviving **species** of tuatara, *Sphenodon punctatus* and *S. guntheri*, are very rare animals, living only on a few isolated small islands off New Zealand. These species used to have much broader ranges, but they were widely extirpated from the mainland and many islands of New Zealand after European colonization. The catastrophic decline of tuataras was mainly due to predation of adults and eggs by introduced **mammals**, such as **cats**, foxes, **rats**, and others. It is significant that these predators do not occur on the few, closely protected islands where the world's last tuataras live.

The initial survival of tuataras was due to the early isolation of New Zealand from other continents, where ecologically more capable types of reptiles, **birds**, and mammals evolved and came to dominate **animal** communities. In other parts of Gondwanaland (the southern **continent** which New Zealand was originally a part of) tuataras and their relatives were incapable of competing with the better adapted creatures, and so became extinct. Tuataras are truly extraordinary survivors, relics from a bygone era of the evolution of animal forms on **Earth**. However, tuataras are now critically endangered as a re-

sult of recent influences by humans. The continued, though precarious, survival of tuataras now depends on their strict and perpetual protection on a few small islands.

Bill Freedman

Tuber

A tuber is a swollen, underground storage **organ** that develops on the roots of certain **species** of plants. Some types of tubers are highly nutritious, mostly because of their **energy** content in the form of starch.

Agricultural species of plants that develop edible tubers include the white **potato** (*Solanum tuberosum*), **sweet potato** (*Ipomoea batatas*), tapioca or cassava (*Manihot esculenta*), and **yam** (*Dioscorea batatas*).

The white potato is the most important and best-known of the agricultural tubers. The potato is a native of the Andean plateau of **South America**. In this species, tubers develop at the end of roots that emerge from underground stems, known as stolons. Potato tubers have stem buds known as "eyes" which can sprout and grow new, aboveground stems. It appears that potatoes have been cultivated by indigenous peoples of the Andean plateau for at least 6,500 years.

The varieties of potatoes that are most commonly cultivated in modern, industrial agriculture typically develop rather large tubers with white centers. However, there is a great diversity of other varieties of potatoes, especially in their native Andean range. These varieties commonly form relatively small tubers with brown, red, yellow, or purple skin, and white or darker-colored interiors.

Potato tubers are very nutritious, especially as a source of starch. However, they also contain about 2% protein and valuable **minerals** and vitamins, especially vitamins B and C. Interestingly, potato foliage is poisonous because of its content of a toxic **alkaloid** known as solanin. This chemical also occurs in green sprouts of the eyes of the tubers, which is why these should be excised and not eaten.

The potato was first discovered by Europeans in the mid-1530s when Spanish conquistadors observed its cultivation in Peru. The potato subsequently became a commonly cultivated food in **Europe**, initially as a food for **livestock**. Around the beginning of the seventeenth century, people also began to commonly eat the potato. Its high productivity and favorable nutritional qualities are believed to have been important in allowing population growth in Europe during the next several centuries. However, at the beginning of the nineteenth century a new **disease** of potatoes occurred. The potato blight, caused by the fungus *Phytophthora infestans*, became a recurrent disaster, and widespread famine resulted. In some regions such as Ireland, poor people ate little else but potatoes, and the blight caused **mass** starvation which to some degree was alleviated by massive immigration to **North America**. The potato blight is still an important disease. However, this disease is now controlled by growing resistant varieties of potatoes by managing the environment to make it less favorable to the fungus and by the use of fungicides.

See also Nightshade.

Bill Freedman

Tuberculosis

Tuberculosis is a **disease** caused by the bacterium *Mycobacterium tuberculosis*. The **organism** infects the lungs and causes a debilitating condition that historically was known as consumption. In the 1970s, scientists considered tuberculosis as largely defeated following the widespread use of **antibiotics**. Today, multi-drug resistant *Mycobacterium tuberculosis* has developed, and tuberculosis has reemerged as a worldwide public health problem.

Tuberculosis is not a new disease. Indeed, examinations of Egyptian mummies that are over 4,000 years old have shown symptoms of tuberculosis. Hippocrates, the father of medicine, described a malady in 460 B.C. that he termed *pthisis*. His description makes it clear that he was describing tuberculosis.

The term consumption reflected the progression of tuberculosis. Patients became lethargic, weak, and seemed to waste away. The role of **bacteria** in tuberculosis became clear in 1882, when Robert Koch discovered a staining technique that allowed bacteria to be visible using a light **microscope**. In Koch's time, and even into the middle of the twentieth century, the main treatment for tuberculosis was the isolation of patients in facilities called sanitoriums. Here, patients spent much of their time exposed to the dry, fresh outdoor air, which lessened the spread of the bacteria.

Tuberculosis is infectious, and is spread from person to person via inhaling contaminated droplets in the air. Often, someone can be infected with *Mycobacterium tuberculosis* yet not feel ill. This is also called latent tuberculosis. The infected person can spread the germ to others, however, by coughing or breathing. An active **infection** can appear at a later time.

In the twentieth century, specific chemical treatments for tuberculosis were developed. French bacteriologists Leon Calmette and Camille Guérin found a way to grow *Mycobacterium tuberculosis* so that its ability to cause disease was weakened. This weakened, or attenuated, microbe eventually formed the basis of a **vaccine**. The bacille Calmette-Guérin (BCG) vaccine, which was first given to people in 1921, is still in use today.

Unfortunately, the potential of BCG has not been realized because the vaccine is protective against the form of tuberculosis that occurs in children rather than in adults. Also, because samples of the original vaccine were kept, the development of numerous formulations of BCG have made it difficult to establish which of the many variations of the vaccine is effective.

One variant of BCG, called Evans BCG, was withdrawn from use in July 2002 because of concerns that it was not protective against infection. Every vaccine, particularly those that use attenuated **microorganisms**, carries some risk. In the case of Evans BCG, the risks of its use were deemed to be greater than any benefit resulting from its use.

Beginning in the 1940s, antibiotics were discovered and chemically synthesized. An antibiotic called streptomycin was an immediate success against the tuberculosis bacteria. Unfortunately, resistance to the antibiotic soon developed. A succession of antibiotics was able to keep tuberculosis at bay for some decades. However, beginning in the 1980s, the number of cases of tuberculosis once again began to rise.

Some 30 million people around the world died of tuberculosis in the 1990s. The U.S. Department of Health and Human Services predicts tuberculosis, the "forgotten plague," will spread further by the year 2005. In 1990 there were 7,537,000 TB cases worldwide. That number is expected to rise by almost 60%, to 11,875,000 cases by 2005.

Part of this increase has paralleled the emergence and spread of Acquired Immunodeficiency Syndrome (**AIDS**). AIDS devastates the **immune system**, which gives other infections, including tuberculosis, a better chance to flourish. Another factor contributing to the spread of tuberculosis is the emergence of strains of *Mycobacterium tuberculosis* that are resistant to many antibiotics.

See also Epidemiology.

Resources

Books

Reichman, L.B., and J.H. Tanne. *Time Bomb: The Global Epidemic of Multi-Drug Resistant Tuberculosis.* New York: McGraw-Hill, 2001.

Periodicals

Starke, J.R., et al. "The Role of BCG Vaccine in the Prevention and Control of Tuberculosis in the United States: A Joint Statement by the Advisory Council for the Elimination of Tuberculosis and the Advisory Committee on Immunization Practices." *Morbidity and Mortality Weekly Report* 45 (April 1996): 1–18.

Organizations

Centers for Disease Control and Prevention. Division of Tuberculosis Elimination. 1600 Clifton Road, NE Mailstop E-10, Atlanta, GA 30333. (404) 639–8135. <http://www.cdc.gov/nchstp/tb/>.

Tulip *see* **Lily family (Liliaceae)**

Tumbleweed

The true tumbleweeds are various **species** of herbaceous plants in the **amaranth family (Amaranthaceae)**. These are usually annual plants that develop a spherical, bush-shaped **biomass**. At the end of the growing season when their small **seeds** are ripe, the tumbleweeds wither and detach from their base and are blown about by winds, scattering their seeds widely over the surface of the ground. Therefore, the tumbling habit of these plants is an **adaptation** to extensive dispersal of their ripe seeds.

One common species of tumbleweed is *Amaranthus graecizans*. This annual **plant** is native to semi-deserts but is now a common weed of recently disturbed land in agricultural and urban areas. This plant has a whitish stem, green leaves, and numerous small, greenish flowers. *Amaranthus albus*, also known as tumble pigweed, is a closely related tumbleweed.

Other unrelated species of plants also tend to tumble to disperse their seeds. Two examples in the goosefoot family (Chenopodiaceae) are the Russian-thistle or Russian tumbleweed (*Salsola kali*) and the Russian pigweed (*Axyris amaranthoides*). These are both **introduced species** and can be important weeds. The winged pigweed or tumbleweed (*Cycloloma atriplicifolium*) is a related native species of arid habitats of the West which has also become weedy in open, disturbed habitats. Another species with a tumbling habit is the tumble mustard (*Sisymbrium altissimum*), in the **mustard family (Brassicaceae)**. Tumble panic-grass (*Panicum capillare*, family Poaceae) produces a large, bushy inflorescence that often detaches and blows about at the end of the growing season.

The "tumbling tumbleweeds" have become a romanticized element of the landscape of the American West through the pervasive influence of their images in songs and movies. However, many of these very capable and opportunistic tumbleweeds are also extremely widespread as weeds. As such, the tumbleweeds are taking advantage of many types of disturbed habitats that are created on the landscape by humans and their activities.

Tumor

A tumor (also known as a neoplasm) is an abnormal **tissue** growth. Neoplasm means "new formation." Tumors can be either malignant (cancerous) or nonmalignant (benign) but either type may require therapy to remove it or reduce its size. In either case the tumor's growth is unregulated by normal body control mechanisms. Usually the growth is not beneficial to the **organ** in which it is developing and may be harmful.

It is not known what triggers this abnormal growth. Normally cells are generated at a **rate** needed to replace those that die or are needed for an individual's growth and development. Muscle cells are added as one grows as are bone cells and others. Genetic controls modulate the formation of any given cells. The process of some cells becoming muscle cells, some becoming nerve cells, and so on is called **cell** differentiation. Tumor formation is an abnormality in cell differentiation.

A benign tumor is a well-defined growth with smooth boundaries. This type of tumor simply grows in diameter. A benign growth compresses adjacent tissues as it grows. A malignant tumor usually has irregular boundaries and invades the surrounding tissue. This **cancer** also sheds cells that travel through the bloodstream implanting themselves elsewhere in the body and starting new tumor growth. This process is called metastasis.

It is important that the physician determine which kind of tumor is present when one is discovered. In some cases this is not a simple matter. It is difficult to determine whether the growth is benign without taking a sample of it and studying the tissue under the **microscope**. This sampling is called a biopsy. Biopsy tissue can be frozen quickly, sliced thinly, and observed without staining (this is called a frozen section); or it can be sliced, stained with dyes, and observed under the microscope. Cancer tissue is distinctly different from benign.

A benign tumor can be lethal if it compresses the surrounding tissue against an immovable obstacle. A benign **brain** tumor compresses brain tissue against the skull or the bony floor of the cranium and results in paralysis, loss of **hearing** or sight, dizziness, and/or loss of control of the extremities. A tumor growing in the abdomen can compress the intestine and interfere with di-

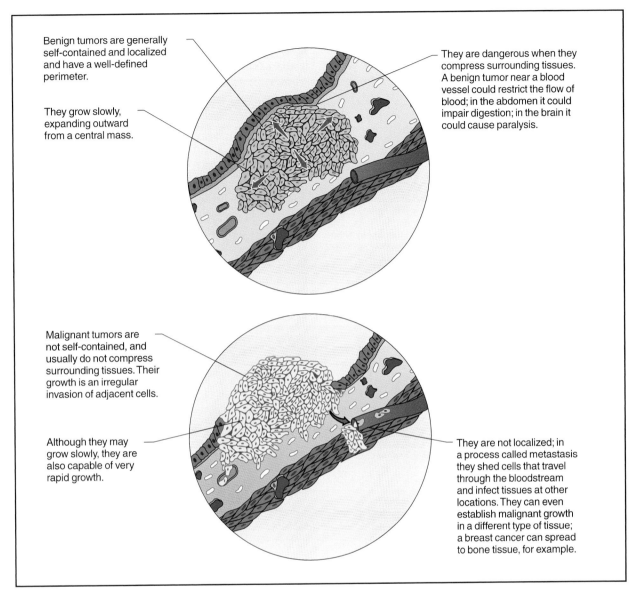

Benign tumors are generally self-contained and localized and have a well-defined perimeter.

They grow slowly, expanding outward from a central mass.

They are dangerous when they compress surrounding tissues. A benign tumor near a blood vessel could restrict the flow of blood; in the abdomen it could impair digestion; in the brain it could cause paralysis.

Malignant tumors are not self-contained, and usually do not compress surrounding tissues. Their growth is an irregular invasion of adjacent cells.

Although they may grow slowly, they are also capable of very rapid growth.

They are not localized; in a process called metastasis they shed cells that travel through the bloodstream and infect tissues at other locations. They can even establish malignant growth in a different type of tissue; a breast cancer can spread to bone tissue, for example.

A comparison of benign and malignant tumor characteristics. *Illustration by Hans & Cassidy. Courtesy of Gale Group.*

gestion. It also can prevent proper liver or pancreatic function. The benign tumor usually grows at a relatively slow pace and may stop growing for a time when it reaches a certain size.

A cancer may grow quite rapidly or slowly, but usually is irregular in shape. It invades the neighboring tissue instead of pressing it aside. Most importantly, a cancer sheds cells, that is, metastasizes, so that new cancer growths can spring up in areas distant from the original cancer. The cancerous cells also can establish a cancer in tissue that is different from the original cancer. A breast cancer could spread to bone tissue or to liver.

A benign tumor can be removed surgically if it is in a location that a surgeon can reach. A tumor growing in an unreachable area of the brain can be treated using **radiation**. It can also be treated by inserting thin probes through the brain tissue into the tumor and circulating liquid **nitrogen** through the probe to freeze the tumor. This operation is called cryosurgery.

A malignancy requires steps to remove it but consideration must be given to the possibility that the tumor has begun to metastasize. The main or primary tumor may be removed surgically but if the tumor has been growing for some time the patient also may require treatment with powerful drugs to kill any stray cells. This treatment is called chemotherapy. Chemotherapy allows the antitumor drug to be circulated throughout the body to counter any small tumor growths.

KEY TERMS

. . . .

Biopsy—The surgical removal of a small part of a tumor. The excised tissue is studied under the microscope to determine whether it is benign or malignant.

Chemotherapy—Use of powerful drugs to kill cancer cells in the human body.

Differentiation—The process by which cells take the form of a given type of tissue. That is, basic cells become muscle cells, neurons, stomach lining, kidney, or other cell types.

Metastasis—Spreading of a cancerous growth by shedding cells that grow in other locales.

Resources

Periodicals

Carpi, J. "Progress Against Cancer." *American Health* 13 (October 1994): 15-16+.

DeVita, E. "Conquering Cancer." *American Health* 13 (November 1994): 66-73.

Xu, D., et al. "Downregulation of Telomerase Reverse Transcriptase mRNA Expression by Wild Type in Human Tumor Cells." *Oncogene* 19 (26 October 2000): 53.

Larry Blaser

Tuna

Tuna are large, fast-swimming **bony fish** (teleosts of the family Scombridae) found in waters of the world's oceans. All **species** of tuna are economically important, usually supporting a large commercial fishery, and sometimes a local sport fishery. The largest species is the bluefin tuna (*Thunnus thynnus*), which can achieve a length of 13 ft (4 m) and weigh up to 1,760 lb (800 kg). Populations of bluefin tuna occur in temperate and warm waters throughout the world. Smaller species of tunas include the albacore tuna (*T. alalunga*), yellowfin tuna (*T. albacares*), skipjack tuna (*Euthynnus pelamis*), bonitos (*Sarda* spp.), frigate mackerels (*Auxis* spp.), Spanish mackerels (*Scomberomorus* spp.), and the wahoo (*Acanthocybium* spp.).

Biology of tuna

A remarkable aspect of the **physiology** of tuna in the genus *Thunnus* is their ability to maintain a body temperature significantly warmer than that of the ambient seawater. For example, the bluefin tuna can maintain a core body temperature of 75-95°F (24-35°C), even in **water** as cold as 43°F (6°C). However, unlike typical **endothermic** creatures such as **mammals** and **birds**, the body temperatures of tuna are not held constant within a relatively narrow range.

The endothermy of tuna is achieved by conserving the **heat** generated through normal body **metabolism**. This is accomplished through the action of an intertwined meshwork of **veins** and **arteries**, known as the rete mirable (meaning "wonderful net"), located in the periphery of the body. The rete mirable reclaims much of the heat in the venous **blood**, and transfers it to arterial blood through the action of a counter-current exchange system. This **heat transfer** slows down the **rate** of cooling of the tuna at the body surface, and thereby allows the **animal** to maintain a warmer core temperature. Higher body temperatures allow tuna to use their muscles more efficiently, and therefore swim more quickly with relatively little expenditure of **energy**. Higher temperatures of the body core may also contribute to the rapid and efficient digestion and absorption of food.

Tuna are fast swimmers, reaching speeds of up to 56 mph (90 kph) in the case of the large bluefin tuna. Moreover, tuna are well adapted for cruising great distances at a relatively brisk speed. However, tuna can also accelerate quickly while predating on other **fish**, or to avoid their own predators. A tuna's body is relatively elongated and fusiform, that is, tapering at both ends. Their fins are of a size and position designed to minimize drag, so that maximum speeds can be achieved and maintained with relatively small expenditures of energy. The major source of forward thrust while swimming is side-to-side movements of the caudal, or tail fin. Tunas also have relatively large median fins (especially the dorsal, or top fin), which is an **adaptation** to minimizing the drag associated with sideways slippage of the body during swimming.

Tuna fisheries

Tuna support large commercial fisheries wherever they are abundant, and are thus economically important fish. However, as with any fishery, stocks of tuna can be easily exhausted through excessive harvesting. Indeed, most if not all stocks of tuna have been significantly degraded by overharvesting. This problem can be illustrated by the case of the fishery for bluefin tuna in the western Atlantic Ocean.

The bluefin tuna is a very large, fast-swimming fish which undertakes regular trans-oceanic migrations. Bluefin tuna are extremely valuable since they are eagerly sought for sale as a delicacy in Japanese sushi restau-

A tuna. © Lanceau Y. Jacana, National Audubon Society Collection/Photo Reasearchers, Inc. Reproduced by permission.

rants. During the early 1990s, a prime bluefin tuna caught in **North America** could be sold for $30,000 (U.S. currency) at the wharf, and then for at least $60,000 at an auction in Tokyo. The tuna meat might then be sold for about $350 per pound, as prepared sushi in restaurants. This is equivalent to about $230,000-$385,000 per fish, depending on its weight.

Because bluefin tuna are so enormously valuable, they have been exploited intensively, and their populations are declining rapidly. For example, in 1975 there were an estimated 150,000 bluefin tuna in the western Atlantic Ocean, but by the early 1990s this number had decreased by 90% through excessive harvesting, to only 22,000 animals. This resource collapse occurred even though the fishery was regulated by an international agency, the Atlantic Commission for the Conservation of Atlantic Tunas. The problem was that the managers of the commission consistently ignored the advice of their resource scientists, and set the allowable catches higher than was recommended, or was prudent. In addition, there was substantial unregulated pirate fishing by ships flying the flags of nations that are not members of the commission. As of 1999, this situation has not significantly improved.

The regulated and non-regulated overfishing were both ultimately caused by greed, and a desire to reap large, short-term profits. This was done without significant regard for the sustainability of the enterprise, or of the natural resource of bluefin tuna.

Some of the smaller species of tuna are caught using a type of net called a purse seine, set around a school of fish. The net initially floats vertically, with one side buoyed at the surface. Once the purse seine is set around a group of fish, the deeper side of the net is closed using a drawstring-like apparatus, trapping the fish inside the "purse." Unfortunately, purse seines also trap other species, including dolphins and porpoises that often associate with schools of tuna in some regions. In fact, in some fishing sectors boat captains deliberately set their nets around groups of these marine mammals, because they know that schools of tuna are generally found beneath them. From the 1960s to the late 1980s, huge numbers of some species of dolphins and porpoises were killed in purse seines set for tuna, perhaps 200,000 of these marine mammals each year. More recently, the dolphin kill rate has declined to about 100,000 per year. This non-target fishing mortality has significantly depleted the populations of dolphins and porpoises in some regions.

KEY TERMS

. . . .

Counter-current exchange—An exchange of heat or respiratory gases between two fluids moving across each other in different directions. In the rete mirable of tunas, the fluids are contained in veins and arteries, and the transfer of heat occurs across the walls of these vessels.

Endothermy—Refers to animals that maintain their body temperatures within a range substantially warmer than their ambient environment. The source of heat is from internal metabolism.

Overharvesting—The unsustainable exploitation of a potentially renewable, biological resource. In such a case, the harvesting rate exceeds the rate of regeneration, so the quantity of the resource diminishes over time, sometimes to commercial or even biological extinction.

Large tuna often have a significant **contamination** of their flesh with mercury. This commonly occurs to a degree that exceeds the maximum acceptable concentration of mercury in fish intended for human consumption, that is, 0.5 ppm (parts per million, on a fresh weight basis). The contamination of tuna and other large, oceanic fish by mercury is apparently a natural phenomenon. One study found no difference in the mercury concentrations of modern tuna, and animals collected between 1878 and 1909 and stored in a museum.

The World Conservation Union (IUCN) considers the southern bluefin tuna (*Thunnus maccoyii*), bigeye tuna (*Thunnus obesus*), and Monterrey Spanish **mackerel** (*Scomberomorus concolor*) to be threatened species. The southern bluefin tuna is listed as critically endangered.

See also Drift net.

Resources

Books

Freedman, B. *Environmental Ecology.* San Diego: Academic Press, 1994.
Harris, C. L. *Concepts in Zoology.* New York: Harper-Collins, 1992.
Nelson, J.S. *Fishes of the World.* 2nd ed. New York: Wiley, 1984.
Scott, W. B., and M. G. Scott. *Atlantic Fishes of Canada.* Toronto: University of Toronto Press, 1988.

Periodicals

Safina, C. "Bluefin Tuna in the West Atlantic: Negligent Management and the Making of an Endangered Species." *Conservation Biology* 7: 229-34.

Bill Freedman

Tundra

Tundra is a generic name for a low-growing **ecosystem** found in climatically stressed environments with short and cool growing **seasons**. Latitudinal tundra occurs in the Arctic and to a much lesser extent in the Antarctic, where the environments are characterized by cool, short growing seasons. Altitudinal tundra occurs under a similar climatic regime, but at the tops of **mountains**.

After **temperature**, the second most-important environmental factor affecting most tundra communities is moisture. Under wet conditions, sedge and grass-dominated meadow communities develop, while moist conditions favor a vegetation dominated by dwarf shrubs and herbaceous **species**, and dry sites have cushion plants and **lichens**. The vegetation of arctic and alpine tundras share many structural characteristics, most genera, and some species. However, there are important environmental differences between these two tundras, with the alpine type being subject to much larger variations of daily temperature during the growing season, as well as more intense inputs of solar **radiation** during the day. In contrast, arctic tundra can experience continuous exposure to the sun's rays, with 24-hour days for an extended period during the growing season.

Arctic tundra

Arctic tundra occurs in the northern most parts of the Northern Hemisphere, intergrading across the latitudinal tree-line with the boreal forest to the south. In **North America**, arctic tundra occurs on all non-glaciated regions of the islands of northern Canada and on Greenland, as a northern fringe of Alaska and continental Canada, and penetrating to relatively southern latitudes in the vicinity of Hudson Bay, which acts as a climate-influencing extension of the cold Arctic Ocean. (An ecological analogue of arctic tundra occurs on the southernmost islands of the Southern Ocean and as a narrow fringe of parts of **Antarctica**. However, compared with arctic tundra, antarctic tundras are less well developed and have few species of plants.) The distribution of arctic tundra is determined by climate, generally occurring where the annual **precipitation** is less than about 2 in (50 cm) each year, and the average annual temperature is colder than 23°F (-5°C). Much of the arctic tundra covers permanently frozen ground, or **permafrost**, above which only the active layer—comprised of the surface 1.6 ft (0.5 m) or so—thaws during the growing season. Arctic tundra experiences continuous inputs of solar radiation during much of the growing season, a condition that can last for more than two months, depending on latitude. The incessant insolation during this time allows the vegetation to be relatively productive, as long as the availabilities of moisture and **nutrients** are not excessively constrained. During the arctic winter, when plants are dormant but some animals remain active, there is continuous night for several months, and extremely cold conditions.

There are several categories of arctic tundra vegetation. The low-arctic tundra occurs in southern areas. On wet sites, the low-arctic tundra of North America develops as relatively productive wet meadows dominated by a tussock-forming cottongrass (*Eriophorum vaginatum*), while better-drained sites are dominated by shrubs such as willow (e.g, *Salix glauca*) and birch (e.g., *Betula glandulosa*), growing to about 1.6 ft (0.5 m) in height. These sites also support herbaceous plants, such as arctic lupin (*Lupinus arcticus*).

The high-arctic tundra is less productive and the plants are of lower stature. In North America, poorly drained wet meadows are dominated by graminoid plants, especially **sedges** (e.g., *Carex membranacea* and *C. stans*) and cottongrass (*Eriophorum angustifolium*). Drier sites are typically dominated by dwarf shrubs such as arctic willow (*Salix arctica*), arctic heather (*Cassiope tetragona*), mountain cranberry (*Vaccinium vitis-idaea*), and arctic bilberry (*V. uliginosum*), along with herbaceous plants such as lousewort (*Pedicularis* spp.), **grasses** (such as *Arctagrostis latifolia*), mosses, and lichens.

The polar **desert** is a very sparse tundra that occurs where climatic extremes of temperature and moisture availability allow only an incomplete cover of vegetation to develop. Such sites typically have a sparse cover of lichens and a few species of vascular plants, including cushion plants such as arctic avens (*Dryas integrifolia*) and purple saxifrage (*Saxifraga oppositifolia*), along with a few grasses and herbs. However, within the polar desert landscape there are occasional places that, because of their topography and drainage patterns, are relatively warm and moist throughout the growing season. These more moderate places are called high-arctic oases. They sustain a relatively lush growth of vegetation and, if the oasis is large enough, relatively large populations of animals.

The arctic tundra sustains only a few species of resident animals that remain active throughout the year. In the high-arctic tundra of North America, resident **birds** include the raven (*Corvus corax*) and rock ptarmigan (*Lagopus mutus*). Resident **mammals** include Peary **caribou** (*Rangifer tarandus pearyi*), muskox (*Ovibos moschatus*), arctic fox (*Alopex lagopus*), collared lemming (*Dicrostonyx torquatus groenlandicus*), and arctic hare (*Lepus arcticus*). However, there is a much larger number of seasonally abundant animals. These include **insects**, some of which can be very abundant during the

Rocky tundra on Bear Island in the Barents Sea.
Photograph by E.R. Degginger. National Audubon Society Collection/Photo Researchers, Inc. Reproduced by permission.

growing season, including dense populations of **mosquitoes**. Migratory species of birds include **finches** such as snow bunting (*Plectrophenax nivalis*), Lapland longspur (*Calcarius lapponicus*), and hoary redpoll (*Acanthis hornemanni*); shorebirds such as Baird's sandpiper (*Calidris bairdii*) and red knot (*Calidris canutus*); waterfowl such as oldsquaw duck (*Clangula hyemalis*) and greater snow goose (*Chen caerulescens atlantica*); and larids such as arctic tern (*Sterna paradisaea*), glaucous gull (*Larus hyperboreus*), and parasitic jaeger (*Stercorarius parasiticus*).

Alpine tundra

Alpine tundra occurs in climatically stressed environments at high mountainous altitudes. Alpine tundra can even occur on mountaintops at tropical latitudes, although this vegetation type is much more common in temperate regions. Compared with the arctic tundra, alpine environments have much larger daily variations of temperature and solar radiation during the growing season. Because of the thinness of the atmosphere at high altitude, alpine tundra is also subject to large inputs of ultraviolet radiation, which can be an important biological stressor. Because the skies are often clear at high altitude, the surface cools very quickly at night, so that frost can be a daily occurrence during the growing season.

In general, alpine tundra is considerably richer in **plant** species than arctic tundra. At temperate latitudes, this occurs because alpine environments were not regionally obliterated by glacial **ice** during the most recent glaciations, so the component species were able to endure this period of intense climatic stress. This survival was made possible by the occurrence of non-glaciated refugia where plants could survive on some mountaintops (these are called nunataks). In addition, as the climate deteriorated, alpine tundra could migrate to lower-altitude, non-glaciated parts of mountainous regions, as the tree-line moved downwards. In contrast, almost all of the arctic tundra was destroyed by the extensive continental **glaciers** that covered northern regions during the most recent **ice ages**. It is believed that after deglaciation the arctic tundra was re-established by a northward **migration** of some of the plant species of alpine tundra. However, because only some species were capable of undertaking the extensive migrations that were necessary, the arctic tundra is relatively poor in species, compared with temperate alpine tundras.

There are relatively few animals that only occur in alpine tundra and not in other types of ecosystems. In North America some of the characteristic mammals of alpine tundra include a small relative of rabbits called the pika (*Ochotona princeps*), and the hoary marmot (*Marmota caligata*). The gray-crowned rosy finch (*Leucostichte tephrocotis*), horned lark (*Eremophila alpestris*), and water pipit (*Anthus spinoletta*) all breed in alpine tundra of North America, but also in arctic tundra (the lark also breeds in other open habitats, such as **prairie** and fields).

Resources

Books

Barbour, M.G., and W.D. Billings. *North American Terrestrial Vegetation.* Cambridge: Cambridge University Press, 1988.

Barbour, M.G., J.H. Burk, and W.D. Pitts. *Terrestrial Plant Ecology, 2nd ed.* Don Mills, Ontario: Benjamin/Cummings Pub. Co., 1987.

Bill Freedman

Tungsten *see* **Element, chemical**

Tunneling

Tunneling, also known as the tunnel effect, is a quantum mechanical phenomenon by which a tiny particle can penetrate a barrier that it could not, by any classical or obvious means, pass. Though seemingly miraculous, the effect does have some intuitive characteristics. For instance, thin barriers allow more particles to tunnel than do thick ones, and low barriers permit more tunneling than do high ones.

Tunneling does not generally show itself in the macroscopic world. It only starts to become a factor for microscopic items. **Atoms** can tunnel, as can electrons, but things such as tennis balls and **grapes**, easily seen

with the naked **eye**, will not. For microscopic particles, the barrier heights are described in terms of **energy** instead of distance, but for conceptual purposes there is little difference.

It is important to note that the effect can only be understood with the aid of **quantum mechanics**. Classical mechanics, the system pioneered by Isaac Newton that stood unchallenged until the early twentieth century, has no way of explaining tunneling. In the Newtonian philosophy, all particles, even the tiniest of microscopic particles, can be located precisely. Any uncertainty is seen as the result of an imperfect measuring device or a sloppy scientist. In addition, each microscopic particle is considered to be like a tiny pebble and there can be nothing wave-like about it.

The quantum view of the universe is fundamentally different from the Newtonian view. Each particle is said to have both corpuscular (pebble-like) *and* wave-like properties. Furthermore, a quantum particle cannot, in general, be located precisely. It has a built-in uncertainty that cannot be taken away by the best of measuring instruments. For these reasons, a particle in quantum mechanics is often treated as a wave function. This is another way of saying that the particle is akin to a small bundle of waves.

Representing a quantum particle as such a bundle has two advantages. For one, it reveals that the particle is, in some sense, blurry and can never be exactly pinned down. It exists over a range of space, not a specific point. For another, the wave function format allows particles to exhibit wave-like properties (see **interference** and **diffraction**). Tunneling is a one of these wave-like properties. By the more general name of "barrier penetration," it is a well-established characteristic of waves. **Light** waves, for instance, have long been observed to overcome daunting optical barriers.

In a fundamental sense, the quantum mechanical explanation of tunneling can be illustrated by an analogy. If we roll a ball very slowly toward, say, a cement speed bump, we confidently say that it will not surpass the barrier and predict that it will roll back toward us. However, if a very modest **ocean** swell approaches an offshore sandbar, we are not as sure of the results and rightly so. The character of the wave, even if diminished in size, often makes its way past the sandbar. Similarly, treating a microscopic particle with the mathematical model of a ball clearly tells us tunneling is impossible, while using the mathematical model of a wave just as clearly states that particles will always have a chance to tunnel.

History

Some scattering experiments of the early twentieth century paved the way for tunneling. Ernest Rutherford, the pioneer of the scattering method, came across a paradox in a series of **uranium** experiments in 1910. Scattering involves the probing of a microscopic object by bombarding it with other particles and then keenly observing how the particles are, literally, scattered by the object in question. Specifically, Rutherford tried to pinpoint where the bombarding particles were scattered and how fast they were moving.

An example using tennis balls gives one a simple and accurate picture of a scattering experiment. Say a statue exists in a dark courtyard and we want information about the statue without being able to see it. Mainly, we want to know what its shape is (e.g., a horse or a person) and from what materials it is made (e.g., granite or clay). To learn more about the statue, we stand at the edge of the courtyard and toss tennis balls into the dark area, hoping to hit the statue. After many thousands of tennis balls, we've learned a lot. If only a few of the tennis balls have hit anything, we know the statue is probably quite small. Also, the directions that tennis balls were scattered are important. If most of them come straight back to us, we can ascertain that the statue has a large, flat front like a wall. Moreover, by observing the speed of the scattered tennis balls, we can get an idea of how hard the statue is. Atomic scattering set a similar scene for Rutherford, but he was interested in the nuclei of atoms and he preferred alpha particles to tennis balls.

Through a series of remarkable scattering experiments, Rutherford, working in conjunction with his students Hans Geiger and Ernest Marsden, had accurately mapped the insides, or nucleus, of the uranium atom. Uranium was of great interest at the time because of its radioactive properties. The summation of Rutherford's results for Uranium came in the form of a potential diagram. Potential refers to electric potential energy, so his diagram mapped out an energy barrier. Any particle that escaped the nucleus (i.e., any particle of nuclear **radiation**) would have to overcome this energy obstacle before it left. Rutherford estimated the top of this barrier to be at least about nine units of energy high. However, when observing the occasional radiated **alpha particle** (see radiation), he found that it had only four units of energy. The paradox was born. How could a particle with so little energy overcome a barrier of such height? Such a problem can be compared to a man walking next to a huge baseball stadium and suddenly seeing a baseball floating toward him, as if gently tossed. Surely any ball hit out of the stadium would have been moving fast enough to at least sting his palm.

The question lingered for 18 years until George Gamow, assisted by his colleagues Edward Uhler Condon and Ronald Gurney, proposed a solution. In 1928, quantum mechanics was gaining credibility, and the

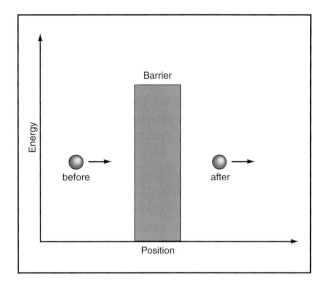

Figure 1. A particle before and after tunneling. It approaches from the left with far less energy than it would need to pass over the energy barrier. *Illustration by Hans & Cassidy. Courtesy of Gale Group.*

three physicists performed a relatively simple calculation, treating the alpha particle as a quantum mechanical wave function. In essence, they analyzed the problem from the viewpoint that an alpha particle was not located precisely at any given spot, but rather that its existence was spread out, like a wave. Their explanation proposed that the alpha particles tunneled out of uranium's energy barrier, and it fit Rutherford's observations perfectly. The acceptance and practical application of tunneling theory had begun.

Applications

One of the first applications of tunneling was an **atomic clock** based on the tunneling **frequency** of the **nitrogen** atom in an **ammonia** (chemical formula NH_3) **molecule**. The **rate** at which the nitrogen atom tunnels back and forth across the energy barrier presented by the **hydrogen** atoms is so reliable and so easily measured that it was used as the timing mechanism in one of the earliest atomic clocks.

A current and quickly advancing application of tunneling is Scanning Tunneling Microscopy (abbreviated STM). This technique can render high-resolution images, including individual atoms, that accurately map the surface of a material. As with many high-tech tools, its operation is fairly simple in principle, while its actual construction is quite challenging.

The working part of a tunneling **microscope** is an incredibly sharp **metal** tip. This tip is electrically charged and held near the surface of an object (known as

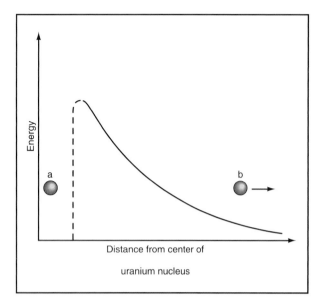

Figure 2. Rutherford's potential energy map (dark curve) and the theoretical other side of the barrier (dashed curve), which serves to contain the pieces of the nucleus; (a) shows an alpha particle as part of the nucleus and (b) shows its troublesome appearance outside the uranium atom. *Illustration by Hans & Cassidy. Courtesy of Gale Group.*

the sample) that is to be imaged. The energy barrier in this case is the gap between the tip and the sample. When the tip gets sufficiently close to the sample surface, the energy barrier becomes thin enough that a noticeable number of electrons begin to tunnel from the tip to the object. Classically, the technique could never work because the electrons would not pass from the tip to the sample until the two actually touched. The number of tunneling electrons, measured by incredibly sensitive equipment, can eventually yield enough information to create a picture of the sample surface.

Another application of tunneling has resulted in the tunnel **diode**. The tunnel diode is a small electronic switch and, by incorporating **electron** tunneling, it can process electronic signals much faster than any ordinary physical switch. At peak performance, it can switch on and then off again ten billion times in a single second.

Resources

Books

Hewitt, Paul. *Conceptual Physics.* New York: Prentice Hall, 2001.

Meriam, J.L., and L.G. Kraige. *Engineering Mechanics, Dynamics.* 5th ed. New York: John Wiley & Sons, 2002.

Wiesendanger, Roland, and Hans-Joachim Goentherodt, eds. *Scanning Tunneling Microscopy I.* New York: Springer-Verlag, 1993.

Brandon Brown

KEY TERMS

Electron—One of the fundamental particles of the universe; carries a negative charge and has very little mass.

Energy barrier—An obstacle analogous to a physical wall, where any object that passes must either possess an energy greater than the barrier's or tunnel through the barrier.

Macroscopic—Not needing a microscope to be seen, readily observed by the human eye.

Potential—Closely related to potential energy, which is known as the "energy of position" or the energy that a body possesses due to its circumstances as opposed to its motion.

Quantum mechanics—The theory that has been developed from Max Planck's quantum principle to describe the physics of the very small. The quantum principle basically states that energy only comes in certain indivisible amounts designated as quanta. Any physical interaction in which energy is exchanged can only exchange integral numbers of quanta.

Scattering—An experimental technique by which an item of interest is studied by inducing other particles to collide with it.

Wave function—A useful mathematical construct commonly employed in quantum mechanics to represent both a particle's wavelike characteristics and its uncertainty in location.

Turacos

Turacos, or touracos, are 18 **species** of sub-Saharan **birds** that make up the family Musophagidae, in the order Cuculiformes, which also includes the **cuckoos**, anis, coucals, and roadrunner. The usual **habitat** of turacos is dense tropical **forests** or forest edges. Turacos do not migrate, although they may move locally.

Turacos are medium- to large-sized birds, with a body length of 1.2-2.5 ft (38-76 cm). They have short, rounded wings, long, broad tails, and small, crested heads. Their bills are short, stout, and slightly hooked at the tip of the upper mandible. The outer toe is highly dexterous, and can be turned and used in a forward or backward position.

Turacos have a soft, thick plumage, colored in hues of green, brown, blue, or gray, with patches of white and red. The feather pigments of turacos are rather unusual. Most birds achieve a green coloration using a combina-

Turacos lay two to three eggs in a frail, arboreal nest made of twigs. These are incubated by the female, but both parents share in the feeding and care of the downy young.

The giant blue plantain-eater (*Corythaeola cristata*) is the biggest turaco, achieving a length of 2.5 ft (76 cm).

Bill Freedman

Turbine

A turbine is any of various rotary machines that convert the kinetic **energy** in a stream of fluid (gas or liquid) into mechanical energy by passing the stream through a system of fixed and moving fans or blades. Turbines are simple but powerful machines that embody Newton's third law of **motion** which states that for every action there is an equal and opposite reaction. They are classified according to the driving fluid they use: steam, gas, **water**, and **wind**. Today, different types of turbines generate **electricity**, power ships and submarines, and propel jet **aircraft**.

History

The idea of using naturally moving water or air to help do **work** is an ancient one. Waterwheels and windmills are the best examples of ancient mankind's ability to capture some of nature's energy and put it to work. The Romans were grinding corn with a **waterwheel** as early as 70 B.C., and modern-type windmills were being used in Persia around A.D. 700. Both are ancestors of the turbine. Both have large surfaces (paddles, buckets, or a sail) at their wheel edges that are struck by moving wind or water which forces the wheel to turn. It was through the turning of this large central wheel, which could turn other smaller wheels, that mechanical energy was obtained and work, like grinding corn or operating a pump, could be accomplished.

The most ancient of these methods was the undershot wheel or paddle wheel. On these old waterwheels, only the very lowest part of the wheel was submerged beneath a moving body of water, and the entire wheel was turned as the river flowed past it, pushing against its paddles. This was a prototype for what came to be called an impulse turbine, which is one that is driven by the **force** of a fluid directly striking it. The undershot waterwheel was followed during medieval times by the overshot wheel. This first made its appearance in Germany around the middle of the twelveth century and became the prototype for the modern reaction turbine. Contrasted to the impulse turbine whose energy source is kinetic energy (something striking something else and giving it

A Ross's turaco. © *Anthony Mercieca, National Audubon Society Collection/ Photo Researchers, Inc. Reproduced by permission*

tion of yellow and melano-blue pigments, but in turacos an actual, green pigment known as turacoverdin is present in the feathers. Turacos also produce a unique red copper-containing pigment known as turacin.

Turacos are arboreal birds. They are weak, undulating fliers, and they commonly run along branches through **tree** foliage, suggesting the movements of a small mammal, such as a squirrel. These birds eat **fruits**, **seeds**, buds, and **invertebrates**. The family name, Musophagidae, translates from the Greek as **banana** eater, although this is a misnomer, because these birds are not actually known to feed on bananas or plantains.

Turacos are somewhat gregarious, congregating in small, noisy groups. **Courtship** includes fluttering displays to reveal bright patches of coloration, along with raising of the conspicuous crest on the head, and energetic bobbing of the tail. These displays are accompanied by loud calls, and feeding of the female with fruit offered by the male.

some of its energy), the energy source for an overshot wheel (or reaction turbine) is known as potential energy. This is because it is the weight of the water acting under gravity that is used to turn the wheel. Renaissance engineers studied the waterwheel and realized that the action of water on a wheel with blades would be much more effective if the entire wheel were somehow enclosed in a kind of chamber. They knew very well that only a small amount of the water pushing or falling on a wheel blade or paddle actually strikes it, and that much of the energy contained in the onrushing water is lost or never actually captured. Enclosing the wheel and channeling the water through this chamber would result in a machine of greater efficiency and power. They were hampered, however, by a lack of any theoretical understanding of hydraulics as well as by a lack of precision **machine tools** with which they could carefully build things. Both of these problems were resolved to some degree in the eighteenth century, and one of the earliest examples of a reaction turbine was built in 1750 by the German mathematician and naturalist Johann Andres von Segner (1704-1777). In his system, the moving water entered a cylindrical box containing the shaft of a runner or rotor and flowed out through tangential openings, acting with its weight on the inclined vanes of the wheel.

A really efficient water turbine was now within reach it appeared, and a prize was offered in France by the Societe d'Encouragement pour l'Industrie Nationale. The prize was won by the French **mining** engineer Claude Burdin (1778-1873), who published his results in 1828. It was in this publication that Burdin coined the word "turbine" which he took from the Latin "turbo" meaning a whirling or spinning top. It was Burdin's student, Benoit Fourneyron (1801-1867), who improved and developed his master's work and who is considered to be the inventor of the modern hydraulic turbine. Fourneyron built a six-horsepower turbine and later went on to build larger machines that worked under higher pressures and delivered more horsepower. His main contribution was his addition of a distributor which guided the water flow so that it acted with the greatest efficiency on the blades of the wheel. His was a reaction type turbine, since water entering through the vanes of the distributor (that was fitted inside the blades) then acted on the blades of the wheel. Following Fourneyron's first turbine, which happened to be a hydraulic or water turbine, other turbines were developed that used the energy of a different material like gas or steam. Although these different types of turbines have different means of operation and certainly different histories, they still embody the basic characteristics of a turbine. They all spin, or receive their energy from some form of a moving fluid, and they all convert it into mechanical energy.

Types of turbines

While turbines can be classed as either impulse or reaction according to the way they function, there are four broad types of turbines categorized according to the fluid that supplies the driving force: steam, gas, water, or wind. Steam, water, and wind turbines are all used to generate electricity, and gas turbines are most often used by jet aircraft for propulsion. The steam turbine is mainly used by power plants that burn either **fossil fuels** or use nuclear energy to drive generators for consumer electricity. Steam turbines also power submarines and ships. The water or hydraulic turbine is used almost exclusively in hydroelectric plants to power an electric **generator** which then produces electric power for homes, offices, and factories. Wind turbines are the least common, but Scotland now uses the vertical machines called Darrieus turbines whose giant, bow-shaped blades look like huge egg beaters to generate electricity via the wind. The gas turbine is primarily used by jet aircraft.

Steam turbines transform the thermal energy stored in steam into mechanical work. The earliest steam turbine was also the earliest known **steam engine**. During the first century A.D., the Greek mathematician and engineer, Hero of Alexandria, built what was basically a novelty and produced no useful work, but was nonetheless the first steam turbine. It consisted of a small, hollow **sphere** with two nozzles or bent tubes sticking out of it. The sphere was attached to a boiler which produced steam. As the steam escaped from the sphere's hollow tubes, the sphere itself would rotate on its axis and continue to whirl. This was in principle a reaction steam turbine because the force of the escaping steam itself provided the thrust to make it spin. Steam was not considered in any type of turbine context again until the Italian Giovanni Branca published a work in 1629, in which he suggested the principle of the impulse steam turbine. In his book he details that it would be simple to convert the linear motion of a cylinder into the rotary motion needed for work by directing a jet of steam onto the vanes of a wheel, like water against a waterwheel. It is not known if he ever built such an engine.

Despite the advances made in understanding and managing steam that were gained in the eighteenth century, the steam turbine could not be built until the precision and strength of machining and materials had reached a certain level. In 1884, English engineer Charles Algernon Parsons (1854-1931) produced the first practical steam turbine engine. Although designed for the production of electric power, it was soon applied to marine propulsion and drove a ship named *Turbinia* in 1887. The spectacular speed and performance of this great ship opened a new era of steam propulsion at sea. Parsons overcame several major **engineering** difficulties involving stress, vibration, and balancing and truly deserves the title of father of the modern steam turbine. Besides their use at sea, steam turbines went on to generate an overwhelming proportion of the electricity used in the twentieth century. Today, the bulk of our electricity is generated by power stations using steam turbines. The steam is produced by the burning of fossil fuels (**coal** or gas) or by the use of nuclear energy. Most agree that steam turbines are still evolving and will play a considerable role in the generation of electrical power for some time to come.

Water or hydraulic turbines are identified with **dams** and the generation of hydroelectric power. When a turbine is operated by rapidly flowing or falling water, it is called an impulse turbine. The huge hydro-electric plant at Niagara Falls that was built at the end of the nineteenth century is this type of turbine. Water conditions usually determine what type of turbine is needed, and impulse water turbines require a constant flow of water to operate efficiently. Two aspects of this water flow is critical, its **volume** and its head. Water head is the distance water must fall before it strikes the turbine's wheel. With a sufficient volume and head like Niagara, the impulse turbine can have its wheel or rotor mounted on either a vertical or a horizontal shaft. The ends of the turbine's blades act like cup-shaped buckets, and as the water is directed at them at very high speeds by jets, the blades turn. As might be expected, most hydraulic turbines are of the reaction type since they are best suited to low-head situations. Here, the turbine is underwater and is turned by both the weight and speed of its flow. Its shaft is vertical and has either spirally curved blades or ones that resemble a ship's propeller. Unlike impulse turbines which achieve **rotation** by the **acceleration** of water from the supply nozzles, reaction turbines work because of the acceleration of water in the rotor or runner. Both then transform the energy from the rushing water into mechanical energy.

Wind turbines are the least common or significant of all turbine types, and many technical texts do not even mention them. Unlike waterwheels which directly led to the hydraulic turbine, the windmill has for the most part not evolved as a significant source of modern energy. As with the noted Darrieus turbines in Scotland however, wind turbines do exist and have proved useful in areas of high, continuous winds. Wind turbine clusters generate electricity in the Tehachapi Mountains near Barstow, California, as well as in certain areas of Hawaii and New Hampshire.

The best-known use for gas turbines is for jet engines. Gas turbines utilize hot gases as their names implies, and they are the newest type of turbine engine. Their gases are produced by the burning of some type of fuel, like kerosene. Air is then drawn into the front of the

Gas turbine manufacturing in South Carolina. *Photograph by Brownie Harris. Stock Market. Reproduced by permission.*

turbine and passed through a compressor where the compressed air is mixed with fuel in a **combustion** chamber and is burned. This produces hot gases that expand and therefore rush through the turbine rotors, causing them to spin. This spinning can be used to power an electric generator or a pump, but in the case of a jet aircraft, the hot expanding gases are sent out at very high speed from the rear nozzle of the engine, producing thrust which then pushes the engine and the aircraft forward. Gas turbines attain temperatures higher than those of a steam turbine (the hotter a gas turbine is, the more efficiently it runs) and consequently cannot be built with ordinary metals.

Turbine engines are an example of an idea that could not be put into practice until technology had accomplished certain advances. Probably the most important technical advance was the widespread introduction of **steel** and its alloys that occurred during the second half of the nineteenth century. The popularity and use of certain types of turbine engines rises and falls as needs, priorities, and situations change. A good example is the

KEY TERMS

. . . .

Hydroelectric power—Electric power derived from generators that are driven by hydraulic or water turbine engines.

Impulse turbine—The force of a fastmoving fluid striking the blades that makes the rotor spin.

Kinetic energy—That part of the energy of a body that it possesses as a result of its motion.

Mechanical energy—Energy in the form of mechanical power.

Reaction turbine—The rotor turns primarily as a result of the weight or pressure of a fluid on the blades.

use of steam turbines for ship propulsion. After dominating sea travel for many years, steam turbines declined after the 1973 oil embargo because the fuel to make steam became prohibitively expensive. Diesels moved in to take their place since they required less fuel. Diesels can only use liquid fuel however, and as oil becomes scarcer in the next century, steam turbines for ships may again be the choice, since they can be driven by coal-burning boilers.

See also Alternative energy sources; Jet engine.

Resources

Books

Gunston, Bill. *The Development of Jet and Turbine Aero Engines.* 2nd ed. New York: Haynes Publishing, 1998.

IEEE Guide for the Operation and Maintenance of Turbine Generators. Institute of Electrical and Electronics Engineers, 1990.

Meriam, J.L., and L.G. Kraige. *Engineering Mechanics, Dynamics.* 5th ed. New York: John Wiley & Sons, 2002.

Leonard C. Bruno

Turbot *see* **Flatfish**

Turbulence

Turbulence is the formation of eddies in a fluid (liquid or gas). It is produced whenever a fluid (under certain conditions) is in contact with a solid and there is relative **motion** between them; for example: when **wind**

flows past a building or past a mountain; when the **ocean** flows past an **island**; when a baseball flies by; when a jet plane moves in the stratosphere; or when a river flows past a bridge pier. In all these cases, eddies form behind the obstacle (i.e., "downstream"), and eventually are carried away by the main body of fluid.

Historical overview

Turbulence has long been observed, but its scientific study began with the work of William John Macquorn Rankine (1820-1872); later, Osborne Reynolds (1842-1912) defined the number bearing his name, and Ludwig Prandtl (1875-1953) put forth the limiting-layer hypothesis.

Today, the study of turbulence, experimental and theoretical, continues; but an agreement between both approaches is still in the future.

The Reynolds number

There is a number—called Reynolds number—whose values indicate clearly whether the motion of a fluid in a certain region is turbulent or not. It is defined as:

$$R = \frac{\text{Inertial force}}{\text{frictional force}} = \frac{vd}{\eta}$$

Whether an obstacle carries any eddies, and whether these are released into the flow downstream depends upon the speed of the incoming fluid, the size of the obstacle and the internal **friction** (**viscosity**) of the fluid, according to Karen J. Heywood writing in *Physics Education*. Just like a solid body, a parcel of liquid or gas has mass, and therefore inertia. Inertial **force** is the amount of force required to stop a body that is moving along steadily with its own inertia (for example, the force to stop a charging rhino moving towards you would be greater than that needed to stop a hummingbird at the same speed). The inertial force necessary to stop a parcel of **water** that occupies a unit **volume** (1 m^3, say) is proportional to the square of its speed (v^2) divided by a length typical of the obstacle (d)—say, the diameter of a stone in a river. All fluids, as they flow, present friction between their different parts. This property is called viscosity. Liquids are more viscous than gases, and, among liquids, corn syrup is much more viscous than water. The viscous force (internal friction) working on an object of diameter d moving through a fluid at speed v is proportional to $h.d/d\,2$, where h is the viscosity **coefficient** of the fluid. Dividing the expression for the inertial force by that for the frictional force, we obtain R as given by the above equation.

Formation of eddies

If we place an obstacle, e.g., a **sphere**, at rest in the midst of a fluid stream, our physical intuition suggests

Turbulence in the water below Niagara Falls. *JLM Visuals. Reproduced by permission.*

KEY TERMS

Fluid—A piece of matter that flows; i.e., is deformed under the action of even the weakest forces.

Inertia—The tendency of an object in motion to remain in motion, and the tendency of an object at rest to remain at rest.

Viscosity—The internal friction within a fluid that makes it resist flow.

that the part of the fluid which is really in contact with the obstacle must be at rest. However, photographs taken in different laboratories appeared not to support this idea. In order to explain what was happening, Ludwig Prandtl (1875-1953) introduced the hypothesis of the "limiting layer," according to which in the immediate neighborhood of the obstacle there is a very thin layer of fluid whose **velocity** parallel to the surface of the object brows very rapidly, from zero at the surface itself to the velocity of the main body of fluid far from the object.

This limiting layer is very thin upstream, but broadens downstream, i.e., behind the obstacle.

Inside the downstream limit layer the fluid begins to move backwards and in circles, until the eddies form when the Reynolds number is $R = 5$. As the fluid velocity grows, bringing R to a value of 70, the limit layer broadens still more downstream, forming what is called a "von Karman vortex street" after Theodor von Karman (born 1881). These eddies finally leave the vicinity of the obstacle and float away with the main fluid current. If R keeps increasing to values of 1000 or 2500, eddies become more frequent and the vortex street broadens still more, finally breaking up and forming a "turbulent wake." At this stage the motion of fluid particles is chaotic and varies in **time**.

See also Fluid dynamics.

Resources

Periodicals

Heywood, Karen J. "Fluid Flows in the Environment: An Introduction." *Physics Education* 28 (1993): 43.

Simon, R.A. "La Hidrodinamica." *Charlas de Fisica* 9 (1992): 43.

Raul A. Simon

Turkeys

Turkeys are relatively large, powerful, ground-feeding, North American **birds** with colorful, featherless heads, classified in the family Phasianidae. The original range of the common turkey (*Meleagris gallopavo*) was from extreme southern Ontario to Mexico, but it now occupies a much smaller area. The second **species** in this family is the ocellated turkey (*Agriocharis ocellata*), which occurs in southern Mexico, Guatemala, and Belize.

Turkeys are economically important birds. They are widely hunted in the wild, and are intensively reared on farms. The populations of wild turkeys are now greatly reduced, and much of their natural **habitat** has been destroyed, but many millions of these birds occur in captivity.

Biology of turkeys

Wild turkeys have a rather dark plumage, with some degree of iridescence. Turkeys have an featherless head, with brightly colored naked skin which is blue and red in the common turkey, and blue and orange in the ocellated turkey.

Turkeys are sexually dimorphic. Male turkeys (called "toms") are relatively colorful and large, with a body length in the common turkey of up to 4 ft (1.2 m), and a weight of up to 20 lb (9 kg). Turkeys have powerful legs, and male birds have a sharp spur on the back of the foot that can inflict serious wounds during combat with other males or possibly when fending off a **predator**.

The beak of male common turkeys is adorned by a wattle, which is a long, red, pendulous appendage that develops from tissues over the base of the upper mandible. During **courtship** displays the wattle is extended to a droopy length that is several times that of the beak. Male common turkeys also develop a fat-rich growth on their breast prior to the breeding season. This **tissue** helps to sustain the male turkeys during this intensive period of the year, when their frequent, time-consuming, aggressive encounters with other males do not allow them to feed regularly. The constant preoccupation with displaying, mating, and fighting during the breeding period is hard on the toms, and they can be quite emaciated by the time this season has passed.

Wild turkeys mostly occur in forested and shrubby habitats, often with open glades. Turkeys forage on the

Wild turkeys (*Meleagris gallopavo*) at the Arizona Sonora Desert Museum, Arizona. *Photograph by Robert J. Huffman. Field Mark Publications. Reproduced by permission.*

ground in small groups, and spend the night roosting in trees. Turkeys are mostly herbivorous birds, eating a wide range of **plant** foods, although they also eat **insects** as they are encountered. Hard **tree** fruits such as acorns and other nuts, known collectively as "mast," is an important food that is gleaned from the forestfloor. These hard **seeds** are ground with small stones and other grit in the powerful gizzard of turkeys, so that the nutritious matter can be digested and assimilated.

Turkeys are polygamous, meaning that a male bird will mate with as many females as possible. Male turkeys court females by elaborately spreading their fan-like tail feathers, and by other visual displays, in which the wattle figures prominently. These displays are given while the tom struts proudly about, making loud "gobbling" noises. Male turkeys are extremely aggressive amongst themselves during the breeding season, and well-matched toms may fight to the death over access to females.

Female turkeys are alone responsible for building the ground nest, brooding the eight to 15 eggs, and raising the young. Turkey chicks are precocious, leaving the nest within a day of being born, and following the female about and feeding themselves. Turkeys are gregarious after the breeding season, forming flocks that forage and roost together.

Turkeys and humans

Because of their large size and mild-tasting flesh, turkeys have long been hunted by humans as food and for sport. Until recently, wild common turkeys were badly overhunted in **North America**. This caused the wild populations of turkeys to decline over large areas, a resource collapse that was especially intense during the nineteenth century. Turkey populations were also badly damaged wherever there were extensive conversions of their forest habitat into agriculture, a change that has occurred over widespread regions.

Today, common turkeys do not occupy much of their former range, and they generally occur as isolated populations in fragmented habitats. However, turkeys have been re-introduced to many areas from which they were eliminated, and also to some regions to which they were not native. These introductions, coupled with controls over hunting pressures, have allowed substantial increases in the populations of wild turkeys over much of their North American range.

It is not known when the common turkey was first domesticated, but this had already been accomplished by indigenous peoples of Mexico long before the Spanish conquest. The first turkeys viewed by Europeans were apparently those domestic birds, some of which were taken to **Europe** for display and cultivation as a novel and tasty food from the New World. The turkeys that are raised intensively today are derived from Mexican wild turkeys. Most domestic turkeys are white, although some varieties are black. Domestic turkeys have been artificially selected to have large amounts of meat, especially on the breast.

If the turkey came from the Americas, how did this bird receive its common name, which implies a Turkish origin? During the sixteenth century, when the domestic turkey was first introduced to England, the bird was thought to resemble the helmeted **guinea fowl** (*Numida meleagris*). This species had been kept domestically since the fourth century B.C., but had disappeared from Europe after the collapse of the Roman Empire. During the fifteenth and sixteenth centuries, Portuguese traders re-introduced domestic guinea fowl to Europe, using birds that had been obtained in the region of Turkey. The common name in England of these re-introduced guinea fowl was "turkey," and this name was transferred to the superficially similar domestic turkey of the Americas when it was introduced somewhat later on.

Bill Freedman

Turner syndrome

The identification of Turner syndrome

Turner syndrome (also referred to as gonadal dysgenesis) is a rare genetic disorder that only affects females. Approximately one in 2,000-5,000 females in the general population are affected with Turner syndrome. This is a large percentage considering that about 99% of pregnancies with fetuses affected with Turner syndrome spontaneously abort, usually during the first trimester of pregnancy. It is also estimated that 10% of all spontaneously aborted fetuses result from Turner syndrome.

Genetic defects

First described by Dr. Henry Turner in 1938, Turner syndrome was later found to be due to a loss of genetic material in one of the two X-chromosomes. There are several ways in which Turner syndrome arises. Females have two X-chromosomes, while males have an X-chromosome and a Y-chromosome. While a Y-chromosome contains genes required for development of testicles in males, both X-chromosomes are required for normal ovarian development in females. The majority of Turner syndrome cases are caused by a sporadic event during a specific stage of cellular division, where an X-chromosome is lost. Normally, when sex cells divide, an equal amount of genetic material is divided into each **cell**. During **fertilization** of the egg, if the sex cell void of the X-chromosome is fertilized by a sex cell that is missing the other X-chromosome, the result is only one sex **chromosome**. If the fetus survives, the baby will be born with Turner syndrome. A fertilized egg with only one Y-chromosome is incompatible with life.

Turner syndrome can also result from the loss of a single X-chromosome after fertilization, sometime during embryonic development. This particular condition is referred to as mosaicism, with clinical manifestations being proportional to the percentage of cells missing the X-chromosome. Finally, Turner syndrome can result due to a defective X-chromosome such as large deletion. The clinical consequences vary depending on the nature of the structural abnormality of the X-chromosome. In the case X-chromosome deletions, where affected individuals are fertile, there is potential for a recurrence risk in pregnancies from an affected female. Otherwise, there is typically little recurrence risk (if any) in subsequent pregnancies or in individuals with Turner syndrome since the majority of these individuals are infertile.

Clinical manifestations of Turner syndrome

There are a broad **spectrum** of clinical manifestations associated with Turner syndrome that can involve anything from major **heart** defects to minor defects in **tissue** development. Some affected individuals only manifest a few clinical features, while others have many abnormalities consistent with the disorder. The majority of individuals with Turner syndrome have short stature and loss of ovarian function. Other disorders include **learning** difficulties, skeletal abnormalities (e.g., webbed neck, low posterior hair line), lymphedema (swelling of a part of the body due to an obstruction or deficiency of the lymphatic drainage system), heart and kidney abnormalities, **infertility**, **obesity**, formation of keloids (thick scars), and thyroid gland dysfunction (hypothyroidism).

Loss of gene function and developmental consequences

Short stature is usually present in females with Turner syndrome. This is partially due to a loss of the SHOX **gene**, which encodes a protein important for long bone growth. The height in adults with Turner syndrome ranges from 143-145 cm (approximately 4 ft 8 in). Treatment using **growth hormones** during early childhood development can increase growth by a few inches in some cases. The loss of the X-chromosome genes may also be related to the intrauterine growth retardation, a gradual decline in growth **rate** during childhood, and the absence of a pubertal growth spurt. Females with Turner

syndrome have abnormal body proportions characterized by markedly shortened lower extremities.

Lost X-chromosome genes that are involved in the regulation of ovarian development and function results in a failure of individuals with Turner syndrome to enter into **puberty** at a normal age. Although 10% of females with Turner syndrome will go through puberty spontaneously, most will require hormone therapy for development of secondary sexual characteristics and menstruation. Without hormonal intervention, most teenagers that undergo partial breast development and menstruate spontaneously will eventually cease further development and menstruation. A few pregnancies have been reported and most likely occur prior to ovarian failure. The time of initiation of therapy varies with each female but usually begins no later than 15 years of age. Various estrogenic and progestational agents and schedules have been used as hormone therapy to maintain their secondary sexual development and prevent **osteoporosis** (bone degradation) later in life. Although most women with Turner syndrome do not have functional ovaries, pregnancy may be possible through *in vitro* fertilization (assisted reproductive technology).

Renal abnormalities occur in 1/3 to 1/4 of females with Turner syndrome. The most common abnormality is a horseshoe kidney. Cardiac abnormalities are also common, with the coarctation of the aorta being the most common defect. This is a condition that results from a severe constriction a major **blood** vessel in the heart, can be treated with **surgery**, and occurs in 5-10% of affected children. Turner syndrome females generally have normal intelligence, however, most may exhibit learning disabilities, especially with regard to spatial **perception**, visual-motor coordination, and **mathematics**. As a result, the nonverbal IQ in Turner syndrome tends to be lower, with a relatively normal verbal IQ. Females with Turner syndrome may also be socially immature for their age and may need support in developing independence and social relationships.

See also Gene mutation; Genetic disorders; Genetic engineering; Genetics; Skeletal system.

Resources

Books

Nussbaum, R.L., Roderick R. McInnes, and Huntington F. Willard. *Genetics in Medicine.* Philadelphia: Saunders, 2001.
Rimoin, D.L. *Emery and Rimoin's Principles and Practice of Medical Genetics.* London; New York: Churchill Livingstone, 2002.

Organizations

Turner Syndrome Society. 14450 TC Jester, Suite 260, Houston TX 77014, (800) 365-9944. <http://www.turner-syndrome-us.org/>

Bryan Cobb

Turtles

Turtles are familiar, four-legged **reptiles** whose body is enclosed within a bony shell. Turtles constitute the reptilian order Testudines. The 257 living **species** inhabit all continents except **Antarctica**, plus many islands, and there are marine turtles in all tropical and temperate oceans.

History and fossil record

Turtles first appear in the fossil record of the Triassic period, from about 215 million years ago. This gives them an older fossil history than any other living kind of four-legged **animal**. Turtles were already present when the first dinosaurs appeared, and they shared the ancient seas with ichthyosaurs, watched pterosaurs soar overhead, and saw the first small, furry **mammals**. Undergo-

ing relatively little change themselves, turtles witnessed the **evolution** of **birds** from feathered dinosaurs, and they were present as some of the early mammals evolved into elephants, whales, **bats**, and even human beings.

The earliest known fossil turtles (*Proganochelys*) are from late Triassic (Norian) sedimentary deposits in Germany. These ancestral turtles had a 3 ft (1m) shell length, and were terrestrial or marsh-dwelling animals. Like modern turtles, the shell of the fossil animals was composed of a rounded upper half (or carapace) and a flattened lower one (plastron). The carapace of the earliest fossil turtles incorporated 10 vertebrae, their associated ribs, and additional bone between the ribs. The skull was solidly constructed, without temporal openings, and the jaws were toothless and presumably beaked as in modern turtles. There were, however, small teeth in the palate, which are not found among living turtles. The eight cervi-

The endangered Kemp's Ridley sea turtle (*Lepidochelys kempii*). *Photograph by Tom McHugh. The National Audubon Society Collection/Photo Researchers, Inc. Reproduced by permission.*

cal (neck) vertebrae were primitive, in that they were not modified to allow the head to withdraw into the shell.

Turtles are fairly large animals, and have a great deal of bone. In addition, they often occur in aquatic or marshy habitats where their bones are likely to be buried and preserved. Consequently, their fossils are found relatively frequently in ancient sedimentary deposits from the Jurassic and younger eras. Many fossils are found in Cretaceous deposits of **North America**.

Morphology

Turtles have an unmistakable appearance, with a head, tail, and four legs projecting from a broad bony shell. The domed upper carapace and flattened lower plastron serve to protect the torso and its organs. In most land-dwelling and amphibious turtles the head, limbs, and tail can be withdrawn inside this shell if danger threatens. A horny beak (like that of a bird) covers the jaws. The head, legs, and tail are covered by horny scales, and the feet have horny nails. The bony shell is covered with an epidermal layer of scutes in most turtles, occurring in a regular pattern that may be diagnostic of the species. Only the leatherback sea turtle and the soft-shell **freshwater** turtles lack these horny plates.

The internal **anatomy** of turtles is rather typical of vertebrate animals, with two lungs, a reptilian three-chambered **heart** and associated **circulatory system**, and an unremarkable **digestive system** with an esophagus, stomach, small and large intestines, and an associated liver. As with other reptiles, the digestive, urinary, and reproductive systems vent to the outside through a chamber known as the cloaca, which also encloses the penis of male animals. The **nervous system** comprises a well-developed **brain**, a spinal cord enclosed in a vertebral column, and peripheral nerves. The senses of **vision** (including **color** vision) and **hearing** are acute.

Species of turtles have an extreme size range. Adult American mud turtles (*Kinosternon subrubum*) are less than 5.0 in (12.5 cm) long, while the gigantic leatherback sea turtle (*Dermochelys coriacea*) can attain a length of more than 6 ft (183 cm) and weight up to 1,500 lb (680 kg).

Ecology

Most modern species of turtles are semi-aquatic, living in such habitats as ponds, swamps, and marshes. Several species are marine. In fact, turtles have diversified into species that are specialized in various ways. The sea turtles, for example, are ocean-dwelling animals that fly through the **water** using their paddle-like forelimbs, emerging on land only to lay their eggs. Others turtles,

such as the soft shells, are river- and lake-dwellers, and are flattened like a pancake to hide on sandy or muddy bottom **habitat**. Still others, such as the tortoises, are strictly land-dwellers, with a high domed shell, elephant-like feet, and ranging into **grasslands** and semi-desert habitats.

Many turtles are omnivores, eating both plants and animals, but others are more specialized in their food habits. The giant tortoises and some of the sea turtles are vegetarian as adults, although their young may eat **invertebrates** or small **vertebrates**. Some river turtles, such as the map turtles (*Graptemys* spp.), are specialists that feed only on **snails** and clams. Softshell turtles (*Trionyx* spp.) are mainly fish-eaters, while snapping turtles (*Chelydra serpentina*) will eat any animal they can subdue.

Behavior and life history

Most species of turtles mate in the spring or early summer, when the males actively search for receptive females. **Courtship** may include interesting **behavior**, such as that of the male red-eared turtle (*Trachemys scripta*), which swims backward in front of the female, while stroking her head and neck with his greatly elongated front claws. The males of some tortoises make noises during courtship or mating. Smaller species, such as the Mediterranean tortoise (*Testudo*), cluck like chickens, whereas the giant Galápagos tortoise (*Geochelone*) bellows.

All turtles lay eggs, which vary in shape from cylindrical to spherical. Smaller species may lay only two or three eggs in a clutch, while sea turtles may lay three or more clutches of 100-150 eggs in a year. Most turtles dig a nesting cavity with their hind feet, lay their eggs inside, and cover the entrance, leaving the eggs to be incubated by the **heat** of the **sun**. Typically, the eggs hatch in 60-90 days.

Most turtles are long-lived. The eastern box turtle (*Terrapene carolina*) may live 100 years, and giant tortoises have been reported to live for more than 150 years. Studies of the vigor of populations of turtles must take the age structure into account. For example, although populations of giant tortoises on certain Galápagos Islands have numerous large individuals, they may nevertheless be endangered if no young are being produced because of excessive predation by introduced mammals.

Classification

The turtles are separated into two major groups (subclasses) that can be readily identified by the way they retract their head into their shell.

Side-neck turtles

The "side-neck" turtles (Pleurodira) fold their neck into a lateral S-shape, so when the head is retracted one side is tucked between the shells. This is a relatively small group of mainly pond-dwelling animals found in **South America**, **Africa** (including Madagascar), and **Australia**. Both of the constituent families (Pelomedusidae and Chelidae) occur in South America. The pelomedusids are also found in Africa, Madagascar, and on some islands of the Indian Ocean, while the chelids occur otherwise in Australia and New Guinea.

One of the best-known side-necks is the bizarre-looking matamata (*Chelus fimbriatus*) of South America, which lies in wait on the bottom of ponds or **rivers** and until a **fish** comes near, then suddenly opens its mouth and expands its throat to "suck in" its **prey**.

Hidden-neck turtles

The "hidden-neck" turtles (Cryptodira) retract their head with the neck in a vertical S-shape, appearing to pull the head directly into the shell with nothing showing but the snout. This is a larger and much more diversified group, and includes the pond turtles, the land-dwelling tortoises, and the large sea turtles. Members of this group are found throughout the temperate and tropical regions of the world.

Three of the more primitive families are found mainly in North America, with some ranging also into Central and South America. The snapping turtles comprise two genera, each with a single species. The common snapper (*Chelydra serpentina*) is best known, and ranges from Canada to Ecuador. It is a large turtle, with a shell length up to 18 in (47 cm), and a long tail. The even larger alligator snapper (*Macrochelys temminckii*) can exceed 24 in (66 cm) in shell length. It feeds mainly on fish that it attracts with a worm-like "bait" on its tongue. Both species lay 20-80 spherical eggs in a flask-shaped hole dug into a sandy bank.

The mud and musk turtles include about 25 small- to medium-sized species found from Canada to Brazil. Most of these are carnivorous, feeding on **insects**, worms, and other small animals. They lay only a few (two to 10) elongate, brittle-shelled eggs.

The highly aquatic Central American river turtle (*Dermatemys mawii*) is a large species, with a shell length of up to 25 in (65 cm). It is seldom found more than a few feet from water, is herbivorous, and highly prized as food by people living in its range.

The pond turtles (Emydidae) and tortoises (Testudinidae) comprise the largest numbers of turtle species, and occur throughout the tropical and temperate regions of the world, other than Australasia. The American pond turtles are closely related to the pond turtles of southern **Asia**, as are the European turtles.

The pelagic sea-turtles comprise only a few genera. They are extremely large, conspicuous animals, and once occurred in huge numbers. Sea turtles spend almost all of their lives in the open sea, but must come to land to lay their eggs. They nest on certain tropical and subtropical beaches, and return year-after-year to these same places. Once the nesting beaches were in remote locations, but no longer; Miami Beach, for example, was once an important nesting place.

The largest of the sea-turtles (and the biggest living turtle) is the giant leatherback. It lacks the horny plates that cover the shells of most turtles, and instead has a smooth leathery covering. It largely feeds on **jellyfish** and is the fastest swimmer of all turtles. It has a thick fatty layer under the skin that helps to retain body heat, as well as a heat-exchanging circulatory mechanism that conserves heat generated by muscular effort. These allow this species to range into cool waters during the northern summer, when jellyfish are abundant there.

Turtles and humans

Turtles as food

Large tortoises have long been used by humans as a source of meat. Some species have become extinct because of over-hunting for this purpose. The only surviving giant tortoises live on islands that were relatively recently discovered by people, such as Aldabra in the Indian Ocean, and the Galápagos Islands in the Pacific. The giant tortoises of the Galápagos Islands were hunted by whalers because they could be kept alive for months in the hold of a ship, providing a source of fresh meat during the long whaling season. Female tortoises were preferred for this purpose, because the mature males were too heavy to carry. As a result, some islands were left only with large male tortoises. Predation of young tortoises by **goats** and **rats** introduced from the ships contributed to the population decline, and most of the giant tortoises are now endangered.

According to accounts of sailors of the sixteenth and seventeenth centuries, sea turtles used to occur in great flotillas in regions such as the Caribbean. However, all species of sea turtles, but especially the green turtle (*Chelonia mydas*), were (and are) hunted for their meat. In addition, the hawksbill (*Eretmochelys imbricata*) was killed for its beautiful "tortoise shell," which can be made into combs and ornaments (these are now illegal in the United States). The sea turtles are most vulnerable on their nesting beaches, where people and other predators may easily take the eggs and the female turtles. In spite of **conservation** measures initiated by many countries, sea turtle populations are continuing to decline throughout their range.

The American salt-water terrapin (*Malaclemmys terrapin*) has also been eaten in large numbers, and declined precipitously in abundance. Fortunately, the initiation of conservation measures resulted in the survivors increasing to a greater abundance today.

Even the snapping turtle, one of the most common turtles in North America, has been over-exploited as a source of food. Turtle soup from these animals has been especially popular in Philadelphia, and for several decades a major soup company used thousands of turtles per year to supply the commercial demand, and others were used by restaurant chefs.

Captive turtles

Zoological parks have helped to save some turtle species from almost certain **extinction**. Some of these are large or brightly colored tortoises that provide good public exhibits. There is also a huge trade in turtles as pets. Turtle farmers in Louisiana and elsewhere have provided millions of baby red-eared sliders (*Trachemys scripta*) to the pet trade. However, some turtle farms became contaminated with **disease** microorganisms such as *Salmonella*, and the sale of baby turtles was made illegal in many states.

Some rare species of turtles and tortoises are being bred and reared commercially, and can bring prices of more than $2,000 each. Europeans and Americans are major purchasers of captive-bred turtles. However, wild-caught animals are also being illegally sold, and this is an extremely serious risk to the survival of rare species.

The future of turtles

In addition to the hazards mentioned above, almost all species of turtles are suffering serious losses of their habitat because of the actions of humans. For example, the sand-hill habitat of the endangered Florida gopher tortoise (*Gopherus polyphemus*) is prime land for development into residential areas and shopping malls. Similarly, the semi-desert habitat of the **desert** tortoise (*Gopherus agassizii*) in California and Arizona is being damaged by motorcycles and off-road vehicles. And all over the world, **wetlands** are being dredged or drained for various reasons, so that valuable habitat for turtles and other **wildlife** is being destroyed. The **automobile** is another important killer of turtles. In the United States, many thousands are run over each year while trying to cross highways. There is no question that most species of turtles have been severely depleted in abundance, and are continuing to decline.

Resources

Books

Ernst, C. H., and R. W. Barbour. *Turtles of the United States.* Lexington: University Press of Kentucky, 1972.

Ernst, C. H., and R. W. Barbour. *Turtles of the World.* Washington D.C.: Smithsonian Institution Press, 1989.

Minton, S. A., Jr. and M. R. Minton. *Giant Reptiles.* New York: Scribner's Sons, 1973.

Obst, F. J. Turtles, *Tortoises and Terrapins.* New York: St. Martin's Press, 1988.

Herndon G. Dowling

Tyndall effect *see* **Colloid**

Typhoid fever

Typhoid fever is a severe **infection** causing a sustained high fever, and caused by the **bacteria** *Salmonella typhi*. *S. typhi* is in the same tribe of bacteria as the type spread by chicken and eggs, commonly known as "*Salmonella* poisoning," or **food poisoning**. *S. typhi* bacteria, however, do not multiply directly in food, as do the *Salmonella* responsible for food poisoning, nor does it have vomiting and diarrhea as the most prominent symptoms. Instead, persistently high fever is the hallmark of infection with *Salmonella typhi*.

How *Salmonella typhi* is spread

S. typhi bacteria are passed into the stool and urine of infected patients, and may continue to be present in the stool of asymptomatic carriers (individuals who have recovered from the symptoms of the **disease**, but continue to carry the bacteria). This carrier state occurs in about 3% of all individuals recovered from typhoid fever.

The disease is passed between humans, then, through poor hygiene, such as deficient hand washing after toileting. Individuals who are carriers of the disease and who handle food can be the source of **epidemic** spread of typhoid. One such individual is the source for the expression "Typhoid Mary," a name given to someone with whom others wish to avoid all contact. The real "Typhoid Mary" was a cook named Mary Mallon (1855-1938) who lived in New York City around 1900. She was a carrier of typhoid and was the cause of at least 53 outbreaks of typhoid fever.

Typhoid fever is a particularly difficult problem in parts of the world with less-than-adequate sanitation practices. In the United States, many patients who become afflicted with typhoid fever have recently returned from travel to another country, where typhoid is much more prevalent, such as Mexico, Peru, Chile, India, and Pakistan.

Progression and symptomatology

To cause disease, the *S. typhi* bacteria must be ingested. This often occurs when a carrier does not wash hands sufficiently well after defecation, and then serves food to others. In countries where open sewage is accessible to **flies**, the **insects** land on the sewage, pick up the bacteria, and then land on food to be eaten by humans.

Ingested bacteria head down the gastrointestinal tract, where they are taken in by cells called mononuclear phagocytes. These phagocytes usually serve to engulf and kill invading bacteria and viruses. However, in the case of *S. typhi*, the bacteria survive ingestion by the phagocytes, and multiply within these cells. This period of time, during which the bacteria are multiplying within the phagocytes, is the 10-14 day incubation period. When huge numbers of bacteria fill an individual phagocyte, the bacteria are discharged out of the **cell** and into the bloodstream, where their presence begins to cause symptoms.

The presence of increasingly large numbers of bacteria in the bloodstream (called bacteremia) is responsible for an increasingly high fever, which lasts throughout the four to eight weeks of the disease, in untreated individuals. Other symptoms include constipation (initially), extreme fatigue, headache, a rash across the abdomen known as "rose spots," and joint **pain**.

The bacteria move from the bloodstream into certain tissues of the body, including the gallbladder and lymph **tissue** of the intestine (called Peyer's patches). The tissue's inflammatory response to this invasion causes symptoms ranging from **inflammation** of the gallbladder (cholecystitis) to intestinal bleeding and actual perforation of the intestine. Perforation of the intestine refers to an actual hole occurring in the wall of the intestine, with leakage of intestinal contents into the abdominal cavity. This causes severe irritation and inflammation of the lining of the abdominal cavity, called peritonitis, which is frequently a fatal outcome of typhoid fever.

Other complications of typhoid fever include liver and spleen enlargement (sometimes so extreme that the spleen ruptures), **anemia** (low red **blood** cell count due to blood loss from the intestinal bleeding), joint infections (especially frequent in patients with **sickle cell anemia** and **immune system** disorders), **pneumonia** (due to a superimposed infection, usually by *Streptococcus pneumoniae*), **heart** infections, **meningitis**, and infections of the **brain** (causing confusion and even **coma**). Untreated typhoid fever may take several months to resolve fully.

Diagnosis

Samples of a patient's stool, urine, blood, and bone marrow can all be used to culture (grow) the *S. typhi* bacteria in a laboratory for identification under a **microscope**. These types of cultures are the most accurate methods of **diagnosis**.

Treatment

Chloramphenicol is the most effective drug treatment for *S. typhi*, and symptoms begin to improve slightly after only 24-48 hours of receiving the medication. Another drug, ceftriaxone, has been used recently, and is also extremely effective, lowering fever fairly quickly.

Carriers of *S. typhi* must be treated even when asymptomatic, as they are responsible for the majority of new cases of typhoid fever. Eliminating the carrier state is actually a fairly difficult task, and requires treatment with one or even two different medications for four to six weeks. In the case of a carrier with gall stones, **surgery** may need to be performed to remove the gall bladder, because the *S. typhi* bacteria are often housed in the gall bladder, where they may survive despite antibiotic treatment.

Prevention

Hygienic sewage disposal systems in a community, as well as hygienic personal practices, are the most important factors in preventing typhoid fever. For travelers who expect to go to countries where *S. typhi* is a known public health problem, immunizations are available. Some of these immunizations provide only short term protection (for a few months), while others may be protective for several years. Immunizations which provide a longer period of protection, with fewer side effects from the **vaccine** itself, are being developed.

Resources

Books

Berkow, Robert, and Andrew J. Fletcher. *The Merck Manual of Diagnosis and Therapy.* Rahway, NJ: Merck Research Laboratories, 1992.

KEY TERMS

Asymptomatic—A state in which an individual experiences no symptoms of a disease.

Bacteremia—Bacteria in the blood.

Carrier—An individual who has a particular bacteria present within his/her body, and can pass this bacteria on to others, but who displays no symptoms of infection.

Epidemic—A situation in which a particular infection is experienced by a very large percentage of the people in a given community within a given time frame.

Mononuclear phagocytes—A type of cell of the human immune system which is responsible for ingesting bacteria, viruses, and other foreign matter, thus removing potentially harmful substances from the bloodstream.

Cormican, M.G., and M.A. Pfaller. "Molecular Pathology of Infectious Diseases." In *Clinical Diagnosis and Management by Laboratory Methods* 20th ed. Philadelphia: W. B. Saunders, 2001.

Isselbacher, Kurt J., et al. *Harrison's Principles of Internal Medicine.* New York: McGraw Hill, 1994.

Kobayashi, G., Patrick R. Murray, Ken Rosenthal, and Michael Pfaller. *Medical Microbiology.* St. Louis: Mosby, 2003.

Rosalyn Carson-DeWitt

Typhoon *see* **Tropical cyclone**

Typhus

Typhus is a **disease** caused by a group of **bacteria** called *Rickettsia*. Three forms of typhus are recognized: **epidemic** typhus, a serious disease that is fatal if not treated promptly; rat-flea or **endemic** typhus, a milder form of the disease; and scrub typhus, another fatal form. The *Rickettsia* species of bacteria that cause all three forms of typhus are transmitted by **insects**. The bacteria that cause epidemic typhus, for instance, are transmitted by the human body louse; the bacteria that cause endemic typhus are transmitted by the Oriental rat flea; and bacteria causing scrub typhus are transmitted by chiggers.

Characteristics of typhus

Typhus takes its name from the Greek word "typhos," meaning smoke, which accurately describes the mental state of infected persons. Typhus is marked by a severe stupor and delirium, as well as headache, chills, and fever. A rash appears within four to seven days after the onset of the disease. The rash starts on the trunk and spreads to the extremities. In milder forms of typhus, such as endemic typhus, the disease symptoms are not severe. In epidemic and scrub typhus, however, the symptoms are extreme, and death can result from complications such as **stroke**, renal failure, and circulatory disturbances. Fatality can be avoided in these forms of typhus with the prompt administration of **antibiotics**.

Epidemic typhus

Epidemic typhus is a disease that has played an important role in history. Because typhus is transmitted by the human body louse, epidemics of this disease break out when humans are in close contact with each other under conditions in which the same clothing is worn for long periods of time. Cold climates also favor typhus epidemics, as people will be more likely to wear heavy clothing in colder conditions. Typhus seems to be a disease of war, poverty, and famine. In fact, according to one researcher, Napoleon's retreat from Moscow in the early nineteenth century was caused by a louse. During World War I, more than three million Russians died of typhus, and during the Vietnam war, sporadic epidemics killed many American soldiers.

Epidemic typhus is caused by *Rickettsia prowazekii*. Humans play a role in the life cycle of the bacteria. **Lice** become infected with the bacteria by biting an infected human; these infected lice then bite other humans. A distinguishing feature of typhus disease transmission is that the louse bite itself does not transmit the bacteria. The feces of the lice are infected with bacteria; when a person scratches a louse bite, the lice feces that have been deposited on the skin are introduced into the bloodstream.

If not treated promptly, typhus is fatal. Interestingly, a person who has had epidemic typhus can experience a relapse of the disease years after they have been cured of their **infection**. Called Brill-Zinsser disease, after the researchers who discovered it, the relapse is usually a milder form of typhus, which is treated with antibiotics. However, a person with Brill-Zinsser disease can infect lice, which can in turn infect other humans. Controlling Brill-Zinsser relapses is important in stopping epidemics of typhus before they start, especially in areas where lice infestation is prominent.

Endemic typhus

Endemic typhus is caused by *R. typhi*. These bacteria are transmitted by the Oriental rat flea, an insect that lives on small **rodents**. Endemic typhus (sometimes called murine typhus or rat-flea typhus) is found worldwide. The symptoms of endemic typhus are mild compared to those of epidemic typhus. In fact, many people do not seek treatment for their symptoms, as the rash that accompanies the disease may be short-lived. Deaths from endemic typhus have been documented, however; these deaths usually occur in the elderly and in people who are already sick with other diseases.

Scrub typhus

Scrub typhus is caused by *R. tsutsugamushi*, which is transmitted by chiggers. The term "scrub typhus" comes from the observation that the disease is found in habitats with scrub vegetation, but the name is somewhat of a misnomer. Scrub typhus is found in beach areas, savannas, tropical rainforests, and deserts—in short, anywhere chiggers live. Scientists studying scrub typhus label a **habitat** that contains all the elements that might prompt an outbreak of the disease a "scrub typhus island." A scrub typhus **island** contains chiggers, **rats**, vegetation that will sustain the chiggers, and, of course, a reservoir of *R. tsutsugamushi*. Scrub typhus islands are common in the geographic area that includes **Australia**, Japan, Korea, India, and Vietnam.

The rash that occurs in scrub typhus sometimes includes a lesion called an eschar. An eschar is a sore that develops around the chigger bite. Scrub typhus symptoms of fever, rash, and chills may evolve into stupor, **pneumonia**, and circulatory failure if antibiotic treatment is not administered. Scrub typhus, like epidemic typhus, is fatal if not treated.

Prevention of typhus

Prevention of typhus outbreaks takes a two-pronged approach. Eliminating the carriers and reservoirs of *Rickettsia* is an important step in prevention. Spraying with **insecticides**, rodent control measures, and treating **soil** with insect-repellent chemicals have all been used successfully to prevent typhus outbreaks. In scrub typhus islands, cutting down vegetation has been shown to lessen the incidence of scrub typhus. The second preventative prong is protecting the body from insect bites. Wearing heavy clothing when venturing into potentially insect-laden areas is one way to protect against insect bites; applying insect repellent to the skin is another. Proper personal hygiene, such as frequent bathing and changing of clothes, will eliminate human body lice and

---

KEY TERMS

Brill-Zinsser disease—A relapse of typhus symptoms experienced by persons who have had epidemic typhus. Symptoms are usually milder than those experienced with the first bout of typhus.

Endemic typhus—A relatively mild form of typhus that is transmitted by the Oriental rat flea.

Epidemic typhus—A form of typhus that is transmitted by the human body louse and can be fatal if not promptly treated.

Scrub typhus—A form of typhus that is transmitted by chiggers and can be fatal if not promptly treated.

---

thus prevent epidemic typhus. A typhus **vaccine** is also available; however, this vaccine only lessens the severity and shortens the course of the disease, and does not protect against infection.

Resources

Books

Cormican, M. G., and M. A. Pfaller. "Molecular Pathology of Infectious Diseases." In *Clinical Diagnosis and Management by Laboratory Methods*. 20th ed. Philadelphia: W. B. Saunders, 2001.

Hardy, Anne. *The Epidemic Streets: Infectious Diseases and the Rise of Preventive Medicine, 1956-1900*. New York: Oxford University Press, 1993.

Joklik, Wolfgang, et al. *Zinsser Microbiology*. 20th ed. Norwalk, CT: Appleton and Lange, 1992.

Periodicals

Dumler, Stephen J., et al. "Clinical and Laboratory Features of Murine Typhus in South Texas, 1980 through 1987." *The Journal of the American Medical Association* 266 (September 11, 1991): 1365-70.

Green, Cornelia R., and Ira Gleiberman. "Brill-Zinsser: Still With Us." *The Journal of the American Medical Association* 264 (October 10, 1990): 1811-12.

Kathleen Scogna

Tyrannosaurus rex

Tyrannosaurus rex or *T. rex*, is easily the most famous of the Tryannosaurids (tyrant reptile). Despite its popularity, *T. rex* appears to have had a limited range in **North America** and **Asia**, and existed for a relatively short period of **time**. *T. rex* appeared during the late Cretaceous Period, about 85 million to 65 million years ago. This was toward the end of the Mesozoic Era or the Age of **Reptiles**, and just prior to the dinosaurs' **extinction**. Many people grow up believing that *T. rex* was the "King of the dinosaurs," as implied by the addition of *rex* after its name. However, it was neither the largest **dinosaur** of its time nor the most intelligent, and at least one paleontologist argues that although it was a teropod or meat-eating dinosaur, *T. rex* was not even a **predator**.

But there is no doubt that it was big. Based on fossils founds, *T. rex* weighed 5-6 tons, stood about 15 ft (4 m) tall, was 20-46 ft (6.5-15 m) long, and had 6 in (15 cm) long, sharp, serrated teeth. It is currently believed that *T. rex* was the largest terrestrial **carnivore** of all time.

Its skull was one unified whole, solidly constructed, with no moving parts except at the joint of the jaw. The compartments in the skull and in the lower jaw that housed the muscles were enlarged more than in any other predator. Its snout was sharply pinched to clear its **field** of vision. Its eyes faced forward to provide some overlap between visual fields from the right and left eyes, permitting stereoscopic vision.

This lizard-hipped dinosaur walked upright on two powerful hind legs, which ended in birdlike feet with three forward-pointing toes with large claws. These were its weapons. **Evolution** shortened its torso for balance and speed. Some say the beast was surprisingly slender-limbed, graceful, and fast, able to attack other slower plant-eating dinosaurs such as the *Triceratops*. Although it possessed two small and muscular forelimbs, many paleontologists believe they were of little practical use.

At least one paleontologist, Jack Horner, of the Museum of the Rockies (Bozeman, MT), recently raised the question as to whether or not *T. rex* was an effective hunter, given its small eyes, small arms, and relatively slow gait. Horner theorizes that *T. rex* scavenged its food from other animals' kills instead of killing its own. He supports this theory citing the beast's large olfactory lobes, which would be able to smell dead animals from a great distance, and its powerful legs with the thigh about the size of the calf as in humans, which were built for walking long stretches. As with any theory, there are many arguments refuting this hypothesis. Dr. Kenneth Carpenter from the Denver Museum of Natural History found a healed *T. rex* tooth mark on the tail of a hadrosaur (a duck-billed dinosaur), proving to many that *T. rex* was an active predator. Other experts say that the small eyes do not necessarily imply poor vision, the forearms were not needed for predation, and its gait, in spite of its bulk, was far from slow.

A coprolite (fossilized feces) believed to be from a *T. rex* was found in Saskatchewan, Canada by a team led by Karen Chin. This 65 million-year-old specimen con-

tained chunks of bones believed to be from the head frill of a Triceratops, an herbivorous (plant-eating) dinosaur eaten by the *T. rex*. The coprolite is a whitish-green rock that is 17 in (44 cm) long, 6 in (15 cm) high and 5 in (13 cm) wide. This fossil provides evidence that *T. rex* crushed bones before swallowing them, since the bones in this coprolite were broken up.

Resources

Books

Bakker, Robert T. *The Dinosaur Heresies*. William Morrow and Company, Inc. 1986.

Lambert, David, the Diagram Group. *The Field Guide to Prehistoric Life*. Facts On File Publications. 1985.

Norman, David. *The Illustrated Encyclopedia of Dinosaurs*. Crescent Books, 1985.

Palmer, Douglas. *The Marshall Illustrated Encyclopedia of Dinosaurs & Prehistoric Animals: A Comprehensive Color Guide to over 500 Species*. New York: Todtri, 2002.

Tweedie, Michael. *The World of Dinosaurs*. William Morrow and Company, Inc., 1977.

Periodicals

Chin, K., G.M. Erickson, et al. "A King-sized Theropod Coprolite." *Nature* 393 (June 18): 680.

Laurie Toupin

Tyrant flycatchers

The tyrant flycatchers are a large family of perching **birds**, containing 367 **species**, and making up the family Tyrannidae in the order Passeriformes. Tyrant flycatchers only breed in the Americas, from the northern boreal forest of Canada, through the rest of **North America**, Central America, and to **South America** as far south as Patagonia. Rarely, individual tyrant flycatchers may occur in coastal **Europe**, but they would have been blown there by a windstorm.

Species of tyrant flycatchers occur in a great diversity of habitats, ranging from sparsely treed prairies to savannas, and **forests** and woodlands of all types. Species that occur in strongly seasonal, temperate climates are migratory, spending their non-breeding season in the tropics or subtropics.

The species in the Tyrannidae are a highly variable group. The range of body lengths is from 3-9 in (8-23 cm)—not including the long tail of some species. Their wings are relatively long and pointed in species that pursue their **prey** in the air, or short and rounded in species that glean **arthropods** from foliage. The tail is usually square-backed or forked, but is very long in some species.

The head of tyrant flycatchers is relatively large, and the beak is stout, somewhat flattened, and has a hook at the tip of the upper mandible. However, fly-catching species have a relatively large bill, while gleaners have a small, more-pointed beak. The fly-catching species also have stiff bristles, known as rictal bristles, around the base of their mouth. The feet are small and weak, and only used for perching.

The plumage of tyrant flycatchers is typically a rather plain, olive-green, brown, or gray, with a lighter belly. Some species, however, can be quite brightly colored, and may have bright hues of yellow or red. Except for the brighter-colored species, the plumage of male and female birds does not differ.

Tyrant flycatchers typically feed by sitting upright at a well-vantaged, prominent perch, from which they can scan the local environment for flying **insects**, or for insects or spiders on the ground or on foliage. When likely prey items are observed, they are caught in the beak after a brief sally. Often the bird returns to the same perch. This foraging strategy is known, quite appropriately, as "fly-catching." Almost all of the North American flycatchers fly-catch for their living, but many species of tropical forests glean their prey from foliage and other surfaces. Some species add fruit to their diet, and some of the larger species will prey on **mice** and small lizards, which are caught on the ground.

Tyrant flycatchers are solitary, and do not form flocks. Tyrant flycatchers are strongly territorial during their breeding season, and some hyper-aggressive species, such as kingbirds, even drive other species away from the proximity of their territory. Tyrant flycatchers have distinctive calls, but the song is not very well developed in most species.

Most species of tyrant flycatchers build cup-shaped nests in trees or shrubs. The clutch size ranges from two to six eggs, with northern species having larger clutches than tropical ones. The eggs are usually incubated by the female. However, the male flycatcher assists with the care and feeding of the young, downy birds.

North American species of tyrant flycatchers

A total of 31 species of tyrant flycatchers breed regularly in the United States or Canada. All of these are migratory, spending their non-breeding season in Central and South America. Many of the species of tyrant flycatchers are remarkably similar in appearance, and they can be extremely difficult to identify to species, even for experienced birders. This problem is especially acute during the spring and autumn migrations, when the birds

A least flycatcher (*Empidonax minimus*). *Photograph by Robert J. Huffman. Field Mark Publications. Reproduced by permission.*

are not necessarily in their typical, breeding **habitat**, and are not singing.

Some of the North American species, and also species elsewhere, are virtually impossible to tell apart by **color** or morphology. However, the species occupy different sorts of habitats and niches, have different songs, do not interbreed, and are reproductively isolated. These sorts of difficult-to-distinguish species are known to evolutionary biologists as sibling species.

In North America, the best examples of sibling species are two types in the so-called Traill's flycatcher (*Empidonax traillii*) group, which was shown by detailed **field** studies to be composed of two, morphologically identical, species. The alder flycatcher (*E. alnorum*) has a relatively northern distribution, breeding from central Alaska and northern Canada south of the **tundra**, to southern British Columbia, Michigan, New York, New England, and Appalachia as far south as Maryland. This species breeds in wet alder and willow thickets, bogs, and regenerating burns and cut-overs, and its song sounds like: "*fee-bee-ow.*" The willow flycatcher (*E. traillii*) breeds further to the south through most of the continental United States, in shrubbery along grassy **lake** edges and streams, and its song sounds like: "*fitz-bew.*"

Actually, the difficulties do not end with these two sibling species. The alder and willow flycatchers and the Acadian flycatcher (*Empidonax virescens*) of the eastern United States are the same size, and they have a very similar coloration, as does the almost imperceptibly smaller least flycatcher (*E. minimus*). As in true sibling species, these four types breed in distinctively different habitats, they have different songs, and they do not interbreed. Biologically, therefore, they are different species, even if frustrated bird-watchers cannot always tell who is what during the migrations of the birds, when they do not occur in typical habitat, and do not usually sing.

KEY TERMS

. . . .

Sibling species—Pairs or groups of very closely related species that cannot be distinguished morphologically, and may even occur in the same region. However, they do not hybridize, they utilize different habitats, and may be different in other respects.

Other tyrant flycatchers are considerably easier to deal with. One of the more familiar species over much of the **continent** is the eastern kingbird (*Tyrannus tyrannus*), a slaty-backed, white breasted species of open country. The western kingbird (*T. verticalis*) has a more southwestern distribution.

The great-crested flycatcher (*Myiarchus crinitus*) occurs in southeastern North America, and is a relatively large, olive-backed bird, with a rufous tail and a yellow belly. The ash-throated flycatcher (*M. cinerascens*) occurs in the western United States.

The eastern phoebe (*Sayornis phoebe*) is a small, gray-backed species that ranges widely in eastern North America south of the tundra. This species was named after its call, which sounds like: "*fee-bee*." Say's phoebe (*S. saya*) has a more-western distribution. Another species that says its name is the eastern **wood** peewee (*Contopus virens*) of southeastern North America, whose call sounds like: "*pee-awee*."

The olive-sided flycatcher (*Nuttallornis borealis*) is another widespread species, whose call sounds like: "*whip-three-wheers*," although most birders actually learn it as: "*quick-three-beers*."

All of the above species are lively and interesting birds, but they are not strikingly colored or patterned. However, some of the more southern species of tyrant flycatchers are quite gaudy. The scissor-tailed flycatcher (*Muscivora forficata*) has a light-gray back, with a pink-orange wash on the flanks, and a spectacular, forked tail that is about two-times as long as the bird's body. The kiskadee flycatcher (*Pitangus sulphuratus*) has a bold, black-and-white pattern on its face, and a bright-yellow belly. However, the most spectacular of the North American species is the vermilion flycatcher (*Pyrocephalus rubinus*), in which the male is garbed in a dark-charcoal black and tail, but has a fiery, vermilion breast and head.

Resources

Books

Ehrlich, P., D. Dobkin, and D. Wheye. *The Birders Handbook.* New York: Simon and Schuster, 1989.

Farrand, J., ed. *The Audubon Society Master Guide to Birding.* New York: Knopf, 1983.

Forshaw, Joseph. *Encyclopedia of Birds.* New York: Academic Press, 1998.

Bill Freedman

Ulcers

An ulcer is a sore that develops in the lining of the stomach or the duodenum, the short section of small intestine that leads away from the stomach, or on the surface of the skin as a result of **infection** with **bacteria**. An ulcer in the stomach is called a gastric ulcer; an ulcer in the duodenum is called a duodenal ulcer; and an ulcer on the skin is called a decubitus ulcer.

Gastric and duodenal ulcers

Until recently, the sole cause of gastric and duodenal ulcers was thought to be overproduction of stomach acid due to prolonged **stress**, smoking, or poor eating habits. However, in 1992, researchers confirmed that many ulcers are caused by infection with bacteria capable of living in the highly acidic environment of the stomach. This revolution in ulcer research has radically changed the way ulcers are diagnosed and treated. For instance, instead of treating ulcer patients with acid-reducing drugs for months or even years, patients with the bacterially-caused ulcers take a week-long course of **antibiotics**. And in contrast to the 50–95% relapse **rate** with conventional treatment, the new antibiotic treatment has reduced the recurrence rate to 20%.

Cause of gastric and duodenal ulcers

Until recently, excess stomach acid was believed to be the cause of ulcers. Hydrochloric acid (HCL) is normally produced in the stomach to help break down food. HCL is secreted from special cells in the stomach lining, is mixed with the stomach contents, and initiates the preliminary digestion of **proteins** in the stomach. From the stomach, the partially digested food moves into the duodenum, where more digestion takes place. But before the partially broken down food moves from the stomach to duodenum, the acid must be neutralized. If it is not, the acidic food will irritate the sensitive duodenum. Sodium bicarbonate—the active ingredient in baking soda—is released from other cells in the stomach lining and neutralizes the acid in the partially digested food before it moves into the duodenum.

Sometimes, however, the acid is not neutralized effectively, and the duodenum is irritated by the acidic food. In this case, a duodenal ulcer may develop. Sometimes the lining of the stomach itself cannot tolerate the high levels of acid that are released within the stomach. In this case, a gastric ulcer lining may result. In either of these cases, the ulcers can be traced to a sensitivity to acid or to its overproduction.

But acid overproduction or sensitivity are not the only causes of ulcers. A bacterium called *Heliobacter pylori*, discovered and named in 1982, has been shown to cause ulcers by colonizing the lining of the stomach. These bacteria can survive in the stomach's highly acidic environment because they have an **enzyme** that neutralizes acid. Scientists now believe that most—60%—of all ulcers diagnosed throughout the world can be traced to the *H. pylori* bacterium.

Symptoms of gastric and duodenal ulcers

The classic symptom of an ulcer is stomach **pain**. Usually the pain is sharp or burning. Patients commonly note that the pain is more intense when the stomach is empty. Eating can sometimes relieve the pain of an ulcer because excess acid is neutralized by food being introduced into the stomach and duodenum.

If the ulcer is severe enough, it may perforate, or "punch through," the lining of the stomach. If this perforation occurs, stomach contents may leak into the body cavity, causing infection. The patient may also bleed internally, which may lead to shock. A perforated ulcer is extremely serious. **Blood** in the stool or vomiting blood are signs that require immediate medical attention.

Treatment of gastric and duodenal ulcers

Before 1992, most ulcers were treated with a regimen of diet and medication. Patients were advised to

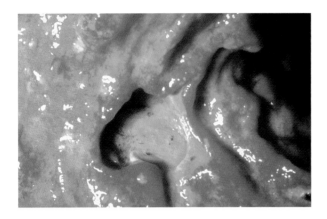

A clinical photograph of a large duodenal ulcer after surgical resection (removal). The ulcer is the prominent triangular crater at center. © Dopamine-CNRI, National Audubon Society Collection/Photo Researchers, Inc. Reproduced by permission.

take over-the-counter antacids and to control their intake of irritating foods and substances, such as **alcohol**, **caffeine**, and fried foods. In the 1980s, drug researchers developed sophisticated medications that target the stomach's acid production mechanism. The patient must take these medications for at least a month, and sometimes years, to suppress the stomach's acid secretion. These drugs only treat the symptoms of ulcers; they do not cure them. Unfortunately, most ulcers recurred despite these state-of-the-art drugs.

Since the discovery of *Heliobacter pylori*, new treatments for ulcers that target the bacteria have been implemented in ulcer patients, with good results. Tests can confirm whether or not a patient has *Heliobacter pylori*. In one of these tests, a tube with a tiny camera on the end is snaked through the patient's esophagus into the stomach and duodenum. An instrument can be passed through the tube to pinch a bit of the intestinal lining. If *Heliobacter pylori* bacteria are found in the sample, the patient is put on a course of antibiotic drugs that kill the bacteria, effectively curing the ulcer. In addition, researchers have found that bismuth subsalicylate—the active ingredient in the over the counter medication Pepto Bismol—is also effective against these bacteria. Some evidence suggests that a medication regiment combining antibiotics and bismuth subsalicylate may be the best treatment for bacterial ulcers.

Prevention of gastric and duodenal ulcers

Since the discovery of *H. pylori*, some researchers have suggested that bacterial ulcers may be prevented with a **vaccine** given early in childhood. Research has already begun into this kind of vaccine; however, it is unlikely that an *H. pylori* vaccine will be available within the next few years.

Evidence also suggests that *H. pylori* infection is highest in areas with poor sanitation facilities, suggesting that the bacteria may be transmitted—like many other human pathogens—by drinking fecally contaminated **water**. Researchers are currently working on this question, as well as studying transmission routes in the United States. Interestingly, *H. pylori* has been found in dental plaque, which may explain why the United States, despite its excellent sanitation facilities, has large numbers of people with *H. pylori* infection: it may be transmitted by kissing or other oral contact.

Despite the revolution brought about by *H. pylori*, ulcers caused by acid overproduction still represent about 40% of all diagnosed ulcers. But instead of tracing the acid overproduction to nerves, physicians are now digging deeper for the actual cause of the excess acid. Cigarette smoking has long been linked to ulcers. Smoking causes acid to be secreted into the stomach, and if the stomach does not have adequate defenses, the acid secretion, over time, can lead to an ulcer. Aspirin intake is also another culprit. Aspirin irritates the lining of the stomach and may set the stage for an ulcer.

Prevention of both kinds of ulcers is a matter of maintaining a healthy lifestyle. Good hygiene habits and the avoidance of cigarettes and excess aspirin may keep the stomach lining free of ulcers. In the future, a vaccine may entirely eliminate the cause of most ulcers, but until that time, lifestyle still plays the major role in avoiding the pain of ulcers.

Decubitus ulcers

Ulcers on the skin are caused by an infection with certain kinds of bacteria called the Enterocci and the Streptococci. These bacteria invade the skin tissues and multiply, causing the ulcer, or sore, to erupt on the skin surface. Experts believe that many people with bacteria-related ulcers acquired these infections during a stay in a hospital. When bacteria are transmitted within hospital settings, the infection is described as nosocomial. Skin ulcers caused by bacteria are treated with antibiotics.

Others skin ulcers are caused by constant **pressure** against the skin that does not allow air circulation. For instance, people who are bedridden for long periods of time frequently develop ulcers on the back, buttocks, and backs of the legs. If these ulcers are not treated promptly, they can quickly become infected with bacteria, and deep wounds can result. Pressure ulcers can be avoided in the bedridden if patients are turned periodically throughout the day so that all the surfaces of the body are exposed to air.

Still another type of skin ulcer primarily affects people with diabetes. One of the complications of diabetes is neuropathy, a condition in which nerve endings become irritated. The nerves may eventually die and the area in which the nerves are located becomes anesthetized. Diabetic patients typically experience neuropathy in the feet. If they injure their feet, the neuropathy may prevent them from feeling any pain from the injury. The injury worsens until a full-blown ulcer develops. People with diabetes are encouraged to examine their feet daily for signs of injury and to seek prompt care for any foot injury, even minor injuries.

Kathleen Scogna

Ultracentrifuge

An ultracentrifuge is a mechanical device that separates substances of different densities by spinning them very fast. It greatly reduces the **time** it would take to separate substances that would eventually separate if left alone. The first successful **centrifuge** was invented in 1883 by Swedish engineer Carl de Laval. It was used to separate cream from milk. Forty years later, another Swede, chemist Theodor Svedberg, invented the ultracentrifuge.

Centrifuges use centrifugal **force**, the force directed outward from a something spinning in a **circle**, to separate particles. You can feel the effects of centrifugal force when you swing a rope with a weight tied to one end above your head. The faster you swing the rope, the more centrifugal force you create on the weight. A washing machine is a type of centrifuge during its spin cycle. It spins **water** out of wet clothes using centrifugal forces far less powerful than those created by an ultracentrifuge.

In an ultracentrifuge, samples are placed in a container holding closed, narrow tubes like test tubes. It

spins them so fast that the centrifugal forces created can be more than one-half million times greater than the force of gravity. The tubes are suspended horizontally while they are spinning and heavier, denser particles, or those with high specific gravity, travel farther in the outstretched tubes than lighter, less dense particles or those with lower specific gravity. (Specific gravity is the **mass** of a substance divided by the mass of an equal **volume** of distilled water at 4°F [-16°C]. It is a way to compare objects based on how much mass they have packed into the space they occupy.) One common use of ultracentrifuges is to separate mixtures of different sized molecules. They are also used to separate and determine the relative sizes and densities of microscopic particles such as parts of cells. Ultracentrifuges are so powerful, for example, that they can separate two groups of molecules that differ only by having different types of **nitrogen** in their structures, nitrogen-14 versus nitrogen-15. Nitrogen-15 differs from nitrogen-14 by having one more **neutron** in its atomic nucleus.

Ultrasonics

Ultrasonics or ultrasound, derived from the Latin words "ultra," meaning beyond, and "sonic," meaning sound, is a term used to describe **sound waves** that vibrate more rapidly than the human **ear** can detect.

Sound waves travel as concentric hollow spheres. The surfaces of the spheres are compressed air molecules, and the spaces between the spheres are expansions of the air molecules through which the sound waves travel. Sound waves are thus a series of compressions and expansions in the medium surrounding them. Although we are used to thinking of sound waves as traveling through air, they may also propagate through other media.

The technical name for one expansion and one compression is a cycle. Thus, a vibration **rate** of 50 cycles per second produces 50 expansions and 50 compressions each second. The term frequency designates the number of cycles per unit of **time** that a sound wave vibrates. One cycle per second is called a hertz and is abbreviated Hz. Other useful units of scale in ultrasonics are kilohertz (kHz), which represents 1,000 Hz; and megahertz (MHz), representing 1,000,000 Hz or 1,000 kHz.

Most people can only detect frequencies of sound that fall between 16 and 16,000 Hz. Ultrasonics has come to describe sound waves with frequencies greater than 16,000 Hz, or 16 kHz. Some **insects** can produce ultrasound with frequencies as high as 40 kHz. Small an-

TABLE 1. VELOCITY OF SOUND IN VARIOUS MEDIA[a]

Material	Velocity (ft/sec)
Sea water	5023
Distilled water	4908
Chloroform	3237
Dry air at 0°C	1086
Hydrogen at 0°C	4212
Brick	11,972
Clay rock	11,414
Cork	1640
Paraffin	4264
Tallow	1279
Polystyrene	3018
Fused silica	18,893
Aluminum	16,400
Gold	6658
Silver	8790
Concrete	12,000
Stainless steel	16,400

[a] All measurements at 25°C (room temperature) unless otherwise indicated.

imals such as **cats** and dogs hear frequencies of up to 30 kHz; and **bats** are known to detect frequencies of up to 100 kHz.

A sound wave that causes compressions and expansions of the molecules in the medium surrounding it as it propagates is called a longitudinal wave. The distance from one compression to the next is known as the wave-length of the sound wave. Sound waves with long wave-lengths pass over small objects in much the same way that **ocean** waves pass over small objects. Sound waves with short wavelengths, on the other hand, tend to be diffracted or scattered by objects comparable to them in size.

The propagation velocity of a sound wave is obtained by multiplying the **frequency** of the sound wave by its wavelength. Thus, if the wavelength and frequency of the sound wave in a given medium are known, its **velocity** can also be calculated. The sound velocities in a variety of materials are shown in Table 1.

As ultrasonic waves tend to have very high frequencies, it follows that they also have very short wavelengths. As a result, ultrasonic waves can be focused in narrow, straight beams.

How ultrasonic waves are generated

In order to duplicate ultrasonic frequencies, humans have harnessed the electrical properties of materials. When a specially cut piezoelectric quartz **crystal** is compressed, the crystal becomes electrically charged and an **electric current** is generated: the greater the **pressure**, the greater the electric current. If the crystal is suddenly stretched rather than being compressed, the direction of the current will reverse itself. Alternately compressing and stretching the crystal has the effect of producing an alternating current. It follows that by applying an alternating current that matches the natural frequency of the crystal, the crystal can be made to expand and contract with the alternating current. When such a current is applied to the crystal, ultrasonic waves are produced.

Depending on which way the crystal is cut, the waves can be focused along the direction of ultrasound propagation or at right angles to the direction of propagation. Waves that travel along the direction of propagation are called longitudinal waves; as noted above, these waves travel in the direction in which molecules in the surrounding medium move back and forth. Waves that travel at right angles to the propagation direction are called transverse waves; the molecules in the surrounding medium move up and down with respect to the direction that the waves propagate. Ultrasound waves can also propagate as surface waves; in this case, molecules in the surrounding medium experience up-and-down **motion** as well as expanding and contracting motion.

In most applications, ultrasonic waves are generated by a **transducer** that includes a piezoelectric crystal that converts electrical **energy** (electric current) to mechanical energy (sound waves). These sound waves are reflected and return to the transducer as echoes and are converted back to electrical signals by the same transducer or by a separate one. Alternately, one can generate ultrasonic waves by means of magnetostriction (from *magneto,* meaning magnetic, and *strictio,* meaning drawing together.) In this case an **iron** or nickel element is magnetized to change its dimensions, thereby producing ultrasonic waves. Ultrasound may also be produced by a whistle or siren-type **generator**. In this method, gas or liquid streams are passed through a resonant cavity or reflector with the result that ultrasonic vibrations characteristic of the particular gas or liquid are produced.

Applications

The number of applications for ultrasound seems to be limited only by the human imagination. There are literally dozens of ways that people have already found to make use of ultrasound.

Coagulation

Ultrasound has been used to bind, or coagulate, solid or liquid particles that are present in dust, mist, or smoke into larger clumps. The technique is used in a process called ultrasonic scrubbing, by which particulate matter is coagulated in smokestacks before it pollutes the atmosphere. Coagulation has also been used at airports to disperse **fog** and mist.

Humidification

In ultrasonic humidification, **water** is reduced to a fine spray by means of ultrasonic vibrations. The water droplets are propelled into a chamber where they are mixed with air, and a mist of air and water leaves the humidifier and enters the room to be humidified.

Ultrasonic dispersion

Two liquids that do not ordinarily mix, i.e., oil and water, can be combined as a liquid by exposing a **solution** of the two to very high frequency sound waves. Such mixtures are called *dispersions*. With this technique, alloys of **aluminum** and **lead**, iron and lead, and aluminum and cadmium can be mixed as liquids—and kept mixed—until they solidify. This technique is known as ultrasonic dispersion. It is also used to produce stable and consistent photographic emulsions.

Milk homogenization and pasteurization

Ultrasonic waves can be used to break up **fat** globules in milk, so that the fat mixes with the milk (homogenization). In addition, pasteurization, the removal of harmful **bacteria** and **microorganisms**, is sometimes done ultrasonically.

Ultrasonic cleaning

Ultrasound is routinely used to clean, process, and degrease **metal** parts, precision machinery, and fabrics. The technique has found heavy use in the automotive, **aircraft**, and **electronics** industries, as well as for cleaning optical, dental, surgical, and other precision instruments. Fabrics can be laundered using ultrasound because the ultrasonic vibrations break down the attraction between dirt particles and fabrics, literally shaking the dirt loose. The principle by which ultrasonic cleaning is accomplished is known as *cavitation*. In cavitation, ultrasonic waves produce microscopic bubbles that collapse, sending out many tiny shock waves. These shock waves loosen the dirt and other contaminants on metals, **plastics**, or **ceramics**. The frequencies used in ultrasonic cleaning range from 15 kHz to 2 MHz.

Welding

Intense ultrasonic vibrations can be used to locally **heat** and weld two materials together. This technique works well with both plastics and metals. Thus, metal wire leads can be connected to semiconductor devices, or thermoplastic films sealed using ultrasound to locally heat, melt, and fuse the materials' surfaces. When used to bond metals to plastics, ultrasonic waves create an even flow of molten plastic at the point of contact. When the liquid plastic solidifies, cohesive bonding takes place.

Drilling

By attaching an ultrasonic impact grinder to a magnetostrictive transducer and using an abrasive liquid, holes of practically any shape can be drilled in hard, brittle materials such as tungsten carbide or precious stones. The actual cutting or drilling is done by feeding an abrasive material, frequently silicon carbide or aluminum oxide, to the cutting area.

Soldering

In ultrasonic soldering, high frequency vibrations are used to produce microscopic bubbles in molten solder. This process removes the metal oxides from the joint or surface to be soldered, and eliminates the need for flux.

Nondestructive testing

When used as flaw detectors, ultrasonic devices locate defects in materials and bounce back images of the defects, thus revealing their shapes and locations. Nondestructive testing neither damages the object being tested, nor harms the person performing the test. Metals, glasses, ceramics, liquids, plastics, and rubbers can be evaluated by this technique. Nondestructive testing of forged parts

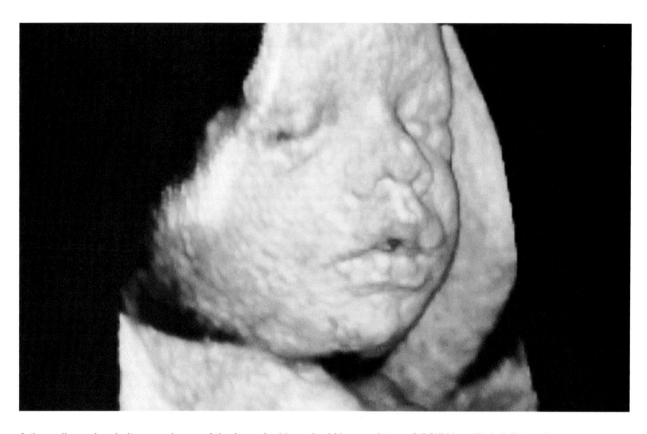

A three-dimensional ultrasound scan of the face of a 30-week-old human fetus. © BSIP/Kretz Technik/Photo Researchers. Reproduced by permission.

is now a standard manufacturing practice. The technique is used to detect **corrosion** in metal parts. It is also used to measure the thickness of many materials (with accuracy of up to 0.0001 in [0.00025 cm] for metals, and 0.001 in [0.0025 cm] for plastics), including **concrete** structures. Farmers have even used ultrasound to measure the fat layers on their cattle prior to sending them to market.

Scientific research

Ultrasound has been used to investigate the physical properties of materials, to determine the molecular weights of liquid polymers, to investigate the associated states of water, and to induce and speed up **chemical reactions**. Ultrasound has also been used to break up high **molecular weight** polymers, thereby making possible the creation of new plastic materials. Agricultural research indicates that **seeds** have been found to germinate more rapidly and to give higher yields after they have been subjected to ultrasound.

Medicine

Perhaps in no other **field** has there been such an explosion of ultrasound applications as in medicine. Ultrasound has been used in the following applications:

(1) to photograph body organs and bones. Body parts as small as 0.004 in (0.1 mm) may be imaged using ultrasound. **Heart** examinations may be performed to locate tumors, valve diseases, and accumulation of fluids. Pregnancies may be detected as early as five weeks after conception, and fetal size and development is monitored throughout pregnancy and delivery using ultrasonic imaging.

(2) to measure the rate and direction of **blood** flow using the principle that the frequency of sound changes as it travels toward an observer, but decreases as it moves away. This phenomenon, known as the **Doppler effect**, accounts for why the pitch of a train whistle, for example, becomes higher as a train first approaches, then becomes lower as it passes people standing on a station platform. Doctors can determine the direction of blood flow in the body by observing increases or decreases in the frequency of the ultrasonic measurements.

(3) to detect tumors in the body and to distinguish between malignant tumors and healthy **tissue**. Ultrasound is also employed by oncologists to destroy malignant tumors and inclusions, eliminating the need for **surgery**. **Cancer** cells are destroyed using ultrasound to produce microscopic bubbles that collapse and send out

KEY TERMS

. . . .

Cycle—One wave expansion and compression.

Hertz—A unit of measurement for frequency, abbreviated Hz. One hertz is one cycle per second.

Kilohertz (kHz)—One thousand hertz.

Megahertz (MHz)—One thousand kilohertz.

Piezoelectric—A material that becomes electrically charged when compressed, generating an electric current.

Transducer—An electronic device used to generate ultrasound.

Ultrasound—Another term for ultrasonic waves; sometimes reserved for medical applications.

Wavelength—The distance between two consecutive crests or troughs in a wave.

intense shock waves (cavitation effect). The same technique is used to destroy gallstones and kidney stones.

(4) to view living cells without damaging them. Ultrasonic microscopes can be used to image cellular structures to within 0.2 microns (two-thousandths of a millimeter). Ultrasonic methods are also used to locate foreign objects in the **eye** during surgery and in routine eye examinations, and to measure the depth of burns in burn patients. This technique affords an accuracy of 0.05-0.1 in (0.1-0.2 mm).

(5) to relieve muscle strain. Ultrasonic heat has been used to treat **arthritis**, bursitis, myelitis, neuralgia, malignancy, lumbago, rheumatism, arthritis, sciatica, sinitis, and post-operative **pain**.

(6) to clean teeth by means of ultrasonic prophylaxis units operating at 25 kHz.

Electronic eavesdropping

Conversations can be overheard without using microphones by directing ultrasonic waves at the window of the room being monitored. Sounds in the room cause the window to vibrate; the **speech** vibrations produce characteristic changes in the ultrasonic waves that are reflected back into the monitor. A transducer can be used to convert the reflected vibrations to electrical signals that can be reconstructed as audible sounds.

Detection devices

Ultrasound has been used to detect undersea naval vessels, to measure the depth of the ocean floor, and to locate schools of **fish**. When used in these ways, ultrasound is usually referred to as sonar, an acronym for SOund NAvigation and Ranging. The frequencies used in most sonar systems range from 5 to 50 kHz. Ultrasonic detectors also measure chemical fluid levels automatically in tanks and containers, as well as in the fuel tanks of aircraft. In addition, ultrasonic devices are used in burglar alarms. When an intruder trips an ultrasonic alarm, a signal can be relayed to the local police station, and the burglar apprehended.

Radio

Radio talk shows routinely use ultrasonic delay lines to monitor and cut off abusive callers before their comments are aired during radio talk shows. The ultrasonic delay line bounces the voice signal back and forth between two transducers until it has been monitored, then releases it for broadcast.

See also Acoustics; Solder and soldering iron.

Resources

Books

Knight, David C. *Silent Sound: The World of Ultrasonics.* New York: Morrow, 1980.

Randall Frost

Ultraviolet astronomy

Ultraviolet astronomy is the study of astronomical objects in the ultraviolet portion of the **electromagnetic spectrum**. Because Earth's atmosphere prevents ultraviolet **radiation** from reaching its surface, ground-based observatories cannot observe in the ultraviolet. Only with the advent of space-based telescopes has this area of **astronomy** become available for research. Ultraviolet radiation has a shorter wavelength and more **energy** than visual radiation, and much of ultraviolet astronomy therefore centers on energetic processes in stars and galaxies. Hot regions of stellar atmospheres, for example, invisible to optical telescopes, reveal a wealth of information to the ultraviolet **telescope**. The crowded, violent regions at the centers of some galaxies are also prime targets for ultraviolet telescopes.

Ultraviolet radiation

We often refer to electromagnetic radiation in terms of its wavelength, the distance from one peak of a **light** wave to the next peak. A convenient unit of wavelength

ROSAT (Roentgensatellit) satellite prior to its launch on June 1, 1990. This German/United Kingdom/NASA satellite is capable of detecting both x rays and extreme ultraviolet (EUV) light. © *Dornier Space/Science Photo Library, National Audubon Society Collection/Photo Researchers, Inc. Reproduced by permission.*

is the Angstrom Å. One Angstrom equals one 10 billionth of a meter.

Visual light, the light our eyes are sensitive to, has wavelengths from about 4,000-7,000 Angstroms. Beyond the visual is infrared light—we cannot see it, but we can feel it as **heat**. On the short wavelength side of the visual part of the **spectrum** is the ultraviolet. Ultraviolet (often just called UV) light has wavelengths from 100-4,000 Angstroms.

Earth's atmosphere is opaque to UV light, meaning UV radiation cannot penetrate it. This is fortunate for us, since UV light is what causes sunburn and in sufficiently large doses, skin **cancer**. Optical telescopes cannot see wavelengths much shorter than 3,600 Angstroms, and to observe UV radiation from astronomical objects it is therefore necessary to go above the atmosphere. Orbiting, space-based telescopes are needed, and only in the past few decades have they been available.

Ultraviolet observatories

Astronomers have developed many different kinds of telescopes besides the familiar optical instruments. **Radio**, infrared, ultraviolet, x-ray, and gamma-ray telescopes all have unique design requirements to maximize their efficiency in the part of the spectrum they are intended to study.

Like gamma-ray and x–ray telescopes, UV telescopes have only been possible in the era of spaceflight, and the longest lived and most important of these so far has been the *International Ultraviolet Explorer.* Launched in 1978, IUE was designed to observe the UV sky for five years. Instead, the telescope was not shut down until September 30, 1996, and took tens of thousands of spectra of stars, nebulae, and galaxies.

IUE was a joint project of United States and European **space** agencies, and was operated for 16 hours each day at the Goddard Space Flight Center in Greenbelt, MD, and for eight hours each day at the Villafranca Satellite Tracking Station in Spain. Astronomers around the world used IUE for their research, and it has been one of the most productive missions in the history of spaceflight.

Despite its glowing track record, IUE had some important limitations. Its primary mirror was only 17 in (45 cm) in diameter, and IUE therefore could not observe very faint objects. Also, its instrumentation was developed in the 1970s and was not as technologically advanced as that available in the 1980s and 1990s. For this reason, a new generation of UV observatories was designed and built.

On June 7, 1992, the Extreme Ultraviolet Explorer (EUVE) was launched. This **satellite** was designed to extend the spectral coverage of IUE, which only went down to 1,100 Å. The EUVE telescope observed at wavelengths as short as 70 Å, and extended the observing capability of space-based observatories throughout the UV.

On June 24, 1999, the Far Ultraviolet Spectroscopic Explorer (FUSE) was launched. This satellite is also designed to look father into the ultraviolet—i.e., to shorter wavelengths—than IUE, observing at wavelengths from 900-1,200 Å. With FUSE, astronomers will explore conditions in the Universe as they existed only shortly after the big bang, in addition to myriad studies of high-energy processes in stars and galaxies.

These observatories—IUE, EUVE, and FUSE—are what NASA calls "Explorer-class" missions. These are smaller, less ambitious and expensive projects, designed to perform a specific task. This is in contrast to the "Great Observatories" such as the **Hubble Space Telescope** (HST), which includes a UV instrument called the Goddard High Resolution Spectrograph (GHRS). The GHRS can observe the same part of the spectrum as IUE, but the 8.5 ft (2.6 m) mirror of the HST is much larger than the 16 in (45 cm) mirror of IUE, and GHRS can observe much fainter objects than IUE.

Research with UV telescopes

UV telescopes reveal a wealth of information about hot and energetic processes in astronomical objects. This

KEY TERMS

. . . .

EUVE—The Extreme Ultraviolet Explorer, launched in 1992. The EUVE telescope observes the short-wavelength end of the UV, from 70-760 Angstroms.

GHRS—The Goddard High Resolution Spectrograph. An ultraviolet instrument that is part of the Hubble Space Telescope, GHRS extends the capabilities of IUE owing to its more modern instrumentation and the HST's large mirror.

IUE—The International Ultraviolet Explorer. Launched in 1978, IUE operated for nearly two decades despite its original five-year design lifetime. IUE was one of the most successful and productive of space-based observatories, and has been used to observe nearly every kind of astronomical object, from planets and comets to stars, nebulae, and galaxies.

UV radiation—Radiation with wavelengths between 100-4,000 Angstroms. UV radiation causes sunburn and, in sufficient doses, skin cancer. Fortunately for us, Earth's atmosphere prevents most UV radiation from reaching the ground, but this also means that the ultraviolet radiation from astronomical objects can only be studied by telescopes orbiting above the atmosphere.

is because the hotter an object is, the more energy it radiates at short wavelengths. UV radiation has shorter wavelengths than visual light, so hot objects are brighter in the UV than in the visual. For example, a hot **star** like Rigel (the blue-white star that forms Orion's left foot) emits much more UV radiation than the **Sun**.

UV telescopes have greatly enhanced our understanding of the stars. It is well-known that the **temperature** rises in the outer atmospheres of stars like the Sun, but the causes of this temperature rise are poorly understood. Because the atmospheres get very hot, they emit much of their radiation in the UV, and until the launch of IUE in 1978, the nature of these hot atmospheres was largely unknown. UV telescopes have also been used to study winds from hot stars, stars that are still in the process of forming, and hot, dead stars that **orbit** other stars, drawing matter off them and heating it until it emits large amounts of UV and x–ray radiation.

Another place where hot, high-energy conditions prevail is at the center of galaxies. The so-called active galaxies have intense high-energy sources at their centers. These galaxies often have huge jets of hot, high-en-

ergy material streaming out of them. A hypothesized source of the intense energy generation is an enormous **black hole** at the galactic center. IUE and other UV telescopes have been used to study galactic centers in an effort to understand the processes occurring in the crowded, violent environments thought to prevail there.

See also Galaxy.

Resources

Books

Bacon, Dennis Henry, and Percy Seymour. *A Mechanical History of the Universe.* London: Philip Wilson Publishing, Ltd., 2003.

Kaufmann, W. *Discovering the Universe.* 2nd ed. Freeman, 1991.

Mark, Hans, Maureen Salkin, and Ahmed Yousef, eds. *Encyclopedia of Space Science & Technology.* New York: John Wiley & Sons, 2001.

Jeffrey C. Hall

Ultraviolet radiation *see* **Electromagnetic spectrum**

Unconformity

An unconformity is a widespread surface separating **rocks** above and below, which represents a gap in the rock record. Unconformities occur when either **erosion** wears away rocks, or rock deposits never form. Therefore, a **time** gap exists between when the rocks below the unconformity formed and when those above it formed.

Unconformities are classified as three types. The most easily recognized are angular unconformities, which show horizontal layers of **sedimentary rock** lying on tilted layers of sedimentary rock. The upper layers may not be perfectly horizontal, but they do not lie **parallel** to the lower layers. The second type of unconformities are disconformities, which lie between parallel layers of sedimentary rock. The third type are nonconformities, which divide sedimentary layers from metamorphic and intrusive (cooled inside the **earth**) **igneous rocks**. Common to all three, erosion causes them to form, and younger rocks sit on older rocks.

Angular unconformities

Four basic steps create angular unconformities. In step one, sediment weathered from land and carried to the sea accumulates on the sea floor and over millions of years turns to rock layers. Then the collision of plates, giant sections of the earth's crust that constantly shift, lift and tilt the layers until the layers rise above **sea level** and then **weather** and erode. They erode for millions of years until the edges of the tilted layers become a flattened **plane** (a "peneplain" is a broad land surface flattened by erosion). Finally, in step four, sea level rises or land sinks. Sediments wash down, forming new horizontal layers that cover the submerged, tilted layers. These four steps could take hundreds of millions of years to complete.

The Colorado River at the Grand Canyon exposed one of the best angular unconformities in the world. From even miles away on the rim of the canyon, observers can see tilted layers of rock truncated roughly 550 million years ago by a horizontal sedimentary layer called the Tapeats Sandstone. The inner gorge of the Grand Canyon provides a great example of angular unconformity formation. Except, instead of four steps, the rocks tell of seven: (1) Over two billion years ago, layers of sediment accumulated and turned to rock. (2) Around two billion years ago, plate collisions lifted **mountains** and turned the sedimentary rocks into the Vishnu Schist, a **metamorphic rock**. (3) A half a billion years later, the mountains eroded into a peneplain. (4) The land subsided or sea level rose to deposit new layers (known as the Grand Canyon Series) on the old Vishnu Schist. (5) New plate collisions tilted and uplifted the Grand Canyon Series. (6) Erosion truncated the tilted Series and created another peneplain. The erosional episode lasted almost a billion years. (7) The land subsided eventually and the Tapeats Sandstone accumulated on the tilted Grand Canyon Series. In some places in the canyon, the Tapeats lies not on the angled Series but directly on the metamorphic Vishnu Schist—making this a nonconformity.

Another famous angular nonconformity is Scotland's Siccar Point, a site which played a part in the development of modern **geology**. In the eighteenth century, most people believed the earth to be only 6,000 years old, a figure arrived at earlier by Bishop Ussher, a prominent theologian who added the ages of Biblical characters and thus concluded the world was created in 4004 B.C. Scottish scientist James Hutton, however, realized that thousand-year-old Roman ruins in Great Britain were barely touched by **weathering** and erosion. He therefore wondered how long it takes for whole mountains, like those in Scotland, to wear down.

The angular unconformity he discovered at Siccar Point in Scotland provided dramatic evidence for his time "expansion." He saw nearly horizontal sandstone resting on nearly vertical graywacke (a sedimentary rock similar to sandstone) and marveled at how long it took to deposit the graywacke, tilt it, erode it, and then lay sandstone across it. As his friend, John Playfair, wrote "The mind seemed to grow giddy by looking so far into the abyss of time."

Disconformity

From a mile away, or perhaps from a few hundred feet away, disconformities can hide. The layers appear regular, all parallel. However, between the layers, a disconformity can lie. Not until a geologist closely examines the fossils in the layers for the presence or absence of certain organisms can he or she recognize the gap in time—an erosional period when sediment accumulation or deposition halted or perhaps when anything laid down washed away.

As with an angular unconformity, disconformities form in steps. In step one, sediments collect on an **ocean** floor (or perhaps on the bed of a large **lake**). They compact and become rock layers. In the second phase, sea level falls or the sea floor rises to expose the layers to weathering and erosion. The main difference in the formation of disconformities and angular unconformities lies in this second step. As the layers of the future disconformity rise above sea level, they remain horizontal—no tilting occurs. If they tilt in this step, they later form an angular unconformity. Then, in step three, the land subsides or sea level rises, and new sediments collect on the older, still horizontal, layers.

Back in the Grand Canyon, where our story paused roughly 550 million years ago, the Tapeats Sandstone draped across the Grand Canyon Series and the Vishnu Schist to form the Great Unconformity. At least two more layers of sediments—the Bright Angel Shale and the Muav Limestone—collected on the Tapeats over the next 50 million years. Other layers may have been formed, too, but disappeared when the land rose and eroded for the next 80 million years. When the area again sank beneath the sea, the Temple Butte Formation, 80 million years younger than the Muav, accumulated on the disconformity. The cycle of deposition, **uplift**, erosion, **subsidence**, and more deposition repeated at least four times from 550 to 250 million years ago.

Nonconformities

Nonconformities separate sedimentary rock layers from metamorphic rock layers and from intrusive igneous rock (like granite). In a step-by-step process similar to the other two unconformities, sediment accumulates and becomes rock. Then plate collisions deform these layers and change them into metamorphic rocks. Associated with this mountain-building, molten rock often squeezes upward into the metamorphic rock frac-

KEY TERMS

KEY TERMS

. . . .

Angular unconformity—An unconformity, or gap, in the rock record, where horizontal rock layers overlie tilted layers.

Disconformity—An unconformity, or gap, in the rock record, situated between parallel rock layers.

Nonconformity—An unconformity, or gap, in the rock record, where sedimentary rocks overlie metamorphic or igneous rocks.

tured by the deformities and solidifies, forming igneous rock (usually granite). In phase three, the mountains erode to a peneplain. Then, finally, new layers collect over the flattened metamorphic and igneous rocks. As described earlier, the Tapeats Sandstone layer in the Grand Canyon forms an angular unconformity where it overlies the tilted Grand Canyon Series and a nonconformity where it rests on the Vishnu Schist.

An ongoing process

As rocks continue to wear away, more unconformities appear. As road crews cut through mountains, they expose unconformities for the speeding motorist as well as the geologist to enjoy. However, these new exposures and the mountains that contain them will erode flat. The Appalachians, the Himalayas, the Alps, the Rockies, even the Grand Canyon, will die their slow erosional deaths as nature levels the continents, which may then subside beneath the seas. However, more sediment will soon accumulate, which will uplift and erode, and so on into eternity—an unbroken cycle of geologic processes.

See also Geologic time.

Resources

Books

Baars, Donald L. *The Colorado Plateau: A Geologic History.* Albuquerque: University of New Mexico Press, 1983.

Chronic, Halka. *Pages of Stone: Geology of Western National Parks and Monuments.* Vol. 4. *Grand Canyon and the Plateau Country.* Seattle: The Mountaineers, 1988.

Chronic, Halka. *Roadside Geology of Arizona.* Missoula, MT: Mountain Press, 1984.

Dixon, Dougal, and Raymond L. Bernor, ed. *The Practical Geologist.* New York: Simon and Schuster, 1992.

Harris, Ann G., Esther Tuttle, and S. D. Tuttle. *Geology of National Parks.* 4th ed. Dubuque: Kendall/Hunt Publishing Co., 1990.

McPhee, John. *Basin and Range.* New York: Farrar, Strauss & Giroux, 1982.

Underwater exploration

Underwater exploration is the relatively recent process of investigating the depths of the sea to understand its physical and chemical characteristics and to learn about the life forms that inhabit this realm. Underwater exploration near the surface and near the shore is an ancient form of earning a livelihood and enjoying the pleasures of the **water**; but deep-sea exploration is a recent phenomenon (compared to many other sciences) because technological developments have been essential to the survival of human beings in deeper water. Alternatively, these developments have eliminated the need for humans to journey to these depths.

History

The very earliest "explorations" of the sea depended on human endurance, that is, the depth a person could sustain a dive. Our ancient ancestors certainly explored the near shore. The Polynesians dived from their sea-going outrigger canoes, but the depth they could explore was limited to relatively shallow water. The women who dive for pearls in and near Japan and the Greeks who dive for **sponges** have achieved phenomenal endurance records (presumably in ancient as well as modern times) for holding their breath, but diving for pearl-bearing oysters or for sponges requires perseverance for searching not for depth.

Scientific study of the **physics** of the deep sea began when the French mathematician, astronomer, and scientist Pierre Simon de Laplace (1749-1827) used only tidal motions along the shores of West **Africa** and Brazil to calculate the average depth of the Atlantic Ocean. He estimated this average to be 13,000 ft (3,962 m), which scientists later proved with soundings over the ocean to be relatively accurate. Investigations of the sea bottom were begun when submarines were manufactured, and soundings were used to lay **submarine** cables.

Nineteenth and twentieth century technology has caused an explosion in the exact sciences. The captains of sailing vessels made precise ships' logs in the early nineteenth century that proved valuable in early **oceanography**. These were compiled by Matthew Fontaine Maury (1806-1873), who set documentation standards later followed by many international congresses on oceanography and other sciences of the sea. The expeditions of Captain James Cook (1728-1779) and the polar explorers (notably Sir James Ross [1800-1862] who explored the North Pole with Sir William Parry [1790-1855] as well as the Antarctic Region and his uncle Sir John Ross [1777-1856] who was also an explorer of the North Pole) added more information about oceanic surfaces.

In the mid-1800s, Norwegian scientists proved life exists in the deep sea when they recovered a stalked crinoid from a depth of 10,200 ft (3,109 m). In 1870, the British began the first expedition strictly to explore the deep ocean. The H.M.S. *Challenger* expedition left England at Christmastime in 1872 and spent four years conducting oceanographic studies in the oceans of the world, returning to England in May 1876. The ship's crew was under the command of Sir George Nares, and Sir Charles Wyville Thomson (1830-1882) was the chief scientist on board. The crew is credited with discovering 715 new genera and 4,417 new **species** of marine organisms. At about the same time, the German ship the *S.M.S. Gazelle* made observations of southern waters including the South Atlantic, South Pacific, and Indian Oceans. The U.S.S. *Tuscarora* cruised the North Pacific to make soundings for the trans-Pacific cable line and recorded many other scientific observations along with the soundings.

Oceanography

Oceanography is literally the science of mapping the floor, **geometry**, and configuration of large bodies of water. The history of deep-sea exploration began with practical applications of oceanography, such as the laying of undersea cables, and was extended by natural and scientific curiosity. Aspects of the condition of the oceans studied by oceanographers include relief of the sea floor, volumes of ocean basins and numerous subareas, character of the ocean surface including atmospheric effects, transportation and properties of sediments found in marine environments (as well as their origins, such as land, volcanic, organic, and inorganic sources), the **chemistry** of sea water (including the gas content), physical properties of sea water like **density** and **pressure**, characteristics of **ice** and **icebergs**, and biological oceanography (including **plankton**, **bacteria**, and **plant nutrients** as well as more familiar plants and animals).

Based on the surface map of the world, the oceans cover 71% of the globe. In the past century, oceanography has dramatically changed our understanding of the importance of the sea to land dwellers. Not only have people become more environmentally aware, but we are more knowledgeable of the vastness of Earth's seas.

Oceanography led to the development of a number of instruments that are used to chart the bottom of the sea; some of these are also used in undersea exploration for other purposes. Sounding devices were the first key oceanographic tool. The first sounding weight, the Baillie sounding machine, was used on the *Challenger* expedition and consisted of a large weight dropped to the sea floor. When the weight hit the bottom, the line was pulled taut and the depth measurement was read from the line. The Baillie sounding machine also had a tube below the weight that drove into the sea-floor sediments. Samples could be retrieved in this fashion. Early explorations also collected samples from the seabed using dredges (that were pulled along the sea floor) and an assortment of scoops. These tools collected **soil**, rock, some plant life, and other biological specimens.

Weight-sounding techniques were replaced after World War II by echo-sounding that uses sounds or acoustic impulses from ships on the ocean surface to measure **reflections** of the **sound waves** off the bottom. The time lapse of the sound wave's return to the ship indicates the depth, although early uses of echo-sounding were often in error if the device was not properly calibrated for the density of **saltwater**.

Instrumentation

Oceanographers also use drilling and coring techniques for sampling the seabed. The gravity corer replaced the sampling device on the Baillie sounding machine with an open and weighted tube that is triggered to release as soon as sediments are encountered. It then drills into the sea floor to up to about 33 ft (10 m). When the corer is extracted and brought on board ship, the core can be extruded, and layers in the sediments are logged by a geologist specializing in ocean sediments. Some specially equipped ocean drilling rigs are able to retrieve core samples from greater depths (as much as 4,900 ft or 1,500 m), and samples from the drilling of test wells for oil and gas extraction and the foundations for offshore oil platforms are also examined by oceanographers and other specialists.

Other oceanographic instruments include flow meters for measuring the **velocity** of deep-sea **currents**, seismographs for detecting **earthquake** activity far from land-based equipment, pressure meters that measure pressure beneath the ocean with depth, and thermometers. These instruments are usually attached to sounding devices because their measurements with respect to the depth to the sea floor are important. Research vessels carry these instruments, but the instruments can also be tethered to buoys and left at sea. The research ships themselves are precise, highly equipped floating laboratories with sophisticated navigation systems including links to global positioning **satellite** (GPS) systems and positioning systems that use computers in the ship's controls to keep it in a fixed location at sea. A sonar beacon seated on the ocean floor usually provided the point of orientation for the ship's fixed position. A variety of **television**, video, and still cameras and audio detection equipment is also standard for research vessels.

Satellite technology has greatly advanced the science of oceanography. One of the techniques, known as satellite altimetry, utilizes **radar** to measure the distance from an orbiting satellite to the ocean surface. While usually considered smooth and spherical, the surface of Earth's oceans actually exhibits a multitude of broad dimples and bulges that reflect the topography of the ocean floor. The uneven surface of the ocean is due to localized gravity effects from **mountains** and depression at the bottom of the ocean. Although the relief of these prominences is greatly subdued when compared to the ocean floor, their extent is sufficient to be quantified by means of satellite altimetry which has an astounding vertical resolution of 1 in (0.03m). The altimetry data provided by the US Navy's *Geosat* and European Space Agency's *ERS-1* satellites permit the construction of topographic maps of the world's ocean basins. This is particularly important in deep, remote portions of the basins where little depth information is available.

Modern surface mapping techniques of similar resolution would require approximately 125 years and several hundred million dollars to complete.

The technique has a wide variety of applications. Navigation of ships, submarines, and even **aircraft** are frequently affected by local gravitational variations. The information provided by satellite altimetry allows the variations to be accounted for and the course corrections applied. The topographic information permits the identification of subsurface controls of ocean currents and favorable fishing locations. Geologists utilize the information to investigate various aspects of plate tectonic theory, identify and study subsurface volcanoes, locate potential **petroleum** reserves, and even measure the structural characteristics of Earth's oceanic crust.

Diving tools and techniques

Diving suits and devices to help divers stay longer underwater were invented and tested as early as the fourth century B.C. Aristotle mentions artificial breathing devices for divers, and Alexander the Great supposedly dove in a primitive version of a diving bell. The first practical diving bell was invented in 1717 by Edmund Halley, the British astronomer for whom the comet is named. It had a wooden chamber with an open bottom and **glass** in the top or ceiling for **light**. Leather tubes supplied air to the occupants, and the air was furnished through casks lowered into the water as they were needed. As water flowed into the casks, it forced the air out through the tubes, in a simple form of compressed air. **Steel** chambers similar to Halley's invention are still used today for some types of underwater construction, except that the compressed air is supplied from tanks.

Individual diving suits to protect divers and let them move freely were first tried in the seventeenth century. In 1819, the first successful diving suit was invented by Augustus Siebe (1788-1872), an inventor of German and British extraction. He used the principle of the diving bell in fitting the diver's head into a **metal** helmet that was attached to a leather jacket. Air was pumped into the helmet through a hose. The system was not watertight, but the forced-air pressure kept the water below the diver's nose and mouth. Siebe followed his invention with several improvements, the last of which was made in 1830. The modern diving suit fully encloses the diver in a suit of rubberized fabric and a helmet. The unit is airtight, and the diver can regulate both air pressure and buoyancy with valves on the helmet. Diving suits for greater depths include weighted shoes, **lead** plates for the back and chest, and a communications line linked to a **telephone** at the surface. For still greater depths, metal suits with special airtight joints help divers withstand the higher water pressures. Air pressure within these suits can be properly regulated so, in fact, the suits for greater depths impose less physical stress on the diver than those for shallower waters. Self-contained underwater breathing apparatus (SCUBA) supports both skin divers and divers with gear for deeper water and eliminates the troublesome supply hoses.

Work underwater is done with special equipment that is also pneumatically powered (powered by compressed air). Drills, wrenches, and other tools require supplied air for power although standard cutting and **welding** torches can be used underwater. Electrically powered lights are needed at depth because light only penetrates a few yards (meters) in some waters. Underwater stations for working and habitation have been tested; depending on depth, different air supplies using mixtures of **oxygen** and helium or **hydrogen** instead of **nitrogen** are needed to prevent fatal bubbles in the **blood** stream of the diver. Divers adapt to the underwater world in stations no more than 328 ft (100 m) deep, but they can work for shorter periods of time at depths of 1,300-1,500 ft (approximately 400-650 m) in flexible suits. Underwater habitats or stations are supplied air and power by stationary surface craft.

Deep-sea submersible vessels

The tool that made true exploration of the deepest waters of the seas possible is the deep-sea submersible vessel or vehicle, simply called a submersible. The submersible is a miniature submarine, but submarines are not submersibles. Submarines are fully contained quarters for human occupancy and for machines, usually with a military purpose, that can survive at depth for an extended length of time. Some nuclear-powered submarines stay submerged for months, carry food and fresh water for crews of over 100 persons, purify air for breathing, and perform specific tasks related to warfare, espionage, and research. While they also have highly sophisticated equipment, including sounding devices, pressure and **temperature** meters, and elaborate navigation and power systems, these are used for different purposes than the instruments on a research ship or submersible.

Submersibles are designed to dive to much greater depths than submarines. Because of the tremendous pressures in the deep ocean realms, they are built for strength, survival of two or three human occupants (if any), and specific research tasks. They do not carry stores of food or water, and oxygen is furnished from limited on board storage tanks or piped in from the support vessel at the surface.

Early submersibles were called bathyscaphs from the Greek roots for deep and boat, bathyspheres meaning deep-diving spheres, or diving saucers. The **bathysphere** was a steel diving chamber suspended from a host ship at the surface on a steel cable and a separate telephone cable. The bathyscaph also had a steel diving **sphere**, but it was suspended beneath a football-shaped blimp that carried gasoline to keep the craft afloat until the crew wanted to make the bathyscaph descend. For descent, the gasoline was released and replaced with seawater. Diving saucers were a specialty of the French; Jacques Yves-Cousteau designed an early diving saucer called the *Soucoupe* (the French word for saucer) that was unique in using hydrojets to maneuver in the water. Later saucer-like vessels, the *Deepstar 4000* and the *Cyana*, also made landmark explorations into the underwater world. The *Cyana* was used in 1974 in the pioneering exploration of the Mid-Atlantic Ridge and its deep rift valley.

A number of countries around the world operate submersibles through their oceanic research organizations. Manned submersibles have descended to over 20,000 ft (6,000 m) deep; one of these, the *Argo*, was used by Dr. Robert D. Ballard to locate the wreck of the H.M.S. *Titanic* in 1985. After the *Titanic's* location was discovered using a manned submersible, a smaller, unmanned robot submersible named *Jason* ventured into the wreck to photograph its interior. Most submersibles carry still cameras, television systems, and special lighting systems to provide light for **photography**. All of these are designed and built specifically for the deep-ocean environment and its severely limiting hardships. Many submersibles are also equipped with mechanical manipulators (arms and scoops) that can collect samples from the sea floor, biological specimens, and oddities such as debris from the *Titanic*.

Deep-sea pioneers

Charles William Beebe and Otis Barton

Charles William Beebe (1877-1952) was the designer of the first practical bathysphere. Beebe, part scientist and part showman, never completed his degree at Columbia University in New York: instead, he became a curator at a zoo, tracked rare species of **birds** in South and Central America, and climbed volcanoes before becoming interested in underwater exploration. In 1934, he and Otis Barton made a record-setting descent to 3,028 ft (923 m) below the waters off the Bermuda Islands. Barton was a far different character, a virtual recluse who had been born to an extraordinarily wealthy family and was interested in the ocean from his youth. Barton invented his own deep diving helmet and weighted himself down with **rocks** to explore Massachusetts waters before teaming up with Beebe.

By 1926, Beebe was famous as an adventurer; Barton contacted him and showed him detailed designs for a steel sphere that would serve as a capsule for carrying two passengers beneath the sea. Two oxygen tanks in the sphere carried eight hours worth of air, trays of absorbents collected **carbon dioxide** and moisture, and panes of quartz that had been pressure-tested were fitted into the sphere as windows. Conditions were so primitive that Beebe and Barton carried small palm-leaf fans to circulate air in the chamber. A steam-powered winch on the host ship hoisted the bathysphere to the surface on a steel cable, and another cable carried two wires for telephone communications with the surface and two for an electric searchlight mounted inside the sphere and aimed through a window. Beebe wore headphones during the dives and described observations by telephone to an assistant onboard the surface craft.

During their historic 1934 dive, the captain of their crew allowed the bathysphere to stay at its greatest depth for only three minutes before beginning the surface ascent. Beebe described eerie and extravagant undersea creatures as well as great water voids with no apparent life. For years, he was condemned for deceiving the public until the observations and photographs made by others verified his observations, and Beebe was officially credited with discovery of hundreds of new life forms.

Auguste and Jacques Piccard

Swiss physicist Auguste Piccard (1884-1962) had twin fascinations, the atmosphere above Earth's surface and the sea below. He was world-famous as an inventor (who collaborated with Albert Einstein, among others), balloonist, and adventurer, and, at the Chicago World's Fair in 1933, his hydrogen-filled **balloon** was displayed next to Beebe's bathysphere. This led to a meeting of the two like minds, and, in 1937, Piccard began building his bathyscaph with its gasoline-filled float and suspended chamber or gondola of spherical steel. Largely supported in his atmospheric explorations by the Belgian organization Fonds National de la Recherche Scientifique (FNRS),

Piccard asked them to back him in building the bathy-scaph, named *FNRS-2* (his atmospheric exploration balloon had been named *FNRS-1*). His research was suspended for the duration of World War II, but, in 1948, Piccard and his son Jacques (1922-) reached a new record depth of 4,500 ft (1,500 m). Jacques was educated in Trieste, Italy, and, in 1953, the Piccards in a new Swiss/Italian bathy-scaphe named *Trieste* engaged the French/Belgian *FNRS* in a battle to beat the Piccards' last depth record.

In September 1953, the Piccards set the new record of 10,390 ft (over 3,100 m), they were limited only by the depth of the Mediterranean Sea. The U.S. Navy joined the race in 1957 and wanted to purchase the *Trieste* for test dives and further attempts at world records in the Pacific Ocean off the coast near San Diego. The ultimate objective was a dive into Challenger Deep, the deepest "hole" in the world's oceans in the Mariana Trench near Guam where the Pacific forms "Mount Everest in reverse," a 35,800-ft-deep (over 11,400-m-deep) chasm discovered in 1949 by the H.M.S. *Challenger II* research ship. The "Big Dive" was scheduled for January 23, 1960, and Jacques Piccard was selected by the Navy as half of the two-man crew with Lieutenant Don Walsh. After descending at the speed of an **elevator** and having their fragile craft buffeted by thermoclines (differences in ocean temperatures), Piccard and Walsh reached the deepest known point on **Earth**. With the depth race over, the oceans were open to more thorough scientific exploration.

Maurice Ewing

Maurice Ewing (1906-1974) was a professor of **geology** at Lehigh University in Lehigh, Pennsylvania. He had used seismic reflection, a technique for bouncing mini earthquake waves generated by **explosives** off surfaces and measuring their reflections, to locate deep oil and gas reserves in Texas. Different types of rock and other materials reflect seismic waves of different wavelengths. He was approached about applying the same method over the ocean to map the **continental shelf**, the border of any **continent** at the point where it drops steeply to deep ocean. In 1934, Ewing began a study of the continental shelf off the coast of Virginia. In 1940, Ewing went to the Woods Hole Oceanographic Institute to learn about the sea, and, during World War II, he performed secret research for the Navy and worked with Allyn Vine and John Worzel to develop the first underwater cameras. He was a leader in developing techniques for sampling soil from the sea floor and in investigating the Mid-Atlantic Ridge and discovered the great rift that divides this ridge. For 40 years, these and Ewing's other pioneering techniques were used to establish depths, bottom characteristics, and conditions below the sea floor, not just along the Continental shelf but over the deepest oceans.

Sylvia A. Earle

Sylvia A. Earle (1935-) extended public awareness of our need to preserve the environment from beyond the shore to the deepest ocean. She spent 40 years working as a marine scientist, assisting government agencies, writing, lecturing, and establishing records for diving and exploring her ocean world. In 1968, she joined a submarine crew on a Smithsonian program for exploring the ocean and fell in love with its challenges and habitats. In 1970, she led a team of women scientists in the Tektite II Project in which the team lived underwater for two weeks to help develop techniques for survival in confined circumstances that might be used in the space program. The media dubbed these women the "aquababes," and Earle learned the power to educate through media coverage. She set her first deep diving record in 1979 and the experience so intrigued her that she and British engineer Graham Hawkes built a deep-water submersible called the *Deep Rover* and later the *Phantom*, a remotely operated vehicle (ROV). In 1990, she was named the chief scientist of the National Oceanic and Atmospheric Administration (NOAA), the first woman to hold that post. Earle continues to campaign for "sea change" and popular support for the ocean environment.

Allyn Vine

The deep-sea submersibles that provide so many stunning images from the depth of the ocean are Allyn Vine's (1941-1994) work. Vine had worked with Maurice Ewing at Lehigh University and on the *Atlantis*, the Woods Hole Oceanographic Institution's research vessel. In the 1940s, there were about 45 ocean research vessels around the world, but all of them had the same capabilities with limited ability to explore the greatest depths of the ocean. Vine obtained funds from the Navy's research department to design and build a deep sea submersible, a miniature submarine that could withstand the tremendous water pressures at depth, hold a crew of only two or three, powered by golf-cart batteries, and controlled by a mother ship at the surface, and host a number of cameras, sampling devices, and instruments. The passenger ship on the submersible was fully detachable; if the main craft could not rise to the surface, the passenger ship would. In 1964, the first submersible called *Alvin*, for the first two letters from Allyn and the first three from Vine, was ready for a deep-dive test. The *Alvin* was successfully certified on her first deep dive. In 1994, after thousands of improvements, she celebrated her thirtieth birthday and 2,772 dives in the name of scientific research.

Vine's early experience, in the company of Ewing, was with the Navy during World War II in testing and improving the bathythermograph (BT), a device that measures temperature differences with depth in sea water. Because temperature and density in water are directly related, enemy submarines could hide from detection by sonar from the surface by hiding in dense water. Vine's improvement of the BT helped the Navy capture and destroy enemy subs but also helped its own subs find the most efficient hiding places.

Robert D. Ballard

Robert Ballard (1942-) is best known as the discoverer of the wreckage of the H.M.S. *Titanic*, the legendary ocean liner that in theory could not be sunk, but crashed to the ocean floor on her maiden voyage in 1912, taking over 1,500 lives with her. But Ballard is a geologist and oceanographer with many other astounding achievements to his credit. He was the first to take a submersible on a dive of the Mid-Atlantic Ridge, and, in an exploration of the volcanic sea floor around the Galápagos Islands, he discovered new life forms around **hydrothermal vents** at depths thought impossible for life. He investigated the sunken nuclear submarines the *Thresher* and the *Scorpion*, but finding the *Titanic* was a dream. An avid researcher, author, and writer of technical papers, Ballard used the fame that came with the discovery of *Titanic* to launch the JASON Project to educate schoolchildren about undersea explorations; through satellite links, the students can view the findings of submersibles as they work and even help manipulate it. The JASON Project was named for the *Jason* robot or ROV (remotely operated vehicle) that Ballard used to photograph the interior of the *Titanic*; Ballard describes the *Jason* ROV as "a tethered eyeball."

Jacques-Yves Cousteau and the Calypso

For his immeasurable contribution to oceanography and the preservation of the wealth of the seas, Jacques-Yves Cousteau (1910-1997), a former French sailor, deserves special mention. After his education at the French Naval School at Brest, Cousteau served as a gunnery officer and became fascinated with the depths of the sea. During and immediately following his Navy career, Cousteau dived underwater extensively himself, experimented with diving equipment, and created the improvements he needed. His underwater inventions were many, but the most notable is the aqualung or SCUBA (Self-Contained Underwater Breathing Apparatus), which he and Émile Gagnan (a French engineer) designed in 1943. The aqualung consists of a face mask, a pressure-regulating valve, and an attached cylinder of compressed air that enables a trained diver to stay underwater for several hours. For the first time, an individual could go beyond his own breathing limitations in exploring the sea. In the 1940s, he was named captain of the *Ingçnieur Elie Mon-*

nier, the world's first marine research vessel and the pride of the French Navy.

In 1950, Cousteau obtained indefinite leave from the Navy to devote himself to underwater exploration (he was to retire from the Navy with the rank of corvette captain in 1957). He needed a research vessel himself and found one in the *Calypso*, a former minesweeper that had been built for the British Navy in World War II and served as a ferryboat around the **island** of Malta after the war. The ship was extensively remodeled to work as a floating laboratory, and funding for this was provided by British brewery heir Noel Guinness. Accommodations were overhauled, sophisticated navigation and exploratory instrumentation was installed, and a "false nose" or underwater observation chamber was constructed on the tip of the ship's prow in a metal cage. Rigging and facilities for diving equipment were installed.

Aside from pure oceanography, *Calypso* was equipped to study and monitor patterns of biological populations, behavior of coastal and marine animals, the shapes and operations of a coral reef, the effects of undersea instruments, and special diving conditions and equipment performance. Her other assignments included topography, **weather**, **acoustics**, geology, chemistry, physics, and **geophysics**. Other private sources, the French Navy, manufacturers, and even donations from school children kept *Calypso* constantly moving about the world's oceans, making discoveries that benefited and educated the world. In 1951, Cousteau put *Calypso* to sea with his wife and two sons (among others) as crew. Operating from a base in Toulon on the French Mediterranean, and under the administration of the Campagnes Oceanographiques Franaises (COF) or French Oceanographic Expeditions (a nonprofit organization), *Calypso* began her voyages of discovery.

Cousteau brought *Calypso's* voyages into many families' living rooms thanks to his other skills as an underwater photographer, maker of documentary films, and author. Cousteau learned underwater photography and deep-sea photography at the feet of a master; in 1953, he began working with Doctor Harold Edgerton, known as "Papa Flash," who had pioneered deep-sea cameras and the use of strobe lights for flash as an inventor and electrical engineer at the Massachusetts Institute of Technology. Cousteau and Edgerton developed a sonar device to trigger a flash near the sea floor and a sled-like device for mounting cameras. By separating the cameras from the flash sources, the pair took some of the most remarkable deep-sea photographs ever seen. Cousteau's first film debuted in 1943. He made full-length films, documentary shorts, and many made-for-television films. Two of these, *The Silent World* (1956) and *World Without Sun* (1966) won Cousteau Academy Awards for best documentary feature. His best-known books may be those in the series called *The Undersea World of Jacques Cousteau*. The books, films, and television programs interested many children in the mysteries of the underwater world and help expand the environmental movement beyond the confines of land.

In the 1960s, Cousteau started a series of experiments in building underwater habitats where people could work and live. These concepts were abandoned because of economics, but, again, they awakened the public's interest in the compatibility of man and the underwater world. He turned more strongly toward environmental interests in the 1970s and started the Cousteau Society for marine **conservation** before his death in 1977.

Key findings in underwater exploration

The invention and improvement of the submersible from about 1930 to the early 1970s opened the possibility of vastly improving our understanding of the extremes of the deep. This world is so enormous and full of mysteries that nearly every submersible dive introduces new life forms or discoveries leading to greater knowledge of the mechanics of our **planet**. Some of the landmark studies involving submersibles are the 1974 exploration of the Mid-Atlantic Ridge, the 1979-1980 study of the rift valley near the Galápagos Islands off the coast of Ecuador, and the 1985 discovery of the wreck of the *Titanic*.

In 1974, a French-American team of scientists explored the great rift in the Mid-Atlantic Ridge by using occupied submersibles and a collection of support ships. The FAMOUS Project (French-American Mid-Ocean Undersea Study) used the American submersible *Alvin*, the French diving saucer *Cyana*, and the French bathyscaphe *Archimde* to dive into the rift south of the Azores Islands, where geologists believe two great plates of Earth's crust, the Eurasian Plate and the North American Plate, are pulling away from each other allowing **magma** (molten rock) to flow into the rift and the sea floor to enlarge or spread. The research ship the *Glomar Challenger* was the sea-level base for the submersibles and was assisted by a small flotilla of support ships. Samples of solidified but geologically young magma were collected by manipulators on the submersibles, and over 5,200 photographs were taken in this region. Exploration would have been impossible without the submersibles; in some places, the edges of the *Cyana*, which was about 7 ft (2.1 m) in diameter, nearly touched both sides of the ridge and hovered over the depths of the rift that are far greater than the highest mountains on Earth's surface. Analysis of the findings from the FAMOUS Project proved that the central fissure of the rift valley is widening by about 1 in (2.5 cm) per year and added substan-

tially to both proving and helping scientists understand **plate tectonics** (the motions of the massive plates comprising Earth's crust) and sea-floor spreading (the separations of those plates beneath the sea where new crustal material is made).

The 1979-1980 study of the Galápagos Rift was begun as a further study of sea-floor spreading but found it occurring in a very different environment. Mexican, French, and American scientists united efforts and discovered expanses of hydrothermal vents, which are chimney-like growths on the seabed that discharge hot springs of mineral-rich water. The water temperature of these vents is about 570°F (300°C), and the vent chimneys are about 12 ft (3.7 m) in diameter and 30 ft (9 m) tall. The smokey plumes of dissolved metals form deposits laden with nickel, **copper**, **uranium**, cadmium, and chromium; and the ecological community supported by the hot springs is rich in plants and animals that would have remained hidden without the camera eyes of submersibles. The hydrothermal vents and their surrounding communities proved that the deep sea is neither the barren abyss nor the realm of sea monsters of popular imagination.

Deep seas, the final frontier

Future exploration of the oceans of the world parallels our exploration of outer space in many ways. To increase understanding of both, technologies are being merged in creative ways. In the Arctic Ocean Basin, a submarine and sophisticated acoustics are combined to measure water temperature. The U.S.S. *Hawkbill*, a nuclear attack submarine operated by the U.S. Navy, launches acoustical probes that measure the density, salt content, and temperature of seawater along the path of the sound. The *Hawkbill* is part of a research platform for SCICEX 99, the fifth year of a working relationship between the Navy, the National Science Foundation, and other federal departments interested in the relationship among the atmosphere, oceans, and climate. The acoustic tests performed by the *Hawkbill* show that, after shrinking for four years running, the Arctic ice cap may be building again; the pattern of shrinkage and growth enters into our understanding of the importance of the ice cap.

Exciting underwater finds like the discovery of the *Titanic* have led to a burst of shipwreck hunts. The *Titanic* adventure proved that the technology exists to find any lost vessel anywhere, and all parties from historians to gold grabbers are looking for Spanish galleons, passenger liners, Roman vessels, and historic ships for their cargo and the answer to questions about their fate. Ethical and legal questions have arisen over control of ship-

KEY TERMS

Bathyscaph—A deep-sea exploration vehicle or submersible consisting of a ballast-filled float (resembling a blimp or balloon) with a spherical metal gondola for carrying occupants and equipment suspended below it.

Bathysphere—A deep-sea exploration vehicle or submersible consisting of a sphere that carries a crew and equipment and is lowered to the sea floor on a cable.

Bathythermograph (BT)—An instrument for measuring the differences in temperature in sea water depths.

Black smokers—Hydrothermal vents on the sea floor that emit black clouds of hot, mineral-rich water much like a chimney belches black smoke.

Continental shelf—A relatively shallow, gently sloping, submarine area at the edges of continents and large islands, extending from the shoreline to the continental slope.

Diving bell—An enclosed device for carrying a single diver exploring relatively shallow waters; re-

placed by submersibles except in some work situations.

Global Positioning System (GPS)—A system of satellites whose signals can be used to locate objects on Earth (including below sea level) very precisely.

Hydrothermal vent—An opening of Earth's crust on the sea floor where hot springs bearing mineral-rich waters are emitted. Hydrothermal vents are important sources of minerals and warmth for species of life not found in other environments.

Magma—The molten rock from the core of Earth that emerges on the surface through volcanic eruption and sea-floor spreading. When magma cools, it forms igneous rock.

Oceanography—The science of measuring the ocean.

Plate tectonics—The theory now widely accepted that the crust of Earth is composed of about 12 giant plates that form the land masses and sea floors and that grind slowly past each other, caus-

wrecks; apart from monetary value, their contents are historically and scientifically important. The United Nations Economic, Scientific, and Cultural Organization (UNESCO) has drafted a treaty establishing the limits of a nation's cultural underwater heritage offshore, which may help regulate the hot underwater marketplace. Even television rights for photographing discovered wrecks is highly contested.

Similarly, the underwater riches that occur naturally as mineral deposits are being mined at shallow depths, but the rights for deep sea **minerals** are contested. The "black smokers" or hydrothermal vents in the Mid-Atlantic and other rift zones belch minerals like smoke, but these minerals include gold, lead, and silver. Undersea craters off the coast of Japan were discovered in 1998 and are thought to have over $2 billion in mineral riches on and near them. Deep-sea submersibles are an expensive ($1 million per month at sea) but available tool for harvesting the minerals, but these vents also support exotic life forms, including tube worms, anemones, and giant clams that are not found in any other Earth environment. Just as archaeologists are contesting shipwreck hunters over historical disasters, marine biologists are trying to compete with the **mining** industry in preserving nature's secret treasure trove.

Pure observation to further our knowledge of the underwater world is also progressing, thanks to technology. Off shore near New Jersey, the Long-Term Ecosystem Observatory has been constructed to record a **battery** of measurements of physical, chemical, and biological state of the sea. Complex instrument packages along with instrument-bearing torpedoes and surface vessels transmit, collect, and convert a variety of signals into information about the ocean. An underwater **habitat** named *Aquarius* is sited off the Florida coast about 60 ft (20 m) under water. Aquanauts including Sylvia Earle are studying coral reefs that indicate the health of near-shore waters but also the deep ocean. The Monterey Bay Aquarium uses two remotely operated vehicles (ROVs) for similar purposes of probing the characteristics and life forms in the deep canyon under the Monterey (California) Bay.

Despite the huge technological leap into deep waters in the past century, there are other creative ways of exploring underwater. A team from the Smithsonian Institution is using natural enemies to its advantage. To attempt to film giant **squid** in their natural environment, the Smithsonian is using a "crittercam," a video camera attached to an **animal** to pursue and film this elusive creature. The sperm whale preys on the giant squid, and, using a suction cup to mount a small video camera on the whale's back,

ing earthquakes, mountain-building, and other large-scale geologic occurrences.

Remotely operated vehicle (ROV)—A deep-sea submersible that carries equipment only (no human occupants) and can be remotely operated from a surface ship.

Rift valley—A large, deep valley, either on the land surface or beneath the sea, created by the movement of two plates composing Earth's crust away from each other. The Mid-Atlantic Ridge, the Marianas Trench, and the Galápagos Rift are examples of submarine rift valleys.

Sea-floor spreading—The part of plate tectonics that describes the movement of the edges of two of the plates forming Earth's crust away from each other under the ocean. Sea-floor spreading results in the formation of new submarine surfaces.

Sediment—Soil and rock particles that wash off land surfaces and flow with water and gravity toward the sea. On the sea floor, sediment can build

up into thick layers. When it compresses under its weight, sedimentary rock is formed.

Self-contained underwater breathing apparatus (SCUBA)—Also called an aqualung. The mask, mouthpiece, valves, and oxygen or compressed air tank that can be worn by a diver to sustain breathing for periods up to several hours under water.

Sounding—The process of using dropped weights (weight sounding), sound waves (sonar), or seismic waves artificially induced by man-made explosions to produce waves that, when reflected back to their source, can be used to measure distances and the densities of the materials through which the waves pass.

Submersible—Deep-sea exploration vehicles that carry two or three human occupants, cameras, and other equipment to relatively great depths in the ocean. Submersibles can also carry equipment only and be remotely operated.

Thermocline—A difference in temperature in sea water or in the atmosphere.

scientists hope to obtain candid shots of the squid. The whales are not expected to return the camera for processing; instead, the camera films for three hours, then releases the suction on the cup, and floats to the surface.

Similarly, scientists at McMurdo Station, **Antarctica** have attached cameras to Weddell **seals** to study the **ecology** of fishes living beneath the sea ice. In this case, wild seals are captured and fitted with photographic and other sensing equipment. The seals are taken to an isolated area of sea ice with no natural breathing holes. A hole is drilled into the ice and the seals are allowed to hunt freely. Because there are no other options, the seals must return to the artificial hole for breathing. The equipment allows the scientists to monitor the activities of the seals and environment in which they and their **prey** exist. Once the information is collected, the equipment is removed from the seal and it is released at the location it was captured.

See also Abyssal plain; Seamounts.

Resources

Books

Ballard, Robert D. *Eternal Darkness: A Personal History of Deep-sea Exploration.* Princeton, NJ: Princeton University Press, 2000.

Cousteau, Jacques, and Alexis Sivirine. *Jacques Cousteau's Calypso.* New York: Harry N. Abrams, Incorporated, 1983.

Cousteau, Jacques-Yves. *The Ocean World of Jacques Cousteau: Guide to the Sea.* New York: The World Publishing Company, 1974.

Cousteau, Jacques-Yves. *The Ocean World of Jacques Cousteau: Inner and Outer Space.* New York: The World Publishing Company, 1974.

Kunzig, Robert. *The Restless Sea: Exploring the World Beneath the Waves.* New York: W. W. Norton & Company, 1999.

Polking, Kirk. *Oceanographers and Explorers of the Sea.* Springfield, NJ: Enslow Publishing, Inc., 1999.

Periodicals

Broad, William J., and Blake Edgar. "The Hot Dive: In the Chilly Depths of the Pacific, Explorers Find the Unexpected: Superheated Towers of Rock." *Earth* 6, no. 4 (August 1997).

Epstein, Aaron. "As the Sea Hunt Intensifies, So Does Legal, Ethical Debate Over Control." *Knight-Ridder/Tribune News Service* (December 30, 1998).

Fischman, Josh. "In Search of the Elusive Megaplume." *Discover* 20, no. 3 (March 1999): 108.

Foley, Don. "Creatures of the Twilight Zone." *Popular Science* 253, no. 3 (September 1998): 50.

Guynup, Sharon. "Undersea Riches." *Science World* 56, no. 3 (October 4, 1999): 16.

Hui, Li. "New Robotic Vessel Extends Deep-Ocean Exploration." *Science* 278, no. 5344 (December 5, 1997): 1,705.

Spotts, Peter N. "Ice Station Pieces Together Arctic's Global Sway." *The Christian Science Monitor* (May 20, 1999).

Stiefel, Chana Freiman. "Science On the Sea Floor." *Science World* 53, no. 1 (March 7, 1997): 8.

Stone, Richard. "Researchers Ready for the Plunge Into Deep Water." *Science* 283, no. 5405: 929.

Vanderkam, Laura R. "Divers Adapt To the Deep." *Insight on the News* 14, no. 35 (September 21, 1998): 41.

Wiley, John P. Jr. "Wiring the Jersey Coast." *Smithsonian* (October 1, 1998): 22.

Young, Catherine. "Scientists Use Crittercam To Search For Sea's Most Elusive Creature." *Insight on the News* 13, no. 16 (May 5, 1997): 39.

Other

Billings, Alistair W. "Through Wiser Eyes: Scientists Let Weddell Seals 'Do the Walkin'." Cyber Diver News Network. January 19, 2002 [cited January 20, 2003]. <http://www.cdnn.info/eco/e020119/ e020119.html>.

Sandwell, David T., and Walter H. F. Smith. "Exploring the Ocean Basins with Satellite Altimeter Data." January 17, 2003 [cited January 20, 2003]. <http://www.ngdc.noaa.gov/mgg/bathymetry/predicted/explore.HTML>.

Gillian S. Holmes

Ungulates

Ungulates are large grazing animals whose toenails have become enlarged into hooves. There are two orders of ungulates: Perissodactyla and Artiodactyla.

Animals in the order Artiodactyla have an even number of toes (usually two) that form a cloven hoof. This order is relatively diverse, containing 82 genera and several hundred **species**. There are nine families in this order, the most familiar of which are the **pigs** (Suidae), **peccaries** (Tayasuidae), **hippopotamuses** (Hippopotamidae), **camels** (Camelidae), **deer** (Cervidae), giraffes (Giraffidae), **sheep**, cattle, antelopes (Bovidae), and **pronghorn** antelopes (Antilocapridae).

Animals in the order Perissodactyla have an odd number of toes (usually one) that form a single, large hoof. Examples of this order include **horses** (family Equidae), **tapirs** (Tapiridae), and rhinos (Rhinocerotidae), together comprising six extant genera and 16 species.

At the time of the European discovery of **North America**, the native **fauna** of ungulates included **bison** (*Bison bison*), pronghorn antelope (*Antilocapra americana*), collared peccary (*Tayassu tayacu*), muskox (*Ovibos moschatus*), mountain goat (*Oreanmnos americanus*), and mountain and Dall sheep (*Ovis canadensis*

and *O. dalli*). The North American Cervidae includes white-tailed deer and mule deer (*Odocoileus virginianus* and *O. hemionus*), wapiti or elk (*Cervus elaphus*), **moose** (*Alces alces*), and **caribou** (*Rangifer tarandus*).

About 8,000-10,000 years before Europeans first came to America, North America supported a substantially larger number of ungulate species, including 10 species of horses, four species of camels, a species of cow, two additional species of bison, and the **saiga antelope** (*Saiga tatarica*). There were also other large, now-extinct **mammals**, including four species of elephants, such as the mastodon (*Mammut americanum*) and mammoth (*Mammuthus primigenius*), a giant ground sloth (*Gryptotherium listai*), and large predators such as the sabertooth cat (*Smilodon fatalis*) and the American lion (*Panthera leo atrox*). These large mammals disappeared during a great wave of extinctions that occurred at the end of the last **ice** age (about 8-12 thousand years ago). These extinctions may have been caused by overhunting by the first human inhabitants of North America, migrants from **Asia** who colonized the **continent** at about that time.

Some species of ungulates that have been domesticated are important in agriculture and sometimes as draft animals. The most abundant domesticated ungulates are sheep (*Ovis aries*), **goats** (*Capra hircus*), cows (*Bos taurus*), zebu cows (*B. indica*), pigs (*Sus scrofa*), horses (*Equus caballus*), camels (*Camelus dromedarius* and *C. bactrianus*), and **water** buffalo (*Bubalus bubalis*).

Many of the wild species of ungulates have recently become endangered and some have become extinct as a result of human influences. The most important of the human activities that endanger **wildlife** are the **habitat** losses associated with extensive conversions of natural ecosystems into agricultural or urban lands, and overhunting for the meat, hide, horns, or antlers of these large animals. It is critical that these human influences be rigorously controlled, if there is to be room on **Earth** to sustain all living species of ungulates. Natural ecosystems would be severely impoverished if only domesticated species of ungulates used in agriculture were to survive, along with the few species tolerant of habitats created by humans.

Bill Freedman

Uniformitarianism

Uniformitarianism is commonly oversimplified where stated in geological textbooks as "the present is a guide to interpreting the past" (or words to that effect). This explanation, however, is not correct about the true meaning of uniformitarianism. In order to understand uniformitarianism, one must examine its roots in the Enlightenment era (c. 1750–1850) and how the term has been distorted in meaning since that time.

Geology is an historical science, yet the phenomena and processes studied by geologists operated under nonhistorical natural systems that are independent of the time in which they operated. It is clear from the insights of one of geology's founding fathers of the Enlightenment era, James Hutton (1726–1797), that he understood this fact very well. In *Theory of the Earth* (1795), he stated: "In examining things present, we have data from which to reason with regard to what has been; and, from what has actually been, we have data for concluding with regard to that which is to happen thereafter." With his book, Hutton popularized the notion of "examining things present...with regard to what has been," but gave the concept no specific name. Hutton did not use the term uniformitarianism and used the word "uniformity" only rarely.

Charles Lyell (1797–1875), one of geology's founding fathers from later in the Enlightenment era, wrote about the subject **matter** of uniformitarianism (but did not use that specific term) in his widely read text, *Principles of Geology* (1830). Partly in response to strident criticism that his notions about geology did not conform to Biblical edicts about supernatural catastrophic events, Lyell developed a much more radical and extreme view of the subject matter of the "uniformity of nature." Careful reading of what Lyell laid out in his discussion of the "uniformity of nature" shows that he embraced both the concept of Hutton, which can be summarized as a uniformity of known causes or processes throughout time, and his own separate view that there must be a uniformity of process rates. The latter, more radical aspect of Lyell's "uniformity of nature" was intended to be a statement of general principle to counter the catastrophist interpretations of the past set forth by geologists of the day who were more inclined to look to the scriptures for their geological interpretations. In Lyell's view, a strong notion of uniformity of rates precluded divine (i.e., catastrophic) intervention.

In 1837, the name uniformitarianism was coined by William Whewell (1794–1866) as a term meant to convey Hutton's sense of order and regularity in the operation of nature and Lyell's sense that there was a uniformity of rates of geological processes through time. It is Whewell's definition that became the most common definition of uniformitarianism.

Lyell's work was influential, and he succeeded in imbuing generations of geologists with the notion of a dual foundation for "uniformity of nature." This dual foundation encompassed both uniformity of causes and uniformity of intensity. The former view is more commonly called actualism, and the latter, gradualism. In large part, the presence of Lyell's strongly defended gradualism succeeded in freeing nineteenth century geology from the firm grasp of Biblical preconception and allowed it to develop as a legitimate science.

One of the most elegant statements about actualism was made by John Playfair in his book, *Illustrations of the Huttonian Theory* (1802). He said: "Amid all the revolutions of the globe the economy of Nature has been uniform, and her laws are the only thing that have resisted the general movement. The **rivers** and the **rocks**, the seas, and the continents have been changed in all their parts; but the laws which describe those changes, and the rules to which they are subject, have remained invariably the same." Actualism is not unique to geology, as it is really a basic and broad scientific concept of many fields. Even though Playfair mentions laws, it is, of course, nature itself that is constant, not laws that have been written by people in order to try to predict nature.

The other side of Lyell's "uniformity of nature," i.e., gradualism, has no such elegant prose behind it. It has been referred to in inglorious terms by some of the leading minds of our time as "false and stifling to hypothesis formation," "a blatant lie," and "a superfluous term...best confined to the past history of geology." In other words, gradualism is no longer considered a valid idea.

Because uniformitarianism has this historical component of uniformity of process rates (i.e., gradualism), many writers have advocated its elimination from the geological vocabulary. Others argue that should be retained, but with careful notation about its historical meaning. Some writers ignore this historical debate and continue to tout the term uniformitarianism as the most basic principle of geology. The range of misguided meanings of this term from some recent geology texts includes definitions that span the gamut from something near the nineteenth century meaning to the assumption that the **Earth** is very old, to the logical method of geologic investigation.

Careful analysis of geological texts and recent scientific articles shows that there are at least 12 basic fallacies about uniformitarianism, (such as those explained by University of Wisconsin Geology Professor James H. Shea), which are perpetuated by some writers. These are:

• Uniformitarianism is unique to geology.

• Uniformitarianism was first discussed by James Hutton.

• Uniformitarianism was named by Lyell, who gave us its modern meaning.

• Uniformitarianism is the same as actualism, and should be re-named actualism.

• Uniformitarianism holds that only processes that are currently active could have occurred in the geologic past.

• Uniformitarianism holds that rates and intensities of geologic processes are constant through time.

• Uniformitarianism holds that only non-catastrophic, or gradual processes have operated during geologic time.

• Uniformitarianism holds that Earth's conditions have changed little over geologic time.

• Uniformitarianism holds that Earth is very old.

• Uniformitarianism is a testable hypothesis, theory, or law.

• Uniformitarianism applies to the past only as far back as present conditions have existed on Earth's surface.

• Uniformitarianism holds only that the governing laws of nature are constant through **space** and **geologic time**.

Through historical analysis of uniformitarianism, one is able to see how these twelve common conceptions are false and misleading. Most scientists argue that uniformitarianism should be kept in its proper historical perspective in the future, and that a more specific term like actualism might supplant uniformitarianism in places where the word is meant to convey strictly the modern concept of uniformity of causes.

See also Stratigraphy (archeology); Stratigraphy.

Resources

Books

Hancock, P.L., and B.J. Skinner, eds. *The Oxford Companion to the Earth.* New York: Oxford University Press, 2000.

Periodicals

Gould, S.J. "Is Uniformitarianism Necessary?" *American Journal of Science* 263 (1965): 223–28.

Gould, S.J. "Reply to C.R. Longwell's Criticism of 'Is Uniformitarianism Necessary?'" *American Journal of Science* 263 (1965): 919–21.

Shea, J.H. "Twelve Fallacies of Uniformitarianism." *Geology* (September 1982): 457.

David T. King, Jr.

Units and standards

A unit of measurement is some specific quantity that has been chosen as the standard against which other measurements of the same kind are made. For example,
the meter is the unit of measurement for length in the **metric system**. When an object is said to be 4 m long, that means that the object is four times as long as the unit standard (1 m).

The term "standard" refers to the physical object on which the unit of measurement is based. For example, for many years the standard used in measuring length in the metric system was the distance between two scratches on a platinum-iridium bar kept at the Bureau of Standards in Sèvres, France. A standard serves as a model against which other measuring devices of the same kind are made. The meter stick in your classroom or home is thought to be exactly 1 m long because it was made from a permanent model kept at the manufacturing plant that was originally copied from the standard meter in France.

All measurements consist of two parts: a **scalar** (numerical) quantity and the unit designation. In the measurement 8.5 m, the scalar quantity is 8.5 and the unit designation is meters.

History

The need for units and standards developed at a point in human history when people needed to know how much of something they were buying, selling, or exchanging. A farmer might want to sell a bushel of **wheat**, for example, for 10 dollars, but he or she could do so only if the unit "bushel" was known to potential buyers. Furthermore, the unit "bushel" had to have the same meaning for everyone who used the term.

The measuring system that most Americans know best is the British system, with units including the foot, yard, second, pound, and gallon. The British system grew up informally and in a disorganized way over many centuries. The first units of measurement probably came into use shortly after 1215. These units were tied to easily obtained or produced standards. The yard, for example, was defined as the distance from King Henry II's nose to the thumb of his outstretched hand.

The British system of measurement consists of a complex, irrational collection of units whose only advantage is its familiarity. As an example of the problems it poses, the British system has three different units known as the quart. These are the British quart, the United States dry quart, and the United States liquid quart. The exact size of each of these quarts differs.

In addition, a number of different units are in use for specific purposes. Among the units of **volume** in use in the British system, (in addition to those mentioned above) are the bag, barrel (of which there are three types—British and United States dry, United States liquid, and United States **petroleum**), bushel, butt, cord, drachm, firkin, gill, hogshead, kilderkin, last, noggin, peck, perch, pint, and quarter.

The metric system

In an effort to bring some rationality to systems of measurement, the French National Assembly established a committee in 1790 to propose a new system of measurement, with new units and new standards. That system has come to be known as the metric system and is now the only system of measurement used by all scientists and in every country of the world except the United States and the Myanmar Republic. The units of measurement chosen for the metric system were the gram (abbreviated g) for **mass**, the liter (l) for volume, the meter (m) for length, and the second (s) for time.

A specific standard was chosen for each of these basic units. The meter was originally defined as one ten-millionth the distance from the north pole to the equator along the prime meridian. As a definition, this standard is perfectly acceptable, but it has one major disadvantage: a person who wants to make a meter stick would have difficulty using that standard to construct a meter stick of his or her own.

As a result, new and more suitable standards were selected over time. One improvement was to construct the platinum-iridium bar standard mentioned above. Manufacturers of measuring devices could ask for copies of the fundamental standard kept in France and then make their own copies from those. As you can imagine, the more copies of copies that had to be made, the less accurate the final measuring device would be.

The most recent standard adopted for the meter solves this problem. In 1983, the international Conference on Weights and Measures defined the meter as the distance that **light** travels in 1/299,792,458 second. The standard is useful because it depends on the most accurate physical measurement known—the second—and because anyone in the world is able, given the proper equipment, to determine the true length of a meter.

Le Système International d'Unités (the SI system)

In 1960, the metric system was modified somewhat with the adoption of new units of measurement. The modification was given the name of Le Système International d'Unités, or the International System of Units—more commonly known as the SI system.

Nine fundamental units make up the SI system. These are the meter (abbreviated m) for length, the kilogram (kg) for mass, the second (s) for time, the ampere (A) for **electric current**, the Kelvin (K) for **tempera-**ture, the candela (cd) for light intensity, the **mole** (mol) for quantity of a substance, the radian (rad) for **plane** angles, and the steradian (sr) for solid angles.

Derived units

Many physical phenomena are measured in units that are derived from SI units. As an example, frequency is measured in a unit known as the hertz (Hz). The hertz is the number of vibrations made by a wave in a second. It can be expressed in terms of the basic SI unit as s^{-1}. **Pressure** is another derived unit. Pressure is defined as the **force** per unit area. In the metric system, the unit of pressure is the Pascal (Pa) and can be expressed as kilograms per meter per second squared, or $kg/m \; x^2$. Even units that appear to have little or no relationship to the nine fundamental units can, nonetheless, be expressed in these terms. The absorbed dose, for example, indicates that amount of **radiation** received by a person or object. In the metric system, the unit for this measurement is the gray. One gray can be defined in terms of the fundamental units as meters squared per second squared, or $m^2 \; x \; s^2$.

Many other commonly used units can also be expressed in terms of the nine fundamental units. Some of the most familiar are the units for area (square meter: m^2), volume (cubic meter: m^3), **velocity** (meters per second: m/s), concentration (**moles** per cubic meter: mol/m^3), **density** (kilogram per cubic meter: kg/m^3), luminance (candela per square meter: cd/m^2), and magnetic **field** strength (amperes per meter: A/m).

A set of prefixes is available that makes it possible to use the fundamental SI units to express larger or smaller amounts of the same quantity. Among the most commonly used prefixes are milli- (m) for one-thousandth, centi- (c) for one-hundredth, micro- (æ) for one-millionth, kilo- (k) for one thousand times, and mega- (M) for one million times. Thus, any volume can be expressed by using some combination of the fundamental unit (liter) and the appropriate prefix. One million liters, using this system, would be a megaliter (ML) and one millionth of a liter, a microliter (æL).

Natural units

One characteristic of all of the above units is that they have been selected arbitrarily. The committee that established the metric system could, for example, have defined the meter as one one-hundredth the distance between Paris and Sèvres. It was completely free to choose any standard it wished.

Some measurements, however, suggested "natural" units. In the field of **electricity**, for example, the charge

KEY TERMS

· ·

British system—A collection of measuring units that has developed haphazardly over many centuries and is now used almost exclusively in the United States and for certain specialized types of measurements.

Derived units—Units of measurements that can be obtained by multiplying or dividing various combinations of the nine basic SI units.

Metric system—A system of measurement developed in France in the 1790s.

Natural units—Units of measurement that are based on some obvious natural standard, such as the mass of an electron.

SI system—An abbreviation for Le Système International d'Unités, a system of weights and measures adopted in 1960 by the General Conference on Weights and Measures.

British and metric systems. To convert from the pound to the kilogram, for example, it is necessary to multiply the given quantity (in pounds) by the factor 0.45359237. A conversion in the reverse direction, from kilograms to pounds, involves multiplying the given quantity (in kilograms) by the factor 2.2046226. Other relevant conversion factors are 1 inch = 2.54 centimeters and 1 yard = 0.9144 meter.

Resources

Books

Adams, Herbert F.R. *SI Metric Units: An Introduction.* Toronto: McGraw-Hill Ryerson, 1974.

Jerrard, H.G., and D.B. McNeil. *A Dictionary of Scientific Units: Including Dimensionless Numbers and Scales.* London: Chapman and Hall, 1980.

Nelson, Robert A. *SI: The International System of Units.* Stony Brook, NY: American Association of Physics Teachers, 1982.

David E. Newton

carried by a single **electron** would appear to be a natural unit of measurement. That quantity is known as the elementary charge (e) and has the value of $1.6021892 \times 10^{-19}$ coulomb. Other natural units of measurement include the speed of light (c: 2.99792458×10^8 m/s), the Planck constant (6.626176×10^{-34} joule per hertz), the mass of an electron (m_e: $0.9109534 \times 10^{-30}$ kg), and the mass of a **proton** (m_p: $1.6726485 \times 10^{-27}$ kg). As you can see, each of these natural units can be expressed in terms of SI units, but they are often used as basic units in specialized fields of science.

Unit conversions between systems

For many years, an effort has been made to have the metric system, including SI units, adopted worldwide. As early as 1866, the United States Congress legalized the use of the metric system. More than a hundred years later, in 1976, the Congress adopted the Metric Conversion Act, declaring it the policy of the nation to increase the use of the metric system in the United States.

In fact, little progress has been made in that direction. Indeed, elements of the British system of measurement continue in use for specialized purposes throughout the world. All flight navigation, for example, is expressed in terms of feet, not meters. As a consequence, it is still necessary for an educated person to be able to convert from one system of measurement to the other.

In 1959, English-speaking countries around the world met to adopt standard conversion factors between

Uplift

Uplift is the process by which the earth's surface slowly rises either due to increasing upward **force** applied from below or decreasing downward force (weight) from above.

During uplift, land, as well as the sea floor, rises. The outer shell of the **earth**, the crust, divides into moving sections called plates. Uplift, forming **mountains** and plateaus, usually results as these plates crash into each other over millions of years. Although the plates move at roughly the speed fingernails grow, their **motion** still has a tremendous impact on the earth, since plates can be as big as a **continent** or the Pacific Ocean. Sometimes mountains rise from crust separating rather than colliding. The study of these moving plates is **plate tectonics**.

Collision between two pieces of continents lifts the tallest mountains. When India, formerly a large **island**, slammed into the south side of **Asia** around 55 million years ago, the Himalayas uplifted. When **Africa** and **Europe** smashed into **North America** 300 million years ago, the Appalachians uplifted. As the western edge of the westward-moving North America smacked into various islands over the past 200 million years, the Rockies uplifted. In these uplifts, pieces of ocean floor trapped between the approaching continents rise as well. Limestone, a rock formed on the sea floor, composes the summit of Mt. Everest, over 26,000 ft (8,000 m) above **sea level**.

Stratified rock in Glacier National Park, Montana. The rocky crust has uplifted in isostatic rebound from the weight of glaciers. *JLM Visuals. Reproduced by permission.*

Uplift also results when sea floor crust collides with continental crust or with other pieces of sea floor crust. Volcanic mountains (Andes, Cascades) or volcanic islands (Indonesia, Japan, Aleutians) result from sea floor colliding with and diving beneath a continent or another sea floor. As the sea floor crust descends, some of the sediment on the floor scrapes off the plunging crust and piles up to form a ridge called an accretionary wedge. For example, as the Indo-Australian plate dove beneath the Eurasian plate at Indonesia, the scraped and folded sediments form the Java Ridge off the west coast of the volcanic island of Java.

Sea floor also uplifts along mid-ocean ridges where crust separated as **magma** (melted rock) from inside the earth tries to reach the surface. The magma that rises from below these ridges lifts the ocean floor. Mid-ocean ridges circle the earth like the seams on a giant baseball. If the magma rises similarly beneath a continent, the land will bulge and eventually crack, potentially tearing the continent in half and creating a new ocean, as at the Red Sea.

Although compression creates most uplift, the **Basin** and Range region of western North America re-

sulted from a combination of collision and then extension. A map of North America illustrates a series of **parallel** north-south oriented mountain ranges separated by north-south trending valleys (basins) extending from Nevada and Utah down into Mexico. These mountains formed when first the crust arched from the collision of North America with a piece of ocean floor, and then later the crust began to separate. As the top of the arches cracked (similar to how the top of a bent piece of clay cracks along its upper surface when bent), pieces of the crust dropped down to form the valleys, and other pieces formed the mountain ranges.

Finally, when a huge weight is removed from the crust, the crust will slowly rise up in a process called isostatic rebound. During an **ice** age, when **glaciers** up to 1.9 mi (3 km) thick cover continents, the weight of that ice pushes down on the crust, causing it to sink or subside. When the ice melts, the crust uplifts just as a raft in a pool rises when the swimmer gets off it. Scandinavia still responds to glaciers melting 10,000 years ago by uplifting by as much as a centimeter per year.

See also Volcano.

Upwelling

Upwellings are a flow to the surface of deep, cold, nutrient-rich waters from greater depths in the **ocean**. The most extensive upwellings are associated with persistent coastal **currents** that draw surface **water** away from or along the coast to be replaced by a surfaceward flow of deeper waters. The most famous of these sorts of regional upwellings are found off the west coast of southern **South America** and in parts of the Antarctic Ocean.

Extensive upwellings can also develop where large currents are moving in opposite directions. This occurs in parts of the Pacific Ocean where the equatorial current in the Southern Hemisphere tends to move in a southerly direction, while that in the Northern Hemisphere moves to the north. Where these divergent equatorial currents flow beside each other, they develop extensive upwellings of deep waters.

More local upwellings can be caused when currents encounter surface or subsurface obstructions to their flow which can also **force** deeper water to the surface. Upwellings also develop on the leeward (downflow) side of islands that obstruct the passage of a prevalent current.

Upwellings are most common in high-latitude regions of the Arctic and Antarctic Oceans, along the equator, and in certain coastal locations on the eastern sides of oceans. The most important of the latter types of upwellings are associated with the California current, the Peru current off South America, and the Benguala current off west **Africa**.

Upwelling waters are relatively rich in inorganic **nutrients** such as nitrate and phosphate. As a result, upwelling waters can sustain a large productivity of **phytoplankton** when they reach the surface of the ocean where there is ample sunlight to support **photosynthesis**. The relatively great primary productivity of phytoplankton can in turn support a large productivity of **zooplankton**, which can sustain a great abundance of **fish**, seabirds, and marine **mammals**. Because of their intrinsic fertility, upwelling waters are much more ecologically productive than the open ocean, which is generally highly deficient in nutrients.

Some of the world's largest fisheries are associated with extensive upwellings. These include the cold-water fisheries of certain regions of the Antarctic and Arctic Oceans and the temperate fisheries off Peru, Chile, California, and west Africa. The world's largest and most productive populations of seabirds and marine mammals also occur in those upwelling-driven marine ecosystems.

Uranium

Uranium is the metallic chemical element with an **atomic number** of 92. Its symbol is U, **atomic weight** is 238.0, and specific gravity is 18.95. It melts at 2,071.4°F (1,133°C) and boils at 6,904.4°F (3,818°C). Natural uranium consists of three isotopes of **mass** numbers 234 (0.00054%), 235 (0.711%) and 238 (99.275%). All are radioactive.

History and applications

With the exception of tiny amounts of neptunium, uranium is the heaviest element found on Earth—that is, the element with the highest atomic number and atomic weight. It has held that distinction ever since it was first recognized as an element by the German chemist Martin H. Klaproth in 1789, who named it uranium in honor of the new **planet** that had recently been discovered: **Uranus**. Until 1896, when Henri Becquerel discovered radioactivity, uranium remained a dull, uninteresting **metal** that found occasional use in making yellow **glass**. But then it acquired the distinction of being one of only two known elements that possessed the mysterious property of being radioactive. (The other element was thorium.) When **nuclear fission** was discovered in 1938, uranium suddenly became the most fateful element in the **periodic table**. Because of its ability to undergo nuclear fission with the release of huge amounts of **energy**, it became a brand-new source of power, which people would use for both peaceful and destructive purposes.

Aside from its nuclear properties of radioactivity and fission, uranium is literally dull; freshly cut uranium metal is silvery white, but it soon develops a dull gray **color** in air because of a thin coating of black uranium oxide.

In spite of its radioactivity, uranium has a few useful applications because it is so heavy. Having a **density** of 19.0 grams per cubic centimeter, it is almost as dense as gold (19.3) and platinum (21.5). But it is much cheaper for two reasons: it is much more plentiful on **Earth** (40 or 50 times as abundant as silver) and it is a byproduct of the **nuclear power** industry after the very valuable uranium-235 **isotope** has been removed. It therefore finds some military uses in which a lot of weight is needed in a small space, such as for counterweights in **aircraft** control systems, ballast for missile reentry vehicles, and shielding against **radiation**.

Chemically, uranium is a member of the actinide series of elements, which runs from atomic number 89 (actinium) to atomic number 103 (lawrencium). Those with atomic numbers higher than uranium's 92 are the transuranium elements. Uranium's most important features lie not in its **chemistry**, but in its radioactivity and its ability to undergo nuclear fission.

Uranium's radioactivity

Although uranium is indeed radioactive—the discovery of radioactivity occurred during a study of uranium's properties—it has a very long **half-life**, which means that it emits its radiations at a rather leisurely pace. Also, it emits mostly alpha particles, which do not travel very far through the air and will not even penetrate the skin. Its radiations are therefore not very harmful, and uranium and its compounds can be handled with a reasonable amount of care, like any other highly poisonous chemicals.

The half-life of the most abundant uranium isotope, uranium-238, is 4.47×10^9 years, or about 4.5 billion years, which happens to be equal to the age of Earth as a planet. This fact allows scientists to use the disintegration of uranium as a sort of clock to determine the ages of **rocks** and other geological features of Earth.

Uranium-238 is the "parent" atom of a series of radioactive isotopes that we find associated with it in uranium ores. Through radioactive disintegrations, the uranium has been producing these "daughter" isotopes ever since the **ore** was laid down where we find it today. Uranium-235, which has a half-life of 7.04×10^8 (700 million) years, is the parent of another radioactive series. Among the daughters in these two series are various radioactive isotopes of radium, **radon**, and other elements. Both series of disintegrations proceed by producing consecutive radioactive isotopes until they **wind** up as stable isotopes of lead. Thus, the uranium isotopes are slowly turning into lead at a steady **rate** that is well-known from their half-lives. By measuring the relative amounts of uranium and lead isotopes in a uranium-containing rock, scientists can calculate how old it is.

One of the disintegration products of uranium is radium, the element of atomic number 88. Radium was discovered by Marie Curie (1867-1934), who isolated it from a uranium ore. Another important disintegration product of uranium-238 is radon-222, which has a half-life of 3.8 days. Radon is a gas (a rare gas) that can diffuse out of uranium in the ground and seep into people's houses. Radon can cause lung **cancer** because when inhaled, it can emit alpha particles directly in the lung, where they can do the most damage. Testing houses for radon gas has become an important precaution ever since this hazard was uncovered only 10 or 15 years ago.

The fission of uranium

By far the most important characteristic of uranium is that it undergoes the nuclear reaction called fission. Urani-

Nine-and-a-half pound (4.3 kg) button of uranium-235, which will be manufactured into a nuclear weapons component. *United States Department of Energy.*

um-235, which is only about 0.7% of all uranium **atoms**, is the isotope that fissions most readily. For use in nuclear reactors, natural uranium is enriched in the 235 isotope (that is, the percentage of the 235 isotope is increased) by gaseous **diffusion**. In this process the uranium is converted into the gaseous compound uranium hexafluoride, UF_6, and allowed to diffuse through a series of porous barriers. Those molecules which contain atoms of the slightly lighter uranium-235 isotope diffuse slightly faster and therefore separate themselves from the heavier, more slowly moving molecules that contain uranium-238.

See also Element, chemical.

Resources

Books

Emsley, John. *Nature's Building Blocks: An A-Z Guide to the Elements.* Oxford: Oxford University Press, 2002.

Greenwood, N.N., and A. Earnshaw. *Chemistry of the Elements.* 2nd ed. Oxford: Butterworth-Heinemann Press, 1997.

Kirk-Othmer Encyclopedia of Chemical Technology. 4th ed. Suppl. New York: John Wiley & Sons, 1998.

Lide, D.R., ed. *CRC Handbook of Chemistry and Physics* Boca Raton: CRC Press, 2001.

Robert L. Wolke

Uranus

Uranus is the seventh **planet** from the **Sun**. It has a large size (its diameter is almost four times that of **Earth**) and **mass**, low mean **density**, fairly rapid **rotation**, and well-developed ring (11 components) and **satellite** (15 members) systems. The planet has a strong magnetic **field** with a large tilt (58.6°) to its rotation axis and offset (0.3 Uranus radius) from its center. Analysis of the observations made by *Voyager 2* during its flyby of **Neptune** in August 1989 shows that Uranus and Neptune are similar in most of these properties and form a subgroup of the Jovian planets; **Jupiter** and **Saturn**, much larger and more massive, form the other subgroup.

Discovery

William Herschel (1738–1822) fortuitously discovered Uranus in 1781; it was the first planet discovered telescopically. It was found to **orbit** the sun at a mean distance of about 19.2 astronomical units (a.u.) (2,870,000,000 km), about twice as far from the sun as Saturn (9.54 a.u), the most distant planet known before 1781.

Observations from Earth

Knowledge about Uranus came slowly because of its distance. Even when it is closest, Uranus shows a disk of only 4 in (10 cm) in apparent diameter through a **telescope** and is 5.7^m apparent magnitude (barely visible to the unaided **eye** even in the best observing conditions). Herschel discovered Oberon and Titania, the outermost and largest, respectively, satellites of Uranus in 1787. Determination of their orbits around Uranus from observations gave their periods of revolution P and mean distances A from Uranus. This allowed one to determine Uranus' mass from the general form of Kepler's third law; it turned out to be 14.5 Earth masses. Using its radius of 15,873 mi (25,560 km), one calculated Uranus' mean density (its mass divided by its **volume**), which is 1.27 grams/cm^3. This indicated that Uranus is a smaller type of Jovian planet similar to Jupiter and Saturn; they are characterized by large masses and sizes and low mean densities (compared to Earth), and are inferred to consist largely of gases.

The planes of the orbits of Oberon and Titania were expected to lie in or near the **plane** of Uranus' equator, since most other planetary satellites have orbital planes that are in or near the equatorial planes of their planets. When the orbital planes of Oberon and Titania were determined, however, they indicated that the plane of Uranus' equator is almost **perpendicular** to the plane of its orbit around the sun (and also to the ecliptic). This is unlike the other planets, whose equatorial planes are tilted by at most 30° to the planes of their orbits around the sun. This implied that Uranus' axis (and poles) of rotation lie almost in its orbital plane. This conclusion has been confirmed by observations of Uranus' rotation from the *Voyager 2* spacecraft in 1986. This is the first of several interesting characteristics we have discovered for Uranus, and gives Uranus interesting **seasons** during its year (period of revolution around the sun), which is 84.1 Earth years long. The Uranian seasons will be discussed in more detail below. The cause of this unusual orientation of Uranus' rotation axis is still unknown and is now the subject of considerable speculation and theoretical research. One theory is that the orientation of its rotation axis was produced by the collision of an Earth-sized body with Uranus near the end of its formation.

Unexplained perturbations of Uranus' orbit in the early nineteenth century led to the prediction of the existence of a still more distant large planet, resulting in the discovery of Neptune in 1846. Neptune is the most distant (30.06 a.u. mean distance from the sun) Jovian planet, and it has several properties (mass, size, rotation period, rings, and magnetic field) like those of Uranus.

Three more satellites of Uranus, closer to it than Titania, were discovered during the 105 years after Neptune's discovery. They are, in order of closeness to Uranus, Umbriel and Ariel, discovered in 1851 by Lassell (1799–1880), and Miranda, discovered in 1948 by G. P. Kuiper (1905–1973). These discoveries showed that Uranus has a satellite system comparable to those of Jupiter and Saturn, although Titania, its largest (980 mi [1,580 km] diameter) and most massive satellite, and the slightly smaller Oberon, are comparable in size and mass to Saturn's satellites Iapetus and Rhea rather than to its much larger satellite Titan and Jupiter's four Galilean satellites. All other satellites of Uranus are smaller and less massive.

The best telescopic observations of Uranus from Earth's surface show a small, featureless, bluish green disk. Spectroscopic observations show that this **color** is produced by the absorption of sunlight by methane gas in its atmosphere; this gas is also present in the atmospheres of Jupiter and Saturn. Observations of occultations (similar to **eclipses**) or stars by Uranus indicated that Uranus' atmosphere is mostly composed of molecular **hydrogen** and helium, which are also the main components of the atmospheres of Jupiter and Saturn.

Observations at infrared wavelengths (that are longer than those of red **light**), where planets radiate away most of their **heat energy**, show that Uranus radiates at most only slightly more infrared radiative energy than its atmosphere absorbs from sunlight. Any excess energy originating from Uranus' interior can be attributed to the decay of radioactive elements, which also produces much of the heating in **Earth's interior**. This is not true for the other Jovian planets, which all emit as much as twice as much infrared energy as their atmospheres absorb from sunlight; this requires another internal energy source for them, which is possibly continuing gravitational contraction.

One of the last major discoveries about the Uranus system from Earth-based observations was made on March 10, 1977, during observations of Uranus' occultation of the **star** SAO 158657, when J. L. Elliot's (1943–) group and other observers noticed unexpected dimming of the star's light before the occultation and again after it. These dimmings were correctly identified with the existence of several faint, thin rings orbiting Uranus well inside Miranda's orbit which were hitherto undetected; unlike Saturn's rings, they are too faint to be directly observed from Earth's surface by ordinary methods. Uranus' rings have been observed several times since then during stellar occultations and by the *Voyager 2* spacecraft. The rings are very dark; their albedos (the fraction of the light that falls on them which they reflect) are only about 0.05.

A second major discovery made by Earth-based infrared observations was the detection of **water** ice on the surfaces of some of Uranus' satellites.

Let us now return to the seasons of Uranus. The fact that Uranus' rotation axis lies almost in the plane of its orbit around the sun means that at some season its south pole will be pointed nearly at the sun, and nearly all of its southern hemisphere will be in continuous sunlight (early southern hemisphere summer), while nearly all of its northern hemisphere will be in continuous night (early northern hemisphere winter). These seasons occurred in 1901 and in late 1985, and will occur next in 2069. The sun was last above Uranus' equator in December 1965, when it rose at Uranus' south pole and set at the north pole; the sun will then shine continuously on Uranus' south pole for the next 41.6 years until mid-2007, when it will again be above Uranus' equator and will set at the south pole and rise at the north pole, which will be in continuous sunlight for the next 42.5 years while the south pole will be in continuous night. The north pole will point closest to the sun in early 2030 (early northern hemisphere summer), and the sun will set there and rise at the south pole in 2050. Calculations show that a horizontal unit surface area, say a square meter, at either pole will receive over a Uranian year about 1.5 times the sunlight that the same surface would receive at Uranus' equator over a Uranian year (84.1 Earth years).

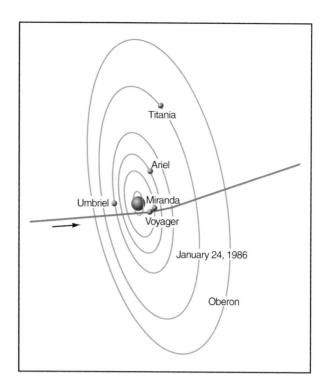

Figure 1. Trajectory of *Voyager 2* through the Uranus system in January 1986. *Illustration by Hans & Cassidy. Courtesy of Gale Group.*

Results from the flyby of the *Voyager 2* spacecraft

The *Voyager 2* spacecraft was launched from Earth on August 20, 1977. As it flew by Jupiter in July 1979, it was accelerated toward Saturn which, in turn, accelerated *Voyager 2* toward Uranus during the August 1981 flyby. *Voyager 2* flew by Uranus on a hyperbolic orbit, passing it at a minimum distance of 66,447 mi (107,000 km) from the center of Uranus on January 24, 1986. The observations that *Voyager 2* made of the Uranus system from November 4, 1985 to February 26, 1986 added immensely to our knowledge about it. Because Uranus' rotation axis was pointed less than 8° from the Sun then, and because *Voyager 2* was approaching Uranus along a path that made about a 35° **angle** with the line from the Sun, *Voyager 2* passed through Uranus' ring and satellite system much like a bullet passing through a "bull's eye target" and could not pass fairly close to any more than two of Uranus's satellites at most.

The satellites closely approached by *Voyager 2* were Miranda and Ariel; the spacecraft passed by these two at distances of 16,146 mi (26,000 km) and 17,388 mi (28,000 km), respectively.

The main discoveries made by *Voyager 2* during its encounter with Uranus are the following:

Uranus's magnetic field

Like Earth and the other Jovian planets, Uranus has a strong magnetic field which arises in its interior. Evidence for Uranus' magnetic field and **magnetosphere** (the region of **space** where the planet's magnetic field is dominant over the interplanetary field) was not found until January 22, 1986, two days before closest approach to Uranus, when **radio** noise from charged particles trapped in its magnetosphere was detected. *Voyager 2* crossed into Uranus' magnetosphere on January 24 and remained inside it for 45 hours. Uranus' magnetic field was found to be quite strong but very unusual. First, Uranus' magnetic poles were found to be 58.6° from its poles of rotation, which is much greater than the tilts of the magnetic fields to the poles of rotation found for Earth (11°), Jupiter (9.6°), and Saturn (0°). Second, the center of Uranus' magnetic field was found to be offset from its center of mass by 0.3 of Uranus' radius, a much greater offset than those found for the above-named planets. One effect of this magnetic field offset is that the magnetic field strength at the cloud level in Uranus' atmosphere is expected to vary by factors of five to 10 depending on Uranian **latitude and longitude**. **Radiation** belts of charged particles trapped in Uranus' magnetosphere were detected. They consist mainly of low energy protons and electrons; very few heavy ions are detected. The particle densities in these radiation belts are low compared with those densities in the radiation belts of Earth and Jupiter, possibly because the large tilt of Uranus' magnetic field to the interplanetary magnetic field allows the **solar wind** to make convective sweeps of particles out of Uranus' radiation belts fairly frequently.

Uranus' rotation

The fact that Uranus' magnetic field is tilted to its rotation axis and is offset from its center causes fluctuations of its magnetic field that are associated with the rotation of Uranus' interior. From measurements of these fluctuations by *Voyager 2*, the rotation period of Uranus' interior was found to be 17 hours 14 minutes. This is the first accurate rotation period for Uranus; earlier attempts in the last 100 years to determine its rotation period from spectroscopic and photometric observations gave very diverse, conflicting, and, as we now know, incorrect results. Other somewhat different rotation periods found from *Voyager 2* observations of cloud features in Uranus' atmosphere, which range from 16 to 17.5 hours, are caused by winds in Uranus' atmosphere.

Atmospheric temperature

In the earlier discussion of Uranus' seasons, it was mentioned that a unit surface at the poles will receive about 1.5 times as much sunlight as the same surface would on Uranus equator over a Uranian year. Based on this, one might expect Uranus' south polar region, which at the time of the *Voyager 2* flyby had been in continuous sunlight for the order of 20 years, to be warmer than its equatorial region, which would be warmer than the north polar region, which had been in prolonged night. *Voyager 2's* infrared instruments did not observe this; at the level of **clouds** in Uranus' atmosphere, the **temperature** seemed to be the same, about -346°F (-210°C), from the south pole across the equator to the north pole. The most evident temperature change at this level was a 34°F (1°C) decrease centered at about 30° south latitude. This surprising observation shows the great capacity of the enormously thick atmosphere of Uranus (and those of the other Jovian planets) to absorb and transport away almost all the sunlight energy from a region sunlit continuously for decades.

Uranus' atmosphere

By solar and stellar occultations and visual, infrared, and radio observations from *Voyager 2*, the structure and circulation of Uranus' atmosphere has been mapped from just below the cloud layers to its exosphere. The main components of the atmosphere are hydrogen and helium, the most abundant elements in the universe. Methane comprises 1–2% of the observable troposphere. Water vapor and **ammonia** are inferred to be important components of the atmosphere below the clouds, but they have not been detected in the observable part of the troposphere because it is too cold and freezes them out.

Uranus' upper atmosphere is dominated by hydrogen, mainly molecular. Methane and other hydrocarbons are nearly all frozen out by the underlying cold atmosphere, especially the lower stratosphere (at 428°F [220°C] temperature). Molecular hydrogen is broken down into atomic hydrogen mainly by the absorption of ultraviolet radiation from the sun, and some atomic hydrogen is ionized into free protons and electrons, forming an ionosphere. The temperature of the upper atmosphere increases to 482°F (250°C) at 497 mi (800 km) above the cloud layers, and continues to increase to over 932°F (500°C) at 3,105 mi (5,000 km) above the clouds. Atomic hydrogen becomes the main component of Uranus' upper atmosphere above 4,658 mi (7,500 km) above the clouds, forming a hydrogen thermal corona in its exosphere that extends at least 15,525 mi (25,000 km) above the clouds (extending through the zone of the rings from 9,936 mi [16,000 km] to 16,146 mi [26,000 km] above the clouds). The source of most of the heating of Uranus' upper atmosphere is still unknown as is also true for the heating of the upper atmospheres of Earth, Jupiter, and of Saturn.

Uranus's internal structure

Evidence indicates that Uranus may have a silicate rock core (perhaps rich in **iron** and **magnesium**), which is 4,800 km in diameter (approximately 40% of the planet's mass). The mantle is likely **ice** or ice-rock mixture (water ice, methane ice, ammonia ice) that may be molten in part (perhaps evidence of convention produced in the magnetic field). Above the mantle is the lower atmosphere, which consists of molecular (gaseous) hydrogen, helium, and traces of other gasses (approximately 10% of planet's mass). Finally, the upper atmosphere is methane with cloud layers of ammonia or water ice. The magnetic field discovered and mapped by *Voyager 2* implies a field generating region in Uranus' interior which extends out to 0.7 of Uranus radius from the center, and that part of Uranus' interior is a fluid and has a high internal temperature.

Uranus's rings

Uranus' ring particles are dark grey to black and form rings about 1–37 mi (2–60 km) wide and less than one kilometer thick. The rings are in the equatorial plane of Uranus, which is tilted at 98 degrees to the typical planetary attitude of the **solar system** (i.e., the plane of the orbit of the inner eight planets). This suggests that the rings formed after the planet was titled.

There is a ring hierarchy about Uranus. The inner eight rings are very thin and have no known shepard satellites. In an outward order, these rings are known as Delta, Gamma, Eta, Beta, Alpha, 4-ring, 5-ring, and 6-ring. The two outer rings have variable thickness and have shepard satellites (Cordelia on the inner edge and Ophelia on the outer edge). The outer rings are URI and Epsilon.

The *Voyager 2* observations also indicate that the rings contain a large fraction of large particles; the average particle size in the rings was calculated to be between 8 and 28 in (20-70 cm). An appreciable amount of micron-sized dust seems to be distributed throughout the ring system. However, this dust is probably transitory; atmospheric drag on the dust particles from Uranus' hydrogen corona that was mentioned above is expected to decelerate them and cause them to **spiral** into the denser layers of Uranus' atmosphere after, at most, a few thousand years. Uranus has 10 known rings that are nearly circular and lie in or nearly in the plane of Uranus' equator. The Epsilon ring is not only the most distant ring from Uranus, but it is also the most elliptical and the widest one.

Uranian satellites

Uranus' satellites all lie in the equatorial plane (like the rings). There are several groups of satellites. The inner satellites, of which there are ten, are irregular dark

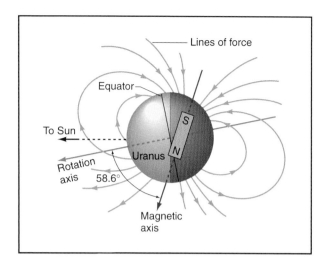

Figure 2. Diagram of offset and tilted magnetic field of Uranus. *Illustration by Hans & Cassidy. Courtesy of Gale Group.*

objects (which may mean they are carbonaceous rock or are methane ice bodies coated with **carbon** material) under 93 mi (150 km) in diameter. The inner satellites are Cordelia, Ophelia, Bianca, Cressida, Desdemona, Juliet, Portia, Rosalind, Belinda, and Puck. The outer satellites are all rather large satellites (292–982 mi [470–1580 km] in diameter) that are locked in 1:1 spin-orbit couples with Uranus. The outer satellites (Miranda, Ariel, Umbriel, Titania, and Oberon) are all spherical objects of water ice surrounding rock.

Voyager 2 discovered 10 satellites which all orbit Uranus closer to it than Miranda and are all smaller than Miranda. Surface **albedo** could be found for only Puck and Cordelia, which are 0.08 and 0.07, respectively, indicating that they are somewhat brighter than the ring particles. The other eight newly found satellites seem to be dark like Puck, Cordelia, and the rings. Cordelia and Ophelia, the two satellites closest to Uranus, seem to serve as "shepherd satellites" for the Epsilon ring, keeping its particles in the ring by their gravitational perturbations on them, thereby increasing this ring's orbital stability. Gravitational perturbations produced by several other satellites near the rings may make the rings more stable. Ophelia is slightly more and Cordelia slightly less than two Uranus radii from Uranus' center; this raises the possibility that the rings were formed by satellites inside Uranus' Roche limit that were torn to pieces by collisions or by tidal forces produced in them by Uranus.

Observations of Miranda and other satellites

Since *Voyager 2* passed fairly close to Miranda, it was possible to determine Miranda's mass from its perturbation

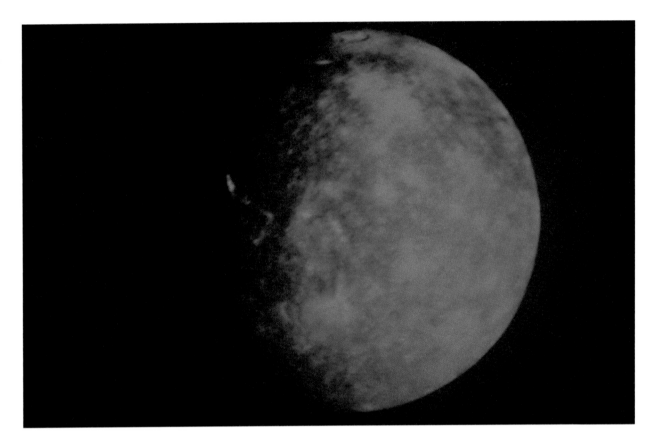

The surface of Titania, the largest of the Uranian satellites at 980 mi (1580 km) in diameter, is marked by both impact craters and past geologic activity. The deep fault valley visible in this image near the terminator (day-night boundary) is the result of at least one episode of tectonic activity, and the basin-like feature near the top of the image appears to be the result of a heavy impact on Titania's surface. *U.S. National Aeronautics and Space Administration (NASA).*

of *Voyager 2's* hyperbolic orbit past Uranus found from an analysis of the *Voyager 2* radio data. Ariel's mass was then determined from its perturbations of the orbits of Miranda and *Voyager 2*. *Voyager 2* did not approach Umbriel, Titania, and Oberon closely enough for reliable determination of their masses from perturbations of its flyby orbit. Instead their masses are determined from their perturbations of Ariel's orbit and each other's orbits using both *Voyager 2* observations and Earth-based observations made over many years. Accurate radii were found for all five satellites from *Voyager 2* images; this allowed the calculation of mean densities for these satellites. Their mean densities, which range from 1.20 grams/cm^3 for Miranda to 1.69 grams/cm^3 for Titania, are compatible with their interiors being largely composed of water ice, which was detected earlier on their surfaces. All five satellites were found to be tidally locked to Uranus, as predicted by theory, so their rotation periods are the same as their orbital periods of revolution. Their rotation axes have become aligned nearly parallel to Uranus' rotation axis, so that only their southern hemispheres could be imaged. *Voyager 2* infrared measurements near the south poles of Miranda and Ariel gave tem-

peratures of -304.6°F (-187°C) and -308.2°F (-189°C), respectively, considerably warmer than the cloud layer of Uranus' atmosphere, but understandable for a surface that has been in continuous sunlight for about 20 years. None of the satellites show an appreciable atmosphere.

Voyager 2 obtained detailed images of parts of the sunlit surface of all five previously known satellites and also of Puck. Bright and dark (albedo) regions, craters with or without bright ray systems around them, **mountains**, cliffs, scarps, valleys, canyons, graben, faults, and other geological features are clearly seen on these images. Maps of the parts of the satellite surfaces that have been imaged have been made, and names have been assigned to many surface features. Summaries of the surface features of the six satellites with imaged surface features are given below. The images with the best resolution obtained were Ariel and Miranda because *Voyager 2* flew closest to them.

Oberon

Oberon, Uranus' most distant and second largest satellite, has extensive, heavily cratered terrain interrupted by canyons (rift valleys) and scarps. Some craters are surrounded by bright ray systems; others show **dark matter** on their floors. A prominent feature on Oberon's limb is a large mountain 7 mi (11 km) high and 28 mi (45 km) wide. Oberon shows the least evidence from modification of its cratered terrain by geologic activity of any Uranian satellite. Image resolution is poor due to Oberon's distance.

Titania

Titania, Uranus' largest and most massive satellite, shows similar geological features to those found on Oberon. Heavily cratered plains are the most extensive surfaces found here. Oberon also shows a global rift valley network related to global **tectonics**. Image resolution is somewhat better for Titania than Oberon, since *Voyager 2* was closer to Titania. A prominent system of canyons and scarps is the most noticeable class of features on Titania's surface; shadows indicate some of them may be as deep as 3.7 mi (6 km). Moderately cratered and smooth plains are also observed, which indicate resurfacing. These features indicate that more geologic activity occurred on Titania than on Oberon after the end of the heavy bombardment (crater formation) phase of their histories. Here the resurfacing material should be liquid or icy water, ammonia, or methane, perhaps mixed with rocky material, rather than terrestrial type volcanic lava. Impact and/or internal heating may have melted or softened these ices, which then erupted and flowed over the satellite surfaces, resurfacing them. This phenomenon is called cryovolcanism (sometimes "water volcanism"), and it may have been important in the geological histories of many of the satellites of the Jovian planets.

Umbriel

With an average albedo of 0.21, Umbriel is the darkest of Uranus' five largest satellites. It also has a more uniform albedo than Titania and Oberon, showing only a few striking albedo features on the part of its surface imaged by *Voyager 2*. Umbriel also has extensive, heavily cratered terrain and several groups of canyons, scarps, and lineaments which seem to be of more recent origin than the craters which they cut. Evidence is also seen of resurfacing in and near several craters, indicating cryovolcanism. Much of its surface may have been resurfaced with dark material of uncertain origin (cryovolcanic, dark ejecta from a crater, or from an external source).

Ariel

Ariel is similar in size and mass to Umbriel, but whereas Umbriel is the darkest of Uranus' five largest

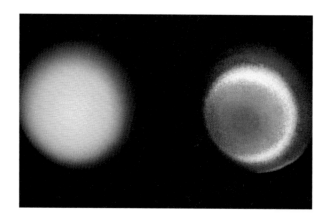

Unlike the other gas giants, Uranus reveals few atmospheric features in visible light. However, latitudinal atmospheric bands do exist, as can be seen in the enhanced image on the right; because Uranus has such an inclined axis of rotation (which is pointing almost directly down into this image), atmospheric bands that cross the surface of the planet (in the same way that atmospheric bands cross Jupiter's surface) appear as concentric circles in the photo. *U.S. National Aeronautics and Space Administration (NASA).*

satellites, Ariel is the brightest, with a 0.40 average albedo. Ariel's surface seems to have evolved more like Titania's surface than that of Umbriel; it shows global-scale faulting (canyons, scarps, and lineaments) and resurfacing by cryovolcanism, but on a more extensive scale than occurred on Titania. Ariel's most heavily cratered plains show less cratering than the most cratered units on Titania and Oberon. Evidence has been found for the extrusion of ice into Ariel's surface, filling part of a valley and partially burying an **impact crater**.

Miranda

Miranda shows the most interesting evidence for geologic activity and surface modification among Uranus' satellites. In addition to old cratered plains, canyons, scarps, lineaments, and valleys, the three coronae Arden, Elsinore, and Inverness exist on its surface; they are its most prominent surface features. They are large, lightly cratered regions of up to 186 mi (300 km) or more extent; Arden shows banded regions, while Elsinore and Inverness show numerous grooves and ridges. A cliff near Miranda's south pole may be as high as 12 mi (20 km). One plausible model for Miranda's **evolution** hypothesizes the following sequence of events. First, Miranda accreted from several smaller bodies near Uranus and nearly in its equatorial plane; impacts during accretion formed the old cratered plains. Some of the canyon systems were formed near the end of accretion. Next, a large body impacted Miranda, forming the Arden **basin**. Debris was scattered over and ejected from Miranda, and cryovolcanism flooded the basin, forming Arden Corona. Then the Inverness basin was flooded, forming Inverness Corona. The last main cryovolcanic activity formed Elsinore Corona. This left Miranda's surface early in its present state, with only a few more recent craters added since then.

Puck

Puck's surface was also imaged by *Voyager 2*, revealing a cratered surface that is considerably darker (0.08 average albedo) than that for any of the five larger satellites. The *Voyager 2* images also show that Puck is almost spherical. The three largest impact craters on Puck are named Bogle, Lob, and Butz.

See also Planetary atmospheres; Space probe.

Resources

Books

Beatty, J. Kelly, Carolyn Collins Petersen, and Andrew L. Chaikin. *The New Solar System.* Cambridge: Cambridge Univ. Press, 1999.

Bergstrahlh, Jay T., Ellis D. Miner, and Mildred S. Matthews, eds. *Uranus.* Tucson: University of Arizona Press, 1991.

de Pater, Imke and Jack J. Lissauer. *Planetary Sciences* Cambridge, UK: Cambridge University Press, 2001.

Morrison, D., and Tobias Owen. *The Planetary System.* 3rd ed. Addison-Wesley Publishing, 2002.

Taylor, F.W. *The Cambridge Photographic Guide to the Planets.* Cambridge University Press, 2002.

Periodicals

Beatty, J.K. "*Voyager 2*'s Triumph." *Sky & Telescope* 72, no. 4 (October 1986): 336-342.

Cuzzi, Jeffrey, and Larry Esposito. "The Rings of Uranus." *Scientific American* (July 1987): 52-66.

Dowling, T. "Big Blue: Twin Worlds of Uranus and Neptune." *Astronomy* 18, no. 10 (October 1990): 42-53.

Other

Arnett, B. SEDS, University of Arizona. "The Nine Planets, a Multimedia Tour of the Solar System." (November 6, 2002) [cited February 8, 2003]. <http://seds.lpl.arizona.edu/nineplanets/nineplanets/nineplanets.html>.

Frederick R. West
David T. King, Jr.

Urea

Urea is a white, crystalline solid also known as carbamide. It is highly soluble in **water** and is the major **molecule** used by **mammals** and **amphibians** as a means of excreting nitrogenous waste (which generally comes from **proteins**). It is used in making **fertilizers** (where it serves as source of **nitrogen**) and in cattle feed,

where it also raises the nitrogen levels. Urea is also used in the manufacturing of **barbiturates** and in manufacturing some **plastics** such as urethanes. Urea melts at 271°F (133°C). Solutions of urea in water are slightly basic. The formula of urea is shown in Figure 1.

The NH_2 groups are derived from nitrogen containing portions of proteins. Urea was first isolated from urine in 1773 by Hillaire-Malin Rouelle. In 1828 urea became the first organic (**carbon** based) molecule to be synthesized from inorganic components. This was accomplished by Freidrich Wolher by heating ammonium cyanate, forming urea. This synthesis began the decline of a "vital force" theory which held that only living things could make organic compounds.

Urea and metabolism

Urea is the final product of the **metabolism** of amino acids (the building blocks of proteins) in mammals, amphibians, and **turtles**. In the liver, **ammonia** reacts with **carbon dioxide** and through a series of seven steps that are controlled by enzymes (protein catalysts that speed up specific reactions), urea is produced. Each molecule of urea is "built" from two ammonia molecules (NH_3) and one carbon dioxide (CO_2) molecule. Ammonia itself is toxic so many animals have developed metabolic steps that take the ammonia formed and convert it into less toxic and easily dealt with molecules. The extremely high **solubility** of urea in water (35 oz [1,000 g] of urea dissolve in a gal [l] of water) make it ideal for eliminating nitrogen-based waste products. The **concentration** of urea rises in many kidney diseases, and **blood** urea concentration is often used to monitor kidney function. The urea produced by the body is excreted in urine. A healthy adult will excrete about 0.9 oz (25 g) of urea per day. Upon standing the urea in urine will decompose to carbon dioxide and ammonia, accounting for the "ammonia" smell of old urine. In many animals and in some vegetarians, cloudy urine is common. This is the result of a precipitate (insoluble compound) formed from urea and **calcium** or **magnesium** ions. Medically urea is used as a diuretic, or substance that promotes water loss through urination. Urea-containing creams are used on wounds.

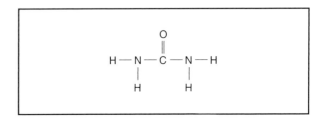

Figure 1. Structural formula of urea. *Illustration by Hans & Cassidy. Courtesy of Gale Group.*

KEY TERMS

. . . .

Barbiturates—A group of drugs widely used as tranquilizers and anti-anxiety medications. One of the basic molecules of this class is barbituric acid, which is made directly from urea.

Diuretic—A substance that increases water loss through urination.

Resin—A type of organic polymer. A polymer is a large molecule made of repeating basic units known as monomers.

Urea and industry

Industrially, urea is used in the manufacturing of fertilizers (as a source of nitrogen) and in the synthesis of some barbiturates and in the **petroleum** industry to help separate straight chain and branched hydrocarbons in petroleum. Urea is widely used in the production of many plastics and **resins**. One of the most common of these is a resin formed by the reaction of urea and formaldehyde. Urea-formaldehyde resins are used in the adhesives industry (urea-formaldehyde resins are used in making laminated woods), in textile finishes, and surface coatings for plastics. These resins are widely used since they accept dyes easily. Over 661,000 lb (300,000 kg) of urea-formaldehyde resins are produced annually in the United States. Urea is also used as a stabilizer for many **explosives**, allowing greater control over the reactions. Urea is made commercially by reacting ammonia and carbon dioxide under high **pressure**.

See also Excretory system.

Resources

Books

Atkins, P.W. *Molecules.* W. H. Freeman, 1987.

Louis Gotlib

Urinary system *see* **Excretory system**

Urology

Urology is the branch of medicine that deals with the urinary tract in females and with the urogenital tract in males. In both sexes, the urinary tract consists of the kidneys, ureters, bladder, and urethra. In males, additional structures such as the prostate gland are included in the urogenital system.

The problems with which a urologist deals tend to fall into three general categories: **infection**, **cancer**, and stone formation. Cystitis is any infection of the urinary tract. The condition often appears to be centered in the urinary bladder, but is usually associated with infections of other parts of the urinary system. Cystitis is accompanied by frequent and painful urination. It is treated relatively effectively with **antibiotics**, although other urinary tract problems with which it is associated may require other treatments.

Enlargement of the prostate gland is now one of the most common disorders among males, especially older males. In some cases, the condition is benign and is primarily a matter of inconvenience for men who find that urination becomes more difficult and more frequent. Non-cancerous enlargement of the prostate is known as benign prostatic hyperplasia (BPH). BPH can be treated surgically by the removal of excess fatty **tissue**, although a number of urologists now recommend the use of a newly approved drug known as Proscar as a way of shrinking the enlarged gland.

Cancer of the prostate has become one of the leading causes of death among older males in the United States and other parts of the developed world. Some physicians hypothesize that carcinogenic substances in the environment accumulate in the prostate, one of the fattiest organs in the body, and induce tumors. Prostate cancer can be treated with the same surgical techniques used with BPH, although the complete removal of the testicles may be recommended if the cancer has metastasized.

Kidney stones form when certain chemicals that normally dissolve in urine begin to precipitate out and form stones ranging from microscopic particles to marble-size structures. In the majority of cases, the stones are expelled from the urinary system without incident. In some cases, however, they may become lodged in various parts of the system: along the ureter, in the bladder, or in the prostate, for example. When this happens, the stone may cut into tissue and cause extreme **pain**.

Stones can be removed in a number of ways: surgically, with drugs that dissolve the stony material, or with ultrasound therapy. In the last of these treatments, high frequency **sound waves** are used to break apart a stone, allowing the smaller fragments to be carried away in urine.

See also Surgery.

Vaccine

A vaccine is a medical preparation providing immunity from a vaccine specific **disease**. Vaccines generally consist of a weakened (attenuated) or killed antigens, associated with a particular disease that are capable of stimulating the body to make specific antibodies to that disease. Vaccines use a variety of different substances ranging from dead **microorganisms** to genetically engineered antigens to defend the body against potentially harmful microorganisms. Effective vaccines change the **immune system** by promoting the development of antibodies that can quickly and effectively attack a disease causing microorganism when it enters the body, preventing disease development.

The development of vaccines against diseases ranging from polio and **smallpox** to **tetanus** and measles is considered among one of the great accomplishments of medical science. Contemporary researchers are continually attempting to develop new vaccinations against such diseases as Acquired Immune Deficiency Syndrome (**AIDS**), **cancer**, **influenza**, and other diseases.

Physicians have long observed that individuals who were exposed to an infectious disease and survived were somehow protected against that disease in the future. Prior to the invention of vaccines, however, infectious diseases swept through towns, villages, and cities with a horrifying vengeance.

The first effective vaccine was developed against smallpox, an international peril that killed thousands of its victims and left thousands of others permanently disfigured. The disease was so common in ancient China that newborns were not named until they survived the disease. The development of the vaccine in the late 1700s followed centuries of innovative efforts to fight smallpox.

The ancient Chinese were the first to develop an effective measure against smallpox. A snuff made from powdered smallpox scabs was blown into the nostrils of uninfected individuals. Some individuals died from the therapy; however, in most cases, the mild **infection** produced offered protection from later, more serious infection.

By the late 1600s, some European peasants employed a similar method of immunizing themselves against smallpox. In a practice referred to as "buying the smallpox," peasants in Poland, Scotland, and Denmark reportedly injected the smallpox **virus** into the skin to obtain immunity. At the time, conventional medical doctors in **Europe** relied solely on isolation and quarantine of people with the disease.

Changes in these practices took place, in part, through the vigorous effort of Lady Mary Wortley Montague, the wife of the British ambassador to Turkey in the early 1700s. Montague said the Turks injected a preparation of smallpox scabs into the **veins** of susceptible individuals. Those injected generally developed a mild case of smallpox from which they recovered rapidly, Montague wrote.

Upon her return to Great Britain, Montague helped convince King George I to allow trials of the technique on inmates in Newgate Prison. Success of the trials cleared the way for variolation, or the direct injection of smallpox, to become accepted medical practice in England until a vaccination was developed later in the century. Variolation also was credited with protecting United States soldiers from smallpox during the Revolutionary War.

Doubts remained about the practice. Individuals were known to die after receiving the smallpox injections.

The next leap in the battle against smallpox occurred when Edward Jenner (1749–1823) acted on a hunch. Jenner, a country physician and an orphan since the age of five, observed that people who were in contact with cows often developed cowpox, which caused pox but was not life threatening. Those people did not develop smallpox. In 1796, Jenner decided to test his hypothesis that cowpox could be used to protect humans against smallpox. Jenner injected a healthy eight-year-old boy with cowpox obtained from a milkmaid's sore. The boy was moderately ill and recovered. Jenner then injected the boy twice with the smallpox virus, and the boy did not get sick.

Jenner's discovery launched a new era in medicine, one in which the intricacies of the immune system would become increasingly important. Contemporary knowledge suggests that cowpox was similar enough to smallpox that the antigen included in the vaccine stimulated an immune response to smallpox. Exposure to cowpox antigen transformed the boy's immune system, generating cells that would remember the original antigen. The **smallpox vaccine**, like the many others that would follow, carved a protective pattern in the immune system, one that conditioned the immune system to move faster and more efficiently against future infection by smallpox.

The term vaccination, taken from the Latin for cow (*vacca*) was developed by Louis Pasteur (1822–1895) a century later to define Jenner's discovery. The term also drew from the word vaccinia, the virus drawn from cowpox and developed in the laboratory for use in the smallpox vaccine. In spite of Jenner's successful report, critics questioned the wisdom of using the vaccine, with some worrying that people injected with cowpox would develop **animal** characteristics, such as women growing animal hair. Nonetheless, the vaccine gained popularity, and replaced the more risky direct inoculation with smallpox. In 1979, following a major cooperative effort between nations and several international organizations, world health authorities declared smallpox the only infectious disease to be completely eliminated.

The concerns expressed by Jenner's contemporaries about the side effects of vaccines would continue to follow the pioneers of vaccine development. Virtually all vaccinations continue to have side effects, with some of these effects due to the inherent nature of the vaccine, some due to the potential for impurities in a manufactured product, and some due to the potential for human error in administering the vaccine.

Virtually all vaccines would also continue to attract intense public interest. This was demonstrated in 1885 when Louis Pasteur (1822–1895) saved the life of Joseph Meister, a nine year old who had been attacked by a rabid dog. Pasteur's series of experimental **rabies** vaccinations on the boy proved the effectiveness of the new vaccine.

Until development of the rabies vaccine, Pasteur had been criticized by the public, though his great discoveries included the development of the **food preservation** process called pasteurization. With the discovery of a rabies vaccine, Pasteur became an honored figure. In France, his birthday declared a national holiday, and streets renamed after him.

Pasteur's rabies vaccine, the first human vaccine created in a laboratory, was made of an extract gathered from the spinal cords of rabies-infected rabbits. The live virus was weakened by drying over potash. The new vaccination was far from perfect, causing occasional fatalities and temporary paralysis. Individuals had to be injected 14 to 21 times.

The rabies vaccine has been refined many times. In the 1950s, a vaccine grown in duck embryos replaced the use of live virus, and in 1980, a vaccine developed in cultured human cells was produced. In 1998, the newest vaccine technology—genetically engineered vaccines—was applied to rabies. The new **DNA vaccine** cost a fraction of the regular vaccine. While only a few people die of rabies each year in the United States, more than 40,000 die worldwide, particularly in **Asia** and **Africa**. The less expensive vaccine will make vaccination far more available to people in less developed nations.

The story of the most celebrated vaccine in modern times, the polio vaccine, is one of discovery and revision. While the viruses that cause polio appear to have been present for centuries, the disease emerged to an unusual extent in the early 1900s. At the peak of the **epidemic**, in 1952, polio killed 3,000 Americans and 58,000 new cases of polio were reported. The crippling disease caused an epidemic of fear and illness as Americans—and the world—searched for an explanation of how the disease worked and how to protect their families.

The creation of a vaccine for **poliomyelitis** by Jonas Salk (1914–1995) in 1955 concluded decades of a drive to find a cure. The Salk vaccine, a killed virus type, contained the three types of polio virus which had been identified in the 1940s.

In 1955, the first year the vaccine was distributed, disaster struck. Dozens of cases were reported in individuals who had received the vaccine or had contact with individuals who had been vaccinated. The culprit was an impure batch of vaccine that had not been completely inactivated. By the end of the incident, more than 200 cases had developed and 11 people had died.

Production problems with the Salk vaccine were overcome following the 1955 disaster. Then in 1961, an oral polio vaccine developed by Albert B. Sabin (1906–1993) was licensed in the United States. The continuing controversy over the virtues of the Sabin and Salk vaccines is a reminder of the many complexities in evaluating the risks versus the benefits of vaccines.

The Sabin vaccine, which used weakened, live polio virus, quickly overtook the Salk vaccine in popularity in the United States, and is currently administered to all healthy children. Because it is taken orally, the Sabin vaccine is more convenient and less expensive to administer than the Salk vaccine.

Advocates of the Salk vaccine, which is still used extensively in Canada and many other countries, contend that it is safer than the Sabin oral vaccine. No individuals have developed polio from the Salk vaccine since the 1955 incident. In contrast, the Sabin vaccine has a very small but significant **rate** of complications, including the development of polio. However, there has not been one new case of polio in the United States since 1975, or in the Western Hemisphere since 1991. Though polio has not been completely eradicated, there were only 144 confirmed cases worldwide in 1999.

Effective vaccines have limited many of the life-threatening infectious diseases. In the United States, children starting kindergarten are required to be immunized against polio, **diphtheria**, tetanus, and several other diseases. Other vaccinations are used only by populations at risk, individuals exposed to disease, or when exposure to a disease is likely to occur due to travel to an area where the disease is common. These include influenza, **yellow fever**, typhoid, **cholera**, and **hepatitis** A and B.

The influenza virus is one of the more problematic diseases because the viruses constantly change, making development of vaccines difficult. Scientists grapple with predicting what particular influenza strain will predominate in a given year. When the prediction is accurate, the vaccine is effective. When they are not, the vaccine is often of little help.

The classic methods for producing vaccines use biological products obtained directly from a virus or a **bacteria**. Depending on the vaccination, the virus or bacteria is either used in a weakened form, as in the Sabin oral polio vaccine; killed, as in the Salk polio vaccine; or taken apart so that a piece of the microorganism can be used. For example, the vaccine for *Streptococcus pneumoniae* uses bacterial polysaccharides, carbohydrates found in bacteria which contain large numbers of monosaccharides, a simple sugar. These classical methods vary in safety and efficiency. In general, vaccines that use live bacterial or viral products are extremely effective when they work, but carry a greater risk of causing disease. This is most threatening to individuals whose immune systems are weakened, such as individuals with **leukemia**. Children with leukemia are advised not to take the oral polio vaccine because they are at greater risk of developing the disease. Vaccines which do not include a live virus or bacteria tend to be safer, but their protection may not be as great.

The classical types of vaccines are all limited in their dependence on biological products, which often must be kept cold, may have a limited life, and can be difficult to produce. The development of recombinant vaccines—those using chromosomal parts (or DNA) from a different organism—has generated hope for a new generation of man-made vaccines. The hepatitis B vaccine, one of the first recombinant vaccines to be approved for human use, is made using recombinant **yeast** cells genetically engineered to include the **gene** coding for the hepatitis B antigen. Because the vaccine contains the antigen, it is capable of stimulating antibody production against hepatitis B without the risk that live hepatitis B vaccine carries by introducing the virus into the **blood** stream. To mitigate dangers of side effects, researchers and clinicians are preparing vaccines that consist of synthetic peptides representing only a small part of the antigen present in the enitre **organism** are currently in development.

As medical knowledge has increased—particularly in the **field** of DNA vaccines—researchers have set their sights on a wealth of possible new vaccines for cancer, melanoma, AIDS, influenza, and numerous others. Since 1980, many improved vaccines have been approved, including several genetically engineered (recombinant) types which first developed during an experiment in 1990. These recombinant vaccines involve the use of so-called "naked DNA." Microscopic portions of a viruses' DNA are injected into the patient. The patient's own cells then adopt that DNA, which is then duplicated when the **cell** divides, becoming part of each new cell. Researchers have reported success using this method in laboratory trials against influenza and **malaria**. These DNA vaccines work from inside the cell, not just from the cell's surface, as other vaccines do, allowing a stronger cell-mediated fight against the disease. Also, because the influenza virus constantly changes its surface **proteins**, the immune system or vaccines cannot change quickly enough to fight each new strain. However, DNA vaccines work on a core protein, which researchers believe should not be affected by these surface changes.

Since the emergence of AIDS in the early 1980s, a worldwide search against the disease has resulted in clinical trials for more than 25 experimental vaccines. These range from whole-inactivated viruses to genetically engineered types. Some have focused on a therapeutic approach to help infected individuals to fend off further illness by stimulating components of the immune system; others have genetically engineered a protein on the surface of HIV to prompt immune response against the virus; and yet others attempted to protect uninfected individuals. The challenges in developing a protective vaccine include the fact that HIV appears to have multiple viral strains and mutates quickly.

In January, 1999, a promising study was reported in Science magazine of a new AIDS vaccine created by injecting a healthy cell with DNA from a protein in the AIDS virus that is involved in the infection process. This cell was then injected with genetic material from cells involved in the immune response. Once injected into the individual, this vaccine "catches the AIDS virus in the act," exposing it to the immune system and triggering an immune response. This discovery offers considerable hope for development of an effective vaccine. As of June 2002, a proven vaccine for AIDS had not yet been proven in clinical trials.

Stimulating the immune system is also considered key by many researchers seeking a vaccine for cancer. Currently numerous clinical trials for cancer vaccines are in progress, with researchers developing experimental vaccines against cancer of the breast, colon, and lung, among other areas. Promising studies of vaccines made from the patient's own **tumor** cells and genetically engineered vaccines have been reported. Other experimental techniques attempt to penetrate the body in ways that could stimulate vigorous immune responses. These include using bacteria or viruses, both known to be efficient travelers in the body, as carriers of vaccine antigens. Such bacteria or viruses would be treated or engineered to make them incapable of causing illness.

Current research also focuses on developing better vaccines. The Children's Vaccine Initiative, supported by the World Health Organization, the United Nation's Children's Fund, and other organizations, are working diligently to make vaccines easier to distribute in developing countries. Although more than 80% of the world's children were immunized by 1990, no new vaccines have been introduced extensively since then. More than four million people, mostly children, die needlessly every year from preventable diseases. Annually, measles kills 1.1 million children worldwide; **whooping cough** (pertussis) kills 350,000; hepatitis B 800,000; Haemophilus influenzae type b (Hib) 500,000; tetanus 500,000; rubella 300,000; and yellow fever 30,000. Another eight million die from diseases for which vaccines are still being developed. These include pneumococcal **pneumonia** (1.2 million); acute respiratory virus infections (400,000), malaria (two million); AIDS (2.3 million); and rotavirus (800,000). In August, 1998, the Food and Drug Administration approved the first vaccine to prevent rotavirus—a severe diarrhea and vomiting infection.

The measles epidemic of 1989 was a graphic display of the failure of many Americans to be properly immunized. A total of 18,000 people were infected, including 41 children who died after developing measles, an infectious, viral illness whose complications include pneumonia and **encephalitis**. The epidemic was particularly troubling because an effective, safe vaccine against measles has been widely distributed in the United States since the late 1960s. By 1991, the number of new

measles cases had started to decrease, but health officials warned that measles remained a threat.

This outbreak reflected the limited reach of vaccination programs. Only 15% of the children between the ages of 16 and 59 months who developed measles between 1989 and 1991 had received the recommended measles vaccination. In many cases parent's erroneously reasoned that they could avoid even the minimal risk of vaccine side effects "because all other children were vaccinated."

Nearly all children are immunized properly by the time they start school. However, very young children are far less likely to receive the proper vaccinations. Problems behind the lack of immunization range from the limited health care received by many Americans to the increasing cost of vaccinations. Health experts also contend that keeping up with a vaccine schedule, which requires repeated visits, may be too challenging for Americans who do not have a regular doctor or health provider.

Internationally, the challenge of vaccinating large numbers of people has also proven to be immense. Also, the reluctance of some parents to vaccinate their children due to potential side effects has limited vaccination use. Parents in the United States and several European countries have balked at vaccinating their children with the pertussis vaccine due to the development of neurological complications in a small number of children given the vaccine. Because of incomplete immunization, whooping cough remains common in the United States, with 30,000 cases and about 25 deaths due to complications annually. One response to such concerns has been testing in the United States of a new pertussis vaccine that has fewer side effects.

Researchers look to **genetic engineering**, gene discovery, and other innovative technologies to produce new vaccines.

See also Immunology.

Resources

Periodicals

Borchers. A.T., Keen, C.L.Y. Shoenfeld, J. Silva Jr. M.E. Gershwin. "Vaccines, Viruses, and Voodoo." *Journal of Investigational Allergology and Clinical Immunology* 12 (2002) 155-68.

Dodd, D. "Benefits of Combination Vaccines: Effective Vaccination on a Simplified Schedule." *Americue Journal of Managed Care.* 9 Suppl. 1 (2003) S6-12.

Machiels, J.P., N. van Baren, M. Marchand. "Peptide-based Cancer Vaccines." *Seminars in Oncology* 29 (2002) 494-502.

Mwau, M., A.J. McMichael. "A Review of Vaccines for HIV Prevention." *Journal of Genetic Medicine* 5 (2003) 3-10.

Sibley, C.L. "Smallpox Vaccination Revisited." *American Journal of Nursing.* 102 (2002) 26-32.

Vacuum

Vacuum is a term that describes conditions where the **pressure** is lower than that of the atmosphere. A sealed container is said to be "under vacuum" in this case whereas it is "pressurized" when the pressure is higher than atmosphere. In a vacuum, it becomes necessary to define pressure microscopically. This means that the pressure, or **force** per unit area, is determined by the number of collisions between the **atoms** or molecules present and the walls of the container.

The first experiments involving vacuum date back to 1644 when Evangelista Torricelli worked with columns of mercury, leading to the first **barometer** (a device for measuring pressure). The famous experiment of Otto von Guericke in 1654 demonstrated the astounding force of vacuum when he evacuated the **volume** formed by a pair of joined hemispheres and attached each end to a team of **horses** that were unable to pull the hemispheres apart.

In order to create a vacuum, some kind of pump is needed. Simple mechanical pumps create a pressure difference, or suction force, which can be sufficient to pump **water**, for example. The most common use of vacuum, the vacuum cleaner, is simply a chamber and hose which are continuously evacuated by a fan. More sophisticated vacuum pumps must be sealed to prevent air from leaking back into the pumping volume too quickly. These pumps increase in complexity as better vacuum is needed. Pumps can generally be grouped into two categories: dynamic pumps, using mechanical or turbo-molecular action, and static pumps, using electrical ionization or low **temperature** (cryogenic) condensation.

Vacuum is important for research and industry, especially for manufacturing. Many industrial processes require vacuum either to be efficient or to be possible at all. Vacuum can be used for the prevention of **chemical reactions**, such as clotting in **blood plasma** or the removal of water in the process of freeze drying. Vacuum is also necessary for the prevention of particle collisions with background gas, in a **television** picture tube for example. For the fabrication of integrated **electronics**, it is very important to avoid impurities on a microscopic scale. It is only with excellent vacuum that such conditions can be obtained.

See also Atmospheric pressure; Vacuum tube.

Vacuum tube

A vacuum tube is a hollow **glass** cylinder containing a positive electrode and a **negative** electrode between which is conducted in a full or partial **vacuum**. A grid between these electrodes controls the flow of **electricity**.

The hollow cylinder of a vacuum tube contains a filament, typically tungsten coated with another **metal**. When the filament is sufficiently heated by an **electric current**, it emits electrons. This filament, or electrode, which emits electrons is known as a **cathode** and has a negative charge. Because it has a negative charge, it attracts electrons, thus nullifying the process. Therefore, free electrons must be supplied to the cathode. This is usually done by connecting the cathode to the negative terminal of a **generator** or **battery**. The other electrode, known as an **anode**, has a positive charge. The electrons move from the cathode to the anode, resulting in a one-way current within the tube.

History

In 1884, Thomas Edison, while working on his **incandescent light** bulb, inserted a metal plate between glowing filaments. He observed that electricity would flow from the positive side of the filament to the plate, but not from the negative. He did not understand why this was so and treated this effect (now known as the Edison effect) as a curiosity. Unwittingly, he had created the first **diode**.

Later, John Ambrose Fleming of England, one of Edison's former assistants, became involved in designing a **radio** transmitter for Guglielmo Marconi. In 1904 Fleming realized that the diode had the ability to convert alternating current (AC) into direct current (DC), and incorporated it into his very efficient radio wave detector. Fleming called his device the thermionic valve because it used **heat** to control the flow of electricity just as a valve controls the flow of **water**. In the United States the invention became known as a vacuum tube.

In Germany, Arthur Wehnelt, who also worked with thermionic **emission**, had applied for a patent in January 1904 for a tube that converted AC into DC. However, he neglected to mention the use of the device in radio wave detection and was unable to sell his invention for that purpose after Fleming applied for his own patent.

Lee de Forest (1873-1961) improved on Fleming's valve by adding a third element in 1906, thus inventing the triode. This made an even better radio wave detector but, like Edison, he did not realize the full potential of his invention; his device, called an audion, created an electrical current that could be amplified considerably.

In 1912 Edwin Howard Armstrong realized what de Fest had wrought. He used the triode to invent a regenerative circuit that not only received radio signals, it amplified them to such a degree they could be sent to a loudspeaker and heard without the use of headphones.

KEY TERMS

Amplitude—The farthest an object can get from its resting point, as in the highest position a pendulum reaches in its swing.

Filament—A fine wire heated to a high temperature and, thus, emitting electrons.

Kinetic energy—The energy possessed by an object due to the object's movement; for example, the energy in a baseball when it flies through the sky after being struck by a bat.

Semiconductor—A solid whose conductivity varies between that of a conductor (like a metal) at high temperatures and that of an insulator (such as rubber) at low temperatures.

Tungsten—A metal which makes a good conductor and has a high melting point.

Diodes were usually made of two concentric cylinders, one inside the other. The cathode emitted electrons and the anode collected them. Fleming's thermionic valve operated at a **temperature** of 4,532°F (2,500°C), generating a considerable amount of heat. Deforest placed a grid between the cathode and anode. The electrons passed through the triode's grid, inducing a larger current to flow.

These early vacuum tubes were called soft valves. The vacuum was not the best and some air remained within the tube, shortening its lifespan. Langmuir devised a more efficient vacuum pump in 1915; with a better vacuum, the tubes lasted longer and were more stable. The improved tubes were called hard valves and their operating temperature dropped to 3,632°F (2,000°C). In 1922 the temperature was reduced yet again, to 1,832°F (1,000°C), with the introduction of new elements. Indirect heating improved tube efficiency.

Triodes were limited to low frequencies of less than one megahertz. In 1927 American physicist Albert Wallace Hull (1880-1966) invented the tetrode to eliminate high-frequency **oscillations** and improve the frequency range. A year later the pentode, which improved performance at low voltage, was developed and became the most commonly used valve.

Over the course of years, a variety of vacuum tubes came into use. Low-voltage/low-power tubes were used in radio receivers as well as early digital computers. Photo tubes were used in sound equipment, making it possible to record and retrieve audio from **motion** picture film. The cathode-ray tube focused an **electron** beam, leading to the invention of oscilloscopes, televisions, and cameras. Microwave tubes were used in **radar**, early **space** communication, and microwave ovens. Storage tubes, which could store and retrieve data, were essential in the advancement of computers.

Despite its numerous advantages, the vacuum tube had many drawbacks. It was extremely fragile, had a limited life, was fairly large, and required a lot of power to operate its heating element. The successor to the vacuum tube, the **transistor**, invented by Walter Houser Brattain, John Bardeen, and William Shockley in 1948, had none of these drawbacks. After 1960 the small, lightweight, low-voltage transistors became commercially available and replaced vacuum tubes in most applications, but with the creation of microscopic vacuum tubes (microtubes) in the 1990s, vacuum tubes are again being used in electronic devices.

See also Cathode ray tube.

Resources

Books

Collins, A. Frederick. "Vacuum Tubes." *The Radio Amateur's Handbook.* Revised by Robert Herzberg. New York: Harper & Row, 1983.

Moyer, James A., and John F. Wostrel. *Radio Receiving and Television Tubes: Including Applications for Distant Control of Industrial Processes and Precision Measurements.* New York: McGraw-Hill, 1936.

Oldfield, R. L. "Electron Tubes." *Radio-Television & Basic Electronics.* Chicago: American Technical Society. 1960.

Stollberg, Robert, and Faith Fitch Hill. "Electrons in a Vacuum." *Physics Fundamentals and Frontiers.* Rev. ed. Boston: Houghton Mifflin Co, 1980.

Periodicals

"Cold Cathodes: Vacuum Microelectronics Enter the Flat-display Race." *Scientific American* 263 (October 1990): 127-128.

Goodman, Billy. "Return of the Vacuum Tube." *Discover* 11 (March 1990): 55-57.

Valence

Valence refers to a number assigned to elements that reflects their ability to react with other elements and the type of reactions the element will undergo. The term valence is derived from the Latin word for strength and can reflect an element's strength or affinity for certain types of reactions.

The electrons in an atom are located at different **energy** levels. The electrons in the highest energy level are called valence electrons. In accord with the octet rule—

and to become more energetically stable—atoms gain, lose, or share valence electrons in an effort to obtain a noble gas configuration in their outer shell. The configuration of electrons in an atom's outer shell determines its ability and affinity to enter into **chemical reactions**.

The valence number of an element can be determined by using a few simple rules relating to an element's location on the **periodic table**. In ionic compounds (formed between charged **atoms** or groups of atoms called ions) the valence of an atom is the number of electrons that atom will gain or lose to obtain a full outer shell. In group one of the periodic table, elements are assigned a valence number of 1. A valence number of 1 means that an element will generally react to lose one **electron** to obtain a full outer shell. Group two elements are assigned a valence number of 2. A valence number of 2 means that a group two element will generally react to lose two electrons to obtain a full outer shell. Group 17 elements are assigned a valence number of **negative** one (-1). A valence number of -1 means that a group two element will generally react to gain one electron to obtain a noble gas electron configuration. Reflecting an inability to react with other elements, Nobel gases, already maintaining a stable arrangement of electrons, are assigned a valence of zero (O).

The term valence can also refer to the charge or oxidation number on an atom. In **magnesium** atoms (Mg^{+2}) the valence is +2. An atom or ion with a charge of +2 is said to be divalent.

In covalent compounds the valence of an atom may be less obvious. In this case it is the number of bonds formed, that is, whether the bonds are single, double, or triple bonds. A **carbon** atom with two single bonds and one double bond carries a valence of four (4). In **water** (H_2O), the valence of **oxygen** is 2 and the valence of **hydrogen** is 1. In both cases the valence number gives an indication of the number of bonds each atom forms.

Valence bond theory is similar to molecular orbital theory in that it is concerned with the formation of covalent bonds. Valence bond theory describes bonds in term of interactions between outer orbitals and hybridized orbitals to explain the formation of compounds.

Valence Shell Electron Pair Repulsion (VSEPR) theory is one of the favored models to explain covalent bonds. This theory states that molecules will be shaped so as to minimize the repulsion that takes place between valence electrons. Because they are all negatively charged, valence shell electrons repel one another. VSEPR theory states that the atoms of a **molecule** will arrange themselves and assume a shape around a central atom so as to minimize repulsion between valence electrons.

See also Atomic models; Atomic number; Atomic theory; Chemical bond; Chemistry.

Van Allen belts

Radiation belts are enormous populations of energetic, electrically charged particles—principally protons and electrons—trapped in the external magnetic **field** of a **planet**. Durable radiation belts exist at the planets **Earth**, **Jupiter**, **Saturn**, **Uranus**, and **Neptune** but not at Mercury, **Venus**, or **Mars**.

Discovery of the radiation belts of Earth

The first successfully launched American **satellite** of Earth was *Explorer I*. It was propelled into **orbit** from Cape Canaveral, Florida, on January 31, 1958, by a four-stage combination of rockets developed by the United States Army Ballistic Missile Agency and the Jet Propulsion Laboratory (JPL). The principal scientific instrument within the payload was a Geiger tube radiation detector developed by Professor James A. Van Allen and graduate student George H. Ludwig of the University of Iowa. The instrument's intended purpose was a comprehensive survey of **cosmic ray** intensity above Earth's atmosphere.

The launch of an improved version of the radiation instrument was attempted on March 5 on *Explorer II*, but an orbit was not achieved because the fourth stage rocket failed to ignite. On March 26, the launch of the improved instrument on *Explorer III*, including the first magnetic tape recorder ever flown in **space**, was successful.

The in-flight data from the **radiation detectors** on *Explorer I* and *Explorer III* revealed that there are enormous numbers of energetic, electrically charged particles trapped in the external magnetic field of Earth. This discovery was promptly confirmed and extended later in 1958 with additional space flights by the Iowa group and others, including Soviet investigators on *Sputnik III*. In subsequent years the continuing study of this phenomenon has included the efforts of over 1000 scientists in at least 20 different countries.

Description

The populations of energetic particles trapped in **Earth's magnetic field** have come to be known as radiation belts because the doughnut-shaped regions within which they are confined encircle Earth like huge belts. There are two distinct belts: an inner one whose lower boundary is at an altitude of about 250 mi (402 km) and whose less-well defined outer boundary is at a radial distance of about 10,000 mi (16,100 km) and an outer one which extends outward from 10,000 mi (16,100 km) to over 50,000 mi (80,500 km). Both belts encircle Earth in longitude and have the greatest concentration of trapped particles at its magnetic equatorial **plane**. The concentration diminishes with increasing latitude north and south of the magnetic equator and falls to nearly zero over the north and south polar caps at latitudes greater than about 67°. Each trapped particle spirals around a magnetic line of **force**, oscillates between magnetic "mirror" points in northern and southern hemispheres, and drifts slowly in longitude. This defines a doughnut-shaped region.

The principal source of particles in the outer belt is **solar wind**, a hot ionized gas that flows outward from the **Sun** through interplanetary space. Some of the electrons, protons, and other ions in the impinging solar **wind** are injected into Earth's external magnetic field. These electrons, protons, and ions subsequently diffuse inward and are accelerated to greater energies by natural fluctuations of electrical and magnetic fields induced by the varying solar wind. The solar wind also sweeps back the outer portion of Earth's magnetic field to produce a long wake called the *magnetotail* extending on the night side of Earth to a distance of about 4,000,000 mi (6,440,000 km) far beyond the orbit of the **Moon**. In the **heart** of the outer belt, typical intensities of electrons with energies greater than 1,000,000 **electron** volts (1.0 MeV) (the most penetrating component there) are 2,000,000 particles per square inch per second. The numbers of particles of lesser **energy** and lesser penetration are much greater. The outer belt exhibits marked variations on time scales of hours, weeks, and months.

In the heart of the relatively stable inner belt, typical intensities of protons of energy exceeding 30,000,000 electron volts (30 MeV) are 120,000 particles per square inch per second. The residence times of such protons are many years. Penetrating particles in the inner belt are attributable to neutrons from reactions of cosmic rays in the gas of the upper atmosphere. A small fraction of such (uncharged) neutrons decay into protons, electrons, and neutrons as they move outward. At the points of decay the electrically charged protons and electrons are injected into trapped orbits. A radiation belt of this type would be created around a magnetized planet by cosmic rays even in the absence of the solar wind, though no such example has been found.

The fate of trapped particles is loss into the atmosphere or outward into space. A quasi-equilibrium population of any specified **species** of trapped particle is achieved when losses are equal to sources.

Radiation belts are part of a more complex system called the *magnetosphere* which also contains large populations of relatively low energy ionized gas (**plasma**). This gas plays a central role in the overall physical dynamics of the system.

Artificial radiation belts

Nine artificial radiation belts of Earth were produced during the period 1958-62 by the injection of electrons from radioactive nuclei produced by United States and USSR nuclear bomb bursts at high altitudes. These experiments made important contributions to understanding the natural belts. Since 1962 such high altitude bursts have been prohibited by international treaty.

Related geophysical effects

The aurorae, in both northern and southern latitudes (northern and southern lights) at typical but widely fluctuating magnetic latitudes of about 67°, are one of the visible, widely observed manifestations of magnetospheric phenomena. Aurorae are not a direct product of the outer radiation belt but share the solar wind as the primary agent for their creation. Magnetic storms detected by the fluctuations of sensitive magnetic compasses are attributable to electrical **currents** in the outer radiation belt.

Limitations on space flight

The inner radiation belt imposes an altitude ceiling on the region around Earth within which orbital flights of humans and animals can be conducted without exposing the occupants to excessive or fatal radiation exposures. Prolonged flights of human crews above an altitude of about 250 mi (402 km) are unsafe though rapid traversals of the radiation belts requiring only a few hours (as in the Apollo missions to the Moon) result in moderate exposures.

Two common misperceptions

Contrary to some common statements, trapped particles are *not* radioactive. Rather they are mainly ordinary electrons and protons such as those accelerated in high energy **physics** laboratories. Radiation belts do *not* shield Earth's surface, though the magnetic field deflects some cosmic rays away from Earth. The atmosphere acts as an effective shield against many solar and other radiations that impinge on it.

Radiation belts of other planets

In 1958-59 **radio** astronomers discovered that the planet Jupiter has an enormous radiation belt of high energy electrons. This discovery provided a powerful impe-

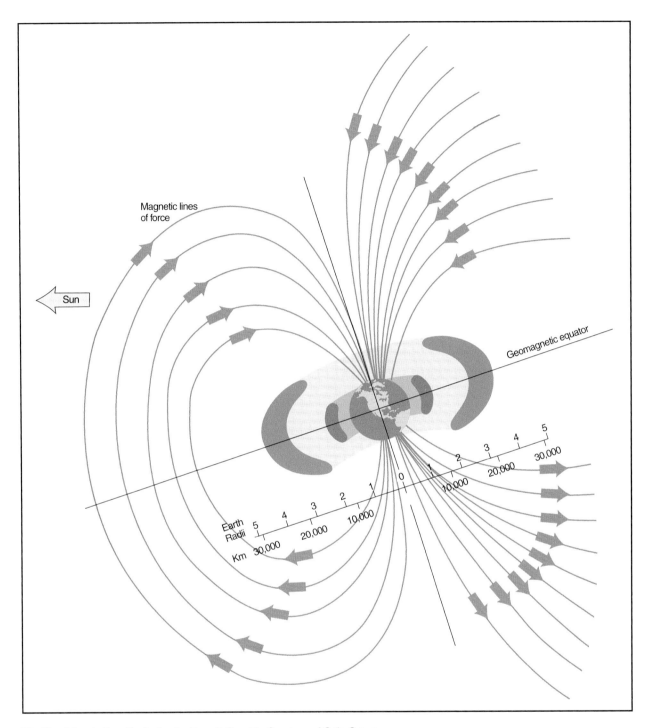

The Van Allen belts. *Illustration by Hans & Cassidy. Courtesy of Gale Group.*

tus for the investigation of Jupiter and the other planets by scientifically instrumented spacecraft.

Beginning in 1962 in situ investigations of the planets have been conducted by American, Soviet, and European spacecraft. The radiation belt of Jupiter has been explored in detail. Enormous radiation belts of Saturn, Uranus, and Neptune also have been discovered and investigated. It has been found that Venus, Mars, and Mercury have no durable radiation belts, but there are significant magnetospheric effects at these planets because of the obstacles that they present to the flow of the solar wind. Such effects have also been observed at the Moon and at three **comets**.

A durable radiation belt at a planet can only exist if that planet is strongly magnetized so that energetic electrically charged particles can be trapped durably in its external magnetic field. Earth, Jupiter, Saturn, Uranus, and Neptune meet this condition by virtue of electrical currents circulating in their interiors to produce huge electromagnets.

Venus, Mars, Mercury, the Moon, comets, and asteroids are insufficiently magnetized to retain radiation belts. It is likely that the planet **Pluto** is also in this group.

Pulsars and other distant astrophysical objects have radiation belts.

See also Magnetosphere.

Resources

Books

Introduction to Astronomy and Astrophysics. 4th ed. New York: Harcourt Brace, 1997.

Smolin, Lee. *The Life of the Cosmos.* Oxford: Oxford University Press, 1999.

Van Allen, James A. "Magnetospheres, Cosmic Rays, and the Interplanetary Medium" *The New Solar System.* 3rd ed. Cambridge: Sky Publishing Corporation and Cambridge University Press, 1990.

Van Allen, James A. *Origins of Magnetospheric Physics.* Washington, DC: Smithsonian Institution Press, 1983.

Periodicals

Van Allen, James A. "Radiation Belts Around the Earth." *Scientific American* 200 (March 1958): 39-47.

James A. Van Allen

Van de graaf accelerator *see* **Accelerators**

Van der Waals forces

Van der Waals forces are weak attractive forces between electrically neutral **atoms** or molecules. They are much weaker than the electrostatic forces which bind charged atoms or molecules (ions) of opposite sign or the covalent forces that bond neighboring atoms by sharing electrons. These forces develop because the rapid shifting of electrons within molecules causes some parts of the **molecule** to become momentarily charged, either positively or negatively. For this reason, weak, transient forces of attraction can develop between particles that are actually neutral. The magnitude of the forces is dependent on the distance between neighboring molecules. Van der Walls forces cause gas molecules to condense first to a liquid and finally to a solid as the gas is cooled.

The forces are named for the Dutch physicist Johannes Diederik van der Waals (1837-1930). The discovery of these forces evolved from van der Waals's research on the mathematical equations describing the gaseous and liquid **states of matter**. These equations are generally known as gas laws and relate the **temperature**, pressures and **volume** of gases. Originally derived for an idealized gas, these equations assumed that gas molecules had zero volume and that there were no attractive forces between them. In 1881, van der Waals proposed an empirical gas law which included two parameters to account for molecular size and attraction. This more accurate model was the first serious attempt to formulate gas laws for real gases, and it earned van der Waals the Nobel Prize for **physics** in 1910.

See also Gases, properties of.

Vanadium *see* **Element, chemical**
Vanilla *see* **Orchid family**

Vapor pressure

Definition

Vapor **pressure** is a **force** exerted by the gaseous phase of a two phase—gas/liquid or gas/solid system.

All liquids and solids have vapor pressure at all temperatures except at **absolute zero**, -459°F (-273°C). The pressure of the vapor that is formed above its liquid or solid is called the vapor pressure. If a substance is in an enclosed place the two phase system will arrive to an equilibrium state. This equilibrium state is a dynamic, balanced condition with no change of either phase. The pressure of the vapor measured at equilibrium state is the equilibrium vapor pressure. This pressure is a fraction of the total pressure, which is equal to 760 mm Hg at **sea level**. For a given substance, vapor pressure is constant under isothermal and isobarometric conditions, but its value depends on the **temperature**, pressure, and on the nature of the substance. As temperature increases so does the vapor pressure. At a constant temperature and pressure existing inter-molecular forces of the substance are the determining factors of the vapor pressure. The molecules of polar liquids and solids are held together with relatively large inter-molecular forces (e.g., dipole-dipole forces and **hydrogen** bounding). Polar compounds such as **water**, **acetic acid,** and ethyl alcohol have low vapor pressure at a given temperature. Non-polar liquids like **ether**, hexane, and **benzene** or solids like naphthalene have relatively small intermolecular forces (no hydrogen

TABLE 1. EQUILIBRIUM VAPOR PRESSURE OF WATER			
TEMPERATURE °C	WATER VAPOR PRESSURE mm Hg	TEMPERATURE °C	WATER VAPOR PRESSURE mm Hg
0	4.6	60	149
10	9.2	70	237
20	17.5	80	355
30	31.8	90	526
40	55	100	760
50	93		

bounding or dipole-dipole forces). These substances have relatively high vapor pressure and are known as volatile substances. However, it should be noted that substances of high **molecular weight** evaporate more slowly than similar substances of low molecular weight.

The atmosphere has considerable water vapor in it; noted in the **weather** report as relative **humidity**. This relative humidity can be calculated by the equation below.

$$\% \text{ Relative Humidity} = \frac{\text{Actual partial pressure of water vapor}}{\text{Equilibrium vapor pressure of water at existing temperature}} \times 100$$

The water vapor present in the air is temperature, geography, and weather dependent. Many living systems, including humans, are effected by humidity. On a cold, wintry day the air is dry due to the very low water vapor pressure (as low as 4 mm Hg) in the air. On a hot, humid summer day the humidity can be above 40 mm Hg, in which is close to the equilibrium vapor pressure resulting ~ 90% relative humidity. People use devices like humidifiers and dehumidifiers to compensate for such extreme conditions and keep the relative humidity level around 55-60%.

The equilibrium vapor pressure of water at different temperatures is given in Table 1.

Jeanette Vass

Variable

A variable is a mathematical symbol which is used to represent a member of a given set and is typically de-noted by a letter such as x, y, or z. The idea of a variable, invented during the late sixteenth century, is characteristic of modern **mathematics** and was not widely used in ancient times. Since a variable reflects a quantity which can take on different values, its use has become a critical part of nearly all disciplines which use mathematical models to represent the real world.

The idea of using letters to represent variables was first suggested by the sixteenth century mathematician François Viète (1540-1603) in his work, *In artem analyticam isagoge* (1591). Although his notations barely resemble our modern day symbolism, they were an important step in the development of the concept of using a letter to represent a changing value in a mathematical equation. His ideas were further developed in the decades that followed. Rene Descartes (1596-1650) is generally credited for making standard, the use of the letters x, y, and z for variables.

Characteristics of a variable

A variable is often denoted by a letter in an algebraic expression and represents a value which can be changed or varied. For example, in the expression x + 2, the letter x is a real variable and can take on the value of any real number. If x is 4 then the expression has a value of 6 because 4 + 2 = 6. Similarly, if x is 10 the expression has a value of 12. The number 2 in this expression is known as a constant because it never changes. Generally, a constant can be any number or letter in an equation whose value does not change.

In an equation, the value of a variable is often not given and is therefore called an unknown. In the equation y + 7 = 12, the letter y is an unknown variable and it represents some number. The value of the unknown which makes the equation true is called the solution or root of the equation. In this example, the solution of the equation is y = 5 because 5 + 7 = 12. Often, there is more than one solution to an equation so the unknown variable is equal to all of these values. The solution to the equation $x^2 = 4$ is both 2 and -2 because each of these values make the equation true.

Variables in a function

Some algebraic equations, known as functions, represent relationships between two variables. In these functions, the value of one variable is said to depend on the value of the other. For instance, the sales tax on a pair of gym shoes depends on the price of the shoes. The **distance** a car travels in a given time depends on its speed. In these examples, the sales tax and the distance travelled are called dependent variables because their value depends on the value of the other variable in the **function**. This variable, known as the independent variable, is represented by the price of the gym shoes and the speed of the car.

Using variables to represent unknowns was an important part of the development of **algebra**. Variables have distinct advantages over the rhetorical (written out) algebra of the ancient Greeks. They allow mathematical ideas to be communicated clearly and briefly. The equation $2x^2 + y = 6$ is much clearer than the equivalent phrase "two times some number times itself, plus some other number is equal to six." Variables also make mathematics more generally applicable. For instance, the area of a certain square with sides of 2 cm is 4 cm². The area of another square with 3 cm sides is 9 cm². By representing the side of any square with the variable s, the area of any square can be represented by s^2.

Although any letter or character can represent any variable, over time, mathematicians and scientists have used certain letters to represent certain values. The letters x, y, and z are the most commonly used variables to represent unknown values in polynomial equations. The letter r is often used to represent the radius of a **circle** and the character q is used to signify an unknown **angle**. Other commonly used variables include t to represent time, s to represent speed, and p to represent **pressure**.

See also Solution of equation.

Resources

Books

Bittinger, Marvin L., and Davic Ellenbogen. *Intermediate Algebra: Concepts and Applications.* 6th ed. Reading, MA: Addison-Wesley Publishing, 2001.

Paulos, John Allen. *Beyond Numeracy.* New York: Alfred A. Knopf, Inc., 1991.

KEY TERMS

. . . .

Algebraic expression—A symbolic representation of a mathematical statement made up of numbers, letters, and operations.

Constant—A part of an algebraic expression which does not change, such as a number.

Dependent variable—A variable in a function whose value depends on the value of another variable in the function.

Function—A mathematical relationship between two or more variables.

Independent variable—A variable in a function whose value determines the value of the dependent variable.

Rhetorical algebra—A method of communicating mathematical ideas by using words to express relationships between values.

Solution of the equation—The value of a variable that makes an equation true.

Unknown—A term used to describe a variable whose value is not evident.

Weisstein, Eric W. *The CRC Concise Encyclopedia of Mathematics.* New York: CRC Press, 1998.

Perry Romanowski

Variable stars

Most people regard the stars as constant and unchanging. A character in one of Shakespeare's plays refers to a friend "as constant as the Pole Star." While Shakespeare was probably referring to the constant position of the Pole Star, he did not know about the **precession of the equinoxes**. Or, if he was referring to the constant **light** of the Pole Star (Polaris), he was in error there also. Astronomers now know that the light from Polaris varies, visible only with a **telescope**. Polaris is an example of what astronomers call a variable star. Originally, the term was used only for stars that vary a great deal in brightness, but now it is used to cover a wide range of stars. In fact, all stars probably vary in one way or another. The **Sun**, for example, has very slight variations in its **energy** output during the life of one of its sunspot cycles. Here, we will restrict ourselves to stars that are much more variable than the Sun.

Most people today do not observe the stars closely enough to note any of their brightness changes. Ancient people with darker skies did observe some stars changing in brightness. The Arabs noted that a bright **star** in the **constellation** Perseus dropped to about half its normal brightness for two hours every three days. They named this star Algol meaning the demon star. We know today that Algol is an example of an eclipsing **binary star** system. The change in brightness results when a dimmer companion star moves in front of the brighter star, blocking part of the light of the brighter star.

But the change in brightness of most variable stars is due to changes taking place within a single star. These stars are known as intrinsic variable stars. Astronomers have identified several classes of intrinsic variable stars. Some variable stars are classed as periodic, meaning that they go through regular changes in brightness every few hours or within a few days or weeks. This group includes such stars as the cepheid variables and RR Lyrae variables. Other stars are only roughly periodic or non-periodic in their brightness changes and are known as semiregular and irregular variables, respectively. The variable stars that are most apt to capture the attention of the public are called **nova** and **supernova** stars. While these are not often seen, they sometimes become bright enough so that they are easy to observe without the use of a telescope.

Cepheid variables are so named because one of the first of this class studied is found in the constellation Cepheus. These stars are all supergiant stars and are many times larger than the Sun. These stars vary because of pulsations within the stars themselves. They go through a cycle of expansion (brightening) and contraction (dimming). Over a period of many weeks the brighter cepheids may vary by a **factor** of 300% in their energy output. Polaris is an example of a dimmer cepheid and it varies only about 7% over a period of four days. A general pattern of brightness variations of cepheid variables was discovered by a Harvard astronomer, Henrietta Leavitt, early in the twentieth century. She discovered that the brighter cepheids had longer periods of brightness variation and the shorter period cepheids were generally dimmer. This period-luminosity relationship later made it possible for Edwin Hubble to demonstrate that there are galaxies other than our own **Milky Way galaxy**.

They are called the RR Lyrae stars because, when studying this class, one of the first stars was found in the constellation Lyra. These stars have very short periods of light variation (usually less than one day) and they are dimmer than the cepheids. They are usually found in large spherical collections of stars called globular clusters. We now know that globular clusters uniformly surround our galaxy. In the 1920s, Harlow Shapley used RR Lyrae variables to map the distribution of globular clusters and concluded that the Sun was not in the center of our galaxy as was generally believed at the time. He found that our Sun and **solar system** are located about 30,000 light years from our galactic center.

Some stars vary because they have large spots on their surface. The amount of area covered by these spots changes with time and this effects the output of energy from their surfaces. The Sun is one example of a spotted star. The study of how other spotted stars change their energy output has lead another Harvard astronomer, Sallie Baliunas, to suggest that there may be a link between the number of spots on the sun and the climate patterns on the **earth**. Records suggest that when the spots are at a minimum for a long period of time, as they were in the late 1600s, there was a significant change in our climate.

Ancient astronomers noted that once in awhile stars would appear where they had not been seen before. These were called nova stars from the Latin word for "new." Later it was noted that some of these stars were much brighter than other nova and the name supernova was coined. Nova are usually noticed only by astronomers but historical records indicate that some of the supernova were so bright that they could be seen by ordinary people.

Astronomers now believe that nova always occur in double-star systems. Over a long period of time, one of the members of the pair evolves and becomes a **white dwarf** star. Later, the other member of the pair evolves to become a giant star and a portion of its outer layer is drawn onto the surface of the white dwarf. A temporary nuclear explosion, similar to a **hydrogen** bomb, takes place.

A supernova, on the other hand, usually results when a single, massive star evolves until it has used up most of its hydrogen fuel. The star then collapses on itself with a tremendous explosion, leaving behind small but massive remnants that become either a **neutron star** or a **black hole**. If an observer is located in the right orientation, the neutronstar may be observed as a object called a **pulsar**. These objects are some of the most unusual variable stars. They are thought to be highly magnetic, rotating objects only a few miles in diameter. Pulsars release their electromagnetic energy in regular bursts as frequently as a fraction of a second up to five seconds.

In a few seconds a supernova can produce as much energy as 10 billion Suns. It becomes a brilliant star and even can be seen in the daytime. In our own galaxy we have seen only a few in all of recorded history. The Chinese reported the appearance of supernovas in A.D. 1006, 1054, and 1181. The July 4, 1054 supernova was so spectacular that there are rock carvings recording the

KEY TERMS

. . . .

Black hole—A supermassive object with such a strong gravitational field that nothing, not even light, can escape it.

Cepheid variable star—Stars that belong to a class of supergiant pulsating stars.

Eclipsing binary star system—A double star system in which the plane of their mutual orbit is nearly parallel to our line of sight. As one star passes in front of the other star it periodically dims the other.

Intrinsic variable star—Any star that changes its energy output because of changes within the star or on the surface of the star.

Light year—A light year is equal to the distance that light would travel in one year and equals about 6 trillion miles or 9 trillion kilometers.

Neutronstar—The result of the collapse of a supernova star. It is believed that they are almost completely composed of neutrons, hence their high densities.

Nova—A star that has a temporary increase in energy output on the order of a thousand times its normal energy output.

Period-luminosity relationship—An empirical rela-

tionship exhibited by cepheid variables between the variation in their energy output and the time period for this variation.

Pulsars—Small, dense, rapidly rotating neutron stars whose magnetic fields direct their energy output in much the same way as a lighthouse been that sweeps over the observer.

RR Lyrae variable stars—A class of giant pulsating stars who have periods of about a day.

Sunspot—Cooler and darker areas on the surface of the sun. They appear dark only because they are cooler than the surrounding surface. Sunspots appear and disappear in cycles of approximately 11 years.

Supernova—The final collapse stage of a supergiant star.

Variable star—Any star that has a variation in energy output.

White dwarf—A star that has used up all of its thermonuclear energy sources and has collapsed gravitationally to the equilibrium against further collapse that is maintained by a degenerate electron gas.

event by Native Americans of the American Southwest. Later, in 1572 and 1604, two other supernova were observed and recorded in **Europe**, and these are the last seen in our galaxy. Most astronomers believe that we are overdue for a supernova in our own galaxy. Life on any **planet** near a supernova would be instantly vaporized, but we should not worry. The Sun is not massive enough to become a supernova. The closest likely candidate for a supernova is the supergiant star Betalguese in Orion, and it is located nearly 500 light years away. At this distance, there would likely not be any terrestrial effects if the star did become a supernova.

See also Stellar evolution.

Resources

Books

Abell, George, David Morrison, and Sydney Wolff. *Exploration of the Universe.* Sixth edition. Philadelphia, PA: Saunders College Publishing, 1993.

Bacon, Dennis Henry, and Percy Seymour. *A Mechanical History of the Universe.* London: Philip Wilson Publishing, Ltd., 2003.

Hartman, William. *The Cosmic Voyage.* Belmont, CA: Wadsworth Publishing, 1992.

Hoff, Darrel, Linda Kelsey, and John Neff. *Activities in Astronomy.* 4th ed. Dubuque: Kendall/ Hunt Publishers, 1995.

Pasachoff, Jay. *Astronomy: From the Earth to the Universe.* 4th ed. Philadelphia: Saunders College Publishing, 1991.

Zeilik, Michael. *Astronomy: The Evolving Universe.* 6th ed. New York: John Wiley & Sons, Inc., 1991.

Periodicals

Wheeler, J., and K. Nomoto. "How Stars Explode." *American Scientist* 73 (1985): 240.

Darrel B. Hoff

Variance

Variance is a mathematical expression of how data points are spread across a data set. Such expressions are known as measures of dispersion since they indicate how values are dispersed throughout a population. The variance is the average or **mean** of the squares of the **distance** each data point in a set is from the mean of all the data points in the set. Mathematically, variance is represented as σ^2, according to the equation: $\sigma^2 = [(x_1 - \mu)^2 +$

$(x_2-\mu)^2 + (x_3-\mu)^2 +...(x_n-\mu)^2]/n$; where $x_{1,2,3,.....n}$ are the values of specific variables; μ is the mean, or average, of all the values; and n is the total number of values. Variance is commonly replaced in applications by its **square root**, which is known as the standard deviation or σ.

Variance is one of several measures of dispersion which are used to evaluate the spread of a distribution of numbers. Such measures are important because they provide ways of obtaining information about data sets without considering all of the elements of the data individually.

To understand variance, one must first understand something about other measures of dispersion. One measure of dispersion is the "average of deviations." This value is equal to the average, for a set of numbers, of the differences between each number and the set's mean. The mean (also known as the average) is simply the sum of the numbers in a given set divided by the number of entries in the set. For the set of eight test scores: 7 + 25 + 36 + 44 + 59 + 71 + 85 + 97, the mean is 53. The deviation from the mean for any given value is that value minus the value of the mean. For example, the first number in the set, 7, has a deviation from the mean of -46; the second number, 25, has a deviation from the mean of -28; and so on. However, the sum of these deviations across the entire data set will be equal to 0 (since by definition the mean is the "middle" value with all other values being above or below it.) A measure that will show how much deviation is involved without having these deviations add up to **zero** would be more useful in evaluating data. Such a nonzero sum can be obtained by adding the absolute values of the deviations. This average is the absolute mean deviation. However, for reasons that will not be dealt with here, even this expression has limited application.

A still more informative measure of dispersion can be obtained by squaring the deviations from the mean, adding them, and dividing by the number of scores; this value is known as the average squared deviation or "variance." For example, in the series of test scores cited above, the variance can be calculated as follows:

$$\delta^2 = [(x_1-\mu)^2+(x_2-\mu)^2+(x-\mu)^2+...(x_n-\mu)^2]/n$$
$$\delta^2 = [(7-53)^2+(25-53)^2+(36-53)^2+(44-53)^2+$$
$$(59-53)^2+(71-53)^2+(85-53)^2+(97-53)^2]/8$$
$$\delta^2 = [(-46)^2+(-28)^2+(-17)^2+(-9)^2+(7)^2+$$
$$(18)^2+(32)^2+(44)^2]/8$$
$$\delta^2 = [2116+784+289+81+49+324+ 1024+1936]/8$$
$$\delta^2 = 6603/8$$
$$\delta^2 = 825.38$$

Theoretically, the value of σ^2 should relate valuable information regarding the spread of data. However, in order for this concept to be applied in practical situations (we can-

KEY TERMS

. . . .

Absolute deviation from the mean—The sum of the absolute values of the deviations from the mean.

Average deviation from the mean—For a set of numbers, the average of the differences between each number and the set's mean value.

Measure of dispersion—A mathematical expression which provides information about how data points are spread across a data set without having to consider all of the points individually.

Standard deviation—The square root of the variance.

not talk about squared test scores) we may elect to use the square root of the variance. This value is called the standard deviation of the scores. For this series of test scores the standard deviation is the square root of 825.38 or 28.73. In general, a small standard deviation indicates that the data are clustered closely around the mean; a large standard deviation shows that the data are more spread apart.

While modern computerization reduces the need for laborious statistical calculations, it is still necessary to understand and interpret the concept of variance and its daughter, standard deviation, in order to digest the statistical significance of data. For example, teachers must be thoroughly familiar with these statistical tools in order to properly interpret test data.

See also Set theory; Statistics.

Resources

Books

Dunham, William. *Journey Through Genius.* New York: John Wiley & Sons Inc., 1990.

Facade, Harold P., and Kenneth B. Cummins. *The Teaching of Mathematics from Counting to Calculus.* Columbus, OH: Charles E. Merrill Publishing Co., 1970.

Lloyd, G.E.R. *Early Greek Science: Thales to Aristotle.* New York: W.W. Norton and Company, 1970.

Randy Schueller

Varicella zoster virus

Varicella zoster virus is a member of the alphaherpesvirus group and is the cause of both **chickenpox** (also known as varicella) and **shingles** (herpes zoster).

The **virus** is surrounded by a covering, or envelope, that is made of **lipid**. As such, the envelope dissolves readily in solvents such as **alcohol**. Wiping surfaces with alcohol is thus an effective means of inactivating the virus and preventing spread of chickenpox. Inside the lipid envelope is a protein shell that houses the deoxyribonucleic acid.

Varicella zoster virus is related to herpes simplex viruses types 1 and 2. Indeed, **nucleic acid** analysis has revealed that the genetic material of the three viruses is highly similar, both in the genes present and in the arrangement of the genes.

Chickenpox is the result of a person's first **infection** with the virus. Typically, chickenpox occurs most often in children. Recovery is usually complete within a week or two and immunity to another bout of chickenpox is usually, but not always, life-long. Treatment for chickenpox is available. Fortunately for adults, a **vaccine** to chickenpox exists for those who have not contracted chickenpox in their childhood.

Naturally acquired immunity to chickenpox does not prevent someone from contracting shingles years, even decades later. Shingles occurs in between 10% and 20% of those who have had chickenpox. In the United States, upwards of 800,000 people are afflicted with shingles each year. The annual number of shingles sufferers worldwide is in the millions. The **disease** occurs most commonly in those who are over 50 years of age.

As the symptoms of chickenpox fade, varicella zoster virus is not eliminated from the body. Rather, the virus lies dormant in nerve **tissue**, particularly in the face and the body. The roots of sensory nerves in the spinal cord are also a site of virus **hibernation**. The virus is stirred to replicate by triggers that are as yet unclear. Impairment of the **immune system** seems to be involved, whether from immunodeficiency diseases or from cancers, the effect of drugs, or a generalized debilitation of the body with age. Whatever forces of the immune system that normally operate to hold the hibernating virus in check are abrogated.

Reactivation of the virus causes **pain** and a rash in the region that is served by the affected nerves. The affected areas are referred to as dermatomes. These areas appear as a rash or blistering of the skin. This can be quite painful during the one to two weeks they persist. Other complications can develop. For example, shingles on the face can lead to an **eye** infection causing temporary or even permanent blindness. A condition of muscle weakness or paralysis, known as **Guillian-Barre syndrome**, can last for months after a bout of shingles. Another condition known as post herpetic neuralgia can ex-

tend the pain of shingles long after the visible symptoms have abated.

See also Childhood diseases; Immunology.

Variola virus

Variola virus (or variola major **virus**) is the virus that causes **smallpox**. The virus is one of the members of the poxvirus group (Family Poxviridae). The virus particle is **brick** shaped and contains a double strand of deoxyribonucleic acid. The variola virus is among the most dangerous of all the potential biological weapons. At the time of smallpox eradication approximately one third of patients died—usually within a period of two to three weeks following appearance of symptoms.

Variola virus infects only humans. The virus can be easily transmitted from person to person via the air. Inhalation of only a few virus particles is sufficient to establish an **infection**. Transmission of the virus is also possible if items such as contaminated linen are handled.

The origin of the variola virus in not clear. However, the similarity of the virus and cowpox virus has prompted the suggestion that the variola virus is a mutated version of the cowpox virus. The **mutation** allowed the virus to infect humans. If such a mutation did occur, then the adoption of farming activities by people, instead of the formally nomadic existence, would have been a selective **pressure** for a virus to adopt the capability to infect humans.

Vaccination to prevent infection with the variola virus is long established.

In the late 1990s, a resolution was passed at the World Health Assembly that the remaining stocks of variola virus be destroyed, to prevent the re-emergence of smallpox and the misuse of the virus as a biological weapon. At the time only two high-security laboratories were thought to contain variola virus stock (Centers for Disease Control and Prevention in Atlanta, Georgia, and the Russian State Centre for Research on Virology and Biotechnology, Koltsovo, Russia). However, this decision was postponed until 2002, and now the United States government has indicated its unwillingness to comply with the resolution for security issues related to potential **bioterrorism**. Destruction of the stocks of variola virus would deprive countries of the material needed to prepare **vaccine** in the event of the deliberate use of the virus as a biological weapon. This scenario has gained more credence in the past decade, as terrorist groups have demonstrated the resolve to use biological weapons, including smallpox. In addition, intelligence agencies in several Western European countries issued opinions that additional stocks of the variola virus exist in other than the previously authorized locations.

See also Viral genetics.

Vegetables

The word vegetable is not scientifically defined by botanists. Rather, the plants and **plant** parts that are considered to be vegetables have been specified by a broad consensus among farmers, grocers, and consumers.

In general, vegetables are plant tissues that are eaten as a substantial part of the main course of a meal. In contrast, **fruits** have a culinary definition as relatively sweet, often uncooked plant parts that are eaten as desserts or snacks.

Many vegetables are above-ground, leafy tissues or stems of herbaceous plants, for example: cabbage, lettuce, and celery. However, the common understanding of vegetables also includes certain plant parts that are botanical fruits but are normally cooked before being eaten, such as the tomato, bell **pepper**, cucumber, and squash. Certain below-ground plant parts are also considered to be vegetables; for example: onions, garlic, and carrots, even though other below-ground tissues such as potatoes and yams are not.

Globally, the production of the most important types of vegetables are as follows: tomatoes, 45 million tons; various types of cabbage, 30 million tons; onions and garlic, 20 million tons; cucumber and squashes, 15 million tons; bell and chili peppers, six million tons; peas and beans, eight million tons; and carrots, eight million tons.

Vegetables derived from stems, petioles, or foliage

The cabbage is a **species** of mustard, originally native to Eurasia. Various agricultural varieties of the cabbage (*Brassica oleracea*) have been developed and are eaten cooked or sometimes raw. The familiar head and savoy cabbages are modifications of the leafy, pre-flowering rosette of the original mustard and can be green or a rich-red **color**. These cabbages are sometimes served raw, for example, finely chopped and dressed with oil and mayonnaise as coleslaw. Alternatively, these cabbages may be cooked and perhaps used as wrappers of cabbage rolls. Cabbage may also be slightly fermented and pickled, a food known as sauerkraut. Brussels sprouts are small, leafy rosettes that develop abundantly on an upright stem of another variety of cabbage and are served cooked.

Lettuce (*Lactuca sativa*) is another head-forming, leafy vegetable, native to southern **Europe** and usually served raw in salads. Endive (*Cichorium endivia*) is a similar related species that is also served raw as a salad green.

The celery (*Apium graveolens*) develops thick, fleshy, light-green petioles which are eaten raw or cooked. **Rhubarb** (*Rheum rhaponticum*) also develops tart, nutritious petioles which are cooked, sweetened with sugar or fruit, and eaten. The leaves of this species are poisonous.

The dark-green leaves of **spinach** (*Spinacea oleracea*) are an iron-rich vegetable, served steamed or raw.

The young shoots of the asparagus (*Asparagus officinalis*) are collected in the early springtime and are cooked as a tasty vegetable.

The ostrich fern (*Matteucia struthiopteris*) is a species of **North America**. This plant is not normally cultivated, but it can develop prolific stands in moist bottomlands where its young shoots, known as "fiddleheads," are collected in the early springtime. These are served as a steamed vegetable.

The leafy stem of the leek (*Allium porrum*) is served as a cooked vegetable. The foliage of chives (*A. schoenoprasum*) is usually served raw in salads or as a garnish to other cooked vegetables.

Vegetables derived from fruits or flowers

The cauliflower and broccoli are additional varieties of the cabbage that are derived from modified versions of the inflorescence and are nutritious vegetables. The edible part of the globe artichoke (*Cynara scolymus*) is the pre-flowering inflorescence of the plant which is prepared by boiling or steaming. The fleshy bracts are eaten by drawing them between the teeth to remove the starchy tissues, or the bractless "heart" is served as an unusual vegetable.

The fruits of many species in the pea family are served as cooked vegetables, sometimes intact in their pod. Some of the more common species are the French bean (*Phaseolus vulgaris*), the broad bean (*Vicia faba*), the scarlet runner bean (*Phaseolus multiflorus*), the garden pea (*Pisum sativum*), and the lentil (*Lens esculenta*).

Several species in the **tomato family** develop fleshy fruits that are used as vegetables. The red, plump fruits of the tomato (*Lycopersicum esculentum*), originally native to **South America**, are served raw or cooked. The larger fruits of the closely related eggplant or aubergine (*Solanum melongena*) are served cooked. The fruits of the red pepper, bell pepper, capsicum, or chili (*Capsicum annuum*) occur in diverse, domesticated varieties and are served cooked or raw. In Hungary, a red powder called paprika is made from dried capsicum fruits.

The elongate, fleshy fruits of several species in the cucumber family are also served as vegetables. The cucumber (*Cucumis sativus*) is usually served pickled in vinegar, raw, or cooked. The pumpkin, squash, or vegetable marrow (*Cucurbita pepo*) are usually served cooked, or their **seeds** may be extracted, roasted, and eaten.

Vegetables derived from below-ground tissues

The carrot (*Daucus carota*) is a biennial plant native to Eurasia. The carrot grows a large, starchy, cone-shaped tap-root during its first year of growth. As in many biennials, the large root is intended to support the great metabolic demands of the development of the flowering structure during the second and final year of the life of the plant. In carrots, the root develops a rich, orange color because of the presence of large concentrations of the pigment carotene, a precursor of **vitamin** A when eaten and metabolized by animals. Carrots are eaten raw or cooked. In North America and elsewhere, Queen Anne's lace is a weedy variety of carrots that has escaped from cultivation and is a common plant of disturbed places.

The parsnip (*Pastinaca sativa*) is another biennial plant, native to Europe. This plant is also cultivated for its conical tap-root which is usually eaten cooked.

The **beet** (*Beta vulgaris*) is another biennial plant that is grown for its large root. The beet root develops an intensely red-purple color due to the presence of large concentrations of anthocyanin pigments. Beet root is served cooked or is pickled in vinegar. The foliage of this plant is sometimes served as a steamed, leafy vegetable.

The radish (*Raphanus sativus*) is an annual plant that develops a large, below-ground storage **organ** which is anatomically a hypocotyl (the **tissue** between the true stem and the root). Radishes are red-colored on the outside and white inside. A biennial cultivar of the radish develops a large root which can be red, white, or brown colored. Radishes have a pungent flavor due to the presence of mustard oils and are usually served raw.

The Jerusalem artichoke (*Helianthus tuberosus*) is a perennial relative of the sunflower, native to North America. This plant produces tubers on its below-ground stems which are cooked and eaten.

The onion (*Allium cepa*) is a member of the lily family. The below-ground bulbs of this plant are served cooked or raw in salads. The garlic (*A. sativum*) also develops a strong-tasting bulb which is served raw or cooked and is mostly used as a flavoring.

KEY TERMS

Cultivar—A distinct variety of a plant that has been bred for particular agricultural or culinary attributes. Cultivars are not sufficiently distinct in the genetic sense to be considered to be subspecies.

Eat your vegetables

Most vegetables are highly nutritious foods, very rich in **carbohydrate** energy as well as vitamins, **minerals**, fiber, and to a lesser degree, protein. A balanced diversity of vegetables is an important component of any healthy diet, and these plant products should be eaten regularly.

See also Composite family; Legumes; Nightshade; Sweet potato; Tuber.

Resources

Books

Hvass, E. *Plants That Serve and Feed Us.* New York: Hippocrene Books, 1975.

Klein, R.M. *The Green World. An Introduction to Plants and People.* New York: Harper and Row, 1987.

Bill Freedman

Veins

Veins are vessels designed to collect and return **blood**, including deoxygenated hemoglobin, from tissues to the **heart**. In humans, veins and the venous vascular system can be divided in to three separate systems depending on anatomical relationships and function. Initially, veins can be divided into systemic and pulmonary systems. The veins that drain the heart, comprising the coronary venous system, may be described as an independent venous system, or be considered a subset of the systemic vascular system. The systemic veins transport venous blood—deoxygenated when compared with arterial blood—from the body to the heart. The pulmonary veins return freshly oxygenated blood from the lungs to the heart so that it may be pumped into the systemic arterial system.

Veins can also be described by their anatomical position. Deep veins run in organs or **connective tissue** that supports organs, muscle, or bone. Superficial veins are those that drain the outer skin and fascia.

In contrast to **arteries**, veins often run a more convoluted course, with frequent branching and fusions with other veins (anastomoses) that make the tracing of the venous system less straightforward than mapping the arterial system. In addition, there are reservoirs or pools (sinus) that collect venous return from multiple sources. Many veins contain valves that assure a unidirectional (one way) flow of venous blood toward the heart.

The systemic venous system can be roughly divided into groups depending on the region they drain, and the vessel through which they return blood to the heart.

The first systemic venous group consists of veins that drain the head, neck, thorax, and upper limbs. These veins ultimately return blood to the heart through the superior vena cava.

Veins that drain the abdomen, pelvis, and lower limbs return blood through the inferior vena cava. Both the superior and the inferior vena cava return deoxygenated blood to the right atrium of the heart. The coronary sinus collects blood from a number of cardiac veins before returning blood to the right atrium near the point where the inferior vena cava enters the right atrium.

The pulmonary veins return blood oxygenated in the lungs to the left atrium. There are four major pulmonary veins, each lung being drained by a pair of pulmonary veins. Akin to the drainage of a land **basin** from streams into a larger river system, smaller venules arise from the lung alveolar capillary bed, the venules fuse to form single veins that separately drain isolated lobes of the lung. The veins from the upper and middle lobes of the three lobed right lung fuse to create a pair of veins—a superior and inferior pulmonary vein that separately transport blood to the left atrium.

At a microscopic or histological level, veins have thinner walls than do arteries. They are more elastic and capable of a wider range of lower **pressure** volume transformations. The **elasticity** is a result of the fact that veins have far less subendothelial connective **tissue** in their vascular walls. In addition the tunica media and tunica adventitia are often indistinguishable layers or are poorly developed when compared with arterial linings.

Venules drain **capillaries** and capillary beds. The venules ultimately fuse (coalesce) into veins that, as they increase in size, also increase in organization and differentiation of their vascular walls. In general, the larger the vein the more likely it is to be invested or surrounded with smooth muscle tissue. The values with the venous system are formed from a cusp forming multiple folding of the tunica intima. Valves are generally absent from the largest veins and the pulmonary veins.

Veins also serve in fluid uptake and can receive lymph fluid from lymphatic vessels. The major lymphatic

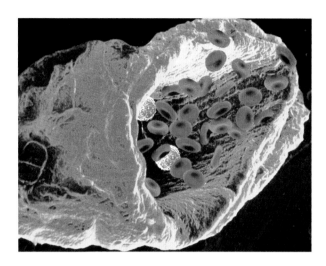

Scanning electron micrograph (SEM) showing red and white blood cells flowing through saphenous vein. ©Image Shop/Phototake. Reproduced by permission.

duct, for example, drains into the fused vein formed from the fusion of the subclavian and left internal jugular vein.

See also Anatomy; Blood gas analysis; Blood supply; Circulatory system; Heart diseases; Stroke.

Velocity

Velocity is the time **rate** of change of the position of a body. Mathematically, velocity is a vector quantity having direction as well as magnitude. Speed, on the other hand, is a **scalar** quantity which has only magnitude. The magnitude of velocity is expressed in units such as miles per hour or meters per second when describing **motion** along a straight or curved path. A body which is rotating about an axis has angular velocity. Angular velocity is also a vector quantity and is expressed as units of angular **rotation** per unit of time such as revolutions per minute or radians per second.

Venus

Venus is the second **planet** from the **Sun** and is in some superficial geological features like **Earth**. In many important geochemical features, however, it is quite different. Next to the Sun and **Moon**, Venus can be the brightest object in the sky. At its most brilliant, Venus is sixteen times brighter than Sirius (α Canis Majoris), the brightest **star** in the night sky. The extreme brilliance of

Venus is partly due to its occasional closeness to Earth, and partly due to it having a highly reflective atmosphere.

Basic properties

Venus orbits the Sun on a near circular **orbit** at mean **distance** of 0.723 Astronomical Units (AU). At aphelion the planet is at a maximum distance of 0.728 AU from the Sun, while at perihelion Venus it is 0.718 AU away. Coming as near as 0.272 AU, no other planet can approach the earth as close as Venus does.

Since it circles the Sun in an orbit smaller than that of Earth's, Venus is never very far away from the Sun for terrestrial viewers. The greatest angular distance between the Sun and Venus, as seen from Earth, is 47 degrees. This means that even under the most favorable of conditions Venus will set at most three hours after the Sun, or rise no earlier than three hours before the Sun. Observed since the most ancient of times, Venus is often called the "morning star," if it rises before dawn or the "evening star" if it sets after the Sun. The Greek philosopher Homer referred to Venus in his *Illiad* as "the most beautiful star set in the sky."

The time required for Venus to complete one orbit about the Sun, its sidereal period, is 224.701 days, whereas the time for Venus to repeat alignments with respect to the Sun and Earth is 584 days. The best times for viewing Venus are when it is near greatest eastern, or greatest western elongation. Greatest western elongation follows about five months after greatest eastern elongation, and greatest eastern elongation's repeat about every 19 months.

As it circles the Sun, Venus shows phases just like the Moon, and the planet Mercury. The phase changes of Venus were first recorded by Galileo Galilei in 1610. Galileo's observations of Venusian phase changes were important since they strengthened his belief in the heliocentric model of the **solar system** which had been proposed by Copernicus in his now famous book, *De Revolutionibus*, published in 1543. Galileo reasoned that if Venus circled Earth, as the then accepted geocentric model of the Universe decreed, it would not show the full range of phases that were observed. Likewise, the correlation between angular size and phase would also be different to that observed if Venus orbited Earth.

Transits of Venus across the disk of the Sun, as seen from Earth, are not common. If a transit is to occur, however, it will take place in either June or December, when Earth is at the line along which Venus's orbit cuts the ecliptic—the line of nodes for Venus. A very precise geometrical alignment of Earth, Venus and Sun is required for a Venusian transit to take place. The last Venusian transit occurred on December 6, 1882. The next transit will take place on June 8, 2004. In general, Venusian transits are seen in pairs. Each of the transits in a pair are eight years apart, and the pair-cycle repeats on an alternating cycle of 121.5 and 105.5 years.

The rotation rate of Venus

The Venusian atmosphere is both optically thick and highly reflective. The upper cloud deck has an **albedo** of 0.76, meaning that it reflects 76% of the sunlight that falls on it. In addition, the low-lying Venusian **clouds** are so dense that they completely obscure any optical view of the planet's surface. Not being able to monitor variations in surface detail has meant that astronomers have only recently discovered the true **rotation** rate of Venus.

While the Venusian atmosphere presents an impenetrable shroud against optical observations, it is transparent to **radio** and microwave **radiation**. Using the giant radio **telescope** at the Arecibo Observatory in Puerto Rico, astronomers were able to bounce microwave signals off the surface of Venus. By analyzing the Doppler shift in the returned signals the astronomers were then able to determine the planet's rotation rate. The results were a complete surprise.

The microwave measurement of Venus's rotation rate showed that the planet was spinning on its axis in the opposite sense to which it orbits the Sun. If one could stand on the surface of Venus, sunrise would be in the west and sunset would be in the east, the exact opposite to that seen on Earth. When viewed from above, all the planets in the solar system orbit the Sun in a counter clockwise direction. Most of the planets also rotate about their spin axes in a counter clockwise sense. The only exceptions to this rule are the planets Venus, **Uranus**, and **Pluto**. When a planet spins on its axis in the opposite sense to its orbital **motion**, the rotation is said to be retrograde.

Venus takes 243.01 days to spin once on its axis. With this slow rotation rate it actually takes 18.3 days longer for Venus to spin once on its axis than it takes for the planet to orbit the Sun. The Venusian day, that is the time from one noon to the next, is 116.8 terrestrial days long. A curious relationship exists between the length of the Venusian day and the planet's syndic period. The syndic period of Venus, that is, the time for the planet to repeat the same alignment with respect to Earth and Sun, is 584 days, and this is five times the Venusian day (584 = 5 × 116.8). It is not known if this result is just a coincidence, or the action of some subtle orbital interaction. The practical consequence of the relationship is that, should a terrestrial observer make two observations of Venus that are 584 days apart, then they will "see" the same side of the planet turned towards Earth.

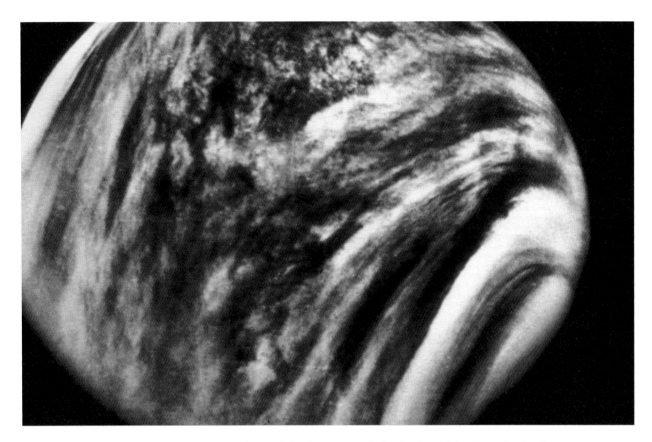

The surface of Venus is obscured by a dense layer of clouds composed of sulfuric acid droplets. The cloud cover creates a greenhouse effect that results in surface temperatures over 860°F (460°C), higher than that of Mercury. *U.S. National Aeronautics and Space Administration (NASA).*

Venusian surface detail

At is closest approach the planet Venus can have an angular diameter just slightly larger than 1/60th of a degree. This angular size translates to a physical diameter of 7,520 mi (12,104 km), making the planet about 95% the size of Earth. Given the near similar sizes of Earth and Venus, it might be expected that the surface of Venus has been shaped by the same geological processes that operate on Earth. This expectation is only partly true.

While **radar** maps of the surface of Venus were constructed during the 1970s, the most detailed topographic maps obtained to date are those from the *Magellan* spacecraft mission. The *Magellan* spacecraft was placed into Venusian orbit by NASA in 1990, and a powerful on-board imaging radar system was used to map the entire Venusian surface to a resolution of a few hundred meters.

The *Magellan* radar data showed that Venus is remarkably flat, and that some 80% of the planet's surface is covered by smooth volcanic plains, the result of many lava out-flows. The altitude map constructed from *Magellan* data has revealed the existence of two large continent-like features on Venus. These features are known as Ishtar Terra (named after the Babylonian Goddess of love), and Aphrodite Terra (named after the Greek Goddess of love). Ishtar terra, which measures some 621 mi (1,000 km) by 931 mi (1,500 km), lies in Venus's northern hemisphere, and has the form of a high plateau ringed with **mountains**. The largest mountain in the region, Maxwell Montes, rises to a height of 7 mi (11 km). Aphrodite Terra is situated just to the south of the Venusian equator and is some 9,936 mi (16,000 km) long by 1,242 mi (2,000 km) wide. It is a region dominated by mountainous highlands and several large volcanoes.

Venus has three major terrains, based mainly upon elevation (above mean planet radius). The lowlands are rolling volcanic plains (~ 60% of the planet) with under 1,600 ft (500 m) of relief. The uplands are intermediate in elevation and represent a transition between lowlands and highlands. Elevations range from 0 to 1.2 mi (0 to 2 km) in the uplands. The highlands have up to 3 mi (5 km) relief and are quite mountainous in some places. The highlands have compression ridges and fractured **rocks** and comprise about 15% of the planet's surface.

Given the similarity in size of Venus and Earth, geologists had speculated on the possibility that **plate tectonics**, which is the primary agent for re-shaping Earth's surface, might operate on Venus. The *Magellan* probe found no evidence, however, for large scale tectonic activity on Venus. The reasons underlying the absence of any large-scale tectonic activity on Venus are presently unclear, but it may be indicative that the planet has a thinner and weaker **lithosphere** than Earth. The ridges and folds that cover many of the plain regions of Venus is indicative, however, of a certain amount of local tectonic activity on the planet.

The *Magellan* maps revealed many large craters on the Venusian surface. There is an apparent cut-off for craters with diameters less than a few kilometers. This is a selection effect imposed by the dense Venusian atmosphere. The atmosphere is in fact such a good filter of incoming meteoroids, that only those objects larger than a few kilometers in diameter survive their passage through the atmosphere, with sufficient **mass**, to produce a crater at the Venusian surface.

The relative age of different regions on a planet's surface can be gauged by counting the number of craters that appear per unit area. This method of crater counting has proved very useful for dating the various regions on our Moon. The essential idea being exploited in crater count dating is that, assuming the cratering rate is the same over the whole planet, if one region has fewer craters per unit area than another, it implies that some re-surfacing, e.g., by a lava flow, has taken place in the region with fewer craters. The re-surfacing has in effect erased the older craters and re-set the cratering clock.

The number of craters in the Venusian plains is typically about 15% of the crater counts for the lunar maria. This observation indicates that the Venusian plains have an age roughly equivalent to 15% the age of the lunar maria. From the lunar rock samples that were returned from the Apollo Moon landing missions we know that the lunar maria are about 3.2 billion years old, and consequently the likely age of the Venusian plains is about 500 million years.

The main agent for re-surfacing the Venusian plains is believed to be aperiodic but widespread volcanism. Certainly, many (apparently) extinct volcano's were mapped by *Magellan* during its five-year survey. Some lava flow regions observed by *Magellan* are believed to be no more than ten millions years old, and they may be much younger.

Some of the more remarkable surface features discovered by *Magellan* were the pancake-shaped volcanoes. These flat-topped, circular volcanoes are unique to Venus, and it is thought that they are probably formed through the surface extrusion of a very thick and viscous lava.

In recognition of Venus being named in honor of the Greek Goddess Aphrodites, whom the Romans called Venus, the International Astronomical Union assigns only female names to the planet's surface features. Craters, for example, can be named after any famous women; linear features are named after Goddesses of War, while plains are named in honor of mythological heroines.

Venusian surface processes

Processes affecting Venus include impact cratering, lava flows, solid-state creep (viscous flow of rocks at the surface due to high temperatures and pressures), and eolian (**wind**) effects. The latter include deflation (blowing away fine particles), wind **erosion** (forming yardangs), and wind deposition (formation of streaks, transverse dunes, and wind-shadow dunes). **Weathering** on the surface is due to interaction between **carbon dioxide** and **sulfur dioxide** at high **temperature** with silicate rocks. At elevations above about 2.2 mi (3.5 km) above mean planet radius, weathering seems to favor formation of iron-sulfur compounds and below that level, iron-oxide and calcium-sulfate compounds, tend to form. The latter was inferred from the nature of the reflected radar signal off surficial materials.

Venusian internal structure

Venus has no natural satellites and consequently its mass has only been determined through the gravitational effect that the planet has on passing **space** probes. A mass equivalent to 82% that of Earth's, or 4.9×10^{24} kg, has been found for Venus. The bulk **density** of Venus is 5240 kg/m^3, slightly smaller than that of Earth's.

The similarity between the mass, radius, and bulk density of Venus and Earth suggests that the two planets have similar internal structure. Venus most probably has, therefore, a thin rocky crust, a large iron- and magnesium-silicate mantle, and an inner nickel-iron **alloy** core ($\sim 25\%$ of the planet's mass).

One Venusian anomaly that defies present-day theory relates to the planet's magnetic field, or more correctly to the complete lack of any detectable magnetic field. It is believed that the **earth's magnetic field** is created by a dynamo effect that operates in its hot, liquid nickel-iron alloy core. If, as has been previously argued, Venus has an internal structure similar to that of Earth, why does it not have a similar magnetic field? The answer may lie with the slow Venusian rotation rate. One of the key ingredients of the dynamo theory is that the conducting, liquid core is rotating. Since Venus rotates much more slowly than Earth, by a multiplicative factor of 1/243, it may be that the dynamo effect cannot operate in the planet's core.

The Venusian atmosphere

While Venus is often referred to as Earth's twin on the basis that the two planets have similar physical characteristics (radius, mass, density, composition, etc), it is far from being Earth's twin when atmospheric characteristics are compared.

The many spacecraft that have flown past, or landed on, the Venusian surface have found that the uppermost cloud tops, which obscure Earth-based observers' view of the planet, are about 40 mi (65 km) above the surface. For comparison, on Earth, the highest clouds are about 10 mi (16 km high.) Observations taken of ultraviolet wavelengths reveal that the upper Venusian clouds follow a jet stream-like pattern and circle the planet once every four days, or so. The circulation **velocity** of the upper cloud deck is much greater than the rotation rate of the planet, and it is believed that this is the result of extensive atmospheric **convection** driven by solar heating.

The upper cloud deck is about 3 mi (5 km) thick. At about 31 mi (50 km) altitude there is a second much more dense cloud deck. Below about 18 mi (30 km) in altitude the Venusian atmosphere is clear of clouds. The upper cloud deck has been found to contain substantial amounts of **sulfur**, which give the clouds their dark yellow to yellow-orange **color**. The lower cloud deck has been found to contain large concentrations of sulfur dioxide, hydrogen sulfide compounds, and droplets of **sulfuric acid**. It has been suggested that the presence of atmospheric sulfides is indicative of very recent volcanic activity on the planet's surface.

The greenhouse effect

Astronomers began to suspect that the surface of Venus was a decidedly inhospitable place when radio telescope measurements, made in the 1950s, indicated surface temperatures as high as 750K (891°F; 477°C). It is believed that a **greenhouse effect** is responsible for maintaining the high surface temperature on Venus.

A greenhouse effect occurs whenever incoming sunlight warms the planetary surface, but the atmospheric gases do not allow the infrared radiation emitted by the heated surface to escape back into space. The net result of the atmospheric trapping of infrared radiation is that the atmosphere heats up, and the surface temperature continues to rise.

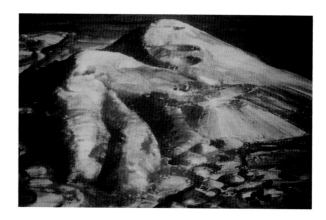

An artist's impression of one of the three continent-sized highland regions of Venus, Beta Regio, which consists of two huge shield volcanoes rising above the plains. Both are about 2.5 mi (4 km) high, and appear to be on a north-south fault line connecting them with other possibly volcanic features in the south. They have smooth surfaces and are shaped like Hawaiian volcanoes. The Russian spacecrafts *Veneras 9* and *10* landed directly east of Beta Regio and determined that the surface rock was volcanic basalt. *U.S. National Aeronautics and Space Administration (NASA).*

That a strong greenhouse effect can operate at Venus is not surprising given that its primary atmospheric component is carbon dioxide. This gas has long been recognized as a problematic "greenhouse" gas on Earth.

Building a spacecraft to land on the Venusian surface has proved to be a major **engineering** challenge. Not only must a lander be able to operate at temperatures that exceed the melting point of **lead**, but it must also withstand an **atmospheric pressure** some 90 times greater than that experienced at **sea level** on Earth. The **pressure** exerted by the Venusian atmosphere, at the planet's surface, is equivalent to that exerted by a 0.6 mi (1 km) column of **water** in Earth's oceans. The first of only four spacecraft to soft-land, and successfully transmit images of the surface of Venus back to Earth, was the former Soviet Union-built spacecraft *Venera 7*. The lander, which set-down on August 17, 1970, managed to transmit data for 23 minutes.

Impact craters on Venus

Venus' atmosphere protects the surface from smaller objects that would otherwise impact the surface if the atmosphere were thinner. Smaller objects, especially those under 0.6 mi (1 km) in diameter are largely broken up in the atmosphere and do not directly impact the surface. For this reason and due to Venus' high volcanic resurfacing rate, rather few impact craters on Venus are known (less than 1000). Impact craters on Venus are frequently attended by flows of impact ejecta that look much like lava flows. This is due to the effect of the thick atmos-

KEY TERMS

Albedo—The fraction of sunlight that a surface reflects. An albedo of zero indicates complete absorption, while an albedo of unity indicates total reflection.

Doppler effect—The apparent change in the wavelength of a signal due to the relative motion of the source and the observer.

Dynamo effect—A model for the generation of planetary magnetic fields: the circulation of conducting fluids within a planet's hot, liquid inner-core results in the generation of a magnetic field.

Greenhouse effect—The phenomenon that occurs when gases in a planet's atmosphere capture radiant energy radiated from a planet's surface thereby raising the temperature of the atmosphere and the planet it surrounds.

Lithosphere—The solid outer layer, or crust, of a planet's mantle.

Mantle—The major portion of a terrestrial planet's interior, made of plastic rock.

Retrograde rotation—Axial spin that is directed in the opposite sense to that of the orbital motion.

Tectonic activity—The theory of crustal motion.

phere on ejecta behavior during impact. Meade Patera is the largest **impact crater** basin on Venus at 174 mi (280 km) in diameter. There are only six known multi-ring impact basins on Venus, of which Meade is one. There are also large "splotches" (radar bright spots) about 6–44 mi (10–70 km) in diameter thought to be due to the effect of near-surface atmospheric detonation of incoming objects.

Venus geologic history

Venus has two alternative histories, depending upon how one views the age of the extensive resurfacing volcanic lavas. In the "catastrophic" model of Venus history, there was a huge resurfacing event in Venus' history about 200 to 700 million years ago, probably due to rather sudden solidification of the interior of the planet. Vast volcanic features of about the same age favor this interpretation. In the "gradualistic" model of Venus history, global resurfacing has occurred gradually over the whole of Venus' history and the rate of this activity has been rather high so no older surface areas still exist. The mechanism for this is **random** and continuous volcanic activity. This model does not explain the lack of magnetic field (explained in the other model by the internal so-

lidification event). Future studies of Venus will hopefully help us understand which of these (or perhaps some other) model is correct about Venus' past.

See also Planetary atmospheres.

Resources

Books

Beatty, J. Kelly, Carolyn Collins Petersen, and Andrew L. Chaikin. *The New Solar System.* Cambridge: Cambridge University Press, 1999.

de Pater, Imke, and Jack J. Lissauer. *Planetary Sciences.* Cambridge, UK: Cambridge University Press, 2001.

Morrison, D., and Tobias Owen. *The Planetary System.* 3rd ed. Addison-Wesley Publishing, 2002.

Taylor, F.W. *The Cambridge Photographic Guide to the Planets.* Cambridge University Press, 2002.

Periodicals

Goldman, Stuart. "Venus Unveiled." *Sky and Telescope* 83 (March 1992).

Luhmann, Janet, J. Pollack, and C. Lawrence. "The Pioneer Mission to Venus." *Scientific American* (April 1994).

Paul, Jeffrey. "Venus in 3-D." *Sky & Telescope* 86 (August 1993).

Saunders, Stephen. "The Surface of Venus." *Scientific American* (December 1990).

Other

Arnett, B. SEDS, University of Arizona. "The Nine Planets, A Multimedia Tour of the Solar System." November 6, 2002 [cited February 8, 2003] <http://seds.lpl.arizona.edu/nineplanets/nineplanets/nineplanets.html>.

Martin Beech
David T. King, Jr.

Venus fly-trap *see* **Carnivorous plants**

Verbena family (Verbenaceae)

The verbena or vervain family (Verbenaceae) is a diverse group of about 3,000 **species** of plants, most of which occur in the tropics.

Plants in this family can be herbs, shrubs, trees, or lianas (tropical vines). The leaves are usually simple, arranged alternately on the stem, which is often square-sided. The flowers are small, but often occur in attractive inflorescences.

Some species of trees in the verbena family are extremely valuable for the production of lumber.

Tropical hardwoods in the verbena family

Teak (*Tectona grandis*) is one of the world's most prized species of tropical hardwood. Teak is a large **tree** of mature, tropical **forests** of South and Southeast **Asia**, and can grow as tall as 131 ft (40 m). Teak lumber can vary in **color** from light to brownish yellow, or a deep chocolate-brown. Lumber made from teak is heavy, strong, durable, resistant to splitting and cracking, and highly resistant to damages associated with immersion in **water**. Teak **wood** contains an aromatic, resinous oil that makes the wood feel slightly greasy to the touch, and helps to make it almost invulnerable to **termites** and highly resistant to wood-rotting **fungi**. Teak is valued for the manufacture of durable decking and trim on boats, and for making flooring, panelling, and fine furniture.

Teak is harvested from tropical forests wherever it occurs in Asia. Typically, teak trees are girdled and stripped of their lower **bark**, and then left standing for two years prior to felling. This allows the trees to dry somewhat before they are cut down, so the logs will be lighter and can be more easily dragged out of the forest. After the teak trees are felled they are sectioned into manageable-sized logs. These are then transported out of the forest using elephants or mechanical skidders, often to a river, on which the logs are floated to the coast for processing into lumber or veneer.

Unfortunately, teak occurring in natural forests is rarely harvested on a sustainable basis, and the resources of this extremely valuable tropical hardwood are being rapidly mined. Today and more so into the future, much of the teak available in commerce must be grown in plantations established for the production of this precious wood.

Other tropical species of tree in the Verbena family are also valuable as sources of hardwood lumber. These include species of *Petitea*, *Premna*, and *Vitex celebica*. Zither wood is a specialized material derived from *Citharexylum* spp. of Central and **South America**, and used to manufacture musical instruments.

The tropical trees *Lippia citriodora* and *Vitex agnus-castus* are useful as a source of natural oils, known as oil of verbena.

Ornamental species

Some species in the Verbena family are cultivated for their showy flowers. The most common garden verbenas in North American gardens are *Verbena hortensis* and *V. hybrida*, both frequently used as bedding plants. Two native species with showy flowers, the large-flowered verbena (*Verbena canadensis*) and small-flowered verbena (*V. bipinnatifida*), are commonly grown in gardens, and are often crossed with other verbenas to develop new varieties for **horticulture**. Purple mulberry (*Callicarpa purpurea*) is an attractive Asian shrub that is sometimes cultivated in **North America**.

Several vines and shrubs in the genus *Clerodendrum* are sometimes grown as ornamentals in temperate areas, including the bleeding-heart (*C. thomsoniae*) and the pagoda **flower** (*C. paniculatum*). The shrub known as the **lilac** chaste tree (*Vitex agnus-castus*) is sometimes cultivated for its attractive, blue or white flowers. The beauty-berry (*Callicarpa americana*) is sometimes grown for its ornamental **fruits**.

Attractive species of *Lantana* are often cultivated as greenhouse and bedding plants. Unfortunately, some species of *Lantana* have escaped from gardens in the tropics, and in many places these have become serious weeds of pastures because **livestock** can be poisoned by eating this **plant**. Some horticultural species in the verbena family have also become naturalized as weeds in North America.

North American species

A number of species of wildflowers in the verbena family occur naturally in North America, or have been introduced from elsewhere and have spread to natural habitats.

One of the more familiar native species of verbenas in North America is the blue vervain or wild hyssop (*Verbena hastata*), a common plant of moist, temperate habitats. The French or Bermuda mulberry (*Callicarpa americana*) is a native shrub of moist thickets in southern parts of North America.

The European vervain or berbine (*Verbena officinalis*) is a common, **introduced species** in North America, and is sometimes an important weed.

Resources

Books

Hartmann, H.T., A.M. Kofranek, V.E. Rubatzky, and W.J. Flocker. *Plant Science. Growth, Development, and Utilization of Cultivated Plants.* Englewood Cliffs, NJ: Prentice-Hall, 1988.

Woodland, D.W. *Contemporary Plant.* Heigh *Systematics.* Englewood Cliffs, NJ: Prentice-Hall, 1991.

Bill Freedman

Vertebrate paleontology *see* **Paleontology**

Vertebrates

Vertebrates are animals classified in the subphylum Vertebrata, phylum Chordata. Vertebrates share a number of features. They all have an internal skeleton of bone and/or cartilage, which includes a bony cranium surrounding the **brain** and a bony vertebral column enclosing the spinal cord. Vertebrates are all covered by a skin composed of dermal and superficial epidermal layers of scales, feathers or fur, a ventral **heart**, formed red and white **blood** cells, a liver, pancreas, kidney, and a number of other internal organs. The most advanced vertebrates also have jaws, teeth, limbs or fins, and an internal skeletal structure with pelvic and pectoral girdles, and thoracic lungs.

Eight classes of vertebrates are recognized. These are listed below, in the order of their first appearance in the fossil record:

(1) The class Agnatha is a group of jawless, fish-like animals with poorly developed fins, which first appeared more than 500 million years ago, during the late Cambrian. The 75 surviving **species** include the jawless **lampreys and hagfishes**.

(2) The class Placodermi is an extinct group of bony-plated aquatic animals. The placoderms were primitive, jawless, fish-like creatures, whose head was heavily armored by an external shield of bony plates. These creatures were most abundant during the Devonian period, some 413-365 million years ago.

(3) The class Chondrichthyes includes about 800 living species of **sharks**, **rays**, and rat fishes, all of which have a cartilaginous skeleton, true jaws, and a number of other distinctive characters.

(4) The class Osteichthyes includes some 20,000 species of true fishes, with a bony skeleton, a sutured skull, teeth fused to the jaws, lobed or rayed fins, and a number of other distinguishing features.

(5) The class Amphibia includes some 3,500 living species of **frogs**, **toads**, **salamanders**, **newts**, and **caecilians**, all of which have four limbs (making them tetrapods), a moist glandular skin, external **fertilization**, and a complex life cycle.

(6) The class Reptilia are four-legged, tailed animals, with dermal scales, internal fertilization, amniotic eggs, and direct development. Living **reptiles** include about 6,200 species of **crocodiles**, **turtles**, lizards, **snakes**, and tuataras. Important extinct groups of reptiles include the dinosaurs, hadrosaurs, ichthyosaurs, pterosaurs, and plesiosaurs.

(7) The class Aves, the **birds**, is a diverse group of about 8,800 species of warm-blooded (or homoiothermic) tetrapods whose forelimbs are specialized for flight (although some species are secondarily flightless). Birds have a characteristic covering of feathers, a beak which lacks teeth, and reproduce by laying eggs.

(8) The class Mammalia includes more than 4,000 species of homoiothermic tetrapods, with epidermal hair and female mammary **glands** for suckling the young. All give **birth** to young, although a very few, primitive species reproduce by laying eggs.

Vertebrates are the most complex of Earth's **animal** life forms. The earliest vertebrates were marine, jawless, fish-like creatures that probably fed on **algae**, small animals, and decaying organic **matter**. The **evolution** of jaws allowed a more complex exploitation of ecological opportunities, including the pursuit of a predatory life style. The evolution of limbs and the complex life cycle of **amphibians** allowed the adults to exploit moist terrestrial habitats as well as aquatic habits. The subsequent evolution of internal fertilization and the self-contained, amniotic eggs of reptiles, birds, and **mammals** allowed reproduction on land, and led to fully terrestrial forms. Birds and mammals further advanced vertebrate adaptations to terrestrial environments through their complex anatomical, physiological, and behavioral adaptations, and this has allowed them to extensively exploit all of Earth's habitable environments.

See also Cartilaginous fish; Chordates; Fish; Invertebrates.

Bill Freedman

Video recording

The term "video recording" refers to storing a video signal (information designed to specify a moving image) in a recording medium such as magnetic tape, optical disc, or computer memory. Video signals have much larger bandwidths ≅65 MHz) than do audio signals (≅20 kHz), and thus involve a more complex recording and playback technology.

Basic principles of video recording

Magnetic tape is still the most common method of storing video signals, whether analog or digital. In analog magnetic recording, a thin layer of metallic material (e.g., **iron** oxide) on some moving substrate (e.g., a tape being wound from one reel to another, or a rotating disc) is magnetized under the control of an oscillating electrical signal (the video signal). The video signal is controls passed to a recording head, which consists of a coil of wire wound around a core made of ferrite (iron-based) material. When this signal is passed through the recording-head core, which is Ω-shaped, a magnetic field arcs across the gap in the Ω.

As the video signal goes through a positive-negative oscillation, the polarity of the two ends of the core changes, reversing of the direction of the magnetic flux. The intensity of the video signal determines the strength of the magnetic flux. This flux impresses a magnetized area on the flexible tape or other magnetic medium, which is moving past the recording head. That is, the field produced by the recording head forces **atoms** in the medium's coating to shift their alignment; this alignment remains fixed even after the recording head is no longer in the vicinity, producing a weak, permanent magnetic field on the surface of the medium.

As the video signal oscillates, a linear series of such magnetized areas are produced on the recording medium, the magnetic-field directions and strengths of these areas correspond to the polarity (positive or negative) and strength of the original video signal at each moment. These magnetized areas comprise the recording of the analog video signal.

In order for the video signal to be recorded properly, the medium (usually a tape) has to move at a constant and sufficient speed across the end gap of the head. This leads to magnetization of the tape according to the signal content at each moment of time.

Although a digital video signal has a very different electrical structure—a series of sudden flips between a high level and a low level, rather than a smoothly varying level—recording of a digital signal on a magnetic medium works much the same way as for an analog signal. The major difference is that a digital magnetic recording consists of a series of discrete microregions of tape (or disc surface), each one of uniform field strength, rather than a smoothly varying continuum of magnetized particles.

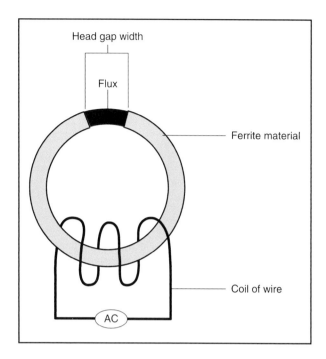

Figure 1. A schematic diagram of a recording head.
Illustration by Hans & Cassidy. Courtesy of Gale Group.

Recording techniques

The baseband (original) frequencies contained in a video signal lie between 10 Hz and 5 MHz. For a given bandwidth, assuming a simple, linear **motion** of the tape over a stationary recording head, the speed of the tape would be given by:

Speed of tape (meters per second) = width of gap (meters) × 2 × **frequency** (Hertz).

For a 5-MHz bandwidth and gap width of 1×10^{-6} meters, the speed of the tape required would thus be 10 m per second, and 36,000 meters (over a mile) of tape would be needed to record a one-hour program. In practice, a linear tape speed for recording video signals of 24 mm per second can be achieved using various techniques. Some of the recording techniques used in the past and others currently in use are discussed in the following sections.

Transverse recording

The transverse recording technique is based upon the concept of **rotation** of the head simultaneous with transverse movement of the tape over the head. The head rotates at a speed of 14,400 revolutions per minute, recording a track that zigzags along the tape and gives an effective writing speed of 38 meters per second. In this method, a single image is divided into 16 segments. All these segments are then recorded linearly onto the magnetic tape, in

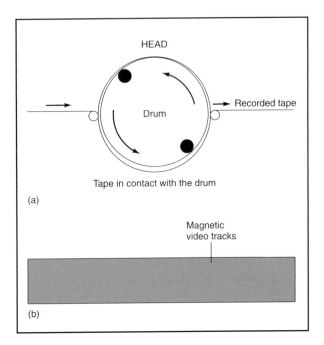

Figure 2. (A) A schematic representation of the rotating drum and moving tape. (B) A schematic diagram of a magnetic tape that has recorded signals. *Illustration by Hans & Cassidy. Courtesy of Gale Group.*

parallel. This requires a great deal of horizontal synchronization while reproducing the video signal.

Helical recording

Helical recording enables the linear speed of the tape itself to be reduced while increasing the writing speed. Instead of a single recording head, two heads are set diametrically into a small rotating drum. The magnetic tape wraps around the drum as it moves forward, thus both the head and the tape are moving in the same direction. This drum is tilted at an angle, which causes the heads to traverse the magnetic tape in slanted tracks. Again a *track length* that is much longer than the *tape length* is achieved.

For maximum utilization of the magnetic tape, at least two heads are essential. The two heads are set in the drum so that their gaps are at an angle of 6° plus or minus from the "zero" position. The "zero" position is defined as the right angle to the direction of the rotation. This angle is called the azimuthal angle, and this type of recording is also referred to as azimuth recording. The plus and minus sign in the angles ensure that the two heads identify their own tracks while reproducing the video signals. Unlike transverse recording, the picture field is divided into two segments and each segment is recorded by each head. Thus in one rotation of the drum one picture field is written completely.

The width of these magnetic tracks is 0.049 mm and the total width of the tape is 12.65 mm. In addition to the video tracks, the tape has two other tracks, the sound track and a control track for synchronizing tape speeds. The latter two tracks are stored in a linear fashion.

Frequency modulation

The wide bandwidth of a video signal poses a problem, due to the way that the inductive impedance of the recording head (i.e., resistance of the recording head to rapid changes in current flow) rises with the increase in the frequency. For normal recording a thousand times more head voltage will be required for a 5-MHz signal than for a 30-Hz signal. To avoid this problem, the wide-width luminance (brightness) signal is not recorded directly, but is instead recorded using a process called frequency modulation (FM), where the original signal is used to modulate (vary) the frequency of a high-frequency carrier signal. This effectively increases the **ratio** of the lowest and highest frequencies but does not reduce the bandwidth. FM signals give a better signal-to-noise performance and are less sensitive to unwanted **interference**.

Video systems

Video-recording systems are dependent, in their mechanical and electrical details, on the format of the **television** signals to be recorded. These signals vary in different parts of the world. For example, electrical-power standards vary from region to region, with two of the most common power frequencies being 50 Hz and 60 Hz. In order to match the frequency of the power supply, the television signals have been adapted to these standards. In countries with 60-Hz power, like the United States, Canada, and Japan, 30 video frames per second are transmitted, while in countries with 50-Hz power, like **Australia**, India, and some European countries, 25 frames per second are transmitted.

A T.V. picture consists of a series of dots. For a black and white picture the dots are black, gray, and white; in a **color** picture, they are usually red, green, and blue. These dots are usually very small and, if viewed from more than a meter or so away, invisible to the human **eye**. A series of these dots are synchronized in the form of horizontal lines and vertical lines, forming the image. The structure of the image lines also determine the bandwidth of the television signal, and, as with electrical power, the number of horizontal lines varies from country to country. Countries with 60-Hz power use 525 horizontal lines, while others use 625 lines. There is no world standard yet for these horizontal lines and the number of frames for transmission. This has led to the development of various incompatible video recording systems. There are three major types of video recording: NTSC (National Television System Committee); PAL (phase alternating line); and SECAM (from the French for "sequential color with memory"). The major differences in the three systems are:

Horizontal lines: NTSC = 525, PAL = 625, SECAM = 625. Fields per second: NTSC = 60, PAL = 50, SECAM = 50. Frames per second NTSC = 30, PAL = 25, SECAM = 25

Besides these major differences, there are variations in their subcarrier frequency, luminance, and chrominance bandwidths. (PAL and SECAM differ in these variables, not in their arrangemtns of lines, fields, and frames per second.) There are also variants of these video systems, the differences between these minor variants being mainly in FM bandwidth.

Video formats

The video signal is recorded using the helical scanning technique. Use of different azimuthal angles for recording leads to different video formats.

VHS format

This is one of the most commonly used video formats. The azimuthal angle is +6 or -6 degrees. The writing speed is usually 4.85 m per second, while the linear speed of the tape is 23.99 mm per second. The video track width is 0.049 mm and the actual tape width is 12.65 mm.

Betamax format

The azimuthal angle for this format is +7 and -7 degrees. The linear speed of the tapes is 18.7 mm per second slower than the VHS tape speed, although the writing speed is 5.83 m per second. The video track width is 0.0328 mm on a tape 12.7 mm wide.

Video-8 format

This format uses azimuthal angles of +10 and -10 degrees. The standard video tracks are 0.0344 mm wide on a tape that is 8 mm wide. The writing speed is 3.12 m per second with a linear tape speed of 20.051 mm per second. The drum size used in this format is smaller than those used in VHS or Betamax format.

VHS-C format

VHS-C stands for VHS compact. This format is widely used in video recording cameras and is fully compatible with the standard VHS format. The tape width is same as in VHS, but the drum is 41.33 mm instead of 62 mm in VHS. The other difference is that VHS-C uses four-head helical scanning in order to produce the same magnetic pattern in VHS.

Digital recording

Most of the current video recording devices are still analog recorders, but this is changing rapidly as hand-held digital video cameras, digital video discs (DVDs), and other digital video technologies capture increasingly large segments of the consumer market. The broadcast television market is also shifting rapidly to digital signal standards. **Digital recording** requires high-density recording technology due to the large bandwidth—125–270 million bits per second—but has many advantages over analog recording, including greater reliability, low cost (given recent advances in digital-signal-processing technologies), higher resolution, and greater color accuracy. Digital video has the additional virtue of transferability, as it may be recorded on any medium capable of storing digital data: computer hard drive, digital videotape, optical disc, or other. Given contemporary standards for memory and processing speed in affordable desktop computers, both professional and amateur video users can now upload digital video into working computer memory and edit it at will. There is little doubt that analog television signals, both for broadcast and recording, will be a thing of the past within some 10 or 20 years; indeed, in May, 1997 the U.S. Federal Communications Commission (FCC) mandated that U.S. broadcasters begin to phase out NTSC in favor of digital television.

Unfortunately, there is presently even more global disparity among digital television signal types than among analog signal types. The most pervasive—used on DVDs, streaming video on the Internet, and in broadcast—is probably that which exploits the type of data compression termed MPEG. This acronym is itself compression of "motion-JPEG," where JPEG stands for Joint Photographic Experts Group, the name of the body that designed this compression **algorithm**. Data compression enables the number of bits in a video frame to be reduced by as much as 75% without (hopefully) compromising the image quality. Image compression can, however, degrade image quality if misapplied. MPEG is actually two standards: MPEG-1 for low-quality video (e.g., streaming video on the Internet), while MPEG-2 is for broadcast-quality video.

Resources

Books

Marsh, Ken. *Independent Video: A complete Guide to the Physics, Operation, and Application of the New Television for the Student, the Artist, and for Community TV.* Straight Arrow Books, 2001.

Other

"Digital Television Frequently Asked Questions." U.S. Federal Communications Commission, 2002 [cited February 7, 2003]. <http://www.fcc.gov/mb/policy/dtv/>.

Satyam Priyadarshy

Violet family (Violaceae)

The violet family (Violaceae) includes about 900 **species** of plants. Species in this family occur in all parts of the world, but are mostly in the temperate zones, and at high altitude in the tropics. The largest group in the family is the genus containing violets and pansies (*Viola* spp.), with about 500 species.

Most species in this family are annual or perennial herbs. The leaves are simple, commonly heart-shaped, and are alternately arranged on the stem, or arise from a basal **rhizome**. In most species the flowers are irregular, that is, they are composed of two complimentary halfs. The flowers of most species have both female (pistillate) and male (staminate) organs. Most species **flower** in the spring and early summer and have relatively large and showy flowers, sometimes grouped into an inflorescence. They typically produce fragrance and **nectar** to attract insect pollinators. Later in the growing season, some species also develop self-fertilized flowers that do not fully open, an unusual trait known as cleistogamy. The fruit is a many-seeded berry or capsule.

The major commercial value of species in the violet family is horticultural. One species is important in perfumery, and a few are used as medicinals.

Species native to North America

Most species of violets native to **North America** are wildflowers of the spring and early summer. Blue-colored violets are relatively common, with some of the more widespread species including the wooly blue violet (*Viola sororia*), northern blue violet (*V. septentrionalis*), New England blue violet (*V. novae-angliae*), western blue violet (*V. retusa*), **prairie** or larkspur violet (*V. pedatifida*), and marsh violet (*V. palustris*).

Some of the more common white-colored violets include the large-leaved white violet (*Viola incognita*), kidney-leaved violet (*V. renifolia*), sweet white violet (*V. blanda*), and northern white violet (*V. pallens*). Yellow-colored violets include the hairy yellow violet (*Viola pubescens*), smoothish yellow violet (*V. eriocarpa*), and round-leaved yellow violet (*V. rotundifolia*).

The green violet (*Hybanthus concolor*) is a species found in moist **forests** of eastern North America.

Ornamental violets

Many species of violets and pansies and their diverse hybrids are grown in gardens as ornamentals, particularly as bedding plants. Most commonly cultivated are the so-called garden pansies (*Viola x wittrockiana*), a **hybrid** complex that is largely based on the European pansy (*V. tricolor*). Ornamental pansies are now available in diverse floral colors, including solid and mixed hues of blue, purple, red, yellow, and white. Some pansy varieties develop quite large flowers.

The English or sweet violet (*Viola odorata*) is also commonly cultivated as an ornamental **plant**, and it sometimes escapes from cultivation to become a minor weed of North American lawns. Less commonly cultivated species include the horned violet (*V. cornuta*) and the alpine violet (*V. labradorica*).

Other uses of violets

The sweet violet has been cultivated in large quantities in southern France for the production of a fragrant oil from its flowers, known as oil-of-violets. This oil is used in the mixing of perfumes and other scents. The yield of one tonne of fresh flowers is only 28-43 g (.98-1.51 oz) of oil-of-violets. Interestingly, this light-green oil only has a faint scent when in its concentrated, distilled state, but when diluted to about 1:5000 its odor becomes very strong. The scent of violets can now been synthesized by chemists, so natural oil-of-violets is now rare.

Sometimes the flowers of pansies are served as an attractive, edible garnish to well-presented, epicurean foods. Pansies are also sometimes candied as an exotic confectionary.

Some minor organic medicines are prepared from several species in the Violaceae, mostly for use as emetics. These medicinal plants include *Anchietea salutaris*, *Corynostylis hybanthus*, and *Hybanthus ipecacuanha*.

Resources

Books

Conger, R.H.M., and G.D. Hill. *Agricultural Plants.* 2nd ed. Cambridge, UK: Cambridge University Press,1991.

Hartmann, H.T., A.M. Kofranek, V.E. Rubatzky, and W.J. Flocker. *Plant Science: Growth, Development, and Utilization of Cultivated Plants.* Englewood Cliffs, NJ: Prentice-Hall,1988.

Judd, Walter S., Christopher Campbell, Elizabeth A. Kellogg, Michael J. Donoghue, and Peter Stevens. *Plant Systematics: A Phylogenetic Approach.* 2nd ed. with CD-ROM. Suderland, MD: Sinauer, 2002.

Bill Freedman

Vipers

Vipers are **snakes** in the family Viperidae, a group of short-tailed, (usually) stout-bodied snakes with long fangs at the front of the mouth, sited on a short jawbone that can be rotated to bring the fangs from their resting position parallel with the palate to an erect position for striking.

This efficient venom delivery system allows vipers to eat large (and sometimes dangerous) animals without a struggle that might expose them to harm. Vipers make a swift strike in which the long hollow (hypodermic needle-like) fangs inject a strong venom deep into the prey's body. The snakes then wait until the **animal** dies, tracking it down if necessary, and then calmly swallowing it. The venom also has the effect of initiating digestion even before the **prey** is swallowed. Many vipers do not find it necessary to eat more than once a month. The venom of these snakes are diverse, being adapted to quickly kill the preferred prey animals of each **species**.

Vipers are an old and diverse group. They are generally divided into the Old World "true" vipers (Viperinae) and the pit vipers (Crotalinae), which are found in **Asia** and the Americas. One strange viper, (*Azemiops*) from

A Pope's pit viper. *Photograph by Tom McHugh. Photo Researchers. Reproduced by permission.*

southern China and northern Myanmar (Burma) of unknown relationships, is placed in its own subfamily, Azemiopinae.

Old World vipers

True vipers (family Viperinae) are found in the Old World and lack facial pits; this distinguishes them from the pit vipers of the Americas. **Africa** is the home of 30 of the 45 species of Old World vipers. Among the African vipers are the Gaboon viper (*Bitis gabonicus*) and the **rhinoceros** viper (*B. nasicornis*) that attain lengths of almost 6 ft (2 m) and have fangs that may be 2 in (5 cm) in length. There are green tree-vipers (*Atheris*), desert-dwelling sandvipers (*Cerastes*), and even a little-known worm-eating, burrowing viper (*Adorhinos*).

The deserts of north Africa and south Asia are home to a number of sandvipers, one of which is the notorious carpet viper (*Echis carinatus*). Although an adult of this snake may be little more than 1 ft (30 cm) in length, its bite is highly toxic and is responsible for many deaths, particularly in the **desert** regions of Pakistan and western India.

The small European viper or adder (*Vipera berus*), by contrast, is not dangerous to humans. Its bite has been described as "little worse than a bee-sting," and the few reported deaths are apparently due to over-treatment. This viper is notable in that it is one of very few snakes that ranges above the Arctic Circle in Sweden and Norway.

Some large vipers of the genus *Vipera* range from the eastern Mediterranean eastward through southern Asia. The ill-tempered and highly venomous Russell's viper *(Vipera russelii)* follows the **rats** into the **rice** fields when the fields are drained for harvesting. It is the major cause of fatal snake bites (killing perhaps 10,000 people annually) in Burma (Myanmar) and Thailand.

Pitvipers

The pitvipers are easily identified by the loreal pit, a heat-sensitive receptor that lies between the nostril and the **eye** on either side of the face. With this receptor the snakes can detect and accurately strike a warm-blooded prey animal in absolute darkness, guided by the infra-red (**heat**) rays that the prey animal produces. The Viperinae lack this heat-sensitive pit.

The pitvipers range from eastern **Europe** to the East Indies and Japan, and from Canada to Argentina in the Americas. Although the 145 species of New World pit vipers are much more diverse than the Old World vipers, most are terrestrial, with a few (usually green) arboreal species. The best-known pit vipers are the rattlesnakes (*Crotalus* and *Sistrurus*) which are found only in the

Americas and whose modified tail skin vibrates to produce a rattling, warning sound. Most rattlesnakes are North American, ranging from southern Canada to Panama, with the greatest number inhabiting the dry regions of the southwestern United States and northern Mexico. However, a small tropical group (*Crotalus durissus* and its relatives) ranges as far as southern Brazil.

There are about 60 species of tropical pitvipers (*Bothrops* and relatives) that range from coastal Mexico southward to the Patagonian plains of Argentina. Most of these are relatives of the large terrestrial South American "fer-de-lance" *Bothrops atrox,* but Central America has a more diverse assemblage composed of several genera, some of which are treevipers.

The largest of all vipers is the bushmaster (*Lachesis muta*) of northern **South America** and Panama, which attains a reported length of 12 ft (4 m). Like the large African vipers, it has very long fangs (2 in/5 cm or more) and a large supply of venom. Very few people are bitten by these snakes, however, because bushmasters live in forested areas and are active only at night.

Asia has a large contingent of pitvipers (including green, arboreal species), similar to the American *Bothrops,* that are placed in the genus *Trimeresurus*. Some of its members are so similar to some of the American species that there is some doubt that they should be in different genera.

The **water** moccasins (genus *Agkistrodon*) are found both in **North America** and in Asia. The copperhead (*Agkistrodon contortrix*) is common in the **forests** of the eastern United States, and is responsible for most of the venomous snakebites in that region. Fortunately, its venom is not highly toxic to humans, and almost no one dies from these bites.

See also Reptiles.

Resources

Books

Campbell, J. A., and E. D. Brodie Jr., eds. *Biology of the Pitvipers.* Tyler, Texas: Selva, 1992.

Cogger, Harold G., David Kirshner, and Richard Zweifel. *Encyclopedia of Reptiles and Amphibians.* 2nd ed. San Diego: Academic Press, 1998.

Greene, Harry W., Patricia Fogdon, and Michael Fogden. *Snakes: The Evolution of Mystery in Nature.* Berkeley: University of California Press, 1997.

Klauber, L.M. *Rattlesnakes: Their Habits, Life Histories, and Influences on Mankind.* Berkeley: University of California Press, 1982.

Russell, F.E. *Snake Venom Poisoning.* Philadelphia: Lippincott, 1980.

Herndon G. Dowling

Viral genetics

Viral genetics, the study of the genetic mechanisms that operate during the life cycle of viruses, utilizes biophysical, biological, and genetic analyses to study the viral **genome** and its variation. The **virus** genome consists of only one type of **nucleic acid**, which could be a single or double stranded DNA or RNA. Single stranded RNA viruses could contain positive-sense (+RNA), which serves directly as mRNA or negative-sense RNA (-RNA) that must use an RNA polymerase to synthesize a complementary positive strand to serve as mRNA. Viruses are obligate **parasites** that are completely dependant on the host **cell** for the replication and transcription of their genomes as well as the translation of the mRNA transcripts into **proteins**. Viral proteins usually have a structural function, making up a shell around the genome, but may contain some enzymes that are necessary for the virus replication and life cycle in the host cell. Both bacterial virus (bacteriophages) and **animal** viruses play an important role as tools in molecular and cellular **biology** research.

Viruses are classified in two families depending on whether they have RNA or DNA genomes and whether these genomes are double or single stranded. Further subdivision into types takes into account whether the genome consists of a single RNA **molecule** or many molecules as in the case of segmented viruses. Four types of bacteriophages are widely used in biochemical and genetic research. These are the T phages, the temperate phages typified by **bacteriophage** lambda, the small DNA phages like M13, and the RNA phages. Animal viruses are subdivided in many classes and types. Class I viruses contain a single molecule of double stranded DNA and are exemplified by adenovirus, simian virus 40 (SV40), herpes viruses, and human papillomaviruses. Class II viruses are also called parvoviruses and are made of single stranded DNA that is copied in to double stranded DNA before transcription in the host cell. Class III viruses are double stranded RNA viruses that have segmented genomes which means that they contain 10-12 separate double stranded RNA molecules. The negative strands serve as template for mRNA synthesis. Class IV viruses, typified by poliovirus, have single plus strand genomic RNA that serves as the mRNA. Class V viruses contain a single negative strand RNA which serves as the template for the production of mRNA by specific virus enzymes. Class VI viruses are also known as retroviruses and contain double stranded RNA genome. These viruses have an **enzyme** called reverse transcriptase that can both copy minus strand DNA from genomic RNA catalyze the synthesis of a complementary plus DNA strand. The resulting double stranded DNA is integrated in the host **chromosome** and is transcribed by the host's own machinery. The resulting transcripts are either used to synthesize proteins or produce new viral particles. These new viruses are released by budding, usually without killing the host cell. Both HIV and HTLV viruses belong to this class of viruses.

Virus genetics is studied by either investigating genome mutations or exchange of genetic material during the life cycle of the virus. The **frequency** and types of genetic variations in the virus are influenced by the nature of the viral genome and its structure. Especially important are the type of the nucleic acid that influence the potential for the viral genome to integrate in the host, and the segmentation that influence exchange of genetic information through assortment and recombination.

Mutations in the virus genome could either occur spontaneously or be induced by physical and chemical means. Spontaneous mutations that arise naturally as a result of viral replication are either due to a defect in the genome replication machinery or to the incorporation of an analogous base instead of the normal one. Induced virus mutants are obtained by either using chemical mutants like nitrous oxide that acts directly on bases and modify them or by incorporating already modified bases in the virus genome by adding these bases as substrates during virus replication. Physical agents such as ultra-violet **light** and **x rays** can also be used in inducing mutations. Genotypically, the induced mutations are usually point mutations, deletions, and rarely insertions. The phenotype of the induced mutants is usually varied. Some mutants are conditional lethal mutants. These could differ from the **wild type** virus by being sensitive to high or low **temperature**. A low temperature mutant would for example grow at 88°F (31°C) but not at 100°F (38°C), while the wild type will grow at both temperatures. A mutant could also be obtained that grows better at elevated temperatures than the wild type virus. These mutants are called hot mutants and may be more dangerous for the host because fever, which usually slows the growth of wild type virus, is ineffective in controlling them. Other mutants that are usually generated are those that show drug resistance, enzyme deficiency or an altered pathogenicity or host range. Some of these mutants cause milder symptoms compared to the parental virulent virus and usually have potential in **vaccine** development as exemplified by some types of **influenza** vaccines.

Besides **mutation**, new genetic variants of viruses also arise through exchange of genetic material by recombination and reassortment. Classical recombination involves breaking of covalent bonds within the virus nucleic acid and exchange of some DNA segments followed by rejoining of the DNA break. This type of recombination is almost exclusively reserved to DNA viruses and retroviruses. RNA viruses that do not have a DNA phase rarely use this mechanism. Recombination usually enables a virus to pick up genetic material from similar viruses and even from unrelated viruses and the eukaryotic host cells. Exchange of genetic material with the host is especially common with retroviruses. Reassortment is a non-classical kind of recombination that occurs if two variants of a segmented virus infect the same cell. The resulting progeny virions may get some segments from one parent and some from the other. All known segmented viruses that infect humans are RNA viruses. The process of reassortment is very efficient in the exchange of genetic material and is used in the generation of viral vaccines especially in the case of influenza live vaccines. The ability of viruses to exchange genetic information through recombination is the basis for virus-based vectors in **recombinant DNA** technology and hold great promises in the development of **gene therapy**. Viruses are attractive as vectors in **gene** therapy because they can be targeted to specific tissues in the organs that the virus usually infect and because viruses do not need special chemical reagents called transfectants that are used to target a plasmid vector to the genome of the host.

Genetic variants generated through mutations, recombination or reassortment could interact with each other if they infected the same host cell and prevent the appearance of any phen of any phenotype. This phenomenon, where each mutant provides the missing function of the other while both are still genotypically mutant, is known as complementation. It is used as an efficient tool to determine if mutations are in a unique or in different genes and to reveal the minimum number of genes affecting a function. Temperature sensitive mutants that have the same mutation in the same gene will for example not be able to complement each other. It is important to distinguish complementation reactivation where a higher dose of inactivated mutants will be reactivated and infect a cell because these inactivated viruses cooperate in a poorly understood process. This reactivation probably involves both a complementation step that allows defective viruses to replicate and a recombination step resulting in new genotypes and sometimes regeneration of the wild type. The viruses that need complementation to achieve an infectious cycle are usually referred to as defective mutants and the complementing virus is the helper virus. In some cases, the defective virus may interfere with and reduce the infectivity of the helper virus by competing with it for some factors that are involved in the viral life cycle. These defective viruses called "defective interfering" are sometimes involved in modulating natural infections. Different wild type viruses that infect the same cell may exchange coat components without any exchange of genetic material. This phenomenon, known as phenotypic mixing is usually restricted to related viruses and may change both the morphology of the packaged virus and the tropism or **tissue** specificity of these infectious agents.

See also Archaeogenetics; Epidemiology; Genetic engineering; Genetic identification of microorganisms; Immunology; Medical genetics; Mendelian genetics; Microbial genetics; Molecular biology; Organelles and subcellular genetics.

Resources

Books

Beurton, Peter, Raphael Falk, and Hans-Jörg Rheinberger., eds. *The Concept of the Gene in Development and Evolution.* Cambridge, UK: Cambridge University Press, 2000.

Coffin, J.M., S.H. Hughes, and H.E. Varmus. *Retroviruses.* Cold Spring Harbor, NY: Cold Spring Harbor Press, 1997.

Flint, S.J., et al. *Principles of Virology: Molecular Biology, Pathogenesis, and Control.* Washington: American Society for Microbiology, 1999.

Lodish, H., et al. *Molecular Cell Biology.* 4th ed. New York: W. H. Freeman & Co., 2000.

Richman, D.D., and R.J. Whitley. *Clinical Virology.* 2nd ed. Washington: American Society for Microbiology, 2002.

Periodicals

Buchschacher, G.L., Jr. "Introduction to Retroviruses and Retroviral Vectors." *Somatic Cell and Molecular Genetics* no. 26 (1-6) (November 2001) :1-11.

Bonhoeffer S., P. Sniegowski. "Virus Evolution: the Importance of Being Erroneous." *Nature,* 28, no. 420 (6914) (November 2002): 367, 369.

Abdel Hakim Nasr

Vireos

Vireos are 44 **species** of small arboreal **birds** that comprise the family Vireonidae, in the order Passeriformes. As it is considered here, the Vireonidae is an assembly of three sub-families: the true vireos or Vireoninae, the shrike vireos or Vireolaniinae, and the **pepper** shrikes or Cyclarhinae. It should be pointed out, however, that some taxonomic treatments consider these to be separate families.

The vireos only occur in the Americas. Species of vireos range from the temperate **forests** of Canada, through the rest of **North America**, Central America, and south to northern Argentina. The northern, temperate species are all migratory, breeding in the northern parts of their biological range, but spending the nonbreeding season in tropical and subtropical forests. Vireos occur in all types of tropical and temperate forests, and in shrubby habitats as well.

Vireos are small birds, ranging in body length from 3.9-7.1 in (10-18 cm). The bill is relatively heavy for a small bird, and the upper mandible has a hook at the tip. Depending on the species, the wings are either long and pointed, or short and rounded, while the legs and feet are short but strong. The plumage of vireos is plain, generally olive-green or gray on the back and wings, and lighter colored on the throat and belly. Species may have **eye** rings, eye stripes, wing bars, and other diagnostically useful markings.

Vireos glean foliage and branches for their food of **insects** and spiders, and they may also eat small **fruits**. Compared with other types of foliage-gleaning birds, vireos are rather sluggish and deliberate in their movements. Vireos generally occur as solitary birds, or in family groups. They are aggressively territorial during the breeding season. The territory is demarcated and defended by loud, melodious songs, consisting of multisyllabic, persistently repeated phrases.

The nest is cup-like, or is an open, pendulous, baglike structure woven of **plant** fibers, usually located in a horizontal fork of a branch. The clutch size is two to five. The male helps with incubation, and both sexes cooperate in rearing the young birds.

North American species of vireos

A total of 12 species of vireos breed regularly in the United States or Canada. All of these are in the genus *Vireo*, and all are migratory, spending their nonbreeding season in Mexico or further south in Central America, or in the case of the red-eyed vireos, in Amazonia.

The red-eyed vireo (*Vireo olivaceus*) is a widespread species, occurring in deciduous forests over much of the **continent**, except for parts of the southwestern United States. The red-eyed vireo is an abundant species, and is one of the most commonly netted birds at mist-netting sites in eastern North America, where bird **migration** is studied.

The warbling vireo (*Vireo gilvus*) breeds in much of temperate North America and south into Mexico. The warbling vireo is a secretive bird, and males can be so

A red-eyed vireo (*Vireo olivaceus*). *JLM Visuals. Reproduced by permission.*

confident in their camouflage that they will sing from the nest while incubating their clutch.

The solitary or blue-headed vireo (*Vireo solitarius*) breeds in mixed hardwood-conifer forests through much of north temperate North America. This is one of the only species of vireo whose wintering range commonly includes the southern United States.

The white-eyed vireo (*Vireo griseus*) is an abundant species of moist deciduous forests and forest edges in the eastern United States. The yellow-throated vireo (*Vireo flavifrons*) has a bright yellow throat and breast, and is perhaps the most attractive of the North American species. This relatively uncommon species breeds in deciduous forests throughout the eastern United States.

Vireos elsewhere

Most species of true vireos occur in subtropical, tropical, and montane forests of Central and **South America**, but these species are too numerous and diverse to deal with here in any detail. One representative is the ashy-headed greenlet (*Hylophilus pectoralis*), a species found in the tropical forests of Guyana, Surinam, Brazil, and Bolivia.

The shrike vireos are relatively stout, tropical birds with a heavy, hooked bill. The chestnut-sided shrike vireo (*Vireolanius melitophrys*) occurs in tropical rainforests of Mexico and Guatemala.

The pepper shrikes also are more stout than the true vireos, and have a laterally compressed, hooked bill. The rufous-browed pepper shrike (*Cyclarhis gujanensis*) occurs in open forests from Mexico to Argentina.

Vireos and people

In spite of their small size, vireos are economically important. Sightings of these birds are avidly sought by

KEY TERMS

· ·

Nest parasite—A species that lays its eggs in the nests of other species. The host raises the parasitic egg, and usually does not raise any of its own babies.

birdwatchers. Birding is a nonconsumptive field sport, and is increasing rapidly in popularity. Birding and related activities such as bird feeding have large economic impacts, and give great aesthetic pleasure to many people.

Unfortunately, the populations of many species of birds that are the targets of these activities, including vireos, are declining greatly because of human activities. This is especially true of many species native to North America.

Vireos and other birds that share their **habitat** are at risk from changes occurring in both their breeding and wintering ranges. There have been tremendous decreases in the areas of mature forests that most vireos require for breeding in North America. The vireos are affected directly by these losses of area of their essential habitat, as well as by indirect effects associated with the fragmentation of much of the remaining habitat into small woodlots.

Small, isolated, habitat "islands" are highly influenced by their proximity to edges with younger habitat. This circumstance exposes vireos and other birds of the forest-interior to a greater intensity of predation, and to the disastrous effects of nest-parasitism by the brown-headed cowbird (*Molothrus ater*). Many ornithologists believe that these factors are causing large declines in the populations of numerous species of migratory forest birds, including many of the vireos.

Resources

Books

Ehrlich, P., D. Dobkin, and D. Wheye. *The Birders Handbook.* New York: Simon and Schuster, 1989.

Forshaw, Joseph. *Encyclopedia of Birds.* New York: Academic Press, 1998.

Sibley, David Allen. *The Sibley Guide to Birds.* New York: Knopf, 2000.

Bill Freedman

Virtual particles

Virtual particles are **subatomic particles** that form out of "nothing" (**vacuum** fields conceptually analogous

to lines of **force** between magnetic poles) for extremely short periods of **time** and then disappear again. Such particles permeate **space**, mediate particle decay, and mediate the exchange of the fundamental forces (electromagnetic, weak, strong, and—in accord with quantum theory—gravititational forces). Virtual particles are real and have measurable effects, but the same uncertainty principle that allows them to come into existence dictates that they cannot be directly observed.

Heisenberg's uncertainty principle, which explains the virtual particle phenomenon, is most commonly stated as follows: It is impossible to exactly and simultaneously measure both the **momentum** and position of a particle. There is always an uncertainty in momentum and an uncertainty in position. More importantly, these two uncertainties cannot be reduced to **zero** together.

One consequence of Heisenberg's uncertainty principle is that the **energy** and duration of a particle are also characterized by complementary uncertainties. There is always, at every point in space and time, even in a perfect vacuum, an uncertainty in energy and an uncertainty in duration, and these two complementary uncertainties cannot be reduced to zero simultaneously.

The meaning of Heisenberg's uncertainty principle is that "something" can arise from "nothing" *if* the "something" returns to the "nothing" after a very short time—an **interval** too short in which to be observed. These micro-violations of energy **conservation** are not only allowed to happen, they do, and so "empty" space is seething with particle-antiparticle pairs that come into being and then annihilate each other again after a very short interval. Although these particles cannot be observed individually, their existence can be demonstrated.

Normally, a **metal** plate experiences a **storm** of fleeting impacts from virtual particles on both of its surfaces; this "vacuum pressure" is equal on both sides of the plate, and so cancels out. If, however, two parallel metal plates are too closely spaced to allow the formation of relatively large virtual particles between them, the vacuum **pressure** between the plates is less than that on their outer surfaces, and they experience a net force pushing them together. This force is termed the Casimir effect after Dutch physicist Hendrik Casimir (1909–2000), who predicted its existence in 1948, and was experimentally measured in 1997.

The Casimir effect is only one manifestation of the reality of virtual particles. Virtual particles also mediate the exchange of all forces between particles. For example, when an **electron** experiences electrical repulsion from another electron (electrons are negatively charged, and like charges repel), it is actually exchanging virtual photons with that other electron. Higher-energy virtual photons are only allowed by the uncertainty principle to exist for shorter periods of time, as shown by the uncertainty equation, and thus cannot travel as far as lower-energy virtual photons; this explains why the electric force is stronger at short distances. (In fact, all the basic forces—electric, strong, weak, and gravitational—diminish with distance for this reason. Gravity, however, has not been satisfactorily integrated with the equations that describe the other three forces.)

A third role for virtual particles is in decay mediation. When an unstable subatomic particle decays (i.e., breaks down into two or more other subatomic particles), it does so by first taking the form of a virtual particle. The virtual particle then completes the decay process. In some cases, the intermediate virtual particle has *more mass* than the initial particle or the final set of decay products; this does not violate the conservation of **mass** because the intermediate particle is virtual, that is, exists for such a short period of time that it falls within the uncertainty bounds prescribed for the system's energy by the **Heisenberg uncertainty principle**.

This list of phenomena does not describe all the properties of virtual particles, but does indicate their prevalence.

See also Quantum mechanics.

Resources

Books

Barnett, R. Michael, Henry Mühry, and Helen R. Quinn. *The Charm of Strange Quarks.* New York: Springer-Verlag, 2000.
Ne'eman, Yuval, and Yoram Kirsh. *The Particle Hunters.* Cambridge, UK: Cambridge University Press, 1996.

Other

Lambrecht, Astrid. "The Casimir Effect: A Force From Nothing." PhysicsWeb. September 2002 [cited February 14, 2003]. <http://physicsweb.org/article/world/15/9/6>.

Larry Gilman

Virtual reality

Virtual reality is a product of the evolution of the computer from an instrument that merely received input from a user to a machine that can adapt to the user's cues to create an almost lifelike experience.

The term virtual reality was coined in 1989 by Jaron Lanier. Others have described the concept as "artificial reality," "cyberspace," and "virtual worlds."

Virtual reality combines state-of-the-art imaging with computer technology to allow users to experience a three-dimensional simulated environment. It is this environment that was called cyberspace in a novel by Canadian science-fiction writer William Gibson.

Cyberspace is interactive. In other words, the user can alter the appearance of the image or the nature of the scene. This interactive medium incorporates powerful computers with video displays, sensors, electronic headsets, and gloves. With these tools, users can both see and manipulate a phantom environment that appears real. Virtual reality tools under development include a whole body suit, which, like diving into **water**, would totally immerse the user in a virtual world. Although virtual reality has been popularized as a new form of entertainment, it has applications in business, industry, and medicine.

The origin of virtual reality

The concept of virtual reality dates back to World War II. Then, piloting training for combat missions had need of realistic flight simulators. The technology of the day was insufficient to produce much beyond a rudimentary simulation.

By the 1960s, technology advanced to a point where virtual reality became possible. In 1966, Ivan Sutherland conducted experiments with the first head-mounted three-dimensional displays at the Massachusetts Institute of Technology's Lincoln Laboratory. Although the headset was extremely cumbersome, the user was able to view a computer-generated three-dimensional cube floating in space and, by moving his or her head, inspect various aspects of the cube and determine its dimensions. Sutherland built the first fully functional head-mounted display unit in 1970.

Myron Krueger also worked on the infant science of virtual reality, first at the University of Utah and later at the University of Connecticut. His "artificial realities" used both computers and video systems. VIDEOPLACE was first exhibited in 1975 at the Milwaukee Art Center. Using video displays, computer graphics, and position-sensing technologies, Kreuger was able to create a virtual environment in darkened rooms containing large video screens. People in the room could see their own computer-generated silhouettes and follow their movements in the virtual world projected onto the screen. In addition, people in two different rooms could see each others' silhouettes and interact in the same virtual world.

As is the case with other technological advance, much of the initial development of virtual reality was funded by the military. By 1972, the General Electric Corporation had built one of the first computerized flight simulators, using three screens surrounding the training cockpit to provide a 180-degree field of view that simulated flying conditions. In 1979, virtual reality technology was incorporated into a head-mounted display developed by the McDonnell-Douglas Corporation. Three years later, Thomas Furness III, who had created visual displays for the military since 1966, developed the prototype Visually Coupled Airborne Systems Simulator. Donning a specialized oversized helmet, pilots were presented for the first time with an abstract view of flying conditions instead of a reality-based image. Since they were unable to see anything but the computerized cockpit's field of view, pilots became totally immersed in the graphic representation.

While scientists like Sutherland and Furness concentrated on the visual components of virtual reality, Frederick Brooks began experimenting with tactile feedback, or the sense of **touch**, in the early 1970s at the University of North Carolina. However, it was not until 1986 that the computer industry developed the tools to simulate tactile experience (i.e., sensing by touch). Brooks was able to develop his GROPE-III system, which used a specialized remote manipulator based on a device that mimicked arm motions to handle radioactive substances. Specifically, the GROPE-III system generated stereoscopic images of molecules and protein structures that could be felt and manipulated as though they existed in the physical world.

Components of virtual reality

Essentially, virtual reality systems consist of the computer and software—known as the reality engine—input sensors, and output sensors. The input sensors are the equipment to computer enthusiasts, and include the keyboard, mouse, knobs, and joysticks. Output devices include the printer and the video display monitor. In addition, virtual reality input and output devices include the head- and ear-mounted equipment mentioned above, and gloves for controlling the virtual world. Finally, the fourth sensory component is the user, who both directs and reacts to the chosen environment.

The reality engine

The reality engine employs both computer hardware and software to create the virtual world. Reality engines are based largely on the same components that make up a personal computer (PC), although much more computing power is required for the reality engine than what is available in a standard PC.

One reason for the increased computing power is the complexity of the hardware and software necessary to create a world that appears real. The images created by the computer and software are extremely complex, compared to the relatively simple line-based graphics associated with computer games. Virtual reality images are

made with thousands of dots called pixels (or picture elements). The more pixels per given amount of area, the higher the quality of the image. Hence, an image will be more realistic. Creating realistic images that can be manipulated is known as "realization." These images can be either opaque, in which all the viewer sees is the virtual world, or see-through, in which the virtual image is projected or superimposed onto the outer world.

The reality engine is also involved in bringing sound to the virtual world. Sound enriches the virtual world. For example, in a flight simulator, the experience of soaring through the air in a simulated cockpit is more realistic if the user hears the roar of the engines. Sound also enhances participation in the virtual world by providing the user with audio cues. For example, the user may be directed to look for another virtual airplane flying overhead.

To incorporate the total experience provided by the sight and sound cues, the reality engine can use what is known as haptic enhancement. Haptic enhancement utilizes the participant's other senses of touch and **pressure** in the virtual world. Haptic enhancement is a complex process, and the hardware and software that are required increase the cost of the system tremendously. To date haptic enhancement is used mainly military and research applications.

Headsets

Head-mounted display (HMD) units use a small screen or a pair of screens (one for each **eye**) that are worn in a helmet or a pair of glasses. The HMD allows viewers to look at an image from various angles or change their field of view by simply moving their heads. In contrast, a movie is a passive experience, where the view of the audience is controlled by the position of the camera that recorded the scene.

HMD units usually employ **cathode ray tube** (CRT) or liquid crystal display (**LCD**) technology. The optical systems in CRTs reflect an image onto the viewer's eye, creating an image of very clear and realistic image. CRT images can be semi-reflective. This means that the user can experience the virtual world while still being able to see the outside world. This permits the user to operate another machine or device while viewing the virtual world.

LCD technology has lagged behind CRT in picture quality. LCD monitors display two slightly different images to each eye. The **brain** processes and merges the images into a single three-dimensional view. However LCD systems have the advantages of being slimmer, lighter, and less expensive than CRT systems. Thus, LCD is better suited to home entertainment. As the

A virtual reality system. *Photograph by Thomas Ernstein. Bilderberg/Stock Market. Reproduced by permission.*

image quality improves, LCDs will find a lucrative **niche** in the home entertainment market.

Audio units

Sound effects in virtual reality rely on a prerecorded sound set. This aspect of the virtual reality experience is less prone to alteration.

The audio portion of virtual reality is transmitted through small speakers placed over each **ear**. Audio cues may include voices, singing, the sound of bubbling water, thud-like noises of colliding objects—in short, any sound that can be recorded.

While the sounds themselves cannot be changed from a recording, the presentation of the sounds to the user can be changed. Three-dimensional (omnidirectional) sound further enhances the virtual reality experience. Sound that seems to come from above, below, or either side provides audio cues that mimic how sounds are heard in the real world (e.g., footsteps approaching or a plane flying overhead). Three-dimensional sound is achieved through the use of complex filtering devices. This technology must take into account the delay in the detection of sound by the ear that is furthest away from

the source of the sound (interaural time difference) and the tendency of one ear to hear a sound more loudly than the other ear (interaural amplitude difference).

The most complex human hearing dynamic is called head-related transfer functions (HRTF). HRTF accounts for how the eardrum and inner ear process **sound waves**. Factors that are influential in HRTF include the various frequencies at which the sound waves travel, and how waves are absorbed and reflected by other objects. HRTF audio processing enables the listener to locate a sound source and to focus in on a specific sound out of a multitude of sounds. (i.e., the sound of their name called out in the midst of a noisy party).

Gloves

A popular image of a virtual reality experience shows the user wearing gloves. The gloves allow the user to interact with the virtual world. For example, the user may pick up a virtual block, and, by turning their gloved hands, turn the block over and set it on a virtual table.

Virtual reality gloves are wired with thin fiber-optic cables, or have light-emitting diodes positioned at critical points over the glove's surface. The **optics** detects the amount of **light** passing through the cable in **relation** to the movement of the hand or joint. The computer then analyzes the corresponding information and projects this moving hand into the virtual reality. Magnetic tracking systems are also used to determine where the hand is in space in relation to the virtual scene.

Some gloves use haptic enhancement to provide a sense of touch and feel. In haptic enhancement, the reality engine relays the various sensations of force, **heat**, and texture that are experienced by the user to the **computer software**. The software can use the information to determine an outcome of the user's actions, and relay the outcomes back to the user. For example, if the user closes a hand on a virtual squeeze toy, the software will alter the virtual image to show the toy becoming compressed. To achieve this two-way communication, virtual reality gloves may use either air pressure (such as strategically placed, inflated air pockets in the glove) or vibrating transducers placed next to the skin (such as a voice coil from a stereo speaker or alloys, which change shape through the conduction of electrical currents) to simulate tactile experience.

Tools under development

Many other virtual reality tools are in the phases of research and development. Remote control robotic or manipulation devices are being tested for industry and medicine. Already, **surgery** has been done by a physician located hundreds of miles away from the patient, by means of **robotics** and virtual imaging.

Special wands with sensors, joysticks, and finger sensors such as picks and rings will eventually be as common to virtual reality technology as microwaves are to cooking. The technology to control the virtual world through voice commands is also rapidly advancing.

Perhaps the most impressive technology under development is the whole body suit. These suits would function similarly to the gloves, creating a virtual body that could take a stroll through a virtual world and feel a virtual windstorm.

Applications of virtual reality

The potential for virtual reality as an entertainment medium is apparent. Instead of manipulating computerized images of two boxers or a car race, the virtual playground allows the user to experience the event. Disney World's Epcot Center houses a virtual reality system.

Most entertainment applications of the present day are visually based. Virtual reality will allow players of the future to experience a variety of tactile events. For example, in a simulated boxing match, virtual reality users would bob and weave, and throw, land, and receive punches in return.

Virtual reality also has practical applications in business, manufacturing, and medicine. Already, the National Aeronautics and Space Administration (NASA) has developed a virtual **wind** tunnel to test **aerodynamics** shape. Virtual reality holds promise for discovering the most efficient manufacturing conditions by allowing planners to evaluate the actual physical motions and strength needed to complete a job. For example, the McDonnell-Douglas Corporation is using virtual reality to explore the use of different materials and tools in building the F-18 E/F **aircraft**. The study of people in relation to their environments (ergonomics) may also be revolutionized by trials in cyberspace. Engineers at the Volvo car company use virtual reality to test various designs for the dashboard configuration from the perspective of the user.

In medicine, virtual reality systems are being developed to help surgeons plan and practice delicate surgical procedures. Philip Green, a researcher at SRI International, has developed a telemanipulator, a special remote-controlled robot, to be used in surgery. Such surgery was performed in 2002 by a physician in Halifax, Nova Scotia, on a patient located hundreds of miles away. Using instruments connected to a computer, the operation was performed cyberspace, while the computer sent signals to direct the telemanipulator.

Virtual reality may even have applications in **psychiatry**. For example, someone with acrophobia (a fear of heights) may be treated by having the patient stand atop virtual skyscrapers or soar through the air like a bird.

On the horizon of virtual reality

Virtual reality will no doubt mirror the breath taking pace of development that is the norm for other computerized applications. Thus, what is state of the art now will be commonplace in decades.

Aspects of virtual technology that are just ideas now will become reality soon. For example, technology is being developed to use the retina of the eye as a screen for images that could be transmitted directly to the brain through the optic nerve. Virtual sight would become a replacement for natural sight in those blind people whose optical hardware was intact.

Also under development is technology to allow the remote operation of aircraft and other machines as though the user were actually in the machine. Traveling to France or Greece, including the experience of climbing the Eiffel Tower or basking on a sun-drenched beach,

KEY TERMS

· ·

Pixel—A word used for picture elements, or dots, that make up a computerized image.

Three-dimensional—A visual representation in terms of height, width, and *depth* as opposed to a "flat" image that represents only height and width.

may be as easy as donning a headset and body suit and plugging in.

Like most technological advances, virtual reality has social and psychological ramifications. Critics argue that virtual reality could cause some people to forego emotions and interpersonal relationships for the safe, controllable virtual world. But proponents say advances from the proper applications of this technology—both as a means of interacting with the real world and as an end of facilitating training and entertainment—far outweigh the potential for antisocial abuse.

Resources

Books

Dix, Alan J., Janet E. Finlay, Gregory D. Abowd, et al. *Human-Computer Interaction.* 2nd ed. Upper Saddle River, NJ: Prentice Hall, 1998.

Hsu, Feng-Hsiung. *Behind Deep Blue: Building the Computer That Defeated the World Chess Champion.* Princeton: Princeton University Press, 2002.

Sherman, William, R., and Alan B. Craig. *Understanding Virtual Reality: Interface, Application, and Design.* San Francisco: Morgan Kaufmann Publishers, 2002.

Other

University of Michigan Virtual Reality Laboratory. "Virtual Reality: A Short Introduction." College of Engineering. January 6, 2003 [cited January 17, 2003]. <http://www.vrl.mich.edu/intro/>.

David Petechuk

Virus

A virus is a small, infectious agent that consists of a core of genetic material (either deoxyribonucleic acid [DNA] or ribonucleic acid [RNA]) surrounded by a shell of protein. Viruses cause **disease** by infecting a host **cell** and commandeering the host cell's synthetic capabilities to produce more viruses. The newly made viruses then leave the host cell, sometimes killing it in the process, and proceed to infect other cells within the host. Because

viruses invade cells, no drug therapy has yet been designed to kill viruses. The human **immune system** is the only defense against a viral disease.

Viruses can infect both plants, **bacteria**, and animals. The tobacco mosaic virus, one of the most studied of all viruses, infects tobacco plants. Bacterial viruses, called bacteriophages, infect a variety of bacteria, such as *Escherichia coli,* a bacteria commonly found in the human digestive tract. **Animal** viruses cause a variety of fatal diseases. Acquired Immune Deficiency Syndrome (**AIDS**) is caused by the Human Immunodeficiency Virus (HIV); **hepatitis** and **rabies** are viral diseases; and the so-called hemorrhagic fevers, which are characterized by severe internal bleeding, are caused by filoviruses. Other animal viruses cause some of the most common human diseases. Often, these diseases strike in childhood. Measles, mumps, and **chickenpox** are viral diseases. The common cold and **influenza** are also caused by viruses. Finally, some viruses can cause **cancer** and tumors. One such virus, Human T-cell Leukemia Virus (HTLV), was only recently discovered and its role in the development of a special kind of **leukemia** is still being elucidated.

Although viral structure varies considerably between the different types of viruses, all viruses share some common characteristics. All viruses contain either RNA or DNA surrounded by a protective protein shell called a capsid. Some viruses have a double strand of DNA, others a single strand of DNA. Other viruses have a double strand of RNA or a single strand of RNA. The size of the genetic material of viruses is often quite small. Compared to the 100,000 genes that exist within human DNA, viral genes number from 10 to about 200 genes.

Viruses contain such small amounts of genetic material because the only activity that they perform independently of a host cell is the synthesis of the protein capsid. In order to reproduce, a virus must infect a host cell and take over the host cell's synthetic machinery. This aspect of viruses—that the virus does not appear to be "alive" until it infects a host cell—has led to controversy in describing the nature of viruses. Are they living or non-living? When viruses are not inside a host cell, they do not appear to carry out many of the functions ascribed to living things, such as reproduction, **metabolism**, and movement. When they infect a host cell, they acquire these capabilities. Thus, viruses are both living and non-living. It was once acceptable to describe viruses as agents that exist on the boundary between living and non-living; however, a more accurate description of viruses is that they are either active or inactive, a description that leaves the question of life behind altogether.

The origin of viruses is also controversial. Some viruses, such as the pox viruses, are so complex that they appear to have been derived from some kind of living eukaryote or **prokaryote**. The origin of the poxvirus could therefore resemble that of mitochondria and chloroplasts, organelles within eukaryotic cells which are thought to have once been independent organisms. On the other hand, some viruses are extremely simple in structure, leading to the conclusion that these viruses are derived from cellular genetic material that somehow acquired the capacity to exist independently. This possibility is much more likely for most viruses; however, scientists still believe that the poxvirus is the exception to this scenario.

Structure of viruses

All viruses consist of genetic material surrounded by a capsid, but within the broad range of virus types, variations exist within this basic structure. Studding the envelope of these viruses are protein "spikes." These spikes are clearly visible on some viruses, such as the influenza viruses; on other enveloped viruses, the spikes are extremely difficult to see. The spikes help the virus invade host cells. The influenza virus, for instance, has two types of spikes. One type, composed of hemagglutinin protein (HA), fuses with the host cell **membrane**, allowing the virus particle to enter the cell. The other type of spike, composed of the protein neuraminidase (NA), helps the newly formed virus particles to bud out from the host cell membrane.

The capsid of viruses is relatively simple in structure, owing to the few genes that the virus contains to encode the capsid. Most viral capsids consist of a few repeating protein subunits. The capsid serves two functions: it protects the viral genetic material and it helps the virus introduce itself into the host cell. Many viruses are extremely specific, targeting only certain cells within the **plant** or animal body. HIV, for instance, targets a specific immune cell, the T helper cell. The cold virus targets respiratory cells, leaving the other cells in the body alone. How does a virus "know" which cells to target? The viral capsid has special receptors that match receptors on their targeted host cells. When the virus encounters the correct receptors on a host cell, it "docks" with this host cell and begins the process of **infection** and replication.

Most viruses are rod- or roughly sphere-shaped. Rod-shaped viruses include tobacco mosaic virus and the filoviruses. Although they look like rods under a **microscope**, these viral capsids are actually composed of protein molecules arranged in a helix. Other viruses are shaped somewhat like spheres, although many viruses are not actual spheres. The capsid of the adenovirus, which infects the respiratory tract of animals, consists of 20 triangular faces. This shape is called a icosahedron. HIV is a true **sphere**, as is the influenza virus.

Some viruses are neither rod- or sphere-shaped. The poxviruses are rectangular, looking somewhat like bricks. Parapoxviruses are ovoid. Bacteriophages are the most unusually shaped of all viruses. A **bacteriophage** consists of a head region attached to a sheath. Protruding from the sheath are tail fibers that dock with the host bacterium. The bacteriophage's structure is eminently suited to the way it infects cells. Instead of the entire virus entering the bacterium, the bacteriophage injects its genetic material into the cell, leaving an empty capsid on the surface of the bacterium.

Viral infection

Viruses are obligate intracellular **parasites**, meaning that in order to replicate, they need to be inside a host cell. Viruses lack the machinery and enzymes necessary to reproduce; the only synthetic activity they perform on their own is to synthesize their capsids.

The infection cycle of most viruses follows a basic pattern. Bacteriophages are unusual in that they can infect a bacterium in two ways (although other viruses may replicate in these two ways as well). In the lytic cycle of replication, the bacteriophage destroys the bacterium it infects. In the lysogenic cycle, however, the bacteriophage coexists with its bacterial host, and remains inside the bacterium throughout its life, reproducing only when the bacterium itself reproduces.

An example of a bacteriophage that undergoes lytic replication inside a bacterial host is the T4 bacteriophage which infects *E. coli*. T4 begins the infection cycle by docking with an *E. coli* bacterium. The tail fibers of the bacteriophage make contact with the cell wall of the bacterium, and the bacteriophage then injects its genetic material into the bacterium. Inside the bacterium, the viral genes are transcribed. One of the first products produced from the viral genes is an **enzyme** that destroys the bacterium's own genetic material. Now the virus can proceed in its replication unhampered by the bacterial genes. Parts of new bacteriophages are produced and assembled. The bacterium then bursts, and the new bacteriophages are freed to infect other bacteria. This entire process takes only 20-30 minutes.

In the lysogenic cycle, the bacteriophage reproduces its genetic material but does not destroy the host's genetic material. The bacteriophage called lambda, another *E. coli*-infecting virus, is an example of a bacteriophage that undergoes lysogenic replication within a bacterial host. After the viral DNA has been injected into the bacterial host, it assumes a circular shape. At this point, the replication cycle can either become lytic or lysogenic. In a lysogenic cycle, the circular DNA attaches to the host cell **genome** at a specific place. This combination host-viral genome is

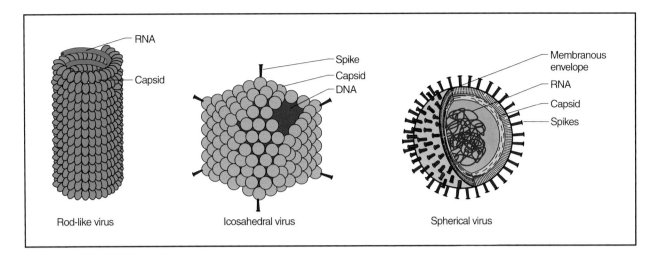

Figure 1. In addition to the capsid, some viruses are surrounded by an envelope, consisting of protein and lipid, which is acquired as the virus exits the host cell. Viruses that have an envelope exit host cells by exocytosis, in which they "bud out" from the host cell. As they bud out, they take pieces of the host cell's plasma membrane with them and use these pieces as the protective envelope. *Illustration by Hans & Cassidy. Courtesy of Gale Group.*

called a prophage. Most of the viral genes within the prophage are repressed by a special repressor protein, so they do not encode the production of new bacteriophages. However, each time the bacterium divides, the viral genes are replicated along with the host genes. The bacterial progeny are thus lysogenically infected with viral genes.

Interestingly, bacteria that contain prophages can be destroyed when the viral DNA is suddenly triggered to undergo lytic replication. **Radiation** and chemicals are often the triggers that initiate lytic replication. Another interesting aspect of prophages is the role they play in human diseases. The bacteria that cause **diphtheria** and **botulism** both harbor viruses. The viral genes encode powerful toxins that have devastating effects on the human body. Without the infecting viruses, these bacteria may well be innocuous. It is the presence of viruses that makes these bacterial diseases so lethal.

Types of viruses

Scientists have classified viruses according to the type of genetic material they contain. Broad categories of viruses include double-stranded DNA viruses, single-stranded DNA viruses, double-stranded RNA viruses, and single stranded RNA viruses. For the description of virus types that follows, however, these categories are not used. Rather, viruses are described by the type of disease they cause.

Poxviruses

Poxviruses are the most complex kind of viruses known. They have large amounts of genetic material and

fibrils anchored to the outside of the viral capsid that assist in attachment to the host cell. Poxviruses contain a double strand of DNA.

Viruses cause a variety of human diseases, including **smallpox** and cowpox. Because of worldwide vaccination efforts, smallpox has virtually disappeared from the world, with the last known case appearing in Somalia in 1977. The only places on **Earth** where smallpox virus currently exists are two labs: the Centers for Disease Control in Atlanta and the Research Institute for Viral Preparation in Moscow. Prior to the eradication efforts begun by the World Health Organization in 1966, smallpox was one of the most devastating of human diseases. In 1707, for instance, an outbreak of smallpox killed 18,000 of Iceland's 50,000 residents. In Boston in 1721, smallpox struck 5,889 of the city's 12,000 inhabitants, killing 15% of those infected.

Edward Jenner (1749-1823) is credited with developing the first successful **vaccine** against a viral disease, and that disease was smallpox. A vaccine works by eliciting an immune response. During this immune response, specific immune cells, called **memory** cells, are produced that remain in the body long after the foreign microbe present in a vaccine has been destroyed. When the body again encounters the same kind of microbe, the memory cells quickly destroy the microbe. Vaccines contain either a live, altered version of a virus or bacteria, or they contain only parts of a virus or bacteria, enough to elicit an immune response.

In 1797, Jenner developed his **smallpox vaccine** by taking pus from a cowpox lesion on the hand of a milkmaid. Cowpox was a common disease of the era, transmitted through contact with an infected cow. Unlike smallpox, however, cowpox is a much milder disease. Using the cowpox pus, he inoculated an eight-year-old boy. Jenner continued his vaccination efforts through his lifetime. Until 1976, children were vaccinated with the smallpox vaccine, called vaccinia. Reactions to the introduction of the vaccine ranged from a mild fever to severe complications, including (although very rarely) death. In 1976, with the eradication of smallpox complete, vaccinia vaccinations for children were discontinued, although vaccinia continues to be used as a carrier for **recombinant DNA** techniques. In these techniques, foreign DNA is inserted in cells. Efforts to produce a vaccine for HIV, for instance, have used vaccinia as the vehicle that carries specific parts of HIV.

Herpesviruses

Herpesviruses are enveloped, double-stranded DNA viruses. Of the more than 50 herpes viruses that exist, only eight cause disease in humans. These include the human herpes virus types 1 and 2 that cause cold sores and genital herpes; human herpes virus 3, or varicella-zoster virus (VZV), that causes chicken pox and **shingles**; cytomegalovirus (CMV), a virus that in some individuals attacks the cells of the **eye** and leads to blindness; human herpes virus 4, or **Epstein-Barr virus** (EBV), which has been implicated in a cancer called Burkitt's lymphoma; and human herpes virus types 6 and 7, newly discovered viruses that infect white **blood** cells. In addition, herpes B virus is a virus that infects **monkeys** and can be transmitted to humans by handling infected monkeys.

Adenoviruses

Adenoviruses are viruses that attack respiratory, intestinal, and eye cells in animals. More than 40 kinds of human adenoviruses have been identified. Adenoviruses contain double-stranded DNA within a 20-faceted capsid.

Adenoviruses that target respiratory cells cause **bronchitis**, **pneumonia**, and **tonsillitis**. Gastrointestinal illnesses caused by adenoviruses are usually characterized by diarrhea and are often accompanied by respiratory symptoms. Some forms of appendicitis are also caused by adenoviruses. Eye illnesses caused by adenoviruses include conjunctivitis, an infection of the eye tissues, as well as a disease called pharyngoconjunctival fever, a disease in which the virus is transmitted in poorly chlorinated swimming pools.

Papoviruses

Human papoviruses include two groups: the papilloma viruses and the polyomaviruses. Human papilloma viruses (HPV) are the smallest double-stranded DNA viruses. They replicate within cells through both the lytic and the lysogenic replication cycles. Because of their lysogenic capabilities, HPV-containing cells can be produced through the replication of those cells that HPV initially infects. In this way, HPV infects epithelial cells, such as the cells of the skin. HPVs cause several kinds of benign (non-cancerous) warts, including plantar warts (those that form on the soles of the feet) and genital warts. However, HPVs have also been implicated in a form of cervical cancer that accounts for 7% of all female cancers.

HPV is believed to contain oncogenes, or genes that encode for growth factors that initiate the uncontrolled growth of cells. This uncontrolled proliferation of cells is called cancer. When the HPV oncogenes within an epithelial cell are activated, they cause the epithelial cell to proliferate. In the cervix (the opening of the uterus), the cell proliferation manifests first as a condition called cervical neoplasia. In this condition, the cervical cells proliferate and begin to crowd together. Eventually, cervical neoplasia can lead to full-blown cancer.

Polyomaviruses are somewhat mysterious viruses. Studies of blood have revealed that 80% of children aged five to nine years have antibodies to these viruses, indicating that they have at some point been exposed to polyomaviruses. However, it is not clear what disease this virus causes. Some evidence exists that a mild respiratory illness is present when the first antibodies to the virus are evident. The only disease that is certainly caused by polyomavirses is called progressive multifocal leukoencephalopathy (PML), a disease in which the virus infects specific **brain** cells called the oligodendrocytes. PML is a debilitating disease that is usually fatal, and is marked by progressive neurological degeneration. It usually occurs in people with suppressed immune systems, such as cancer patients and people with AIDS.

Hepadnaviruses

The hepadnaviruses cause several diseases, including hepatitis B. Hepatitis B is a chronic, debilitating disease of the liver and immune system. The disease is much more serious than hepatitis A for several reasons: it is chronic and long-lasting; it can cause **cirrhosis** and cancer of the liver; and many people who contract the disease become carriers of the virus, able to transmit the virus through body fluids such as blood, semen, and vaginal secretions.

The hepatitis B virus (HBV) infects liver cells and has one of the smallest viral genomes. A double-stranded DNA virus, HBV is able to integrate its genome into the host cell's genome. When this integration occurs, the viral genome is replicated each time the cell divides. Individuals who have integrated HBV into their cells become carriers of the disease. Recently, a vaccine against

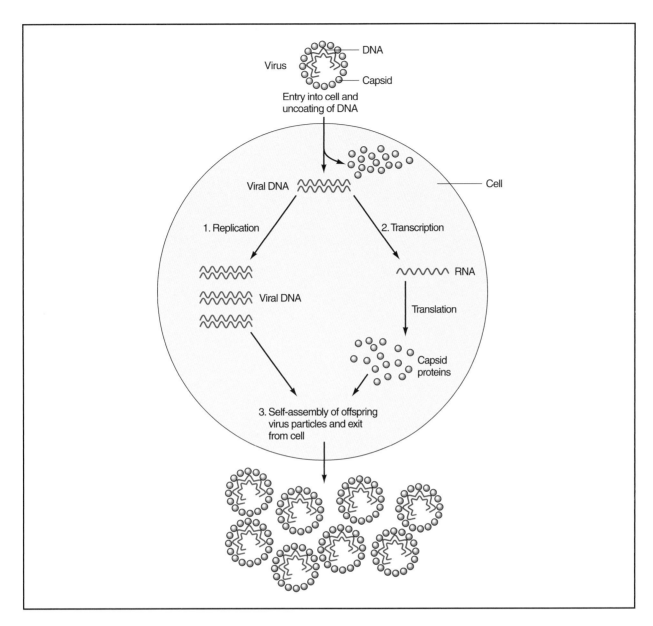

Figure 2. First, the virus docks with the host cell. The genetic material of the virus then enters the host cell and the virus loses its capsid. Once inside the host cell, the viral RNA or DNA takes over the cellular machinery to make more viruses. Viral subunits are produced, which are then assembled to make whole viruses. Finally, the new viruses leave the host cell, either by exocytosis or by destroying the host cell. *Illustration by Hans & Cassidy. Courtesy of Gale Group.*

HBV was developed. The vaccine is especially recommended for health care workers who through exposure to patient's body fluids are at high risk for infection.

Parvoviruses

Parvoviruses are icosahedral, single-stranded DNA viruses that infect a wide variety of **mammals**. Each type of parvovirus has its own host. For instance, one type of parvovirus causes disease in humans; another type causes disease in **cats**; while still another type caus-

es disease in dogs. The disease caused by parvovirus in humans is called erythremia infectiosum, a disease of the red blood cells that is relatively rare except for individuals who have the inherited disorder **sickle cell anemia**. Canine and feline parvovirus infections are fatal, but a vaccine against parvovirus is available for dogs and cats.

Orthomyxoviruses

Orthomyxoviruses cause influenza ("flu"). This highly contagious viral infection can quickly assume

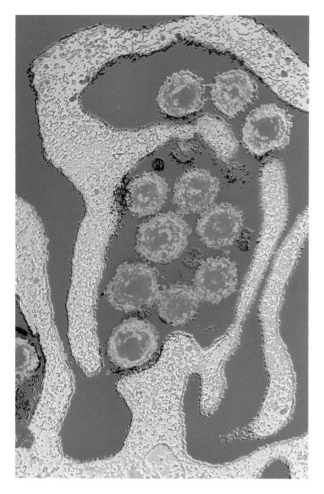

A transmission electron micrograph (TEM) of *Human Herpes Virus type 6 (HHV6)* infecting a human cell. This cell was isolated from an AIDS patient suffering a secondary infection with the virus. *HHV6* is now known to be the cause of the childhood disease *roseola infantum* which produces sudden fever, irritability, and a skin rash. *Photograph by A. B. Dowsett. Science Photo Library, National Audubon Society Collection/ Photo Researchers, Inc. Reproduced by permission.*

epidemic proportions, given the right environmental conditions. An influenza outbreak is considered an epidemic when more than 10% of the population is infected. In 1918, the influenza virus infected 25 million Americans and killed 22 million people worldwide in 18 months. Most people who require hospitalization due to influenza or who die from the infection are elderly individuals, especially those with a pre-existing chronic lung or **heart** condition. The most common complication of influenza is pneumonia.

Influenza viruses are spherical, single-stranded RNA viruses, with visible protein spikes that protrude from the capsid. Three strains of influenza virus—strains A, B, and C—cause illness in humans. Strain C causes a relatively mild illness that usually does not **balloon** into an epidemic.

Strains A and B cause more debilitating illnesses and are the frequent cause of epidemics. These strains undergo frequent genetic **mutation**, so that antibodies that are made in the body against prior strains are ineffective against mutated strains. Strain A is the most common influenza virus; strain B emerges only once every two to four years.

A vaccine against the current strain of influenza is prepared each year. This vaccination should be performed during the fall months, so that antibodies against the virus reach optimum numbers by the winter flu season. For reasons that are not clear, flu vaccinations are only 50-60% effective in children, while in the elderly, they can be 70-90% effective. Therefore, most experts recommend that the elderly, who are most at risk for developing serious complications from flu, receive a flu vaccination every year.

Enteroviruses

The enteroviruses are icosahedral, enveloped, single-stranded RNA viruses. Five types of enteroviruses cause diseases in humans, including the polioviruses (which cause polio) and echoviruses (which cause viral **meningitis**). Enteroviruses have an unusual replication cycle. They first enter the body through the upper respiratory tract and replicate within respiratory cells. Once the virus particles have been replicated, they are shed back into the oral secretions and are then swallowed. In the gastrointestinal tract, more replication takes place, and the viruses then enter the bloodstream. From the bloodstream, the viruses are carried to all parts of the body. However, further replication takes place only in cells for which the virus has an affinity. For example, polioviruses enter the nerve cells in the brain. There, viruses undergo further replication, and symptoms begin to appear.

Polio, the disease caused by polioviruses, is now controlled in the United States and other developed countries by the polio vaccine. However, before the introduction of this vaccine in the late 1950s, polio was one of the most feared human diseases. It causes a widespread paralysis of the muscles, sometimes including those of the **respiratory system**. Prolonged breathing assistance is needed for those individuals in which the respiratory muscles are paralyzed.

Echoviruses, another kind of enterovirus, cause viral meningitis, an **inflammation** of the **nervous system**. This disease is not as serious as the meningitis caused by bacteria, called bacterial meningitis.

Rhinoviruses

Rhinoviruses (from the Latin word meaning "nose") cause the common cold. They come from the same fami-

KEY TERMS

. . . .

Bacteriophage—A virus that infects bacteria.

Capsid—The outer protein coat of a virus.

Deoxyribonucleic acid (DNA)—The genetic material in a cell; in the nucleus, DNA transcribes RNA for the synthesis of proteins.

Envelope—The outermost covering of some viruses; it is composed of lipid and protein acquired from the host cell's plasma membrane as the virus buds out from the cell.

Eukaryote—A cell whose genetic material is carried on chromosomes inside a nucleus encased in a membrane. Eukaryotic cells also have organelles that perform specific metabolic tasks and are supported by a cytoskeleton which runs through the cytoplasm, giving the cell form and shape.

Exocytosis—The process by which a virus "buds out" from the host cell.

Genome—The complete sequence of genes within a cell or virus.

Host cell—The specific cell that a virus targets and infects.

Icosahedron—A 20–sided polyhedron.

Lysogenic cycle—A viral replication cycle in which the virus does not destroy the host cell but co-exists within it.

Lytic cycle—A viral replication cycle in which the virus destroys the host cell.

Obligate intracellular parasite—An organism or agent, such as a virus, that cannot reproduce unless it is inside a cell.

Oncogene—A gene that encodes for growth factors; oncogenes are believed to cause the cancerous growth of cells.

Prokaryote—A type of cell without a true nucleus, such as a bacterium.

Retrovirus—A type of virus that inserts its genetic material into the chromosomes of the cells it infects.

Reverse transcriptase—The enzyme that allows a retrovirus to transcribe DNA from RNA.

Ribonucleic acid—RNA; the molecule translated from DNA in the nucleus that directs protein synthesis in the cytoplasm; it is also the genetic material of many viruses.

ly as the enteroviruses, described above, and thus contain a single strand of RNA enclosed within an enveloped, 20-sided capsid. Interestingly, 113 types of rhinoviruses have been classified. These 113 types differ slightly in the composition of their capsids, so antibodies that are made against one type of rhinovirus are often ineffective against other types of viruses. For this reason, most people are susceptible to colds from season to season.

Paramyxoviruses

These helical, enveloped, single-stranded RNA viruses cause pneumonia, croup, measles, and mumps in children. A vaccine against measles and mumps has greatly reduced the incidence of these diseases in the United States. In addition, a paramyxovirus called respiratory syncytial virus (RSV) causes bronchiolitis (an infection of the bronchioles) and pneumonia.

Flaviviruses

Flaviviruses (from the Latin word meaning "yellow") cause insect-carried diseases including **yellow fever**, an often fatal disease characterized by high fever and internal bleeding. Flaviviruses are single-stranded RNA viruses.

Filoviruses

The two filoviruses, **Ebola virus** and Marburg virus, are perhaps the most lethal of all human viruses. Both cause severe fevers accompanied by internal bleeding, which eventually kills the victim. The fatality **rate** of Marburg is about 60%, while the fatality rate of Ebolavirus is about 90%. Both are transmitted through contact with body fluids. Marburg and Ebola also infect **primates**.

Rhabdoviruses

Rhabdoviruses are bullet-shaped, single-stranded RNA viruses. They are responsible for rabies, a disease that affects dogs, **rodents**, and humans.

Retroviruses

Retroviruses are unique viruses. They are double-stranded RNA viruses that contain an enzyme called reverse transcriptase. Within the host cell, the virus uses reverse transcriptase to make a DNA copy from its RNA genome. In all other organisms, RNA is synthesized from DNA. Cells infected with retroviruses are the only living things that reverse this process.

The first retroviruses discovered were viruses that infect chickens. The Rous sarcoma virus, discovered in the 1950s by Peyton Rous (1879-1970), was also the first virus that was linked to cancer. But it was not until 1980 that the first human **retrovirus** was discovered. Called Human T-cell Leukemia Virus (HTLV), this virus causes a form of leukemia called adult T-cell leukemia. In 1983-4, another human retrovirus, Human Immunodefiency Virus, the virus responsible for AIDS, was discovered independently by two researchers. Both HIV and HTLV are transmitted in body fluids.

See also Childhood diseases; Cold, common; Meningitis; Poliomyelitis.

Resources

Books

Doerfler, Walter, and Petra Bohm, eds. *Virus Strategies: Molecular Biology and Pathenogenesis.* New York: VCH, 1993.

Flint, S.J., et al. *Principles of Virology: Molecular Biology, Pathogenesis, and Control.* Washington: American Society for Microbiology, 1999.

Kurstak, Edouard, ed. *Control of Virus Diseases.* New York: Marcel Dekker, 1993.

Richman, D.D., and R.J. Whitley. *Clinical Virology.* 2nd ed. Washington: American Society for Microbiology, 2002.

Thomas, D. Brian. *Viruses and the Cellular Immune Response.* New York: Marcel Dekker, 1993.

Periodicals

Appleton, Hazel. "Foodborne Viruses." *The Lancet* 336 (December 1990): 1362.

Berns, Kenneth I., and Michael R. Linden. "The Cryptic Life Style of Adeno-associated Virus." *BioEssays* 17 (March 1995): 237.

Bloch, Alan B., et al. "Recovery of Hepatitis A Virus from a Water Supply Responsible for a Common Source Outbreak of Hepatitis A." *American Journal of Public Health* 80 (April 1990): 428.

Cimons, M. "New Prospects on the HIV Vaccine Scene." *ASM News* no. 68 (January 2002): 19-22.

Dutton, Gail. "Biotechnology Counters Bioterrorism." *Genetic Engineering News* no. 21 (December 2000): 1-22ff.

Dybul, M., T.-K. Chun, C. Yoder, et al. "Short-cycle Structured Intermittent Treatment of Chronic HIV Infection with Highly Effective Antiretroviral Therapy: Effects on Virologic, Immunologic, and Toxicity Parameters." *Proceedings of the National Academy of Sciences* no. 98 (18 December 2001): 15161-15166.

Ganem, Don. "Oncogenic Viruses: Of Marmots and Men." *Nature* 347 (September 20, 1990): 230.

Slater, P. E., et al. "Poliomyelitis Outbreak in Israel in 1988: A Report with Two Commentaries." *The Lancet*: 335 (May 19, 1992): 1192.

Zur Hausen, Harald. "Viruses in Human Cancer." *Science* 254 (November 22, 1991): 1167.

Kathleen Scogna

Viscosity

The viscosity of a fluid is a measure of its resistance to continuous deformation caused by sliding or shearing forces. Imagine a fluid between two flat plates; one plate is stationary and the other is being moved by a **force** at a constant **velocity** parallel to the first plate. The applied force per unit area of the plate is called the shear stress. The applied shear stress keeps the plate in **motion** and, when the plate velocity is steady, this shear stress is in equilibrium with the frictional and drag forces within the fluid. The shear stress is proportional to the speed of the plate and inversely proportional to the distance between the plates. The proportionality factor between the shear stress and the velocity difference between the plates is defined as the **coefficient** of viscosity or simply the viscosity of the fluid. Thick fluids such as tar or honey have a high viscosity; thin fluids such as **water** or **alcohol** have a low viscosity.

In general, viscosity is a function of **temperature** and **pressure**; however, in some fluids viscosity is dependent on the **rate** of shear and time. When brushed on (sheared) quickly, fluids such as paint have a low viscosity and flow easily. After paint is applied, only the slow and steady pull of its weight causes it to flow; at this slow shear rate the viscosity of paint is high and it resists the tendency to flow or sag. Fluids that behave in this manner are called non-Newtonian fluids. Other examples are liquid **plastics** and mud. For gases and non-polymeric liquids like water, viscosity is independent of the fluid's shear stress and history. These are called Newtonian fluids. In the case of gases, the viscosity increases with temperature because of the increased molecular activity at

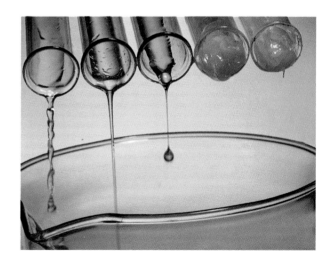

Liquids of different viscosities. © *Yoav Levy/Phototake NYC. Reproduced by permission.*

higher temperatures. Liquids, conversely, generally show decreasing viscosity with increasing temperature.

The flow of liquids in pipes, the performance of oil-lubricated bearings in engines or oil-filled automotive shock absorbers, and the air resistance on a moving car or airplane are all dependent on the viscosity of the fluids involved.

Vision

Vision is sight, the act of seeing with the eyes. In humans, sight conveys more information to the **brain** than either **hearing**, **touch**, **taste**, or **smell**, and contributes enormously to **memory** and other requirements for our normal, everyday functioning. Because we see objects with two eyes at the same time, human vision is **binocular**, and therefore *stereoscopic*. Vision begins when **light** enters the **eye**, stimulating photoreceptor cells in the retina called rods and cones. The retina forms the inner lining of each eye and functions in many ways like film in a camera. The photoreceptor cells produce electrical impulses which they transmit to adjoining nerve cells (neurons), which converge at the optic nerve at the back of the retina. The visual information coded as electrical impulses travels along nerve tracts to reach each visual cortex in the posterior of the brain's left and right hemispheres. Each eye conveys a slightly different, two-dimensional (flat) image to the brain, which has the amazing ability to decode and interpret these images into a clear, colorful, three-dimensional view of the world.

Our 3-D view of the world

Because our eyes are separated by about 2.6 in (6.5 cm), each eye has a slightly different horizontal view. This phenomenon is called "binocular displacement." The visual images reaching the retina of each eye is a two-dimensional flat image. In normal binocular vision, the blending of these two images into one single image is called *stereopsis*, which produces a three-dimensional view (one with a sense of depth), and allows the brain to accurately judge an object's depth and distance in space in **relation** to ourselves and other objects.

Depth and distance **perception** is also available without binocular displacement and is called *monocular stereopsis*. Even with one eye closed, a car close to us will appear much larger than the same sized car a mile down the road, or two **rails** of a railway line appear to draw closer together the further they run off into the distance. The ability to unconsciously and instantaneously assess depth and distance enables us to move about in

space without continually bumping into objects or stumbling over steps.

Ocular dominance

Studies strongly indicate there is a *critical period* during which normal development of the visual system takes place—a period when environmental information is permanently encoded within the brain. Although the exact time frame of the critical period is not clear, it is believed that by age six or seven years, visual maturation is complete. **Animal** studies show that if one eye is completely covered during the entire critical period, neurons in the visual pathway and brain connected to the covered eye do not develop normally. When that eye is finally uncovered, only neurons relating to the eye that was not covered function in the visual process. This is an example of "ocular dominance," when cells activated by one eye dominate over the cells of the other.

Memory

Just as vision plays an important role in memory, memory plays an important role in vision. The brain accurately stores an incredible amount of visual data which it draws upon every time the eyes look at something. For example, imagine fishing in a quiet stream and the **cork** on your line begins to bob up and down in the **water**. Although you cannot see under the water, your brain—from previous knowledge—remembers that a **fish** tugging at the worm on the hook will cause the floater to bob and tells you to pull the line in.

Electrochemical messengers

The entire visual pathway—from the retina to the visual cortex—is paved with millions of neurons. From the time light enters the eye until the brain forms a visual image, vision relies upon the process of electrochemical communication between neurons. Each **neuron** has a **cell** body with branching fibers called dendrites and a single long, cylindrical fiber called an axon. When a neuron is stimulated it sends chemicals called neurotransmitters, which causes the release of electrical impulses along the axon. The point where information passes from one cell to the next is a gap called a **synapse**, and neurotransmitters affect the transmission of electrical impulses on to an adjacent cell. This synaptic transmission of impulses is repeated until the message reaches the appropriate location in the brain. In the retina, approximately 125 million rods and cones transmit information to approximately one million ganglion cells. This means that many rods and cones must converge onto one single cell. At the same time, however, information from each single rod and cone "diverges" on to more than one ganglion cell. This complicated phenomenon of convergence and divergence occurs along the entire optic pathway. The brain must transform all this stimulation into useful information and respond to it by sending messages back to the eye and other parts of the brain before we can see.

Our eyes adapt to an incredible range of light intensities—from the glare of sunlight on glistening snow to the glow of moonlight on rippling water. Although the pupil regulates to some degree the amount of light entering the eye, it is the rods and cones which allow our vision to adapt to such extremes. Rod vision begins in dim light at the level of darkness and responds for up to five orders of intensity. Cones function in bright light and are responsible for **color** vision and visual activity.

When light hits the surface of an object, it is either absorbed, reflected, or passes through—as it does through clear **glass**. The amount of pigment in an object helps us determine its color. The amount of light absorbed by an object is determined by the amount of pigment, or color, contained in that object. The more heavily pigmented the object, the darker it appears because it absorbs more light. A sparsely pigmented object, which absorbs very little light and reflects a lot back, appears lighter.

Color vision

Human color perception is dependent on three conditions. First, whether we have normal color vision; second, whether an object reflects or absorbs light; and third, whether the source of light transmits wavelengths within the visible **spectrum**. Rods contain only one pigment which is sensitive to very dim light, and which facilitates night vision but not color. Cones are activated by bright light and allow us to see colors and fine detail. There are three types of cones that contain different pigments which absorbs wavelengths in the short (S), middle (M), or long (L) ranges. Cones are often labeled blue, green, and red, because they detect wavelengths in those color spectrums. The peak wavelength absorption of the S (blue) cone is approximately 430 nm; the M (green) cone 530 nm; and the L (red) cone 560 nm.

The range of detectable wavelengths for all three types of cones overlap, and two of them—the L and M cones—respond to all wavelengths in the visible spectrum. Most of the light we see consists of a mixture of all visible wavelengths which results in "white" light, like that of sunshine. However, cone overlap and the amount of stimulation they receive from varying wavelengths produces a fabulous range of vivid colors and gentle hues present in normal color vision. Approximately 8% of all human males experience abnormal color vision, or **color blindness**.

Actually, we do not "see" colors at all. A **leaf**, for example, appears green because it absorbs long- and short-wavelengths but reflects those in the middle (green) range, stimulating the M cones to transmit electrochemical messages to the brain which interprets the signals as the color green.

Optic pathway

Only about 10% of the light which enters the eye actually reaches the photoreceptors in the retina. This is because light must pass first through the cornea, pupil, **lens**, aqueous and vitreous humors (the liquid and gel-like fluids inside the eye) then through the **blood** vessels of the lining of the eye and then through two layers of nerve cells (ganglion and bipolar cells in the retina).

Visual field

The entire scene projected onto the retinas of both eyes is called the "visual field."

Optic chiasma

Synaptic transmission of impulses from retinal cells follows the optic nerve to the *optic chiasma*, an x-shaped junction in the brain where half the fibers from each eye cross to the other side of the brain. This means that some visual information from the right half of each retina (from the left visual field) travels to the right visual cortex, and visual information from the left half of each retina (from the right visual field) travels to the left visual cortex. Information from the right half of our environment is processed in the left hemisphere of the brain, and vice versa. Damage to the optic pathway or visual cortex in the left brain—perhaps from a stroke—can cause complete loss of the right visual field. This means only information entering the eye from the left side of our environment is processed, even though information still enters the eye from both visual fields.

Visual cortex

Each visual cortex is about 2 in (5 cm) square and contains about 200 million nerve cells which respond to very elaborate stimuli. In **primates**, there are about 20 different visual areas in the visual cortex, the largest being the primary, or *striate*, cortex. The striate cortex sends information to an adjacent area which in turn transmits to at least three other areas about the size of postage stamps. Each of these areas then relays the information to several other remote areas called *accessory optic nuclei*. It is thought that the accessory optic nuclei plays a role in coordinating movement between the head and eyes so images remain focused on the retina when the head moves.

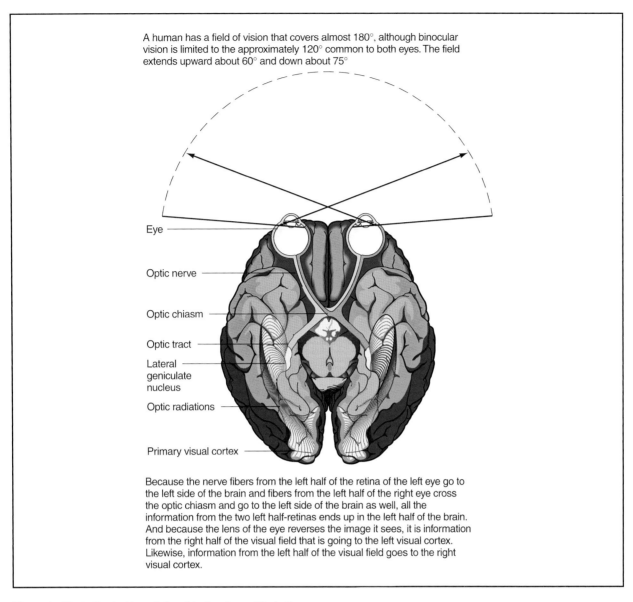

A human has a field of vision that covers almost 180°, although binocular vision is limited to the approximately 120° common to both eyes. The field extends upward about 60° and down about 75°

Eye

Optic nerve

Optic chiasm

Optic tract

Lateral geniculate nucleus

Optic radiations

Primary visual cortex

Because the nerve fibers from the left half of the retina of the left eye go to the left side of the brain and fibers from the left half of the right eye cross the optic chiasm and go to the left side of the brain as well, all the information from the two left half-retinas ends up in the left half of the brain. And because the lens of the eye reverses the image it sees, it is information from the right half of the visual field that is going to the left visual cortex. Likewise, information from the left half of the visual field goes to the right visual cortex.

Figure 1. *Illustration by Hans & Cassidy. Courtesy of Gale Group.*

Visual acuity

Visual acuity, keenness of sight and the ability to distinguish small objects, develops rapidly in infants between the age of three and six months and decreases rapidly as people approach middle age. Good visual acuity is often called 20/20 vision. Optometrists test visual acuity when we have our eyes examined, and poor acuity is often correctable with glasses or contact lenses. As with every other aspect of vision, visual acuity is highly complex, and is influenced by many factors.

Retinal eccentricity

The area of the retina on which light is focused influences visual acuity, which is sharpest when the object is projected directly onto the central *fovea*—a tiny indentation at the back of the retina comprised entirely of cones. Acuity decreases rapidly toward the retina's periphery. It was initially believed this was because cones decrease in number moving out from the retina, disappearing altogether at the retina's periphery where only rods exist. However, recent studies indicate it may result from the decreasing **density** of ganglion cells toward the retina's periphery.

Luminance

Luminance is the intensity of light reflecting off an object, and influences visual acuity. Dim light activates only rods, and visual acuity is poor. As luminance in-

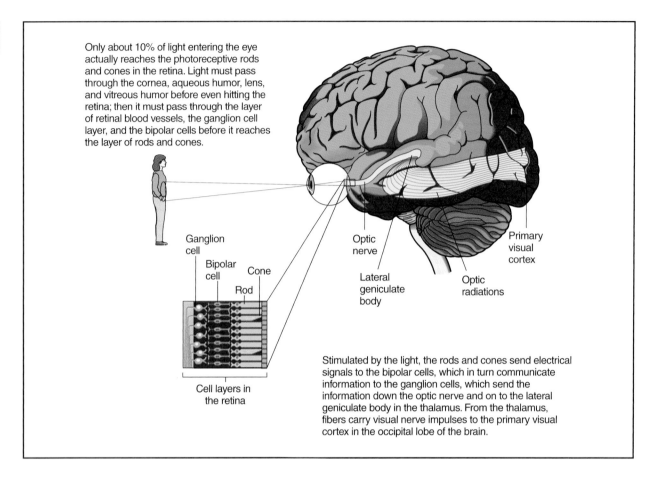

Only about 10% of light entering the eye actually reaches the photoreceptive rods and cones in the retina. Light must pass through the cornea, aqueous humor, lens, and vitreous humor before even hitting the retina; then it must pass through the layer of retinal blood vessels, the ganglion cell layer, and the bipolar cells before it reaches the layer of rods and cones.

Ganglion cell

Bipolar cell

Cone

Rod

Cell layers in the retina

Optic nerve

Lateral geniculate body

Optic radiations

Primary visual cortex

Stimulated by the light, the rods and cones send electrical signals to the bipolar cells, which in turn communicate information to the ganglion cells, which send the information down the optic nerve and on to the lateral geniculate body in the thalamus. From the thalamus, fibers carry visual nerve impulses to the primary visual cortex in the occipital lobe of the brain.

Figure 2. If you were to draw a line down the center of this scene, what you observe left of that line is your "left visual field," and to the right is your "right visual field." Images from the left visual field project to the right half of each retina, and images from the right visual field project to the left half of each retina. *Illustration by Hans & Cassidy. Courtesy of Gale Group.*

creases, more cones become active and acuity levels rise sharply. Pupil size also affects acuity. When the pupil expands, it allows more light into the eye. However, because light is then projected onto a wider area of the retina, optical irregularities can occur. A very narrow pupil can reduce acuity because it greatly reduces retinal luminance. Optimal acuity seems to occur with an intermediate pupil size, but the optimum size varies depending on the degree of external luminance. The difference in luminance reflected by each object in an image produces varying degrees of light, dark, or color. Contrast between a white page and black letters enables us to read. The greater the contrast, the more acute the visual image.

Accommodation

Accommodation is the eye's ability to adjust its focus to bring about clear, sharp images of both far and near objects. Accommodation begins to decline around age 20 and is so diminished by the mid-fifties that sharp close-up vision is seldom possible without corrective lenses. This condition, called *presbyopia*, is the most common vision problem in the world.

Common visual problems

Strabismus

Strabismus is seeing two images of a single object. Strabismus results from a lack of parallelism of the visual axes of the eyes. In one form (cross-eyes) one or both eyes turn inward toward the nose. In another form, wall-eyes, one or both eyes turn outward. A person with strabismus does not usually see a double image—particularly if onset was at a young age and remained untreated. This is because the brain suppresses the image from the weaker eye, and neurons associated with the dominant eye (ocular dominance) take over.

While the causes of strabismus are not fully understood, it appears to be hereditary, often obvious soon after **birth**. In many cases, strabismus is correctable.

KEY TERMS

. . . .

Accommodation—Changes in the curvature of the eye lens to form sharp retinal images of near and far objects.

Cones—Photoreceptors for daylight and color vision are found in three types, each type detecting visible wavelengths in either the short, medium, or long, (blue, green, or red) spectrum.

Ganglion cells—Neurons in the retina whose axons form the optic nerves.

Ocular dominance—Cells in the striate cortex which respond more to input from one eye than from the other.

Optic pathway—The neuronal pathway leading from the eye to the visual cortex. It includes the eye, optic nerve, optic chiasm, optic tract, geniculate nucleus, optic radiations, and striate cortex.

Rods—Photoreceptors which allow vision in dim light but do not facilitate color.

Stereopsis—The blending of two different images into one single image, resulting in a three-dimensional image.

Suppression—A "blocking out" by the brain of unwanted images from one or both eyes. Prolonged, abnormal suppression will result in underdevelopment of neurons in the visual pathway.

Synapse—Junction between cells where the exchange of electrical or chemical information takes place.

Visual acuity—Keenness of sight and the ability to focus sharply on small objects.

Visual field—The entire image seen with both eyes, divided into the left and right visual fields.

However, the critical period (probably to age six or seven years) involved in normal neuronal development of vision makes it necessary that the problem be detected and treated as early as possible.

Amblyopia

Amblyopia, or lazy eye, is the most common visual problem associated with strabismus. Amblyopia involves severely impaired visual acuity, and is the result of suppression and ocular dominance; it affects an estimated four million people in the United States alone. One study suggests it causes blindness in more people under 45 years of age than any other ocular **disease** and injury combined.

Other common visual problems

Slight irregularities in the shape or structure of the eyeball, lens, or cornea cause imperfectly focused images on the retina. Resulting visual distortions include *presbyopia* (far-sightedness, or the inability to focus on close objects), *myopia* (near-sightedness, in which distant objects appear out of focus), and *astigmatism* (which causes distorted visual images). All of these distortions can usually be rectified with corrective lenses.

Valuable vision

Our memory and mental processes rely heavily on sight. There are more neurons in the **nervous system** dedicated to vision than to any other of the five senses, indicating vision's importance in our lives. The almost immediate interaction between the eye and the brain in producing vision makes even the most intricate computer program pale in comparison. Although we seldom pause to imagine life without sight, vision is the most precious of all our senses. Without it, our relationship to the world about us, and our ability to interact with our environment, would diminish immeasurably.

See also Blindness and visual impairments; Brain; Color; Depth perception; Vision disorders.

Resources

Books

Hart, William M., Jr., ed. *Adler's Physiology of the Eye.* St. Louis: Mosby Year Book, 1992.

Hubel, David H. *Eye, Brain, and Vision.* New York: Scientific American Library, 1988.

Leibovic, K. N., ed. *Science of Vision.* New York: Springer-Verlag, 1990.

Lent, Roberto, ed. *The Visual System-From Genesis to Maturity.* Boston: Birkhauser, 1992.

Moller, Aage R. *Sensory Systems: Anatomy and Physiology.* New York: Academic Press, 2002.

von Noorden, Gunter K. *Binocular Vision and Ocular Motility-Theory and Management of Strabismus.* St. Louis: The C.V. Mosby Company, 1990.

Marie L. Thompson

Vision disorders

Vision disorders are irregularities or abnormalities either of the **eye**, visual pathway, or **brain**, which affect one's ability to see. In healthy vision, visual acuity—often referred to as "20/20 vision"—develops rapidly by three to six months of age and generally decreases rapidly as people approach 45. Poor visual acuity is often correctable with glasses or contact lenses. However, many other factors affect human's ability to see—some preventable or correctable and others not. Vision disorders may manifest from refractive errors, defective eye muscles, cataracts, **lens** displacement, glaucoma, fundus conditions, **color** vision deficits, eyelid conditions, orbital diseases, eye injuries, and optic nerve and visual pathway damage.

Refractive errors

When parallel rays of **light** enter the eye they are refracted (bent) by the crystalline lens and projected into the eye. The healthy eye adjusts for distance, focusing each image perfectly on a minute hollow in the retina at the back of the eye called the fovea. In refractive errors, the image either falls short of the fovea or lands behind it. This causes blurred **vision** as a result of inadequate adjustment of the eye, irregular axial length (distance from the front to the back of the eyeball), or incorrect curvature of the cornea. These problems can usually be fixed with corrective lenses or **surgery**, which adjusts the shape of the cornea.

Hyperopia/presbyopia

When the eye's axial length is shorter than normal, close objects appear blurry. This is called hyperopia or presbyopia, commonly termed long-sightedness. It may be latent—meaning the eye can compensate through accommodation, the ability to adjust; or it may be absolute—in which case correction requires convex or positive lenses. Around middle-age, the eye's ability to accommodate deteriorates, which is why almost all people over 45 or 50 years of age require glasses for close-up vision. Irregular curvature of the cornea or lens will produce the same deficit.

Myopia

When the axial length is longer than normal, distant objects appear blurry. This error is called myopia, or short-sightedness. Myopia can also be caused by irregular curvature of the cornea or lens, and is correctable with concave or negative lenses.

Astigmatism

Astigmatism, irregular curvature of the cornea, causes blurred vision of objects both near and far. It may be regular, which means light rays fall on the retina in two different areas, or irregular, resulting from corneal damage which projects light onto many different points on the retina. The former is correctable with lenses, the latter is not. Sometimes the faulty component is the crystalline lens and in these instances the error is called lenticular astigmatism.

Other refractive errors

While the above refractive errors may be **congenital** (originating at **birth**) or occur gradually, fast changes in the eyes' focusing ability should be investigated immediately by an ophthalmologist. Some causes may be senile cataracts (described below); diabetes mellitus—in which fluctuating **blood** sugar levels cause corresponding fluctuations in focus; external **pressure** from lumps or cysts on the eyelids distorting the shape of the cornea; subluxation (partial dislocation) of the crystalline lens, causing hyperopia when it moves backward and myopia when it moves forward; and keratoconus, the thinning of the cornea allowing irregular curvature of the eyeball.

Strabismus

Strabismus comes from the Greek *strabismos* meaning twisted, and results from a lack of parallelism of the visual axes of the eyes. Often, the cause of strabismus is not known, but it appears to be hereditary and is usually obvious soon after birth. Cosmetically, strabismus causes a squint—convergence or divergence of one or both eyes from the parallel line. Sometimes the term "cross-eyed" is used to describe one or both eyes turning toward the nose, and "wall-eyed" when one or both eyes turn outward. It may be concomitant (nonparalytic), in which the divergence or convergence remains the same no matter what way the eyes turn; or noncomitant (paralytic), which means the deviation is more noticeable when the eyes look in one direction than in another. Strabismus is often correctable if treated before the age of four or five years.

Nonparalytic strabismus

Nonparalytic strabismus is thought to be due to underdeveloped **binocular** reflexes within the brain resulting in diplopia, the projection of a double image of a single object to the brain. The brain does not see both images, however—particularly if onset is at a young age and remains untreated—because the visual system develops an adaptive sensory mechanisms to deal with the confusion. This mechanism is called suppression, in which the brain suppresses the image from the weaker eye and neurons associated with the dominant eye take over.

The most common problem associated with suppression is strabismic amblyopia, or lazy eye, which affects more than four million people in the United States and causes blindness in more people under 45 years of age than any other ocular **disease** and injury combined. If it is present at birth or occurs within the first few months, vision never develops in the affected eye. This is called amblyopia of arrest. When the problem appears during the first two or three years, it is termed suppression amblyopia.

Paralytic strabismus

Paralytic strabismus is caused by some form of **interference** in the transmission of motor impulses from the brain to the eye muscles. This paralysis results in limited eye movement, particularly when the eyes turn in the direction of the paralyzed muscle. When it occurs a year or more after birth, it is usually the result of diseases such as **meningitis**, **encephalitis**, and multiple sclerosis and is termed acquired paralytic strabismus. Some congenital forms are evident at birth, perhaps caused by developmental abnormalities in the neural pathway or head injury during birth.

Cataracts

A cataract is the clouding or opacity of the crystalline lens which alters the amount of light entering the eye. Developmental cataracts are relatively minor and harmless, primarily present at birth or developing in early childhood, but which can occur later in life. No reduction in vision is noticeable with this type, which are thought to be hereditary. Congenital cataracts are present at birth and may appear as a white or gray pupil; they create a squint like that apparent in strabismus and cause visual impairment. In most instances, the cause is unknown; however, diseases and disorders contracted by the child's mother during pregnancy—such as rubella, syphilis, **diabetes mellitus**, or **birth defects** as a result of **Down syndrome**, are thought to be causative. Systemic disease-associated cataracts develop as a result of such disorders as diabetes, high doses of steroids over long periods, and a dysfunctioning thyroid. Some cataracts develop secondary to ocular diseases like keratitis (**inflammation** of the cornea), iritis (inflammation of the iris), **radiation** (either from radiation therapy or the **sun**), or from trauma, such as intraocular surgery or penetration of the eye.

The most common cataracts are senile cataracts, a phenomenon of aging that occurs in almost all people over 65 years of age. Sometimes they develop in younger people whose **nutrition** is poor and are then termed presenile cataracts. Both types are probably caused by changes in or loss of **enzyme** production, which synthesizes **nutrients** that in turn feed the layer of cells on the lens surface. Senile cataracts grow slowly over months

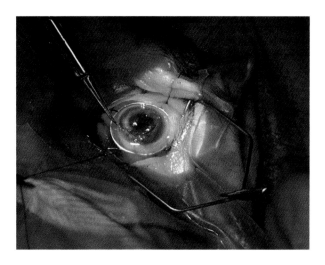

Suturing of a donor's cornea onto the eye of a transplant recipient. The transplant replaces the patient's own damaged cornea, which was impairing his vision. *Photograph by Chet Szymecki. © Chet Szymecki/Phototake. Reproduced by permission.*

or years, cause no **pain**, usually affect both eyes, and gradually reduce visual acuity. Symptoms include two or more images from one eye, dark spots in the center of vision, extreme sensitivity to the glare of bright lights or sunshine, and reduced color vision. If not removed surgically, they eventually cause blindness.

Lens displacement

In a normal eye, the crystalline lens fits snugly behind the iris. Either complete (dislocated) or partial (subluxated) lens displacement causes the iris to wobble as the lens is no longer supporting it, or the edge of the lens may be visible through the pupil. If the lens moves forward into the anterior (frontal) chamber or backward into the posterior (rear) chamber, interrupted flow of fluid may cause glaucoma. Immediate rectification is necessary to prevent blindness in the affected eye. Posterior subluxation will also cause cataracts, and—because the lens moves about—the patient experiences rapid and frequent changes in visual acuity.

Glaucoma

Glaucoma is a serious condition which, left undiagnosed and treated, will cause permanent blindness. Glaucoma is the unrelieved increase of intraocular pressure which acts like air in a tire. In a normal eye, pressure is maintained by the continuous production of aqueous humor, its movement through the pupil into the anterior chamber, and its drainage out of the chamber. When drainage is decreased or prevented, pressure inside the chamber increases.

Chronic simple glaucoma

Because this type occurs gradually and there is no pain or evident loss of visual acuity in the early stages, it causes more blindness word-wide than any other disease. It affects perhaps as many as 2% of the population over the age of 40 years, with the highest incidence after 60 years. This disease is inherited by approximately 10% of all children whose parents are known to be affected. Symptoms are gradual loss of close-up vision but occurring more quickly than presbyopia, and partial loss of visual field, noticeable because the patient frequently bumps into objects.

Acute glaucoma

Acute glaucoma may happen in a matter of hours and causes very fast vision loss, severe pain, and a bloodshot and watery eye. Colored haloes around lights may be visible prior to onset. It primarily affects one eye at a time, usually begins in the evening, and often after a physical or emotional trauma. It is most common in people over the age of 40 years, particularly in the 60-70 year age group. Immediate medical intervention is imperative to prevent damage to the optic nerve and subsequent blindness.

Secondary glaucomas

These are complications commonly caused by other eye diseases such as iritis, injury, blockage to the retinal vein, or long-term use of steroid eye drops for other disorders.

Fundus disorders

The fundus is the lining inside the eye which includes the retina and its blood vessels, the macula, and the optic disc. Viewed through an ophthalmoscope—a special instrument for examining the interior of the eye—the fundus appears a reddish brown in color, the macula in the center of the retina appears as a darker red spot, and the optic disc located toward the nose side of center a lighter red. Many general health and vision-related abnormalities can be detected by examining the fundus. Congenital myopia appears as a pale crescent **moon** in the vicinity of the optic disc, while cloudy nerve fibers—which should be the same color as the retina—appear as a whitish patch. Both these conditions are harmless. Others, however, are not.

Vascular conditions

Blood vessels in the fundus affected by **hypertension** (high blood pressure) are narrow and appear quite bright. Rapid-onset or severe hypertension, which can cause strokes, may first be diagnosed when a patient seeks an eye exam for blurred vision caused by the disorder. **Arteriosclerosis** (hardening of the **arteries**) is a normal process of aging and poses no danger to vision. Central retinal artery occlusion and central retinal vein occlusion can cause sudden loss of vision in the affected eye when the **blood supply** to the retina is obstructed for longer than two hours, or if hemorrhaging occurs within the retina. Vision is then severely limited to light only and damage is virtually irreversible. Causes include hypertension, diabetes mellitus, narrowing of a carotid artery (the primary arteries supplying blood to the head), and blood disorders such as **anemia**, **leukemia**, and sickle-cell anemia.

Degeneration of the macula

This is primarily an age-related disorder resulting in deterioration and often distortion of vision. It is one of the leading causes of blindness in developed countries. There is no treatment; however, if detected early, **laser** treatment may prevent further degeneration.

Detached retina

This condition manifests with the sudden appearance of bright, floating specks and flashes of light in the affected eye with a "shadow" or "curtain" appearing later which may gradually cover the entire visual field. It occurs when part of the retina falls away from the outer layer of cells to which it is normally attached. Breaks or tears in the retina also cause similar symptoms. Although most common in people over 50 years of age, especially those with a high degree of myopia, detachment can also occur in younger people, particularly following trauma. Surgery is the only form of treatment for detachment but **laser surgery** for a tear or break will sometimes prevent full detachment.

Fundus inflammation and tumors

The choroid is the layer of cells over which the retina lies. Inflammation of this layer is visible through the retina and manifests as rapid onset of blurred vision—usually only in one eye—and haziness and spots over the entire field of vision. It may be caused by a parasitic protozoan **infection** in the bloodstream; larvae of dog tapeworms (the eggs of which may be accidentally ingested through contact with dog feces); a seriously compromised immune system—from **AIDS**, for example; a fungus infection called Histoplasmosis; or syphilis passed on to unborn babies.

Benign and malignant tumors

Benign melanomas produce no symptoms and usually cause no loss of vision. Malignant melanomas of the

KEY TERMS

. . . .

Astigmatism—Irregular curvature of the cornea causing distorted images.

Cataract—Eye disease characterized by the development of a cloudy layer in the lens of the eye.

Diplopia—Double image.

Dystrophy—Atrophy, deterioration.

Fundus—Layers lining the inside of the eye.

Glaucoma—A disease of the eye in which increased pressure within the eyeball can cause gradual loss of vision.

Hyperopia/presbyopia—Far-sighted; close objects are out of focus.

Keratoconus—Thinning of the cornea.

Moypia—Near-sighted; distant objects out of focus.

Stabismus—Non-parallel eye axes.

choroid occur primarily around middle-age, producing symptoms similar to those caused by a detached retina—which can occur secondary to the **tumor** itself—and may result in spread of **cancer** to the bloodstream. When localized treatment with radiation therapy is not effective, the removal of the entire eye is recommended. Retinoblastoma usually affects children within the first two years of life, is closely related to astrocytoma—a tumor of the brain—and is the only primary tumor which occurs on the retina. It manifests as a "white pupil" or squint and if treatments with radiation therapy are ineffective, the affected eye is removed.

Retinal dystrophies

Retinitis pigmentosa and macular dystrophy are degenerative disorders which appear to be hereditary, generally affect young children and adolescents, and cause night blindness, tunnel vision, slow deterioration of central vision, and increasing loss of sight. As yet, effective treatment has not been found.

Other causes

Many people, especially males, experience **color blindness**, varying degrees of which range from inability to differentiate between red and green to total inability to see one or more colors. Damage to any area in the visual pathway—including the optic nerve, optic tract, optic chiasm, optic radiations, or visual cortex—will cause vision deficits or loss. Although not vision

disorders themselves, diseases, disorders, and damage of the eye lid, eye, and bony **orbit** (eye socket) may—if severe and left untreated—result in reduced, impaired, or lost vision.

See also Blindness and visual impairments; Radial keratotomy.

Resources

Books

Bankes, James L. Kennerley. *Clinical Ophthalmology—A Text and Colour Atlas.* London: Churchill Livingstone, 1994.

Chawla, Hector Bryson. *Ophthalmology.* London: Churchill Livingstone, 1993.

Stein, Harold A., Bernard J. Slatt, and Raymond M. Stein. *Ophthalmic Terminology-Speller and Vocabulary Builder.* St. Louis: Mosby, 1987.

Marie L. Thompson

Vitamin

Vitamins are organic molecules that are needed in small amounts in the diet. They are frequently molecules that bind in the active site of an **enzyme** and thereby alter its structure in a way that permits it to react more readily. Vitamins serve nearly the same role in all forms of life and many are essential in the **metabolism** of all living organisms. They are synthesized by plants and micro-organisms and the absolute requirement for vitamins in the diet of higher animals is the result of the loss of this biosynthetic capability during **evolution**. The biosynthetic abilities and thus the dietary requirement of different **species** vary. For example, ascorbic acid (vitamin C) is a vitamin only for **primates** and a few other animals, such as the guinea pig, but most other animals can synthesize it, so for them it is not a vitamin. Certain vitamins can be synthesized from provitamins obtained from the diet. Some of the vitamin requirements of humans and higher animals are supplied by the intestinal **flora**, for example most of the vitamin K required by humans is provided in this way.

Several diseases resulting from vitamin deficiencies were prevalent until the last century and sailors on long sea voyages, where fresh **vegetables** were not readily available, were often victims. In the Orient, the **disease** beriberi was rampant and millions died of its associated polyneuritis. The condition could be relieved by feeding the patients **rice** polishings. The founder of the vitamin concept was Lumin (1853–1937). During subsequent decades, the importance of accessory food factors for normal growth and development was gradually recog-

ESSENTIAL VITAMINS

Vitamin	What It Does For The Body
Vitamin A (Beta Carotene)	Promotes growth and repair of body tissues; reduces susceptibility to infections; aids in bone and teeth formation; maintains smooth skin
Vitamin B-1 (Thiamin)	Promotes growth and muscle tone; aids in the proper functioning of the muscles, heart, and nervous system; assists in digestion of carbohydrates
Vitamin B-2 (Riboflavin)	Maintains good vision and healthy skin, hair, and nails; assists in formation of antibodies and red blood cells; aids in carbohydrate, fat, and protein metabolism
Vitamin B-3 (Niacinamide)	Reduces cholesterol levels in the blood; maintains healthy skin, tongue, and digestive system; improves blood circulation; increases energy
Vitamin B-5	Fortifies white blood cells; helps the body's resistance to stress; builds cells
Vitamin B-6 (Pyridoxine)	Aids in the synthesis and breakdown of amino acids and the metabolism of fats and carbohydrates; supports the central nervous system; maintains healthy skin
Vitamin B-12 (Cobalamin)	Promotes growth in children; prevents anemia by regenerating red blood cells; aids in the metabolism of carbohydrates, fats, and proteins; maintains healthy nervous system
Biotin	Aids in the metabolism of proteins and fats; promotes healthy skin
Choline	Helps the liver eliminate toxins
Folic Acid (Folate, Folacin)	Promotes the growth and reproduction of body cells; aids in the formation of red blood cells and bone marrow
Vitamin C (Ascorbic Acid)	One of the major antioxidants; essential for healthy teeth, gums, and bones; helps to heal wounds, fractures, and scar tissue; builds resistance to infections; assists in the prevention and treatment of the common cold; prevents scurvy
Vitamin D	Improves the absorption of calcium and phosphorous (essential in the formation of healthy bones and teeth) maintains nervous system
Vitamin E	A major antioxidant; supplies oxygen to blood; provides nourishment to cells; prevents blood clots; slows cellular aging
Vitamin K (Menadione)	Prevents internal bleeding; reduces heavy menstrual flow

Stanley Publishing. Reproduced by permission.

nized. The Polish biochemist, Casimir Funk formulated the vitamin theory in 1912 and proposed that several common diseases such as beriberi, pellagra, rickets and scurvy resulted from lack in the diet of essential **nutrients**. It was Funk who suggested the name "vitamin" for these accessory factors, from the Latin *vita* + amine, the "amine" reflecting the fact that the first of these factors to be studied, vitamin B_1, contained **nitrogen**.

The metabolic role of vitamins is largely catalytic. Most vitamins serve as coenzymes and prosthetic groups of enzymes. For most of these, the nature of the biocatalytic function has been elucidated. Vitamin D, however, acts as a regulator of bone metabolism and is thus has an activity similar to **hormones**. As a component of the visual pigments, vitamin A acts as a prosthetic group, however, it is not known whether it is associated with catalytic **proteins** in its other functions. Nicotinamide and riboflavin are constituents of the hydrogen-transferring enzymes, such as those in the respiratory **electron** transport chain. Biotin, folic acid, pantothenic acid, pyridoxine, cobalamin and thiamine are coenzymes, or precursors of coenzymes, of group transfer reactions. The low daily requirements for vitamins reflect their catalytic and/or regulatory roles. Thus vitamins are nutritionally quite different from **fat**, **carbohydrate**, or protein, which are required in the diet in considerable quantities as substrates of **tissue** synthesis and **energy** metabolism.

Vitamins can be grouped according to whether they are soluble in **water** or polar solvents. The water-soluble vitamins are ascorbic acid, the vitamin B series (thiamain, B_1, riboflavin, B_2, pyridoxine, B_6, cobalamin, B_{12},), folic acid, niacin and pantothenic acid. Ascorbate, the ionised form of ascorbic acid, is essential in the prevention of scurvy and acts as a reducing agent (an antioxidant). It serves, for example, in the hydroxylation of proline residues in **collagen**. The vitamin B series are components of coenzymes. For example, riboflavin (vitamin B_2) is a precurser of FAD, and pantothenate is a component of coenzmye A. Vitamin B_1 (thiamine) was found to cure beriberi.

Much is known about the molecular actions of the fat-soluble vitamins, which are designated by the letters A, D, E and K. Vitamin K, which is required for normal **blood** clotting, participates in the carboxylation of γ-carboxyglutamate, which makes it a much stronger chelator of Ca^{2+}. Vitamin A (retinol) is the precurser of retinal, the **light** sensitive group in rhodopsin and other visual pigments. A deficiency of this vitamin leads to night blindness. Furthermore it is required for growth by young animals. Retinoic acid, which contains a terminal carboxylate in place of the **alcohol** terminus of retinal, activates the transcription of specific genes that mediate growth and development. The metabolism of **calcium** and **phosphorus** is regulated by a hormone derived from vitamin D. A deficiency of vitamin D impairs bone formation in growing animals and causes the disease rickets. **Infertility** in **rats** is a consequence of vitamin E (α-tocopherol) deficiency and this vitamin also protects unsaturated **membrane** lipids from oxidation.

Most vitamins were purified between 1920 and 1950. The last one was vitamin B_{12}, in 1948, whose chemical structure was elucidated by A. R. Todd in 1955. Chemical syntheses are known for all vitamins.

See also Biochemistry; Malnutrition; Nutrient deficiency diseases; Nutrition.

Resources

Books

Guyton, Arthur C., and John E. Hall. *Textbook of Medical Physiology.* 10th ed. Philadelphia: W.B. Saunders Co., 2000.

Thibodeau, Gary A., and Kevin T. Patton. *Anatomy & Physiology.* 5th ed. St. Louis: Mosby, 2002.

Ulene, A. *Dr. Art Ulene's Complete Guide to Vitamins, Minerals and Herbs.* New York: Avery, 2000.

Periodicals

Mulinare, Joseph A. "Vitamin and Mineral Supplement Use in the United States: Results from the Third National Health and Nutrition Examination Survey." *Archives of Family Medicine* 9 (2000): 258.

Judyth Sassoon

Viviparity

Viviparity is a form of reproduction found in most **mammals** and in several other **species**. Viviparous animals give **birth** to living young that have been nourished in close contact with their mothers' bodies. Humans, dogs, and **cats** are viviparous animals. Viviparous animals differ from egg-laying animals, such as **birds** and most **reptiles**. Egg-laying, or **oviparous**, animals obtain all nourishment as they develop from the yolk and the protein-rich albumen, or "white," in the egg itself, not from direct contact with the mother, as is the case with viviparous young.

The offspring of both viviparous and oviparous animals develop from fertilized eggs, but the eggs of viviparous animals lack a hard outer covering or shell like the chicken egg. Viviparous young grow in the adult female until they are able to survive on their own outside her body. In many cases, the developing fetuses of viviparous animals are connected to a placenta in the mother's body. The placenta is a special membranous **organ** with a rich **blood supply** that lines the uterus in pregnant mammals. It provides nourishment to the fetus through a supply line called an umbilical cord. The time between **fertilization** and birth of viviparous animals is called the gestation period.

All mammals except the **platypus** and the echidnas are viviparous. Only these two unusual mammals, called montremes, lay eggs. Some **snakes**, such as the Garter snake, are viviparous. So are some lizards and even a few **insects**. Ocean **perch**, some **sharks**, and a few popular aquarium **fish**, guppies, and mollies are also viviparous.

Although certain snakes give birth to live young, they are not viviparous. These snakes hatch from eggs which never leave the body of the parent snake. Because these young snakes hatch from eggs, and do not receive nourishment directly from the mother's body, this type of reproduction is called ovoviviparity. It is considered a more primitive form of reproduction than viviparity.

Even some plants, such as the mangrove and the tiger lily, are described as viviparous because they produce **seeds** that germinate, or sprout, before they become detached from the parent **plant**.

See also Embryo and embryonic development.

Vivisection

Vivisection originally meant the dissection of a live **animal**, usually for the purpose of teaching or research. Historically, the word came also to mean the use of a live animal in any experiment. Vivisection, especially in its broader meaning, is a time-tested tool that has helped humans understand how the bodies of animals function, how **disease** alters that function, and how such diseases can be treated. However, changing attitudes toward animals and a more cynical outlook toward the biomedical enterprise have caused some people to question the continued usefulness and morality of using animals for this purpose.

An ancient history

The practice of true vivisection dates back to ancient times. Around 500 B.C., one of the earliest known vivisectionists, Akmaeon of Croton, discovered that the optic nerve is necessary for **vision** by cutting it in living animals. One of the most well known—and controversial— early vivisectionists was Galen of Pergamon, physician to Roman Emperor Marcus Aurelius. Galen, who lived in the second century A.D., is remembered today for his pioneering use of vivisection of animals to understand health and disease in the human body. But Galen was also a poor scientist, failing to identify such major bodily functions as the circulation of the **blood**. An unquestioning adherence to Galen's false beliefs in succeeding generations of physicians was undoubtedly a major hindrance to medical progress in **Europe**.

Real progress in medical knowledge began again with the experiments of the Italian physicians Andreas Vesalius and his student, Realdo Colombo, in the sixteenth century. They pioneered the use of vivisection to correct and expand, rather than merely to confirm, Galen's science. In the early seventeenth century, English physician William Harvey used vivisection to discover the circulation of the blood and to debunk many of Galen's other beliefs.

But this century also saw the beginnings of an antivivisectionist movement. Physician Jean Riolan Jr. in France and Irish physician Edmund O'Meara both argued that the painful and violent deaths suffered by vivisected animals—remember, there was no **anesthesia** yet—were putting the animals into an unnatural state that could lead to faulty assumptions about the functioning of a healthy animal.

Also in this century, vivisection received an important philosophical boost from the French philosopher René Descartes. Descartes believed that the mind and the body are separate entities, and that animals differ from humans in that they have bodies but no true minds. As such, animals were morally no different from machines, and so vivisection was not morally wrong. Descartes even went so far as to say that animals did not feel real **pain** (a belief that is sometimes still repeated today, although few believe it to be true), although he stressed that vivisection was primarily defensible because it helped humans, not because hurting animals was right. Unfortunately, some of Descartes's later followers lost this fine distinction, and were known for their gratuitous cruelty to animals.

Battle lines are drawn

Around the turn of the eighteenth century, both vivisectionists and their critics gained **momentum**. English physicist Robert Boyle's experiments with animals in a **vacuum** chamber were hailed at the time for their contribution to the understanding of breathing and the function of the lungs. But also in England, the literary figures Joseph Addison, Alexander Pope, and Samuel Johnson all condemned the practice of vivisection and live animal experiments as cruel. Johnson suggested that medical students who learned to become insensitive to the suffering of animals in experiments would also become insensitive to human pain—an argument that continues to this day.

But it was the nineteenth century that saw many of today's attitudes forming. The medical advances of this era, derived at least in part from animal experiments, were profound, including widespread use of vaccination (first discovered at the end of the eighteenth century); the understanding of **microorganisms** as a cause of disease; the use of sterile procedures in **surgery**; the discovery of the **diphtheria** antitoxin, which could be used to save patients with that often fatal disease; and the discovery of anesthesia.

This last discovery, however, did not have as great an effect on the vivisection debate as might have been expected. For one thing, antivivisectionists argued that researchers were not using anesthesia as often as they should; they also began to argue that anesthesia was only being given at low levels, to prevent the animal from struggling, rather than the higher levels necessary to control pain. Again, these arguments are repeated today.

A major confrontation between pro-vivisection physicians and antivivisectionists came to a head in turn of the century America. This conflict was sometimes bitter, calling to mind more recent events in the debate. Antivivisectionists were accused of using out of date and misleading pictures of animals in experimental devices to garner sympathy for their cause; they also planted spies in laboratories to expose vivisectionist practices. The fight came to a head at the U.S. Senate hearings of 1896 and 1900, where testimony by some of the world's greatest medical researchers led to the defeat of a bill to regulate vivisection in the District of Columbia.

The debate today

Today the biomedical establishment and antivivisectionists are again locked in a struggle over the appropriateness of animal experimentation. The biomedical research enterprise has become a huge endeavor in the last century; in addition, the pharmaceutical, agricultural, and cosmetics industries use many animals each year to test new products and procedures. A new wrinkle in the debate is scientists' ability to genetically engineer animals—especially mice—to either lack or to contain genes related to human disease.

Present United States law does not require detailed regulation or recording of experiments involving animals except for dogs, **cats**, and **primates**, so it is difficult to estimate how many animals are involved in total. In 1991, 108,000 dogs and 35,000 cats were used in biomedical experiments in the United States. By comparison, in England, where detailed records are kept of all animal experiments, the total number of animals used in 1990 was about 3.2 million, with roughly 16,000 being dogs and cats; so we know that the total number of animals used in experiments in the United States must be very large.

Sometimes the debate today may seem a contest between scientists who oppose all regulation and animal extremists who would deny that animal research has helped human beings at all. In 1975 animal rights activist Peter Singer's book *Animal Liberation* suggested that animal research is wrong not merely because it is cruel, but because animals have the same rights as humans not to suffer. This book fueled a movement that has sometimes been violent, including groups such as the Animal Liberation Front that have threatened researchers' lives and broken into and vandalized laboratories. On the other side, occasionally one may hear a scientist voicing the Cartesian opinion that animals can't feel pain, and that the goal of medical knowledge justifies any treatment of animals.

But others have voiced more moderate views that may represent a more realistic goal for the future. Animal welfare activists seek to institute safeguards against cruelty in laboratories without banning animal research. The famous English primatologist Jane Goodall, while recognizing the importance of primate research for human health, has argued that our growing understanding of animal intelligence demands more humane conditions in laboratories—especially for highly intelligent animals such as **chimpanzees**.

Both scientists and laypeople have played an important role in minimizing animal pain in research by serving on the now-required animal welfare committees at institutions engaging in government-funded research—although the political power of such committees has been claimed to vary between different institutions. And the growing use of computer simulations, tissues grown in the laboratory, and other alternatives to animal use is being hailed for its ethical as well as scientific value.

But in the end, it seems unlikely that animal research will either be abolished or will continue without further regulation. Research alternatives, while helpful, must be compared to the real thing in order to be validated: for instance, the safe testing of drugs will continue to require large numbers of animals, even when alternatives are also used. On the other hand, many animal welfare activists note that instances of cruelty and broken regulations continue to be discovered at some laboratories, and that vast numbers of animal experiments in the cosmetics and agricultural industries are hardly regulated at all.

In the end, the question may be as enduring as human beings' ability to view the same event in very different ways. The value of animal experimentation to human health and knowledge is not seriously in doubt. But past "scientific" beliefs—such as that animals cannot feel pain; that an animal rendered motionless by anesthesia cannot feel pain; and that higher animals such as dogs and primates cannot feel **anxiety** and fear—have been overturned by increased scientific understanding.

Resources

Books

Orlans, F. Barbara. *In the Name of Science: Issues in Responsible Animal Experimentation.* Oxford: Oxford University Press, 1993.

Paton, William. *Man and Mouse: Animals in Medical Research.* Oxford: Oxford University Press, 1993.

Rupke, Nicolaas A., ed. *Vivisection in Historical Perspective.* Kent, England: Croom Helm Ltd., 1987.

Periodicals

Goodall, Jane. "A Plea for the Chimps." *The New York Times Magazine* (May 17, 1987): 108-120.

Other

"The New Research Environment." Video recording. Washington, DC: The Foundation for Biomedical Research, in cooperation with the Association of American Medical Colleges and Johns Hopkins University School of Medicine, 1987.

Kenneth B. Chiacchia

Volatility

Volatility is the ease with which a substance is converted to the gaseous, or vapor, state. The term is usually used to describe the speed with which a liquid evaporates, but it can also apply to the process of a solid changing to a gas, known as sublimation. Liquids that boil at low temperatures, such as gasoline, are volatile liquids, while liquids that boil at higher temperatures, such as **water**, are less volatile or nonvolatile. Extremely volatile substances have such low boiling points that they exist as gases at room **temperature**, such as **oxygen** gas.

Several factors affect the volatility of a liquid. In general, larger molecules are less volatile than smaller ones among molecules that share similar composition and construction. For example, ethyl alcohol is more volatile and evaporates more readily than a larger **alcohol** such as decanol. Almost every chemical is more volatile at higher temperatures than at lower ones. Water, which evaporates faster on a hot day than on a cold one, is a good example of this. Volatility is also affected by **atmospheric pressure**. Liquids are more volatile at higher altitudes, where atmospheric **pressure** is less. Some substances that are liquids at **sea level** will evaporate very quickly on top of a high mountain, assuming no change in other factors such as temperature.

Some compounds, such as water, are extremely non-volatile; often this is because of strong chemical bonds between the molecules—the most common of which are **hydrogen** bonds—that resist the tendencies of individual molecules to enter the gaseous state.

Volcanic arcs *see* **Plate tectonics**

Volcanic island arc *see* **Plate tectonics**

Volcano

A volcano is an opening in Earth's surface through which molten rock, hot gases, and **rocks** are ejected. Volcanoes create new land and islands. They can also produce economically important mineral deposits, fertile soils, and beautiful landscapes. However, volcanoes can also destroy lives and property. Therefore, they constitute significant geologic hazards in many parts of the world.

With regard to the hazard that they present, volcanoes can be classified as active, dormant, or extinct. Active volcanoes are those that have erupted within recorded history. Dormant volcanoes are those that have not erupted during recorded history but may erupt again, whereas extinct volcanoes are those for which there is little or no chance of future eruptions.

Where volcanoes develop

Most of the volcanoes on **Earth** are located along the boundaries between lithospheric plates, which can be convergent (subduction zones) or divergent (mid **ocean** ridges). The chain of volcanoes along the Pacific Rim, often referred to as the Ring of Fire, is an example of subduction zone volcanism. Iceland, in contrast, is a volcanic **island** straddling the Mid-Atlantic Ridge (a divergent plate boundary). Although they are not as numerous as plate boundary volcanoes, intraplate volcanoes can occur where plates pass over mantle hot spots or along continental rift zones where plates are being pulled apart. The Hawaiian Islands, for example, were formed as the Pacific Plate slowly passed over a magma-generating **hot spot** within the mantle.

Oceanic ridges are chains of volcanoes located along the boundary between two diverging oceanic plates. New oceanic crust is formed along the ridges as two oceanic plates move apart. Where the **rate** of plate formation is rapid, older crust is quickly pushed out of the way and only small volcanic vents form. Where the spreading rate is slower, volcanic eruptions may form large volcanoes.

Subduction zones are areas in which oceanic plates are overridden by continental plates, forcing the heavier oceanic crust deep enough to be melted and recycled as second generation **magma**. Because the size of Earth is constant, the amount of oceanic crust consumed by subduction zones must be approximately equal the amount produced along mid-ocean ridges. The volcanoes of the western coast of North and **South America**, Indonesia, the Philippines, Japan, Kamchatka, and Alaska all sit atop subduction zones and collectively comprise the Ring of Fire.

Some volcanoes form above hot spots far from the edges of tectonic plates, and are known as intraplate volcanoes. A hot spot is an **upwelling** of magma from far beneath the Earth's crust, caused by a disturbance at the boundary between the solid mantle and the liquid outer core of **Earth's interior**. Hot spots are, compared to mid-ocean ridges or subduction zones, relatively small and isolated features. Well known intraplate hot spot volcanoes include the Hawaiian Islands, the Jemez Mountains and Capulin volcanic field of New Mexico, and the volcanoes that produced volcanic rocks throughout the Yellowstone region of Wyoming.

Hot spots provide information about the rates and directions of movement of tectonic plates over **geologic time** scales. When magma from a hot spot rises from the lower mantle and through the lithospheric plate, a volcano forms on the surface. The plate continues to move over the stationary hot spot and eventually produces a chain of volcanoes that increase in age away from the hot spot and show the direction of plate movement. The Hawaiian Islands mark the track of a hot spot, over which the Pacific Plate has moved in a northwesterly direction, toward Japan. Likewise, the volcanic rocks of the Snake River Plain in southern Idaho recorded the progress of the North American plate as it moved over the Yellowstone hot spots.

The origin of magma

All magma forms through melting of pre-existing rock. Generally, this occurs in one of two ways: (1) by **convection** of rock upwards through the mantle until it melts, or (2) by melting rock at a subduction zone. Mantle convection occurs because deep within the earth, **radioactive decay** raises the **temperature** of rock, making it expand. This expansion lowers the rock's **density**, causing it to rise, or convect. As the rock rises through the mantle, the surrounding **pressure** decreases and eventually the convecting rock melts as a result. Geologists call this pressure-relief melting. The magma moves upward and erupts to form either an oceanic ridge or a hot spot volcano. At subduction zones, volatile compounds (especially **water**) escape from the subducting plate and lower the melting temperature of the overlying mantle rocks. This triggers melting and magma forms as a result.

Magma comes from a variety of sources and may have a complicated history. For example, as magma rises in the mantle and crust, it undergoes a process known as fractional crystallization. Each mineral in a rock has its own crystallization (or melting) temperature. Because different **minerals** crystallize at different temperatures, certain minerals form from magma earlier than others. This produces a magma with a composition different

The July 22, 1980, eruption of Mount St. Helens in southern Washington. *JLM Visuals. Reproduced by permission.*

from that when the minerals first began to crystallize. Therefore, the minerals that crystallize later, and the rocks that they form, will be of a different composition than those that form earlier. Fractional crystallization is thought to be one way of producing rocks of different compositions from the same magma. Partial melting and magma **contamination** are also important.

If a rock is not exposed to a high enough temperature to melt all of its minerals, only some minerals will melt. This is known as partial melting. If a rock melts only partially, the magma produced will have a different chemical composition than the rock from which the magma originated. As magma rises toward the earth's surface it may also cause rocks in the overlying crust to partially melt, contaminating the magma with molten rock of a different composition. The composition of magma therefore depends on many factors, including original magma composition resulting from partial melting, fractional crystallization, and magma contamination.

Volcanic rocks produced from partially melted continental crust usually appear red, brown, or gray in **color** and are known as felsic rocks. Felsic rocks such as rhyolite are rich in the minerals feldspar and quartz, both of which contain abundant silica. Lava formed by melting of mantle rocks contains abundant iron- and magnesium-rich minerals, which are poorer in silica than quartz and feldspar, and produces mafic volcanic rocks such as basalt. Lava with a chemical composition that falls between these two extremes is said to be of intermediate composition. Andesite is an example of a volcanic rock of this type.

Types of volcanic eruptions

Lava flows are streams of molten rock that flow onto Earth's surface from a vent or fissure, and most commonly have a basaltic composition. Two common types of basaltic lava have been distinguished in the Hawaiian language. The Hawaiian words adopted into English to describe these different lava rocks are aa (pronounced AH-ah), also called blocky lava, and pahoehoe (pronounced pa-HOY-hoy) or ropey lava. Aa forms from viscous and slow-moving, and aa flows are characterized by an irregular, jagged appearance. Pahoehoe flows, in contrast, are characterized by smooth, wavy surfaces. **Submarine** eruptions of basalt form large lobes known as pillows, and are commonly referred to as pillow basalts.

Pyroclastic ("fiery fragment") deposits are the result of explosive eruptions. Explosive eruptions occur when magma containing water or gases (or magma that has been in contact with ground water) rises near enough to the surface that the pressure exerted by the rock above it can no longer keep the magma from boiling. The result is an explosive eruption of pyroclastic debris. Volcanic dust, ash, cinders, and blocks are collectively known as tephra. Ash from pyroclastic eruptions can cover large areas, thinning with distance from the volcano. The rock produced by a volcanic ash fall is known as tuff.

An ash flow, or pyroclastic flow, is a dense body of ash, superheated gases, and rock that moves as a fluid from an erupting volcano, crossing the landscape and filling valleys with the fluid mixture. This material deflates as it cools and produces a rock known as ignimbrite, or welded tuff. Ignimbrites can cover hundreds of square kilometers of landscape, such as the Mitchell Mesa Tuff of West Texas. Pyroclastic flows from a prehistoric eruption of Taupo, a volcano in New Zealand, produced ignimbrite deposits that covered the tops of hills hundreds of meters tall.

A pyroclastic surge is a kind of pyroclastic flow that occurs when magma encounters **groundwater** close to Earth's surface. Also called a nuee ardente, a French phrase meaning "glowing cloud," this was the kind of eruption that destroyed the city of St. Pierre, on the Caribbean island of Martinique, in 1902. The volcano that is formed by this kind of eruption is called a maar or tuff ring.

Controls on the explosivity of volcanic eruptions

Generally, the **viscosity** of a magma controls the type and violence of eruptions. Viscosity is the resistance of a fluid to flow. The more viscous the magma, the more explosive its eruption is likely to be. Very viscous magmas tend to resist eruption, and so gas pressure builds within the magma pipe leading to the volcanic vent. By the time sufficient pressure builds to displace a viscous magma, the **force** released by the eruption will be much greater than for a fluid magma. This leads to explosive eruptions. The most important controls on viscosity are the silica

content of the magma and its temperature. Basaltic (mafic) lavas are very fluid due to their low silica content. Conversely, rhyolitic lavas are very viscous due to their high silica content. Magma and lava viscosity is also a function of temperature; therefore, a the viscosity of a lava flow will increase as its temperature decreases.

Volatile substances are elements or compounds—hydrogen sulfide, water, **carbon dioxide**, **radon**, and other gasses—that escape during eruptions. The Latin root for volatile means "winged." Volatile compounds in magma can cause violent explosive eruptions.

Different kinds of volcanic structures

As new rock forms around a volcanic vent, the resulting volcano takes shape according to the kind of erupted material, which is in turn related to lava composition. The most common types of volcanoes, from largest to smallest, are the shield volcano, composite volcano, and cinder cone.

A shield volcano is a very large, broad, and low profile volcano consisting of layers of basaltic rocks. Shield volcanoes most commonly form in the middle of oceanic plates or in continental rifts, which are areas in which the continental crust is being pulled apart. The shape of shield volcanoes resembles the round shields used by warriors of ancient times to protect themselves in battle. The tallest individual mountain on Earth—the island of Hawaii—is a shield volcano. This volcanic island slopes gently down from the 13,796 ft (4205 m) summit of Mauna Kea to the ocean abyss more than 30,000 ft (9 km) below. Kenya's Mt. Kilimanjaro, the tallest mountain in **Africa**, is a shield volcano, as is Olympus Mons, the tallest mountain on **Mars**. Mountains very much like shield volcanoes have been mapped on **Venus** by the Magellan spacecraft **radar** mapping expedition. Shield volcanoes can produce large amounts of lava and bury large areas, but are not known for for violent explosive eruptions.

A composite volcano, or stratovolcano, is a large, steep-sided andesitic volcano made of alternating sequences of lava and pyroclastic debris. Composite volcanoes are most commonly located along plate boundaries. Japan's Mt. Fuji is a composite volcano, as are Mount St. Helens in Washington, Mt. Ararat in the Caucasus, and Popocatepetl, near Mexico City. Composite volcanoes can grow over millions of years and then collapse in a cataclysmic event, forming a large volcanic crater known as a caldera.

A cinder cone is a small, steep-sided volcano made of pyroclastic material, with lava composition ranging from basaltic to rhyolitic. Cinder cone eruptions occur as an incandescent liquid solidifies in midair and falls into a heap. Landslides sculpt the sides of the still-hot rock

A volcanic cinder cone and lava flow in northern Arizona. *Photograph by Tom Bean. Stock Market. Reproduced by permission.*

pile, forming such cinder cones as Mexico's Paricutin, New Mexico's Mt. Capulin, and Arizona's Sunset Crater.

Some mafic eruptions occur through cracks known as fissures. Instead of building a mountain in one place, fissure eruptions cover broad areas with basaltic lava flows known as flood basalts. A fissure eruption occurred in Iceland in 1783, and the resulting environmental catastrophe wiped out one-fifth of its population. Fissure eruptions have:

• Filled the Rio Grande Rift with hundreds of feet (100 m) of volcanic rock at Taos, New Mexico.

• Constructed the Columbia River Plateau in Washington and Oregon.

• Covered southern India with approximately 240,000 cubic mi (1 million cubic km) of basalt 65 million years ago, forming the Deccan Traps.

• Formed the Siberian Flood Basalt plateau in northern Russia 250 million years ago, the largest flood basalt in the world.

Some of these basaltic eruptions happened at the same time great extinctions occurred. No conclusive evidence has been found for a connection between fissure eruptions and global **extinction**, but there is no doubt that the gas and **heat** released could affect environmental conditions for the worse all over the planet's surface.

Calderas

The most violent large volcanic eruption is the collapse of a composite volcano. This normally happens on the active margins of tectonic plates, that is, at subduction zones or along a continental rift valley (where a **continent** is breaking apart). The process is part of the **evolution** of a composite volcano, which starts with a reservoir of molten rock, several miles wide and under high pressure. This magma rises in the earth's crust and forces its way to the surface. A composite volcano is born in **clouds** of ash, supersonic steam explosions filling the air with hot rock, ash, and various gases.

After a series of eruptions, perhaps over millions of years, the volcano forms a mountain of lava and pyroclastic material as much as 2-3 mi (3-4 km) high. Eventually, there is one last eruption of ash and pyroclastic flows. The magma begins to boil, gas bubbles expand the magma to many times its original **volume**, and it explodes upward. The magma chamber rapidly empties its contents onto the landscape above and the volcano collapses into the void, forming a depression known as a caldera.

Volcanic catastrophes

According to the United States Geological Survey, between 50 and 60 volcanoes erupt each year, usually in sparsely populated areas. In the past 500 years more than 200,000 people have died as a result of volcanic activity. Between 1900 and 1986, volcanoes directly or indirectly killed an average of 845 people each year. The 1980 eruption of Mount St. Helens, the last volcanic eruption within the contiguous United States, killed 57 people.

During the 1980s and 1990s the science of volcanology greatly improved our knowledge of volcano behavior. Consequently, volcanologists can now warn civil authorities when eruptions are likely to occur. Mt. Pinatubo, a composite volcano in the Philippine Islands, erupted explosively in 1991. In 1997, a pyroclastic flow burst from the Sourfriere Hills volcano on the island of Montserrat in the Caribbean. Lives were lost in both of these disasters; however, many more lives were saved by timely evacuation of the populations flanking the volcanoes.

Mt. Mazama, which existed in what is now southern Oregon, erupted and collapsed nearly 7000 years ago, creating Crater Lake. The story of the eruption evolved into a Native American myth, the Battle of Llao and Skell, that was eventually translated to English. Twentieth century geologists found the myth to be an accurate description of the development of a caldera nearly 250 generations ago.

An ash fall from Mt. Vesuvius buried the Roman city of Pompeii in A.D. 79. The volcano struck down the people where they lived, and preserved the shapes of their bodies where they fell in the ash. Pompeii is a remarkable volcanic disaster because much of the city was preserved along with the names, portraits, writings, and even graffiti of those who lived there. Plaster casts were made of them as the city was excavated in the late 1700s. The nearby city of Herculaneum was covered by a pyroclastic flow that destroyed it in seconds.

Kuwae, in Melanesia, erupted in 1453. A legendary chieftain hastily assembled the population, moving them to safety in the last-minute. The eruption destroyed all remaining life on the island, and split it into several sections. Decades after the caldera eruption, the people returned to the new archipelago of which Epi and Tongoa are the principal islands. Eruptions of this size affect **weather** and climate worldwide, and cause peculiar optical effects in the atmosphere. Ash in the stratosphere causes the sky to appear strangely colored and dims the sunlight. Some geologists speculate that the eruption's optical effects in 1453

may have filled the defenders of Constantinople with superstitious dread, hastening the city's demise.

The Laki fissure eruption of 1783 produced huge volumes of fluorine gas, which poisoned the grass that fed Icelanders' flocks. Approximately 229,000 animals died as a result, and 10,000 Icelanders subsequently starved to death, reducing the population by one-fifth. Benjamin Franklin observed the blue haze that covered **Europe**, and deduced that the **pollution** from the eruption must have caused the abnormally cold winter that year.

Tambora, an Indonesian composite volcano, erupted and collapsed in 1815, chilling the world for the next year. The explosion destroyed the volcano's summit and filled the stratosphere with volcanic dust, significantly decreasing the amount of sunlight that reached Earth's surface. Tambora's collapse killed 10,000 people, and another 80,000 starved to death as a result of crop losses.

Krakatau, another Indonesian composite volcano, erupted and collapsed in 1883, causing worldwide cooling similar to Tambora. The collapse was heard by people 2,500 mi (4,000 km) away. Tsunamis killed 36,000 people in coastal Java and Sumatra. The atmospheric effects of this eruption in the equatorial latitudes included brilliant green sunrises, sunsets, and moonrises, followed by blue sunlight throughout the day.

Volcanologists know of even larger caldera eruptions in the geologic past than those in this list. Likewise, there are many volcanoes that have the potential to erupt catastrophically in the near future. Some of them are located in the world's most populous areas: Seattle, Washington; Guadalajara, Mexico; the Bay of Naples, Italy; and Rabaul, Papua New Guinea.

Volcanic benefits

Through geologic time, volcanic eruptions have shaped Earth's environment in many ways. The eruption of volcanoes built the continents on which we live. We owe much of the composition of our atmosphere to volcanic eruptions on the early Earth. Our oceans formed from water expelled during these same eruptions. Some of the world's richest farmland draws its fertility from minerals provided by nearby volcanoes. Volcanoes, particularly collapsed calderas, can develop geothermal systems when groundwater or rainwater seep into the volcano. Geothermally powered electric generating stations provide **electricity** in Iceland, Italy, and New Zealand. Volcanic processes are also responsible for precious and non-precious mineral deposits, which are formed when minerals precipitate from geothermal waters circulating in the rocks beneath and around volcanoes.

KEY TERMS

. . . .

Ash fall—A layer of volcanic ash that falls from an erupted ash cloud.

Cinder cone—A small, steep-sided volcano made of pyroclastic material. A cinder cone is an accumulation of loose volcanic material that erupts as a liquid, and cools into cinders in the air, falling to the ground in a heap.

Composite volcano—A large, steep-sided volcano made of alternating sequences of lava and pyroclastic debris. Sometimes called a stratovolcano.

Convection current—The motion of a fluid that rises as it is heated and sinks as it cools, moving in a circular path.

Felsic—A term applied to light-colored igneous rocks, such as rhyolite, that are rich in silica. Felsic rocks are rich in the minerals feldspar and quartz.

Fissure—A crack through which lava erupts onto Earth's surface

Hot spot—An upwelling of magma from beneath the earth's crust, caused by a disturbance at the boundary between the solid mantle and the liquid outer core. This upwelling is not related to the convection currents associated with oceanic ridges, although some hot spots do occur there.

Lava—Molten rock erupted onto Earth's surface.

Mafic—A term applied to dark-colored igneous rocks, such as basalt, that are poor in silica and contain large amounts of the iron and magnesium.

Magma—Molten rock beneath Earth's surface.

Oceanic ridge system—A long (40,000 mi; 64,000 km) crack in the earth's crust where new ocean crust is continuously forming, causing ocean basins to grow wider.

Pyroclastic flow—A fast moving body of pyroclastic material from an erupting volcano. It moves as a fluid, in some cases covering thousands of square kilometers.

Pyroclastic material—Volcanic debris formed by solidification of erupted lava in air; includes dust, ash, cinders, and blocks of rock.

Shield volcano—A broad, low profile volcano consisting of layers of basaltic rock, typically formed in the middle of oceanic plates or on continental rifts.

Silica—Any of the mineral forms of silicon dioxide.

Subduction zone—A boundary between tectonic plates in which a dense oceanic plate is forced beneath a less dense continental plate.

Viscosity—The internal friction within a fluid that makes it resist flow.

Volatile—Readily able to form a vapor at a relatively low temperature.

See also Asthenosphere; Boiling point; Catastrophism; Geology; Geophysics; Lithosphere; Planetary geology; Plate tectonics; Seamounts; Tectonics.

Resources

Books

De Boer, Jelle Zeilinga, and Donald T. Sanders. *Volcanoes in Human History: The Far-Reaching Effects of Major Eruptions.* Princeton, NJ: Princeton University Press, 2001.

Harris, Stephen. *Fire Mountains of the West.* Missoula: Mountain Press Publishing Co., 1988.

Periodicals

Gore, Rick. "Cascadia: Living On Fire." *National Geographic* (May 1998): 6-37.

Heiken, Grant. "Will Vesuvius Erupt? Three Million People Need to Know." *Science* (November 26, 1999): 1685

McClintock, Jack. "Under the Volcano." *Discover* (November 1999): 82.

Mothes, Patricia. "Waiting for the Eruption: Tungurahua Volcano, Ecuador." *Geotimes* (March 2000): 26-27.

Williams, A.R. "Under the Volcano: Montserrat." *National Geographic* (July 1997): 58-73.

Other

U.S. Forest Service. "Mount St. Helens National Volcanic Monument." January 13, 2003 [cited February 9, 2003]. <http://www.fs.fed.us/gpnf/mshnvm/>.

U.S. Geological Survey. "U.S. Geological Survey Volcano Hazards Program." January 17, 2003 [cited February 9, 2003]. <http://volcanoes.usgs.gov/>.

Clay Harris

Voles

Voles are small mouse-like **mammals** in the family Muridae, order Rodentia. Other members of this family include the **gerbils, hamsters, lemmings, rats,** and **mice**. Voles occur in a wide range of open, often grassy habitats, such as alpine and arctic **tundra**, prairies, savannas, and pastures and other types of agricultural fields.

Voles have a body length of about 3-5 in (8-12 cm), and typically weigh 1.1-1.8 oz (30-50 g), with mature males being slightly larger than females. Voles have a stout, plump body, a small head, blunt nose, small eyes, short ears, a short neck, and a short, stubby tail. Their pelage is dense, and is typically colored brownish or grayish.

Voles are herbivores, eating a wide range of **plant** tissues, especially the shoots and rhizomes of **grasses** and **sedges**. Voles typically live in tunnels and runways that they dig in the ground and surface litter, equipped with numerous resting dens and food-storage areas. These tunnels are often lived in by communal groups of voles. The runways are kept free of obstructions, and the voles are intimately familiar with the twists and turns of these paths. Consequently, voles can move along their runways much faster than they can run along the ground surface. This is an important advantage when these small animals are attempting to flee from a **predator**.

Voles remain active throughout the year. During the winter, voles develop tunnels beneath the snow at the snow-ground interface, and they feed on rhizomes and food that they have stored from the previous autumn.

Populations of voles often irrupt in a cyclic fashion, with regularly occurring years of great abundance punctuated by longer periods of **time** during which these small animals may be rather scarce. Some **species** of voles have enormous potential for population growth because of their intrinsic fertility. For example, females of the common vole (*Microtus arvalis*) of **Europe** can become sexually mature before they are two weeks old, and while they are still suckling on their mother's milk! Potentially, a female vole can give **birth** to her first litter of four to seven young when she is only five weeks old. One captive common vole bore 33 litters, and had a total of 127 young. Because these animals breed more or less continuously during the growing season, and sometimes even in the winter, their population growth rates are potentially enormous.

What prevents an unmitigated growth of voles is mortality associated with predation, coupled with unfavorable environmental conditions, such as the availability of adequate food. In most cases, it appears that the most important ecological factors that allow rapid population growth in voles is the availability of an abundance of food, associated perhaps with relatively favorable growing conditions for one or several years.

Voles are always an important **prey** item for many species of predators. However, when voles are abundant, they are avidly hunted by virtually all small predators, including **weasels**, foxes, **owls**, and **hawks**, and even some larger animals such as wolves and **bears**.

Voles are sometimes considered to be **pests**, because they can cause serious damages to **crops** growing in fields, and to stored grains and other foods. These dam-

ages are caused by the actual eating of foods, as well as by the **contamination** of stored grains with vole feces, which can render the crops unsalable.

North American voles

There are about 50 species of voles. Most voles are included in the genus *Microtus*. The meadow vole (*M. pennsylvanicus*) is the most common and widespread species of vole in **North America**, occurring through most of Canada south of the high arctic tundra, and in most of the northern United States. The meadow vole is a familiar species of fields, wet meadows, and disturbed **forests**. Because of its wide distribution and periodic irruptions of abundance, the meadow vole is both ecologically important as a component of ecological food webs, and economically important as an occasional pest.

The tundra vole (*Microtus oeconomus*) is another widespread species, occurring in arctic tundra and open boreal forests of Alaska, Yukon, and Eurasia. The woodland vole (*M. pinetorum*) is widespread in forests of southeastern North America. The montane vole (*M. montanus*) occurs in alpine **grasslands** and tundras of the **mountains** of the western United States. The **prairie** vole (*M. ochrogaster*) occurs in the grasslands of the interior of the **continent**, while the rock vole (*M. chrotorrhinus*) occurs in boulder slopes and other rocky places in northeastern North America. The chestnut-cheeked vole (*M. xanthognathus*) occurs in local populations in the boreal forest of northwestern North America. The long-tailed vole (*M. longicaudus*), singing vole (*M. miurus*), Townsend's vole (*M. townsendii*), Richardson's vole (*M. richardsoni*), and California vole (*M. californicus*) are all species of coniferous forests of western North America.

The heather vole (*Phenacomys intermedius*) occurs widely in boreal and northern temperate forests of Canada and the northwestern United States. This species is very similar in appearance to the meadow vole, and was overlooked as a distinct species by most field biologists until the 1950s, when reliable, diagnostic characters were discovered (these involve the shape of the cheek teeth).

The northern red-backed vole (*Clethrionomys rutilus*) occurs in tundra and open boreal forests of northwestern Canada and Alaska, and also in Siberia and eastern Scandinavia. There is also a disjunct population in coniferous rain-forests of Oregon and Washington. Gapper's red-backed vole (*C. gapperi*) occurs more widely throughout temperate North America, reaching as far south as the mountains of Arizona.

The sagebrush vole (*Lagurus curtatus*) occurs in high altitude sagebrush steppes and semi-deserts of southwestern North America. Richardson's water vole

(*Arvicola richarsoni*) is a species of alpine meadows and streams of the Rocky Mountains.

See also Rodents.

Resources

Books

MacDonald, David, and Sasha Norris, eds. *Encyclopedia of Mammals.* New York: Facts on File, 2001.

Wilson, D. E., and D. Reeder. *Mammal Species of the World.* 2nd ed. Washington, DC: Smithsonian Institution Press, 1993.

Bill Freedman

Voltage *see* **Electric circuit**

Volume

Volume is the amount of space occupied by an object or a material. Volume is said to be a derived unit, since the volume of an object can be known from other measurements. In order to find the volume of a rectangular box, for example, one only needs to know the length, width, and depth of the box. Then the volume can be calculated from the formula, $V = l \times w \times d$.

Volume of most physical objects is a function of two other factors, **temperature** and **pressure**. In general, the volume of an object increases with an increase in temperature and decreases with an increase in pressure. Some exceptions exist to this general rule. For example, when **water** is heated from a temperature of 32°F (0°C) to 39°F (4°C), it decreases in volume. Above 39°F (4°C), however, further heating of water results in an increase in volume that is more characteristic of **matter**.

Units of volume

The term unit volume refers to the volume of one something: one quart, one milliliter, or one cubic inch, for example. Every measuring system that exists defines a unit volume for that system. Then, when one speaks about the volume of an object in that system, what he or she means is how many times that unit volume is contained within the object. If the volume of a **glass** of water is said to be 35.6 cubic inches, for example, what is meant is that 35.6 cubic inch unit volumes could be placed into that glass.

Mathematically, volume would seem to be a simple extension of the concept of area, but it is actually more complicated. The volume of simple figures with **integral** sides is found by determining the number of unit cubes that fit into the figure. When this idea is extended to include all possible positive **real numbers**, however, paradoxes of volume occur. It theoretically is possible to take a solid figure apart into a few pieces and reassemble it so that it has a different volume.

The units in which volume is measured depend on a variety of factors, such as the system of measurement being used and the type of material being measured. For example, volume in the British system of measurement may be measured in barrels, bushels, drams, gills, pecks, teaspoons, or other units. Each of these units may have more than one meaning, depending on the material being measured. For example, the precise size of a "barrel" ranges anywhere from 31 to 42 gallons, depending on federal and state statutes. The more standard units used in the British system, however, are the cubic inch or cubic foot and the gallon.

Variability in the basic units also exists. For example, the "quart" differs in size depending on whether it is being used to measure a liquid or dry volume and whether it is a measurement made in the British or customary U.S. system. As an example, 1 customary liquid quart is equivalent to 57.75 cubic inches, while 1 customary dry quart is equivalent to 67.201 cubic inches. In contrast, 1 British quart is equivalent to 69.354 cubic inches.

The basic unit of volume in the international system (often called the **metric system**) is the liter (abbreviated as l), although the cubic centimeter (cc or cm³) and milliliter (ml) are also widely used as units for measuring volume. The fundamental relationship between units in the two systems is given by the fact that 1 U.S. liquid quart is equivalent to 0.946 L or, conversely, 1 liter is equivalent to 1.057 customary liquid quarts.

The volume of solids

The volume of solids is relatively less affected by pressure and temperature changes than is that of liquids or gases. For example, heating a liter of **iron** from 32°F (0°C) to 212°F (100°C) causes an increase in volume of less than 1%, and heating a liter of water through the same temperature range causes an increase in volume of less than 5%. But heating a liter of air from 32°F (0°C) to 212°F (100°C) causes an increase in volume of nearly 140%.

The volume of a solid object can be determined in one of two general ways, depending on whether or not a mathematical formula can be written for the object. For example, the volume of a cube can be determined if one knows the length of one side. In such a case, $V = s^3$, or the volume of the cube is equal to the cube of the length of any one side (all sides being equal in length). The volume of a cylinder, on the other hand, is equal to the product of the area of the base multiplied by the altitude of the cylinder. For a right circular cylinder, the volume is equal to the product of the radius of the circular base (r) squared multiplied by the height (h) of the cone and by **pi** (π), or $V = \pi r^2 h$.

Many solid objects have irregular shapes for which no mathematical formula exists. One way to find the volume of such objects is to sub-divide them into recognizable shapes for which formulas do exist (such as many small cubes) and then approximate the total volume by summing the volumes of individual sub-divisions. This method of approximation can become exact by using **calculus**. Another way is to calculate the volume by water displacement, or the displacement of some other liquid.

Suppose, for example, that one wishes to calculate the volume of an irregularly shaped piece of rock. One way to determine that volume is first to add water to some volume-measuring instrument, such as a graduated cylinder. The exact volume of water added to the cylinder is recorded. Then, the object whose volume is to be determined is also added to the cylinder. The water in the cylinder will rise by an amount equivalent to the volume of the object. Thus, the final volume read on the cylinder less the original volume is equal to the volume of the submerged object.

This method is applicable, of course, only if the object is insoluble in water. If the object is soluble in water, then another liquid, such as **alcohol** or cyclohexane, can be substituted for the water.

The volume of liquids and gases

Measuring the volume of a liquid is relatively straight forward. Since liquids take the shape of the container in which they are placed, a liquid whose volume is to be found can simply be poured into a graduated container, that is, a container on which some scale has been etched. Graduated cylinders of various sizes, ranging from 10 ml to 1 l are commonly available in science laboratories for measuring the volumes of liquids. Other devices, such as pipettes and burettes, are available for measuring exact volumes, especially small volumes.

The volume of a liquid is only moderately affected by pressure, but it is often quite sensitive to changes in temperature. For this reason, volume measurements made at temperatures other than ambient temperature are generally so indicated when they are reported, as $V = 35.89$ ml (95°F; 35°C).

The volume of gases is very much influenced by temperature and pressure. Thus, any attempt to measure or report the volume of the gas must always include an indication of the pressure and temperature under which that volume was measured. Indeed, since gases expand

to fill any container into which they are placed, the term volume has meaning for a gas *only* when temperature and pressure are indicated.

David E. Newton

Voyager spacecraft

Twin robotic **space** probes, *Voyager 1* and *Voyager 2*, were launched by the United States in 1977. Their original mission was to fly by **Jupiter** and **Saturn**, but the journey of *Voyager 2* was successfully extended to **Uranus** and **Neptune**. The Voyagers were the most scientifically fruitful space mission ever launched, collecting, among other data, high-quality photographs of four planets and dozens of moons, most of which were previously known only as specks of **light** in telescopes. The Voyagers are still functioning today, returning data to **Earth** as they coast toward interstellar space.

The structural foundation of each Voyager is a puck-shaped frame or "bus" about 6 ft (1.8 m) diameter. To this bus are attached 11 science instruments, communications antennae, computers and a power-generation unit. A high-gain dish **antenna** 12 ft (3.7 m) in diameter is mounted directly onto the bus and devoted to receiving command signals from Earth. A radioiosotope thermoelectric **generator** provides each Voyager with electric power derived from the **heat** given off by several kilograms of plutonium-238. Voyager's cameras and spectrometers are mounted on a scan platform attached to a gearbox at the end of a 10-ft (3-m) boom. The gearbox makes it possible to select

Voyager U.S. National Aeronautics and Space Administration (NASA).

observational targets, within limits, without having to rotate the entire spacecraft. Each Voyager also possesses several computers to handle scientific data, coordinate its own subsystems, and control its position.

Since it was known that the Voyagers would eventually leave the **solar system**, they are also equipped with messages for any extraterrestrial beings that might encounter them, perhaps millions of years hence. These messages are preserved on gold-plated audio discs of the now-obsolete type that encodes sound as sinuous grooves on a surface. The Voyager records include a wide variety of Earthly sounds (e.g., rain, a kiss, the rock-and-roll classic "Johnny B. Goode" by Chuck Berry) and images (e.g., a snowflake, Australian hunters, rush-hour traffic), and are supplied with needles, cartridges, and playback instructions in pictorial form.

Voyager 1 made its closest approach to Jupiter in March, 1979, taking detailed pictures of the **planet** and several of its moons; *Voyager 2* encountered Jupiter in July of the same year. Voyager photographs revealed that the **moon** Io is the most volcanically active body in the solar system, with its interior kneaded to hot liquid by periodic gravitational tugs from the moon Europa and its surface pockmarked by hundreds of volcanoes. Some of these volcanoes squirt liquid **sulfur** compounds at 0.6 miles per second (1 km/sec) up to 190 mi (300 km) above the surface, forming umbrella-shaped plumes that can be easily seen from space. Europa was found to be among the smoothest bodies in the solar system, covered with a network of cracks suggesting a relatively thin layer of **ice** over a watery world-ocean. Ganymede and Callisto, consisting mostly of ice, were found to contain fascinating **geology** of their own. Callisto, thanks to data obtained in 1998 by the **space probe** *Galileo,* is now thought to also possess a world-ocean of **saltwater,** albeit under a thicker crust than Europa's.

Both spacecraft received a gravitational assist from Jupiter that increased their speed and redirected them toward Saturn, which *Voyager 1* reached in November, 1980, and *Voyager 2,* 10 months later. Their observations showed that Saturn's ring structure was more complex and finely divided than was suspected. Close observation of Saturn's moon Titan, the largest **satellite** in the solar system, revealed little because of its dense, hazy atmosphere.

After Saturn, *Voyager 1* proceeded to head out of the solar system, but *Voyager 2* continued on what was termed the Grand Tour—a course that would take it to Neptune and then, with yet another gravitational assist from that planet, to Uranus. A Grand Tour is only available to spacecraft when the outer planets are in a certain alignment; this alignment was present when the Voyagers were launched in the 1970s, but will not recur for another 150 years. In January, 1986, *Voyager 2* swept through the Uranian system of moons, which is oriented at right angles to the **plane** of the ecliptic so that Uranus, with its system of moons, moves as if rolling along its **orbit**. *Voyager 2,* moving along the ecliptic, shot through the Uranian system like a dart through a bull's eye, gathering detailed images of its five previously known large moons—all of which revealed unique geology—and discovering ten new, lesser satellites. In August, 1989, *Voyager 2* encountered Neptune, outermost of the major planets, passing within a mere 3,000 mi (4,800 km) of its north pole. It made thorough observations of Neptune's ring system (similar to Saturn's, but less spectacular) and observed bizarre **nitrogen** geysers on its largest moon, Triton.

For its encounters with Uranus and Neptune, *Voyager 2* was ingeniously reprogrammed to cope with conditions it had not been designed to face. The power yielded by its thermoelectric generators had declined, forcing controllers to dole it out by switching essential systems on and off in an intricate sequence. Additionally, the Sun's light is much dimmer at Uranus and Neptune—the former being four times and the latter six times as far from the **Sun** as is Jupiter—necessitating lengthy camera exposures (over a minute in some cases). In order to keep its instruments steady for such long periods, the slight jerk caused by the onboard tape recorder starting up was compensated for by milliseconds-long steering-rocket blasts.

Voyager 1 and *Voyager 2* continue to relay data on charged particles, plasma, and magnetic fields from approximately 70 and 80 AU (astronomical units) away, respectively, where 1 AU equals the average distance of the Earth from the Sun. They will have enough electrical power and hydrazine steering-rocket propellant to remain functional until about 2020, at which time, after over 40 years of productive science, their mission will cease. *Voyager 1* will be the first human-made object to pass into true interstellar space when it crosses the heliopause (the limit of the solar wind's influence) some time before the year 2013.

See also Astronomy; Cosmology; Galaxy; Interstellar matter; Mars Pathfinder; Planetary atmospheres; Planetary geology; Planetary nebulae; Planetary ring systems.

Resources

Books

Swift, David W. *Voyager Tales: Personal Views of the Grand Tour.* Reston VA: American Institute of Aeronautics and Astronautics, 1997.

Harwood, William, and Stephen P. Maran. *Space Odyssey: Voyaging Through the Cosmos.* Washington, DC: National Geographic, 2001.

Other

National Aeronautics and Space Administration. "Voyager: Celebrating 25 Years of Discovery." Jet Propulsion Laboratory, California Institute of Technology. November 15, 2002 [cited December 30, 2002]. <http://voyager.jpl.nasa.gov/>.

Larry Gilman

Vulcanization

Vulcanization is the process by which rubber molecules (polymers or macromolecules made of repeating units or monomers called isoprene) are cross-linked with each other by heating the liquid rubber with **sulfur**. Cross-linking increases the **elasticity** and the strength of rubber by about ten-fold, but the amount of cross-linking must be controlled to avoid creating a brittle and inelastic substance. The process of vulcanization was discovered accidentally in 1839 by the American inventor Charles Goodyear (1800-1860) when he dropped some rubber containing sulfur onto a hot stove. Goodyear followed up on this discovery and subsequently developed the process of vulcanization. In 1844, Goodyear was issued United States Patent #3644.

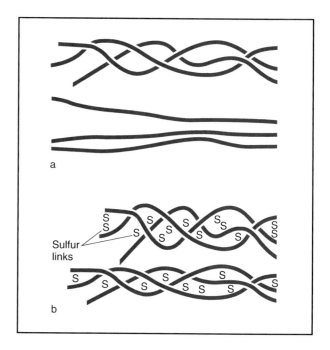

Figure 1. The process of vulcanization. *Illustration by Hans & Cassidy. Courtesy of Gale Group.*

Walker, Jearly. "Why are the First Few Puffs the Hardest When You Blow up a Balloon?" *Scientific American* (December 1989): 136.

Louis Gotlib

Rubber as a natural product

Natural rubber comes from the rubber **tree** (*Hevea brasiliensis*) and is a white, milky liquid called *latex*. Most rubber comes from Malaysia and other nations in East **Asia**. Latex can also be seen as the white fluid in dandelion stalks. The latex from the tree is actually a suspension of rubber particles in **water**. Rubber is a **polymer** (long chain made of repeating units) of isoprene. Natural rubber is relatively reactive, and is especially vulnerable to oxidation.

Vulcanization and properties of vulcanized rubber

In the process of vulcanization, the added sulfur allows some C-H bonds to be broken and replaced by C-S bonds. The process of vulcanization cross-links the chains or polyisoprene to each other. The cross-linked molecules create a three-dimensional network of rubber. Each cross-link is a chain of about eight sulfur **atoms** between two long chains of polyisoprene.

Vulcanized rubber is about 10 times stronger than natural rubber and is also about 10 times more rigid. However, it is still very elastic, which means that is can be stretched reversibly. Polymers that are elastic are sometimes called elastomers. The optimum amount of sulfur to be added to the rubber is about 10% by weight. Adding an excess of sulfur produces a very brittle and inelastic substance called ebonite. Man-made or synthetic rubber can also be vulcanized, and the process is similar.

Figure 1 shows what happens to rubber when the long chains of polyisoprene are cross-linked. In part a, the macromolecules are bent and randomly arranged. In part b, the chains are cross-linked but still randomly arranged. The molecules become aligned when the rubber is stretched. If the individual chains were not cross-linked, each chain could slide freely past each other.

Resources

Periodicals

"Feeling Bad-Latex Sensitivity." *The Economist* 32 (November 14, 1992): 105.

Smith, Emily, T. "Rubber That's So Tough it Goes the Extra Mile." *Business Week* (February 11, 1991): 80.

Vultures

Vultures are large **birds of prey** specialized to scavenge the bodies of dead animals. **Species** of vultures are assigned to two families in the order Falconiformes. The vultures of the Americas include seven species in the family Cathartidae. The vultures of Eurasia and **Africa**, numbering 14 species, are specialized members of the Accipitridae, a family that also includes **hawks** and **eagles**. The Cathartidae and the Accipitridae are not closely related.

Biology of vultures

Although the vultures of the Cathartidae and Accipitridae have evolved from different ancestral stock, the two groups occupy broadly similar ecological niches. As a result, the vultures in these families are highly convergent in many aspects of their **biology** and **ecology**. All vultures have very broad wings with marginated primary feathers at the tips that allow them to soar at great heights and position themselves in thermals, thus flying for hours with little effort. This is a very useful ability for animals with excellent eyesight, because they can scan for carrion over great expanses of terrain. Once most vultures become airborne, it is unusual to see them flap their wings. In fact, most species are rather weak at active flying, and they sometimes appear to be straining their capabilities when they are taking flight from the ground.

All vultures have a hooked beak. The neck muscles and beak of vultures are too weak to tear the tough skin of recently dead, large animals, but they are able to deal with carcasses that have had some time to decompose. Until this happens, only the eyes of the dead **animal**, apparently a delicacy among vultures, can be eaten. Vultures have long, clawed toes, but their feet cannot grasp with much strength. Because of their relatively weak beaks and feet, these **birds** survive almost entirely by scavenging dead animals, and increasingly in many areas, the refuse of human towns and habitations.

Vultures rarely kill anything. They are rather timid when confronted by a living adult animal, even if the creature is obviously unwell, and therefore a potential meal. In such a situation, vultures will typically wait patiently nearby until the animal dies before they begin to eat. However, vultures have been known to attack relatively helpless, recently born, young animals.

Vultures are binge eaters of the first order. Their **prey** of dead animals is not always plentiful, and these birds sometimes must pass a rather long time between meals. However, when a large corpse is available, vultures will avail themselves of it to the fullest possible degree. Under such conditions, vultures can sometimes become so engorged with ingested carrion they are unable to fly. If they are disturbed when thus grounded, vultures must lighten their load so that they can become airborne. They do this by regurgitating their food, representing a case of reluctant avian bulimia.

New World vultures

The vultures of the Cathartidae have perforate nasal septums, which means that when one is looking at a sideways profile of the head of one of these birds, daylight can be clearly seen through their paired nostrils.

It appears that at least some of the American vultures have an excellent sense of **smell**, a very rare and unusual trait among birds. The turkey vulture (*Cathartes aura*), for example, has the largest olfactory system of any bird. There is also a great deal of behavioral evidence that a sense of smell is a significant aid to vultures in finding their food of dead animals. This ability is especially useful for vultures that find some of their fragrant food of carrion in relatively closed-canopied ecosystems, such as shrublands and **forests**.

Most species of vultures have a featherless, naked head. This is likely an **adaptation** that facilitates sanitation, because these birds often have to reach far into a decomposing carcass in order to feed. Most species with naked heads have brightly colored and patterned, warty skin on their neck and head, which is important in species recognition and in courting and aggressive interactions. The male Andean condor (*Vultur gryphus*) has a large, fleshy structure known as a caruncle on the top of its naked head, seemingly bizarre to humans, but undoubtedly most alluring to females of the species.

The **courtship** rituals of American vultures generally involve open-winged strutting on the ground, followed by impressive displays of aerial soaring and gliding. Vultures are monogamous, and both sexes brood the eggs and care for the young. Vulture chicks mature slowly, taking up to six months to leave the nest, in the case of the Andean condor.

Species of New World vultures

The most widespread vulture in the Americas is the turkey vulture (*Cathartes aura*), which ranges from southern Canada to **South America**. This species is migratory in the northern part of its range, often travelling in flocks of several hundred birds. The black vulture (*Coragyps atratus*) is another common species that breeds from southern **North America** to South America. The only other species of vulture in North America is the California condor (*Gymnogyps californianus*), an **endangered species** on the edge of **extinction**.

Other vultures occur in South America. The largest species is the increasingly rare Andean condor, which has a wingspread of 9 ft (3m), the largest of any living bird. This species, which can weigh more than 14 lb (6.5 kg), is also the heaviest flying bird.

A now-extinct American vulture was the very impressive *Terratornis mirabilis*, which had an enormous wing span of at least 16 ft (5 m). This species is known from fossils collected at the Rancho La Brea tar pits of southern California. *Terratornis mirabilis* may have still been around as recently as the end of the most recent glacial epoch, about 10,000-15,000 years ago. This enormous vulture may have become extinct as an indirect consequence of the disappearance at that time of many large species of **mammals**, an event that was likely due to the first colonization of the Americas by very effective human hunters.

New World vultures and man

As scavengers of dead animals, vultures provide an important ecological service. This has long been recognized by some human societies, which have viewed vultures as useful birds because they contribute to cleaning up some types of unpleasant refuse around habitations. However, in some places these views have recently changed and the presence of these scavengers is no longer encouraged. This has happened, for example, in parts of the Andes, where vulture dropping has been shown to contribute to the **contamination** of open reservoirs of drinking **water** with microbial **pathogens**.

Species of vultures have had great cultural significance to various groups of people. The Maya of Central America commonly used a hieroglyphic associated with the king vulture (*Sarcoramphus papa*) to signify the thirteenth day of each month, and this species also appears to have been a religious symbol. The Andean condor was also culturally important to indigenous peoples over large parts of its range.

Unfortunately, because of their feeding habits vultures are regarded with great distaste in some other human cultures. Until rather recently, vultures were erroneously thought to be responsible for spreading some important, contagious diseases of **livestock**, because they fed on the bodies of animals that had died of those maladies. Vultures have sometimes been shot or poi-

Black vultures on a dead alligator in Florida. *JLM Visuals. Reproduced by permission.*

soned in large numbers for this reason. One American farmer claimed to have shot 3,500 black vultures during a single winter, in an attempt to relieve himself and his ranch of this harmless bird, which he perceived as a pest. Fortunately, vultures are rarely treated this way in North America any more, although they are still persecuted in some other areas, usually by a few misguided people.

The California condor is a critically endangered species of vulture. This species was formerly widespread in North America, especially in the west. However, this slowly reproducing species has declined enormously in abundance because of hunting and poisoning, and until recently it only survived in a critically endangered population of a few birds in the San Joachin Valley of southern California. In 1984, only 15 birds survived, and by 1986 there were only three adult male California **condors**. The few surviving wild condors are still under threat from shooting, **lead** poisoning following ingestion of lead bullets in scavenged carrion, and **habitat** degradation.

In 1987, no California condors remained in the wild. The handful of surviving wild condors were caught for use in a captive-breeding program, with the hope that enough birds could be produced to allow the eventual re-introduction of this endangered species into the wild. At that time the total population of the species was only 27 individuals, all in captivity. These last California condors are being used in a program of captive breeding supported by the United States Fish and Wildlife Service, and located at the San Diego Wild Animal Park and the Los Angeles Zoo. Fortunately, this breeding program is showing encouraging signs of success, and in 1991 the first, careful releases of California condors were made to a large ecological reserve created and managed for their benefit in southern California.

The California condor remains a critically endangered species, but there is now guarded optimism for its longer-term prospects of avoiding extinction.

Status

• Andean condor (*Vultur gryphus*). The largest flying bird. Resident of the Andes.

• Turkey vulture (*Cathartes aura*). Highly developed sense of smell. Ranges from southern Canada to Tierra del Fuego. Reportedly declining population on the southern plains. Eggshell thinning has been a widespread problem. Today the populations appear stable.

- Yellow-headed vulture (*Cathartes burrovianus*). Resident of Mexico, Panama, and the lowland areas of South America.

- Greater yellow-headed vulture (*Cathartes melambrotus*). Found in several regions of northern South America.

- Black vulture (*Coragyps atratus*). Resident of middle North America to South America. Winters in North America. Loss of suitable **tree** cavities due to fire control has been a problem, as has eggshell thinning as a result of pesticide use. The population has expanded in the northeastern United States, but declined in the Southeast (possibly due to a loss of nesting sites in hollow trees).

- King vulture (*Sarcorhamphus papa*). Striking bird with pinkish-white plumage and black flight feathers. Some records suggest this bird may once have been a resident of Florida. Today it is a rare resident of tropical America.

- California condor (*Gymnogyps californianus*). Resident of North America. On the verge of extinction, now an endangered species. A captive breeding program introduced in 1987 has met with some success, and hopes now exist that a wild population may again be re-established.

Old World vultures

Some Old World vultures have evolved unusual behaviors that help them with feeding. The lammergeier or bearded vulture (*Gypaetus barbatus*) of Eurasia and Africa is known to drop large bones onto **rocks** in order to break them open so that they can eat the nutritious, internal marrow. The Egyptian vulture (*Neophron percnopterus*) is known to pick up stones in its beak, and throw them at ostrich eggs to break them, again so that the contents can be eaten.

Unlike the New World vultures, the Old World vultures lack a perforate nasal septum, and they are generally heavier billed. The Old World vultures have an ancient evolutionary lineage, extending to at least 30 million years, whereas it appears that the American vultures evolved much more recently.

In regions of India, vultures are given access to human corpses within secluded, walled facilities. This custom would seem to be a cultural acknowledgement of the **continuity** of life and death, and of the relationships of humans with other species and ecosystems.

Old World vultures, unlike their New World counterparts, lack a highly developed sense of smell, and are therefore entirely dependent on their **vision** to find food. Their dependence on their sense of sight has restricted their range to the open country, and has kept them from making their homes in rainforests like those inhabited by

the New World vultures. Representative Old World vultures include the following:

- Lappet-faced vulture (*Torgos tracheliotus*). Largest of the African vultures, and second in size only to the European Black Vulture among Old World vultures.

- Griffon vultures (*Gyps fulvus;* also *G. bengalensis, G. indicus, G. himaleyensis, G. africanus, G. rueppellii, G. coprotheres*). Medium to large-billed resident of Eurasia. Griffins are found chiefly in Spain and the Balkans; in **Asia**, Iran, the Hindu Kush, and parts of the Himalayas.

- Lammergeir or bearded vulture (*Gypaetus barbatus*). A handsome bird, rather unlike many of the vultures that many consider repellant. A threatened species in **Europe** (1989), this vulture maintains healthy populations in the Middle East, Central Asia, and the African highlands. Strychnine used in bait has led to its decline in Europe.

- Egyptian vulture (*Neophron pecnopterus*). Small vulture. Resident of Africa, southern Europe, the Middle East, southwestern and central Asia, the Himalayas, and India.

- Hooded vulture (*Necrosyrtes monachus*). Small vulture. Resident of Africa, south of the Sahara.

- Palm-nut vulture (*Gypohierax angolensis*). Feeds on **crabs**, molluscs, fish, and the husks of the Oil Palm. Resident of Africa.

- Indian black vulture (*Sarcogyps calvus*). Resident of India, the Himalayas, Burma, Thailand, the Malay **Peninsula**, and southern Vietnam.

- European black vulture (*Aegypius monachus*). Resident of southern Europe, Afghanistan, Tibet, the Himalayas, and China. Small numbers.

- White-headed vulture (*Trigonoceps occipitalis*). Resident of southern Africa, south of the Sahara. Scarce.

Resources

Books

Ehrlich, Paul R., David S. Dobkin, and Darryl Wheye. *The Birder's Handbook.* New York: Simon & Schuster Inc., 1988.

Ehrlich, P. R., D. S. Dobkin, and D. Wheye. *Birds in Jeopardy.* Stanford, CA: Stanford University Press, 1992.

Forshaw, Joseph. *Encyclopedia of Birds.* New York: Academic Press, 1998.

Freedman, B. *Environmental Ecology.* 2nd ed. San Diego: Academic Press, 1994.

Peterson, Roger Tory. *North American Birds.* Houghton Miflin Interactive (CD-ROM). Somerville, MA: Houghton Miflin, 1995.

Bill Freedman
Randall Frost

VX agent

VX nerve agent (O-ethyl S-[2-diisoproylamino-ethyl] methylphsophonothioate) is one of the most toxic substances ever developed. Like other nerve agents, it is an organophosphate. Although it is often called a nerve gas, VX is usually a clear, odorless, tasteless liquid. A tiny amount of VX, about 10 mg, absorbed through the skin, eyes, or ingested is fatal, and death usually occurs within an hour of exposure. VX poisons by binding to the **enzyme** cholinesterase and inactivates it. As a result, the chemical signals passed between nerve cells are transmitted uncontrollably. Symptoms of VX poisoning include constriction of the pupils, headache, runny nose, and nasal congestion, chest tightness, giddiness, **anxiety,** and nausea, eventually progressing to convulsions and respiratory failure. VX poisoning can be treated immediately with two antidotes: atropine and pralidoxome chloride. Because of its extreme toxicity, VX is considered a weapon of **mass** destruction.

VX poisoning

Chemical signals are transmitted between nerve cells by means of small molecules called neurotransmitters. One of the most common neurotransmitters in the central and peripheral **nervous system** is **acetylcholine**. Under normal conditions, acetylcholine is released from the terminal axon of one nerve **cell**, crosses the synaptic cleft between nerve cells, and binds with a receptor on the **membrane** of the post-synaptic nerve cell. Then, the enzyme cholinesterase binds to acetylcholine and inactivates it. This completes the chemical signaling between nerve cells.

When the VX nerve agent is present in the nervous system, it inactivates the enzyme cholinesterase. As a result, the receptor on the post-synaptic nerve cell is indefinitely stimulated by acetylcholine. In addition, the pre-synaptic nerve cell continues to release acetylcholine. Nervous signals are never completed and the nervous system is eventually destroyed.

VX poisoning can occur by exposure to the eyes or skin, inhalation, or ingestion. Symptoms occur within minutes. Autonomic nervous system symptoms include constricted pupils, reduced **vision** and other visual effects, drooling, sweating, diarrhea, nausea, vomiting, and abdominal **pain**. Neuromuscular symptoms are twitching, weakness, paralysis, and eventually respiratory failure. Symptoms affecting the central nervous system are headache, confusion, **depression**, convulsions, **coma**, respiratory depression, and respiratory arrest.

Treatment of VX poisoning

Two antidotes exist for VX poisoning: atropine and pralidoxime chloride, also called 2-PAM. Atropine blocks one type of acetylcholine receptor on the post-synaptic nerve cell membrane. This prevents acetyl-choline that is in the synaptic cleft from binding to the receptor. Pralidoxime chloride prevents VX from binding to cholinesterase. Together, these drugs have been combined in an antidote kit called Mark I.

VX is an extremely toxic material with low **volatility** and therefore, it dissipates very slowly. VX also has **adhesive** properties, which make it difficult to remove from surfaces. These characteristics make a powerful strategic contaminant. For example, military bases contaminated with VX could result in casualties for several weeks if the base continued to be used. In order to counter such tactics by terrorist groups, scientists at the Department of Energy's Idaho National **Engineering** and Environmental Laboratory have recently developed technology to detect VX and to predict its degradation **rate** on **concrete** surfaces.

See also Bioterrorism.

Resources

Books

Haugen, David M., ed. *Biological and Chemical Weapons.* San Diego: Greenhaven Press, Inc., 2001.

Seagrave, Sterling. *Yellow Rain: A Journey Through the Terror of Chemical Warfare.* New York: M. Evans and Company, Inc., 1981.

Sifton, David W., ed. *PDR Guide to Biological and Chemical Warfare Response.* Montvale, NJ: Thompson/Physician's Desk Reference, 2002.

Wise, David. *Cassidy's Run: The Secret Spy War over Nerve Gas.* New York: Random House, Inc., 2000.

Other

Chemical Weapons: Nerve Agents [cited February 11, 2003]. <http://faculty.washington.edu/chudler/weap.html>.

United States Army. Chemical Agent Fact Sheet: VX [cited February 11, 2003]. <http://www.sbccom.army.mil/services/edu/vx.htm>.

Material Safety Data Sheet: Lethal Nerve Agent VX [cited February 11, 2003]. <http://www.ilpi.com/msds/vx.html>.

Juli Berwald

W

W and Z bosons *see* **Subatomic particles**

Wagtails and pipits

Wagtails and pipits are 48 **species** of terrestrial **birds** that make up the family Motacillidae. Species in this group occur on all of the continents but **Antarctica**. The usual **habitat** of these birds is deserts and semideserts, prairies, tundras, shores, and cultivated fields. Many species are migratory, with northern species travelling to the tropics to spend their nonbreeding season, and alpine species moving to lower-altitude valleys and lowlands.

Wagtails and pipits are slender birds, with a body length of 5.1-8.7 in (13-22 cm). They have pointed wings, a long tail, long, slim legs, rather long toes and claws, and a thin, short, straight beak.

Pipits are typically colored in streaked or mottled browns, and the sexes are similar. Wagtails are brighter colored, with bold patterns of yellow, black, and white, and males being brighter than females. The tail of both types is commonly edged with white feathers, and it is wagged frequently as the bird walks, particularly by the well-named wagtails. The flight of these birds is undulating.

These birds hunt on the ground, by walking and searching for their food of small **insects**, other **invertebrates**, and **seeds**, especially during the winter.

Courtship in wagtails and pipits includes songflights, which feature the male making a rapid ascent, and then undertaking a slower, fluttering, tinkling descent. Wagtails and pipits nest on the ground in a small cupshaped nest woven of **plant** fibers. The clutch size can range from two to seven, with high-latitude species having a larger number of eggs, and tropical species fewer. Both sexes of wagtails incubate the eggs, but in the pipits only the females do. Both sexes of pipits and wagtails cooperate in feeding the young.

The most widespread species in **North America** is the water pipit (*Anthus spinoletta*). This bird nests in alpine and arctic tundras. It spends its nonbreeding season in prairies and agricultural fields of the southern United States and Central America, generally occurring as seed-eating flocks of various size. The water pipit also breeds in arctic and alpine tundras of Eurasia.

The Sprague's pipit (*A. spragueii*) breeds in the mixed-grass and short-grass prairies of central North America, and winters in Arizona, Texas, and Louisiana, and as far south as southwestern Mexico.

The yellow wagtail (*Motacilla flava*) breeds in western Alaska and the northern Yukon, and much more widely in northern Eurasia. This species has a number of well-defined subspecies, and these have created some confusion among taxonomists, who in the past have treated certain of these subspecies as full species. The type occurring in Alaska is a subspecies of northeastern Siberia and the Beringean region, known as *Motacilla flava tschutschensis*. The white wagtail (*M. alba*) is another species that breeds widely in northern Eurasia, while also breeding in Greenland and western, coastal Alaska.

Bill Freedman

Walkingsticks

Walkingsticks are **insects** with a long, thin body, lengthy delicate legs and a brown-green **color** which gives them a striking resemblance to a twig. Walkingsticks are in the family Phasmidae in the order Orthoptera, which also includes the **grasshoppers** and **crickets**. There are almost 2,500 **species** of walkingsticks (phasmids), ranging in size from 1 in (2.5 cm) to 1 ft (30 cm), the largest species occurring in the tropics. Some phasmids resemble leaves (rather than twigs) and are called **leaf** insects.

A walkingstick. *Photograph. © Art Wolfe/The National Audubon Society Collection/Photo Researchers, Inc. Reproduced by permission.*

All species of phasmids are herbivorous and are found on or near vegetation. Often, a phasmid will look like the **plant** on which it is feeding, an **adaptation** known as protective resemblance, which blends these insects with their surroundings and protects them from predation. Walkingsticks not only resemble twigs, they also walk with an odd, rocking **motion** which makes them appear to be a stick being blown in the **wind**. If phasmids are attacked by a **predator** then delicate legs break off so the insect can escape. Some species of phasmids when attacked spray a foul-smelling fluid from **glands** on their thorax. In large numbers, phasmids can cause extensive defoliation damage to trees, but these insects are otherwise harmless.

After copulation, female phasmids simply lay their eggs while feeding, scattering them on the ground. Some species project their eggs with a flick of the abdomen while others place their eggs in a protected crack or crevice. The eggs usually hatch within two years. Thus, the majority of walkingsticks are usually found every second year. Young phasmids (nymphs) resemble adult insects, and grow in a series of stages (instors) to the adult stage, a type of development known as incomplete **metamorphosis**.

Wallabies *see* **Kangaroos and wallabies**

Walnut family

The walnut family contains about 60 **species** of trees in the family Juglandaceae, divided among seven genera. North American representatives are the walnuts (*Juglans* spp.) and hickories (*Carya* spp.). All of these species produce edible nuts and useful **wood**, and some are cultivated in orchards for the production of these **crops**.

Biology of walnuts

Species in the walnut family are woody plants that develop as trees, and mostly grow in angiosperm-dominated **forests** in temperate and subtropical climates. The range of most species is the Northern Hemisphere, although a few species penetrate to the Andes of **South America** and the southwest Pacific.

The wood of trees in the walnut family is strong and resilient and is highly prized as lumber. The twigs have a chambered pith which is visible in a longitudinal cross-section.

Species in the walnut family have prominent, hairy buds and seasonally deciduous, compound leaves which are shed in the autumn. The flowers are small and greenish. The staminate flowers occur in catkins, while pistillate flowers occur individually. The ripe fruit of the hickories and walnuts is properly termed a drupe in which their large, hard-coated seed is encased in a leathery case. The **fruits** of walnuts and hickories are edible.

An interesting characteristic of the black walnut and some related species is their ability to apparently poison the **soil** in their immediate vicinity. Few plants are able to grow beneath a large walnut **tree**, an observation attributed to the presence of a toxic **alkaloid** known as juglone. Even walnut seedlings cannot normally grow beneath a parent tree. The production of phytotoxic chemicals by plants for use in a type of **chemical warfare** with other competing plants is a form of allelopathy. Black walnut is often cited by ecologists as an archetypal, allelopathic species.

Species of walnuts

Six species of walnuts occur in **North America**. The black walnut (*Juglans nigra*) and butternut (*J. cinerea*) are widespread in eastern North America. The little walnut (*J. microcarpa*) and Arizona walnut (*J. major*) range into Texas and Arizona and south into Mexico. The California walnut (*J. californica*) and Hinds walnut (*J. hindsii*) have relatively localized distributions in southern California. The English walnut (*J. regia*) is native to **Europe** and **Asia** but has been widely planted in North America.

Twelve native species of hickories occur in North America. The most famous species is the pecan (*Carya illinoensis*). This species occurs naturally throughout the central United States and south through eastern Mexico, but is now cultivated more widely throughout the eastern United States. Other species that are widespread in

Pecans growing on a tree (above). Ripened pecans showing nut meat (below). *JLM Visuals. Reproduced by permission.*

southeastern North America are the shagbark hickory (*C. ovata*), mockernut hickory (*C. tomentosa*), pignut hickory (*C. glabra*), and bitternut hickory (*C. cordiformis*).

Economic importance

Various species of walnuts and hickories are economically important trees for both their wood and their edible fruits which may be gathered in the wild but are now mostly grown in plantations.

The wood of black, English, and other walnuts is close-grained, dark-brown colored, and very strong. Walnut wood is used to manufacture lumber and veneers for fine furnitures and cabinets, and it is sometimes carved into components for artisanal furniture. A well-formed tree of black walnut with a good grain and solid core can be worth more than $12,000 as raw material for fine lumber or veneer. Because of this enormous per-tree value, walnut trees are sometimes illegally "rustled" from private or public property to be sold in a black market.

KEY TERMS

· ·

Compound leaf—A leaf in which the blade is separated into several or many smaller units, called leaflets, arranged along a central petiole or stalk known as a rachis.

Drupe—A fruit which has a fleshy outer layer, and a hard inner layer which encloses a single seed. A cherry is a typical example, but the fruit of a walnut is also a drupe.

Nut—A generic term for a dry, one-seeded fruit with a hard coat which is usually quite difficult to open.

Pith—A large-celled tissue that is found inside of the roots or stems of certain species. Members of the walnut family have a chambered pith in which the tissue is separated into discrete zones of solid tissue and air chambers.

Hickories also provide an excellent hard wood, used to manufacture fine furniture and wooden baseball bats.

The best-known edible fruits harvested from species in the walnut family are those of the European or English walnut (*Juglans regia*), the black walnut (*J. nigra*), the pecan (*Carya illinoensis*), and the hickory (*Carya ovata*). The first three of these species are commonly grown in plantations established for the production of their fruits. When they reach a large size, the walnuts may be harvested from the plantations for their extremely valuable wood. However, this is not done for pecans because their wood does not have qualities that are as desirable as those of large walnut trees.

The most important use of the fruits of walnuts and hickories is directly for eating. However, fresh walnut **seeds** contain about 50% of their weight as oil, which can be expressed from these fruits and used as an edible oil or to manufacture **soap**, perfume, cosmetics, or paint.

Walnuts have sometimes been used as minor folk medicines. The inner **bark** of the black walnut can be used as a laxative, while the fruit rind has been used to treat intestinal **parasites**, **ulcers**, and syphilis. An infusion of boiled leaves has been used to get rid of bedbugs.

The doctrine of signatures was a medicinal theory that developed in Europe during the Middle Ages (about 500 to 1,500 years ago), but also occurred independently in some other cultures. This theory held that the potential usefulness of plants for medical purposes was revealed through the growth form of the **plant** or its parts. For example, a similarity between the form of the plant or its parts and some component of the human **anatomy** was commonly thought to reveal a signature of usefulness. When the hard, outer shell of a walnut is removed, the seed looks superficially like a human **brain**, viewed from above with the top of the skull removed. Consequently, it was believed that walnuts were somehow useful for the treatments of insanity and headaches.

Resources

Books

Judd, Walter S., Christopher Campbell, Elizabeth A. Kellogg, Michael J. Donoghue, and Peter Stevens. *Plant Systematics: A Phylogenetic Approach.* 2nd ed. with CD-ROM. Suderland, MD: Sinauer, 2002.

Klein, R.M. *The Green World. An Introduction to Plants and People.* New York: Harper and Row, 1987.

Bill Freedman

Walruses

The walrus (*Odobenus rosmarus*) is one of the largest **seals**, order Pinnipedia. Although similar in many respects to other seals, particularly the eared seals, the walrus is sufficiently different to merit being placed in its own family, Odobenidae. It is the sole member of that family, with only a few subspecies. The genus name is derived from the Greek words for "tooth" and "I walk"—thus, "tooth walker," which refers to the walrus's use of its tusks to haul out on **ice** floes. The **species** name is derived from the Scandinavian word for walrus.

Walruses are found along the coast of the Arctic Ocean, where they spend most of their time on the seasonal pack ice. (If ice is unavailable, they will gather on land.) They have bred as far south as Nova Scotia, the Aleutians, and the White Sea, and stray individuals have been recorded in more temperate waters, including coastal Massachusetts, Ireland, southern England, France, and even northern Spain.

Among the seals, the walrus is second in size only to the **elephant** seal. It varies in size according to location; the smallest walruses are found in Hudson Bay, the largest in the Bering and Chukchi Seas. Hudson Bay males average 9.5 ft (2.9 m) in length and weigh about 1,750 lb (795 kg), and the females are about 8.2 ft (2.5 m) long and weigh 1,250 lb (565 kg). In the Bering Sea population, the adult males are about 10.5 ft (3.2 m) long and weigh 2,670 lb (1,210 kg) and the females are 8.9 ft (2.7 m) long and 1,830 lb (830 kg).

It is impossible to mistake a walrus for anything else. Its distinctive tusks are present in both sexes. Tusk

Walruses (*Odobenus rosmarus*) on Round Island, Alaska. *Photograph by Charles Krebs. Stock Market. Reproduced by permission.*

length, like body size, varies among populations, but can exceed 39 in (100 cm) in males and 32 in (80 cm) in females. The tusks are derived from the canine teeth, and they grow throughout life. The tusk is almost purely dentine (ivory); this has made the walrus a target of hunting. Although fearsome weapons, the tusks are mostly used as ice axes, when the **animal** hauls itself out of the **water** onto an ice floe. They are also used as a social signal, much like the antlers of **deer**. The size and shape of tusks differs between males and females, so it is likely that walruses use these clues to determine sex and age.

The body of a walrus is massive. The head, which appears to be set directly on the animal's trunk, is round and the skull is thick. A walrus may use its head as a battering ram to break through ice up to 8 in (20 cm) thick. The deeply folded skin is 1.2-1.6 in (3-4 cm) thick, and is underlain by a 4 in (10 cm) layer of blubber. In females and young males, the skin is covered with coarse hair about 0.4 in (1 cm) long, which gives them a soft, velvety look. Older males are nearly hairless, and their thick, folded skin is bare on their neck and shoulders. When an old male hauls himself out of the Arctic water, he may be nearly white; but as the **sun** warms him, his skin will turn a rosy shade of pink. The skin is particular-

ly thick on the male's neck, where it can be up to 2.4 in (6 cm). Males also have a pair of pouches extending from the pharynx. When inflated, these pouches produce a distinctive bell-like sound, and help the male float.

The walrus has four flippers, each with five digits. The flippers are thick and cartilaginous. All the flippers have a bare, warty sole that provides good traction on ice. Like the eared seals, the walrus can rotate the flippers forward and use them to walk. In the water, however, the walrus propels itself almost entirely with its hind flippers, using the fore-flippers only to steer. They swim at an average speed of about 4.3 mph (7 km/h), and can reach 22 mph (35 km/h). However, they are generally not great migrators, and seem to prefer to hitch rides on ice floes that drift with the current. Dives last about 2-10 minutes; the depth is typically 32-164 ft (10-50 m), but the deepest known dive was 262 ft (80 m).

The eyes are small, and the external ears are mere folds of skin. While these senses may not be particularly good, there is an excellent sense of **touch** using the distinctive moustache, used to find food. The moustache, found in both sexes, is composed of about 450 sensitive whiskers.

The whiskers help the walrus detect the bottom-dwelling organisms on which it feeds. Bivalve **mollusks** such as clams comprise the bulk of the typical diet. Once a walrus noses a mollusk out of the mud (an old wives' tale suggests that walrus use their tusks to dig up mollusks), it sucks the creature out and spits the shell back into the mud. Thousands of mollusks may make up a meal. The diet also includes **crabs**, **octopus**, **sea cucumbers**, polychaete worms, tunicates, **fish**, and occasionally, **birds**. Some individuals have been known to hunt other seals and even whales.

Walrus are gregarious animals. They seem to enjoy nothing more than sunbathing in great agglomerations called haul-outs. The haul-outs are generally all-male or all-female during the non-breeding season, and within them several thousand walruses may lie close together. Within a haul-out there is a social hierarchy, with the largest animals with the largest tusks at the top. A dominant animal need just show its tusks to a subordinate and the latter will generally move along; sometimes there is violence, but generally there is no lasting injury.

Mating occurs during January and February, in the water, and the female can delay implantation of the fertilized egg until July. The calf is born 10-11 months later, about 140 lb (63 kg) in weight, 45 in (113 cm) in length, and tuskless. Mothers and calves share a strong bond, and the mother is very protective. Unlike in the eared seals, walrus mothers do not leave their calf for long periods. However, females will "babysit" each other's young, and even adopt orphans. Calves remain with their mother for two years. Females are sexually mature between ages six or seven; males mature later, at around eight to 10, but they generally cannot compete for females until they reach their full growth at around age 15. Females are full-grown between 10-12. Walrus can live about 40 years.

Walrus have been hunted for thousands of years by aboriginal Arctic peoples, who used all parts of the animal—skin, **fat**, meat, and bones—for clothing, fuel, food, tools, and boat-skins. Then Europeans realized the walrus was an easy-to-kill source of oil, ivory, and skins. By the mid-1800s, the Atlantic subspecies (*O. r. rosmarus*) had vanished at southern locations they had traditionally used, such as Sable Island and the Magdalen Islands off eastern Canada. This subspecies has never recovered from the onslaught; about 30,000 North Atlantic walruses remain, but they occur much farther to the north than in former times. During the nineteenth century, western hunters turned their attention to the Pacific subspecies (*O. r. divergens*), and tens of thousands of those animals were killed. On the Pribilof Islands alone, hunters took 35,280 lb (16,000 kg) of walrus ivory. The walrus of the Laptev Sea, *O. r. laptevi*, is considered rare, as it numbers only between 4,000-5,000 animals.

Although both the United States and Russia have prohibited hunting except by native peoples, some conservationists contend that this "subsistence" hunting is now primarily commercial. Poaching has increased since an international moratorium on international trade of elephant ivory was enacted (walrus ivory is a good substitute for many purposes). Between poaching and the legal killing of 10,000-15,000 walruses in the eastern and western Arctic each year, the population of all walruses is likely to decrease greatly.

Other than human beings, walruses have few natural enemies; they include polar **bears**, but walruses are quite good at fending off these predators.

F. C. Nicholson

Wandering jew *see* **Spiderwort family**

Warblers

Warblers are small, perching **song birds** with a large number of **species** distributed throughout the world. There are two families of warblers, one in the New World and one in the Old World. The New World warblers (family Parulidae) comprise 113 species that occur throughout the Americas. The Old World warblers (family Silviidae) occur in Eurasia, **Africa**, and **Australia**, and include some 325 species. Although the warblers in these two families are not closely related, having evolved from different ancestral stocks, they are of rather similar general appearance, and many species in both families are accomplished singers. These are the main reasons for their shared common name.

In addition, **birds** in both families of warblers are very active hunters of **insects**, spiders, and other small **invertebrates**. Most species of warblers are tropical or subtropical. However, many migratory species of warblers breed in forested and shrubby habitats at higher latitudes, and spend the non-breeding season in warmer latitudes. These migratory warblers can be regarded as essentially tropical birds that breed and raise their young in the **forests** of higher latitudes in order to take advantage of the seasonal abundance of **arthropods**.

The Old World and American warblers bear a superficial resemblance to each other. This is largely because of convergent evolutionary influences, resulting from the fact that these unrelated birds occupy rather similar niches in their ecosystems—that of small, arthropod-eating birds. One difference between the families is the occurrence of 10 primary wing feathers in the Silviidae, compared with nine in the Parulidae. The American warblers, particularly male birds, tend to be much more brightly colored than the Old World warblers.

American warblers

The American warblers are small, brightly colored, insectivorous birds. Most species of American warblers are resident in tropical forests, but some migrate to northern ecosystems to breed.

At least 52 species of American warblers breed north of Mexico. Some of these species have very wide ranges, occurring over much of the United States and Canada. The yellow warbler (*Dendroica petechia*), for example, is a common species found in shrubby habitats from the northern low arctic through all but the southernmost deciduous forests and western deserts. The common yellowthroat (*Geothlypis trichas*) is similarly widespread in marshes and shrubby habitats, from the subarctic through the southern states and into Mexico. The yellow-breasted chat (*Icteria virens*) is the largest species of warbler in **North America**, with a body length of 15 in (38 cm). This species also ranges widely, from southern Canada through most of the United States. The yellow-rumped warbler (*Dendroica coronata*) is one of the most familiar species, because it breeds in many types of **conifer** forests, and winters throughout much of the southern United States and down into Mexico and elsewhere in Central America.

Male birds of these and many of the other North American warblers are very colorful. The prothonotary or golden swamp warbler (*Protonotaria citrea*) occurs in hardwood swamps and riparian forests, and has a brilliant yellow plumage, offset by its gray-blue wings. The northern parula (*Parula americana*) occurs in eastern hardwood forests, and has a bright blue back, and a yellow breast with a black-and-red band running across. The male blackburnian warbler (*Dendroica fusca*) of eastern spruce-fir forests has an orange throat framed by black and white markings. The red-faced warbler (*Cardellina rubrifrons*) of southern Arizona and Mexico has a crimson face and throat. The American redstart (*Setophaga ruticilla*) is a widespread forest species with a black body and bright, orange patches on its flanks and tail.

American warblers are a diverse component of the community of birds breeding in most forests of the Americas. In some places in North America, 10 or more species may breed in the same stand. Although all of these birds feed on small arthropods, they segregate ecologically by feeding in different parts of the **tree** canopy, on the surface of tree **bark**, or on the forest floor.

A male black-throated green warbler (*Dendroica virens*) at Metropolitan Beach Metropark, Michigan. Usually found high in trees, this warbler was in the grass feeding on grounded insects after a cold snap. *Photograph by Robert J. Huffman. Field Mark Publications. Reproduced by permission.*

Old World warblers

About one-half of the Old World warblers are resident or short-distance migrants that breed in tropical and sub-tropical ecosystems of Africa. Africa is also the wintering ground for many of the other species of Old World warblers, which migrate to more northern latitudes to breed. Various species in this family are widespread throughout Eurasia, breeding as far as the northern **tundra**, but wintering in the tropics of Africa or **Asia**. Other migratory species breed in more temperate habitats.

Only a few species of Silviid warblers breed in North America. The ruby-crowned kinglet (*Regulus calendula*) and golden-crowned kinglet (*R. satrapa*) are widespread and common birds of northern conifer forests. The blue-gray gnatcatcher (*Polioptila caerula*) is more southern in distribution, and prefers broad-leaf forests. The black-tailed gnatcatcher (*P. melanura*) occurs in **desert** scrub. These are all small, very active, insect- and spider-hunting birds. The gnatcatcher was named after it habit of catching small insects while hovering, or through brief, aerial pursuits.

Like other songbirds, male warblers sing to proclaim their breeding territory, and to advertise their availability as a mate to females. Some of the Silviid warblers of Eurasia are very difficult to tell apart on the basis of plumage, but they have different **habitat** preferences and distinctive songs.

Conservation of warblers

Warblers are small, often colorful birds, and there are many species of both the Parulid and Silviid warblers. As a result, these birds are among the most avidly sought-after sightings by bird-watchers. Birding is a non-con-

KEY TERMS

. . . .

Convergence—An evolutionary pattern by which unrelated species that fill similar ecological niches tend to develop similar morphologies and behavior. Convergence occurs in response to similar selection pressures.

Nest parasite—A bird that lays its eggs in the nests of other species. The host raises the parasitic egg, and often fails to successfully raise any of its own babies.

sumptive field sport rapidly increasing in popularity. Birding, in conjunction with related activities such as bird-feeding, has a very large economic impact, and gives great aesthetic enjoyment to many people. Unfortunately, populations of many of the small birds that are the object of these activities, including warblers, are becoming increasingly at risk from a number of human activities.

In spite of their small size, migrating Old World warblers are commonly hunted as food in the Mediterranean region. These days, they are mostly captured with nets or using sticky perches from which the birds cannot extricate themselves. Large numbers of migrating warblers are caught in these ways, and are eaten locally or are offered for sale as a delicacy at markets. Some of the catch, generally pickled, is even traded internationally. The use of warblers in this way is a quite uncontrolled, free-for-all exploitation, and represents a significant risk to the populations of these tiny birds.

One species of North American warbler apparently became extinct around the 1950s or 1960s. The Bachman's warbler (*Vermivora bachmanii*) used to breed in mature broad-leaf forests of the southeastern United States. This species has not been seen for decades even though there still seems to be sufficient breeding habitat, including intact stands where Bachman's warbler used to successfully breed. The **extinction** of this species was probably caused by the conversion of its wintering habitat on the **island** of Cuba to agriculture. Once the surviving numbers of Bachman's warbler decreased to below a critical threshold of abundance, potential mates were probably unable to locate each other in their relatively large breeding range, and the population collapsed to extinction.

Many other species of North American warblers are at risk from decreases in the area of their wintering and/or breeding habitat. This concern is especially acute for those warblers whose habitat is mature forest. Losses of natural habitat directly decrease the populations of birds that can be sustained on the landscape. In addition much

of the remaining breeding habitat is fragmented into small woodlots. This means that much of the remaining forest habitat is ecologically influenced by proximity to an edge with younger habitat. This factor appears to expose warblers and other forest birds to more intense predation, and to the debilitating effects of nest-parasitism by the brown-headed cowbird (*Molothrus ater*). Many North American ornithologists feel that these factors are causing large declines in the populations of numerous species of migratory forest birds, including many species of warblers. The same environmental problem, along with hunting, is affecting populations of warblers and other small birds in **Europe** and elsewhere.

When patience and birding skills allow their close observation, warblers can be wonderfully charismatic. Unfortunately, like so many other creatures native to North America and other parts of the rapidly changing world, many species of warblers are at great risk from the environmental changes being caused by humans and their activities.

Resources

Books

Ehrlich, P.R., D.S. Dobkin, and D. Wheye. *Birds in Jeopardy.* Stanford, CA: Stanford University Press, 1992.

Forshaw, Joseph. *Encyclopedia of Birds.* New York: Academic Press, 1998.

Bill Freedman

Wasps

Wasps are slim-waisted, stinging **insects** in the order Hymenoptera. There are two main groups of wasps: the solitary wasps are relatively small **parasites** of other **arthropods**, while the social wasps are larger and live in colonies. Some other groups of tinier hymenopterans are also commonly known as wasps.

Wasps are familiar insects to most people, and they are good insects to know about because wasps can deliver a very painful sting when they feel threatened or are agitated. It is less well-known that many of the tinier **species** of wasps provide a very valuable service to humans because these parasites and predators can be quite effective at reducing populations of injurious species of insects.

Biology of wasps

Wasps have a complete **metamorphosis** with four stages in their **life history**: egg, larva, pupa, and adult.

Adult wasps have four sparsely veined, membranous wings, and most species have their abdomen joined to the thorax across a very narrow waist. Wasps have chewing mouth parts, useful for masticating their food and for pulping **wood** to manufacture the **paper** that wasps often use to build the cell-like walls of their nests. The ovipositor of female wasps is modified into a stinger, located at the end of their abdomen. This **organ** can deliver a dose of venom to kill or paralyze other insects which may be eaten or used as a living provision for young wasps. The social wasps will also aggressively sting large animals if they feel that their nest is threatened.

There are males and females of most species of wasps. However, as with most species in the order Hymenoptera, males are normally produced from nonfertilized eggs in a developmental process known as **parthenogenesis**. Female wasps develop from fertilized eggs and are usually much more abundant than males. In some species, male wasps are not even known to occur.

Solitary wasps are relatively small insects that build their nests in burrows in the ground or out of mud on an exposed surface. The nest is then provisioned with a insect or spider that has been paralyzed by stinging and upon which one or more eggs are laid. The **prey** serves as a living but immobile food for the developing larvae of the wasp. Although quite small, parasitic wasps can be rather abundant, and they can exert a substantial measure of control over the populations of their prey species.

In contrast, the social wasps are relatively large insects that live in colonies of various size. Vespid wasps develop colonies with three castes: queens, drones, and workers. The drones are relatively short-lived males and serve only to fertilize the queens. The queens are long-lived wasps, and their major function is to initiate a colony and then spend their life laying eggs. Once a colony is established, the eggs and young are tended by workers which are nonfertile female wasps that can be very numerous in large colonies. Social wasps cooperatively feed their developing young on a continuous basis, often with chewed-up insects and other animal-derived foods.

Important groups of North American wasps

The most familiar wasps to most people are the relatively large social species, such as hornets, yellow jackets, and potter wasps, all in the family Vespidae. These wasps are brightly colored, have yellow-and-black or white-and-black stripes on their abdomen, and buzz audibly when flying. Adults of these species catch insects as prey, and they also feed on **nectar** and soft **fruits**. Vespid wasps build nests out of paper, made from the **cellulose** fibers of well-chewed wood. These wasps sometimes attack people who have stepped on their nests

This female wasp of the family Mutillidae mimics the appearance of dangerous species of ants to ward off predators. *© George Bernard, National Audubon Society Collection/Photo Researchers, Inc. Reproduced by permission.*

or are too close for the wasps' comfort. The stings of these large wasps, often delivered in multiple doses, can be very painful and often cause a substantial swelling of the surrounding **tissue**. Some people develop allergies to the stings of wasps (and **bees**), and fatalities can be caused if these hypersensitive individuals are stung.

The name yellow jacket is applied to various ground-nesting species in the genus *Vespula*, including *V. pennsylvanica* in western **North America** and *V. maculifrons* in the east. The closely related bald-faced hornet (*V. maculata*) is a widespread and abundant species in the United States and Canada. The polistes wasp (*Polistes fuscatus*) builds paper nests that are suspended from **tree** limbs or the eaves of roofs. The potter or mud-dauber wasp (*Eumenes fraterna*) makes clay nests suspended from branches of trees and shrubs.

The spider wasps (family Pompilidae) build their nests in the ground and provision them with paralyzed spiders. One of the better-known species is the tarantula wasp (*Pepsis mildei*) which is famous for its skills at hunting and subduing tarantula spiders which are much larger than the wasp. Virtually all tarantulas that are located by a tarantula wasp become living pantries for the young of these efficient predators.

Chalcid wasps are various, minute-bodied species of parasitic wasps in the superfamily Chalcidoidea made up of several families. Adult chalcid wasps feed on nectar, **plant** juices, or honeydew, a sweet secretion of **aphids**. The young wasps, however, are reared in the bodies of arthropods, usually eventually killing the host. Some species of chalcid wasps are bred in captivity in enormous numbers and are then released into fields or orchards in an attempt to achieve a measure of biological control over important insect **pests**. For example, the trichogramma wasp (*Trichogramma minutum*), only 0.1 in (2.5 mm) long, will parasitize more than 200 species of insects. This useful wasp has been captive-reared and released to reduce populations of bollworms of **cotton**, corn earworms, and **spruce** budworms in **conifer forests**.

Resources

Books

Arnett, Ross H. *American Insects.* New York: CRC Publishing, 2000.

Borror, D. J., C.J. Triplehorn, and N. Johnson. *An Introduction to the Study of Insects.* New York: Saunders, 1989.

Carde, Ring, and Vincent H. Resh, eds. *Encyclopedia of Insects.* San Diego: Academic Press, 2003.

Ito, Y. *Behaviour and Social Evolution of Wasps.* Oxford, England: Oxford University Press, 1993.

Ross, K.G., and R.W. Matthews, eds. *The Social Biology of Wasps.* Ithaca, NY: Cornell University Press, 1991.

Wilson, E. O. *The Insect Societies.* Cambridge, MA: Harvard University Press, 1976.

Bill Freedman

Waste management

Waste management is the handling of discarded materials. **Recycling** and **composting**, which transform waste into useful products, are forms of waste management. The management of waste also includes disposal, such as landfilling.

Waste can be almost anything, including food, leaves, newspapers, bottles, construction debris, chemicals from a factory, candy wrappers, disposable diapers, old cars, or radioactive materials. People have always produced waste, but as industry and technology have evolved and the human population has grown, waste management has become increasingly complex.

A primary objective of waste management today is to protect the public and the environment from potentially harmful effects of waste. Some waste materials are normally safe, but can become hazardous if not managed properly. For example, 1 gal (3.75 l) of used motor oil can potentially contaminate one million gallons (3,790,000 l) of drinking **water**.

Every individual, business or organization must make decisions and take some responsibility regarding the management of their waste. On a larger scale, government agencies at the local, state, and federal levels enact and enforce regulations governing waste management. These agencies also educate the public about proper waste management. In addition, local government agencies may provide disposal or recycling services, or they may hire or authorize private companies to perform those functions.

History of waste management

Throughout history, there have been four basic methods of managing waste: dumping it; burning it; finding another use for it (reuse and recycling); and not creating the waste in the first place (waste prevention). How those four methods are utilized depends on the wastes being managed. Municipal solid waste is different than industrial, agricultural, or **mining** waste. Hazardous waste is a category that should be handled separately, although it sometimes is generated with the other types.

The rapid increase in government regulation since about the early 1970s is a phenomenon that has had an enormous impact on all forms of waste management. This has been especially true in the United States, but it has also occurred in many other countries.

Municipal solid waste

Municipal solid waste (MSW) is what most people think of as garbage, refuse, or trash. It is generated by households, businesses (other than heavy industry), and institutions, such as schools and hospitals. However, MSW does not include toilet wastes or other liquid wastes from these sources, which are commonly handled through public **sewage treatment** systems.

The first humans did not worry much about waste management. They simply left their garbage where it dropped. However, as permanent communities developed, people began to dispose their waste in designated dumping areas. The use of such "open dumps" for garbage is still common in many parts of the world. Open dumps have major disadvantages, however, especially in

A household hazardous waste disposal day sponsored by the city of Livonia, Michigan. Residents are asked to bring toxic materials, such as paint, petroleum products, insecticides, and antifreeze, to a central location where they are combined and placed in barrels for disposal. *Photograph by Robert J. Huffman. Field Mark Publications. Reproduced by permission.*

heavily populated areas. Toxic chemicals can filter down through a dump and contaminate **groundwater**. The liquid that filters through a dump or **landfill** is called leachate. Dumps may also generate methane, a flammable and explosive gas produced when organic wastes decompose under **anaerobic** (oxygen-poor) conditions.

The landfill, also known as the "sanitary landfill," was invented in England in the 1920s. At a landfill, the garbage is compacted and covered at the end of every day with several inches of **soil**. Landfilling became common in the United States in the 1940s. By the late 1950s, it was the dominant method for disposing municipal solid waste in the nation.

Early landfills had significant problems with leachate and methane, but those have largely been resolved at facilities built since about the early 1970s. Well-engineered landfills are lined with several feet of clay and with thick plastic sheets. Leachate is collected at the bottom, drained through pipes, and processed. Methane gas is also safely piped out of many landfills.

The dumping of waste does not just take place on land. **Ocean** dumping, in which barges carry garbage out to sea, was once used as a disposal method by some United States coastal cities and is still practiced by some nations. Sewage sludge, or waste material from sewage treatment, was dumped at sea in huge quantities by New York City as recently as 1992, but this is now prohibited in the United States Also called biosolids, sewage sludge is not generally considered solid waste, but it is sometimes composted with organic municipal solid waste.

Burning has a long history in municipal solid waste management. Some American cities began to burn their garbage in the late nineteenth century in devices called cremators. These were not very efficient, however, and cities went back to dumping and other methods. In the 1930s and 1940s, many cities built new types of more-efficient garbage burners known as incinerators. The early incinerators were rather dirty in terms of their emissions of air pollutants, and beginning in the 1950s they were gradually shut down.

However, beginning in the 1970s waste burning enjoyed another revival. These newer incinerators, many of which are still in operation, are called "resource recovery" or "waste-to-energy" plants. In addition to burning garbage, they produce **heat** or **electricity** that can be used in nearby buildings or residences, or sold to a utility. Many local governments became interested in waste-to-energy plants following the **energy** crisis in 1973. However, since the mid-1980s, it became difficult to find locations to build these facilities, mainly because of public opposition focused on air-quality issues.

Another problem with **incineration** is that it generates ash, which must be landfilled. Incinerators usually reduce the **volume** of garbage by 70–90%. The remainder of the incinerated MSW comes out as ash that often contains high concentrations of toxic substances.

Municipal solid waste will likely always be landfilled or burned to some extent. In the past 25 years, however, non-disposal methods such as waste prevention and recycling have become more common. Because of public concerns and the high costs of landfilling and burning (especially to build new facilities), local governments want to reduce the amount of waste that must be disposed in these ways.

Even the earliest civilizations recycled some items before they became garbage. For example, broken pottery was often ground up and used to make new pottery. Recycling has taken many forms. One unusual type of recycling was common in large United States cities from about 1900 to 1930. This involved so-called reduction plants, where food waste, dead **horses**, and other dead animals were cooked in large vats to produce grease and fertilizer. A more familiar, and certainly less unappealing, type of recycling took place during World War II, when all scrap **metal** was fervently collected to help the war effort. Modern-day recycling has had two recent "booms," from about 1969 to 1974, and from the late 1980s until the present.

Reuse and repair are the earliest forms of waste prevention, which is also known as waste reduction. When tools, clothes and other necessities were scarce, people naturally repaired them again and again. When they were beyond repair, people tried to find other uses for them. People are again realizing the value of reuse, and are increasingly doing this with furniture, clothing, and disused building materials. In fact, some of these reuse activities are becoming quite fashionable.

Refillable soft drink bottles are another example of reuse. These are still the norm in many countries, but have become increasingly rare in the United States. One form of waste prevention, called source reduction, is a reduction in the quantity or the toxicity of the material used for a product or packaging.

Agricultural, mining, and industrial waste

Municipal solid waste is a relatively small part of the overall waste generated in the United States. More than 95% of the total 4.5 billion tons of solid waste generated in the United States each year is agricultural, mining, or industrial waste.

These wastes do not receive nearly as much attention as municipal solid waste, because most people do not have direct experience with them. Also, agricultural and mining wastes, which make up 88% of the overall total of solid waste, are largely handled at the places they are generated, that is, in the fields or at remote mining sites.

Mining nearly always generates substantial waste, whether the material being mined is **coal**, clay, **sand**, gravel, building stone, or metallic **ore**. Early mining concentrated on the richest lodes of **minerals**. Because modern methods of mining are more efficient, they can extract the desired minerals from veins that are less rich. However, much more waste is produced in the process.

Many of the **plant** and **animal** wastes generated by agriculture remain in the fields or rangelands. These wastes can be beneficial because they return organic **matter** and **nutrients** to the soil. However, modern techniques of raising large numbers of animals in small areas generate huge volumes of animal waste, or manure. Waste in such concentrated quantities must be managed carefully, or it can contaminate groundwater or surface water.

Industrial wastes that are not hazardous have traditionally been sent to landfills or incinerators. The rising cost of disposal has prompted many companies to seek alternative methods for handling these wastes, such as waste prevention and recycling. Often a manufacturing plant can reclaim certain waste materials by feeding them back into the production process.

Hazardous waste

Hazardous wastes are materials considered harmful or potentially harmful to human health or the environment. Wastes may be deemed hazardous because they are poisonous, flammable, or corrosive, or because they react with other substances in a dangerous way.

Industrial operations have produced large quantities of hazardous waste for hundreds of years. Some hazardous wastes, such as mercury and dioxins, may be released as gases or vapors. Many hazardous industrial wastes are in liquid form. One of the greatest risks is that these wastes will contaminate water supplies.

An estimated 60% of all hazardous industrial waste in the United States is disposed using a method called deep-well injection. With this technique, liquid wastes are injected through a well into an impervious rock formation that keeps the waste isolated from groundwater and surface water. Other methods of underground burial are also used to dispose hazardous industrial waste and other types of dangerous material.

Hazardous wastes are also disposed at specially designed landfills and incinerators. A controversial issue in international relations is the export of hazardous waste, usually from relatively wealthy industrialized countries, to poorer developing nations. Such exports often take place with the stated intent of recycling, but many of the wastes end up being dumped.

Pesticides used in farming may contaminate agricultural waste. Because of the enormous volumes of pesticides used in agriculture, the proper handling of unused pesticides is a daunting challenge for waste managers. Certain mining techniques also utilize toxic chemicals. Piles of mining and metal-processing waste, known as waste rock and tailings, may contain hazardous substances. Because of a reaction with the **oxygen** in the air, large amounts of toxic acids may form in waste rock and tailings and leach into surface waters.

Hazardous wastes also come from common household products that contain toxic chemicals. Examples include drain cleaner, pesticides, glue, paint, paint thinner, air freshener, detergent, and nail polish. Until about the early 1970s, most people dumped these products in their domestic garbage. However, local waste managers do not want hazardous domestic wastes in with the regular garbage. They also do not want residents to pour leftover household chemicals down the drain, since municipal sewage treatment plants are not properly equipped to treat them.

The trend during the 1980s and 1990s was for local governments to open facilities where residents could take their household hazardous wastes, or to sponsor periodic collection events for those materials. City and county governments in the United States conduct more than 800 of these events each year.

Modern practices

Three main, inter-related factors influence the way wastes are handled today: government regulation, cost, and public attitudes. Industry and local governments must comply with increasingly strict federal and state regulations for landfills and incinerators. Partly because those regulations have driven up the costs of disposal, it has become critical for local governments, industry and businesses of all sizes to find the lowest-cost waste management options.

Public attitudes also play a pivotal role in decisions about waste management. Virtually every proposed new landfill or waste-to-energy plant is opposed by people who live near the site. Public officials and planners refer to this reaction as NIMBY, which stands for "Not In My BackYard." If an opposition group becomes vocal or powerful enough, a city or county council is not likely to approve a proposed waste-disposal project. The public also wields considerable clout with businesses. Recycling and waste prevention initiatives enjoy strong public support.

About 19% of United States municipal solid waste was recycled or composted in 1994, 10% was incinerated, and 71% was landfilled. The recycling **rate** is expected to rise to 25% by the year 2000.

Waste prevention

Preventing or reducing waste is typically the least expensive method for managing waste. Waste prevention may also reduce the amount of resources needed to manufacture or package a product.

For example, most roll-on deodorants once came in a plastic bottle, which was inside a box. Beginning about 1992, deodorant manufacturers redesigned the bottle so that it would not tip-over easily on store shelves, which eliminated the need for the box as packaging. This is the type of waste prevention called source reduction. It can save businesses money, while also reducing waste.

Waste prevention includes many different practices that result in using fewer materials or products, or using materials that are less toxic. For example, a chain of clothing stores can ship its products to its stores in reusable garment bags, instead of disposable plastic bags. Manufacturers of household batteries can reduce the amount of mercury in their batteries. In an office, employees can copy documents on both sides of a sheet of **paper**, instead of just one side. A family can use cloth instead of paper napkins.

Composting grass clippings and **tree** leaves at home, rather than having them picked up for disposal or municipal composting, is another form of waste prevention. A resident can leave grass clippings on the lawn after mowing (this is known as grass-cycling), or can compost leaves and grass in a backyard composting bin, or use them as a mulch in the garden.

Waste prevention is preferable over recycling or municipal composting programs, because it does not require transportation, processing, and administration. However, waste prevention does have limitations. It will never eliminate waste; it just reduces the amount that has to be recycled or disposed. Waste prevention also is extremely difficult for a government to measure, since waste that is prevented never really existed in the first place. The lack of good data can make it hard for governments to justify spending money on education programs in support of waste prevention. Even though waste prevention is less expensive than other forms of waste management in the long run, local governments and businesses may need to spend substantial amounts over the short term, to provide education about waste prevention or to make changes in operating procedures so that less waste is produced. Waste prevention can also be a valuable tool for managing industrial and hazardous wastes, since disposal of those materials is particularly expensive and heavily regulated.

Farmers can use natural or alternative methods of pest control to replace some or all of their use of pesticides, which in turn reduces the amount of hazardous wastes produced. Many government agencies strongly encourage this. A program started by the province of Ontario, Canada, aims to reduce agricultural pesticide use by 50%.

Recycling and composting

Recycling is a simple concept: using disused (or waste) material to make a new product. In practice, however, recycling is far from simple. Recycling consists of three essential elements: collection of the waste materials, also known as secondary materials or recyclables; processing those materials and manufacturing them into new products; and the marketing and sale of those new products. Dozens of different materials can be recycled, including **glass** bottles, **aluminum** cans, **steel** cans, plastic bottles, many types of paper, used motor oil, car batteries, and scrap metal. For each material, the collection, processing, and marketing needs can be quite different.

When the current recycling boom began in the late 1980s, markets for the recyclables were not sufficiently considered. A result was that some recyclable materials were collected in large quantities but could not be sold, and some ended up going to landfills. Today, the development of recycling markets is a high priority. "Close the loop" is a catch-phrase in recycling education; it means that true recycling (i.e., the recycling loop) has not taken place until the new product is purchased and used.

To boost recycling markets, many local and state governments now require that their own agencies purchase and use products made from recycled materials. In a major step forward for recycling, President Bill Clinton issued an executive order in 1993 requiring the federal government to use more recycled products.

Many managers of government recycling programs feel that manufacturers should take more responsibility for the disposal of their products and packaging, rather than letting municipalities bear the brunt of the disposal costs. An innovative and controversial law in Germany requires manufacturers to set up collection and recycling programs for disused packaging of their products.

The high cost of government-created recycling programs is often criticized. Supporters of recycling argue it is still less expensive than landfilling or incineration, when all costs are considered. Another concern about recycling is that the recycling process itself may generate hazardous wastes that must be treated and disposed.

Recycling of construction and demolition (C&D) debris is one of the growth areas for recycling. Although C&D debris is not normally considered a type of municipal solid waste, millions of tons of it have gone to municipal landfills over the years. If this material is separated at the construction or demolition site into separate piles of **concrete**, **wood**, and steel, it can usually be recycled.

Composting is considered either a form of recycling, or a close relative. Composting occurs when organic waste, such as yard waste, food waste, and paper, is broken down by microbial processes. The resulting material, known as compost, can be used by landscapers and gardeners to improve the fertility of their soil.

Yard waste, primarily grass clippings and tree leaves, makes up about one-fifth of the weight of municipal solid waste. Some states do not allow this waste to be disposed. These yard-waste bans have resulted in rapid growth for municipal composting programs. In these programs, yard waste is collected by trucks (separately from garbage and recyclables) and taken to a composting plant, where it is chopped up, heaped, and regularly turned until it becomes compost.

Waste from food-processing plants and produce trimmings from grocery stores are composted in some parts of the country. Residential food waste is the next frontier for composting. The city of Halifax, Canada, collects food waste from households and composts it in large, central facilities. For more details on recycling and composting, see the entries on those topics.

Biological treatment, a technique for handling hazardous wastes, could be called a high-tech form of composting. Like composting, biological treatment employs microbes to break down wastes through a series of metabolic reactions. Many substances that are toxic, carcinogenic (cancer-causing), or undesirable in the environment for other reasons can be rendered harmless through this method.

Extensive research on biological treatment is in progress. **Genetic engineering**, a controversial branch of **biology** dealing with the modification of genetic codes, is closely linked with biological treatment, and could produce significant advances in this field.

New developments in disposal

Waste management became a particularly expensive proposition during the 1990s, especially for disposal. Consequently, waste managers constantly seek innovations that will improve efficiency and reduce costs. Several new ideas in land-filling involve the reclamation of useful resources from wastes.

For example, instead of just burning or releasing the methane gas that is generated within solid-waste landfills, some operators collect this gas, and then use it to produce power locally or sell it as fuel. At a few landfills, managers have experimented with a bold but relatively untested concept known as landfill mining. This involves digging up an existing landfill to recover recyclable materials, and sometimes to re-bury the garbage more efficiently. Landfill mining has been criticized as costly and impractical, but some operators believe it can save money under certain circumstances.

In the high-tech world of incineration, new designs and concepts are constantly being tried. One waste-to-energy technology for solid waste being introduced to the United States is called fluidized-bed incineration. About 40% of incinerators in Japan use this technology, which is designed to have lower emissions of some air pollutants than conventional incinerators.

A 1994 United States Supreme Court ruling could increase the cost of incineration significantly. The Court ruled that some ash produced by municipal solid-waste incinerators must be treated as a hazardous waste, because of high levels of toxic substances such as **lead** and cadmium. This means that incinerator ash now has to be tested, and part or all of the material may have to go to a hazardous waste landfill rather than a standard landfill.

A much smaller type of incinerator is used at many hospitals to burn medical wastes, such as **blood**, surgical waste, syringes, and laboratory waste. The safety of these medical waste incinerators has become a major issue in some communities. A study by the Environmental Protection Agency released in 1994 found that medical waste incinerators were leading sources of **dioxin** emissions into the air. The same study warned that dioxins, which can be formed by the burning of certain chemical compounds, pose a high risk of causing **cancer** and other health hazards in humans.

Trends for the twenty-first century

Even with the tremendous increase in the regulation of waste since the early 1970s, many problems still exist with hazardous and other wastes. Citizens often express their concerns about this issue. However, governments have a limited amount of money to spend on environmental protection. As the twenty-first century begins, governments and industry strive to continue to improve the efficiency of recycling and disposal. Governments in the United States and elsewhere will also rely more on

waste prevention strategies, because of their lower cost and greater environmental benefit.

The greatest impetus for waste prevention will likely come from the public. More and more citizens will come to understand that pesticides, excessive packaging, and the use of disposable rather than durable items have important environmental costs. Through the growth of the information society, knowledge about these and other environmental issues will increase. This should result in a continuing **evolution** towards more efficient and environmentally sensitive waste management.

See also Emission; Pollution.

Resources

Books

American Water Works Association. *Water Quality and Treatment.* 5th ed. Denver: American Water Works Association, 1999.

Hirschhorn, Joel, and Kirsten Oldenburg. *Prosperity Without Pollution: The Prevention Strategy for Industry and Consumers.* New York: Van Nostrand Reinhold, 1991.

Miller, E. Willard, and Ruby Miller. *Environmental Hazards: Toxic Waste and Hazardous Material: A Reference Handbook.* Santa Barbara: ABC-CLIO, 1991.

Rathje, William, and Cullen Murphy. *Rubbish!: The Archaeology of Garbage.* New York: HarperCollins Publishers, 1992.

KEY TERMS

Composting—The process by which organic waste, such as yard waste, food waste, and paper, is broken down by microorganisms and turned into a useful product for improving soil.

Hazardous wastes—Wastes that are poisonous, flammable, or corrosive, or that react with other substances in a dangerous way.

Incineration—The burning of solid waste as a disposal method.

Landfilling—A land disposal method for solid waste, in which the garbage is covered every day with several inches of soil.

Recycling—The use of disused (or waste) materials, also known as secondary materials or recyclables, to produce new products.

Source reduction—Reduction in the quantity or toxicity of material used for a product or packaging; this is a form of waste prevention.

Waste prevention—A waste management method that involves preventing waste from being created, or reducing waste.

Periodicals

Nakamura. "Input-Output Analysis Of Waste Management." *Journal Of Industrial Ecology* 6, no. 1 (2002): 39-63.

Other

The League of Women Voters Education Fund. *The Garbage Primer.* New York: Lyons & Burford, 1993.

Tom Watson

Waste, toxic

Toxins are poisonous materials that interfere with vital metabolic processes to sicken or kill living organisms. Toxins can be either general poisons that kill many types of cells and organisms, or they can be extremely specific in their target and mode of action. Some are extremely reactive and can be lethal even in very dilute concentrations. **Ricin**, for instance, is a protein found in castor beans, and is one of the most toxic organic compounds known. Three hundred picograms (trillionths of a gram) injected intravenously is enough to kill an average mouse. That means that a few teaspoonsful of this substance, if divided and delivered in individual doses, could potentially kill all the **mice** in the world. Put another way, an amount of supertoxins that is invisible to the naked **eye**, if delivered in the right way, could be lethal.

An important principle of **toxicology** (the study of poisons) is that "the dose makes the poison." This dictum, first pronounced by the German physician Paracelsus in the sixteenth century, means that almost everything is dangerous at some level. Even compounds like table **salt** and **water** that are essential parts of our diet in reasonable amounts could make you very sick or even kill you if ingested in excess. Contrarily, even the most toxic compounds generally have a threshold level below which they are effectively harmless. Toxicity depends on the amount, time, mode of delivery of the toxin, as well as age and physiological state of the target **organism**. Among the most dangerous toxins are carcinogens (cause **cancer**), mutagens (genetic damage), teratogens (**birth defects**), and neurotoxins (nerve damage). Not all toxins are organic compounds. Many metals, such as **lead**, mercury, cadmium, and chromium, are highly poisonous as are elements such as arsenic and selenium, and **minerals** such as **asbestos**. Many people also assume that all human-made chemicals are poisonous, while all natural materials must be benign and wholesome. This is far from the truth. Many synthetic, industrial chemicals are relatively innocuous while perfectly natural materials, like some of those mentioned above, are extremely dangerous.

Toxic wastes, as their name implies, are unwanted materials known to be fatal to humans or laboratory animals at low doses or that are carcinogenic, mutagenic, teratogenic, or neurotoxic to humans or other life forms. Radioactive materials are considered especially dangerous, and their use and disposal is tightly regulated. Modern societies produce, use, and discard a vast array of toxic chemical substances. According to the Environmental Protection Agency (EPA), the United States generates about 265 million metric tons of officially classified hazardous and toxic wastes each year. This amounts to about one ton per year for every individual in the United States. Fortunately, most of this material is stored, recycled, converted to non-hazardous forms, or otherwise disposed of safely. Shockingly, however, at least 40 million metric tons (22 billion lb) of toxic and hazardous chemicals are released each year in the United States into the air, water, or land by unsound or illegal disposal methods. This represents an immediate health hazard to many people who live close to these disposal sites, and it may well represent a long-term health and ecological hazard to all of us. Scientists are discovering that persistent chemicals such as **pesticides**, dioxins, **polychlorinated biphenyls (PCBs)**, and mercury can be carried over long distances and accumulate to levels that appear to be causing worrisome health effects in **wildlife** and human populations thousands of miles and dozens of years from their original source.

The preferred hierarchy of **waste management** is to reduce, reuse, recycle, detoxify, and—only as a last resort—store safely. Reducing waste amounts means not making it in the first place. Often we can find alternative products or industrial processes that avoid creating a particular waste. Reuse means using a material for some other purpose or process. What is one person's unwanted waste can be a valuable resource for someone else. **Recycling** and detoxification involve chemical, biological, or physical treatments to change toxins into harmless forms that could be used in beneficial ways. Storage of toxic wastes requires specialized facilities in which materials are isolated from the environment by secure **metal** containers, impermeable plastic liners, compacted clay cushions, and other coverings that prevent materials from ever escaping. Permanent, secure waste disposal sites are both very expensive to construct as well as difficult to site and maintain.

Water

Water is the most abundant liquid on **Earth**. It covers more than 70% of the earth's surface. Including the

4282

clouds (which are, of course, also water), it makes our entire **planet** look blue and white from space.

The earth's supply of water is constantly being recycled. It is evaporated from the oceans by the **sun** and is given off by the **forests**. The vapor condenses into clouds, which rain out onto the land. The land water runs off into the lakes and **rivers**, which then run back to the seas, and the cycle is complete. The total amount of water on Earth, in the form of oceans, lakes, rivers, clouds, polar **ice**, etc. is 1.5×10^{18} (one-and-a-half billion) tons, occupying a total **volume** of 8.7 million cubic miles.

It is impossible to overstate the importance of water to almost every process on Earth, from the life processes of the lowest **bacteria** to the shaping of continents. Water is the most familiar of all chemical compounds known to humans. It is essential to all living things, **plant** and **animal**. We drink it, we wash with it, we play in it and we cook in it. In fact, we ourselves are more than half water.

We never see absolutely pure water because it dissolves so many substances. If we want pure water we have to prepare it laboriously by such means as **distillation**, **ion exchange**, and reverse **osmosis**. Moving water even dissolves rock slightly, to form caves and to wear away **mountains**. All of the water on Earth, therefore, is in the form of solutions. The dissolved substances change the properties of water from what they would be in absolutely pure water. They affect its freezing point and its **boiling point**, among many other physical and chemical properties. The dissolved or suspended substances in water can be in the form of ions, molecules, or larger particles. For drinking water, bacteria must also be killed.

What is water?

Water is an odorless, tasteless, transparent liquid that appears colorless but is actually very pale blue. The **color** is obvious in large quantities of water such as lakes and oceans, but it can even be seen in a full bathtub. It is a single chemical compound whose molecules consist of two **hydrogen atoms** attached to one **oxygen** atom. The chemical formula of this compound is therefore H_2O. The two hydrogen atoms are attached to the oxygen atom in such a way as to make an angle-shaped **molecule**.

The **angle** isn't 90°, however, but 104.5°—close to a right angle, but a little wider.

The formula H_2O means that no matter how much water we are talking about, it always contains exactly twice as many hydrogen atoms as oxygen atoms. Considering that a hydrogen atom weighs only about one-

sixteenth as much as an oxygen atom, most of the weight in water is due to oxygen: 88.8% of the weight is oxygen and 11.2% is hydrogen. That goes for everything from a single molecule to a **lake**.

Water can be made (synthesized) from hydrogen and oxygen, both of which are gases. When these two gases are mixed, however, they do not react unless the reaction is started with a flame or spark. Then they react with explosive violence. The tremendous **energy** that is released is a signal that hydrogen and oxygen are very eager to become water. Another way of saying that is that water is an extremely stable compound, compared with loose molecules of hydrogen and oxygen. It is hard to break water molecules apart into its components.

Not only that, but water molecules are stuck quite tightly to each other—at least when compared with similar compounds. The molecule-to-molecule stickiness is caused mainly by the fact that the water molecule is a **dipole**, because the oxygen atom pulls electrons away from the hydrogen atoms, giving the oxygen corner of the molecule a slight **negative** charge and the two hydrogen ends a slight positive charge. The negative part of one water molecule then attracts the positive parts of others like a magnet, although they can still slide around over each other as molecules do in any liquid. Water molecules stick to each other also by hydrogen bonds.

An unusual liquid

The strong attractions that water molecules have for each other are responsible for many of water's highly unusual properties, as compared with other liquids of about the same **molecular weight**. Among these are:

1. Its unusually high boiling point (if it were similar to the other liquids, it would be a gas at room **temperature**).

2. Its high **heat** of vaporization (the amount of heat it takes to change the liquid to a gas).

3. Its high heat of fusion (the amount of heat it takes to melt solid ice).

4. Its high **heat capacity** (the amount of heat it takes to raise its temperature by a certain amount).

5. Its world-champion rank among liquids as a solvent (it has been called the universal solvent because it dissolves so many different substances).

6. The low **density** (the lightness) of ice, which makes it float on the surface of liquid water. As water is cooled to make ice, it gets slightly denser, like all liquids. But at 39.2°F (4°C) it reaches its maximum density; when cooled below that temperature, it gets less dense until it reaches 32°F (0°C), at which time it freezes and

suddenly decreases to 91.7% of the density of the water. Being less dense than the water, the ice floats.

The normal boiling point of water is 212°F (100°C), and its freezing point is 32°F (0°C). In fact, **zero** and 100 degrees on the Celsius scale are defined as the freezing and boiling points of water. Water is also the standard by which many other quantities are measured. For example, the density of a material is often expressed as its specific gravity or specific weight: how many times denser it is than water.

In pure water, one out of every 555 million molecules is broken down—dissociated—into a hydrogen ion and a hydroxide ion:

$$H_2O \longleftrightarrow H^+ + OH^-$$

These ions are enough to make water a slight conductor of **electricity**. That is why water is dangerous when there is electricity around. The slight **dissociation** of water is responsible for the acid and base balances of all of the **chemical reactions** that take place in water, and that includes almost all the chemical reactions that take place anywhere, including those in the human body. Acid-base balance is probably the most important single factor that affects chemical reactions.

See also Groundwater; Hydrologic cycle; Hydrology; Irrigation; Precipitation; States of matter; Water conservation; Water pollution.

Resources

Books

American Water Works Association. *Water Quality and Treatment.* 5th ed. Denver: American Water Works Association, 1999.

Hancock, P.L., and B.J. Skinner eds., *The Oxford Companion to the Earth.* Oxford: Oxford University Press, 2000.

Herschy, Reginald, and Rhodes Fairbridge, eds. *Encyclopedia of Hydrology and Water Resources.* Boston: Kluwer Academic Publishing, 1998.

Lide, D.R., ed. *CRC Handbook of Chemistry and Physics.* Boca Raton: CRC Press, 2001.

McConnell, Robert, and Daniel Abel. *Environmental Issues: Measuring, Analyzing, Evaluating.* 2nd ed. Englewood Cliffs, NJ: Prentice Hall, 2002.

Oxtoby, David W., et al. *The Principles of Modern Chemistry.* 5th ed. Pacific Grove, CA: Brooks/Cole, 2002.

Walton, Alan J. *The Three Phases of Matter.* Oxford: Oxford University Press, 1996.

Periodicals

Malin, M.C., and K.S. Edgett. "Evidence for Recent Groundwater Seepage and Surface Runoff on Mars." *Science* no. 288 (2000): 2330-2335.

Wieczorek, Gerald F., et al. "Unusual July 10, 1996, Rock Fall at Happy Isles, Yosemite National Park, California." *Bulletin of the Geological Society of America* 112, no. 1 (January 2000): 75-85.

Robert L. Wolke

Water bears

Water bears or tartigrades are about 500 **species** of tiny aquatic invertebrate animals in the phylum Tartigrada, including about 90 species in **North America**. Water bears have a very widespread distribution, occurring in moist habitats from the Arctic to the Antarctic and on mountains as high as 19,680 ft (6,000 m).

Water bears have roughly cylindrical dark-colored bodies with four body segments and four pairs of stumpy legs each tipped with tiny claws. The awkward, pawing locomotion of these animals was thought to vaguely resemble that of slow-moving **bears** and hence the common name of these creatures which was given in the nineteenth century. The mouthparts of water bears are adapted for piercing, and they have a muscular pharynx for sucking the juices of mosses, liverworts, and **algae**. Water bears do not have active circulatory or respiratory systems; materials move about within their tiny body by simple processes such as **diffusion**, while respiratory gases diffuse across the surface of the body.

Water bears are typically found in moist films on mosses, **lichens**, **angiosperm** plants with a rosette growth form, and **plant** litter. Water bears can utilize this micro-aquatic **habitat** because they are only 0.002-0.05 in (0.05-1.2 mm) long. Water bears are highly tolerant of the desiccation that frequently afflicts these sorts of habitats because they have an ability to encapsulate into a shrivelled, spherical state until moist conditions again return. Water bears may also be collected from debris or mud in more typical shallow-water habitats, but they are seldom abundant in ponds or lakes. Some species occur in sandy and pebbly marine habitats above the zone of most vigorous wave action.

Water bears have separate sexes, but males are quite uncommon. Usually males are only relatively abundant during the winter and spring. Some species only produce females, reproducing by a process known as **parthenogenesis** which does not require sex. There are four to 12 post-hatching developmental stages, each beginning with a molt of the cuticle.

Water bears are interesting little creatures which have fascinated zoologists ever since the discovery of **microscopy** first made these animals visible several centuries ago.

Bill Freedman

Water conservation

The **hydrosphere** refers to that portion of the **earth** that is made of **water**, including all oceans, lakes, **rivers**, streams, **glaciers**, and underground water. Less than 3% of the water of Earth is **freshwater**, an amount that includes **polar ice caps**, glaciers, **groundwater**, surface water of rivers and freshwater lakes, and even atmospheric water. However, the amount of freshwater useable by people and other members of the **biosphere** is less than 0.5% of the total (this is water in rivers and lakes, and in the ground). This relatively small amount of available freshwater is recycled and purified by the action of processes within the **hydrologic cycle**, including **evaporation**, condensation, **precipitation**, and percolation through the ground. All life depends on the availability of freshwater.

Of all the freshwater used directly by humans, agricultural **irrigation** accounts for about 70% of the total. The remainder is used for industrial and domestic purposes. However, these proportions vary widely due to the climatic and economic conditions of the particular locality. Within this century, one third of the countries situated in areas of water scarcity may encounter severe water shortages. By 2025, two thirds of the world's population is likely to live in areas of moderate or severe water shortage. The need for more effective **conservation** of the limited supplies of water that are available for use by people and required by natural ecosystems will intensify as water stress grows.

Freshwater resources

Available freshwater resources are either groundwater or surface water (rivers and lakes). Water that flows on the surface of the land is surface runoff. The relationship among surface runoff, precipitation, evaporation, and percolation is summarized in the following equation:

Surface runoff = precipitation − (evaporation + percolation)

When surface runoff resulting from rainfall or snowmelt is confined to a relatively narrow, well-defined channel, it is called a river or stream.

Groundwater is that water that has percolated downward through the **soil** and is present within porous spaces in soil and **bedrock**. It has been estimated that the global groundwater resource is equivalent to about 34 times the **volume** of all surface waters (i.e., rivers and lakes) of the world. This resource is present nearly everywhere and has the additional advantages of typically needing no storage or treatment. Utilization does require the construction of a well, sometimes presenting a problem in the most needy locations.

Water utilization efficiency is measured by the **ratio** of water withdrawal and its subsequent consumption. Water withdrawal is water pumped from rivers, reservoirs, or groundwater wells, and is then transported for use. Water consumption is water that is withdrawn and actually used for some specific purpose. It is then returned to the environment through evaporation, **transpiration**, discharge to a river or **lake**, or in some other way.

Water consumption

Water consumption varies greatly among regions due to differences in economic development. The average municipal use in the United States is about 150 gal (568 l) per person per day, though the **rate** can be higher than 350 gal (1324 l) in some locations. This includes home use for bathing, waste disposal, and gardening, as well as institutional and commercial usage. Per capita (per person) water usage in **Asia** is only 22 gal (85 l) per day, and just 12 gal (47 l) in **Africa**.

According to the World Health Organization (WHO) of the United Nations, people have a minimum water requirement of about 5 gal (20 l) per person per day. This is the minimum amount needed for physiological rehydration, cooking, washing, and other subsistence requirements. However, the WHO estimates that nearly two billion people consume contaminated water. This carries a significant risk of developing such water-borne diseases as **cholera**, **dysentery**, polio, or typhoid, which kill about 25 million people per year. Both conservation and sanitation are obvious necessities in meeting the huge demand for freshwater.

Because irrigation accounts for 70% of the water used by humans worldwide, achieving a better efficiency of agricultural use is a logical step in advancing water conservation. This can be accomplished by lining water delivery systems with **concrete** or other impervious materials to minimize loss by leaking during transport, and by using drip-irrigation systems to minimize losses by evaporation. Drip-irrigation systems have been successfully used on fruit trees, certain row-crops, and horticultural plants. Conservation can also be accomplished by improving the efficiency of utilization of water by **crops**, including the cultivation of plants that are less demanding of moisture.

Efficient water utilization efforts

Subsurface irrigation is an emerging technology with high water-utilization efficiency. Subsurface irrigation uses a drip-irrigation tubing buried 6–8 in (15–20 cm) underground, with a spacing of 12–24 in (30–60 cm) between parallel lines. The tubing contains drip outlets that deliver water and **nutrients** within the root zone at a desired rate. In addition to water conservation, subsurface irrigation has other advantages that overhead sprinklers do not: minimal over watering, fewer **disease** and aeration problems, less runoff and **erosion**, fewer weeds, and better protection from vandalism. However, this system is relatively expensive to install. In California, subsurface irrigation has been used on fruit trees, field crops, and lawns, and has achieved water-use savings of about 50%. However, this methodology can, in arid environments, lead to the buildup of soil salinity levels, damaging plants and reducing crop yields. Balancing the water needs of the **plant** with maintenance of soil quality is an important component of water conservation measures. Technologically advanced irrigation systems now incorporate climate-based controls. These systems utilize meteorological information to determine the need for irrigation and modify the length and duration of irrigation to match the plant's requirements. Though these systems are currently used primarily on large-scale applications, development of economical models for the small-scale user is underway.

Xeriscaping, or the cultivation of plants requiring little water, is an especially suitable horticultural practice for conserving water in regions with a dry, hot climate. For example, over much of the southwestern United States, more than 50% of the domestic water consumption may be used to irrigate lawns and other horticultural plants that are intolerant of **drought**. Xeriscaping uses plants such as cacti, succulents, and shrubs of semi-desert **habitat** (such as trailing rosemary *Rosemarinus officinale* and rock rose *Cistus cobariensis*), which are well-adapted to a hot, dry climate and need little water.

Water conservation can also be advanced by improving other domestic uses of water. One simple conservation practice is to install ultra-low-flush (ULF) toilets and low-flow showerheads in homes and other buildings. A ULF toilet uses only 1.6 gal (6.1 l) per flush, compared to 5–7 gal by a standard toilet. Replacing a standard toilet with an ULF saves about 30–40 gal (114–151 l) of water per day, equivalent to 10,000–16,000 gal (37,850–60,560 l) per year. More recently, advanced toilets and urinals requiring no water have been developed and are beginning to be utilized on a limited basis.

Another way to conserve the freshwater supply is to desalinize seawater. Desalinization is the removal of salts and other impurities from seawater by either **distillation** or reverse **osmosis** (RO), and this method being increasingly used to provide high-quality water for drinking, cooking, and other domestic uses. In 1993, the world production of desalinated water was about 3.5 billion gallons per day (13 billion liters), most of which was produced in Saudi Arabia and other nations of the Gulf of Arabia, where **energy** costs are relatively low (the cost of desalinated water is highly sensitive to the cost of energy). Desalinization is also practiced in California and Florida, where the cost is about three dollars per thousand gallons, which is four to five times the cost paid for domestic water by typical urban consumers in the United States, and more than 100 times the cost paid by farmers for water for irrigation.

Widespread recognition of the importance of reusing water has begun to change traditional water use methods. As the value of water increases, users are willing employ methods that may increase the initial cost of a project, with the hope of regaining those costs through water savings in the future. One of the first of these reuse applications was the irrigation of golf courses and landscaping. In many areas, treated wastewater is diverted from its normal disposal path to be reused in irrigation. This has gained in popularity and is also utilized in small artificial ponds for decorative purposes. Greywater systems capture water that drains from sinks, tubs, laundry, and dishwashers for reuse in irrigation. Greywater systems do not incorporate toilet wastes because of the potential health threat. Dual plumbing is required for such a system and some treatment is required prior to reuse. Though home construction costs are obviously increased by including a greywater system, many have become dedicated believers in the benefits of water reuse, while others question the economic benefit of small-scale systems. The widespread application of greywater systems has, however, been hampered by codes and laws that make such systems illegal in many locations.

Economic incentives for water conservation

As the availability of water becomes more restricted, the costs to both the provider and consumer are increased. In a situation unique to the water supply industry, providers are frequently placed in the position of trying to convince consumers to use less of the commodity that they supply. Most large water providers have departments dedicated to education of the public with regard to conservation. In general, these education efforts have been largely ineffective and conservation of freshwater resources has been best achieved through economic incentives. Water providers frequently provide rebates for those consumers that are willing to change from older technology to newer, such as low-flush toilets and modern washing machines, convert to water efficient landscaping, or otherwise demonstrate lower water usage. Greatest effect has been achieved through tiered pricing. In this pricing structure, users are charged higher rates for each successive unit, or block, of water used. The rate structure penalizes heavy users with greatly increased rates. This technique has been shown to be highly effective in reducing overall usage. In Tucson, Arizona, an increasing tiered price structure resulted in decreased usage of 26% over a three-year period. Additionally, some communities have implemented the use of water conservation monitors and water waste hotlines to penalize those that continue to waste the resource. Many communities currently limit the type and size of landscaping, the **time** and nature of outdoor water use, and in extreme cases, have completely banned outdoor water use during crisis periods.

Throughout history, the availability of water has been a vital factor in the rise and fall of human cultures. This is largely because water is a **limiting factor** for the **carrying capacity** for human activities in any region. It is crucial that humans learn to live within the limits of available natural resources, including the supply of fresh water. Because the supply of usable water is finite, the consumption per person must be reduced in regions that are using this resource excessively.

See also Desalination; Water pollution.

Resources

Books

Buzzelli, B. *How to Get Water Smart: Products and Practices for Saving Water in the Nineties.* Santa Barbara, CA: Terra Firma Publishing, 1991.

Clarke, R. *Water: The International Crisis.* Cambridge: MIT Press, 1993.

Keller, Edward. *Environmental Geology.* Upper Saddle River, NJ: Prentice-Hall, Inc., 2000.

Morrison, Jason I., Sandra L. Postel, and Peter H. Gleick. *The Sustainable Use of Water in the Lower Colorado River Basin.* Oakland, CA: Pacific Institute for Studies in Development, Environment, and Security, 1996.

Postel, Sandra L. *Last Oasis: Facing Water Scarcity.* W.W. Norton and Co., 1997

van der Leeden, Frits, Fred L. Troise, and David K. Todd. *The Water Encyclopedia.* Chelsea: Lewis Publishers, Inc., 1990.

Vickers, Amy. *Handbook of Water Use and Conservation.* Amherst, MA: Waterplow Press, 2001.

Yudelman. M., et al. *New Vegetative Approaches to Soil and Water Conservation.* Washington, DC: World Wildlife Fund, 1990.

Periodicals

Graves, William, ed. "Water: The Power, Promise, and Turmoil of North America's Fresh Water." Special Edition *National Geographic Special Edition* (November 1993): 1–119.

Postel, Sandra L. "Plug the Leak, Save the City." *International Wildlife.* 23 (January-February 1993): 38–41.

Reisner, Marc. "Unleash the Rivers." *Time Magazine Special Edition* (April-May 2000): 66–71.

Other

California Urban Water Conservation Council. "H₂OUSE: Water Saver Home" 2002 [cited October 20, 2002]. <http://www.h2ouse.org/>.

"Greywater: What It Is, Ways To Treat It, Ways To Use It." 2000 [cited October 20, 2002]. <http://www.greywater.com/>.

International Food Policy Research Institute. "Domestic Water Supply, Hygiene, And Sanitation." October 2001 [cited October 20, 2002] <http:// www.ifpri.cgiar.org/2020/focus/focus09/focus09_03.htm.

National Wildlife Federation. "Population, Water & Wildlife: Finding a Balance." 2001 [cited October 20, 2002]. <http://www.nwf.org/nwfWebAdmin/binaryVault/PWWReport.pdf>.

United Nations Department of Economic and Social Affairs,Division for Sustainable Development. "Facts About Water." 2002 [cited October 20, 2002]. <http://www.johannesburgsummit.org/html/media_info/press releases_factsheets/wssd4_water.pdf>.

United States Environmental Protection Agency. "How to Conserve Water and Use It Effectively." June 7, 2002 [cited October 20, 2002]. <http://www.epa.gov/water/you/chap3.html>.

United States Environmental Protection Agency. "How We Use Water In These United States." June 7, 2002 [cited October 20, 2002]. <http://www.epa.gov/water/you/chap1.html>.

United States Geological Survey. "Thirsty? How 'bout a Cool, Refreshing Cup of Seawater?" June 12, 2001 [cited October 20, 2002]. <http://ga.water.usgs.gov/edu/drinkseawater.html>.

David Goings

KEY TERMS

Drip irrigation—A method of irrigation utilizing small, low-flow emitters that are located at or above the plant root zone. Designed to reduce the quantity of water lost to evaporation.

Grey water—Used wash water collected from sinks, laundry, etc. that is reused for irrigation. Grey water does not include toilet wastes.

Per capita usage—The amount used by one person in a given amount of time.

Reverse osmosis—A process for purification of water in which water is forced through a semipermeable membrane, retaining most ions while transmitting the water.

Tiered pricing—A system of pricing in which unit quantities of a commodity are priced with increasingly higher rates, such that, higher rates of usage result in rapidly increasing costs for the consumer.

Water lilies

The water lily, yellow water lily, lotus, and several other aquatic plants are about 60 **species** of aquatic herbs that make up the family Nymphaeaceae. These plants occur in shallow, fresh waterbodies from the boreal to the tropical zones. The usual habitats of these plants are ponds and shallow **water** around **lake** edges, as well as slowly-flowing pools and stagnant backwaters in streams and **rivers**.

Water lilies are perennial, herbaceous plants. Their green foliage dies back each year at the end of the growing season, but the **plant** perennates itself by issuing new growth from a long-lived **rhizome** occurring in the surface sediment. The foliage of most species in the family is comprised of simple, glossy, dark-green leaves, which float on the water surface. The solitary, perfect (or bisexual) flowers are held just above the water surface, and are large and showy, with numerous petals and a strong fragrance, making them highly attractive to pollinating **insects**. In most species, the flowers open at dawn, and close at dusk for the night.

On breezy days, the leaves and flowers of water lilies appear to dance lightly on the water surface. This superficial aesthetic led to the choosing of the family name of these plants (Nymphaeaceae) by the famous Swedish naturalist Carolus Linnaeus (1707-1778), who likened the water lilies to frolicking nymphs.

Species of water lilies

Species of water lilies are prominent in many shallow-water habitats, from the boreal zones to the tropics. The genera are described below, with particular reference to species occurring in **North America**.

There are about 40 species of water lily (*Nymphaea* spp.). The white water lilies (*Nymphaea odorata* and *Nymphaea tuberosa*) are widespread in North America, and have large, roundish leaves with a triangular cleft at one end, at the point of which the petiole attaches. The large flowers are colored white, or rarely pink. The blue water lily (*Nymphaea elegans*) occurs in the southern United States and down south into Latin America, and has bluish or pale violet flowers.

The are about 10 species of yellow water lilies or spatterdock (*Nuphar* spp.), including the widely dispersed North American species *Nuphar microphyllum*, *N. variegatum*, and *N. advena*. The floating leaves of these plants are rather oblong in shape, with a basal cleft, at the apex of which the petiole attaches. The flowers are greenish on the outside, and are bright-yellow or sometimes reddish on the inside.

The water-shield (*Brasenia schreberi*) occurs widely in ponds and shallow waters along lake and pond shores in North America, Central and northern **South America**, and Eurasia. The petiole joins the oblong, floating leaves at their center, a morphology known as peltate. The flowers are relatively small, and are colored a dull red.

There are about seven species of fanwort or water-shield, including *Cabomba caroliniana*, a widespread species of pools and quiet streams over much of North America. This species has dense, oppositely arranged, finely dissected, submersed foliage, as well as small, alternately arranged, floating leaves. The six-petalled flowers are relatively small, and are colored white or lavender.

The lotus lily or water chinquapin (*Nelumbo lutea*) occurs in scattered populations in North America. The roundish leaves float on the water surface, or are held slightly above, and the petiole is attached at the middle. The flowers are pale-yellow in **color**. The Oriental sacred lotus (*N. nucifera*) and sacred lotus (*N. nelumbo*) are native species in **Asia**, and are widely cultivated ornamentals there, and sometimes in North America and **Europe**.

The largest-leafed species in the Nymphaeaceae is the royal water lily (*Victoria amazonica*) of tropical South America, whose floating leaves can be larger than one-meter across, and can support the weight of a child.

Ecological and economic importance

Species in the water lily family are important components of the plant communities of most **freshwater** lakes, ponds, and other shallow-water habitats. They provide food for many types of herbivorous animals, and a **habitat** substrate for others, such as the long-toed **birds** known as "lily-trotters" or **jacanas** (family Jacanidae, including *Jacana spinosa* of Central and South America).

Species of water lilies provide a beautiful aesthetic to aquatic habitats, which is greatly appreciated by many people. The sacred lotuses (*Nelumbo nucifera* and *N. nelumbo*) are especially important in this regard in a number of cultures. This is particularly true in India, China, Japan, and elsewhere in Asia, where sacred lotuses are featured prominently in horticultural plantings in many gardens and parks, in paintings and other visual arts, in architectural motifs and decorations, and as symbolism in literature.

Several other species in the water lily family are of minor economic importance as horticultural plants, because of the pleasing aesthetics of their floating leaves, as well as their attractive flowers. Various species of water lilies and spatterdocks are commonly planted in gardens which have shallow ponds incorporated into their design. A water lily native to North America, *Nymphaea odorata*, is commercially available in rose-hued flowers, as well as the wild-type white color.

Another minor use of some species is in the production of food for fishes grown in tropical aquaculture. Water lilies growing in commercial **fish** ponds are eaten

KEY TERMS

Hydrophyte—A perennial plant that is adapted to growing in permanently aquatic habitats.

Perfect—In the botanical sense, this refers to flowers that are bisexual, containing both male and female reproductive parts.

as a food by certain herbivorous fish, and thereby contribute to the productivity of the agricultural **ecosystem**.

Some people eat the **seeds** of *Nymphaea*, *Nelumbo*, and *Victoria*, but this is a relatively minor use of the plants.

Resources

Books

Hochkiss, N. *Common Marsh, Underwater, and Floating-leafed Plants of the United States and Canada.* New York: Dover Publications, 1972.

Judd, Walter S., Christopher Campbell, Elizabeth A. Kellogg, Michael J. Donoghue, and Peter Stevens. *Plant Systematics: A Phylogenetic Approach.* 2nd ed. with CD-ROM. Suderland, MD: Sinauer, 2002.

Klein, R.M. *The Green World. An Introduction to Plants and People.* New York: Harper and Row, 1987.

Bill Freedman

Water microbiology

Water microbiology is concerned with the **microorganisms** that live in **water**, or can be transported from one **habitat** to another by water.

Water can support the growth of many types of microorganisms. This can be advantageous. For example, the chemical activities of certain strains of yeasts provide us with beer and bread. As well, the growth of some **bacteria** in contaminated water can help digest the poisons from the water.

However, the presence of other **disease** causing microbes in water is unhealthy and even life threatening. For example, bacteria that live in the intestinal tracts of humans and other warm blooded animals, such as *Escherichia coli*, *Salmonella*, *Shigella*, and *Vibrio*, can contaminate water if feces enters the water. **Contamination** of drinking water with a type of *Escherichia coli* known as O157:H7 can be fatal. The contamination of the municipal water supply of Walkerton, Ontario, Cana-

da in the summer of 2000 by strain O157:H7 sickened 2,000 people and killed seven people.

The intestinal tract of warm-blooded animals also contains viruses that can contaminate water and cause disease. Examples include rotavirus, enteroviruses, and coxsackievirus.

Another group of microbes of concern in water microbiology are **protozoa**. The two protozoa of the most concern are *Giardia* and *Cryptosporidium*. They live normally in the intestinal tract of animals such as beaver and **deer**. *Giardia* and *Cryptosporidium* form dormant and hardy forms called cysts during their life cycles. The cyst forms are resistant to **chlorine**, which is the most popular form of drinking water disinfection, and can pass through the filters used in many **water treatment** plants. If ingested in drinking water they can cause debilitating and prolonged diarrhea in humans, and can be life threatening to those people with impaired immune systems. *Cryptosporidium* contamination of the drinking water of Milwaukee, Wisconsin with in 1993 sickened more than 400,000 people and killed 47 people.

Many microorganisms are found naturally in fresh and **saltwater.** These include bacteria, cyanobacteria, protozoa, **algae**, and tiny animals such as rotifers. These can be important in the food chain that forms the basis of life in the water. For example, the microbes called cyanobacteria can convert the **energy** of the **sun** into the energy it needs to live. The plentiful numbers of these organisms in turn are used as food for other life. The algae that thrive in water is also an important food source for other forms of life.

A variety of microorganisms live in fresh water. The region of a water body near the shoreline (the littoral zone) is well lighted, shallow, and warmer than other regions of the water. Photosynthetic algae and bacteria that use **light** as energy thrive in this zone. Further away from the shore is the limnitic zone. Photosynthetic microbes also live here. As the water deepens, temperatures become colder and the **oxygen** concentration and light in the water decrease. Now, microbes that require oxygen do not thrive. Instead, purple and green **sulfur** bacteria, which can grow without oxygen, dominate. Finally, at the bottom of fresh waters (the benthic zone), few microbes survive. Bacteria that can survive in the absence of oxygen and sunlight, such as methane producing bacteria, thrive.

Saltwater presents a different environment to microorganisms. The higher **salt** concentration, higher **pH**, and lower **nutrients**, relative to **freshwater,** are lethal to many microorganisms. But, salt loving (halophilic) bacteria abound near the surface, and some bacteria that also live in freshwater are plentiful (i.e., *Pseudomonas* and *Vibrio*). Also, in 2001, researchers demonstrated that the ancient form of microbial life known as **archaebacteria** is one of the dominant forms of life in the **ocean**. The role of archaebacteria in the ocean food chain is not yet known, but must be of vital importance.

Another microorganism found in saltwater are a type of algae known as dinoflagellates. The rapid growth and multiplication of dinoflagellates can turn the water red. This "red tide" depletes the water of nutrients and oxygen, which can cause many **fish** to die. As well, humans can become ill by eating contaminated fish.

Water can also be an ideal means of transporting microorganisms from one place to another. For example, the water that is carried in the hulls of ships to stabilize the vessels during their ocean voyages is now known to be a means of transporting microorganisms around the globe. One of these organisms, a bacterium called *Vibrio cholerae*, causes life threatening diarrhea in humans.

Drinking water is usually treated to minimize the risk of microbial contamination. The importance of drinking water treatment has been known for centuries. For example, in pre-Christian times the storage of drinking water in jugs made of **metal** was practiced. Now, the anti-bacterial effect of some metals is known. Similarly, the boiling of drinking water, as a means of protection of water has long been known.

Chemicals such as chlorine or chlorine derivatives has been a popular means of killing bacteria such as *Escherichia coli* in water since the early decades of the twentieth century. Other bacteria-killing treatments that are increasingly becoming popular include the use of a gas called **ozone** and the disabling of the microbe's genetic material by the use of ultraviolet light. Microbes can also be physically excluded form the water by passing the water through a filter. Modern filters have holes in them that are so tiny that even particles as miniscule as viruses can be trapped.

An important aspect of water microbiology, particularly for drinking water, is the testing of the water to ensure that it is safe to drink. Water quality testing can de done in several ways. One popular test measures the turbidity of the water. Turbidity gives an indication of the amount of suspended material in the water. Typically, if material such as **soil** is present in the water then microorganisms will also be present. The presence of particles even as small as bacteria and viruses can decrease the clarity of the water. Turbidity is a quick way of indicating if water quality is deteriorating, and so if action should be taken to correct the water problem.

In many countries, water microbiology is also the subject of legislation. Regulations specify how often water sources are sampled, how the sampling is done, how the analysis will be performed, what microbes are

detected, and the acceptable limits for the target microorganisms in the water sample. Testing for microbes that cause disease (i.e., *Salmonella typhymurium* and *Vibrio cholerae*) can be expensive and, if the bacteria are present in low numbers, they may escape detection. Instead, other more numerous bacteria provide an indication of fecal **pollution** of the water. *Escherichia coli* has been used as an indicator of fecal pollution for decades. The bacterium is present in the intestinal tract in huge numbers, and is more numerous than the disease-causing bacteria and viruses. The chances of detecting *Escherichia coli* is better than detecting the actual disease causing microorganisms. *Escherichia coli* also had the advantage of not being capable of growing and reproducing in the water (except in the warm and food-laden waters of tropical countries). Thus, the presence of the bacterium in water is indicative of recent fecal pollution. Finally, *Escherichia coli* can be detected easily and inexpensively.

See also Chlorination; Oil spills; Sewage treatment; Water pollution.

Resources

Books

Chapelle, F.H. *Ground Water Microbiology and Geochemistry.* New York: John Wiley & Sons, 2000.

Madigan, M.M., J. Martinko, and J. Parker. *Brock Biology of Microorganisms.* 8th ed Upper Saddle River, NJ: Prentice-Hall, 2000.

Periodicals

Karner, M.B., E.F. DeLong, and D.M. Karl. "Archae Dominance in the Mesopelagic Zone of the Pacific Ocean." *Nature* 409 (January 2001): 507–510.

Ruiz, G.M., T.K. Rawlings, F.C. Dobbs, et al. "Global Spread of Microorganisms by Ships." *Nature* 406 (November 2000): 49.

Brian Hoyle

Water pollution

Any physical, biological, or chemical change in **water** quality that adversely affects living organisms or makes water unsuitable for desired uses can be considered **pollution**.

Often, however, a change that adversely affects one **organism** may be advantageous to another. Conversely, antibiotic designed for use at one site, might pose a pollution threat to non-target or beneficial downstream **microorganisms** and ultimately other life forms.

The United States Environmental Protection Agency (EPA) oversees National Primary Drinking Water Regulations (NPDWRs or primary standards) that are legally enforceable standards regarding water contained in public water systems. Primary standards are intended to promote and protect public health by setting limits for levels of contaminants in drinking water. Agents of pollution or contaminants are divided into categories of microorganisms, disinfectants, disinfection byproducts, organic chemicals, inorganic chemicals, and radionuclides.

Nutrients that stimulate growth of **bacteria** and other oxygen-consuming decomposers in a river or **lake**, for example, are good for the bacteria but can be lethal to game **fish** populations. Similarly, warming of waters by industrial discharges may be deadly for some **species** but may create optimal conditions for others. Whether the quality of the water has suffered depends on your perspective. There are natural sources of water **contamination**, such as arsenic springs, oil seeps, and sedimentation from **desert erosion**, but most environmental scientists restrict their focus on water pollution to factors caused by human actions and that detract from conditions and uses that humans consider desirable.

Water **pollution control** regulations usually distinguish between point and nonpoint pollution sources. Factories, power plants, **sewage treatment** facilities, underground mines and oil wells, for example, are classified as point sources because they release pollution from specific locations, such as drain pipes, ditches, or sewer outfalls. These individual, easily identifiable sources are relatively easy to monitor and regulate. Their unwanted contents can be diverted and treated before discharge. In contrast, nonpoint pollution sources are scattered or diffuse, having no specific location where they originate or discharge into water bodies. Some nonpoint sources include runoff from farm fields, feedlots, lawns, gardens, golf courses, construction sites, logging areas, roads, streets, and parking lots. Whereas point sources often are fairly uniform and predictable, nonpoint runoff often is highly irregular. The first heavy rainfall after a dry period, for example, may flush high concentrations of oil, gasoline, rubber, and trash off city streets, while subsequent runoff may have much lower levels of these contaminants. The irregular timing of these events, as well as their multiple sources, scattered location, and lack of specific ownership make them much more difficult to monitor, regulate, and treat than point sources.

Among the most important categories of water pollutants are sediment, infectious agents, toxins, **oxygen** demanding wastes, **plant** nutrients, and thermal changes. Sediment (dirt, **soil**, insoluble solids) and trash make up the largest **volume** and most visible type of water pollution in most **rivers** and lakes. Rivers have always carried silt, **sand**, and gravel down to the oceans but human-

Garbage washed up on the shore of the Hudson River.
Yoav Levy/Phototake NYC. Reproduced with permission.

caused erosion now probably rivals the effects of geologic forces. Worldwide, erosion from croplands, **forests**, grazing lands, and construction sites is estimated to add some 75 billion tons of sediment each year to rivers and lakes. This sediment smothers gravel beds in which fish lay their eggs. It fills lakes and reservoirs, obstructs shipping channels, clogs hydroelectric turbines, and makes drinking water purification more costly. The most serious water pollutant in terms of human health worldwide is pathogenic (disease-causing) organisms. Among the most deadly waterborne diseases are **cholera**, **dysentery**, polio, infectious **hepatitis**, and schistosomiasis. Together, these diseases probably cause at least two billion new cases of **disease** each year and kill somewhere between six and eight million people. The largest source of infectious agents in water is untreated or insufficiently treated human and **animal** waste. The United Nations estimates that about half the world's population has inadequate sanitation and that at least one billion people lack access to clean drinking water.

Toxins are poisonous chemicals that interfere with basic cellular **metabolism** (the **enzyme** reactions that make life possible). Among some important toxins found in water are metals (**lead**, mercury, cadmium, nickel), inorganic elements (selenium, arsenic), acids, salts, and organic chemicals such as **pesticides**, solvents, and industrial wastes. Some of these materials are so toxic that exposure to extremely low levels (perhaps even parts per billion) can be dangerous. Others, while not usually found in toxic concentrations in most water bodies, can be taken up by living organisms, altered into more toxic forms, stored, and concentrated to dangerous levels through food chains. For example, fish in lakes and rivers in many parts of the United States have accumulated mercury (released mainly by power plants, waste disposal, and industrial processes) to levels that are considered a threat to human health for those who eat fish on a regular basis.

The United States continues to work toward a goal of making all surface waters "fishable and swimmable." Investments in sewage treatment, regulation of toxic waste disposal and factory effluents, and other forms of pollution control have resulted in significant water quality increases many areas. Nearly 90% of all the river miles and lake acres that are assessed for water quality in the United States fully or partly support their designed uses. Lake Erie, for instance, which was widely described in the 1970s as being "dead," now has much cleaner water and more healthy fish populations than would ever have been thought possible 25 years ago. Unfortunately, surface waters in developing countries have not experienced similar progress in pollution control. In most developing countries, only a tiny fraction of human wastes are treated before being dumped into rivers, lakes, or the **ocean**. In consequence, water pollution levels often are appalling. In India, for example, two-thirds of all surface waters are considered dangerous to human health.

See also Waste, toxic.

Water table *see* **Groundwater**

Water treatment

Water is treated to make it safe to drink and to use for other purposes, such as to spray on agricultural plants. Water that contains domestic and industrial waste is often required to be treated to lessen or remove the contaminants prior to the discharge of the water into a river, **lake**, or **ocean**.

Some industrial processes require water that is free of impurities and **microorganisms**. One example is the water used in the manufacture of pharmaceuticals. The preparation of a medicine using contaminated water could be disastrous for the patient.

The need for treatment of drinking water is becoming more urgent, even in developed countries. The increasing populations of developing countries are encroaching more on previously undisturbed watersheds. As the **watershed** quality deteriorates, the ability of the watershed to naturally purify the water flowing through it is lessened. As well, the increasing use of chemicals is contaminating **groundwater**. Watersheds that were pristine only a few decades ago are now under threat.

The treatment of water for drinking is also referred to as water purification. Purification typically involves several steps. These are designed to remove objects from the water, particularly if the water is from a surface source like a river or a lake, and also to treat the water to minimize the risk from microorganisms.

The physical removal of objects like sticks and leaves is the first step in drinking water treatment. The water is filtered and then passed into a settling tank. As the name implies, the tank allows **sand** and grit to settle out on the bottom. Even smaller material is next removed in a step called coagulation. Here, a chemical called alum is added. The alum forms globs that attach to **bacteria**, silt, and other materials. The globs subsequently sink to the bottom of the holding tank.

Water can then be treated in several ways. It can be pumped through a filter that has much smaller holes in it than the filter designed to remove large objects. The holes or pores of the filter are so small that particles as small as viruses, bacteria, and **protozoa** cannot pass through to the other side of the filter. The **filtration** is intended to mimic the movement of water down from the surface through the **soil** and rock layers.

Filtration has become an important way to clear water of protozoa such as Cryptosporidium and Giardia. These organisms, which are typical residents of wild animals like the beaver, are resistant to the traditional chemical treatment of water. Chemical treatment utilizes **chlorine** to kill susceptible microorganisms. The process of killing the microbes is referred to as disinfection.

Chlorination disinfection has the advantage that a residual amount of the chemical remains in the water as the water passes through the pipelines on its way to the tap. This property, and the efficiency of killing by chlorine and chlorine-containing compounds, has made chlorination the most popular drinking water treatment method for over 50 years. However, the method is not without drawbacks. In particular, chlorine by-products can form in the presence of organic material. These by-products, which are known as trihalomethanes, have been linked to health problems in humans. There is concern that the long-term ingestion of trihalomethanes can be harmful to health. Increasingly, the use of chlorine dioxide, which does not form trihalomethanes, or alternatives to chlorination, either alone or as secondary treatment that permit the chlorine **concentration** to be lowered, are being used.

Other means of disinfection that are becoming increasingly popular include the use of **ozone** and ultraviolet **light**. Home-based ultraviolet systems that sterilize the water just prior to the tap are becoming popular.

Wastewater treatment includes domestic and industrial waters. Domestic water commonly includes water flushed down toilets and the "gray" water from bathing and dish washing. Industrial water is water that has been used in production processes. Such water can contain chemicals that are toxic or foul smelling. Processing of domestic and industrial wastewater is necessary to remove the noxious compounds and microorganisms, or reduce the amounts of these items to acceptable levels, before the water is discharged into another body of water. Increasingly, the treatment of wastewater is a legal requirement.

Like the treatment of drinking water, wastewater treatment is a multi-stage process. Initially, a pre-treatment step filters out or grinds up objects such as sticks, rags and bottles that would clog equipment further on in the process.

The primary treatment step allows materials to either settle to the bottom or, in the case of liquids such as grease or oil that do not mix with water, to float to the surface. The surface waste is skimmed off. The clarified water passes on to the secondary treatment.

Secondary treatment uses microorganisms to digest organic material in the water. This can be done in one of three ways. The first method is called the fixed film system. This was developed in the mid-nineteenth century. The film is a film of microorganisms that has grown on **rocks**, sand, or plastic. In the case of a film on a flat support such as a plastic sheet, as in a typical domestic septic field, the wastewater can be flowed over the microbial film. As the water slowly passes over the film, the bacteria in the film digest the impurities in the water. Alternatively, the fixed film can be positioned on an arm, which can slowly sweep through the wastewater.

A third version of secondary treatment is called the suspended film. Microorganisms are suspended in the wastewater. Over **time**, the microbes clump together and settle out as sludge. The sludge can then be removed. Some of the sludge is added back to the wastewater to keep the digestion process going. This cycle can be repeated on the same **volume** of water, in order to digest most of the impurities.

A water treatment plant in Orange County, California. *Photograph by Alan Towse. © Ecoscene/CORBIS. Reproduced by permission.*

The sludge that is collected can be subsequently used as compost, or can be digested by the bacteria, which produce methane that can be collected for use as a fuel and power source.

A forth version of secondary treatment is a lagoon. Wastewater is added to a lagoon and the sewage is degraded over the next few months. The **algae** and bacteria that are normal residents of the lagoon will use compounds such as **phosphorus** and **nitrogen** as food sources. Bacteria will produce **carbon dioxide** that is used by algae. The resulting algal activity produces **oxygen** that stimulates growth of the bacteria. This cycle of microbiological activity can continue until the organic **matter** in the water is consumed.

The final treatment step removes or neutralizes bacteria and other microorganisms. This step involves the use of a disinfectant like chlorine, or the use of filters, ozone, or ultraviolet light. **Neutralization** of the disinfectant chemical might be necessary prior to the flow of the treated water into a river, stream, lake, or other body of water. For example, chlorine can be removed by a reaction with **sulfur dioxide**.

Within the past several decades, the use of treatments that rely on the presence of living material such as plants to treat wastewater has become more popular. These systems, which are known as "living machines," can produce water that meets the requirements of purity for drinking water.

See also Hard water; Rivers; Sewage treatment.

Resources

Books

American Water Works Association. *Water Quality and Treatment.* 5th ed. Denver: American Water Works Association, 1999.

Droste, R. L. *Theory and Practice of Water and Wastewater Treatment.* New York: John Wiley & Sons, 1996.

Brian Hoyle

Waterbuck

Waterbucks belong to the large family of bovids, plant-eating hooved animals with horns and a four-chambered stomach for extracting **nutrients** from a diet of grass or foliage. These ruminants regurgitate food that is rechewed (chewing the cud). Domestic cattle, also of the bovid family, chew their cud.

Description

Waterbucks belong to a subfamily of bovids called Reduncinae, which also includes kobs, reedbucks, and lechwes. Waterbucks are the largest animals in this subfamily, with males weighing more than 500 lb (227 kg) and standing more than 4 ft (1.2 m) at the shoulders. Female waterbucks are slightly smaller. Only male waterbucks have horns, which are V-shaped, tapered, ridged, and curve backward, then slightly forward at the tips.

Two subspecies of waterbuck are recognized: the common waterbuck *Kobus ellipsiprymnus ellipsiprymaus* and the defassa waterbuck *K. e. defassa.* Some scientists consider the defassa waterbuck to be a separate **species**. One distinguishing marking between the two is their coloration. Defassa waterbucks are silvery-gray or reddish to dark brown with a streak of white on their rump. Common waterbucks have a distinctive white ring around the rump, the lower half of the legs are black, and they have white markings at the throat, an eyebrow line, and a white snout. Males tend to darken with age.

Waterbucks lack scent **glands**, but possess a greasy coat with skin glands that secrete a smelly, musk odor that can be detected as far away as 1600 ft (488 m) away. The greasy, shaggy coat serves to protect waterbucks from the damp habitats. Waterbucks do not run rapidly, moving mostly at a trot.

Habitat and population

Waterbucks have even less tolerance to dehydration than cattle. Therefore, they are usually found close to **water** in regions with ample rainfall.

Waterbucks prefer to eat the protein-rich **grasses**, herbs, and foliage found in valleys and areas where rain water drains. Waterbucks are found in the central and south-central African countries of Kenya, Tanzania, Rwanda, Uganda, Zambia, Zimbabwe, and northern parts of South **Africa**.

In areas where the different subspecies of waterbuck overlap, many **hybrid** specimens have been seen. In a national park in Kenya the population **density** of waterbucks has been observed to reach as many as 250 animals per sq km while the average is 30 per sq km. These numbers are much higher than their typical density outside the refuge where they average about four animals per sq km. In areas where there is a good water supply, waterbucks tend to live in more dense populations.

Social relationships

The importance of proximity to water plays an important role in the social **behavior** of waterbucks. Not only do they need the water for drinking, but they also use water to fend off predators, for they are good swimmers. **Competition** to be near water is reflected among male relationships. Male waterbucks are territorial, but they will tolerate several other males within their territory. These so-called satellite bulls are submissive to the dominant male and will help him defend their territory from other males.

Females have loose, come-and-go relationships, congregating in female herds with as many as 70 animals, including dependent young waterbucks. Female herds may have a home range as large as 1500 acres (608 hectares), while bachelor groups have a range of 250 acres (101 hectares). Young males form bachelor herds of five to 40 young bulls and there is evidence of a hierarchy among them. These are also loose social groups with members coming and going. Young males join these groups at the age of nine months until they are six years old, their age of maturity.

Female waterbucks conceive after the age of three, reproducing about once a year after an eight-month gestation period. The mother stays with her calf at night, but during the day wanders away from her young. The calf is nursed three times a day for the first two to four weeks and the calf is weaned at six to eight months. Mothers

KEY TERMS

. . . .

Coloration—The overall color of an animal, plus distinctive color markings on various parts of the body.

Family—A large group of animals classified together because they share common characteristics of physical structure and consequent behavior.

Herbivore—An animal that only eats plant foods.

Hybrid animals—Animals that are crosses between species or subspecies that have already been classified.

Regurgitation—The casting up of food that has only been partly digested in a stomach chamber of cud-chewing animals.

Satellite bulls—Submissive bulls that the dominant bull tolerates within his territory and that help defend it from other intruders.

with young offspring tend to remain in woodland areas as a protection from predators.

See also Antelopes and gazelles.

Resources

Books

Estes, Richard D. *Behavior Guide to African Mammals.* Berkeley: University of California, 1991.

Estes, Richard D. *The Safari Companion.* Post Mills, Vermont: Chelsea Green, 1993.

Grmzimek, Bernhard. *Encyclopedia of Mammals.* New York: McGraw-Hill, 1990.

Haltenorth, T., and H. Diller. *A Field Guide to the Mammals of Africa.* London: Collins, 1992.

MacDonald, David, and Sasha Norris, eds. *Encyclopedia of Mammals.* New York: Facts on File, 2001.

Nowak, R.M., ed. *Walker's Mammals of the World.* 5th ed. Baltimore: Johns Hopkins University Press, 1991.

Vita Richman

Watermelon *see* **Gourd family (Cucurbitaceae)**

Watershed

A watershed refers to land that is drained by an interconnected system of rivulets, streams, **rivers**, lakes, and groundwater. **Water** from a watershed eventually drains into a common destination. Both rain and snow contribute to the watershed. Coastal watersheds that begin as rivulets often end as large rivers that empty into a **lake** or an **ocean**. Watersheds can range in size from just a few square miles to many hundreds, even millions of square miles.

A catchment or drainage **basin**, which is the total area of land that drains into a water body, is usually a topographically delineated area that is drained by a stream system. River basins are large watersheds that contribute to water flow in a river. The watershed of a lake is the total land area that drains into the lake. In addition to being hydrologic units, watersheds are useful units of land for planning and managing multiple natural resources. By using the watershed as a planning unit, management activities and their effects can be determined for the land area that is directly affected by management. The hydrologic effects of land management downstream can be evaluated as well. Sometimes **land use** and management can alter the quantity and quality of water that flows to downstream communities. By considering a watershed, many of these environmental effects can be taken into consideration.

Watersheds are important as habitats for many creatures, and as a source of drinking and recreational water for many communities. As well, because one watershed can often by connected to another watershed that lies "downstream," the environmental quality of one watershed can affect other watersheds. As more communities rely on watersheds for their drinking water, the preservation of watersheds is becoming more urgent.

To function properly, a watershed needs to be maintained in a fairly undisturbed state, especially near watercourses. This undisturbed **habitat** helps to keep unwanted pollutants and excess **soil** and runoff from reaching the water course. The preservation of watershed habitats is recognized as a priority by local, regional, and national governments.

See also Hydrologic cycle; Water pollution.

Resources

Books

Grossman, E. *Watershed: The Undamming of America.* Boulder: Counterpoint Press, 2002.

Organizations

Center for Watershed Protection. 8391 Main Street, Ellicott, MD, 213043-4605. (410) 461–8323. <http//www.cwp.org/>.

United States Environmental Protection Agency, Office of Wetlands, Oceans, and Watersheds (4501T). 1200 Pennsylvania Avenue, NW, Washington, DC 20460. <http://www.epa.gov/owow/>.

Waterwheel

The waterwheel is considered the first rotor mechanism in which an outside **force** creates power to spin a shaft. The Greeks are said to have first developed the waterwheel, using it to raise **water** from **rivers**. Polls or pots were attached around the circumference of a large wheel; then oxen would walk in a circle round a vertical shaft connected through a simple gear to the horizontal shaft of the waterwheel. When possible, the current of a fast-flowing river would do the **work** of the oxen. By Roman times, ancient engineers realized that the spinning shaft could be used as a power source to turn millstones. However, few were built for this purpose due to the abundance of slaves to grind grain into flour.

After the fall of the Roman Empire, waterwheels spread throughout Medieval **Europe** largely through the efforts of monks who introduced the technology to landowners, many of whom had lost workers to **disease** and war. The first factories—textile mills—were powered by water. Other early uses of waterwheels included sawmills for sawing lumber and gristmills for grinding grain. Many large cities owe their existence to power drawn from nearby rivers, but this power source was not very dependable due to floods and droughts that changed the amount of water flowing in the river.

Several types of waterwheels evolved over the centuries. The two most used were the overshot wheel, which involved water falling on the paddles from above, and the undershot wheel, which was placed directly in the water and worked especially well in swift streams. A more unusual and less frequently used type was the reaction wheel in which water was dropped into a hollow tube from a great height. The tube had hollow spouts attached to it, and the water would spray out the spouts, causing the tube to rotate much like today's lawn sprinklers. The tidal wheel required an enclosed pond with a gate. As the tide rose, water flowed into the pond until the gate was closed, effectively trapping the water. Later, the gate was opened, allowing water to flow out and turn a wheel for power.

By the late eighteenth century, more efficient and powerful steam engines began to replace waterwheels as Europe's primary source of power. However, the principle of the waterwheel lives on in today's water turbines.

The waterwheel of an old mill at Stone Mountain Park, Georgia. *Photograph by Chris Hamilton. Stock Market. Reproduced by permission.*

Wave motion

A wave is nothing more than a disturbance that moves from place to place in some medium, carrying **energy** with it. Since the behavior of waves is so closely related to the concept of **oscillations**, that is a good place to start.

There are many examples of simple oscillations, but a very good one is that of an object attached to the end of a spring. Assume that the other end is held fixed, perhaps by a clamp. Suppose the spring hangs vertically and slowly lowers the object until it becomes stationary. The spring is now stretched enough for its upward pull to balance the weight of the object, which at that location is in equilibrium. Now disturb the object by lifting it a short distance above that point and letting go. The object then begins to oscillate vertically as it first falls, until the spring stops it and pulls it back upward to the original position, then it falls again, etc. The energy in the oscillating **motion** came from the original disturbance, which in this case moved the object to a position above the equilibrium point. At that instance, the object was at its maximum displacement whose size is the amplitude. The larger the amplitude, the greater the energy in the motion.

If we hang a duplicate object and spring side-by-side with the first one, but without the two being in contact in any way. If we disturb the first object, it oscillates just as before and the second object remains stationary at its equilibrium position. However, the situation becomes different if we connect the two objects with a rubber band and then disturb only the first object. It begins to oscillate as before, but soon the second object starts to oscillate also. The rubber band allows energy to transfer to the second object, which will move with the same **frequency** as the first oscillation. We can make the experiment more complicated if we use a large number of springs hanging in a row, with each object connected to the one before and after it with rubber bands. If only the first object is disturbed, the oscillation will pass its energy through all the springs. After the energy has been transferred to the next few springs, the first spring will become still with its object back at its equilibrium point. This demonstrates exactly how a disturbance can move as a wave through a medium. In this example, the medium is composed of the objects on springs, which act as coupled oscillators.

The way that a disturbance is transferred from one part of a medium to another in the spring model is very similar to the way that waves move in **water**, air, or a guitar string. We can think of the water, perhaps in a bathtub, as being composed of a great number of H_2O molecules lying very close side-by-side. If the water has been undisturbed for a long **time**, its surface will be calm and the molecules will be relatively stationary (of course, they are not completely stationary, but their motion is microscopic). Now suppose we disturb the water by tapping it with a finger. This produces a disturbance as many molecules are forced downward, the opposite of what happened in our spring example. We have all seen a wave move on the surface of the water away from the position of our tap, and this is simply the disturbance being transferred from one **molecule** to another. Just as before, the larger the original displacement, the greater the energy in the disturbance and the bigger the energy that the wave carries. You might ask how the transfer of energy takes place. Actually, the molecules of water all exert some force on their neighbors which are quite close. This holds them together and provides the same effect as the rubber bands.

The wave we just made is an example of a traveling wave since the disturbance moves from place to place in the medium. It can also be classified as a transverse wave because the direction of the disturbance (vertical in this case) is perpendicular to the direction that the wave travels (horizontally). There are also longitudinal waves,

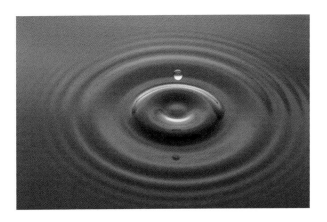

Transverse waves produced by a water droplet penetrating the surface of a body of liquid. © Martin Dohrn/Science Photo Library, National Audubon Society Collection/Photo Researchers, Inc. Reproduced with permission.

like those which carry the energy of sound in air, in which the direction of the disturbance is the same as that of the wave motion. This is easy to visualize if you have ever seen a speaker move when the volume is turned up very loud. The sudden movements of the speaker compress the nearby air and that disturbance moves in the same direction toward you and your ears.

A repeated pattern of individual waves, a wave train, often occurs. A wave train can be produced in water by tapping the surface with a specific rhythm, or frequency. A complementary characteristic of a wave train is the wavelength. Suppose a friend taps the water with a specific frequency and you take a picture of the resulting waves. In effect, this lets you "freeze" the wave train in time and examine it. You will notice that there is a constant distance between the individual waves; this is the wavelength. The frequency tells how often the wave train repeats itself in time and the inverse of the wavelength tells how often the wave train repeats itself in **space**. Multiplying the numerical values of the frequency and the wavelength gives the speed at which the waves move.

Waves have many interesting properties. They can reflect from surfaces and refract, or change their direction, when they pass from one medium into another. If these properties seem familiar, that is because we are accustomed to **light** behaving in exactly this way. Obviously light reflects, and an example of refraction is the bending of light when it passes from water into air. For this reason, when you look into water, objects appear to be at different locations than their real positions. Light is therefore considered in these instances to behave as though it was a wave produced by moving disturbances of electric and magnetic fields.

Waves can also combine, or interfere. For example, two waves can reach a particular point at just the right

KEY TERMS

Frequency—The number of cycles, or repetitions, of an oscillating motion which occur per second.

Medium—A material in which a disturbance (oscillation) at one location can transfer its energy to nearby locations, for example from one molecule of water to another.

Wave train—A repeated pattern of individual waves which are produced with a specific rhythm, or frequency.

time for both to disturb the medium in the same way (such as if two water waves both try to lift the surface at the same time). This is constructive **interference**. Likewise, destructive interference happens when the disturbances of different waves cancel. Interference can also lead to standing waves which appear to be stationary—the medium is still disturbed, but the disturbances are oscillating in place. This can only occur within confined regions, like a bathtub or a guitar string (fixed at both ends). For just the right wavelengths, traveling wave trains and their **reflections** off the boundaries can interfere to produce a wave that appears stationary.

See also Acoustics; Fluid dynamics.

Resources

Books

Clark, J.O.E. *Matter and Energy: Physics in Action.* New York: Oxford University Press, 1994.

Ehrlich, R. *Turning the World Inside Out, and 174 Other Simple Physics Demonstrations.* Princeton, NJ: Princeton University Press, 1990.

Epstein, L.C. *Thinking Physics: Practical Lessons in Critical Thinking, Second Edition.* San Francisco: Insight Press, 1994.

Gough, W., et al. *Vibrations and Waves.* 2nd ed. Englewood Cliffs, NJ: Prentice Hall, 1995.

James J. Carroll

Waxbills

Waxbills are 107 **species** of finch-like **birds** that make up the family Estrilidae. Species of waxbills occur in the tropics of **Africa**, South and Southeast **Asia**, New Guinea, **Australia**, and many islands of the South Pacific. Their usual habitats are **grasslands**, marshes, savannas, forest edges, and disturbed **forests**. Waxbills are sedentary, nonmigratory birds.

Waxbills are small birds, ranging in body length from 3.2-5.9 in (8-15 cm). The bill is short, stout, conical-shaped, and pointed, and adapted to eating **seeds**. The plumage of male birds is generally brightly colored, with contrasting patterns that can include hues of red, blue, purple, green, yellow, black, or white. Female birds have a much more inconspicuous coloration.

Waxbills forage on the ground for seeds, **fruits**, and small **invertebrates**. Species that occur in open, seasonally dry habitats are gregarious during the drier times of the year, while species occurring in more closed moist habitats defend their territories throughout the year.

All waxbills are territorial during their breeding season, which for many species typically begins immediately after the seasonal rains begin. Their songs are weak hisses, buzzes, and chatterings. Waxbills construct bulky, domed nests of grassy fibers, often lined with feathers on the inside. Waxbills lay four to 10 eggs, which are incubated by both sexes, who also share duties in caring for the young.

Several species of waxbills have been domesticated, and are kept in captivity as cagebirds. These include the zebra finch (*Taeniopygia guttata*) of Australia, the striated finch or white-rumped munia (*Lonchura striata*) of South and Southeast Asia, and the Java finch (*Padda oryzivora*), native to Java and Bali in Indonesia, but widely introduced elsewhere in Indochina and Southeast Asia.

The Gouldian or rainbow finch (*Peophila gouldiae*) is a particularly gaudy bird of northern Australia, which has a yellow bill, black face, purple breast, yellow belly, and green back. The white-crowned mannikin (*Lonchura nevermanni*) is a white-headed, two-toned brown-bodied species that lives within grassy savannas in southern New Guinea. The red-headed parrot finch (*Erythrura cyaneovirens*) has a red head, blue breast, green belly and back, and burgundy tail. The orange-eyed pytilia or red-faced waxbill (*Pytilia afra*) is a red-faced, gray-bodied species of East Africa.

Some species of waxbills are regarded as serious **pests** of agriculture. For example, the Java finch is a pest in parts of Southeast Asia, because of the quantities of ripened **rice** that this flock consumes.

See also Finches.

Bill Freedman

Waxwings

Waxwings are medium-sized, fruit-eating, perching **birds** found in northern Eurasia and **North America** that

The bohemian waxwing (*B. garrulus*) breeds in coniferous and mixedwood **forests** and muskegs of northwestern North America, as well as in northern Eurasia, where it is known simply as the waxwing. However, during the non-breeding season the bohemian waxwing aggregates into flocks, which forage widely for berries far to the south of the breeding range, and as far east as the Atlantic coast.

The Phainopepla

Phainopepla nitens is a red-eyed, glossy-black bird that occurs in the southwestern United States and central Mexico. The *Phainopepla* is found in semi-arid scrub, and is rather more insectivorous than the waxwings. The *Phainopepla* also travels in small flocks during the non-breeding season.

The waxwings are rather irregular in both their breeding and wintering abundances. Waxwings sometimes occur in unusually large numbers outside of their usual range in some winters, probably as a result of poor berry **crops** in their normal wintering **habitat**. During such irruptive events of abundance, waxwings often occur in large flocks in cities, avidly feeding on berry-laden, urban trees and shrubs such as junipers and mountain-ash. These winter occurrences of flocks of bohemian and cedar waxwings are unpredictable pleasures that are relished by bird watchers.

Weak acids and bases *see* **Acids and bases**

Weak electrolyte *see* **Electrolyte**

Cedar waxwing (*Bombycilla cedrorum*). *Photograph by Robert J. Huffman. Field Mark Publications. Reproduced by permission.*

are included in the family Bombycillidae. Waxwings have a crest on the top of their head, and have soft, sleek, often shiny plumage. The secondary feathers often have a soft, wax-like appendage at the tip, from which the common name of these birds was derived.

Waxwings are largely fruit-eating birds, although they also eat **insects**, especially during the breeding season. These birds usually catch their insect **prey** by fly-catching, which involves aerial sallies from an observation **perch** to catch insects on the wing.

The waxwing family includes eight **species** worldwide. Three species in this family are familiar to North Americans.

The cedar waxwing (*Bombycilla cedrorum*) breeds in a wide range of habitats throughout the range of northern coniferous and broad **leaf** forests of Canada and the northern United States. The cedar waxwing winters in the southern United States and Central America.

Weasels

Weasels, ermines, and stoats are various **species** of small carnivores in the family Mustelidae, which also includes the **otters**, **badgers**, martens, minks, **skunks**, and **wolverine**. Species of weasels occur in **North America**, northern **South America**, northern **Africa**, **Europe**, **Asia**, and Southeast Asia.

Weasels have a long, lithe, almost serpentine body, and short legs. This body shape is highly adaptive for pursuing their **prey** of small **mammals** through small holes and along narrow passages. Weasels are very active, inquisitive animals, and they can run remarkably quickly and nimbly when they are chasing their prey.

Weasels have a very soft and dense fur, and although these animals are quite small, they are commonly trapped as furbearers. Northern species of weasels develop a thick, white coat in the wintertime, and these espe-

A long-tailed weasel (*Mustela frenata*). *Photograph by Judd Cooney. Phototake NYC. Reproduced by permission.*

cially valuable furs are known as ermine. Ermine is mostly used as a fine trim for coats and robes, or to make neck-pieces and stoles.

All of the weasels are terrestrial animals, occurring in a wide range of habitats, including **tundra**, various types of **forests**, and **grasslands**. Weasels are voracious carnivores, and they are capable of subduing animals substantially larger than themselves. They will often climb trees to hunt **squirrels**.

When they are faced with a super-abundance of food, weasels sometimes go on a lustful killing spree. A weasel in a chicken-house, for example, will kill a much larger number of **birds** than it could ever hope to eat. However, in more natural circumstances, these animals will attempt to cache their excess food. Weasels that learn how to kill chickens can be quite a problem, but they can be selectively killed. There is no need to kill weasels indiscriminately as perceived **pests**, especially considering the large numbers of small **rodents** that they kill, thereby providing a valuable service to farmers.

Weasels are generally solitary animals. However, the young stay with their mother until they learn to hunt for themselves. Young weasels engage in rough-and-tumble play, important in **learning** some of the physical

skills that are necessary in hunting. Weasels are usually most active at night, although they often hunt during the day as well.

Species of weasels

The ermine or stoat (*Mustela erminea*) is a circumboreal species, occurring widely in conifer-dominated boreal forests and tundras in northern North America, Europe, and Asia. The ermine is a ferocious **predator**. Although only some are 11.8 in (30 cm) long and weigh no more than 3.5 oz (100 g) (this is the weight of an adult male; females are about half as heavy), this **carnivore** can hunt and subdue animals as large as rabbits and hares weighing several kilograms. The winter coat of northern populations of this weasel is a well-camouflaged white, except for the black-tipped tail, while the summer pelage is tawny brown above, and yellow-white beneath.

The ermine has delayed implantation. This is characterized by mating in the summer, but the fertilized embryos remaining dormant in the uterus after they are fertilized, and not implanting and developing as embryos until three to four weeks prior to **birth**, which occurs the following spring or early summer. In this species the total post-fertilization gestation period is 200-340 days.

KEY TERMS

. . . .

Carnivore—A flesh-eating animal.

Circumboreal—A biogeographic distribution that includes the conifer-dominated boreal forests of both North America and Eurasia.

The long-tailed weasel (*Mustela frenata*) occurs from the southern half of North America to northern South America. Like the ermine, the long-tailed weasel has a white coat in winter, and delayed implantation.

The least, dwarf, or pygmy weasel (*Mustela rixosa*) of North America is the smallest of any of the predatory mammals (that is, order Carnivora). This species only attains a length (body plus tail) of 4.9 in (12.5 cm), and a weight of 1.47 oz (42 g). Because of its diminutive size, this species feeds mostly on **mice**, **voles**, and other small mammals. Also because of the small size of the least weasel, it must feed voraciously in order to maintain its weight and body **temperature**. This is because very small animals have a large **ratio** of body surface to **mass**, and they therefore lose **heat** quickly. In fact, the least weasel must eat an amount of small mammals equal to more than one-half of its own body weight each day. Like many other species of weasel, the dwarf weasel has a white coat in winter, and a brown coat in summer.

The common or Old World weasel (*Mustela nivalis*) is closely related to the least weasel, and these are sometimes considered to be geographic variants of the same species. The Old World weasel is a diminutive species that occurs in forests of Europe, northern and central Asia, and northern Africa. This species does not display delayed implantation. The Old World weasel has been introduced to New Zealand in a misguided attempt to control introduced **rats** and mice, but it has caused great damages through its depredations on native species of birds.

The alpine weasel (*M. altaica*) occurs in montane forests and alpine tundra of **mountains** in Asia. The yellow-bellied weasel (*M. kathiah*) and Siberian weasel (*M. sibirica*) are additional Asian species. The Java weasel (*M. lutreolina*) and bare-footed weasel (*M. nudipes*) are tropical-forest species of Southeast Asia.

Other species in the genus *Mustela* are relatively large in comparison with the true weasels. These include the rare and endangered black-footed ferret (*M. nigripes*) of the shortgrass prairies of western North America, the polecat or ferret (*M. putorius*) of northern Africa and Eurasia, the **mink** (*M. vison*) of North America, the Eurasian mink (*M. lutreola*) of Europe and Asia, and a tropical weasel (*M. africana*) of northern South America.

Resources

Books

Grzimek, B. (ed.). *Grzimek's Encyclopedia of Mammals.* London: McGraw Hill, 1990.

King, C. *The Natural History of Weasels and Stoats.* London: Academic Press, 1989.

Nowak, R.M. (ed.). *Walker's Mammals of the World.* 5th ed. Baltimore: Johns Hopkins University Press, 1991.

Wilson, D.E., and D. Reeder. *Mammal Species of the World.* 2nd ed. Washington, DC: Smithsonian Institution Press, 1993.

Bill Freedman

Weather

Weather can be defined as the condition of the atmosphere at any given time and place. Weather conditions are determined by six major factors: air **temperature**, air **pressure**, **humidity** of the air, amount and kind of cloud cover, amount and kind of **precipitation**, and speed and direction of the **wind**. Weather condition patterns for any one region or for the whole **planet** can be charted on a weather map containing information about all six of these factors. This information often can be used to produce a weather forecast, a prediction of weather conditions at some future time for some given region.

The study of weather is known as **meteorology**. No exact date can be given for the beginnings of this science, since humans have studied weather conditions for many centuries. Indeed, the word meteorology itself goes back to *Meterologica*, a book written by the Greek natural philosopher Aristotle in about 340 B.C. Many scholars date the rise of modern meteorology to the work of a Norwegian father and son team, Vilhelm and Jakob Bjerknes. The Bjerknes's were the first to develop the concept of masses of air moving across the earth's surface, affecting weather conditions as they moved. They also created the first widespread system for measuring weather conditions throughout their native Norway.

The six factors determining weather conditions result from the interaction of four basic physical elements: the **Sun**, the earth's atmosphere, the **earth** itself, and natural landforms on the earth's surface.

Solar energy

The driving force behind all meteorological changes taking place on the earth is solar **energy**. Each minute, the outer portions of the earth's atmosphere receive an average of 2 calories/sq cm. This value is known as the solar constant. Although the solar constant changes over very long periods of time, it does not vary enough to affect the general nature of the earth's weather over short periods of time.

The solar energy reaching the outer atmosphere may experience a variety of fates. Thirty **percent** of all solar energy is lost to **space** by means of scattering and by reflection off **clouds** and the earth's surface. Another 19% is absorbed by gases in the atmosphere and by clouds. About a quarter of it (25%) reaches the earth's surface directly; another quarter (26%) eventually reaches the surface after being scattered by gases in the atmosphere.

An important factor in determining the fate of solar **radiation** is its wavelength. Shorter wavelengths tend to be absorbed by gases in the atmosphere (especially **oxygen** and **ozone**) while radiation of longer wavelengths tends to be transmitted to the earth's surface.

Solar radiation that reaches the earth's surface is absorbed to varying degrees, depending on the kind of material on which it falls. Since darker colors and rougher surfaces absorb radiation better than lighter colors and smoother surfaces, **soil** tends to absorb more solar radiation than **water**.

Solar energy that reaches the earth's surface is re-radiated back to the atmosphere as **heat**, also referred to as infrared radiation. Infrared radiation consists of much longer wavelengths. This re-radiated energy is likely to be absorbed by certain gases in the atmosphere such as **carbon dioxide** and nitrous oxide. This absorption process, the **greenhouse effect**, is responsible for maintaining the planet's annual average temperature.

Humidity, clouds, and precipitation

The absorption of solar energy by the earth's surface and its atmosphere is directly responsible for most of the major factors making up weather patterns. For example, when the water in oceans, lakes, **rivers**, streams, and other bodies of water is warmed, it tends to evaporate and move upward into the atmosphere. The amount of moisture found in the air at any one time and place is called the humidity. Humans are very sensitive to this characteristic of weather.

Water that has evaporated from the earth's surface (or escaped from plants through the process of **transpiration**) rises to an altitude in the atmosphere at which the air around it is cold enough to cause condensation. When moisture condenses into tiny water droplets or tiny **ice** crystals, clouds are formed.

Clouds are an important factor in the development of weather patterns. They tend to reflect sunlight back into space. Thus, an accumulation of cloud cover may contribute to a decrease in heat retained in the atmosphere.

Clouds are also the breeding grounds for various types of precipitation. As water droplets or ice crystals collide with each other, they coalesce, and form larger particles. Eventually, the particles become large enough and heavy enough to overcome upward drafts in the air and fall to the earth as precipitation. The form of precipitation that occurs (rain, snow, sleet, hail, etc.) depends on the atmospheric conditions through which the water or ice falls.

Atmospheric pressure and winds

Solar energy also is directly responsible for the development of differing atmospheric pressures at various locations on the planet's surface, and the winds that result from these differences. Since the earth's surface is different in **color** and texture from place to place, some locations will be heated more intensely by solar radiation than others. Warm places usually heat the air above them, setting **convection** currents into movement that carry masses of air upward into the upper atmosphere. Those same convection currents then carry other masses of air downward from the upper atmosphere toward the earth's surface.

In regions where warm air moves upward, the **atmospheric pressure** tends to be low; downward air movements are associated with higher atmospheric pressures. These higher or lower atmospheric pressures can be measured by a **barometer**. Barometers provide information not only about current pressures but about possible future weather patterns as well.

The existence of areas with different atmospheric pressures accounts for the movement of air, which we call wind. Wind is simply the movement of air from a region of high pressure to one of lower pressure.

Terrestrial characteristics

If the Sun provides the energy by which weather patterns can develop, certain features of the earth itself determine the precise forms in which those patterns may be exhibited. One example has already been provided above. Earth's surface is highly variable, ranging from oceans to deserts to cultivated land to urbanized areas. The way solar energy is absorbed and reflected from each of these regions is different, accounting for variations in local weather patterns.

Other characteristics of the planet account for more significant variations in weather patterns. These characteristics include such features as the tilt of Earth on its axis in **relation** to its **plane** of revolution, and the variations in Earth's distance from the Sun.

The fact that Earth's axis is tilted at an **angle** of 23 1/2° to the plane of its **orbit** means that the planet is heated unevenly by the Sun. During the summer, sunlight reaching the Northern Hemisphere strikes more nearly at right angles than it does in the Southern Hemisphere. In the winter, the situation is reversed.

The elliptical shape of the earth's orbit around the Sun also affects weather conditions. At certain times of the year the planet is closer to the Sun than at others. This variation means that the amount of solar energy reaching the outer atmosphere will vary from month to month depending on Earth's location in its path around the Sun.

Even **Earth's rotation** on its own axis influences weather patterns. If Earth did not rotate, air movements on the planet would probably be relatively simple: air heated along the equator would rise into the upper atmosphere, travel northward toward the poles, be cooled, and then return to the earth's surface at the poles.

Earth's rotation causes the deflection of these theoretically simple air movements. Instead of a single overall equator-to-poles air movement, global winds are broken up into smaller cells. In one cell warm air rises above the equator, moves northward in upper altitudes, is cooled, and returns to the earth in the regions around 30° north and south latitude. A second cell consists of air that moves upward in the regions around 60° north and south latitude, across the upper atmosphere, and then downward at about 30° north and south latitude. The final cell contains winds traveling upward at 60° north and south latitude and then downward again at the poles.

Topographic factors

Irregularities on Earth's surface also affect weather. A mountain range can dramatically affect the movement of approaching air masses. Suppose that a **mass** of warm moisture-filled air is forced to ascend one side of a mountain range. As the air is pushed upward it cools off and moisture begins to condense out, first in the form of clouds then as precipitation. This side of the mountain range will experience high rates of precipitation.

As the air mass continues over the top of the mountain range, it does so without its moisture. The winds that sweep down the far side of the range will tend to be warm and dry. The term orographic is used to describe changes in weather patterns like these induced by mountain ranges.

Weather and climate

The terms weather and climate often are used in conjunction with each other, but they refer to quite different phenomena. Weather involves atmospheric conditions that currently prevail or that exist over a relatively short period of time. Climate refers to the average weath-

KEY TERMS
. . . .

Humidity—The amount of water vapor contained in the air.

Meteorology—The study of the earth's atmosphere and the changes that take place within it.

Orographic—A term referring to effects produced when air moves across a mountain range.

Solar constant—The rate at which solar energy strikes the outermost layer of the earth's atmosphere.

Solar energy—Any form of electromagnetic radiation that is emitted by the Sun.

Topography—The detailed surface features of an area.

er pattern for a region (or for the whole planet) over a much longer period of time (at least three decades according to some authorities).

Changes in weather patterns are easily observed. It may rain today and be clear tomorrow. Changes in climate patterns are much more difficult to detect. If the summer of 1997 is unusually hot, there is no way of knowing if that fact is part of a general trend towards warmer weather or a single variation that will not appear again for some time.

See also Atmospheric temperature; Seasons; Weather forecasting; Weather mapping.

Resources

Books

Ahrens, C. David, Rachel Alvelais, and Nina Horne. *Essentials of Meteorology: An Invitation to the Atmosphere.* Belmont, CA: Brooks/Cole, 2000.

Bramwell, Martyn. *Weather.* New York: Franklin Watts, 1994.

Danielson, Eric W., James Levin, and Elliot Abrams. *Meteorology.* 2nd ed. with CD-ROM. Columbus: McGraw-Hill Science/Engineering/Math, 2002.

Lutgens, Frederick K., Edward J. Tarbuck, and Dennis Tasa. *The Atmosphere: An Intorduction to Meteorology.* 8th ed. New York: Prentice-Hall, 2000.

Lynott, Robert E. *How Weather Works and Why.* Gadfly Press, 1994.

Watts, Alan. *The Weather Handbook.* Dobbs Ferry: Sheridan House, 1994.

Williams, Jack. *The Weather Book.* New York: Vintage Books, 1997.

Periodicals

"Temperature And Rainfall Tables: July 2002." *Journal of Meteorology* 27, no. 273 (2002): 362.

"Weather Extremes: July 2002." *Journal Of Meteorology* 27 no. 273 (2002): 361.

David E. Newton

Weather forecasting

Weather forecasting is the attempt by meteorologists to predict the state of the atmosphere at some future time and the **weather** conditions that may be expected. Weather forecasting is the single most important practical reason for the existence of **meteorology** as a science. It is obvious that knowing the future of the weather can be important for individuals and organizations. Accurate weather forecasts can tell a farmer when the best time to plant is; an airport control tower what information to send to planes that are landing and taking off; and residents of a coastal region when a hurricane might strike.

Humans have been looking for ways to forecast the weather for centuries. The Greek natural philosopher Theophrastus wrote a book, *Book of Signs*, in about 300 B.C. listing more than 200 ways of knowing when to expect rain, **wind**, fair conditions, and other kinds of weather.

Scientifically based weather forecasting was not possible until meteorologists were able to collect data about current weather conditions from a relatively widespread system of observing stations and organize that data in a timely fashion. By the 1930s these conditions had been met. Vilhelm and Jacob Bjerknes developed a weather station network in the 1920s that allowed for the collection of regional weather data. The weather data collected by the network could be transmitted nearly instantaneously by use of the **telegraph**, invented in the 1830s by Samuel F. B. Morse. The age of scientific forecasting, also referred to as synoptic forecasting, was under way.

The National Weather Service

In the United States weather forecasting is the responsibility of the National Weather Service (NWS), a division of the National Oceanic and Atmospheric Administration (NOAA) of the Department of Commerce. NWS maintains more than 400 field offices and observatories in all 50 states and overseas. The future modernized structure of the NWS will include 116 weather forecast offices (WFO) and 13 river forecast centers, all collocated with WFOs. WFOs also collect data from ships at sea all over the world and from meteorological satellites circling **Earth**. Each year the Service collects nearly four million pieces of information about atmospheric conditions from these sources.

The information collected by WFOs is used in the weather forecasting work of NWS. The data is processed by nine National Centers for Environmental Prediction (NCEP). Each center has a specific weather-related responsibility: seven of the centers focus on weather prediction—the Aviation Weather Center, the Climate Prediction Center, the Hydrometeorological Prediction Center, the Marine Prediction Center, the Space Environment Center, the Storm Prediction Center, and the Tropical Prediction Center—while the other two centers develop and run complex computer models of the atmosphere and provide support to the other centers—the Environmental Prediction Center and NCEP Central Operations. Severe weather systems such as thunderstorms and hurricanes are monitored at the National Storm Prediction Center in Norman, Oklahoma. Hurricane watches and warnings are issued by the Tropical Prediction Center in Miami, Florida, (serving the Atlantic, Caribbean, Gulf of Mexico, and eastern Pacific Ocean) and by the Forecast Office in and Honolulu, Hawaii, (serving the central Pacific). WFOs, other government agencies, and private meteorological services rely on NCEP's information, and many of the weather forecasts in the **paper**, and on **radio** and **television**, originate at NCEP.

Global weather data are collected at more than 1,000 observation points around the world and then sent to central stations maintained by the World Meteorological Organization, a division of the United Nations. Global data also is sent to NWS's NCEPs for analysis and publication.

Types of weather forecasts

The less one knows about the way the atmosphere works the simpler weather forecasting appears to be. For example if **clouds** appear in the sky and a light rain begins to fall one might predict that rain will continue throughout the day. This type of weather forecast is known as a persistent forecast. A persistent forecast assumes the weather over a particular geographic area simply will continue into the future.

The validity of persistent forecasting lasts for a few hours, but not much longer because weather conditions result from a complex interaction of many factors that still are not well understood and that may change very rapidly.

A somewhat more reliable approach to weather forecasting is known as the steady-state or trend method. This method is based on the knowledge that weather conditions are strongly influenced by the movement of air masses which often can be charted quite accurately. A weather map might show that a cold front is moving across the great plains of the United States from west to east with an average speed of 10 mph (16 kph). It might be reasonable to predict that the front would reach a place 100 mi (1,609 km) to the east in a matter of 10 hours. Since characteristic types of weather often are associated with cold fronts it then might be reasonable to predict the weather at locations east of the front with some degree of confidence.

A similar approach to forecasting is called the analogue method because it uses analogies between existing weather maps and similar maps from the past. For example suppose a weather map for December 10, 1996, is found to be almost identical with a weather map for January 8, 1993. Since the weather for the earlier date is already known it might be reasonable to predict similar weather patterns for the later date.

Yet another form of weather forecasting makes use of statistical probability. In some locations on Earth's surface one can safely predict the weather because a consistent pattern has already been established. In parts of Peru it rains no more than a few inches per century. A weather forecaster in this region might feel confident that he or she could predict clear skies for tomorrow with a 99.9% chance of being correct.

Long-range forecasting

The complexity of atmosphere conditions is reflected in the fact that none of the forecasting methods outlined above is dependable for more than a few days at best. This reality does not prevent meteorologists from attempting to make long-term forecasts. These forecasts might predict the weather a few weeks, a few months, or even a year in advance. One of the best known (although not necessarily the most accurate) of long-term forecasts is found in the annual edition of the *Farmer's Almanac*.

The basis for long-range forecasting is a statistical analysis of weather conditions over an area in the past. For example a forecaster might determine that the average snow fall in December in Grand Rapids, Michigan, over the past 30 years had been 15.8 in (40.1 cm). A reasonable way to try estimating next year's snowfall in Grand Rapids would be to assume that it might be close to 15.8 inches (40.1 cm).

Today this kind of statistical data is augmented by studies of global conditions such as winds in the upper atmosphere and ocean temperatures. If a forecaster knows that the **jet stream** over Canada has been diverted southward from its normal flow for a period of months, that change might alter **precipitation** patterns over Grand Rapids over the next few months.

Numerical weather prediction

The term numerical weather prediction is something of a misnomer since all forms of forecasting make use of numerical data like **temperature**, **atmospheric pressure**, and **humidity**. More precisely numerical weather prediction refers to forecasts that are obtained by using complex mathematical calculations carried out with high-speed computers.

Numerical weather prediction is based on mathematical models of the atmosphere. A mathematical model is a system of equations that attempt to describe the properties of the atmosphere and changes that may take place within it. These equations can be written because the gases which comprise the atmosphere obey the same physical and chemical laws that gases on Earth's surface follow. For example, Charles' law says that when a gas is heated it tends to expand. This law applies to gases in the atmosphere as it does to gases in a laboratory.

The technical problem that meteorologists face is that atmospheric gases are influenced by many different physical and chemical factors at the same time. A gas that expands according to Charles' law may also be decomposing because of chemical forces acting on it. How can anyone make use of all the different chemical and physical laws operating in the atmosphere to come up with a forecast of future atmospheric conditions?

The role of computers in weather forecasting

The answer is that no human can solve such a problem. The **mathematics** involved are too complex. The task is not too much for computer, however. Computers can perform a series of calculations in a few hours that would take a meteorologist his or her whole lifetime to finish.

In numerical weather predicting meteorologists select a group of equations that describe the conditions of the atmosphere as completely as possible for any one location at any one time. This set of equations can never be complete because even a computer is limited as to the number of calculations it can complete in a reasonable time. Thus, meteorologists pick out the factors they think are most important in influencing the development of atmospheric conditions. These equations are fed into the computer. After a certain period of time, the computer will print out the changes that might be expected if atmospheric gases behave according to the scientific laws to which they are subject. From this printout a meteorologist can make a forecast of the weather in an area in the future.

The accuracy of numerical weather predictions depend primarily on two factors. First, the more data that is available to a computer the more accurate its results. Second, the faster the speed of the computer the more calculations it can perform and the more accurate its report will be. In the period from 1955 (when computers were first used in weather forecasting) to the current

KEY TERMS

Analogue method of forecasting—A prediction of future weather conditions based on the assumption that current conditions will produce weather patterns similar to those observed in the past.

Cold front—The leading edge of an advancing mass of cold air.

Hurricane warning—A notice issued when a hurricane has been observed either visually or on a radar screen.

Hurricane watch—A notice to the general public that a hurricane may be expected within a particular area.

Long-term forecast—A prediction of weather conditions over a matter of weeks, months, or a year.

Mathematical model—A system of equations that attempts to describe the properties of the atmosphere and changes that may take place within it.

Numerical forecast—A prediction of future weather patterns obtained by using high speed computers to carry out complex mathematical calculations derived from mathematical models of the atmosphere.

Percent skill—The likelihood that a weather forecast will be better than a pure chance prediction.

Persistent forecast—A prediction of weather conditions based on the assumption that the weather over a particular geographic area will remain constant over the near future.

Short-term forecast—A prediction for weather conditions over a matter of hours or days.

Statistical probability forecast—A prediction of future weather conditions based on an analysis of the likelihood of various conditions having occurred in the past.

Steady-state forecast, Trend method—A prediction of weather conditions based on the movement of air masses over a given geographical area at about the same direction and approximately the same speed as they have been moving.

Synopic forecasting—Scientifically based forecasts derived from the rapid collection and analysis of weather data from as extensive an area as possible.

time, the **percent** skill of forecasts has improved from about 30% to more than 60%. The percent skill measure was invented to describe the likelihood that a weather forecast will be better than pure chance.

Accuracy of weather forecasts

Weather forecasters have long been the subject of jokes, probably as much today as they were in Theophrastus's time. One reason for this is that there is no standard measure of a "correct" weather forecast. Suppose that a forecaster predicts heavy rain for your area tomorrow. Does a rainfall of 1 in (2.5 cm) prove that prediction correct? Or a rainfall of 1.5 in (1 cm)? Or a rainfall of 5 in (13 cm)?

Forecast accuracy also is difficult to judge since the average person's expectations probably have increased as the percent skill of forecasts also has increased. A hundred years ago, few people would have expected to have much idea as to what the weather would be like 24 hours in the future. Today, a good next-day forecast often is possible.

In general it is probably safe to say that the shorter the time period and the more limited the geographic area involved, the more accurate a forecast is likely to

be. For periods of less than a day, a forecast covering an area of 100 sq mi (259 sq km) is likely to be quite dependable. Predictions about weather patterns six months from now for the state of California are likely to be much less reliable.

See also Air masses and fronts; Atmosphere observation; Atmospheric circulation; Atmospheric temperature; Global climate; Weather mapping.

Resources

Books

Danielson, Eric W., James Levin, and Elliot Abrams. *Meteorology.* 2nd ed. with CD-ROM. Columbus: McGraw-Hill Science/Engineering/Math, 2002.

Hodgson, Michael, and Devin Wick. *Basic Essentials: Weather Forecasting.* 2nd ed. Guilford, CT: Globe Pequot Press, 1999.

Lutgens, Frederick K., and Edward J. Tarbuck. *The Atmosphere: An Introduction to Meteorology.* 8th ed. New York: Prentice-Hall, 2000.

Lynott, Robert E. *How Weather Works and Why.* Gadfly Press, 1994.

Periodicals

"Boundary-Layer Meteorology." *Boundary-Layer Meteorology* 105, no. 3-3 (2002): 515-520.

Lee, Thomas. "Eleventh AMS Conference on Satellite Meteo-rology and Oceanography." *Bulletin of the American Meteorological Society* 83, no. 11 (2002): 1645-1648.

Spellman, Greg. "Experiences Teaching Meteorology To Adults." *Journal Of Meteorology* 27 no. 268 (2002): 133-137.

Other

The National Weather Service [cited 2003]. <http://www.nws.noaa.gov>.

David E. Newton

Weather mapping

Weather mapping is the process of representing existing **weather** patterns and their future development and movement on a map. The process is only possible if two conditions are met. First, current weather conditions must be available at a number of relatively widely distributed stations. Second, reports of those conditions must be transmitted to a central collecting station within a short time period.

History

Weather maps first came in use in the last third of the nineteenth century when the invention of the **telegraph** made the transmission of weather data from far-flung observing points possible. In 1870 the newly created National Weather Service produced weather maps for limited regions of the United States.

In Norway, Vilhelm and Jakob Bjerknes developed one of the most successful early weather mapping systems. The Bjerknes convinced the Norwegian government to set up weather stations in strategically important areas throughout their country. Data from these stations was telegraphed to Bergen, where it was assembled and used to produce some of the most complete weather maps available at the time.

Data collection and transmission

Today, data on weather conditions around the world are collected by more than 10,000 individuals stations, hundreds of ships at sea, and a variety of instruments traveling through Earth's atmosphere. These data are transmitted regularly four times a day: usually at 0:00 hours, 6:00 hours, 12:00 hours, and 18:00 hours (Greenwich Mean Time). The data are used by national weather services to develop weather maps and forecasts for their own regions. Overseeing these individuals forecasts is the World Meteorological Organization (WMO), an international organization consisting of more than 130 nations. WMO is responsible for ensuring that all stations follow standard collection and transmission procedures and for the exchange of weather information among member nations.

The data sent from an observation station to the central collecting point are transmitted by a standard code consisting of number blocks. Meteorologists understand the significance of each block of numbers and translate them into specific weather conditions such as barometric **pressure**, **temperature**, and **wind** speed.

Constructing the weather map

Many different kinds of weather maps exist. Synoptic maps show current weather conditions, while prognostic maps show weather predictions for some time in the future. Some weather maps are complex and contain a great deal of detailed information. Others are simpler and provide only general patterns and trends. Maps can be sub-divided into those that summarize weather close to Earth's surface (surface charts) and those that describe weather at upper altitudes.

One very detailed type of map makes use of the station model. The reporting station is indicated with a small circle and the data received from that station is arranged around the circle in a predetermined pattern. Among the kinds of data plotted in the station model are cloud cover, wind direction, visibility in miles, present weather, barometric pressure, current air temperature, cloud types, **dew point**, and **precipitation**.

Most of these variables can be represented by numbers. Visibility might be indicated as 1/2 for 0.5 mi (0.8 km) and current temperature as 22 for 22°C (71°F). Other data are represented by standard symbols. Wind direction and speed are indicated by a one-edged arrow. The number of feathers on the arrow indicate the wind speed and the orientation of the arrow indicates the wind direction. A single full feather represents a wind speed of 8-12 knots; a double feather, a speed of 18-22 knots; and so on. A variety of symbols represent current weather conditions, such as * for intermittent snow fall and +, for intermittent drizzle.

The daily weather map

The weather map that appears in daily newspapers can be used to predict with some degree of **accuracy** conditions in the next few days. It usually does not include as much station information as the more detailed maps described above. Instead the major features of the daily weather map include **isobars** and high and low pressure areas.

Isobars connect locations with the same barometric pressure. The pressure described by each isobar is often indicated at one end, the other, or both, in inches, mil-

KEY TERMS

Barometric pressure—Air pressure; the force exerted by a column of air at any given point.

Cloud cover—The portion of the sky that is covered by clouds at any given time and place.

Dew point—The temperature to which air must be cooled for it to become saturated.

Isobar—A line on a weather map connecting points of equal atmospheric pressure.

Surface chart—A map that shows weather conditions at and just above Earth's surface.

Visibility—The distance to which an observer can see at any given location.

libars, or other unit. Isobars often enclose regions of high or low pressure indicated on the map as H or L.

The outer edge of a concentric series of isobars marks a front. The nature of the front is indicated by means of solid triangles, solid half-circles, or a combination of the two. An isobar with solid triangles attached represents a cold front; one with solid half-circles, a warm front; one with triangles and half-circles on opposite sides, a stationary front; and one with triangles and half-circles on the same side, an occluded front. The daily weather map may also include simplified symbols that indicate weather at a station as a T enclosed in a circle for thunderstorms, F enclosed in a circle for **fog**, and Z in a circle for freezing rain.

See also Air masses and fronts; Atmosphere observation; Atmospheric circulation; Atmospheric pressure; Atmospheric temperature; Clouds; Weather forecasting.

Resources

Books

Danielson, Eric W., James Levin, and Elliot Abrams. *Meteorology*. 2nd ed. with CD-ROM. Columbus: McGraw-Hill Science/Engineering/Math, 2002.

Lutgens, Frederick K., Edward J. Tarbuck, and Dennis Tasa. *The Atmosphere: An Intorduction to Meteorology*. 8th ed. New York: Prentice-Hall, 2000.

McNeill, Robert C. *Understanding the Weather*. Las Vegas: Arbor Publishers, 1991.

Moran, Joseph M., and Michael D. Morgan. *Meteorology: The Atmosphere and the Science of Weather*. New York: Macmillan, 1994.

Periodicals

"Boundary-Layer Meteorology." *Boundary-Layer Meteorology* 105, no. 3-3 (2002): 515-520.

David E. Newton

Weather modification

The term weather modification refers to any deliberate effort on the part of humans to influence **weather** patterns for some desirable purpose. Probably the most familiar example of weather modification is the seeding of **clouds**, most often done in order to increase the amount of **precipitation** during periods of **drought**.

The earliest scientific programs on weather modification date to the 1940s when Vincent J. Schaefer carried out experiments on cloud seeding. A half century later, the science of weather modification is still in its infancy with many questions surrounding the most effective way of bringing about the changes desired in any particular setting. The major types of weather modification that are currently in use or under study include cloud seeding, frost prevention, **fog** and cloud dispersal, hurricane modification, hail suppression, and lighting suppression.

Cloud seeding

A cloud is a large **mass** of **water** droplets and **ice** crystals. Precipitation normally occurs in a cloud only when ice crystals grow large enough to fall to **Earth** as rain, snow, hail, or some other form of precipitation. When conditions do not favor the growth of ice crystals, moisture remains suspended in the clouds, and precipitation does not occur.

The general goal of cloud seeding is to find some way of converting the supercooled droplets of liquid water in a cloud to ice crystals. Supercooled water is water that remains in a liquid state even below its freezing point. The two substances most commonly used to transform water droplets to ice crystals are dry ice (solid **carbon dioxide**) and silver iodide.

The ability of dry ice to trigger the condensation of supercooled water droplets was discovered accidentally in 1946 by Schaefer. He had planned to use a block of dry ice to cool a container of moist air, but discovered that the dry ice actually initiated the formation of ice crystals in the container. Shortly after Schaefer's research, the ability of silver iodide to produce similar results was also discovered.

Methods of cloud seeding

Any technique of cloud seeding depends on the release of millions of tiny particles of dry ice or silver iodide into a cloud. One way of accomplishing that goal is to ignite solid silver iodide in burners on the ground. The smoke thus formed consists of many tiny particles of the compound which are then carried upward into a cloud.

A more efficient way of seeding a cloud is to drop the seeding agent from an airplane onto the top of the cloud. If silver iodide is used, it can be released from flares attached to the wing tips of the **aircraft**. If dry ice is used, it is first pulverized into a fine powder and then sprayed onto the cloud.

Effectiveness of cloud seeding

A number of large experiments have been conducted to determine the effectiveness of cloud seeding as a way of increasing precipitation. The Atmospheric Water Resources Program of the Bureau of Reclamation, for example, has supported about a dozen research projects on cloud seeding. In one of these projects, the Colorado River Basin Project, cloud seeding was thought to have produced an increase of about 30% in the amount of snow falling in Colorado's San Juan Mountains. Experts estimated the value of this additional water for farmers and other consumers at about $100 million.

Other agencies at federal, state, and county level have also supported cloud research projects. A group of counties in Kansas, for example, annually join together to support cloud seeding experiments during periods of unusually low precipitation.

A great deal has been learned in the last 50 years about the conditions under which cloud seeding is most likely to be effective. Scientists have discovered, for example, that the optimal **rate** for seeding with dry ice is 0.17 oz (5 g) for each 0.62 mi (1 km) of cloud surface and 0.89 oz (25 g) per 0.62 mi (1 km) for silver iodide. Still, precise evaluations of the effectiveness of various forms of cloud seeding are often difficult because of the inherent uncertainty about most weather patterns such as cloud formation and dissipation and precipitation rates.

Fog and cloud dispersal

The techniques of cloud seeding can also be used for a second purpose, the removal of clouds and fog. This goal is desirable, as an example, in regions around an airport where prolonged fog can bring air travel to a halt, at great economic cost. The use of dry ice as a seeding agent can cause water droplets in fog to condense on ice crystals, after which they precipitate out of the air. In the process, fog banks and clouds may disappear.

Some dramatic results have been achieved with commercial fog dispersion at airports. In some cases, areas of a few square kilometers have been cleared in a matter of hours by seeding with dry ice.

Hail and lighting suppression

The federal government has been interested to a greater or less extent in programs of hail and lighting suppression for more than two decades. Hail suppression is of interest because of the devastating effects that this form of precipitation can have on **crops**, while lighting suppression is of importance because of the many forest fires it causes.

The principle underlying most hail suppression research is that hail can be prevented if the atmosphere is flooded with nuclei on which moisture can condense and freeze. The more nuclei present, the argument goes, the less likely large pieces of ice (hail) are to form. Results of research on hail suppression thus far have not been especially encouraging.

Research on lighting suppression, while of considerable economic value, has so far not received a great deal of attention. One suggestion has been to seed thunderheads with very small **aluminum** fibers in an attempt to dissipate electrical charges in a cloud. Early research in the late 1960s and early 1970s produced some promising results, but relatively little work is now being done in the field.

Hurricane modification

The control of hurricanes would be another area with very significant economic and human benefit. Some researchers have suggested that cloud seeding techniques might be a way of dissipating the **energy** stored in a hurricane. Proposals have been made for the seeding of both the hurricane center (its "eye") and the high **velocity** winds that surround it.

Project Stormfury is an on-again, off-again project of the federal government to test theories of weather modification. The project seeded Hurricane Ester in 1961, Hurricane Beulah in 1963, and Hurricane Debbie in 1969, among others. At its most effective, the seeding process appears to have reduced hurricane winds by as much as 30%. Hurricane modification research continues in the early 1990s, but at a relatively modest level.

Frost prevention

The most promising field of weather modification may well be in the prevention of frost. Frost is such a devastating event for grape growers, citrus farmers, and other fields of agriculture that extensive efforts have been made to develop fog prevention systems. The two general principles that underlie most of these systems have been to increase the **temperature** of air near the ground, where frost forms, and to reduce the amount of **heat** lost at night, when frost formation usually occurs.

Some of the specific techniques used to accomplish these goals include heaters (to warm air), **wind** machines (to insure mixing of air), sprinkling systems (to provide water which will release heat when it freezes), and smudge pots (to release heat).

KEY TERMS
· ·

Cloud—A large mass of condensed moisture consisting of water droplets and/or ice crystals.

Cloud seeding—The introduction of particles of (usually) dry ice or silver iodide with the hope of increasing precipitation from the cloud.

Dry ice—Solid carbon dioxide.

Fog—A cloud whose base lies at or near the earth's surface.

Hail—A form of precipitation consisting of relatively large masses of ice.

Precipitation—Any form of solid or liquid water that reaches the ground from the atmosphere.

Supercooled—Water than exists in a liquid state at temperatures below 32°F (0°C).

Wartime applications

Some defense experts have suggested the use of weather modification techniques as a military weapon. The claim has been made, for example, that the United States used cloud seeding during the Vietnam War. The hope was that increased rainfall would make the movement of personnel and material along the Ho Chi Minh trail more difficult.

Social and ethical issues

The use of weather modification techniques is often surrounded by controversy. An increase of precipitation over an area might be of benefit to some individuals in the area, but a disadvantage to others. For example, suppose that the owner of a private ski resort wants to have clouds seeded in order to increase snowfall over his or her property. If that effort is successful, the ski area benefits, economically. But other individuals and businesses in the area might suffer from this change in the weather. As an example, the county or state might have to pay more to keep roads and highways clear of the additional snow.

See also Tropical cyclone.

Resources

Books

Danielson, Eric W., James Levin, and Elliot Abrams. *Meteorology*. 2nd ed. with CD-ROM. Columbus: McGraw-Hill Science/Engineering/Math, 2002.

Fleagle, R.G., et al. *Weather Modification in the Public Interest*. Seattle: University of Washington Press, 1974.

Lutgens, Frederick K., Edward J. Tarbuck, and Dennis Tasa. *The Atmosphere: An Intorduction to Meteorology*. 8th ed. New York: Prentice-Hall, 2000.

Newton, David E. *Science and Social Issues*. Portland: J. Weston Walch, Publishers, 1992.

David E. Newton

Weathering

Weathering is the process by which **rocks** and **minerals** are broken down into simpler materials by means of physical (mechanical), chemical, and biological processes. Weathering is an extremely important phenomenon for the human **species** since it is the mechanism by which one of the planet's most important natural resources—soil—is formed.

The exact way in which weathering occurs in any particular situation depends primarily on two factors: the type of rock and the environmental conditions to which the rock is exposed. For example, rocky formations along a seacoast are likely to be exposed to the mechanical action of waves and **tides**. But rocks buried underground are more likely to be attacked by **chemical reactions** made possible by **water** that runs through them.

Physical (mechanical) weathering

During physical weathering, a large piece of rock is broken down into smaller and smaller pieces. This process can come about as the result of a number of natural processes. For example, the **force** of gravity may cause a large boulder to break loose from the top of a mountain and fall. When the boulder hits solid ground, it may break apart into many smaller pieces.

Ground movements can also result in physical weathering. As overlying rocks and soils are removed by natural or human-caused forces, underlying rocks may work their way to Earth's surface. As **pressure** on these rocks is relieved, they may begin to expand outward, often forming an flaky appearance known as exfoliation.

Abrasion can also cause physical weathering. Imagine a **wind storm** blowing across a broad expanse of sandy **desert**. Tiny particles of **sand** are carried along by the wind, a current of air that acts like sandpaper on rocks that stand in its pathway. The wind-sandpaper scours off pieces of grit and sand from these rocks, contributing to their physical weathering.

Temperature and moisture

In many locations, changes in **temperature** and moisture content of the environment cause significant physical weathering. When rock is warmed, it expands; when it cools, it contracts. In some regions, rocks are

Bryce Canyon National Park, Utah, where weathering caused the exposed rock to erode into a system of canyons. *Photograph by Richard Frear. National Park Service.*

heated to relatively high temperatures during the day and then cooled to much lower temperatures during the night. The constant expansion and contraction of the rocks may result in pieces being broken off.

This effect is likely to be more pronounced if water is present. Suppose that water fills the cracks in a rock during the day. At night, if the temperature drops far enough, that water will freeze and form veins of **ice** in the rock. But water expands as it freezes. Therefore, the veins of ice are likely to break apart pieces of the rock, a process that is repeated day after day and night after night when temperatures follow the pattern described above.

Chemical weathering

Chemical weathering is the process by which changes take place in the very chemical structure of rocks themselves. Chemical weathering represents a second stage of rock disintegration in which small pieces of rock produced by physical weathering are then further broken apart by chemical processes.

Three chemical reactions in particular are effective in bringing about the weathering of a rock: acid reactions, **hydrolysis**, and oxidation. Acids form readily in the **soil**. One of the most common such reactions occurs when **carbon dioxide** in the air reacts with water to form a weak acid, carbonic acid. Carbonic acid has the ability to attack many kinds of rocks, changing them into other forms. For example, when carbonic acid reacts with limestone, it produces calcium bicarbonate, which is partially soluble in water. Caves are formed when underground water containing carbonic acid travels through blocks of limestone, dissolves out the limestone, and leaves empty pockets (caves) behind.

Acids produced by human activities can also produce chemical weathering. For example, the conversion of metallic ores to the pure metals often results in the formation of **sulfur dioxide**. When sulfur dioxide combines with water, it forms the weak acid sulfurous acid and, eventually, the stronger acid **sulfuric acid**. Both of these acids are capable of attacking certain kinds of rocks in much the way that carbonic acid does.

Hydrolysis is a chemical reaction by which a compound reacts with water to form one or more new substances. A number of rock-forming minerals readily undergo hydrolysis, especially in acidic conditions. For ex-

Abrasion—The mechanism by which one material rubs against another material, sometimes producing weathering in the process.

Chemical reaction—Any change that takes place in which one substance is changed into one or more new substances.

Exfoliation—The process by which skinlike layers form on the outer surface of a rock and, in some cases, eventually peel off.

Oxidation—The chemical reaction by which a substance reacts with oxygen.

ample, the common mineral feldspar will undergo hydrolysis to produce a clay-type mineral known as kaolinite and silicic acid. Both of these new compounds are much more soluble in water than is feldspar. Hydrolysis of the mineral results, therefore, in the degradation of any rocks in which it may occur.

Oxidation occurs when the metallic part of a mineral reacts with **oxygen** in the air (or from some other source) to produce a new substance that is different in structure or more soluble than the original mineral. The spectacular red, orange, and yellow **color** of certain natural rock formations—such as those in Utah's Bryce Canyon—are an indication that an oxide of **iron** has been produced during the chemical weathering of the rock formations.

Biological weathering

The presence of living organisms can also cause weathering. Imagine that a seed falls into a small crevice in a rock and begins to germinate. As the **plant** continues to grow and send down roots, it will work its way into the rock and eventually make the crevice grow in size. Eventually the plant's roots may actually tear the rock apart.

Rates of weathering

The **rate** at which rocks disintegrate depends both on the type of rock involved and the external forces to which the rock is exposed. As an example, sandstone tends to **weather** rather easily, while granite is quite resistant to weathering. The presence of moisture, high temperatures, large temperature variations, and air movement also tend to increase the rate at which weathering takes place. Human activities can also affect the rate of weathering. For example, large quantities of gaseous oxides are produced by electrical power generating plants.

When these oxides react with water vapor in the air, they form "acid rain." When **acid rain** falls to the earth's surface, it may attack rocky materials in much the same way that natural acids like carbonic acid do.

Resources

Books

Hamblin, W.K., and E.H. Christiansen. *Earth's Dynamic Systems.* 9th ed. Upper Saddle River: Prentice Hall, 2001.
Skinner, Brian J., and Stephen C. Porter. *The Dynamic Earth: An Introduction to Physical Geology.* 4th ed. John Wiley & Sons, 2000.

David E. Newton

Weaver finches

Weaver finches are a relatively large family of 156 **species** of perching **birds**, comprising the family Ploceidae. Weaver finches are native to **Africa**, Madagascar, Eurasia, and Malaysia. This group is richest in species in Africa. However, some species have been widely introduced outside of their natural range.

Species of weaver finches occur in a wide range of terrestrial habitats, including semi-deserts, **grasslands**, savannas, and various types of **forests**. Weaver finches do not migrate, although during times of **drought** and food shortage, they may undertake wanderings covering hundreds of kilometers.

Weaver finches are relatively small birds, most with a body length of 3.9-9.8 in (10-25 cm)—not including the very long tail of some African species. Weaver finches are rather stout-bodied, and they have a short, pointed, conical, seed-eating bill. The coloration and patterns of their plumage are highly variable among species, and are quite attractive in some cases.

Weaver finches mostly eat **seeds**, but they also eat other small **fruits**, succulent foliage, and **insects**.

Most species of weaver finches are gregarious, occurring in flocks during the nonbreeding season, and often nesting in colonial groups. Their calls are harsh and repeated chirps, buzzes, and chattering, and not very musical.

Nesting and breeding systems are extremely variable among the weaver finches. Nesting strategies range from large, woven colonial nests to individually woven nests. Some species are polygynous, in which a male breeds with as many females as possible, and helps little with the incubation of eggs or care of the young birds. One species, the cuckoo weaver (*Anomalospiza imberbis*), is a parasitic breeder, laying its eggs in the nests of other

A yellow and black village weaver. *Photograph by Nigel Dennis. The National Audubon Society Collection/Photo Researchers, Inc. Reproduced by permission.*

Polygyny—A breeding system in which a male attempts to breed with as many females as possible. In avian polygyny, the female usually incubates the eggs and raises the babies.

Sexual selection—This is a type of natural selection in which anatomical or behavioral traits may be favored because they confer some advantage in courtship or another aspect of breeding. For example, the bright coloration, long tail, and elaborate displays of male pheasants have resulted from sexual selection by females, who apparently favor extreme expressions of these traits in their mates.

species, which then raise the parasitic baby. In most species, however, a one-family domed nest is constructed, the clutch size is two to eight, the female or both sexes incubate the eggs, and both parents care for the young.

Species of weaver finches

The typical weaver finches are in the genus *Passer*, including two species that commonly nest in cities and around farms in many regions—the house sparrow (*Passer domesticus*) and the Eurasian tree sparrow (*P. montanum*). Both of these species have been introduced far beyond their original, natural ranges. This includes **North America**, where the house sparrow in particular is a common bird in cities. In fact, the house sparrow is now one of the world's most widely distributed species of land birds.

The village weaver (*Ploceus cucullatus*) of African savannas is a colonial nester, with many pairs of birds building individual, pendulous nests from the same tree. The social weaver (*Philetarus socius*) is also a colonial nester. However, this species builds an aggregate, apartment-like nest, comprised of large numbers of arboreal hanging nests constructed immediately adjacent to each

other, so that the finished **mass** looks rather like a haystack. The entrance holes to the individual compartments are on the underside of the mass, so that entry requires a brief hover.

The snow finch (*Montifringilla nivalis*) is a species existing in alpine **tundra** in **Europe** and **Asia**.

One of the most spectacular weaver finches is the paradise widowbird or whydah (*Steganura paradisea*), which breeds in savannas of tropical Africa. Males of this polygynous species achieve a length of 13.4 in (34 cm), of which about 3/4 is due to their fabulously long tail, comprised of three or four over-developed, black feathers. This attractive bird also has a black face and back, a chestnut breast, a yellow nape, and a whitish belly. Female paradise widowbirds are relatively drab, brownish-and-whitish birds.

Field studies have shown that paradise widowbird males with the longest tail are more successful in attracting females. In part, this was demonstrated by clipping the tail of some individuals. These truncated fellows were then considerably less fortunate in their love life than males who had not been tampered with, or had their tail cut, and then glued back on. This is an example of sexual **selection**, in which traits that may be detrimental in some respects, for example, in foraging or escaping from predators, may nevertheless be selected for because they enhance reproductive success. In this case, the extraordinarily long tail of male paradise weaverbirds is favored by sexual selection, because a long tail has an irresistible appeal to females.

Some other weaver finches also have long tails, for example, the queen whydah (*Vidua regia*), and the pin-tailed weaverbird (*Vidua macroura*), both of tropical Africa.

Conflicts with humans

Some species of weaver finches occur in large numbers in urban and agricultural areas, where for various reasons they may be regarded as **pests**. The house sparrow and the Eurasian tree sparrow are most important in this respect.

The world's most important avian pest is probably the quelea (*Quelea quelea*) of Africa, which eats large quantities of ripe grains in places where it is abundant. This bird roosts communally in huge numbers, where it is sometimes sprayed with an organophosphate pesticide. It has been estimated that as many as one billion of these weaver finches are killed in this way each year.

See also Finches.

Resources

Books

Forshaw, Joseph. *Encyclopedia of Birds.* New York: Academic Press, 1998.
Perrins, C.M., ed. *The Birds of the Western Palaearctic. Vol. VIII. Crows to Finches.* Oxford, England: Oxford University Press, 1994.

Bill Freedman

Weevils

Weevils (Curculionidae) comprise a very large group of **insects** that are closely related to **beetles** (order Coleoptera); more than 40,000 **species** are recognized worldwide, ranging in size from 0.2-2 in (0.5-5 cm). A weevil is easily distinguished from a beetle by its extended head, which forms a rostrum, and long, segmented antennae that are clubbed at the end and are usually bent in an elbow fashion near the base. Most species have a dull colouration—commonly brown, gray, or buff—but others are more brilliantly decorated with tinges of reddish brown, pale green, and blue. The body of most weevils is covered with a **mass** of tiny scales that give these insects an iridescent **color**. Adults usually have two pairs of wings, although several species are wingless and the elytra (wing cases) are fused together. Males are often smaller than females.

All weevils are herbivorous and feed off a wide range of plants. The small mouth is located at the tip of the rostrum. The chewing mouth parts, or mandibles, move in a horizontal manner in most species. In some, the rostrum is used as a boring tool, primarily to create a tiny hole for egg laying, or to reach the inner tissues of a **plant** stem. Adult weevils lay their eggs either directly on or inserted within their food plants. The larvae do not possess legs, and most feed within roots, stems, or **seeds** of a wide range of plants. Here they receive some degree of protection from predators such as **birds**, small **mammals**, and parasitic insects. When the larvae are fully grown they develop a cocoon and pupate; some root-living species pupate in the **soil**, but the majority appear to remain within the plant itself.

In view of their **life history**, weevils are of considerable economic importance, particularly in **relation** to agriculture. No part of a plant, from the roots to the seeds, is safe from the attacks of one or more species of weevil. Although it is mainly the larval stages that cause the most damage, the adults too can be quite destructive. Some species, such as the grain weevil (*Sitophilus granarius*), are serious **pests** of stored grains. Others, such as the cotton boll weevils (*Anthonomus barbirostris*), are responsible for widescale destruction of stored **cotton** in the United States, into which it was introduced in the 1890s from Latin America.

Welding

Welding is a group of processes used to join nonmetallic and metallic materials, by applying **heat**, **pressure**, or a combination of both. Most welding procedures require heat, although some procedures require only extreme pressure (cold welding). The welding process chosen to join materials together depends upon the mechanical, physical, and chemical properties of the materials to be joined, and the use for which the product is intended. The welding processes most commonly used today are resistance welding, gas welding, and arc welding. Special welding processes used include electrogas, electroslag, plasma arc, submerged arc welding, underwater, **electron** beam, **laser** beam, ultrasonic, **friction** welding, thermit, brazing, and soldering.

In industry, welding usually refers to joining metals, although materials such as **plastics** or **ceramics** are welded. Thermoplastics, such as polyvinylchloride, polyethylene, polypropylene, and acrylics can be welded. Like **metal**, plastics are welded with localized heat. New welding processes have been developed as new metals, alloys, plastics, and ceramics have been created. Welding is a means of construction, and a method for maintenance and repair. Various welding processes are used in numerous industries, such as **aircraft**, automotive, **mining**, nuclear, railroad, shipping, building construction, tool-making, and farm equipment. Welding by robots is one of the more common—and spectacular—applications of robots in industry.

When welding, wearing protective clothing is necessary to avoid injury from sparks, metal fragments, flames, and ultraviolet and infrared rays. Different welding processes require specific clothing. Clothing should be flame resistant, hair and skin should be covered, and special goggles must be worn. Sometimes leather clothing and helmets are recommended, as are **steel** toed boots. The work area should be properly ventilated; some welding procedures are required to be done in specially vented areas or booths to avoid toxic fumes.

Tests have been devised to inspect welds for flaws and defects. There are two types of testing, nondestructive and destructive. Often, a visual inspection is all that is needed but to test for internal or extremely small defects, other methods are necessary. Some nondestructive methods include air pressure leak tests, and ultrasonic, x ray, magnetic particle, and liquid penetrant inspections. Nondestructive tests do not damage the weld. Destructive tests are used to test the physical properties of the weld. Usually a test piece is removed from the weld, or a sample weld is made and then tested, completely destroying the weld. Some examples of destructive tests are tensile, hardness, bend, impact, pressure, and fillet testing.

Evolution of welding

The oldest type of welding is forge welding, a process that dates to 2,000 B.C. Forge welding is a pressure-heat procedure used by blacksmiths and artisans to form metal into specific shapes, and to join metals such as **copper** and bronze together. Toward the beginning of the twentieth century, several new welding techniques were developed. The discovery of acetylene gas in 1836 by Edmund Davy led to oxyacetylene welding. Resistance welding was invented in 1877 by the British-born American electrical engineer Elihu Thomson (1853-1937). The **electric arc**, discovered by Sir Humphry Davy, was first used for welding by Auguste de Meritens in 1881. In the United States, C. L. Coffin received a patent for a bare metal electrode arc welding process in 1892. As the arc metal welding process was developed and improved, welding replaced riveted joints as a method of joining pieces of metal. In 1918, the first all-welded ship was launched, and in 1920 the first all welded building was constructed. Electric arc processes were used extensively during the post-World War I period. During World War II, inert gas welding was developed, and the gas shielded welding process was developed in 1948. Today, there are around 40 welding processes in use. Some newer welding processes include electron beam welding, laser beam welding, and solid state procedures such as friction and ultrasonic welding.

Welding methods

A weld is defined as a blend or coalescence of two or metals (or nonmetals) by heating them until they reach a critical **temperature** and flow together. Upon cooling, the metal becomes hard. The piece of metal to be welded is called the base metal, workpiece, or work. The edges of the base metal are often specially prepared for welding by, for example, machining, shearing, or gouging. There are five basic weld joints: butt, lap, corner, "T," and edge. The American Welding Society has developed a system of symbols that are added to mechanical drawings, to convey precisely how a welding site should be prepared, what type of weld should be made, and any other considerations.

Fusion welding, a heat process that sometimes requires the use of a filler metal, uses either **electricity** (arc welding) or gas (gas welding) as its source of heat **energy**. Solid state processes, such as friction welding and ultrasonic welding, weld metals at a temperature below their melting points, without the addition of a filler metal. Pressure is always used to achieve a weld with this method. When most metals are heated, a reaction takes place between the base metal and the surrounding atmosphere. For example, some metals oxidize when melted, which can interfere with the quality of the weld. Other common atmospheric contaminants are **nitrogen** and **hydrogen**. To control this problem, fluxes and inert gases are used to rid the welding area of impurities, and to protect the area from the atmospheric gases by displacing the surrounding air from the weld site. Welding is done mechanically or manually with welding guns or torches, and can also be performed by robots.

Arc welding

The electric arc used in welding processes is created between a covered or bare metal electrode and the base metal or workpiece. With shielded metal arc welding (SMAC), an **electric circuit** is set up between the welding machine (AC or DC continuous power source), the workpiece, the electrical cables, the electrode holder, electrode, and ground wire. To strike an arc, the electrode must be touching the base metal; this is usually done by scratching or pecking the base metal with the electrode. As the electricity begins flowing, the electrode is held away from base metal, creating a gap. The electrical current flows across the gap, resulting in an arc. The intense heat from the arc melts the workpiece and the electrode, which contains metal powder that, when melted, becomes the filler metal. The covering or coating on the metal electrode is a flux material that melts, which removes impurities from the weld and sometimes creates a gas that shields the area from at-

KEY TERMS

Acetylene—Colorless fuel gas. When burned with oxygen, acetylene produces one of the highest flame temperatures.

Arc—A stream of bright light or sparks formed as a strong electric current jumps from one conductor to another. In welding, an arc is formed when an electrode connected to a power supply touches the base metal.

Base metal—Metal to be welded or joined. It is also called workpiece or work.

Brazing—Process of welding in which two base metals are joined with a filler metal heated to above 800°F (427°C), but below the melting point of the base metals. The piecework is grooved, the brazing rod is melted, and the molten filler metal flows into the grooves via "capillary action."

Chemical properties—The way in which a material reacts in a given environment. Some examples are oxidation resistance and corrosion resistance.

Electrode—Terminal point to which electricity is brought to produce the arc for welding. Some electrodes are melted and become part of the weld.

Electron beam welding—Process in which a fo-cused beam of electrons heats and fuses the material being welded.

Filler metal—Metal or alloy added to the base metal to make welded, brazed, or soldered joints.

Flux—A material to facilitate melting and the removal of unwanted contaminants.

Laser beam welding—Welding process that uses the energy of a laser beam to fuse materials.

Mechanical properties—The way a material reacts under loads or forces, such as hardness, brittleness, ductility, and toughness.

Physical properties—The characteristics used to describe or identify a metal, such as color, melting temperature, or density.

Resistance—The property of a material to oppose the passage of an electric current. In welding, metal "resists" the electrical current and heats up.

Shielding gas—A gas that is used to guard the weld from surrounding air contamination.

Soldering—A group of welding processes that join materials by heating a filler metal or solder to around 800°F (427°C), which is below the melting point of the base metal.

---

mospheric **contamination**. Essentially, the electrode and its shielding control the mechanical, chemical, and electrical characteristics of the weld. The heat—6,000–9,000°F (3,351–4,982°C)—of the electric arc brings the base metal and the consumable electrode to molten state, within a matter of seconds.

Gas metal arc welding (GMAW or MIG) is similar to SMAW. A direct current (DC) is always used with this process and there is a gas supply apparatus. A consumable electrode is housed within a nozzle that supplies an inert shielding gas such as helium or argon. GMAW has several advantages over SMAW. With GMAW, welding speed is faster, no slag is produced, there is deeper penetration, and the electrode wires are continuously fed so that longer welds can be made. A type of arc welding that does not use a consumable electrode is gas tungsten arc welding (GTAW or TIG). An arc is produced between the base metal and a tungsten electrode, a shielding gas is used, and there must be a **water** supply to cool off the torch. Instead of a consumable electrode, a metal rod or welding rod is used to provide filler metal, if required. This type of welding is also called heliarc welding.

Gas welding

Gas welding, also called oxyfuel gas welding (OFW), refers to a group of welding processes that use gas as the source of heat energy. Oxyacetylene and oxy-hydrogen are two types of fuel gases used. Acetylene is commonly used for welding because, when combined with **oxygen**, the flame temperature can reach 5,600°F (3,093°C), the highest temperature produced by any fuel gas-oxygen combination. A filler metal rod may or may not be used with this type of welding process. The fuel gas and oxygen are contained in separate pressurized tanks or cylinders. Specially designed hoses run from the gas cylinders and connect to the welding torch. The welding torch has valves that control the amount of incoming gases, and a mixing chamber where the gases are mixed. The blended gases flow to the tip of the torch where the flame is ignited by a torch lighter, usually a flint and steel sparklighter. After the flame is lit, it must be adjusted until the correct balance of gases is achieved (a neutral flame). Oxyfuel gas welding was the primary welding process during the first part of the twentieth century. As newer methods of welding and new materials were developed, other, more suitable, welding processes replaced the oxyfuel gas process. Currently, the oxyacetylene process is used for braze welding, brazing, and soldering.

Resistance welding

Resistance, or spot, welding (RW) is a process where two or more layers or pieces of metal, stacked together, are welded together by a combination of pressure and heat. An electrical current, along with an appropriate amount of pressure, is applied to the area or spot of the desired weld. When the electricity flows through the metal, it heats up, due to the metal's resistance to the flow of electricity. When two or more metals are touching, the heat flows through from one piece to the next. The greatest amount of heat is generated at the spot where the two metals are touching. As the temperature reaches the critical point where the metals melt, a weld is created. Pressure is applied through air pressure, hydraulic pressure, or mechanical leverage. Electrodes used for spot welding are not consumable, and can be manufactured into specific shapes. Some electrodes are shaped like wheels for seam welding.

See also Metallurgy; Solder and soldering iron.

Resources

Books

Althouse, Andrew D., et al. *Modern Welding.* South Holland, IL: Goodheart-Willcox, 1988.

Bowditch, W., and K. Bowditch. *Welding Technology Fundamentals.* South Holland, IL: Goodheart-Willcox, 1992.

Jeffus, L., and H. Johnson. *Welding, Principles and Applications.* 2nd ed. Albany, NY: Delmar, 1988.

Christine Miner Minderovic

West Nile virus

West Nile virus is a member of the family of viruses that is called Flaviviridae. The **virus** is similar to other members of this viral family, is passed to humans from **birds** by the bite of a mosquito, and is capable of causing **disease** in humans. For example, another mosquito borne flavivirus causes St. Louis **encephalitis** in humans.

West Nile virus is **endemic** in **Africa**. Periodic outbreaks have occurred in **Europe** for decades. The virus has become more prominent in Europe and **North America** in the past decade. The virus was first discovered in North America in September 1999, during an investigation of an encephalitis outbreak in New York City.

Encephalitis is a swelling of the **brain**. The malady occurs in domestic animals such as **horses**, dogs, **cats**, wild animals, and wild birds. The virus can be transferred from an infected bird to humans by the bite of a mosquito. When transferred to a human, the viral **infection** produces encephalitis and **inflammation** of nerve cells of the spinal cord (**meningitis**).

In 1999, 62 cases of the disease were reported in New York City. Seven people died. The following spring, 21 more cases occurred, and two of the people died. It is thought that infected mosquitos that survived the winter were responsible for the renewed outbreak. In 1999 and 2000, the West Nile virus was confined to the northeastern region of the United States. However, since then, the virus has spread to most of the continental United States (42 of the 48 states) and Canada. In the summer of 2001, dead birds infected with the virus were found as far north as Toronto, Canada, as far south as the northern portion of Florida and Louisiana, and as far west as Milwaukee, Wisconsin. An encephalitis outbreak that occurred in Louisiana garnered a great deal of media attention and led to fears of a mass outbreak of encephalitis. This has not occurred. As of November 2002, there are 3,559 confirmed cases of encephalitis and 211 deaths in the United States. Most of these cases occurred during 2002.

Infected migratory birds may have aided the rapid spread of the virus. As well, the presence of the virus or infected **mosquitoes** in transported equipment, luggage, and on people is thought to be a factor in the spread.

The symptoms of infection begin three to 15 days following the bite of an infected mosquito. Most people experience only mild symptoms that mimic the flu (i.e., fever, headache, body aches). Others will also develop a mild rash or have swollen lymph **glands**.

In about 3–15% of those who are infected with the virus, the infection is more serious. This is particularly the case for the elderly or those whose immune systems are not functioning properly. These are the people who are at risk for developing encephalitis or meningitis. Symptoms appear suddenly, and include a severe headache, high fever, stiff neck, vomiting, confusion, and loss of consciousness. Even after recovery, a person may have muscle weakness and brain related complications for a long time.

The origin of the virus dates back to 1937, when the virus was isolated from a woman in the West Nile District of Uganda. This is the basis for the name of the virus. The disease causing nature of the virus for humans was discovered in the 1950s, and in animals during the 1960s. It is unclear whether the virus radiated out from Uganda, or whether it has long been present in North America and was previously undetected. However, the pattern of detection suggested that the virus spread globally from one region.

A Culex mosquito, a major carrier of West Nile virus in the southern United States. *AP/Wide World Photos. Reproduced by permission.*

The mosquito is responsible for spread of the West Nile virus. Many **species** of mosquito can become infected with the virus. Three species in particular have been most commonly associated with outbreaks. These species are *Culex pipiens*, *Culex restuans*, and *Culex quinquefasciatus*.

Mosquitoes acquire the virus when they obtain a **blood** meal from an infected **animal** or a bird. The virus resides in the salivary glands of the mosquito and can be passed into a human from which the mosquito subsequently obtains a blood meal. A period of time, thought to be approximately two weeks, must pass before a mosquito is able to transmit the virus to a human or animal.

The number of bird species involved in transmission of the West Nile virus is not known. It is known that over 70 species of birds can be infected with the virus, and that the infection kills many crows, blue jays, magpies, and ravens. The recent outbreaks in North America have been associated mainly with crows.

The virus enters the host's bloodstream and, by a mechanism that is not yet known, is able to cross the barrier between the blood and the brain. Multiplication of the virus in brain **tissue** disrupts the cells, causing the

nervous system to malfunction and the brain tissue to become inflamed.

Despite the publicity of public fear surrounding the spread of the West Nile virus in North America, the chance of acquiring West Nile virus via a mosquito bite is small. The available data from surveys of mosquito populations indicates that less than 1% of mosquitoes carry the virus, even in areas where the virus is known to be present.

Because the mosquito will not survive cold northerly winters in the wild, the spread of the virus in colder climates is slow. However, if mosquitoes can find protection from the winter, such as in buildings or sewer pipelines, than over wintering of the mosquitoes can occur. Indeed, in New York, infections diminished in the winters of 1999 and 2000, but increased in the spring.

Another climatic factor may be hot and dry spring and summer **seasons**. The dry conditions limit the open **water** supplies that are frequented by mosquitoes. The large mosquito populations that gather at the available water supplies make it easier for the virus to become established in large numbers of the **insects**.

The mosquito to human route of infection is the only route to have been confirmed thus far. While ticks can be infected with the virus, a tick-borne outbreak of the disease has not been documented in humans. Person to person contact cannot occur. Even exchange of body fluids like saliva from an infected to an uninfected person will not transmit the virus. The claim that the virus can be transmitted by sexual contact has been refuted.

In 2002, several people became ill after receiving blood from a donor who was subsequently found to be infected with West Nile virus. This has sparked concern that the virus can be transferred via blood donation. Currently, blood is not routinely screened for the presence of the virus.

Currently no **vaccine** to the West Nile virus exists for humans, although researchers have recently reported on the success of a weakened West Nile virus-dengue virus construct in stimulating an immune response in experimental animals. A vaccine for horses has been developed, but is not widely available yet. As of 2002, prevention of infection consists of the use of chemical mosquito repellent and insecticide, the use of protective clothing, and simply avoiding being outside at places or times when mosquitoes are typically present.

See also Microorganisms; Zoonoses.

Resources

Books

Despommier, D. *West Nile Story.* New York: Apple Trees Productions, LLC, 2001.

White, D. J., and D. L. Morse. *West Nile Virus: Detection, Surveillance, and Control.* Baltimore: Johns Hopkins University Press, 2002.

Periodicals

Pletnev, A. G., R. Putnak, J. Speicher, et al. "West Nile Virus/ Dengue Type 4 Virus Chimeras that are Reduced in Neurovirulence and Peripheral Virulence Without Loss of Immunogenicity or Protective Efficacy." *Proceedings of the National Academy of Sciences* 99 (March 2002): 3036–3041.

Organizations

Centers for Disease Control and Prevention, Division of Vector-Borne Infectious Diseases. P.O. Box 2087, Fort Collins, CO 80522. (888) 246–2675. <http://www.cdc. gov/ncidod/dvbid/westnile/>.

Health Canada. 0904A Brooke Claxton Bldg., Tunney's Pasture, Ottawa, ON K1A 0K9. (613) 957–2991. <http:// www.hc-sc.gc.ca/pphb-dgspsp/publicat/info/wnv_e.html.>

Brian Hoyle

Wetlands

Wetlands are low-lying, depressional ecosystems that are permanently or periodically saturated with water at or close to the surface. The vegetation of wetlands must be adapted to the physical and chemical stresses associated with flooded substrates. The most common types of wetlands are swamps, marshes, shallow open waters, and mires, the latter consisting of peat-accumulating fens and bogs. Wetlands vary greatly in their productivity, mostly because of intrinsic differences in the **rate** of supply of **nutrients**. Wetlands provide important **habitat** for a wide variety of plants and animals. However, wetlands are rapidly disappearing because they are being drained and in-filled for agricultural, urbanization, and industrial purposes. Wetlands are also being degraded by nutrient loading, which causes **eutrophication**, and by **pollution** associated with inputs of toxic chemicals and organic materials. Losses of wetlands and the **biodiversity** that they support are an extremely important aspect of the environmental crisis.

Types of wetlands

Wetlands can be characterized on the basis of their **hydrology**, morphology, water **chemistry**, and vegetation. All of these factors can vary regionally and locally, depending on the climate, character of the surrounding **watershed**, and the **species** that are present (that is, the biogeographic region). The major kinds of wetlands are described below, with an emphasis on North American types.

Swamps

Swamps are forested or shrub-dominated wetlands, usually associated with low-lying, periodically or permanently flooded areas around streams and **rivers**. Water flows through swamps, although the movement can sometimes be imperceptible. In southeastern **North America**, alluvial and floodplain swamp **forests** are sometimes extensive, and are typically dominated by such **tree** species as bald cypress (*Taxodium distichum*), water tupelo (*Nyssa sylvatica*), swamp tupelo (*N. sylvatica*), and eastern white cedar (*Chamaecyparis thyoides*). More northern temperate swamps are usually dominated by red maple (*Acer rubrum*), silver maple (*Acer saccharinum*), American **elm** (*Ulmus americana*), and green or swamp ash (*Fraxinus pennsylvanica*). **Freshwater** tropical swamps can support much-more diverse species of **angiosperm** trees, while tropical mangrove swamps support only a few tree species that are tolerant of the **brackish** water.

Swamps provide habitat for numerous species of animals, many of which have a specific requirement for this type of habitat. For example, swamps of bald cypress in southeastern North America used to support the now-extinct American ivory-billed woodpecker (*Campephilus principalis principalis*), and they still provide habitat for the pileated woodpecker (*Dryocopus pileatus*), red-shouldered hawk (*Buteo lineatus*), prothonotary warbler (*Protonotaria citrea*), Carolina wren (*Thryothorus ludovicianus*), and many other small **birds**. These swamps also provide nesting habitat for wood duck (*Aix sponsa*) and for colonies of wading birds such as **herons** and egrets (e.g., great blue heron, *Ardea herodias*, and common egret, *Casmerodius albus*) and wood stork (*Mycteria americana*). Cypress swamps also support **mammals**, including swamp rabbit (*Sylvilagus aquaticus*), white-tailed **deer** (*Odocoileus virginianus*), and panther (*Felis concolor*), along with many species of **amphibians** and **reptiles**, including the American alligator (*Alligator mississippiensis*).

Marshes

Marshes are a relatively productive wetland in which the vegetation is dominated by tall, emergent, graminoid (that is, grass-like) plants. Typical plants of North American marshes include **cattails** (e.g., *Typha latifolia*), reeds (e.g., *Phragmites communis*), bulrushes (e.g., *Scirpus validus*), and saw-grass (*Cladium jamaicense*). Marshes dominated by these plants are relatively productive, because they have access to nutrients dissolved in their slowly flowing water. Wet meadows are less productive types of marshes, and are dominated by shorter graminoid plants such as **sedges** (*Carex* spp.) or a grass known as blue-joint (e.g., *Calamagrostis* *canadensis*). **Salt** marshes are brackish because they are periodically inundated by oceanic water. Temperate salt marshes are dominated by species of cordgrass (e.g., *Spartina alterniflora*).

Because they are rather productive, marshes can support relatively large populations of certain mammals, such as **muskrat** (*Ondatra zibethicus*). Birds can also be abundant in marshes. This is true of large, extensive marshes, and also of relatively small, fringing marshes around bodies of open water, such as lakes and ponds. For example, small ponds are common in the prairies of North America, where they are called "potholes." The marshy borders of potholes have historically provided important breeding habitat for most of the continent's surface-feeding **ducks** (these are known as "dabbling" ducks), such as mallard (*Anas platyrhynchos*), pintail (*Anas acuta*), and blue-winged teal (*Anas discors*). Unfortunately, most **prairie** potholes have been drained or filled to provide land for agriculture. This ecological conversion has increased the importance of the remaining potholes as habitat for declining populations of ducks, other animals, and native plants. Consequently, further losses of this habitat type are vigorously resisted by the **conservation** community, even though agricultural interests continue to encourage the drainage of these important wetlands.

Farther to the north, extensive salt marshes and freshwater fringing marshes in boreal and sub-arctic regions provide important breeding habitat for **geese**, especially snow goose (*Chen caerulescens*) and Canada goose (*Branta canadensis*). Historically, these migratory waterfowl wintered in extensive temperate marshes farther to the south, but now many of these birds spend much of the winter foraging for unharvested grain in agricultural fields.

Shallow open water

This is a heterogenous wetland type, transitional from deeper open-water habitats such as lakes, and more completely vegetated wetlands such as marshes. Shallow, open-water wetlands are known locally by names such as ponds, sloughs, and potholes. These are small bodies of surface water, less than about 7 ft (2 m) in depth, and free of emergent plants, but often having floating-leaved vegetation. These wetlands and their fringing marshes can support relatively large populations of waterfowl, amphibians, and other animals. For example, the open-water portions of prairie potholes provide habitat for populations of diving ducks, such as lesser scaup (*Aythya affinis*), canvasback (*Aythya valisneria*), and redhead (*Aythya americana*), along with coot (*Fulica americana*), western grebe (*Aechmorphorus occiden-*

talis), and other species of birds. Beaver (*Castor canadensis*) can also be abundant in shallow, open-water wetlands, and in fact they often create such habitat by damming streams.

Fens

Fens are a type of mire, or a peat-accumulating wetland, and are most commonly found in boreal and sub-arctic regions. Because surface water percolates slowly through fens (that is, they are minerotrophic), they are relatively well supplied with nutrients and alkalinity, and are only slightly acidic. Consequently, fens are relatively productive, and they are non-acidic or only slightly acidic. Fens have a **plant** community dominated by short-statured graminoids, especially sedges and **rushes**, some shrubs, and species of peat-moss (*Sphagnum* spp.) that do not prefer acidic habitats. Because they are not very productive, fens support relatively small populations of animals, although their associated open-water habitat may be used by breeding ducks, geese, and **cranes**.

Bogs

Bogs are another type of peat-accumulating wetland, but they only receive very small inputs of nutrients and alkalinity, entirely from atmospheric deposition associated with rain, snow, and particulates (that is, bogs are ombrotrophic, or "fed from the clouds"). As a result of their nutrient-poor status, bogs are highly unproductive and acidic, with a **pH** less than about 4.5 and as low as 3.5. The surface of the most ombrotrophic bogs is often raised above the level of the surrounding terrain, occurring as blanket- or raised-bogs, which can have peat depths of more than 30-50 ft (10-15 m). The vegetation of bogs is typically dominated by acid-loving species of peat-moss (these are known as acidophilous species, and they are different *Sphagnum* spp. than the ones found in fens), along with various shrub species in the **heath family (Ericaceae)**. Because they are so acidic and unproductive, not many animals breed in bogs or their associated, brown-water pools.

Wetland ecology

Wetlands are dynamic ecosystems, transitional between terrestrial and aquatic habitats. Over **time**, most wetlands gradually in-fill because of the cumulative deposition of sediment and peat. Consequently, wetlands are most numerous in places where geological forces, such as glaciation or the **migration** of oxbow rivers, periodically create conditions that are favorable to their formation.

The ecological conditions of wetlands are, of course, dominated by the influences of permanent or

Sunset on a wetland near Lake Erie in Crane Creek State Park. During low water periods, mud flats are exposed, providing nesting platforms for waterfowl like the Canada goose in the photo. *Photograph by Robert J. Huffman. Field Mark Publications. Reproduced by permission.*

temporary waterlogging. Clearly, the availability of water to sustain plant growth is not a problem in wetlands, as it is in many terrestrial ecosystems. However, waterlogged **soil** or sediment are usually lacking in **oxygen**, a factor that inhibits **respiration** by plant roots. To cope with this stressful environmental condition, some plants have evolved specific adaptations to supply oxygen to their roots. Many herbaceous plants, such as cattails and bulrushes, have spongy, air-filled stem and root tissues, called aerenchyma, which helps to facilitate the transport of oxygen to underwater tissues. Some trees, such as bald cypress and black mangrove (*Avicennia nitida*), have specialized woody structures called pneumatophores, that extend from roots into the air, and have extensive intercellular spaces that are useful in supplying oxygen to below-water tissues.

The **anaerobic** nature of wetland substrates also causes other chemical changes that can pose important problems for plants, by affecting their **nutrition** and exposing roots to toxic chemicals. For example, access to certain nutrients can be difficult under anaerobic conditions. This is because the nutrients may not be present in a chemical form that is easy for roots to assimilate, or

because roots cannot sustain the oxygen-demanding respiratory demands required for the active uptake of nutrient ions. Anaerobic conditions also encourage the solubilization of certain potentially toxic metals, such as manganese. In addition, anaerobic **metabolism** within root tissues can lead to excessive accumulations of alcohols, possibly causing toxicity. In general, wetland plants are well-adapted to these conditions, although they may nevertheless be physiologically stressed if these factors are severe enough.

Wetland hydrology is, of course, a highly variable character. Some wetlands are permanently flooded, while others are only waterlogged some of the time, usually seasonally. These dynamics are highly influential on the types of plants that can occur in particular wetlands, and on the communities that they develop. Tolerance of permanent or frequent **flooding**, as occurs, for example, in salt marshes, mangroves, and some swamps, requires highly adapted species of plants. In comparison, wetlands that are only occasionally flooded are, in some respects, a more ephemeral transitional between truly aquatic and terrestrial environments. In these situations, plants must only be tolerant of the stresses of sporadic events of flooding, while growing relatively freely when the water recedes and the soil is drier.

Another highly influential environmental factor in wetlands is the supply of plant nutrients. In general, wetlands that are well supplied with **phosphorus** (in the form of phosphate), and to a lesser degree **nitrogen** (as nitrate or ammonium), sustain relatively high rates of plant productivity, and consequently large populations of animals. This is commonly the case for marshes, which are among the most productive natural ecosystems on **Earth**. In contrast, wetlands with restricted supplies of nutrients, such as ombrotrophic bogs, sustain only small productivities of plants and animals.

Losses of wetlands

All wetlands have great intrinsic value as natural ecosystems, and they all support species of plants and animals that occur nowhere else. Consequently, wetlands have great value in terms of biodiversity.

Sometimes, the biodiversity-related importance of particular types of wetlands is a matter of their relative abundance, in the regional context. For example, although bogs and fens can be extremely abundant in boreal and sub-arctic regions of northern North America, these types of wetlands are uncommon farther south, where they usually occur as relict, post-glacial ecosystems. In these southern regions, the few bogs and fens that occur have great conservation value as scarce and unusual ecosystems, and because most of their species of plants and **arthropods** are regionally or locally rare. As such, any proposals to "develop" these wetlands into agricultural or urbanized lands are highly controversial, because these conversions would cause an irretrievable loss of natural values.

Wetlands also provide essential habitat for species of birds and mammals that are hunted, and this gives them economic value. Waterfowl such as ducks and geese occur primarily in marshes and swamps. During the past century the populations of some of these hunted waterfowl were greatly decreased, as a combined effect of overhunting and loss of natural habitats. Consequently, there are now substantial efforts to regulate hunting, and to preserve or enhance the marshes and swamps that are required as habitat by these birds. Some species of waterfowl are responding well to these conservation measures, and their populations are increasing.

Wetlands are also important because they offer other ecological goods and services, in addition to those previously described. For example, wetlands maintain some control over hydrology, helping to prevent extremes of water flow. This service moderates the risks of flooding caused by heavy rain or the spring flush of snowmelt in northern regions. It also helps to extend supplies of water for drinking or **irrigation** longer into the drier **seasons** of the year. Wetlands also provide important services by cleansing the water that flows through them of pollutants, including nutrients and toxic chemicals, such as metals and certain **pesticides**. Furthermore, wetlands are useful in protecting shorelines from **erosion**, controlling sedimentation, and providing essential habitat for **fish**, birds, and other **wildlife**. Wetlands have good aesthetics, and this also contributes to their value as an ecological resource.

Unfortunately, wetlands are being rapidly lost in most of the world. The most important causes of the destruction of wetlands are drainage and in-filling to provide dry land for agriculture, urbanization, and industrialization. Wetlands are also sometimes used as convenient places for the disposal of mine tailings, municipal solid wastes, and sewage. In some cases, wetlands are degraded or lost because economically useful products can be mined from them, especially peat from bogs, and wood from forested swamps. Wetlands are also degraded if they are subjected to large inputs of nutrients through the runoff of agricultural **fertilizers** or by sewage dumping. These nutrient inputs can cause eutrophication, with a consequent loss of the original ecological values of the wetland.

All of these disturbances, stresses, and ecological conversions result in net losses of wetlands. The ecological consequences include endangerment of natural wetland ecosystems, endangerment of their species of plants and animals, and the loss of many important services that wetlands can provide. The loss of wetlands is an important environmental issue, which can only be resolved by protection of those wetlands that still survive, and in some areas where the losses have been especially severe, by the active restoration of wetlands.

The protection and conservation of wetlands is an important activity of many governments and private organizations. In the United States and Canada, wetlands are among the highest-priority natural habitats for protection by governments at all levels (national, state or provincial, and local). In addition, non-governmental organizations such as the World Wildlife Fund, The Nature Conservancy, the Nature Conservancy of Canada, and Ducks Unlimited have made the conservation and protection of wetlands a high priority in their activities. Internationally, the *Convention on Wetlands of International Importance, Especially as Waterfowl Habitat* (also known as the Ramsar Convention, after the city in Iran where it was negotiated) is an intergovernmental treaty that provides a framework for worldwide cooperation in the conservation of wetlands. The activities of all of these agencies are important and useful, but much more needs to be done to give wetlands and their species the degree of protection that they require.

KEY TERMS

· ·

Alkalinity—The amount of alkali in a solution. In fresh water, alkalinity is mainly associated with bicarbonates, carbonates, and hydroxides, and it is generally measured by titration with acid to a fixed end point.

Anaerobic—Environments in which oxygen is not present, or only present in a very small concentration.

Eutrophication—An aquatic ecosystem process by which increased productivity results from an increase in the rate of nutrient input. Excessive eutrophication and its symptoms are regarded as a type of ecological degradation.

Graminoid—A generic term for plants with a grass-like growth form, such as grasses (Poaceae), sedges (Cyperaceae), rushes (Juncaceae), and cattails (Typhaceae).

Hydrology—The study of the distribution, movement, and physical-chemical properties of water in Earth's atmosphere, surface, and near-surface crust.

Minerotrophic—This refers to wetlands that receive much of their nutrient supply as substances dissolved in water draining from a part of the watershed that is higher in altitude.

Ombrotrophic—This refers to wetlands with no input of nutrients from ground water or surface water, so that all of the nutrient supply arrives from the atmosphere with precipitation and dust.

Watershed—The expanse of terrain from which water flows into a wetland, waterbody, or stream.

See also Alluvial systems; Biodiversity.

Resources

Books

Barbour, M.G., and W.D. Billings. *North American Terrestrial Vegetation.* Cambridge: Cambridge University Press, 1988.

Barbour, M.G., J.H. Burk, and W.D. Pitts. *Terrestrial Plant Ecology, 2nd ed.* Ontario: Benjamin/Cummings Pub. Co., Don Mills, 1987.

Finlayson, M., and M. Moser. *Wetlands.* New York: Facts on File, 1991.

Hamblin, W.K., and E.H. Christiansen. *Earth's Dynamic Systems.* 9th ed. Upper Saddle River: Prentice Hall, 2001.

Hancock, P. L., and B. J. Skinner, eds. *The Oxford Companion to the Earth.* Oxford: Oxford University Press, 2000.

Mitsch, W.J., and J.G. Gosselink. *Wetlands.* John Wiley and Sons, 1997

Sobel, Jack. *Marine Reserves: A Guide to Science, Design, and Use.* Washington, DC: Island Press, 2003.

Vileisis, A. *Discovering the Unknown Landscape: A History of America's Wetlands.* Island Press, 1997

Periodicals

"Satellite Remote Sensing of Wetlands." *Wetlands Ecology And Management* 10, no. 5-5 (2002): 381-402.

Bill Freedman

Whales *see* **Cetaceans**

Wheat

Wheat is one of the oldest and most important cereal **crops**. Wheat is grown for its grain, which is ground into flour used to make breads and pastas. Wheat consists of approximately 20 **species** in the genus *Triticum* of the grass family (Poaceae). The most important wheats are: *Triticum aestivum,* used to make bread; *T. durum,* used to make pasta; and *T. compactum,* used to make softer cakes, crackers, cookies, and pastries.

Wheat plants have slender leaves and, in most varieties, long hollow stems. Each stem is topped by a single head or spike, which is an aggregation of 20-100 individual **flower** clusters called spikelets. Each flower cluster may contain up to six flowers, and each fertilized flower produces a single, edible grain.

About 13% of the **mass** of the ripe grain is formed by the fused layers of the fruit wall, seed coat, and aleurone. These layers, known as the bran, are nutritionally important because they contain fiber, some protein, and the vitamins thiamine, riboflavin, niacin, and **vitamin** A. About 84% of the grain is endosperm, a food-storage **tissue** that consists mainly of starch. The embryo, or germ, represents only 2.5% of the grain and contains most of the oil and protein. Because wheat provides a balance of several vitamins, starch, **proteins**, and oils, it is an excellent source of **nutrition**. Furthermore, because of its small **water** content, generally about 12%, wheat grains are easily transported and stored, and are resistant to microbial spoilage.

Wheat was one of the first plants to be domesticated. Along with **barley** (*Hordeum vulgare*), wheat provided the agronomic basis for development of the earliest civilizations in Mesopotamia, and has been found at archaeological sites in the region dating back to 7000 B.C. The cultivation of wheat gradually spread by trade from

Mesopotamia to other parts of the world, becoming established in India by 3000 B.C., and in **Europe** by 2000 B.C. Wheat did not occur in the New World until brought there by the Spanish in 1520. Wheat was introduced by settlers to the United States in the early 1600s. Nowadays, wheat is grown on every arable **continent**, and it is the most important food for people living in temperate regions of the world (it is replaced by **rice** in the tropics).

By far the most important wheat is the common bread wheat (*Triticum aestivum*). This species evolved by a series of natural hybridizations followed by **chromosome** doubling (polyploidy) about 6,000 years ago in the Middle East, where its wild relatives still occur. The first hybridization is believed to have been between a primitive einkorn wheat (*Triticum urartu*) and an unknown species of wild goat grass (related to *Triticum speltoides*), each with 14 chromosomes. The resulting **hybrid** doubled its chromosome number and became fertile, producing emmer wheat (*Triticum turgidum*) with 28 chromosomes. Emmer wheat then hybridized with another wild species of goat grass (*Triticum tauschii*), which had 14 chromosomes, and the hybrid doubled its chromosome number to produce bread wheat with 42 chromosomes. Some of this hybridization and chromosome doubling has been duplicated experimentally in modern times, to yield an "artificial" wheat that resembles certain cultivars of bread wheat.

Winter wheats are planted in the fall and germinate before winter. The seedlings can survive cold winter temperatures—in fact, the low temperatures are needed for proper growth and development of the grain. The seedlings start growing again in the spring as soon as the frost is out of the ground, and by late spring the mature plants are ready for harvest. In contrast, spring wheats are planted in the spring and harvested in the fall.

For thousands of years, wheat was laboriously harvested using a sickle, and then threshed, or beaten, to separate the grains from the heads and flower parts (chaff). In the first half of the 1800s, the reaper was developed, which mechanized cutting and greatly reduced the amount of labor required. Nowadays, cutting the standing plants, threshing the heads, separating the grain from the chaff, cleaning the grain, and discharging it into bags, are all combined in a large, self-propelled machine called a combine.

About 25% of the world's farmland is devoted to wheat cultivation. This is more than is used for any other crop. About 566 million acres (229 million hectares) are sown to produce 478 million U.S. tons (527 million metric tons) of wheat. The world's greatest wheat producing countries are: the United States, China, Ukraine, India, Canada, **Australia**, Turkey, and Pakistan. The major

KEY TERMS

Aleurone—The protein-rich, outer layer of the endosperm in cereal grains.

Bran—The fused fruit wall, seed coat, and aleurone layer of a cereal grain. It is usually removed during the milling process.

Cereal—A member of the grass family with edible grains, such as oats, barley, rye, maize, and wheat.

Combine—A self-propelled, tractor-like machine that combines the separate functions of cutting, threshing, cleaning, and bagging grains.

Endosperm—The food-storage tissue of a seed of a flowering plant. It consists mostly of starch.

Germ—The embryo of a grain.

Grain—The dry, one-seeded fruit of a grass, differing from other one-seeded fruits (such as nuts and achenes) by having the fruit wall fused to the seed. Known botanically as a *caryopsis*.

Leavened bread—Wheat bread that has been made to raise by the action of its dough trapping carbon dioxide gas produced by yeast, baking soda, or baking powder.

Wheat flour—The powder that results from grinding grains of wheat.

White flour—Wheat flour made from grains having the bran and embryo removed.

Whole-wheat flour—Flour made from milling the whole grain; that is, grain with the endosperm, bran, and embryo left intact.

wheat-exporting countries are the United States, Australia, Canada, and Argentina. Nearly all wheat-growing countries have breeding programs to improve the races of wheat adapted to local conditions. These programs strive to enhance qualities such as yield, **disease** and insect resistance, and nutrient content of the grain.

If wheat grains are eaten whole or ground whole by traditional stone grinding, all of the **nutrients** are retained. Modern milling methods, however, remove the germ and bran, thereby eliminating most of the proteins, oils, and nearly all vitamins. The resulting white flour stores longer and tastes good but is nutritionally impoverished. For this reason, white flour is often artificially enriched with vitamins to improve its nutritional value.

Wheat flour is primarily used to bake bread. Wheat flour is especially suited for this purpose because it contains two proteins, glutenin and gliadin (collectively known as gluten), that make a sticky, elastic dough. During baking, the dough traps bubbles of **carbon dioxide** produced by **yeast** or by chemical leaveners such as baking powder or baking soda. The trapped bubbles cause the bread to rise. (The holes you see in sliced bread were formed by trapped gas bubbles.) Other cereal grains, such as rice, barley, corn, and oats, do not contain glutenin and gliadin and therefore are not suitable for making leavened bread.

Durum wheat (*Triticum durum*) does not rise as well as bread wheat because of the different protein composition of its grain. Therefore, durum wheat is used for making spaghetti, macaroni, and other kinds of pasta. Other products made from wheat include bulgur, which is prepared by cooking, dehydrating, and peeling wheat. Wheat germ, which is removed in the milling process, has a large content of vitamin E and protein, and is often added to or sprinkled on other foods as a nutritional supplement. Entire wheat grains, either rolled or puffed, are often used in breakfast cereals. In addition, starch and gluten are extracted from wheat grains. The starch is used in laundering and in making a sweet syrup. The gluten is used in making **monosodium glutamate (MSG)**, a flavor enhancer of cooked foods, especially commonly used in oriental cooking. Wheat is the most important grain fed to poultry. However, wheat is not generally fed to **livestock**, because it is more expensive than other suitable grains such as maize, barley, and oats.

See also Grasses.

Robbin C. Moran

Whisk fern

The whisk fern (*Psilotum* spp., family Psilotaceae) splays its leafless, whisk-like branches upward, and is a living fossil from the time before the dinosaurs. It can grow as an epiphyte in moist climates or as a terrestrial **plant** in drier areas. Found in the tropics from around the world, the whisk fern is descended from the first vascular land plants, the Rhyniophytes, which appeared about 400 million years ago. It is not a true fern, unlike the popular Boston fern, but both the whisk fern and true **ferns** are ancient plants when compared to the flowering plants or angiosperms.

The leaved genus *Tmesipteis* (family Tmesipteridaceae) and *Psilotum* are the only representatives of the division Psilophyta (order Psilopsida). The principal use-

fulness of *Psilotum* to humans lies in their limited decorative use, and in scientific study as a living example of a very ancient land plant.

The primitive nature of the whisk fern is underscored by its having flagellated sperm, unlike the more advanced flowering plants, the angiosperms. The simple branched stems of *Psilotum* recalls the structure of the rhyniophytes, and the whisk fern is unique among living vascular plants in its lack of roots and leaves.

In place of roots the whisk fern has rhizomes, that is, modified underground stems. In *Psilotum nudum* the **rhizome** occurs with a mutualistic fungus in a type of **mycorrhiza** useful for obtaining necessary **nutrients**. Because *Psilotum* is without leaves, the interior parts of the stem conduct food and **water**, known as the vascular cylinder. In *Psilotum* the vascular cylinder lacks a central part made of large, open-looking cells, called pith. The lack of these cells defines the type of vascular cylinder known as a protostele.

The lack of **seeds** in the reproductive cycle of the whisk fern is another example of its ancient evolutionary origins. In place of the pollen and ovule of angiosperms, *Psilotum* has multicellular male and female gametophytes, and the whisk fern has spores which give rise to the gametophytes.

The gametophyte is the stage of the plant life cycle which has a haploid complement of chromosomes (1n). The gametophytes of flowering plants are extremely reduced in size. The pollen grain and the seven-celled ovule are hidden within the unpollinated ovary. However, in ancient plants such as the whisk fern, the gametophyte is relatively large. The gametophyte of *Psilotum* even has vascular **tissue** and a distinct area of food—and water—conducting tissues, unlike the gametophytes of more ancient plants, such as **moss** and liverworts. The cigar-shaped gametophytes also grow underground, unlike the gametophytes of many other plants, where they are nourished by an endophytic fungus. Scientists have now learned how to germinate the spores of some **species** of *Psilotum* in the laboratory, allowing for a more complete study of their gametophytes.

White dwarf

During the first quarter of this century astronomers found that the brightest **star** in the sky, Sirius, was orbited by a much fainter companion. Analysis of the **orbit** yielded a **mass** for the companion similar to that of the **sun** while an analysis of its **light** suggested that its size was approximately the same as Earth's. Further observation revealed that these small massive stars are reasonably common but had gone undetected because they are so faint. The stars of this unique class are called white dwarfs. They range in mass from perhaps a third to just under one and a half times the mass of the sun. We now know that these stars have exhausted their supply of nuclear fuel, which would enable them to shine like the sun and other ordinary stars. Under the weight of their own **matter**, they have collapsed to roughly a hundredth the size of the sun. A fundamental law of **quantum mechanics** (i.e., the Pauli exclusion principle) limits the ability of the crushing gravity to pack electrons into an ever decreasing **volume**. The **pressure** of these highly packed electrons balances the weight of the overlying stellar material, stopping any further collapse. Since the **electron** pressure results from a simple packing of the electrons in a particular volume, there is no need for them to be heated by **nuclear fusion**, as is the case in normal stars (the nuclear fusion produces the pressure to balance the weight of overlying material). We call such densely packed matter electron-degenerate. A teaspoon of such matter would weigh 40 tons (36.3 tonnes) on **Earth**. Thus, white dwarfs are stable stars which slowly cool off, becoming dark stellar cinders.

Since two thirds of all stars occur in **binary star** systems where the components orbit one another, it is not surprising that many white dwarfs are found in binary systems. As the companion of the white dwarf ages it will expand and begin to lose matter to the gravitational pull of the white dwarf. This results in the injection of hydrogen-rich matter from the normal companion into the outer layers of the white dwarf. After a certain amount of this matter accumulates, it will undergo a nuclear fusion explosion which will blow off the outer ten-thousandth of the white dwarf. The **energy** released during this explosion will cause the system to brighten perhaps a million times or more, forming what we call a **nova**. The extent of the explosion critically depends on aspects of the donor star as well as the white dwarf. Smaller explosions may be called dwarf novae. This material will be blown clear of the system and the expanding shell will become visible to later observers.

There is a limit in mass to which the white dwarf can grow because there is a limit to the extent that electrons can resist the increased gravitational forces. Depending on the chemical composition of the white dwarf, this limit occurs when it reaches between 1.2 and 1.4 times the mass of the sun. In binary systems where donated mass from the companion forces the white dwarf beyond this limit, the white dwarf will collapse by a factor of one thousand to become a **neutron star**. The cir-

cumstances of the collapse may be so violent as to result in an explosion called a Type I **Supernova**. Such an explosion is far more violent than a nova and may result in the disruption of the binary system.

See also Stellar evolution.

White-eyes

White-eyes are 85 **species** of small, perching songbirds that constitute the family Zosteropidae. White-eyes occur in sub-Saharan **Africa**, South **Asia**, Japan, Southeast Asia, New Guinea, **Australia**, New Zealand, and many other islands in the Pacific and Indian Oceans.

White-eyes are arboreal **birds**, occurring in a wide range of forest types, including mangroves, lowland **forests**, and montane forests. They also occur in brushy habitats, especially at higher altitude on **mountains**.

White-eyes are small birds, ranging in body length from 2.8-5.5 in (7-14 cm). They have a slender, pointed bill, with the upper mandible slightly downward curved. The wings are rounded, and the tail square backed. The plumage is subdued in coloration, similar among the various species, and is most commonly brown, olive-green, grey, or yellow, the latter usually occurring on the throat or belly. Most species have a white-feathered eye-ring, from which these birds are given their common name. The sexes are similar in plumage.

White-eyes forage actively in trees and shrubs for **insects** and spiders. White-eyes also feed on **fruits**, **nectar**, and pollen. Some species poke holes into soft, larger fruits, and feed on the pulp and juice using specialized, brushlike structures at the end of their tongue.

White-eyes commonly occur in flocks, which can be quite large during the nonbreeding season. They may also be part of mixed-species foraging flocks with other insectivorous birds.

White-eyes are territorial during the breeding season, when they declare their territories with soft, warbling songs. They build a deep, cup-shaped nest of woven fibers and spider webs, located in a horizontal fork of a branch. The clutch size is one to five. The eggs are incubated by both parents, who also share the feeding and care of the nestlings.

The genus Zosterops accounts for 62 of the 85 species in the family Zosteropidae. The oriental white-eye (Z. palpebrosa) occurs from Afghanistan to south China, and south into Indonesia. The black-capped white-eye (Z. atricapilla) occurs in the tropical forests of

Sumatra, Borneo, peninsular Malaya, and Thailand in Southeast Asia.

The cinnamon white-eye (*Hypocryptadius cinnamomeus*) is a brown-backed species, with a dark **eye** but lacking in a white eye-ring, that breeds in the Philippines.

Bill Freedman

Whooping cough

Whooping cough is a highly contagious **disease** caused by the **bacteria** *Bordatella pertussis*. It is characterized by classic paroxysms (spasms) of uncontrollable coughing, followed by a sharp intake of air which creates the characteristic "whoop" of the disease name.

B. pertussis is uniquely a human pathogen, meaning that it neither causes disease in other animals, nor survives in humans without resulting in disease. It exists worldwide as a disease-causing agent, and causes epidemics cyclically in all locations.

B. pertussis causes its most severe symptoms by attacking specifically those cells in the respiratory tract which have cilia. Cilia are small, hair-like projections which beat constantly, and serve to constantly sweep the respiratory tract clean of such debris as mucus, bacteria, viruses, and dead cells. When *B. pertussis* interferes with this normal, janitorial function, mucus and cellular debris accumulate and cause constant irritation to the respiratory tract, triggering the cough **reflex** and increasing further mucus production.

Children under the age of two, particularly infants, are most at risk for serious **infection**, although the disease can occur at any age. However, once an individual has been exposed to *B. pertussis*, subsequent exposures result in mild illness similar to the common cold, and thus usually not identifiable as resulting from *B. pertussis*.

Symptoms and progression of whooping cough

Whooping cough has four somewhat overlapping stages: incubation, catarrhal stage, paroxysmal stage, and convalescent stage.

An individual usually acquires *B. pertussis* by inhaling droplets infected with the bacteria, coughed into the air by an individual already suffering from whooping cough symptoms. Incubation is the period of seven to 14 days after exposure to *B. pertussis*, and during which the bacteria penetrate the lining tissues of the entire respiratory tract.

The catarrhal stage is often mistaken for an exceedingly heavy cold. The patient has teary eyes, sneezing, fatigue, poor appetite, and a very runny nose. This stage lasts about 10-14 days.

The paroxysmal stage, lasting two to four weeks, is heralded by the development of the characteristic whooping cough. Spasms of uncontrollable coughing, the "whooping" sound of the sharp inspiration of air, and vomiting are hallmarks of this stage. The whoop is believed to occur due to **inflammation** and mucous which narrow the breathing tubes, causing the patient to struggle to get air in, and resulting in intense exhaustion. The paroxysms can be caused by overactivity, feeding, crying, or even overhearing someone else cough.

The mucus which is produced during the paroxysmal stage is thicker and more difficult to clear than the waterier mucus of the catarrhal stage, and the patient becomes increasingly exhausted while attempting to cough clear the respiratory tract. Severely ill children may have great difficulty maintaining the normal level of **oxygen** in their systems, and may appear somewhat blue after a paroxysm of coughing due to the low oxygen content of their **blood**. Such children may also suffer from encephalopathy, a swelling and degeneration of the **brain** which is believed to be caused both by lack of oxygen to the brain during paroxysms, and also by bleeding into the brain caused by increased **pressure** during coughing. Seizures may result from decreased oxygen to the brain. Some children have such greatly increased abdominal pressure during coughing, that hernias result (hernias are the abnormal protrusion of a loop of intestine through a weaker area of muscle). Another complicating factor during this phase is the development of **pneumonia** from infection with another bacterial agent, which takes hold due to the patient's weakened condition.

If the patient survives the paroxysmal stage, recovery occurs gradually during the convalescent stage, and takes about three to four weeks. Spasms of coughing may continue to occur over a period of months, especially when a patient contracts a cold or any other respiratory infection.

Children who die of pertussis infection usually have one or more of three conditions present: 1) severe pneumonia, perhaps with accompanying encephalopathy; 2) extreme weight loss, weakness, and metabolic abnormalities due to persistent vomiting during paroxysms of coughing; 3) other pre-existing conditions, so that the patient is already in a relatively weak, vulnerable state (such conditions may include low **birth** weight, poor **nutrition**, infection with the measles **virus**, presence of other respiratory or gastrointestinal infections or diseases).

Diagnosis

Diagnosis based just on symptomatology is not particularly accurate, as the catarrhal stage may appear to be a heavy cold, a case of the flu, or **bronchitis**. Other viruses and **tuberculosis** infections cause symptoms similar to those found during the paroxysmal stage. The presence of a pertussis-like cough along with an increase of certain specific white blood cells (lymphocytes) is suggestive of *B. pertussis* infection, although it could occur with other pertussis-like viruses. The most accurate method of diagnosis is to culture (grow on a laboratory plate) the organisms obtained from swabbing mucus out of the nasopharynx (the breathing tube continuous with the nose). *B. pertussis* can then be identified during microscopic examination of the culture.

Treatment

Treatment with the antibiotic erythromycin is helpful only at very early stages of whooping cough: during incubation and early in the catarrhal stage. After the cilia, and the cells bearing those cilia, are damaged, the process cannot be reversed. Such a patient will experience the full progression of whooping cough symptoms, which will only abate when the old, damaged lining cells of the respiratory tract are replaced over time with new, healthy, cilia-bearing cells. However, treatment with erythromycin is still recommended to decrease the likelihood of *B. pertussis* spreading. In fact, all members of the household in which a patient with whooping cough lives should be treated with erythromycin to prevent spread of *B. pertussis* throughout the community.

Other treatment is supportive, and includes careful monitoring of fluids, rest in a quiet, dark room to decrease paroxysms, and suctioning of mucus.

Prevention

The mainstay of prevention lies in the **mass** immunization program which begins in the United States when an infant is two months old. The pertussis **vaccine**, most often given as one immunization together with **diphtheria** and **tetanus**, has greatly reduced the incidence of whooping cough. Unfortunately, there has been some concern about serious neurologic side effects from the vaccine itself. This concern led huge numbers of parents in England, Japan, and Sweden to avoid immunizing their children, which in turn led to major epidemics of disease in those countries. Multiple carefully constructed research studies, however, have disproved pertussis vaccine as the cause of neurologic damage.

See also Childhood diseases; Respiratory diseases; Respiratory system.

KEY TERMS

. . . .

Cilia—Tiny, hair-like projections from a cell. In the respiratory tract, cilia beat constantly in order to move mucus and debris up and out of the respiratory tree, in order to protect the lung from infection or irritation by foreign bodies.

Encephalopathy—Any abnormality in the structure or function of the brain.

Pathogen—A disease causing agent, such as a bacteria, virus, fungus, etc.

Resources

Books

Berkow, Robert, and Andrew J. Fletcher. *The Merck Manual of Diagnosis and Therapy.* Rahway, NJ: Merck Research Laboratories, 1992.

Kobayashi, G., Patrick R. Murray, Ken Rosenthal, and Michael Pfaller. *Medical Microbiology.* St. Louis: Mosby, 2003.

Krugman, Saul, et al. *Infectious Diseases of Children.* St. Louis: Mosby-Year Book, Inc., 1992.

Rosalyn Carson-DeWitt

Wild type

In **genetics**, the specific types of genes (**alleles**) carried by individuals in any population comprise that individual's genotype. The actual expression of those genes produces a set of observable characteristics (phenotype). In any population of organisms, the wild type (also often printed in a hyphenated form as "wild-type") represents the most common genotype. With many organisms, alleles that are not a part of that genotype are often considered mutant alleles. The designation of wild type is based upon a quantitative (numerical) representation or estimation of the norm (normal) or standard in a population.

For example, one of the first descriptions of a wild-type **gene** was made with reference to the *Drosophila* fruit fly. In early studies of genetic traits of *Drosophila*, the American geneticist Thomas Hunt Morgan (1866-1945) noted a white-eyed fly in an isolated breeding population of red-eyed *Drosophila* **flies** (the flies were isolated in a bottle). Because the vast majority of *Drosophila* have red eyes, Morgan considered the white-eyed fly a mutant and termed the gene for red eyes in *Drosophila* the wild-type gene.

Outside of strict reference to genotype or phenotype, the term wild type is also used to denote the natural state of an **organism**, or the natural life cycle of an organism. When wild type is used to describe an entire organism, the sub-population of most prevalent phenotypes with the population is often referred to as the wild-type strain.

A genetic complementation test is used to determine the location and nature of mutations. Essentially, a complementation test looks for restoration of the wild-type phenotype in a mating between organisms with mutant genes. Complementation testing also determines the capability of mutants to act independently to supply the genetic information needed to result in the expression of a wild-type phenotype. For example, when two mutations affect the same gene, and neither **mutation** is capable of generating a wild-type phenotype, if these mutations are combined in the same **cell** the resulting strain must have a mutant phenotype. On the other hand, if the mutations affect different genes, so that each is able to generate some of the gene products required to produce a wild-type phenotype, then between the two genes the sum of the two gene products might still be able to generate a wild-type phenotype.

Geneticists use a variety of symbols and type scripts (capitals, italics, etc.) to denote wild-type alleles of a gene. One method commonly used indicates a wild type gene by the presence of a plus sign (+). Most often, this symbol is used as a superscript next to the notation for the allele. For example, the notation Pax1$^+$ denotes the wild type allele of a Pax1 gene in **mice** that is the most prevalent allele for the gene. In contrast, when an organism undergoes a mutation that reverts the gene back to the wild type the plus sign is associated with a superscripted allele symbol. Addition reversions are usually identified by numbers preceding the allele in question. Geneticists also often use the letter "w" to denote the wild type gene. In the case of *Drosophila* the allele for red eyes is often designated by the letter "w" or the plus sign

A revertant is a mutation that restores the phenotype to the wild type (most prevalent form). In a true revertant, the original mutation itself is mutated back to the original wild type. With pseudo-revertants, or with pseudo reversions, the original mutation remains while another mutation that takes place within the same gene restores the wild-type phenotype. In the case of *Drosophilae*, a revertant would restore red eyes to the fly regardless whether it was a true revertant or a pseudo-revertant.

Initial forms of **gene therapy** were essentially gene replacement therapies that sought to introduce complete copies of the relevant wild-type gene into the organism having a genetic **disease**. The theory was that a wild-type gene, introduced via an appropriate agent (vector), might allow for wild-type (normal) gene expression. In the case of an **enzyme** deficiency, for example, such an introduction of the wild-type gene for the enzyme would allow the cell to produce the otherwise deficient enzyme.

See also Genetic disorders; Genetic engineering; Genotype and phenotype; Mutagenesis.

Wildfire

Wildfire is a periodic ecological disturbance, associated with the rapid **combustion** of much of the **biomass** of an **ecosystem**. Once ignited by **lightning** or by humans, the biomass oxidizes as an uncontrolled blaze, until the fire either runs out of fuel or is quenched. Wildfire is best known as a force affecting **forests**, although **savanna**, chaparral, **prairie**, and **tundra** also burn. A large wildfire can kill mature trees over an extensive area, after which a process of ecological recovery ensures, called secondary **succession**. Fire can be an important **factor** affecting the nature of ecological communities. In the absence of wildfire or other catastrophic disturbances, relatively stable, climax communities tend to develop on the landscape, the nature of which is determined by climate, **soil**, and the participating biota. However, intervening wildfires can arrest this process, so that the climax or other late-successional communities are not reached.

The nature of wildfire

Wildfire is especially frequent in ecosystems that experience seasonal **drought**, for example, boreal forests, temperate pine forests, tall-grass prairie, chaparral, and savannah. Wildfires can be very extensive, and in aggregate they affect tremendous areas of landscape each year. For example, an average of about 8 million acres (3 million ha) of forest burns each year in Canada, and in some years it exceeds 25 million acres (10 million ha). Most of those fires are started by humans, although most of the actual burned area is ignited naturally by lightning. The natural fires predominantly affect northern, non-commercial forests, where most fires are not actively quenched by humans, so that individual burns can exceed 2.5 million acres (one million ha) in area. However, even vigorously fought fires can be enormous, as was the case of the famous Yellowstone fires of 1988, which burned more than 1.2 million acres (0.5 million ha), including 45% of Yellowstone National Park. Even moist tropical **rainforest** will occasionally burn, as happened over more than 7.4 million acres (3.0 million ha) of Borneo during relatively dry conditions in 1982-1983.

Fire causes a number of changes in soil quality. Depending on the intensity of the burn, much of the organic

Labels in illustration:
Mature trees

Climax forest

1.

2.
Ash layer
Buried seeds
Deposition of minerals

3.
Species from other sites invading

Buried seeds sprouting
Roots regenerating

4.
Recovered forest

A climax forest (1) destroyed by wildfire (2) and its recovery (3, 4). *Illustration by Hans & Cassidy. Courtesy of Gale Group.*

matter and litter of the forest floor may be consumed, and mineral soil may be exposed. The combustion of organic matter results in a large **emission** of **carbon dioxide** to the atmosphere, along with gaseous oxides of **nitrogen** derived from the oxidation of organic nitrogen, and sooty particulates. The layer of ash that deposits onto the soil surface is of a basic quality, so soil acidity is temporarily decreased after a fire. The ash also supplies large quantities of certain **nutrients**, especially **calcium**, **magnesium**, potassium, and **phosphorus**, some of which leaches from the site. Often, post-fire soils are relatively fertile for several years as a net effect of these physical and chemical changes, and **plant** growth can be rather lush until the intensity of **competition** increases when the canopy closes again.

Post-fire succession

When an ecosystem is disrupted by a wildfire, it can quickly suffer an intensive mortality of its dominant **species**, along with disruptions of its physical ecological structure and other damages. However, except in the case of rare, extremely intense fires, some plants survive the disturbance, and these can contribute to the post-fire regeneration that immediately begins.

Plant species vary greatly in the strategies they have evolved to survive wildfire, and to regenerate afterwards. Often, the below-ground tissues of certain plants can survive the fire even though the above-ground biomass was killed by combustion or scorching, and the regeneration may then occur through stump or root-sprouting. In **North America**, trembling aspen (*Populus tremuloides*), other woody plants, and many understorey herbs commonly display this sort of survival and regeneration strategy.

Other plants may survive the fire as long-lived **seeds** that are buried in the forest floor, and are stimulated to germinate by post-fire environmental conditions. Species such as pin cherry (*Prunus pensylvanica*) and red raspberry (*Rubus strigosus*) can regenerate vigorously from this so-called buried seedbank. A few conifers maintain their seedbank in persistent, aerial cones, which are stimulated to open by the **heat** of the burn, so that seeds are released to the fire-prepared seedbed immediately afterwards. These fire-adapted **tree** species often form even-aged stands after fire, as is the case of knobcone pine (*Pinus attenuata*) in the southwestern United States, and jack pine (*Pinus banksiana*) farther to the north.

In other cases, species may invade the burned site, by dispersing from unburned communities nearby. Species with light, windblown seeds are especially efficient at colonizing burned sites, as is the case for white birch (*Betula papyrifera*), fireweed (*Epilobium angustifolium*), and various species in the aster family, such as goldenrod (e.g., *Solidago rugosa*).

The post-fire ecological recovery that is manifest in the regenerating vegetation is a type of secondary succession. In the absence of another wildfire, or some other catastrophic disturbance of the stand, the post-fire secondary succession often restores an ecosystem similar to the one present prior to the fire.

If the return **frequency** of natural wildfire is shorter than the **time** required for a climax ecosystem to develop, then disturbance by fire may be important in maintaining the land in an earlier successional stage. For example, most of the region of North America that supported a tall-grass prairie was climatically suitable for the development of an oak (*Quercus* spp.) dominated forest. It was only the periodic burning of the prairie that prevented the encroachment of shrubs and trees, and maintained the natural prairie. Today, tall-grass prairie is an endangered ecosystem, because most of its original area has been converted to agricultural purposes. To maintain the **ecological integrity** of the few, small remnants of tall-grass prairie that remain in protected areas, these

KEY TERMS

Secondary succession—A succession that follows any disturbance that is not so intense as to eliminate the regenerative capabilities of the biota. Secondary succession occurs on soils that have been modified biologically, and on sites where plants survived the disturbance. In contrast, primary succession occurs on a bare substrate that has not previously been influenced by organisms.

Succession—A process of ecological change, involving the progressive replacement of earlier communities with others over time, and generally beginning with the disturbance of a previous type of ecosystem.

areas must be deliberately burned to prevent them from successionally turning into forest.

Even stands of the giant **sequoia** (*Sequoiadendron giganteum*) appear to require periodic ground fires of a particular intensity if that community type is to be sustained over the longer term. Fire helps to maintain stands of this species, by reducing fuel loads and thereby preventing catastrophic crown fires that could kill mature trees, and by optimizing recruitment of sequoia seedlings.

Management of fires

Sometimes, to achieve particular ecological objectives, fire may be used as a tool in ecosystem management. The use of prescribed burns to maintain prairie was previously described, but similar practices have also been used to manage other ecological communities, and even some species. For example, prescribed burning is an essential component of the management strategy used to maintain an appropriate **habitat** of jack pine required by Kirtland's warbler (*Dendroica kirtlandii*), an endangered bird that only nests in northern Michigan. Prescribed burning is also used in **forestry** in some regions, to reduce the quantities of slash left after logging operations, to prepare a suitable seedbed for particular species of trees, or to prevent large build-ups of fuel that could lead to a more catastrophic wildfire.

To protect stands of timber that are important commercially or for other reasons, many agencies actively engage in fire protection activities. Fire protection is usually achieved by attempting to prevent humans from starting uncontrolled blazes, by using prescribed burns to prevent dangerous accumulations of large quantities of fuel, and by quenching fires that are accidentally or deliberately ignited. However, fire is a natural, ecological force, and even the greatest efforts of humans are not always capable of preventing or quenching large fires. This fact is occasionally brought to our attention when uncontrollable conflagrations destroy homes, commercial timber, or forest in protected areas such as parks.

See also Disturbance, ecological.

Resources

Books

Barbour, M.G., et al. *Terrestrial Plant Ecology, 2nd ed.* Don Mills, Ont.: Benjamin/Cummings Pub. Co, 1987.

Periodicals

Christensen, N.L., et al. "Interpreting the Yellowstone Fires of 1988." *BioScience* 39 (1989): 678-685.

Bill Freedman

Wildlife

It was once customary to consider all undomesticated **species** of vertebrate animals as wildlife. **Birds** and **mammals** still receive the greatest public interest and concern, consistently higher than those expressed for **reptiles** and **amphibians**. Most concern over fishes results from interest in sport and commercial value. The tendency in recent years has been to include more life-forms under the category of wildlife. Thus, **mollusks, insects**, and plants are all now represented on national and international lists of threatened and **endangered species**.

People find many reasons to value wildlife. Virtually everyone appreciates the aesthetic value of natural beauty or artistic appeal present in **animal** life. Giant **pandas**, bald **eagles**, and infant harp **seals** are familiar examples of wildlife with outstanding aesthetic value. Wild species offer recreational value, the most common examples of which are sport hunting and bird watching.

Less obvious, perhaps, is ecological value, resulting from the role an **individual** species plays within an **ecosystem**. Alligators, for example, create depressions in swamps and marshes. During periods of droughts, these "alligator holes" offer critical refuge to water-dependent life-forms. Educational and scientific values are those that serve in teaching and **learning** about **biology** and scientific principles.

Wildlife also has utilitarian value which results from its practical uses. Examples of utilitarian value range from genetic reservoirs for crop and **livestock** improvement to diverse biomedical and pharmaceutical uses. A related category, commercial value, includes such familiar examples as the sale of furs and hunting leases.

Shifts in human lifestyle have been accompanied by changes in attitudes toward wildlife. Societies of hunter-gatherers depend directly on wild species for food, as many plains Indian tribes did on the **bison**. But as people shift from hunting and gathering to agriculture, wildlife comes to be viewed as more of a threat because of potential crop or livestock damage. In modern developed nations, people's lives are based less on rural ways of life and more on business and industry in cities. Urbanites rarely if ever feel threatened economically by wild animals. They have the leisure **time** and mobility to visit wildlife refuges or parks, where they appreciate seeing native wildlife as a unique, aesthetic experience. They also sense that wildlife is in decline and therefore favor greater protection.

The most obvious threat to wildlife is that of direct exploitation, often related to commercial use. Exploitation helped bring about the **extinction** of the passenger pigeon (*Ectopistes migratorius*), the great auk, Stellar's sea cow, and the sea **mink**, as well as the near extinction of the American bison. In the late nineteenth and early twentieth century, state and federal laws were passed to help curb exploitation. These were successful for the most part, and they continue to play a crucial role in wildlife management.

Introductions of exotic species represent another threat to wildlife. Insular or island-dwelling species of wildlife are especially vulnerable to the impacts of exotic plants and animals. Beginning in the seventeenth century, sailors deliberately placed **goats** and **pigs** on **ocean** islands, intending to use their descendants as food on future voyages. As the exotic populations grew, the native vegetation proved unable to cope, creating drastic **habitat** changes. Other species, such as **rats**, **mice**, and **cats**, jumped ship and devastated island-dwelling birds, which had evolved in the absence of mammalian predators and had few or no defenses.

Pollution is yet another threat to wildlife. Bald eagles, ospreys, peregrine **falcons** (*Falco peregrinus*), and brown **pelicans** (*Pelecanus occidentalis*) experienced serious and sudden population declines in the 1950s and 60s. Studies showed that these **fish** eaters were ingesting heavy doses of **pesticides**, including DDT. The pesticides left the shells of their eggs so thin that they cracked under the weight of incubating parents, and numbers declined due to reproductive failure. Populations of these birds in the United States rebounded after regulatory laws curbed the use of these pesticides. However, thousands of other chemicals still enter the air, **water**, and **soil** every year, and the effects of most of them on wildlife are unknown.

By far the most critical threat to wildlife is habitat alteration. Unfortunately, it is also more subtle than direct exploitation, and thus often escapes public attention. As the twenty-first century approaches, human activities are altering some of the most biologically rich habitats in the world on a scale unprecedented in history. Tropical rainforests, for example, originally covered only about 7% of the earth's land surface, yet they are thought to contain half the planet's wild species. Other rich habitats undergoing rapid changes include tropical dry **forests** and coral reefs. As extensive areas of natural habitat are irrevocably changed, many of the native species that once occurred there will become extinct, even those with no commercial value.

Any species of wild animal has a set of habitat requirements. These begin with food requirements, adequate amounts of available food for each season. Cover requirements are structural components that are used for nesting, roosting, or watching, or that offer protection from severe **weather** or predators. Water is habitat requirement that affects wildlife directly, by providing drinking water, and indirectly, by influencing local vegetation. The final habitat requirement is space. Biologists can now calculate a minimum area requirement to sustain a particular species of a given size.

Even in the absence of human activities, populations of wild animals change as a result of variations in **birth** and death rates. When a population is sparse relative to the number that can be supported by local habitat conditions, birth rates tend to be high. In such circumstances, natural mortality, including predation, **disease**, and starvation, tends to be low. As populations increase, birth rates decline and death rates rise. These trends continue until the population reaches **carrying capacity**, the number of animals of a particular species that can be sustained within a given area.

Carrying capacity, though, is difficult to define in practice. Variations in winter severity or in summer rainfall can, between years, alter the carrying capacity for a particular area. In addition, carrying capacity changes as forests grows older, **grasslands** mature, or **wetlands** fill in through natural siltation. Despite these limitations, the concept of carrying capacity illustrates an important biological principle: living wild animals cannot be stockpiled beyond the practical limits that local habitat conditions can support.

While all populations vary, some undergo extreme fluctuations. When they occur regularly, such fluctuation are called cycles. Cyclical populations fall into two categories, the three- to four-year cycle typical of **lemmings** and **voles**, and the eight- to 11-year cycle known in snowshoe hares and lynx of the Western Hemisphere. The mechanisms that keep cycles going are complex and not completely understood, but the existence of cycles is wide-ly accepted. Extreme populations fluctuations that occur at irregular intervals are called population irruptions. Local populations of **deer** tend to be irruptive, suddenly showing substantial changes at unpredictable intervals.

There are more species of wild plants and animals in tropical rain forests than on arctic tundras. Such patterns illustrate variations in the complexities of life-forms due to climatic conditions. Measurement of **biodiversity** usually focuses on a particular group of organisms such as trees or birds. The most basic indication of species richness is the total number of species in a certain area or habitat. A measure of species evenness is more valuable because it indicates the relative abundance of each species. As diversity includes richness and evenness, an area with many species uniformly distributed would have a high overall diversity.

Biodiversity is important to wildlife **conservation** as well as to basic **ecology**. Comparisons of species diversity patterns indicate the extent to which natural conditions have been affected by human activities. They also help establish priorities for acquiring new protected areas.

Resources

Books

Caughley, G. *Analysis of Vertebrate Populations.* New York: John Wiley & Sons, 1977.

Kellert, S. *Trends in Animal Use and Perception in 20th Century America.* Washington, DC: U.S. Department of the Interior, Fish and Wildlife Service, 1981.

Matthiessen, P. *Wildlife in America.* 2nd ed. New York: Viking Books, 1987.

McCullough, D. *The George Reserve Deer Herd.* Ann Arbor, MI: University of Michigan Press, 1979.

James H. Shaw

Wildlife trade (illegal)

Many **endangered species**, or their body parts, are extremely valuable for one reason or another. In some cases, they are avidly sought by public zoos or botanical gardens, or by private collectors, who may be willing to pay large sums of money for living or dead specimens to add to their collection. In other cases, parts of an **animal** or **plant** may be valuable. This can result in **species** being killed for their precious fur, ivory, horn, or internal organs.

Much of the harvesting and trade in **wildlife** is legal, and not a threat to species that are widespread and abundant. In many other cases, however, endangered species are being illegally harvested because of the vast sums of money that can be made. The illegal trade in wildlife involves a well-organized chain of commerce, which includes: the hunters (also known as poachers), the buyers of the living animal or plant or its body parts, the traders exporting or importing the goods, the manufacturers of consumer products, and ultimately the consumers. Each of these actors plays a crucial role in the illegal drama of wildlife trade. This trade is greatly increasing the risk of **extinction** of many endangered species of plants and animals.

The trade in wildlife

The international commerce in wildlife has a value of about $20 billion per year. In recent years, this economic activity has involved about five million wild **birds**, 32 thousand **primates**, 12 million orchids, 11 million cacti, and huge numbers of other kinds of organisms. Most of this trade is legal, but a great deal is not, and involves an organized network of poaching, smuggling, and illicit sales.

Much of the wildlife trade involves the sale of living organisms for public zoological or botanical collections, or as private pets. This affects millions of wild-collected animals and plants each year, including endangered species. There is also an enormous trade in the parts of animals and plants. For example, the facial horns of rhinoceroses are extremely valuable in eastern **Asia** as an ingredient in traditional medicine, and in Yemen for manufacturing into dagger handles. Bile from the gall bladder of bears is also a precious material in traditional Asian medicine, as are the bones of tigers, and the roots of wild **ginseng**. Another example is **elephant** ivory, which is valued for use in artisanal crafts in many countries. Other valuable products of endangered species include rare furs, a fact that has threatened large **cats** such as the tiger, cheetah, leopard, and jaguar. Many species of plants and animals have become endangered because of excessive hunting and trade of their valuable body parts.

Monitoring and regulating the international trade in endangered species

The Convention on International Trade in Endangered Species, often referred to by its acronym CITES, is a treaty that since 1973 has committed 145 signatory nations to preventing or controlling the international trade of endangered species. CITES was established in 1973 under the auspices of the United Nations Environment Program (UNEP). The goal of CITES is to monitor and regulate the international trade in endangered species. For these purposes, the **conservation** status of species (that is, as being endangered, vulnerable, or rare) is designated by the International Union for the Conservation of Nature (IUCN). The actual international trade of

species-at-risk is monitored by the "Traffic" network of the World Wildlife Fund (WWF) and the IUCN. The headquarters of CITES, IUCN, and WWF are all located in Switzerland. In addition, the World Conservation Monitoring Center (WCMC), located in England, publishes a series of so-called "red books" that summarize the status and commerce of about 60,000 species of plants and 2,000 of animals.

CITES and its partners regulate or monitor the international trade of about 639 species of **mammals**, 1,557 birds, 464 **reptiles**, 81 **amphibians**, 36 **fish**, 2,070 **invertebrates**, and 25,660 plants. In most of these cases, the international trade is only monitored. However, in 821 cases involving species threatened with extinction, any international trade is banned. A few examples of species for which trade is not allowed include endangered hyacinth macaws, rhinoceroses, tigers, sea **turtles**, and certain rare orchids.

The United States is a member of CITES. Some of its responsibilities under the treaty are to monitor and report on its international trade of all species dealt with by WCMC. The United States also has its own legislation governing the domestic trade in endangered species: the Endangered Species Act of 1973. The Fish and Wildlife Service has the responsibility of monitoring and policing any illegal trade in wildlife, both domestic and international.

Resources

Books

Blair, C.B. *Endangered Species: Must They Disappear?* Information Plus Publishers, 1996.

Fitzgerald, S. *International Wildlife Trade: Whose Business Is It?* World Wildlife Fund, 1990.

Henley, G. *International Wildlife Trade: A CITES Sourcebook.* Island Press, 1994.

Periodicals

Li, Y.M. "Illegal Wildlife Trade In The Himalayan Region Of China." *Biodiversity and Conservation* 9, no. 7 (2000): 901-918.

Martin, E. "Wildlife for Sale." *Biologist* 47, no. 1 (2001): 27-30.

"Profile: Ian Redmond: An 11th-Hour Rescue for Great Apes?" *Science* 297 no. 5590 (2002): 2203.

Other

United States Fish and Wildlife Service. 1849 C St., NW, Washington, DC 20240. [cited 2003]. <http://www.fws.org>.

World Conservation Monitoring Center (WCMC). 219 Huntingdon Road, Cambridge CB3 0DL, United Kingdom. [cited 2003]. <http://www.wcmc.org.uk>.

World Wildlife Fund (WWF). Avenue du Mont-Blanc, CH-1196, Gland, Switzerland. [cited 2003]. <http://www.panda.org>.

Bill Freedman

Willow family (Salicaceae)

Willows are a diverse group of about 300 **species** of woody **angiosperm** plants in the genus *Salix*, family Salicaceae. Willows are widely dispersed and occur on all continents except **Antarctica**, but they are most diverse in cooler regions of the Northern Hemisphere.

All willows are woody plants, but the species vary greatly in size. Some species of willows are trees that can grow taller than 49 ft (15 m), while others are dwarf shrubs of the **tundra** that never get any taller than a few centimeters.

Biology of willows

Willows have simple, slender leaves, alternately arranged on the twigs, and with toothed or entire margins. The foliage of willows is seasonally deciduous, being shed in the autumn. Willow plants are dioecious, meaning that particular individuals bear either male or female flowers but not both. Both types of flowers usually produce **nectar** so that **pollination** is by **insects**. The flowers of willows are arranged in elongate inflorescences, known as catkins. The **fruits** are a capsule, containing tiny **seeds** with tufted hairs that make them aerodynamically buoyant so that they can be dispersed widely by the **wind**.

Willows are rather fast growing woody plants, but they are relatively short lived. Some species of willows sprout prolifically, and they may form dense thickets in moist, recently disturbed habitats. Most willows are relatively easy to cultivate from stem cuttings.

The usual **habitat** of willows is moist places, often beside streams, **rivers**, lakes, and other surface waters.

Species of willows

More than 100 species of willows are native to **North America**. Most of these are shrubs or dwarf shrubs, but about forty species reach **tree** size. Willow species commonly hybridize with each other and this,

along with their relatively great richness of species, can make some of the willows difficult to identify.

Some of the more common species of willow that can attain the size of trees include the following: the black willow (*Salix nigra*) is a widespread tree in low-lying and riparian habitats in the eastern United States and southern Ontario; the peachleaf willow (*S. amygdaloides*) is also widespread in central North America; the Pacific willow (*S. lasiandra*) is another tree-sized species, occurring widely from southern California to central Alaska.

Shrub-sized species of willows are richer in species and include the following: The sandbar willow (*S. interior*) occurs widely in eastern and central North America and through the boreal forest to central Alaska. This species often forms thickets in flat, moist, alluvial habitats. The arroyo willow (*S. lasiolepis*) occurs in moist canyons and along streams in the western United States. The Mackenzie willow (*S. mackenzieana*) is a northwestern species. The coastal plain willow (*S. carolineana*) is widespread in the southeastern United States. The Bebb willow (*S. bebbiana*) is a common shrub of boreal and cool-temperate regions. The feltleaf willow (*S. alaxensis*) is widespread in the northeastern boreal forest.

Many species of willows are dwarf shrubs, occurring in alpine and arctic tundra. The stems of these tiny willows grow horizontally along the ground, and in some cases they never rise any higher than several centimeters above the ground surface. The most widespread of the dwarf willows is the arctic willow (*S. arctica*). This species occurs throughout much of the tundra of North America, Greenland, and Eurasia, as far north as the limits of land. Another arctic species is the reticulated willow (*S. reticulata*).

Economic and ecological importance of willows

Many species of willows are important ecologically. Willows are often species of early **succession**, and they are important in the early and middle stages of successional recovery after disturbance. Willows are commonly an important browse of **mammals** such as **deer**, **moose**, rabbits, hares, and other species, especially during the winter when herbaceous forage is not very available.

Tree-sized willows are sometimes used for lumber. The black willow is the only species used much for this purpose in North America. Because its **wood** is not very strong, it is generally used to manufacture boxes and similar goods.

Because willows can be so productive, there has been research into the cultivation of tree-sized willows in

Willow in bloom. *Photograph by Robert J. Huffman. Field Mark Publications. Reproduced by permission.*

plantations for use as a **biomass** fuel. This use of willows as a source of renewable **energy** may prove to be important in the future. The willow biomass can be burned directly, or it can be chemically converted into more easily portable liquid fuels such as **alcohol** or a synthetic, petroleum-like mixture which can be manufactured under **heat** and **pressure**.

Because willows grow quickly and are so easy to propagate using stem cuttings, they are often used to vegetate stream banks to help prevent **erosion** and sometimes to re-vegetate other types of disturbed lands.

Willows have long had some use in folk medicine. Many cultures are known to have chewed willow twigs to relieve **pain** and fever. The original source from which salicylic acid was extracted was the **bark** of the white willow (*S. alba*) of **Europe**. This chemical is used to manufacture **acetylsalicylic acid** or ASA (sometimes known as aspirin), an economically important analgesic useful for treating pain, fever, and **inflammation**.

Willows may be an important source of nectar for **bees** in the early spring, a time when few other species of insect-pollinated plants are flowering. Willow honey may be a locally significant product in some areas.

Willow twigs are rather flexible and have been used to weave baskets, for caning, and to make woven fences and other lattices.

Some species of willows have good aesthetics and are utilized in **horticulture**. One of the best known species for this purpose is the weeping willow (*Salix babylonica*), a beautiful, pendulous tree. This species is native to northern China. However, the weeping willow was considered to be so beautiful by the famous Swedish botanist Carl von Linnaeus, who gave the species its scientific name, that he decided that it must have been present in the biblical Garden of Babylon; hence, the origin of the geographically inaccurate, scientific binomial of the weeping willow.

The weeping willow has been widely introduced to North America as an ornamental tree. Other non-native species that are commonly used in horticulture include the crack willow (*S. fragilis*) of Eurasia and the white willow (*S. alba*) and basket willow (*S. viminalis*) of Europe. Some of these species have escaped from cultivation and have become locally invasive in natural habitats.

Wild willows also have pleasant aesthetics. Most famous in this sense are the several species known as "pussy willows," especially the pussy willow (*Salix dis-*

KEY TERMS

Browse—A food consisting of the foliage, twigs, and flowers of woody plants.

Catkin—An elongate, spikelike cluster of unisexual flowers, often drooping at maturity. Catkins are the floral type of the willow family.

Dioecious—Plants in which male and female flowers occur on separate plants.

color). These species produce large, attractive catkins in the early springtime. In fact, stems of these species can be collected in the late winter before they have bloomed and placed in **water** in a vase. In a short time, the pussy-willow stems will bloom indoors to pleasantly herald the arrival of spring.

Resources

Books

Judd, Walter S., Christopher Campbell, Elizabeth A. Kellogg, Michael J. Donoghue, and Peter Stevens. *Plant Systematics: A Phylogenetic Approach.* 2nd ed. with CD-ROM. Suderland, MD: Sinauer, 2002.

Klein, R.M. *The Green World. An Introduction to Plants and People.* New York: Harper and Row, 1987.

Bill Freedman

Willy-willy *see* **Tropical cyclone**

Wind

The term wind refers to any flow of air relative to the Earth's surface in a roughly horizontal direction. Breezes that blow back and forth from a body of **water** to adjacent land areas—on-shore and off-shore breezes—are examples of wind.

The ultimate cause of Earth's winds is solar **energy**. When sunlight strikes Earth's surface, it heats that surface differently. Newly turned **soil**, for example, absorbs more **heat** than does snow.

Uneven heating of Earth's surface, in turn, causes differences in air **pressure** at various locations. On a **weather** map, these pressure differences can be found by locating **isobars**, lines that connect points of equal pressure. The pressure at two points on two different isobars will be different. A pressure gradient is said to exist between these two points. It is this pressure gradient that provides the **force** that drives air from one point to the other, causing wind to blow from one point to the other. The magnitude of the winds blowing between any two points is determined by the pressure gradient between those two points.

The Coriolis effect and wind direction

In an ideal situation, one could draw the direction of winds blowing over an area simply by looking at the isobars on a weather map. But the **earth** is not an ideal situation. At least two important factors affect the direction in which winds actually blow: the **Coriolis effect** and **friction**. The Coriolis effect is a pseudoforce that appears to be operating on any moving object situated on a rotating body, such as a stream of air traveling on the surface of the rotating **planet**. The effect of the Coriolis force is to deflect winds from the straight-forward direction that we might expect them to take simply from an examination of isobars. In the Northern Hemisphere, the Coriolis effect tends to deflect winds to the right and in the Southern Hemisphere, it tends to drive winds to the left.

Imagine how the Coriolis effect will determine the movement of winds in the Northern Hemisphere. Suppose that air initially begins to move from west to east as a result of pressure gradient forces. At once, the Coriolis effect will begin to drive the stream of air to the right, that is, to the south. The actual path followed by the wind, then, is a compromise between the pressure gradient force and the Coriolis force. Since each of these forces can range widely in value, the precise movement of wind in any one case is also variable.

At some point, the two forces driving the wind are likely to come into balance. At that point, the wind begins to move in a straight line that is perpendicular to the direction of the two forces. Such a wind is known as a geostrophic wind.

Friction and wind movement

The picture described above applies to winds that blow in the upper atmosphere. At distances of more than a kilometer or so above the ground, pressure gradient and Coriolis forces are the only factors affecting the movement of winds. Thus, air movements eventually reach an equilibrium point between pressure gradient forces and the Coriolis force, and geostrophic winds blow parallel to the isobars on a weather map.

Such is not the case near ground level, however. An additional factor affecting air movements near the earth's surface is friction. As winds pass over the earth's surface, they encounter surface irregularities and slow down. The decrease in wind speed means that the Coriolis effect acting on the winds also decreases. Since the pressure gradient force remains constant, the wind direction is driven more strongly toward the lower air pressure. Instead of developing into geostrophic winds, as is the case in the upper atmosphere, the winds tend to curve inward towards the center of a low pressure area or to spiral outward away from the center of a high pressure area.

Friction effects vary significantly with the nature of the terrain over which the wind is blowing. On very hilly land, winds may be deflected by 30 degrees or more, while on flat lands, the effects may be nearly negligible.

Local winds

In many locations, wind patterns exist that are not easily explained by the general principles outlined above. In most cases, unusual topographic or geographic features are responsible for such winds, known as local winds. **Land and sea breezes** are typical of such winds. Because water heats up and cools down more slowly than does dry land, the air along a shoreline is alternately warmer over the water and cooler over the land, and vice versa. These differences account for the fact that winds tend to blow offshore during the evening and on-shore during the day.

The presence of **mountains** and valleys also produces specialized types of local winds. For example, Southern Californians are familiar with the warm, dry Santa Ana winds that regularly sweep down out of the San Gabriel and San Bernadino Mountains, through the San Fernando Valley, and into the Los Angeles **Basin**, often bringing with them widespread and devastating wildfires.

See also Atmospheric circulation; Atmospheric pressure.

Resources

Books

Ahrens, C. Donald. *Meteorology Today.* 2nd ed. St. Paul: West Publishing Company, 1985.

Battan, Louis J. *Fundamentals of Meteorology.* Englewood Cliffs, NJ: Prentice-Hall, Inc., 1979.

Holton, James R. *An Introduction to Dynamic Meteorology.* 2nd ed. New York: Academic Press, 1979.

Lutgens, Frederick K., and Edward J. Tarbuck. *The Atmosphere: An Introduction to Meteorology.* 4th ed. Englewood Cliffs, NJ: Prentice Hall, 1989.

David E. Newton

Wind chill

Wind chill is the **temperature** felt by humans as a result of air blowing over exposed skin. The temperature

WIND CHILL EQUIVALENT TEMPERATURE TABLE

DRY BULB TEMPERATURE (°F)

WIND VELOCITY (MPH)	45	40	35	30	25	20	15	10	5	0	−5	−10	−15	−20	−25	−30	−35	−40	−45	
4	45	40	35	30	25	20	15	10	5	0	−5	−10	−15	−20	−25	−30	−35	−40	−45	4
5	43	37	32	27	22	16	11	6	0	−5	−10	−15	−21	−26	−31	−36	−42	−47	−52	5
10	34	28	22	16	10	3	−3	−9	−15	−22	−27	−34	−40	−46	−52	−58	−64	−71	−77	10
15	29	23	16	9	2	−5	−11	−18	−25	−31	−38	−45	−51	−58	−65	−72	−78	−85	−92	15
20	26	19	12	4	−3	−10	−17	−24	−31	−39	−46	−53	−60	−67	−74	−81	−88	−95	−103	20
25	23	16	8	1	−7	−15	−22	−29	−36	−44	−51	−59	−66	−74	−81	−88	−96	−103	−110	25
30	21	13	6	−2	−10	−18	−25	−33	−41	−49	−56	−64	−71	−79	−86	−93	−107	−109	−116	30
35	20	12	4	−4	−12	−20	−27	−35	−43	−52	−58	−67	−74	−82	−89	−97	−105	−113	−120	35
40	19	11	3	−5	−13	−21	−29	−37	−45	−53	−60	−68	−76	−84	−92	−100	−107	−115	−123	40
45	18	10	2	−6	−14	−22	−30	−38	−46	−54	−62	−70	−78	−85	−93	−102	−109	−117	−125	45

VERY COLD — BITTER COLD — EXTREME COLD

Table 1. *Illustration by Hans & Cassidy. Courtesy of Gale Group.*

that humans actually feel, called the *sensible temperature*, can be quite different from the temperature measured in the same location with a **thermometer**. The reason for such differences is that the human body constantly gives off and absorbs **heat** in a variety of ways. For example, when a person perspires, **evaporation** of moisture from the skin removes heat from the body, and one feels cooler than the true temperature would indicate.

In still air, skin is normally covered with a thin layer of warm molecules that insulates the body and produces a sensible temperature somewhat higher than the air around it. When the **wind** begins to blow, that layer of molecules is swept away, and body heat is lost to the surrounding atmosphere. An individual begins to feel colder than would be expected from a thermometer reading at the same location. The faster the wind blows, the more rapidly heat is lost and the colder the temperature appears to be.

The National Weather Service has published a wind chill chart that shows the relationship among actual temperature, wind speed, and wind chill factor, or the temperature felt by a person at the given wind speed. (See Table 1.) According to this chart, a wind speed of 4 MPH (6 km/h) or less results in no observable change in temperature sensed. At a wind speed of 17 MPH (30 km/h) and a temperature of 32°F (0°C), however, the perceived temperature is 7°F (-14°C).

The colder the temperature, the more strongly the wind chill factor is felt. At a wind speed of 31 mi/h (50 km/h), for example, the perceived temperature at 32°F (0°C) is 7°F (-14°C), but at -40°F (-40°C), the perceived temperature is -112°F (-80°C).

Wind energy *see* **Alternative energy sources**

Wind shear

Wind shear is the difference in speed or direction between two layers of air in the atmosphere. Wind shear may occur in either a vertical or horizontal orientation. An example of the former situation is the case in which one layer of air in the atmosphere is traveling from the west at a speed of 31 mph (50 kph) while a second layer above it is traveling in the same direction at a speed of 6.2 mph (10 kph). The **friction** that occurs at the boundary of these two air currents is a manifestation of wind shear.

An example of horizontal wind shear occurs in the **jet stream** where one section of air moves more rapidly than other sections on either side of it. In this case, the wind shear line lies at the same altitude as various cur-

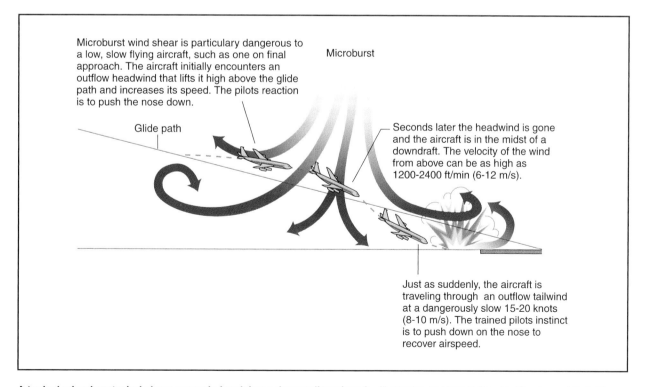

A typical microburst wind shear scenario involving a descending aircraft. *Illustration by Hans & Cassidy. Courtesy of Gale Group.*

rents in the jet stream, but at different horizontal distances from the jet stream's center.

Wind shear is a crucial factor in the development of other atmospheric phenomena. For example, as the difference between adjacent wind currents increases, the wind shear also increases. At some point, the boundary between currents may break apart and for eddies that can develop into clear air **turbulence** or, in more drastic circumstances, tornadoes and other violent storms.

Wind shear has been implicated in a number of disasters resulting in property damage and/or loss of human life. The phenomenon is known as a microburst, a strong localization down draft (down burst) which, when it when reaches the ground, continues as an expanding outflow. For example, it is associated with the movement of two streams of air at high rates of speed in opposite directions. An airplane that attempts to fly through a microburst passes through the wind shear at the boundary of these two air streams. The plane feels, in rapid **succession**, an additional lift from headwinds and then a sudden loss of lift from tailwinds. In such a case, a pilot may not be able to maintain control of the **aircraft** in time to prevent a crash.

See also Atmospheric circulation; Wind.

Wintergreen

Wintergreen is the common name for an evergreen shrub belonging to the genus *Gaultheria* and heath family Ericaceae. *Gaultheria procumbens* is native to eastern **North America** and grows wild in sandy, wooded areas or shady clearings. This shrub grows 4-6 in (10-15 cm) high with creeping stems. Stalks grow from the stems and have elliptical, shiny green leaves and leathery leaves. During mid-summer, white, drooping, bell-shaped flowers grow; in the fall, wintergreen has bright red edible berries. The berries, called deerberries or checkerberries, contain several **seeds** and remain on the plants into the winter season.

Some plants of the genus *Chimaphila*, related to *Gaultheria*, are also called wintergreens. One **plant** of interest is pipsissewa (*C. umbelata*), which comes from the Cree word pipisisikweu, meaning "[the juice] breaks it into small pieces." It has been speculated that the Cree used this type of wintergreen as a remedy for kidney stones.

Wintergreen is also called teaberry or checkerberry. The whole plant, particularly the leaves, is a source of the volatile oil that has made wintergreen's spicy, sweet taste very popular. Wintergreen oil is used to flavor gum, candy, toothpaste, mouthwash, and birch beer (a carbonated soft drink). The oil is also used in topical antiseptics and liniments. The active element of wintergreen is methyl salicylate, which is a **derivative** of salicylic acid (an important ingredient of aspirin). American Indians of the eastern woodlands made poultices from wintergreen leaves as a remedy for muscle and joint aches, inflammations, and toothaches. The Indians also taught early settlers how to make wintergreen tea for sore throats, nausea, and fevers.

During the early colonization days of North America, large amounts of wintergreen were gathered, dried, and transported to distilleries where the oil was extracted from the leaves. Today chemical factories make synthetic oil of wintergreen in the form of methyl salicylate. The leaves of natural wintergreen can be chewed; at first they will taste sweet, but the taste turns bitter very quickly. One should be careful when using pure wintergreen oil, as it can be irritating to the skin and poisonous if ingested internally. Many drugstores no longer sell pure wintergreen oil, as it is illegal to sell it in many local jurisdictions.

Christine Miner Minderovic

Wire chambers see **Particle detectors**

Wolverine

The wolverine (*Gulo gulo*) is an uncommonly large member of the weasel family (Mustelidae) that occurs in the subarctic and boreal **forests** of **North America** and northern Eurasia. The wolverine is famous for its aggressive, combative nature and its remarkable strength. Although the wolverine is only a medium-sized **animal**, it can dominate much larger animals during an aggressive encounter at a food source. For example, wolverines are capable of driving away a bear, or a group of wolves, from an animal carcass. If the movement of a **moose** or **deer** is hampered by snow, it can be killed by a single, much smaller wolverine.

Wolverines typically have a body length of 5.5-8.5 ft (2-2.5 m) plus a tail of about 8-12 in (20-30 cm). Adult animals weigh 30-60 lb (14-28 kg). Female wolverines are typically smaller than males. Wolverines have a long, dense, rather lustrous fur, usually dark brown or black. They generally have lighter brown-colored bands on each side of the body, extending from the shoulders to the rump.

Wolverines are rather omnivorous in their feeding habits. They are primarily eaters of carrion, that is, an-

A wolverine in the snow. *Photograph by Tom Brakefield. Stock Market. Reproduced by permission.*

imals that have died of natural causes, or have been killed by other **species** of predators. The wolverine has powerful jaws and teeth, and it can crush the bones of dead animals to extract the nutritious marrow. Their food of carrion is mostly located by smell, as wolverines have rather poor **vision**. Wolverines sometimes kill their own **prey**. However, this only happens opportunistically, because these animals are rather slow moving and often cannot run down a healthy prey animal. Wolverines will also eat bird eggs, insect larvae, and berries when these foods are abundant. Wolverines are famous for their avaricious appetite, as is reflected in some of their alternative common names, such as glutton.

Adult wolverines are solitary animals, and they are active throughout the year. The territories of male wolverines can be very large, as much as thousands of square kilometers in extent, but shared with several resident females. Wolverines move over their territories with a persistent, loping gait, but they are also adept at climbing trees in pursuit of prey.

The fur of the wolverine is very highly regarded by northern peoples for use in the ruff around the hood of parkas. Wolverine fur is relatively effective at repelling the moisture emitted during breathing, so it does not frost up to the degree of other furs.

Wolverines are indicators of wilderness and of ecosystems that are relatively unaffected by humans. Unfortunately, trapping has eliminated wolverines from much of their natural range, and they are becoming increasingly rare in those ranges where they still manage to hang on. Currently, wolverines are classified as a "vulnerable" species, meaning that the species is not critically endangered or otherwise endangered, but is facing a high risk of **extinction** in the wild in the medium-term future. The affected wolverine populations include those of Canada, the United States, Finland, Norway, Russia, and Sweden. It is critical that this special animal of wild landscapes be allowed to survive in its remaining territories, and perhaps be reintroduced to parts of its former range where suitable **habitat** still remains.

Wolves *see* **Canines**

Wombats

Wombats are thickset, bear-like, Australian **marsupials** (order Marsupalial). They dig burrows, are about the size of a small dog, and have perpetually growing teeth (like placental **rodents**). Wombats are members of the family Vombatidae, which includes three **species**. The critically endangered Queensland or northern hairy-nosed wombat (*Lasiorhinus krefftii*) is limited to a small area of Queensland National Park, where the surviving animals (as few as 40 individuals) live in burrows in an old riverbed. The southern hairy-nosed wombat (*L. latifrons*) lives in dry **grasslands** of southern **Australia**, such as the Nullabor Plain along the Great Australia Bight. The hairy-nosed wombats have fine white hairs on a fairly large nose, while the nose of the common wombat (*Vombatus ursinus*) is leathery and bare. The common wombat lives in **forests** on mountainsides, and has characteristic, rounded ears.

Wombats can weigh more than 80 lb (40 kg) and can be 3 ft (1 m) long. They are gray or brownish gray, and are darker on the back than on the belly. The fur of the common wombat is coarse, while that of the hairy-nosed wombats is soft and silky. The feet of wombats have bare, leathery pads that withstand the energetic digging by their broad, flat, shovel-like claws (five on the front feet, four on the back).

In Australia and nearby islands, wombats occupy the ecological **niche** of a burrow-living plant-eater, similar to that of the woodchuck in **North America**. Wombats have chisel-like teeth that must grow continuously because

A southern hairy-nosed wombat. *Photograph by Terry Whittaker. The National Audubon Society Collection/Photo Researchers, Inc. Reproduced by permission.*

they are eroded by constant chewing on coarse **grasses** and roots. Wombats are quick and efficient at digging, which they do by lying on their side and using their heavy-clawed front feet to dig and their hind feet to push the **soil** backward. Wombats often lie near the entrance to their burrow and bask in the **sun**, feeding at night.

Hairy-nosed wombats dig systems of interconnecting tunnels, while the common wombat is likely to inhabit a limited system of only two or three tunnels. Common wombats establish a territory of up to 60 acres (25 hectares), but do not demand exclusivity, for it often overlaps with that of another wombat. Common wombats may even share the same burrow, though they occupy it at different times of day.

Wombats have a poorly developed cooling system, and if they get too hot they can die. They usually avoid high temperatures by retreating to their burrows, which provide cooler places in the summer and a warmer spot in winter. The hairy-nosed wombats rarely drink, getting all their **water** from the plants they eat. Their food is digested very slowly in order to get the most nourishment from it.

Wombats are solitary animals, except during the mating season, and then only if there is sufficient food available. Males fight for mating rights and females fight off males during **courtship**, so any pair that mates is usually bloody and scarred. After a gestation of only three or four weeks, a single offspring is born, which is carried in a rear-facing pouch. The wombat's pouch is held closed by a strong muscle. As the tiny wombat grows and begins to move around, this pouch muscle relaxes, and the infant wombat may be seen peering out from between its mother's hind legs. It takes at least five months for the eyes to open and the fur to grow in. The young wombat does not leave the pouch until it is about a year old. Wombats make amiable pets and have lived in captivity for over 25 years.

Resources

Books

Lavine, Sigmund A. *Wonders of Marsupials.* New York: Dodd, Mead & Co., 1978.

Lyne, Gordon. *Marsupials and Monotremes of Australia.* New York: Taplinger Publishing, 1967.

Triggs, Barbara. *Wombats.* Boston: Houghton Mifflin, 1991.

Jean F. Blashfield

Wood

Wood, also known as *secondary xylem*, is a composite of tissues found in trees. Secondary xylem is composed primarily of cells, called vessel elements in angiosperms, or of slightly different cells in gymnosperms called *tracheids*. These cells of secondary xylem, along with specialized cells of a type called *parenchyma*, are made by a meristematic **tissue** called the *vascular cambium*. As the vascular cambium generates new cells, secondary xylem accumulates on its inside, and the **tree** increases in diameter.

Newly made vessel elements and tracheids are **water** conduits from the roots of plants to their leaves. When first made, vessel elements and tracheids are alive but once they mature and become functional, they die. The functional vessel elements or tracheids occur in a few cell layers behind the vascular cambium, in a water-conducting section of the secondary xylem known as *sapwood*.

The parenchyma are made by the vascular cambium along with the vessels or tracheids, and are located at certain points along the perimeter of the vascular cambium. As the tree expands through growth, these narrow columns of parenchyma cells, called xylem rays, become longer, and ultimately extend from the vascular cambium to very near the center of the tree trunk. The function of xylem rays is to transfer aqueous material horizontally along the diameter of the tree, at a right angle to the flow of water in vessel elements and tracheids. The parenchyma cells of the xylem rays are alive in their mature, functional state.

As newer vessel elements or tracheids are made, older ones become buried under successive layers of more recently formed xylem. As the tree gets progressively larger in diameter, older secondary xylem tissues no longer conduct water. After this happens, these non-conducting cells are used to store waste products, such as **resins**. The xylem rays function to conduct wastes from actively functioning cells near the vascular cambium, to the non-functioning xylem cells. This waste-filled secondary xylem is called *heartwood*. By the time that a tree is larger than about 4-8 in (10-20 cm) in diameter, most of its **biomass** is composed of heartwood. New sapwood is created during each growing season but, within two to three years these cells become part of the heartwood. It is the heartwood of trees that is harvested to manufacture the lumber and **paper** used by people.

Wood of different **species** varies in **density** and strength, due to the size and density of the vessel elements or tracheids in the secondary xylem. For example, heartwood of the Brazilian ironwood (*Caesalpinia ferrea*) has very tiny vessel elements and is extremely dense. At the opposite extreme, the heartwood of balsa (*Ochroma pyramidale*) has very large vessel elements, and is correspondingly light in density. The wood of typical gymnosperms is generally soft and light in density,

KEY TERMS

· ·

Meristem—A cluster of similar, undifferentiated dendrochronology plant cells that produce cells which do differentiate, and become mature tissues.

Parenchyma—A non-vascular tissue composed of large, thin-walled cells that may differ in size, shape, and structure of cell wall.

Tracheids—Thick walled, lignified elements of xylem which have no perforations on the cross-walls of adjoining cells. Instead, water is transferred among tracheid cells through holes in the side of the cell walls known as bordered pits.

Vascular cambium—Undifferentiated plant tissue which gives rise to phloem and xylem.

Vessel elements—Thick-walled, lignified elements of xylem that have perforated or missing end walls. The relatively large openings in the cross-walls between adjoining cells allow a continuous, vertical transport of water.

Xylem—Plant tissue that transports water and minerals upward from the roots.

because tracheids do not fit together as closely as the vessel elements in the xylem of most angiosperms.

The size of tracheids and vessel elements also varies within a single tree, according to the season of the year that they were laid down during growth. In spring, when air temperatures are cool and **soil** moisture is typically plentiful, the vascular cambium of trees makes large diameter xylem cells. As the moisture wanes and temperatures increase in the summer, the vascular cambium makes smaller diameter cells. In the winter, no new cells are made, because of the cold temperatures. This cycle repeats itself every year and makes visible growth rings in the tree (except in the tropics). These rings are evident because spring wood, with larger diameter cells, is relatively dark in appearance, while summer wood is lighter in **color**. This annual repetition of differing cell sizes in growth rings is useful in ecological studies through *dendrochronology*. Because the size of vessel elements or tracheids is dependent on both air **temperature** and water, dendrochronologists can determine past periods of **drought**, flood, and unseasonal cold or **heat**, by studying variations in the width of growth rings.

Vessel elements and tracheids differ in length between angiosperms and gymnosperms. Although the length of **individual** cells makes little difference in the ability of trees to conduct water, the length of cells is of

great importance to the paper industry. The length of these cells corresponds to the fiber length of pulp that is to turned into paper, and influences the quality of paper that can be produced. Short fibers make fine grade papers, while longer fibers make coarser grade papers.

Resources

Books

Bell, P. R. *Green Plants, Their Origin and Diversity.* Portland: Dioscorides Press, 1992.

Fahn, A. *Plant Anatomy.* 5th ed. New York: Pergamon Press, 1990.

Lewington, A. *Plants for People.* New York: Oxford University Press, 1990.

Stephen R. Johnson

Woodpeckers

Woodpeckers are **birds** in the family Picidae, which includes about 200 **species** of true woodpeckers, **wrynecks**, as well as the diminutive piculets. Woodpeckers are widespread in the world's forested areas, occurring everywhere but **Australia**, New Zealand, New Guinea, Madagascar, and **Antarctica**. Birds in the woodpecker family range in size from the relatively enormous imperial woodpecker (*Campephilus imperialis*) of Mexico, with a body length of 21.7 in (55 cm) and weight of 1.1 lb (550 g), to tropical piculets only 3.2 in (8 cm) long.

Instinctive behavior

Woodpeckers spend a great deal of their time pecking **wood** and chiseling **bark** off trees. They do this for several reasons—to search for their principal food of wood-boring **arthropods**, to excavate their nesting and roosting cavities, and to proclaim their territory and impress potential mates by loud drumming. Woodpeckers accomplish these tasks by hammering vigorously at softer, fungal-rotted parts of living and dead trees, using their chisel-shaped bill. In their territorial drumming, however, woodpeckers tend to choose more resonant, unrotted trees. Some woodpeckers habitually use **telephone** poles and tin roofs as their sonorous drumming posts. Both males and females engage in drumming displays, which may be supplemented by loud, raucous, laughing calls. Woodpeckers use their cavities for both nesting and roosting. Although both sexes participate in brooding the eggs and young, only the male bird spends the night in the nesting cavity.

Physical adaptions

Woodpeckers have a number of adaptations that permit the vigorous hammering of wood without damaging the bird. Their skull is thick-walled and the **brain** is cushioned by absorbent **tissue**, which helps withstand the physical shocks of their head blows. The tongue of the woodpecker is long, barbed, and sticky to help extract **insects** from crevices, and the **organ** is supported by an extended hyoid bone and its muscles. The bill of woodpeckers is stout and pointed, and it grows continuously because of the wear to which it is subjected. As an **adaptation** for gripping vertical bark surfaces, woodpeckers have feet in which two toes point forward and two backward. The stiff, downward-propping tail feathers of woodpeckers also provide mechanical support while they are pecking.

Most woodpeckers live in **forests**, eating arthropods in or on trees, but a few species occur in more open habitats, where they often forage on the ground for arthropods. Some species are at least partly herbivorous, seasonally eating soft **fruits** and nuts. Many species of woodpeckers are migratory, while others are resident throughout the year in or near their territories. All of the true woodpeckers nest in cavities that they excavate in the soft, rotted interior of living or dead trees. However, some birds in the family use natural cavities or are secondary users of the abandoned excavations of other birds.

Woodpeckers in North America

About 21 species of woodpeckers regularly breed in **North America**. The largest species is the 18 in (46 cm) American ivory-billed woodpecker (*Campephilus principalis*) of the southeastern United States, although this species is rare and may even be extinct.

The pileated woodpecker (*Dryocopus pileatus*) is another large species, with a body length of 15 in (38 cm). This species is still widespread, although uncommon throughout its range.

The northern flicker (*Colaptes auratus*) occurs very widely across North America. The yellow-shafted flicker is a subspecies (*C. a. borealis*) with bright yellow underwing feathers and is predominant in the eastern and northern range, while the red-shafted flicker (*C. a. cafer*) is southwestern in distribution. Flickers often feed on the ground, eating **ants** and other insects.

The yellow-bellied sapsucker (*Sphyrapicus varius*) drills a horizontal series of holes in trees, which then ooze sugary sap that attracts and ensnares insects. They are later eaten by the sapsucker, as is some of the sap. The acorn woodpecker (*Melanerpes formicivorus*) occurs in oak forests of the southwestern United States. This species collects and caches acorns for future consumption, storing them in small holes that it excavates in **tree** bark. The acorn woodpecker lives in social groups of four to 10 closely related individuals. These cooperat-

ing birds engage in communal defense of a breeding territory, and they collect and store their acorns together.

The hairy and downy woodpeckers (*Picoides villosus* and *P. pubescens*) are the most widespread species in North America, occurring in almost every forest. The downy woodpecker is more abundant and familiar, often occurring in suburban environments. Both species will feed on suet and peanut butter at feeders.

Woodpeckers and humans

Woodpeckers have sometimes been regarded as **pests**. Sapsuckers occasionally cause damage when their horizontal rows of drillings girdle trees and prevent the free flow of sap and **water**. Pileated woodpeckers can damage wooden utility poles, sometimes requiring their premature replacement. Overall, woodpeckers provide more benefit than detriment to humans because they feed on injurious insects, provide nesting cavities for a wide range of other species of **wildlife**, and have positive aesthetics for birdwatchers and other people who enjoy sightings of these interesting and personable birds.

The populations of some species of woodpeckers have decreased greatly as a result of human activities. The American ivory-billed woodpecker may never have been very abundant in the North American part of its range, and it quickly declined when its preferred **habitat** of bottom land forests of **angiosperm** trees and swamps of cypress were cleared for agriculture or harvested for lumber. This species has not been seen in North America since the 1940s, and it is probably extirpated. The subspecies known as the Cuban ivory-billed woodpecker (*Campephilus principalis bairdii*) is also critically endangered, as is the closely related Imperial woodpecker of Mexico.

The red-cockaded woodpecker (*Picoides borealis*) occurs in old-growth pineforests in the southeastern United States. This species breeds colonially, and has a relatively complex social system, involving clan-helpers that aid in the rearing of broods. There have been large reductions in the pine forests that satisfy the relatively stringent habitat requirements of the red-cockaded woodpecker, because these ecosystems have been converted to agricultural uses, plantation forests, and residential developments. The diminished populations of red-cockaded woodpeckers are now extremely vulnerable to further losses of their habitat through human activities or because of natural disturbances such as **wildfire** and hurricanes, and the species is listed as endangered. To prevent the **extinction** of this species, it is necessary to protect suitable habitat, and to manage these protected areas sustainably. In other areas of suitable habitat lacking protection, it is necessary to greatly restrict the types

A male Gila woodpecker (*Melanerpes uropygialis*). *JLM Visuals. Reproduced by permission.*

of **forestry** that are permitted in the vicinity of known colonies of this **endangered species**.

Beyond the specific case of the endangered red-cockaded woodpecker, intensive forest management poses a more general risk to woodpeckers. This happens because the birds require a forest habitat that contains standing dead trees (snags), in which they can excavate cavities, feed, and display. Forestry tends to greatly reduce the numbers of snags in the forest, because dead trees can pose a tree-fall hazard to workers, and because they take up space without contributing to the economically productive forest resource. This is especially true of forestry plantations, where large snags may not be present at all, therefore depriving woodpeckers of an opportunity to utilize these industrial forests. One of the sensible accommodations that will have to be made by foresters to encourage woodpeckers (and the many other species of birds and **mammals** that utilize dead wood in forests) will be the provision of snags in managed forests to allow the native animals to sustain breeding popula-

tions. This would mean the integrated management of the land for both forest products and for woodpeckers and other species of wildlife.

Status

• Acorn woodpecker (*Melanerpes formicivorus*). Population appears stable.

• Black-backed woodpecker (*Picoides arcticus*). Local populations rise and fall with changes in the food supply, but the overall population appears stable.

• Downy woodpecker (*Picoides pubescens*). A common and widespread woodpecker. The population appears stable. Does not nest in birdhouses.

• Gila woodpecker (*Melanerpes uropygialis*). The population in California declined in the twentieth century, but this species remains abundant in Arizona.

• Golden-fronted woodpecker (*Melanerpes aurifrons*). This bird was once shot by railroad personnel because it was considered a telephone pole pest (soft pine is much easier to drill holes in than mesquite), and many were shot in Texas in the early 1900s. Today, the population appears stable.

• Great spotted woodpecker (*Dendrocopos major*). An Alaskan stray.

• Hairy woodpecker (*Picoides villosus*). Has declined in some areas due to loss of nesting sites. **Starlings** and house sparrows sometimes take over the nesting cavities.

• Ivory-billed woodpecker (*Campephilus principalis*). Almost certainly extinct. The last confirmed sitings in the United States were in the 1950s; reports persisted, however, of sitings in Cuba into the 1980s.

• Ladder-backed woodpecker (*Picoides scalaris*). Indications are that there has been a slight decline in number in recent years, but the population today appears stable.

• Lewis's woodpecker (*Melanerpes lewis*). Found erratically, so populations have been hard to monitor. There is some indication, however, that populations have declined in recent years. This woodpecker is sometimes considered an orchard pest.

• Nuttall's woodpecker (*Picoides nuttallii*). Population appears stable.

• Pileated woodpecker (*Drycopus pileatus*). The population declined in the East in the eighteenth and nineteenth centuries due to **deforestation**. This woodpecker has made a gradual comeback since 1900, becoming once again common in some areas.

• Red-bellied woodpecker (*Melanerpes carolinus*). The population in the North declined over the first half of the twentieth century, but this trend has recently re-

KEY TERMS

. . . .

Cavity nester—A bird that builds its nest in a hollow in a tree. Woodpeckers excavate their own avities, but other species of birds and some mammals use natural cavities, or excavations created and abandoned by other species, especially woodpeckers.

Integrated management—A management system that focuses on more than a single economic resource. In forestry, integrated management might be designed to enhance the resource of lumber and pulpwood, as well as the needs of hunted animals such as deer, non-hunted animals such as song birds and woodpeckers, the aesthetics of landscapes, and other values.

Snag—An erect but dead tree.

versed itself. The overall population appears stable, and may actually be increasing.

- Red-cockaded woodpecker (*Picoides borealis*). Endangered. The total population is estimated at less than 10,000, with many local groups facing extinction. The cause of this woodpecker's decline has been the suppression of natural fires and overcutting of the pine forests in the Southeast.

- Red-headed woodpecker (*Melanerpes erythrocephalus*). The population has been in decline for a number of years, probably due to loss of nesting sites and **competition** with starlings for nest cavities. This woodpecker avoids birdhouses.

- Strickland's woodpecker (*Picoides striclandi*). The population in the United States appears stable.

- Three-toed woodpecker (*Picoides tridactylus*). Local variations in population. Although usually uncommon, this woodpecker may become abundant during periods of heavy insect infestation. Appears stable in remote northern range.

- White-headed woodpecker (*Picoides albolarvatus*). Population appears stable.

- Northern flicker (*Colaptes auratus*). Abundant and widespread, but there may have been some decline in population. The flicker competes at a disadvantage with the starling for new nesting sites. The yellow- and red-shafted subspecies appear stable.

- Williamson's sapsucker (*Sphyrapicus thyoideus*). Population appears stable.

- Red-breasted sapsucker (*Sphyrapicus ruber*). Although the population may have declined due to cutting of forests of the Northwest, this species is still fairly numerous.

- Yellow-bellied sapsucker (*Sphyrapicus varius*). Although this bird has disappeared from some traditional southern nesting areas, it is still fairly numerous.

- Red-naped sapsucker (*Sphyrapicus nuchalis*). Population appears stable.

Resources

Books

Bent, A.C. *Life History of North American Woodpeckers (Deluxe Edition).* Indiana University Press, 1992.

Brooke, M., and T. Birkhead. *The Cambridge Encyclopedia of Ornithology.* Cambridge, UK: Cambridge University Press, 1991.

Ehrlich, Paul R., David S. Dobkin, and Darryl Wheye. *The Birder's Handbook.* New York: Simon & Schuster Inc., 1988.

Forshaw, Joseph. *Encyclopedia of Birds.* New York: Academic Press, 1998.

Godfrey, W.E. *The Birds of Canada.* Toronto: University of Toronto Press, 1986.

Sibley, David Allen. *The Sibley Guide to Birds.* New York: Knopf, 2000.

Bill Freedman
Randall Frost

Woolly mammoth

The woolly mammoth (*Mammuthus primigenius*) was a large mammal that coexisted with early humans. It became extinct at the end of the last ice age, about 10,000 years ago. One of four **species** of mammoths, woolly mammoths were abundant on the cold **tundra** that extended beyond the glaciated **ice** fields of **Europe**, **Asia**, and **North America**. It is unclear whether the **extinction** of the woolly mammoth was a result of climatic warming at the end of the last ice age, leading to the loss of mammoth **habitat**, over-hunting by human predators, or a combination of both.

The woolly mammoth belonged to the same family as modern elephants. Standing approximately 11 ft (4 m) tall and weighing 6-8 tons, this **animal** was well-adapted to the cold tundra, especially compared to the other species of mammoths. The woolly mammoth was so-named because of the thick, long brown fur that covered its entire body, including the ears and trunk. Beneath its fur was a wool undercoat, and layers of **fat** provided additional insulation from the cold. Its extremities (ears, feet, trunk, and tail) were small in comparison to those of other mam-

Woolly mammoth. *Jonathan Blair/Corbis Corporation. Reproduced by permission.*

moth species, to minimize the loss of body **heat** through its surface. As an **herbivore** feeding on coarse tundra vegetation, the woolly mammoth had huge specialized teeth and a lower jaw that swung back and forward to shred plants. The woolly mammoth may have used its long, curved ivory tusks to scrape snow from the plants.

Three lines of elephants, the African and Asian elephants and the mammoths, evolved during the Miocene Epoch, about 24 million years ago. Mammoths first appeared in **Africa** and then spread through Europe and Asia. The woolly mammoth was one of two mammoth species that migrated across the land bridge from Siberia to North America, less than one million years ago. Numerous fossilized bones and teeth of the woolly mammoth have been found across the northern United States and southern Canada. Three woolly mammoths, along with numerous Columbian mammoths, have been found at the Hot Springs sinkhole in South Dakota, where they were trapped and died 26,000 years ago. Bodies of woolly mammoths also have been found preserved in ice, with their last meal of tundra plants still in their stomachs.

In 1999, a 23-ton block of ice, containing a fully-preserved 23,000 year-old male woolly mammoth, was excavated from the **permafrost** of the Siberian tundra.

Scientists plan to move the block to an ice **cave**, where it will be carefully thawed using hair dryers. This specimen will provide new information about the woolly mammoth and, potentially, woolly mammoth DNA for use in cloning experiments.

Margaret Alic

Work

Who is doing more work: a weight lifter holding up, but not moving, a 200 lb (91 kg) barbell, or an office worker lifting a pen? The weight lifter is certainly exerting more effort, and many people would say he is doing more work. To a physicist, however, the office worker is doing more work as long as the weight lifter does not actually move the barbell. The weight lifter does a considerable amount of work lifting the barbell in the first place, but not in holding it up.

The term work has a very specific meaning in **physics** that is different from the everyday use of the term. In physics, the amount of work is the distance an object is moved times the amount of **force** applied in the direction of the **motion**. If the force is not parallel to the direction of motion, the force must be multiplied by the cosine of the **angle** between the force and the direction of motion to get the component of the force parallel to the motion.

If the force applied in the direction of motion is **zero**, then the work done is zero regardless of the amount of motion. Likewise, if the distance moved is zero, then the work done is zero regardless of the force applied. Any number multiplied by zero is still zero. In the above example, the weight lifter is exerting a large force, but as long as he does not actually move the weight he is doing zero work, just exerting a lot of effort. The office worker does not need to exert much force to lift the pen, but the force is not zero. So lifting and moving the pen is more work than supporting but not moving the weight. Now think about the weight lifter actually lifting the weight. There is a large force required to lift the weight, and it moves several feet. The weight lifter is now doing quite a bit of work. To do work you must actually move something. Just exerting a force, no **matter** how large, is not enough.

See also Energy.

Wren-warblers

The wren-warblers or Australian **warblers** are 83 **species** of **birds** that constitute the family Maluridae. These are nonmigratory birds, occurring in New Guinea, **Australia**, New Zealand, and nearby islands. Their usual **habitat** is **forests**, shrublands, and heaths.

Wren-warblers are small birds, with a body length of 3.9-7.9 in (10 to 20 cm), including the long, cocked tail that many species have. Their wings are short and rounded, and the bill is small and weak. The males of most species are brightly colored in contrasting patterns of blue, red, brown, black, or white. Other species, however, are more drab in coloration, and the sexes do not differ.

Wren-warblers are gregarious, and they hunt in small flocks for their **prey** of **insects** and other small **invertebrates** in canopy foliage. Some species also eat **seeds**.

Wren-warblers are loosely territorial during their breeding season. The males of most species are good singers, and they defend their individual territories in this way. However, the fairy **wrens** (*Malurus* spp.) often breed in a social group, in which one male is dominant, and is the most handsomely colored. One or more other males assist with his breeding effort, and if anything happens to the dominant individual, another will quickly moult to a brighter plumage and assume the central role.

The nest of wren-warblers is a dome-shaped structure with a side entrance. The clutch size is two to five eggs, which are incubated by the female. Both sexes rear the young birds.

The fairy wrens (*Malurus* spp.) are especially lovely. The variegated wren (*Malurus lamberti*) occurs in southeastern Australia, and has a sky blue cap and tail, a violet throat, and chestnut-and-black wings. The superb blue wren (*M. cyaneus*) has a similar range, and a violet throat and tail, light blue cap and face, and a brown back.

The emu wren (*Stipiturus malachurus*) breeds in southern Australia and Tasmania. This species has a **chestnut** cap, a bright blue throat, and a light brown body.

Bill Freedman

Wrens

Wrens are 63 **species** of small, restless perching **birds** in the family Troglodytidae. Species of wrens are most diverse in **North America** and **South America**, although one species, the winter wren, breeds widely in **Europe**, **Asia**, and North **Africa**. Wrens occur in a wide range of habitats, including semidesert, **prairie**, **savanna**, **forests**, and **wetlands**. Species of wrens breed from the boreal zone to the humid tropics.

Wrens are small, stout birds, ranging in body length from 3.9-8.7 in (10 to 22 cm). They have short, rounded wings, and long, strong legs, feet, and claws, and they hold their tail cocked upwards. Their bill is rather long, slender, pointed, and downward curved. Wrens are relatively dull colored, commonly in gray, brown, or rufous hues, patterned with white or black bars, mottles, or spots, and often a white belly. The sexes do not differ in coloration, and juveniles are similar to adults.

Wrens are active birds, often chattering and flitting about in dense undergrowth or shrubbery in search of their food of **insects** and other **invertebrates**. However, wrens are furtive animals and do not often emerge from dense cover, so that in spite of their bustling activity, they are not frequently seen. Wrens roost in concealed nestlike structures at night. During cold **weather**, wrens may roost together in huddled social groups.

Wrens are territorial. Males proclaim and defend their breeding territory using a rapidly phrased, melodious song. The nest may be placed in a hollow cavity, or it may be constructed as a dome-shaped structure of **plant** fibers and twigs, usually placed on or close to the ground. The clutch size ranges from two to 11, with larger numbers of eggs being laid by birds of temperate ecosystems, and smaller

A marsh wren (*Cistothorus palustris*) perched in cattail stalks at Stoney Point, Ontario, Canada. *Photograph by Robert J. Huffman. Field Mark Publications. Reproduced by permission.*

clutches by wrens breeding in tropical habitats. The female incubates the eggs, but males help with raising the brood. Some species of wrens breeding in boreal and temperate habitats are commonly polygynous, particularly in situations where the territory of the male is of high quality.

Species of wrens

Ten species of wrens breed regularly in North America.

The winter wren (*Troglodytes troglodytes*) breeds in moist, conifer-dominated forests, and winters in the eastern United States and western coastal forests. This species is the widest-ranging of the wrens, breeding extensively in North America, and also in Eurasia and North Africa, where it is known as the common wren. The winter wren breeds in a wide range of habitats, from boreal, **conifer** forests on offshore islands, to densely shrubby suburban gardens and parks.

The house wren (*Troglodytes aedon*) is a widespread and familiar species, breeding through much of southern Canada and extensively through the United States, except for parts of the southeast. The house wren winters as far south as southern Mexico and the Gulf Coast of the southern states. This species will often accept a nest box located in a shrubby **habitat**, and in this way can be lured to breed in suburban gardens.

Bewick's wren (*Thryomanes bewickii*) is a relatively common breeder in the western United States and south to Mexico, and is less abundant in the eastern states.

The Carolina wren (*Thryothorus ludovicianus*) is a relatively abundant breeding species in southern Ontario and most of the eastern United States. This species is partial to thick, brushy habitats in open forests, along forest edges, and in parks and gardens.

The long-billed marsh wren (*Cistothorus palustris*) breeds abundantly in its habitat of tall marshes with bulrushes, **cattails**, and reeds across much of central and southern North America. This species winters in the southern United States and Central America. The sedge wren or short-billed marsh wren (*Cistothorus platensis*) is a less common species in its range in central-eastern North America, and breeds in shorter wet meadows and fens dominated by **sedges**. This species winters in coastal marshes of the southern Atlantic states and Gulf of Mexico.

KEY TERMS

. . . .

Polygynous—A breeding system in which a male will attempt to breed with as many females as possible. In birds, the female of a polygynous species usually incubates the eggs and raises the babies.

The rock wren (*Salpinctes obsoletus*) breeds in semiarid rocky habitats through the western United States to Costa Rica in Central America.

The cactus wren (*Campylorhynchus brunneicapillus*) is the largest of the North American wrens, achieving a length of almost 6.7 in (17 cm). This species breeds in deserts with thorny shrubs and large **cactus** plants, especially saguaro, from the southwestern United States through Central America.

Wrens and humans

Some species of wrens in North America have suffered greatly from habitat losses associated with human activity. Other stressors have also been important, including the use of **pesticides** in agriculture, **forestry**, and in the shrubby parks and gardens in which some wrens breed.

The San Clemente Bewick's wren (*Thryomanes bewickii leucophrys*) was a resident breeder on San Clemente Island off southern California. This subspecies of the Bewick's wren became extinct through severe habitat damages that were caused by introduced populations of **goats** and **sheep**. These are generalized herbivores that essentially devoured the limited habitat of the San Clemente Bewick's wren, and that of other native local species of plants and animals.

Other populations of wrens have also declined in many places in North America. The leading causes of these changes are habitat losses associated with the conversion of natural ecosystems into land-uses associated with agriculture and housing, and to a lesser degree, with forestry. Pesticide use is also important in some cases.

Resources

Books

Ehrlich, P.R., D.S. Dobkin, and D. Wheye. *Birds in Jeopardy.* Stanford, CA: Stanford University Press, 1992.
Forshaw, Joseph. *Encyclopedia of Birds.* New York: Academic Press, 1998.
Sibley, David Allen. *The Sibley Guide to Birds.* New York: Knopf, 2000.

Bill Freedman

Wrynecks

Wrynecks are two **species** of small, woodpecker-like **birds**. Wrynecks are in the family Picidae, which also includes the **woodpeckers** and piculets. However, the distinctively different wrynecks are in their own subfamily, the Junginae. Wrynecks received their common name from their habit of twisting their head and neck when disturbed.

The plumage of wrynecks is a mottled and cryptic brown, grey, and black. Wrynecks are somewhat less specialized feeders than the true woodpeckers. They lack the stiff propping tail feathers of the woodpeckers, do not climb vertical tree-trunks, and do not drill holes in **bark** and **wood**. Wrynecks do, however, nest in cavities in trees, although they do not excavate these for themselves.

Wrynecks forage on the ground for their food of **ants** and other small **invertebrates**. Their usual **habitat** is angiosperm-dominated **forests**.

The wryneck (*Jynx torquilla*) breeds widely in forests of Eurasia and north **Africa**, and migrates to sub-Saharan Africa and tropical **Asia**. The African wryneck (*J. ruficollis*) occurs in forests in Africa.

Bill Freedman

X-ray astronomy

At the high-energy end of the **electromagnetic spectrum**, **x rays** provide a unique window on some of the hottest and most violent objects in the universe. Since the discovery of extra-solar x-ray sources in 1962, scientists have investigated a large number of phenomena which emit x rays. With each new space mission, more sources and more details of the structure of the x-ray universe have been gleaned.

Background

Although they are among the most energetic of the electromagnetic **spectrum**, and thus provide a window on some of the most violent processes in the universe, x rays are not able to penetrate Earth's atmosphere; they are absorbed at about 62 mi (100 km) above the surface. Thus, only with the advent of rocket and **satellite astronomy** have astronomers been able to study the wide-ranging phenomena which produce x rays. The highest **energy** x rays have also been studied by balloons high in Earth's atmosphere, but there are far fewer photons at these energies than at the lower energies that can be observed above the atmosphere.

X rays are also difficult to bring to a focus, since their energies are so high. Therefore, an important breakthrough in x-ray astronomy was the advent of imaging telescopes, replacing instruments which could only crudely tell in which direction an x-ray source was located. The telescopes with which we are most familiar, consisting of lenses or **mirrors** that capture **light** arriving at normal incidence (perpendicular to the surface) won't work in the x-ray region of the spectrum, since the x rays pass through unchanged or are absorbed by the **optics**. Instead, x-ray astronomers use grazing incidence telescopes, in which the light from the source strikes mirrors at angles of only a few degrees, skipping like stones over the surface of **water**. By combining two mirrors, the energy can be focused onto a detector in order to provide a sharp image of the source.

History

Although the **temperature** of the sun's surface is about 6,000K (10,341°F; 5,727°C), by the 1930s there was evidence that the outer regions of the solar atmosphere were much hotter, meaning that they could be a source of x rays. At that time there was no way to verify this prediction, however. After World War II, when captured V-2 rockets allowed scientists to place instruments outside the protective atmosphere for the first time, a number of experiments were able to show that the **Sun** did indeed produce x rays.

The strongest early evidence came in 1948, when x-ray detectors registered x rays were coming from the direction of the Sun. Further investigations showed that the total x-ray output of the Sun was only a tiny fraction of the total energy generated. Because the total x-ray output was so small, despite the fact that the Sun is so close in terms of interstellar distances, many believed that no other sources would be found.

In 1962, a rocket was sent up to look for x rays from the **Moon**, which was theorized to generate x rays due to **solar wind** bombardment. No **emission** was detected from the Moon, but in a surprising discovery, the detector registered an x-ray source in the direction of the **constellation** Scorpio, along with a diffuse background coming from all directions; the source was called Scorpius X-1.

Since that time, a large number of rocket and Earth-orbiting satellites have discovered tens of thousands of x-ray sources in the sky, many of which are many orders of magnitude brighter than the Sun. The Crab Nebula, for instance, produces approximately 2,000 times more energy in the x-ray region of the spectrum than the Sun does over all wavelengths. Thus we now know that the Sun is relatively quiet as far as x-ray sources go.

The x-ray universe

A wide variety of x-ray sources have been seen since the first extrasolar identification in 1962. A few of the most interesting types of sources are:

The Sun. A number of x-ray satellites have monitored the Sun. Solar flares produce enhancements in its x-ray output.

Stars. Many stars, particularly those with coronae or rapid stellar winds, emit x rays from their outer layers.

Comets. Astronomers have detected x-ray emission from 10 different comets since the phenomenon was first discovered in 1996 with Comet Hyakutake.

Scientists believe that x rays are generated by some sort of interaction between the solar **wind** and the comet's atmosphere, ionosphere, or **atoms** within the nucleus.

Groups of galaxies in hot **clouds**. Bright x-radiation is seen emanating from clusters of galaxies, which, due to their enormous gravitational pull, trap gas in the region. This gas is very hot, and there is a large amount of it. It thus glows in the x-ray region.

X-ray background. The sky is not dark in the x-ray region of the sky like it is in the visible. The diffuse background which was detected in the rocket flight described above is still not understood, although some believe it may be the result of many individual, unresolved sources.

X-ray binaries. These are close binary stars in which gas from one **star** falls onto its companion, heats up, and gives off x rays. This is especially bright when the companion is a compact stellar remnant such as a **neutron star** or **black hole**, because the enormous gravitational field compresses and heats the incoming gas, causing it to glow at x-ray wavelengths.

Supernova remnants. Explosions of stars, or supernovae, show traces of the heavy elements that are formed there when their x-ray spectra are examined.

Quasars and **active galactic nuclei**. These are among the most energetic objects in the universe, and they emit enormous quantities of **radiation** at x-ray wavelengths. It is thought that the ultimate source of this energy is a supermassive black hole, surrounded by an **accretion disk** of in-falling gas that is heated to many millions of degrees.

X-ray missions

Among the largest and most productive x-ray missions were Uhuru (1970), which catalogued 339 x-ray sources; Einstein (also known as HEAO-2, 1978-1981); and EXOSAT (1983-1986). In addition, there have been many smaller-scale observations.

The more recent missions, such as the German ROSAT (Röntgensatellit), launched in 1990, contain very sophisticated instrumentation, including detectors and grazing incidence telescopes, which can pinpoint the location of an x-ray source to very high accuracy, and take x-ray pictures to show the shape and distribution of

KEY TERMS

Grazing incidence telescope—A telescope design in which the incoming radiation strikes the mirrors at very small angles.

Spectrum—A display of the intensity of radiation versus wavelength.

the source. This is an important improvement over early missions, which often were not able to determine the exact location of the x-ray sources, making it difficult to correlate the source with an object that could be detected in another wavelength region.

Recent missions have also been able to measure the x-ray spectrum, or strength of the radiation in different energy bands. This allows the identification of particular elements in the source. ROSAT identified more than 50,000 x-ray sources during its survey phase, when it scanned the sky for six months. It also finally succeeded in detecting x rays from the Moon, nearly 30 years after the first attempt to do so.

NASA launched the Advanced X ray **Astrophysics** Facility (AXAF), named the Chandra X-ray Observatory. Designed with a resolution 25 times better than any preceding x-ray **telescope**, CXO passes around the **earth** in an elliptical **orbit**, studying black holes, supernovas, and **dark matter** and in an attempt to increase our understanding of the origin and **evolution** of the universe.

Resources

Books

Tucker, Wallace, and Riccardo Giacconi. *The X-ray Universe.* Cambridge: Harvard University Press, 1985.

Periodicals

Beatty, J. Kelly. "ROSAT and the X-ray Universe." *Sky & Telescope* (August 1990):128.

Margon, Bruce. "Exploring the High-Energy Universe." *Sky & Telescope* (December 1991): 607.

Van den Heuvel, Edward P.J., and Jan van Paradijs. "X-ray Binaries." *Scientific American* (November 1993): 64.

David Sahnow

X-ray crystallography

X-ray crystallography is a laboratory technique used for the study of the internal structure of crystalline materials. More specifically known as x-ray **diffraction**, the technique is based on the **interference** pattern produced as **x rays** pass through the three-dimensional, repeating pattern of **atoms** within a **crystal** lattice. The characteristic interference patterns produced are reflective of the molecular structure of the sample. X-ray diffraction has enabled the measurement of distances between planes of atoms and the determination of the arrangement of atoms within the lattice. Once the characteristic pattern for a substance has been identified, x-ray diffraction may also be used to identify an unknown sample of that same material by matching the diffraction pattern of the unknown to the appropriate known pattern. Prior to the discovery of x-ray diffraction, crystallographers had no means by which to measure the internal positions of atoms within crystals and could only hypothesize as to the internal structure based upon external and optical features. X-ray diffraction has allowed crystallographers to demonstrate the orderly internal structure of crystals and has profoundly affected science since the inception of the technique.

In 1895, x rays were discovered by German physicist Wilhelm Conrad Roentgen (1845–1923) while experimenting with **cathode** rays. In 1912, German physicist Max von Laue (1879–1960) suggested that x rays interacting with a crystal could produce a distinctive interference pattern. His hypothesis, for which he was awarded the 1914 Nobel Prize, proved to be correct. The procedure demonstrated the internal order of atoms within a crystal and was the origin of x-ray crystallography. In 1914, the father and son team of English physicists, William Henry Bragg (1862–1942) and William Lawrence Bragg (1890–1971) refined the analysis of crystalline structure with x-ray diffraction, determined the atomic structure of a simple inorganic substance, common **salt** (NaCl), and deciphered the mathematical relationships between crystal structure and the associated diffraction pattern. They were jointly awarded the Nobel Prize in 1915; the younger Bragg was the youngest-ever Nobel laureate at age 25.

Crystalline substances have an ordered three-dimensional arrangement with a particular spacing of atoms. When x rays strike the atoms within the crystal, the atoms absorb and reemit the **energy** from the x rays in the form of spherical wave fronts emanating from each atom. The waves traveling outward from each atom interact with other waves in the processes known as constructive and destructive interference. In some directions, the waves cancel each other and little energy remains; in other directions the energy is reinforced and a zone of increased energy exists. The resulting pattern of constructive and destructive interference is known as a diffraction pattern. The patterns are controlled by the spacing of atoms within the matrix and are unique to that substance.

In its most basic form, a diffractometer consists of three main components; a source of x rays and the means to direct the beam to the sample, a sample holder, and a method for collecting the resultant **radiation** and recording the diffraction pattern. In the Laue method, a single crystal is placed in the x-ray beam and the diffraction pattern is captured on photographic film. The crystal is stationary and the method allows for the study of **symmetry** within the crystal structure. The rotational method of diffraction is similar to the Laue method in that a single, well-formed crystal is used. As the name suggests, however, the crystal is rotated about one axis, allowing the collection of a greater quantity of diffraction data. The difficulties associated with obtaining and orienting well-formed crystals eventually led to the development of the powder method of x-ray crystallography. In this case, the sample is ground to a powder and the diffracted energy from all of the atomic planes within the material are measured simultaneously.

On modern diffractometers, electronic detectors linked to chart recorders have replaced photographic film. The information provided by each of these methods is quite similar, but the automated electronic system has a number of advantages. These advantages include the ability to read the data values directly from the chart without the need for careful measurements, the intensity of the energy peaks is clearly visible on the chart, no need for film developing, and rapid data collection.

X-ray crystallography was initially used to investigate the structure of **minerals**, confirming and refining the crystallographic descriptions. Use of the technique was expanded to the investigation of metals, alloys, and inorganic and organic chemical substances. More recently, biomedical research has utilized the technique for the investigation of the structure and dynamics of **proteins**, nucleic acids, and other biological molecules. Research into microelectronics and semiconductors, as well as pharmaceutical research, continue to rely on the qualities of x-ray crystallography.

See also Electromagnetic spectrum; Mineralogy.

Resources

Books

Bragg, William L. *The Crystalline State: A General Survey.* London: G. Bell and Sons Ltd., 1949.

Clegg, William, ed. *Crystal Structure: Principles and Practice.* New York: Oxford University Press, 2001.

Hammond, Christopher. *The Basics of Crystallography and Diffraction.* New York: Oxford University Press, 2001.

Other

Rupp, Bernhard. "Crystallography 101." [cited January 14, 2003]. <http://www-structure.llnl.gov/Xray/101index.html>.

KEY TERMS

· ·

Crystal—A solid, homogeneous body composed of a single element or compound having a fixed and regular internal atomic arrangement that may be expressed by external planar faces.

Crystal lattice—The ordered, three-dimensional arrangement of atoms in a crystal.

Crystallography—The study of crystals, including their growth, structure, properties, and classification.

Diffraction—The process by which the direction of wave motion is modified by bending around an obstacle.

Diffractometer—The laboratory instrument used for x-ray crystallographic analysis.

Interference—The effect two sets of electromagnetic waves have on each other, and the combined pattern which may be detected as formed by this interaction.

X ray—Electromagnetic radiation of very short wavelength, and very high energy.

Weiss, Manfred S. "X-ray Crystallography." 1998 [cited January 14, 2003]. <http://www. imb-jena.de/www_sbx/manfred/zteach/page1.html>.

David B. Goings

X rays

X rays are electromagnetic waves with wavelengths covering a fairly broad range from about 3×10^{-8} ft (10^{-8} m) to 3×10^{-11} ft (10^{-11} m). There is no sharp boundary between x rays and ultraviolet **light** on the long wavelength side of this range. Similarly, on the short wavelength side, x rays blend into that portion of the **electromagnetic spectrum** called gamma rays which have even shorter wavelengths. X rays have wavelengths much shorter than visible light, which occurs between 1.2×10^{-6} and 2.1×10^{-6} ft (4×10^{-7} and 7×10^{-7} m), and they also behave quite differently. They are invisible, are able to penetrate substantial thicknesses of **matter**, and can ionize matter. Since the time of their discovery in 1895 they have been an extremely important tool in the physical and biological sciences and the fields of medicine and **engineering**.

History

X rays were discovered in Germany in 1895 by Wilhelm Roentgen (1845-1923) quite by accident while he was studying the conduction of **electricity** through gases at low **pressure**. The discovery was made when these mysterious "X" rays were observed to light up a fluorescent screen a few meters from the source. Roentgen soon found that these rays were quite penetrating and was actually able to insert his hand between the source and the screen and see on the screen the faint shadow of the bones in his hand. This indicated that more dense materials such as bone absorbed more x rays than less dense material such as human flesh. He soon found that photographic plates were sensitive to x rays and was able to make the first crude x-ray photographs.

Roentgen had been experimenting with what was called a **cathode** ray discharge tube, i.e., a partially evacuated **glass** tube with **metal** electrodes at each end. When a high electrical voltage was applied between the electrodes a discharge took place in the tube. One effect of the discharge was to produce electrons which acquired high velocities as they were attracted to the positive electrode. When they hit this metal electrode the x rays were produced. It was not until 1913 in the United States that W. D. Coolidge (1873-1975) invented the x-ray tube similar to those still used today. Coolidge removed as much air from the tube as possible and used a hot tungsten filament as the source of electrons. This permitted more careful experiments in which the high voltage applied to the tube and the **rate** at which electrons hit the target could be varied independently.

Mechanisms for x-ray production

The intensity of x rays from an x-ray tube varies with wavelength. A diagram of the wavelength **spectrum** from an x-ray tube shows several sharp peaks superimposed on what appears to be a continuous distribution. The peaks and the continuous region are produced by two quite different mechanisms. The continuous spectrum is produced by the incident electrons as they strike and enter the metal target. They are attracted by the positively charged nuclei of the **atoms** in the target and are suddenly deflected. Electromagnetic theory tells us that when electric charges are accelerated they radiate. Similarly, in the **antenna** of a **radio** or TV transmitter, electrons are made to oscillate rapidly back and forth, but in that case the accelerations are not as large and the wavelengths are much larger being measured in meters or centimeters.

A completely different mechanism produces the sharp peaks in the x-ray spectrum. These peaks are at very specific wavelengths and occur at different wavelengths for different targets. This **radiation** is produced when an incident **electron** knocks an electron out of one of the inner or low **energy** levels of the atom. An electron in a higher energy level falls into the vacant level and in the process an x ray is given off. The energy of this x ray is equal to the difference in energy between these two levels. Characteristic x rays have wavelengths ranging from about 10^{-11} m for **uranium** to 2.5×10^{-8} m for **lithium**.

Measuring x-ray wavelengths

Development of the x-ray tube greatly speeded up the detailed study of x rays, the origin and nature of which had finally been discovered by 1912, with the help of the suggestion by the German scientist Max von Laue (1879-1960) that x rays could be diffracted by three-dimensional crystals and thus must be electromagnetic radiation similar to visible light. This new approach was necessary because the wavelengths of x rays are so small that the **diffraction** gratings used for visible light will not work because the lines on the grating cannot be made with small enough spacings. A "natural" three-dimensional grating in the form of a single **crystal** of a material such as **sodium chloride** (**salt**) or calcite works very well since the spacing of the atoms in the crystal is roughly the same as the x-ray wavelengths of interest. Intrigued by von Laue's discovery of x-ray diffraction, English physicist W. L. Bragg (1890-1971), working with his father W. H. Bragg (1862-1942), began a series of experiments that culminated in the invention of the x-ray spectrometer in 1913. This device allowed the Braggs to examine the structure of certain crystals, laying the foundation for the science of **x-ray crystallography**. Ultimately, the development of the x-ray spectrometer led to important advances in atomic **physics** and an improved understanding of the **periodic table**.

In 1913, the English physicist H. G. J. Moseley (1887-1915) used a Bragg-type spectrometer to look at the characteristic x rays from many of the elements. Taking advantage of the regular decrease in wavelength of characteristic x rays as one looks at successively heavier elements, he discovered that it was possible to tell one element from another by looking at the characteristic x rays. He found that elements should be listed in the periodic table in terms of their **atomic number**, not the **atomic weight** as had previously been done. He determined, for example, that cobalt should come before nickel even though cobalt had a larger atomic weight. He was also able to predict the existence of several elements, such as scandium and promethium, which were then unknown but later discovered.

Detection of x rays

In order for the Bragg spectrometer to be useful, the reflected rays must be detected. The detection of x rays is always based on their ability to eject electrons from atoms, i.e., to ionize matter. Thus the detector in the first Bragg spectrometer was an ionization chamber which collected the **electric charge** produced by x rays as they impacted a gas filled chamber. This ionization chamber was soon replaced by the familiar Geiger counter, developed in the study of radioactivity which was discovered at about the same time as x rays.

Applications of x rays

The uses of x rays in the fields of medicine and **dentistry** have been extremely important. X-ray photographs utilize the fact that portions of the body such as bones and teeth with higher **density** are less transparent to x rays than other parts of the human body. X rays are widely used for diagnostic purposes in these fields. Examples might include the observation of the broken bones and torn ligaments of football players, the detection of breast **cancer** in women, or the discovery of cavities and impacted wisdom teeth.

Since x rays can be produced with energies sufficient to ionize the atoms making up human **tissue**, it is not surprising that x rays can be used to kill these cells. This is just what is done in some types of cancer therapy in which the radiation is directed against the malignancy in the hope of destroying it while doing minimal damage to nearby normal tissue. Unfortunately, too much exposure of normal tissue to x rays can cause the development of cancer, a fact that was learned too late for many of the early workers in this field. For this reason, great care is taken by physicians and dentists when taking x rays of any type to be sure that the exposure to the rest of the body is kept at an absolute minimum.

A relatively new technique for using x rays in the field of medicine is called **computerized axial tomography**, producing what are called CAT scans. These scans produce a cross-sectional picture of a part of the body which is much sharper than a normal x ray. This is because a normal x ray, taken through the body, often shows organs and body parts superimposed on one another. To produce a CAT scan, a narrow beam of x rays is sent through the region of interest from many differentangles and a computer is used to reconstruct the cross-sectional picture of that region.

Moseley found that various natural elements can be identified by measuring the energy of their characteristic x rays. This fact makes a useful form of elemental analysis possible. If x rays of sufficient energy impact a sample of unknown composition, electrons will be knocked out of the atoms of the various elements in the sample and characteristic x rays will be given off by these atoms. Measurement of the energy of these x rays permits a determination of the elements present in the sample. This technique is known as x-ray **fluorescence** analysis. It is often used by chemists to perform a nondestructive elemental analysis and by law enforcement agencies when it is necessary to know what elements are present in a sample of hair or **blood** or some other material being used as evidence in a criminal investigation.

X rays are used in business and industry in many other ways. For example, x-ray pictures of whole engines or engine parts can be taken to look for defects in a nondestructive manner. Similarly, sections of pipe lines for oil or **natural gas** can be examined for cracks or defective welds. Airlines also use x-ray detectors to check the baggage of passengers for guns or other illegal objects.

In recent years an interesting new source of x rays has been developed called synchrotron radiation. Many particle **accelerators** accelerate charged particles such as electrons or protons by giving them repeated small increases in energy as they move in a circular path in the accelerator. A circular ring of magnets keeps the particles in this circular path. Any object moving in a circular path experiences an **acceleration** toward the center of the circle, so the charged particles moving in these paths must radiate and therefore lose energy. Many years ago, the builders of accelerators for research in nuclear physics considered this energy loss a nuisance, but gradually scientists realized that accelerators could be built to take advantage of the fact that this radiation could be made very intense. Electrons turn out to be the best particle for use in these machines, called electron synchrotrons, and now accelerators are built for the sole purpose of producing this radiation which can be adjusted to produce radiation anywhere from the visible region up to the x ray region. This synchrotron radiation, from which very intense beams at nearly one wavelength can be produced, is extremely useful in **learning** about the arrangement of atoms in various compounds of interest to biologists, chemists, and physicists.

One of the more important commercial applications of synchrotron radiation is in the field of x-ray **lithography**, used in the **electronics** industry in the manufacture of high density integrated circuits. The **integrated circuit** chips are made by etching successive layers of electric circuitry into a wafer of semiconducting material such as silicon. The details of the circuitry are defined by coating the wafer with a light sensitive substance called a photoresist and shining light on the coated surface through a stencil like mask. The pattern of the electric circuits is cut into the mask and the exposed photoresist can easily be washed away leaving the circuit outlines in the remaining photoresist. The size of the circuit elements is limited by the wavelength of the light-the shorter the wavelength the smaller the circuit

KEY TERMS

Bragg x-ray spectrometer—A device using a single crystal with regularly spaced atoms to measure the wavelengths of x rays.

Continuous x-ray spectrum—The x rays produced by the electrons in an x-ray tube as they strike the target and are suddenly deflected. A broad range of wavelengths is produced.

Synchrotron radiation—Electromagnetic radiation from electron accelerators called synchrotrons that can range from the visible region to the x-ray region.

X-ray fluorescence analysis—A method of detecting the presence of various elements in an unknown sample by observing the characteristic x rays given off by the sample when excited by sufficiently energetic x rays.

X-ray tube—Evacuated tube in which electrons moving at high velocities are made to hit a metal target producing x rays.

elements. If x rays are used instead of light, the circuits on the wafer can be made much smaller and many more elements can be put on a wafer of a given size, permitting the manufacture of smaller electronic devices such as computers.

Resources

Books

Dyson, Norman A. *X Rays in Atomic and Nuclear Physics.* White Plains, NY: Longman, 1973.

Hewitt, Paul. *Conceptual Physics.* Englewood Cliffs, NJ: Prentice Hall, 2001.

Young, Hugh. *University Physics.* Reading, MA: Addison-Wesley, 1999.

Robert L. Stearns

Xenogamy

When used by botanists and **plant** breeders, xenogamy (also called outbreeding) generally refers to a form of cross-pollination. Xenogamy is also a term more broadly used in **genetics** to describe the union of genetically unrelated organisms within the same **species**. In all cases, xenogamy promotes genetic diversity and thus, also enhances the overall fitness of a species.

In some circumstances, xenogamy and outbreeding are also referred to as crossbreeding. Regardless of the exact terminology, the core concept involves an increase in genetic variability. With crossbreeding, genetically dissimilar or unrelated animals from the same breed can be crossed in a process known as outcrossing. True crossbreeding exists when, for example, differing breeds of cattle are allowed to mate and produce offspring. Extreme xenogamy exists with species crossing (a mating between organisms from two different species).

Induced xenogamy and crossbreeding are often attempts by scientists to genetically combine desirable traits from two differing species or breeds in order to produce offspring with more desirable characteristics. Successful crossbreeding, whether in plants or animals, usually results in increased **hybrid** vigor (a set of characteristics that can include increased fertility, faster growth rates, increased immunological tolerance, greater strength, and/or other desired characteristics of a hybrid species).

In addition to induced xenogamy, outbreeding is a fundamental part of natural **selection** and, by producing new and varied genetic combinations, an essential element of **evolution**. As such, outbreeding is an important tool in the continued survival and evolution of a species.

In terms of genes and **alleles**, xenogamy promotes genetic variability and vitality within a breeding population by reducing homozygosity (the state of being homozygous). Organisms that are homozygous carry identical alleles on both chromosomes of a pair of homologous chromosomes.

For example, using traditional notation to designate dominant and recessive alleles (e.g., "T" for tall stems and "t" for short stems in a particular plant species), a homozygous plant with a pair of corresponding chromosomes would be designated as either "TT" (two "T" alleles) or "tt" (two "t" alleles). With regard to phenotype (the outward expression of genotype), "TT" homozygous plants should normally produce tall-stemmed plants. The "tt" genotype should—under normal environmental conditions—result in a short stemmed plant. If the "T" allele is dominant, tall stemmed plants would be the normal expected phenotypic expression of "Tt" or "tT" genotypes.

Because homozygotes contain identical alleles on their **chromosome**, in the absence of **mutation** they can only produce gametes (sex cells) that contain those same alleles. For example, an **organism** with a homozygous "TT" genotype can contribute only a "T" allele to its off-spring. Such organisms, when mated with other homozygotes, breed true with regard to a particular trait (e.g., stem length). In many cases, xenogamy allows the reintroduction of alleles—or the introduction of new alleles—into a population.

By uniting differing genotypes, xenogamy allows increases in genetic variability that, in turn, exert measurable influences on the **frequency** of genes, types of alleles, and traits within a population.

Although such simple examples as above serve to illustrate broad genetic principles, the degenerate nature of the genetic code (i.e., there are multiple codes that convey the same genetic instructions) means that homozygosity more specifically means that the products of the instructions contained in the homozygous genes are similar enough to produce identical visible expression (identical phenotypic expression). Because there are multiple codes that represent multiple **sequences** of bases in the nucleic acids (deoxyribonucleic acid [DNA] and ribonucleic acid [RNA]) to convey essentially identical instructions for the construction of **proteins**, multiple instructions can result in the formation of the same protein. As a result, although an **individual** may be homozygous for a particular **gene**, this does not necessarily mean that the base sequence found in those genes is identical.

In plant species, there are several natural mechanisms that can result in xenogamy, including self-incompatibility. With self-incompatibility, there is an inability on the part of sex cells (gametes) from the same species of plants to produce a viable embryo. With such species, it is usually the case that pollen, unable to induce **fertilization** on its own stigma, is able to successfully grow on the stigma of other plants of the same species. The process involving the transfer of pollen to a foreign stigma is termed allogamy. Regardless of the exact mechanism, self-incompatibility mechanisms promote xenogamy (outcrossing) and heterozygosity while acting to prevent inbreeding. Self-incompatibility is often the result of incomplete nuclear or genetic fusion. The failure of the fusion processes are usually traced to a single genetic **locus** (S-locus) that exists as multiple alleles.

See also Evolutionary mechanisms; Genetic engineering; Genetically modified foods and organisms; Genotype and phenotype.

Xenon see **Rare gases**
Xylem see **Plant**

Yak

The yaks are members of the family Bovidae (oxen), order Artiodactyla, which also includes the domestic cattle and existing wild cattle **species** such as the aurochs, the gaur or seladang, and the koupray. The generally accepted species name for yak is *Bos grunniens*, and it seems to have an affinity to **bison**, which belong to the same genus *Bos*. Like some other *Bos* species the yak is a large, massive **animal** with stout limbs and a long tail, usually tufted at the tip. Wild yak males measure 10.5 ft (3.25 m) in head and body. The shoulder height is over 6.5 ft (2 m), and they weigh 1,800-2,200 lb (820-1,000 kg). Females are smaller, and weigh about one third as much.

Both sexes bear horns which are black and positioned quite far apart on each extremity of the top of the skull. In the male, they are larger, curving up and then down. They are approximately oval in cross-section. There are no suborbital, inguinal, or interdigital **glands** in yaks. The body is covered with long hair, blackish brown in **color**; it hangs down almost to the ground like a fringe around the lower part of the shoulders, the sides, the flanks, and the thighs. Wild yaks live on the Tibetan plateau where temperatures in winter may drop to -40°F (-40°C). The long, thick coat protects the yak from these frigid temperatures.

Domesticated yaks are smaller, have weaker horns, and may vary greatly in color: red, mottled brown, or black.

Wild yaks inhabit desolate steppes at up to 20,000 ft (6,100 m) above **sea level**. They are expert climbers, sure-footed and sturdy. During the relatively warmer months of August and September, wild yaks remain in high areas with permanent snow, but spend the rest of the year at lower elevations. They feed chiefly on grass, herbs, and **lichens**.

The females with young congregate in large herds up to 1,000 animals. The males live alone most of the year, or in groups of not more than 12. The bulls join the herd and fight for the females when mating season begins in September. Only during that time do wild yaks make a strange grunting sound, which domesticated yaks emit all year round. After a gestation period of nine months, a single calf is born, usually in June, with births occurring every other year. The newborn becomes independent after a year, and reaches full size at six to eight years of age. Maximum life span for the wild yak is estimated at 25 years. Domesticated yaks may give **birth** every year. Wild yaks are classified as endangered by IUCN-The World Conservation Union. They are officially protected in China, but their numbers have declined due to uncontrolled hunting.

Yaks were probably domesticated during the first millennium B.C., and now they are found in association with people in the high plateaus and **mountains** of Central **Asia**. They are docile and powerful, and they are surely the most useful of domesticated animals at elevation above 6,500 ft (2,000 m). The serve as mounts, beasts of burden, for milk and meat, and are also sheared for their wool. There are about 12 million domesticated yaks, but perhaps no more than a few hundred wild ones.

Yam

Yams are any of the 10 economically important **species** of *Dioscorea*, a genus in the monocotyledonous family Dioscoriaceae. These species, all tropical in their origin, are cultivated for their edible tubers (enlarged, fleshy, usually underground storage stems). In the United States, the name *yam* is often misapplied to the **sweet potato** (*Ipomea batatas*).

Yams are herbaceous plants whose stems twine up and around bushes, trees, or poles. Depending on the species of yam, stems twine either clockwise or counterclockwise. The stems bear stalked, palmately veined leaves that are simple and entire, although a few species have three-lobed leaves. All yams have a dioecious lifestyle, which means that the staminate and pistillate flowers are borne on separate plants. The flowers are inconspicuous, being only 1/8 in (2-4 mm) long and whitish or greenish. The **fruits** produced from the flowers are three-angled and contain winged **seeds**. Some cultivars of yam, however, rarely **flower** or set seed.

In commonly cultivated yams, the tubers lie underground and are one (rarely two or three) per **plant**. These tubers resemble huge, elongated potatoes, typically growing 2-6 ft long (0.6-2 m) and weighing 11-33 lb (5-15 kg). A thin skin protects their outer surface, and on the inside they are filled with starch which can be white or yellow depending on the species. One cultivated yam (*Dioscorea bulbifera*) bears small tubers along the aerial stems in the **leaf** axils (the angle between the stem and leaf stalk).

There are two centers of yam cultivation worldwide. The first is the high rainfall region of western **Africa**, from the Ivory Coast to Cameroon. Here the most important species are the white yam (*Dioscorea rotundata*) and the yellow yam (*D. cayenensis)*, named for the **color** of their tuber's flesh. The second center is Vietnam, Cambodia, Laos, and neighboring regions where the most commonly cultivated species is the Asiatic yam (*D. alata*). Secondary areas of yam cultivation are the West Indies, Pacific islands, and southeastern United States (from Louisiana to Georgia). Most yam species originated in **Asia** and Africa; only one, the cush-cush yam (*D. trifida*), is native to the New World.

The world production of yams amounts to about 22 million tons (20 million metric tons) per year, of which two-thirds comes from tropical West Africa. Yams are to tropical West Africans what wheaten bread is to North Americans and Europeans. In tropical west-Africa, many social and religious festivals are associated with planting and harvesting yams.

Yams are propagated from cuttings of the **tuber**. Because the plants climb, they are provided with poles or trellises for support. It generally takes seven to 10 months before the tubers can be harvested, and this must be done by hand because mechanical harvesters tend to damage the tubers. Yams store better than most tropical tuber **crops** and this is one reason why they are widely grown. Before eating, yams are usually peeled and then either boiled, roasted, or fried. In Africa yams are usually prepared as *fufu* or *four-fou*, made from peeling, cutting, and boiling the tuber, and then pounding it into a gelatinous dough. It is served with soups or stews or cooked raw in palm oil. Nutritionally, the yams are equivalent to the common **potato**, containing 80-90% carbohydrates, 5-8% protein, and about 3.5% **minerals**. Yam production is now declining because cassava (*Manihot utilissima*) and sweet potatoes (*Ipomea batatas*)—sources of starch that

are easier to cultivate—are increasingly being used. Yams are not fed to **livestock** because they are more expensive than other kinds of **animal** feed.

Yams are a source of steroids and alkaloids-chemicals that are extremely active physiologically in vertebrate animals. The most important yam steroid is diosgenin used in the production of birth-control pills. Alkaloids from yams have been used to kill **fish** and to poison darts and arrows for hunting. Some yams are poisonous to humans because of their high **alkaloid** content, and their tubers must be boiled before eating to remove the toxins.

Yeast

Yeasts are single-celled **fungi**, belonging mainly to the Ascomycetes, that serve as nutrient recyclers in nature, but are also important in industry, **biotechnology**, and as the agents of **disease** in humans. The term yeast is generically used in reference to many **species** of single-celled, budding fungi, including *Saccharomyces*—used in baking and **brewing**—and *Candida*—an infectious yeast common in people with compromised immune systems, such as **AIDS** victims.

Life cycle

Yeasts secrete enzymes that break down carbohydrates (through **fermentation**) to yield **carbon dioxide** and **alcohol**. The source of carbohydrates are either living hosts or non-living hosts such as rotting vegetation, or the moist body cavities of animals. Yeasts are considered by some scientists to be closely related to the **algae**, lacking only in photosynthetic capability—perhaps as a result of an evolutionary trend toward a lifestyle dependent upon host **nutrition**. Ecologically yeasts are decomposers that secrete enzymes which dismantle the complex **carbon** compounds of **plant cell** walls and **animal** tissues, which they convert to sugars for their own growth and sustenance. Yeast reproduction may involve sexual **spore** production or asexual budding, dependent upon surrounding conditions. Though yeasts are highly tolerant of environmental variations in **temperature** and acidity, they thrive in warm and moist places high in **oxygen** and low in carbon dioxide. Whether or not they reproduce through asexual budding depends on the favorability of surrounding conditions: when times are good, yeast clones are produced by budding. In times of environmental stress, yeasts produce spores which are capable of withstanding periods of environmental hardship—perhaps even to lie dormant, until conditions improve and the mingling of genes can take place with the spore of another yeast. This rare version of yeast reproduction provides for genetic

Yeast (*Anthrocobia muelleri*). *Carolina Biological Supply./Phototake NYC. Reproduced by permission.*

variation when conditions demand it, though budding is the predominant mode of yeast reproduction.

The importance of yeast for humans

People have been using yeast in bread baking for centuries, but suffering from its scourges for much longer. The biochemical by-products of yeast sugar metabolism—carbon dioxide and alcohol—are essential in baking and brewing. Bakers yeast (*Saccharomyces cerevisiae*), when added to baker's dough, produces carbon dioxide pockets that make bread rise. Brewer's yeast, another strain of *Saccharomyces*, takes advantage of the yeasts characteristic of switching to **anaerobic** fermentation when deprived of oxygen to produce alcohol as a by-product of incomplete sugar breakdown. Yeasts that occur naturally on the skins of **grapes** also play a vital role in fermentation—converting the sugars of grapes into alcohol for wine production.

Yeasts also comprise some of the natural microbial **flora** residing in and upon animals, including humans. The yeast species *Candida albicans* is perhaps the most notorious of the yeast inhabitants of the human body, responsible for the affliction Candidiasis, which may take many forms. Through normal health and hygiene, *Candida* is held in check by the populous and benign bacterial residents of our skin and mucous membranes. But in instances of compromised health, *Candida albicans* can result in skin sores (such as Thrush), urogenital tract infections (such as vaginitis), and internally, endocarditis (**heart** muscle **infection**), **inflammation** of the spleen, liver, kidneys and lungs. Victims of AIDS are particularly susceptible to Candidiasis.

Biotechnology and yeast

Yeasts are rising stars in the toolbox of biotechnologists. Early in the development of biotechnology, which

used cells as recipients of transplanted genes, **bacteria** were the **organism** of choice. However, limitations involving differences between bacterial cells and our own have relegated bacteria to a second place behind yeasts. Both yeast cells and human cells are eukaryotic—possessing a nucleus and membrane-bound organelles such as **ribosomes** and mitochondria. As biotechnology techniques have progressed over the last decade, yeasts have come to the fore as host cells for human **gene** implantation and as potential surrogate cells for housing human chromosomes. Yeasts have also been subjected to the alteration of their own genes, as biotechnologists attempt to develop strains of yeast more efficient in the **metabolism** of sugars for food and industrial applications.

By studying yeast **genetics**, scientists hope to gain insight into how the genes of all eukaryotic cells, including our own, function. In 1993 the first yeast **chromosome** was completely mapped, and the function of each of the 182 genes has yielded insight into such vital genetic processes as **mutation** repair, **enzyme** production, and cellular division regulation. These findings may yield insights into such fundamental human health issues as **cancer** and the aging process. Yeasts have also been transformed through **genetic engineering** to produce the first genetically engineered **vaccine** for **hepatitis** B. **Insulin** and hemoglobin are also being commercially manufactured by yeast cells which have been re-coded by inserted human genes. Yeasts are a hopeful host cell for housing the entire human **genome** in an international effort to decipher the gene library of human cells.

Resources

Books

Davis, Joel. *Mapping the Code: The Human Genome Project and the Choices of Modern Science.* Wiley, 1990.

Gross, Cynthia S. *The New Biotechnology: Putting Microbes to Work.* New York: Lerner, 1988.

Hanlin, Richard, and Miguel Ulloa. *Atlas of Introductory Mycology.* 2nd Ed. H. W. Wilson Co., 1988.

Krasner, R.I. *The Microbial Challenge: Human-Microbe Interactions.* Washington: American Society for Microbiology, 2002.

Sagan, Dorion. *Garden of Microbial Delights: A Practical Guide to the Subvisible World.* Harcourt Brace Jovanovich, 1988.

Stephens, R. B., ed. *Mycology Guidebook.* Mycology Guidebook Committee, Mycological Society of America, University of Washington Press, 1981.

Periodicals

Chan, K.C., A. Csikasz-Nagy, et al. "Kinetic Analysis of a Molecular Model of the Budding Yeast Cell Cycle". *Molecular Biology of the Cell* 11 (2000): 369-391.

Hoffman, M. "Yeast Biology Enters a Surprising New Phase." *Science* (March 20, 1992): 1210-1511.

Jeffrey Weld

Yellow fever

Yellow fever is a severe illness that causes outbreaks of **epidemic** proportions throughout **Africa** and tropical America. The first written evidence of such an epidemic dates back to a 1648 outbreak in the Yucatan Peninsula in Mexico. Since that time, much has been learned about the interesting transmission patterns of this devastating illness.

How yellow fever is spread

Many of the common illnesses in the United States, including the common cold, diarrhea, and **influenza** are spread via direct passage of the causative **virus** between human beings. Yellow fever, however, cannot be passed from one human being to another. Rather, the virus responsible for yellow fever is transmitted through an intermediate vector—a mosquito—which carries the virus from one host to another. The hosts of yellow fever include both humans and **monkeys**. The cycle begins when an infected monkey is bitten by a tree-hole breeding mosquito. This mosquito acquires the virus, and can pass the virus to any number of other monkeys which it may bite. When a human is bitten by such a mosquito, the human may acquire the virus. In the case of South American yellow fever, the infected human may return to the city, where an urban mosquito (*Aedes aegypti*) serves as a viral vector, spreading the **infection** between humans.

The host-vector-host cycle of yellow fever was first described by Walter Reed (1851-1902), the military surgeon for whom the Walter Reed Medical Center in Washington, D.C., is named. Reed was commissioned by the United States government to study yellow fever transmission. Reed's discovery of the mosquito as an intermediate vector led to improved control over the spread of the **disease**, ultimately allowing the building of the Panama Canal in an area prone to yellow fever epidemics.

Clinical course of yellow fever

Once a mosquito has transmitted the yellow fever virus to a human, the likelihood of symptomatic disease development is about 5-20%. Some infections may be warded off by the host's **immune system**; others may be subclinical, meaning they lack the severity of symptoms that would usually result in the identification of infection.

After a human host has received the yellow fever virus, there are five distinct stages through which a classic yellow fever infection evolves. These have been termed the periods of incubation, invasion, remission, intoxication, and convalescence.

The incubation period is the amount of time between the introduction of the virus into the host and the development of symptoms. For yellow fever this period is three to six days. During this time there are generally no symptoms identifiable to the host.

The period of invasion lasts two to five days; it begins with an abrupt onset of symptoms, consisting of fever and chills, intense headache, lower backache, muscle aches, nausea, and extreme exhaustion. The patient's tongue shows a characteristic white furry coating in the center, surrounded by beefy red margins. While most other infections that cause an elevation in **temperature** also cause an increase in **heart rate**, yellow fever produces an unusual symptom, called Faget's sign—the simultaneous occurrence of a high fever with a slowed heart rate. Throughout the period of invasion, there are still viruses circulating in the patient's **blood** stream, so continued viral transmission through mosquito vectors is possible.

The next phase is called the period of remission. The fever falls, and symptoms decrease in severity for a period of several hours to several days. In some patients, this signals the end of the disease; in other patients, this proves to be only the calm before the **storm**.

The period of intoxication is the most severe and potentially fatal phase of the illness. During this time, lasting three to nine days, a type of degeneration, or **tissue** breakdown, of the internal organs—specifically the kidneys, liver, and heart—occurs. This fatty degeneration results in what is considered the classic triad of yellow fever symptoms: **jaundice**, black vomit, and the release of protein into the urine. Jaundice causes the skin and the whites of the patient's eyes to take on a distinctive yellow tint. This yellow **color** is due to liver damage, resulting in the accumulation of bilirubin normally processed by a healthy liver. The liver damage also results in a tendency toward bleeding; the patient's vomit appears black due to the presence of blood. Protein, which is normally kept out of the urine by healthy and intact kidneys, appears in the urine due to disruption of the kidneys' composition by fatty degeneration.

Patients who survive the period of intoxication enter into a relatively short period of convalescence, and recover with no long-term deficits related to the yellow fever infection. Further, infection with the yellow fever virus results in lifelong immunity against repeated infection with the virus.

Five to 10 **percent** of all diagnosed cases of yellow fever are fatal. The occurrence of jaundice during a yellow fever infection is an extremely grave predictor, with 20-50% of these patients dying from their infection. Death may occur due to hemorrhaging (massive bleeding), often following a lapse into a comatose state.

Diagnosis

The **diagnosis** of yellow fever is made through examining the blood using various techniques in order to demonstrate the presence of either yellow fever viral antigens (the part of the virus that initiates the patient's immune response) or specific antibodies (the cells produced by the patient's immune system that are specifically directed against the yellow fever virus). Yellow fever is strongly suspected when Faget's sign is present, or when the classic triad of symptoms is noted.

Treatment

Medical management of yellow fever infection is directed toward relief of symptomatology. No active antiviral treatment currently exists. Fevers and **pain** should be relieved with acetaminophen; neither aspirin nor ibuprofen should be used, because either one could exacerbate the bleeding tendency already present. Dehydration, due to fluid loss both from fever and bleeding, must be carefully avoided. The risk of bleeding into the stomach can be decreased through the administration of antacids and other medications. Hemorrhages may require blood transfusions, and kidney failure may require **dialysis**, a process that allows the work of the kidneys in clearing the blood of potentially toxic substances to be taken over by a machine that is outside of the body.

Degeneration—Breakdown of tissue.

Faget's sign—The simultaneous occurrence of a high fever with a slowed heart rate.

Host—The organism, such as a monkey or human, in which another organism, such as a virus or bacteria, is living.

Vector—Any agent, living or otherwise, that carries and transmits parasites and diseases.

Prevention

A very safe, highly effective yellow fever **vaccine** exists, with about 95% of vaccine recipients acquiring long-term immunity to the yellow fever flavivirus. Careful measures to decrease mosquito populations in both urban areas and jungle areas in which humans are working, along with programs to vaccinate all people living in such areas, are necessary to avoid massive yellow fever outbreaks.

See also Mosquitoes.

Resources

Books

Andreoli, Thomas E., et al. *Cecil Essentials of Medicine.* Philadelphia: W. B. Saunders, 1993.

Berkow, Robert, and Andrew J. Fletcher. *The Merck Manual of Diagnosis and Therapy.* Rahway, NJ: Merck Research Laboratories, 1992.

Isselbacher, Kurt J., et al. *Harrison's Principles of Internal Medicine.* New York: McGraw Hill, 1994.

Kobayashi, G., Patrick R. Murray, Ken Rosenthal, and Michael Pfaller. *Medical Microbiology.* St. Louis: Mosby, 2003.

Mandell, Douglas, et al. *Principles and Practice of Infectious Diseases.* New York: Churchill Livingstone, 1995.

Rosalyn Carson-DeWitt

Yew

Yews are various **species** of tree- or shrub-sized, woody plants that comprise the **conifer** family, Taxaceae. All yews are in the genus *Taxus*, and about seven species are known, occurring in moist, temperate forest habitats in **North America**, **Europe**, **Asia**, and **Africa**.

Biology of yews

Yews have dark-green colored, rather soft, evergreen, needle-like leaves. Yews usually have separate male and female plants, although sometimes these flowers will develop on separate branches of the same **plant**. The male flowers are small and inconspicuous, and develop as globular bodies in **leaf** axils. The female flowers of yews develop in the springtime, and they occur singly and naked in the axis of a leaf. The fruit is a relatively large, hard seed, encased in a bright-red or scarlet, fleshy, cup-like structure known as an aril. The arils are sought by **birds** as food, which eat and disperse the **seeds**. The seeds of yews are poisonous to people and other **mammals**, although the fleshy arils are not. The symptoms of poisoning by yew seeds or foliage are dilation of the pupils, **pain**, and vomiting, leading to **coma** and death if the dose is large enough.

The **wood** of yews is dense, strong, flexible, and resistant to decay. The **bark** is thin, fibrous, scaly, and dark colored. The young twigs of yews are generally greenish in **color**.

Yews are slow growing and highly tolerant of shading. As a result, yews can survive in the shade beneath a closed forest canopy. This can be especially true beneath **angiosperm** trees, which allow the yews to grow relatively freely during the spring and autumn when the trees are in a leafless condition. However, yews also do well beneath conifer trees, as long as occasional gaps in the canopy allow exposure to light as brief sunflecks during the day.

Yews have a dense foliage and so cast a rather deep shade. As a result of this shade, as well as a toxic quality of the yew's leaf litter, few plants will grow beneath a closed canopy of yew **tree** or shrubs.

Species of yew

Three species of yew grow naturally in North America. The Pacific or western yew (*Taxus brevifolia*) is the only species that reaches the size of a small tree, typically 19-39 ft (6-12 m) tall, but as tall as 75 ft (23 m). The western yew is a species of the sub-canopy of conifer rain **forests** of the Pacific coast, ranging from central California to southern Alaska. The species also occurs on the relatively moist, western slopes of the Rocky Mountains in Washington and southern British Columbia. The western yew is relatively widespread in mature, conifer-dominated rainforests, but it is rarely abundant.

The Florida yew (*T. floridana*) is a rare, small tree that occurs in northern Florida. The Canadian yew (*T. canadensis*), also known as ground hemlock or poison hemlock, is relatively widely distributed in coniferous and mixed-wood forests of cool-temperate regions of eastern North America. The European or English yew (*T. baccata*) of Europe and Asia is a relatively tall species, which can grow to almost 65 ft (20 m) in height.

Economic and ecological importance of yews

Yew wood is very tough and elastic, a consequence of the structural qualities of its elongated, water-conducting cells known as tracheids. The wood of yews is of minor commercial importance. However, it is prized for certain uses in which strength and flexibility are required. During medieval times in western Europe, the wood of European yew trees was favored for the manufacture of bows. Today, bows are mostly made of synthetic materials, but some traditional archers still like to use bows made from yew. Canoe paddles are also sometimes carved from yew wood.

Yews are very popular shrubs in **horticulture**, probably being used more commonly than any other types of conifers. Various species are cultivated, but the European yew is the most common, and it is available in a number of cultivated varieties, or cultivars. It is quite easy to propagate desired cultivars of yews because these plants will root from stem cuttings, a relatively unusual trait for a conifer.

Yews are commonly used to accent the lines of buildings, or to create interesting entrances around doors. Very attractive hedges can also be developed using dense plantings of yews, which may be sheared to achieve a desired shape.

Sometimes, the dense foliage of taller yews is sheared in interesting ways to create a special visual effect. For example, yews can be trimmed to develop globular or cubed shapes, or to look like animals of various sorts. Yews are best sheared in the late summer.

Other yews that are commonly used in horticulture are the Japanese yew (*T. cuspidata*) and the columnar-shaped *T. hicksi* and *T. hilli*, all originally from Asia. These and some other species of yews have been widely introduced into North America and elsewhere as attractive species in horticulture. However, there is no indication that any of these non-native yews have escaped from cultivation and become invasive **pests** of natural habitats.

Medicinal uses of yews

Yews have widely been recognized as toxic to **livestock** and humans. In smaller doses, yew has been used as a minor folk medicine in some parts of its range. Example of the medicinal uses of yews include the induction of menstruation, and the treatment of **arthritis**, kidney **disease**, scurvy, **tuberculosis**, and other ailments. However, in recent years, yews have become famous for their use in the treatment of several deadly cancers.

In particular, the dark-brown or purple bark of western yew has been found to contain relatively large concentrations of an **alkaloid** known as taxol. Taxol has been demonstrated as being an effective treatment against advanced cases of ovarian and breast **cancer**, two deadly diseases. The use of taxol for these purposes is now approved by the U.S. Food and Drug Administration and by regulatory agencies in other countries. This recently developed use of western yew has led to an essentially insatiable demand for its bark.

The concentrations of taxol in bark are variable, ranging from only one part per million (ppm) to 690 ppm. The concentration of taxol in foliage of western yew is generally smaller, ranging from 12 to 80 ppm. At the present time, taxol is most efficiently extracted from yew bark. Unfortunately, it takes about 14 kg of yew bark to yield only 1 g of taxol, equivalent to three to 12 yew trees per cancer patient treated. Clearly, very large amounts of yew bark must be collected in order to have sufficient material to satisfy the medical and research demands.

Yew is not a commercial species in **forestry**, and up until about 1989 the species was routinely trashed and usually burned after logging. Today, however, the bark of western yew is routinely stripped from the survivors of the logging aftermath. In addition, the bark is now widely stripped from living western yews in unlogged forests, including old-growth rainforests of the Pacific coast, where this species tends to be most abundant. Sometimes, this process is rather inefficient, and in some cases of poaching, the bark has only been taken from the lower part of taller yew trees, where it can be easily reached. Unfortunately, the removal of its bark usually kills the yews by destroying its vascular system, so bark stripping is a rather wasteful use of the plant. A system has been designed and implemented to regulate the harvest of yew bark, but it is difficult to enforce over extensive areas, and a great deal of poaching occurs.

The enthusiastic collection of western yew bark, while contributing to the successful treatment of some cancer cases, threatens to exhaust the very resource upon which this treatment depends, as it is extremely wasteful and does not aim to conserve the species for future use. This dilemma may be resolved if pharmaceutical biochemists develop an economical method of synthesizing taxol in the laboratory, making the extraction of the alkaloid from western yew bark unnecessary. Alternatively, means could be found to economically extract the taxol found in relatively small concentrations in other parts of the yew, especially the foliage. Another possibility is to establish managed plantations of western yew for the specific purpose of obtaining taxol. This is actually being done, but the western yew is slow growing, and it will take some years before the plantations can be economically harvested.

KEY TERMS

Aril—A fleshy, often brightly colored covering that partially encases a seed. The aril is edible and is intended to encourage an animal to eat the fruit and thereby disperse the seed.

Cultivar—A distinct variety of a plant that has been bred for particular, agricultural or culinary attributes. Cultivars are not sufficiently distinct in the genetic sense to be considered to be subspecies.

Part per million (ppm)—A unit of concentration, equivalent to 0.0001%, or one milligram in a kilogram, or one milliliter in a liter.

Sunfleck—A transient patch of sunlight that travels over the forest floor as the Sun arcs overhead during the day.

Taxol—An alkaloid chemical that can be extracted from the bark and other tissues of yews and is active in the treatment of human ovarian and breast cancers.

So far, these relatively sustainable approaches have not been developed to the point where **pressure** on wild western yews can be relieved, and this plant is being rapidly mined from its natural habitats. Because the western yew is not naturally abundant, its populations will rapidly become depleted, and the species will become endangered in the wild.

Resources

Books

Brockman, C.F. *Trees of North America*. Golden Press, New York: 1968.

Periodicals

Daly, D. "Tree of Life." *Audubon* 70 (1992): 76-85.

Joyce, C. "Taxol: Search for a Cancer Drug." *BioScience* 43 (1993): 133-36.

Bill Freedman

Yttrium

Yttrium is not itself a rare **earth** element; however, its history is closely tied to that of the rare earths, and its chemical properties are similar to those of the members of that family. It also occurs in close association in nature with the rare earths.

Yttrium was the first new element to be identified in the complex mineral called ytteriteytterite (now known as gadolinitegadolinite), discovered in 1787. Johan Gadolin analyzed the dense black mineral and realized that it contained a new substance. That substance was further analyzed by the Swedish chemist Anders Gustav Ekeberg in 1799 and given the name of yttriayttria. Over the next 12 years, yttria was shown to contain nine other elements in addition to yttrium itself. An impure form of the element was produced by Friedrich Wöhler in 1828.

Yttrium's **atomic number** is 39, its **atomic weight**, 88.9059, and its chemical symbol, Y. The element is a silvery **metal** with a melting point of 2,771.6°F (1,522°C) and a **boiling point** of about 6,053°F (3,345°C). It is a relatively active metal that decomposes cold **water** slowly and boiling water rapidly.

Yttrium metal turnings may ignite spontaneously in air.

Yttrium is used in alloys to decrease grain size or add strength. Its greatest use, in the form of yttrium oxide, is in **television** phosphors. When doped with erbium, the phosphors produce a red glow. Synthetic garnets containing yttrium are very hard and have been used as gemstones that are similar to diamonds. The garnets are also used in microwave filters and in lasers. Compounds containing yttrium have been shown to become superconducting at relatively high temperatures. Such uses could conceivably become the most important application of the element in the future.

Y2K

As the end of the 1990s approached, the world became preoccupied with the coming of the Year 2000, nicknamed "Y2K" (Y for year and 2 times K, a standard designation for a thousand). Some were superstitious about the turning of year numbers to 2000, but many focused on a predicted technological problem commonly known as the"Y2K Glitch" or the "millennium bug" feared to cause computers and computer-assisted devices to malfunction. Computer experts predicted that older computers and several large data systems—including those controlling bank transactions, transportation networks, and government data—were not equipped to process dates beyond December 31, 1999. Though the problem was discovered decades earlier, the "Y2K Glitch" did not become widely known until the mid-1990s when media sources began reporting on the computer bug and its possible consequences. Some technological experts and lay-persons feared widespread computer failures, especially in regions dependent on older technology. Most experts, companies,

and governments however implemented detailed programs to fix the suspected problem in their computer systems well before January 1, 2000. As the date recognition problem was relatively simple to correct, such fixes were highly successful, and the "millennium bug" was averted. Though its effects were not largely realized, the "Y2K Bug" did highlight the pervasiveness of computer systems and their importance in everyday commerce and industry.

The millennium

In the modern era, the word millennium was appropriated to signify the largest metric division of calendar years on the French Republican calendar. The Republican calendar, implemented in France during the French Revolution (1789–1804) utilized metric divisions (groups of 10) to divide hours, days, weeks, months, and years. The use of the term millennium to describe a 1,000-year-long span of **time** remains.

Landmarks on the calendar

There are many types of **calendars** in the world with different systems of measuring time, distinct origins (religious or cultural), and various chronologies (starting events and dates). The Gregorian or Christian calendar (decreed by Pope Gregory XIII in 1582) has become an accepted international standard. This calendar begins with the **birth** of Christ; the years before Christ (B.C.) increase back in time from the year of Christ's birth. The years since Christ's birth (in 1 B.C.) are "the year(s) of Our Lord" in Latin, *anno domini* or A.D. and have grown larger in number from one forward. No year **zero** separates B.C. from A.D. As another example, the Hebrew calendar uses 3761 B.C., which was the year of the creation of the world (according to Old Testament interpretations), as its beginning (or 1 A.M., *anno mundi*, "the year of the world").

In the past, landmark centuries have also had great significance. When the first millennium was reached in A.D. 1000, in Medieval **Europe**, various events were predicted, from the return of Christ to a solar eclipse, from a plague to the obliteration of the world. Others chose to forgo doomsday predictions in favor of celebration.

Similarly, as the year 2000 was commonly thought to begin the third millennium, some superstitious persons, numerologists, and others attached unusual significance to a simple tick of the clock. Technically, the new millennium did not begin until January 1, 2001 (because the Christian calendar begins with the year A.D. 1). While this event was marked by many, it did not have the mass appeal of the turning of the year 2000.

The millennium bug and its origins

The year 2000, Y2K, did have another association. Some feared the so-called Y2K bug or millennium bug would be devastating. The Y2K bug was a fault built into computers. It came about because early developers of computer programs were uncertain that computers would have a future. Grace Murray Hopper and Robert Berner created COBOL (COmmon Business-Oriented Language), a standard programming language that began one of the keystones of **computer languages** and software. One of their purposes was to keep the language simple so anyone could use it.

To cut corners wherever possible, they built in standardized dates with two digits each for the day, month, and year, as in 061458 for June 14, 1958. This short form could also mean 1558 or 2058. Berner accidentally solved the problem without recognizing it in the late 1950s when the Mormons asked him to adapt the computer's storage capability to the massive genealogical library in Salt Lake City. Genealogical records embrace centuries, so Berner created a "picture clause" in COBOL that allowed years to be written with four digits. When others adapted COBOL, they ignored Berner's saving device. Notably, IBM made the two-digit-year formula part of its System/360 computers, which were as pervasive in the 1960s as Windows systems and software are today.

By the mid-1970s, organizations and programmers were beginning to recognize the potential obstacle. The Pentagon promised adjustments for the coming century beginning in 1974, but the transition reportedly proceeded slowly. Programmers began experimenting with plugging 2000-plus dates into their systems and software and found they did not compute. It was not until 1995, however, that Congress, the media, and the public all seemed to "discover" the problem. As of 1999, 1.2 trillion lines of computer code still needed fixed, and up to $600 billion were reportedly spent to reprogram and test vulnerable computers.

The Y2K phenomenon was put to an early test on September 9, 1999. Computer shorthand used to include the messages "0000" at the start of a file and "9999" to end or delete a file. Those predicting the worst for Y2K also expected computer files to disappear on September 9, 1999. Few incidents were reported in the days following 9999.

The potential for disaster

The U.S. Government prepared for Y2K for years, but the actual progress made by the dawning of 2000 was uncertain. The credit card giants Visa, American Express, and MasterCard invested millions of dollars in solving the problem together, assisted merchants who carry their cards, and time-proofed the VeriFone system that autho-

KEY TERMS

Bug—An error in software that causes a computer to operate incorrectly, produce wrong results, or shut down. Synonym: glitch.

Embedded system—A computer chip or processor built into an appliance or mechanical device that is not a computer itself.

Millennium—A time period of 1000 years.

Millennium bug or Y2K bug—A software or computer programming error in which 00 as the year is read as a mistake or a symbol for the year 1900 instead of 2000, causing the computer to reject the information or stop operating.

rizes transactions. Among the gloomiest predictions were worries that air traffic control systems would fail, major utilities would be unable to supply **water** and power, and banks would not acknowledge mortgage payments or social security checks. The heritage of the original COBOL programming was passed down through generations of computers to the personal computer or PC, and computer chips that control the workings of some of our simplest devices like coffeemakers that display the date also control automobiles, heavy machinery, building security systems, and space satellites. Such built-in processors are called embedded systems, and those that are date-dependent were not expected to function properly or to shut down as 1999 rolled over to 2000.

Realities of compliance

Companies in the United States and many other countries encouraged businesses to share compliance information regarding the potential Y2K problem. The Year 2000 Readiness and Disclosure Act made businesses liable if they provided consumers with inaccurate information regarding their Y2K readiness. Businesses typically did not provide free upgrades to make computers and other products Y2K-compliant, but they did furnish information on their Web sites, including corrections that could be downloaded.

Perhaps the greatest fears were focused on developing countries. Pacific Rim and Western countries publicized Y2K's associations well, but, according to the World Bank, many key systems and emerging economies were at risk in the Third World. Fewer and larger computers control a range of systems, the systems of several countries are often interdependent, emergency management is short-handed for day-to-day problems aside from

Y2K, and funds were scarce for buying emergency help and repairs.

Y2K: The aftermath

As the year 2000 became a reality, most all major computer systems were fully Y2K compliant. There were no widespread failures, and industries from airlines to power plants, functioned normally. Even in developing regions, the predicted effects of the millennium bug, for the most part, were successfully prevented. Newspapers and pundits pronounced the Y2K bug a complete bust. So much money and so many technological hours were spent anticipating the Y2K bug, and protecting against it, that the possible scope of its effects will remain unknown.

Resources

Periodicals

Teresi, Dick. "Zero." *The Atlantic Monthly* (July 1997): 88.
Woodward, Kenneth L. "Uh-oh, Maybe We Missed the Big Day." *Newsweek* (August 11, 1997): 51.

Other

United States General Services Administration, National Y2K Clearinghouse. [cited October 18, 2002] <http://www.y2k.gov>.
United States Senate Special Committee on the Year 2000 Technology Problem. *Y2K Aftermath: Crisis Averted, Final Committee Report.* February, 29, 2000.

Gillian S. Holmes

Z

Zebras

Zebras are members of the horse family (Equidae) that inhabit tropical **grasslands** (savannas) in much of sub-Saharan **Africa**. Three of the seven **species** of equids are zebras. Zebras are herd-living social **ungulates** (hoofed **mammals**) recognized by a black-and-white (or cream or yellowish) striped coat, short erect mane, and a tail averaging about 18 in (0.5 m) long. The body length of a zebra is about 6-8.5 ft (2-2.6 m) long, and body weight can reach 770 lb (350 kg), with males slightly bigger than females.

There are three species and several subspecies of zebra. The common, Burchell's, or plains zebra *(E. burchelli)* lives throughout much of eastern and southern Africa, and is the best-studied species. Grévy's zebra *(Equus grevyi)* is found in Somalia, Kenya, and Ethiopia, and is the largest of the wild **horses**, characterized by large ears, narrow stripes, and a thick neck. The third species is the mountain zebra *(E. zebra)* found in the hill country of Angola, Namibia, and western South Africa.

The stripes of the zebra camouflage the **animal** in the waving **grasses** of the African grasslands, especially at dusk and dawn. Zebras rest in herds in open ground where they can see predators approaching, rather than lying down in the grass. When a herd of zebras is being chased by predators such as lions or hyenas, their stripes make it difficult for predators to visually latch on to a specific animal to attack. Other suggestions are that the stripes identify individual zebras as part of a group and also initiates mating **behavior**.

The common zebra

The common or Burchell's zebra stands about 50-52 in (1.3 m) tall at the shoulder. It has the widest stripes of all zebras, and the stripes continue down to the belly of the animal. The stripes of Burchell's zebra become fainter and almost disappear in southern populations.

There are several subspecies of the common zebra, distinguishable by the pattern of stripes on the rump. Grant's zebra (*E. burchelli granti*) has wide rump stripes, the Chapmann's or Damaraland zebra has tan shadow stripes between the black stripes, and Selous's zebra has only faint shadow stripes between the black stripes. A subspecies known as the "true" Burchelli's zebra had reddish brown stripes, and became extinct in the early 1900s.

Herds of common plains zebras are actually a collection of family groups. Each family group is headed by a dominant male stallion and one female is dominant over the other seven or eight females and their young.

When the group is feeding or drinking, one animal stands guard. If a **predator** approaches, such as a troop of wild dogs or lions, the stallion zebra moves to the back of the fleeing herd and ensures that no single animal falls behind or gets separated and becomes vulnerable to attack. Zebras that become isolated call attention to their plight by making a harsh sound like a combination bark and bray, which draws the other zebras back to protect it.

The zebra's large, rounded ears turn in every direction and are able to pick up the slightest sound. The zebra's primary defense is running, which it can do at speeds approaching 40 mph (64 kph), though for only a short time. Given the chance, a zebra fights primarily by kicking with its hind hooves. However, an attacking predator is likely to grasp the animal's neck with one quick leap that prevents the zebra from acting.

After a young mare mates for the first time, she will stay with the same family group for the rest of her life. Males that have not yet developed herds—or that have been ousted from their herds by other males—live in male-only bands (bachelor herds), which they leave to find mates at about six years old. If stallions try to take mares from an existing family group, they will have to fight that group's stallion.

Gestation lasts approximately one year, with only one offspring produced. The young are born with disproportionately long legs and are ready to run with the herd within a few minutes. The brown-striped coat of a new foal gradually changes to the familiar black and white.

Mountain zebras

There are two subspecies of mountain zebras in South Africa which live on rough, rocky highlands, but once also occurred in large numbers in the grassy lowland plains. The Cape mountain zebra of South Africa is the smallest zebra, standing about 4 ft (1.2 m) at the shoulder and weighing about 600 lb (272 kg). The stripes of the Cape mountain zebra are slightly wider and shorter than those of the other subspecies, Hartmann's mountain zebra of Namibia. The Cape mountain zebra has a dewlap under the lower jaw, which other zebras do not have.

Herds of the common zebra readily mix with herds of wildebeest, but Cape mountain zebras tend to keep apart from other animals. Hartmann's zebras have the ability to locate **water** in apparently dry valleys.

Grévy's zebra

Grévy's zebra is quite different in both appearance and behavior from the common zebra. It stands close to 5 ft (1.5 m) tall at the shoulder and weighs about 850 lb (400 kg). Its stripes are very narrow, do not cross over the lower back as they do in the Cape mountain zebra.

Grévy's zebras live in semidesert country and do not form herds. Instead, only the mother foals remain together, while the males and the young females drift off. Individual stallions have remarkably large breeding territories, and any female found within the territory belongs to the male. A stallion will tolerate other males in his territory only if they do not try to mate with the mares. The gestation period of the Grévy's zebra is 12 1/2 months, one month longer than other species of zebra.

The recently extinct zebra-like quagga (*E. quagga*) looked like a combination of wild ass and zebra and had a reddish coat and stripes only on the head, neck, and shoulders. The name quagga is derived from the odd bray of this species, which was described as "kwa-ha." The quagga lived throughout South Africa but was killed by nineteenth-century settlers for its meat and hides. The last quagga died in an Amsterdam zoo in 1883.

The populations of the common zebra are high, especially in the Serengeti Plains of Tanzania and Kenya, where zebras run with the huge wildebeest herds on their annual migrations.

Grévy's zebra and the Cape mountain zebra are both endangered. There are probably fewer than 15,000 Grévy's zebras in Ethiopia, Somalia, and Kenya, in three

---

KEY TERMS

Dewlap—A loose fold of skin that hangs from the neck.

Equid—Any member of family Equidae, including horses, zebras, and asses.

Gestation—The period of carrying developing offspring in the uterus after conception; pregnancy.

---

separate populations. A major herd of Grévy's zebra is protected in a game reserve in Kenya. There are less than 200 Cape mountain zebras which live in Mountain Zebra National Park in South Africa, and this small number makes the long-term survival of this species doubtful.

Resources

Books

Arnold, Caroline. *Zebra.* New York: William Morrow & Co., 1987.

Duncan, P., ed. *Zebras, Horses and Asses: An Action Plan for the Conservation of Wild Equids.* Island Press, 1992.

MacDonald, David, and Sasha Norris, eds. *Encyclopedia of Mammals.* New York: Facts on File, 2001.

Scuro, Vincent. *Wonders of Zebras.* New York: Dodd, Mead And Co., 1983.

Stidworthy, John. *Mammals: The Large Plant-Eaters.* Encyclopedia of the Animal World. New York: Facts On File, 1988.

Willis, Terri. *Serengeti Plain.* Wonders of the World series. Austin, TX: Raintree Steck-Vaughan Publishers, 1994.

Zebras. Zoobooks. San Diego: Wildlife Education, Ltd., 1989.

Zero

Zero is often equated with "nothing," but that is not a good analogy. Zero can be the absence of a quality, but it can also be a starting point, such as 0° on a **temperature** scale. In a mathematical system, zero is the *additive identity*. It is a number which can be added to any given number to yield a sum equal to the given number. Symbolically, it is a number 0, such that $a + 0 = a$ for any number a.

In the Hindi-Arabic numeration system, zero is used as a placeholder as well as a number. The number 205 is distinguished from 25 by having a 0 in the tens place. This can be interpreted as no tens, but the early use of 0 in this way was more to show that 2 was in the hundreds place than to show no tens.

Zero is used in some ways that take it beyond ordinary **addition** and **multiplication**. One use is as an **ex-**

ponent. In an exponential function such as $y = 10^x$ the exponent is not limited to the counting numbers. One of its possible values is 0. If 10^0 is to obey the rule for exponents—$10^m = 10^{m+0} = 10^m \times 10^0$—$10^0$ must equal 1. This is true not only for 10^0 but for any a^0, where a is any **positive number**. That is, $a^0 = 1$.

Another curious use of zero is the expression 0!. Ordinarily n! is the product $1 \times 2 \times 3 \times ... \times n$, of all the **integers** from 1 to n. In a formula such as n!/r!(n - r)!, which represents the number of different combinations of things which can be chosen from n things r at a time, 0! can occur. If 0! is assigned the value 1, the formula works. This happens in other instances as well.

The symbol for zero does not appear before about A.D. 800, when it appears in connection with Hindu-Arabic base-10 numerals. In these numerals it functions as a place holder. The Mayans also used a zero in writing their base-20 numerals. It was a symbol which looked something like an **eye**, and it acted as a place holder.

The reason that the symbol appeared so late in history is that the number systems used by the Greeks, Romans, Chinese, Egyptians, and others did not need it. For example, one can write the Roman numeral for 1056 as MLVI. No zero as a place holder is needed. The Babylonians did have a place-value system with their base-60 numerals, and a symbol for zero would have eliminated some of the ambiguity that shows up in their clay tablets, but was probably overlooked because, within each place, the numbers from 1 to 59 were represented with wedge-shaped tallies. In a tally system all that is required to represent zero is the absence of a tally. Sometimes Babylonians did use a dot or a space as a placeholder, but failed to see that this could be a number of its own.

The word zero appears to be a much metamorphosed translation of the Hindu word "sunya," meaning void or empty.

Zero also has the property $a \times 0 = 0$ for any number a. This property is a consequence of zero's additive property.

In ordinary **arithmetic** the statement ab = 0 implies that a, b, or both are equal to 0; that is, the only way for a product to equal zero is for one or more of its factors to equal zero. This property is used when one solves equations such as (x - 2)(x + 3) = 0 by setting each **factor** equal to zero.

The multiplicative property of zero is also used in the argument for not allowing zero to be used as a divisor or a denominator. The law which defines a/b is (a/b)b = a. If one substitutes 0 for b, the result is (a/0)0 = a, which forces a to be 0. But even when a is 0, the law allows 0/0 to be any number, which is intolerable.

Zero sometimes appears in disguise. In even-and-odd arithmetic we have "even plus odd equals odd," "odd times odd equals odd," and so on. The various combinations can be listed in the tables

+	even	odd		x	even	odd
even	even	odd		even	even	even
odd	odd	even		odd	even	odd

Is there a zero? Is there an element 0 such that 0 + a = a for either of the possible values of a? The top line in the addition table says that there is. "Even" is such an element. Does $0 \times a = 0$ for both values of a? The top line of the multiplication table says that it does. Does ab = 0 imply that one or both of the factors is 0? Only if both factors are odd, is the product odd; so, yes, it does.

Thus this miniature arithmetic has a zero, and it is "even."

Another arithmetic is clock arithmetic. In this arithmetic 3 is three hours past 12; 3 + 7 is 10 hours past 12; and 3 + 12 is 15 hours past 12. But on a clock, every 12 hours the hands return to their original position; so 15 hours past 12 is the same as three hours past 12. For any a, a + 12 = a. [In number-theory symbolism this would be written a + 12 intergral a (mod 12).] So in clock arithmetic, 12 behaves like 0 in ordinary arithmetic.

It also multiplies like 0. Twelve 3-hour periods equal 36 hours, which the hands show as 12. Twelve periods of a hours each leave the hands at 12 for any a (a is limited to whole numbers in clock arithmetic), so $12 \times a = 12$.

Thus, in clock arithmetic 12 does n0t look like zero, but it behaves like zero. It could be called 0, and on a digital 24-hour clock, where the number 24 behaves like 0, 24 is called 0. The next number after 23:59:59 is 0:0:0.

In this arithmetic, unlike ordinary arithmetic, the law "ab = 12 if and only if a, b, or both equal 12" does not hold. The "if" part does, but not the "only if." Six times 2 is 12, but neither 6 nor 2 is 12. Three times 8 is 12 (the hands go around twice, passing 12 once and ending at 12), but neither 3 nor 8 is 12. Thus in clock arithmetic there can be two numbers, neither of them zero, whose product is zero. Such numbers are called divisors of zero. This happens because we use 12-hour (or 24-hour) clocks. If we used 11-hour clocks, it would not.

See also Numeration systems.

Resources

Books

Clawson, Calvin C. *The Mathematical Traveler: Exploring the Grand History of Numbers.* Cambridge, MA: Perseus Publishing, 2003.

Gelfond, A.O. *Transcendental and Algebraic Numbers.* Dover Publications, 2003.

Gullberg, Jan, and Peter Hilton. *Mathematics: From the Birth of Numbers.* W.W. Norton & Company, 1997.

Stopple, Jeffrey. *A Primer of Analytic Number Theory: From Pythagoras to Riemann.* Cambridge: Cambridge University Press, 2003.

J. Paul Moulton

Zinc *see* **Element, chemical**

Zirconium *see* **Element, chemical**

Zodiacal light

The term zodiacal light is used to describe a faint, glowing band of **light** that occasionally appears near the eastern or western horizon, which is caused by reflection of sunlight from tiny dust particles in the **solar system**.

Our solar system can be pictured as a huge disk, with the planets, their moons, and asteroids mostly orbiting in or near the same **plane**. If you look at the sky some evening when the **Moon** and two or three planets are visible, you will see they fall more or less in a line across the sky; this imaginary line, which is the plane of Earth's **orbit** projected against the sky, is typically called the ecliptic, but also is sometimes called the zodiac.

Planets and moons are not the only things in the solar system; they're just the most visible. The ecliptic is filled with tiny particles of dust left over from the formation of the solar system. The solar system formed from a vast, rotating cloud of gas and dust termed the solar nebula. As this cloud contracted, its **rotation rate** increased. The faster rotation flattened the cloud into a disk, much as a blob of pizza dough becomes a disk when the baker tosses it spinning into the air. Most of this spinning disk of gas and dust condensed into the conglomerations of **matter** that are now the **Sun** and planets, but some was left behind, floating free in the space between the planets.

Today, this remnant dust is still there. It is far too sparse and cold to be directly detectable, but it does reflect light. As the Sun's **radiation** streams through the solar system, a small amount of it is reflected by these dust particles. Some is reflected toward **Earth**.

As a result, we occasionally see a thin band of light, called the zodiacal light, extending upward from the eastern or western horizon along the plane of the ecliptic, or zodiac. The light appears in a band because the dust particles reflecting it lie mostly in the plane of the ecliptic. The effect is much the same as when we shine a flashlight beam through a cloud of chalk dust in the air: we can see the faint, reflected light from the particles. The zodiacal light typically appears in deep dusk after sunset or before sunrise, because it is at those times of night that the Sun is favorably placed to illuminate the intraplanetary dust. Zodiacal light is also easier to see in the fall and winter in the northern hemisphere, when the zodiac is highest in the sky and the plane of the ecliptic is oriented more nearly perpendicular to the horizon than in summer. The ghostly zodiacal light is not visible every night, and it is easily washed out by moonlight or the glow of a nearby town or city. But from a dark viewing site under the right conditions, it is easily visible and makes a splendid addition to the array of wonders above.

Jeffrey Hall

Zoonoses

Zoonoses are diseases caused in humans by **bacteria**, viruses, **parasites**, and **fungi** that have been transmitted from animals, **reptiles**, or **birds** to people.

Because many of the **microorganisms** that cause zoonotic **disease** are normal inhabitants of domestic animals and birds, people who are involved in agriculture and those who work in food processing plants can be at risk for **infection**. Prevention of such infections is actively studied. Understanding the host factors that contribute to immunity from disease and which factors promote the establishment of disease is essential if zoonotic infections are to be successfully prevented.

Humans can develop zoonotic diseases by different routes, depending on the microorganism. For example, some bacteria can cause infection following entry through a cut in the skin. Another common method of disease transmission is by the inhalation of bacteria, viruses, or fungi. A third route is via the ingestion of improperly cooked food or **water** contaminated with the fecal material.

A classic example of a zoonotic disease is **yellow fever**. The discovery of the disease and of its origin came during the construction of the Panama Canal during the first decade of the twentieth century. The construction of the canal took humans into previously unexplored regions of the Central American jungle. This brought the workers into contact with animals and microorganisms that they had not been in contact with before. One of the microbes produced the serious illness that came to be called yellow fever.

Another example occurred beginning in the 1970s. Then, the clearing of the Amazonian rain forest to provide agricultural land accelerated. Woodcutters became ill with infections caused by the Mayaro and Oropouche viruses. Another final example dates back only a decade ago. In the mid 1990s, a rapidly developing and often-fatal lung infection in the Southwestern United States was found to be caused by the Hanta **virus** that can be transmitted from **rodents** to humans. A final example is the viral zoonoses called **West Nile virus**. This virus, which can infect birds such as crows and blue jays, can be transmitted to humans by a mosquito that feeds on the bird and then subsequently feeds on a human.

There are a variety of zoonotic diseases that are caused by the movement of bacteria from an **animal**, bird, or insect host to humans. Examples include Tularemia, which is caused by *Francisella tulerensis*, Leptospirosis (*Leptospiras spp.*), **Lyme disease** (*Borrelia burgdorferi*), Chlaydiosis (*Chlamydia psittaci*), Salmonellosis (*Salmonella spp.*), **Brucellosis** (*Brucella melitensis, suis, and abortus*), Q-fever (*Coxiella burnetti*), and Campylobacteriosis (*Campylobacter jejuni*).

Fungi also produce zoonoses. An example is Aspergillosis, which is caused by *Aspergillus fumigatus*.

Two protozoan zoonoses have emerged in accelerating numbers in the past two decades. One protozoan is called Giardia. The debilitating diarrhea that is commonly dubbed "beaver fever" is caused by drinking water that is contaminated with *Giardia lamblia*. *Cryptosporidium parvum* causes an equally debilitating diarrheal malady called cryptosporidiosis.

Giardia and Cryptosporidium live naturally in many wild animals. The loss of natural **habitat** with increasing human development has brought the animals and their microbial cargo into more frequent contact with people, and human infections are the result.

Human encroachment is also fueling the emergence of fatal viral hemorrhagic fevers such as Ebola and Rift Valley fever. The zoonotic origin of these agents is not definitively known. However, the available evidence strongly supports the idea that primate populations harbor the viruses, and that periodically transmission of the viruses from **primates** to humans occurs.

Beginning in 1986, an outbreak of bovine spongiform encephalopathy (BSE) among cattle in the United Kingdom led to the eventual slaughter of over 150,000 domestic animals. Ominously, scientists have evidence that links the "mad cow" disease in the domestic animals with the **brain** degeneration in humans known as Creutzfeld-Jacob disease. Although not conclusively established, there is the possibility that the disease-causing agent of Creutzfeld-Jacob disease was passed to some of the people when they ate infected beef.

The increasing incidence of these and other zoonotic diseases may be in part due to the increased ease of global travel. Microorganisms are global residents, which can easily move along with people and cargo. New combinations of microorganisms and susceptible human populations have and continue to arise. As of July 2003, the virus responsible for **severe acute respiratory syndrome** (SARS) was thought to be zoonotic in origin, but further research is needed to clarify origin and transmission modes.

See also Chimpanzees; Prions.

Resources

Books

Hugh-Jones, M.E., H.V. Hagstad, and W.T. Hubbert. *Zoonoses: Recognition, Control and Prevention*. Ames: Iowa State University press, 2000.

Organizations

Centers for Disease Control and Prevention, National Center for Infectious Diseases, Rabies Section, MSG-33. 1600 Clifton Road, Atlanta, GA 30333. October 23, 2001 [cited November 11, 2002] <http://www.cdc.gov/ncidod/dvrd/rabies/bats_&_rabies/bats&.htm>.

Los Angeles County Department of Animal Services. 3834 S. Western Ave., Room 238, Los Angeles, CA 90062. (323) 730–3723. [cited November 11, 2002]. <http://www.la publichealth.org/vet/guides/vetzooman.htm>.

World Health Organization. Avenue Appia 20, 1211 Geneva 27, Switzerland. (+41 22) 791 21 11. January 2002 [cited November 11, 2002] <http://www.who.int/health-topics/zoonoses.htm>.

Brian Hoyle

Zooplankton

Zooplankton are small animals that occur in the **water** column of either marine and **freshwater** ecosystems. Zooplankton are a diverse group defined on the basis of their size and function, rather than on their taxonomic affinities.

Most **species** in the zooplankton community fall into three major groups—Crustacea, Rotifers, and Protozoas. Crustaceans are generally the most abundant, especially those in the order Cladocera (waterfleas), and the class Copepoda (the **copepods**), particularly the orders Calanoida and Cyclopoida. Cladocerans are typically most abundant in freshwater, with common genera including *Daphnia* and *Bosmina*. Commonly observed genera of marine calanoid copepods include *Calanus*, *Pseudocalanus*, and *Diaptomus*, while abundant cyclopoid copepods include *Cyclops* and *Mesocyclops*. Other crustaceans in the zooplankton include species of opossum shrimps (order

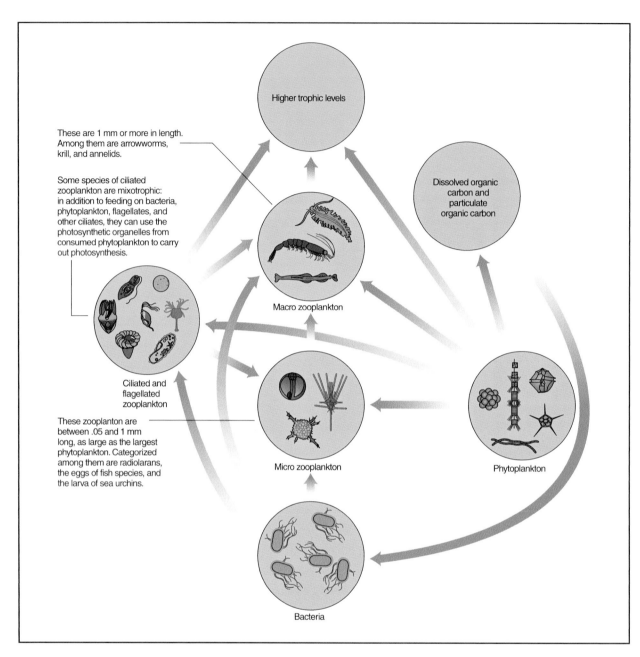

These are 1 mm or more in length.
Among them are arrowworms,
krill, and annelids.

Some species of ciliated
zooplankton are mixotrophic:
in addition to feeding on bacteria,
phytoplankton, flagellates, and
other ciliates, they can use the
photosynthetic organelles from
consumed phytoplankton to carry
out photosynthesis.

Higher trophic levels

Dissolved organic
carbon and
particulate
organic carbon

Macro zooplankton

Ciliated and
flagellated
zooplankton

These zooplanton are
between .05 and 1 mm
long, as large as the largest
phytoplankton. Categorized
among them are radiolarans,
the eggs of fish species, and
the larva of sea urchins.

Micro zooplankton

Phytoplankton

Bacteria

The planktonic ecosystem; arrows show the movement of biomass through the food chain. *Illustration by Hans & Cassidy. Courtesy of Gale Group.*

Mysidacea), amphipods (order Amphipoda), and fairy **shrimp** (order Anostraca). Rotifers (phylum Rotifera) are also found in the zooplankton, as are protozoans (kingdom **Protista**). **Insects** may also be important, especially in fresh waters close to the shoreline.

Most zooplankton are secondary consumers, that is, they are herbivores that graze on **phytoplankton**, or on unicellular or colonial **algae** suspended in the water column. The productivity of the zooplankton community is ultimately limited by the productivity of the small algae

upon which they feed. There are times when the **biomass** of the zooplankton at any given time may be similar to, or even exceed, that of the phytoplankton. This occurs because the animals of the zooplankton are relatively long-lived compared with the algal cells upon which they feed, so the turnover of their biomass is much less rapid. Some members of the zooplankton are detritivores, feeding on suspended organic detritus. Some species of zooplankton are predators, feeding on other species of zooplankton, and some spend part of their lives as **parasites** of larger animals, such as **fish**.

Zooplankton are very important in the food webs of open-water ecosystems, in both marine and fresh waters. Zooplankton are eaten by relatively small fish (called planktivorous fish), which are then eaten by larger fish. Zooplankton are an important link in the transfer of **energy** from the algae (the primary producers) to the ecologically and economically important fish community (the consumers).

Species of zooplankton vary in their susceptibility to environmental stressors, such as exposure to toxic chemicals, acidification of the water, **eutrophication** and **oxygen** depletion, or changes in **temperature**. As a result, the species assemblages (or communities) of the zooplankton are indicators of environmental quality and ecological change.

GENERAL INDEX

References to individual volumes are listed in **boldface**; numbers following a colon refer to page numbers.
A **boldface** page number indicates the main essay for a topic.
An *italicized* page number indicates a photo, figure (f), illustration, or table (t).

A

A horizons, **3**:2006–2007
AA (Alcoholics Anonymous), **1**:112
Aardvarks, **1**:*1*, **1**:1–2
Aardwolfs, **3**:2069
Aargus pheasants, **5**:3038
AAT (Anglo-Australian telescope), **6**:*3984*
Abaca, **1**:429
Abacus, **1**:*2*, **1**:2–3, **1**:*3*, **1**:672
Abbé, Ernst, **4**:2553
Abbey, Edward, **2**:1017
Abbott, Patrick, **1**:102
Abdominal hernias, **3**:1978, **3**:*1979*
Abegg, Richard, **2**:821
Abel, Niels Henrik, **3**:1883
Abelam, P. H., **4**:3022
Abelian groups, **3**:1883
Abelson, Phillip, **2**:1429, **6**:3881
Abetalipoproteinemia, **4**:2510
Abies balsamea. See Balsam firs
Abies concolor. See Firs
Abies spp., **3**:1615
AbioCor® artificial heart, **1**:302
Abiogenesis. *See* Spontaneous generation
ABMs (Antiballistic missiles), **5**:3475
Abnormal psychology, **5**:3276
ABO antigens, **1**:227
Abocets, **6**:3848–3849
Aboriginal tribes, **1**:*265*
Abortion, **2**:1046–1047
Abrasive disc cutoff saws, **4**:2399
Abrasives, **1**:3–5, **1**:*3t*, **1**:*4t*
Abscesses, **1**:*5*, **1**:5–6
Absolute dating, **2**:1155
Absolute humidity, **3**:2039, **3**:2040
Absolute maxima, **4**:2477
Absolute minima, **4**:2477
Absolute zero, **1**:6–7, **2**:1116, **2**:1381
Absorption lines, **3**:1984–1985
Absorption spectra, **6**:3758
Abundance, in biological communities, **1**:515
Abyssal plains, **1**:7, **4**:2851
Abyssopelagic zones, **4**:2854
AC. *See* Alternating current

AC motors, **2**:1374–1375
Acacias, **4**:2312
Acadian flycatchers, **6**:4151
Acanthaster planci. See Crown-of-thorns starfishes
Acanthodoras spinosissimus. See Talking catfish
Acanthophis antarcticus. See Death adders
Acanthophis pyrrhus. See Death adders
Acari, **4**:2595
Acceleration, **1**:7–9, **1**:11–12, **1**:425, **4**:2649
Accelerators, **1**:9–13, **1**:*10*, **1**:*11*, **5**:3356
Accidental hypothermia, **3**:2076–2077
Accipiters, **3**:1922–1923, **3**:1925
Accipitridae, **1**:550–551, **6**:4261
Accommodation, **2**:1197–1198, **6**:4240
Accretion disks, **1**:13, **2**:1538
Accumulators. *See* Storage cells
Accuracy, **1**:13–14
ACE inhibitors, for hypertension, **3**:2075
Acer negundo. See Box-elders
Acer rubrum. See Red maples
Acer saccharum. See Sugar maples
Acetaldehyde, **1**:111–112, **2**:892
Acetaminophen, **1**:159, **1**:181, **1**:288
Acetic acids, **1**:14
Acetic anhydride, **2**:1518
Acetins, **3**:1834–1835
Acetone, **1**:15
Acetylcholine, **1**:15–16, **1**:152, **6**:4265
Acetylcholine esterase (AChE), **3**:2140
Acetylene, **3**:2049
Acetylsalicylic acid, **1**:16–17, **1**:180, **3**:1971
AChE (Acetylcholine esterase), **3**:2140
Achenbach Child Behavior Rating Scales, **1**:377
Acheulean tools, **3**:1916, **3**:1917
Achondrite meteorites, **4**:2538
Achondroplasia, **1**:558
Achra zapota. See Sapodilla trees
Acid-base indicators, **3**:2112–2113
Acid fog, **3**:1650
Acid mine drainage, **1**:535, **3**:1926
Acid rain, **1**:18–25, **1**:*21*, **1**:*23*, **5**:3218
from aerosols, **1**:61

from agrochemicals, **1**:80
lake fertility and, **4**:2263
from non-point sources, **4**:2788
sulfur cycle, **6**:3898–3899
transport of, **1**:92–93
Acidification bioremediation, **1**:534–535
Acidophiles, **1**:412
Acids, **1**:25–28, **1**:*27t*
acetic, **1**:14
benzoic, **1**:485–486, **2**:1523
boric, **1**:597
carboxylic, **1**:724–725
cirtric, **5**:3426
citric, **1**:25, **2**:893, **2**:894
definitions, **2**:972
dissociation of, **2**:1255–1256
glutamic, **4**:2630
linoleic, **3**:1579
nitric, **4**:2773
nucleic, **2**:1190, **4**:2822
organic, **1**:27–28
oxalic, **4**:2924
stearic, **6**:3821–3822
sulfuric, **1**:26, **6**:3900
tartaric, **6**:3956–3957, **6**:*3956f*
weathering and, **6**:4311
See also specific acids and types of acids
Acinonyx jubatus. See Cheetahs
Acne, **1**:28–29, **1**:*29*
Acomys spp. *See* Spiny mice
Aconcagua, **5**:3720
Acorn barnacles, **1**:440
Acorn woodpeckers, **6**:4345, **6**:4346
Acorn worms, **1**:29–30
See also Tongue worms
Acorns, **4**:2843
Acoustics, **1**:30–34, **1**:*31f*
Acquired immune deficiency syndrome. *See* AIDS
Acquired resistance, bacterial, **1**:416
Acquired traits, **4**:2264
Acridotheres cristatellus. See Crested mynahs
Acridotheres spp., **4**:2696
Acridotheres tristis. See Common mynahs
Acromegaly, **3**:1887–1888

Algebra, **1:**121–124
  addition in, **1:**50
  analytic geometry and, **1:**183, **1:**184
  Boolean, **1:**595–597, **1:***596*
  fundamental theorem of, **3:**1712
  of inequalities, **3:**2122–2123
  linear, **4:**2346–2348, **4:**2478
Algebraic functions, **3:**1711–1712
  See also Functions
Algebraic invariance, **3:**2182
Algebraic numbers, **2:**1086
Algorithms, **1:**49, **1:**124–125, **2:**1121
Alhazen (Arab mathematician), **2:**942,
  4:2336
Alice in Wonderland, **4:**2678
Aliphatic hydrocarbons, **3:**2047–2049
Alkali metals, **1:**125–126, **2:**1426,
  4:3019
Alkaline earth metals, **1:**126–128,
  4:2518
Alkaline fuel cells, **3:**1709
Alkaline phosphatase (ALP),
  4:2511–2512
Alkaline zinc/manganese dioxide cells,
  1:459
Alkaloids, **1:**128–130, **2:**1129
Alkanes, **1:**131, **3:**2048
Alkaptonuria, **4:**2509
Alkenes, **1:**131, **3:**2048–2049
Alkyl bromides, **3:**1903
Alkyl chlorides, **3:**1903
Alkyl fluorides, **3:**1902
Alkyl group, **1:**131–132, **5:**3039
Alkyl iodides, **3:**1903
Alkyl radicals, **1:**131
Alkynes, **3:**2049
Allard, H. A., **1:**516
Allegheny Mountains, **4:**2793
Alleles, **1:**132, **1:**708, **4:**2492
Allen, John, **1:**539
Allenopithecus spp., **3:**1890
Allen's bushbabies, **4:**2376
Allergic rhinitis, **5:**3431
Allergies, **1:**132–135, **1:***133,* **1:**235,
  5:3430
Alligator gars, **3:**1735
Alligator mississipiensis, **2:**1104
Alligator sinensis, **2:**1104
Alligator snapper turtles, **6:**4145
Alligators, **2:**1104
Allium cepa. See Onions
Allium porrum. See Leeks
Allium sativum. See Garlic
Allium schoenoprasum. See Chives
Allium spp., **4:**2341
Allosaurus, **2:**1246
Allosteric enzymes, **4:**2514
Allotropes, **1:**135–136
Alloys, **1:**136, **4:**2521
  magnesium, **4:**2409
  steel, **6:**3826
Allspice, **4:**2698
Allspice, Carolina, **1:**130
Alluvial mining, **4:**2581

Alluvial sediments, **6:**3863
Alluvial systems, **1:**136–139, **1:***137*
The Almagest, **6:**4095
Alnus rubra. See Red alders
Alnus spp. *See* Alders
Alopochen aegyptiacus. See Egyptian
  geese
ALOS (Advanced Land Observing
  Satellite), **5:**3411
Alosa pseudoharengus. See Alewives
Alosa sapidissima. See American shad
Alouattinae. See Howler monkeys
ALP (Alkaline phosphatase),
  4:2511–2512
Alpacas, **1:**688
Alper, Ticvah, **5:**3237
Alpha amanitin, **4:**2694
Alpha-beta blockers, for hypertension,
  3:2075
Alpha and beta radioemissions, **5:**3344
Alpha helix, **5:**3260
Alpha-hydroxypropanoic acid. *See* Lac-
  tic acid
Alpha particles, **1:**139–140
Alpha receptors, **1:**487
Alpha rhythms, **1:**511
Alpha satellite DNA, **3:**2028
Alpher, Ralph, **2:**1072, **2:**1431
Alpine glaciers, **3:**1823
Alpine marmots, **4:**2447–2448
Alpine tundra, **1:**528, **6:**4131–4132
Alps, **2:**1533
Alterations, chromosomal, **2:**870–873,
  2:*873*
Alternate leaves. *See* Spiral phyllotaxy
Alternate personalities, **4:**2666
Alternating current (AC), **1:**703–704,
  2:1372, **2:**1374–1375, **2:**1383,
  2:1388–1389, **3:**1762
Alternative energy sources, **1:**140–143,
  1:*142*
Alternative gene splicing, **3:**1753–1754
Alternative medicine, **1:**143–148,
  3:1971
  See also Herbal medicine
Alternia spp., **4:**2694
Altimeters, **1:**442
Altman, Sidney, **5:**3454
Altocumulus clouds, **2:**906
Altostratus clouds, **2:**906
Altricial birds, **1:**547
Altruism, **1:**148–149
Alumina fibers, **2:**980
Aluminum, **1:**149–151
  in Australia, **1:**386
  leaching of, **4:**2295
  pollutants, **5:**3558
  production of, **2:**1396–1397
Aluminum/air cells, **1:**459
Aluminum hydroxide, **1:**151–152
Aluminum oxide, **1:**3, **1:**4, **1:**805–806
Aluminum oxide fibers, **2:**980
Alvarez, Luis Walter, **2:**1564, **4:**2460
Alvi, Abass, **1:**41

Alvin, **4:***2856,* **6:**4167, **6:**4169
Alzheimer, Alois, **1:**152
Alzheimer disease, **1:**152–154, **1:***153,*
  2:1177–1179, **4:**2749
Alzheimer's Disease and Related Dis-
  orders Association, **2:**1177
AM frequency, **5:**3334, **5:**3335, **5:**3342
Amanita muscaria. See Fly agaric
Amanita muscaria formosa. See Ameri-
  can fly agaric
Alpha amanitin, **4:**2694
Amaranth family, **1:***154,* **1:**154–155,
  1:*155,* **6:**4126
Amaranthaceae. *See* Amaranth family
Amaryllidaceae. *See* Amaryllis family
Amaryllis family, **1:** 155–157, **1:***156*
Amazon basin, **2:***1205,* **5:**3718, **5:**3722
Amazon Cone, **2:**1038
Amazon Delta, **1:**139
Amazon rainforest, **5:**3722
Amazon River, **5:**3460
Amazona viridigenalis. See Red-
  crowned parrots
Amblyncus spp. *See* Marine iguanas
Amblyopia, **6:**4241, **6:**4242–4243
Ambros, Victor, **5:**3465
Ambrosia artemesiifolia. See Ragweeds
Ambrosia trifida. See Ragweeds
Ambystoma mexicanum. See Axolotyl
Ambystoma tigrinum. See Tiger sala-
  manders
Amebic dysentery, **2:**1301–1302,
  4:2973
Amelung, John F., **3:**1827
American Academy of Family Physi-
  cians, **6:**3866
American alligators, **2:**1104
American avocets, **6:**3849
American badgers, **1:**419
American basswoods, **1:**451
American beavers, **1:**465–467, **1:***467,*
  5:3442
American bison, **1:**560–561, **1:***561,*
  4:2360, **6:**4332
American bitterns, **1:**562, **1:***563,*
  3:1981–1982
American black bears, **1:**463–464
American Board of Physical Medicine
  and Rehabilitation, **5:**3396
American buffaloes. *See* American
  bison
American bullfrogs, **3:**1704
American Chemical Society (ACS),
  2:1428, **2:**1429
American chestnuts, **1:**469–471, **2:**834
American Conference of Governmental
  Industrial Hygienists, **4:**2713
American coots, **5:**3360
American Council of Pediatrics, **1:**251
American crocodiles, **2:***1102,* **2:**1104
American Dietetic Association (ADA),
  4:2835
American eels, **6:**4116, **6:**4117
American elk, **5:**3442

American fly agaric, **3:**_1715_
American ginseng, **3:**1818
American goldfinches, **3:**1613
American Heart Association
  on blood pressure, **3:**2072
  on marijuana, **4:**2442
  on newborn status, **1:**251
  on stress, **6:**3866
American hornbeams, **1:**544–545
American Iron and Steel Institute,
  3:2196
American kestrels, **3:**_1576_, **3:**1577,
  5:3369–3370
American lobsters, **4:**_2362_
American mahogany, **5:**3365
American minks, **4:**2582, **4:**_2583_
American oil palms, **4:**2959
American Orchardist, **5:**3118
American oystercatchers, **4:**2932
American pelicans, **4:**3001, **4:**3002
American persimmons, **2:**1334
American pikas, **4:**2258
American plaice, **3:**1621
American Psychiatric Association,
  4:2665, **5:**3273
American Psychological Association,
  5:3276
American robins, **5:**3467,
  6:4025–4026, **6:**4027
American shad, **3:**1983
American shrew-moles, **4:**2618
American Sign Language, **3:**2032
American Society of Anesthetists,
  1:193
American Society for Laser Medicine
  and Surgery, Inc., **4:**2286
American Standard Code for Informa-
  tion Interchange (ASCII), **1:157–158**
American sycamores, **5:**3097–3098
American system (Manufacturing). _See_
  Uniformity system
American Telephone and Telegraph,
  4:2651
American toads, **1:**173, **6:**4040,
  6:4041, **6:**_4041_
American tree sparrows, **5:**3743
American vultures. _See_ Condors
American Welding Society, **6:**4315
American white pelicans, **4:**3001
American woodcocks, **5:**3508
Americium, **2:**1430
Ames, Bruce, **1:**158, **4:**2687
Ames test, **1:158**
Amethystine pythons, **5:**3293
Amia spp., **1:**604
Amicable numbers, **1:158–159**
Amides, **1:159–160**
Amino acid racimization, **2:**1155
Amino acids, **1:160–163**, **1:**161_f,_
  1:724, **4:**2607
  deficiencies, **4:**2834
  formation of, **4:**2569–2570
  Murchison meteorite and, **4:**2669

Ammodramus caudacuta. See Sharp-
  tailed sparrows
Ammodramus savannarum. See
  Grasshopper sparrows
Ammodramusl econteii. See LeConte's
  sparrows
Ammonia, **1:163–165**, **3:**2056
  molecular shape, **4:**2611
  nitrogen and, **4:**2775, **4:**2776,
    4:2778, **4:**2780
  in urea, **6:**4186
Ammonification, **1:165–167**
Ammonium nitrate-fuel oil (ANFO) ex-
  plosives, **2:**1562, **2:**1563
Ammonoids, **3:**1684
Amnesia, **1:167–168**, **4:**2490
Amniocentesis, **1:**_168_, **1:168–169**,
  2:1281, **5:**3446
Amniotic sac, **1:**553
Amoco Cadiz, **4:**2859, **4:**2860
Amoebas, **1:169–170**, **5:**3266, **5:**3271
Amorphous silicon, **5:**3074–3075
Ampère, André Marie, **2:**1378, **2:**1388,
  2:1406, **4:**2417
Ampère's law, **2:**1373
Amphetamines, **1:170–173**, **1:**437
Amphibians, **1:173–175**, **1:**_174_, **6:**4214
  metamorphosis in, **4:**2532–2533
  paleontology, **4:**2950–2951
  See also Herpetology; specific types
    of amphibians
Amphispiza bilineata. See Black-throat-
  ed sparrows
Amphiuma laterale. See Blue-spotted
  salamanders
Amphiuma maculatum. See Spotted
  salamanders
Amphiuma means. See Two-toed am-
  phiumas
Amphiuma opacum. See Marbled sala-
  manders
Amphiuma tridactylum. See Three-toed
  amphiumas
Amplifiers, **1:**_175_, **1:175–176**, **2:**1413
Amputation, **1:176–177**
Amsterdam albatrosses, **1:**104
Amtrak, **4:**2467
Amygdala, **1:**611
Amyloid plaques, **1:**152, **1:**153
Amyloidosis, **4:**2511
An-Nafud, **1:**315
Anabaena spp., **3:**1588
Anabolism, **1:177–179**, **1:**751–752,
  4:2513–2515
Anacardiaceae. _See_ Cashew family
Anacardium occidentale. See Cashew
  family
Anacondas, green, **1:**586, **1:**_587_
Anaerobic, **1:**166, **1:179**
Anaerobic bacteria, **1:**412, **1:**413,
  5:3066
Anaerobic glycolysis, **3:**1837
Anaerobic respiration, **1:**179, **1:**796,
  4:2256, **4:**2515

Analemma, **1:179–180**
Analgesia, **1:180–182**, **4:**2639–2640
  See also specific analgesics
Analog computers, **2:990–991**
Analog recording, **2:**1238, **2:**1239
Analog signals, **1:182–183**
Analysis
  blood gas, **1:579–580**
  blotting, **1:581–582**
  chi-square, **2:834–836**
  DNA, **1:**581, **3:**1669, **3:**_1751_
  flame, **3:1618–1619**
  genetic, **3:**1764
  genome, **3:**1783
  geochemical, **3:1787–1788**
  neutron activation, **3:**1670
  proteins, **1:**582
  ribonucleic acid, **1:**582
Analytic geometry, **1:183–186**
Analytical chemistry, **1:**36, **2:**832
Analytical engines, **1:**673, **2:**991, **2:**996
Ananus comosus. See Pineapples
Anaphase, **4:**2597
Anaphylaxis, **1:186–187**
Anas platyrhynchos. See Mallard ducks
Anatomy, **1:187–190**
  See also specific plants and animals
Anatomy, comparative, **1:190**
Anatosaurus, **2:**1247
Anchiceratops, **2:**1247
Anchovies, **1:190–191**
Andaman Sea, **1:**317
Andean avocets, **6:**3848
Andean bears. _See_ Spectacled bears
Andean condors, **2:**1006, **2:**1007,
  6:4262, **6:**4263
Anderson, Carl, **1:**363, **2:**1408, **6:**3875
Anderson, W. French, **1:**42, **3:**1759,
  3:1769
Andes Mountains, **2:**1314,
  5:3717–3718, **5:**3719, **5:**3720
Andesite, **3:**2089
Andre, Nicholas, **4:**2902
Andreesen, Marc, **3:**2175
Andrenid bees, **1:**472
Andrew, Thomas, **5:**3117
Andrias davidianus. See Asiatic sala-
  manders
Andrias japonicus. See Japanese sala-
  manders
Andromeda Galaxy, **3:**_1723_
Anemia, **1:191–192**, **1:**390, **3:**1960
  See also Sickle cell anemia
Aneroid barometers, **1:**354, **1:**441,
  1:_442_
Anesthesia, **1:193–196**, **1:**_194_, **1:**555,
  2:852
  in dental care, **2:**1185–1186
  novocain, **4:**2796–2797
Anethum graveolus. See Dill
Aneuploidy, **2:**870–872, **2:**_872_, **3:**1764,
  3:1766
Aneurisms, **1:**_196_, **1:196**
Anfinsen, Christian, **5:**3261, **5:**3454

Arctic tundra, **1**:528, **6**:4130–4131
Arctium minus. See Burdock
Arctocebus calabarensis. See Angwantibos
Arctocephalus townsendi. See Guadalupe fur seals
Arctonyx collaris. See Hog-badgers
Ardea herodias. See Great blue herons
Arden Corona, **6**:4185
Ardeotis kori. See Kori bustards
Area landfill method, **4**:2270
Area scales, **1**:746
Areas (Geometry), **1**:676, **3**:1802, **3**:*1802*
Areca catechu. See Betel palms
Arenaria interpres. See Ruddy turnstones
Arenaviruses, **3**:1963
Areolar connective tissue, **2**:1015
Areoles, **1**:661
Aretaeus, **3**:1940
Arethusa bulbosa. See Dragon's-mouth orchids
Arfwedson, August, **4**:2351
Argand, Jean Robert, **2**:973–974
Argentina, **2**:1244, **5**:3720
Argentine fire ants, **3**:2183–2184
Argentine hemorrhagic fever, **3**:1964
Argillaceous stone, **6**:3851
Argo, **5**:3063, **6**:4166
Argon, **5**:3371, **5**:3372
Argon lasers, **1**:*767,* **4**:2282, **4**:2286
Argos, **3**:1831
Argusianus argus. See Aargus pheasants
Ariane 4, **5**:*3473*
Ariel, **6**:4180, **6**:4181, **6**:4184, **6**:4185
Arisaema triphyllum. See Jack-in-the-pulpit
Aristarchus, **3**:1955, **6**:4095
Aristotle, **1**:214
　on abiogenesis, **6**:3778
　on aging, **3**:1810
　botany and, **1**:598
　cetaceans and, **1**:808
　on contraception, **2**:1042
　embryology and, **2**:1449–1450
　geophysics and, **3**:1804–1805
　gravity and gravitation, **3**:1866
　marine life and, **4**:2855
　on menopause, **4**:2496
　on meteorology, **4**:2534
　observation analysis, **1**:521
　physics and, **5**:3079
　on plant diseases, **5**:3122
　solar system and comets, **2**:958
　on solids, **2**:1121
　underwater exploration and, **6**:4165
Arithmetic, **1**:277–279, **4**:2471–2472
　clock, **6**:4373
　with complex numbers, **2**:973
　fundamental theorem of, **3**:1712
　integers in, **3**:2147–2148
　modular, **4**:2601
　place-value systems, **4**:2829
　See also Algebra

Arithmetic growth and decay, **3**:1885
Arkwright, Richard, **1**:393, **3**:2120
d'Arlincourt, Ludovic, **3**:1585
Armadillos, **1**:279–280, **1**:*280,* **4**:2322
Armatures, **3**:1762
Armillaria bulbosa, **3**:1717
Armorheads. *See* Boarfish
Arms (Artificial). *See* Prosthetics
Armstrong, Edwin Howard, **2**:1411, **5**:3334, **6**:4193
Armstrong, Neil A., **4**:*2634,* **5**:3737
Arnold-Chiari malformations, **6**:3766
Arnold of Villanova, **1**:720
Aromatherapy, **3**:1971
Aromatic aldehydes, **1**:115
Aromatic hydrocarbons, **3**:2049–2050, **5**:3179–3180
Aromatic plants. *See* specific plants
Arousal disorders (Parasomnias), **5**:3656
ARPANET (Advanced Research Projects Agency Network), **3**:2174
Arrays, **3**:1609, **3**:*1610,* **6**:3984
d'Arrest, Heinrich, **5**:3099
Arrhenius, Svante August, **1**:25, **1**:720, **2**:972
Arrhythmias, **2**:1392
Arrow worms, **1**:280–281, **1**:*281*
Arrowgrass, **1**:281
Arrowroot, **1**:281–282
ARS (Artificial silicon retina), **1**:307–308
Arsenic, **2**:1423, **6**:4060
Art, ice age, **3**:2035
ART (Assisted reproductive techniques), **3**:2129–2130
　See also specific techniques
Arteries, **1**:282–283, **1**:*283,* **2**:888
Arteriograms, **1**:198
Arteriosclerosis, **1**:*283,* **1**:283–286, **1**:*284,* **2**:888–889, **6**:4244
　aneurisms from, **1**:196
　coronary heart disease and, **1**:282
　gangrene and, **3**:1732
　hypertension and, **3**:2073
Arthoboric acid. *See* Boric acid
Arthritis, **1**:*286,* **1**:286–288, **1**:*287,* **4**:2904
Arthroplasty, **5**:3252, **5**:3253
Arthropods, **1**:288–289, **4**:2949, **4**:2974
Arthroscopic surgery, **1**:289–290, **4**:2903–2904
Artichokes, **2**:*975,* **6**:4205, **6**:4206
Articulate brachiopods, **1**:605–606, **1**:*606*
Artifacts, **1**:290–292
　bronze, **1**:270
　metal, **1**:272
　shipwrecks, **4**:2716–2719
Artificial biospheres, **1**:537
Artificial blood, **1**:576–577
Artificial chromosomes
　bacterial, **3**:2037, **3**:2038
　human, **3**:2026–2028, **3**:2037

Artificial eyes, **1**:306–308, **1**:*307,* **5**:3251–3252
Artificial fibers, **1**:292–301, **1**:*293t–298t,* **1**:*298t–300t,* **1**:*300*
Artificial hearts, **1**:*301,* **1**:301–302
Artificial insemination, **1**:203, **3**:1593, **3**:2127–2128
Artificial intelligence, **1**:303–306, **1**:394, **4**:3012
Artificial kidneys, **4**:2910
Artificial limbs. *See* Prosthetics
Artificial organs, **6**:4086
　See also specific organs
Artificial retina component chip (ARCC), **1**:*307,* **1**:307–308
Artificial silicon retina (ARS), **1**:307–308
Artificial vision, **1**:*307,* **1**:307–308
Artiodactyla, **6**:4172
Artist's fungi, **4**:2677
Artocarpus spp., **4**:2664
Arum family, **1**:308–310, **1**:*309*
Aryl bromides, **3**:1903
Aryl chlorides, **3**:1903
Aryl fluorides, **3**:1902, **3**:1903
Aryl iodides, **3**:1903
A.S.A. (Angle-side-angle) congruence, **2**:1009, **2**:*1009*
Asbestos, **1**:310–312, **3**:2118, **4**:2531, **4**:2713
Ascariasis, **6**:4110
Ascaris lumbricoides. See Nematodes
ASCC (Automatic Sequence Controlled Calculators), **1**:674
ASCII (American Standard Code for Information Interchange), **1**:157–158
Asclepias syriaca. See Milkweeds
Ascomycota, **3**:1715–1716
Ascorbic acid. *See* Vitamin C
ASDIC (British Anti-Submarine Detection and Investigation Committee), **5**:3713
Aserinsky, Eugene, **5**:3652
Asexual reproduction, **1**:312–314, **1**:*313*
Ash, volcanic, **6**:4252
Asia, **1**:314–319, **1**:*317,* **5**:3190
Asian cobras, **2**:1364
Asian elephants, **2**:1432–1438, **2**:*1435*
Asian fairy-bluebirds, **4**:2900
Asian flu, **3**:2133
Asian mahogany, **5**:3365
Asian monsoons, **4**:2633–2634
Asian reticulated pythons, **5**:3294
Asian rock pythons, **5**:3294
Asian small-clawed otters, **4**:2917–2918
Asian tapirs, **6**:3952
Asiatic black bears, **1**:464
Asiatic salamanders, **5**:3496
Asiatic wild asses, **1**:322
Asilidae, **6**:4120
Asklepiades of Prusa, **2**:821
Aspens, **3**:2114, **6**:4331

B

Bridges, C. B., **2:**1285

Bridges (Architecture), **1:616–619,**
1:_617,_ **1:**_618,_ **1:**642–643, **2:**1325

Brinell hardness test, **6:**3828

Bristle-thighed curlews, **2:**1130

Bristlecone pines, **5:**3093, **5:**3583

Bristletails, **1:619–620**

Britain. _See_ England

British Anti-Submarine Detection and
Investigation Committee (ASDIC),
5:3713

British Gloster Meteor, **1:**97

British Imperial System of Units,
4:2540, **6:**4174–4175, **6:**4176

British soldiers (Lichens), **4:**_2330_

Brittle stars, **1:620–621**

Broad-billed rollers, **5:**3481

Broad fish tapeworms, **3:**1623–1624

Broad-headed skinks, **5:**3646

Broad-spectrum insecticides,
3:2139–2141, **3:**2152

Broad-winged hawks, **1:**658, **3:**1925

Broadcast spraying, of herbicides,
3:1974–1975

Broca, Pierre Paul, **6:**3773

Broca's area, **1:**610

Broccoli, **6:**4205

Broglie, Louis-Victor de, **1:**362–363,
1:370, **1:**589, **4:**2553, **5:**3305

Brome-grass, **3:**1861

Bromeliaceae. _See_ Bromeliad family

Bromeliad family, **1:**_621,_ **1:621–623,**
1:_622_

Bromides, **3:**1903

Bromine, **3:**1911, **3:**1912

Bromus tectorum. See Brome-grass

Bronchioles, **1:**_327_

Bronchitis, **1:623,** **5:**3431

Bronchodilators, **1:**326–328, **5:**3432

Brønsted, J. N., **1:**26, **2:**972

Brønsted-Lowry definition, **1:**26–27

Brontosaurus. _See_ Apatosaurus

Bronze, **2:**1055

Bronze age, **1:**270

Brookesia spp., **1:**815

Brooks, Frederick, **6:**4226

Broussonetia papyrifera. See Paper
mulberries

Brown, R. Hanbury, **3:**2161

Brown, Robert, **1:**625, **3:**1637

Brown algae. _See_ Kelps

Brown anoles, **1:**_209_

Brown bears, **1:**461, **1:**462–463, **1:**_481_

Brown boobies, **1:**593–594

Brown capuchins, **1:**710–711

Brown dwarfs, **1:623–625,** **2:**1568,
6:3833

Brown fat, **3:**1989

Brown-headed cowbirds, **1:**_569,_
5:3442, **6:**4274

Brown-headed spider monkeys, **4:**2760

Brown howler monkeys, **4:**2760

Brown hyenas, **3:**2068–2069

Brown pelicans, **4:**3001, **4:**3002,
4:_3002,_ **4:**3025, **6:**4332

Brown rats. _See_ Norway rats

Brown towhees, **5:**3744

Brown tree snakes, **2:**1460

Brownian motion, **1:**625, **2:**937

Browsers, Web. _See_ Web browsers

Brucella spp., **1:**625–626

Brucellosis, **1:**520, **1:625–627**

Brundtland Report, **2:**1018, **6:**3925

Brunet, Michael, **3:**2033–2034

Brunhes, Bernard, **4:**2946

Brush, Charles, **1:**259

Brush-tailed porcupines, **5:**3194

Brush-tailed rat-kangaroos. _See_ Bet-
tongs

Brush turkeys. _See_ Moundbuilders

Brushtail possums, **4:**3034

Brussels sprouts, **6:**4205

Bryce Canyon National Park, Utah,
6:_4311_

Bryophytes, **1:627–630,** **4:**2642–2644,
5:3112

See also Hornworts; Liverworts;
Mosses

Bryozoa. _See_ Moss animals

BSE (Bovine spongiform encephalopathy), **6:**4375

BSOC (Bit-Serial Optical Computer),
2:992

BSRMBs (Battlefield short range ballistic missiles), **1:**422

BT (Bathythermographs), **6:**4167–4168

Bt corn, **2:**1067

Bubal hartebeests, **3:**1920

Bubalus bubalis. See Water buffaloes

Bubble chambers, **4:**2987–2988

Bubo bubo. See Eagle owls

Bubonic plague, **1:630–632,** **1:**_631,_
3:1626

Bubulcus ibis. See Cattle egrets

Buchner, Eduard, **1:**502, **2:**1494,
3:1586

Buchner, Hans, **1:**502

Buckeyes, **3:**2013–2014

Buckland, William, **3:**1794

Buckminsterfullerenes, **1:632–634,**
1:_633_

Buckthorns, **1:634**

Buckwheat, **1:634–635,** **1:**_635_

Buckyballs, **4:**2703

Budgerigars, **4:**2979

Buds and budding, **1:635–636,** **1:**_636_

Buffalo gnats, **6:**4120

Buffalo gourds, **3:**1847

Buffaloes, **1:**765, **1:**_766,_ **4:**2360

Buffers, **1:636–638**

Buffon, Georges Louis Leclerc de,
3:1794

Bufo americanus. See American toads

Bufo boreas. See Boreal toads

Bufo marinus. See Cane toads

Bugeranus carunculatus, **2:**1093

Bugs (Internet), **3:**2172–2173

"Build absolutely nothing near anybody
or anything" syndrome, **3:**2111

Building design, **1:638–644**

Building materials, **1:**642–643

Bulbuls, **1:644–645**

Bulimia, **1:**46, **2:**1329–1330

Bullfinches, **3:**1614

Bullfrogs, **1:**173, **3:**1704

Bullhead sharks, **5:**3611

Bullheads, **1:**754, **1:**_754_

Bulls, Longhorn, **1:**_764_

Bumble bees, **1:**472

Bunsen, Robert, **1:**645, **3:**1618, **4:**2409

Bunsen burners, **1:645**

Buntings, **5:3742–3745,** **5:**_3744_

Bunyaviruses, **3:**1963

Buoyancy, principle of, **1:**426,
1:645–647, **1:**_646,_ **3:**1641

Buoyant force, **1:**645

Burbank, Luther, **5:**3118

Burbidge, Geoffery, **2:**1431

Burbidge, Margaret, **2:**1431

Burchell's zebras. _See_ Common zebras

Burckhardt, Gottlieb, **5:**3281

Burdin, Claude, **6:**4136

Burdock, **5:**3572

Bureau of Land Management, **3:**2184

Bureau of Reclamation, **2:**1490

Buret, Frederic, **5:**3606

Burets, **1:647**

Burgdorfer, Willy, **4:**2382

Burgess Shale, **3:**1584, **3:**1685

Burial grounds
archaeology excavation and,
1:269–270
Indian mounds, **1:**266
mounds, **4:**2655–2656

Burial metamorphism, **4:**2528

Burial mounds, **4:**2655–2656

Burkitt's lymphoma, **4:**2388

Burnett salmon. _See_ Australian lungfish

Burning
controlled, **6:**4331–4332
prescribed, **5:**3225–3226, **5:**_3226_

Burns, **1:647–649,** **1:**_648,_ **5:**3329

Burns, Jack, **6:**3909

Burrowing owls, **4:**2923

Burst and Transient Source Experiment
(BATSE), **3:**1731, **3:**1732

Bursts, gamma ray, **3:**1730–1731,
3:1731–1732

Burton, Robert, **4:**2496

Bush, George W., **3:**2030, **4:**2819

Bush, Vannevar, **2:**990

Bush pigs, **5:**3088

Bushbabies, **4:**2374–2376

Bushmasters, **6:**4220

Bushnell, David, **6:**3880

Bushveld Complex, **5:**3215

Business land uses, **4:**2267

Bussy, Antoine, **4:**2409

Bustards, **1:649–650,** **1:**_650_

Butane. _See_ Hydrocarbons

Buteo buteo. See Common buzzards

D

incandescent lamps and, **3:**2107
isolation of calcium, **1:**669–670
isolation of chlorine, **2:**856
isolation of sodium, **5:**3678
magnesium, **4:**2409
nitrous oxide and, **1:**193
on ytterbia, **4:**2279
Dawkins, Richard, **5:**3578
Dawn man, **5:**3090
Dawn redwoods, **6:**3929
Dawson, Charles, **5:**3090
Days, sidereal, **1:**780
DC. *See* Direct current
DDT (Dichlorodiphenyl-
trichloroethane), **1:**81–82, **1:**526,
2:854, **2:1158–1162**, **2:***1159,* **3:**1653,
3:1903, **4:**3025
bald eagles, **5:**3370
bioaccumulation and, **1:**499
damage from, **3:**2139–2140
grebes and, **3:**1874
peregrine falcons and, **4:**3014
wildlife, **6:**4332
De Chamant, M. Dubois, **2:**1185
de Colmar, Charles Xavier Thomas,
1:673
De Forest, Lee, **6:**4193–4194
de Gardanne, C. P. L., **4:**2496
De La Rue, **3:**2107
de Méré, Chevalier, **5:**3240
de Meritens, Auguste, **6:**4315
De Moivre, Abraham, **2:**973
De Moivre's theorem, **2:**974
De Morgan, Augustus, **3:**1798
De Re Metallica, **4:**2580, **4:**2886,
5:3476
Dead of Night, **6:**3816
Dead Sea, **1:**315, **5:**3069
Deadly nightshade, **3:**1910, **4:**2771,
4:2772
Deafness, **2:1162–1163**, **2:***1163*
Dean, H. Trendley, **3:**1646
Death
aging and, **1:***73,* **1:***73–74*
of cells, **1:788–789**
of comets, **2:**961
Death adders, **2:**1365
Death Valley, California, **5:**3319
DeBakey, Michael, **1:**302
DeBary, Heinrich Anton, **5:**3122
Debye, Peter, **2:**1253
Decay
exponential, **3:**1902
radioactive, **3:**1794, **3:**1902, **6:**4225
DeChateau, M., **2:**1185
Decibels, **3:**1931, **4:**2370–2371
Deciduous forests, **3:***1677*
cold, **3:**1676
temperate, **1:**528, **3:**1676
Decimal fractions, **1:**50, **2:1163**
Deciphering matrices, **4:**2474
Declinations, **1:**772, **1:**778–779
Decomposer food web, **3:**1653
Decomposers, **1:**718, **3:**1988

Decomposition, **1:**165–166, **2:***1164,*
2:*1164***–1165**, **4:**2270–2271
Decomposition reactions, **2:**826
Decongestants, **5:**3431
See also Antihistamines
Decubitus ulcers, **6:**4154–4155
Deep-bed filters, **3:**1610
Deep Rover, **6:**4167
Deep-sea diving, **1:**451–452
Deep-sea exploration, **6:**4163–4171
Deep-sea jellyfish, **3:**2215–2216
Deep-sea submersible vessels. *See* Sub-
mersibles
Deep water currents, **2:**1131–1132
Deep well injection, **3:**1929
Deepstar 4000, **6:**4166
Deer, **2:1165–1168**, **2:***1167,* **3:**2183,
4:2360
Deer flies, **6:**4120
Deer mice, **2:1168–1169**, **4:**2543
Deer ticks, **4:**2382–2384
Deere, John, **1:**75
Definite integrals, **1:***678*–679, **3:**2149
Deflation, **2:**1206
Deforestation, **1:**508, **2:1169–1172**,
2:*1170,* **3:**1678
agent orange, **1:**70–72
ecological economics, **2:**1342,
2:1345
endangered species and, **2:**1461
old growth, **2:**1260
population, **5:**3191
slash-and-burn agriculture,
5:3649–3650
Degree (Algebra), **2:1172**
Degrees of freedom, chi-square distrib-
utions, **2:***835,* **2:**836
Dehydration, food, **3:**1659–1660
Dehydroepiandrosterone (DHEA),
1:54, **2:1172–1174**
Deimos, **4:**2452
Deinonychus, **2:**1246
Deisenhofer, Johann, **5:**3066
Delaroche, Paul, **5:**3059
Delayed sleep phase syndrome, **5:**3656
Delesse, Achilles, **4:**2945–2946
Deletions (Genetics), **4:**2688
Delisle, Jean Baptiste Louis Romé,
2:1121
Delta-9-tetrahydrocannabinol (THC),
3:1966
Deltas, **1:**138, **2:1174–1177**, **2:***1176*
alluvial systems and, **1:***137,*
1:138–139
sedimentary basins and, **1:**448–449
Demagnetization, adiabatic, **2:**1118
Demand paging, **2:**995
Dementia, **2:1177–1181**
in Alzheimer disease, **1:**152
Parkinson disease, **4:**2977
Democritus, **1:**360, **1:**366, **1:**375, **5:**3079
Demodulation, **5:**3335
Demographics, **3:**2031, **3:**2121
Demoiselle cranes, **2:**1094

DeMoyleyns, Frederick, **3:**2107
Dempster, A. J., **4:**2465
Dendrimers, **3:**1758
Dendrite formation, **3:**2168
Dendroapsis polylepis. See Black mam-
bas
Dendrochronology, **1:***1155,* **1:**271–272,
2:1156
Dendroica kirtlandii. See Kirtland's
warblers
Dendroica virens. See Black-throated
green warblers
Dendrolagus spp. *See* Tree kangaroos
Dengue fever, **2:1181**
Denis, Jean-Baptiste, **1:**580
Denisyuk, Yuri, **3:**2000
Denitrification, **2:1181–1183**, **4:**2780
Denominators, **3:**1691
Dense connective tissue, **2:**1015
Density, **2:1183**
See also specific substances
Density currents, **2:**1131–1132
Dentistry, **2:1183–1188**, **2:***1185,*
4:2796–2797, **6:**4357
Denver, Colorado, air pollution in, **1:***92*
Deoxyglucose technique, **5:**3276
Deoxyribonucleic acid. *See* DNA
Depo-medroxyprogesterone acetate
(DMPA), **2:**1046
Deposits, **2:1195**
Depression (Psychology), **1:**242,
2:1195–1196
See also Manic depression
Depth perception, **2:1196–1199**,
2:*1197,* **4:**3010, **6:**4237
Derivatives, **2:1199–1202**, **2:**1200*f,*
2:1200*t,* **2:**1201*f,* **3:**2149
See also specific derivatives
Derived unis, **6:**4175
Dermatemys mawii. See Central Ameri-
can river turtles
Dermatomyositis, **1:**389
Dermestid beetles, **1:**478
Dermis, **3:**2156
Derrick, Godfrey, **2:**1093
Derrick cranes, **2:**1093
Derricks, **4:**2862
Derris. *See* Rotenone
DES (Diethylstilbestrol), **2:1226–1228**,
5:3418
Desalination, **2:1202**
reverse osmosis, **4:**2910
solar system, **5:***3073*
Desargues' theorem, **5:**3244–3245
Descartes, René, **1:**183, **1:**184, **5:**3361,
6:3771
on Cartesian coordinate system,
1:740, **3:**1853
on functions, **3:**1711
idealistic view of, **2:**933
Kepler's postulates and, **2:**1121
on momentum, **4:**2621
nativist views of, **2:**1004
on variable notations, **6:**4199

punctuated equilibrium,
5:3290–3291
stellar, 3:1985*f,* 3:1987,
6:3829–3833
symbiosis, 6:3936
taxonomy, 6:3962–3964
universe, 1:492–493, 6:3815–3816
See also Adaptation; Selection
Ewing, Maurice, 6:4167
Ex situ conservation. *See* Captive
breeding and reintroduction
Excavation methods, 1:*267,*
1:268–270, 2:1549–1552, 2:*1550,*
4:2657
See also Archaeology
Exclusion principle, Pauli, 1:35,
2:1552–1553
defined, 1:363, 5:3305, 6:4326
quarks and, 5:3309, 6:3876
stars and, 4:2755
Sun and, 4:2756
Excretory system, **2:1553–1556**
Exercises, **2:1556–1558**
Exhaust systems, automotive, 1:399
Exit pupil, 1:497
Exocrine glands, 2:1558–1559, 3:1826
Exons, 3:1753
Exoskeletons, 3:2155
Exothermic reactions, 2:826
Exotic species, 3:2182–2183
Expanding Universe. *See* Big bang theory
Expected frequency, in chi-square test,
2:835–836, 2:836*t*
Expendable launch vehicles (ELVs),
5:3726, 5:3741
Experiment (Locomotive), 6:4064
Expert systems, 1:304–305
Exploration
balloons and, 1:427
underwater, **6:4163–4172**
"Explorer-class" missions, 6:4160
Explorer spacecraft, 6:4195
Explosives, 2:1559–1563, 2:*1561*
See also specific explosives
Exponential decay, 3:1902
Exponential functions, 3:1712
Exponents, 2:1563–1564, 4:2369–2371
Exposed plutons, 4:2660
Exposure therapy, for phobias, 5:3042
Extended radio galaxies, 3:1725
Extinctions, **2:1564–1566**
ammonoids, 3:1684
Archaeoceti, 1:808
aurochs, 1:763
Australian frogs, 4:2921–2922
bennettites, 1:483–484
biodiversity and, 1:506
bubal hartebeests, 3:1920
cave bears, 1:461
comet, 2:961
Cooper's sandpiper, 5:3509
of dinosaurs, 3:2227
Genyornis spp., 3:1629

giant beavers, 1:465
heath hens, 3:1885, 5:3207
humans and, 1:506–507
Lake Nakuru hartebeests, 3:1920
mylodon listai sloths, 5:3660
New Zealand moas, 3:1629
passenger pigeons, 5:3085–3086,
6:4332
Phalacrocorax perspicillatus,
2:1066
pig-footed bandicoots, 1:429, 1:*430*
Rhyniopsida, 3:1590
sea minks, 4:2582–2583
skinks, 5:3647
tarpans, 3:2014–2015
Teratornis incredibilis, 2:1006
See also Mass extinctions
Extracorporeal shock wave (ESWL)
lithotripsy, 4:2354–2356, 4:*2355*
Extractive metallurgy. *See* Chemical
metallurgy
Extragalactic astronomy, 3:1723
Extrasolar planets, 1:331, 1:339,
2:1566–1568, 5:3099
Extraterrestrial organic processes,
4:2669
Extreme Ultraviolet Explorer (EUVE),
6:4160
Extremophiles, 3:1803
Extrusive rocks, 3:2089
Exxon Valdez, 1:534, 2:1489, 4:2859,
4:2860
Eye diseases. *See* Vision disorders
Eyebalm, 3:1841–1842
Eyebrowed thrushes, 6:4026
Eyeroot, 3:1841–1842
Eyes, **2:1568–1571**, 2:*1569*
albinism, 1:105
artificial, 1:306–308, 1:*307,*
5:3251–3252
color and, 1:202
corneal transplants, 6:4082
depth perception, 2:1196–1199,
2:*1197*
laser surgery, 4:2285–2286
night vision, 4:2768–2769
perception, 4:3008–3012
radial keratotomy, 5:3322–3324,
5:*3323*
See also Vision
Eyre Peninsula of South Australia,
1:382

F

F-numbers, 4:2319
Faber, Joseph, 6:3945
Fabrici, Girolamo, 2:1450
Fabrics, bleaching, 5:3686–3687
Fabry, Charles, 3:2160
Fabry-Perot interferometers, 3:2160,
3:*2160*

Face bricks, 1:615
Facelifts, 5:3129
Facies, metamorphic, 4:2526–2528,
4:2529
Facioscapulohumeral muscular dystrophy, 4:2738–2739
Facsimile machines, 3:1585, 3:*1585*
Factor VIII, 1:580
See also Hemophilia
Factorials, 3:1574
Factories, in Industrial Revolution,
3:2121–2122
Factors, **3:1573–1574**
Faculae, 6:4052
FADH$_2$ (Reduced flavin adenine dinucleotide), 5:3426
Fagaceae. *See* Beech family
Fagopyrum esculentum. See Buckwheat
Fagus spp., 1:469, 1:471
Fahrenheit, Daniel Gabriel, 4:2501,
6:3990, 6:4020
Faint Object Spectrograph (FOS),
3:2024, 3:2025
Falco columbarius. See Merlins
Falco femoralis. See Aplomado falcons
Falco mexicanus. See Prairie falcons
Falco peregrinus. See Peregrine falcons
Falco puctatus. See Mauritius kestrels
Falco rusticolus. See Gyrfalcons
Falco sparverius. See American
kestrels
Falconiformes, 5:3369–3371
Falconry, 3:1576
Falcons, **3:1574–1577**, 5:3369, 5:3370
American kestrels, 3:*1576,* 3:1577
Aplomado, 3:1577
crested caracaras, 3:1577
gyrfalcons, 3:1575–1576, 3:1577
Merlins, 3:1577
prairie, 3:1577, 5:3209
See also Peregrine falcons
Falkland Islands, 1:212, 5:3720
Falling bodies, law of, 3:1867
Fallopian tubes, 3:2128, 5:3416–3417
Fallopio, Gabriele, 2:1043
Fallout, radioactive, 1:92, 4:2821,
5:3345–3346
Falls (Water). *See* Waterfalls
False backgrounds, in motion pictures,
4:2653
False gavials, 2:*1103,* 2:1104
Familial hypercholesterolemia, 3:1759
Family therapy, for schizophrenia,
5:3534
Famine, Great Irish, 5:3200–3201
FAMOUS Project (French-American
Mid-Ocean Undersea Study), 6:4169
Fan Lei, 1:583
Fans
on hovercraft, 3:2022
turbojet, 3:2218
Fantails, 4:2622
FAO (United Nations Food and Agriculture Organization), 2:1170, 5:3118

G

H

H-Bombs, **2:**1561
HAART (Highly active antiretroviral therapy), **1:**87
Haber, Fritz, **1:** 63
Haber-Bosch process, **1:**163, **1:**164, **3:**2056
Habitats, **3:1901**
    carrying capacity, **1:**739–740
    critical, **2:1100–1101**
    ecological economics, **2:**1342, **2:**1345
    endemic species, **2:**1462–1463
    extinction and, **2:**1564
    herbicides and, **3:**1975–1976
    protected areas, **5:**3257–3259
    restoration ecology, **5:**3440–3443
    underwater, **6:**4170
    See also specific habitats, plants and animals
Habituation, **1:**482
Hack saws, **4:**2399
Hackworth, Timothy, **6:**4064
HACs (Human artificial chromosomes), **3:2026–2028**
Hadalpelagic zones, **4:**2854
Hadfield, Robert, **6:**3824
Hadley, George, **1:**347, **1:**348, **1:**349
Hadley, John, **5:**3602
Hadley cells, **1:**348, **1:**349, **3:**1828, **5:**3102
Hadrons, **6:**3793
Hadrosaurs, **2:**1247
Haeckel, Ernst, **2:**1211, **5:**3265
Haematopus. See Oystercatchers
Haematopus bachmani. See Black oystercatchers
Haematopus fuliginosus. See Sooty oystercatchers
Haematopus ostralegus. See Musselpeckers
Haematopus palliatus. See American oystercatchers
Haematoxylon campechianum. See Logwood
Haemophilius influensae, **4:**2281
Haemophilus ducreyi, **5:**3608
Hafele, J. C., **5:**3409
Hafnium, **2:**1424
Hagfishes, **3:1901–1902, 4:2264–2266**
Haggerty, B. M., **4:**2460
Hahn, Otto, **2:**1429, **4:**2797, **4:**2817
Hahnemann, Samuel, **1:**147, **3:**1971
Haig, David, **3:**2103
Haig Hypothesis, **3:**2103–2104
Hail, **5:**3217, **6:**4029–4030, **6:**4309
Hairy sakis, **4:**2759
Haitian solenodons, **3:***2142*
Haldane, J. B. S., **1:**505, **2:**1496, **2:**1539, **4:**2897, **4:**2898
Hale, Alan, **2:**956
Hale, George Ellery, **6:**3834, **6:**3901

Hale telescope, **6:**3984
Hales, Stephen, **1:**598
Half-lives, **3:1902**
    See also specific elements
Haliaeetus leucocephalus. See Bald eagles
Halibut, **3:**1621
Halichoerus gryptus. See Gray seals
Halide lamps, **1:**260
Halides, organic, **3:1902–1904**
Halite. *See* Sodium chloride
Halitosis, **2:**1187
Hall, Asaph, **5:**3515
Hall, Charles Martin, **1:**149, **4:**2520
Hall, E. H., **3:**1904
Hall effect, **3:1904–1905**, **4:**2419
Hall-Héroult process, **1:**149
Haller, Albrecht von, **2:**1450
Halley, Edmond, **1:**451, **2:**958, **2:**1076, **3:**1905, **4:**2762, **6:**4165
Halley's comet, **1:**335, **2:**958, **2:***959,* **3:1905–1906**, **3:***1906,* **4:**2762, **4:**2881
Hallucinations, **5:**3280
Hallucinogenic mushrooms, **3:**1909, **4:**2678, **4:**2694
Hallucinogens, **3:1906–1910**
    See also specific hallucinogens
Halobacterium, **5:**3069
Haloes, **1:**352–353
Halogenated hydrocarbons, **3:1910–1911**
Halogens, **3:1911–1913**
    See also Periodic table of elements
Halons, ozone depletion and, **4:**2938
Halophiles, **1:**412
Halosaurs, **3:1914**
Halothane, **1:**194
Hamadryas baboons, **1:**405–406
Hamilton, William D., **1:**148
Hammer, Adam, **3:**1943
Hammer-headed fruit bats, **1:**455
Hammersley Mountain Range, **1:**383, **1:**386
Hammond, George, **5:**3395
Hammond, William, **2:**918
Hams Caves, Spain, **1:***769*
Hamsters, **3:1914–1915**, **3:***1915*
Hand tools, **3:1915–1917**
"Handedness," of chemicals, **4:**2616
Hands (Artificial). *See* Prosthetics
Hansen's disease. *See* Leprosy
Hantavirus infections, **3:1917–1918**, **3:**1964
Hantavirus pulmonary syndrome, **3:**1917, **3:**1918
Hantaviruses, **3:**1917
Hanuman langurs, **4:**2276, **4:**2277–2278
Haplothrips mali. See Black hunter thrips
Haptic system, **4:**3008, **6:**4226, **6:**4228
Harbor seals, **5:**3553
Hard valves, **6:**4194
Hard water, **3:1918–1919**

Hardening, of concrete, **2:**1003
Harder, Delmar S., **1:**393
Hardness
    abrasion and, **1:**5
    Brinell test, **6:**3828
    of minerals, **4:**2573, **4:**2578, **4:**2601–2602
Hardy-Weinberg theorum, **1:**132, **2:**1540
Hares, **4:**2257–2259
Hargreaves, James, **3:**2120
Hari Rud River, **1:**318
Harmonics, **1:**30, **3:1919**
Harnesses, **1:**76
Harp seals, **5:**3553
Harpactes erythrocephalus. See Redheaded trogons
Harpacticoida, **2:**1053
Harpagomantis discolor. See South African flower dwelling mantids
*Harpia harpyja*f. *See* Harpy eagles
Harpooning, of ocean sunfish, **4:**2853
Harpy eagles, **2:***1308,* **2:**1309
Harriers, **3:**1924
Harrington, Joseph, **1:**665
Harris, Rollin, **2:**990
Harrison, Ed, **2:**1077
Harrison, John, **4:**2290
Harrison, Michael R., **5:**3222
Harrison Act, **2:**920
Hartebeests, **3:1919–1921**, **3:***1920*
Harvest mice, **4:***2543*
Harvesting machines, **1:**76–79
Harvey, William, **1:**521, **3:**1808, **3:**1935, **3:**1941, **6:**3917, **6:**4248
Hashimoto's thyroiditis, **1:**390
Hausdorf, Felix, **3:**1690
Hausdorf dimension, **3:**1690
HAV (Hepatitis A virus), **3:**1967
Havestmen, **1:***257,* **1:**258
Hawaiian-Emperor island chain, **3:**2203
Hawaiian Islands, **3:**2022, **4:**3028, **6:**4250–4252
Hawaiian stilts, **6:**3849
Hawfinches, **3:**1614
Hawk moths, **4:**2648
Hawkbill, **6:**4169
Hawkes, Graham, **6:**4167
Hawkesbee, Francis, **2:**1418
Hawking, Stephen, **1:**568, **2:1538**, **5:**3405
Hawking radiation, **1:**568
Hawkins, Gerald, **2:**1340
Hawkins, Richard, **4:**2832
Hawks, **3:1921–1925**, **3:***1923,* **5:**3369, **5:**3370
    See also Buzzards
Hayford, John Fillmore, **3:**2209
Hayford-Bowie concept, **3:**2209
Hayward fault, **2:**1324
Hazard, Cyril, **5:**3310
Hazard zoning flood plains, **3:**1635

pneumonia, **5:**3156
retroviruses, **1:**578–579, **5:**3444, **6:**4236
T cells, **1:**578–579
therapies for, **1:**88–89
vaccines, **5:**3609–3610
See also AIDS
HLA (Human leukocyte antigens), **1:**227, **4:**2992, **5:**3450
HMD (Head-mounted display) units, **6:**4226–4227
H.M.S. *Beagle,* **2:**1544
H.M.S. *Challenger,* **4:**2856, **6:**4164
H.M.S. *Eagle,* **6:**3880
H.M.S. *Titanic. See* Titanic
HMSN (Hereditary motor and sensory neuropathy), **4:**2739
hnRNA, **5:**3464–3465, **5:**3466
Hoary marmots, **4:**2447, **4:***2447*
Hoatzins, **3:2011–2013**, **3:***2012*
Hodgkin's cells, **3:***1997*
Hodgkin's disease, **3:1996–1998**, **3:***1997,* **4:**2388
Hoes, **1:**75
Hoff, Ted, **1:**675
Hoffmann, Klaus, **1:**516
Hoffmann, Paul Erich, **5:**3606
Hoffman's two-toed sloths. *See* Two-toed sloths
Hofmann, Albert, **3:**1907
Hog-badgers, **1:**420
Hog-nosed skunks, **5:**3648
Hogbacks, **4:**2275
Hokkaido, **1:**316
Holbrook, Stephen R., **5:**3465
Holbrookia maculata. See Lesser earless lizards
Holland, John, **6:**3880
Holland, tulips in, **4:**2341
Hollerith, Herman, **2:**991
Holley, Robert W., **2:**927
Holly family, **3:1998–2000**, **3:***1999*
Hollywood, **4:**2651
Holograms, **3:2000–2002**, **3:***2001*
Holographic interferometry, **3:**2161
Holography, **2:**1229, **3:2000–2002**
Holometabola, **4:**2532
Holstein calves, **1:***552*
Homarus americanus. See Northern lobsters
Homarus vulgaris. See European lobsters
Homeopathic Pharmacopoeia of the United States (HPUS), **3:**1971
Homeopathy, **1:**147, **3:**1971
Homeostasis, **3:2002–2004**, **3:**2003*f,* **5:**3081, **5:**3082–3083
Homer, **6:**4208
Hominids, **3:**2032–2034, **5:**3230
Homo erectus, **3:**2034
Homo ergaster, **3:**2034
Homo habilis, **3:**2034
Homo sapiens neanderthalensis. See Neanderthal man

Homo sapiens sapiens, **3:**2034–2035
Homogeneous catalysis, **1:**752
Homogenization, **6:**4157
Homozygosity, **6:**4359
Honey bees, **1:**472, **1:**473
Honey dews, **3:**1846
Honeycreepers, **2:**1460, **2:**1462, **3:2004–2005**
Honeyeaters, **3:2005–2006**
Honeypot ants, **1:**246
Hong Kong flu, **3:**2133
Hong River, **1:**318
Honshu, **1:**316
Hooded mergansers, **5:**3442
Hooded parakeets, **4:***2980*
Hooke, Robert, **2:**958, **2:**1121, **2:**1146, **2:**1366, **3:**1808
on crystal structure, **4:**2575
on microscopy, **4:**2552
Hooker Chemical Company, **3:**1927
Hooker process, **5:**3686
Hooke's law, **2:**1366
Hookworms, **6:**4110–4111
Hoolock gibbons, **3:***1813,* **3:**1814
Hoopoes, **3:2006**
Hoover, Herbert, **4:**2580
Hoover, J. Edgar, **3:**1668
Hoover, Lou Henry, **4:**2580
Hoover dam, **2:***1150*
Hope, James, **3:**1942
Hopewell burial mouonds, **4:**2656
Hopper, Grace Murray, **2:**993, **6:**4369
Hordeum sativum. See Six-rowed barley
Hordeum vulgare. See Barley
Horizon coordinates, **1:**771
Horizon problem, **2:**1079
Horizons, soil, **3:2006–2007**, **5:**3688–3689
Horizontal band saws, **4:**2399
Horizontal disease transmission. *See* Autosomal recessive diseases
Horizontal drills, **4:**2398
Horizontal milling machines, **4:**2395
Horizontal plain shapers, **4:**2398
Hormonal regulation, **4:2919–2920**
Hormone replacement therapy (HRT), **4:**2497
Hormone therapy, **1:**697
Hormones, **3:2007–2011**
adrenal glands, **1:53–55**
aging and, **1:**72
anterior pituitary, **5:**3416
cardiac cycle and, **3:**1938
diethylstilbestrol, **5:**3418
endocrine, **2:**1464–1465
fetal, **1:**552
growth, **3:1886–1888**, **5:**3285, **5:**3286
in metabolism, **4:**2514–2515
in metamorphosis, **4:**2532, **4:**2533
mitosis and, **1:**791
plants, **5:**3114
sex change, **5:**3600–3601
Hornbeams, **1:**544

Hornbills, **3:2011–2013**, **3:***2012*
Horned larks, **4:**2280
Horned screamers, **5:**3540
Horner, Jack, **6:**4149–4150
Hornworts, **1:**628
Horse chestnuts, **3:***2013,* **3:2013–2014**
Horse flies, **3:**1628, **6:**4120
Horsehair worms, **3:2014**
Horsehead Nebula, **6:**3801–3802
Horses, **3:2014–2016**, **3:***2015,* **4:**2360
Horseshoe crabs, **3:2016–2018**, **3:***2017*
Horsetails, **3:2018–2019**, **3:***2019*
Horsley, Victor, **4:**2745
Horticulture, **3:2019–2021**, **3:***2020*
See also specific plants
Horvitz, Robert, **1:**789
Hot-dip galvanizing, **4:**2525
Hot extrusion, **4:**2524
Hot spots, **3:2021–2022**, **4:**2794, **6:**4251
Hot working, **4:**2524
Houdini, Harry, **1:**511
Hour angles, **1:**772
House, Royal E., **6:**3972
House-dust mites, **4:***2595*
House finches, **3:**1612
House mice, **4:**2543–2544
House wrens, **6:**4350
Houseflies, **3:***1627,* **6:**4120–4121
Household hazardous wastes, **3:**1927
Household laundering, **1:**571
Houston, Edwin, **1:**259
Housz, Jan Ingen, **1:**719
Hover flies, **3:**1628
Hovercraft, **3:2022–2023**
Hovey, C. M., **5:**3118
Howard, Albert, **2:**982
Howard, Luke, **2:**905
Howe, Elias, **3:**2121
Howler monkeys, **4:**2759–2760
Hoyle, Fred, **1:**491, **2:**1073, **2:**1078, **2:**1431, **6:**3816–3817, **6:**3838
HPLC (High performance liquid chromatography), **2:**868
HPUS (Homeopathic Pharmacopoeia of the United States), **3:**1971
HPV (Human papilloma viruses), **6:**4232
HRS (High Resolution Spectrograph), **3:**2024, **3:**2025
HRT. *See* Hormone replacement therapy
HST. *See* Hubble Space Telescope
HTLV (Human T-cell leukemia virus), **5:**3444, **6:**4236
HTML (HyperText Markup Language), **3:**2175
HTTP (HyperText Transfer Protocol), **3:**2174–2175
Huanacos. *See* Guanacos
Huang Hai, **1:**316
Huang Ma, **3:**1971
Huang River, **1:**316
Hubbard, William, **5:**3103

Instincts, **3:2145–2146**
Institutional land uses, **4:**2267
Instrumental conditioning. *See* Operant conditioning
Instrumental methods, of quantitative analysis, **5:**3300*t*, **5:**3301
Instrumentation, for underwater exploration, **6:**4164–4165
Insulin, **3:**1768, **3:**2010, **3:2147**
    diabetes mellitus and, **2:**1214, **2:**1215–1216
    pancreas and, **2:**1236
Insulin-dependent diabetes. *See* Type 1 diabetes
Insulin-like growth factor 2 (IGF2), **3:**2103
Integers, **1:**278, **2:**1085*t*, **3:**2131, **3:2147–2148**, **3:**2148*t*
Integral calculus, **1:**676, **1:**677–678
Integrals, **1:***678,* **1:**678–679, **3:2148–2150**
Integrated amplifiers, **1:**175–176
Integrated circuits (ICs), **2:**1412, **3:2150–2151**, **3:***2151,* **4:**2352–2353, **6:**4077, **6:**4358
Integrated Molecular Analysis of Gene Expression (IMAGE), **3:**2037
Integrated pest management (IPM), **3:**2141, **3:2151–2154**, **3:***2153,* **4:**3026
Integumentary system, **3:2154–2157**, **3:***2155,* **4:**2895
Intelligence
    artificial, **1:303–306**
    homo sapiens, **3:**2032
    See also specific species
Intelligent computer-assisted instruction (ICAI), **1:**306
Intelsat VI, **5:***3512*
Intensity (Loudness), **3:**1931
Intentional hypothermia, **3:**2076
Interchangeable parts, **4:**2461–2462
Intercontinental range ballistic missiles (ICBMs), **1:**422, **1:**423, **1:**424
Interfaces, CD-ROM, **2:**967–968
Interference (Psychology), **4:**2490
Interference (Waves), **2:**941, **3:***2157,* **3:2157–2158**
Interferometry, **3:2158–2162**, **3:**2158*f*, **3:**2159*f*, **3:**2160*f*, **3:**2161*f*, **4:**2284
Interferon-tau, **3:**2162, **3:**2163
Interferons, **3:2162–2164**, **3:***2163*
Interleukin-2, **1:**207
Intermediate filaments, **1:**786–787
Intermediate range ballistic missiles (IRBMs), **1:**422
Internal clocks, **1:**516–517
Internal combustion engines, **3:2164–2167**
    automotive, **1:**397–398
    diesel, **2:***1224,* **2:**1224–1226, **2:***1225*
    for farm machinery, **1:**78
    See also Combustion

International Acupuncture Training Center (Shanghai College of Traditional Chinese Medicine), **1:**39
International Astronomical Union (IAU), **1:**323, **4:**2506
International Board for Plant Genetic Resources, **5:**3118
International Bureau of Weights and Measures, **4:**2541
International Business Machine Corporation (IBM), **1:**674
International cooperation, **2:**1018–1019
    See also International Space Station; Treaties
International Date Line, **6:**4034
International diamond trade, **2:**1221
International Geophysical Year (IGY), **1:**215
International Harvester Company, **1:**77
International Olympic Committee (IOC), **3:**2234–2236
International Polar Year, **1:**215
International Space Station (ISS), **3:2167–2170**, **3:***2169,* **4:**2592–2593, **5:**3734, **5:**3738, **5:**3739, **5:**3741
International Standard of Units. *See* Le Système International d'Unités (SI)
International System for Human Cytogenetic Nomenclature (ISCN), **3:**2233–2234
International Tsunami Warning System (ITWS), **6:**4122–4123
International Ultraviolet Explorer (IUE), **3:2170–2172**, **6:**4160
International Union for the Conservation of Nature (IUCN), **5:**3257–3258
International Union of Pure and Applied Chemistry (IUPAC), **2:**1428, **2:**1430, **4:**3019, **4:**3023–3024
International William Fetal Medicine and Surgery Society, **5:**3222
Internet, **3:2173–2176**
    addiction, **1:**47
    communications, **3:**2174
    tracking, **3:2172–2173**
Internet Architecture Board (IAB), **3:**2174
Internet file transfers, **3:2172–2173**
Internet Network Information Center (InterNIC), **3:**2174
InterNIC. *See* Internet Network Information Center
Interphase, in cell division, **1:**791
Interpreter program, **2:**994
Interspecific competition, **2:**969
Interstellar gas, **3:**2176–2177, **4:**2567
Interstellar matter, **3:2176–2178**
Interstitial-cell-stimulating hormones (ICSH), **5:**3414
Intervals, **3:2178–2179**
Interventional radiology, **5:**3358
Intestines, **2:**862, **2:**1234
Intracranial pressure, **4:**2494–2495

Intracytoplasmic sperm injection (ICSI), **3:**1729, **3:**2104
Intraspecific competition, **2:**969
Intrauterine devices (IUDs), **2:**1044
Intravenous feeding, **4:**2427–2428
Introduced species, **2:**1459–1460, **3:2179–2181**, **3:***2180,* **3:**2182–2185
Introduction To the Study of Disease, **2:**1525
Introns, **3:**1753, **3:**1754
Intrusive rocks, **3:**2089, **4:**2408
Invariants, **3:2181–2182**
Invasive species, **3:2182–2185**
Invasive Species Council, **3:**2184
Inventions or Devices (Bourne), **6:**3880
Inverness Corona, **6:**4185–4186
Invertebrate paleontology, **4:**2948–2950
Invertebrates, **1:**574, **3:2185**, **3:***2186*
    brain, **1:**608
    circulation, **2:**886
    culture, **1:**584
    integumentary system, **3:**2155
Inverted topography, **4:**2275, **4:**2661
Involuntary muscles. *See* Smooth muscles
Io, **3:**2224, **3:***2225,* **6:**4259
IOC. *See* International Olympic Committee
Iodides, **3:**1903
Iodine, **3:**1911, **3:**1912
Iodine-129, **3:**1796
Ion beams, focused, **3:1649–1650**
Ion exchange, **3:***2186,* **3:2186–2189**, **3:**2187*t*, **3:**2188*f*, **4:**2279–2280
Ion exchangers, **3:**1918, **3:**1919
Ionic bonds, **2:**821, **4:**2573, **4:**2574–2575, **4:**2616–2617
Ionic compounds, **2:**1122, **2:**1123–1124, **2:**1124*t*, **3:**1679, **4:**2609
Ionization smoke detectors, **1:**140
Ionizing radiation, **3:**2192, **3:2193**, **4:**2687, **4:**2688, **5:**3329, **5:**3347
    See also Food irradiation
Ionosphere, **1:**344
Ions and ionization, **1:**360, **3:**1649, **3:2189–2193**, **3:**2190*t*, **3:**2191*t*
    freshwater, **3:**1698, **3:**1699
    Lewis structures of, **4:**2326–2328
    photochemistry, **5:**3051
    plasma, **5:**3127
IPM. *See* Integrated pest management
Ipomoea batatas. See Sweet potatoes
Ipswich sparrows, **5:**3743
Iquique, **5:**3720
IRA. *See* Infrared Astronomical Satellite
Iran, **1:**314
Iran-Iraq War, **2:**828
Iraq, **1:**314
IRAS (Infrared Astronomical Satellite), **3:**2135

L

M

MEMS (Micro electrical mechanical systems), **4**:2557–2558

Mendel, Gregor Johann, **1**:540, **1**:598, **2**:880, **2**:1190, **2**:1539, **3**:1751, **3**:1774, **4**:2492, **5**:3117, **5**:3173
  mutations, **4**:2689

Mendeléev, Dmitri Ivanovitch, **1**:367, **2**:1423, **2**:1426, **2**:1430, **2**:1553, **4**:3017, **4**:3018, **4**:3019, **4**:3020

Mendelevium, **2**:1430

Mendelian disorders, **3**:1763

Mendelian genetics, **4**:*2492,* **4:2492–2493**, **4**:*2493*

Mengele, Josef, **2**:1263–1264

Menhadens, Atlantic, **3**:1983

Meninges, **1**:609, **4**:2493

Meningitis, **4:2493–2496**

Meningomyelocele, **3**:2052–2053

Menopause, **2**:1226, **3**:1898, **4:2496–2498**, **4**:2500, **5**:3417

Menstrual cycles, **3**:1898, **3**:2010–2011, **4:2498–2500**, **4**:*2499,* **5**:3286

Menstruation, **4**:2499–2500, **5**:3416

Mental disorders. *See* specific mental disorders

Menten, Maud, **2**:1494

Mentha spp., **4**:2590

Menthaceae. *See* Mint family

Menura alberti. See Prince Albert lyre-birds

Menville, C. F., **4**:2496

Mephitis mephitis. See Skunks

Mercalli scale, **2**:1324

Mercator projections, **4**:2289

Mercuric oxide/zinc cells, **1**:458

Mercurous chloride, **4:2500–2501**

Mercury barometers, **1**:354, **1**:441–442

Mercury (Element), **2**:1424, **4:2501–2502**, **6**:4060
  biomagnification, **1**:525–526
  poisoning, **4**:2501–2502
  from solid waste incineration, **3**:2111

Mercury (Planet), **1**:775, **4:2502–2508**, **4**:2503*f,* **4**:*2504,* **4**:*2505,* **4**:*2506,* **5**:3702
  atmosphere, **5**:3100, **5**:3101, **5**:3102
  orbit of, **5**:3402
  space probes, **5**:3728
  spacecraft, **5**:3736

Mercury vapor lamps, **1**:260

Mergansers, **2**:1288, **2**:1290

Merganthaler, Ottmar, **5**:3234

Merginae. *See* Mergansers

Meridians. *See* Latitudes and longi-tudes

Merino sheep, **4**:2359, **5**:3616, **5**:3618

Meriones unguiculatus, **3**:1806–1807

Merlins, **3**:1577

Meropidae. *See* Bee-eaters

Merops apiaster. See European bee-eaters

Merops ornatus. See Rainbow-birds

Merops persicus. See Blue-cheeked bee-eaters

Merrifield, Bruce, **5**:3454

Mersenne prime numbers, **4**:3014

Mesabi Mountains, **4**:2792

Mesas, **4**:2275, **4**:2660–2661

Mesazoa, **4:2509**

Mescal buttons, **1**:664

Meselson, Matthew, **2**:1264

Mesmerism, **1**:145

Mesocricetus auratus. See Golden hamsters

Mesons, **6**:3879

Mesopelagic zones, **4**:2854

Mesophiles, **1**:411

Mesophyll, **4**:2299

Mesoscopic systems, **4:2508**

Mesosphere, **1**:344, **1**:355

Mesotrophic lakes, **4**:2262

Mesquite, **4**:2312

Messenger ribonucleic acid. *See* mRNA

Messerschmitt 262, **1**:97

Metabolic disorders, **3**:1765–1766, **4:2509–2513**

Metabolism, **4:2513–2517**
  adenosine diphosphate and, **1**:50
  aerobic, **1:55–56**, **1**:*56*
  antimetabolites and, **1**:238
  crassulacean-acid, **1**:661
  dieting and, **2**:1330
  drug, **4**:3036
  lactic acid in, **4**:2256
  of molds, **4**:2602
  obesity and, **4**:2846
  urea and, **6**:4186
  See also Adenosine triphosphate

Metabolites, **4**:2513

Metal alloys. *See* Alloys

Metal chlorides, **3**:2058

Metal fatigue, **4**:2519, **6**:3828

Metal forming, **4**:2525

Metal molds, **4**:2523–2524

Metal pollution, **1**:534

Metal production, **4:2519–2521**, **4**:*2520*

Metal sulfides, **6**:3898

Metal tools, **3**:1916

Metalcutting. *See* Machining

Metallic bonds, **2**:823, **4**:2573, **4**:2575, **4**:2617

Metallic coatings, **4**:2525–2526

Metallic crystals, **2**:1122–1123

Metallic elements, **4**:3018, **4**:3019, **4**:3022
  See also Metals

Metallic oxides, **4**:2520–2521

Metallizing, **4**:2525

Metallo-organic chemical vapor deposi-tion (MOCVD), **4**:2703

Metalloporphyrin, **2**:*1050,* **2**:1051

Metallurgy, **1**:270, **4:2521–2526**, **4**:2522*t*

Metals, **2**:1122, **4:2517–2519**
  corrosion and, **2**:1072
  earth, **1:126–128**, **2**:1426, **4**:2418
  heavy, **1**:34

ligands and, **4**:2332, **4**:*2332,* **4**:2335
  misch, **4**:2280
  precious, **5:3213–3216**, **5**:*3214*
  qualitative analysis of, **5**:3297
  transition, **2**:972, **2**:1427, **4**:2518, **4**:3019
  See also specific metals and types of metals

Metamorphic facies, **4**:2526–2528, **4**:2529

Metamorphic grade, **4:2526–2528**, **4**:*2527*

Metamorphic ores, **4**:2887

Metamorphic rocks, **4**:2526–2527, **4**:*2527,* **4:2528–2529**, **5**:3478

Metamorphism, **4**:2526–2528, **4**:2528–2529, **4:2529–2531**

Metamorphosis, **4:2531–2534**, **4**:*2532*

Metaphase, **4**:2597

Metasequoia glyptostroboides. See Dawn redwoods

Metchnikoff, Elie, **3**:2099

Meteor Crater, **3**:*2102*

Meteor showers, **4**:2537–2538

Meteorite craters. *See* Craters

Meteorites, **2**:1564–1565, **3**:1796, **3**:2085, **4:2535–2539**, **4**:2536*f,* **4**:2537*f,* **4**:*2538*
  astroblemes and, **1**:331
  Henbury craters, **1**:383
  from Mars, **4**:2450–2451
  Murchison, **4:2669–2670**

Meteorological satellites, **1**:346, **2**:1287, **4**:2534–2535

Meteorology, **1**:346, **2**:1317, **4:2534–2535**
  atmosphere circulation, **1**:347–350, **1**:*348*
  drought, **2**:1285–1287, **2**:1286
  See also Weather

Meteors, **1**:330, **4:2535–2539**, **4**:2536*f,* **4**:2537*f,* **4**:*2538*

Meterologica, **6**:4301

Meters, **4**:2541, **4**:*2541,* **6**:4175

Methamphetamines, **1**:170, **1**:171

Methane, **1**:143, **1**:164, **1**:179, **1**:509, **3**:1832, **3**:1876, **3**:2048, **4**:2271, **4**:2539

Methanol, **1**:109, **1**:110, **1**:509, **4**:2539–2540

Methyl alcohol, **3**:2056

Methyl ether, **4**:*2615,* **4**:2616

Methyl group, **4:2539–2540**

Methyl mercury, **3**:1718–1719, **4**:2502

Methyl red, **3**:2112

Methylene chloride, **3**:1903

Metric Conversion Act of 1976, **6**:4176

Metric system, **4:2540–2542**, **4**:*2541,* **6**:4175

Metula, **4**:2859

Mexico City earthquakes, **2**:1326

Meyerhof, Otto, **1**:714, **3**:1586

MFOs (Mixed-function oxidases), **2**:1029

Namibia, **2:***1295*

Nanoscale systems. *See* Mesoscopic systems

Nanostructures, **4:**2508

Nanotechnology, **2:**1266, **4:**2508, **4:2703–2706,** **4:**2705*f*

Naphthalene, **3:***2049,* **3:**2050

Napier, John, **4:**2371

Napo River, **5:**3719

Narcisseae, **1:**156

Narcolepsy, **1:**170

Narcotic blockers, **1:**181–**1:**182

Narcotics, **4:2707**

   See also specific narcotics

Nares, George, **6:**4164

Narina trogons, **6:**4101

Narlikar, Jayant, **6:**3817

NASA. *See* National Aeronautics and Space Administration

Nasua nasua. See Ringtailed coatis

National Acid Precipitation Act of 1980, **2:**1488

National Aeronautics and Space Administration (NASA)
   on airborne infrared astronomy, **3:**2135
   Chandra X-ray Observatory, **6:**3833
   Hubble Space Telescope, **3:**2023–2026
   International Space Station and, **3:**2168
   International Ultraviolet Explorer, **3:**2170
   on jet stream, **3:**2219
   Kepler telescope, **2:**1568
   manned spacecraft, **5:**3734, **5:**3735
   Mars Pathfinder, **4:**2452–2453
   multistage rockets, **5:**3474
   Murchison meteorite, **4:**2669
   near-Earth asteroids, **4:**2588
   Radarsat, **5:**3411
   SETI, **5:**3590
   space shuttles, **5:**3730, **5:**3732, **5:**3741
   telescopes and, **6:**3985

National Center for Supercomputing Applications, **3:**2175

National Centers for Environmental Prediction (NCEP), **6:**4304

National Commission for the Protection of Human Subjects of Biomedical and Behavioral Research, **5:**3283

National Environmental Policy Act, **2:**1488, **2:**1491

National Eye Institute, **5:**3324

National Forest Management Act, **2:**1488

National Human Genome Research Institute (NHGRI), **3:**2037

National Institute for Neurological and Communicative Disorders and Stroke, **2:**1177

National Institute of Mental Health (NIMH), **4:**2667

National Institute of Standards and Technology (NIST), **1:**358, **1:**683, **2:**1514–1516

National Institutes of Health (NIH), **1:**39
   Human Genome Project, **3:**2036, **3:**2037
   on marijuana, **4:**2441

National monuments, **5:**3257

National Oceanic and Atmospheric Administration (NOAA), **2:**1360, **6:**4304

National Organ Procurement and Transplantation Network, **6:**4085

National Overflights Act, **4:**2784

National Park Service (U.S.). *See* U.S. National Park Service

National parks, **5:**3257–3259

National Primary Drinking Water Regulations (NPDWRs), **6:**4290–4291

National Radio Astronomy Observatory, **5:**3339

National Research and Education Network (NREN), **3:**2174

National Science Foundation (NSF), **2:**1348, **3:**2174

National Storm Prediction Center, **6:**4304

National Tay-Sachs and Allied Diseases Association, **6:**3967

National Television System Committee (NTSC) video recording, **6:**4217

National Weather Service, **6:**4339

Native American Grave Protection and Repatriation Act (NAGPRA), **1:**269–270

Native American Religious Freedom Act of 1978, **1:**269

Native Americans
   archaeology excavation and, **1:**269–270
   earthern mounds, **4:**2655–2657
   goldenseal and, **3:**1842
   petroglyphs by, **4:**3029

Native elements, **4:**2577–2578

Natural composites, **2:**979

Natural disasters. *See* specific types of disasters

Natural family planning, **2:**1044

Natural fibers, **4:2707–2714,** **4:***2708,* **4:**2709*t,* **4:**2710*t,* **4:**2711*t,* **4:**2712*t,* **4:**2713*t*

Natural gas, **1:**140, **4:**2539, **4:2714–2716,** **4:**2714*t,* **4:***2715*
   Australia, **1:**386
   liquefied, **3:**1736

Natural language processing, **1:**305

Natural logarithms, **4:**2370

Natural numbers, **1:**49, **1:**278, **3:**2131, **4:2716**

Natural resins, **5:**3420*t*

Natural resources, **5:3424**
   See also specific resources

Natural rubber, **6:**4261

Natural selection. *See* Selection

Natural units, **6:**4175–4176

Nature Conservancy, **1:**507, **2:**1461, **2:**1566

Naturopathy, **1:**144

Nauss, Lee, **1:**41

Nautical archaeology, **4:2716–2719**

Nautilus, **6:**3881

Naval Research Laboratory, **5:**3339

Navigation
   aircraft, **1:**98–99
   global positioning systems, **3:1830–1831,** **4:**2374
   inertial, **3:**2123–2125
   LORAN, **3:**2071–2072, **4:**2373–2374
   of migratory animals, **4:**2564
   satellite, **3:**1831

Navigation Satellite for Time and Ranging (Navstar), **3:**1831

Navigation weirs, **4:**2364

Navistar, **1:**77

Navstar (Navigation Satellite for Time and Ranging), **3:**1831

NCEP (National Centers for Environmental Prediction), **6:**4304

Nd:YAG lasers, **4:**2282, **4:**2284, **4:**2285

Nd:YLF lasers, **4:***2283*

Neanderthal man, **3:**2034

Near Earth Asteroid Rendezous-Shoemaker. *See* NEAR-Shoemaker

Near-Earth Asteroid Rendezvous (NEAR). *See* NEAR spacecraft

Near-Earth asteroids (NEAs), **1:**777, **4:**2587–2589, **4:**2587*t,* **4:**2720

NEAR-Earth Object Hazard Index, **4:2719–2720**

Near Infrared Camera and Multi-Object Spectrometer (NICMOS), **3:**2025–2026

NEAR-Shoemaker, **4:**2588–2589

NEAR spacecraft, **5:**3704, **5:**3728

NEAs. *See* Near-Earth asteroids

Nebulae. *See* Planetary nebulae

Nebulizers, **2:***1143*

Necromancer, **2:**999

Nectars, **3:**2040–2042, **4:2720–2721**

Necturus maculosus. See Mudpuppies

Needham, John, **6:**3779

Needle-nailed galagos, **4:**2376

Ne'eman, Yuval, **5:**3308

Negative feedback, **3:**2002–2003, **3:***2003*

Negative numbers, **1:**278

Negative pressure ventilators, **5:**3429–3430

Negative selection, **4:**2547–2548

Negatives (Mathematics), **4:2721–2722**

Negev desert, **1:**315

Neisser, Albert, **5:**3606

Neisseria gonorrhoeae, **5:**3606

Neisseria meningitidis, **4:**2494, **4:**2495

Nelson, David, **4:**2698

Nelumbo lutea. See Lotus water lilies

O

S

T

U.S. Fish and Wildlife Service, **2:**1290, **2:**1458, **6:**4334
　biodiversity and, **1:**507
　California condors, **6:**4263
　on invasive species, **3:**2184
U.S. Food and Drug Administration (FDA)
　on butylated hydroxyanisole, **1:**657
　on butylated hydroxytoluene, **1:**657
　on cyclamate, **2:**1138–1139
　esters, **2:**1516
　on interferons, **3:**2163
　on laser surgery, **4:**2288
　on lithium, **4:**2351
　on medical marijuana, **4:**2442
　pharmaceuticals, **2:**1277
　on vaccines, **6:**4192
U.S. Geological Survey, **2:**1322
　on petroleum, **4:**3031
　volcanoes, **6:**4254
U.S. military bases, hazardous wastes and, **3:**1926
U.S. National Aeronautics and Space Adminstration. *See* NASA
U.S. National Park Service, **2:**1490, **3:**2184
U.S. Naval Observatory, **6:**4035
U.S. Post Office, **1:**431
Units and standards, **4:**2540, **4:**2612, **6:4174–4176**
Units of concentration, **2:**1002
Universal milling machines, **4:**2395
Universal Product Code (UPC), **1:***431*
Universal time, **6:**4034–4035
Universe
　age of, **1:69–70**, **1:**491, **2:**1077
　distance measurement in, **4:**2338
　geocentric, **3:***1786*
　heliocentric theory and, **3:**1957–1959
　See also Big bang theory
Unix. *See* Linux
UNOCD (United Nations Conference on Desertification), **2:**1204
Unsaturated fats, **3:**2056
Unsaturated fatty acids, **3:**1579
Unsaturated hydrocarbons. *See* Alkenes
Up quarks, **5:**3308–3309
Upatnieks, J., **3:**2000
UPC (Universal Product Code), **1:***431*
Uplift, **6:4176–4177**, **6:***4177*
Upper respiratory infection. *See* Colds, common
Upright drills, **4:**2398
UPS. *See* Ultraviolet photoelectron spectroscopy
Upslope fog, **3:**1651
Upupa epops. See Hoopoes
Upwellings, **1:**531, **6:4178**
Ural Mountains, **2:**1314
Uranium, **1:**34, **2:**1421, **3:**2210, **6:4178–4179**, **6:***4179*
　early discoveries with, **1:**368
　mining, **5:**3359

nuclear energy, **2:**1473, **4:**2754, **4:**2797–2801, **4:**2812, **4:**2817
　ore, **1:**386
　radioactivity of, **4:**3022
Uranium-235, **6:**4179, **6:***4179*
Uranium-238, **5:**3343, **6:**4179
Uranium isotopes, **1:**140
Uranium series dating, **2:**1156
Uranus, **1:**774, **5:**3701, **5:**3702, **6:4180–4186**, **6:**4181*f,* **6:**4183*f,* **6:***4184*, **6:***4185*
　atmosphere, **5:**3100, **5:**3101, **5:**3103
　discovery of, **1:**336
　radiation belts of, **6:**4197–4198
　rings of, **5:**3109, **5:**3110
　Voyager 2 and, **6:**4260
Urban-industrial techno-ecosystems, **1:**532
Urban planning, **4:**2268–2269
Urbanization, in Industrial Revolution, **3:**2121
Urea, **1:**166, **6:4186–4187**, **6:**4186*f*
Urea-formaldehyde resins, **6:**4187
Urey, Harold, **2:**824, **2:**1209, **2:**1210, **3:**2056, **4:**2569, **4:**2669
Urinary system, **1:**187, **2:**1554–1555, **4:**2896, **5:**3222
Urology, **6:4187**
Ursidae. *See* Bears
Ursus americanus. See American black bears
Ursus arctos. See Brown bears
Ursus arctos horibilis. See Grizzly bears
Ursus spelaeus. See Cave bears
Uruguay, **5:**3720
Uruguay River, **5:**3720
USDA. *See* U.S. Department of Agriculture
Usherwood, William, **2:**959
U.S.S. *Hawkbill,* **6:**4169
U.S.S. *Nautilus,* **6:**3881
U.S.S. *Skate,* **6:**3881
U.S.S. *Tuscarora,* **6:**4164
Ussher, Bishop, **6:**4162
Ussher, James, **3:**1793
USSR. *See* Soviet Union
Ustilago myadis, **5:**3493
Ustilago violacea, **5:**3493
Uterus, **3:**2128, **5:**3417
UV radiation. *See* Ultraviolet radiation
Uvalas, **5:**3637

V

V-2s, **1:**424
Vaccinations
　antibodies and, **1:**227–228
　tetanus, **6:**4003
Vaccines, **2:**839, **3:**1667, **6:4189–4192**
　AIDS, **1:**87, **1:**89, **6:**4191–4192
　AIDS/HIV, **5:**3609–3610

chicken pox and, **2:**838, **2:**840
cholera, **2:**863
for diphtheria, **2:**1252
DNA, **2:1267**, **6:**4190, **6:**4191–4192
flu, **3:**2133, **6:**4191, **6:**4234
foot and mouth disease, **3:**1667
hepatitis B, **6:**4191, **6:**4233
for leprosy, **4:**2323
malarial, **4:**2426
measles, **2:**840
polio, **2:**842, **6:**4190–4191, **6:**4234
to prevent meningitis, **4:**2495
rabies, **5:**3315
tuberculosis, **6:**4125
whooping cough, **6:**4328
yellow fever, **3:**1964–1965, **6:**4365
Vaccinium spp., **3:**1951
Vacuoles, **1:**788, **5:**3115
Vacuum drying, **3:**1659
Vacuum filters, **3:**1610–1611
Vacuum tubes, **6:4193–4194**
Vacuums, **6:4193**
Vadose zone, **3:**1882
Vagina, **5:**3417
Valence, **6:4194–4195**
Valence bond theory, **2:***1050,* **2:**1051
Valence electrons, **4:**2326, **4:***2327,* **6:**4194–4195
Valence Shell Electron Pair Repulsion (VSEPR), **4:**2610–2611, **6:**4195
Valence shells, **4:**3020
Valenstein, Elliot, **5:**3283
Valhalla, **3:**2225
Validity (Testing), **5:**3278
Valium. *See* Diazepam
Valles Marinaris, **4:**2449
Valley fog, **3:**1650
Valleys, **4:**2275
Vallisneria americana. See Tape-grass
Valproate, for manic depression, **4:**2435
Valproic acid compounds, **1:**231
Valued ecosystem components (VECs), **2:**1492
Vampire bats, **1:**453, **1:**456
Van Allen, James A., **6:**4195
Van Allen belts, **2:**1321, **5:**3706, **6:4195–4198**, **6:***4197*
Van de Graaff, Robert Jemison, **1:**9, **2:**1419
Van de Graaff accelerators, **1:**9
Van de Graaff generators, **2:***1419*
van de Kamp, Peter, **2:**1566
Van der Waals, Diederik, **6:**4198
Van der Waals equation of state, **3:**1740
Van der Waals forces, **2:**823, **4:**2575, **6:4198**
van Leyden, Lucas, **2:**1483
Van Niel, Cornelius, **5:**3066, **5:**3265
Vanadium, **2:**1425
Vane, John R., **1:**181
Vanilla fragrans. See Vanilla orchids
Vanilla orchids, **4:**2884

W

X

Y

Z